AutoCAD 14 Instructor

James A. Leach
University of Louisville

IRWIN

GRAPHICS
SERIES

Boston   Burr Ridge, IL   Dubuque, IA   Madison, WI   New York   San Francisco   St. Louis
Bangkok   Bogotá   Caracas   Lisbon   London   Madrid
Mexico City   Milan   New Delhi   Seoul   Singapore   Sydney   Taipei   Toronto

The Irwin Graphics Series

Providing you with the highest quality textbooks that meet your changing needs requires feedback, improvement, and revision. The team of authors and McGraw-Hill are committed to this effort. We invite you to become part of our team by offering your wishes, suggestions, and comments for future editions and new products and texts.

Please mail or fax your comments to: Jim Leach
c/o McGraw-Hill
1333 Burr Ridge Parkway
Burr Ridge, IL 60521
fax 630-789-6946

TITLES IN THE IRWIN GRAPHICS SERIES INCLUDE:

Technical Graphics Communication, 2e
Bertoline, Wiebe, Miller, and Mohler, 1997

Fundamentals of Graphics Communication, 2e
Bertoline, Wiebe, and Miller, 1998

Engineering Graphics Communication
Bertoline, Wiebe, Miller, and Nasman, 1995

Problems for Engineering Graphics Communication and Technical Graphics Communication Workbook #1, *1995*

Problems for Engineering Graphics Communication and Technical Graphics Communication Workbook #2, *1995*

AutoCAD Instructor Release 12
James A. Leach, 1995

AutoCAD Companion Release 12
James A. Leach, 1995

AutoCAD 13 Instructor
James A. Leach, 1996

AutoCAD 13 Companion
James A. Leach, 1996

AutoCAD 14 Instructor
James A. Leach, 1998

AutoCAD 14 Companion
James A. Leach, 1998

The Companion for CADKEY 97
John Cherng, 1998

Hands-On CADKEY: A Guide to Versions 5, 6, and 7
Timothy J. Sexton, 1996

Modeling with AutoCAD Designer
Dobek and Ranschaert, 1996

Engineering Design and Visualization Workbook
Dennis Stevenson, 1995

Graphics Interactive CD-ROM
Dennis Lieu, 1997

IRWIN
GRAPHICS
SERIES

To Nicholas—the future is yours.

"You can accomplish anything
you make up your mind to. . ."

Norman G. Leach

DEDICATION

WCB/McGraw-Hill

A Division of The McGraw·Hill Companies

AutoCAD 14 Instructor

This book is printed on acid-free paper.

4   5   6   7   8   9   0   CSI/CSI   1   0

ISBN 0-256-26602-6

Vice president and editorial director: *Kevin T. Kane*
Publisher: *Tom Casson*
Executive editor: *Elizabeth A. Jones*
Senior developmental editor: *Kelley Butcher*
Marketing manager: *John T. Wannemacher*
Senior project manager: *Beth Cigler*
Production supervisor: *Heather Burbridge*
Designer: *Jennifer McQueen Hollingsworth*
Compositor: *Interactive Composition Corporation*
Printer: *Courier Stoughton, Inc.*

Library of Congress Cataloging-in-Publication Data

Leach, James A.
    AutoCAD 14 instructor / James A. Leach
       p.   cm.
    Includes index.
    ISBN 0-256-26602-6
    1. Computer graphics.   2. AutoCAD (Computer file)   I. Title
  T385.L38264   1998
  620'.0042'02855369--dc21   97-49114

http://www.mhhe.com

PREFACE

ABOUT THIS BOOK

This Book is Your AutoCAD 14 Instructor
The objective of *AutoCAD 14 Instructor* is to provide you the best possible printed medium for learning AutoCAD, whether you are a professional or student learning AutoCAD on your own or whether you are attending an instructor-led course.

Complete Coverage
AutoCAD 14 Instructor is written to instruct you in the full range of AutoCAD Release 14 features. All commands, system variables, and features within AutoCAD are covered. This text can be used for a two-, three-, or four-course sequence.

Graphically Oriented
Because *AutoCAD 14 Instructor* discusses concepts that are graphical by nature, many illustrations (approximately **1550**) are used to communicate the concepts, commands, and applications.

Pedagogical Progression
AutoCAD 14 Instructor is presented in a pedagogical format by delivering the fundamental concepts first, then moving toward the more advanced and specialized features of AutoCAD. The book begins with small pieces of information explained in a simple form and then builds on that experience to deliver more complex ideas, requiring a synthesis of earlier concepts. The chapter exercises follow the same progression, beginning with a simple tutorial approach and ending with more challenging problems requiring a synthesis of earlier exercises.

Easy Upgrade from Release 12 or 13
AutoCAD 14 Instructor is helpful if you are already an AutoCAD user but upgrading from Release 12 or 13. All Release 13 and Release 14 commands, concepts, features, and variables are denoted by a "R13" or "R14" vertical bar on the edges of the pages. In this way, the book provides a useful reference to all AutoCAD topics, but allows you to easily locate the Release 13 and Release 14 features.

Valuable Reference Guide
AutoCAD 14 Instructor is structured to be used as a reference guide to AutoCAD. Several important tables, lists, and variable settings are "tabbed" on the edge of the page for easy access. Every command throughout the book is given with a "command table" listing the possible methods of invoking the command. A <u>complete index</u> gives an alphabetical listing of all AutoCAD commands, command options, system variables, and concepts discussed.

For Professionals and Students in Diverse Areas
AutoCAD 14 Instructor is written for professionals and students in the fields of engineering, architecture, design, construction, manufacturing, and any other field that has a use for AutoCAD. Applications and examples from many fields are given throughout the text. The applications and examples are not intended to have an inclination toward a particular field. Instead, applications to a particular field are used when they best explain an idea or use of a command.

www.mhhe.com/leach
Please visit our web page at the above address. Beginning early 1998, ancillary material is available for reading or download. Questions for each chapter (25 true-false, multiple choice, and written answer) are available for review and testing. Additional drawing problems <u>specifically</u> for architectural, mechanical engineering, and other engineering applications are available. Solutions for drawing problems and questions can be downloaded by requesting a password at the web site.

Have Fun

I predict you will have a positive experience learning AutoCAD. Although learning AutoCAD is not a trivial endeavor, you will have fun learning this exciting technology. In fact, I predict that more than once in your learning experience you will say to yourself, "Cool!" (or something to that effect).

James A. Leach

ABOUT THE AUTHOR

James A. Leach (B.I.D., M.Ed.) is an associate professor of engineering graphics at the University of Louisville. He began teaching AutoCAD at Auburn University early in 1984 using Version 1.4, the first version of AutoCAD to operate on IBM personal computers. Jim is currently the director and primary instructor at the AutoCAD Training Center (ATC) at the University of Louisville, one of the first fifteen centers to be authorized by Autodesk, having been established in 1985.

In his 21 years of teaching Engineering Graphics and AutoCAD courses, Jim has published numerous journal and magazine articles, drawing workbooks, and textbooks about AutoCAD and engineering graphics instruction. He has designed CAD facilities and written AutoCAD-related course materials for Auburn University, University of Louisville, the ATC at the University of Louisville, and several two-year and community colleges. Jim is the author of six AutoCAD textbooks published by Richard D. Irwin and McGraw-Hill.

CONTRIBUTING AUTHORS

Steven H. Baldock is an engineer at a consulting firm in Louisville and operates a CAD consulting firm, Infinity Computer Enterprises (ICE). Steve is an Autodesk Certified Instructor and teaches several courses at the University of Louisville AutoCAD Training Center. He has ten years experience using AutoCAD in architectural, civil, and structural design applications. Steve has degrees in engineering, computer science, and mathematics. Steve Baldock prepared material for several sections of *AutoCAD 14 Instructor,* such as xrefs, groups, text, multiline drawing and editing, and dimensioning. Steve also created several hundred figures used in *AutoCAD 14 Instructor.*

Michael E. Beall is the owner of Computer Aided Management and Planning in Shelbyville, Kentucky. Michael offers contract services and professional training on AutoCAD as well as CAP products from Sweets Group, a division of McGraw-Hill. He has recently co-authored *AutoCAD 14 Fundamentals,* and was a contributing author to *Inside AutoCAD 14* from New Riders Publishing. Other efforts include co-authoring *AutoCAD Release 13 for Beginners* and *Inside AutoCAD LT for Windows 95.* He was also author of the ATC certified courseware *AutoCAD Release 13 for the Professional: Level I,* all from New Riders Publishing. Mr. Beall has been presenting CAD training seminars to architects and engineers since 1982 and is currently an Autodesk Certified Instructor (ACI) at the University of Louisville ATC. He is also a presenter for the *Mastering Today's AutoCAD* seminar series from At.A.Glance, Inc., an organization founded by Hugh Bathurst (www.awarenesslearning.com). Mr. Beall received a Bachelor of Architecture degree from the University of Cincinnati. Michael Beall assisted with several topics in *AutoCAD 14 Instructor,* including the geometric calculator, slides and scripts, layer filters, toolbar customization, and point filters. Contact Michael at 502.633.3994 or MBEALL_CAMP@compuserve.com.

Bruce Duffy has been an AutoCAD trainer and consultant for 10 years. He has aided many Fortune 500 companies in the United States with training, curriculum development, and productivity assessment/training. Bruce works primarily as an instructor at the University of Louisville Speed Scientific School and Authorized AutoCAD Training Center. He consults and provides training for national and international clients. Bruce Duffy contributed much of the material for Chapters 31 and 32 of *AutoCAD 14 Instructor*.

Patrick McCuistion, Ph.D., is an assistant professor of Industrial Technology at Ohio University. Dr. McCuistion taught three years at Texas A&M University and previously worked in various engineering design, drafting, and checking positions at several manufacturing industries. He has provided instruction in geometric dimensioning and tolerancing to many industry, military, and educational institutions, as well as prepared several articles and presentations on the topic. Dr. McCuistion is an active member in several ANSI subcommittees, including Y14.5 Dimensioning and Tolerancing, Y14.3 Multiview and Section View Drawings, Y14.35 Revisions, Y14.36 Surface Texture, and B89.3.6 Functional Gages. Dr. McCuistion contributed the material on geometric dimensioning and tolerancing for *AutoCAD 14 Instructor*.

Brian Shelly is a CAD specialist at a consulting engineering firm in Louisville, Kentucky. He began using AutoCAD version 2.0 for drafting, civil, and structural design applications. Brian has over ten years experience programming in AutoLISP and customizing menus for architectural, mapping, civil, structural, power and controls, and piping design applications. He develops presentations, training materials, and supports and manages various CAD packages under multiple operating systems. Brian has an associate degree in Civil Engineering Technology and a Bachelor of Science degree in Construction Technology from Purdue University. Brian contributed material for Chapter 46, Menu Customization.

ACKNOWLEDGMENTS

I want to thank all of the contributing authors for their assistance in writing *AutoCAD 14 Instructor*. Without their help, this text could not have been as application-specific nor could it have been completed in the short time frame. I especially want to thank Steven H. Baldock for his valuable input and hard work on earlier editions of this book.

I am very grateful to Gary Bertoline for his foresight in conceiving the Irwin Graphics Series and for including my efforts in it.

I would like to give thanks to the excellent editorial and production group at WCB/McGraw-Hill who gave their talents and support during this project, especially Betsy Jones, Kelley Butcher, and Beth Cigler.

A special thanks goes to Karen Collins and Bryan Chamberlain of The Cobb Group for all the file conversions and last-minute jobs. Barry Bergin of Interactive Composition Corporation and Karen and Bryan deserve much credit for the layout and design of *AutoCAD 14 Instructor*. They were instrumental in fulfilling my objective of providing the most direct and readable format for conveying concepts.

Doug Clark and Kristine Clayton deserve credit for the creation of the "command tables" used throughout the text. Thanks for meeting the deadlines.

I appreciate the time and attention given by Mike Anderson in quality checking the Chapter Exercises with a fine-tooth comb.

I also acknowledge: my colleague and friend, Robert A. Matthews, for his support of this project and for doing his job well; Charles Grantham of Contemporary Publishing Company of Raleigh, Inc., for generosity and consultation; and Speed Scientific School Dean's Office for support and encouragement.

Special thanks, once again, to my wife, Donna, for the many hours of copy editing required to produce this and the other texts.

TRADEMARK AND COPYRIGHT ACKNOWLEDGMENTS

The object used in the Wireframe Modeling Tutorial, Chapter 36 appears courtesy of James H. Earle, *Graphics for Engineers, Third Edition,* (pg. 120), ©1992 by Addison-Wesley Publishing Company, Inc. Reprinted by permission of the publisher.

The following drawings used for Chapter Exercises appear courtesy of James A. Leach, *Problems in Engineering Graphics Fundamentals, Series A* and *Series B,* ©1984 and 1985 by Contemporary Publishing Company of Raleigh, Inc.: Gasket A, Gasket B, Pulley, Holder, Angle Brace, Saddle, V-Block, Bar Guide, Cam Shaft, Bearing, Cylinder, Support Bracket, Corner Brace, and Adjustable Mount. Reprinted or redrawn by permission of the publisher.

The following are registered U.S. trademarks (®) of Autodesk, Inc.: 3D Studio, 3D Studio MAX, Advanced Modeling Extension, AME, ATC, AutoCAD, Autodesk LT, Autodesk, Autodesk Animator, AutoLISP, AutoShade, AutoSurf, AutoVision, Heidi, and Mechanical Desktop. The following are currently pending U.S. trademarks (™) of Autodesk, Inc.: Autodesk Device Interface, AutoFlix, AutoSnap, DXF, FLI, FLIC, Kinetix, and ObjectARX. All AutoCAD Release 13 and Release 14 sample drawings appearing throughout the text are reprinted courtesy of Autodesk, Inc. Windows 95, Windows NT, Notepad, WordPad, Excel, and MS-DOS are registered trademarks of Microsoft Corporation. Corel WordPerfect is a registered trademark of Corel, Inc. Norton Editor is a registered trademark of S. Reifel & Company. City Blueprint, Country Blueprint, EuroRoman, EuroRoman-oblique, PanRoman, SuperFrench, Romantic, Romantic-bold, Sans Serif, Sans Serif-bold, Sans Serif-oblique, Sans Serif-BoldOblique, Technic Technic-light, and Technic-bold are Type 1 fonts, copyright 1992 P. B. Payne.

LEGEND

The following special treatment of characters and fonts in the textual content is intended to assist you in translating the meaning of words or sentences in *AutoCAD 14 Instructor.*

Underline	Emphasis of a word or an idea.
Helvetica font	An AutoCAD prompt appearing on the screen at the command line or in a text window.
Italic (Upper and Lower)	An AutoCAD command, option, menu, toolbar, or dialog box name.
UPPER CASE	A file name.
UPPER CASE ITALIC	An AutoCAD system variable or a drawing aid (*OSNAP, SNAP, GRID, ORTHO*).

Anything in **Bold** represents user input:

Bold   What you should <u>type</u> or press on the keyboard.

Bold Italic   An AutoCAD <u>command</u> that you should type or <u>menu item</u> that you should select.

BOLD UPPER CASE   A <u>file name</u> that you should type.

BOLD UPPER CASE ITALIC   A <u>system variable</u> that you should type.

PICK   Move the cursor to the indicated position on the screen and press the <u>select</u> button (button #1 or left mouse button).

TABLE OF CONTENTS

COMMAND
TABLE INDEX

Command Name or Variable Name	Pull-down Menu	COMMAND (TYPE)	ALIAS (TYPE)	Short-cut	Screen (side) Menu	Tablet Menu	Chapter in This Text
(VARIABLE NAME)	*Tools Inquiry > Set Variable*	*(VARIABLE NAME)*	...	...	*TOOLS 1 Setvar*	*U,10*	
3DARRAY	*Modify 3D Operations > 3D Array*	*3DARRAY*	*3A*	...	*MODIFY2 3Darray*	*W,20*	Chapter 38
3DFACE	*Draw Surfaces > 3D Face*	*3DFACE*	*3F*	...	*DRAW 2 SURFACES 3Dface*	*M,8*	Chapter 40
3DMESH	*Draw Surfaces > 3D Mesh*	*3DMESH*	...	...	*DRAW 2 SURFACES 3Dmesh:*	...	Chapter 40
3DPOLY	*Draw 3D Polyline*	*3D POLY*	*3P*	...	*DRAW 1 3dpoly*	*O,10*	Chapter 37
3DSIN	*Insert 3D Studio...*	*3DSIN*	...	...	*INSERT 3DSin*	...	Chapter 32
3DSOUT	*File Export...*	*3DSOUT*	...	...	*FILE Export*	...	Chapter 32
ACISIN	*Insert ACIS Solid...*	*ACISIN*	...	...	*INSERT ACISin*	...	Chapters 32, 39
ACISOUT	*File Export... *.sat*	*ACISOUT*	...	...	*FILE Export *.sat*	...	Chapters 32, 39
AI_MESH	*Draw Surfaces > 3D Surfaces... Mesh*	*AI_MESH*	...	...	*DRAW 2 SURFACES 3Dobjec: Mesh*	...	Chapter 40
ALIASEDIT	*Bonus Tools> Command Alias Editor*	*ALIASEDIT*	...	...	...	...	Chapter 47
ALIGN	*Modify 3D Operations > Align*	*ALIGN*	*AL*	...	*MODIFY2 Align*	*X,14*	Chapters 16, 38
AMECONVERT	...	*AMECONVERT*	...	...	...	...	Chapter 39
ARC	*Draw Arc >*	*ARC*	*A*	...	*DRAW 1 Arc*	*R,10*	Chapter 8
ARCTEXT	*Bonus Text> Arc Aligned Text*	*ARCTEXT*	...	...	...	...	Chapter 47
AREA	*Tools Inquiry > Area*	*AREA*	*AA*	...	*TOOLS 1 Area*	*T,7*	Chapter 17
ARRAY	*Modify Array*	*ARRAY*	*AR*	...	*MODIFY1 Array*	*V,18*	Chapter 10
ATTACHURL	...	*ATTACHURL*	...	...	...	...	Chapter 43
ATTDEF and DDATTDEF	*Draw Block > Define Attributes...*	*ATTDEF or DDATTDEF*	*-AT or AT*	...	*DRAW2 Ddattdef*	...	Chapter 22
ATTDISP	*View Display > Attribute Display*	*ATTDISP*	...	...	*VIEW 2 Attdisp*	*L,1*	Chapter 22

Command Name or Variable Name	Pull-down Menu	COMMAND (TYPE)	ALIAS (TYPE)	Short-cut	Screen (side) Menu	Tablet Menu	Chapter in This Text
ATTEDIT	*Modify Object > Attribute > Global*	*ATTEDIT*	-ATE	...	*MODIFY1 Attedit*	...	Chapter 22
ATTEXT and DDATEXT	...	*ATTEXT or DDATTEXT*	...	...	...	...	Chapter 22
ATTREDEF	...	*ATTREDEF*	...	...	...	...	Chapter 22
AUDIT	*File Drawing Utilities > Audit*	*AUDIT*	...	...	*FILE Audit*	Y,24	Chapter 44
BACKGROUND	*View Render > Background...*	*BACKGROUND*	...	...	*VIEW 2 Backgrnd*	Q,2	Chapter 41
BASE	*Draw Block > Base*	*BASE*	...	...	*DRAW 2 Base*	...	Chapter 21
BEXTEND	*Bonus Modify> Extend to Block Entities*	*BEXTEND*	...	...	...	...	Chapter 47
BHATCH	*Draw Hatch*	*BHATCH*	BH or H	...	*DRAW 2 Bhatch*	P,9	Chapter 26
BLIPMODE	*Tools Drawing Aids...*	*BLIPMODE*	...	...	...	...	Chapter 44
BLOCK and BMAKE	*Draw Block > Make...*	*BLOCK or BMAKE*	-B or B	...	*DRAW2 Bmake*	N,9	Chapter 21
BMAKE and BLOCK	*Draw Block > Make...*	*BMAKE or BLOCK*	B or -B	...	*DRAW2 Bmake*	N,9	Chapter 21
BMPOUT	*File Export...*	*BMPOUT*	...	...	*FILE Export*	...	Chapter 32
BONUSPOPUP	*Bonus Tools> Popup Menu*	*BONUSPOPUP*	...	...	...	...	Chapter 47
BOUNDARY	*Draw Boundary...*	*BOUNDARY or -BOUNDARY*	BO or -BO	...	*DRAW 2 Boundary*	Q,9	Chapter 15
BOX	*Draw Solids > Box*	*BOX*	...	...	*DRAW 2 SOLIDS Box*	J,7	Chapter 38
BREAK	*Modify Break*	*BREAK*	BR	...	*MODIFY2 Break*	W,17	Chapter 9
BROWSER	*Help Connect to Internet*	*BROWSER*	...	...	...	...	Chapter 43
BTRIM	*Bonus Modify> Trim Block Entities*	*BTRIM*	...	...	...	...	Chapter 47
BURST	*Bonus Text> Explode Attributes to..*	*BURST*	...	...	...	...	Chapter 47
CAL	...	*CAL*	...	...	...	...	Chapter 44

Command Name or Variable Name	Pull-down Menu	COMMAND (TYPE)	ALIAS (TYPE)	Short-cut	Screen (side) Menu	Tablet Menu	Chapter in This Text
CELTSCALE	...	*CELTSCALE*	...	...	...	...	Chapter 12
CHAMFER	*Modify Chamfer*	*CHAMFER*	*CHA*	...	*MODIFY2 Chamfer*	*W,18*	Chapters 10, 38
CHANGE	...	*CHANGE*	*-CH*	...	...	...	Chapter 16
CHPROP and DDCHPROP	*Modify Properties...*	*CHPROP or DDCHPROP*	*CH*	...	*MODIFY1 Modify*	*Y,14*	Chapter 16
CHT	*Bonus Text> Change Text*	*CHT*	...	...	...	...	Chapter 47
CIRCLE	*Draw Circle >*	*CIRCLE*	*C*	...	*DRAW 1 Circle*	*J,9*	Chapter 8
CLIPIT	*Bonus Modify> Extended Clip*	*CLIPIT*	...	...	...	...	Chapter 47
COLOR and DDCOLOR	*Format Color*	*COLOR or DDCOLOR*	*COL*	...	*FORMAT Ddcolor*	*U,4*	Chapter 12
CONE	*Draw Solids > Cone*	*CONE*	...	...	*DRAW 2 SOLIDS Cone*	*M,7*	Chapter 38
CONVERT	...	*CONVERT*	...	...	...	...	Chapters 8, 26
CONVERTPLINES	*Bonus Tools> Pline Converter*	*CONVERTPLINES*	...	...	...	...	Chapter 47
COPY	*Modify Copy*	*COPY*	*CO or CP*		*MODIFY1 Copy*	*V,15*	Chapters 1, 4, 10
COPYCLIP	*Edit Copy*	*COPYCLIP*	...	*Ctrl+C*	*EDIT Copyclip*	*T,14*	Chapter 31
COPYLINK	*Edit Copy Link*	*COPYLINK*	...	...	*EDIT Copylink*	...	Chapter 31
CUTCLIP	*Edit Cut*	*CUTCLIP*	...	*Ctrl+X*	*EDIT Cutclip*	*T,13*	Chapter 31
CYLINDER	*Draw Solids > Cylinder*	*CYLINDER*	...	...	*DRAW 2 SOLIDS Cylinder*	*L,7*	Chapter 38
DBLIST	...	*DBLIST*	...	...	...	...	Chapter 17
DDATEXT and ATTEXT	...	*ATTEXT or DDATTEXT*	...	...	...	...	Chapter 22
DDATTDEF and ATTDEF	*Draw Block > Define Attributes...*	*DDATTDEF or ATTDEF*	*AT or -AT*	...	*DRAW 2 Ddattdef*	...	Chapter 22
DDATTE	*Modify Object > Attribute > Single...*	*DDATTE*	*ATE*	...	*MODIFY1 Ddatte*	*Y,20*	Chapter 22
DDCHPROP and CHANGE	*Modify Properties...*	*DDCHPROP or CHPROP*	*CH or -CH*	...	*MODIFY1 Modify*	*Y,14*	Chapters 12, 16
DDCOLOR and COLOR	*Format Color*	*DDCOLOR or COLOR*	*COL*	...	*FORMAT Ddcolor*	*U,4*	Chapter 12

Command Name or Variable Name	Pull-down Menu	COMMAND (TYPE)	ALIAS (TYPE)	Short-cut	Screen (side) Menu	Tablet Menu	Chapter in This Text
DDEDIT	*Modify Object > Text…*	*DDEDIT*	*ED*	…	*MODIFY1 Ddedit*	*Y,21*	Chapter 18
DDIM	*Dimension Style…*	*DDIM*	*D*	…	*DIMNSION Ddim*	*Y,5*	Chapter 29
DDINSERT and INSERT	*Insert Block…*	*DDINSERT or INSERT*	*I or -I*	…	*INSERT Ddinsert*	*T,5*	Chapter 21
DDMODIFY	*Modify Properties…*	*DDMODIFY*	*MO*	…	*MODIFY1 Modify*	*Y,14*	Chapters 12, 16, 18, 28
DDPTYPE	*Format Point Style…*	*DDPTYPE*	…	…	*DRAW 2 Point Ddptype:*	*U,1*	Chapter 8
DDRENAME and RENAME	*Format Rename…*	*DDRENAME or RENAME*	*REN or -REN*	…	*FORMAT Ddrename*	*V,1*	Chapter 21
DDSELECT	*Tools Selection…*	*DDSELECT*	*SE*	…	*TOOLS 2 Ddselect*	*X,9*	Chapter 20
DDUCS	*Tools UCS > Named UCS…*	*DDUCS*	*UC*	…	*TOOLS 2 Dducs*	*W,8*	Chapter 36
DDUCSP	*Tools UCS > Preset UCS…*	*DDUCSP*	*UCP*	…	*TOOLS 2 Dducsp*	*W,9*	Chapter 36
DDUNITS and UNITS	*Format Units…*	*DDUNITS or UNITS*	*UN or -UN*	…	*FORMAT Ddunits*	*V,4*	Chapter 6
DDVIEW and VIEW	*View Names Views…*	*DDVIEW or VIEW*	*V or -V*	…	*VIEW 1 Ddview*	*M,5*	Chapter 11
DDVPOINT	*View 3D Viewpoint > Select…*	*DDVPOINT*	*VP*	…	*VIEW 1 Ddvpoint*	*N,5*	Chapter 35
DETACHURL	…	*DETACHURL*	…	…	…	…	Chapter 43
DIMALIGNED	*Dimension Aligned*	*DIMALIGNED*	*DIMALI or DAL*	…	*DIMNSION Aligned*	*W,4*	Chapter 28
DIMANGULAR	*Dimension Angular*	*DIMANGULAR*	*DIMANG or DAN*	…	*DIMNSION Angular*	*X,3*	Chapter 28
DIMBASELINE	*Dimension Baseline*	*DIMBASELINE*	*DIMBASE or DBA*	…	*DIMNSION Baseline*	*W,2*	Chapter 28
DIMCENTER	*Dimension Center Mark*	*DIMCENTER*	*DCE*	…	*DIMNSION Center*	*X,2*	Chapter 28
DIMCONTINUE	*Dimension Continue*	*DIMCONTINUE*	*DIMCONT or DCO*	…	*DIMNSION Continue*	*W,1*	Chapter 28
DIMDIAMETER	*Dimension Diameter*	*DIMDIAMETER*	*DIMDIA or DDI*	…	*DIMNSION Diameter*	*X,4*	Chapter 28
DIMEDIT	*Dimension Oblique*	*DIMEDIT*	*DIMED or DED*	…	*DIMNSION Dimedit*	*Y,1*	Chapter 28
DIMEX	*Bonus Tools> Dimstyle Export…*	*DIMEX*	…	…	…	…	Chapter 47
DIMIM	*Bonus Tools> Dimstyle Import…*	*DIMIM*	…	…	…	…	Chapter 47
DIMLINEAR	*Dimension Linear*	*DIMLINEAR*	*DIMLIN or DLI*	…	*DIMNSION Linear*	*W,5*	Chapter 28

Command Name or Variable Name	Pull-down Menu	COMMAND (TYPE)	ALIAS (TYPE)	Short-cut	Screen (side) Menu	Tablet Menu	Chapter in This Text
DIMORDINATE	*Dimension Ordinate*	*DIMORDINATE*	*DIMORD or DOR*	...	*DIMNSION Ordinate*	W,3	Chapter 28
DIMOVERRIDE	*Dimension Override*	*DIMOVERRIDE*	*DIMOVER or DOV*	...	...	Y,4	Chapter 29
DIMRADIUS	*Dimension Radius*	*DIMRADIUS*	*DIMRAD or DRA*	...	*DIMNSION Radius*	X,5	Chapter 28
DIMSTYLE	*Dimension Update*	*DIMSTYLE*	*DIMSTY or DST*	...	*DIMNSION Dimstyle*	Y,3	Chapter 29
DIMTEDIT	*Dimension Align Text >*	*DIMTEDIT*	*DIMTED*	...	*DIMNSION Dimtedit*	Y,2	Chapter 28
DISTANCE	*Tools Inquiry > Distance*	*DIST*	*DI*	...	*TOOLS 1 Dist*	T,8	Chapter 17
DIVIDE	*Draw Point > Divide*	*DIVIDE*	*DIV*	...	*DRAW 2 Divide*	V,13	Chapter 15
DONUT	*Draw Donut*	*DONUT*	*DO*	...	*DRAW 1 Donut*	K,9	Chapter 15
DRAGMODE	...	*DRAGMODE*	...	...	...	...	Chapter 44
DRAWORDER	*Tools Display Order >*	*DRAWORDER*	*DR*	...	*TOOLS 1 Drawordr*	T,9	Chapters 26, 32
DSVIEWER	*View Aerial View*	*DSVIEWER*	*AV*	...	*VIEW 1 Dsviewer*	K,2	Chapter 11
DTEXT	*Draw Text > Single Line Text...*	*DTEXT*	*DT*	...	*DRAW 2 Dtext*	K,8	Chapter 18
DVIEW	*View 3D Dynamic View*	*DVIEW*	*DV*	...	*VIEW 1 Dview*	R,5	Chapter 35
DWFOUT	*File Export...*	*DWFOUT*	...	...	*FILE Export*	...	Chapters 32, 43
DXBIN	*Insert Drawing Exchange...*	*DXBIN*	...	...	*INSERT DXBin*	...	Chapter 32
DXFIN	*Insert Drawing Exchange...*	*DXFIN*	...	...	...	...	Chapter 32
DXFOUT	*File Export...*	*DXFOUT*	...	...	*FILE Export*	...	Chapter 32
EDGE	*Draw Surfaces > Edge*	*EDGE*	...	...	*DRAW 2 SURFACES Edge:*	...	Chapter 40
EDGESURF	*Draw Surfaces > Edge Surface*	*EDGESURF*	...	...	*DRAW 2 SURFACES Edgsurf:*	R,8	Chapter 40
ELEVATION	...	*ELEVATION*	...	...	...	...	Chapter 40
ELLIPSE	*Draw Ellipse*	*ELLIPSE*	*EL*	...	*DRAW 1 Ellipse*	M,9	Chapters 15, 25
ERASE	*Modify Erase*	*ERASE*	*E*	...	*MODIFY1 Erase*	V,14	Chapters 4, 9
EXCHPROP	*Bonus Modify> Extended Change Properties*	*EXCHPROP*	...	...	...	...	Chapter 47

Command Name or Variable Name	Pull-down Menu	COMMAND (TYPE)	ALIAS (TYPE)	Short-cut	Screen (side) Menu	Tablet Menu	Chapter in This Text
EXIT	File Exit	EXIT	…	…	…	Y,25	Chapter 2
EXPLODE	Modify Explode	EXPLODE	X	…	MODIFY2 Explode	Y,22	Chapters 16, 21
EXPORT	File Export…	EXPORT	EXP	…	FILE Export	…	Chapter 32
EXTEND	Modify Extend	EXTEND	EX	…	MODIFY2 Extend	W,16	Chapter 9
EXTRIM	Bonus Modify> Cookie Cutter Trim	EXTRIM	…	…	…	…	Chapter 47
EXTRUDE	Draw Solids > Extrude	EXTRUDE	EXT	…	DRAW 2 SOLIDS Extrude	P,7	Chapter 38
FILL	…	FILL	…	…	…	…	Chapter 44
FILLET	Modify Fillet	FILLET	F	…	MODIFY2 Fillet	W,19	Chapter 10, 38
FILTER	…	FILTER	FI	…	ASSIST Filters	…	Chapter 20
FIND	Bonus Text> Find and Replace Text	FIND	…	…	…	…	Chapter 47
FOG	View Render > Fog…	FOG	…	…	VIEW 2 Fog	P,2	Chapter 41
GATTE	Bonus Text> Global Attribute Edit	GATTE	…	…	…	…	Chapter 47
GETSEL	Bonus Tools> Get Selection Set	GETSEL	…	…	…	…	Chapter 47
GRID	Tools Drawing Aids…	GRID	…	F7 or Ctrl+G	TOOLS 2 Ddrmodes	W,10	Chapter 6
GRIPS	Tools Grips…	GRIPS or DDGRIPS	GR	…	TOOLS 2 Ddgrips	X,10	Chapter 23
GROUP	Tools Object Group…	GROUP or -GROUP	G or -G	…	TOOLS 2 Group	X,8	Chapter 20
HATCH	…	HATCH	-H	…	…	…	Chapter 26
HATCHEDIT	Modify Object > Hatchedit	HATCHEDIT	HE	…	MODIFY1 Hatchedt	Y,16	Chapter 26
HELP	Help	HELP	?	F1	HELP Help	Y,7	Chapter 5
HIDE	View Hide	HIDE	HI	…	VIEW 2 Hide	M,2	Chapter 35
ID	Tools Inquiry > Id Point	ID	…	…	TOOLS 1 ID	U,9	Chapter 17
IMAGE and -IMAGE	Insert Raster Image…	IMAGE	IM or -IM	…	INSERT Image	T,3	Chapter 32

Command Name or Variable Name	Pull-down Menu	COMMAND (TYPE)	ALIAS (TYPE)	Short-cut	Screen (side) Menu	Tablet Menu	Chapter in This Text
IMAGEADJUST	*Modify Object> Image> Adjust*	*IMAGEADJUST*	*IAD*	...	*MODIFY1 Imageadj*	*X,20*	Chapter 32
IMAGEATTACH	*Insert Raster Image... Attach...*	*IMAGEATTACH*	*IAT*	...	*INSERT Image Attach...*	...	Chapter 32
IMAGECLIP	*Modify Object> Image Clip*	*IMAGECLIP*	*ICL*	...	*MODIFY1 Imageclp*	*X,22*	Chapter 32
IMAGEFRAME	*Modify Object> Image> Frame*	*IMAGEFRAME*	...	...	*MODIFY1 Imagefrm*	...	Chapter 32
IMAGEQUALITY	*Modify Object> Image> Quality*	*IMAGEQUALITY*	...	...	*MODIFY1 Imagequa*	...	Chapter 32
IMPORT	...	*IMPORT*	*IMP*	...	...	*T,2*	Chapter 32
INDEXCTL	*File SaveAs... Options... Index type*	*INDEXCTL*	...	...	*FILE Saveas Options... Index type*	...	Chapter 30
INETCFG	...	*INETCFG*	...	...	...	...	Chapter 43
INETHELP	...	*INETHELP*	...	...	...	...	Chapter 43
INETLOCATION	...	*INETLOCATION*	...	...	...	...	Chapter 43
INSERT and DDINSERT	*Insert Block...*	*INSERT or DDINSERT*	*-I or I*	...	*INSERT Ddinsert*	*T,5*	Chapter 21
INSERTOBJ	*Insert OLE Objects...*	*INSERTOBJ*	...	...	*INSERT Insertob*	*T,1*	Chapter 31
INSERTURL	...	*INSERTURL*	...	...	...	...	Chapter 43
INTERFERE	*Draw Solids > Interference*	*INTERFERE*	*INF*	...	*DRAW 2 SOLIDS Interfer*	...	Chapter 39
INTERSECT	*Modify Boolean > Intersect*	*INTERSECT*	*IN*	...	*MODIFY2 Intrsect*	*X,17*	Chapters 16, 38
LAYCUR	*Bonus Layers> Change to Current*	*LAYCUR*	...	...	...	...	Chapter 47
LAYER	*Format Layer...*	*LAYER or -LAYER*	*LA or -LA*	...	*FORMAT Layer*	*U,5*	Chapter 12
LAYFRZ	*Bonus Layers> Layer Freeze*	*LAYFRZ*	...	...	...	...	Chapter 47
LAYISO	*Bonus Layers> Layers Isolate*	*LAYISO*	...	...	...	...	Chapter 47

Command Name or Variable Name	Pull-down Menu	COMMAND (TYPE)	ALIAS (TYPE)	Short-cut	Screen (side) Menu	Tablet Menu	Chapter in This Text
LAYLCK	*Bonus Layers> Layer Lock*	*LAYLCK*	...	...	...	...	Chapter 47
LAYMCH	*Bonus Layers> Layer Match*	*LAYMCH*	...	...	...	...	Chapter 47
LAYOFF	*Bonus Layers> Layer Off*	*LAYOFF*	...	...	...	...	Chapter 47
LAYON	*Bonus Layers> Turn All Layers On*	*LAYON*	...	...	...	...	Chapter 47
LAYTHW	*Bonus Layers> Thaw All Layers*	*LAYTHW*	...	...	...	...	Chapter 47
LAYULK	*Bonus Layers> Layer Unlock*	*LAYULK*	...	...	...	...	Chapter 47
LEADER	*Dimension Leader*	*LEADER*	*LEAD or LE*	...	*DIMNSION Leader*	*R,7*	Chapter 28
LENGTHEN	*Modify Lengthen*	*LENGTHEN*	*LEN*	...	*MODIFY2 Lengthen*	*W,14*	Chapter 9
LIGHT	*View Render > Light...*	*LIGHT*	...	...	*VIEW2 Light*	*O,1*	Chapter 41
LIMITS	*Format Drawing Limits*	*LIMITS*	...	...	*FORMAT Limits*	*V,2*	Chapter 6
LINE	*Draw Line*	*LINE*	*L*	...	*DRAW1 Line*	*J,10*	Chapters 3, 8
LINETYPE	*Format Linetype*	*LINETYPE or -LINETYPE*	*LT or -LT*	...	*FORMAT Linetype*	*U,3*	Chapter 12
LIST	*Tools Inquiry > List*	*LIST*	*LS or LI*	...	*TOOLS 1 List*	*U,8*	Chapter 17
LISTURL	...	*LISTURL*	...	...	...	...	Chapter 43
LMAN	*Bonus Layers> Layer Manager...*	*LMAN*	...	...	...	...	Chapter 47
LOGFILEON and LOGFILEOFF	*Tools Preferences... General*	*LOGFILEON or LOGFILEOFF*	...	...	*TOOLS 2 Prefernc General*	...	Chapter 45
LSEDIT	*View Render > Landscape Edit...*	*LSEDIT*	...	...	*VIEW 2 Lsedit*	...	Chapter 41
LSLIB	*View Render > Landscape Library...*	*LSLIB*	...	...	*VIEW 2 Lslib*	...	Chapter 41
LSNEW	*View Render > Landscape New...*	*LSNEW*	...	...	*VIEW 2 Lsnew*	...	Chapter 41

Command Name or Variable Name	Pull-down Menu	COMMAND (TYPE)	ALIAS (TYPE)	Short-cut	Screen (side) Menu	Tablet Menu	Chapter in This Text
LTSCALE	Format Linetype... Details >>	LTSCALE	LTS	...	FORMAT Linetype Details >>	...	Chapter 12
MASSPROP	Tools Inquiry > Mass Properties	MASSPROP	...	...	TOOLS 1 Massprop	U,7	Chapter 39
MATCHPROP	Modify Match Properties	MATCHPROP or PAINTER	MA	...	MODIFY1 Matchprp	Y,15	Chapters 12, 15
MATLIB	View Render > Materials Library...	MATLIB	...	...	VIEW 2 Matlib	Q,1	Chapter 41
MEASURE	Draw Point > Measure	MEASURE	ME	...	DRAW 2 Measure	V,12	Chapter 15
MENU	...	MENU	...	...	...	...	Chapter 46
MENULOAD	Tools Customize Menus...	MENULOAD	...	...	TOOLS 2 Menuload	Y,9	Chapters 45, 46
MENUUNLOAD	Tools Customize Menus...	MENUUNLOAD	...	...	TOOLS 2 Menuload	Y,9	
MINSERT	...	MINSERT	...	...	...	...	Chapter 21
MIRROR	Modify Mirror	MIRROR	MI	...	MODIFY1 Mirror	V,16	Chapter 10
MIRROR3D	Modify 3D Operations > Mirror 3D	MIRROR3D	...	...	MODIFY2 Mirror3D	W,21	Chapter 38
MLEDIT	Modify Object > Multiline	MLEDIT or -MLEDIT	...	...	MODIFY1 Mledit	Y,19	Chapter 16
MLINE	Draw Multiline	MLINE	ML	...	DRAW 1 Mline	M,10	Chapter 15
MLSTYLE	Format Multiline Style...	MLSTYLE	...	...	DRAW 1 Mline Mlstyle:	V,5	Chapter 15
MOCORO	Bonus Modify> Move Copy Rotate	MOCORO	...	...	...	...	Chapter 47
MOVE	Modify Move	MOVE	M	...	MODIFY2 Move	V,19	Chapters 4, 9, 38
MPEDIT	Bonus Modify> Multiple Pedit	MPEDIT	...	...	...	...	Chapter 47
MSLIDE	...	MSLIDE	...	...	...	...	Chapter 44
MSPACE	View Model Space (Floating)	MSPACE	MS	...	VIEW 1 Mspace	L,4	Chapter 19
MSTRETCH	Bonus Modify> Multiple Entity Stretch	MSTRETCH	...	...	...	...	Chapter 47

Command Name or Variable Name	Pull-down Menu	COMMAND (TYPE)	ALIAS (TYPE)	Short-cut	Screen (side) Menu	Tablet Menu	Chapter in This Text
MTEXT	*Draw* *Text >* *Multiline Text...*	*MTEXT or* *-MTEXT*	*T or* *-T or* *MT*	...	*DRAW 2* *Mtext*	*J,8*	Chapter 18
MULTIPLE	...	*MULTIPLE*	...	...	...	...	Chapter 44
MVIEW	*View* *Floating Viewports >*	*MVIEW*	*MV*	...	*VIEW 1* *Mview*	*M,4*	Chapter 19
MVSETUP	...	*MVSETUP*	...	...	...	...	Chapters 33, 42
NCOPY	*Bonus* *Modify>* *Copy Nested Entities*	*NCOPY*	...	...	...	...	Chapter 47
NEW	*File* *New...*	*NEW*	...	*Ctrl+N*	*FILE* *New*	*T,24*	Chapter 2
OBJECT SNAPS (SINGLE POINT)	...	*END, MID, etc.* *(first three letters)*	...	...	**** *(asterisks)*	*T,16 –* *U,22*	Chapter 7
OBJECTS SNAPS (RUNNING)	*Tools* *Object Snap* *Settings...*	*OSNAP or* *-OSNAP*	*OS or* *-OS*	*Ctrl+F*	*TOOLS 2* *Ddosnap*	...	Chapter 7
OFFSET	*Modify* *Offset*	*OFFSET*	*O*	...	*MODIFY1* *Offset*	*V,17*	Chapters 10, 27
OLELINKS	*Edit* *OLE Links...*	*OLELINKS*	...	...	*EDIT* *OLElinks*	...	Chapter 31
OOPS	...	*OOPS*	...	...	*MODIFY1* *Erase* *Oops:*	...	Chapter 5
OPEN	*File* *Open...*	*OPEN*	...	*Ctrl+O*	*FILE* *Open*	*T,25*	Chapter 2
OPENURL	...	*OPENURL*	...	...	...	...	Chapter 43
PACK	*Bonus* *Tools>* *Pack 'n Go...*	*PACK*	...	...	...	...	Chapter 47
PAN	*View* *Pan*	*PAN or* *-PAN*	*P or* *-P*	...	*VIEW 1* *Pan*	*N,11–* *P,11*	Chapter 11
PASTECLIP	*Edit* *Paste*	*PASTECLIP*	...	*Ctrl+V*	*EDIT* *Pasteclp*	*U,13*	Chapter 31
PASTESPEC	*Edit* *Paste special...*	*PASTESPEC*	...	...	*EDIT* *Pastespe*	...	Chapter 31
PEDIT	*Modify* *Object >* *Polyline*	*PEDIT*	*PE*	...	*MODIFY1* *Pedit*	*Y,17*	Chapter 16
PFACE	...	*PFACE*	...	...	...	...	Chapter 40
PLAN	*View* *3D Viewpoint >* *Plan View >*	*PLAN*	...	...	*VIEW 1* *Plan*	*N,3*	Chapter 35
PLINE	*Draw* *Polyline*	*PLINE*	*PL*	...	*DRAW 1* *Pline*	*N,10*	Chapter 8
PLOT	*File* *Print*	*PLOT or* *PRINT*	...	*Ctrl+P*	*FILE* *Plot*	*W,25*	Chapter 14

Command Name or Variable Name	Pull-down Menu	COMMAND (TYPE)	ALIAS (TYPE)	Short-cut	Screen (side) Menu	Tablet Menu	Chapter in This Text
POINT	Draw Point > Single or Multiple Point	POINT	PO	...	DRAW 2 Point	O,9	Chapter 8
POINT FILTERS	...	(during command) .X, .Y, .XY, etc.	...	...	ASSIST .x, .y, .xy, etc.	...	Chapter 15
POLYGON	Draw Polygon	POLYGON	POL	...	DRAW 1 Polygon	P,10	Chapter 15
PREVIEW	File Print Preview	PREVIEW	PRE	...	...	X,24	Chapter 14
PROJECTNAME	Tools Preferences... Files Project files...	PROJECTNAME	...	...	TOOLS2 Preferenc Files Project files..	...	Chapter 30
PSDRAG	...	PSDRAG	...	...	...	...	Chapter 32
PSFILL	...	PSFILL	...	...	...	...	Chapter 32
PSIN	Insert Encapsulated Postscript	PSIN	...	...	INSERT PSin	...	Chapter 32
PSLTSCALE	Format Linetype... Details>> Use paper space units...	PSLTSCALE	...	...	FORMAT Linetype Details>> Use p.s. units	...	Chapter 33
PSOUT	File Export...	PSOUT	...	...	FILE Export	...	Chapter 32
PSPACE	View Paper Space	PSPACE	PS	...	VIEW 1 Pspace	L,5	Chapter 19
PSQUALITY	...	PSQUALITY	...	...	...	...	Chapter 32
PURGE	File Drawing Utilities > Purge >	PURGE	PU	...	FILE Purge	X,25	Chapter 21
QLATTACH	Bonus Draw> Leader Tools> Attach Leader to...	QLATTACH	...	...	...	...	Chapter 47
QLEADER	Bonus Draw> Leader Tools> Quick Leader	QLEADER	...	...	...	...	Chapter 47
QSAVE	...	QSAVE	...	Ctrl+S	FILE Qsave	...	Chapter 2
QTEXT	...	QTEXT	...	...	...	...	Chapter 18
QUIT	...	QUIT	EXIT	...	FILE Quit	...	Chapter 2
RAY	Draw Ray	RAY	...	...	DRAW 1 Ray	K,10	Chapters 15, 27

Command Name or Variable Name	Pull-down Menu	COMMAND (TYPE)	ALIAS (TYPE)	Short-cut	Screen (side) Menu	Tablet Menu	Chapter in This Text
RECOVER	*File Drawing Utilities > Recover...*	RECOVER	...	...	*FILE Recover*	...	Chapter 2
RECTANG	*Draw Rectangle*	RETANG	REC	...	*DRAW 1 Rectang*	Q,10	Chapter 15
REDO	*Edit Redo*	REDO	...	Crtl+Y	*EDIT Redo*	U,12	Chapter 5
REDRAW	*View Redraw*	REDRAW	R	...	*VIEW 1 Redraw*	...	Chapter 5
REGEN	*View Regen*	REGEN	RE	...	*VIEW 1 Regen*	J,1	Chapter 5
REGENAUTO	...	REGENAUTO	...	...	...	...	Chapter 44
REGION	*Draw Region*	REGION	REG	...	*DRAW 2 Region*	R,9	Chapter 15
REINIT	...	REINIT	...	...	...	...	Chapter 44
RENAME and DDRENAME	*Format Rename...*	RENAME or DDRENAME	-REN or REN	...	*FORMAT Ddrename*	V,1	Chapter 21
RENDER	*View Render > Render...*	RENDER	RR	...	*VIEW 2 Render*	M,1	Chapter 41
REPLAY	*Tools Display Image > View...*	REPLAY	...	...	*TOOLS 1 Replay*	V,8	Chapter 41
REVCLOUD	*Bonus Draw> Revision Cloud*	REVCLOUD	...	...	...	...	Chapter 47
REVOLVE	*Draw Solids > Revolve*	REVOLVE	REV	...	*DRAW 2 SOLIDS Revolve*	Q,7	Chapter 38
REVSURF	*Draw Surfaces > Revolved Surface*	REVSURF	...	...	*DRAW 2 SURFACES Revsurf:*	O,8	Chapter 40
RMAT	*View Render > Materials...*	RMAT	...	...	*VIEW 2 Rmat*	P,1	Chapter 41
ROTATE	*Modify Rotate*	ROTATE	RO	...	*MODIFY2 Rotate*	V,20	Chapter 9
ROTATE3D	*Modify 3D Operations > Rotate 3D*	ROTATE3D	...	...	*MODIFY2 Rotate3D*	W,22	Chapter 38
RPREF	*View Render > Preferences...*	RPREF	RPR	...	*VIEW 2 Rpref*	R,2	Chapter 41
RULESURF	*Draw Surfaces > Ruled Surface*	RULESURF	...	...	*DRAW 2 SURFACES Rulsurf:*	Q,8	Chapter 40
SAVE	*File Save As*	SAVE	...	Ctrl+S	...	U,24 – U,25	Chapter 2

Command Name or Variable Name	Pull-down Menu	COMMAND (TYPE)	ALIAS (TYPE)	Short-cut	Screen (side) Menu	Tablet Menu	Chapter in This Text
SAVEAS	File Save As...	SAVEAS	...	...	FILE Saveas	V,24	Chapter 2
SAVEIMG	Tools Display Image> Save...	SAVEIMG	...	...	TOOLS1 Saveimg	...	Chapters 32, 41
SAVETIME	Tools Preferences... General	SAVETIME	...	...	TOOLS 2 Prefernc General	...	Chapter 2
SAVEURL	...	SAVEURL	...	...	...	...	Chapter 43
SCALE	Modify Scale	SCALE	SC	...	MODIFY2 Scale	V,21	Chapter 9
SCENE	View Render > Scene...	SCENE	...	...	VIEW 2 Scene	N,1	Chapter 41
SCRIPT	Tools Run Script...	SCRIPT	SCR	...	TOOLS 1 Script	V,9	Chapter 44
SECTION	Draw Solids > Section	SECTION	SEC	...	DRAW 2 SOLIDS Section	...	Chapter 39
SELECT	...	SELECT	...	...	...	...	Chapter 4
SELECTURL	...	SELECTURL	...	...	...	...	Chapter 43
SETUV	View Render > Mapping...	SETUV	...	...	VIEW 2 Mapping	R,1	Chapter 41
SETVAR	Tools Inquiry > Set Variable	SETVAR	SET	...	TOOLS 1 Setvar	U,10	Chapter 17
SHADE	View Shade >	SHADE	SHA	...	VIEW 2 Shade	N,2	Chapter 35
SHELL	...	SHELL	...	...	...	...	Chapter 2
SKETCH	...	SKETCH	...	...	...	...	Chapter 15
SLICE	Draw Solids > Slice	SLICE	SL	...	DRAW 2 SOLIDS Slice	...	Chapter 39
SNAP	Tools Drawing Aids...	SNAP	SN	F9 or Ctrl+B	TOOLS 2 Ddrmodes	W,10	Chapters 6, 25
SOLDRAW	Draw Solids > Setup > Drawing	SOLDRAW	...	...	DRAW 2 SOLIDS Soldraw	...	Chapter 42
SOLID	Draw Surfaces > 2D Solid	SOLID	SO	...	DRAW 2 SURFACES Solid:	L,8	Chapter 15
SOLPROF	Draw Solids > Setup > Profile	SOLPROF	...	...	DRAW 2 SOLIDS Solprof	...	Chapter 42

Command Name or Variable Name	Pull-down Menu	COMMAND (TYPE)	ALIAS (TYPE)	Short-cut	Screen (side) Menu	Tablet Menu	Chapter in This Text
SOLVIEW	*Draw Solids > Setup > View*	*SOLVIEW*	...	...	*DRAW 2 SOLIDS Solview*	...	Chapter 42
SPELL	*Tools Spelling*	*SPELL*	*SP*	...	*TOOLS 1 Spell*	*T,10*	Chapter 18
SPHERE	*Draw Solids > Sphere*	*SPHERE*	...	...	*DRAW 2 SOLIDS Sphere*	*K,7*	Chapter 38
SPLFRAME	...	*SPLFRAME*	...	...	...	...	Chapter 40
SPLINE	*Draw Spline*	*SPLINE*	*SPL*	...	*DRAW 1 Spline*	*L,9*	Chapter 15
SPLINEDIT	*Modify Object > Splinedit*	*SPLINEDIT*	*SPE*	...	*MODIFY1 Splinedit*	*Y,18*	Chapter 16
STATS	*View Render > Statistics...*	*STATS*	...	...	*VIEW 2 Stats*	...	Chapter 41
STATUS	*Tools Inquiry > Status*	*STATUS*	...	...	*TOOLS 1 Status*	...	Chapter 17
STLOUT	*File Export... *.stl*	*STLOUT*	...	...	*FILE Export *.stl*	...	Chapter 39
STRETCH	*Modify Stretch*	*STRETCH*	*S*	...	*MODIFY2 Stretch*	*V,22*	Chapter 9
STYLE	*Format Text Style...*	*STYLE or -STYLE*	*ST*	...	*DRAW 2 Dtext Style:*	*U,2*	Chapter 18
SUBTRACT	*Modify Boolean > Subtract*	*SUBTRACT*	*SU*	...	*MODIFY2 Subtract*	*X,16*	Chapters 16, 38
SURFTYPE	...	*SURFTYPE*	...	...	...	...	Chapter 40
SYSVDLG	*Bonus Tools> System Variable Editor*	*SYSVDLG*	...	...	...	...	Chapter 47
TABSURF	*Draw Surfaces > Tabulated Surface*	*TABSURF*	...	...	*DRAW 2 SURFACES Tabsurf:*	*P,8*	Chapter 40
TEXT	...	*TEXT*	...	...	...	...	Chapter 18
TEXTFILL	*File Print Text Fill*	*TEXTFILL*	...	...	*FILE Plot Text Fill*	...	Chapter 18
TEXTFIT	*Bonus Text> Text Fit*	*TEXTFIT*	...	...	...	...	Chapter 47
TEXTMASK	*Bonus Text> Text Mask*	*TEXTMASK*	...	...	...	...	Chapter 47

Command Name or Variable Name	Pull-down Menu	COMMAND (TYPE)	ALIAS (TYPE)	Short-cut	Screen (side) Menu	Tablet Menu	Chapter in This Text
THICKNESS	Format Thickness	THICKNESS	TH	...	FORMAT Thicknes	V,3	Chapter 40
TILEMODE	View Model Space (Tiled)	TILEMODE	TM or TI	...	VIEW 1 Tilemode	L,3	Chapter 19
TIME	Tools Inquiry > Time	TIME	...	...	TOOLS 1 Time	...	Chapter 17
TOLERANCE	Dimension Tolerance...	TOLERANCE	TOL	...	DIMNSION Toleranc	X,1	Chapter 28
TOOLBAR	View Toolbar...	TOOLBAR	TO	...	VIEW 2 Toolbar	R,3	Chapter 45
TORUS	Draw Solids > Torus	TORUS	TOR	...	DRAW 2 SOLIDS Torus	O,7	Chapter 38
TRACKING	...	(during command) TRACKING	TK	...	**** (asterisks) Tracking	T,15	Chapter 15
TRANSPARENCY	Modify Object> Image> Transparency	TRANSPARENCY	...	...	MODIFY1 Transpar	...	
TRIM	Modify Trim	TRIM	TR	...	MODIFY2 Trim	W,15	Chapter 9
TXTEXP	Bonus Text> Explode Text	TXTEXP	...	...	...	...	Chapter 47
U	Edit Undo	U	...	Crtl+Z	EDIT Undo	T,12	
UCS	Tools UCS >	UCS	...	...	TOOLS 2 UCS	W,7	Chapter 36
UCSFOLLOW	...	UCSFOLLOW	...	...	...	...	Chapter 36
UCSICON	View Display > UCS Icon >	UCSICON	...	...	VIEW 2 UCSicon	L,2	Chapters 11, 34
UNDO	...	UNDO	...	...	ASSIST Undo	...	Chapter 5
UNION	Modify Boolean > Union	UNION	UNI	...	MODIFY2 Union	X,15	Chapters 16, 38
UNITS and DDUNITS	Format Units...	UNITS or DDUNITS	-UN or UN	...	FORMAT Ddunits	V,4	Chapter 6
UPDATE	...	UPDATE	...	...	DIMNSION Update	...	Chapter 29
VIEW and DDVIEW	View Named Views...	VIEW or DDVIEW	-V or V	...	VIEW 1 Ddview	M,5	Chapter 11
VIEWRES	Tools Preferences... Performance	VIEWRES	...	...	...	...	Chapter 11

Command Name or Variable Name	Pull-down Menu	COMMAND (TYPE)	ALIAS (TYPE)	Short-cut	Screen (side) Menu	Tablet Menu	Chapter in This Text
VISRETAIN	*Format* *Layer...* *Details>>* *Retain changes to xref-* *dependent layers*	*VISRETAIN*	...	...	*FORMAT* *Layer* *Details>>* *Retain* *changes to...*	...	Chapter 30
VPLAYER	...	*VPLAYER*	...	...	...	...	Chapter 33
VPOINT	*View* *3D Viewpoint >*	*VPOINT*	*-VP*	...	*VIEW 1* *Vpoint*	*N,4*	Chapter 35
VPORTS	*View* *Tiled Viewports >*	*VPORTS*	...	...	*VIEW 1* *Vports*	*M,3*	Chapters 19, 35
VSLIDE	...	*VSLIDE*	...	...	...	...	Chapter 44
WBLOCK	*File* *Export...*	*WBLOCK*	*W*	...	*FILE* *Export*	*W,24*	Chapter 21
WEDGE	*Draw* *Solids >* *Wedge*	*WEDGE*	*WE*	...	*DRAW 2* *SOLIDS* *Wedge*	*N,7*	Chapter 38
WIPEOUT	*Bonus* *Draw>* *Wipeout*	*WIPEOUT*	...	...	...	...	Chapter 47
XATTACH	*Insert* *External Reference* *Attach...*	*XATTACH*	*XA*	...	*INSERT* *Xref* *Attach...*	...	Chapter 30
XBIND and *-XBIND*	*Modify* *Object >* *External Reference >*	*XBIND or* *-XBIND*	*XB or* *-XB*	...	*MODIFY1* *Xbind*	*X,19*	Chapter 30
XCLIP	*Modify* *Object>* *Clip*	*XCLIP*	*XC*	...	*MODIFY1* *Xclip*	*X,18*	
XCLIPFRAME	*Modify* *Object>* *External Reference>* *Frame*	*XCLIPFRAME*	...	...	...	...	Chapter 30
XDATA	*Bonus* *Tools>* *Xdata Attachment*	*XDATA*	...	...	...	...	Chapter 47
XDLIST	*Bonus* *Tools>* *List Entity Xdata*	*XDLIST*	...	...	...	...	Chapter 47
XLINE	*Draw* *Construction Line*	*XLINE*	*XL*	...	*DRAW 1* *Xline*	*L,10*	Chapters 15, 27
XLIST	*Bonus* *Tools>* *List Xref/Block* *Entities*	*XLIST*	...	...	...	...	Chapter 47
XLOADCTL	*Tools* *Preferences...* *Performance* *External reference file* *demand load:*	*XLOADCTL*	...	...	*TOOLS2* *Preferenc* *Performance* *Ext.* *reference...*	...	Chapter 30

Command Name or Variable Name	Pull-down Menu	COMMAND (TYPE)	ALIAS (TYPE)	Short-cut	Screen (side) Menu	Tablet Menu	Chapter in This Text
XLOADPATH	*Tools* *Preferences…* *Files* *Temporary ext. reference file location*	*XLOADPATH*	…	…	*TOOLS2* *Preferenc* *Files* *Temporary ext. refs…*	…	Chapter 30
XPLODE	*Bonus* *Modify>* *Extended Explode*	*XPLODE*	…	…	…	…	Chapter 47
XREF and *-XREF*	*Insert* *External Reference…*	*XREF or* *-XREF*	*XR or* *-XR*	…	*INSERT* *Xref*	*T,4*	Chapter 30
XREFCTL	…	*XREFCTL*	…	…	…	…	Chapter 30
ZOOM	*View* *Zoom >*	*ZOOM*	*Z*	…	*VIEW 1* *Zoom*	*K,11– M,11 or J,2 – K,5*	Chapter 11

INTRODUCTION

WHAT IS CAD?

CAD is an acronym for Computer-Aided Design or Computer-Aided Drafting. CAD allows you to accomplish design and drafting activities using a computer. A CAD software package, such as AutoCAD, enables you to create designs and generate drawings to document those designs.

Design is a broad field involving the process of making an idea into a real product or system. The design process requires repeated refinement of an idea or ideas until a solution results—a manufactured product or constructed system. Traditionally, design involves the use of sketches, drawings, renderings, 2-dimensional and 3-dimensional models, prototypes, testing, analysis, and documentation. Drafting is generally known as the production of drawings that are used to document a design for manufacturing or construction or to archive the design.

CAD is a <u>tool</u> that can be used for design and drafting activities. CAD can be used to make "rough" idea drawings, although it is more suited to creating accurate finished drawings and renderings. CAD can be used to create a 2-dimensional or 3-dimensional computer model of the product or system for further analysis and testing by other computer programs. In addition, CAD can be used to supply manufacturing equipment such as lathes, mills, laser cutters, or rapid prototyping equipment with numerical data to manufacture a product. CAD is also used to create the 2-dimensional documentation drawings for communicating and archiving the design.

The tangible result of CAD activity is usually a drawing generated by a plotter or printer but can be a rendering of a model or numerical data for use with another software package or manufacturing device. Regardless of the purpose for using CAD, the resulting drawing or model is stored in a CAD file. The file consists of numeric data in binary form usually saved to a magnetic or optical device such as a diskette, hard disk, tape, or CD.

WHY SHOULD YOU USE CAD?

Although there are other methods used for design and drafting activities, CAD offers the following advantages over other methods in many cases:

1.   Accuracy
2.   Productivity for repetitive operations
3.   Sharing the CAD file with other software programs

Accuracy

Since CAD technology is based on computers, it offers great accuracy compared to older "manual" methods of drafting and design. When you draw with a CAD system, the graphical elements, such as lines, arcs, and circles, are stored in the CAD file as numeric data. CAD systems store that numeric data with great precision. For example, AutoCAD stores values with fourteen significant digits. The value 1, for example, is stored in scientific notation as the equivalent of 1.0000000000000. This precision provides you with the ability to create designs and drawings that are 100% accurate for almost every case.

Productivity for Repetitive Operations

It may be faster to create a simple "rough" drawing, such as a sketch by hand (pencil and paper), than it would by using a CAD system. However, for larger and more complex drawings, particularly those involving similar shapes or repetitive operations, CAD methods are very efficient. Any kind of shape or operation accomplished with the CAD system can be easily duplicated since it is stored in a CAD file. In short, it may take some time to set up the first drawing and create some of the initial geometry, but any of the existing geometry or drawing setups can be easily duplicated in the current drawing or for new drawings.

Likewise, making changes to a CAD file (known as editing) is generally much faster than making changes to a traditional manual drawing. Since all the graphical elements in a CAD drawing are stored, only the affected components of the design or drawing need to be altered, and the drawing can be plotted or printed again or converted to other formats.

As CAD and the associated technology advance and software becomes more interconnected, more productive developments are available. For example, it is possible to make a change to a 3-dimensional model that automatically causes a related change in the linked 2-dimensional engineering drawing. One of the main advantages of these technological advances is productivity.

Sharing the CAD File with Other Software Programs

Of course, CAD is not the only form of industrial activity that is making technological advances. Most industries use computer software to increase capability and productivity. Since software is written using digital information and may be written for the same or similar computer operating systems, it is possible and desirable to make software programs with the ability to share data or even interconnect, possibly appearing simultaneously on one screen.

For example, word processing programs can generate text that can be imported into a drawing file, or a drawing can be created and imported into a text file as an illustration. (This book is a result of that capability.) A drawing created with a CAD system such as AutoCAD can be exported to a finite element analysis program that can read the computer model and compute and analyze stresses. CAD files can be dynamically "linked" to spreadsheets or databases in such a way that changing a value in a spreadsheet or text in a database can automatically make the related change in the drawing, or vice versa.

Another advance in CAD technology is the automatic creation and interconnectivity of a 2-dimensional drawing and a 3-dimensional model in one CAD file. With this tool, you can design a 3-dimensional model and have the 2-dimensional drawings automatically generated. The resulting set has bi-directional associativity; that is, a change in either the 2-dimensional drawings or the 3-dimensional model is automatically updated in the other.

CAD, however, may not be the best tool for every design related activity. For example, CAD may help develop ideas but probably won't replace the idea sketch, at least not with present technology. A 3-dimensional CAD model can save much time and expense for some analysis and testing but cannot replace the "feel" of an actual model, at least not until virtual reality technology is developed and refined.

With everything considered, CAD offers many opportunities for increased accuracy, productivity, and interconnectivity. Considering the speed at which this technology is advancing, many more opportunities are rapidly obtainable. However, we need to start with the basics. Beginning by learning to create an AutoCAD drawing is a good start.

WHY USE AutoCAD?

CAD systems are available for a number of computer platforms: laptops, personal computers (PCs), workstations, and mainframes. AutoCAD, offered to the public in late 1982, was one of the first PC-based CAD software products. Since that time, it has grown to be the world leader in market share for all CAD products. Autodesk, the manufacturer of AutoCAD, is the world's leading supplier of PC design software and multimedia tools. At the time of this writing, Autodesk is the fifth largest software producer in the world and has three million customers in more than 150 countries.

Learning AutoCAD offers a number of advantages to you. Since AutoCAD is the most widely used CAD software, using it gives you the highest probability of being able to share CAD files and related data and information with others.

As a student, learning AutoCAD, as opposed to learning another CAD software product, gives you a higher probability of using your skills in industry. Likewise, there are more employers who use AutoCAD than any other single CAD system. In addition, learning AutoCAD as a first CAD system gives you a good foundation for learning other CAD packages because many concepts and commands introduced by AutoCAD are utilized by other systems. In some cases, AutoCAD features become industry standards. The .DXF file format, for example, was introduced by Autodesk and has become an industry standard for CAD file conversion between systems.

As a professional, using AutoCAD gives you the highest possibility that you can share CAD files and related data with your colleagues, vendors, and clients. Compatibility of hardware and software is an important issue in industry. Maintaining compatible hardware and software allows you the highest probability for sharing data and information with others as well as offering you flexibility in experimenting with and utilizing the latest technological advancements. AutoCAD provides you with the greatest compatibility in the CAD domain.

This introduction is not intended as a selling point but to remind you of the importance and potential of the task you are about to undertake. If you are a professional or a student, you have most likely already made up your mind that you want to learn to use AutoCAD as a design or drafting tool. If you have made up your mind, then you can accomplish anything. Let's begin.

GETTING STARTED

Chapter Objectives

After completing this chapter you should:

1. understand how the X, Y, Z coordinate system is used to define the location of drawing elements in digital format in a CAD drawing file;

2. understand why you should create drawings full size in the actual units with CAD;

3. be able to start AutoCAD to begin drawing;

4. recognize the areas of the AutoCAD Drawing Editor and know the function of each;

5. be able to use the five methods of entering commands;

6. be able to turn on and off the *SNAP*, *GRID*, and *ORTHO* drawing aids;

7. know how to customize the AutoCAD for Windows screen to your preferences.

CONCEPTS

Coordinate Systems

Any location in a drawing, such as the endpoint of a line, can be described in X, Y, and Z coordinate values (Cartesian coordinates). If a line is drawn on a sheet of paper, for example, its endpoints can be charted by giving the distance over and up from the lower-left corner of the sheet (Fig. 1-1).

These distances, or values, can be expressed as X and Y coordinates; X is the horizontal distance from the lower left corner (origin) and Y is the vertical distance from that origin. In a three-dimensional coordinate system, the third dimension, Z, is measured from the origin in a direction perpendicular to the plane defined by X and Y.

Two-dimensional (2D) and three-dimensional (3D) CAD systems use coordinate values to define the location of drawing elements such as lines and circles (called objects in AutoCAD).

In a 2D drawing, a line is defined by the X and Y coordinate values for its two endpoints (Fig 1-2).

In a 3D drawing, a line can be created and defined by specifying X, Y, and Z coordinate values (Fig. 1-3). Coordinate values are always expressed by the X value first separated by a comma, then Y, then Z.

Figure 1-1

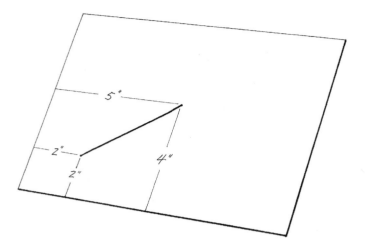

Figure 1-2

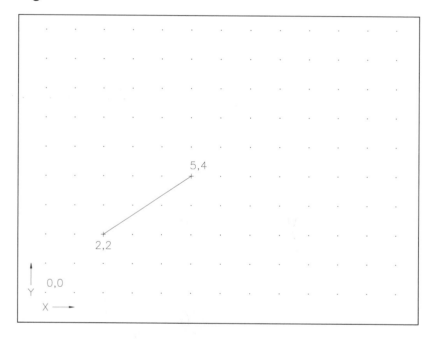

Figure 1-3

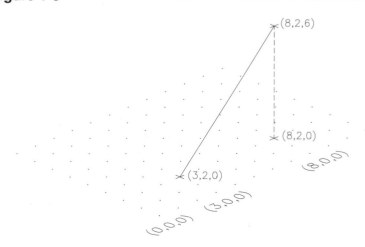

The CAD Database

A CAD (Computer-Aided Design) file, which is the electronically stored version of the drawing, keeps data in binary digital form. These digits describe coordinate values for all of the endpoints, center points, radii, vertices, etc. for all the objects composing the drawing, along with another code that describes the kinds of objects (line, circle, arc, ellipse, etc.). Figure 1-4 shows part of an AutoCAD DXF (Drawing Interchange Format) file giving numeric data defining lines and other objects. Knowing that a CAD system stores drawings by keeping coordinate data helps you understand the input that is required to create objects and how to translate the meaning of prompts on the screen.

Figure 1-4

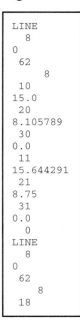

```
LINE
    8
 0
    62
         8
    10
 15.0
    20
 8.105789
    30
 0.0
    11
 15.644291
    21
 8.75
    31
 0.0
     0
LINE
    8
 0
    62
         8
    18
```

Angles in AutoCAD

Figure 1-5

Angles in AutoCAD are measured in a <u>counter-clockwise direction</u>. Angle 0 is positioned in a positive X direction, that is, horizontally from left to right. Therefore, 90 degrees is in a positive Y direction, or straight up; 180 degrees is in a negative X direction, or to the left; and 270 degrees is in a negative Y direction, or straight down (Fig. 1-5).

The position and direction of measuring angles in AutoCAD can be changed; however, the defaults listed here are used in most cases.

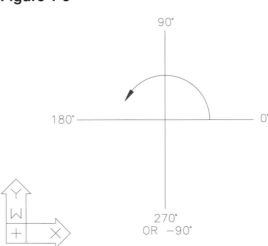

Draw True Size

Figure 1-6

When creating a drawing with pencil and paper tools, you must first determine a scale to use so the drawing will be proportional to the actual object and will fit on the sheet (Fig. 1-6). However, when creating a drawing on a CAD system, there is no fixed size drawing area. The number of drawing units that appear on the screen is variable and is assigned to fit the application.

The CAD drawing is not scaled until it is physically transferred to a fixed size sheet of paper by plotter or printer.

The rule for creating CAD drawings is that the drawing should be created <u>true size</u> using real-world units. The user specifies what units are to be used (architectural, engineering, etc.) and then specifies what size drawing area is needed (in X and Y values) to draw the necessary geometry.

Figure 1-7

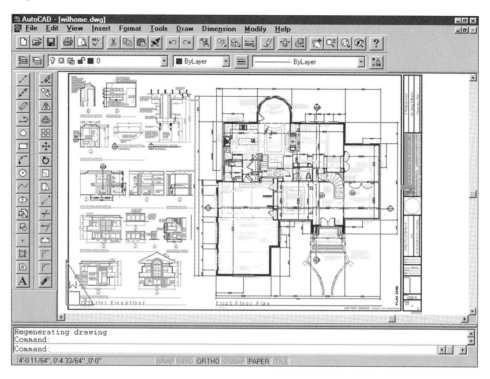

Whatever the specified size of the drawing area, it can be displayed on the screen in its entirety (Fig. 1-7) or as only a portion of the drawing area (Fig. 1-8).

Figure 1-8

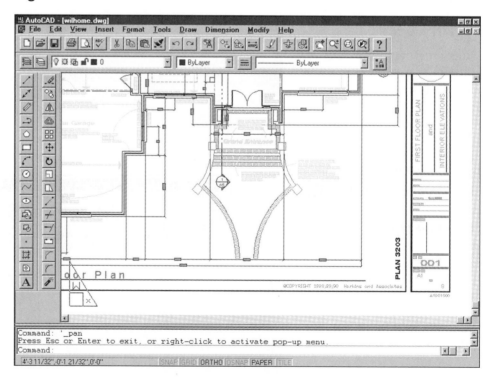

Plot to Scale

As long as a drawing exists as a CAD file or is visible on the screen, it is considered a virtual, full-sized object. Only when the CAD drawing is transferred to paper by a plotter or printer is it converted (usually reduced) to a size that will fit on a sheet. A CAD drawing can be automatically scaled to fit on the sheet regardless of sheet size; however, this action results in a plotted drawing that is not to an accepted scale (not to a regular proportion of the real object). Usually it is desirable to plot a drawing so that the resulting drawing is a proportion of the actual object size. The scale to enter as the plot scale (Fig. 1-9) is simply the proportion of the <u>plotted drawing</u> size to the <u>actual object</u>.

Figure 1-9

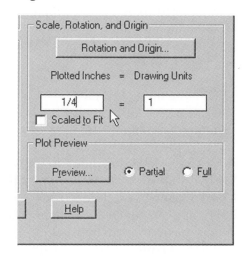

STARTING AutoCAD

Assuming that AutoCAD has been installed and configured properly for your system, you are ready to begin using AutoCAD.

To start AutoCAD for Windows 95 or Windows NT, locate the "AutoCAD R14" shortcut icon on the desktop (Fig. 1-10). Double-clicking on the icon (point the arrow and press the left mouse button quickly two times) opens AutoCAD Release 14.

If you cannot locate a shortcut icon on the desktop, press the "Start" button, highlight "Programs," and search for "AutoCAD R14" in the menu. From the list that appears, select "AutoCAD R14."

To exit AutoCAD, either select *Exit* from the *Files* pull-down menu or click on the "X" in the very upper right corner of the AutoCAD screen.

Figure 1-10

AutoCAD R14

THE AutoCAD DRAWING EDITOR

The *Start Up* Dialog Box

When you start AutoCAD Release 14 for the first time, the *Start Up* dialog box appears (Fig. 1-11). This dialog box includes several tools to help you open an existing drawing or set up a new drawing, including providing access to drawing templates and a setup "Wizard." If you want to <u>start AutoCAD with the default settings</u>, you can select either of the following options:

Figure 1-11

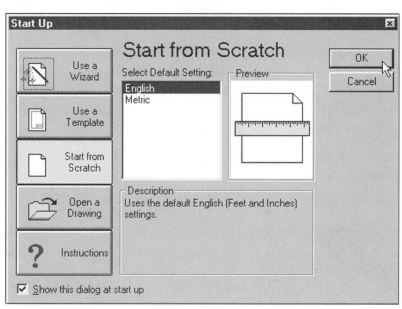

1. Select the *Start from Scratch* button on the left. Text appears in the center portion of the box displaying a choice of *English* or *Metric*. Select *English,* then press *OK*.

2. Select the *Cancel* button. Because AutoCAD's default settings are already set for use with English units, choosing *Cancel* (when starting AutoCAD) accomplishes the same action as selecting an *English* setup.

NOTE: The *Start Up* dialog box and *Create New Drawing* dialog box tools are explained fully in Chapter 6 (Basic Drawing Setup) and Chapter 13 (Advanced Drawing Setup). For the examples and exercises in Chapters 1-5, use the first method above to begin a new drawing.

After specifying your choice in the *Start Up* dialog box or the *Create New Drawing* dialog box, the Drawing Editor appears on the screen and allows you to immediately begin drawing. Figure 1-12 displays the drawing editor in AutoCAD R14 for Windows 95 and Window NT 4.0.

Figure 1-12

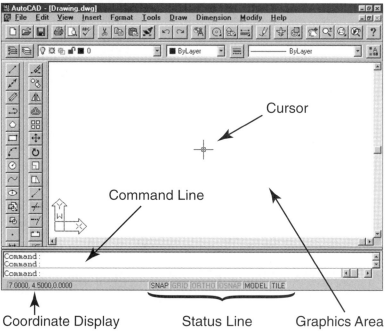

Graphics Area

The large central area of the screen is the Graphics area. It displays the lines, circles, and other objects you draw that will make up the drawing. The cursor is the intersection of the crosshairs (vertical and horizontal lines that follow the mouse or puck movements). The default size of the graphics area is 12 units (X or horizontal) by 9 units (Y or vertical). This usable drawing area (12 x 9) is called the drawing *Limits* and can be changed to any size to fit the application. As you move the cursor, you will notice the numbers in the Coordinate Display change (Fig. 1-12, bottom left).

Command Line

The Command line consists of the three text lines at the bottom of the screen (by default) and is the most important area other than the drawing itself (see Fig. 1-12). Any command that is entered or any prompt that AutoCAD issues appears here. The Command line is always visible and gives the current state of drawing activity. You should develop the habit of glancing at the Command line while you work in AutoCAD. In AutoCAD R14, the command line can be set to display any number of lines and/or moved to another location (see Customizing the AutoCAD Screen).

Pressing the F2 key opens a text window displaying the command history. This window is like an expanded Command line because it displays many more than just the 3 text lines that normally appear at the bottom of the screen. See AutoCAD Text Window, later this chapter.

R14

Toolbars

AutoCAD Release 14 provides a variety of <u>toolbars</u> (Fig. 1-13). Each toolbar contains a number of icon buttons (tools) that can be PICKed to invoke commands for drawing or editing objects (lines, arcs, circles, etc.) or for managing files and other functions. The <u>Standard toolbar</u> is the row of icons nearest the top of the screen. The Standard toolbar contains many standard icons used in other Windows applications (like *New, Open, Save, Print, Cut, Paste,* etc.) and other icons for AutoCAD-specific functions (like *Zoom, Pan,* and *Redraw*). The <u>Object Properties toolbar</u>, located beneath the standard toolbar, is used for managing properties of objects, such as *Layers* and *Linetypes*. The <u>Draw and Modify toolbars</u> also appear (by default)

Figure 1-13

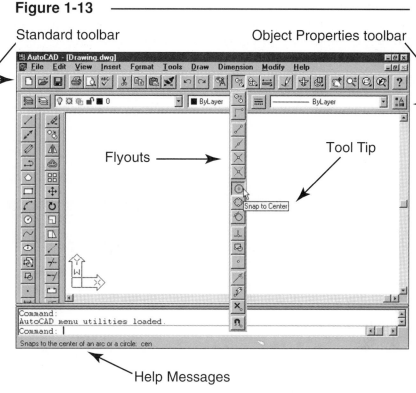

Standard toolbar   Object Properties toolbar

Flyouts

Tool Tip

Help Messages

when you first use AutoCAD. As shown in Figure 1-13, the Draw and Modify toolbars (side) and the Standard and Object Properties toolbars (top) are <u>docked</u>. Many other toolbars are available and can be made to float, resize, or dock (see Customizing the AutoCAD Screen).

If you place the pointer on an any icon and wait momentarily, a <u>Tool Tip</u> and a <u>Help message</u> appear. Tool Tips pop out by the pointer and give the command name (Fig. 1-13). The Help message appears at the bottom of the screen, giving a short description of the function. <u>Flyouts</u> are groups of related icons that pop out in a row or column when one of the group is selected. PICKing any icon that has a small black triangle in its lower-right corner causes the related icons to fly out.

Pull-down Menus

The pull-down menu bar is at the top of the screen just under the title bar (Fig. 1-14). Selecting any of the words in the menu bar activates, or pulls down, the respective menu. Selecting a word appearing with an arrow activates a cascading menu with other options. Selecting a word with ellipsis (. . .) activates a dialog box (see Dialog Boxes). Words in the pull-down menus are not necessarily the same as the formal command names used when typing commands. Menus can be canceled by pressing Escape or PICKing in the graphics area. The pull-down menus <u>do not contain</u> all of the AutoCAD commands and variables but contain the most commonly used ones.

Figure 1-14

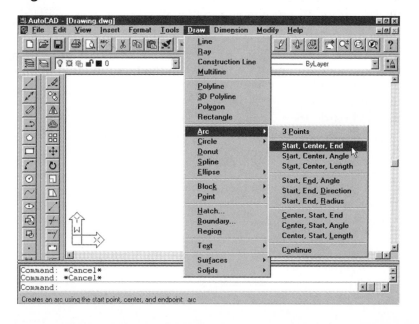

NOTE: Because the words that appear in the pull-down menus are <u>not always the same</u> as the formal commands you would type to invoke a command, AutoCAD can be confusing to learn. The Help message that appears just below the Command line (when a menu is pulled down or the pointer rests on an icon button) can be instrumental in avoiding this confusion. The Help message gives a description of the command followed by colon (:), then the <u>formal command name</u> (see Figures 1-13 and 1-14).

Screen (Side) Menu

The screen menu does not appear by default in Release 14, but can be made to appear by selecting *Preferences…* from the *Tools* pull-down menu. This selection causes the *Preferences* dialog box to appear (Fig. 1-15). Select the *Display* tab and check the *Display AutoCAD screen menu in drawing window* option. Keep in mind that this menu <u>is not needed</u> if you prefer to use the icon buttons, pull-down menus, or keyboard to enter commands.

Figure 1-15 ————————————

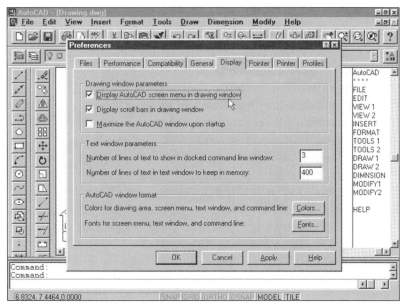

Almost all of the AutoCAD commands can be accessed through the menu system located at the right side of the screen (see Fig. 1-15). The screen menu has a tree structure; that is, other menus and commands are accessed by branching out from the root or top-level menu (Fig. 1-16). Commands (in upper and lower case) or menus (capital letters only) are selected by moving the cursor to the desired position until the word is highlighted and then pressing the PICK button. The root menu (top level) is accessed from any level of the structure by selecting the word "AutoCAD" at the top of any menu.

The commands in these menus are generally the formal command names that can also be typed at the keyboard. (Some commands names are truncated in the screen menu because only eight characters can be displayed.)

If commands are invoked by typing, pull-downs, or toolbars, the screen menu automatically changes to the current command.

Figure 1-16 ————————————

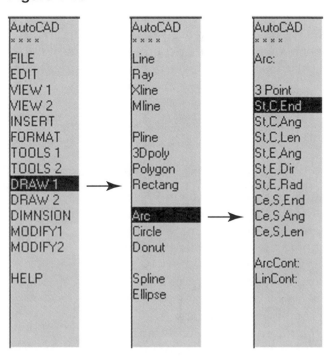

Dialog Boxes

Dialog boxes provide an interface for controlling complex commands or a group of related commands. Depending on the command, the dialog boxes allow you to select among multiple options and sometimes give a preview of the effect of selections. The *Layer & Linetype Properties* dialog box (Fig. 1-17) gives complete control of layer colors, linetypes, and visibility.

Dialog boxes can be invoked by typing a command, selecting an icon button, or PICKing from the menus. For example, typing *Layer*, PICKing the *Layers* button, or selecting *Layer* from the *Format* pull-down menu causes the *Layer & Linetype Properties* dialog box to appear. In the pull-down menu, all commands that invoke a dialog box end with ellipsis points (…). If you type, sometimes you may have to type a command that begins with "DD" to invoke a dialog box; for example, typing *Ddchprop* invokes the *Change Properties* dialog box.

Figure 1-17

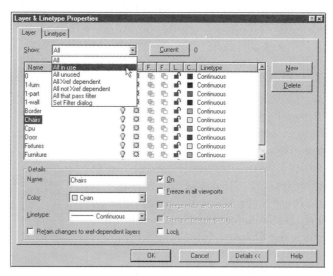

The basic element or smallest component of a dialog box is called a <u>tile</u>. Several types of tiles and the resulting actions of tile selection are listed below.

Button	Resembles a push button and triggers some type of action
Edit box	Allows typing or editing of a single line of text
Image tile	A button that displays a graphical image
List box	A list of text strings from which one or more can be selected
Drop-down list	A text string that drops down to display a list of selections
Radio button	A group of buttons, only one of which can be turned on at a time
Checkbox	A checkbox for turning a feature on or off (displays a check mark when on)

You can change dialog box border colors (for all applications) with the Windows Control Panel. The AutoCAD screen colors can also be customized (see Customizing the AutoCAD Screen, this chapter). See File Dialog Box Functions in Chapter 2 for more information on dialog boxes that are used to access files.

New dialog boxes can be created and customized. Programming the configuration of tiles and resulting action of tile selection requires use of Dialog Control Language (DCL) to configure the tiles and AutoLISP programs to control the action of tile selection.

Status Line

The Status line is a set of informative words or symbols that gives the status of the drawing aids. The Status line appears at the very bottom of the screen (Fig. 1-12). The following drawing aids can be toggled on or off by double-clicking (quickly pressing the left mouse button twice) on the desired word or by using Function keys or Ctrl key sequences. The following drawing aids are explained here or in following chapters:

SNAP, GRID, ORTHO, OSNAP, TILE

Coordinate Display (*Coords*)

The Coordinate Display is located in the lower-left corner of the AutoCAD screen (Fig. 1-18). The Coordinate Display (*Coords*) displays the current position of the cursor in one of two possible formats explained below. This display can be very helpful when you draw because it can give the X, Y, and Z coordinate position of the cursor or give the cursor's distance and angle from the last point established. The format of *Coords* is controlled by toggling the F6 key, pressing Ctrl+D, or double-clicking on the *Coords* display (numbers) at the bottom left of the screen. *Coords* can also be toggled off.

Figure 1-18

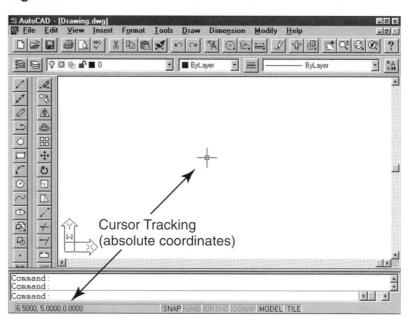

Cursor tracking
When *Coords* is in this position, the values display the current location of the cursor in absolute (X, Y, and Z) coordinates (Fig. 1-18).

Relative polar display
This display is only possible if a draw or edit command is in use. The values give the distance and angle of the "rubberband" line from the last point established (Fig. 1-19).

Figure 1-19

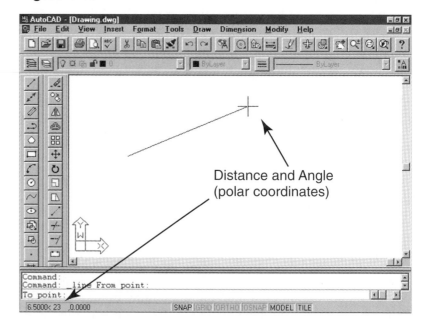

Digitizing Tablet Menu

If you have a digitizing tablet, the AutoCAD commands are available by making the desired selection from the AutoCAD digitizing tablet menu (Fig. 1-20). The open area located slightly to the right of center is called the Screen Pointing area. Locating the digitizing puck there makes the cursor appear on the screen. Locating the puck at any other location allows you to select a command. The icons on the tablet menu are identical to the toolbar icons. Similar to the format of the toolbars, commands are located in groups such as Draw, Edit, View, Dimension, etc. The tablet menu has the columns numbered along the top and the rows lettered along the left side. The location of each command by column and row is given in the "command tables" in his book. For example, the *Line* command can be found at *10,J*.

Figure 1-20 ───

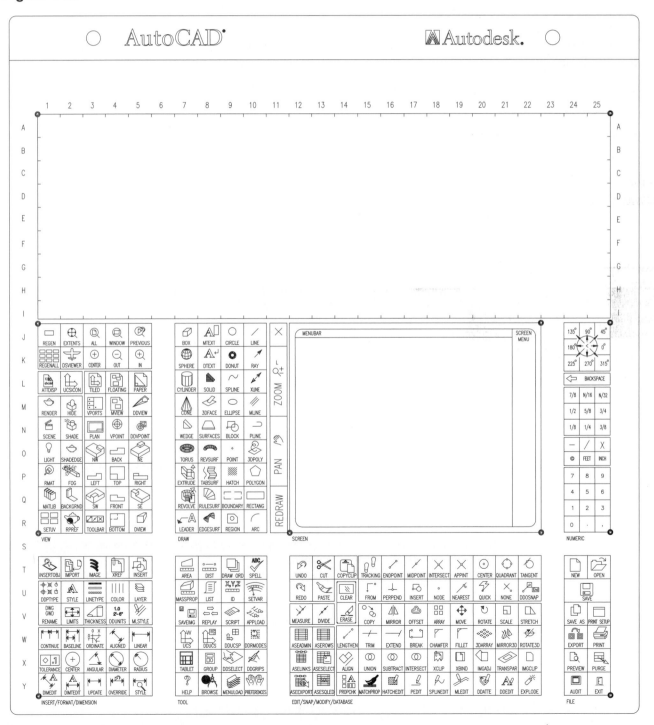

COMMAND ENTRY

Methods for Entering Commands

There are five possible methods for entering commands in AutoCAD depending on your *Preferences* setting (for the screen menu) and availability of a digitizing tablet. Generally, <u>any one</u> of the five methods can be used to invoke a particular command.

1. **Toolbars** — Select the command or dialog box by PICKing an icon (tool) from a toolbar.
2. **Pull-down menu** — Select the command or dialog box from a pull-down menu.
3. **Screen (side) menu** — Select the command or dialog box from the screen (side) menu.
4. **Keyboard** — Type the command name, command alias, or accelerator (Ctrl) keys at the keyboard.*
5. **Tablet menu** — Select the command from the digitizing tablet menu (if available).

*A command alias is a one- or two-letter shortcut. Accelerator keys use the Ctrl key plus another key and are used for utility operations. The command aliases and accelerator keys are given in the command tables (see below).

All five methods of entering commands accomplish the same goals; however, one method may offer a slightly different option or advantage to another depending on the command used. A few considerations are listed below.

- Commands invoked by any method automatically change the screen menu to display the same command.
- Typing commands requires that no other commands are currently in use; therefore, the Esc (Escape) key should be used to cancel active commands before typing.
- Pull-downs are easily visible but do <u>not</u> contain all of the commands.
- The screen menus contain almost <u>all</u> AutoCAD commands; however, locating a particular command in the menu structure may take a few seconds.

All menus, including the digitizing tablet, can be customized by editing the ACAD.MNU file, and command aliases can be changed or added in the ACAD.PGP file so that command entry can be designed to your preference (see Chapter 45, Basic Customization, and Chapter 46, Menu Customization).

Using the "Command Tables" in This Book to Locate a Particular Command

Command tables, like the one below, are used throughout this book to show the possible methods for entering a particular command. The table shows the icon used in the toolbars and digitizing tablet, gives the selections to make for the pull-down and screen (side) menus, gives the correct spelling for entering commands and command aliases at the keyboard, and gives the command location (column, row) on the digitizing tablet menu. This example uses the *Copy* command.

COPY

Pull-down Menu	COMMAND (TYPE)	ALIAS (TYPE)	Short-cut	Screen (side) Menu	Tablet Menu
Modify *Copy*	*COPY*	*CO*	...	*MODIFY1* *Copy*	*V,15*

Mouse and Digitizing Puck Buttons

Depending on the type of mouse or digitizing puck used for cursor control, a different number of buttons is available. In any case, the buttons perform the following tasks:

#1 (left mouse)	**PICK**	Used to select commands or point to locations on screen.
#2 (right mouse)	**Enter**	Generally, performs the same action as the Enter or Return key on the keyboard. When some dialog boxes are visible, right-clicking produces a menu of choices affecting the active dialog box.
#3	*OSNAP*	Activates the cursor *OSNAP* menu.
#4	**Cancel**	Cancels a command.

Function Keys

Several function keys are usable with AutoCAD. They offer a quick method of turning on or off (toggling) drawing aids.

F1	*Help*	Opens a help window providing written explanations on commands and variables.
F2	*Flipscreen*	Activates a text window showing the previous command line activity (command history).
F3	*Osnap Toggle*	If Running Osnaps are set, toggling this tile temporarily turns the Running Osnaps off so that a point can be picked without using Osnaps. If no Running Osnaps are set, F3 produces the *Osnap Settings* dialog box (discussed in Chapter 7).
F4	*Tablet*	Turns the *TABMODE* variable on or off. If *TABMODE* is on, the digitizing tablet can be used to digitize an existing paper drawing into AutoCAD.
F5	*Isoplane*	When using an *Isometric* style *SNAP* and *GRID* setting, toggles the cursor (with *ORTHO* on) to draw on one of three isometric planes.
F6	*Coords*	Toggles the Coordinate Display between cursor tracking mode and off. If used transparently (during a command in operation), displays a polar coordinate format.
F7	*GRID*	Turns the *GRID* on or off (see Drawing Aids).
F8	*ORTHO*	Turns *ORTHO* on or off (see Drawing Aids).
F9	*SNAP*	Turns *SNAP* on or off (see Drawing Aids).
F10	*Status*	Turns the Status line on or off.

Control Key Sequences (Accelerator Keys)

Several Control key sequences (holding down the Ctrl key and pressing another key simultaneously) invoke regular AutoCAD commands or produce special functions. The first six key sequences (Drawing Aids) have the same duties as Function keys F3 through F9.

Drawing Aids

Ctrl+F	(F3)	*Osnap Toggle*	If Running Osnaps are set, pressing Ctrl+F temporarily turns the Running Osnaps off so that a point can be PICKed without using Osnaps. If there are no Running Object Snaps set, Ctrl+F produces the *Osnap Settings* dialog box. This dialog box is used to turn on and off Running Object Snaps (discussed in Chapter 7).
Ctrl+T	(F4)	*Tablet*	Turns the *TABMODE* variable on or off. If *TABMODE* is on, the digitizing tablet can be used to digitize an existing paper drawing into AutoCAD.
Ctrl+E	(F5)	*Isoplane*	When using an *Isometric* style *SNAP* and *GRID* setting, toggles the cursor (with *ORTHO* on) to draw on one of three isometric planes.
Ctrl+D	(F6)	*Coords*	Toggles the coordinate display between cursor tracking mode and off. If used during a command operation, can be toggled to a polar coordinate format.

R14

Ctrl+G	(F7)	*GRID*	Turns the *GRID* on or off (see Drawing Aids).
Ctrl+L	(F8)	*ORTHO*	Turns *ORTHO* on or off (see Drawing Aids).
Crtl+B	(F9)	*SNAP*	Turns *SNAP* on or off (see Drawing Aids).

Windows Copy, Cut, Paste (see Chapter 31)

Ctrl+C	*Copyclip*	Copies the highlighted objects to the Windows clipboard.
Ctrl+X	*Cutclip*	Cuts the highlighted objects from the drawing and copies them to the Windows clipboard.
Ctrl+V	*Pasteclip*	Pastes the clipboard contents into the AutoCAD drawing as a Block Reference.

File Operations (see Chapter 2)

Ctrl+O	*Open*	Invokes the *Open* command to open an existing drawing.
Ctrl+N	*New*	Invokes the *New* command to start a new drawing.
Ctrl+S	*Qsave*	Performs a quick save or produces the *SaveAs* dialog box if the file is not yet named.

Other Control Key Sequences

Ctrl+Z	*Undo*	Undoes the last command (see Chapter 5).
Ctrl+Y	*Redo*	Invokes the *Redo* command (see Chapter 5).
Ctrl+P	*Print/Plot*	Produces the *Print/Plot Configuration* dialog box for creating and controlling prints and plots (see Chapter 14).
Ctrl+A	*Group*	Toggles selectable *Groups* on or off (see Chapter 20).
Ctrl+J	*Enter*	Has the same function as Enter.
Ctrl+K	*PICKADD*	Changes the setting (0 or 1) of the *PICKADD* variable (see Chapter 20).

Special Key Functions

Esc	The Escape key cancels a command, menu, or dialog box or interrupts processing of plotting or hatching.
Space bar	In AutoCAD, the space bar performs the same action as the Enter key or #2 button. Only when you are entering text into a drawing does the space bar create a space.
Enter	If Enter, Spacebar, or #2 button is pressed when no command is in use (the open Command: prompt is visible), the last command used is invoked again.
Up Arrow	The up arrow key can be used to recall previous command line entries, similar to the DOSKEY feature. Pressing the up arrow places the previous command at the Command: prompt, but does not execute it. You can use the up arrow repeatedly to find a previously used command, then press Enter to execute it.
Down Arrow	After the up arrow is used to recall previously used commands in reverse order, the down arrow cycles back through those commands in forward order. The up arrow and down arrow keys are used together to locate particular command line entries.

Drawing Aids

This section gives a brief introduction to AutoCAD's Drawing Aids. For a full explanation of the related commands and options, see Chapter 6.

SNAP (**F9** or **Ctrl+B**)

SNAP is a function that forces the cursor to "snap" to a regular interval (.5 units by default), which aids in drawing geometry accurate to equal interval lengths. The *Snap* command or *Drawing Aids* dialog box allows you to specify any value for the *SNAP* interval. In Figure 1-21, the *SNAP* is set to .125 (note the values in the coordinate display).

GRID (**F7** or **Ctrl+G**)

A drawing aid called *GRID* can be used to give a visual reference of units of length. The *GRID* default value is .5 units. The *Grid* command or *Drawing Aids* dialog box allows you to change the interval to any value. The *GRID* is not part of the geometry and is not plotted. Figure 1-21 displays a *GRID* of 1.0. *SNAP* and *GRID* are independent functions—they can be turned on or off independently. However, you can force the *GRID* to have the same interval as *SNAP* by entering a *GRID* value of 0 or you can use a proportion of *SNAP* by entering a *GRID* value followed by an "X".

Figure 1-21

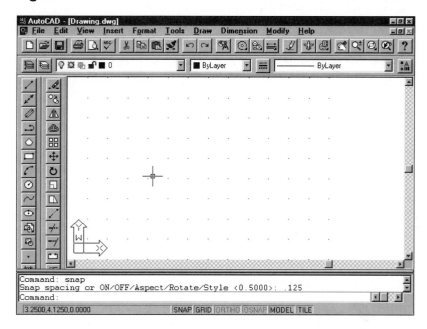

ORTHO (**F8** or **Ctrl+L**)

If *ORTHO* is on, lines are forced to an orthogonal alignment (horizontal or vertical) when drawing (Fig. 1-22). *ORTHO* is often helpful since so many drawings are composed mainly of horizontal and vertical lines. *ORTHO* can only be turned on or off.

Figure 1-22

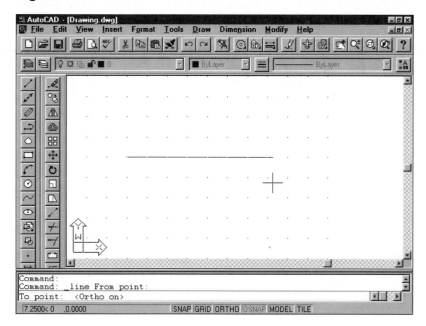

AutoCAD Text Window (F2)

Pressing the F2 key activates the *AutoCAD Text Window,* sometimes called the Command History. Here you can see the text activity that occurred at the command line—kind of an "expanded" command line. Press F2 again to close the text window.

The *Edit* pull-down menu in this text window provides several options. If you highlight text in the window (Fig. 1-23), you can then *Paste to Cmdline* (command line), *Copy* it to another program such as a word processor, *Copy History* (entire command history) to another program, or *Paste* text into the window. The *Preferences* option invokes the *Preferences* dialog box (discussed later).

Figure 1-23

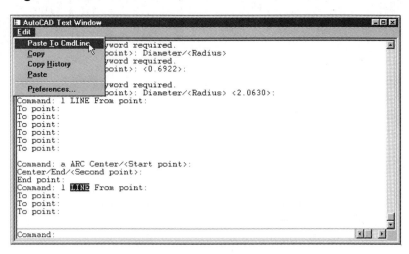

COMMAND ENTRY METHODS PRACTICE

Start AutoCAD. If the *Start Up* dialog box or *Create New Drawing* dialog box appears, select *Start from Scratch,* choose *English* as the default setting, and click the *OK* button. Invoke the *Line* command using each of the command entry methods as follows.

1.   Type the command

STEPS	COMMAND PROMPT	PERFORM ACTION	COMMENTS
1.		press **Escape** if another command is in use	
2.	Command:	type *Line* and press **Enter**	
3.	LINE From point:	**PICK** any point	a "rubberband" line appears
4.	To point:	**PICK** any point	another "rubberband" line appears
5.	To point:	press **Enter**	to complete command

2.   Type the command alias

STEPS	COMMAND PROMPT	PERFORM ACTION	COMMENTS
1.		press **Escape** if another command is in use	
2.	Command:	type *L* and press **Enter**	
3.	LINE From point:	**PICK** any point	a "rubberband" line appears
4.	To point:	**PICK** any point	another "rubberband" line appears
5.	To point:	press **Enter**	to complete command

3. **Pull-down menu**

STEPS	COMMAND PROMPT	PERFORM ACTION	COMMENTS
1.	Command:	select the *Draw* menu from the menu bar on top	
2.	Command:	select *Line*	menu disappears
3.	LINE From point:	**PICK** any point	a "rubberband" line appears
4.	To point:	**PICK** any point	another "rubberband" line appears
5.	To point:	press **Enter**	to complete command

4. **Screen menu** (if activated on your setup)

STEPS	COMMAND PROMPT	PERFORM ACTION	COMMENTS
1.	Command:	select **AutoCAD** from screen menu on the side	only if menu is not at root level
2.	Command:	select *DRAW1* from root screen menu	menu changes to *DRAW 1*
3.	Command:	select *Line*	
4.	LINE From point:	**PICK** any point	a "rubberband" line appears
5.	To point:	**PICK** any point	another "rubberband" line appears
6.	To point:	press **Enter**	to complete command

5. **Toolbars**

STEPS	COMMAND PROMPT	PERFORM ACTION	COMMENTS
1.	Command:	select the *Line* icon from the Draw toolbar (on the left side of the screen)	the *Line* tool should be located at the top of the toolbar
2.	LINE From point:	**PICK** any point	a "rubberband" line appears
3.	To point:	**PICK** any point	another "rubberband" line appears
4.	To point:	press **Enter**	to complete command

6. **Digitizing tablet menu** (if available)

STEPS	COMMAND PROMPT	PERFORM ACTION	COMMENTS
1.	Command:	select *LINE* on the digitizing tablet menu	located at **10,J**
2.	LINE From point:	**PICK** any point	a "rubberband" line appears
3.	To point:	**PICK** any point	another "rubberband" line appears
4.	To point:	press **Enter**	to complete command

When you are finished practicing, use the *Files* pull-down menu and select *Exit* to exit AutoCAD. You do not have to "Save Changes."

CUSTOMIZING THE AutoCAD SCREEN

Toolbars

Many toolbars are available, each with a group of related commands for specialized functions. For example, when you are ready to dimension a drawing, you can activate the Dimension toolbar for efficiency. Selecting the *View* pull-down menu, then *Toolbars...* displays the list of possible selections in the *Toolbars* dialog box (Fig. 1-24). Making a selection in the dialog box (PICKing in the checkbox) activates that toolbar. You may instead type in the *Toolbar* command or right-click on any tool (icon button) to invoke the *Toolbars* dialog box. Toolbars can be removed from the screen by clicking once on the "X" symbol in the upper-right of a floating toolbar.

Figure 1-24

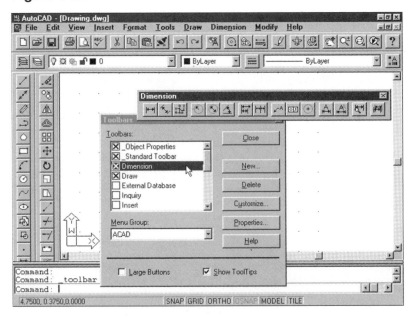

By default, the Object Properties and Standard toolbars (top) and the Draw and Modify toolbars (side) are docked, whereas toolbars that are newly activated are floating (see Fig. 1-24). A floating toolbar can be easily moved to any location on the screen if it obstructs an important area of a drawing. Placing the pointer in the title background allows you to move the toolbar by holding down the left button and dragging it to a new location (Fig. 1-25). Floating toolbars can also be resized by placing the pointer on the narrow border until a two-way arrow appears, then dragging left, right, up, or down (Fig 1-26).

Figure 1-25

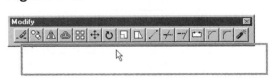

Figure 1-26

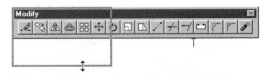

R14

R13

A floating toolbar can be docked against any border (right, left, top, bottom) by dragging it to the desired location (Fig. 1-27). Several toolbars can be stacked in a docked position. By the same method, docked toolbars can be dragged back onto the graphics area. The Object Properties and Standard toolbars can be moved onto the graphics area or docked on another border, although it is wise to keep these toolbars in their standard position. The Object Properties toolbar only displays its full options when located in a horizontal position.

Figure 1-27

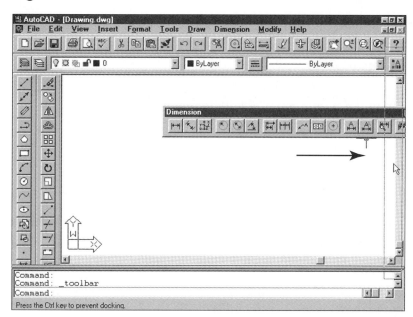

Holding down the **Ctrl** key while dragging a toolbar near a border of the drawing window prevents the toolbar from docking. This feature enables you to float a toolbar anywhere on the screen. New toolbars can be created and existing toolbars can be customized. Typically, you would create toolbars to include groups of related commands that you use most frequently or need for special activities. See Customizing Toolbars in Chapter 45.

Command Line

The Command line, normally located near the bottom of the screen, can be resized to display more or fewer lines of text. Moving the pointer to the border between the graphics screen and the Command line until two-way arrows appear allows you to slide the border up or down (Fig. 1-28). A <u>minimum of two lines</u> of text is recommended. The Command line text window can also be moved to any location on the screen by pointing to the border, holding down the left button, and dragging to the new position.

Figure 1-28

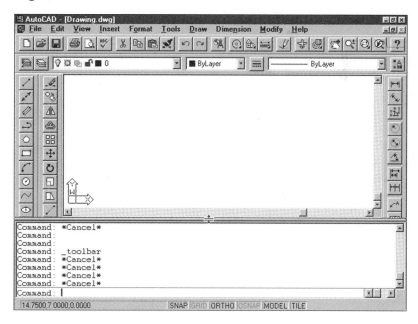

Preferences

Fonts, colors, and other features of the AutoCAD drawing editor can be customized to your liking by using the *Preferences* command and selecting the *Display* tab. Typing *Preferences* or selecting *Preferences…* from the *Tools* pull-down menu activates the dialog box shown in Figure 1-29. The screen menu can also be activated through this dialog box (first checkbox in the dialog box).

Figure 1-29

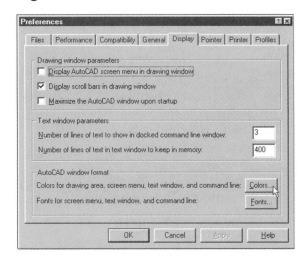

Selecting the *Color…* tile provides a dialog box for customizing the screen colors (Fig. 1-30).

The cursor size (crosshair length) can also be adjusted to your preference using the *Preferences* dialog box. Select the *Pointer* tab and enter the desired percentage value in the *Cursor Size* edit box. The default cursor size (cursor length) is 5%. Previous releases of AutoCAD displayed the cursor completely across the screen. This display can be achieved in Release 14 by entering a percentage value of 100 (Fig. 1-31).

Figure 1-30

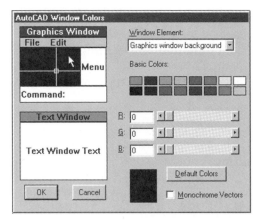

All changes made to the Windows screen by any of the options discussed in this section are automatically saved for the next drawing session. The changes are saved in the system registry. However, if you are working in a school laboratory, it is likely that the computer systems are set up to present the same screen defaults each time you start AutoCAD.

Figure 1-31

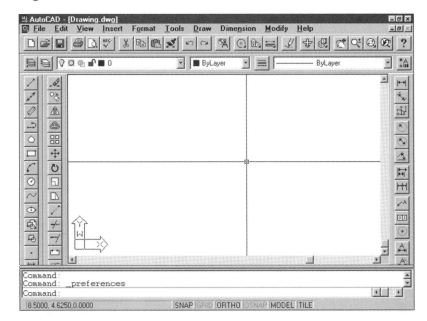

CHAPTER EXERCISES

1. **Starting and Exiting AutoCAD**

   Start AutoCAD by clicking the "14" shortcut icon or "AutoCAD 14" from the Programs menu. If the *Start Up* dialog box or *Create New Drawing* dialog box appears, select *Start from Scratch,* choose *English* as the default setting, and click the *OK* button. Draw a *Line.* Exit AutoCAD by selecting the *Exit* option from the *Files* pull-down menu. Answer *No* to the "Save Changes to Drawing.dwg?" prompt. Repeat these steps until you are confident with the procedure.

2. **Using Drawing Aids**

   Start AutoCAD. Turn on and off each of the following modes:

   SNAP, GRID, ORTHO

3. **Understanding Coordinates**

   Begin drawing a *Line* by PICKing a "From point:." Toggle *Coords* to display each of the <u>three</u> formats. PICK several other points at the "To point:" prompt. Pay particular attention to the coordinate values displayed for each point and visualize the relationship between that point and coordinate 0,0 (absolute value) or the last point established (relative polar value). Finish the command by pressing Enter.

4. **Using *Flipscreen***

   Use *Flipscreen* (**F2**) to toggle between the text screen or window and the graphics screen.

5. **Drawing with Drawing Aids**

   Draw four *Lines* using <u>each</u> Drawing Aid: ***GRID, SNAP, ORTHO.*** Toggle on and off each of the drawing aids one at a time for each set of four *Lines.* Next, draw *Lines* using combinations of the Drawing Aids, particularly ***GRID + SNAP*** and ***GRID + SNAP + ORTHO.***

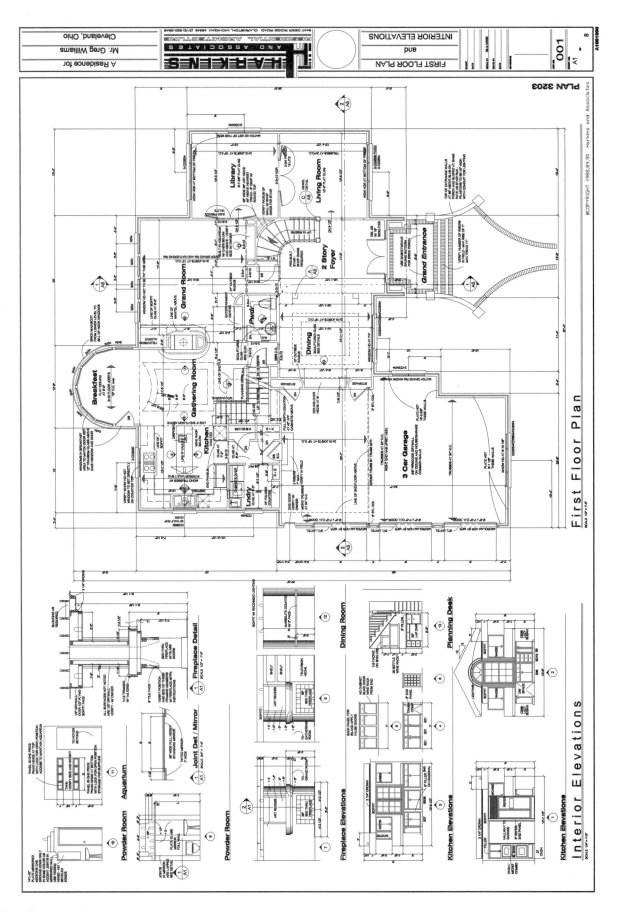

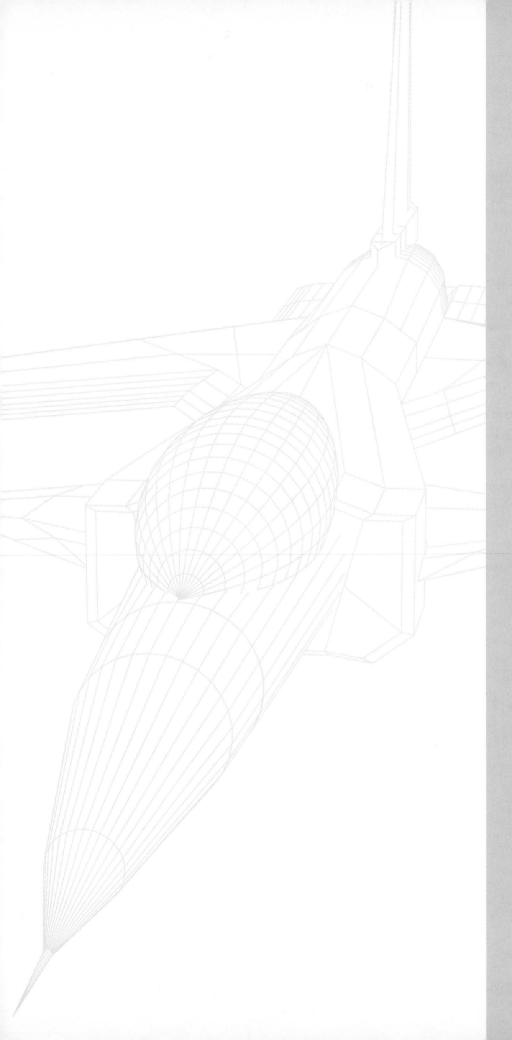

2

WORKING WITH FILES

Chapter Objectives

After completing this chapter you should be able to:

1. name drawing files;

2. use file-related dialog boxes;

3. create *New* drawings;

4. *Open* existing drawings;

5. *Save* drawings to disk;

6. use *SaveAs* to save a drawing under a different name, path, and/or format;

7. *Exit AutoCAD;*

8. practice good file management techniques;

9. use the Windows right-click shortcut menus in file dialog boxes.

AutoCAD DRAWING FILES

Naming Drawing Files

What is a drawing file? A CAD drawing file is the electronically stored data form of a drawing. The computer's hard disk is the principal magnetic storage device used for saving and restoring CAD drawing files. Diskettes, compact disks (CDs), networks, and the Internet are used to transport files from one computer to another, as in the case of transferring CAD files among clients, consultants, or vendors in industry. The AutoCAD commands used for saving drawings to and restoring drawings from files are explained in this chapter.

An AutoCAD drawing file has a name that you assign and a file extension of ".DWG." An example of an AutoCAD drawing file is:

PART-024.DWG

filename   extension

The file name you assign must be compliant with the Windows 95 or Windows NT (FAT) filename conventions; that is, it can have a <u>maximum</u> of 256 alphanumeric characters. File names and directory names (folders) can be in UPPER CASE, Title Case, or lower case letters. Characters such as _ - $ # ( ) ^ and spaces can be used in names, but other characters such as \ / : * ? < > | are not allowed. AutoCAD automatically appends the extension of .DWG to all AutoCAD-created drawing files.

NOTE: The chapter exercises and other examples in this book list file names in UPPER CASE letters for easy recognition. The file names and directory (folder) names on your system may appear as UPPER CASE, Title Case, or lower case.

Beginning and Saving an AutoCAD Drawing

When you start AutoCAD, the drawing editor appears and allows you to begin drawing even before using any file commands. If you choose, you can assign a name for the drawing when you begin with the *New* command. As you draw, you should develop the habit of saving the drawing periodically (about every 15 or 20 minutes) using *Save*. *Save* stores the drawing in its most current state to disk.

The typical drawing session would involve using *New* to begin a new drawing or using *Open* to open an existing drawing. Alternately, the *Start Up* dialog box that appears when starting AutoCAD would be used to begin a new drawing or open an existing one. *Save* would be used periodically, and *Exit* would be used for the final save and to end the session.

Figure 2-1

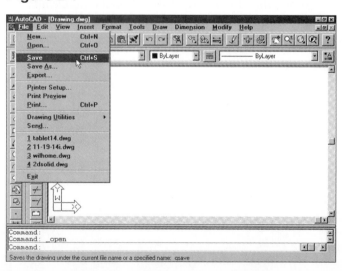

Accessing File Commands

Proper use of the file-related commands covered in this chapter allows you to manage your AutoCAD drawing files in a safe and efficient manner. Although the file-related commands can be invoked by several methods, they are easily accessible via the first pull-down menu option, *File* (Fig. 2-1). Most of the selections from this pull-down menu invoke dialog boxes for selection or specification of file names.

The Standard toolbar at the top of the AutoCAD screen has tools (icon buttons) for *New, Open,* and *Save.* File commands can also be entered at the keyboard by typing the formal command name, the command alias, or using the Ctrl keys sequences. File commands and related dialog boxes are also available from the *File* screen menu or by selection from the digitizing tablet menu.

File Dialog Box Functions

There are many dialog boxes appearing in AutoCAD that help you manage files. All of these dialog boxes operate in a similar manner. A few guidelines will help you use them. The *Save Drawing As* dialog box for Windows (Fig. 2-2) is used as an example.

Figure 2-2

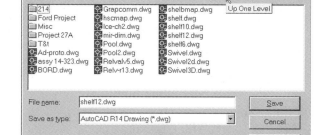

- The top of the box gives the title describing the action to be performed. It is <u>very important</u> to glance at the title before acting, especially when saving or deleting files.
- The desired file can be selected by PICKing it, then PICKing *OK*. Double-clicking on the selection accomplishes the same action. File names can also be typed in the *File name:* edit box.
- Every file name has an extension (three letters following the period) called the <u>type.</u> File types can be selected from the *Save as Type:* or *Files of Type:* section of the dialog boxes, or the desired file extension can be entered in the *File name:* edit box.
- The current folder (directory) is listed in the box near the top of the dialog box. You can select another folder (directory) or drive by using the drop-down list to display the current path or by selecting the *Up One Level* icon to the right of the list (Fig. 2-3). (Rest your pointer on an icon momentarily to make the tool tip appear.)
- A new folder (subdirectory) can be created within the current folder (directory) by selecting the *New Folder* icon.
- The last two icons to the right of the row near the top of the dialog box toggle the listing of files to a *List* or to show *Details*. The *List* option only displays file names (Fig. 2-2), whereas the *Details* option gives file-related information such as file size, file type, and time and date last modified (Fig. 2-3). The detailed list can also be sorted alphabetically by name, by file size, alphabetically by file type, or chronologically by time and date. Do this by clicking the *Name, Type, Size,* or *Modified* tiles immediately above the list. Double-clicking on one of the tiles reverses the order of the list.
- You can resize the width of a column (*Name, Size, Type,* etc.) by moving the pointer to the "crack" between two columns, then sliding the double arrows that appear in either direction.
- Scroll bars appear at the right and/or bottom if the list is longer than one page. PICKing the arrows moves the list one character or line at a time, or sliding the button causes motion at any rate.
- As an alternative, the file name (and path) can be typed in the edit box labeled *File name:*. Make sure the name is highlighted or the blinking cursor appears in the box (move the pointer and PICK) before typing.

Figure 2-3

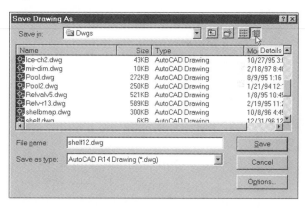

Windows Right-Click Shortcut Menus

AutoCAD for Windows 95 or Windows NT utilizes the right-click shortcut menus that operate with software products running on the Windows 95 and NT operating systems. Activate the shortcut menus by pressing the right mouse button (right-clicking) <u>inside the file list area</u> of a dialog box. There are two menus that give you additional file management capabilities.

Right-Click, No Files Highlighted
When no files are highlighted, right-clicking produces a menu for file list display and file management options (Fig. 2-4). The menu choices are as follows:

Figure 2-4

View	Shows *Large Icons* or *Small Icons* next to the file names, displays a *List* of file names only (see Fig. 2-2), or displays all file *Details* (see Fig. 2-3).
Arrange Icons	Sorts the files *by Name, by Type* (file extension), *by Size,* or *by Date*.
Paste	If the *Copy* or *Cut* function was previously used (see Right-Click, File Highlighted, Fig. 2-5), *Paste* can be used to place the copied file into the displayed folder.
Paste Shortcut	Use this option to place a *Shortcut* (to open a file) into the current folder.
Undo Rename	If the *Rename* function was previously used (see Right-Click, File Highlighted, Fig. 2-5), this action restores the original file name.
New	Creates a new *Folder, Shortcut,* or document.
Properties	Displays a dialog box listing properties of the current folder.

Right-Click, File Highlighted
When a file is highlighted, right-clicking produces a menu with options for the selected file (Fig. 2-5). The menu choices are as follows:

Figure 2-5

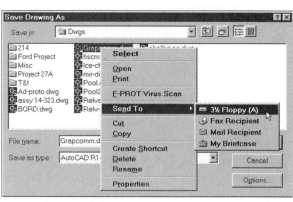

Select	Processes the file (like selecting the *OK* button) according to the dialog box function. For example, if the *SaveAs* dialog box is open, *Select* saves the highlighted file; or if the *Select File* dialog box is active, *Select* opens the drawing.
Open	Opens the application (AutoCAD or other program) and loads the selected file. <u>Do not use this function in AutoCAD unless you want to open a second AutoCAD session.</u>
Print	Sends the selected file to the configured system printer. In AutoCAD, opens the *Print/Plot Configuration* dialog box.
Send To	Copies the selected file to the selected device. <u>This is an easy way to copy a file from your hard drive to a diskette in A: drive.</u>
Cut	This option, in conjunction with *Paste,* allows you to move a file from one location to another.
Copy	In conjunction with *Paste,* allows you to copy the selected file to another location. You can copy the file to the same folder, but Windows renames the file to "Copy of . . .".

Create Shortcut	Use this option to create a *Shortcut* (to open the highlighted file) in the current folder. The Shortcut can be moved to another location using drag and drop.
Delete	Sends the selected file to the Recycle Bin.
Rename	Allows you to rename the selected file. Move your cursor to the highlighted file name, click near the letters you want to change, then type or use the backspace, delete, space, or arrow keys.
Properties	Displays a dialog box listing properties of the selected file.

Specific features of other dialog boxes are explained in the following sections.

THE *START UP* DIALOG BOX

When you start AutoCAD for the first time, the *Start Up* dialog box appears (Fig 2-6). The *Start Up* dialog box may not appear on your system if it has been suppressed (deactivated). It can be suppressed by removing the check in the box at the bottom of the dialog box labeled *Show this dialog at start up*. You can make the dialog box reappear on start up by checking *Show the Start Up dialog box* in the *Compatibility* tab of the *Preferences* dialog box.

Figure 2-6

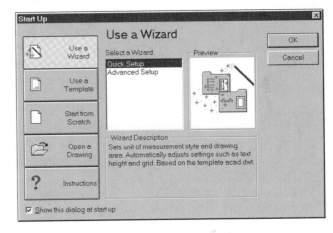

There is no command that invokes the *Start Up* dialog box. However, the *Start Up* dialog box is identical to and <u>combines</u> the functions of the *New* and *Open* commands. Using the *New* command produces the *Create New Drawing* dialog box, which contains the *Use a Wizard, Use a Template,* and *Start from Scratch* options as seen in the *Start Up* dialog box (Fig. 2-6). Using the *Open a Drawing* option in the *Start Up* dialog performs the same action as using the *Open* command.

Because of the similarities of the *Start Up* dialog box and the *New* and *Open* commands, an understanding of the *Start Up* dialog box can be acquired from the following explanation of the *New* command and the *Open* command.

AutoCAD FILE COMMANDS

NEW

Pull-down Menu	COMMAND (TYPE)	ALIAS (TYPE)	Short-cut	Screen (side) Menu	Tablet Menu
File *New...*	*NEW*	...	*Ctrl+N*	*FILE* *New*	*T,24*

The *New* command produces the *Create a New Drawing* dialog box (Fig. 2-7). (Similar functions are available in the *Start Up* dialog box that may appear when you first start AutoCAD.) In this dialog box you can begin a new drawing by choosing *Use a Wizard, Use a Template,* or *Start from Scratch*. Selecting any one of these tiles changes the appearance of the center and right side of the box.

Use a Wizard

You can choose between a *Quick Setup Wizard* and an *Advanced Setup Wizard* (Fig 2-6). The *Quick Setup Wizard* prompts you to select the type of drawing *Units* you want to use and to specify the drawing *Area* (*Limits*). The *Advanced Setup* offers options for units, angular direction and measurement, area, title blocks, and layouts for paper space and model space. Both Wizards also automatically make changes to many size-related system variables. The *Quick Setup Wizard* and the *Advanced Setup Wizard* are discussed in Chapter 13, Advanced Drawing Setup.

Figure 2-7

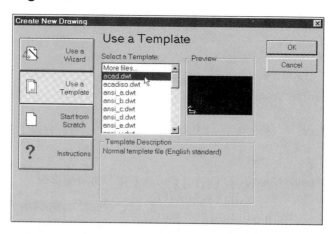

Use a Template

Use this option if you want to use another existing drawing as a template (sometimes known as a "prototype") (Fig. 2-7). A template drawing is one that already has many of the setup steps performed but contains no geometry (graphical objects).

Selecting the ACAD.DWT template begins a new drawing using the traditional AutoCAD default settings for drawing area (called *Limits*) of 12 x 9 units and other default settings. Selecting the ACAD.DWT template accomplishes the same action as using *Start from Scratch, English* option. Other template drawings are discussed in Chapter 13, Advanced Drawing Setup.

Start from Scratch

Choose from either the *English* or *Metric* default settings. Selecting *English* causes AutoCAD to use the default drawing area (called *Limits*) of 12 x 9 units (Fig. 2-8). Choosing the *Metric* option causes AutoCAD to use the metric drawing template, which has *Limits* settings for a A4 metric size sheet (297 x 210 mm).

Figure 2-8

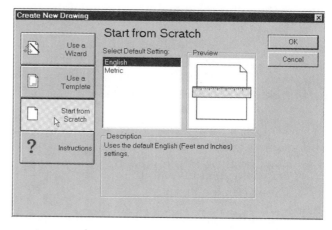

For the purposes of learning AutoCAD starting with the basic principles and commands, it is helpful to use the <u>default AutoCAD settings</u> when beginning a new drawing. Until you read Chapters 6 and 13, you can begin new drawings for completing the exercises and practicing the examples in Chapters 1 through 5 by any of the following methods:

> Selecting the ACAD.DWT template drawing
> Selecting the *English* defaults in the *Start from Scratch* option
> Pressing *Cancel* in the *Setup* dialog box that appears when AutoCAD starts
> Starting AutoCAD after suppressing the *Setup* dialog box

OPEN

Pull-down Menu	COMMAND (TYPE)	ALIAS (TYPE)	Short-cut	Screen (side) Menu	Tablet Menu
File Open...	OPEN	...	Ctrl+O	FILE Open	T,25

Use *Open* to select an existing drawing to be loaded into AutoCAD. Normally, you would open an existing drawing (one that is completed or partially completed) so you could continue drawing or make a print or plot. You can *Open* only one drawing at a time in AutoCAD Release 14 for Windows 95 and NT.

The *Open* command produces the *Select File* dialog box (Fig. 2-9). In this dialog box you can select any drawing from the current directory list. PICKing a drawing name from the list displays a small bitmap image of the drawing in the *Preview* tile. Select the *Open* button or double-click on the file name to open the high-lighted drawing. You could instead type the file name (and path) of the desired drawing in

Figure 2-9

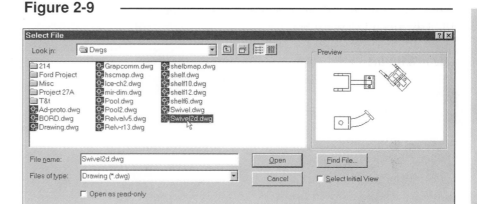

the *File name:* edit box and press Enter, but a preview of the typed entry will not appear. AutoCAD Release 12 or previous DWG files usually do not have bitmap images.

To locate a drawing in another folder (directory) or on another drive on your computer or network, select the drop-down list on top of the dialog box next to *Look in:* or use the *Up one Level* button. Remember that you can also display the file details (size, type, modified) by toggling the *Details* button. You can open a DWG (drawing), DXF (drawing interchange format), or DWT (template) file by selecting from the *Files of Type:* drop-down list.

Selecting the *Find File...* button in the *Select File* dialog box invokes the *Browse/Search* dialog box (Fig. 2-10). This box displays a small bitmap image of all drawings in the search directory. You can open a drawing by double-clicking on its image. The dialog box offers two methods to find files: *Browse* and *Search*.

Figure 2-10

Using the *Browse* tab (Fig. 2-10), you can browse for drawing (DWG, DWT, or DXF) files in any directory. A preview image of all drawings in the selected directory (in the *Directories:* area) is displayed. The preview images can be displayed in a large, medium, or small *Size:*. Use the scroll bar along the bottom to see all drawings in the directory.

The *Search* tab of the *Browse/Search* dialog box enables you to search for files meeting specific criteria that you specify (Fig. 2-11). You can enter file names (including wildcards) in the *Search Pattern* box, and the desired drawing type can be selected from the *File Types* drop-down list. The *Date Filter* helps you search for files by *Date* and *Time* based on when the files were last saved. Any drive or path can be selected using the *Search Location* cluster.

Figure 2-11

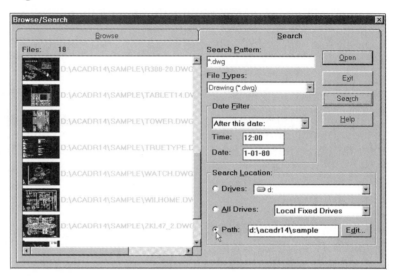

After specifying the search criteria, select the *Search* button. All file names found matching the criteria are displayed with the drawing preview images on the left side of the dialog box. Files can be opened by double-clicking on the name or image tile or by the *Open* button.

NOTE: It is considered <u>poor practice to *Open* a drawing from a diskette in A: drive.</u> Normally, drawings should be copied to a directory on a fixed (hard) drive, then opened from that directory. Opening a drawing from a diskette and performing numerous *Saves* is much slower and less reliable than operating from a hard drive. See other NOTEs at the end of the discussion on *Save* and *SaveAs*.

SAVE

Pull-down Menu	COMMAND (TYPE)	ALIAS (TYPE)	Short-cut	Screen (side) Menu	Tablet Menu
File *Save As*	SAVE	...	Ctrl+S	...	U,24-U,25

The *Save* command is intended to be used periodically during a drawing session (every 15 to 20 minutes is recommended). When *Save* is selected from a menu, the current version of the drawing is saved to disk without interruption of the drawing session. The first time an unnamed drawing is saved, the *Save Drawing As* dialog box (Figs. 2-2 and 2-3) appears, which prompts you for a drawing name. Typically, however, the drawing already has an assigned name, in which case *Save* actually performs a *Qsave* (quick save). A *Qsave* gives no prompts or options, nor does it display a dialog box.

When the file has an assigned name, using *Save* by selecting the command from a menu or icon button automatically performs a quick save (*Qsave*). Therefore, *Qsave* automatically saves the drawing in the same drive and directory from which it was opened or where it was first saved. In contrast, <u>typing *Save* always produces the *Save Drawing As* dialog box,</u> where you can enter a new name and/or path to save the drawing or press Enter to keep the same name and path (see *SaveAs*).

NOTE: If you want to save a drawing directly to a diskette in drive A:, <u>type</u> *Save* and specify drive A:. Even though the *SaveAs* dialog is used, <u>typing *Save*</u> does not reset the path of the current drawing for subsequent saves. (See the NOTE at the end of the explanation of *SaveAs*.) Another useful method for copying a file to a diskette in A: drive is to use the *Send To* option in the shortcut menu that appears if you right-click in the *Save*, *SaveAs*, or *Select File* dialog box (see Windows Right-Click Shortcut Menus and Fig. 2-5).

QSAVE

Pull-down Menu	COMMAND (TYPE)	ALIAS (TYPE)	Short-cut	Screen (side) Menu	Tablet Menu
...	QSAVE	...	Ctrl+S	FILE Qsave	...

Qsave (quick save) is normally invoked automatically when *Save* is used (see *Save*) but can also be typed. *Qsave* saves the drawing under the previously assigned file name. No dialog boxes appear nor are any other inputs required. This is the same as using *Save* (from a menu), assuming the drawing name has been assigned. However, if the drawing has not been named when *Qsave* is invoked, the *Save Drawing As* dialog box appears.

SAVEAS

Pull-down Menu	COMMAND (TYPE)	ALIAS (TYPE)	Short-cut	Screen (side) Menu	Tablet Menu
File Save As...	SAVEAS	...	...	FILE Saveas	V,24

The *SaveAs* command can fulfill four functions:

1. save the drawing file under a new name if desired;
2. save the drawing file to a new path (drive and directory location) if desired;
3. in the case of either 1 or 2, assign the new file name and/or path to the current drawing (change the name and path of the current drawing);
4. save the drawing in a format other than the default Release 14 format (but does not reassign the current drawing format for future saves).

Therefore, assuming a name has previously been assigned, *SaveAs* allows you to save the current drawing under a different name and/or path; but, beware, <u>*SaveAs* sets the current drawing name and/or path to the last one entered.</u> This dialog box is shown in Figures 2-2 and 2-3.

SaveAs can be a benefit when creating two similar drawings. A typical scenario follows. A design engineer wants to make two similar but slightly different design drawings. During construction of the first drawing, the engineer periodically saves under the name DESIGN1 using *Save*. The first drawing is then completed and *Saved*. Instead of starting a *New* drawing, *SaveAs* is used to save the current drawing under the name DESIGN2. *SaveAs* also resets the current drawing name to DESIGN2. The designer then has two separate but identical drawing files on disk which can be further edited to complete the specialized differences. The engineer continues to work on the current drawing DESIGN2.

NOTE: If you want to save the drawing to a diskette in A: drive, do not use *SaveAs*. Invoking *SaveAs* by any method resets the drawing name and path to whatever is entered in the *Save Drawing As* dialog box, so entering A:NAME would set A: as the current drive. This could cause problems because of the speed and reliability of a diskette as opposed to a hard drive. Instead, you should (1) <u>type</u> *Save* (see previous NOTE), or (2) save the drawing to the hard drive (usually C:), then close the drawing (by using *Open*, *New*, or *Exit*). Next, use Windows Explorer or the Windows right-click shortcut menu to copy the drawing file to A:. (See Windows Right-Click Shortcut Menus and Fig. 2-5.)

You can save an AutoCAD Release 14 drawing in several formats other than the default format (*AutoCAD R14 Drawing* *.dwg*). Use the drop-down list at the bottom of the *Save* (or *SaveAs*) dialog box (Fig. 2-12) to save the current drawing as an *AutoCAD R13/LT 95 Drawing, AutoCAD R12/LT2 Drawing*, or a *Drawing Template file* *.dwt*. The current drawing format is not changed for future saves. (If saving to previous versions, especially Release 12 or LT2, some information may be lost when the "downward" conversion is made. Newer objects like *Leader, Xline, Ray, Ellipse, Tolerance, Mline*, OLE objects, *Xref overlay, Mtext, Spline*, 3D solids, solid fills, and raster objects may be converted to the closest matching object in the selected format when possible.)

Figure 2-12

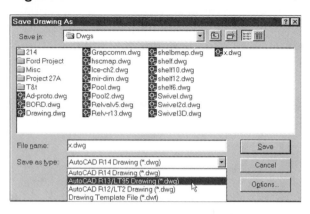

Selecting the *Options…* button produces the *Export Options* dialog box. This is used to select how proxy images of custom objects are saved (see Chapter 30).

EXIT

Pull-down Menu	COMMAND (TYPE)	ALIAS (TYPE)	Short-cut	Screen (side) Menu	Tablet Menu
File Exit	EXIT	…	…	…	Y,25

This is the simplest method to use when you want to exit AutoCAD. If any changes have been made to the drawing since the last *Save, Exit* invokes a dialog box asking if you want to *Save Changes to . . .?* before ending AutoCAD (Fig. 2-13). *Yes* causes AutoCAD to use *Save,* then exit; *No* exits without saving; and *Cancel* does not exit. The *Quit* command could optionally be typed to accomplish the same action as using *Exit*.

Figure 2-13

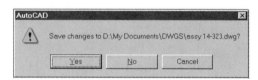

An alternative to using the *Exit* option is to use the standard Windows methods for exiting an application. The two options are (1) select the **"X"** in the extreme upper-right corner of the AutoCAD window, or (2) select the AutoCAD logo (red) in the extreme upper-left corner of the AutoCAD window. Selecting the logo in the upper-left corner produces a pull-down menu allowing you to *Minimize, Maximize,* etc., or to *Close* the window (Fig. 12-14). Using *Close* is the same as using *Exit*.

Figure 2-14

QUIT

Pull-down Menu	COMMAND (TYPE)	ALIAS (TYPE)	Short-cut	Screen (side) Menu	Tablet Menu
…	QUIT	EXIT	…	FILE Quit	…

The *Quit* command accomplishes the same action as *Exit*. *Quit* discontinues (exits) the AutoCAD session and usually produces the dialog box shown in Figure 2-13 if changes have been made since the last *Save*.

RECOVER

Pull-down Menu	COMMAND (TYPE)	ALIAS (TYPE)	Short-cut	Screen (side) Menu	Tablet Menu
File Drawing Utilities > Recover...	RECOVER	...	...	FILE Recover	...

This option is used only in the case that *Open* does not operate on a particular drawing because of damage to the file. Damage to a drawing file can occur from improper exiting of AutoCAD (such as power failure) or from damage to a diskette. *Recover* can usually reassemble the file to a usable state and reload the drawing.

The *Recover* command produces the *Select File* dialog box. Select the drawing to recover. A recovery log is displayed in a text window reporting results of the recover operation (Fig. 2-15).

Figure 2-15

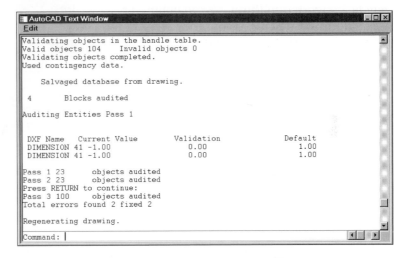

SAVETIME

Pull-down Menu	COMMAND (TYPE)	ALIAS (TYPE)	Short-cut	Screen (side) Menu	Tablet Menu
Tools Preferences... General	SAVETIME	...	...	TOOLS 2 Prefernc General	...

SAVETIME is actually a system variable that controls AutoCAD's automatic save feature. AutoCAD automatically saves the current drawing for you at time intervals that you specify. The default time interval is 120 (minutes)! If you want to use this feature effectively, change the value stored in the *SAVETIME* variable to about 15 or 20 (minutes). A value of 0 disables this feature. The value for *SAVETIME* can also be set in the *General* tab of the *Preferences* dialog box (see Figure 2-18).

When automatic saving occurs, the current drawing is saved under the default name of AUTO.SV$ (sometimes in the Windows/Temp directory, depending on your operating system and setup). The current drawing name is not overwritten. If you want to save the drawing under the name you assigned, *Save* or *Qsave* must be used, or *SaveAs* should be used to save under a new name. The default name and path of the automatic save file can be changed in the *Files* tab of the *Preferences* dialog box under *Menu, Help, Log, and Miscellaneous File Names*.

SHELL

Pull-down Menu	COMMAND (TYPE)	ALIAS (TYPE)	Short-cut	Screen (side) Menu	Tablet Menu
...	*SHELL*	...	...	...	...

If you like to use DOS commands for file management, you can temporarily exit AutoCAD and type commands in a DOS window by using *Shell*. *Shell* automatically switches to the DOS text mode while AutoCAD runs "in the background." The OS command: prompt appears and allows one operating system command before switching back to the AutoCAD Drawing Editor. If you choose to use more than one DOS command, press **Enter** twice after entering **Shell**. The *AutoCAD Shell Active* window appears (Fig. 2-16).

Figure 2-16

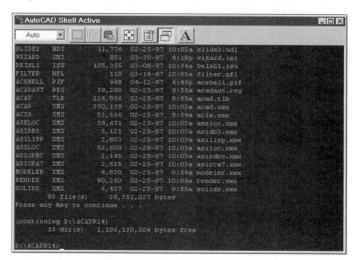

Shell is very handy at times but also very dangerous. You can perform most DOS operations like copying, deleting, or renaming files while using *Shell*. However, it is very important to remember that AutoCAD is running while you are "shelled out." You cannot turn off your computer during *Shell* because AutoCAD is still running. You can type **Exit** or click the **"X"** in the upper-right corner of the *AutoCAD Shell Active* window to return to AutoCAD.

There are many options available for controlling the format of the *AutoCAD Shell Active* window and for specifying memory usage. Using the *Properties* button displays the *AutoCAD Shell Active Properties* dialog box (Fig. 2-17).

Figure 2-17

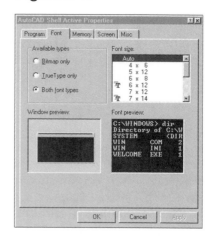

The following tabs are available:

Program   set working directory, shortcut key, window type, etc.
Font   set font size and type (Bitmap, TrueType, or both)
Memory   set how *Shell* uses memory (expended, expanded, protected mode)
Screen   set the window size and performance
Misc   miscellaneous settings and shortcut keys

Whether you use *Shell* or switch to another file management program such as Windows Explorer, there are a few issues to keep in mind while AutoCAD is running. AutoCAD can keep several temporary files open while in operation. These files are only "cleaned up" if you return to AutoCAD and exit AutoCAD properly. Turning off the computer system while "shelled out" can result in loss of the current drawing and having several temporary files left on disk. Do not delete any unrecognizable files (ending with *.AC$ or *.TMP) while using *Shell*. Do not try to return to AutoCAD by selecting the shortcut icon, since that action starts a second AutoCAD session. Do not use the DOS command **chkdsk/f** or **scandisk** while using *Shell*. You can use other DOS commands, but make sure you eventually type **Exit** to return to AutoCAD.

AutoCAD Backup Files

When a drawing is saved, AutoCAD creates a file with a .DWG extension. For example, if you name the drawing PART1, using *Save* creates a file named PART1.DWG. The next time you save, AutoCAD makes a new PART1.DWG and renames the old version to PART1.BAK. One .BAK (backup) file is kept automatically by AutoCAD by default. You can disable the automatic backup function by changing the *ISAVEBAK* system variable to 0 or by accessing the *General* tab in the *Preferences* dialog box (Fig. 2-18).

Figure 2-18

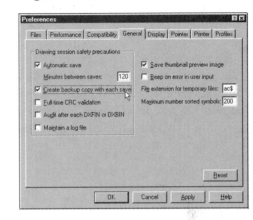

You cannot *Open* a .BAK file. It must be renamed to a .DWG file. Remember that you already have a .DWG file by the same name, so rename the extension <u>and</u> the filename. For example, PART1.BAK could be renamed to PART1OLD.DWG. Use the Windows Explorer, the *Select File* dialog box (use *Open*) with the right-click options, or use *Shell* and the DOS **ren** command to rename the file.

The .BAK files can also be deleted without affecting the .DWG files. The .BAK files accumulate after time, so you may want to periodically delete the unneeded ones to conserve disk space.

AutoCAD Drawing File Management

AutoCAD drawing files should be stored in folders (directories) used exclusively for that purpose. For example, you may use Windows Explorer to create a new folder called "Drawings" or "DWG." This could be a folder in the main C: or D: drive, or could be a subdirectory such as "C:\My Documents\ AutoCAD Drawings." You can create folders for different types of drawings or different projects. For example, you may have a folder named "DWG" with several subdirectories for each project, drawing type, or client, such as "COMPONENTS" and "ASSEMBLIES," or "In Progress" and "Completed," or "TWA," "Ford," and "Dupont." If you are working in an office or school laboratory, a directory structure most likely already has been created for you to use.

It is considered poor practice to create drawings in folders where the AutoCAD system files are kept. Beware, if you install AutoCAD and accept all defaults, when you first use *Save,* the files are saved in the "C:\Program Files\AutoCAD R14" directory.

Safety and organization are important considerations for storage of AutoCAD drawing files. AutoCAD drawings should be saved to designated folders to prevent accidental deletion of system or other important files. If you work with AutoCAD long enough or if you work in an office or laboratory, most likely many drawings are saved on the computer hard drive or network drives. It is imperative that a logical directory structure be maintained so important drawings can be located easily and saved safely.

Using Other File Formats

See Chapter 32 for file commands related to importing and exporting images and exchanging files in other than *.DWG format.

CHAPTER EXERCISES

Start AutoCAD. If the *Start Up* dialog box appears, select *Start from Scratch,* then use the *English* default settings. NOTE: The chapter exercises in this book list file names in UPPER CASE letters for easy recognition. The file names and directory (folder) names on your system may appear as UPPER CASE, Title Case, or lower case.

1. **Create or determine the name of the folder ("working" directory) for opening and saving AutoCAD files on your computer system**

   If you are working in an office or laboratory, a folder (directory) has most likely been created for saving your files. If you have installed AutoCAD yourself and are learning on your home or office system, you should create a folder for saving AutoCAD drawings. It should have a name like "C:\ACADR14\DWG," "C:\My Documents\Dwgs," or "C:\Acadr14\Drawing Files." (HINT: Use the *SaveAs* command. The name of the folder last used for saving files appears at the top of the *SaveAs* dialog box.)

2. *Save* **and name a drawing file**

   Draw 2 vertical *Lines*. Select *Save* from the *Files* pull-down, from the Standard toolbar, or by another method. (The *SaveAs* dialog box appears since a name has not yet been assigned.) Name the drawing **"CH2 VERTICAL."**

3. **Using** *Qsave*

   Draw 2 more vertical *Lines*. Select *Save* again from the menu or by any other method except typing. (Notice that the *Qsave* command appears at the command line since the drawing has already been named.) Draw 2 more vertical *Lines* (a total of six lines now). Select *Save* again.

4. **Start a** *New* **drawing**

   Invoke *New* from the *Files* pull-down, the Standard toolbar, or by any other method. If the *Start Up* dialog box appears, select **Start from Scratch,** then use the **English** default settings. Draw 2 horizontal *Lines*. Use *Save*. Enter **"CH2 HORIZONTAL"** as the name for the drawing. Draw 2 more horizontal *Lines,* but do not *Save*. Continue to exercise 5.

5. *Open* **an existing drawing**

   Use *Open* to open **CH2 VERTICAL.** Notice that AutoCAD first forces you to answer *Yes* or *No* to *Save changes to CH2 HORIZONTAL.DWG?* PICK *Yes* to save the changes. Proceed to *Open* **CH2 VERTICAL.**

6. **Using** *SaveAs*

   Draw 2 inclined (angled) *Lines* in the Ch2 Vertical drawing. Invoke *SaveAs* to save the drawing under a new name. Enter **"CH2 INCLINED"** as the new name. Notice the current drawing name displayed in the AutoCAD title bar (at the top of the screen) is reset to the new name. Draw 2 more inclined *Lines* and *Save*.

7. *Open* **an AutoCAD sample drawing**

   Open a drawing named **"COLORWH.DWG"** usually located in the "C:\Programs\ AutoCAD R14\Sample" directory. Use the *Directories* section of the dialog box to change drive and directory if necessary. Do not *Save* the sample drawing after viewing it. Practice the *Open* command by looking at other sample drawings in the Sample directory, such as **WILHOME.DWG** or **ZKL47_2.DWG.** Be careful not to *Save* the drawings.

8. **Check the *SAVETIME* setting**

   At the command prompt, enter *SAVETIME*. The reported value is the interval (in minutes) between automatic saves. Your system may be set to 120 (default setting). If you are using your own system, change the setting to **15 or 20.**

9. **Find and rename a backup file**

   Use the *Open* command. When the *Select File* dialog box appears, locate the folder where your drawing files are saved, then enter **"*.BAK"** in the *File name:* edit box (* is a wildcard that means "all files"). All of the .BAK files should appear. Search for a backup file that was created the last time you saved **CH2 VERTICAL.DWG,** named **CH2 VERTICAL.BAK. Right-click** on the file name. Select *Rename* from the pop-up menu and rename the file **"CH2 VERT 2.DWG."** Next enter *.DWG in the *File name:* edit box to make all the .dwg files reappear. Highlight **CH2 VERT 2.DWG** and the bitmap image should appear in the preview display. You can now *Open* the file if you wish.

10. **Find and rename the automatic save file**

    If you are familiar with Windows 95 or NT functions, use the "Find File" feature to locate the AutoCAD automatic save file. You <u>may or may not</u> find a file named AUTO?.SV$ (the last automatic save drawing file), depending on the *SAVETIME* interval setting and how long AutoCAD has been used on your computer. (Use the Windows Task Bar *Startup* button, then select *Find Files,* or use Explorer and select *Find Files or Folders...* from the *Tools* pull-down menu.) Then enter **"*.SV$"** in the *Named:* edit box. (*.SV$ lists any AutoCAD automatic save files.)

    Use the *Open* command again. Locate the folder that was indicated in the *Find File* results. Enter **"*.SV$"** in the *File name:* edit box. Using the same technique as in exercise 9, rename the file to **"AUTOSAVE.DWG."** <u>Be extremely careful not to alter any other files in the folder.</u> Highlight the **AUTOSAVE.DWG** file name and the bitmap image should appear in the preview display. You can now *Open* the file if you wish. *Exit* AutoCAD.

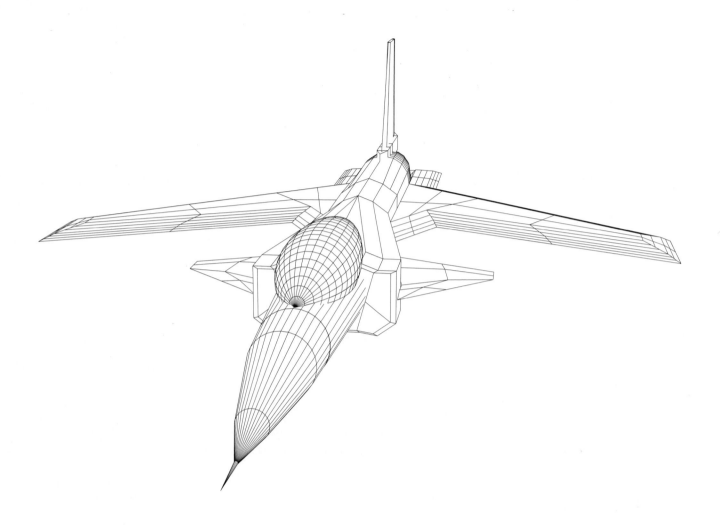

X29.DWG Courtesy Autodesk, Inc. (Release 14 Sample Drawing)

3

DRAW COMMAND CONCEPTS

Chapter Objectives

After completing this chapter you should be able to:

1. recognize drawing objects;

2. draw *Lines* <u>interactively</u>;

3. use *SNAP, GRID,* and *ORTHO* while drawing *Lines* and *Circles* interactively;

4. create *Lines* by specifying <u>absolute</u> coordinates;

5. create *Lines* by specifying <u>relative rectangular</u> coordinates;

6. create *Lines* by specifying <u>relative polar</u> coordinates;

7. create *Lines* by specifying <u>direct distance entry</u> coordinates;

8. create *Circles* by each of the five coordinate entry methods.

AutoCAD OBJECTS

Figure 3-1

The smallest component of a drawing in AutoCAD is called an <u>object</u> (sometimes referred to as an entity). An example of an object is a *Line,* an *Arc,* or a *Circle* (Fig. 3-1). A rectangle created with the *Line* command would contain four objects.

Draw commands <u>create</u> objects. The draw command names are the same as the object names.

Simple objects are *Point, Line, Arc,* and *Circle.*

Complex objects are shapes such as *Polygon, Rectangle, Ellipse, Polyline,* and *Spline* which are created with one command (Fig. 3-2). Even though they appear to have several segments, they are <u>treated</u> by AutoCAD as one object.

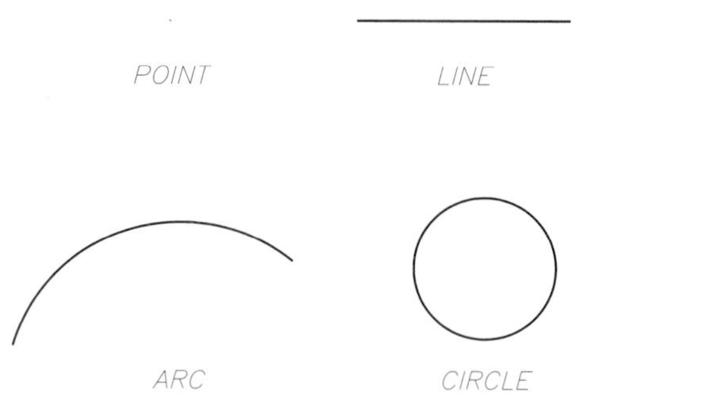

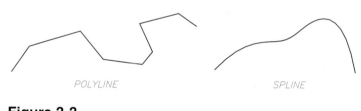

POINT   LINE

ARC   CIRCLE

Figure 3-2

POLYGON   RECTANGLE   ELLIPSE

POLYLINE   SPLINE

It is not always apparent whether a shape is composed of one or more objects. However, if you pick an object with the "pickbox," an object is "highlighted," or shown in a broken line pattern (Fig. 3-3). This high-lighting reveals whether the shape is com-posed of one or several objects.

Figure 3-3

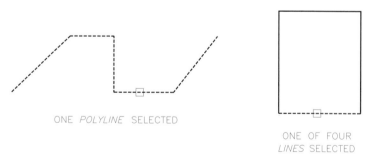

ONE *POLYLINE* SELECTED

ONE OF FOUR
LINES SELECTED

LOCATING THE DRAW COMMANDS

To invoke *Draw* commands, any of the five command entry methods can be used depending on your computer setup.

1. **Toolbars**   Select the icon from the *Draw* toolbar.
2. **Pull-down menu** Select the command from the *Draw* pull-down menu.
3. **Screen menu**   Select the command from the *Draw* screen menu.
4. **Keyboard**   Type the command name, command alias, or accelerator keys at the key-board (a command alias is a one- or two-letter shortcut).
5. **Tablet menu**   Select the icon from the digitizing tablet menu (if available).

For example, a draw command can be activated by PICKing its icon button from the *Draw* toolbar (Fig. 3-4).

Figure 3-4

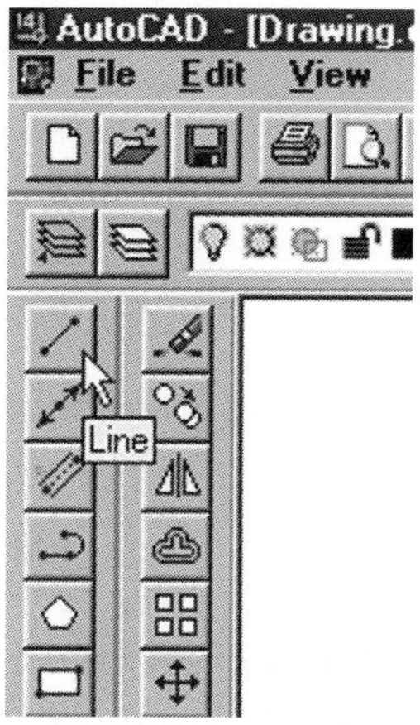

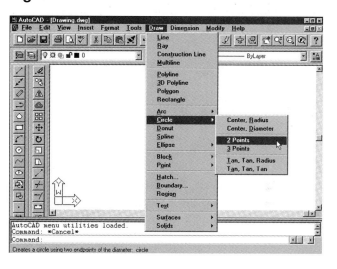

The *Draw* pull-down menu can also be used to select draw commands (Fig. 3-5). Options for a command are found on cascading menus.

Figure 3-5

Additionally, the Screen menu can be used to select draw commands. Two menus contain all the draw commands: *DRAW1* and *DRAW2* (Fig. 3-6).

Figure 3-6

THE FIVE COORDINATE ENTRY METHODS

All drawing commands prompt you to specify points, or locations, in the drawing. For example, the *Line* command prompts you to give the "From point:" and "To point:," expecting you to specify locations for the first and second endpoints of the line. After you specify those points, AutoCAD stores the coordinate values to define the line. A 2-dimensional line in AutoCAD is defined and stored in the database as two sets of X, Y, and Z values (with Z values of 0), one for each endpoint.

There are five ways to specify coordinates; that is, there are five ways to tell AutoCAD the location of points when you draw objects.

1. **Interactive method**   PICK   Use the cursor to select points on the screen.
2. **Absolute coordinates**   X,Y   Type explicit X and Y values relative to the origin at 0,0.
3. **Relative rectangular**   @X,Y   Type explicit X and Y values relative to the last point
   coordinates   (@ means "last point").
4. **Relative polar coordinates** @dist<angle   Type a distance value and angle value relative
                                                          to the last point (< means "angle of").
5. **Direct distance entry**   dist,direction   Type a distance value relative to the last point,
                                                          indicate direction with the cursor, then press Enter.

The interactive method of coordinate entry is simple because you PICK the desired locations with your cursor when AutoCAD prompts for a point. It is generally the easiest to use for most applications. It should be used with *SNAP* or with *Object Snaps* (Chapter 7) for accuracy.

Absolute coordinates are used when AutoCAD prompts for a point and you know the exact coordinates of the desired location. You simply key in the coordinate values at the keyboard (X and Y values are separated by a comma).

Relative rectangular coordinates are similar to absolute values except the X and Y distances are given in relation to the last point, instead of being relative to the origin. AutoCAD interprets the @ symbol as the "last point." Use this type of coordinate entry when you do not know the exact absolute values, but you know the X and Y distances from the last specified point.

Relative polar coordinate entry is used whenever you want to draw a line or specify a point that is at an exact angle with respect to the last point. Interactive coordinate entry is not accurate enough to draw diagonal lines to a specific angle (unless *Object Snap* can be used). AutoCAD interprets the @ symbol as last point and the < symbol as an angular designator for the following value. So, "@ 2<90" means "from the last point, 2 units at an angle of 90 degrees."

Direct distance coordinate entry is a combination of relative polar coordinates and interactive specification because a distance value is entered at the keyboard and the angle is specified by the direction of the cursor movement (from the last point). Direct distance entry is useful primarily for orthogonal (horizontal or vertical) operations by toggling *ORTHO* on. For example, assume you wanted to draw a *Line* 7.5 units in a horizontal (positive X) direction from the last point. Using direct distance entry to establish the "To point:" (second end point), you would turn on *ORTHO* and move the cursor to the right any distance, then type "7.5" and press Enter. Alternately, you could first type the distance value, then move the cursor before pressing Enter. This method is very fast and accurate for horizontal and vertical drawing and editing operations.

DRAWING *LINES* USING THE FIVE COORDINATE ENTRY METHODS

Begin a *New* drawing to complete the *Line* exercises. If the *Start Up* dialog box or *Create New Drawing* dialog box appears, select *Start from Scratch,* choose *English* as the default setting, and click the *OK* button. The *Line* command can be activated by any one of the methods shown in the command table below.

LINE

	Pull-down Menu	COMMAND (TYPE)	ALIAS (TYPE)	Short-cut	Screen (side) Menu	Tablet Menu
	Draw *Line*	*LINE*	*L*	...	DRAW 1 *Line*	*J,10*

Drawing Horizontal Lines

1. Draw a horizontal *Line* of 2 units length starting at point 2,2. Use the <u>interactive</u> method. See Figure 3-7.

Steps	Command Prompt	Perform Action	Comments
1.		turn on *GRID* (**F7**)	grid appears
2.		turn on *SNAP* (**F9**)	"SNAP" appears on Status line
3.	Command:	select or type **Line**	use any method
4.	LINE From point:	**PICK** location **2,2**	watch *Coords*
5.	To point:	**PICK** location **4,2**	watch *Coords*
6.	To point:	press **Enter**	completes command

The preceding steps produce a *Line* as shown in Figure 3-7.

Figure 3-7

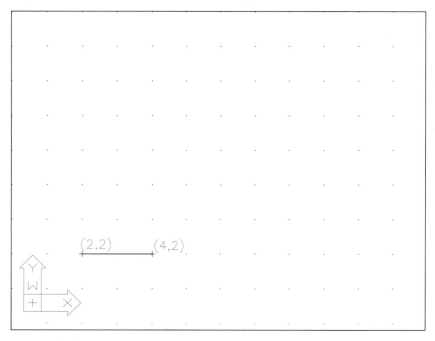

2.   Draw a horizontal *Line* of 2 units length starting at point 2,3 using <u>absolute coordinates</u>.

Steps	Command Prompt	Perform Action	Comments
1.	Command:	select *Line*	use any method
2.	LINE From point:	type **2,3** and press **Enter**	establishes the first endpoint
3.	To point:	type **4,3** and press **Enter**	a *Line* should appear
4.	To point:	press **Enter**	completes command

The above procedure produces the new *Line* above the first *Line* as shown (Fig. 3-8).

Figure 3-8

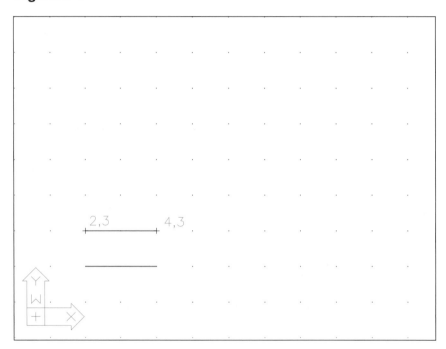

3.   Draw a horizontal *Line* of 2 units length starting at point 2,4 using <u>relative rectangular coordinates</u>.

Steps	Command Prompt	Perform Action	Comments
1.	Command:	select *Line*	use any method
2.	LINE From point:	type **2,4** and press **Enter**	establishes the first endpoint
3.	To point:	type **@2,0** and press **Enter**	@ means "last point"
4.	To point:	press **Enter**	completes command

The new *Line* appears above the previous two as shown in Figure 3-9.

Figure 3-9

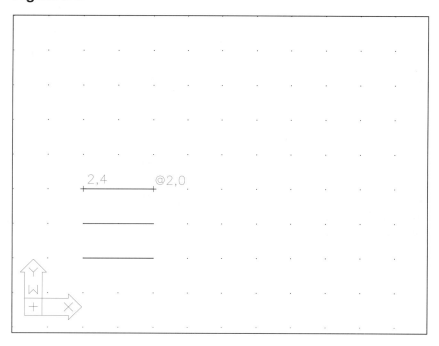

4. Draw a horizontal *Line* of 2 units length starting at point 2,5 using <u>relative polar coordinates</u>.

Steps	Command Prompt	Perform Action	Comments
1.	Command:	select *Line*	use any method
2.	LINE From point:	type **2,5** and press **Enter**	establishes the first endpoint
3.	To point:	type **@2<0** and press **Enter**	@ means "last point" < means "angle of"
4.	To point:	press **Enter**	completes command

The new horizontal *Line* appears above the other three. See Figure 3-10.

Figure 3-10

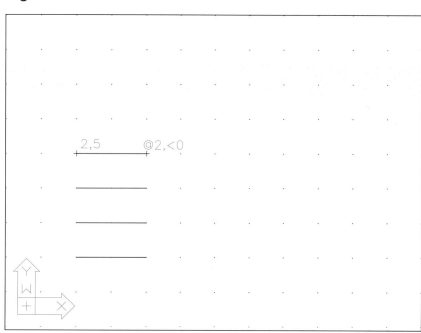

5. Draw a horizontal *Line* of 2 units length starting at point 2,6 using <u>direct distance</u> coordinate entry.

Steps	Command Prompt	Perform Action	Comments
1.	Command:	select *Line*	use any method
2.	LINE From point:	type **2,6** and press **Enter**	establishes the first endpoint
3.	To point:	turn on *ORTHO,* move the cursor to the right, type **2,** then press **Enter**	the cursor movement indicates direction, 2 is the distance
4.	To point:	press **Enter**	completes command

The new line appears above the other four as shown in Figure 3-11.

Figure 3-11

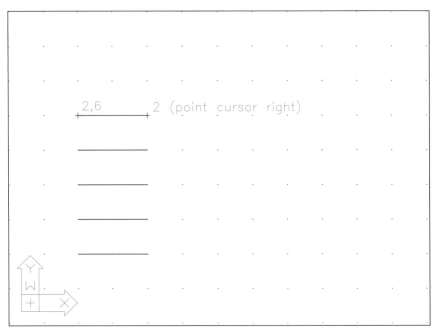

One of these methods may be more favorable than another in a particular situation. The interactive method is fast and easy, assuming that O*SNAP* is used (see Chapter 7) or that *SNAP* and *GRID* are used and set to appropriate values. *SNAP* and *GRID* are used successfully for small drawings where objects have regular interval lengths. Direct distance coordinate entry is fast, easy, and accurate when drawing horizontal or vertical lines. *ORTHO* should be on for most applications of direct distance entry. Direct distance coordinate entry is preferred for horizontal or vertical drawing and editing in cases when O*SNAP* cannot be used (no geometry to O*SNAP* to) and when *SNAP* and *GRID* are not appropriate, such as an architectural or civil engineering application. In these applications, objects are relatively large (lines are long) and are not necessarily drawn to regular intervals; therefore, it is convenient and accurate to enter exact lengths as opposed to interactively PICKing points.

Drawing Vertical Lines

Below are listed the steps for drawing vertical lines using each of the five methods of coordinate entry. The following completed problems should look like those in Figure 3-12.

Figure 3-12

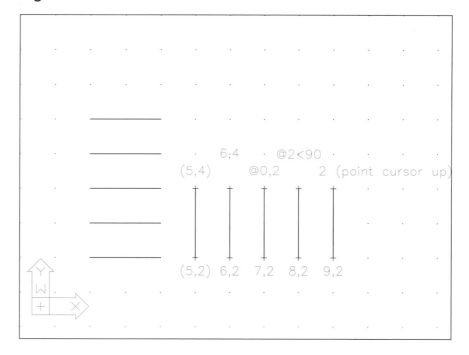

1. Draw a vertical *Line* of 2 units length starting at point 5,2 using the <u>interactive</u> method.

Steps	Command Prompt	Perform Action	Comments
1.		turn on *GRID* (**F7**)	grid appears
2.		turn on *SNAP* (**F9**)	"SNAP" appears on Status line
3.	Command:	select or type **Line**	use any method
4.	LINE From point:	**PICK** location **5,2**	watch *Coords*
5.	To point:	**PICK** location **5,4**	watch *Coords*
6.	To point:	press **Enter**	completes command

2. Draw a vertical *Line* of 2 units length starting at point 6,2 using <u>absolute coordinates</u>.

Steps	Command Prompt	Perform Action	Comments
1.	Command:	select **Line**	use any method
2.	LINE From point:	type **6,2** and press **Enter**	establishes the first endpoint
3.	To point:	type **6,4** and press **Enter**	a *Line* should appear
4.	To point:	press **Enter**	completes command

3. Draw a vertical *Line* of 2 units length starting at point 7,2 using <u>relative rectangular coordinates</u>.

Steps	Command Prompt	Perform Action	Comments
1.	Command:	select *Line*	use any method
2.	LINE From point:	type **7,2** and press **Enter**	establishes the first endpoint
3.	To point:	type **@0,2** and press **Enter**	@ means "last point"
4.	To point:	press **Enter**	completes command

4. Draw a vertical *Line* of 2 units length starting at point 8,2 using <u>relative polar coordinates</u>.

Steps	Command Prompt	Perform Action	Comments
1.	Command:	select *Line*	use any method
2.	LINE From point:	type **8,2** and press **Enter**	establishes the first endpoint
3.	To point:	type **@2<90** and press **Enter**	@ means "last point," < means "angle of"
4.	To point:	press **Enter**	completes command

5. Draw a vertical *Line* of 2 units length starting at point 9,2 using <u>direct distance</u> coordinate entry.

Steps	Command Prompt	Perform Action	Comments
1.	Command:	select *Line*	use any method
2.	LINE From point:	type **9,2** and press **Enter**	establishes the first endpoint
3.	To point:	turn on **ORTHO,** move the cursor upward, type **2,** then press **Enter**	the cursor movement indicates direction, 2 is the distance
4.	To point:	press **Enter**	completes command

The method used for drawing vertical lines depends on the application and the individual, much the same as it would for drawing horizontal lines. Refer to the discussion after the horizontal lines drawing exercises.

Drawing Inclined Lines

Following are listed the steps in drawing <u>inclined lines</u> using each of the five methods of coordinate entry. The following completed problems should look like those in Figure 3-13.

Figure 3-13

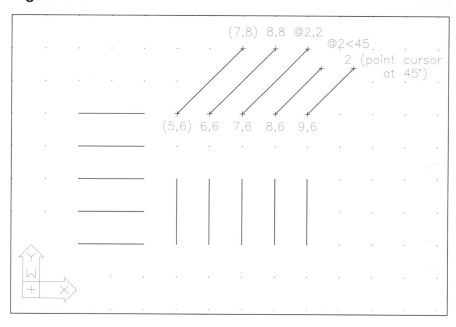

1. Draw an inclined *Line* of from 5,6 to 7,8. Use the <u>interactive</u> method.

Steps	Command Prompt	Perform Action	Comments
1.		turn on *GRID* (**F7**)	grid appears
2.		turn on *SNAP* (**F9**)	"SNAP" appears on Status line
3.		turn off *ORTHO* (**F8**)	in order to draw inclined *Lines*
4.	Command:	select or type *Line*	use any method
5.	LINE From point:	**PICK** location **5,6**	watch *Coords*
6.	To point:	**PICK** location **7,8**	watch *Coords*
7.	To point:	press **Enter**	completes command

2. Draw an inclined *Line* starting at 6,6 and ending at 8,8. Use <u>absolute coordinates</u>.

Steps	Command Prompt	Perform Action	Comments
1.	Command:	select *Line*	use any method
2.	LINE From point:	type **6,6** and press **Enter**	establishes the first endpoint
3.	To point:	type **8,8** and press **Enter**	a *Line* should appear
4.	To point:	press **Enter**	completes command

3.   Draw an inclined *Line* starting at 7,6 and ending 2 units over (in a positive X direction) and 2 units up (in a positive Y direction). Use <u>relative rectangular coordinates</u>.

Steps	Command Prompt	Perform Action	Comments
1.	Command:	select *Line*	use any method
2.	LINE From point:	type **7,6** and press **Enter**	establishes the first endpoint
3.	To point:	type **@2,2** and press **Enter**	@ means "last point"
4.	To point:	press **Enter**	completes command

4.   Draw an inclined *Line* of 2 units length at a 45 degree angle and starting at 8,6. Use <u>relative polar coordinates</u>.

Steps	Command Prompt	Perform Action	Comments
1.	Command:	select *Line*	use any method
2.	LINE From point:	type **8,6** and press **Enter**	establishes the first endpoint
3.	To point:	type **@2<45** and press **Enter**	@ means "last point" < means "angle of"
4.	To point	press **Enter**	completes command

5.   Draw an inclined *Line* of 2 units length at a 45 degree angle starting at point 9,6. Use <u>direct distance</u> coordinate entry.

Steps	Command Prompt	Perform Action	Comments
1.		Turn on *GRID* (**F7**)	grid appears
2.		Turn on *SNAP* (**F9**)	"SNAP" appears on status line
3.	Command:	select *Line*	use any method
4.	LINE From point:	type **9,6** and press **Enter**	establishes the first endpoint
5.	To point:	turn off **ORTHO**, move the cursor up and to the right (the coordinate display must indicate an angle of 45), type **2**, then press **Enter**	the cursor movement indicates direction, 2 is the distance
6.	To point:	press **Enter**	completes command

Method 4, using polar values, is preferred if you are required to draw an inclined line with an exact length and angle. The interactive method and the direct distance coordinate entry method can only be used for drawing inclined lines at 45 degree angle increments (45, 135, etc.) and only when *SNAP* is on and set to equal X and Y intervals. These two methods are <u>not accurate</u> for drawing inclined lines at other than 45 degree increments. Notice that each of the first three methods draws an inclined line of 2.828 (2 * square root of 2) units length since the line is the hypotenuse of a right triangle.

DRAWING *CIRCLES* USING THE FIVE COMMAND ENTRY METHODS

Begin a *New* drawing to complete the *Circle* exercises. If the *Start Up* dialog box or *Create New Drawing* dialog box appears, select *Start from Scratch*, choose *English* as the default setting, and click the *OK* button. The *Circle* command can be invoked by any of the methods shown in the command table.

CIRCLE

Pull-down Menu	COMMAND (TYPE)	ALIAS (TYPE)	Short-cut	Screen (side) Menu	Tablet Menu
Draw *Circle >*	*CIRCLE*	*C*	...	*DRAW 1* *Circle*	*J,9*

Below are listed the steps for drawing *Circles* using the *Center, Radius* method. The circles are to be drawn using each of the four coordinate entry methods and should look like those in Figure 3-14.

Figure 3-14

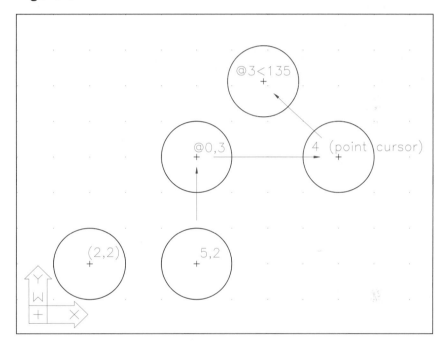

1. Draw a *Circle* of 1 unit radius with the center at point 2,2. Use the <u>interactive</u> method.

Steps	Command Prompt	Perform Action	Comments
1.		turn on *GRID* (**F7**)	grid appears
2.		turn on *SNAP* (**F9**)	"SNAP" appears on Status line
3.		turn on *ORTHO* (**F8**)	"ORTHO" appears on Status line
4.	Command:	select or type **Circle**	use *Center, Radius* method
5.	CIRCLE 3P/2P/TTR/<Center point>:	**PICK** location **2,2**	watch *Coords*
6.	Diameter/<Radius>:	move 1 unit and **PICK**	watch *Coords*

2. Draw a circle of 1 unit radius with the center at point 5,2. Use <u>absolute coordinates</u>.

Steps	Command Prompt	Perform Action	Comments
1.	Command:	select *Circle*	use *Center, Radius* method
2.	CIRCLE 3P/2P/TTR/<Center point>:	type **5,2** and press **Enter**	a *Circle* should appear
3.	Diameter/<Radius>:	type **1** and press **Enter**	the correct *Circle* appears

3. Draw a circle of 1 unit radius with the center 3 units above the last point (previous *Circle* center). Use <u>relative rectangular coordinates</u>.

Steps	Command Prompt	Perform Action	Comments
1.	Command:	select *Circle*	use *Center, Radius* method
2.	CIRCLE 3P/2P/TTR/<Center point>:	type **@0,3** and press **Enter**	a *Circle* should appear
3.	Diameter/<Radius>:	type **1** and press **Enter**	the correct *Circle* appears

(If the new *Circle* is not above the last, type "ID" and enter "5,2." The entered point becomes the last point. Then try again.)

4. Draw a circle of 1 unit radius with the center 4 units to the right of the previous *Circle* center. Use <u>direct distance</u> coordinate entry.

Steps	Command Prompt	Perform Action	Comments
1.		turn on *ORTHO* (**F8**)	"ORTHO" appears on Status line
2.	Command:	select *Circle*	use *Center, Radius* method
3.	CIRCLE 3P/2P/TTR/<Center point>:	type **4,** move the cursor to the right, and press **Enter**	a *Circle* should appear
4.	Diameter/<Radius>:	type **1**, then press **Enter**	the correct *Circle* appears

(If the new *Circle* is not directly to the right of the last, type "ID" and enter "5,5." The entered point becomes the last point. Then try again.)

5. Draw a circle of 1 unit radius with the center 3 units at a 35 degree angle above and to the left of the previous *Circle*. Use <u>relative polar coordinates</u>.

Steps	Command Prompt	Perform Action	Comments
1.	Command:	select *Circle*	use *Center, Radius* method
2.	CIRCLE 3P/2P/TTR/<Center point>:	type **@3<135** and press **Enter**	a *Circle* should appear
3.	Diameter/<Radius>:	type **1** and press **Enter**	the *Circle* appears

(If the new *Circle* is not correctly positioned with respect to the last, type "ID" and enter "9,5." The entered point becomes the last point. Then try again.)

CHAPTER EXERCISES

1. **Start a *New* Drawing**

    Start a *New* drawing. Select the ***Start from Scratch*** option and select ***English*** settings. Next, use the ***Save*** command and save the drawing as "**CH3EX**." Remember to *Save* often as you complete the following exercises. The completed exercise should look like Figure 3-15.

2. **Using interactive coordinate entry**

    Draw a square with sides of 2 units length. Locate the lower-left corner of the square at **2,2**. Use the ***Line*** command with interactive coordinate entry. (HINT: Turn on *SNAP, GRID,* and *ORTHO*.)

3. **Using absolute coordinates**

    Draw another square with 2 unit sides using the ***Line*** command. Enter absolute coordinates. Begin with the lower-left corner of the square at **5,2**.

4. **Using relative rectangular coordinates**

    Draw a third square (with 2 unit sides) using the ***Line*** command. Enter relative rectangular coordinates. Locate the lower-left corner at **8,2**.

5. **Using direct distance coordinate entry**

    Draw a fourth square (with 2 unit sides) beginning at the lower-left corner of **2,5**. Complete the square drawing ***Lines*** using direct distance coordinate entry. Don't forget to turn on ***ORTHO***.

6. **Using relative polar coordinates**

    Draw an equilateral triangle with sides of 2 units. Locate the lower-left corner at **5,5**. Use relative polar coordinates (<u>after</u> establishing the "From point:"). HINT: An equilateral triangle has interior angles of 60 degrees.

7. **Using interactive coordinate entry**

    Draw a *Circle* with a 1 unit <u>radius</u>. Locate the center at **9,6**. Use the interactive method. (Turn on *SNAP* and *GRID*.)

8. **Using relative rectangular, relative polar, or direct distance entry coordinates**

   Draw another *Circle* with a 2 unit <u>diameter.</u> Using any method listed above, locate the center 3 units below the previous *Circle.*

9. *Save* **your drawing**

   Use *Save.* Compare your results with Figure 3-15. When you are finished, *Exit* AutoCAD.

   Figure 3-15 ————————————————————————————

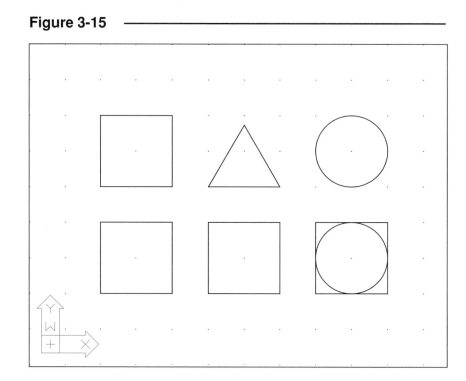

4

SELECTION SETS

Chapter Objectives

After completing this chapter you should:

1. know that Modify commands and many other commands require you to select objects;

2. be able to create a selection set using each of the specification methods;

3. be able to *Erase* objects from the drawing;

4. be able to *Move* objects from one location to another;

5. understand Noun/Verb and Verb/Noun order of command syntax.

MODIFY COMMAND CONCEPTS

Draw commands create objects. Modify commands <u>change existing objects</u> or <u>use existing objects to create new ones</u>. Examples are *Copy* an existing *Circle, Move* a *Line,* or *Erase* a *Circle.*

Since all of the Modify commands use or modify <u>existing</u> objects, you must first select the objects that you want to act on. The process of selecting the objects you want to use is called building a <u>selection set</u>. For example, if you want to *Copy, Erase,* or *Move* several objects in the drawing, you must first select the set of objects that you want to act on.

Remember that any of the five command entry methods (depending on your setup) can be used to invoke Modify commands.

1. **Toolbars**   *Modify* or *ModifyII* toolbar
2. **Pull-down menu**   *Modify* pull-down menu
3. **Screen menu**   *MODIFY1* or *MODIFY2* screen menus
4. **Keyboard**   Type the command name <u>or</u> command alias
5. **Tablet menu**   Select the command icon

All of the Modify commands will be discussed in detail later, but for now we will focus on how to build selection sets.

SELECTION SETS

No matter which of the five methods you use to invoke a Modify command, you must specify a selection set during the command operation. There are two ways you can select objects: you can select the set of objects either (1) immediately before you invoke the command or (2) when the command prompts you to select objects. For example, examine the command syntax that would appear at the command line when the *Erase* command is used (method 2).

> Command: **erase**
> Select Objects:

The "Select Objects:" prompt is your cue to use any of several methods to PICK the objects to erase. As a matter of fact, every Modify command begins with the same "Select Objects:" prompt (unless you selected immediately before invoking the command).

When the "Select Objects:" prompt appears, the "crosshairs" cursor disappears and only a small, square pickbox appears at the cursor (Fig. 4-1).

Figure 4-1

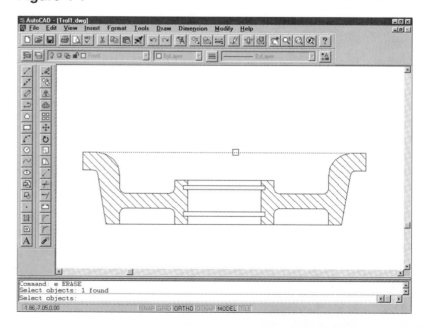

You can PICK objects using only the pickbox or any of several other methods illustrated in this chapter. (Only when you PICK objects at the "Select Objects:" prompt can you use all of the selection methods shown here. If you PICK immediately before the command, called Noun/Verb order, only the *AUto* method can be used. See Noun/Verb Syntax in this chapter.)

When the objects have been selected, they become highlighted (displayed as a broken line), which serves as a visual indication of the current selection set. Press Enter to indicate that you are finished selecting and are ready to proceed with the command.

Accessing Selection Set Options

When the "Select Objects:" prompt appears, you should select objects using the pickbox or one of a variety of other methods. Any method can be used <u>independently</u> or <u>in combination</u> to achieve the desired set of objects because object selection is a <u>cumulative</u> process.

The pickbox is the default option which can automatically be changed to a window or crossing window by PICKing in an open area (PICKing no objects). The other methods can be selected from the screen (side) menu or by typing the capitalized letters shown in the option names following.

If the screen (side) menu is displayed, you can access the object selection options from the *ASSIST* menu. *ASSIST* is displayed at the bottom of all other menus (Fig 4-2).

Figure 4-2

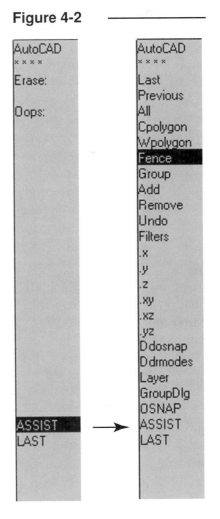

Optionally, you can type the first letter (or two in some cases) of any option explained in the next section. For example, to use the *Fence* option, type *F* at the "Select objects:" prompt. *Window Polygon* (*WP*) and *Crossing Polygon* (*CP*) are the only options that require typing two letters.

In Release 14, there are no icon buttons or pull-down menu selections that are easily available for the object selection options. However, a toolbar can be "customized" to display the object selection options (see Chapter 45, Basic Customization).

Selection Set Options

The options for creating selection sets (PICKing objects) are shown on this and the following pages. Two *Circles* and five *Lines* (as shown in Figure 4-3) are used for every example. In each case, the <u>circles only</u> are selected for editing. If you want to follow along and practice as you read, draw *Circles* and *Lines* in a configuration similar to this. Then use a *Modify* command. (Press Escape to cancel the command after selecting objects.)

Figure 4-3

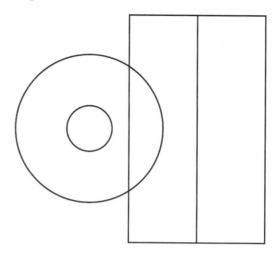

pickbox

This default option is used for selecting <u>one object</u> at a time. Locate the pickbox so that an object crosses through it and **PICK** (Fig. 4-4). You do not have to type or select anything to use this option.

Figure 4-4

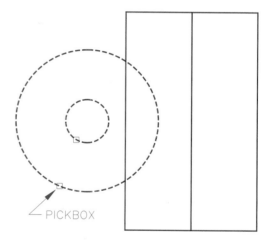

AUto

To use this option, you do not have to type or select anything from the Select Objects menu or toolbars. The pickbox must be positioned in an open area so that no objects cross through it; then **PICK** to start a window. If you drag to the <u>right</u>, a *Window* is created (Fig. 4-5). **PICK** the other corner.

Figure 4-5

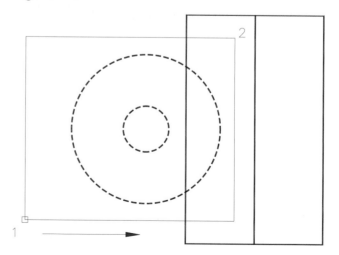

If you drag to the <u>left</u> instead, a *Crossing Window* forms (Fig. 4-6). (See Window and Crossing Window below.)

Figure 4-6

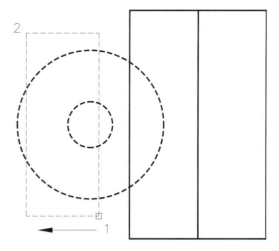

Window
Only objects <u>completely within</u> the *Window* are selected. The *Window* is a solid linetype rectangular box. Select the first and second points (diagonal corners) in any direction as shown in Figure 4-7.

Figure 4-7

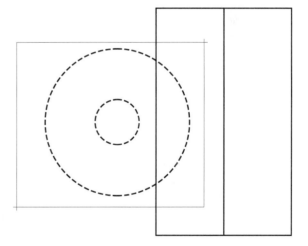

Crossing Window
All objects <u>within and crossing through</u> the window are selected. The *Crossing Window* is displayed as a broken linetype rectangular box (Fig. 4-8). Select two diagonal <u>corners</u>.

Figure 4-8

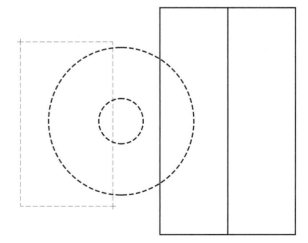

Window Polygon

The *Window Polygon* operates like a *Window*, but the box can be <u>any</u> irregular polygonal shape (Fig. 4-9). You can pick any number of corners to form any shape rather than picking just two corners to form a rectangle as with the *Window* option.

Figure 4-9

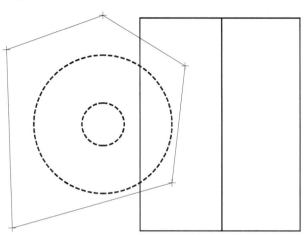

Crossing Polygon

The *Crossing Polygon* operates like a *Crossing Window*, but can have any number of corners like *Window Polygon* (Fig. 4-10).

Figure 4-10

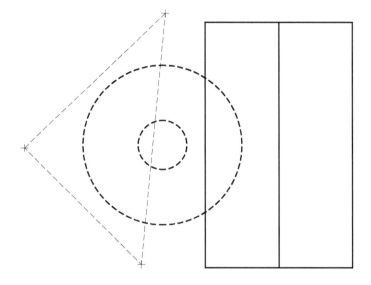

Fence

This option operates like a <u>crossing line</u>. Any objects crossing the *Fence* are selected. The *Fence* can have any number of segments (Fig. 4-11).

Figure 4-11

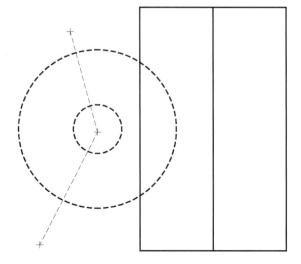

Last

This option automatically finds and selects <u>only</u> the last object <u>created</u>. *Last* does not find the last object modified (with *Move, Stretch,* etc.).

Previous

Previous finds and selects the <u>previous selection set</u>, that is, whatever was selected during the previous command (except after *Erase*). This option allows you to use several editing commands on the same set of objects without having to respecify the set.

ALL

This option selects <u>all objects</u> in the drawing except those on *Frozen* or *Locked* layers (*Layers* are covered in Chapter 12).

Remove

Selecting this option causes AutoCAD to <u>switch</u> to the "Remove objects:" mode. Any selection options used from this time on remove objects from the highlighted set (see Figure 4-12).

Figure 4-12

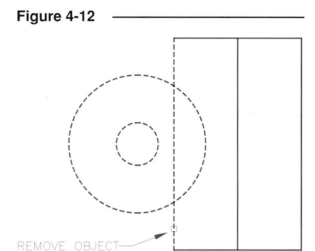

Add

The *Add* option switches back to the default "Select objects:" mode so additional objects can be added to the selection set.

Shift + #1

Holding down the **Shift** key and pressing the **#1** button simultaneously <u>removes</u> objects selected from the highlighted set as shown in Figure 4-12. This method is generally quicker than, but performs the same action as, *Remove*. The advantage here is that the *Add* mode is in effect unless Shift is held down.

REMOVE OBJECT
FROM SELECTION SET

Group

The *Group* option selects groups of objects that were previously specified using the *Group* command. Groups are selection sets to which you can assign a name (see Chapter 20).

Ctrl + #1 (Object Cycling)

Holding down the **Ctrl** key and simultaneously pressing the **PICK** (#1) button causes AutoCAD to cycle through (highlight one at a time) two or more objects that may cross through the pickbox.

For example, if you attempted to PICK an object but other objects were in the pickbox, AutoCAD may not highlight the one you want (Fig. 4-13). In this case, hold down the **Ctrl** key and press the **PICK** button several times (the cursor can be located anywhere on the screen during cycling). All of the objects that passed through the pickbox will be highlighted one at a time. The "<Cycle on>" prompt appears at the command line. When the object you want is highlighted, press Enter and AutoCAD adds the object to the selection set and returns to the "Select objects:" prompt.

Figure 4-13

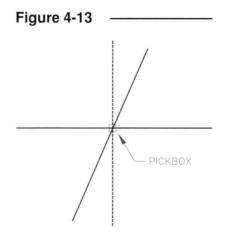

PICKBOX

SELECT

Pull-down Menu	COMMAND (TYPE)	ALIAS (TYPE)	Short-cut	Screen (side) Menu	Tablet Menu
...	*SELECT*	...	...	...	...

The *Select* command can be used to PICK objects to be saved in the selection set buffer for subsequent use with the *Previous* option. Any of the selection methods can be used to PICK the objects.

    Command: **select**
    Select Objects: **PICK** (use any selection option)
    Select Objects: **Enter** (completes the selection process)
    Command:

The selected objects become unhighlighted when you complete the command by pressing Enter. The objects become highlighted again and are used as the selection set if you use the *Previous* selection option in the <u>next</u> editing command.

NOUN/VERB SYNTAX

An object is the <u>noun</u> and a command is the <u>verb</u>. Noun/Verb syntax order means to pick objects (nouns) first, then use an editing command (verb) second. If you select objects first (at the open Command: prompt) and then immediately choose a Modify command, AutoCAD recognizes the selection set and passes through the "Select Objects:" prompt to the next step in the command.

Verb/Noun means to invoke a command and then select objects within the command. For example, if the *Erase* command (verb) is invoked first, AutoCAD then issues the prompt to "Select Objects:"; therefore, objects (nouns) are PICKed second. In the previous examples, and with older versions of AutoCAD, only Verb/Noun syntax order was used.

You can use <u>either</u> order you want (Noun/Verb or Verb/Noun) and AutoCAD automatically understands. If objects are selected first, the selection set is passed to the next editing command used, but if no objects are selected first, the editing command automatically prompts you to "Select Objects:".

If you use Noun/Verb order, you are limited to using only the *AUto* options for object selection (pickbox, window, and crossing window). You can only use the other options (e.g., *Crossing Polygon, Fence, Previous*) if you invoke the desired Modify command first, then select objects when the "Select Objects:" prompt appears.

The *PICKFIRST* variable (a very descriptive name) enables Noun/Verb syntax. The default setting is 1 (*On*). If *PICKFIRST* is set to 0 (*Off*), Noun/Verb syntax is disabled and the selection set can be specified only <u>within</u> the editing commands (Verb/Noun).

Setting *PICKFIRST* to 1 provides two options: Noun/Verb and Verb/Noun. You can use <u>either</u> order you want. If objects are selected first, the selection set is passed to the next editing command, but if no objects are selected first, the editing command prompts you to select objects. See Chapter 20 for a complete explanation of *PICKFIRST* and advanced selection set features.

SELECTION SETS PRACTICE

NOTE: While learning and practicing with the editing commands, it is suggested that *GRIPS* be turned off. This can be accomplished by typing in **GRIPS** and setting the *GRIPS* variable to a value of **0**. The AutoCAD Release 14 default has *GRIPS* on (set to 1). *GRIPS* is covered in Chapter 23.

Begin a *New* drawing to complete the selection set practice exercises. If the *Start Up* dialog box or *Create New Drawing* dialog box appears, select *Start from Scratch,* choose *English* as the default setting, and click the *OK* button. The *Erase, Move,* and *Copy* commands can be activated by any one of the methods shown in the command tables that follow.

Using *Erase*

Erase is the simplest editing command. *Erase* removes objects from the drawing. The only action required is the selection of objects to be erased.

ERASE

Pull-down Menu	COMMAND (TYPE)	ALIAS (TYPE)	Short-cut	Screen (side) Menu	Tablet Menu
Modify *Erase*	*ERASE*	*E*	...	*MODIFY1* *Erase*	V,14

1. Draw several *Lines* and *Circles*. Practice using the object selection options with the *Erase* command. The following sequence uses the pickbox, Window, and Crossing Window.

STEPS	COMMAND PROMPT	PERFORM ACTION	COMMENTS
1.	Command:	type *E* and press **Spacebar**	*E* is the alias for *Erase*, Spacebar can be used like Enter
2.	ERASE Select Objects:	use pickbox to select one or two objects	objects are highlighted
3.	Select Objects:	type *W* or use a *Window*, then select more objects	objects are highlighted
4.	Select Objects:	type *C* or use a *Crossing Window*, then select objects	objects are highlighted
5.	Select Objects:	press **Enter**	objects are erased

2. Draw several more *Lines* and *Circles*. Practice using the *Erase* command with the *AUto Window* and *AUto Crossing Window* options as indicated below.

STEPS	COMMAND PROMPT	PERFORM ACTION	COMMENTS
1.	Command:	select the *Modify* pull-down, then *Erase*	
2.	ERASE Select Objects:	use pickbox to **PICK** an open area, drag *Window* to the <u>right</u> to select objects	objects inside Window are highlighted
3.	Select Objects:	**PICK** an open area, drag *Crossing Window* to the <u>left</u> to Select Objects	objects inside and crossing through Window are highlighted
4.	Select Objects:	press **Enter**	objects are erased

Using *Move*

The *Move* command specifically prompts you to (1) select objects, (2) specify a "basepoint," or point to move <u>from</u>, and (3) specify a "second point of displacement," or point to move <u>to</u>.

MOVE

Pull-down Menu	COMMAND (TYPE)	ALIAS (TYPE)	Short-cut	Screen (side) Menu	Tablet Menu
Modify *Move*	*MOVE*	*M*	...	*MODIFY2* *Move*	*V,19*

1. Draw a *Circle* and two *Lines*. Use the *Move* command to practice selecting objects and to move one *Line* and the *Circle* as indicated in the following table.

STEPS	COMMAND PROMPT	PERFORM ACTION	COMMENTS
1.	Command:	type *M* and press **Spacebar**	*M* is the command alias for *Move*
2.	MOVE Select Objects:	use pickbox to select one *Line* and the *Circle*	objects are highlighted
3.	Select Objects:	press **Spacebar** or **Enter**	
4.	Base point or displacement:	**PICK** near the *Circle* center	basepoint is the handle, or where to move <u>from</u>
5.	Second point of displacement:	**PICK** near the other *Line*	second point is where to move <u>to</u>

2. Use *Move* again to move the *Circle* back to its original position. Select the *Circle* with the *Window* option.

STEPS	COMMAND PROMPT	PERFORM ACTION	COMMENTS
1.	Command:	select the *Modify* pull-down, then *Move*	
2.	MOVE Select Objects:	type *W* and press **Spacebar**	select only the circle; object is highlighted
3.	Select Objects:	press **Spacebar** or **Enter**	
4.	Base point or displacement:	**PICK** near the *Circle* center	basepoint is the handle, or where to move <u>from</u>
5.	Second point of displacement:	**PICK** near the original location	second point is where to move <u>to</u>

Using *Copy*

The *Copy* command is similar to *Move* because you are prompted to (1) select objects, (2) specify a "base-point," or point to copy <u>from,</u> and (3) specify a "second point of displacement," or point to copy <u>to.</u>

COPY

	Pull-down Menu	COMMAND (TYPE)	ALIAS (TYPE)	Short-cut	Screen (side) Menu	Tablet Menu
	Modify *Copy*	*COPY*	*CO*	...	*MODIFY1* *Copy*	*V,15*

1. Using the *Circle* and 2 *Lines* from the previous exercise, use the *Copy* command to practice selecting objects and to make copies of the objects as indicated in the following table.

STEPS	COMMAND PROMPT	PERFORM ACTION	COMMENTS
1.	Command:	type *CO* and press **Enter** or **Spacebar**	*CO* is the command alias for *Copy*
2.	COPY Select objects:	type *F* and press **Enter**	*F* invokes the *Fence* selection option
3.	First fence point:	**PICK** a point near two *Lines*	starts the "fence"
4.	Undo/<Endpoint of line>:	**PICK** a second point across the two *Lines*	lines are highlighted
5.	Undo/<Endpoint of line>:	Press **Enter**	completes *Fence* option
6.	Select objects:	Press **Enter**	completes object selection
7.	<Base point or displacement>/Multiple:	**PICK** between the *Lines*	basepoint is the handle, or where to copy <u>from</u>
8.	Second point of displacement:	enter *@2<45* and press **Enter**	copies of the lines are created 2 units in distance at 45 degrees from the original location

2. Practice removing objects from the selection set by following the steps given in the table below.

STEPS	COMMAND PROMPT	PERFORM ACTION	COMMENTS
1.	Command:	type *CO* and press **Enter** or **Spacebar**	*CO* is the command alias for *Copy*
2.	COPY Select objects:	**PICK** in an open area near the right of your drawing, drag Window to the left to select all objects	all objects within and crossing the Window are highlighted
3.	Select objects:	hold down **Shift** and **PICK** all highlighted objects except one *Circle*	holding down Shift while PICKing removes objects from the selection set
4.	Select objects:	press **Enter**	only one circle is highlighted
5.	<Base point or displacement>Multiple:	**PICK** near the circle's center	basepoint is the handle, or where to copy <u>from</u>
6.	Second point of displacement:	turn on *ORTHO*, move the cursor to the right, type **3** and press **Enter**	a copy of the circle is created 3 units to the right of the original circle

NOUN/VERB Command Syntax Practice

1. Practice using the *Move* command using Noun/Verb syntax order by following the steps in the table below.

STEPS	COMMAND PROMPT	PERFORM ACTION	COMMENTS
1.	Command:	**PICK** one *Circle*	circle becomes highlighted (grips may appear if not disabled)
2.	Command:	type *M* and press **Enter** or **Spacebar**	*M* is the command alias for *Move*
3.	MOVE 1 found Base point or displacement:	**PICK** near the *Circle's* center	command skips the "Select objects" prompt and proceeds
4.	Second point of displacement:	enter **@-3,-3** and press **Enter**	circle is moved -3 X units and -3 Y units from the original location

2. Practice using Noun/Verb syntax order by selecting objects for *Erase*.

STEPS	COMMAND PROMPT	PERFORM ACTION	COMMENTS
1.	Command:	**PICK** one *Line*	line becomes highlighted (grips may appear if not disabled)
2.	Command:	type *E* and press *Enter* or **Spacebar**	*E* is the command alias for *Erase*
3.	ERASE 1 found		command skips the "Select objects" prompt and erases highlighted line

CHAPTER EXERCISES

Open drawing **CH3EX** that you created in Chapter 3 Exercises. Turn off *SNAP* (**F9**) to make object selection easier.

1. **Use the pickbox to select objects**

   Invoke the *Erase* command by any method. Select the lower left square with the pickbox (Fig. 4-14, highlighted). Each *Line* must be selected individually. Press **Enter** to complete *Erase*. Then use the *Oops* command to unerase the square. (Type *Oops*.)

Figure 4-14 ───────────

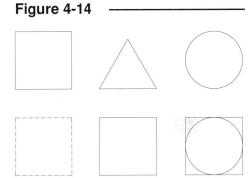

2. **Use the *AUto* Window and *Auto* Crossing Window**

   Invoke *Erase*. Select the center square on the bottom row with the *AUto* Window and select the equilateral triangle with the *AUto* Crossing Window. Press **Enter** to complete the *Erase* as shown in Figure 4-15. Use *Oops* to bring back the objects.

Figure 4-15 ───────────

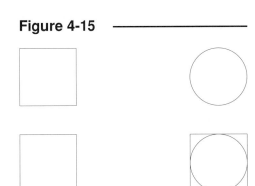

3. **Use the *Fence* selection option**

   Invoke *Erase* again. Use the *Fence* option to select all the vertical *Lines* and the *Circle* from the squares on the bottom row. Complete the *Erase* (see Fig. 4-16). Use ***Oops*** to unerase.

Figure 4-16 —————————

4. **Use the *ALL* option and deselect**

   Use *Erase*. Select all the objects with ***ALL***. Remove the four *Lines* (shown highlighted in Fig. 4-17) from the selection set by pressing **Shift** while PICKing. Complete the *Erase* to leave only the four *Lines*. Finally, use ***Oops***.

Figure 4-17 —————————

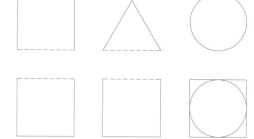

5. **Using Noun/Verb selection**

   <u>Before</u> invoking *Erase,* use the pickbox or *AUto* window to select the triangle. (Make sure no other commands are in use.) <u>Then</u> invoke ***Erase***. The triangle should disappear. Retrieve the triangle with ***Oops***.

6. **Using *Move* with *Wpolygon***

   Invoke the ***Move*** command by any method. Use the **WP** option (*Window Polygon*) to select only the *Lines* comprising the triangle. Turn on *SNAP* and PICK the lower-left corner as the "Base point:". *Move* the triangle up 1 unit. (See Fig. 4-18.)

Figure 4-18 —————————

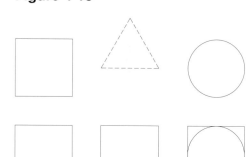

7. **Using *Previous* with *Move***

   Invoke ***Move*** again. At the "Select Objects:" prompt, type **P** for *Previous*. The triangle should highlight. Using the same basepoint, move the triangle back to its original position.

8. ***Exit*** AutoCAD and do <u>not</u> save changes.

5

HELPFUL COMMANDS

Chapter Objectives

After completing this chapter you should be able to:

1. find *Help* for any command or system variable;

2. use *Oops* to unerase objects;

3. use *U* to undo one command or use *Undo* to undo multiple commands;

4. set a *Mark* and use *Undo Back* to undo all commands until the marker is encountered;

5. *Redo* commands that were undone;

6. regenerate the drawing with *Regen*.

CONCEPTS

There are several commands that do not draw or edit objects in AutoCAD, but are intended to assist you in using AutoCAD. These commands are used by experienced AutoCAD users and are particularly helpful to the beginner. The commands, as a group, are not located in any one menu, but are scattered in several menus.

HELP

Pull-down Menu	COMMAND (TYPE)	ALIAS (TYPE)	Short-cut	Screen (side) Menu	Tablet Menu
Help	*HELP*	*?*	*F1*	*HELP* *Help*	*Y,7*

Help gives you an explanation for any AutoCAD command or system variable as well as help for using the menus and toolbars. *Help* displays a window which gives a variety of methods for finding the information that you need. There is even help for using help!

Help can be used two ways: (1) entered as a command at the open Command: prompt or (2) used transparently while a command is currently in use.

1. If the *Help* command is entered at an open Command: prompt (when no other commands are in use), the *Help* window appears (Fig. 5-1).

Figure 5-1

2. When *Help* is used transparently (when a command is in use), it is context sensitive; that is, help on the current command is given automatically. For example, Figure 5-2 displays the window that appears if *Help* is invoked during the *Line* command. (If typing a transparent command, an ' [apostrophe] symbol is typed as a prefix to the command, e.g., **'HELP** or **'?**. If you PICK *Help* from the menus or press **F1**, it is automatically transparent.)

Figure 5-2

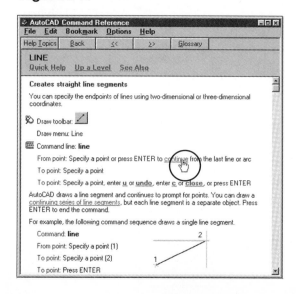

Much of the text that appears in the window can be PICKed to reveal another level of help on that item. This feature, called Hypertext, is activated by moving the pointer to a word (usually underlined and in a green color) or an icon. When the pointer changes to a small hand (Fig. 5-2), click on the item to activate the new information.

If you use *Help* when no commands are in use (at the open Command: prompt), the *Help Topics: AutoCAD Help* dialog box is displayed. There are three tabs in this dialog box described below, *Contents*, *Index*, and *Find*.

Contents

Several levels of the *Contents* tab are available, offering an overwhelming amount of information (Fig. 5-3). Each main level can be opened to reveal a second (chapter level) or third level of information. The main levels are:

Figure 5-3

Using Help	Get instruction on how to use the *Help* system and the on-line documentation.
Tutorial	Several tutorials are given to assist you to learn AutoCAD.
How To. . .	Select from a list of several main topics such as *Set up a Drawing*, *Using Grips*, or *Modify a Dimension*. Another level is available for each of these topics.
Command Reference	The following informative sections are available: *Commands* *System Variables* *Menus* *Toolbars* *Utilities* *Standard Libraries* *AutoCAD Graphical Objects*
User's Guide	Seventeen chapters and three appendices are given, each with several sections explaining concepts and commands for using AutoCAD.
Installation Guide	Eight chapters can be accessed to give information on installing, configuring, and optimizing AutoCAD.
Customization Guide	This section includes information such as how to customize linetypes, shapes, fonts, menus, dialog boxes, and how to write AutoLISP programs.
ActiveX Automation	This section gives information concerning a new programming interface that allows you to manipulate AutoCAD objects from an application controller such as Visual Basic or Excel.

Figure 5-4

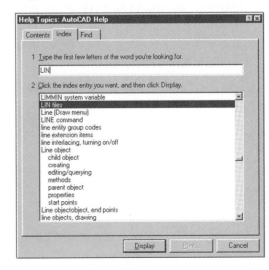

Index

In this section you can type two or three letters of a word that you want information about (Fig. 5-4). As you type, the list below displays the available contents beginning with the letters that you have entered. The word you type can be a command, system variable, term, or concept used in AutoCAD. Once the word you want information on is found, press Enter or click the *Display* button to invoke a new *AutoCAD Help* box.

Find

Find operates like the *Index* function, except you can enter several words describing the topic you want help with (Fig. 5-5). There are three steps to using the *Find* function: (1) type in word or words in the top edit box, (2) select a topic from the list in the center to narrow your search, (3) select a topic from the bottom list and click *Display*. An *AutoCAD Help* box appears.

How To... or To...

Because so much information and so many topics are listed in the *Help* utility, it may take several attempts to locate the information you are looking for. Often your search results in information about programming features or advanced topics that you did not intend to find. Generally, if you are looking for information on how to use AutoCAD to create and edit objects, use any listing that begins with **To...** (do something), or use the **How To...** section of the *Contents* tab. In either case, the resulting *AutoCAD User's Guide* window gives step-by-step information explaining the procedure for accomplishing a task (Fig. 5-6). This window appears near the side of the screen so you can draw or edit in the graphics area while reading the instructions in the Help box. This is generally the most useful help for beginning AutoCAD users.

The simplest way to invoke the *AutoCAD User's Guide* window is by following these steps:

1. Invoke *Help* and select the *Index* tab;
2. Type the command you want help for in the top edit box, <u>then press Enter</u>.
3. The *Topics Found* box appears displaying *AutoCAD User's Guide (AUG)* selections (Fig. 5-7). Make a selection.
4. The *AutoCAD User's Guide* appears on the side of the screen (Fig 5-6).

Figure 5-5

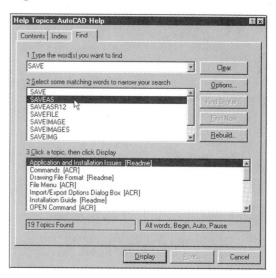

Figure 5-6

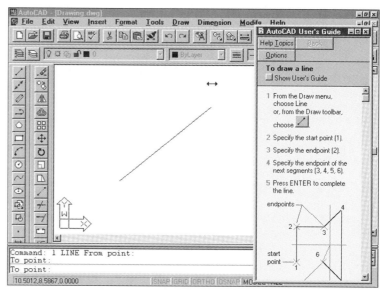

Figure 5-7

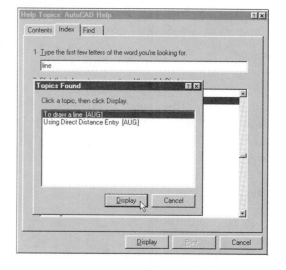

OOPS

Pull-down Menu	COMMAND (TYPE)	ALIAS (TYPE)	Short-cut	Screen (side) Menu	Tablet Menu
...	OOPS	...	...	*MODIFY1* *Erase* *Oops:*	...

The *Oops* command unerases whatever was erased with the <u>last</u> *Erase* command. *Oops* does not have to be used immediately after the *Erase*, but can be used at <u>any time after</u> the *Erase*. *Oops* is typically used after an accidental erase. However, *Erase* could be used intentionally to remove something from the screen temporarily to simplify some other action. For example, you can *Erase* a *Line* to simplify PICKing a group of other objects to *Move* or *Copy*, and then use *Oops* to restore the erased *Line*.

Oops can be used to restore the original set of objects after the *Block* or *Wblock* command is used to combine many objects into one object (explained in Chapter 21).

The *Oops* command is available only from the side menu; there is no icon button or option available in the pull-down menu. *Oops* appears on the side menu only <u>if</u> the *Erase* command is typed or selected. Otherwise, *Oops* must be typed.

U

Pull-down Menu	COMMAND (TYPE)	ALIAS (TYPE)	Short-cut	Screen (side) Menu	Tablet Menu
Edit *Undo*	U	...	Crtl+Z	*EDIT* *Undo*	T,12

The *U* command undoes only the <u>last</u> command. *U* means "undo one command." If used after *Erase*, it unerases whatever was just erased. If used after *Line*, it undoes the group of lines drawn with the last *Line* command. Both *U* and *Undo* do not undo inquiry commands (like *Help*), the *Plot* command, or commands that cause a write-to-disk, such as *Save*.

If you type the letter **U**, select the icon button, or select **Undo** from the pull-down menu or screen (side) menu, only the last command is undone. Typing **Undo** invokes the full *Undo* command. The *Undo* command is explained next.

UNDO

Pull-down Menu	COMMAND (TYPE)	ALIAS (TYPE)	Short-cut	Screen (side) Menu	Tablet Menu
...	UNDO	...	...	*ASSIST* *Undo*	...

The full *Undo* command (as opposed to the *U* command) must be typed. This command has the same effect as the *U* command in that it undoes the previous command(s). However, the *Undo* command allows you to <u>undo multiple commands</u> in reverse chronological order. For example, you can use *Undo* and enter a value of **5** to undo the last 5 commands you used. Entering a value (specifying the number of commands to undo) is the default option of *Undo*. The command syntax is as follows:

```
Command: undo
Auto/Control/BEgin/End/Mark/Back/<Number>:
```

All of the *Undo* options are listed next.

<number>
Enter a value for the number of commands to *Undo*. This is the default option.

Mark
This option sets a marker at that stage of the drawing. The marker is intended to be used by the *Back* option for future *Undo* commands.

Back
This option causes *Undo* to go back to the last marker encountered. Markers are created by the *Mark* option. If a marker is encountered, it is removed. If no marker is encountered, beware, because *Undo* goes back to the <u>beginning of the session</u>. A warning message appears in this case.

BEgin
This option sets the first designator for a group of commands to be treated as one *Undo*.

End
End sets the second designator for the end of a group.

Auto
If *On, Auto* treats each command as one group; for example, several lines drawn with one *Line* command would all be undone with *U*.

Control
This option allows you to disable the *Undo* command or limit it to one undo each time it is used.

REDO

Pull-down Menu	COMMAND (TYPE)	ALIAS (TYPE)	Short-cut	Screen (side) Menu	Tablet Menu
Edit *Redo*	*REDO*	...	*Crtl+Y*	*EDIT* *Redo*	*U,12*

The *Redo* command undoes an *Undo*. *Redo* must be used as the <u>next</u> command after the *Undo*. The result of *Redo* is as if *Undo* was never used. Remember, *U* or *Undo* can be used anytime, but *Redo* has no effect unless used immediately after *U* or *Undo*.

REDRAW

Pull-down Menu	COMMAND (TYPE)	ALIAS (TYPE)	Short-cut	Screen (side) Menu	Tablet Menu
View *Redraw*	*REDRAW*	*R*	...	*VIEW 1* *Redraw*	...

Redraw erases the "blips" from the screen if *BLIPMODE* is on. If *BLIPMODE* is on, small crosses appear on the screen any time you PICK a point during a command such as when selecting objects, drawing *Lines*, etc., or creating windows. The "blips" remain on the screen until a viewing command is used (such as *Zoom*), or a regeneration occurs, or until *Redraw* is used. In AutoCAD Release 14, *BLIPMODE* is off by default, so there is little need to use *Redraw* when *BLIPMODE* is off.

In previous releases of AutoCAD, *BLIPMODE* is on by default. Therefore, if you work with drawings created in previous versions of AutoCAD, the "blips" may appear. The *BLIPMODE* variable setting is stored in the drawing file. *BLIPMODE* can be controlled by typing *BLIPMODE* at the command prompt or using the *Drawing Aids* dialog box from the *Tools* pull-down menu.

Redrawall is needed only when Viewports are being used (see Chapters 19, 33, and 35).

REGEN

Pull-down Menu	COMMAND (TYPE)	ALIAS (TYPE)	Short-cut	Screen (side) Menu	Tablet Menu
View *Regen*	*REGEN*	*RE*	...	*VIEW 1* *Regen*	J,1

The *Regen* command reads the database and redisplays the drawing accordingly. A *Regen* is caused by some commands automatically. Occasionally, the *Regen* command is required to update the drawing to display the latest changes made to some system variables. For example, if you change the style in which *Point* objects appear (displayed as a dot, an "X," or other options), you should then *Regen* in order to display the *Points* according to the new setting. (*Points* are explained in Chapter 8.) *Regenall* is used to regenerate all viewports when several viewports are being used.

CHAPTER EXERCISES

Begin a *New* drawing. If the *Start Up* dialog box or *Create New Drawing* dialog box appears, select **Start from Scratch**, choose **English** as the default setting, and click the **OK** button. Complete the following exercises.

1. *Help*

   Use **Help** by any method to find information on the following commands. Use the *Contents* tab and select the *Command Reference* to locate information on each command. (Use *Help* at the open Command: prompt, not during a command in use.) Read the text screen for each command.

   > *New, Open, Save, SaveAs*
   > *Oops, Undo, U*

2. **Context-sensitive *Help***

   Invoke each of the commands listed below. When you see the first prompt in each command, enter *'Help* or *'?* (transparently) or select **Help** from the menus or Standard toolbar. Read the explanation for each prompt. Select a Hypertext item in each screen (underlined and green in color).

   > *Line, Arc, Circle, Point*

3. *Oops*

   Draw 3 vertical **Lines**. **Erase** one line; then use **Oops** to restore it. Next **Erase** two *Lines*, each with a separate use of the *Erase* command. Use **Oops**. Only the last *Line* is restored. **Erase** the remaining two *Lines*, but select both with a Window. Now use **Oops** to restore both *Lines* (since they were *Erased* at the same time).

4. **Using delayed *Oops***

   Oops can be used at any time, not only immediately after the *Erase*. Draw several horizontal **Lines** near the bottom of the screen. Draw a **Circle** on the **Lines**. Then **Erase** the *Circle*. Use **Move**, select the *Lines* with a Window, and displace the *Lines* to another location above. Now use **Oops** to make the *Circle* reappear.

5. **U**

   Press the letter **U** (make sure no other commands are in use). The *Circle* should disappear (*U* undoes the last command—*Oops*). Do this repeatedly to *Undo* one command at a time until the *Circle* and *Lines* are in their original position (when you first created them).

6. **Undo**

   Use the **Undo** command and select the **Back** option. Answer **Yes** to the warning message. This action should *Undo* everything.

   Draw a vertical **Line**. Next, draw a square with four **Line** segments (all drawn in the same *Line* command). Finally, draw a second vertical **Line**. **Erase** the first *Line*.

   Now <u>type</u> **Undo** and enter a value of **3**. You should have only one *Line* remaining. *Undo* reversed the following three commands:

Erase	The first vertical *Line* was unerased.
Line	The second vertical *Line* was removed.
Line	The four *Lines* comprising the square were removed.

7. **Redo**

   Invoke **Redo** immediately after the *Undo* (from the previous exercise). The three commands are redone. *Redo* must be used immediately after *Undo*.

8. **Exit** AutoCAD and answer **No** to "Save Changes to…?"

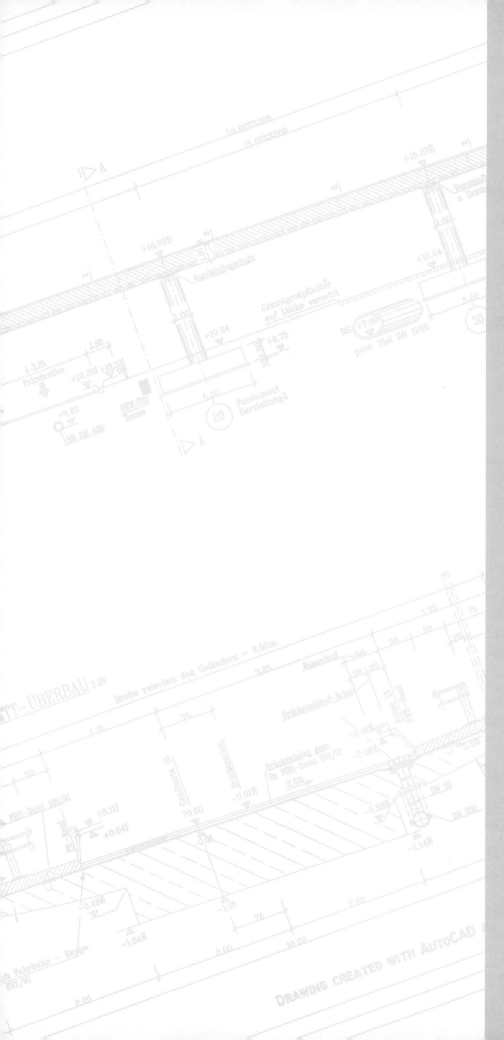

BASIC
DRAWING
SETUP

Chapter Objectives

After completing this chapter you should:

1. know the basic steps for setting up a drawing;

2. be able to specify the desired *Units, Angles* format, and *Precision* for the drawing;

3. be able to specify the drawing *Limits;*

4. know how to specify the *Snap* increment;

5. know how to specify the *Grid* increment;

6. know how to use the *Setup* and *Create New Drawing* dialog boxes to accomplish basic drawing setup.

STEPS FOR BASIC DRAWING SETUP

Assuming the general configuration (dimensions and proportions) of the geometry to be created is known, the following steps are suggested for setting up a drawing:

1. Determine and set the *Units* that are to be used.
2. Determine and set the drawing *Limits;* then *Zoom All.*
3. Set an appropriate *Snap* value.
4. Set an appropriate *Grid* value to be used.

These additional steps for drawing setup are discussed in Chapter 13, Advanced Drawing Setup.

5. Change the *LTSCALE* value based on the new *Limits.*
6. Create the desired *Layers* and assign appropriate *linetype* and *color* settings.
7. Create desired *Text Styles* (optional).
8. Create desired *Dimension Styles* (optional).
9. Create a title block and border (optional).

When you start AutoCAD Release 14 for the first time or use the *New* command to begin a new drawing, the *Start Up* or *Create New Drawing* dialog box appears. Each of these dialog boxes makes available three options for setting up a new drawing. The three options represent three levels of automation/preparation for drawing setup. Only the *Start from Scratch* option is described in this chapter. The *Start from Scratch* option allows you to step through each of the individual commands (listed above) to set up a drawing to your specifications.

It might occur that using the automated *Quick Setup Wizard* or *Advanced Setup Wizard* available in the *Start Up* and *Create New Drawing* dialog boxes would be the fastest and easiest methods to set up a new drawing; however, the automated settings may or may not be exactly what you want or need. <u>Surprisingly, the Wizards do not perform the operations needed to correctly set up a drawing for plotting or printing to a standard scale.</u> As a third alternative, template drawings can be an efficient method for beginning a new drawing since many of the setup steps can be prepared in the template drawing. You must first understand the individual setup commands to be able to create a template drawing or to modify Wizard settings for your needs. The Wizards and template drawings are discussed in detail in Chapter 13.

Setting up a new drawing usually involves both the use of the *Start Up* or *Create New Drawing* dialog box and the individual setup commands listed above and described in the Setup Commands section.

THE *START UP* AND *CREATE NEW DRAWING* DIALOG BOXES

<u>To "manually" achieve the first four steps for drawing setup</u>, use the *Start from Scratch* option of the *Create New Drawing* dialog box.

When you start AutoCAD for the first time, the *Start Up* dialog appears (Fig 6-1). (The *Start Up* dialog box can be suppressed by removing the check in the box at the bottom of the dialog labeled *Show this dialog at start up*. You can make the dialog box reappear on start up by checking *Show the Start Up dialog box* in the *Compatibility* tab of the *Preferences* dialog box.)

Figure 6-1

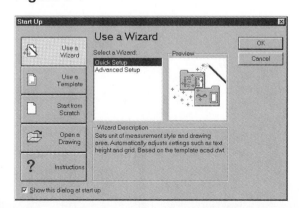

When you use the *New* command, the *Create New Drawing* dialog box appears. This is essentially the same as the *Start Up* dialog box that may appear when you begin AutoCAD, except the *Open a Drawing* option is not included in the *Create New Drawing* dialog box.

The *Start Up* and *Create New Drawing* dialog boxes offer the following three options that are intended to assist you in setting up a drawing.

Use a Wizard

The *Quick Setup Wizard* prompts you to select the type of drawing *Units* you want to use and to specify the drawing *Area* (*Limits*). The *Advanced Setup* offers options for units, angular direction and measurement, area, title blocks, and layouts for paper space and model space. Both Wizards also automatically make changes to many size-related system variables. See Chapter 13, Advanced Drawing Setup, for more information on this option.

Use a Template

Use this option if you want to use an existing template drawing (.DWT file) as a starting point. A template drawing is one that already has many of the setup steps performed but contains no geometry (graphical objects). See Chapter 13, Advanced Drawing Setup, for more information on this option.

Start from Scratch

Use the *Start from Scratch* button and the *English* option (Fig. 6-2) if you want to begin with the traditional AutoCAD default drawing settings, then determine your own system variable values using the individual setup commands such as *Units, Limits, Snap,* and *Grid.* Use the *Metric* option for setting up a metric drawing for plotting on a metric A3 sheet (drawing area of 420mm x 297mm).

Figure 6-2

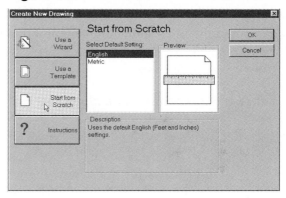

English
Selecting the *English* option causes AutoCAD to use the template drawing ACAD.DWT, which has the traditional AutoCAD default settings (see Table of *Start from Scratch English* Settings). This template drawing is essentially the same drawing used as the default in previous releases of AutoCAD, with settings such as drawing area of 12 x 9 units, etc. As an alternative to selecting the *English* defaults in the *Start from Scratch* option, a drawing can be started using the AutoCAD defaults by any of the following methods:

Selecting the ACAD.DWT template drawing with the *Use a Template* option
Pressing *Cancel* in the *Setup* dialog box that appears when AutoCAD starts
Starting AutoCAD after suppressing the *Setup* dialog box

The following table lists the AutoCAD default settings related to drawing setup. Many of these settings such as *Units, Limits, Snap,* and *Grid* are explained in detail in the following section.

Table of *Start from Scratch English* Settings (ACAD.DWT Defaults)

Related Command	Description	System Variable	Default Setting
Units	drawing units	LUNITS	2 (decimal)
Limits	drawing area	LIMMAX	12.0000,9.0000
Snap	snap increment	SNAPUNIT	.5000
Grid	grid increment	GRIDUNIT	.5000
LTSCALE	linetype scale	LTSCALE	1.0000
DIMSCALE	dimension scale	DIMSCALE	1.0000
Text, Dtext, Mtext	text height	TEXTSIZE	.2000
Hatch	hatch pattern scale	HPSIZE	1.0000

The *English* default setting is recommended if you want to print or plot a drawing to scale on other than a standard A size U.S. architectural or engineering sheet. (See Chapter 13 for information on plotting to scale.)

Metric
Selecting the *Metric* option causes AutoCAD to use the metric template drawing, ACADISO.DWT, in which many of the drawing setups and system variable settings are different from the ACAD.DWT drawing, as shown below.

Table of *Start from Scratch Metric* Settings

Related Command	Description	System Variable	Default Setting
Units	drawing units	LUNITS	2 (decimal)
Limits	drawing area	LIMMAX	420.0000, 297.0000
Snap	snap increment	SNAPUNIT	10.0000
Grid	grid increment	GRIDUNIT	10.0000
LTSCALE	linetype scale	LTSCALE	1.0000
DIMSCALE	dimension scale	DIMSCALE	1.0000
Text, Dtext, Mtext	text height	TEXTSIZE	2.5000
Hatch	hatch pattern scale	HPSIZE	1.0000

The metric drawing setup is intended to be used with ISO linetypes and ISO hatch patterns, which are pre-scaled for these *Limits*, hence the *LTSCALE* and hatch pattern scale of 1. The individual dimensioning variables for arrow size, dimension text size, gaps and extensions, etc. are changed so the dimensions are drawn correctly with a *DIMSCALE* of 1.

The *Metric* option is recommended for plotting to scale on an A3 metric sheet size. If you want to use the *Metric* option for drawing setup and to plot or print on standard U.S. architectural and engineering sheets, the *Limits* should be changed to be proportional to A, B, C, D, or E size sheets. (See Chapter 13 for information on plotting to scale.)

SETUP COMMANDS

UNITS and DDUNITS

Pull-down Menu	COMMAND (TYPE)	ALIAS (TYPE)	Short-cut	Screen (side) Menu	Tablet Menu
Format *Units...*	*UNITS or DDUNITS*	*UN or −UN*	...	*FORMAT* *Ddunits*	*V,4*

The *Units* command allows you to specify the type and precision of linear and angular units as well as the direction and orientation of angles to be used in the drawing. The current setting of *Units* determines the display of values by the coordinates display (*Coords*) and in some dialog boxes. (In previous releases of AutoCAD, the setting for *Units* <u>*Precision*</u> also affected the display of values in dialog boxes, which caused problems in some dialog boxes. *Precision* does not affect display of values in dialog boxes in Release 14.)

You can select *Units ...* from the *Format* pull-down or type *Ddunits* (or command alias *UN*) to invoke the *Units Control* dialog box (Fig. 6-3). Type *Units* (or −*UN*) to produce a text screen (Fig. 6-4).

Figure 6-3 ——————

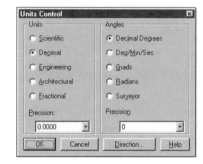

The linear and angular units options are displayed in the dialog box format (Fig. 6-3) and in command line format (Fig. 6-4). The choices for both linear and angular *Units* are shown in the figures.

Figure 6-4 ——————————————

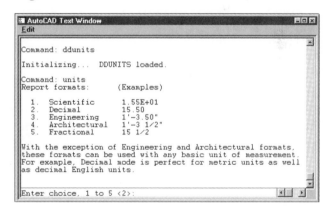

Units	**Format**	
1. Scientific	1.55E + 01	Generic decimal units with an exponent
2. Decimal	15.50	Generic decimal usually used for applications in metric or decimal inches
3. Engineering	1'-3.50"	Explicit feet and decimal inches with notation, one unit equals one inch
4. Architectural	1'-3 1/2"	Explicit feet and fractional inches with notation, one unit equals one inch
5. Fractional	15 1/2	Generic fractional units

Precision

When setting *Units,* you should also set the precision. *Precision* is the number of places to the right of the decimal or the denominator of the smallest fraction to display. The precision is set by making the desired selection from the *Precision* pop-up list in the *Units Control* dialog box (Fig. 6-3) or by keying in the desired selection in command line format.

Precision controls only the <u>display</u> of *Coords*. The <u>actual</u> <u>precision</u> of the drawing database is always the same in AutoCAD, that is, 14 significant digits.

Angles

You can specify a format other than the default (decimal degrees) for expression of angles. Format options for angular display and examples of each are shown in Figure 6-3 (dialog box format).

The orientation of angle 0 can be changed from the default position (east or X positive) to other options by selecting the *Direction* tile in the *Units Control* dialog box. This produces the *Direction Control* dialog box (Fig. 6-5). Alternately, the *Units* command can be typed to select these options in command line format.

The direction of angular measurement can be changed from its default of counter-clockwise to clockwise. The direction of angular measurement affects the direction of positive and negative angles in commands such as *Array Polar, Rotate* and dimension commands that <u>measure</u> angular values but does not change the direction *Arcs* are <u>drawn</u>, which is always counter-clockwise.

Figure 6-5 ⸻

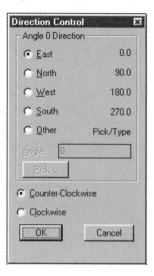

Keyboard Input of *Units* Values

When AutoCAD prompts for a point or a distance, you can respond by entering values at the keyboard. The values can be in <u>any format</u>—integer, decimal, fractional, or scientific—regardless of the format of *Units* selected.

You can <u>type in explicit feet and inch values only if *Architectural* or *Engineering*</u> units have been specified as the drawing units. For this reason, specifying *Units* is the first step in setting up a drawing.

Type in explicit feet or inch values by using the ' (apostrophe) symbol after values representing feet and the " (quote) symbol after values representing inches. If no symbol is used, the values are understood by AutoCAD to be <u>inches</u>.

Feet and inches input <u>cannot</u> contain a blank, so a hyphen (-) must be typed between inches and fractions. For example, with *Architectural* units, key in **6'2-1/2"**, which reads "six feet two and one-half inches." The standard engineering and architectural format for dimensioning, however, places the hyphen between feet and inches (as displayed by the default setting for the *Coords* display).

The *UNITMODE* variable set to **1** changes the display of *Coords* to remind you of the correct format for <u>input</u> of feet and inches (with the hyphen between inches and fractions) rather than displaying the standard format for feet and inch notation (standard format, *UNITMODE* of **0**, is the default setting). If options other than *Architectural* or *Engineering* are used, values are read as generic units.

LIMITS

Pull-down Menu	COMMAND (TYPE)	ALIAS (TYPE)	Short-cut	Screen (side) Menu	Tablet Menu
Format *Drawing Limits*	*LIMITS*	...	...	*FORMAT* *Limits*	*V,2*

The *Limits* command allows you to set the size of the drawing area by specifying the lower left and upper-right corners in X,Y coordinate values.

Command: *limits*
Reset Model space limits
ON/OFF/<Lower left corner> <0,0 or current values>: **X,Y** or **Enter** (Enter an X,Y value or accept the 0,0 default—normally use 0,0 as lower-left corner.)
Upper right corner <12,9>: **X,Y** (Enter new values to change upper-right corner to allow adequate drawing area.)

The default *Limits* values in AutoCAD are 12 and 9; that is, 12 units in the X direction and 9 units in the Y direction (Fig. 6-6). Starting a drawing by any of the following methods (of the *Setup* or *Create New Drawing* dialog boxes) results in *Limits* of 12 x 9:

Selecting the ACAD.DWT template drawing
Pressing *Cancel* in the *Setup* dialog box that appears when AutoCAD starts
Starting AutoCAD after suppressing the *Setup* dialog box
Selecting the *English* defaults in the *Start from Scratch* option
Accepting the default values for the *Area* in the *Quick Setup Wizard*
Accepting the default values for the *Area* in the *Advanced Setup Wizard*

Figure 6-6

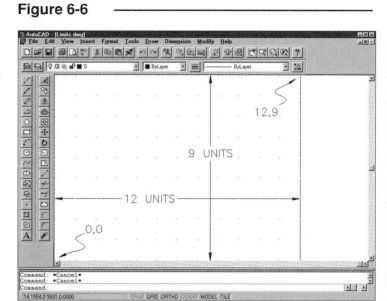

If the *GRID* is turned on, the dots are displayed only over the *Limits*. The AutoCAD screen (default configuration) displays additional area on the right past the *Limits*. The units are generic decimal units that can be used to represent inches, feet, millimeters, miles, or whatever is appropriate for the intended drawing. Typically, however, decimal units are used to represent inches or millimeters. If the default units are used to represent inches, the default drawing size would be 12 by 9 inches.

Remember that when a CAD system is used to create a drawing, the geometry should be drawn <u>full size</u> by specifying dimensions of objects in <u>real-world units</u>. A completed CAD drawing or model is virtually an exact dimensional replica of the actual object. Scaling of the drawing occurs only when plotting or printing the file to an actual fixed-size sheet of paper.

Before beginning to create an AutoCAD drawing, determine the size of the drawing area needed for the intended geometry. After setting *Units*, appropriate *Limits* should be set in order to draw the object or geometry to the <u>real-world size in the actual units</u>. There are no practical maximum or minimum settings for *Limits*.

The X,Y values you enter as *Limits* are understood by AutoCAD as values in the units specified by the *Units* command. For example, if you previously specified *Architectural units*, then the values entered are understood as inches unless the notation for feet (') is given (**240,180** or **20',15'** would define the same coordinate). Remember, you can type in explicit feet and inch values only if *Architectural* or *Engineering* units have been specified as the drawing units.

If you are planning to plot the drawing to scale, *Limits* should be set to a proportion of the <u>sheet size</u> you plan to plot on. For example, setting limits to 22 by 17 (2 times 11 by 8.5) would allow enough room for drawing an object about 20" by 15" and allow plotting at 1/2 size on the 11" x 8.5" sheet. Simply stated, <u>set *Limits* to a proportion of the paper</u>.

ON/OFF

If the *ON* option of *Limits* is used, limits checking is activated. Limits checking prevents you from drawing objects outside of the limits by issuing an outside-limits error. This is similar to drawing "off the paper." Limits checking is *OFF* by default.

Limits also defines the display area for *GRID* as well as the minimum area displayed when a *Zoom All* is used. *Zoom All* forces the full display of the *Limits*. *Zoom All* can be invoked by typing **Z** (command alias) then **A** for the *All* option.

Changing *Limits* does <u>not</u> automatically change the display. As a general rule, you should make a habit of invoking a *Zoom All* <u>immediately following</u> a change in *Limits* to display the area defined by the new limits (Fig. 6-7).

When you <u>reduce</u> *Limits* while *Grid* is *ON*, it is apparent that a change in *Limits* does not automatically change the display. In this case, the area covered by the grid is reduced in size as *Limits* are reduced, yet the display remains unchanged.

Figure 6-7

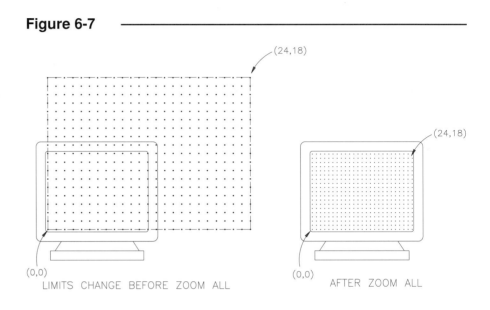

If you are already experimenting with drawing in different *Linetypes*, a change in *Limits* and *Zoom All* affects the display of the hidden and dashed lines. The *LTSCALE* variable controls the spacing of non-continuous lines. The *LTSCALE* is often changed proportionally with changes in *Limits*.

SNAP

Pull-down Menu	COMMAND (TYPE)	ALIAS (TYPE)	Short-cut	Screen (side) Menu	Tablet Menu
Tools *Drawing Aids...*	*SNAP*	*SN*	F9 or Ctrl+B	*TOOLS 2* *Ddrmodes*	*W,10*

SNAP, when activated by pressing **F9** or double-clicking on *SNAP* in the Status line, forces the cursor position to regular increments. This function can be of assistance to you by making it faster and more accurate for creating and editing objects. The default *Snap* setting is **0.5**. The *Snap* command is used to set the value for these invisible snap increments. *Snap* spacing can be set to any value.

Since *Snap* controls the cursor position, the value of *Snap* should be set to the <u>interactive</u> accuracy desired. As a general rule, you specify the *Snap* value to be that of the <u>common</u> dimensional length expected in the drawing. For example, if the common dimensional length in the drawing is 1/2", or you intend for dimensional accuracy of the drawing to be to the nearest 1/2", the *Snap* command is used to change the snap spacing to **1/2** or **.5**. In this way the cursor always "snaps" to .5 increments.

In some cases such as architectural or civil engineering applications, where dimensional increments are small with relation to the size of the geometry, there is no practical <u>interactive</u> accuracy or there may be no <u>common</u> dimensional length; therefore, some users prefer not to use *Snap*.

The *Snap* command is easily typed, displaying the options in command line format. The *Drawing Aids* dialog box (Fig. 6-8) can be invoked by menu selection or by typing *Ddrmodes* or the command alias, *RM*. Enter the desired value for *Snap* in the *X Spacing* box and press *Enter*. The *Y Spacing* value automatically changes to match.

Figure 6-8

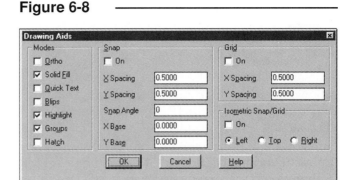

The command line format is as follows:

> Command: **snap**
> Snap spacing or ON/OFF/Aspect/Rotate/Style <current value>: **(value or letter)**
> (enter a value or option)

ON/OFF
Selecting *ON* or *OFF* accomplishes the same action as toggling the **F9** key, pressing **Ctrl+B,** or double-clicking **SNAP** on the Status Line. Typically, *SNAP* should be *ON* for drawing and editing but turned *OFF* to make object selection easier (the cursor moves smoothly to any location with *SNAP OFF*).

Aspect
The *Aspect* option allows specification of unequal X and Y spacing for *SNAP*. This action can also be accomplished in the *Drawing Aids* dialog box by entering different values for *X Spacing* and *Y Spacing*.

Rotate
SNAP can also be *Rotat*ed about any point and set to any angle. When *SNAP* has been rotated, the *GRID, ORTHO,* and "crosshairs" automatically follow this alignment. This action facilitates creating objects oriented at the specified angle, for example, creating an auxiliary view or drawing part of a floor plan at an angle. To accomplish this, use the *R* option in command line format or set *Snap Angle* and *X Base* and *Y Base* (point to rotate about) in the dialog box.

Style
The *Style* option allows switching between a *Standard* snap pattern (the default square or rectangular) and an *Isometric* snap pattern. If using the dialog box, toggle *Isometric Snap/Grid On*. When the *SNAP Style* or *Rotate* angle is changed, the *GRID* automatically aligns with it.

Keep in mind that using the *Quick Setup Wizard* or *Advanced Setup Wizard* causes AutoCAD to automatically calculate *Snap* and *Grid* values for you. In some cases this is helpful, but in other cases this is a hindrance and <u>can cause unintentional errors</u> in geometry creation and editing. Because the Wizard uses a formula to calculate *Snap* and *Grid* values based on the *Area* (*Limits*) settings you specify, the resulting *Snap* and *Grid* can be <u>unusable</u> in some cases and should be changed by using the *Snap* and *Grid* commands or the *Drawing Aids* dialog box. For example, using the Wizard to set up a drawing to plot at a scale of 1=1 on a standard 17 x 11, 18 x 12, or 22 x 17 sheet size results in *Snap* and *Grid* settings of .6, .7, and .9, respectively. (See Chapter 13, Advanced Drawing Setup.)

R14

GRID

Pull-down Menu	COMMAND (TYPE)	ALIAS (TYPE)	Short-cut	Screen (side) Menu	Tablet Menu
Tools *Drawing Aids...*	GRID	...	F7 or Ctrl+G	*TOOLS 2* *Ddrmodes*	*W,10*

GRID is visible on the screen, whereas *SNAP* is invisible. *GRID* is only a <u>visible</u> display of some regular interval. *GRID* and *SNAP* can be <u>independent</u> of each other. In other words, each can have separate spacing settings and the active state of each (*ON, OFF*) can be controlled independently. The *GRID* <u>follows</u> the *SNAP* if *SNAP* is rotated or changed to *Isometric Style*. Although the *GRID* spacing can be different than that of *SNAP*, it can also be forced to follow *SNAP* by using the *Snap* option. The default *GRID* setting is **0.5**.

The *GRID* <u>cannot</u> be plotted. It is <u>not</u> comprised of *Point* objects and therefore is not part of the current drawing. *GRID* is only a visual aid.

Grid can be accessed by command line format (shown below) or set via the *Drawing Aids* dialog box (Fig. 6-8). The dialog box is invoked by menu selection or by typing *Ddrmodes* or *RM*. The dialog box allows only *X Spacing* and *Y Spacing* input for *Grid*.

> Command: **grid**
> Grid spacing(X) or ON/OFF/Snap/Aspect <current value>: **(value or letter)**
> (enter a value or option)

Grid Spacing (X)
If you supply a value for the *Grid spacing*, *GRID* is displayed at that spacing regardless of *SNAP* spacing. If you key in an *X* as a suffix to the value (for example, **2X**), the *GRID* is displayed as that value <u>times</u> the *SNAP* spacing (for example, "2 times" *SNAP*).

ON/OFF
The *ON* and *OFF* options simply make the *GRID* visible or not (like toggling the **F7** key, pressing **Ctrl+G**, or double-clicking **GRID** on the Status line).

Snap
The *Snap* option of the *Grid* command forces the *GRID* spacing to equal that of *SNAP*, even if *SNAP* is subsequently changed.

Aspect
The *Aspect* option of *GRID* allows different X and Y spacing (causing a rectangular rather than a square *GRID*).

Remember that using the *Quick Setup Wizard* or *Advanced Setup Wizard* causes AutoCAD to automatically calculate *Snap* and *Grid* values for you, which can be helpful, but can also be nonproductive in some cases (see Chapter 13, Advanced Drawing Setup). If you want to use values other than the Wizard calculated values, use the *Snap* and *Grid* commands or the *Drawing Aids* dialog box to change them to a more useful increment.

R14

CHAPTER EXERCISES

1. A drawing is to be made to detail a mechanical part. The part is to be manufactured from sheet metal stock; therefore, only one view is needed. The overall dimensions are 18 by 10 inches, accurate to the nearest .125 inch. Complete the steps for drawing setup:

   A. The drawing will be automatically "scaled to fit" the paper (no standard scale).

      1. Begin a *New* drawing. When the *Start Up* or *Create New Drawing* dialog box appears, select *Start from Scratch*. Select the *English* default settings.
      2. *Units* should be *Decimal*. Set the *Precision* to **0.000**.
      3. Set *Limits* in order to draw full size. Make the lower-left corner **0,0** and the upper right at **24,18**. This is a 4 by 3 proportion and should allow space for the part, a title block, border, and dimensions or notes.
      4. *Zoom All*. (Type **Z** for *Zoom;* then type **A** for *All*.)
      5. Set the *GRID* to **1**.
      6. Set *SNAP* to **.125**.
      7. Save this drawing as **CH6EX1A** (to be used again later). (When plotting at a later time, "Scale to Fit" can be specified.)

   B. The drawing will be plotted to scale on engineering A or B size paper (11" by 8.5" or 22" by 17").

      1. Begin a *New* drawing. When the *Start Up* or *Create New Drawing* dialog box appears, select *Start from Scratch*. Select the *English* default settings.
      2. *Units* should be *Decimal*. Set the *Precision* to **0.000**.
      3. Set *Limits* to the paper size (or a proportion thereof), making the lower-left corner **0,0** and the upper-right at **22,17**. This allows space for drawing full size and for a title block, border, and dimensions or notes.
      4. *Zoom All*. (Type **Z** for *Zoom;* then type **A** for *All*.)
      5. Set the *GRID* to **1**.
      6. Set *SNAP* to **.125**.
      7. *Save* this drawing as **CH6EX1B** (to be used again later). (When plotting, a scale of 1=1 can be specified to plot on 22" by 17" paper, or a scale of 1/2=1 can be specified to plot on 11" by 8.5" paper.)

2. A drawing is to be prepared for a house plan. Set up the drawing for a floor plan that is approximately 50' by 30'. Assume the drawing is to be automatically "Scaled to Fit" the sheet (no standard scale).

   A. Begin a *New* drawing. When the *Start Up* or *Create New Drawing* dialog box appears, select *Start from Scratch*. Select the *English* default settings.
   B. Set *Units* to *Architectural*. Set the *Precision* to **0'-0 1/4"**. Each unit equals 1 inch.
   C. Set *Limits* to **0,0** and **80',60'**. Use the **'** (apostrophe) symbol to designate feet. Otherwise, enter **0,0** and **960,720** (size in inch units is: 80x12=960 and 60x12=720).
   D. *Zoom All*. (Type **Z** for *Zoom;* then type **A** for *All*.)
   E. Set *GRID* to **24** (2 feet).
   F. Set *SNAP* to **6** (anything smaller would be hard to PICK).
   G. *Save* this drawing as **CH6EX2**.

3. A multiview drawing of a mechanical part is to be made. The part is 125mm in width, 30mm in height, and 60mm in depth. The plot is to be made on an A3 metric sheet size (420mm x 297mm). The drawing will use ISO linetypes and ISO hatch patterns, so AutoCAD's *Metric* default settings can be used.

    A. Begin a *New* drawing. When the *Start Up* or *Create New Drawing* dialog box appears, select *Start from Scratch*. Select the *Metric* default settings.
    B. *Units* should be *Decimal*. Set the *Precision* to **0.00**.
    C. Calculate the space needed for three views. If *Limits* are set to the sheet size, there should be adequate space for the views. Make sure the lower-left corner is at **0,0** and the upper right is at **420,297**. (Since the *Limits* are set to the sheet size, a plot can be made later at 1:1 scale.)
    D. Change *SNAP* to **5**.
    E. *Save* this drawing as **CH6EX3** (to be used again later).

4. Assume you are working in an office that designs many mechanical parts in metric units. However, the office uses a standard laser jet printer for 11" x 8.5" sheets. Since AutoCAD does not have a setup for metric drawings on non-metric sheets, it would help to carry out the steps for drawing setup and save the drawing as a template to be used later.

    A. Begin a *New* drawing. When the *Start Up* or *Create New Drawing* dialog box appears, select *Start from Scratch*. Select *English* default settings.
    B. Set the *Units Precision* to **0.00**.
    C. Change the *Limits* to match an 11" x 8.5" sheet. Make the lower-left corner **0,0** and the upper right **279,216** (11 x 8.5 times 25.4, approximately). (Since the *Limits* are set to the sheet size, plots can easily be made at 1:1 scale.)
    D. *Zoom All*. (Type **Z** for *Zoom,* then **A** for *All*).
    E. Change the *Snap* to **2**.
    F. Change *Grid* to **10**.
    G. At the command prompt, type *LTSCALE*. Change the value to **12**.
    H. *Save* the drawing as **A-METRIC**.

5. Assume you are commissioned by the local parks and recreation department to provide a layout drawing for a major league sized baseball field. Follow these steps to set up the drawing:

    A. Begin a *New* drawing. When the *Start Up* or *Create New Drawing* dialog box appears, **select** *Start from Scratch,* then *English* default units.
    B. Set the *Units* to *Architectural* and the *Precision* to **1/2"**.
    C. Set *Limits* to an area of **512'** x **384'** (make sure you key in the apostrophe to designate feet).
    D. Use the *Snap* command. Change the *Snap* value to **10'** (don't forget the apostrophe).
    E. Use the *Grid* command and change the value to **20'**. Ensure *SNAP* and *GRID* are on.
    F. *Save* the drawing and assign the name **BALL FIELD CH6**.

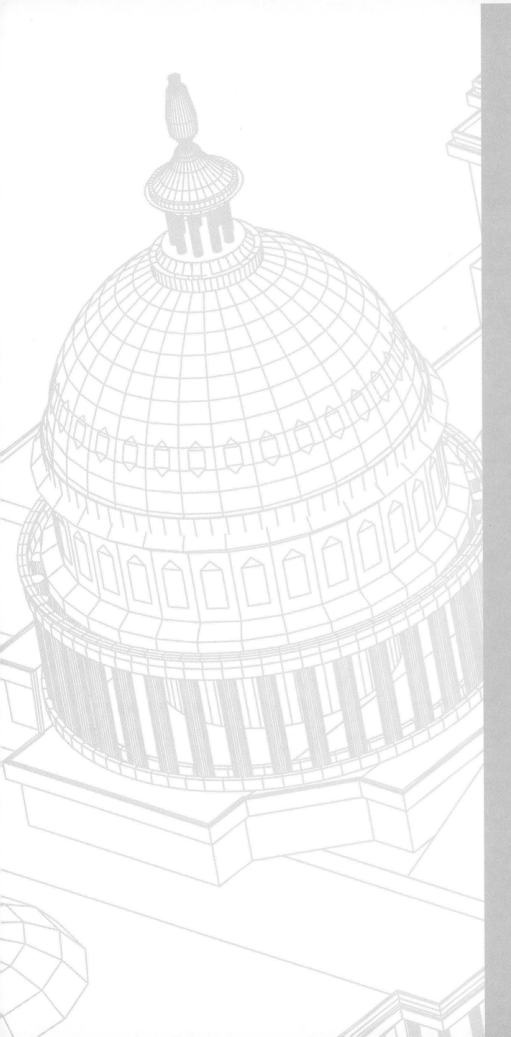

7

OBJECT SNAP

Chapter Objectives

After completing this chapter you should:

1. understand the importance of accuracy in CAD drawings;

2. know the function of each of the Object Snap *(OSNAP)* modes;

3. be able to recognize the AutoSnap Marker symbols;

4. be able to invoke *OSNAP*s for single point selection;

5. be able to operate Running Object Snap modes;

6. know that you can toggle Running Object Snap off to specify points not at object features;

7. know that you can use *OSNAP* any time AutoCAD prompts for a point.

CAD ACCURACY

Because CAD databases store drawings as digital information with great precision (fourteen numeric places in AutoCAD), it is possible, practical, and desirable to create drawings that are 100% accurate; that is, a CAD drawing should be created as an exact dimensional replica of the actual object. For example, lines that appear to connect should actually connect by having the exact coordinate values for the matching line endpoints. Only by employing this precision can dimensions placed in a drawing automatically display the exact intended length, or can a CAD database be used to drive CNC (Computer Numerical Control) machine devices such as milling machines or lathes, or can the CAD database be used for rapid prototyping devices such as Stereo Lithography Apparatus. With CAD/CAM technology (Computer-Aided Design/Computer-Aided Manufacturing), the CAD database defines the configuration and accuracy of the finished part. Accuracy is critical. Therefore, in no case should you create CAD drawings with only visual accuracy such as one might do when sketching using the "eyeball method."

OBJECT SNAP

AutoCAD provides a capability called "Object Snap", or *OSNAP* for short, that enables you to "snap" to existing object endpoints, midpoints, centers, intersections, etc. When an *OSNAP* mode (*Endpoint, Midpoint, Center, Intersection,* etc.) is invoked, you can move the cursor near the desired object feature (endpoint, midpoint, etc.) and AutoCAD locates and calculates the coordinate location of the desired object feature.

Available Object Snap modes are:

> *Center*
> *Endpoint*
> *Insert*
> *Intersection*
> *Midpoint*
> *Nearest*
> *Node* (Point)
> *Perpendicular*
> *Quadrant*
> *Tangent*
> *From*
> *Apparent Intersection* (for 3D use)

Figure 7-1

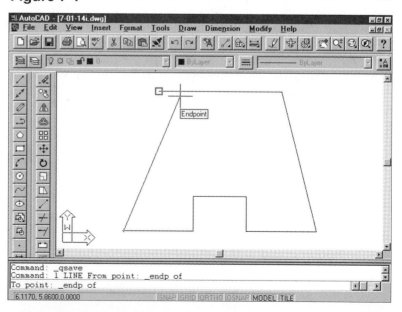

For example, when you want to draw a *Line* and connect its endpoint to an existing *Line,* you can invoke the *Endpoint OSNAP* mode at the "To point:" prompt, then snap exactly to the desired line end by moving the cursor near it and PICKing (Fig. 7-1). *OSNAP*s can be used for any draw or modify operation—whenever AutoCAD prompts for a point (location).

A new Release 14 feature called "AutoSnap" displays a "Snap Marker" indicating the particular object feature (endpoint, midpoint, etc.) when you to move the cursor <u>near</u> an object feature. Each *OSNAP* mode (*Endpoint, Midpoint, Center, Intersection,* etc.) has a distinct symbol (AutoSnap Marker) representing the object feature. This innovation allows you to preview and confirm the snap points before you PICK them (Fig. 7-2). The table in Figure 7-19 gives each Object Snap mode, the related AutoSnap Marker, and the icon button used to activate a single Object Snap selection.

Figure 7-2

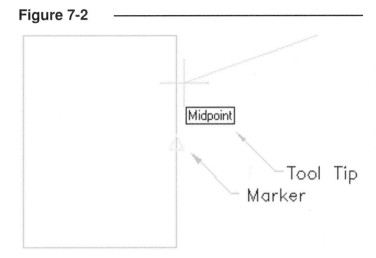

AutoSnap provides two other aids for previewing and confirming *OSNAP* points before you PICK. A "Snap Tip" appears shortly after the Snap Marker appears (hold the cursor still and wait one second). The Snap Tip gives the name of the found *OSNAP* point, such as an *Endpoint, Midpoint, Center,* or *Intersection,* (Fig. 7-2). In addition, a "Magnet" draws the cursor to the snap point if the cursor is within the confines of the Snap Marker. This Magnet feature helps confirm that you have the desired snap point before making the PICK.

A visible target box, or "Aperture," <u>can be displayed</u> at the cursor (invisible by default) whenever an *OSNAP* mode is in effect (Fig. 7-3). The Aperture is a square box larger than the pickbox (default size of 10 pixels square). Technically, this target box (visible or invisible) must be located <u>on an object</u> before a Snap Marker and related Snap Tip appear. In pre-

Figure 7-3

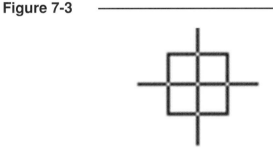

vious versions of AutoCAD, the Aperture is always displayed. Snap Markers, Snap Tips, and the Magnet feature are not available in previous releases; therefore, *OSNAP* selections sometimes result in unexpected snap points.

The settings for the Aperture, Snap Markers, Snap Tips, and Magnet are controlled in the *AutoSnap* tab of the *Osnap Settings* dialog box (discussed later).

The Object Snap modes are explained in the next section. Each mode, and its relation to the AutoCAD objects, is illustrated.

Object Snaps must be activated in order for you to "snap" to the desired object features. Two methods for activating Object Snaps, Single Point Selection and Running Object Snaps, are discussed in the sections following Object Snap Modes. The *OSNAP* modes (*Endpoint, Midpoint, Center, Intersection,* etc.) operate identically for either method.

OBJECT SNAP MODES

AutoCAD provides the following Object Snap Modes.

Center

This *OSNAP* option finds the center of a *Circle, Arc,* or *Donut.* You must PICK the *Circle* <u>object</u>, not where you think the center is.

Figure 7-4 —————————

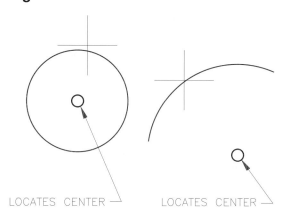

LOCATES CENTER —⏌   LOCATES CENTER —⏌

Endpoint

The *Endpoint* option snaps to the endpoint of a *Line, Pline, Spline,* or *Arc.* PICK the object <u>near</u> the desired end.

Figure 7-5 —————————

LOCATES ENDPOINT —⏋

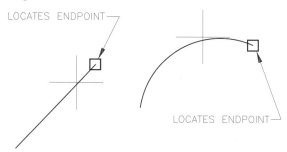

LOCATES ENDPOINT —⏌

Insert

This option locates the insertion point of *Text* or a *Block.* PICK anywhere on the *Block* or line of *Text.*

Figure 7-6 —————————

LOCATES INSERTION POINT —⏋

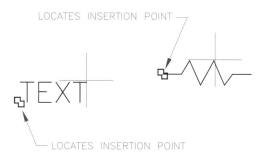

TEXT

⌐ LOCATES INSERTION POINT

Intersection

Using this option causes AutoCAD to calculate and snap to the intersection of any two objects. You can locate the cursor (Aperture) so that <u>both</u> objects pass near (through) it, or you can PICK each object <u>individually</u>.

Figure 7-7 —————————

LOCATES INTERSECTION —⏋

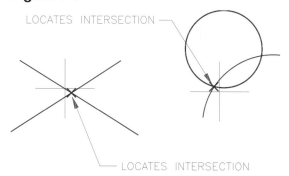

⌐ LOCATES INTERSECTION

Even if the two objects that you PICK do not physically intersect, you can PICK each one individually with the *Intersection* mode and AutoCAD will find the <u>extended</u> intersection.

Figure 7-8

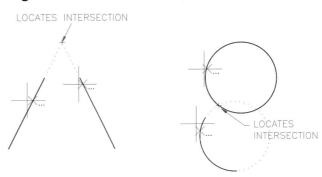

LOCATES INTERSECTION

LOCATES INTERSECTION

R13

Midpoint

The *Midpoint* option snaps to the point of a *Line* or *Arc* that is <u>halfway</u> between the endpoints. PICK anywhere on the object.

Figure 7-9

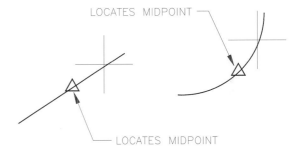

LOCATES MIDPOINT

LOCATES MIDPOINT

Nearest

The *Nearest* option locates the point on an object nearest to the <u>cursor position</u>. Place the cursor center nearest to the desired location, then PICK.

Nearest <u>cannot</u> be used effectively with *ORTHO* because *OSNAPs* override *ORTHO*. In other words, using *Nearest* to locate a "To point:" will not produce an orthogonal line if *ORTHO* is on because *OSNAPs* take priority.

Figure 7-10

LOCATES NEAREST POINT ON OBJECT

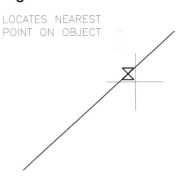

Node

This option snaps to a *Point* object. The *Point* must be within the Aperture (visible or invisible Aperture).

Figure 7-11

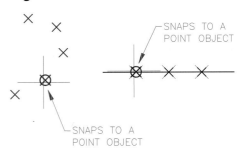

SNAPS TO A POINT OBJECT

SNAPS TO A POINT OBJECT

Perpendicular

Use this option to snap perpendicular to the selected object. PICK anywhere on a *Line* or straight *Pline* segment.

Figure 7-12

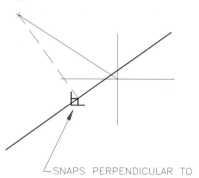

SNAPS PERPENDICULAR TO

Quadrant

The *Quadrant* option snaps to the 0, 90, 180, or 270 degree quadrant of a *Circle*. PICK <u>nearest</u> to the desired *Quadrant*.

Figure 7-13

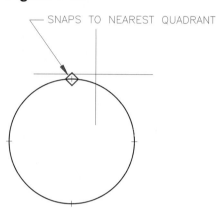

SNAPS TO NEAREST QUADRANT

Tangent

This option calculates and snaps to a tangent point of an *Arc* or *Circle*. PICK the *Arc* or *Circle* as near as possible to the expected *Tangent* point.

Figure 7-14

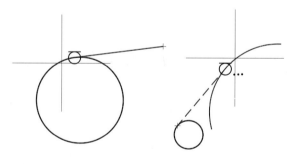

SNAPS TANGENT TO

From

The *From* option is designed to let you snap to a point <u>relative</u> to another point using relative rectangular, relative polar, or direct distance entry coordinates. There are two steps: first select a "Basepoint:" (coordinates or another *OSNAP* may be used); then select an "Offset:" (enter relative rectangular, relative polar, or direct distance entry coordinates).

Figure 7-15

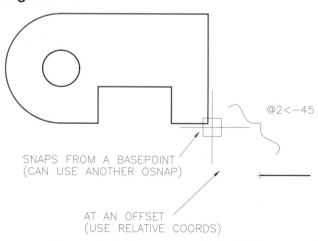

@2<−45

SNAPS FROM A BASEPOINT
(CAN USE ANOTHER OSNAP)

AT AN OFFSET
(USE RELATIVE COORDS)

Apparent intersection

Use this option when you are working with a 3D drawing and want to snap to a point in space where two objects appear to intersect (from your viewpoint) but do not actually physically intersect.

Figure 7-16

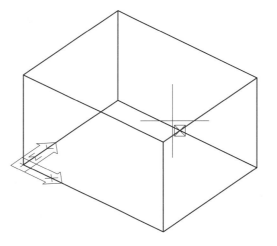

SNAPS TO APPARENT INTERSECTION IN 3D SPACE

Object Snaps must be activated in order for you to "snap" to the desired object features. Two methods for activating Object Snaps, Single Point Selection and Running Object Snaps, are discussed next. The *OSNAP* modes (*Endpoint, Midpoint, Center, Intersection,* etc.) operate identically for either method.

OSNAP SINGLE POINT SELECTION

OBJECT SNAPS (SINGLE POINT)	Pull-down Menu	COMMAND (TYPE)	ALIAS (TYPE)	Short-cut	Screen (side) Menu	Tablet Menu	Cursor Menu (button 3 or shift+button 2)
	...	*END, MID, etc. (first three letters)*	...	...	****(asterisks)*	*T,16 - U,21*	*Endpoint Midpoint, etc.*

Figure 7-17

There are many methods for invoking *OSNAP* modes for single point selection, as shown in the command table. If you prefer to type, enter only the <u>first three letters</u> of the *OSNAP* mode at the "From point: " prompt, "To point:" prompt, or <u>any time AutoCAD prompts for a point.</u>

Object Snaps are available from the Standard toolbar (Fig. 7-17). If desired, a separate *Object Snap* toolbar can be activated to float or dock on the screen by using the *View* pull-down menu and selecting *Toolbars...* , then *Object Snap* (see the docked toolbar on the right, Fig. 7-17). The advantage of invoking the *Object Snap* toolbar is that only one PICK is required for an *OSNAP* mode, whereas the Standard toolbar requires two PICKs because the *OSNAP* icons are on a "flyout."

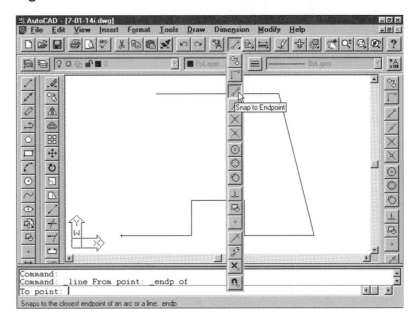

In addition to these options, a special menu called the <u>cursor menu</u> can be used. The cursor menu pops up at the <u>current location</u> of the cursor and replaces the cursor when invoked (Fig. 7-18). This menu is activated as follows:

2-button mouse   press **Shift + #2** (hold down the Shift key while clicking the right mouse button)
3-button mouse   press **#3**
digitizing puck   press **#3**

Figure 7-18 ——————————————

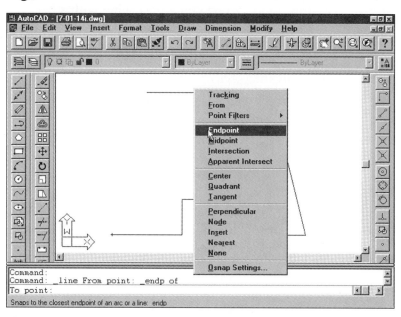

<u>With any of these methods, *OSNAP* modes are active only for selection of a single point</u>. If you want to *OSNAP* to another point, you must select an *OSNAP* mode again for the second point. With this method, the desired *OSNAP* mode is selected <u>transparently</u> (invoked during another command operation) immediately before selecting a point when prompted. In other words, whenever you are prompted for a point (for example, the "From point:" prompt of the *Line* command), select or type an *OSNAP* option. Then PICK near the desired object feature (endpoint, center, etc.) with the cursor. AutoCAD "snaps" to the feature of the object and uses it for the point specification. Using *OSNAP* in this way allows the *OSNAP* mode to operate only for that <u>single point selection</u>.

For example, when using *OSNAP* during the *Line* command, the command line reads as shown:

> Command: **Line**
> From Point: **Endpoint** of (**PICK**)
> To point: **Endpoint** of (**PICK**)
> To point: **Enter** (completes the command)

If you prefer using *OSNAP* for single point specification, it may be helpful to associate the tools (icons) with the Marker that appears on the drawing when you use the *OSNAP* mode. The table in Figure 7-19 displays the *OSNAP* icons and the respective Markers.

Figure 7-19 ——————————————

Object Snaps

Mode	Tool	Marker
Endpoint		□
Midpoint		△
Center		○
Node		⊗
Quadrant		◇
Intersection		✕
Insertion		⊐
Perpendicular		⊥
Tangent		○
Nearest		⊠
Apparent Int		⊠

OSNAP RUNNING MODE

OBJECT SNAPS (RUNNING)	Pull-down Menu	COMMAND (TYPE)	ALIAS (TYPE)	Short-cut	Screen (side) Menu	Tablet Menu	Cursor Menu (button 3 or Shift+button 2)
	Tools *Object Snap* *Settings...*	*OSNAP or* *-OSNAP*	*OS or* *-OS*	*Ctrl+F*	*TOOLS 2* *Ddosnap*	*U,22*	*Osnap* *Settings...*

Running Object Snap

Another method for using Object Snap is called "Running Object Snap" because one or more *OSNAP* modes (*Endpoint, Center, Midpoint*, etc.) can be turned on and kept running indefinitely. This method can obviously be more productive because you do not have to continually invoke an *OSNAP* mode each time you need to use one. For example, suppose you have <u>several</u> *Endpoint*s to connect. It would be most efficient to turn on the running *Endpoint OSNAP* mode and leave it <u>running</u> during the multiple selections. This is faster than continually selecting the *OSNAP* mode each time before you PICK.

You can even have <u>several *OSNAP* modes running at the same time</u>. A common practice is to turn on the *Endpoint, Center, Midpoint* modes simultaneously. In that way, if you move your cursor near a *Circle*, the *Center* Marker appears; if you move the cursor near the end or the middle of a *Line*, the *Endpoint* or the *Midpoint* mode Markers appear.

With versions of AutoCAD previous to Release 14, Running Object Snap is not as useful and is often dangerous because there were no Markers to indicate which *OSNAP* mode would be used for a particular PICK. Therefore, when multiple modes were running simultaneously, users did not know exactly if the selection would find an *Endpoint, Center,* or *Midpoint* until after the PICK. With Release 14, Running Object Snap is much more effective because the AutoSnap <u>Markers</u> make using multiple modes completely predictable.

Accessing Running Object Snap

All features of Running Object Snaps are controlled by the *Osnap Settings* dialog box (Fig. 7-20). This dialog box can be invoked by the following methods (see Command Table above):

1. type the *OSNAP* command;
2. type *OS*, the command alias;
3. select *Object Snap Settings...* from the *Tools* pull-down menu;
4. select *Osnap Settings...* from the bottom of the cursor menu (Shift + #2 button);
5. select the *Object Snap Settings* icon button from the *Object Snap* toolbar;
6. if Running Object Snaps have not yet been set, use an *OSNAP* toggle (F3, Ctrl+F, or double-clicking *OSNAP* at the Status line).

The *Osnap Settings* dialog box has two tabs: *Running Osnap* and *AutoSnap.*

Figure 7-20

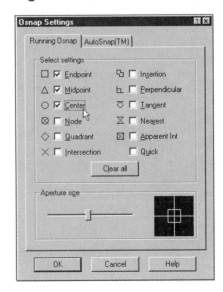

Running Osnap Tab

Use the *Running Osnap* tab to select the desired Object Snap settings. Try using three of four commonly used modes together, such as *Endpoint, Midpoint, Center,* and *Intersection.* The AutoSnap Markers indicate which one of the modes would be used as you move the cursor near different object features. Using similar modes simultaneously, such as *Center, Quadrant,* and *Tangent,* can sometimes lead to difficulties since it requires the cursor to be placed almost in an exact snap spot, or in some cases it may not find one of the modes (*Quadrant* overrides *Center*). In these cases, the tab key can be used to cycle through the options (see Object Snap Cycling).

Quick

This option only operates <u>in addition to</u> other selected *Running Osnap* modes. Using the *Quick* option causes AutoCAD to snap to the first snap point found. In other words, when on, *Quick* finds the quickest of the active options.

Aperture size

Notice that the *Aperture Size* can be adjusted through this dialog box (or through the use of the *Aperture* command). Whether the Aperture is visible or not (see Fig. 7-21, *Display aperture box*), its size determines how close the cursor must be to an object before the Marker appears and the displayed *OSNAP* mode takes effect. In other words, the smaller the Aperture, the closer the cursor must be to the object before a Marker appears and the *OSNAP* mode has an effect, and the larger the Aperture, the further the cursor can be from the object while still having an effect. The Aperture default size is 10 pixels square. The Aperture size influences *Single Point Osnap* selection as well as *Running Osnaps.*

AutoSnap Tab

The *AutoSnap* tab in the *Osnap Settings* dialog box (Fig. 7-21) gives you control over the visual aspects of the following object snap features.

Marker

When this box is checked, an AutoSnap Marker appears when the cursor is moved near an object feature. Each *OSNAP* mode (*Endpoint, Center, Midpoint,* etc.) has a unique Marker (see Fig. 7-19). Normally this box should be checked, especially when using Running Object Snap, because the Markers help confirm when and which *OSNAP* mode is in effect.

Magnet

The Magnet feature causes the cursor to "lock" onto the object feature (*Endpoint, Center, Midpoint,* etc.) when the cursor is within the confines of the Marker. The Magnet helps confirm the exact location that will be snapped to for the subsequent PICK.

Figure 7-21

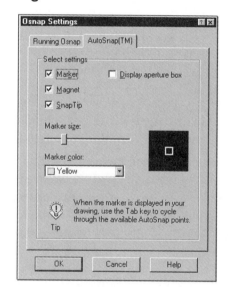

Snap Tip

The Snap Tip (similar to a Tool Tip) is helpful for beginners because it gives a verbal indication of the *OSNAP* mode in effect (*Endpoint, Center, Midpoint,* etc.) when the cursor is near the object feature. Experienced users may want to turn off the Snap Tips once the Marker symbols are learned.

Display aperture box

This checkbox controls the <u>visibility</u> of the Aperture. Whether the Aperture is visible or not, its size determines how close the cursor must be to an object before a Marker appears and the displayed *OSNAP* mode takes effect. (See *Aperture size* in the *Running Osnap* dialog box.)

Marker size

This control determines the size of the Markers. Remember that if the Magnet is on, the cursor "locks" to the object feature when the cursor is within the confines of the Magnet. Therefore, the larger the Marker, the greater the Magnet effective area, and the smaller the Marker, the smaller the Magnet effective area.

Marker color

Use this drop-down box to select a color for the Markers. You may want to change Marker colors if you change the color of the Drawing Editor background. For example, if the background is changed to white, you may want to change the Marker color to dark blue instead of yellow.

Running Object Snap Toggle

Another feature of Release 14 that makes Running Object Snap effective is Osnap Toggle. If you need to PICK a point without using *OSNAP*, use the Osnap Toggle to temporarily override (turn off) the modes. With Running Osnaps temporarily off, you can PICK any point without AutoCAD forcing your selection to an *Endpoint, Center,* or *Midpoint,* etc. When you toggle Running Object Snap on again, AutoCAD remembers which modes were previously set.

The following methods can be used to toggle Running Osnaps on and off:

1. double-click the word *OSNAP* that appears on the Status line (at the bottom of the screen)
2. press **F3**
3. press **Ctrl+F**

None

 The *None OSNAP* option is a Running Osnap <u>override effective for only one PICK</u>. *None* is similar to the Running Object Snap Toggle, except it is effective for one PICK, then Running Osnaps automatically come back on without having to use a toggle. If you have *OSNAP* modes running but want to deactivate them for a <u>single</u> PICK, use *None* in response to "From point:" or other point selection prompt. In other words, using *None* <u>during a draw or edit command</u> overrides any Running Osnaps for that single point selection. *None* can be typed at the command prompt (when prompted for a point) or can be selected from the bottom of the Object Snap toolbar.

Object Snap Cycling

In cases when you have multiple Running Osnaps set, and it is difficult to get the desired AutoSnap Marker to appear, you can use the <u>Tab</u> key to cycle through the possible *OSNAP* modes for the highlighted object. In other words, pressing the Tab key makes AutoCAD highlight the object nearest the cursor, then cycles through the possible snap Markers (for running modes that are set) that affect the object.

For example, when the *Center* and *Quadrant* modes are both set as Running Osnaps, moving the cursor near a *Circle* always makes the *Quadrant* Marker appear but never the *Center* Marker. In this case, pressing the Tab key highlights the *Circle,* then cycles through the four *Quadrant* and one *Center* snap candidates.

TRACKING

Tracking is a new feature in Release 14 that assists you to locate points that are orthogonally aligned with other object features. *Tracking* is used in conjunction with Object Snap. When you are prompted for a point, such as the "From point:" of a *Line* or the center of a *Circle*, you can turn on *Tracking* by one of several methods. With *Tracking* on, you can restrict the point selection to be horizontally or vertically aligned with an *Endpoint, Center, Midpoint,* etc. You can use *Tracking* again to further restrict the point selection to be vertically or horizontally aligned with a second *Endpoint, Center, Midpoint,* etc. In this way, it is easy to find the center of a rectangle by *Tracking* the *Midpoint* of two sides. *Tracking* is explained fully in Chapter 15, Draw Commands II.

OSNAP APPLICATIONS

OSNAP can be used any time AutoCAD prompts you for a point. This means that you can invoke an *OSNAP* mode during any draw or modify command as well as during many other commands. *OSNAP* provides you with the potential to create 100% accurate drawings with AutoCAD. Take advantage of this feature whenever it will improve your drawing precision. Remember, any time you are prompted for a point, use *OSNAP* if it can improve your accuracy.

OSNAP PRACTICE

Single Point Selection Mode

1.   Turn off *SNAP* (**F9**). Draw two (approximately) vertical *Lines*. Follow these steps to draw another *Line* between *Endpoint*s.

STEPS	COMMAND PROMPT	PERFORM ACTION	COMMENTS
1.	Command:	select *Line* by any method	
2.	LINE From point:	type *END* and press **Enter** (or Spacebar)	
3.	endp of	move the cursor near the end of one of the *Lines*	endpoint Marker appears (square box at *Line* end), "Endpoint" Tool Tip may appear
4.		**PICK** while the Marker is visible	"rubberband" line appears
5.	To point:	type *END* and press **Enter** (or Spacebar)	
6.	endp of	move the cursor near the end of the second *Line*	endpoint Marker appears (square box at *Line* end), "Endpoint" Tool Tip may appear
7.		**PICK** while the Marker is visible	*Line* is created between endpoints
8.	To point:	press **Enter**	completes command

2. Draw two **Circles**. Follow these steps to draw a *Line* between the *Centers*.

STEPS	COMMAND PROMPT	PERFORM ACTION	COMMENTS
1.	Command:	select *Line* by any method	
2.	LINE From point:	invoke the cursor menu (press **Shift+#2** button) and select *Center*	
3.	cen of	move the cursor near a *Circle* object (near the *Circle*, <u>not</u> where you think the center is)	the AutoSnap Marker appears (small circle at center), "Center" Tool Tip may appear
4.		**PICK** while the Marker is visible	"rubberband" line appears
5.	To point:	invoke the cursor menu (press **Shift+#2** button) and select *Center*	
6.	cen of	move the cursor near the second *Circle* object (near the *Circle*, <u>not</u> where you think the center is)	the AutoSnap Marker appears (small circle at center), "Center" Tool Tip may appear
7.		**PICK** the other *Circle* while the Marker is visible	*Line* is created between *Circle* centers
8.	To point:	press **Enter**	completes command

3. *Erase* the *Line* only from the previous exercise. Draw another **Line** anywhere, but <u>not</u> attached to the *Circles*. Follow the steps to *Move* the *Line* endpoint to the *Circle* center.

STEPS	COMMAND PROMPT	PERFORM ACTION	COMMENTS
1.	Command:	select *Move* by any method	
2.	MOVE Select objects:	**PICK** the *Line*	the *Line* becomes highlighted
3.	Select objects:	press **Enter**	completes selection set
4.	Base point or displacement:	Select the **Endpoint** icon button from the *Object Snap* toolbar	
5.	endp of	move the cursor near a *Line* endpoint	AutoSnap Marker appears (square box), "Endpoint" Tool Tip may appear
6.		**PICK** while Marker is visible	endpoint becomes the "handle" for *Move*
7.	Second point of displacement:	select **Center** from *Object Snap* toolbar	
8.	cen of	move the cursor near a *Circle*	AutoSnap Marker appears (small circle), "Center" Tool Tip may appear
9.		**PICK** the **Circle** object while Marker is visible	selected *Line* is moved to *Circle* center

Running Object Snap Mode

4.   Draw several *Lines* and *Circles* at random.   To draw several *Lines* to *Endpoints* and *Tangent* to the *Circles*, follow these steps.

STEPS	COMMAND PROMPT	PERFORM ACTION	COMMENTS
1.	Command:	type *OSNAP* or *OS*	*Osnap Settings* dialog box appears
2.		select *Endpoint* and *Tangent* then press *OK*	turns on the Running Osnap modes
3.	Command:	invoke the *Line* command	use any method
4.	LINE From point:	move the cursor <u>near</u> one *Line* endpoint	AutoSnap Marker appears (square box), "Endpoint" Tool Tip may appear
5.		**PICK** while Marker is visible	"rubberband" line appears, connected to endpoint
6.	To point:	move cursor near endpoint of another *Line*	AutoSnap Marker appears (square box), "Endpoint" Tool Tip may appear
7.		**PICK** while Marker is visible	a *Line* is created between endpoints
8.	To point:	move the cursor near a *Circle* object	AutoSnap Marker appears (small circle with tangent line segment), "Tangent" Tool Tip may appear
9.		**PICK** *Circle* while Marker is visible	a *Line* is created *Tangent* to the *Circle*
10.	To point:	press **Enter**	ends *Line* command
11.	Command:	invoke the *Line* command	use any method
12.	LINE From point:	move cursor near a *Circle*	AutoSnap Marker appears (small circle with tangent line segment), "Deferred Tangent" Tool Tip may appear
13.		**PICK** *Circle* while Marker is visible	"rubberband" line does NOT appear
14.	To point:	move the cursor near a second *Circle* object	AutoSnap Marker appears (small circle with tangent line segment), "Deferred Tangent" Tool Tip may appear
15.		**PICK** *Circle* while Marker is visible	a *Line* is created tangent to the two *Circles*
16.	To point:	double-click the word *OSNAP* on the Status line, press **F3** or **Ctrl+F**	temporarily toggles Running Osnaps off

(Continued)

R14

17.		**PICK** a point near a *Line* end	*Line* is created, but does not snap to *Line* endpoint
18.	To point:	double-click the word *OSNAP* on the Status line, press **F3** or **Ctrl+F**	toggles Running Osnaps back on
19.		**PICK** near a *Circle* (when Marker is visible	*Line* is created that snaps tangent to *Circle*
20.	To point:	**Enter**	ends *Line* command
21.	Command:	invoke the cursor menu (press **Shift+#2**) and select *Osnap Settings*...	*Osnap Settings* dialog box appears
22.		Press the ***Clear all*** button, then ***OK***	running *OSNAPS* are turned off

R14

CHAPTER EXERCISES

1. *OSNAP* **Single Point Selection**

   Open **the CH6EX1A** drawing and begin constructing the sheet metal part. Each unit in the drawing represents one inch.

   A. Create four *Circles*. All *Circles* have a radius of **1.685**. The *Circles'* centers are located at **5,5**, **5,13**, **19,5**, and **19,13**.

   B. Draw four *Lines*. The *Lines* (highlighted in Figure 7-22) should be drawn on the outside of the *Circles* by using the ***Quadrant*** *OSNAP* mode as shown for each *Line* endpoint.

   C. Draw two *Lines* from the ***Center*** of the existing *Circles* to form two diagonals as shown in Figure 7-23.

   D. At the ***Intersection*** of the diagonals create a *Circle* with a **3** unit radius.

Figure 7-22

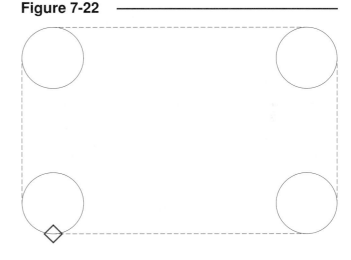

Figure 7-23

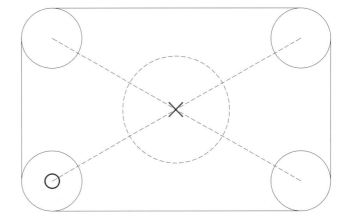

E. Draw two *Lines*, each from the *Intersection* of the diagonals to the *Midpoint* of the vertical *Lines* on each side. Finally, construct four new *Circles* with a radius of .25, each at the *Center* of the existing ones (Fig. 7-24).

F. *SaveAs* **CH7EX1**.

Figure 7-24

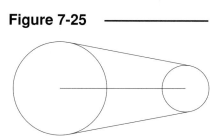

2. *OSNAP* **Single Point Selection**

A multiview drawing of a mechanical part is to be constructed using the A-METRIC drawing. All dimensions are in millimeters, so each unit equals one millimeter.

A. *Open* **A-METRIC** from Chapter 6 Exercises. Draw a *Line* from **60,140** to **140,140**. Create two *Circles* with the centers at the *Endpoints* of the *Line*, one *Circle* having a <u>diameter</u> of **60** and the second *Circle* having a diameter of **30**. Draw two *Lines Tangent* to the *Circles* as shown in Figure 7-25. *SaveAs* **PIVOTARM CH7**.

Figure 7-25

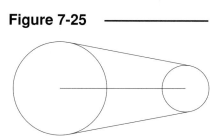

B. Draw a vertical *Line* down from the far left *Quadrant* of the *Circle* on the left. Specify relative polar (**@100<270**) or direct distance entry coordinates to make the *Line* 100 units. Draw a horizontal *Line* **125** units from the last *Endpoint* using relative polar or direct distance entry coordinates. Draw another *Line* between that *Endpoint* and the *Quadrant* of the *Circle* on the right. Finally, draw a horizontal *Line* from point **30,70** and *Perpendicular* to the vertical *Line* on the right.

Figure 7-26

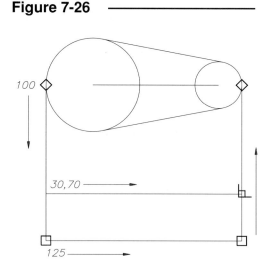

C. Draw two vertical *Lines* from the *Intersections* of the horizontal *Line* and *Circles* and *Perpendicular* to the *Line* at the bottom. Next, draw two *Circles* concentric to the previous two and with diameters of **20** and **10** as shown in Figure 7-27.

Figure 7-27 ────────

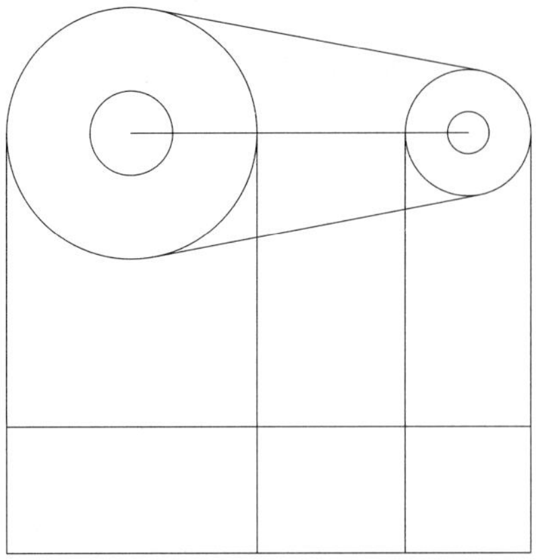

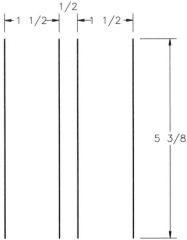

D. Draw four more vertical *Lines* as shown in Figure 7-28. Each *Line* is drawn from the new *Circles'* *Quadrant* and *Perpendicular* to the bottom line. Next, draw a miter *Line* from the *Intersection* of the corner shown to **@150<45**. *Save* the drawing for completion at a later time as another chapter exercise.

Figure 7-28 ────────────────

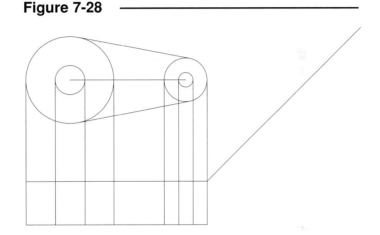

3. **Running Osnap**

Create a cross-sectional view of a door header composed of two 2 x 6 wooden boards and a piece of 1/2″ plywood. (The dimensions of a 2 x 6 are actually 1-1/2″ x 5-3/8″.)

A. Begin a *New* drawing and assign the name **HEADER**. Draw four vertical lines as shown in Figure 7-29.

B. Use the *OSNAP* command or select *Object Snap Settings…* from the *Tools* pull-down menu and turn on the *Endpoint* and *Intersection* modes.

Figure 7-29 ──────────

C. Draw the remaining lines as shown in Figure 7-30 to complete the header cross-section. Don't forget to turn off the running *OSNAP* modes when you are finished by selecting *Clear All* in the *Osnap Settings* dialog box. *Save* the drawing.

Figure 7-30

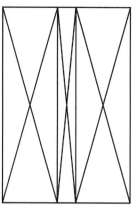

4. **Running Object Snap and *OSNAP* toggle**

Figure 7-31

Assume you are commissioned by the local parks and recreation department to provide a layout drawing for a major league sized baseball field. Lay out the location of bases, infield, and outfield as follows.

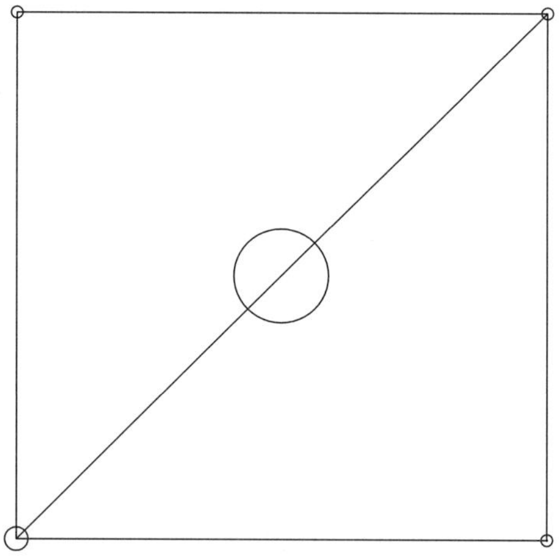

A. *Open* the **BALL FIELD CH6**.**DWG** that you set up in Chapter 6, Exercise 7. Make sure *Limits* are set to 512′,384′, *Snap* is set to 10′, and *Grid* is set to 20′. Ensure *SNAP* and *GRID* are on. Use *SaveAs* to save and rename the drawing to **BALL FIELD CH7**.

B. Begin drawing the baseball diamond by using the *Line* command and using direct distance coordinate entry. **PICK** the "From point:" at **20′,20′** (watch *Coords*). Draw the foul line to first base by turning on *ORTHO*, move the cursor to the right (along the X direction) and enter a value of **90′** (don't forget the apostrophe to indicate feet). At the "To point:" prompt, continue by drawing a vertical *Line* of **90′**. Continue drawing a square with **90′** between bases (see Fig. 7-31).

C. Invoke the *Osnap Settings* dialog box by any method. Turn on the *Endpoint, Midpoint,* and *Center* object snaps.

D. Use the *Circle* command to create a "base" at the lower-right corner of the square. At the "3P/2P/TTR/<Center Point>:" prompt, **PICK** the *Line Endpoint* at the lower-right corner of the square. Enter a *Radius* of **1′** (don't forget the apostrophe). Since Running Object Snaps are on, AutoCAD should display the Marker (square box) at each of the *Line Endpoints* as you move the cursor near; therefore, you can easily "snap" the center of the bases (*Circles*) to the corners of the square. Draw *Circles* of the same *Radius* at second base (upper-right corner), and third base (upper-left corner of the square). Create home plate with a *Circle* of a **2′** *Radius* by the same method (see Fig. 7-31).

E. Draw the pitcher's mound by first drawing a *Line* between home plate and second base. **PICK** the "From point:" at home plate (*Center* or *Endpoint*), then at the "To point:" prompt, PICK the *Center* or the *Endpoint* at second base. Construct a *Circle* of **8′** *Radius* at the *Midpoint* of the newly constructed diagonal line to represent the pitcher's mound (see Fig. 7-31).

F. *Erase* the diagonal *Line* between home plate and second base. Draw the foul lines from first and third base to the outfield. For the first base foul line, construct a **Line** with the "From point:" at the **Endpoint** of the existing first base line or **Center** of the base. Move the cursor (with *ORTHO* on) to the right (X positive) and enter a value of **240'** (don't forget the apostrophe). Press **Enter** to complete the *Line* command. Draw the third base foul line at the same length (in the positive Y direction) by the same method. (see Fig. 7-32).

Figure 7-32

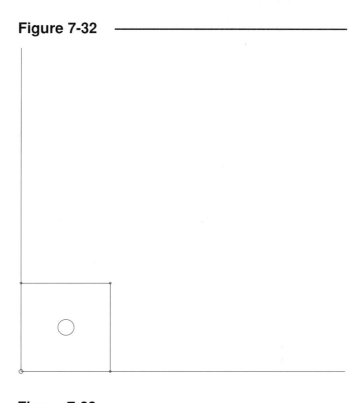

G. Draw the home run fence by using the *Arc* command from the **Draw** pull-down menu. Select the **Start, Center, End** method. **PICK** the end of the first base line (**Endpoint** object snap) for the **Start** of the *Arc* (Fig. 7-33, point 1.), **PICK** the pitcher's mound (**Center** object snap) for the **Center** of the *Arc* (point 2.), and the end of the third base line (**Endpoint** object snap) for the **End** of the *Arc* (point 3.). Don't worry about *ORTHO* in this case because *OSNAP* overrides *ORTHO*.

Figure 7-33

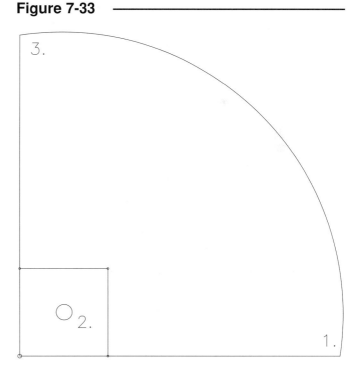

H.   Now turn off **ORTHO**. Construct an Arc to represent the end of the infield. Select the **Arc Start, Center, End** method from the **Draw** pull-down menu. For the **Start** point of the *Arc,* toggle Running Osnaps <u>off</u> by pressing **F3, Ctrl+F,** or double-clicking the word *OSNAP* on the Status line and **PICK** location **140′, 20′, 0′** on the first base line (watch *Coords* and ensure *SNAP* and *GRID* are on). (See Figure 7-34, point 1.) Next, toggle Running Osnaps back on, and **PICK** the **Center** of the pitcher's mound as the **Center** of the *Arc* (point 2.). Third, toggle Running Osnaps off again and **PICK** the **End** point of the *Arc* on the third base line (point 3.).

Figure 7-34

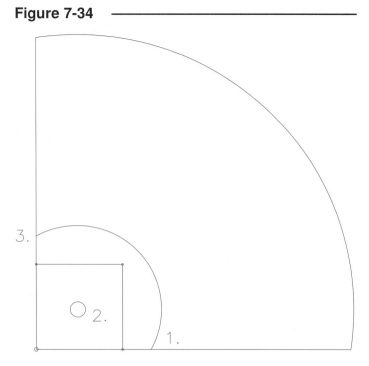

I.   Lastly, the pitcher's mound should be moved to the correct distance from home plate. Type *M* (the command alias for *Move*) or select **Move** from the **Modify** pull-down menu. When prompted to "Select objects:" **PICK** the *Circle* representing pitcher's mound. At the "Basepoint or displacement:" prompt, toggle Running Osnaps <u>on</u> and **PICK** the **Center** of the mound. At the "Second point of displacement:" prompt, enter **@3′2″<235**. This should reposition the pitcher's mound to the regulation distance from home plate (60′-6″). Compare your drawing to Figure 7-34. *Save* the drawing (as BALL FIELD CH7).

8

DRAW
COMMANDS I

Chapter Objectives

After completing this chapter you should:

1. know where to locate and how to invoke the draw commands;

2. be able to draw *Lines;*

3. be able to draw *Circles* by each of the five options;

4. be able to draw *Arcs* by each of the eleven options;

5. be able to create *Point* objects and specify the *Point Style;*

6. be able to create *Plines* with width and combined of line and arc segments.

CONCEPTS

Draw Commands—Simple and Complex

Draw commands create objects. An object is the smallest component of a drawing. The draw commands listed immediately below create simple objects and are discussed in this chapter. Simple objects <u>appear</u> as one entity.

> *Line, Circle, Arc, Point*

Other draw commands create more complex shapes. Complex shapes appear to be composed of several components, but each shape is usually <u>one object</u>. An example of an object that is one entity but usually appears as several segments is listed below and is also covered in this chapter:

> *Pline*

Other draw commands discussed in Chapter 15 (listed below) are a combination of simple and complex shapes:

> *Xline, Ray, Polygon, Rectangle, Donut, Spline, Ellipse, Divide, Mline, Measure, Sketch, Solid, Region, Boundary*

Figure 8-1   —

Draw Command Access

As a review from Chapter 3, Draw Command Basics, remember that any of the five methods can be used to access the draw commands: *Draw* toolbar (Fig. 8-1), *Draw* pull-down menu (Fig. 8-2), *DRAW1* and *DRAW2* screen menus, keyboard entry of the command or alias, and digitizing tablet icons.

Figure 8-2 —————————————

Coordinate Entry

When creating objects with draw commands, AutoCAD always prompts you to indicate points (such as endpoints, centers, radii) to describe the size and location of the objects to be drawn. An example you are familiar with is the *Line* command, where AutoCAD prompts for the "From point:." Indication of these points, called <u>coordinate entry</u> can be accomplished by five formats (for 2D drawings):

1. **Interactive**   **PICK** points on screen with input device
2. **Absolute coordinates**   **X,Y**
3. **Relative rectangular coordinates**   **@X,Y**
4. **Relative polar coordinates**   **@distance<angle**
5. **Direct distance entry**   **dist,direction** (Type a distance value relative to the last point, indicate direction with the cursor, then press Enter.)

Any of these methods can be used <u>whenever</u> AutoCAD prompts you to specify points. (For practice with these methods, see Chapter 3, Draw Command Basics.)

Also keep in mind that you can specify points interactively using *OSNAP* modes as discussed in Chapter 7. *OSNAP* modes can be used <u>whenever</u> AutoCAD prompts you to select points.

COMMANDS

LINE

Pull-down Menu	COMMAND (TYPE)	ALIAS (TYPE)	Short-cut	Screen (side) Menu	Tablet Menu
Draw *Line*	*LINE*	*L*	...	*DRAW 1* *Line*	*J,10*

This is the fundamental drawing command. The *Line* command creates straight line segments; each segment is an object. One or several line segments can be drawn with the *Line* command.

Command: *line*
From point: **PICK** or (**coordinates**) (A point can be designated by interactively selecting with the input device or by entering coordinates. If using the input device, the *Coords* display can be viewed to locate the current cursor position. If entering coordinates, any format is valid.)
To point: **PICK** or (**coordinates**) (Again, device input or keyboard input can be used. If using the input device to select, *ORTHO* (**F8**) can be toggled *ON* to force vertical or horizontal lines.)
To point: **PICK** or (**coordinates**) or **Enter** (Line segments can continually be drawn. Press Enter to complete the command.)
Command:

Figure 8-3

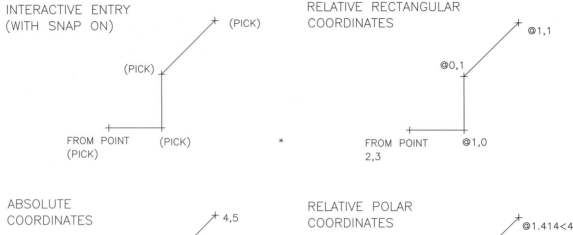

INTERACTIVE ENTRY
(WITH SNAP ON)
(PICK)
(PICK)
FROM POINT   (PICK)
(PICK)

RELATIVE RECTANGULAR
COORDINATES
@1,1
@0,1
FROM POINT   @1,0
2,3

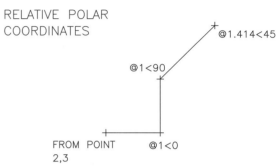

ABSOLUTE
COORDINATES
4,5
3,4
FROM POINT   3,3
2,3

RELATIVE POLAR
COORDINATES
@1.414<45
@1<90
FROM POINT   @1<0
2,3

Figure 8-3 shows four examples of creating the same *Line* segments using different methods of coordinate entry.

Refer to Chapter 3, Draw Command Basics, for examples of drawing vertical, horizontal, and inclined lines using the four formats for coordinate entry.

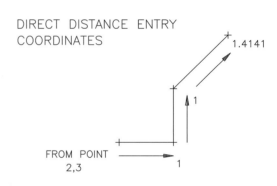

DIRECT DISTANCE ENTRY
COORDINATES
1.4141
1
FROM POINT
2,3
1

CIRCLE

Pull-down Menu	COMMAND (TYPE)	ALIAS (TYPE)	Short-cut	Screen (side) Menu	Tablet Menu
Draw *Circle >*	*CIRCLE*	C	...	**DRAW 1** *Circle*	J,9

The *Circle* command creates one object. Depending on the option selected, you can provide two or three points to define a *Circle*. As with all commands, the command line prompt displays the possible options:

    Command: *circle*
    3P/2P/TTR/<Center point>: **PICK** or (**coordinates**), (**option**). (PICKing or entering coordinates
    designates the center point for the circle. You can enter "3P," "2P" or "TTR" for another option.)

As with many other commands, when typing the command, the default and other options are displayed on the command line. The default option always appears in brackets "<option>:." The other options can be invoked by typing the indicated uppercase letter(s).

All of the options for creating *Circles* are available from the *Draw* pull-down menu and from the *DRAW 1* screen (side) menu. The tool (icon button) for only the *Center, Radius* method is included in the *Draw* toolbar. Tools for the other explicit *Circle* and *Arc* options are available, but only if you customize your own toolbar (See Chapter 45, Basic Customization).

The options, or methods, for drawing *Circles* are listed below. Each figure gives several possibilities for each option, with and without *OSNAPs*.

Center, Radius
Specify a center point, then a radius. (Fig. 8-4).

The *Radius* (or *Diameter*) can be specified by entering values or by indicating a length interactively (PICK two points to specify a length when prompted). As always, points can be specified by PICKing or entering coordinates. Watch *Coords* for coordinate or distance (polar format) display. *OSNAPs* can be used for interactive point specification.

Figure 8-4

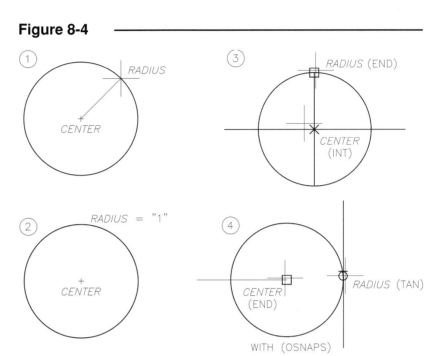

Center, Diameter
Specify the center point, then the diameter (Fig. 8-5).

Figure 8-5

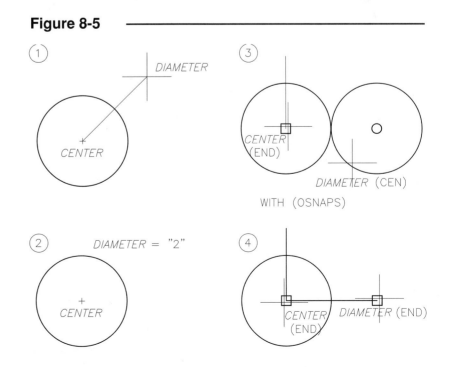

2 Points

The two points specify the location and diameter.

The *Tangent OSNAP*s can be used when selecting points with the *2 Point* and *3 Point* options as shown in Figures 8-6 and 8-7.

Figure 8-6

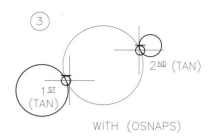

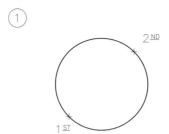

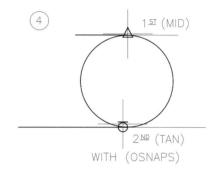

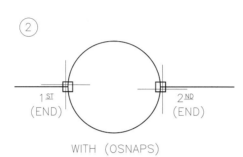

3 Points

The *Circle* passes through all three points specified.

Figure 8-7

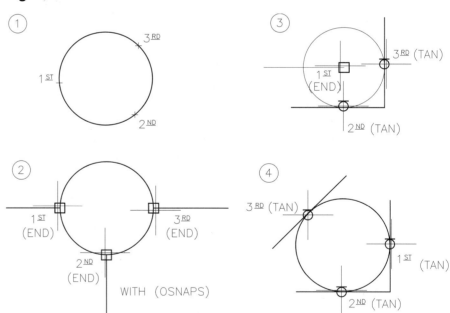

Tangent, Tangent, Radius
Specify two objects for the
Circle to be tangent to; then
specify the radius.

The *TTR* (Tangent, Tangent,
Radius) method is extremely
efficient and productive. The
OSNAP Tangent modes are
automatically invoked (the
aperture is displayed on the
cursor). This is the <u>only</u> draw
command option that auto-
matically calls *OSNAP*s.

Figure 8-8

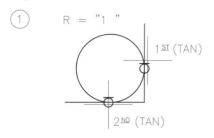
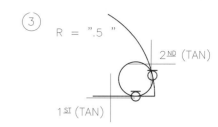
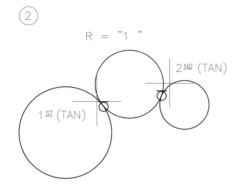
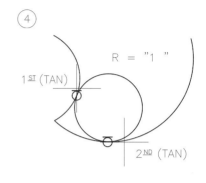

ARC

	COMMAND (TYPE)	ALIAS (TYPE)	Short-cut	Screen (side) Menu	Tablet Menu
Pull-down Menu					
Draw *Arc >*	*ARC*	*A*	...	*DRAW 1* *Arc*	*R,10*

An arc is part of a circle; it is a regular curve of <u>less</u> than 360 degrees. The *Arc* command in AutoCAD
provides eleven options for creating arcs. An *Arc* is one object. *Arcs* are always drawn by default in a
<u>counter-clockwise</u> direction. This occurrence forces you to decide in advance which points should be
designated as *Start* and *End* points (for options requesting those points). For this reason, it is often easier
to create arcs by another method, such as drawing a *Circle* and then using *Trim* or using the *Fillet*
command. (See Use *Arcs* or *Circles*? at the end of this section on *Arcs*.) The *Arc* command prompt is:

> Command: **arc**
> Center/<Start point>: **PICK** or (**coordinates**), or **C** (Interactively select or enter coordinates in any
> format for the start point. Type "C" to use the *Center* option instead.)

The prompts displayed by AutoCAD are different depending on which option is selected. At any time
while using the command, you can select from the options listed on the command line by typing in the
capitalized letter(s) for the desired option.

Alternately, to use a particular option of the *Arc* command, you can select from the *Draw* pull-down
menu or from the *DRAW 1* screen (side) menu. The tool (icon button) for only the *3 Points* method is
included in the *Draw* toolbar. The other tools are available, but only if you customize your own toolbar
(See Chapter 45, Basic Customization). These options require coordinate entry of <u>points in a specific
order</u>.

3Points
Specify three points through which the *Arc* passes (Fig. 8-9).

Figure 8-9

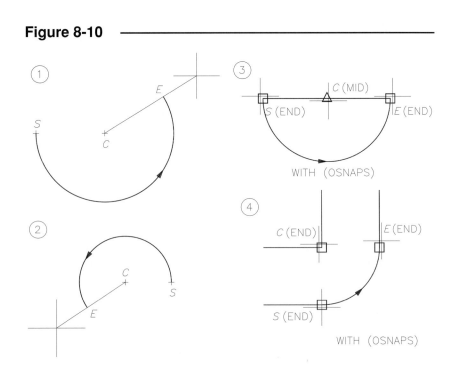

Start, Center, End
The radius is defined by the first two points that you specify (Fig. 8-10).

Figure 8-10

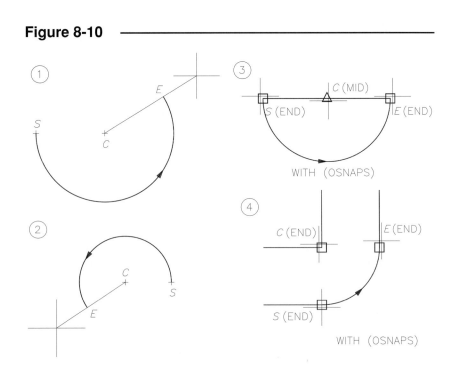

Start, Center, Angle
The angle is the <u>included</u> angle between the sides from the center to the endpoints. A <u>negative</u> angle can be entered to generate an *Arc* in a <u>clockwise</u> direction.

Figure 8-11

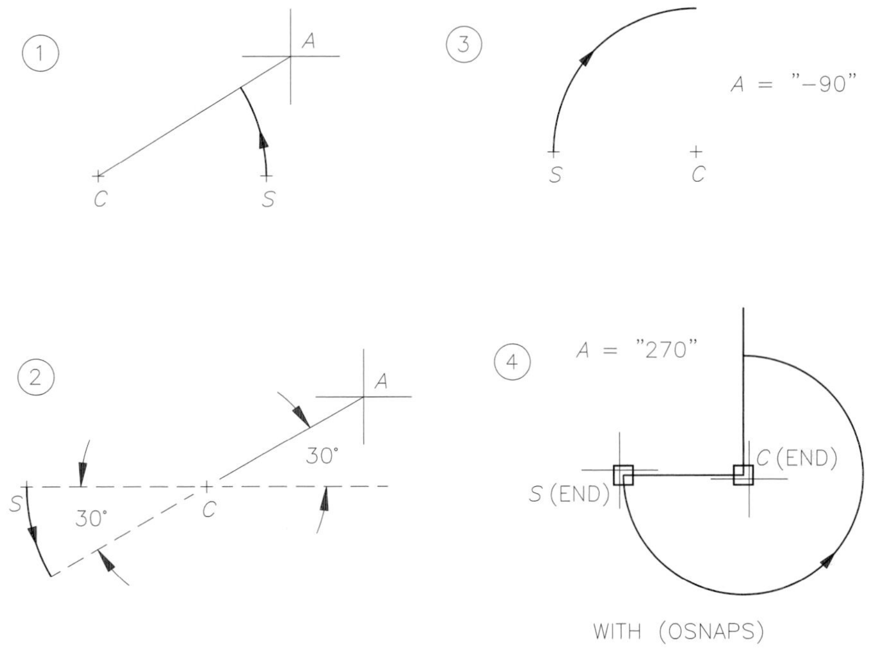

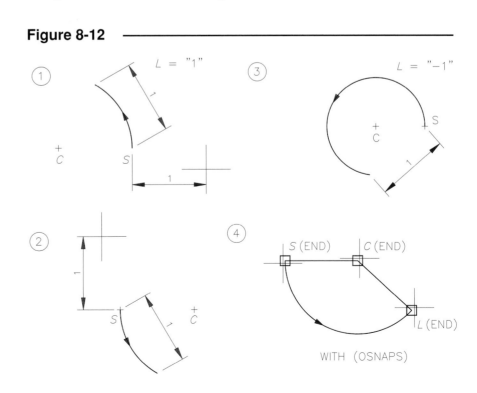

Start, Center, Length
Length means length of chord. The length of chord is between the start and the other point specified. A negative chord length can be entered to generate an *Arc* of 180+ degrees.

Figure 8-12

Start, End, Angle

The included angle is between the sides from the center to the endpoints. Negative angles generate clockwise *Arcs*.

Figure 8-13 ————————————————————————————

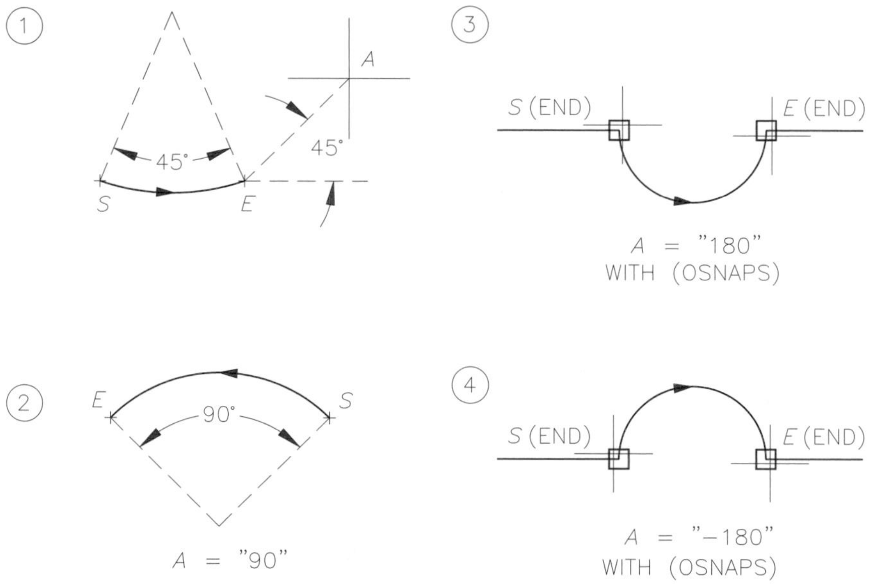

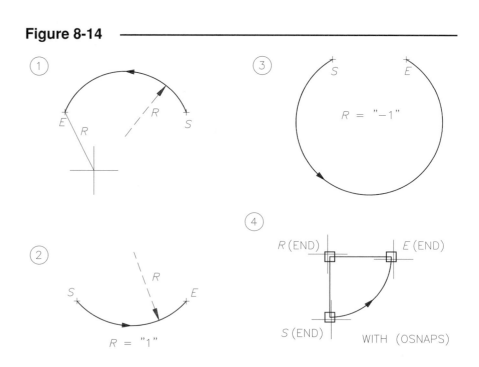

Start, End, Radius

The radius can be PICKed or entered as a value. A negative radius value generates an *Arc* of 180+ degrees.

Figure 8-14 ————————————————————————————

Start, End, Direction

The direction is tangent to the start point.

Figure 8-15

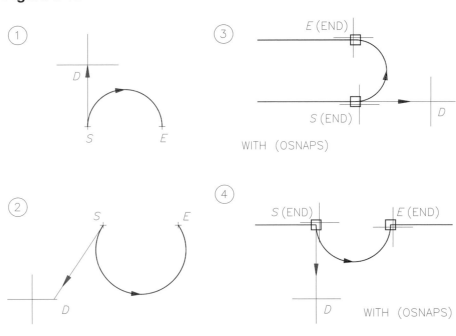

Center, Start, End

This option is like *Start, Center, End* but in a different order.

Figure 8-16

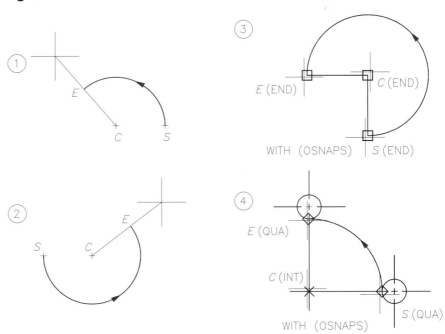

Center, Start, Angle
This option is like *Start, Center, Angle* but in a different order.

Figure 8-17

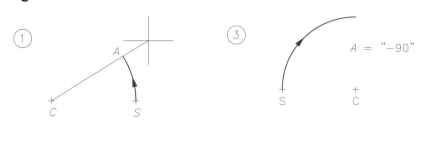

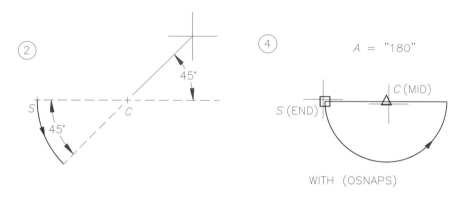

Center, Start, Length
This is similar to the *Start, Center, Length* option but in a different order. *Length* means length of chord.

Figure 8-18

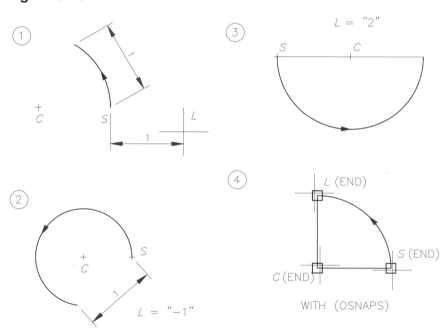

Continue
The new *Arc* continues from
and is tangent to the last
point. The only other point
required is the endpoint of
the *Arc*. This method allows
drawing *Arcs* tangent to the
preceding *Line*
or *Arc*.

Arcs are always created in a
counter-clockwise direction.
This fact must be taken into
consideration when using any
method except the *3-Point*, the
Start, *End*, *Direction*, and the
Continue options. The direction is explicitly specified
with *Start*, *End*, *Direction* and
Continue methods, and direction is irrelevant for *3-Point*
method.

Figure 8-19

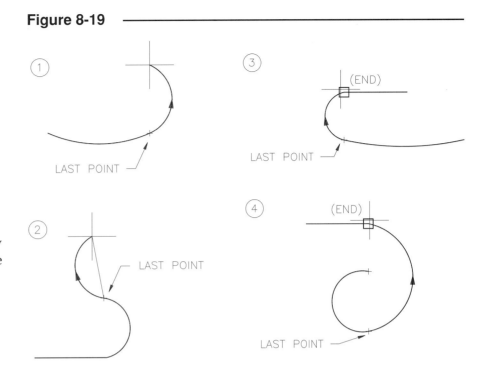

As usual, points can be specified by PICKing or entering coordinates. Watch *Coords* to display coordinate values or distances. *OSNAPs* can be used when PICKing. The *Endpoint, Intersection, Center,
Midpoint*, and *Quadrant OSNAP* options can be used with great effectiveness. The *Tangent OSNAP* option
cannot be used effectively with most of the *Arc* options. The *Radius, Direction, Length*, and *Angle* specifications can be given by entering values or by PICKing with or without *OSNAPs*.

Use *Arcs* or *Circles*?

Although there are sufficient options for drawing *Arcs*, usually it is easier to use the *Circle* command followed by *Trim* to achieve the desired arc. Creating a *Circle* is generally an easier operation than using
Arc because the counter-clockwise direction does not have to be considered. The unwanted portion of
the circle can be *Trimmed* at the *Intersection* of or *Tangent* to the connecting objects using *OSNAP*. The
Fillet command can also be used instead of the *Arc* command to add a fillet (arc) between two existing
objects (see Chapter 10, Modify Commands II).

POINT

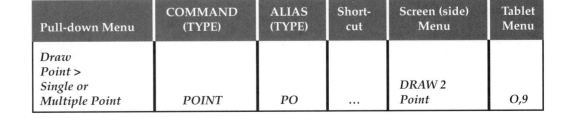

Pull-down Menu	COMMAND (TYPE)	ALIAS (TYPE)	Short-cut	Screen (side) Menu	Tablet Menu
Draw Point > Single or Multiple Point	*POINT*	*PO*	...	DRAW 2 Point	O,9

A *Point* is an object that has no dimension; it only has location. A *Point* is specified by giving only one
coordinate value or by PICKing a location on the screen.

Figure 8-20 compares *Points* to *Line* and *Circle* objects.

Figure 8-20

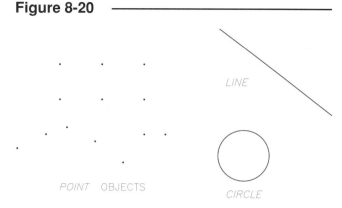

POINT OBJECTS

LINE

CIRCLE

> Command: **point**
> Point: **PICK** or (**coordinates**) (Select a location for the *Point* object.)
> Command:

Points are useful in construction of drawings to locate points of reference for subsequent construction or locational verification. The *Node OSNAP* option is used to snap to *Point* objects.

Points are drawing objects and therefore appear in prints and plots. The default "style" for points is a tiny "dot." The *Point Style* dialog box can be used to define the format you choose for *Point* objects (see *Ddptype* next).

The *Draw* pull-down menu offers the *Single Point* and the *Multiple Point* options. The *Single Point* option creates one *Point*, then returns to the command prompt. This option is the same as using the *Point* command by any other method. Selecting *Multiple Point* continues the *Point* command until you press the Escape key.

If you draw with *BLIPMODE* on, small cross marks appear when you create *Point* objects. Use *Redraw* to make the "blips" disappear to view the *Points* more clearly. This problem may occur if you are editing pre-Release 14 drawings because *BLIPMODE* is on by default (saved in the individual drawing files) and the default *Point* style is a "dot." Therefore, the "blips" conceal the *Points*.

DDPTYPE

Pull-down Menu	COMMAND (TYPE)	ALIAS (TYPE)	Short-cut	Screen (side) Menu	Tablet Menu
Format *Point Style...*	DDPTYPE	...	...	DRAW 2 Point Ddptype:	U,1

The *Point Style* dialog box (Fig. 8-21) is available only through the methods listed in the command table above. This dialog box allows you to define the format for the display of *Point* objects. The selected style is applied immediately to all newly created *Point* objects. The *Point Style* controls the format of *Points* for printing and plotting as well as for the computer display.

Figure 8-21

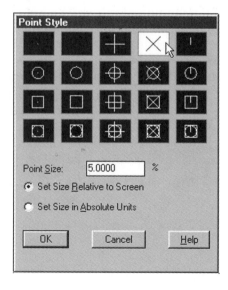

Changing the format of *Point* objects <u>does not automatically update existing *Point* objects</u> in the drawing. A regeneration must occur before existing *Points* are displayed in the new style. Use the *Regen* command immediately after changing the *Point Style* to cause a regeneration of the drawing, which updates the display of *Points* to the new style.

You can set the *Point Size* in *Absolute* units or *Relative to Screen* (default option). The *Relative to Screen* option keeps the *Points* the same size on the display when you *Zoom* in and out, whereas setting *Point Size* in *Absolute Units* gives you control over the size of *Points* for prints and plots. The selected *Point Style* is stored in the *PDMODE* variable.

PLINE

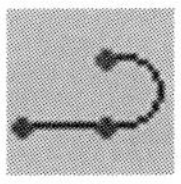

Pull-down Menu	COMMAND (TYPE)	ALIAS (TYPE)	Short-cut	Screen (side) Menu	Tablet Menu
Draw *Polyline*	*PLINE*	*PL*	...	DRAW 1 *Pline*	*N,10*

A *Pline* (or *Polyline*) has special features that make this object more versatile than a *Line*. Three features are most noticeable when first using *Plines*:

1. A *Pline* can have a specified *width*, whereas a *Line* has no width.
2. Several *Pline* segments created with one *Pline* command are treated by AutoCAD as <u>one</u> object, whereas individual line segments created with one use of the *Line* command are individual objects.
3. A *Pline* can contain arc segments.

Figure 8-22 illustrates *Pline* versus *Line* and *Arc* comparisons.

The *Pline* command begins with the same prompt as *Line*; however, <u>after</u> the "From point:" is established, the *Pline* options are accessible.

    Command: **pline**
    From point: **PICK** or (**coordinates**)
    Arc/Close/Halfwidth/Length/Undo/Width/ <Endpoint of Line>: **PICK** or (**coordinates**) or (**letter**)

The options and descriptions follow.

Figure 8-22 ————————————————

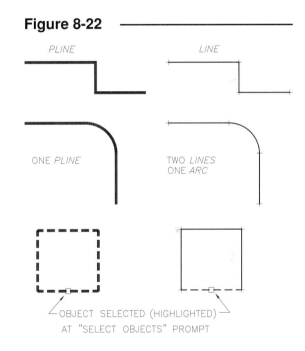

Width
You can use this option to specify starting and ending widths. Width is measured perpendicular to the centerline of the *Pline* segment (Fig. 8-23). *Plines* can be tapered by specifying different starting and ending widths. See NOTE at the end of this section.

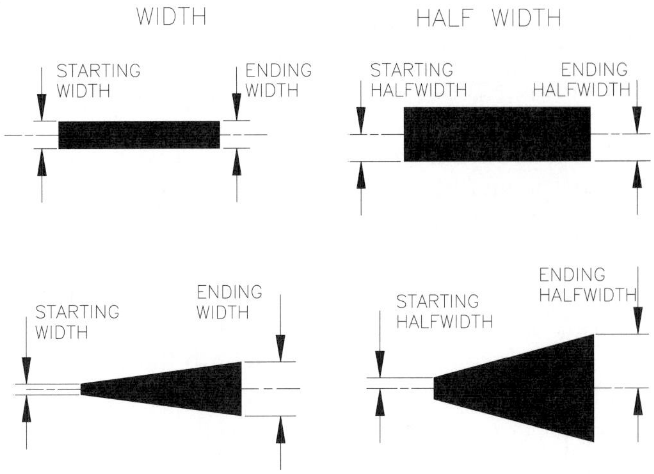

Halfwidth
This option allows specifying half of the *Pline* width. *Plines* can be tapered by specifying different starting and ending widths (Fig. 8-23).

Figure 8-23 ————————————————

Figure 8-24

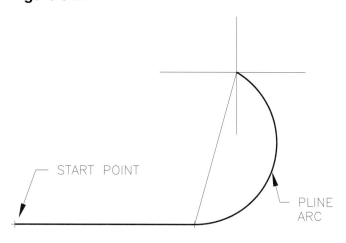

Arc

This option (by default) creates an arc segment in a manner similar to the *Arc Continue* method (Fig. 8-24). Any of several other methods are possible (see *Pline Arc* Segments).

Close

Figure 8-25

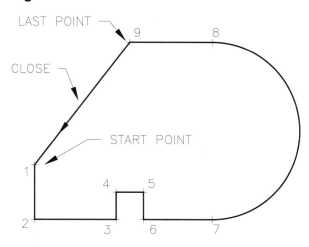

The *Close* option creates the closing segment connecting the first and last points specified with the current *Pline* command as shown in Figure 8-25.

This option can also be used to close a group of connected *Pline* segments into one continuous *Pline*. (A *Pline* closed by PICKing points has a specific start and endpoint.) A *Pline Closed* by this method has special properties if you use *Pedit* for *Pline* editing or if you use the *Fillet* command with the *Pline* option (see *Fillet* in Chapter 10 and *Pedit* in Chapter 16).

Length

Figure 8-26

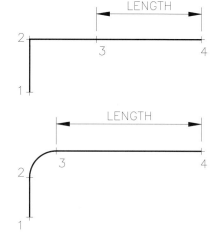

Length draws a *Pline* segment at the same angle as and connected to the previous segment and uses a length that you specify. If the previous segment was an arc, *Length* makes the current segment tangent to the ending direction (Fig. 8-26).

Undo

Use this option to *Undo* the last *Pline* segment. It can be used repeatedly to undo multiple segments.

NOTE: If you change the *Width* of a *Pline*, be sure to respond to <u>both</u> <u>prompts</u> for width ("Starting width:" and "Ending width:") before you draw the first *Pline* segment. It is easy to hastily PICK the endpoint of the *Pline* segment after specifying the "Starting width:" instead of responding with a value (or Enter) for the "Ending width:". In this case (if you PICK at the "Ending Width:" prompt), AutoCAD understands the line length that you interactively specified to be the ending width you want for the next line segment (you can PICK two points in response to the "Starting width:" or "Ending width:" prompts).

Command: **PLINE**
From point:
Current line-width is 0.0000
Arc/Close/Halfwidth/Length/Undo/Width/<Endpoint of line>: **w**
Starting width <0.0000>: **.2**
Ending width <0.2000>: Enter a value or press the Enter key; do not PICK the endpoint of the next *Pline* segment.

Pline Arc **Segments**

When the *Arc* option of *Pline* is selected, the prompt changes to provide the various methods for construction of arcs:

Angle/CEnter/CLose/Direction/Halfwidth/Line/Radius/Second pt/Undo/Width/<Endpoint of Arc>:

Angle
You can draw an arc segment by specifying the included angle (a negative value indicates a clockwise direction for arc generation).

CEnter
This option allows you to specify a specific center point for the arc segment.

CLose
This option closes the *Pline* group with an arc segment.

Direction
Direction allows you to specify an explicit starting direction rather than using the ending direction of the previous segment as a default.

Line
This switches back to the line options of the *Pline* command.

Radius
You can specify an arc radius using this option.

Second pt
Using this option allows specification of a 3-point arc.

Because a shape created with one *Pline* command is <u>one object</u>, manipulation of the shape is generally easier than with several objects. For some applications, one *Pline* shape can have advantages over shapes composed of several objects (see *Offset,* Chapter 10). Editing *Plines* is accomplished by using the *Pedit* command. As an alternative, *Plines* can be "broken" back down into individual objects with *Explode.*

Drawing and editing *Plines* can be somewhat involved. As an alternative, you can draw a shape as you would normally with *Line, Circle, Arc, Trim,* etc., and then <u>convert</u> the shape to one *Pline* object using *Pedit.* (See Chapter 16 for details on converting *Lines* and *Arcs* to *Plines.*)

Lightweight *Polylines*

In AutoCAD Release 14, *Plines* are created and listed as *Lwpolylines*, or "lightweight polylines." In previous releases, complete data for each *Pline* vertex (starting width, ending width, color, linetype, etc.) was stored along with the coordinate values of the vertex, then repeated for each vertex. In Release 14, the data common to all vertices is stored only once and only the coordinate data is stored for each vertex. Because this data structure saves file space, the new *Plines* are known as "lightweight" *Plines*. The differences in the data structure of *Plines* are apparent when the *List* command is used to display coordinates and other information about *Plines* in Release 14 and in previous releases (see *List*, Chapter 17).

Although this feature offers an advantage for users who have drawing applications that utilize many *Plines*, the creation of lightweight *Plines* is transparent to the user in Release 14. The *Pline* command in Release 14 operates identically to its operation in previous releases so there is nothing extra you must do to create a lightweight *Pline* in Release 14.

CONVERT

Pull-down Menu	COMMAND (TYPE)	ALIAS (TYPE)	Short-cut	Screen (side) Menu	Tablet Menu
...	*CONVERT*	...	...	...	...

Plines and hatch patterns in Release 14 have an optimized data structure that reduces file size compared to earlier versions of AutoCAD (see Lightweight *Polylines*). *Convert* transforms *Plines* and associative hatches that were created in Release 13 and previous releases to the optimized Release 14 data structure. The conversion of older drawings that contain a number of *Plines* and hatch patterns can significantly reduce file size and memory requirements.

If you edit drawings that were created with Release 13 or earlier releases, you can use the *Convert* command to convert *Plines* and associative hatches to the optimized Release 14 format. In most cases, however, <u>older *Plines* are automatically converted</u> to the new optimized format when you *Open* a Release 13 or earlier drawing in Release 14. In some cases where *Plines* were created by third-party applications, the *Plines* may require the use of *Convert*. Associative hatch patterns are not automatically converted, therefore *Convert* must be used. (See *Convert*, Chapter 26, for information on associative hatch patterns and command syntax for *Convert*.)

CHAPTER EXERCISES

Create a *New* drawing. When the *Create New Drawing* dialog box appears, select **Start from Scratch** and select **English** settings. Set *SNAP* to **.25** and set *GRID* to **1**. *Save* the drawing as **CH8EX**. For each of the following problems, *Open* **CH8EX**, complete one problem, then use *SaveAs* to give the drawing a new name.

Figure 8-27

1. *Open* **CH8EX**. Create the geometry shown in Figure 8-27. Start the first *Circle* center at point **4,4.5** as shown. Do not copy the dimensions. *SaveAs* **LINK**. (HINT: Locate and draw the two small *Circles* first. Use *Arc*, *Start, Center, End* or *Center, Start, End* for the rounded ends.)

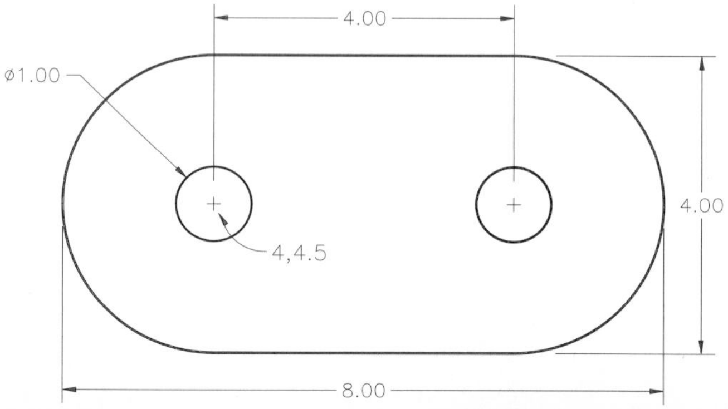

2. *Open* **CH8EX**. Create the geometry as shown in Figure 8-28. Do not copy the dimensions. Assume symmetry about the vertical axis. *SaveAs* **SLOT-PLATE CH8**.

Figure 8-28

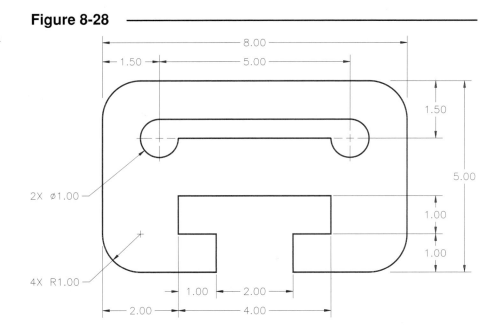

3. *Open* **CH8EX**. Create the shapes shown in Figure 8-29. Do not copy the dimensions. *SaveAs* **CH8EX3**.

Draw the *Lines* at the bottom first, starting at coordinate **1,3**. Then create *Point* objects at **5,7**, **5.4,7**, **5.8,7**, etc. Change the *Point Style* to an X and *Regen*. Use the *NODe OSNAP* mode to draw the inclined *Lines*. Create the *Arc* on top with the *Start, End, Direction* option.

Figure 8-29

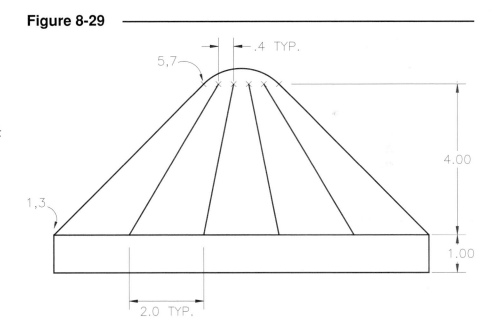

4. ***Open* CH8EX**. Create the shape shown in Figure 8-30. Draw the two horizontal *Lines* and the vertical *Line* first by specifying the endpoints as given. Then create the *Circle* and *Arcs*. *SaveAs* **CH8EX4**.

    HINT: Use the *Circle 2P* method with *Endpoint OSNAP*s. The two upper *Arcs* can be drawn by the *Start, End, Radius* method.

Figure 8-30

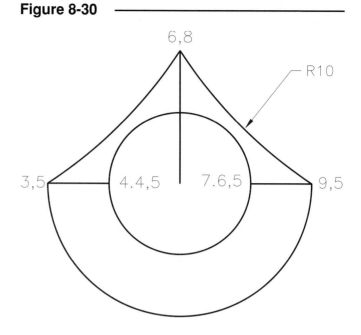

5. ***Open* CH8EX**. Draw the shape shown in Figure 8-31. Assume symmetry along a vertical axis. Start by drawing the two horizontal *Lines* at the base.

    Next, construct the side *Arcs* by the *Start, Center, Angle* method (you can specify a negative angle). The small *Arc* can be drawn by the *3P* method. Use *OSNAP*s when needed (especially for the horizontal *Line* on top and the *Line* along the vertical axis). *SaveAs* **CH8EX5**.

Figure 8-31

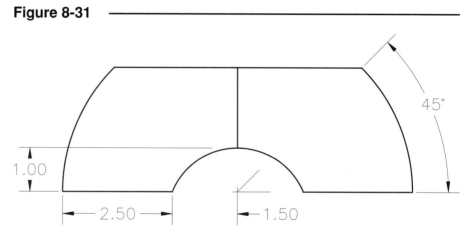

6. **Open CH8EX.**
Complete the geometry
in Figure 8-32. Use the
coordinates to establish
the **Lines.** Draw the
Circles using the
**Tangent, Tangent,
Radius** method.
SaveAs **CH8EX6.**

Figure 8-32

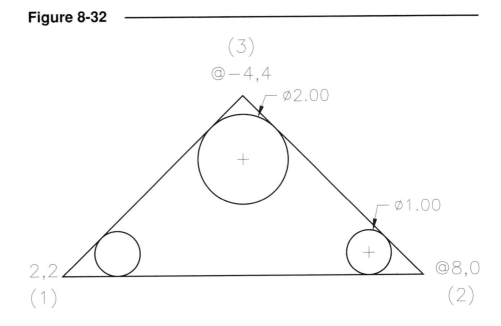

7. **Retaining Wall**

A contractor plans to stake out the edge of a retaining wall located at the bottom of a hill. The following table lists coordinate values based on a survey at the site. Use the drawing you created in Chapter 6 Exercises named **CH6EX2.** *Save* the drawing as **RET-WALL.**

Place **Points** at each coordinate value in order to create a set of data points. Determine the location of one **Arc** and two **Lines** representing the centerline of the retaining wall edge. The centerline of the retaining wall should match the data points as accurately as possible (Fig. 8-33). The data points given in the table are in inch units.

Figure 8-33

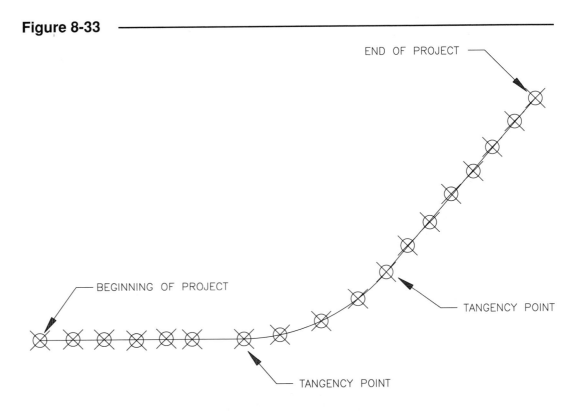

Pt #	X	Y	
1.	60.0000	108.0000	
2.	81.5118	108.5120	
3.	101.4870	108.2560	
4.	122.2305	107.4880	
5.	141.4375	108.5120	
6.	158.1949	108.0000	
7.	192.0000	108.0000	Recommended tangency point
8.	215.3914	110.5185	
9.	242.3905	119.2603	
10.	266.5006	133.6341	
11.	285.1499	151.0284	Recommended tangency point
12.	299.5069	168.1924	
13.	314.2060	183.4111	
14.	328.3813	201.7784	
15.	343.0811	216.4723	
16.	355.6808	232.2157	
17.	370.3805	249.0087	
18.	384.0000	264.0000	

(Hint: Change the *Point Style* to an easily visible format.)

8. *Pline*

Create the shape shown in Figure 8-34. Draw the outside shape with <u>one continuous</u> *Pline* (with 0.00 width). When finished, *SaveAs* **PLINE1**.

Figure 8-34

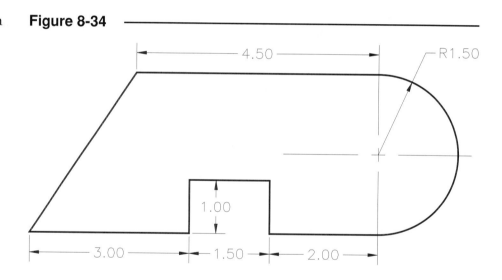

9

MODIFY COMMANDS I

Chapter Objectives

After completing this chapter you should:

1. know where to locate and how to invoke the *Modify* commands;

2. be able to *Erase* objects from the drawing;

3. be able to *Move* objects from a base-point to a second point of displacement;

4. know how to *Rotate* objects about a basepoint;

5. be able to enlarge or reduce objects with *Scale*;

6. be able to *Stretch* selected objects;

7. be able to change the length of *Lines* and *Arcs* with *Lengthen*;

8. be able to *Trim* away parts of objects at cutting edges;

9. be able to *Extend* objects to selected boundary edges;

10. know how to use the four *Break* options.

CONCEPTS

Draw commands are used to create new objects. *Modify* commands are used to change existing objects or to use existing objects to create new and similar objects. The *Modify* commands covered in this chapter only <u>change existing objects</u>. The commands listed below are covered in this chapter, while other *Modify* commands are covered in Chapter 10 and Chapter 16:

  Erase, Move, Rotate, Scale, Stretch, Lengthen, Trim, Extend, and *Break*

Modify commands can be invoked by any of the five command entry methods: toolbar icons, pull-down menus, screen menus, keyboard entry, and digitizing tablet menu. The *Modify* toolbar is visible and docked to the right side of the screen by default in Release 14 (Fig. 9-1).

The digitizing tablet uses the same icons that are displayed for the toolbars.

Figure 9-1

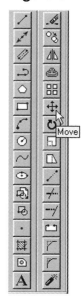

The *Modify* pull-down menu (Fig. 9-2) contains commands that <u>change</u> existing geometry and that use existing geometry to create new but similar objects. All of the *Modify* commands are contained in this single pull-down menu.

Figure 9-2

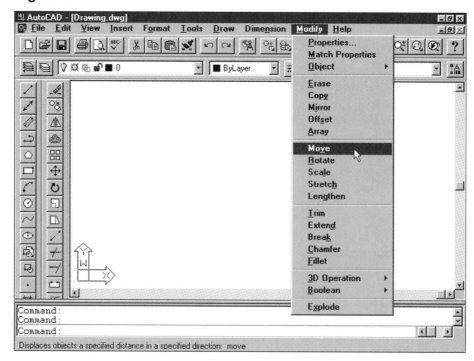

The *MODIFY1* and *MODIFY2* screen menus branch to the individual commands and options (Fig. 9-3).

Since all *Modify* commands affect or use existing geometry, the first step in using any modify command is to construct a selection set (see Chapter 4). This can be done by one of two methods:

1. Invoking the desired command and then creating the selection set in response to the "Select Objects:" prompt (Verb/Noun syntax order) using any of the select object options;

2. Selecting the desired set of objects with the pickbox or *Auto* Window or Crossing Window <u>before</u> invoking the edit command (Noun/Verb syntax order).

Figure 9-3

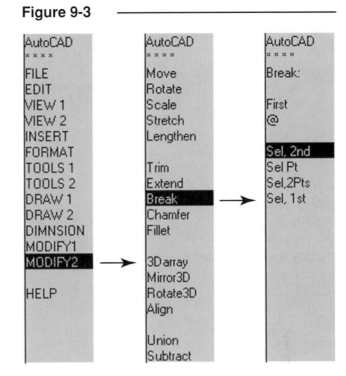

The first method allows use of any of the selection options (*Last, All, WPolygon, Fence,* etc.), while the latter method allows <u>only</u> the use of the pickbox and *Auto* Window and Crossing Window.

COMMANDS

ERASE

Pull-down Menu	COMMAND (TYPE)	ALIAS (TYPE)	Short-cut	Screen (side) Menu	Tablet Menu
Modify *Erase*	ERASE	E	...	*MODIFY1* *Erase*	V,14

The *Erase* command deletes the objects you select from the drawing. Any of the object selection methods can be used to highlight the objects to *Erase*. The only other required action is for you to press *Enter* to cause the erase to take effect.

```
Command: Erase
Select Objects: PICK (Use any object selection method.)
Select Objects: PICK (Continue to select desired objects.)
Select Objects: Enter (Confirms the object selection process and causes Erase to take effect.)
Command:
```

If objects are erased accidentally, *U* can be used immediately following the mistake to undo one step, or *Oops* can be used to bring back into the drawing whatever was *Erased* the last time *Erase* was used (see Chapter 5). If only part of an object should be erased, use *Trim* or *Break*.

MOVE

	COMMAND (TYPE)	ALIAS (TYPE)	Short-cut	Screen (side) Menu	Tablet Menu
Pull-down Menu					
Modify *Move*	MOVE	M	...	*MODIFY2* *Move*	V,19

Move allows you to relocate one or more objects from the existing position in the drawing to any other position you specify. After selecting the objects to *Move,* you must specify the "base-point" and "second point of displacement." You can use any of the five coordinate entry methods to specify these points. Examples are shown in Figure 9-4.

Figure 9-4 ————————————————————

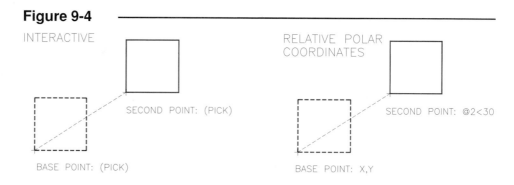

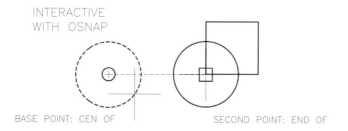

Command: ***move***
Select Objects: **PICK**   (Use any of the object selection methods.)
Select Objects: **PICK**   (Continue to select other desired objects.)
Select Objects: **Enter**   (Press Enter to indicate selection of objects is complete.)
Base point or displacement: **PICK** or **(coordinates)** (This is the point to <u>move from</u>. Select a point to use as a "handle." An *Endpoint* or *Center,* etc., can be used.)
Second point of displacement: **PICK** or **(coordinates)** (This is the point to <u>move to</u>. OSNAPs can also be used here.)
Command:

Keep in mind that *OSNAPs* can be used when PICKing any point. It is often helpful to toggle *ORTHO ON* (**F8**) to force the *Move* in a horizontal or vertical direction.

If you know a specific distance and an angle that the set of objects should be moved, relative rectangular or relative polar coordinates can be used. In the following sequence, relative polar coordinates are used to move objects 2 units in a 30 degree direction (see Fig. 9-4, coordinate entry).

Command: ***move***
Select Objects: **PICK**
Select Objects: **PICK**
Select Objects: **Enter**
Base point or displacement: **X,Y (coordinates)**
Second point of displacement: **@2<30**
Command:

Direct distance coordinate entry can be used effectively with *Move*. For example, assume you wanted to move the right side view of a multiview drawing 20 units to the right using direct distance entry (Fig. 9-5). First, invoke the *Move* command and select all of the objects comprising the right side view in response to the "Select Objects:" prompt. The command sequence is as follows:

Figure 9-5

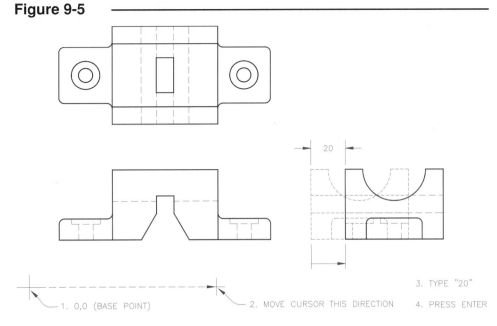

1. 0,0 (BASE POINT)
2. MOVE CURSOR THIS DIRECTION
3. TYPE "20"
4. PRESS ENTER

> Command: **Move**
> Select Objects: **PICK**
> Select Objects: **Enter**
> Base point or displacement: **0,0** (or any value)
> Second point of displacement: **20** (with *ORTHO* on, move cursor to right), then press **Enter**

In response to the "Base point or displacement:" prompt, PICK a point or enter any coordinate pairs or single value. At the "Second point of displacement:" prompt, move the cursor to the right any distance (with *ORTHO* on), then type "20" and press Enter. The value specifies the distance, and the cursor location from the last point indicates the direction of movement.

In the previous example, note that <u>any value</u> can be entered in response to the "Base point or displacement:" prompt. If a single value is entered ("3" for example), AutoCAD recognizes it as direct distance entry. The point designated is 3 units from the last PICK point in the direction specified by wherever the cursor is at the time of entry.

With the relative coordinate entry methods (relative rectangular, relative polar, and direct distance entry), the designated basepoint is of no consequence since the "second point of displacement" is always relative.

R14

ROTATE

Pull-down Menu	COMMAND (TYPE)	ALIAS (TYPE)	Short-cut	Screen (side) Menu	Tablet Menu
Modify *Rotate*	*ROTATE*	*RO*	...	*MODIFY2* *Rotate*	*V,20*

Selected objects can be rotated to any position with this command. After selecting objects to *Rotate*, you select a "basepoint" (a point to rotate about) then specify an angle for rotation.

AutoCAD rotates the selected objects by the increment specified from the original position (Fig. 9-6). For example, specifying a value of **45** would *Rotate* the selected objects 45 degrees counter-clockwise from their current position; a value of **-45** would *Rotate* the objects 45 degrees in a clockwise direction.

Figure 9-6

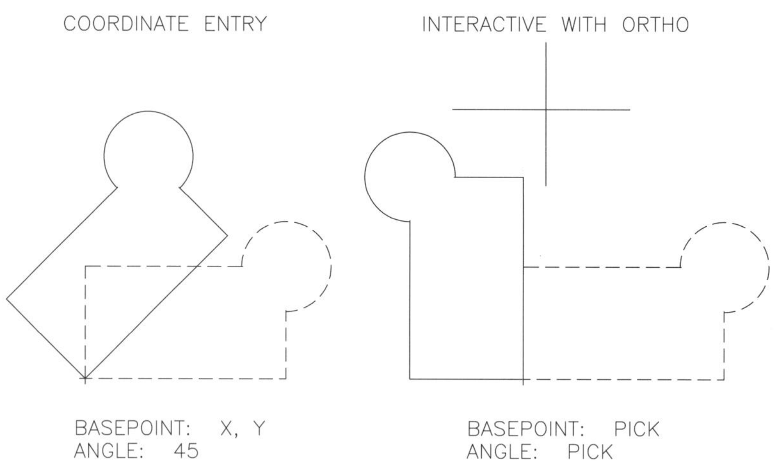

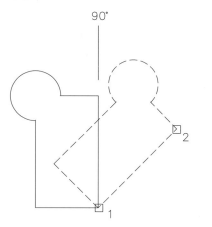

COORDINATE ENTRY   INTERACTIVE WITH ORTHO

BASEPOINT: X, Y   BASEPOINT: PICK
ANGLE: 45   ANGLE: PICK

> Command: **rotate**
> Select Objects: **PICK** or (**coordinates**) (Select the objects to rotate.)
> Select Objects: **Enter** (Indicate completion of the object selection.)
> Base point: **PICK** or (**coordinates**) (Select the point to rotate about.)
> <Rotation angle>Reference: **PICK** or (**value**) or (**coordinates**) (Enter a value for the number of degrees to rotate or interactively rotate the object set.)
> Command:

The basepoint is often selected interactively with *OSNAP*s. When specifying the angle for rotation, a value (incremental angle) can be typed. Alternately, if you want to **PICK** an angle, *Coords* (in the polar format) displays the current angle of the "rubberband" line. Turning on *ORTHO* forces the rotation to a 90 degree increment. The *List* command can be used to report the angle of an existing *Line* or other object (discussed in Chapter 17).

The **Reference** option can be used to specify a vector as the original angle before rotation (Fig. 9-7). This vector can be indicated interactively (*OSNAP*s can be used) or entered as an angle using keyboard entry. Angular values that you enter in response to the "New angle:" prompt are understood by AutoCAD as absolute angles for the *Reference* option only.

Figure 9-7

90°

REFERENCE ANGLE: PICK 1,2
NEW ANGLE: 90°

> Command: **rotate**
> Select Objects: **PICK** or (**coordinates**) (Select the objects to rotate.)
> Select Objects: **Enter** (Indicates completion of object selection.)
> Base point: **PICK** or (**coordinates**) (Select the point to rotate about.)
> <Rotation angle>Reference: **R** (Indicates the *Reference* option.)
> Reference angle<0>: **PICK** or (**value**) (PICK the first point of the vector.)
> Second point: **PICK** (Indicates the second point defining the vector.)
> New angle: **PICK** or (**value**) (Indicates the new angle value.)
> Command:

SCALE

Pull-down Menu	COMMAND (TYPE)	ALIAS (TYPE)	Short-cut	Screen (side) Menu	Tablet Menu
Modify *Scale*	*SCALE*	*SC*	...	*MODIFY2* *Scale*	*V,21*

The *Scale* command is used to increase or decrease the size of objects in a drawing. The *Scale* command does not normally have any relation to plotting a drawing to scale.

After selecting objects to *Scale*, AutoCAD prompts you to select a "Base point:", which is the <u>stationary point</u>. You can then scale the size of the selected objects interactively or enter a scale factor. Using interactive input, you are presented with a "rubberband" line connected to the basepoint. Making the "rubberband" line longer or shorter than 1 unit increases or decreases the scale of the selected objects by that proportion; for example, pulling the "rubberband" line to two units length increases the scale by a factor of two (Fig. 9-8).

Figure 9-8

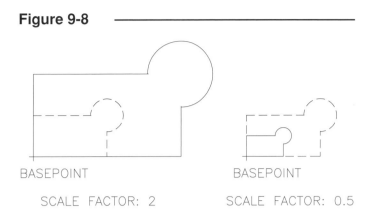

BASEPOINT

SCALE FACTOR: 2

BASEPOINT

SCALE FACTOR: 0.5

> Command: **scale**
> Select Objects: **PICK** or (**coordinates**) (Select the objects to scale.)
> Select Objects: **Enter** (Indicates completion of the object selection.)
> Base point: **PICK** or (**coordinates**) (Select the stationary point.)
> Scale factor<Reference>: **PICK** or (**value**) or (**coordinates**) (Enter a value for the scale factor or interactively scale the set of objects.)
> Command:

It may be desirable in some cases to use the *Reference* option to specify a value or two points to use as the reference length. This length can be indicated interactively (*OSNAP*s can be used) or entered as a value. This length is used for the subsequent reference length that the rubberband uses when interactively scaling. For example, if the reference distance is 2, then the "rubberband" line must be stretched to a length greater than 2 to increase the scale of the selected objects.

Scale normally should <u>not</u> be used to change the scale of an entire drawing in order to plot on a specific size sheet. CAD drawings should be created <u>full size</u> in <u>actual units</u>.

STRETCH

Pull-down Menu	COMMAND (TYPE)	ALIAS (TYPE)	Short-cut	Screen (side) Menu	Tablet Menu
Modify *Stretch*	*STRETCH*	S	...	*MODIFY2* *Stretch*	V,22

Objects can be made longer or shorter with *Stretch*. The power of this command lies in the ability to *Stretch* groups of objects while retaining the connectivity of the group (Fig. 9-9). When *Stretched*, *Lines* and *Plines* become longer or shorter and *Arcs* change radius to become longer or shorter. *Circles* do not stretch; rather, they move if the center is selected within the Crossing Window.

Objects to *Stretch* are not selected by the typical object selection methods but are indicated by a <u>Crossing Window or Crossing Polygon only</u>. The *Crossing Window* or *Polygon* should be created so the objects to *stretch* <u>cross</u> through the window. *Stretch* actually moves the object endpoints that are located within the Crossing Window.

Figure 9-9

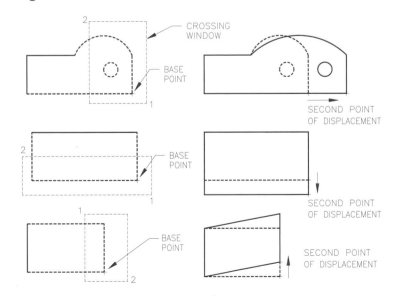

CROSSING WINDOW

BASE POINT

SECOND POINT OF DISPLACEMENT

BASE POINT

SECOND POINT OF DISPLACEMENT

BASE POINT

SECOND POINT OF DISPLACEMENT

Following is the command sequence for *Stretch*.

    Command: **stretch**
    Select object(s) to stretch by crossing-window or -polygon. . .
    Select Objects:
    First Corner: **PICK**
    Other Corner: **PICK**
    Select Objects: **Enter**
    Base point or displacement: **PICK** or (**coordinates**) (Select a point to use as the point to stretch <u>from</u>.)
    Second point of displacement: **PICK** or (**coordinates**) (Select a point to use as the point to stretch <u>to</u>.)
    Command:

Stretch can be used to lengthen one object while shortening another. Application of this ability would be repositioning a door or window on a wall (Fig. 9-10).

In summary, the *Stretch* command <u>stretches objects that cross</u> the selection Window and <u>moves objects that are completely within</u> the selection Window, as shown in Figure 9-10.

Figure 9-10

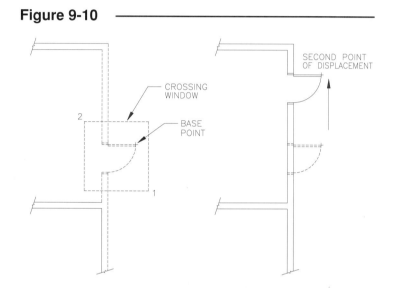

Pull-down Menu	COMMAND (TYPE)	ALIAS (TYPE)	Short-cut	Screen (side) Menu	Tablet Menu
Modify Lengthen	LENGTHEN	LEN	...	MODIFY2 Lengthen	W,14

LENGTHEN

Lengthen changes the length (longer or shorter) of linear objects and arcs. No additional objects are required (as with *Trim* and *Extend*) to make the change in length. Many methods are provided as displayed in the command prompt.

    Command: **lengthen**
    Delta/Percent/Total/DYnamic/<Select object>:

Figure 9-11

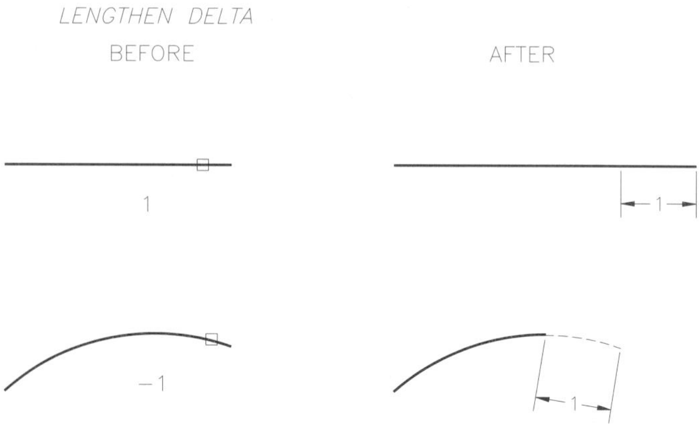

LENGTHEN DELTA

BEFORE   AFTER

Select object
Selecting an object causes AutoCAD to report the current length of that object. If an *Arc* is selected, the included angle is also given.

DElta
Using this option returns the prompt shown next. You can change the current length of an object (including an arc) by any increment that you specify. Entering a positive value increases the length by that amount, while a negative value decreases the current length. The end of the object that you select changes while the other end retains its current endpoint (Fig. 9-11).

Angle/<Enter delta length>: (**value**) or **a**

The *Angle* option allows you to change the <u>included</u> <u>angle</u> of an arc (the length along the curvature of the arc can be changed with the *Delta* option). Enter a positive or negative value (degrees) to add or subtract to the current included angle, then select the end of the object to change (Fig. 9-12).

Figure 9-12 ——————————

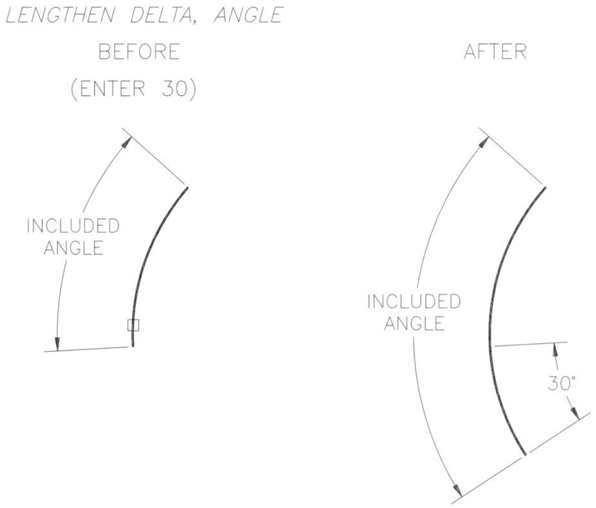

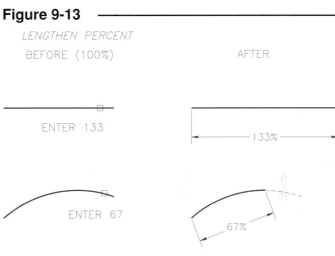

LENGTHEN DELTA, ANGLE

Percent
Use this option if you want to change the length by a percentage of the current total length. For arcs, the percentage applied affects the length and the included angle equally, so there is no *Angle* option. A value of greater than 100 increases the current length, and a value of less than 100 decreases the current length. Negative values are not allowed. The end of the object that you select changes.

Figure 9-13 ——————————

LENGTHEN PERCENT

BEFORE (100%)   AFTER

ENTER 133   133%

ENTER 67   67%

Total
This option lets you specify a value for the new total length. Simply enter the value and select the end of the object to change. The angle option is used to change the total included angle of a selected arc.

Angle/<Enter total length>: (**value**) or **a**

Figure 9-14 ——————————

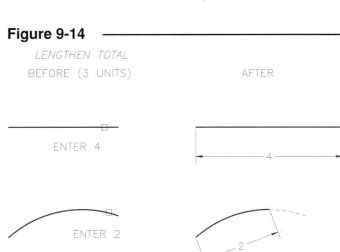

LENGTHEN TOTAL

BEFORE (3 UNITS)   AFTER

ENTER 4   4

ENTER 2   2

DYnamic

This option allows you to change the length of an object by dynamic dragging. Select the end of the object that you want to change. Object snaps can be used.

Figure 9-15

LENGTHEN DYNAMIC

BEFORE   AFTER

TRIM

Pull-down Menu	COMMAND (TYPE)	ALIAS (TYPE)	Short-cut	Screen (side) Menu	Tablet Menu
Modify *Trim*	*TRIM*	*TR*	…	*MODIFY2* *Trim*	*W,15*

The *Trim* command allows you to trim (shorten) the end of an object back to the intersection of another object (Fig. 9-16). The middle section of an object can also be *Trimmed* between two intersecting objects. There are two steps to this command: first, PICK one or more "cutting edges" (existing objects); then PICK the object or objects to *Trim* (portion to remove). The cutting edges are highlighted after selection. Cutting edges themselves can be trimmed if they intersect other cutting edges, but lose their highlight when trimmed.

```
Command: trim
Select cutting edges: (Projmode
= UCS, Edgemode = No extend)
Select Objects: PICK (Select an
object to use as a cutting edge.)
Select Objects: PICK
Select Objects: Enter
<Select object to trim>/Project/Edge/Undo: PICK (Select the end of an object to trim.)
<Select object to trim>/Project/Edge/Undo: PICK
<Select object to trim>/Project/Edge/Undo: Enter
Command:
```

Figure 9-16

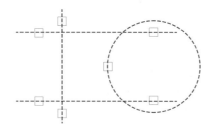
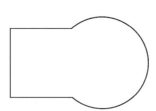

BEFORE *TRIM*   AFTER ALL *TRIMS*

CUTTING EDGES

Edgemode
The *Edgemode* option can be set to *Extend* or *No extend*. In the *Extend* mode, objects that are selected as trimming edges will be <u>imaginarily extended</u> to serve as a cutting edge. In other words, lines used for trimming edges are treated as having infinite length (Fig. 9-17). The *No extend* mode only considers the actual length of the object selected as trimming edges.

Figure 9-17

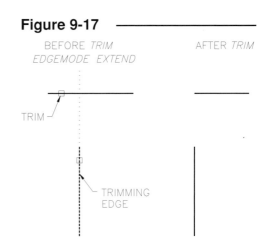

```
Command: trim
Select cutting edges: (Projmode = UCS, Edgemode = No
extend)
Select Objects: PICK
Select Objects: Enter
<Select object to trim>/Project/Edge/Undo: e
Extend/No extend <No extend>: e
<Select object to trim>/Project/Edge/Undo: PICK
<Select object to trim>/Project/Edge/Undo: Enter
Command:
```

Projectmode
The *Projectmode* switch controls how *Trim* and *Extend* operate in 3D space. *Projectmode* affects the projection of the "cutting edge" and "boundary edge." The three options are described here:

```
<Select object to trim>/Project/Edge/Undo: p
None/Ucs/View <UCS>:
```

None
The *None* option does not project the cutting edge. This mode is used for normal 2D drawing when all objects (cutting edges and objects to trim) lie in the current drawing plane. You can use this mode to *Trim* in 3D space when all objects lie in the plane of the current UCS or in planes parallel to it and the objects physically intersect. You can also *Trim* objects not in the UCS if the objects physically intersect, are in the same plane, and are not perpendicular to the UCS. In any case, the <u>objects must physically intersect</u> for trimming to occur.

The other two options (*UCS* and *View*) allow you to trim objects that do not physically intersect in 3D space.

UCS
This option projects the cutting edge perpendicular to the UCS. Any objects crossing the "projected" cutting edge in 3D space can be trimmed. For example, selecting a *Circle* as the cutting edge creates a projected cylinder used for cutting (Fig. 9-18).

Figure 9-18

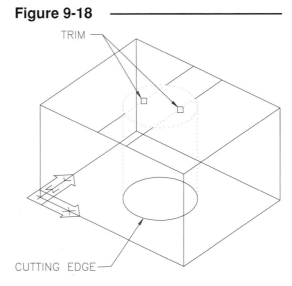

CUTTING EDGE

View

The *View* option allows you to trim objects that <u>appear</u> to intersect from the current viewpoint. The objects do not have to physically intersect in 3D space. The cutting edge is projected perpendicularly to the screen (parallel to the line of sight). This mode is useful for trimming "hidden edges" of a wireframe model to make the surfaces appear opaque (Fig. 9-19).

Figure 9-19

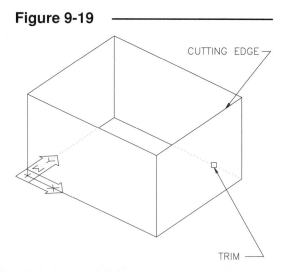

Undo

The *Undo* option allows you to undo the last *Trim* in case of an accidental trim.

EXTEND

Pull-down Menu	COMMAND (TYPE)	ALIAS (TYPE)	Short-cut	Screen (side) Menu	Tablet Menu
Modify *Extend*	*EXTEND*	*EX*	...	*MODIFY2* *Extend*	*W,16*

Extend can be thought of as the opposite of *Trim*. Objects such as *Lines*, *Arcs*, and *Plines* can be *Extended* until intersecting another object called a "boundary edge" (Fig. 9-20). The command first requires selection of <u>existing</u> objects to serve as "boundary edge(s)" which become highlighted; then the objects to extend are selected. Objects extend until, and only if, they eventually intersect a "boundary edge." An *Extended* object acquires a new endpoint at the boundary edge intersection.

Figure 9-20

BEFORE *EXTEND*   AFTER *EXTEND*

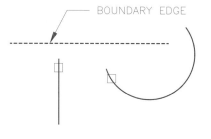
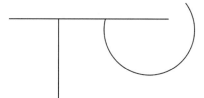

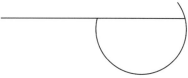

BOUNDARY EDGE

```
Command: extend
Select boundary edges:
(Projmode = UCS, Edgemode
= No extend)
Select Objects: PICK
Select Objects: PICK
Select Objects: Enter
<Select object to extend>/Project/Edge/Undo: PICK  (Select object to extend.)
<Select object to extend>/Project/Edge/Undo: PICK
<Select object to extend>/Project/Edge/Undo: Enter
Command:
```

Edgemode/Projectmode

The *Edgemode* and *Projectmode* switches operate identically to their function with the *Trim* command. Use *Edgemode* with the *Extend* option if you want a boundary edge object to be imaginarily extended (Fig. 9-21).

Figure 9-21

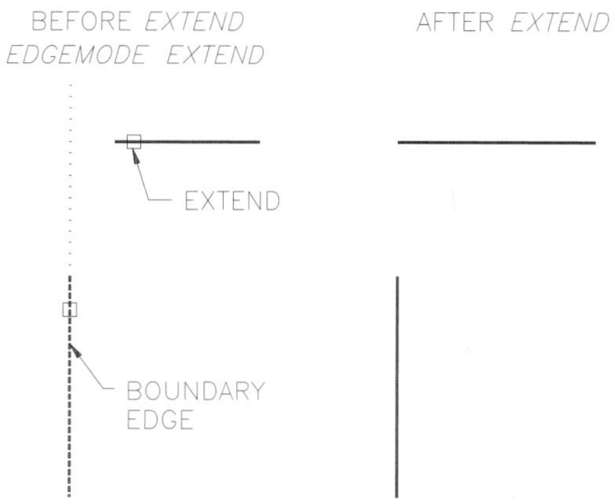

BEFORE *EXTEND*
EDGEMODE EXTEND

AFTER *EXTEND*

EXTEND

BOUNDARY
EDGE

BREAK

Pull-down Menu	COMMAND (TYPE)	ALIAS (TYPE)	Short-cut	Screen (side) Menu	Tablet Menu
Modify *Break*	BREAK	BR	...	*MODIFY2* *Break*	W,17

Break allows you to break a space in an object or break the end off an object. You can think of *Break* as a partial erase. If you choose to break a space in an object, the space is created between two points that you specify (Fig. 9-22). In this case, the *Break* creates two objects from one.

Figure 9-22

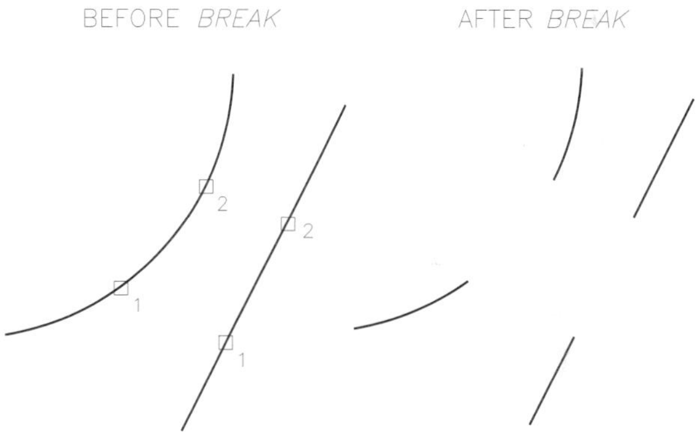

BEFORE *BREAK*

AFTER *BREAK*

If *Break*ing a circle (Fig. 9-23), the break is in a <u>counterclockwise</u> direction from the first to the second point specified.

Figure 9-23

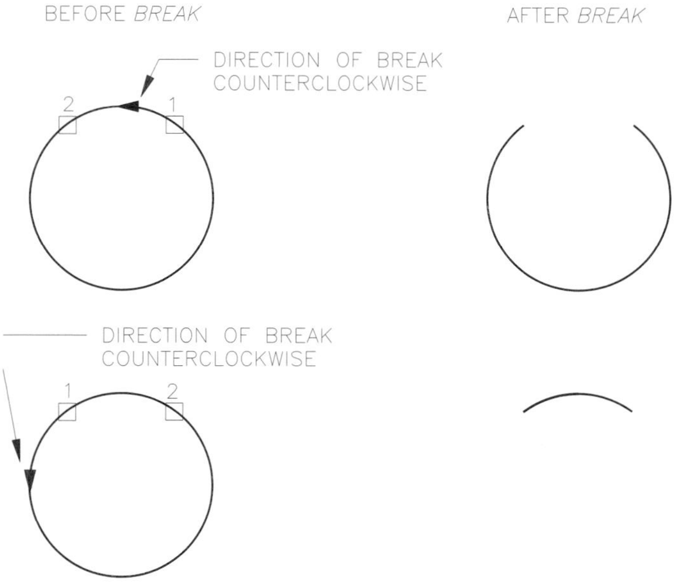

BEFORE *BREAK*

AFTER *BREAK*

DIRECTION OF BREAK
COUNTERCLOCKWISE

DIRECTION OF BREAK
COUNTERCLOCKWISE

If you want to *Break* the end off a *Line* or *Arc*, the first point should be specified at the point of the break and the second point should be <u>just off the end</u> of the *Line* or *Arc* (Fig. 9-24).

Figure 9-24 ——————————————

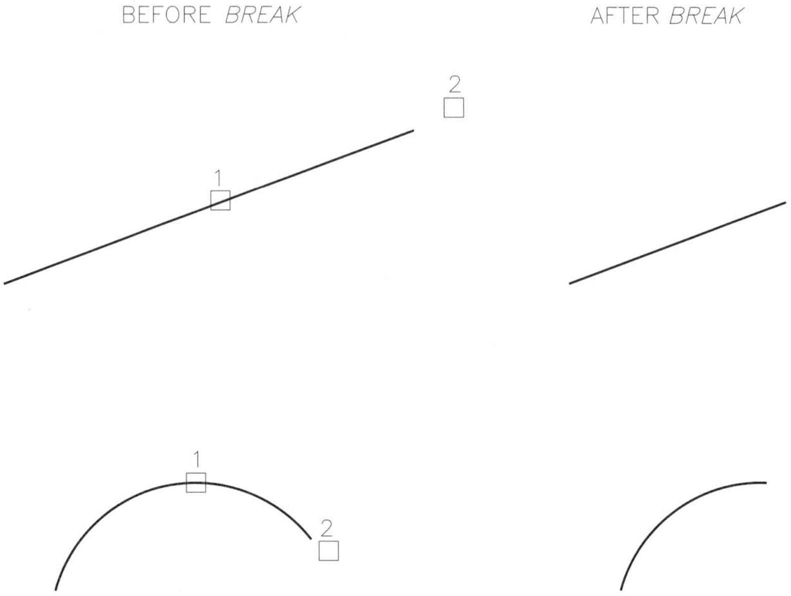

BEFORE *BREAK*   AFTER *BREAK*

There are four options for *Break*. All four options are available <u>only from the screen menu</u>, not from the *Modify* toolbar, *Modify* pull-down menu, or digitizing tablet menu. If you are not using the screen menu, the desired option is selected by keying the letter *F* (for *First point*) or symbol @ (for "last point"). The four options are explained next in detail.

Select, Second

This method (the default method) has two steps: select the object to break; then select the second point of the break. The first point used to select the object is <u>also</u> the first point of the break. (See Figs. 9-22, 9-23, 9-24.)

```
Command: break
Select Objects: PICK or (coordinates) (This is the first point of the break.)
Enter second point (or F for first point): PICK or (coordinates) (This is the second point of the break.)
Command:
```

Select, 2 Points

This method uses the first selection only to indicate the <u>object</u> to *Break*. You then specify the point that is the first point of the *Break*, and the next point specified is the second point of the *Break*. This option can be used with *OSNAP Intersection* to achieve the same results as *Trim*.

Selecting this option from the screen menu automatically sequences through the correct prompts (see Fig. 9-25).

Figure 9-25 ——————————————

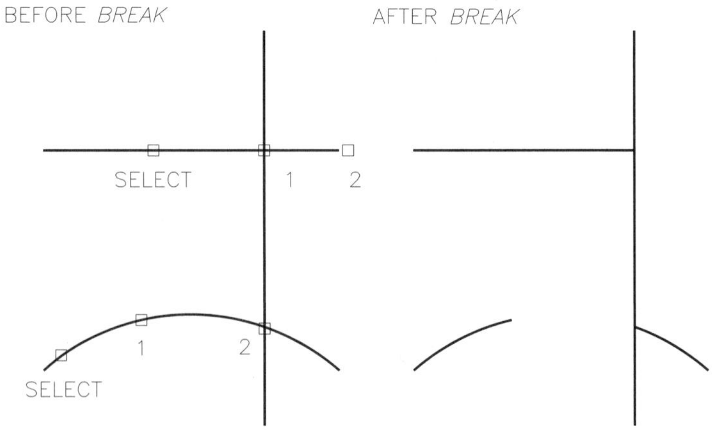

BEFORE *BREAK*   AFTER *BREAK*

SELECT   1   2

SELECT   1   2

If you are typing, the command sequence is as follows:

```
Command: break
Select Objects: PICK or (coordinates) (Select the object to break.)
Enter second point (or F for first point): F (Indicates respecification for the first point.)
```

Enter first point: **PICK** or (**coordinates**) (Select the first point of the break.)
Enter second point: **PICK** or (**coordinates**) (This is the second point of the break.)
Command:

Select Point

This option breaks one object into two separate objects with <u>no space</u> between. You specify only <u>one</u>
point with this method. When you select the object, the point indicates the object <u>and</u> the point of the
break. When prompted for the second point, the @ symbol (translated as "last point") is specified. This
second step is done automatically only if selected from the screen menu. If you are typing the command,
the @ symbol ("last point") must be typed:

Command: ***break***
Select Objects: **PICK** or (**coordinates**) (This PICKs the object and specifies the first point of the break.)
Enter second point (or F for first point): **@** (Indicates the break to be at the last point.)

The resulting object should <u>appear</u> as before;
however, it has been transformed into <u>two</u>
objects with matching endpoints (Fig. 9-26).

Figure 9-26

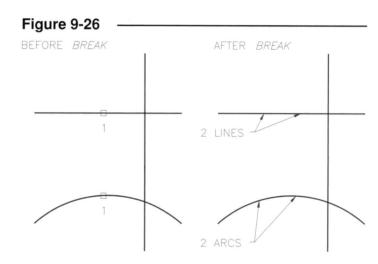

Select, First

This option creates a break with <u>no space,</u>
like the *1 Point* option; however, you can
select the object you want to *Break* first and
the point of the *Break* next (Fig. 9-27).

Figure 9-27

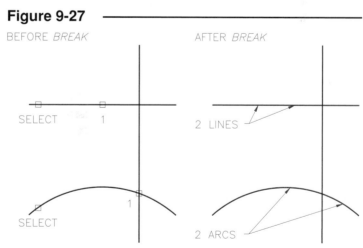

If you are typing the command, the sequence is as follows:

Command: ***break***
Select Objects: **PICK** or (**coordinates**) (Select the object to break.)
Enter second point (or F for first point): ***F*** (Indicates respecification for the first point.)
Enter first point: **PICK** or (**coordinates**) (Select the first point of the break.)
Enter second point: **@** (The second point of the break is the last point.)
Command:

CHAPTER EXERCISES

1. *Move*

Begin a *New* drawing and create the geometry in Figure 9-28 A using *Lines* and *Circles*. If desired, set *SNAP* to **.25** to make drawing easy and accurate.

For practice, turn *SNAP OFF* (**F9**). Use the *Move* command to move the *Circles* and *Lines* into the positions shown in illustration B. *OSNAPs* are required to *Move* the geometry accurately (since *SNAP* is off). *Save* the drawing as **MOVE1**.

Figure 9-28

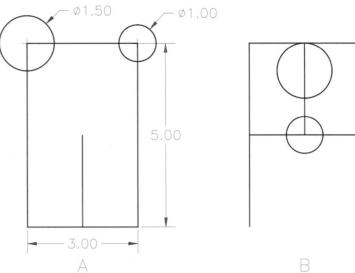

2. *Rotate*

Begin a *New* drawing and create the geometry in Figure 9-29 A.

Rotate the shape into position shown in step B. *SaveAs* **ROTATE1**.

Use the *Reference* option to *Rotate* the box to align with the diagonal *Line* as shown in C. *SaveAs* **ROTATE2**.

Figure 9-29

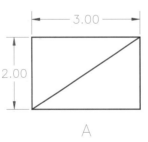
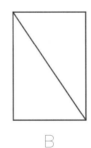
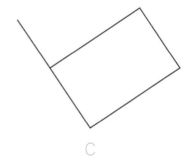

3. *Scale*

Open **ROTATE1** to again use the shape shown in Figure 9-30 B and *SaveAs* **SCALE1**. *Scale* the shape by a factor of **1.5**.

Open **ROTATE2** to again use the shape shown in C. Use the *Reference* option of *Scale* to increase the scale of the three other *Lines* to equal the length of the original diagonal *Line* as shown. (HINT: *OSNAPs* are required to specify the *Reference length* and *New length*.) *SaveAs* **SCALE2**.

Figure 9-30

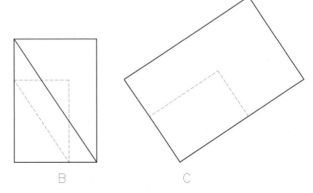

4. *Stretch*

   A design change has been requested. *Open* the **SLOTPLATE CH8** drawing and make the following changes.

   A. The top of the plate (including the slot) must be moved upward. This design change will add 1" to the total height of the Slot Plate. Use *Stretch* to accomplish the change, as shown in Figure 9-31. Draw the Crossing Window as shown.

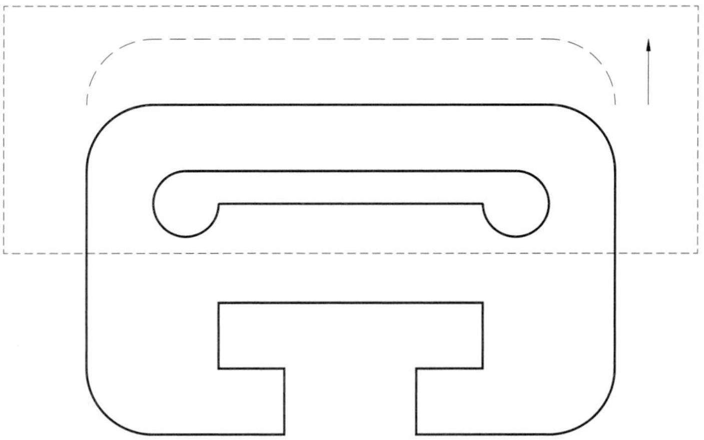

   B. The notch at the bottom of the plate must be adjusted slightly by relocating it .50 units to the right, as shown in Figure 9-32. Draw the Crossing Window as shown. Use *SaveAs* to reassign the name to **SLOTPLATE2**.

Figure 9-31

Figure 9-32

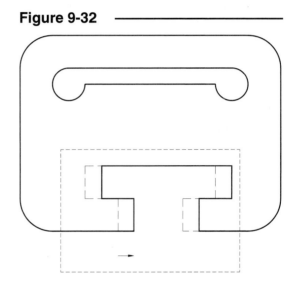

5. *Trim*

   A. Create the shape shown in Figure 9-33 A. *SaveAs* **TRIM-EX**.

   B. Use *Trim* to alter the shape as shown in B. *SaveAs* **TRIM1**.

   C. *Open* **TRIM-EX** to create the shapes shown in C and D using *Trim*. *SaveAs* **TRIM2** and **TRIM3**.

6. *Extend*

   Open each of the drawings created as solutions for Figure 9-33 (**TRIM1**, **TRIM2**, and **TRIM3**). Use *Extend* to return each of the drawings to the original form shown in Figure 9-33 A. *SaveAs* **EXTEND1**, **EXTEND2**, and **EXTEND3**.

Figure 9-33

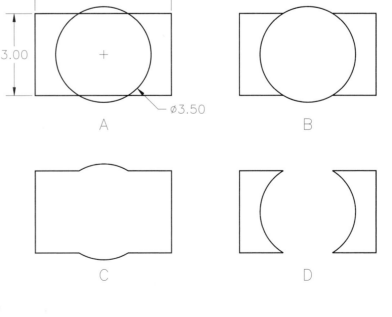

7. Break

A. Create the shape shown in Figure 9-34 A. *SaveAs* **BREAK-EX**.

B. Use *Break* to make the two breaks as shown in B. *SaveAs* **BREAK1.** (HINT: You may have to use *OSNAP*s to create the breaks at the *Intersections* or *Quadrants* as shown.)

C. Open **BREAK-EX** each time to create the shapes shown in C and D with the *Break* command. *SaveAs* **BREAK2** and **BREAK3**.

Figure 9-34

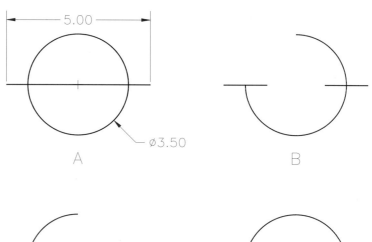

8. Lengthen

Five of your friends went to the horse races, each with $5.00 to bet. Construct a simple bar graph (similar to Fig. 9-35) to illustrate how your friends' wealth compared at the beginning of the day.

Modify the graph with *Lengthen* to report the results of their winnings and losses at the end of the day. The reports were as follows: friend 1 made $1.33 while friend 2 lost $2.40; friend 3 reported a 150% increase and friend 4 brought home 75% of the money; friend 5 ended the day with $7.80 and friend 6 came home with $5.60. Use the *Lengthen* command with the appropriate options to change the line lengths accordingly.

Figure 9-35

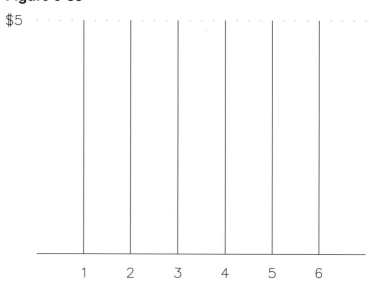

Who won the most? Who lost the most? Who was closest to even? Enhance the graph by adding width to the bars and other improvements as you wish (similar to Fig. 9-36). *SaveAs* **FRIENDS GRAPH.**

Figure 9-36

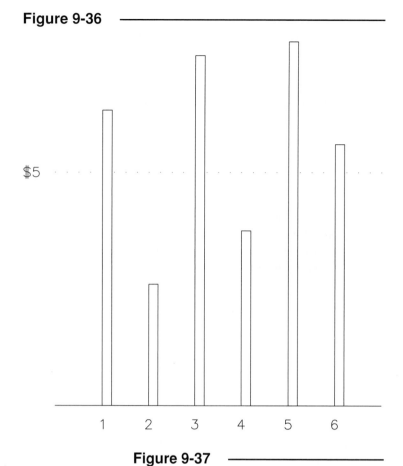

9. *Trim, Extend*

Figure 9-37

A. *Open* the **PIVOTARM CH7** drawing from the Chapter 7 Exercises. Use *Trim* to remove the upper sections of the vertical *Lines* connecting the top and front views. Compare your work to Figure 9-37.

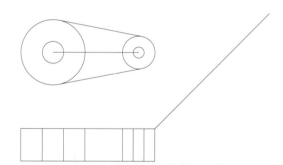

B. Next, draw a horizontal *Line* in the front view between the *Midpoints* of the vertical *Line* on each end of the view, as shown in Figure 9-38. *Erase* the horizontal *Line* in the top view between the *Circle* centers.

Figure 9-38

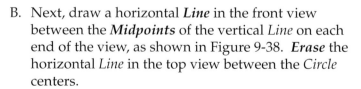

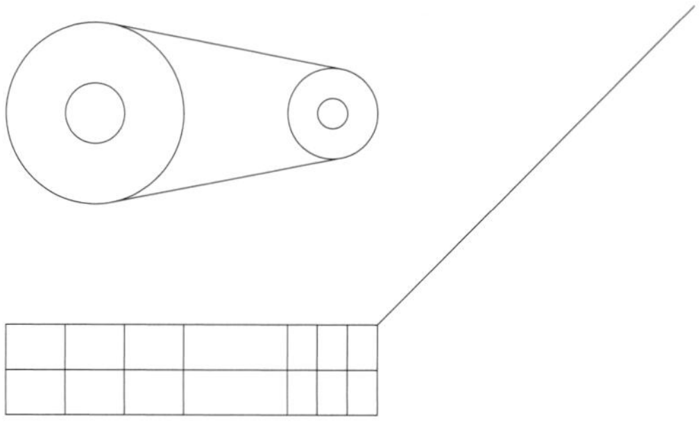

C. Draw vertical *Lines* from the *Endpoints* of the inclined *Line* (side of the object) in the top view down to the bottom *Line* in the front view, as shown in Figure 9-39. Use the two vertical lines as *Cutting edges* for *Trimming* the *Line* in the middle of the front view, as shown highlighted.

Figure 9-39 ———————————

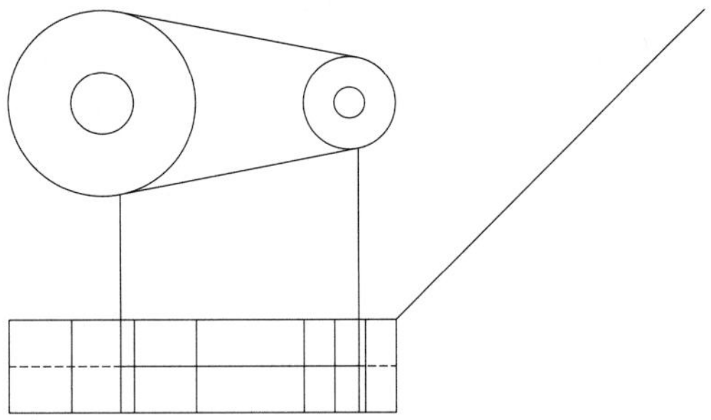

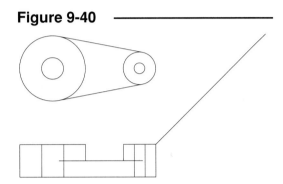

D. Finally, *Erase* the vertical lines used for *Cutting edges* and then use *Trim* to achieve the object as shown in Figure 9-40. Use *SaveAs* and name the drawing **PIVOTARM CH9**.

Figure 9-40 ———————————

10. **GASKETA**

Begin a *New* drawing. Set *Limits* to **0,0** and **8,6**. Set *SNAP* and *GRID* values appropriately. Create the Gasket as shown in Figure 9-41. Construct only the gasket shape, not the dimensions or centerlines. (HINT: Locate and draw the four .5″ diameter *Circles,* then create the concentric .5″ radius arcs as full *Circles,* then *Trim.* Make use of *OSNAPs* and *Trim* whenever applicable.) *Save* the drawing and assign the name **GASKETA.**

Figure 9-41 ———————————

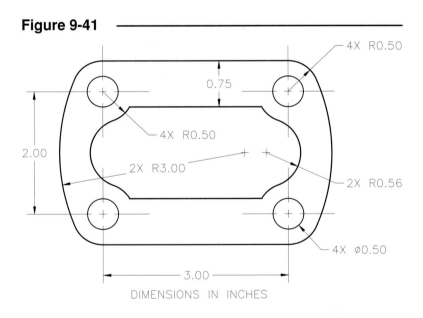

11. **Chemical Process Flow Diagram**

Begin a *New* drawing and recreate the chemical process flow diagram shown in Figure 9-42. Use *Line, Circle, Arc, Trim, Extend, Scale, Break,* and other commands you feel necessary to complete the diagram. Because this is diagrammatic and will not be used for manufacturing, dimensions are not critical, but try to construct the shapes with proportional accuracy. *Save* the drawing as **FLOWDIAG.**

Figure 9-42

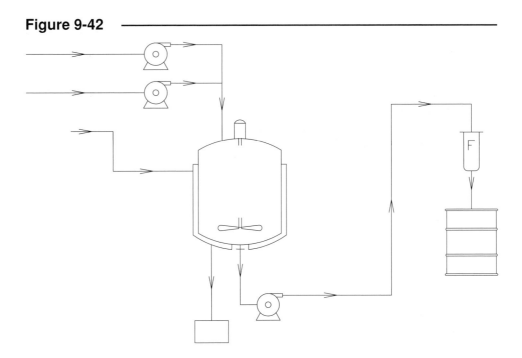

SITE-3D.DWG Courtesy of Autodesk, Inc.

10

MODIFY COMMANDS II

Chapter Objectives

After completing this chapter you should:

1. be able to *Copy* objects;

2. be able to make *Mirror* images of selected objects;

3. know how to create parallel copies of objects with *Offset*;

4. be able to make *rectangular* and *polar Arrays* of existing objects;

5. be able to create a *Fillet* between two objects;

6. be able to create a *Chamfer* between two objects.

CONCEPTS

Modify commands use existing objects. The *Modify* commands covered in Chapter 9 (Modify Commands I) change existing objects. The *Modify* commands covered in this chapter <u>use existing objects to create new and similar objects</u>. For example, the *Copy* command prompts you to select an object (or set of objects), then creates an identical object (or set). The commands covered in this chapter are:

> *Copy, Mirror, Offset, Array, Fillet,* and *Chamfer*

These commands are invoked by the same methods and are found in the same locations in AutoCAD Release 14 as those commands discussed in Chapter 9, Modify Commands I. Please refer to Chapter 9 for information on locating commands in the *Modify* toolbar, *Modify* pull-down menu, and *MODIFY1* and *MODIFY2* screen menus.

COMMANDS

COPY

Pull-down Menu	COMMAND (TYPE)	ALIAS (TYPE)	Short-cut	Screen (side) Menu	Tablet Menu
Modify *Copy*	*COPY*	*CO or CP*	...	*MODIFY1* *Copy*	*V,15*

Copy creates a duplicate set of the selected objects and allows placement of those copies. The *Copy* operation is like the *Move* command, except with *Copy* the original set of objects remains in its original location. You specify a "Base point:" (point to copy <u>from</u>) and a "Second point of displacement:" (point to copy <u>to</u>). See Figure 10-1.

Figure 10-1

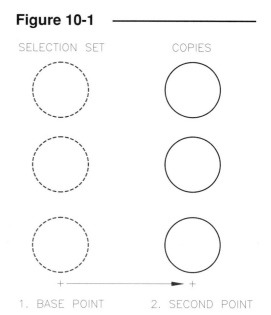

The command syntax for *Copy* is as follows.

    Command: **copy**
    Select Objects: **PICK** (Select objects to be copied.)
    Select Objects: **Enter** (Indicates completion of the
    selection set.)
    <Base point or displacement>Multiple: **PICK** or (**coordi-
    nates**) (This is the point to copy <u>from</u>. Select a point,
    usually on the object, to use as a "handle" or reference point.
    OSNAPs can be used. Coordinates in any format can also
    be entered.)
    Second point of displacement: **PICK** or (**coordinates**)
    (This is the point to copy <u>to</u>. Select a point. *OSNAPs* can be
    used. Coordinates in any format can be entered.)
    Command:

In many applications it is desirable to use *OSNAP* options to PICK the "Base point:" and the "Second point of displacement:" (Fig. 10-2).

Figure 10-2

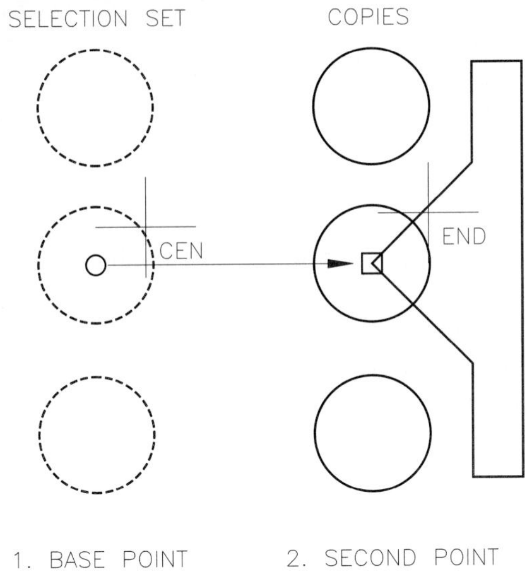

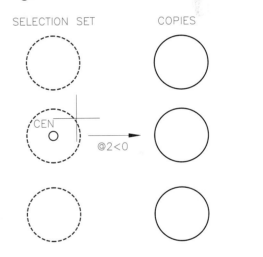

1. BASE POINT   2. SECOND POINT

Alternately, you can PICK the "Base point:" and enter relative rectangular coordinates, relative polar coordinates, or direct distance entry to specify the "Second point of displacement:".

Figure 10-3

SELECTION SET   COPIES

@2<0

1. BASE POINT   2. SECOND POINT

The *Copy* command has a *Multiple* option. The *Multiple* option allows creating and placing multiple copies of the selection set.

```
Command: copy
Select Objects: PICK
Select Objects: Enter
<Base point or displacement>/Multiple: m
Base point: PICK
Second point of displacement: PICK
Second point of displacement: PICK
Second point of displacement: Enter
Command:
```

Figure 10-4

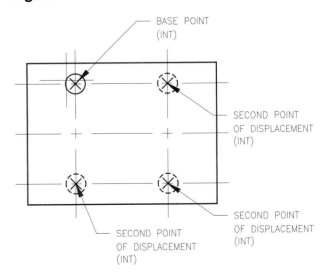

MIRROR

Pull-down Menu	COMMAND (TYPE)	ALIAS (TYPE)	Short-cut	Screen (side) Menu	Tablet Menu
Modify *Mirror*	MIRROR	MI	...	*MODIFY1* *Mirror*	*V,16*

This command creates a mirror image of selected existing objects. You can retain or delete the original objects ("old objects"). After selecting objects, you create two points specifying a "rubberband line," or "mirror line," about which to *Mirror*.

The length of the mirror line is unimportant since it represents a vector or axis (Fig. 10-5).

Figure 10-5

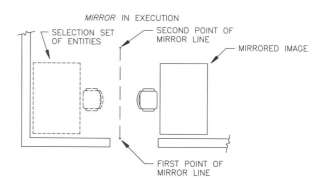

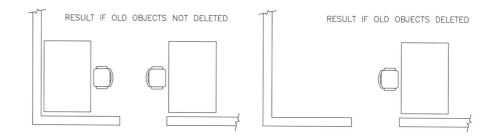

Command: **mirror**
Select Objects: **PICK** (Select object or group of objects to mirror.)
Select Objects: **Enter** (Press Enter to indicate completion of object selection.)
First point of mirror line: **PICK** or (**coordinates**) (Draw first endpoint of line to represent mirror axis by PICKing or entering coordinates.)
Second point of mirror line: **PICK** or (**coordinates**) (Select second point of mirror line by PICKing or entering coordinates.)
Delete old objects? <N> **Enter** or **Y** (Press Enter to yield both sets of objects or enter Y to keep only the mirrored set.)
Command:

If you want to *Mirror* only in a vertical or horizontal direction, toggle *ORTHO* (**F8**) *On* before selecting the "Second point of mirror line."

Mirror can be used to draw the other half of a symmetrical object, thus saving some drawing time (Fig. 10-9).

Figure 10-6

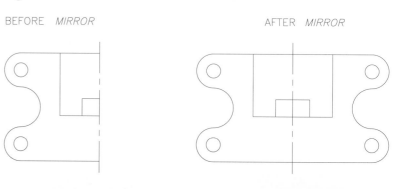

You can control whether <u>text</u> is mirrored by using the *MIRR-TEXT* variable. Type *MIRR-TEXT* at the command prompt and change the value to 0 if you do <u>not</u> want the text to be reflected; otherwise, the default setting of 1 mirrors text along with other selected objects (Fig. 10-7). Dimensions are <u>not</u> affected by the *MIRR-TEXT* variable and therefore are not reflected (that is, if the dimensions are associative). (See Chapter 28 for details on dimensions.)

Figure 10-7

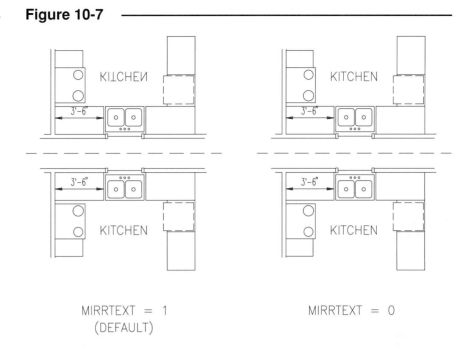

MIRRTEXT = 1
(DEFAULT)

MIRRTEXT = 0

OFFSET

Pull-down Menu	COMMAND (TYPE)	ALIAS (TYPE)	Short-cut	Screen (side) Menu	Tablet Menu
Modify *Offset*	*OFFSET*	*O*	...	*MODIFY1* *Offset*	*V,17*

Offset creates a <u>parallel copy</u> of selected objects. Selected objects can be *Line*s, *Arc*s, *Circle*s, *Pline*s, or other objects. *Offset* is a very useful command that can increase productivity greatly, particularly with *Pline*s.

Depending on the object selected, the resulting *Offset* is drawn differently (Fig. 10-8). *Offset* creates a parallel copy of a *Line* equal in length and perpendicular to the original. *Arc*s and *Circle*s have a concentric *Offset*. *Offsetting* closed *Pline*s or *Spline*s results in a complete parallel shape.

Figure 10-8

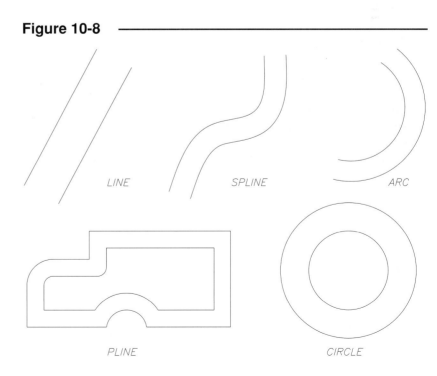

LINE   SPLINE   ARC

PLINE   CIRCLE

Two options are available with *Offset*: (1) *Offset* a specified *distance* and (2) *Offset through* a specified point (Fig. 10-9).

Figure 10-9

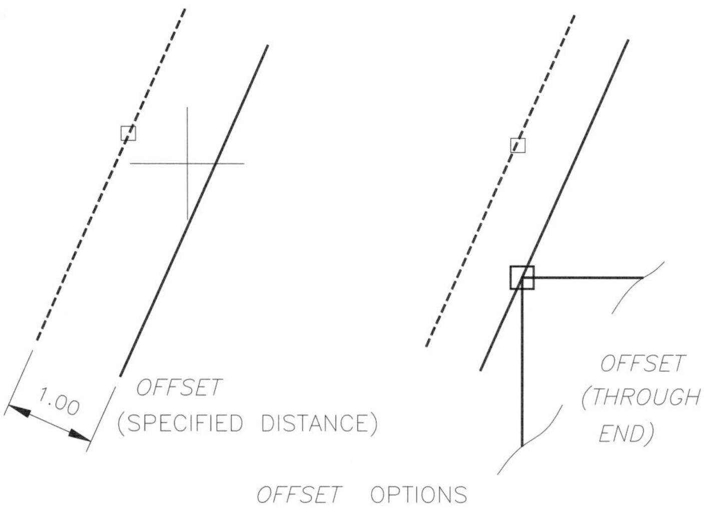

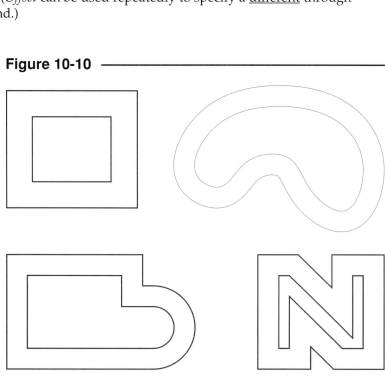

OFFSET (SPECIFIED DISTANCE)

OFFSET (THROUGH END)

OFFSET OPTIONS

Distance
The *Distance* option command sequence is as follows:

    Command: **offset**
    Offset distance or Through <Through>: (**value**) or **PICK** (Indicate the distance to offset by entering a value or interactively picking two points.)
    Select objects to offset: **PICK** (Select an object to make a parallel copy of. Only one object can be selected.)
    Side to offset? **PICK** or (**coordinates**) (Select which side of the selected object for offset object to be drawn.)
    Select objects to offset: **PICK** or **Enter** (*Offset* can be used repeatedly to offset at the <u>same</u> distance as the previously specified value. Pressing Enter completes command.)
    Command:

Through
The command sequence for *Offset through* is as follows:

    Command: **offset**
    Offset distance or Through <Through>: **T**
    Select objects to offset: **PICK** (Select an object to make a parallel copy of. Only one object can be selected.)
    Through point: **PICK** or (**coordinates**) (Select a point for offset object to be drawn through. Coordinates can be entered or point can be selected with or without *OSNAP*s.)
    Select objects to offset: **PICK** or **Enter** (*Offset* can be used repeatedly to specify a <u>different</u> through point. Pressing Enter completes command.)
    Command:

Notice the power of using *Offset* with closed *Pline* or *Spline* shapes (Fig. 10-10). Because a *Pline* or a *Spline* is one object, *Offset* creates one complete smaller or larger "parallel" object. Any closed shape composed of *Lines* and *Arcs* can be converted to one *Pline* object (see *Pedit*, Chapter 16).

Figure 10-10

ARRAY

	Pull-down Menu	COMMAND (TYPE)	ALIAS (TYPE)	Short-cut	Screen (side) Menu	Tablet Menu
	Modify *Array*	*ARRAY*	*AR*	...	*MODIFY1* *Array*	*V,18*

The *Array* command creates either a *Rectangular* or a *Polar* (circular) pattern of existing objects that you select. The pattern could be created from a single object or from a group of objects. *Array* copies a duplicate set of objects for each "item" in the array.

Rectangular

This option creates an *Array* of the selection set in a pattern composed of <u>rows</u> and <u>columns</u>. The command syntax for a rectangular *Array* is given next:

> Command: **array**
> Select Objects: **PICK** (Select objects to be arrayed.)
> Select Objects: **Enter** (Indicates completion of object selection.)
> Rectangular or Polar array (<R>/P): **R** (Indicates rectangular.)
> Number of rows (---) <1>: **(value)** (Enter value for number of rows.)
> Number of columns (||||) <1>: **(value)** (Enter value for columns.)
> Unit cell or distance between rows (---): **(value)** (Enter a value for the distance from any <u>point</u> on one object to the <u>same point</u> on an object in the adjacent row. See Fig. 10-11.)
> Distance between columns (||||): **(value)** (Enter a value for the distance from any *point* on one object to the same point on an object in the adjacent column. See Fig. 10-11.)
> Command:

Figure 10-11

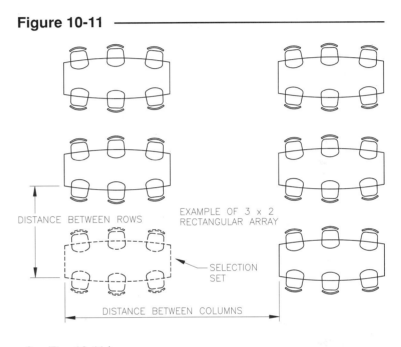

DISTANCE BETWEEN ROWS

EXAMPLE OF 3 x 2 RECTANGULAR ARRAY

SELECTION SET

DISTANCE BETWEEN COLUMNS

The *Array* is generated in a rectangular pattern with the original set of objects as the lower-left "cornerstone." Negative values can be entered (distance between rows and columns) to generate and *Array* in a -X or -Y direction.

If you want to specify distances between cells <u>interactively</u>, use the *Unit cell* method (see Fig. 10-12).

> Unit cell or distance between rows (---): **PICK** (Pick point for first point of cell.)
> Other corner: **PICK** (Pick point for other corner of cell.)

Figure 10-12

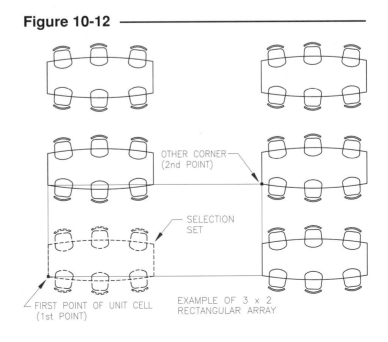

OTHER CORNER (2nd POINT)

SELECTION SET

FIRST POINT OF UNIT CELL (1st POINT)

EXAMPLE OF 3 x 2 RECTANGULAR ARRAY

A rectangular *Array* can be created at an angle by first <u>rotating the *SNAP*</u> to the desired angle and then creating the *Array* as shown in Figure 10-13. See Chapter 6 for information on the *Rotate* option of the *Snap* command.

Figure 10-13

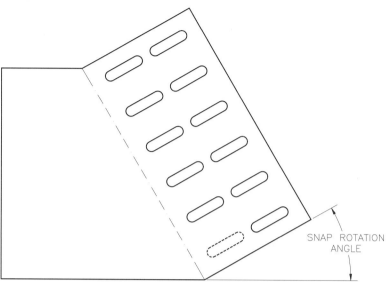

SNAP ROTATION ANGLE

Polar

This option creates a <u>circular</u> pattern of the selection set with any number of copies or "items." The number of items specified <u>includes the original</u> selection set. You also specify the center of array, angle to generate the array through, and orientation of "items:"

Command: **array**
Select Objects: **PICK** (Select objects to be arrayed.)
Select Objects: **Enter** (Indicates object selection completed.)
Rectangular or Polar array (<R>/P): **P** (Indicates polar array.)
Center point of array: **PICK** (Select point for array to be generated around.)
Number of items: (**value**) (Enter value for number of copies <u>including</u> original selection set.)
Angle to fill (+=ccw, -=cw) <360>: **Enter** or (**value**) (Press Enter for full circular array; enter value for less than 360 degree array; enter negative value for clockwise generation of array.)
Rotate objects as they are copied? <Y> **Enter** or **N** (Press Enter for rotation of objects about center; N for keeping objects in original orientation.)

Figure 10-14 illustrates a *Polar Array* created by accepting the default values (360 degrees and "Rotate objects as they are copied").

Figure 10-14

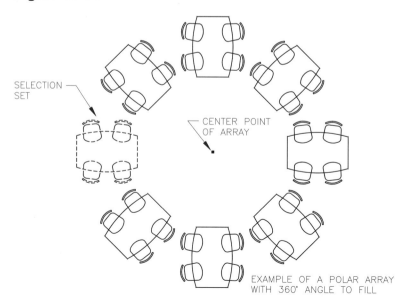

SELECTION SET

CENTER POINT OF ARRAY

EXAMPLE OF A POLAR ARRAY WITH 360° ANGLE TO FILL

Figure 10-15 shows a *Polar Array* through 180 degrees with objects rotated.

In some designs, you may know the <u>angle between objects</u> in a circular pattern rather than the <u>number of objects</u> in the pattern. You can create a *Polar Array* by specifying the "Angle between items:" instead of the "Number of items:" if you choose. This is done by first entering a null response (press **Enter**) to the "Number of items:" prompt. AutoCAD proceeds to prompt you for the "Angle to fill:", then asks for the "Angle between items:"

```
Command: array
Select Objects: PICK
Select Objects: Enter
Rectangular or Polar array (<R>/P): P
Center point of array: PICK
Number of items: Enter (do not specify a value)
Angle to fill (+=ccw, -=cw) <360>: Enter or (value)
Angle between items: (value)
Rotate objects as they are copied? <Y> Enter or N
Command:
```

Figure 10-15

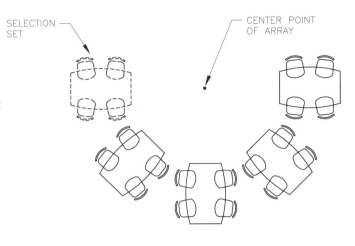

EXAMPLE OF A POLAR ARRAY
WITH 180° ANGLE TO FILL
AND 5 ITEMS TO ARRAY

SELECTION SET

CENTER POINT OF ARRAY

Figure 10-16 illustrates the results of answering *No* to the prompt "Rotate objects as they are rotated?" Note that the "items" remain in the same orientation (not rotated). Also notice the position of the *Array* with relation to the center point of the array.

When a *Polar Array* is created and objects are <u>not</u> rotated, the set of objects may not seem to rotate about the specified center, as shown in Figure 10-16. AutoCAD uses a single reference point for the *Array*—that of the <u>last</u> object in the selection set. If you PICK the objects one at a time (with the pickbox), the <u>last</u> one selected provides the reference point. If you use a Window or other method for selection, the reference point is arbitrary. This usually creates a non-symmetrical circular *Array* about the selected center point. This occurrence is no problem for *Polar Arrays* where the selection set is rotated.

Figure 10-16

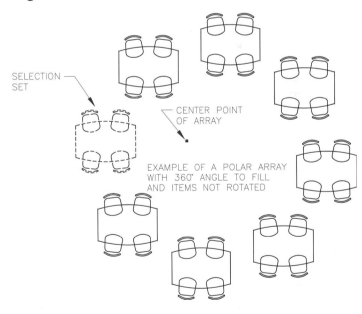

SELECTION SET

CENTER POINT OF ARRAY

EXAMPLE OF A POLAR ARRAY
WITH 360° ANGLE TO FILL
AND ITEMS NOT ROTATED

The solution is to create a *Point* object in the <u>center of the selection set</u> and select it <u>last</u>. This generates the selection set in a circle about the assigned center. (Alternately, you can make the selection set into a *Block* and use the center of the *Block* as the *Insertion* point. See Chapter 21).

FILLET

	Pull-down Menu	COMMAND (TYPE)	ALIAS (TYPE)	Short-cut	Screen (side) Menu	Tablet Menu
	Modify *Fillet*	**FILLET**	*F*	...	*MODIFY2* *Fillet*	*W,19*

The *Fillet* command automatically rounds a sharp corner (intersection of two *Lines, Arcs, Circles,* or *Pline* vertices) with a radius. You only specify the radius and select the objects' ends to be *Filleted*. The objects to fillet do <u>not</u> have to completely intersect but can overlap. You can specify whether or not the objects are automatically extended or trimmed as necessary (Fig. 10-17).

The *Fillet* command is used first to input the desired *radius* (if other than the default 0.5000 value) and a second time to select the objects to *Fillet*.

Figure 10-17 ——————————————

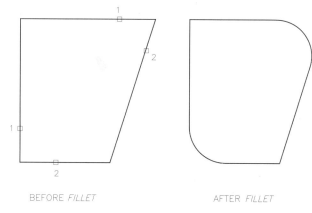

BEFORE *FILLET*   AFTER *FILLET*

```
Command: fillet
(TRIM mode) Current fillet radius=0.5000
Polyline/Radius/Trim/<Select first object>: r (Indicates the radius option.)
Enter fillet radius <0.5000>: (value) or PICK (Enter a value for the desired fillet radius or select two
points to interactively specify the radius.)
Command:
```

No fillet will be drawn at this time. Repeat the *Fillet* command to select objects to fillet.

```
Command: fillet
(TRIM mode) Current fillet radius=0.2500
Polyline/Distances/Trim/<Select first object>: PICK (Select one Line, Arc, or Circle near the point where
fillet should be created.)
Select second object: PICK (Select second object to fillet near the fillet location.)
Command:
```

The fillet is created at the corner selected.

Treatment of *Arcs* and *Circles* with *Fillet* is shown here. Note that the objects to *Fillet* do not have to intersect but can overlap.

Figure 10-18 ——————————————

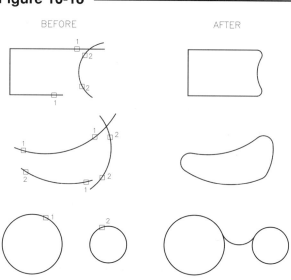

Since trimming and extending are done automatically by *Fillet* when necessary, using *Fillet* with a *radius* of *0* has particular usefulness. Using *Fillet* with a *0 radius* creates clean, sharp corners even if the original objects overlap or do not intersect (Fig. 10-19).

Figure 10-19

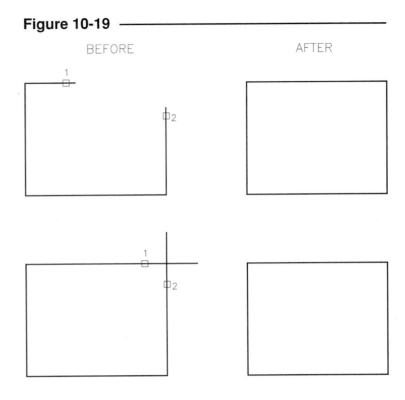

BEFORE   AFTER

If parallel objects are selected, the *Fillet* is automatically created to the correct radius (Fig. 10-20). Therefore, parallel objects can be filleted at any time <u>without specifying a radius value</u>.

Figure 10-20

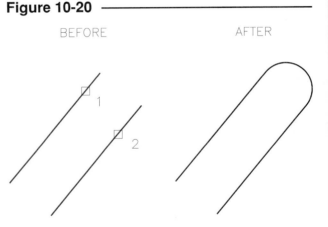

BEFORE   AFTER

Polyline
Fillets can be created on *Polylines* in the same manner as two *Line* objects. Use *Fillet* as you normally would for *Lines*. However, if you want a fillet equal to the specified radius to be added to <u>each vertex</u> of the *Pline* (except the endpoint vertices), use the *Polyline* option; then select anywhere on the *Pline* (Fig. 10-21).

> Command: **fillet**
> Polyline/Distances/<Select first object>: **P**
> (Invokes prompts for the *Polyline* option.)
> Select 2D polyline: **PICK** (Select the desired polyline.)

Figure 10-21

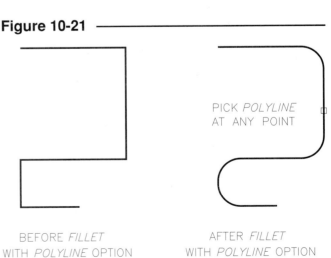

PICK *POLYLINE*
AT ANY POINT

BEFORE *FILLET*
WITH *POLYLINE* OPTION

AFTER *FILLET*
WITH *POLYLINE* OPTION

Closed Plines created by the *Close* option of *Pedit* react differently with *Fillet* than *Plines* connected by PICKing matching endpoints. Figure 10-22 illustrates the effect of *Fillet Polyline* on a *Closed Pline* and on a connected *Pline*.

Figure 10-22

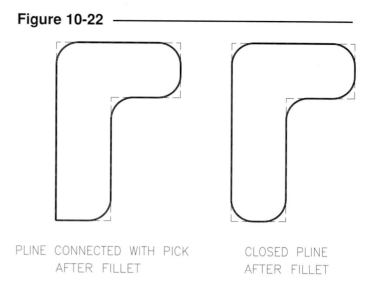

PLINE CONNECTED WITH PICK
AFTER FILLET

CLOSED PLINE
AFTER FILLET

Trim/Notrim

In the previous figures, the objects are shown having been filleted in the *Trim* mode—that is, with automatically trimmed or extended objects to meet the end of the new fillet radius. The *Notrim* mode creates the fillet <u>without</u> any extending or trimming of the involved objects (Fig. 10-23). Note that the command prompt indicates the current mode (as well as the current radius) when *Fillet* is invoked.

Figure 10-23

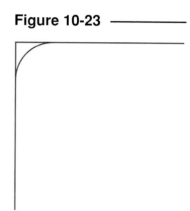

```
Command: fillet
(NOTRIM mode) Current fillet radius=2.000
Polyline/Radius/Trim/<Select first object>: T (Invokes Trim/No trim
option.)
Trim/No trim <Trim>: N (Sets Fillet to No trim.)
Polyline/Radius/Trim/<Select first object>: PICK
Select second object: PICK
Command:
```

FILLET   NOTRIM

The *Notrim* mode is helpful in situations similar to that in Figure 10-24. In this case, if the intersecting *Lines* were trimmed when the first fillet was created, the longer *Line* (highlighted) would have to be redrawn in order to create the second fillet.

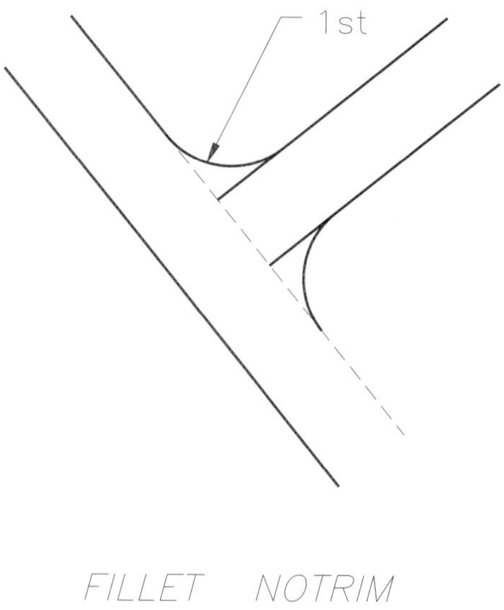

Changing the *Trim/Notrim* option sets the *TRIMMODE* system variable to 0 (*Notrim*) or 1 (*Trim*). This variable controls trimming for both *Fillet* and *Chamfer*, so if you set *Fillet* to *Notrim*, *Chamfer* is also set to *Notrim*.

Figure 10-24

1st

FILLET   NOTRIM

CHAMFER

	COMMAND (TYPE)	ALIAS (TYPE)	Short-cut	Screen (side) Menu	Tablet Menu
Pull-down Menu					
Modify *Chamfer*	*CHAMFER*	*CHA*	...	*MODIFY2* *Chamfer*	*W,18*

Chamfering is a manufacturing process used to replace a sharp corner with an angled surface. In AutoCAD, *Chamfer* is commonly used to change the intersection of two *Lines* or *Plines* by adding an angled line. The *Chamfer* command is similar to *Fillet,* but rather than rounding with a radius or "fillet," an angled line is automatically drawn at the distances (from the existing corner) that you specify.

Chamfers can be created by two methods: *Distance* (specify two distances) or *Angle* (specify a distance and an angle). The current method and the previously specified values are displayed at the Command: prompt along with the options:

> Command: **chamfer**
> (TRIM mode) Current chamfer Dist1 = 0.0000, Dist2 = 0.0000
> Polyline/Distance/Angle/Trim/Method/<Select first line>:

Method
Use this option to indicate which of the two methods you want to use: *Distance* (specify 2 distances) or *Angle* (specify a distance and an angle).

Distance
The *Distance* option is used to specify the two values applied when the *Distance Method* is used to create the chamfer. The values indicate the distances from the corner (intersection of two lines) to each chamfer endpoint (Fig. 10-25). Use the *Chamfer* command once to specify *Distances* and again to draw the chamfer.

Figure 10-25

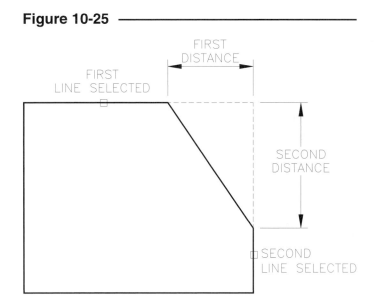

> Command: **chamfer**
> (TRIM mode) Current chamfer Dist1 = 0.0000, Dist2 = 0.0000
> Polyline/Distance/Angle/Trim/Method/<Select first line>: **d** (Indicates the distance option.)
> Enter first chamfer distance <0.0000>: (**value**) or **PICK** (Enter a value for the distance from the existing corner to the endpoint of chamfer on first line or select two points to interactively specify the distance.)
> Enter second chamfer distance <value of first distance>: **Enter, (value)** or **PICK** (Press Enter to use same distance as first distance, enter a value, or PICK as before.)
> Command:

Chamfer with distances of 0 reacts like *Fillet* with a radius of 0; that is, overlapping corners can be automatically trimmed and non-intersecting corners can be automatically extended (using the *Trim* mode).

Angle

The *Angle* option allows you to specify the values that are <u>used for the *Angle Method*</u>. The values specify a distance <u>along the first line</u> and an <u>angle from the first line</u> (Fig. 10-26).

Figure 10-26

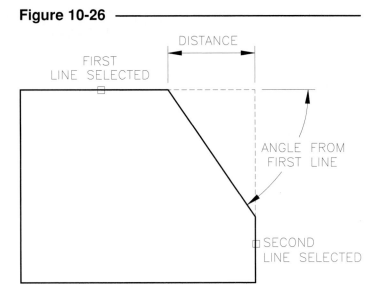

Polyline/Distance/Angle/Trim/Method/<Select first line>: **a**
Enter chamfer length on the first line <1.0000>: (**value**)
Enter chamfer angle from the first line <60.0000>: (**value**)

Trim/Notrim

The two lines selected for chamfering do not have to intersect but can overlap or not connect. The *Trim* setting automatically trims or extends the lines selected for the chamfer (Fig. 10-27), while the *Notrim* setting adds the chamfer without changing the length of the selected lines. These two options are the same as those in the *Fillet* command (see *Fillet*). Changing these options sets the *TRIMMODE* system variable to 0 (*Notrim*) or 1 (*Trim*). The variable controls trimming for both *Fillet* and *Chamfer*.

Figure 10-27

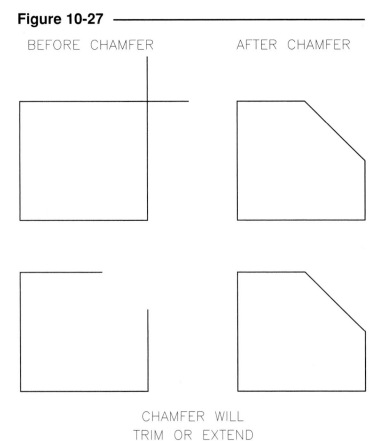

BEFORE CHAMFER   AFTER CHAMFER

CHAMFER WILL
TRIM OR EXTEND

Polyline

The *Polyline* option of *Chamfer* creates chamfers on all vertices of a *Pline*. All vertices of the *Pline* are chamfered with the supplied distances (Fig. 10-28). The first end of the *Pline* that was <u>drawn</u> takes the first distance. Use *Chamfer* without this option if you want to chamfer only one corner of a *Pline*. This is similar to the same option of *Fillet* (see *Fillet*).

Figure 10-28 ——————————

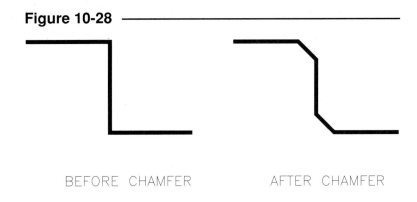

BEFORE CHAMFER   AFTER CHAMFER

A POLYLINE CHAMFER EXECUTES
A CHAMFER AT EACH VERTEX

CHAPTER EXERCISES

1. *Copy*

   Begin a *New* drawing. Set the *Limits* to **24,18**. Set *SNAP* to **.5** and *GRID* to **1**. Create the sheet metal part composed of four *Lines* and one *Circle* as shown in Figure 10-29. The lower-left corner of the part is at coordinate **2,3**.

   Use *Copy* to create two copies of the rectangle and hole in a side-by-side fashion as shown in Figure 10-30. Allow 2 units between the sheet metal layouts. *SaveAs* **PLATES**.

   A. Use *Copy* to create the additional 3 holes equally spaced near the other corners of the part as shown in Figure 10-30 A. *Save* the drawing.

   B. Use *Copy Multiple* to create the hole configuration as shown in Figure 10-30 B. *Save*.

Figure 10-29 ——————————

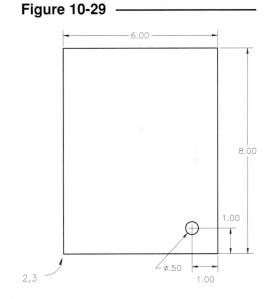

   C. Use *Copy* to create the hole placements as shown in C. Each hole <u>center</u> is **2** units at **125** degrees from the previous one (use relative polar coordinates). *Save* the drawing.

Figure 10-30 ——————————

A   B   C

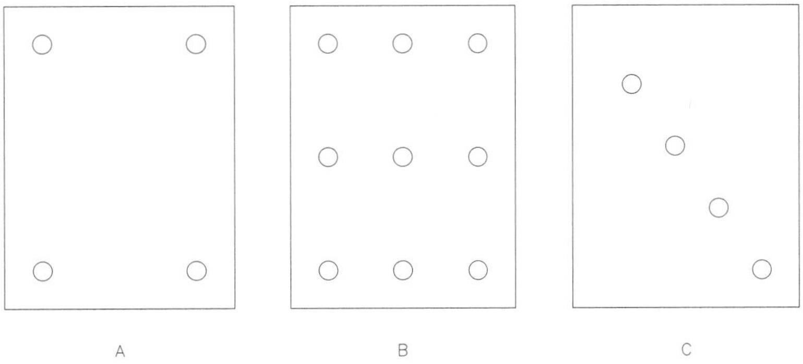

2. *Mirror*

Figure 10-31

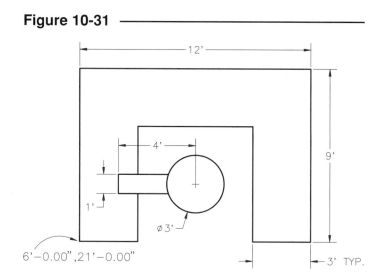

A manufacturing cell is displayed in Figure 10-31. The view is from above, showing a robot <u>centered</u> in a work station. The production line requires 4 cells. Begin by starting a *New* drawing, setting *Units* to *Engineering* and *Limits* to **40′ x 30′**. It may be helpful to set *SNAP* to **6″**. Draw one cell to the dimensions indicated. Begin at the indicated coordinates of the lower-left corner of the cell. *SaveAs* **MANFCELL.**

Use *Mirror* to create the other three manufacturing cells as shown in Figure 10-32. Ensure that there is sufficient space between the cells as indicated. Draw the two horizontal *Lines* representing the walkway as shown. *Save.*

Figure 10-32

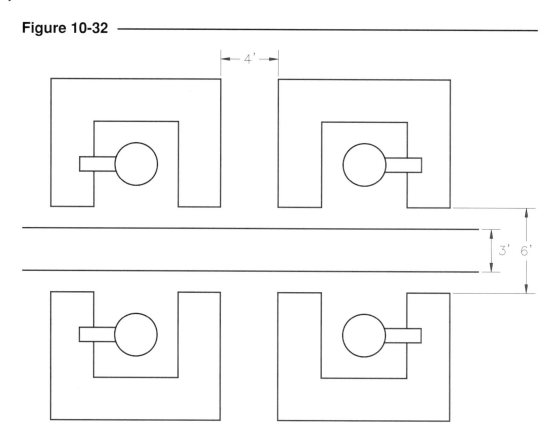

3. *Offset*

   Create the electrical schematic as shown in Figure 10-33. Draw *Lines* and *Plines*, then *Offset* as needed. Use *Point* objects with an appropriate (circular) *Point Style* at each of the connections as shown. Check the *Set Size to Absolute Units* radio button (in the *Point Style* dialog box) and find an appropriate size for your drawing. (Remember to use *Regen* after changing *Point Style* or size.) Because this is a schematic, you can create the symbols by approximating the dimensions. Omit the text. *Save* the drawing as **SCHEMATIC1**. Create a *Plot* and check *Scaled to Fit*.

Figure 10-33 ———

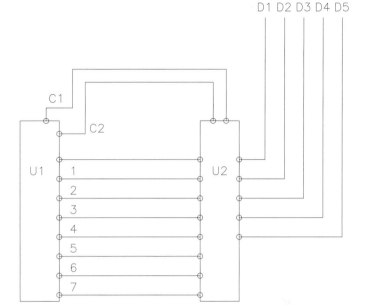

4. *Array, Polar*

   Begin a *New* drawing. Select *Start from Scratch*, *English* defaults. Create the starting geometry for a Flange Plate as shown in Figure 10-34. *SaveAs* **ARRAY**.

Figure 10-34 ———

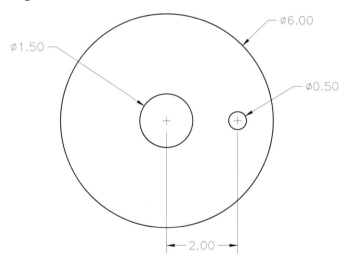

A. Create the *Polar Array* as shown in Figure 10-35 A. *SaveAs* **ARRAY1**.

B. *Open* **ARRAY**. Create the *Polar Array* as shown in Figure 10-35 B. *SaveAs* **ARRAY2**. (HINT: Use a negative angle to generate the *Array* in a clockwise direction.)

Figure 10-35 ———

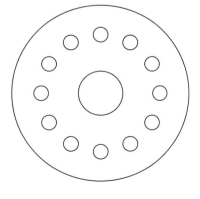
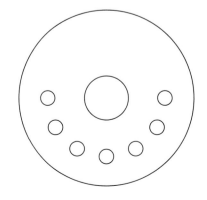

A   B

5. *Array, Rectangular*

Begin a *New* drawing. Select *Start from Scratch, English* defaults. Use *Save* and assign the name **LIBRARY DESKS**. Create the *Array* of study carrels (desks) for the library as shown in Figure 10-36. The room size is 36′ x 27′ (set *Units* and *Limits* accordingly). Each carrel is **30″ x 42″**. Design your own chair. Draw the first carrel (highlighted) at the indicated coordinates. Create the *Rectangular Array* so that the carrels touch side to side and allow a 6′ aisle for walking between carrels (not including chairs).

Figure 10-36

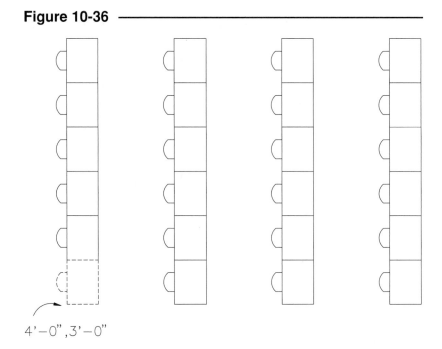

4′−0″,3′−0″

6. *Array, Rectangular*

Create the bolt head with one thread as shown in Figure 10-37 A. Create the small line segment at the end (0.40 in length) as a *Pline* with a *Width* of .02. *Array* the first thread (both crest and root lines, as indicated in B) to create the schematic thread representation. There is **1** row and **10** columns with **.2** units between each. Add the *Lines* around the outside to complete the fastener. *Erase* the *Pline* at the small end of the fastener (Fig. 10-37 B) and replace it with a *Line*. *Save* as **BOLT**.

Figure 10-37

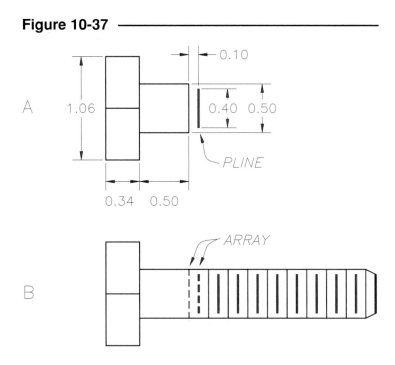

7. *Fillet*

Create the "T" Plate shown in Figure 10-38. Use *Fillet* to create all the fillets and rounds as the last step. When finished, *Save* the drawing as **T-PLATE** and make a plot.

Figure 10-38 ———————————————

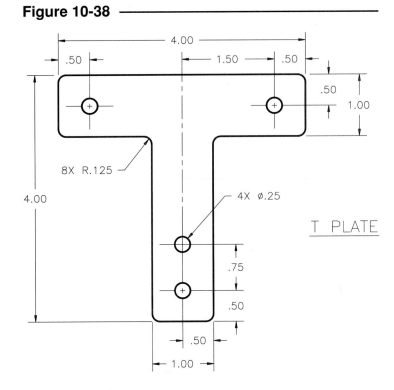

8. *Copy, Fillet*

Create the Gasket shown in Figure 10-39. Use the *Start from Scratch, English* settings when you begin. Include center-lines in your drawing. *Save* the drawing as **GASKETB**.

Figure 10-39 ———————————————

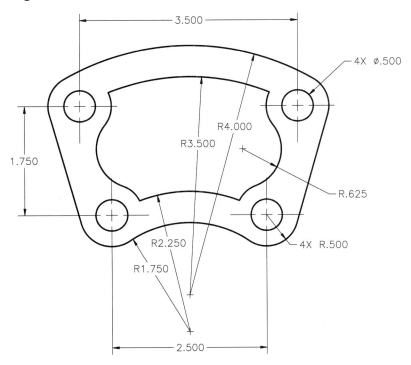

9. *Array*

    Create the perforated plate as shown Figure 10-40. *Save* the drawing as **PERFPLAT**. (HINT: For the *Rectangular Array* in the center, create 100 holes, then *Erase* four holes, one at each corner for a total of 96. The two circular hole patterns contain a total of 40 holes each.)

Figure 10-40

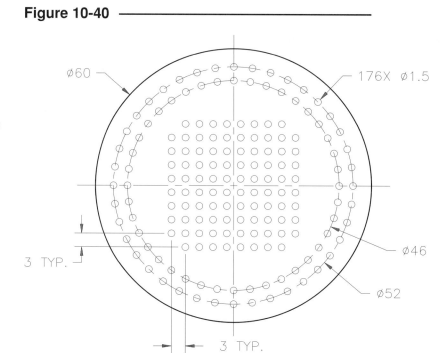

10. *Chamfer*

    Begin a *New* drawing and select *Start from Scratch, English* defaults. Set the *Limits* to **279,216**. Next, type *Zoom* and use the *All* option. Set *Snap* to **2** and *Grid* to **10**. Create the Catch Bracket shown in Figure 10-41. Draw the shape with all vertical and horizontal *Lines*, then use *Chamfer* to create the six chamfers. *Save* the drawing as **CBRACKET** and create a plot.

Figure 10-41

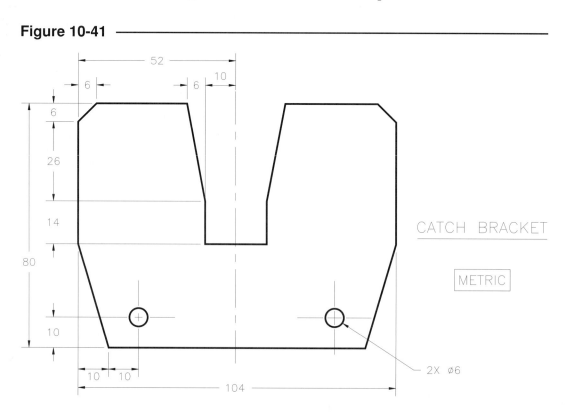

11

VIEWING COMMANDS

Chapter Objectives

After completing this chapter you should:

1. understand the relationship between the drawing objects and the display of those objects;

2. be able to use all of the *Zoom* options to view areas of a drawing;

3. be able to *Pan* the display about your screen;

4. be able to *save* and *restore Views*;

5. know how to use *Viewres* to change the display resolution for curved shapes;

6. be able to use *Aerial View* to *Pan* and *Zoom*.

CONCEPTS

The accepted CAD practice is to draw full size using actual units. Since the drawing is a virtual dimensional replica of the actual object, a drawing could represent a vast area (several hundred feet or even miles) or a small area (only millimeters). The drawing is created full size with the actual units, but it can be displayed at any size on the screen. Consider also that CAD systems provide for a very high degree of dimensional precision, which permits the generation of drawings with great detail and accuracy.

Combining those two CAD capabilities (great precision and drawings representing various areas), a method is needed to view different and detailed segments of the overall drawing area. In AutoCAD the commands that facilitate viewing different areas of a drawing are *Zoom, Pan,* and *View*.

The Viewing commands are found in the *View* pull-down menu (Fig. 11-1).

Figure 11-1

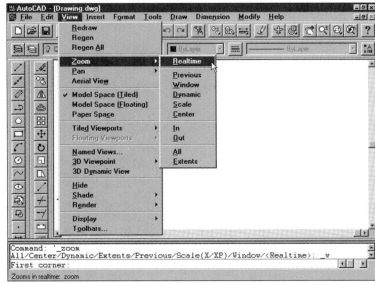

The Standard toolbar contains a group of tools (icon buttons) for the Viewing commands located near the right end of the toolbar (Fig. 11-2). The *Realtime* options of *Pan* and *Zoom, Zoom Previous,* and *Aerial View* each has an icon permanently displayed on the toolbar, whereas the other *Zoom* options are located on flyouts.

Like other commands, the Viewing commands can also be invoked by typing the command or command alias, using the screen menu (*VIEW1*), or using the digitizing tablet menu (if available).

The commands discussed in this chapter are:

    Zoom, Pan, View, and *Dsviewer* (Aerial View)

Figure 11-2

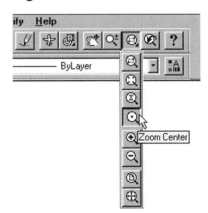

AutoCAD Release 13 Versions C3 and C4 introduced "real-time" panning and zooming (Version C3 for AutoCAD for Windows 3.1 and 3.11, and Version C4 for Windows 95 and Windows NT). Two new commands were introduced: *Rtpan* and *Rtzoom*. However, because the two commands were added after the first version of Release 13, they were not included in the menus. Now with AutoCAD Release 14, the *Realtime* options are the default options for both the *Pan* and *Zoom* commands. *Realtime* options of *Pan* and *Zoom* allow you to <u>interactively</u> change the drawing display by moving the cursor in the drawing area.

COMMANDS

ZOOM

Pull-down Menu	COMMAND (TYPE)	ALIAS (TYPE)	Short-cut	Screen (side) Menu	Tablet Menu
View *Zoom >*	*ZOOM*	*Z*	...	*VIEW 1* *Zoom*	*K,11-M,11* *or J,2-K,5*

To *Zoom in* means to magnify a small segment of a drawing and to *Zoom out* means to display a larger area of the drawing. *Zooming* does <u>not</u> change the <u>size</u> of the drawing; *Zooming* changes only the area that is <u>displayed</u> on the screen. All objects in the drawing have the same dimensions before and after *Zooming*. Only your display of the objects changes.

The *Zoom* options described next can be typed or selected by any method. Unlike most commands, AutoCAD provides a tool (icon button) for each <u>option</u> of the *Zoom* command. If you type *Zoom*, the options can be invoked by typing the first letter of the desired option or pressing Enter for the *Realtime* option. The command prompt appears as follows.

```
Command: zoom
All/Center/Dynamic/Extents/Previous/Scale(X/XP)/Window/<Realtime>:
```

Realtime (RTZOOM)

Realtime is the default option of *Zoom*. If you <u>type</u> *Zoom*, just press Enter to activate the *Realtime* option. Alternately, you can type *Rtzoom* to invoke this option directly.

With the *Realtime* option you can interactively zoom in and out with vertical cursor motion. When you activate the *Realtime* option, the cursor changes to a magnifying glass with a plus (+) and a minus (-) symbol. Move the cursor to any location on the drawing area and hold down the PICK (left) mouse button. Move the cursor up to zoom in and down to zoom out (Fig. 11-3).

Figure 11-3

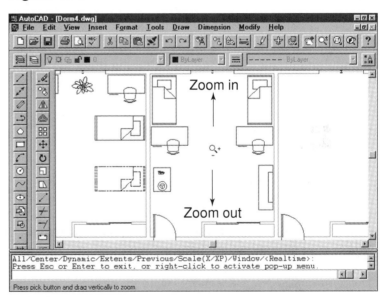

Horizontal movement has no effect. You can zoom in or out repetitively. Pressing Esc or Enter exits *Realtime Zoom* and returns to the Command: prompt.

The current drawing window is used to determine the zooming factor. Moving the cursor half the window height (from the center to the top or bottom) zooms in or out to a zoom factor of 100%. Starting at the bottom or top allows zooming in or out (respectively) at a factor of 200%.

If you have zoomed in or out repeatedly, you can reach a zoom limit. In this case the plus (+) or minus (-) symbol disappears when you press the left mouse button. To zoom further with *Rtzoom*, first type *Regen*.

Pressing the right mouse button (or button #2 on digitizing pucks) produces a small cursor menu with other viewing options (Fig. 11-4). This same menu and its options can also be displayed by right-clicking during *Realtime Pan*. The options of the cursor menu are described below.

Figure 11-4

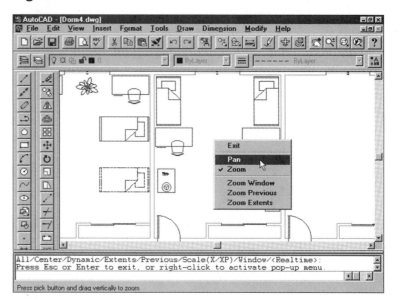

Exit
Select *Exit* to exit *Realtime Pan* or *Zoom* and return to the Command: prompt. The Escape or Enter key can be used to accomplish the same action.

Pan
This selection switches to the *Realtime* option of *Pan*. A check mark appears here if you are currently using *Realtime Pan*. See *Pan* next in this chapter.

Zoom
A check mark appears here if you are currently using *Realtime Zoom*. If you are using the *Realtime* option of *Pan*, check this option to switch to *Realtime Zoom*.

Zoom Window
Select this option if you want to display an area to zoom in to by specifying a *Window*. This feature operates differently but accomplishes the same action as the *Window* option of *Zoom*. With this option, window selection is made with one PICK. In other words, PICK for the first corner, hold down the left button and drag, then release to establish the other corner. With the *Window* option of the *Zoom* command, PICK once for each of two corners. See *Window* next page.

Zoom Previous
Selecting this option automatically displays the area of the drawing that appeared on the screen immediately before using the *Realtime* option of *Zoom* or *Pan*. Using this option successively has no effect on the display. This feature is different from the *Previous* option of the *Zoom* command, in which 10 successive previous views can be displayed.

Zoom Extents
Use this option to display the entire drawing in its largest possible form in the Drawing Editor. This option is identical to the *Extents* option of the *Zoom* command. See *Extents* next page.

Window

 To *Zoom* with a *Window* is to draw a rectangular window around the desired viewing area. You PICK a first and a second corner (diagonally) to form the rectangle. The windowed area is magnified to fill the screen (Fig. 11-5). It is suggested that you draw the window with a 4 x 3 (approximate) proportion to match the screen proportion. If you type *Zoom* or *Z*, *Window* is an <u>automatic</u> option so you can begin selecting the first corner of the window after issuing the *Zoom* command <u>without</u> indicating the *Window* option as a separate step.

Figure 11-5

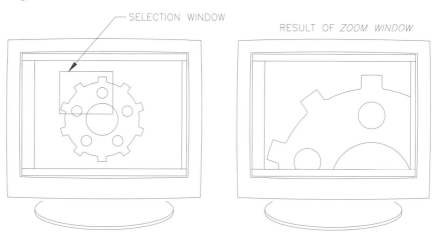

All

 This option displays <u>all of the objects</u> in the drawing <u>and all of the Limits</u>. In Figure 11-6, notice the effects of *Zoom All*, based on the drawing objects and the drawing *Limits*.

Figure 11-6

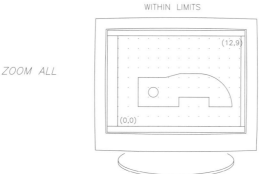

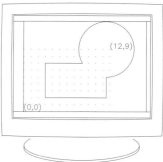

Extents

 This option results in the largest possible display of <u>all of the objects</u>, <u>disregarding</u> the *Limits*. (*Zoom All* includes the *Limits*.) See Figure 11-6.

New in Release 14, invoking *Zoom Extents* causes AutoCAD to execute two steps: (1) a true (pre-

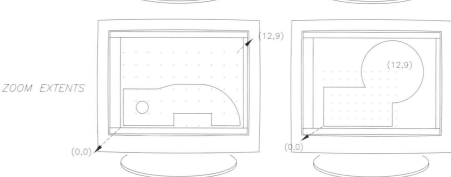

Release 13) *Zoom Extents* is performed which causes the geometry to be maximized so it is "pressed" against the edge of the Drawing Editor window, and (2) a *Zoom Scale .95X* is performed which causes the geometry to be scaled down slightly to fit just within the window (see the *Scale* option). Previous to Release 14, using *Zoom Extents* alone made it difficult to select geometry against the edge of the window, so *Zoom Scale .95X* had to be invoked by the user to make the geometry more accessible for selection.

Scale (X/XP)

This option allows you to enter a scale factor for the desired display. The value that is entered can be relative to the full view (*Limits*) or to the current display (Fig. 11-7). A value of **1**, **2**, or **.5** causes a display that is 1, 2, or .5 times the size of the *Limits*, centered on the current display. A value of **1X**, **2X**, or **.5X** yields a display 1, 2, or .5 times the size of the <u>current display</u>. If you are using paper space and viewing model space, a value of **1XP**, **2XP**, or **.5XP** yields a model space display scaled 1, 2, or .5 times paper space units.

If you type *Zoom* or *Z*, you can enter a scale factor at the *Zoom* command prompt without having to type the letter *S*.

Figure 11-7 ——————————

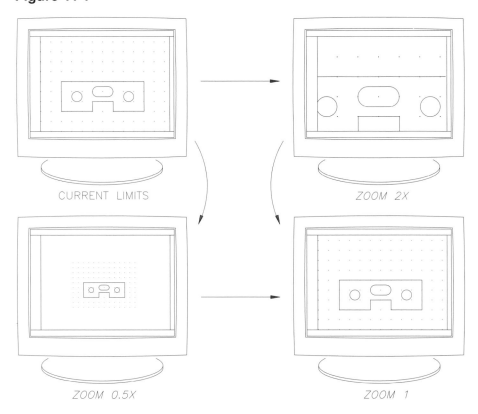

CURRENT LIMITS

ZOOM 2X

ZOOM 0.5X

ZOOM 1

In

Zoom In magnifies the current display by a factor of 2X (2 times the current display). Using this option is the same as entering a *Zoom Scale* of 2X.

Out

Zoom Out makes the current display smaller by a factor of .5X (.5 times the current display). Using this option is the same as entering a *Zoom Scale* of .5X.

Center

First, specify a location as the center of the zoomed area; then specify either a *Magnification factor* (see *Scale X/XP*), a *Height* value for the resulting display, or PICK two points forming a vertical to indicate the height for the resulting display (Fig. 11-8).

Previous

Selecting this option automatically changes to the previous display. AutoCAD saves the previous ten displays changed by *Zoom*, *Pan*, and *View*. You can successively change back through the previous ten displays with this option.

Figure 11-8 ——————————

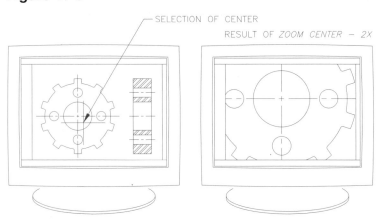

SELECTION OF CENTER

RESULT OF *ZOOM CENTER* — 2X

Dynamic

With this option you can change the display from one windowed area in a drawing to another without using *Zoom All* (to see the entire drawing) as an intermediate step. *Zoom Dynamic* causes the screen to display the drawing *Extents* bounded by a box. The current view or window is bounded by a box in a broken-line pattern. The viewbox is the box with an "X" in the center which can be moved to the desired location (Fig. 11-9). The desired location is selected by pressing **Enter** (not the PICK button as you might expect).

Pressing **PICK** allows you to resize the viewbox (displaying an arrow instead of the "X") to the desired size. Move the mouse or puck left and right to increase and decrease the window size. Press **PICK** again to set the size and make the "X" reappear.

Figure 11-9

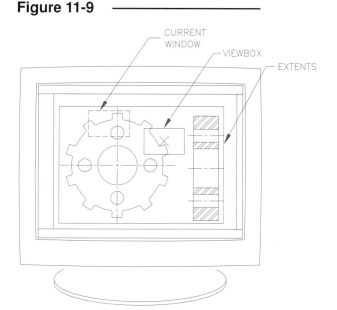

When changing the display of a drawing from one windowed area to another, *Zoom Dynamic* can be a fast method to use because (1) a *Zoom All* is not required as an intermediate step and (2) the viewbox can be moved while AutoCAD is "redrawing" the display.

Zoom Is Transparent

Zoom is a transparent command, meaning it can be invoked while another command is in operation. You can, for example, begin the *Line* command and at the "To point:" prompt 'Zoom with a *Window* to better display the area for selecting the endpoint. This transparent feature is automatically entered if *Zoom* is invoked by the screen, pull-down, or tablet menus or toolbar icons, but if typed it must be prefixed by the ' (apostrophe) symbol, e.g., "To point: 'Zoom." If *Zoom* has been invoked transparently, the >> symbols appear at the command prompt before the listed options as follows:

    LINE from point: PICK
    To point: 'zoom
    >>All/Center/Dynamic/Extents/Previous/Scale(X/XP)/Window/<Realtime>:

PAN

Pull-down Menu	COMMAND (TYPE)	ALIAS (TYPE)	Short-cut	Screen (side) Menu	Tablet Menu
View *Pan*	*PAN or* *-PAN*	*P or -P*	...	*VIEW 1* *Pan*	*N,11-P,11*

Using the *Pan* command by typing or selecting from the tool, screen, or digitizing tablet menu produces the *Realtime* version of *Pan*. The *Point* option is available from the pull-down menu or by typing "-Pan." The other "automatic" *Pan* options (*Left, Right, Up, Down*) can only be selected from the pull-down menu.

The *Pan* command is useful if you want to move (pan) the display area slightly without changing the size of the current view window. With *Pan,* you "drag" the drawing across the screen to display an area outside of the current view in the Drawing Editor.

Realtime (RTPAN)

 Realtime is the default option of *Pan*. It is invoked by selecting *Pan* by any method, including typing *Pan* or *Rtpan*. *Realtime Pan* allows you to interactively pan by "pulling" the drawing across the screen with the cursor motion. After activating the command, the cursor changes to a hand cursor. Move the hand to any location on the drawing, then hold the PICK (left) mouse button down and drag the drawing around on the screen to achieve the desired view (Fig. 11-10). When you release the mouse button, panning is discontinued. You can move the hand to another location and pan again without exiting the command.

Figure 11-10

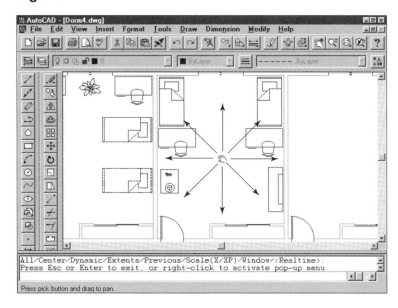

You must press Esc or Enter or use *Exit* from the pop-up cursor menu to exit *Realtime Pan* and return to the Command: prompt. The following command prompt appears during *Realtime Pan*.

Command: **pan**
Press Esc or Enter to exit, or right-click to activate pop-up menu.

If you press the right mouse button (#2 button), a small cursor menu pops up. This is the same menu that appears during *Realtime Zoom* (see Fig. 11-4). This menu provides options for you to *Exit*, *Zoom* or *Pan* (realtime), *Zoom Window*, *Zoom Previous*, or *Zoom Extents*. See *Zoom* earlier in this chapter.

Point

The *Point* option is available only through the pull-down menu or by typing "-*Pan*" (some commands can be typed with a - [hyphen] prefix to invoke the command line version of the command without dialog boxes, etc.). This is the same version of *Pan* that was used in the first shipping of Release 13 and previous versions of AutoCAD. Using this option produces the following command prompt.

Command: **-pan**
Displacement: **PICK** or (**value**)
Second point: **PICK** or (**value**)
Command:

Figure 11-11

The "Displacement:" can be thought of as a "handle," or point to move <u>from</u>, and the "Second point:" as the new location to move <u>to</u> (Fig. 11-11). You can PICK each of these points with the cursor.

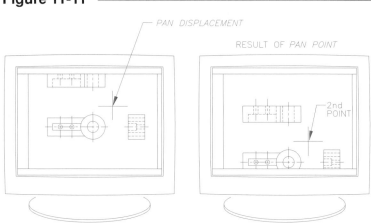

You can also enter coordinate values rather than interactively PICKing points with the cursor. The command syntax is as follows:

<pre>
Command: pan
Pan Displacement: 0,-2
Second point: Enter
Command:
</pre>

Entering coordinate values allows you to *Pan* to a location outside of the current display. If you use the interactive method (PICK points), you can *Pan* only within the range of whatever is visible on the screen.

The following *Pan* options are available from the *View* pull-down menu only.

Left
Automatically pans to the left, equal to about 1/8 of the current drawing area width.

Right
Pans to the right—similar to but opposite of *Left*.

Up
Automatically pans up, equal to about 1/8 of the current drawing area height.

Down
Pans down—similar to but opposite of *Up*.

Pan Is Transparent
Pan, like *Zoom*, can be used as a transparent command. The transparent feature is automatically entered if *Pan* is invoked by the screen, pull-down, or tablet menus or icons. However, if you <u>type</u> *Pan* during another command operation, it must be prefixed by the ' (apostrophe) symbol.

<pre>
Command: LINE
From point: PICK
To point: 'pan
>>Press Esc or Enter to exit, or right-click to activate pop-up menu. Enter
Resuming LINE command.
To point:
</pre>

Scroll Bars

Figure 11-12 ———

You can also use the horizontal and vertical scroll bars directly below and to the right of the graphics area to pan the drawing (Fig. 11-12). These scroll bars pan the drawing in one of three ways: (1) click on the arrows at the ends of the scroll bar, (2) move the thumb wheel, or (3) click inside the scroll bar. The scroll bars can be turned off (removed from the screen) by using the *Display* tab in the *Preferences* dialog box from the *Tools* pull-down menu.

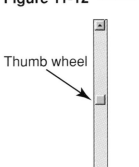

Thumb wheel

VIEW and
DDVIEW

Pull-down Menu	COMMAND (TYPE)	ALIAS (TYPE)	Short-cut	Screen (side) Menu	Tablet Menu
View *Named Views...*	*VIEW or* *DDVIEW*	*V or -V*	...	*VIEW 1* *Ddview*	*M,5*

The *View* command provides options for you to *Save* the current display window and *Restore* it at a later time. For typical applications, you would first *Zoom* in to achieve a desired display, then use *View, Save* to save that display under an assigned name. Later in the drawing session, the named *View* can be *Restored* any number of times. This method is preferred to continually *Zooming* in and out to display the same areas repeatedly. *Restoring* a named *View* requires no regeneration time. A *View* can be sent to the plotter as separate display as long as the view is named and saved. The options of the *View* command are listed below:

?   Displays the list of named views.

Delete   Deletes one or more saved views that you enter.

Restore   Displays the named view you request.

Save   Saves the current display as a *View* with a name you assign.

Window   Allows you to specify a window in the current display and save it as a named *View*.

To *Save* a *View,* first use *Zoom* or *Pan* to achieve the desired display; then invoke the *View* command and the *Save* option as follows:

    Command: **view**
    ?/Delete/Restore/Save/Window: **save**
    View name to save: **name** (Assign a descriptive name for the view.)

For assurance, check to see if the new *View* has been saved by displaying the list by using the *?* option. The *?* option causes a text screen to appear with the list of named (saved) views. Use *Flipscreen* to return to the Drawing Editor after reading the list.

To *Restore* a named *View,* assuming a different area of the drawing is currently displayed, use the *Restore* option and supply the name of the desired *View.*

View can also be used transparently, but only if typed at the keyboard. Remember to preface the command with the ' (apostrophe) symbol, i.e., *'View.*

A feature of saving *Views* is that a named *View* can be plotted by toggling the *View* option in the *Plot Configuration* dialog box. Plotting is discussed in detail in Chapter 14.

Figure 11-13 ————————

Typing the *View* command is generally quick and easy. Selecting *Ddview* from the pull-down or tablet menu invokes the *View Control* dialog box (Fig. 11-13), which is helpful but requires a number of steps to operate. A word of caution when selecting from the list of named views to restore: make sure that you select (1) the desired view <u>and</u> (2) the **Restore** tile, <u>then</u> (3) the **OK** tile.

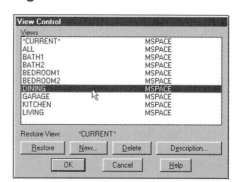

The *View Control* dialog box provides the same options that the *View* command offers. The *New* option causes the *Define New View* dialog box to pop up, giving access to the *Save* option (Fig. 11-14).

Figure 11-14

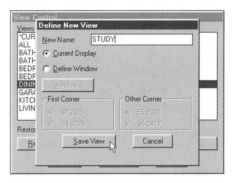

DSVIEWER

Pull-down Menu	COMMAND (TYPE)	ALIAS (TYPE)	Short-cut	Screen (side) Menu	Tablet Menu
View *Aerial View*	*DSVIEWER*	*AV*	...	*VIEW 1* *Dsviewer*	*K,2*

Aerial View is a tool for navigating in a drawing. It can be used to *Zoom* and *Pan* while in another command. Invoking the *Dsviewer* command displays the window shown in Figure 11-15. The options and operations are described next.

Figure 11-15

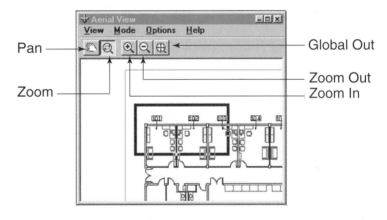

Zoom

This sets the viewer in *Zoom Window* mode. PICKing two corners <u>inside the viewer</u> causes that windowed area to be displayed <u>in the main graphics area</u> (Fig. 11-16).

Figure 11-16

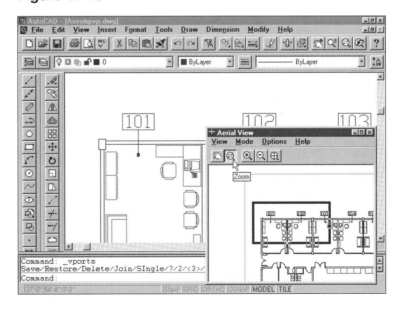

Pan

Selecting this icon or choosing *Pan* from the *View* menu in *Aerial View* sets the viewbox in the *Pan* mode. If you move the viewbox to any location <u>inside the viewer</u>, the selected area of the drawing is displayed in the main graphics area (Fig. 11-17).

Figure 11-17

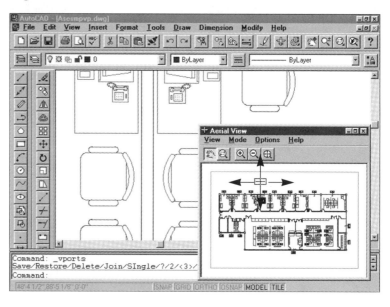

There are two ways to operate the *Pan* option: (1) move the viewbox to a new location and PICK (single-click) to achieve a <u>static display</u> of the viewbox objects in the main graphics area, or (2) hold down the left mouse button and move the viewbox around in the viewer to <u>dynamically display</u> objects in the main graphics area.

This second method is the most unique and useful feature of *Aerial View*. It allows you to dynamically move the viewbox around in viewer while the magnified area is displayed <u>instantaneously and dynamically</u> inside the viewer window. This feature is especially helpful for locating small details in the drawing, like reading text. You can change the size of the viewbox with the *Zoom* option.

Zoom In

PICKing this option causes the <u>display in the viewer</u> to be magnified 2X. This does <u>not</u> affect the display in the main graphics area.

Zoom Out

This option causes the <u>display in the viewer</u> to be reduced to .5X. This does <u>not</u> affect the display in the main graphics area.

Global

Selecting this option will resize the <u>drawing display in the viewer</u> to its maximum size.

VIEWRES

Pull-down Menu	COMMAND (TYPE)	ALIAS (TYPE)	Short-cut	Screen (side) Menu	Tablet Menu
Tools Preferences... Performance	*VIEWRES*	...	...	...	...

Viewres controls the resolution of curved shapes for the <u>screen display</u> only. Its purpose is to speed regeneration time by displaying curved shapes as linear approximations of curves; that is, a curved shape (such as an *Arc, Circle,* or *Ellipse*) appears as several short, straight line segments (Fig. 11-18). The drawing database and the plotted drawing, however, always define a true curve. The range of *Viewres* is from 1–20,000 with 100 being the default. The higher

Figure 11-18 ────────────────

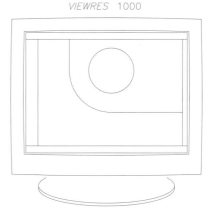

the value (called "circle zoom percentage"), the more accurate the display of curves and the slower the regeneration time. The lower the value, the faster is the regeneration time but the more "jagged" the curves. A value of 500 to 1,000 is suggested for most applications. The command syntax is shown here.

```
Command: Viewres
Do you want fast zooms? <Y> Enter
Enter circle zoom percent (1–20000) <100>: 1000
Command:
```

If you answer *No* to *Do you want fast zooms?*, AutoCAD causes a regeneration after every *Pan, Zoom,* or *View* command. Otherwise, regenerations occur only when *Zooming* extremely far in or out.

UCSICON

Pull-down Menu	COMMAND (TYPE)	ALIAS (TYPE)	Short-cut	Screen (side) Menu	Tablet Menu
View Display > UCS Icon >	*UCSICON*	...	...	*View 2 UCSicon*	*L,2*

The icon that appears in the lower-left corner of the AutoCAD Drawing Editor is the Coordinate System Icon (Fig. 11-19). The icon is sometimes called the "UCS icon" because it automatically orients itself to the new location of a coordinate system that you create, called a "USC" (User Coordinate System).

Figure 11-19 ────

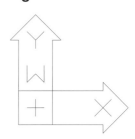

The *Ucsicon* command controls the appearance and positioning of the Coordinate System Icon. It is important to display the icon when working with 3-dimensional drawings so that you can more easily visualize the orientation of the X, Y, and Z coordinate system. However, when you are working with 2-dimensional drawings, it is not necessary to display this icon. (By now you are accustomed to normal orientation of the positive X and Y axes—X positive is to the right and Y positive is up.)

You can use the *Off* option of the *Ucsicon* command to remove the icon from the display. Use *Ucsicon* again with the *On* option to display the icon again. The command prompt is as follows:

```
Command: ucsicon
ON/OFF/All/Noorigin/ORigin <ON>: off
Command:
```

You can also control the visibility of the icon using the *View* pull-down menu. Select *Display*, then *UCS Icon*, then remove the check by the word *On*. Make sure you turn on the Coordinate System Icon when you begin working in 3D.

CHAPTER EXERCISES

1. ***Zoom Extents, Realtime, Window, Previous, Center***

   Open the sample drawing supplied with AutoCAD called **ASESMP.DWG**. If your system is set up with the "Full" installation of AutoCAD, the drawing is located in the **C:\Program Files\ AutoCAD R14\Sample** directory. If you loaded only the "Typical" installation, you can copy the drawing from the AutoCAD R14 CD, Acad\Sample directory. The drawing shows an office building layout (Fig. 11-20). Use *SaveAs* and rename the drawing to **ZOOM TEST** and locate it in your working folder (directory). (When you view sample drawings, it is a good idea to copy them to another name so you do not accidentally change the original drawings.)

   A. First, use ***Zoom Extents*** to display the office drawing as large as possible on your screen. Next, use ***Zoom Realtime*** and zoom in to the center of the layout. You should be able to see rooms 123 and 124 just below the center of your screen. Right-click and use ***Zoom Extents*** from the cursor pop-up menu to display the entire layout again. Press **Esc**, **Enter**, or right-click and select ***Exit*** to exit *Realtime*.

Figure 11-20

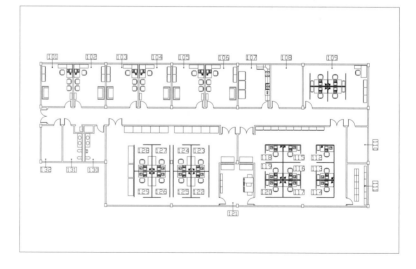

B. Use **Zoom Window** to closely examine room 123. What items in magenta color are on the desk? (HINT: The small solid shape is a digitizer puck.)

Figure 11-21

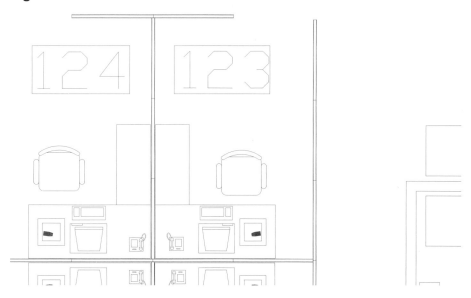

C. Use **Zoom Center**. **PICK** the telephone on the desk next to the computer as the center. Specify a height of **24 (2′)**. You should see a display showing only the telephone. How many keys are on the telephone's number pad (*Point* objects)?

D. Next use **Zoom Previous**. Do you see the same display of rooms 123 and 124 as before (Fig. 11-21)? Use **Zoom Previous** repeatedly until you see the office drawing *Extents*.

2. *Pan Realtime*

A. Using the same drawing as exercise 1 (ZOOM TEST.DWG, originally ASESMP.DWG), use **Zoom Extents** to ensure you can view the entire drawing. Next **Zoom** with a **Window** to an area about the size of two or three rooms.

B. Invoke the **Realtime** option of **Pan**. *Pan* about the drawing and find the kitchen. How many vending machines are in the kitchen (along the left wall)?

3. *Realtime Pan* and *Zoom*

In the following exercise, use **Realtime Pan** and **Zoom** in concert to examine specific details of the office layout.

A. Find the copy machine for the office. What room number is it located in? (HINT: It is centrally located near the majority of the offices.)

B. Assume you were choosing an office for yourself in the office complex. What completely empty office (room number) would you choose, assuming you also wanted to be close to the kitchen? How far is it from your office door to the nearest door to the copy room (direct distance)? HINT: Use the **Distance** command with **Endpoint OSNAP** to select the two indicated points.

C. **Pan** and/or **Zoom** back to the kitchen. How many burners are on the stove? What appliance is located next to the sinks?

4. *Pan Point*

   A. *Zoom* in to your new office (the largest empty office) so that you can see the entire room and room number. Assume you were listening to music on your computer with the door open and calculated it could be heard about 100′ away. Naturally, you are concerned about not bothering the company CEO who has an office in room 101. Use *Pan Point* to see if the CEO's office is within listening range. (HINT: enter **@100′<0** at the "Displacement:" prompt.) Should you turn down the system?

   B. Exit the **ZOOM TEST** drawing, but do not delete it from your working directory because you will use it again in Exercise 6.

5. *Viewres, Zoom Dynamic*

   A. *Open* the **TABLET14** drawing from the **Sample** folder (C:\Program Files\AutoCAD R14\ Sample). Use *SaveAs*, name the new drawing **TABLET TEST**, and locate it in your working folder.

   B. Use *Zoom* with the **Window** option (you can try *Realtime Zoom*, but it may be too slow depending on the speed of your system). Zoom in to the first several command icons located in the middle-left section of the tablet menu. Your display should reveal icons for *Zoom* and other viewing commands. *Pan* or *Zoom* in or out until you see the *Zoom* icons clearly.

   C. Notice how the *Zoom* command icons are shown as polygons instead of circles? Use the *Viewres* command and change the value to **1000**. Do the icons now appear as circles? Use *Zoom Realtime* again. Does the new setting change the speed of zooming? Why do you think *Viewres* was originally set to 100 instead of 1,000? Change *Viewres* back to **100**.

   D. Use *Zoom Dynamic* and try moving the viewbox immediately. Can you move the box before the drawing is completely regenerated? Change the viewbox size so it is approximately equal to the size of one icon. Zoom in to the command in the lower-right corner of the tablet menu. What is the command? What is the command in the lower-left corner? For a drawing of this complexity and file size, which option of *Zoom* is faster and easier—*Realtime* or *Dynamic*?

   E. If you wish, you can delete the **TABLET TEST** drawing from your directory.

6. *View*

   A. *Open* the **ZOOM TEST** drawing that you used in previous exercises. *Zoom* in to the office that you will be moving into (the empty room to the right of the kitchen). Make sure you can see the entire office. Use the *View* command and *Save* the view as **108**. Next, *Zoom* or *Pan* to room 101. *Save* the display as a *View* named **101**.

   B. Use the *View Control* dialog box (type *Ddview* or select *Named Views* from the *View* pull-down menu). *Restore* view **108**.

   C. In order for the CEO to access information on your new computer, a cable must be stretched from your office to room 101. Find out what length of cable is needed to connect the upper-left corner of office 101 to the lower-right corner of office 108. (HINT: Use *Distance* to determine the distance. Type the *View* command <u>transparently</u> [prefix with an apostrophe] during the *Distance* command and use *Running Endpoint OSNAP* to select the two indicated office corners. *Named Views* from the *View* pull-down menu <u>is not automatically transparent</u>.) What length of cable is needed? Do not *Exit* the drawing.

7. *Aerial View*

A. It has been decided to connect several other computers to your office computer. Using a similar technique as in the previous exercise (using *Distance*) determine cable lengths to other locations. Rather than saving a *View* for each location, activate *Aerial View*.

B. Use the *Global* option in *Aerial View* to display the entire office. Next, ensure the **Zoom** button is depressed and make a zoom window into your office (inside *Aerial View*). Your office should appear in the main graphics display area.

C. Ensure that **Endpoint** is checked in the **Running Osnap** dialog box. Activate **Distance** and **PICK** the lower-left corner of your office (in the main graphics display). Next in *Aerial View*, select a *Zoom Window* box around the room 132 (across the hall from 101). What length of cable is needed to connect to the lower-left corner of the room?

D. Use the same technique to connect the lower-left corner of office space 128 to the lower-right corner of room 108. What length of cable is needed?

E. Now that you have finished the exercises with this drawing, you can delete the **ZOOM TEST** drawing from you working directory if you want.

NOTE: If you view and experiment with other sample drawings, remember to immediately use *SaveAs* to create new drawings so the originals are not accidentally changed.

8. *Zoom All, Extents*

Figure 11-22 ——————————————

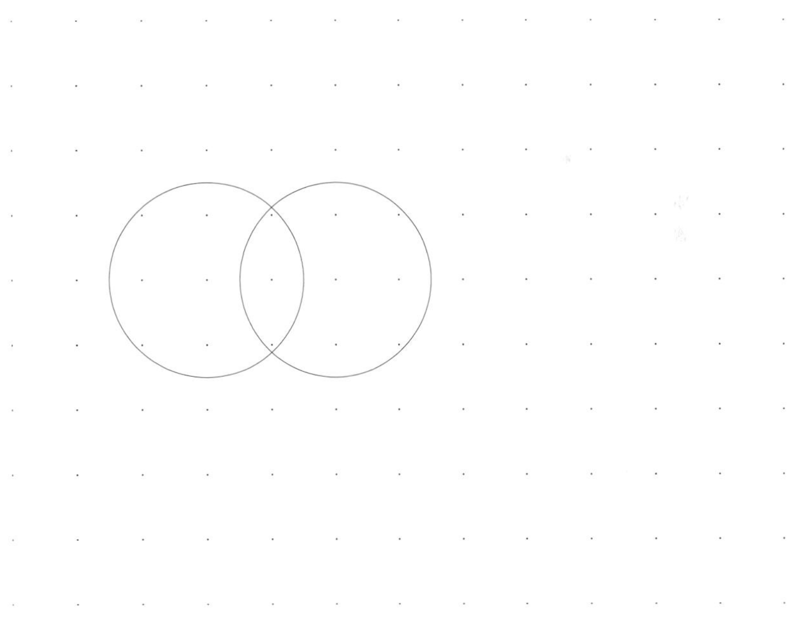

A. Begin a *New* drawing. Turn on the *Snap* (**F9**) and *GRID* (**F7**). Draw two **Circles**, each with a **1.5** unit *radius*. The *Circle* centers are at **3,5** and at **5,5**. See Figure 11-22.

B. Use *Zoom All*. Does the display change? Now use *Zoom Extents*. What happens? Now use *Zoom All* again. Which option <u>always</u> shows all of the *Limits*?

C. Draw a *Circle* with the center at **10,10** and with a *radius* of **5**. Now use *Zoom All*. Notice the *GRID* only appears on the area defined by the *Limits*. Can you move the cursor to 0,0? Now use *Zoom Extents*. What happens? Can you move the cursor to 0,0?

D. *Erase* the large *Circle*. Use *Zoom All*. Can you move the cursor to 0,0? Use *Zoom Extents*. Can you find point 0,0?

E. *Exit* AutoCAD and discard changes.

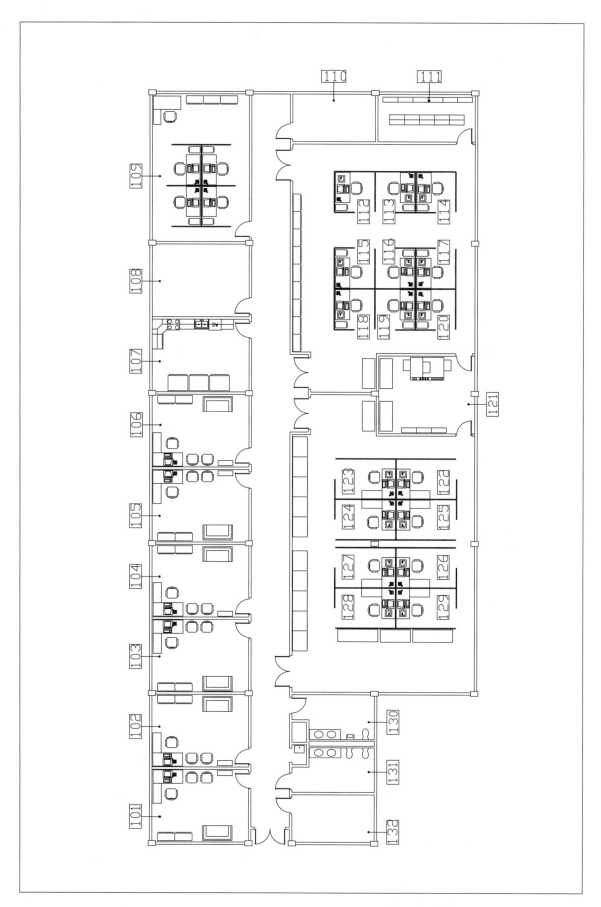

ASESMP.DWG Courtesy, AutoCAD Inc. (Release 14 Sample Drawing)

12

LAYERS, LINETYPES, COLORS, AND OBJECT PROPERTIES

Chapter Objectives

After completing this chapter you should:

1. understand the strategy of grouping related geometry with *Layers*;
2. be able to create *Layers*;
3. be able to assign *Color* and *Linetype* to *Layers;*
4. be able to control a layer's properties and visibility settings (*On, Off, Freeze, Thaw, Lock, Unlock*);
5. be able to select specific layers from a long list using the *Set Layer Filters* dialog box;
6. be able to assign *Color* and *Linetype* to objects;
7. be able to set *LTSCALE* to adjust the scale of linetypes globally;
8. be able to change an object's properties (layer, color, linetype, and linetype scale) with the *Object Properties* toolbar, the *Modify* and *Change Properties* dialog boxes, and with *Match Properties.*

CONCEPTS

In CAD, layers are used to group related objects in a drawing. Objects (*Lines, Circles, Arcs,* etc.) that are created to describe one component, function, or process of a drawing are perceived as related information and, therefore, are typically drawn on one layer. A single CAD drawing is generally composed of several components and, therefore, several layers. Use of layers provides you with a method to control visible features of the components of a drawing. For each layer, you can control its color on the screen, the linetype it will be displayed with, and its visibility setting (on or off). Only visible layers will plot.

Layers in a CAD drawing can be compared to clear overlay sheets on a manual drawing. For example, in a CAD architectural drawing, the floor plan can be drawn on one layer, electrical layout on another, plumbing on a third layer, and HVAC (heating, ventilating, and air conditioning) on a fourth layer (Fig. 12-1). Each layer of a CAD drawing can be assigned a different color, linetype, and visibility setting, similar to the way clear overlay sheets on a manual drawing can be used. Layers can be temporarily turned *OFF* or *ON* to simplify drawing and editing like overlaying or removing the clear sheets. For example, in the architectural CAD drawing, only the floor plan layer can be made visible while creating the electrical layout, but can later be cross-referenced with the HVAC layout by turning its layer on. Final plots can be made of specific layers for the subcontractors and one plot of all layers for the general contractor by controlling the layers' visibility before plotting.

Figure 12-1

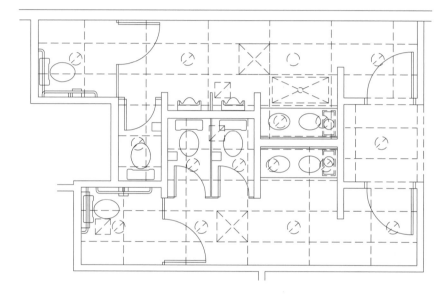

AutoCAD allows you to create a practically unlimited number of layers. You must assign a name to each layer when you create it. The layer names should be descriptive of the information on the layer.

Assigning Colors and Linetypes

There are two strategies for assigning colors and linetypes in a drawing—assign to layers or assign to objects:

<u>Assign colors and linetypes to layers</u>
Usually, layers are assigned a color and a linetype so that all objects drawn on a single layer have the same color and linetype. Assigning colors and linetypes to layers is called *BYLAYER* color and linetype setting. Using the *BYLAYER* method makes it visually apparent which objects are related (on the same layer). All objects on the same layer have the same linetype and color.

<u>Assign colors and linetypes to individual objects</u>
Alternately, you can assign colors and linetypes to specific objects, overriding the layer's color and linetype setting. This method is fast and easy for small drawings and works well when layering schemes are not used or for particular applications. However, using this method makes it difficult to see which layers the objects are located on.

Object Properties

Another way to describe the assignment of color and linetype properties is to consider the concept of Object Properties. Each object has properties such as a layer, a color, and a linetype. The color and linetype for each object can be designated as *BYLAYER* or as a specific color or linetype. Object properties for the two drawing strategies (schemes) are described as follows:

Object Properties	1. *BYLAYER* Drawing Scheme	2. Object-Specific Drawing Scheme
Layer Assignment	Layer name descriptive of geometry on the layer	Layer name descriptive of geometry on the layer
Color Assignment	*BYLAYER*	*Red, Green, Blue, Yellow,* or other specific color setting
Linetype Assignment	*BYLAYER*	*Continuous, Hidden, Center,* or other specific linetype setting

It is recommended that beginners use only one method for assigning colors and linetypes. After gaining some experience, it may be desirable to combine the two methods only for specific applications. Usually, the *BYLAYER* method is learned first, and the object-specific color and linetype assignment method is used when layers are not needed or when complex applications are needed.

(The *BYBLOCK* linetype and color assignment has special applications for *Blocks* and is discussed in Chapter 21.)

Color assignment is of particular importance because the colors (assigned either to layers or to objects) control the pens or line thickness for plotting and printing. In the *Print/Plot Configuration* dialog box (see Chapter 14), *Pen Assignments* are designated based on the object's color on the screen. Each screen color can be assigned to use a separate pen or line thickness for plotting or printing so that all objects that appear in one color will be plotted with the same pen or line thickness.

LAYER, LINETYPE, AND COLOR COMMANDS

LAYER

Pull-down Menu	COMMAND (TYPE)	ALIAS (TYPE)	Short-cut	Screen (side) Menu	Tablet Menu
Format *Layer...*	*LAYER or* *-LAYER*	*LA or -LA*	...	*FORMAT* *Layer*	*U,5*

The easiest way to gain complete layer control is through the *Layer* tab of the *Layer & Linetype Properties* dialog box (Fig. 12-2). The *Layer* tab of the *Layer & Linetype Properties* dialog box is invoked by using the icon button (shown above), typing the *Layer* command or *LA* command alias, or selecting *Layer* from the *Format* pull-down or screen menu.

Figure 12-2

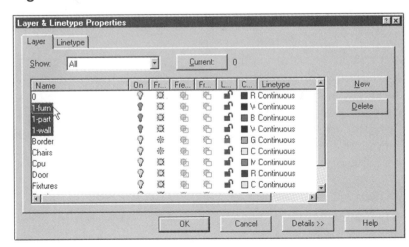

This dialog box allows full control for all layers in a drawing. Layers existing in the drawing appear in the list at the central area (only Layer 0 will appear for new drawings created from standard templates such as ACAD.DWT). New layers can be created by selecting the *New* button near the upper-right corner of the dialog box.

All properties and visibility settings of layers can be controlled by either highlighting the layer name and then selecting tiles such as *Current* or *Delete* or by PICKing one of the icons for a layer such as the light bulb icon (*On, Off*), sun/snowflake icon (*Thaw/Freeze*), *Freeze in the Current Viewport* icon, *Freeze in New Viewports* icon, padlock icon (*Lock/Unlock*), *Color* tile, or the *Linetype*.

Typical Windows 95 dialog box features apply as explained in this paragraph. Multiple layers can be highlighted (see Fig. 12-2) by holding down the Ctrl key while PICKing (to highlight one at a time) or holding down the Shift key while PICKing (to select a range of layers between and including two selected names). Right-clicking in the list area displays a cursor menu allowing you to *Select All* or *Clear All* names in the list (see Fig. 12-4). You can rename a layer by clicking twice <u>slowly</u> on the name (this is the same as PICKing an already highlighted name). The column widths can be changed by moving the pointer to the "crack" between column headings until double arrows appear (Fig. 12-3).

Figure 12-3

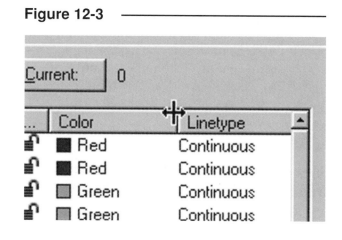

A particularly useful feature is the ability to sort the layers in the list by any one of the headings (*Name, On/Off, Freeze/Thaw, Linetype,* etc.) by clicking on the heading tile. For example, you can sort the list of names in alphabetical order (or reverse order) by clicking once (or twice) on the *Name* heading above the list of names, or you may want to display all of the *Thawed* layers together at the top of the list or all layers of similar colors together.

The *Details* tile near the bottom-right corner of the dialog box displays the details (properties and visibility settings) of the highlighted layer in the list (Fig. 12-4). This area is basically an <u>alternative</u> method of controlling layer properties and visibility settings (instead of using the icons in the list area).

Figure 12-4

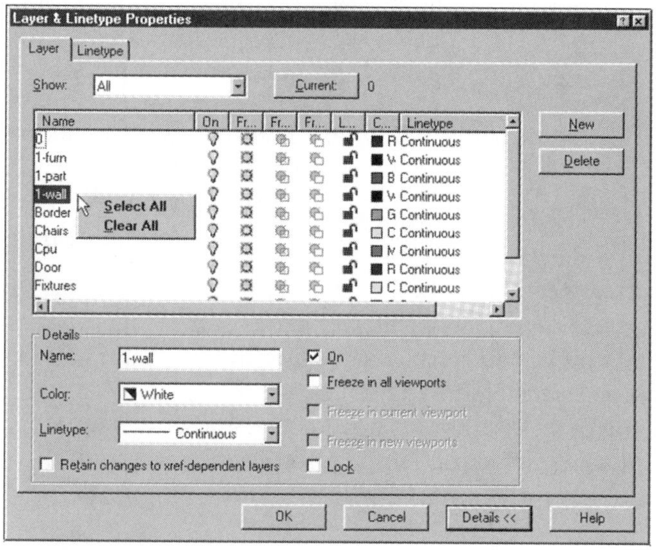

As an alternative to the dialog box, the *Layer* command can be used in command line format by typing *Layer* with a - (hyphen) prefix. The command line format of *Layer* shows all of the available options:

Command: ***-layer***
?/Make/Set/New/ON/OFF/Color/LType/Freeze/Thaw/Lock/Unlock:

Detailed explanations of the options for controlling layer properties and visibility settings, whether using the *Layer* tab of the *Layer & Linetype Properties* dialog box or using *-Layer* in command line format, are listed next.

Current or Set

To *Set* a layer as the *Current* layer is to make it the active drawing layer. Any objects created with draw commands are created on the *Current* layer. You can, however, edit objects on any layer, but draw only on the current layer. Therefore, if you want to draw on the FLOORPLAN layer (for example), use the *Set* or *Current* option. If you want to draw with a certain *Color* or *Linetype,* set the layer with the desired *Color* and *Linetype* as the *Current* layer. Any layer can be made current, but only <u>one layer at a time</u> can be current.

To set the current layer with the *Layer* tab of the *Layer & Linetype Properties* dialog box (Fig. 12-2, 12-4), select the desired layer from the list and then select the *Current* tile. Since only one layer can be current, it may be necessary to "deselect" highlighted layer names from the list until only one is highlighted. Alternately, if you are typing, use the *Set* option of the *-Layer* command to make a layer the current layer.

ON, OFF

If a layer is *ON,* it is visible. Objects on visible layers can be edited or plotted. Layers that are *OFF* are not visible. Objects on layers that are *OFF* will not plot and cannot be edited (unless the *ALL* selection option is used, such as *Erase, All*). It is not advisable to turn the current layer *OFF.*

Freeze, Thaw

Freeze and *Thaw* override *ON* and *OFF. Freeze* is a more protected state than *OFF.* Like being *OFF,* a frozen layer is not visible nor can its objects be edited or plotted. Objects on a frozen layer cannot be accidentally *Erased* with the *All* option. *Freezing* also prevents the layer from being considered when *Regens* occur. *Freezing* unused layers speeds up computing time when working with large and complex drawings. *Thawing* reverses the *Freezing* state. Layers can be *Thawed* and also turned *OFF.* Frozen layers are not visible even though the light bulb icon is on.

Lock, Unlock

Layers that are *Locked* are protected from being edited but are still visible and can be plotted. *Locking* a layer prevents its objects from being changed even though they are visible. Objects on *Locked* layers cannot be selected with the *All* selection option (such as *Erase, All*). Layers can be *Locked* and *OFF.*

Freeze in Current Viewport, Freeze in New Viewports

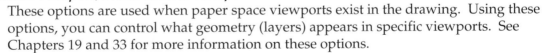 These options are used when paper space viewports exist in the drawing. Using these options, you can control what geometry (layers) appears in specific viewports. See Chapters 19 and 33 for more information on these options.

Color and Linetype **Properties**

Layers have properties of *Color* and *Linetype* such that (generally) an object that is drawn on, or changed to, a specific layer assumes the layer's linetype and color. Using this scheme (*BYLAYER*) enhances your ability to see what geometry is related by layer. It is also possible, however, to assign specific color and linetype to objects that will override the layer's color and linetype (see *Linetype* and *Color* commands and Changing Object Properties).

Color

☐ Green   Selecting one of the small color boxes in the list area of the *Layer* tab of the *Layer & Linetype Properties* dialog box causes the *Select Color* dialog box to pop up (Fig. 12-5). The desired color can then be selected or the name or color number (called the ACI—AutoCAD Color Index) can be typed in the edit box. This action retroactively changes the color assigned to a layer. All objects on the layer with *BYLAYER* setting change to the new layer color. Objects with specific color assigned (not *BYLAYER*) are not affected. Alternately, the *Color* option of the *Layer* command can be typed to enter the color name or ACI number.

The actual number of colors that are available depends on the type of monitor and graphics controller card that are configured. Although most setups allow up to 16 million colors and more, only 256 colors are shown in the standard ACI pallet in AutoCAD.

Figure 12-5

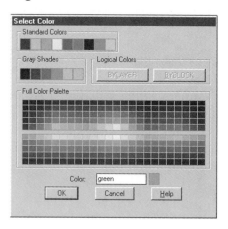

Linetype

To set a layer's linetype, select the *Linetype* (word such as *Continuous* or *Hidden*) in the layer list of the *Layer* tab, which in turn invokes the *Select Linetype* dialog box (Fig. 12-6). Select the desired linetype from the list. Alternately, the *Linetype* option of the *Layer* command can be typed. Similar to changing a layer's color, all objects on the layer with *BYLAYER* linetype assignment are retroactively displayed in the selected layer linetype while non-*BYLAYER* objects remain unchanged.

Figure 12-6

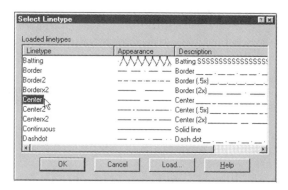

The default AutoCAD template drawing as supplied by Autodesk has only one linetype available (*Continuous*). Before you can use other linetypes in a drawing, you load the linetypes by selecting the *Load* tile (see *Linetype* command) or by using a default template drawing that has the desired linetypes already loaded.

New

The *New* option allows you to make new layers. There is only one layer in the AutoCAD default drawing (ACAD.DWT) as it is provided to you "out of the box." That layer is Layer 0. Layer 0 is a part of every AutoCAD drawing because it cannot be deleted. You can, however, change the *Linetype* and *Color* of Layer 0 from the defaults (continuous linetype and color #7 white). Layer 0 is generally used as a construction layer or for geometry not intended to be included in the final draft of the drawing. Layer 0 has special properties when creating *Blocks* (Chapter 21).

You should create layers for each group of related objects and assign appropriate layer names for that geometry. You can use up to 31 characters (no spaces) for layer names. There is practically no limit to the number of layers that can be created (although 32,767 has been found to be the actual limit). To create layers in the *Layer* tab of the dialog box, select the *New* tile. A new layer named "Layer1" (or other number) then appears in the list with the default color (*White*) and linetype (*Continuous*). The new name initially appears in the rename mode so you can immediately assign a more appropriate and descriptive name for the layer. You can create many new names quickly by typing (renaming) the first layer name, then typing a comma before other names. A comma forces a new "blank" layer name to appear. Colors and linetypes should be assigned as the next step.

If you want to create layers by typing, the *New* and *Make* options of the *Layer* command can be used. *New* allows creation of one or more new layers. *Make* allows creation of one layer (at a time) and sets it as the current layer.

Delete

The *Delete* tile allows you to delete layers. <u>Only layers with no geometry can be deleted</u>. You cannot delete a layer that has objects on it nor can you delete Layer 0, the current layer, or layers that are part of externally referenced (*Xref*) drawings. If you attempt to *Delete* such a layer, accidentally or intentionally, a warning appears (Fig. 12-7).

Figure 12-7

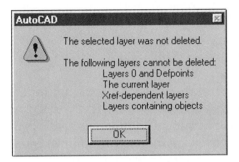

Show

The *Show* drop-down list determines what names appear in the list of layers in the *Layer* tab (Fig. 12-8). The choices are as follows:

Figure 12-8

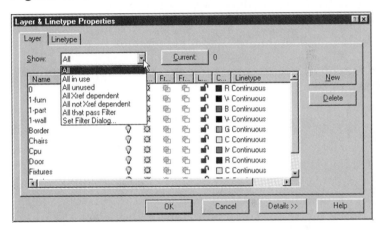

All	shows all layers in the drawing
All in use	shows all layers that have objects on them
All unused	shows only layers without geometry
All Xref dependent	shows only layers that are part of externally referenced drawings
All not Xref	shows all layers except those in externally referenced drawings
All that pass filter	shows the layers that are designated in the *Set Layer Filters* dialog box
Set Filter Dialog. . .	invokes the *Set Layer Filters* dialog box

Set Layer Filters Dialog Box

When working on drawings with a large number of layers, it is time consuming to scroll through the list of layer names in the *Layer* tab of the *Layer & Linetype Properties* dialog box or in the Layer Control drop-down list on the Object Properties toolbar. The *Show:* drop-down list in the *Layer* tab of the *Layer & Linetype Properties* dialog box controls names that appear in the list in that dialog box but not the Layer Control drop-down list. The *Set Layer Filters* dialog box enables you to specify criteria for displaying a selected set of layers in both lists. For example, you may want to list only layer names that are *Thawed*. By using layer filters, you can shorten the layer list by filtering out the layers that do not meet this criterion.

Figure 12-9

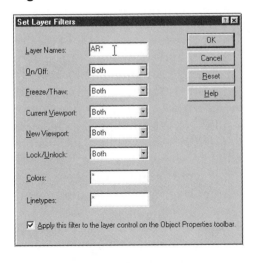

Selecting the *Set Filter dialog...* option from the bottom of the *Show*: list produces the *Set Layer Filters* dialog box (Fig. 12-9). Eight criteria can be specified in the *Set Layer Filters* dialog box. The criteria define <u>what you want to see</u> in the layer listing. When the criteria are set to *Both* or *, all layer names are displayed.

The three boxes displaying a * symbol (asterisk) enable you to enter specific values. These edit boxes allow you to list only layers that match particular names, colors or linetypes. As an example, if you set the *Layer Names* filter to **AR***, only those layers whose names begin with "AR" will be displayed in the layer listing (see Fig. 12-9).

The other criteria are in pairs, enabling you to choose one or "*Both*" of the settings. When a criteria is set to *Both,* you will see layers listed that match both items in the pair, such as layers that are *On* <u>and</u> *Off*. Selecting one of the pair displays only layer names meeting that condition. Selecting *Thawed* from the *Freeze/Thaw* box, for example, causes a display of only layers that that are thawed.

When a layer filter setting is enabled, the *All that pass filter* option is automatically displayed in the *Show*: list in the *Layer* tab of the *Layer & Linetype Properties* dialog box. If you want the filter to also be applied to the Layer Control drop-down list in the Object Properties toolbar, select the checkbox at the bottom of the *Set Layer Filters* dialog box.

Only one layer filter configuration can be applied at a time, and a specific configuration cannot be saved to a file.

Layer Control Drop-down List

The Object Properties toolbar contains a drop-down list for making layer control quick and easy (Fig. 12-10). The window normally displays the current layer's name, visibility setting, and properties. When you pull down the list, all the layers (unless otherwise specified in the *Set Layer Filter* dialog box) and their settings are displayed. Selecting any layer <u>name</u> makes it current. Clicking on any of the visibility/properties icons changes the layers' setting as described previously. Several layers can be changed in one "drop." You cannot change a layer's color or linetype nor can you create new layers using this drop-down list.

Figure 12-10 ───────────

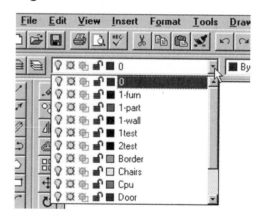

Displaying an Object's Properties and Visibility Settings

When the Layer Control drop-down list in the Object Properties toolbar is in the normal position (not "dropped down"), it can be used to display an <u>object's</u> layer properties and visibility settings. Do this by selecting an object (with the pickbox, Window, or Crossing Window) when no commands are in use. When an object is selected, the Object Properties toolbar displays the layer, color, and linetype <u>of the selected object</u> (Fig. 12-11).

Figure 12-11 ───────────

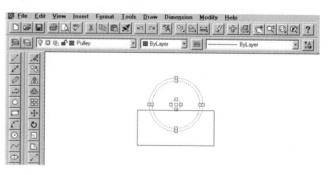

The Layer Control box displays the selected object's layer name and layer settings rather than that of the current layer. The Color Control and Linetype Control boxes in the Object Properties toolbar (just to the right) also temporarily reflect the color and linetype properties of the selected object. Pressing the Escape key causes the list boxes to display the current layer name and settings again, as would normally be displayed. If more than one object is selected and the objects are on different layers or have different color and linetype properties, the list boxes go blank until the objects become unhighlighted or a command is used.

These sections of the Object Properties toolbar have another important new feature that allows you to change a highlighted object's (or set of objects') properties. See Changing Object Properties near the end of this chapter.

Make Object's Layer Current

 A new feature in Release 14 can be used to make a desired layer current simply by selecting <u>any object on the layer</u>. This option is generally faster than using the *Layer & Linetype Properties* dialog box or *-Layer* command. The *Make Object's Layer Current* feature is particularly useful when you want to draw objects on the same layer as other objects you see but are not sure of the layer name. This feature (not a formal command) is only available by selecting the icon on the far-left of the Object Properties toolbar (Fig. 12-11). It cannot be invoked by any other method.

There are two steps in the procedure to make an object's layer current: (1) select the icon and (2) select any object on the desired layer. The selected object's layer becomes current and the new current layer name immediately appears in the Layer Control list box in the Object Properties toolbar.

LINETYPE

Pull-down Menu	COMMAND (TYPE)	ALIAS (TYPE)	Short-cut	Screen (side) Menu	Tablet Menu
Format *Linetype*	*LINETYPE or* *-LINETYPE*	*LT or -LT*	...	*FORMAT* *Linetype*	*U,3*

Invoking this command presents the *Linetype* tab of the *Layer & Linetype Properties* dialog box (Fig. 12-12). Even though this looks similar to the dialog box used for assigning linetypes to layers (shown earlier in Figure 12-6), beware!

When linetypes are selected using the *Linetype* command or *Linetype* tab of the *Layer & Linetype Properties* dialog box, they are <u>assigned to objects</u>—not to layers. That is, selecting a linetype by this manner causes all objects <u>from that time on</u> to be drawn using that linetype, regardless of the layers that they are on (unless the *BYLAYER* type is PICKed). In contrast, selecting linetypes during the *Layer* command (using the *Layer* tab of the *Layer & Linetype Properties* dialog box) results in assignment of linetypes to <u>layers</u>. Remember that using both of these methods for color and linetype assignment in one drawing can be very confusing until you have some experience using both methods.

Figure 12-12

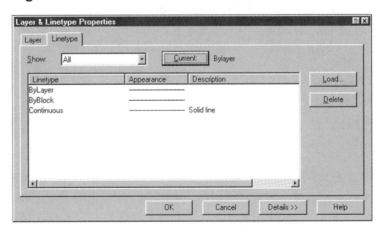

To draw an object in a specific linetype, simply select the linetype from the list and PICK the *OK* tile. That linetype stays in effect (all objects are drawn with that linetype) until another linetype is selected. If you want to draw objects in the layer's assigned linetype, select *BYLAYER*.

Load

The *Linetype* tab of the *Layer & Linetype Properties* dialog box also lets you *Load* linetypes that are not already in the drawing. Just PICK the *Load* button to view and select from the list of available linetypes in the *Load or Reload Linetypes* dialog box (Fig. 12-13). The linetypes are loaded from the ACAD.LIN file. If you want to load all linetypes into the drawing, right-click to display a small cursor menu, then choose *Select All* to highlight all linetypes (see Fig. 12-12). You can also select multiple linetypes by holding down the Ctrl key or select a range by holding down the Shift key. The loaded linetypes are then available for assignment from both the *Linetype* tab and the *Layer* tab of the *Layer & Linetype Properties* dialog box.

AutoCAD supplies numerous linetypes as shown in Figure 12-14. The ACAD_ISO*n*W100 linetypes are intended to be used with metric drawings. For the ACAD_ISO*n*W100 linetypes, *Limits* should be set to metric sheet sizes to accommodate the relatively large linetype spacing. Avoid using both ACAD_ISO*n*W100 linetypes and other linetypes in one drawing due to the difficulty managing the two sets of linetype scales.

Delete

The *Delete* button is useful for deleting any unused linetypes from the drawing. Unused linetypes are those that have been loaded into the drawing but have not been assigned to layers or to objects. Freeing the drawing of unused linetypes should reduce file size slightly.

Typing -*Linetype* produces the command line format of *Linetype*. Although this format of the command does not offer the capabilities of the dialog box version, you can assign linetypes <u>to objects</u> or assign an object to use the layer's linetype by typing -*Linetype* or -*LA* (use the hyphen prefix). This action yields the following prompt:

```
Command: -linetype
?/Create/Load/Set:
```

You can list (*?*), *Load, Set,* or *Create* linetypes with the typed form of the command. Type a question mark (*?*) to display the list of available linetypes. Type *L* to *Load* any linetypes. The *Set* option of *Linetype* accomplishes the same action as selecting from the linetypes in the dialog box. That is, you can set a specific linetype for all subsequent objects regardless of the layer's linetype setting, or you can use *Set* to assign the new objects to use the layer's (*BYLAYER*) linetype.

Figure 12-13

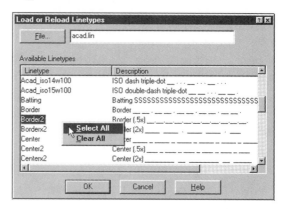

Figure 12-14

BORDER	
BORDER2	
BORDERX2	
CENTER	
CENTER2	
CENTERX2	
CONTINUOUS	
DASHDOT	
DASHDOT2	
DASHDOTX2	
DASHED	
DASHED2	
DASHEDX2	
DIVIDE	
DIVIDE2	
DIVIDEX2	
DOT	
DOT2	
DOTX2	
HIDDEN	
HIDDEN2	
HIDDENX2	
PHANTOM	
PHANTON2	
PHANTOMX2	
ACAD_ISO02W100	
ACAD_ISO03W100	
ACAD_ISO04W100	
ACAD_ISO05W100	
ACAD_ISO06W100	
ACAD_ISO07W100	
ACAD_ISO08W100	
ACAD_ISO09W100	
ACAD_ISO010W100	
ACAD_ISO011W100	
ACAD_ISO012W100	
ACAD_ISO013W100	
ACAD_ISO014W100	
ACAD_ISO015W100	

You can create your own custom linetypes with the *Create* option or by using a text editor. Both simple linetypes (line, dash, and dot combinations) and complex linetypes (including text or other shapes) are possible. See Chapter 45, Basic Customization.

Linetype Control Drop-down List

The Object Properties toolbar contains a drop-down list for selecting linetypes (Fig. 12-15). Although this appears to be quick and easy, you can <u>assign linetypes only to objects</u> by this method unless *BYLAYER* is selected to use the layers' assigned linetypes. Any linetype you select from this list becomes the current object linetype. If you want to select linetypes for layers, make sure this list displays the *BYLAYER* setting, then use the *Layer & Linetype Properties* dialog box to select linetypes for layers.

Keep in mind that the Linetype Control drop-down list as well as the Layer Control and Color Control drop-down lists display the current settings for selected objects (objects selected when no commands are in use). When linetypes are assigned to layers rather than objects, the layer drop-down list reports a *BYLAYER* setting.

Figure 12-15

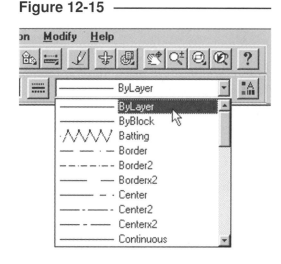

COLOR and
DDCOLOR

	Pull-down Menu	COMMAND (TYPE)	ALIAS (TYPE)	Short-cut	Screen (side) Menu	Tablet Menu
	Format *Color*	*COLOR or* *DDCOLOR*	*COL*	...	*FORMAT* *Ddcolor*	*U,4*

Similar to linetypes, colors can be assigned to layers or to objects. Using the *Color* or *Ddcolor* command assigns a color for all newly created <u>objects</u>, regardless of the layer's color designation (unless the *BYLAYER* color is selected). This color setting <u>overrides</u> the layer color for any newly created objects so that all new objects are drawn with the specified color no matter what layers they are on. This type of color designation prohibits your ability to see which objects are on which layers by their color; however, for some applications object color setting may be desirable. Use the *Layer* tab of the *Layer &Linetype Properties* dialog box to set colors for layers.

Invoking this command by the menus or by typing *Ddcolor* presents the *Select Color* dialog box shown in Figure 12-16. This is essentially the same dialog box used for assigning colors to layers; however, the *BYLAYER* and *BYBLOCK* tiles are accessible. (The buttons are grayed out when this dialog is invoked from the *Layer* tab because, in that case, any setting is a *BYLAYER* setting.) *BYBLOCK* is discussed in Chapter 21.

If you type "*Color,*" the command line format is displayed as follows:

> Command: `color`
> New object color <BYLAYER>: `red` (Specify a color name, number, *BYLAYER* or *BYBLOCK* setting.)
> Command:

Figure 12-16

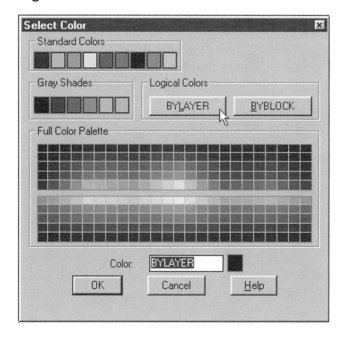

For the example, all new objects drawn will be *red* regardless of the layers' color settings until *Color* or *Ddcolor* is used again to set another color or *BYLAYER* setting.

The current color (whether *BYLAYER* or specific object color) is saved in the *CECOLOR* (Current Entity Color) variable. The current color can be set by changing the value in the *CECOLOR* variable directly at the command prompt (by typing "*CECOLOR*"), by using the *Color* command, *Select Color* dialog box, or by using the Color Control drop-down list in the Object Properties toolbar. The *CECOLOR* variable accepts a string value (such as "red" or "bylayer") or accepts the ACI number (0 through 255).

Color Control Drop-down List

Figure 12-17

When an object-specific color has been set, the top item in the Object Properties toolbar Color Control drop-down list displays the current color (Fig. 12-17). Beware—using this list to select a color assigns an <u>object-specific color unless *BYLAYER* is selected</u>. Any color you select from this list becomes the current object color. Selecting *Other. . .* from the bottom of the list invokes the *Select Color* dialog box (see Figure 12-16). If you want to assign colors to layers, make sure this list displays the *BYLAYER* setting, then use the *Layer* tab of the *Layer & Linetype Properties* dialog box to select colors for layers.

This drop-down list, as well as the others in the Object Properties toolbar, display the current settings for selected objects (objects selected when no commands are in use). A *BYLAYER* setting indicates that properties are assigned to layers rather than objects.

LTSCALE

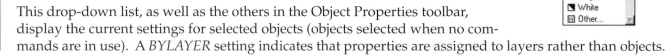

Pull-down Menu	COMMAND (TYPE)	ALIAS (TYPE)	Short-cut	Screen (side) Menu	Tablet Menu
Format *Linetype...* *Details >>*	*LTSCALE*	*LTS*	...	*FORMAT* *Linetype* *Details >>*	...

Hidden, dashed, dotted, and other linetypes that have spaces are called <u>non-continuous</u> linetypes. When drawing objects that have non-continuous linetypes (either *BYLAYER* or object-specific linetype designations), the linetype's dashes or dots are automatically created and spaced. The *LTSCALE* (Linetype Scale) system variable controls the length and spacing of the dashes and/or dots. The value that is specified for *LTSCALE* affects the drawing <u>globally and retroactively</u>. That is, all existing non-continuous lines in the drawing as well as new lines are affected by *LTSCALE*. You can therefore adjust the drawing's linetype scale for all lines at any time with this one command.

If you choose to make the dashes of non-continuous lines smaller and closer together, reduce *LTSCALE*; if you desire larger dashes, increase *LTSCALE*. The *Hidden* linetype is shown in Figure 12-18 at various *LTSCALE* settings. Any positive value can be specified.

Figure 12-18

LTSCALE

4 ───────   ───────   ───────   ───────

2 ── ── ── ── ── ── ── ──

1 ─ ─ ─ ─ ─ ─ ─ ─ ─ ─ ─ ─ ─ ─ ─ ─

0.5 ‐‐‐‐‐‐‐‐‐‐‐‐‐‐‐‐‐‐‐‐‐‐‐‐‐‐‐‐

LTSCALE can be set in the *Linetype* tab of the *Layer & Linetype Properties* dialog box or in command line format. In the *Linetype* tab, select the *Details* button to allow access to the *Global scale factor* edit box (Fig. 12-19). Changing the value in this edit box sets the *LTSCALE* variable. Changing the value in the *Current object scale* edit box sets the linetype scale for the current object only (see *CELTSCALE* below) but does not affect the global linetype scale (*LTSCALE*).

LTSCALE can also be used in the command line format.

Figure 12-19

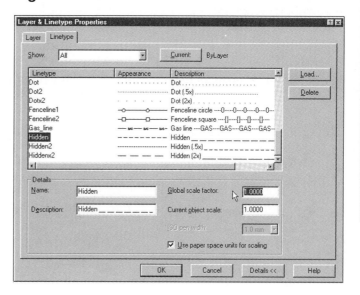

    Command: **LTSCALE**
    New scale factor <1 or current value>:
    (value) (Enter any positive value.)
    Command:

The *LTSCALE* for the default template drawing (ACAD.DWT) is 1. This value represents an appropriate *LTSCALE* for objects drawn within the default *Limits* of 12 x 9.

As a general rule, you should change the *LTSCALE* <u>proportionally</u> when *Limits* are changed (more specifically, when *Limits* are changed to other than the intended plot sheet size). For example, if you increase the drawing area defined by *Limits* by a factor of 2 (to 24 x 18) from the default (12 x 9), you might also change *LTSCALE* proportionally to a value of 2. Since *LTSCALE* is retroactive, it can be changed at a later time or repeatedly adjusted to display the desired spacing of linetypes. (For more information on *LTSCALE*, see Chapters 13 and 14.)

The ACAD_ISO*n*W100 linetypes are intended to be used with metric drawings. Changing *Limits* to metric sheet sizes automatically displays these linetypes with appropriate linetype spacing. For these linetypes only, *LTSCALE* is changed automatically to the value selected in the *ISO Pen Width* box of the *Linetype* tab (see Fig. 12-19). Using both ACAD_ISO*n*W100 linetypes and other linetypes in one drawing is discouraged due to the difficulty managing two sets of linetype scales.

Even though you have some control over the size of spacing for non-continuous lines, you have almost <u>no control</u> over the <u>placement</u> of the dashes for non-continuous lines. For example, the short dashes of centerlines <u>cannot</u> always be controlled to intersect at the centers of a series of circles. The spacing can only be adjusted globally (all lines in the drawing) to reach a compromise. You can also adjust individual objects' linetype scale (*CELTSCALE*) to achieve the desired effect. See *CELTSCALE* below and Chapter 24, Multiview Drawing, for further discussion and suggestions on this subject.

CELTSCALE

Pull-down Menu	COMMAND (TYPE)	ALIAS (TYPE)	Short-cut	Screen (side) Menu	Tablet Menu
Format Linetype... Details>>	*CELTSCALE*	...	...	*FORMAT Linetype Details>>*	...

CELTSCALE stands for Current Entity Linetype Scale. This setting changes the <u>object-specific linetype scale proportional</u> to the *LTSCALE*. The *LTSCALE* value is global and retroactive, whereas *CELTSCALE* sets the linetype scale for all <u>newly created</u> objects and is <u>not retroactive</u>. *CELTSCALE* is object specific.

Using *CELTSCALE* to set an object linetype scale is similar to setting an object color and linetype in that the properties are <u>assigned to the specific object</u>.

For example, if you wanted all non-continuous lines (dashes and spaces) in the drawing to be two times the default size, set *LTSCALE* to 2 (*LTSCALE* is global and retroactive). If you then wanted only say, a select two or three lines to have smaller spacing, change *CELTSCALE* to .5, draw the new lines, and then change *CELTSCALE* back to 1.

You can also use the *Linetype* tab of the *Layer & Linetype Properties* dialog box to set the *CELTSCALE* (see Fig. 12-19). Set the desired *CELTSCALE* by listing the *Details* and changing the value in the *Current object scale* edit box. This setting affects all linetypes for all <u>newly created</u> objects. The linetype scale for individual objects can also be changed retroactively using the method explained in the following paragraph.

The <u>recommended</u> method for adjusting linetypes in the drawing to different scales is <u>not to use</u> *CELTSCALE* but rather the following strategy.

1. Set *LTSCALE* to an appropriate value for the drawing.
2. Create all objects in the drawing with the desired linetypes.
3. Adjust the *LTSCALE* again if necessary to globally alter the linetype scale.
4. Use *Ddmodify*, *Ddchprop*, or *Matchprop* to <u>retroactively</u> change the linetype scale (*CELTSCALE*) <u>for selected objects</u> (see *Ddmodify*, *Ddchprop*, and *Matchprop*).

CHANGING OBJECT PROPERTIES

Often it is desirable to change the properties of an object after it has been created. For example, an object's *Layer* property could be changed. This can be thought of as "moving" the object from one layer to another. When an object is changed from one layer to another, it assumes the new layer's color and linetype, provided the object was originally created with color and linetype assigned *BYLAYER*, as is generally the case. In other words, if an object was created on the wrong layer (possibly with the wrong linetype or color), it could be "moved" to the desired layer, therefore assuming the new layer's linetype and color.

Another example is to change an object's individual linetype scale. In some cases an individual object's linetype scale requires an adjustment to other than the global linetype scale (*LTSCALE*) setting. One of several methods can be used to adjust the object's individual linetype scale (*CELTSCALE*) to an appropriate value.

Several methods can be used to retroactively change properties of selected objects. The Object Properties toolbar, the *Change Properties* (*Ddchprop*) dialog box, the *Modify* (*Ddmodify*) dialog box, and the *Match Properties* command can be used to <u>retroactively</u> change the properties of individual objects. Properties that can be changed by all four methods are *Layer*, *Linetype*, *Color*, and object linetype scale (*CELTSCALE*).

Although some of the commands discussed in this section have additional capabilities, the discussion is limited to changing properties of objects. For this reason, these commands and features are also discussed in Chapter 16, Modify Commands III, and in other chapters.

Object Properties Toolbar

The three drop-down lists in the Object Properties toolbar (when not "dropped down") generally show the current layer, color, and linetype. However, if an object or set of objects is selected, the information in these boxes changes to display the current objects' settings. <u>You can change the selected objects' settings</u> by "dropping down" any of the lists and making another selection.

First, select (highlight) an object when no commands are in use. Use the pickbox (that appears on the crosshairs), Window, or Crossing Window to select the desired object or set of objects. The entries in the three boxes (Layer Control, Color Control, and Linetype Control) then change to display the settings for the <u>selected object or objects</u>. If several objects are selected that have different properties, the boxes display no information (go

Figure 12-20

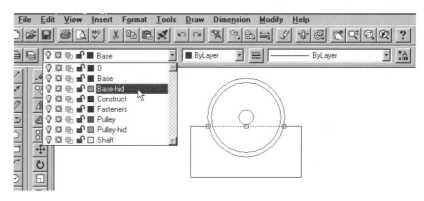

"blank"). Next, use any of the drop-down lists to make another selection (Fig. 12-20). The highlighted object's properties are changed to those selected in the lists. Press the Escape key to complete the process.

Remember that in most cases color and linetype settings are assigned *BYLAYER*. In this type of drawing scheme, to change the linetype or color properties of an object, you would change the object's <u>layer</u> (see Fig. 12-20). If you are using this type of drawing scheme, refrain from using the Color Control and Linetype Control drop-down lists for changing properties.

DDMODIFY

Pull-down Menu	COMMAND (TYPE)	ALIAS (TYPE)	Short-cut	Screen (side) Menu	Tablet Menu
Modify Properties...	*DDMODIFY*	*MO*	...	*MODIFY1 Modify*	*Y,14*

DDCHPROP

Pull-down Menu	COMMAND (TYPE)	ALIAS (TYPE)	Short-cut	Screen (side) Menu	Tablet Menu
Modify Properties...	*DDCHPROP*	*CH*	...	*MODIFY1 Modify*	*Y,14*

These dialog boxes can also be used to change an object's layer, linetype, or color designation retroactively. If you select the *Properties* icon button from the Object Properties toolbar, one of two dialog boxes appears depending on the number of objects selected. <u>If one object is selected</u>, the *Modify* dialog appears (Fig. 12-21), whereas <u>if two or more objects are selected</u>, the *Change Properties* dialog box appears (Fig. 12-22). Alternately, you can type *Ddmodify*, *Ddchprop*, or select *Properties. . .* from the *Modify* pull-down menu.

Figure 12-21

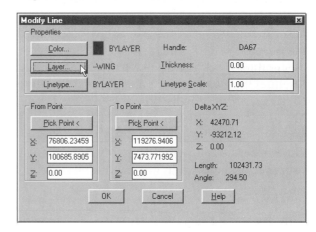

Either of these dialog boxes allows you to change the *Layer, Color, Linetype,* or *Linetype* scale (*CELTSCALE*) of the selected object(s). Selecting *Color* or *Linetype* other than *BYLAYER* sets the <u>individual object's properties</u>. For drawing schemes using the *BYLAYER* color and line-type settings, change the *Layer* to assume the desired linetype and color for the object.

NOTE: The *Linetype Scale* edit box in both of these dialog boxes affects the *CELTSCALE* variable and changes only the <u>selected object's individual linetype scale</u>. It does not affect the drawing *LTSCALE*. This is the <u>recommended method</u> for changing individual object linetype scale rather than setting *CELTSCALE* by command line format.

Figure 12-22 ——————————

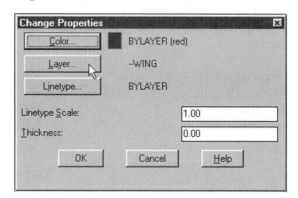

MATCHPROP

Pull-down Menu	COMMAND (TYPE)	ALIAS (TYPE)	Short-cut	Screen (side) Menu	Tablet Menu
Modify *Match Properties*	*MATCHPROP or* *PAINTER*	*MA*	…	*MODIFY1* *Matchprp*	*Y,15*

Matchprop is used to "paint" the properties of one object to another. The process is simple. After invok-ing the command, select the object that has the desired properties (Source Object), then select the object you want to "paint" the properties to (Destination Object). The command prompt is as follows.

```
Command: matchprop
Select Source Object: PICK
Current active settings = color layer ltype ltscale thickness text dim hatch
Settings/<Select Destination Object(s)>: PICK
Settings/<Select Destination Object(s)>: Enter
Command:
```

Several "Destination Objects" can be selected. The "Destination Object(s)" assume all of the properties of the "Source Object" (listed as "Current active settings").

Use the *Settings* option to control which of eight possi-ble properties and other settings are "painted" to the destination objects. At the "Settings/<Select Destination Object(s)>:" prompt, type *S* to display the *Property Settings* dialog box (Fig. 12-23). In the dialog box, designate which of the *Basic Properties* or *Special Properties* are to be painted to the "Destination Objects." The following *Basic Properties* correspond to properties discussed in this chapter:

Figure 12-23 ——————————

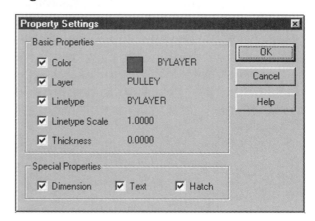

Color	paints the object or *BYLAYER* color
Layer	moves selected objects to Source Object layer
Linetype	paints the object or *BYLAYER* linetype
Linetype Scale	changes the individual object linetype scale (*CELTSCALE*) not global (*LTSCALE*)

The *Special Properties* of the *Property Settings* dialog box are discussed in Chapter 16, Modify Commands III. Also see Chapter 16 for a full explanation of *Change Properties* (*Ddchprop*) and *Modify* (*Ddmodify*).

CHAPTER EXERCISES

1. *Layer* **tab of the** *Layer & Linetype Properties* **dialog box, Layer Control drop-down list, and** *Make Object's Layer Current*

   Open the **WILHOME.DWG** sample drawing located in the C:\Program Files\AutoCAD R14\Sample directory. Activate model space by typing *MS* at the command prompt or by double-clicking the word **PAPER** on the Status line. Next, use *SaveAs* to save the drawing <u>in your working directory</u> as **WILHOME2**.

   A. Invoke the *Layer* command in command line format by typing *-LAYER* or *-LA*. Use the *?* option to yield the list of layers in the drawing. Notice that all the layers are *On* except one. Which layer is *Off*?

   B. Use the **Object Properties** toolbar to indicate the layers of selected items. On the right side of the drawing, **PICK** the yellow lines around the floor plan. What layer are the dimensions on? Press the **Esc** key twice to cancel the selection. Next, **PICK** the cyan (light blue) text around the semi-circular room near the top of the floor plan on the right. What layer is the text on? Press the **Esc** key twice to cancel the selection.

   C. Invoke the *Layer* tab of the *Layer & Linetype Properties* dialog box by selecting the icon button or using the *Format* pull-down menu. Turn *Off* the **Dimensions** layer (click the light bulb icon). Select *OK* to return to the drawing and check the new setting. Are the dimensions displayed? Next, use the **Layer Control** drop-down list to turn the **Dimensions** layer back *On*.

   D. Type *Regen* and count the number of seconds it takes for the drawing to regenerate. Next, use the **Object Properties** toolbar to indicate the name of the layer of the green elevations on the left side of the drawing (**PICK** a green object and examine the layer name that appears). Use the **Layer Control** drop-down list to turn *Off* that layer (light bulb icon) <u>and</u> the layer with the cyan text. Now type *Regen* again and count the number of seconds it takes for the regeneration. Is there any change? Finally, *Freeze* both layers (snowflake/sun icon), then type *Regen* again and count the number of seconds it takes for the regeneration. Is there any difference between *Off* and *Freeze* for regenerations?

   E. Use the *Layer* tab of the *Layer & Linetype Properties* dialog box and **PICK** the *Select All* option (right-click for menu). *Freeze* all layers. Which layer will not *Freeze*? Notice the light bulb icon indicates the layers are still *On* although the layers do not appear in the drawing (*Freeze* overrides *On*). Use the same procedure to select all layers and *Thaw* them.

F.   Use the *Layer* tab of the *Layer & Linetype Properties* dialog box again and drop down the *Show* list. Are there any layers in the drawing that are unused (select *All unused*)? Use the *Show* list again and list *All* layers.

G.   Now we want to *Freeze* all the *Yellow* layers. Using the *Layer* tab of the *Layer & Linetype Properties* dialog box again, sort the list by color. **PICK** the first layer name in the list of the four yellow ones. Now hold down the **Shift** key and **PICK** the last name in the list. All four layers should be selected. *Freeze* all four by selecting any one of the sun/snowflake icons. Select *OK* to return to the drawing.

H.   Assuming you wanted to work only with the layers beginning with AR, you can set the filter to display only those layers. First, select the *Set Filter dialog* in the *Show* list of the *Layer* tab of the *Layer & Linetype Properties* dialog box. In the *Set Layer Filters* dialog box, enter **AR*** in the *Layer Names* edit box and make sure the checkbox at the bottom is set to apply the filter to the Object Properties toolbar. Select *OK*, then examine the list in the *Layer* tab. Now select *OK* in the *Layer & Linetype Properties* dialog box to return to the drawing. Examine the **Layer Control** drop-down list. Does it display only layer names beginning with AR? Return to the *Layer* tab and select *All* in the *Show* list. Select *OK* and examine the **Layer Control** drop-down list again. It should still have the filter applied even though the *Layer* tab shows *All* layers. Use the *Set Layer Filters* dialog box again and enter * in the *Layer Names* edit box to show all layer names in both lists.

I.   Ensure all layers are *Thawed*. Use the *Layer* tab and sort the list alphabetically by *Name*. *Freeze* all layers beginning with **A** and **B**. Select *OK* and then use the drop-down list to display the names. Assuming you were working only with the thawed layers, it would be convenient to display <u>only</u> those names in the drop-down list. Use the *Set Layer Filter* dialog box to set a filter for *Thawed* layers. Ensure the checkbox at the bottom is checked. Return to the drawing and examine the drop-down list. Do only names of *Thawed* layers appear?

J.   Next, you will "move" objects from one layer to another using the **Object Properties** toolbar. Make layer **0** the current layer by using the drop-down list and selecting the name. Use the *Layer* tab and right-click to *Select All* names. **PICK** one icon to *Freeze* all layers (except the current layer). Then select *Clear All* and *Thaw* layer **Gennote**. Select *OK* and return to the drawing. Use a Crossing Window to select all of the objects in the drawing (the small blue grips may appear). The **Object Properties** toolbar should indicate the layer on which the objects reside. Next, list the layers using the drop-down list and **PICK** layer **0**. Wait several seconds. Press **Esc** twice when the drawing regenerates. The objects should be red and on layer 0.

K.   Let's assume you want to draw some additional objects on the same layer as the text and on the dimensions layer. First, *Zoom* in to the upper half of the floor plan on the right. Now select the *Make Object's Layer Current* icon button (on the far-right of the Object Properties toolbar). At the "Select object whose layer will become current" prompt, select a text object (cyan color). The **Gennote** layer name should appear in the Layer Control box as the current layer. Draw a *Line* and it should be on layer Gennote and cyan in color. Next, use *Make Object's Layer Current* again and select a dimension object. The **Dimensions** layer should appear as the current layer. Draw another *Line*. It should be on layer Dimensions and yellow in color. Keep in mind that this method of setting a layer current works especially well when you do not know the layer names or know what objects are on which layers.

L.   Do not save changes to the drawing.

2. *Layer* and *Linetype* tabs of the *Layer &*
   Linetype Properties dialog box

   A. Begin a *New* drawing and use *Save*
   to assign the name **CH12EX2**. Set
   Limits to **11** x **8.5**. Use the *Linetype*
   tab of the *Layer & Linetype* dialog
   box to list the loaded linetypes. Are
   any linetypes already loaded? Right-
   click to *Select All* linetypes then
   PICK the *Load* button. Close the
   dialog box.

   B. Use the *Layer* tab to create 3 new
   layers named **OBJ, HID,** and **CEN**.
   Assign the following colors and linetypes to the layers by clicking on the
   Color and *Linetype* columns:

OBJ	red	continuous
HID	yellow	hidden
CEN	green	center

Figure 12-24

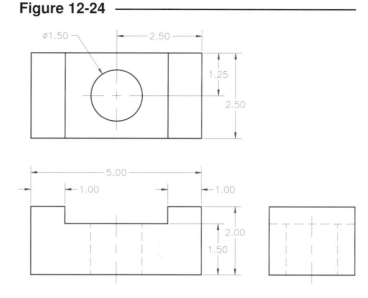

   C. Make the **OBJ** layer *Current* and PICK the *OK* tile. Verify the current layer by looking at the
   Layer Control drop-down list, then draw the visible object *Lines* and the *Circle* only (not the
   dimensions) as shown in Figure 12-24.

   D. When you are finished drawing the visible object lines, create the necessary hidden lines by
   making layer **HID** the *Current* layer, then drawing *Lines*. Notice that you only specify the
   Line endpoints as usual and AutoCAD creates the dashes.

   E. Next, create the centerlines for the
   holes by making layer **CEN** the
   Current layer and drawing *Lines*.
   Make sure the centerlines extend
   <u>slightly past</u> the *Circle* and past the
   horizontal *Lines* defining the hole.
   Save the drawing.

   F. Now create a *New* layer named
   BORDER and draw a border and
   title block of your design on that
   layer. The final drawing should
   appear like that in Figure 12-25.

   G. Open the *Layer & Linetype*
   Properties dialog box again and
   select the *Linetype* tab. Right-click to
   Select All, then select the *Delete*
   button. All unused linetypes should
   be removed from the drawing.

   H. *Save* the drawing.

Figure 12-25

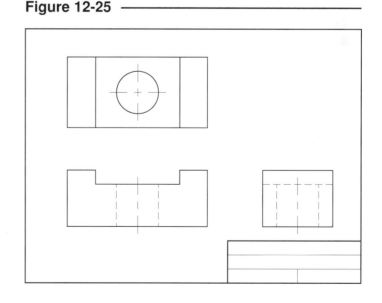

3. **LTSCALE, *Change Properties*   **Figure 12-26**
    dialog box, *Matchprop*

    A. Begin a *New* drawing and
        use *Save* to assign the
        name **CH12EX3**. Create
        the same four *New* layers
        that you made for the pre-
        vious exercise (**OBJ**, **HID**,
        CEN, and **BORDER**) and
        assign the same linetypes
        and colors. Set the *Limits*
        equal to a C size sheet,
        22 x 17.

    B. Create the part shown in
        Figure 12-26. Draw on the
        appropriate layers to
        achieve the desired line-
        types.

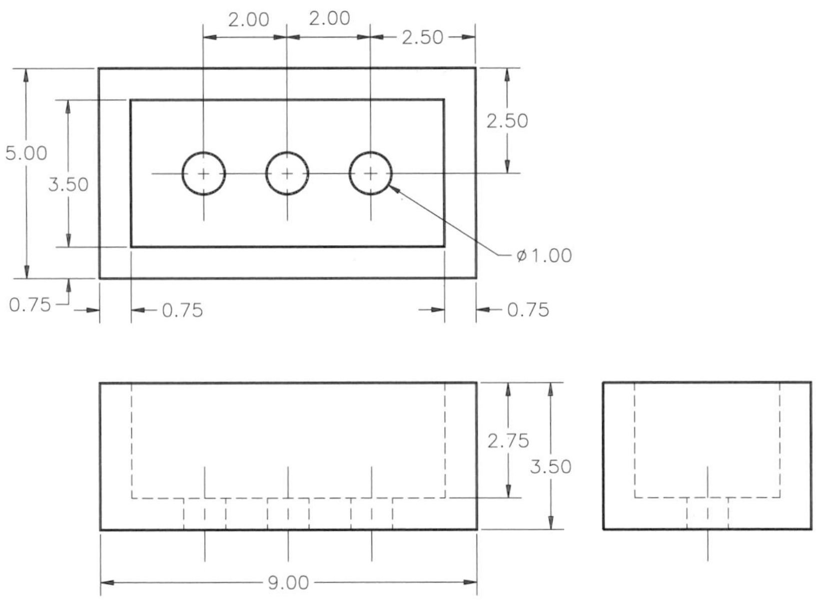

C. Notice that the *Hidden* and *Center* linetype dashes are very small. Use **LTSCALE** to adjust the scale of the non-continuous lines. Since you changed the *Limits* by a factor of slightly less than 2, try using **2** as the **LTSCALE** factor. Notice that all the lines are affected (globally) and that the new *LTSCALE* is retroactive for existing lines as well as for new lines. The dashes appear too long, however, because of the small hidden line segments, so *LTSCALE* should be adjusted. Remember that you cannot control where the multiple short centerline dashes appear for one line segment. Try to reach a compromise between the hidden and center line dashes.

D. The six short vertical hidden lines in the front view and two in the side view should be adjusted to a smaller linetype scale. Invoke the *Change Properties* dialog box by selecting *Properties* from the *Modify* pull-down menu or selecting the *Properties* icon button. Select only the two short lines in the side view. Use the dialog box to retroactively adjust the two individual objects' *Linetype Scale* to **0.6000**.

E. When the new *Linetype Scale* (actually, the *CELTSCALE*) for the two selected lines is adjusted correctly, use *Matchprop* to "paint" the *Linetype Scale* to the several other short lines in the front view. Select **Match Properties** from the *Modify* pull-down menu or select the "paint brush" icon. When prompted for the "Source Object," select one of the two short vertical lines in the side view. When prompted for the "Destination Object(s)," select each of the short lines in the front view. This action should "paint" the *Linetype Scale* to the lines in the front view.

F. When you have the desired linetype scales, *Exit* AutoCAD and *Save Changes*. Keep in mind that the drawing size may appear differently on the screen than it will on a print or plot. Both the plot scale and the paper size should be considered when you set *LTSCALE*. This topic will be discussed further in Chapter 13, Advanced Drawing Setup.

13

ADVANCED DRAWING SETUP

Chapter Objectives

After completing this chapter you should:

1. know the steps for setting up a drawing;

2. be able to determine an appropriate *Limits* setting for the drawing;

3. be able to calculate and apply the "drawing scale factor";

4. know the benefits and shortcomings of the *Quick Setup* and *Advanced Setup Wizards*;

5. be able to access existing and create new template drawings;

6. know what setup steps can be considered for creating template drawings.

CONCEPTS

When you begin a drawing, there are several steps that are typically performed in preparation for creating geometry, such as setting *Units, Limits,* and creating *Layers* with *linetypes* and *colors.* Some of these basic concepts were discussed in Chapters 6 and 12. This chapter discusses setting *Limits* for correct plotting as well as other procedures, such as layer creation and variables settings, that help prepare a drawing for geometry creation.

To correctly set up a drawing for printing or plotting to scale, <u>any two</u> of the following three variables must be known. The third can be determined from the other two.

    Drawing *Limits*
    Print or plot sheet (paper) size
    Print or plot scale

Initially, it would appear that using the automated setup "Wizards" available in the *Start Up* and *Create New Drawing* dialog boxes would be the fastest and easiest method for setting up a drawing. Surprisingly (unlike most other features in AutoCAD Release 14 that work so well), neither the *Quick Setup Wizard* nor the *Advanced Setup Wizard* correctly sets up a drawing for plotting to a standard scale. The reason is because the "wizards" consider only *Limits* and do not consider either sheet size or print/plot scale in their calculations. Therefore, you must have an understanding of the individual commands needed to correctly set up a drawing to scale. In addition, you should know when the setup wizards may and may not be of help to you.

Rather than performing the steps for drawing setup each time you begin, you can use "template drawings." Template drawings have many of the setup steps performed but contain no geometry. AutoCAD provides several template drawings and you can create your own template drawings. A typical engineering, architectural, design, or construction office generally produces drawings that are similar in format and can benefit from creation and use of individualized template drawings. The drawing similarities may be subject of the drawing (geometry), plot scale, sheet size, layering schemes, dimensioning styles, and/or text styles. Template drawing creation and use are also discussed in this chapter.

STEPS FOR DRAWING SETUP

Assuming that you have in mind the general dimensions and proportions of the drawing you want to create, and the drawing will involve using layers, dimensions, and text, the following steps are suggested for setting up a drawing:

1. Determine and set the *Units* that are to be used.
2. Determine and set the drawing *Limits;* then *Zoom All.*
3. Set an appropriate *Snap* value to be used if helpful.
4. Set an appropriate *Grid* value.
5. Change the *LTSCALE* value based on the new *Limits.*
6. Create the desired *Layers* and assign appropriate *linetype* and *color* settings.
7. Create desired *Text Styles* (optional, discussed in Chapter 18).
8. Create desired *Dimension Styles* (optional, discussed in Chapter 29).
9. Create or *Insert* the desired title block and border (optional).

Each of the steps for drawing setup is explained in detail here.

1. Set *Units*

This task is accomplished by using the *Units* command, the *Units Control* dialog box (*Ddunits*), or the *Quick Setup* or *Advanced Setup Wizard.* Set the linear units and precision desired. Set angular units and precision if needed. (See Chapter 6 for details on the *Units* command and see Setup Wizards in this chapter for setting *Units* using the wizards.)

2. Set *Limits*

Before beginning to create an AutoCAD drawing, determine the size of the drawing area needed for the intended geometry. Using the actual *Units,* appropriate *Limits* should be set in order to draw the object or geometry to the <u>real-world size</u>. *Limits* are set with the *Limits* command by specifying the lower-left and upper-right corners of the drawing area. Always *Zoom All* after changing *Limits*. *Limits* can also be set using the *Quick Setup* or *Advanced Setup Wizard*. (See Chapter 6 for details on the *Limits* command and see Setup Wizards in this chapter.)

If you are planning to plot the drawing to scale, *Limits* should be set to a proportion of the sheet size you plan to plot on. For example, if the sheet size is 11" x 8.5", set *Limits* to 11 x 8.5 if you want to plot full size (1"=1"). Setting *Limits* to 22 x 17 (2 times 11 x 8.5) provides 2 times the drawing area and allows plotting at 1/2 size (1/2"=1") on the 11" x 8.5" sheet. Simply stated, set *Limits* to a <u>proportion</u> of the paper size.

Setting *Limits* to the <u>paper size</u> allows plotting at 1=1 scale. Setting *Limits* to a <u>proportion</u> of the sheet size allows plotting at the <u>reciprocal</u> of that proportion. For example, setting *Limits* to 2 times an 11" x 8.5" sheet allows you to plot 1/2 size on that sheet. Or setting *Limits* to 4 times an 11" x 8.5" sheet allows you to plot 1/4 size on that sheet. (Standard paper sizes are given in Chapter 14.)

Drawing Scale Factor

<u>The proportion of the *Limits* to the intended print or plot sheet size is the "drawing scale factor."</u> This factor can be used as a general scale factor for other size-related drawing variables such as *LTSCALE, DIMSCALE* (dimensioning scale), and *Hatch* pattern scales. (Even if you plan to set up a drawing for plotting with Paper Space, the drawing scale factor should be calculated to determine system variable values and the *ZoomXP* scale factor.)

Most size-related AutoCAD drawing variables are set to 1 by default. This means that variables (such as *LTSCALE*) that control

Figure 13-1

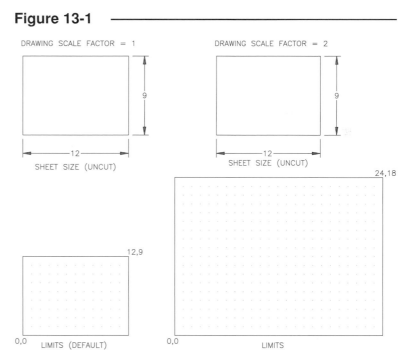

sizing and spacing of objects are set appropriately for creating a drawing plotted full size (1=1). Therefore, when you set *Limits* to the sheet size and print or plot at 1=1, the sizing and spacing of linetypes and other variable-controlled objects are correct. When *Limits* are changed to some proportion of the sheet size, the size-related variables should also be changed proportionally. For example, if you intend to plot on a 12 x 9 sheet and the default *Limits* (12 x 9) are changed by a factor of 2 (to 24 x 18), then 2 becomes the drawing scale factor. Then, as a general rule, the values of variables such as *LTSCALE, DIMSCALE,* and other scales should be multiplied by a factor of 2 (Fig. 13-1). When you then print the 24 x 18 area onto a 12" x 9" sheet, the plot scale would be 1/2, and all sized features would appear correct.

Limits should be set to the paper size or to a proportion of the paper size used for plotting or printing. In many cases, "cut" paper sizes are used based on the 11" x 8.5" module (as opposed to the 12" x 9" module called "uncut sizes"). Assume you plan to print on an 11" x 8.5" sheet. Setting *Limits* to the sheet size provides plotting or printing at 1=1 scale. Changing the *Limits* to a proportion of the sheet size <u>by some multiplier</u> makes that value the drawing scale factor. The <u>reciprocal</u> of the drawing scale factor is the scale for plotting on that sheet (Fig. 13-2).

Since the drawing scale factor (DSF) is the proportion of *Limits* to the sheet size, you can use this formula:

$$DSF = \frac{Limits}{Sheet\ size}$$

Because plot/print scale is the reciprocal of the drawing scale factor:

$$Plot\ scale = \frac{1}{DSF}$$

Figure 13-2

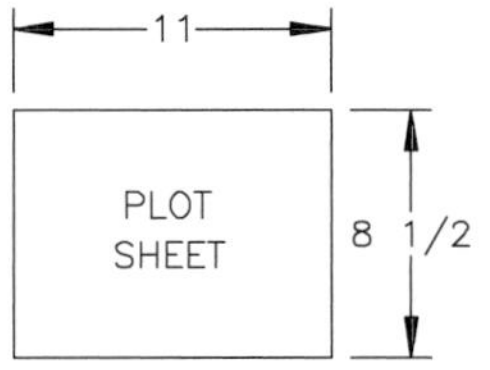

(CUT PAPER SIZE)

LIMITS	SCALE FACTOR	PLOT SCALE	
11 x 8.5	1	1=1	(FULL SIZE)
22 x 17	2	1=2	(HALF SIZE)
44 x 34	4	1=4	(1/4 SIZE)
110 x 85	10	1=10	(1/10 SIZE)

The term "drawing scale factor" is not a variable or command that can be found in AutoCAD or its official documentation. The setup wizards in Release 14 do not calculate or use a drawing scale factor based on sheet size. This concept has been developed by AutoCAD users to set system variables appropriately for printing and plotting drawings to standard scales.

3. Set *Snap*

Use the *Snap* command or *Drawing Aids* dialog box to set an appropriate *Snap* value if *Snap* is useful for your application. The value of *Snap* is dependent on the <u>interactive</u> drawing accuracy that is desired.

In some drawings where the *Limits* are relatively small and the dimensional increments are relatively large, *Snap* can be very useful. However, in other drawings with large *Limits* and small drawing increment lengths or with complex geometry of irregular interval lengths, *Snap* may not be useful and can be turned *Off*. (See Chapter 6 for details on the *Snap* command.)

The accuracy of the drawing and the size of the *Limits* should be considered when setting the *Snap* value. On one hand, to achieve accuracy and detail, you want the *Snap* to be as small of a dimensional increment as would be commonly used in the drawing. On the other hand, depending on the size of the *Limits*, the *Snap* value should be large enough to make interactive selection (PICKing) fast and easy. As a starting point for determining an appropriate *Snap* value, the default *Snap* value can be multiplied by the drawing scale factor.

If you use the *Quick Setup* or *Advanced Setup Wizard*, *Snap* is calculated for you. Because the wizards do not consider sheet size or scale, the resulting value is inappropriate for most applications. Examine the value and enter a new value based on the drawing scale factor.

4. Set *Grid*

The *Grid* value setting is usually set equal to or proportionally larger than that of the *Snap* value. *Grid* should be set to a proportion that is <u>easily visible</u> and to some value representing a regular increment (such as .5, 1, 2, 5, 10, or 20). Setting the value to a proportion of *Snap* gives visual indication of the *Snap* increment. For example, a *Grid* value of **1X, 2X,** or **5X** would give visual display of every 1, 2, or 5 *Snap* increments, respectively. If you are not using *Snap* because only extremely small or irregular interval lengths are needed, you may want to turn *Grid* off. (See Chapter 6 for details on the *Grid* command.)

If you use the *Quick Setup* or *Advanced Setup Wizard, Grid* is also calculated for you but may be inappropriately set. Examine the value and enter a new value based on the drawing scale factor.

5. **Set the *LTSCALE***

A change in *Limits* (then *Zoom All*) affects the display of non-continuous (hidden, dashed, dotted, etc.) lines. The *LTSCALE* variable controls the spacing of non-continuous lines. The drawing scale factor can be used to determine the *LTSCALE* setting. For example, if the default *Limits* (based on sheet size) have been changed by a factor of 2, the drawing scale factor=2; so set the *LTSCALE* to 2. In other words, if the DSF=2 and plot scale=1/2, you would want to double the linetype dashes to appear the correct size in the plot, so *LTSCALE*=2. If you prefer a *LTSCALE* setting of other than 1 for a drawing in the default *Limits,* multiply that value by the drawing scale factor.

American National Standards Institute (ANSI) requires that <u>hidden lines be plotted with 1/8"</u> <u>dashes</u>. An AutoCAD drawing plotted full size (1 unit=1") with the default *LTSCALE* value of 1 plots the *Hidden* linetype with 1/4" dashes. Therefore, an <u>*LTSCALE* of .5 </u>is more appropriate for a drawing plotted full size. (Optionally, you can use the *Hidden2, Center2,* and other *2 linetypes with dash lengths of .5 times normal, then an *LTSCALE* of 1 is correct.) Multiply your *LTSCALE* value by the drawing scale factor when plotting to scale. If individual lines need to be adjusted for linetype scale, use *Modify Properties* (*Ddmodify* or *Ddchprop*) when the drawing is nearly complete. (See Chapter 12 for details on the *LTSCALE* and *CELTSCALE* variables.)

6. **Create *Layers*; Assign *Linetypes* and *Colors***

Using the *Layer* command (*Layer* tab of the *Layer & Linetype Properties* dialog box), create the layers that you anticipate needing. Multiple layers can be created with the *New* option. You can type in several new names separated by commas. Assign a descriptive name for each layer, indicating its type of geometry, part name, or function. Include a *linetype* designator in the layer name if appropriate; for example, PART1-H and PART1-V indicate hidden and visible line layers for PART1.

Once the *Layers* have been created, assign a *Color* and *Linetype* to each layer. *Colors* can be used to give visual relationships among parts of the drawing. Geometry that is intended to be plotted with different pens (pen size or color) or printed with different line widths should be drawn in different screen colors. Use the *Select Linetype* dialog box to load the desired linetypes. You can load only linetypes that you anticipate using or load all linetypes and *Delete* the unused ones when the drawing is completed. (See Chapter 12 for details on creating *Layers, Linetypes,* and *Colors*.)

7. **Create Text Styles**

AutoCAD has only one text style as part of the standard template drawings (ACAD.DWT and ACADISO.DWT) and one or two text styles for most other template drawings. If you desire other *Text Styles,* they are created using the *Style* command or the *Text Style…* option from the *Data* pull-down menu. (See Chapter 18, Creating and Editing Text.) If you desire engineering standard text, create a *Text Style* using the ROMANS.SHX or .TTF font file.

8. **Create Dimension Styles**

If you plan to dimension your drawing, Dimension Styles can be created at this point; however, they are generally created during the dimensioning process. Dimension Styles are names given to groups of dimension variable settings. (See Chapter 29 for information on creating Dimension Styles.)

Although you do not have to create Dimension Styles until you are ready to dimension the geometry, it is helpful to create Dimension Styles as part of a template drawing. If you produce similar drawings repeatedly, your dimensioning techniques are probably similar. Much time can be saved by using a template drawing with previously created Dimension Styles. Several of the AutoCAD-supplied template drawings have prepared Dimension Styles.

9. Create a Title Block and Border

For 2D drawings, it is helpful to insert a title block and border early in the drawing process. This action gives a visual drawing boundary as well as reserves the space needed by the title block.

Since *Limits* are already set to facilitate drawing full size in the actual *Units,* creating a title block and border which will appear on the final plot in the appropriate size is not difficult. A simple method is to use the drawing scale factor to determine the title block size. Multiply the actual size of the title block and border by the drawing scale factor to determine their sizes for the drawing.

One common method is to use the *Insert* command to insert a title block and border as a *Block.* If the *Block* is the actual title block and border size needed for the plot sheet, simply use the drawing scale factor as the *X* and *Y scale factor* during the *Insert* command. (See Chapters 21 and 22 for information on *Block* creation and insertion.)

If you are preparing template drawings, a title block and border can be included as part of a template drawing. Most of the AutoCAD-supplied template drawings have a title block and border included, but in each case the title block and border is in Paper Space. Using Paper Space has many advantages and some disadvantages but is recommended for use only when you have had more experience with AutoCAD. See Chapter 19 for introductory information on Paper Space.

QUICK SETUP WIZARD AND *ADVANCED SETUP* WIZARD

Since the *Quick Setup* and *Advanced Setup Wizards* that are available in the *Start Up* and *Create New Drawing* dialog boxes do not allow input for either the plot/print sheet size or the intended plot scale, <u>they cannot be used effectively for setting up a drawing correctly for plotting or printing to scale.</u> The "Wizards" can be useful in some cases, however, such as setting up a drawing to print or plot on an A size sheet or for some of the initial steps to set up a metric drawing. This section closely examines the *Quick Setup Wizard* and the *Advanced Setup Wizard* to determine their shortcomings and usefulness. See Chapter 6 for coverage of the *Start from Scratch* option in the *Start Up* and *Create New Drawing* dialog boxes. Using template drawings is discussed later in this chapter.

Quick Setup Wizard

Select the *Use a Wizard* button in the *Start Up* or *Create New Drawing* dialog box to present a choice of *Quick Setup* or *Advanced Setup* (Fig. 13-3). The *Quick Setup Wizard* automates the first four steps listed under STEPS FOR DRAWING SETUP on the chapter's first page. Those functions, simply stated, are:

1. *Units*
2. *Limits*
3. *Snap*
4. *Grid*

Figure 13-3 ──────────

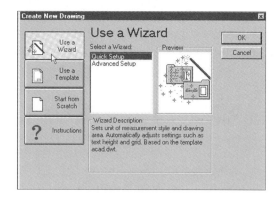

The Wizard prompts you for *Units* and *Area* (*Limits*) and sets the *Grid* and *Snap* values automatically, as well as setting many other size-related system variables, based on your input for the *Limits* values. The *Quick Setup Wizard* uses a multiplier to set the system variables similar to the concept of a drawing scale factor, except the multiplier is based only on the proportion of *Limits* to the default 12 x 9 *Limits* and not to the intended plot sheet size; therefore, the resulting setup is useful only for plotting to scale on an A size (11" x 8.5" or 12" x 9") sheet.

Choosing the *Quick Setup Wizard* invokes the *Quick Setup* dialog box (Fig. 13-4). There are two steps. To proceed through the steps, you can select the *Next* button or select each tab in succession.

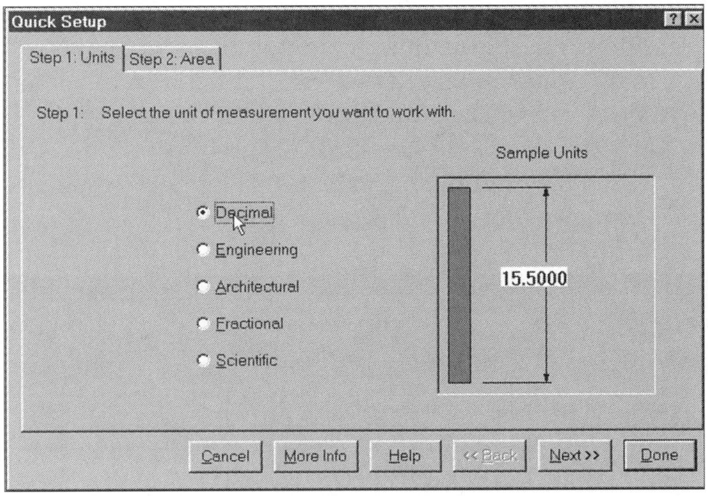

Step 1: Units
Press the desired radio button to display the units you want to use for the drawing. The options are:

Decimal	Use generic decimal units with a precision of 0.0000.
Engineering	Use feet and decimal inches with a precision of 0.0000.
Architectural	Use feet and fractional inches with a precision of 1/16 inch.
Fractional	Use generic fractional units with a precision of 1/16 units.
Scientific	Use generic decimal units showing a precision of 0.0000.

Figure 13-4

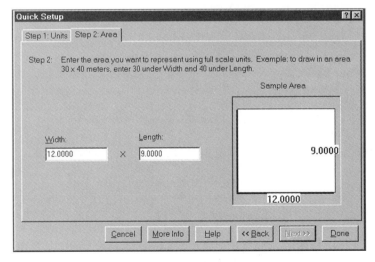

Use *Architectural* or *Engineering* units if you want to specify coordinate input using feet values with the ' (apostrophe) symbol. If you want to set additional parameters for units such as precision or system of angular measurement, use the *Units* or *Ddunits* command. Keep in mind the setting you select in this step changes only the display of units in the coordinate display area of the Status Line (*Coords*) and in dialog boxes but not necessarily for the dimension text format. Select the *Next* button or the *Area* tab after specifying *Units*.

Step 2: Area
Enter two values that constitute the X and Y measurements of the area you want to work in (Fig. 13-5). These values set the *Limits* of the drawing. The first edit box labeled *Width* specifies the X value for *Limits*. The X value is usually the longer of the two measurements for your drawing area and represents the distance across the screen in the X direction or along the long axis of a sheet of paper. The second edit box labeled *Length* specifies the Y value for *Limits* (the Y distance of the drawing area). (Beware: The terms "width" and "length" are misleading since "length" is actually defined as the measurement of something along its greatest dimension. [*The American Heritage Dictionary, Office Edition*, Third Edition, 1994, Dell Publishing.] When setting AutoCAD *Limits*, the Y measurement is generally the shorter of the two measurements.) The two values together specify the upper-right corner of the drawing *Limits*. The Wizard automatically sets *Snap* and *Grid* values for you based on the values you specify for *Area* (*Limits*). The Wizard also calculates settings for most size-related system variables.

Figure 13-5

When you finish the two steps, *Quick Setup Wizard* automatically centers the drawing area (*Limits*) in the drawing editor. If you *Start from Scratch*, change *Limits*, then *Zoom All,* the left edge defined by *Limits* is always against the left edge of the screen.

Table of *Quick Setup Wizard* Settings

The table below gives a few examples (see Table of System Variables Affected by Wizards for full list) of the system variable settings (columns 3, 4, and 5) that result when specific *Area* (*Limits*) values are specified in the *Quick Setup Wizard*. (See Tables of *Limits* Settings, Chapter 14, for a complete list of standard plot scales, paper sizes, and appropriate *Limits* settings.) The area sizes that are used for these examples are proportional to the standard U.S. architectural (based on the 12 x 9 module) and engineering (based on the 11 x 8.5 module) paper sizes. The last column lists the plot scales that are appropriate for the variable settings *Quick Setup Wizard* generates.

Area (*Limits*)	Paper Proportion	*SNAP, GRID*	*LTSCALE, DIMSCALE*	*TEXTSIZE*	Plot Scale for "A" Size Sheet
12 x 9	1 x A architectural	.5	1	.2	1=1
18 x 12	1 x B architectural	.7	1.3333	.2667	3/4=1
24 x 18	1 x C architectural	1	2	.4	1/2=1
36 x 24	1 x D architectural	1	2.667	.5333	3/8=1
48 x 36	1 x E architectural	2	4	.8	1/4=1
12' x 9'	12 x A architectural	6"	12 (1')	2 3/8" (.2')	1"=1'
24' x 18'	24 x A architectural	12 (1')	24 (2')	4 13/16"(.4')	1/2"=1'
48' x 36'	48 x A architectural	24 (2')	48 (4')	9 5/8" (.8')	1/4"=1'
11 x 8.5	1 x A engineering	.5	.9444	.1889	1=1
17 x 11	1 x B engineering	.6	1.2222	.2444	3/4=1
22 x 17	1 x C engineering	.9	1.8889	.3778	1/2=1
34 x 22	1 x D engineering	1	2.4444	.4889	3/8=1
44 x 34	1 x E engineering	2	3.7778	.7556	1/4=1

Note that the *Quick Setup Wizard* generates the *DIMSCALE, LTSCALE,* and *TEXTSIZE* system variable settings (as well as other system variables listed in the Table of System Variables Affected by Wizards) based on the formula

$$N=D(Y/9)$$

where N is the new system variable setting,
       D is the system variable default value,
       Y is the *Area* (*Limits*) Y value, and
       9 is the default *Area* (*Limits*) Y value.

The *Snap* and *Grid* values are rounded to the nearest 1/10 unit when N is under 1 unit and rounded to the nearest integer when N is over 1 unit.

Thus, the *Quick Setup Wizard* calculates appropriate settings (except *Snap* and *Grid* for some cases) for plotting to scale on A size sheets only. This is because the *Quick Setup Wizard* does not consider plot sheet size or plot scale in its calculation. Also notice the *Snap* and *Grid* values for 18 x 12, 17 x 11, and 22 x 17 are inappropriate and should be changed.

In summary, you can use the *Quick Setup Wizard* to automate setting up a drawing to plot to a standard scale on A size sheets. In some cases, you may want to change the *Snap* and *Grid* settings as well as other system variables.

Advanced Setup Wizard

The *Advanced Setup Wizard* performs the same tasks as the *Quick Setup Wizard* with the addition of allowing you to select units precision and other options (normally available in the *Units* dialog box), insert a title block, and enable Paper Space. Selecting *Advanced Setup* produces the dialog box series shown in Figures 13-6 through 13-12. Seven tabs, or "Steps," are involved in the series. To proceed through the steps, you can select the *Next* button or select each tab in succession.

The following list indicates the STEPS FOR DRAWING SETUP on the chapter's first page and the related tabs or "Steps" in the *Advanced Setup Wizard*:

1.	*Units*	*Step 1: Units*
		Step 2: Angle
		Step 3: Angle Measure
		Step 4: Angle Direction
2.	*Limits*	*Step 5: Area*
3.	*Snap*	(automatic)
4.	*Grid*	(automatic)
5.	*Insert Title Block*	*Step 6: Title Block*
		Step 7: Layout

Similar to the function of the *Quick Setup Wizard,* several system variables are changed from the default settings to other values based on your input for the *Limits* values.

Step 1: Units
You can select the units of measurement for the drawing as well as the unit's *Precision*. These are the same options available in the *Units* dialog box (see *Units,* Chapter 6). This is similar to *Step 1* in the *Quick Setup Wizard* but with the addition of *Precision*.

Similar to using the *Units* or *Ddunits* command, your choices in this and the next three dialog boxes determine the display of units for the coordinate display area of the Status line (*Coords*) and in dialog boxes. If you want to use feet units for coordinate input, select *Architectural* or *Engineering*.

Figure 13-6 ————————

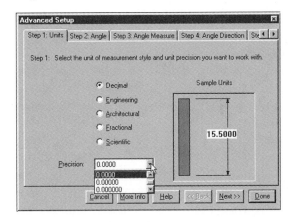

Step 2: Angle
This tab provides for your input of the desired system of angular measurement. Select the drop-down list to select the angular *Precision*.

Figure 13-7 ————————

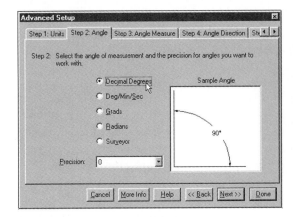

R14

Step 3: *Angle Measure*

This step sets the direction for angle 0. East (X positive) is the AutoCAD default. This has the same function as selecting the *Angle 0 Direction* in the *Units Control* dialog box.

Figure 13-8

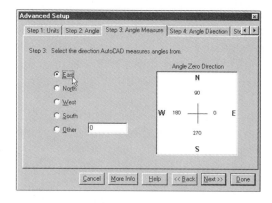

Step 4: *Angle Direction*

Select *Clockwise* if you want to change the AutoCAD default setting for <u>measuring</u> angles. This setting (identical to the *Units Control* dialog box option) affects the direction of positive and negative angles in commands such as *Rotate*, *Array Polar* and dimension commands that measure angles but does not affect the direction *Arcs* are drawn (always counter-clockwise).

Figure 13-9

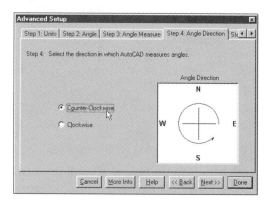

Step 5: *Area*

Enter values to define the upper-right corner for the *Limits* of the drawing. The *Width* refers to the X *Limits* component and the *Length* refers to the Y component. Generally, the *Width* edit box contains the larger of the two values unless you want to set up a vertically oriented drawing area. Even if you plan to insert a title block (*Step 6*) into Model Space (*Step 7*), do not skip this step just because *Limits* will be reset by the insertion. The values you input in this step are used to calculate other system variable settings (see Table of System Variables Affected by Wizards).

If you plan to print or plot to a standard scale, your input for *Area* should be based on the intended plot scale and sheet size. See the Tables of *Limits* Settings in Chapter 14 for appropriate values to use.

Figure 13-10

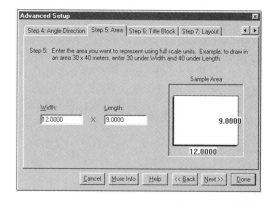

Step 6: *Title Block*

You can select from several AutoCAD-supplied title blocks listed next or select from custom (your own) title blocks by using the *Add* button to select a custom title block in DWG format. The files you select are added to the bottom of the drop-down list. The selected title block is inserted into the drawing as a *Block* object. A new layer is created called TITLE_BLOCK onto which the block is inserted. If you insert the title block into Model Space (*Step 7*), *Limits* are reset to match the title block size.

Figure 13-11

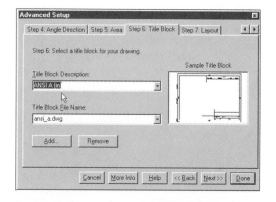

The following list gives a description of the AutoCAD-supplied title blocks. The title blocks match the standard paper sheet sizes.

Title Block	Intended Sheet Size	Sheet Measurement
ANSI A	A size eng. sheet [1]	11″ x 8.5″
ANSI B	B size eng. sheet [1]	17″ x 11″
ANSI C	C size eng. sheet [1]	22″ x 17″
ANSI D	D size eng. sheet [1]	34″ x 22″
ANSI E	E size eng. sheet [1]	44″ x 34″
ANSI V	A size eng. sheet [1]	11″ x 8.5″ vertically oriented
ARCH/ENG	D size arch. sheet	36″ x 24″
DIN A0	A0 size metric sheet [2]	1189 mm x 841 mm
DIN A1	A1 size metric sheet [2]	841 mm x 594 mm
DIN A2	A2 size metric sheet [2]	594 mm x 420 mm
DIN A3	A3 size metric sheet [2]	420 mm x 297 mm
DIN A4	A4 size metric sheet [2]	297 mm x 210 mm
GENERIC D	D size arch. sheet	36″ x 24″
ISO A0	A0 size metric sheet [3]	1188 mm x 840 mm
ISO A1	A1 size metric sheet [3]	840 mm x 594 mm
ISO A2	A2 size metric sheet [3]	594 mm x 420 mm
ISO A3	A3 size metric sheet [3]	420 mm x 297 mm
ISO A4	A4 size metric sheet [3]	297 mm x 210 mm
JIS A0	A0 size metric sheet [4]	1189 mm x 841 mm
JIS A1	A1 size metric sheet [4]	841 mm x 594 mm
JIS A2	A2 size metric sheet [4]	594 mm x 420 mm
JIS A3	A3 size metric sheet [4]	420 mm x 297 mm
JIS A4 (landscape)	A4 size metric sheet [4]	297 mm x 210 mm
JIS A4 (portrait)	A4 size metric sheet [4]	297 mm x 210 mm vertically oriented

[1] American National Standards Institute (ANSI)
[2] Deutsches Institut für Normung (DIN)
[3] International Organization for Standardization (ISO)
[4] Japanese Industrial Standard (JIS)

Inserting the AutoCAD-supplied title blocks into Model Space (*Step 7*) provides for plotting or printing a drawing at 1=1 scale. If you insert the title blocks into Model Space and intend to plot at other than 1=1, you will have to reset *Limits* and use *Scale* to rescale the title block to the appropriate size. If you insert the title blocks into Paper Space in *Step 7*, this procedure is not necessary.

Beginning a drawing with a predefined title block can also be accomplished by using a template drawing (including a title block). See Using and Creating Template Drawings.

Step 7: Layout

In this step you determine if you want to use Paper Space or not. Selecting *No* (not the default in this tab) allows you to work in Model Space, which is the space used for creating geometry and is generally the choice for simple 2D drawings. Select *Yes* if you want to use Paper Space. You should use Paper Space if you need <u>several views</u> (viewports) displaying one drawing or several drawings. The need to (1) plot or print several views of one drawing, (2) create a plot or print of several drawings, or (3) lay out different views of a 3D object are three good reasons for using Paper Space. Selecting the *Yes* option, however, creates only one Paper Space viewport. The viewport is automatically created on layer VIEWPORT. If you have selected a title block (*Step 6*), the title block is inserted into Paper Space.

Figure 13-12

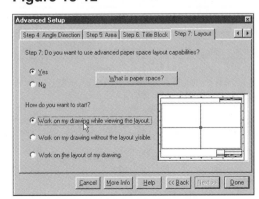

Below is a translation of the three choices for *How do you want to start?*

Option	Meaning
Work on my drawing while viewing the layout.	Tilemode=0, Model Space current
Work on my drawing without the layout visible.	Tilemode=1, Model Space current
Work on the layout of my drawing.	Tilemode=0, Paper Space current

See Chapters 19 and 33 for more information on Paper Space.

Table of System Variables Affected by Wizards

Your choice for *Area* (*Limits*) in both the *Quick Setup* and the *Advanced Setup Wizards* affects the following variables.

AFFECTED SYSTEM VARIABLE	DESCRIPTION
CHAMFERA	First *Chamfer* distance
CHAMFERB	Second *Chamfer* distance
CHAMFERC	*Chamfer* length
DIMSCALE	Overall dimensioning scale factor for size-related features
DIMSTYLE	Current dimensioning style
DONUTID	Default *Donut* inside diameter
DONUTOD	Default *Donut* outside diameter
FILLETRAD	Default *Fillet* radius
GRIDUNIT	*Grid* spacing for current viewport, X and Y
HPSCALE	Default *Hatch* pattern scale factor
HPSPACE	Default *Hatch* pattern line spacing for "U," user-defined simple patterns
LIMMAX	Upper-right drawing *Limits* for the current space
LTSCALE	Global linetype scale factor
OFFSETDIST	Default distance for *Offset*
SNAPUNIT	*Snap* spacing for current viewport, X and Y
TEXTSIZE	The default height for new *Text* objects drawn with the current text style
VIEWCTR	Center of view in current viewport
VIEWSIZE	Height of view in current viewport

All but one of these variables requires numeric input. The value used for each variable is calculated by the formula

$$N=D(Y/9)$$

where N is the new system variable setting,
D is the system variable default value,
Y is the *Area* (*Limits*) Y value, and
9 is the default *Area* (*Limits*) Y value.

The *Snap* and *Grid* values are rounded to the nearest 1/10 unit when N is under 1 unit and are rounded to the nearest integer when N is over 1 unit.

The *DIMSTYLE* variable is set to "STANDARD_WIZARDSCALED" to provide dimensions that are scaled correctly, based on your input for *Area*.

Template Drawings
As an alternative to using *Start from Scratch* to begin a new drawing or using a *Setup Wizard* to set up your drawings, you should consider using template drawings. Template drawings can contain all the features provided by the wizards but can also include many other settings and features.

USING AND CREATING TEMPLATE DRAWINGS

Instead of going through the steps for setup <u>each time</u> you begin a new drawing, you can create one or more template drawings or use one of the AutoCAD-supplied template drawings. AutoCAD Release 14 template drawings are saved as a DWT file format. A template drawing is one which has the initial setup steps (*Units, Limits, Layers, Linetypes, Colors,* etc.) completed and saved, but no geometry has yet been created. Template drawings are used as a <u>template</u> or <u>starting point</u> each time you begin a new drawing. AutoCAD actually makes a copy of the template you select to begin the drawing. The creation and use of template drawings can save hours of preparation.

Using Template Drawings

To use a template drawing, select the *Use a Template* option in the *Create New Drawing* dialog box (Fig. 13-13). This option allows you to select the desired template (DWT) drawing from a list including all templates in the TEMPLATE folder (usually in the C:\Program Files\AutoCAD R14\Template subdirectory). Any template drawings that you create using the *SaveAs* command (as described in the next section) appear in the list of available templates. Select the *More files...* option to browse other folders.

Figure 13-13

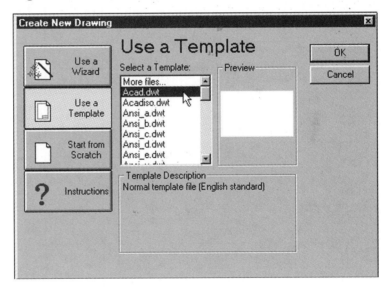

Several template drawings are supplied with AutoCAD Release 14 (see the selections in Fig. 13-13) and are described in the following table. You may notice similarities to the template drawing file names and the title block names that are available in the *Advanced Setup Wizard*. Generally, the template drawing files include the title blocks. In some cases, Dimension Styles have been created but other changes have not been made to the template drawings such as layer setups or loading linetypes.

R14

Table of AutoCAD-Supplied Template Drawing Settings

DWT File	Sheet Size Intended	Limits (Mod. Sp.)	Units, Precn.	Title Block	Paper Space	Layers	Line-types	Dimen. Styles
Acad	—	12, 9	.0000	no	no	Layer 0	no [1]	Standard
Acadiso	—	420, 297	.0000	no	no	Layer 0	no [1]	yes [9]
Ansi_a	11" x 8.5"	11, 8.5	.0000	yes	yes [1]	yes [2]	no [1]	Standard
Ansi_b	17" x 11"	17, 11	.0000	yes	yes [1]	yes [2]	no [1]	Standard
Ansi_c	22" x 17"	22, 17	.0000	yes	yes [1]	yes [2]	no [1]	Standard
Ansi_d	34" x 22"	34, 22	.0000	yes	yes [1]	yes [2]	no [1]	Standard
Ansi_e	44" x 34"	44, 34	.0000	yes	yes [1]	yes [3]	no [1]	Standard
Ansi_v	11" x 8.5" vert	8.5, 11	.0000	yes	yes [1]	yes [2]	no [1]	Standard
Archeng	36" x 24"	36, 24	.0000	yes	yes [1]	yes [3]	no [1]	Standard
Din_a0	1189 x 841 mm	1189, 841	.0000	yes	yes [1]	yes [4]	no [1]	yes [8]
Din_a1	841 x 594 mm	841, 594	.0000	yes	yes [1]	yes [4]	no [1]	yes [8]
Din_a2	594 x 420 mm	594, 420	.0000	yes	yes [1]	yes [4]	no [1]	yes [8]
Din_a3	420 x 297 mm	420, 297	.0000	yes	yes [1]	yes [5]	no [1]	yes [8]
Din_a4	210 x 297 vert.	210, 297	.0000	yes	yes [1]	yes [4]	no [1]	yes [8]
Gs24x36	36" x 24"	34.5, 23	.0000	yes	yes [1]	yes [2]	no [1]	Standard
Iso_a0	1188 x 840 mm	1189, 841	.0000	yes	yes [1]	yes [6]	no [1]	yes [9]
Iso_a1	840 x 594 mm	841, 594	.0000	yes	yes [1]	yes [6]	no [1]	yes [9]
Iso_a2	594 x 420 mm	594, 420	.0000	yes	yes [1]	yes [6]	no [1]	yes [9]
Iso_a3	420 x 297 mm	420, 297	.0000	yes	yes [1]	yes [6]	no [1]	yes [9]
Iso_a4	210 x 297 vert.	210, 297	.0000	yes	yes [1]	yes [6]	no [1]	yes [9]
Jis_a0	1189 x 841 mm	1189, 841	.0000	yes	yes [1]	yes [7]	no [1]	yes [10]
Jis_a1	841 x 594 mm	841, 594	.0000	yes	yes [1]	yes [7]	no [1]	yes [10]
Jis_a2	594 x 420 mm	594, 420	.0000	yes	yes [1]	yes [7]	no [1]	yes [10]
Jis_a3	420 x 297 mm	420, 297	.0000	yes	yes [1]	yes [7]	no [1]	yes [10]
Jis_a4l	297 x 210 mm	297, 210	.0000	yes	yes [1]	yes [7]	no [1]	yes [10]
Jis_a4r	210 x 297 vert.	210, 297	.0000	yes	yes [1]	yes [7]	no [1]	yes [10]

[1] Paper Space is enabled (TILEMODE=0) and title block is in Paper Space.
[2] Layers: 0, TB, and Title_block
[3] Layers: 0 and Title_block
[4] Layers: 0, Rahmen025, Rahmen035, Rahmen05, Rahmen07
[5] Layers: 0, Rahmen025, Rahmen05, Rahmen07
[6] Layers: 0, Frame025, Frame035, Frame050, Frame070, Frame100, Frame140, Frame200, TB
[7] Layers: 0, Defpoints, Object
[8] Dimension Styles: DIN
[9] Dimension Styles: ISO-25, ISO-35, ISO-4, ISO-5, ISO-7, ISO1, ISO1-4, ISO2, Standard
[10] Dimension Styles: JIS

See the Tables of Limits Settings in Chapter 14 for standard plotting scales and intended sheet sizes for the *Limits* settings in these templates.

Creating Template Drawings

To make a template drawing, begin a *New* drawing using the default template drawing (ACAD.DWT) or other template drawing (*.DWT), then make the initial drawing setups. Alternately, *Open* an existing drawing that is set up as you need (layers created, linetypes loaded, *Limits* and other settings for plotting or printing to scale, etc.), and then *Erase* all the geometry. Next, use *SaveAs* to save the drawing under a different descriptive name with a DWT file extension. Do this by selecting the *Drawing Template File (*.dwt)* option in the *Save Drawing As* dialog box (Fig. 13-14). AutoCAD automatically (by default) saves the drawing in the folder where other template (*.DWT) files are found (usually in the C:\Program Files\AutoCAD R14\Template subdirectory).

Figure 13-14

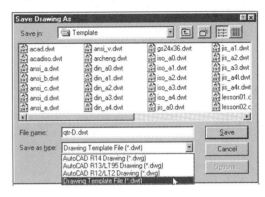

Multiple template drawings can be created, each for a particular situation. For example, you may want to create several templates, each having the setup steps completed but with different *Limits* and with the intention to plot or print each in a different scale or on a different size sheet. Another possibility is to create templates with different layering schemes. There are many possibilities, but the specific settings used depend on your applications (geometry), typical scales, or your plot/print devices.

Typical drawing steps that can be considered for developing template drawings are listed below:

> Set *Units*
> Set *Limits*
> Set *Snap*
> Set *Grid*
> Set *LTSCALE*
> Create *Layers* with color and linetypes assigned
> Create *Text Styles* (see Chapter 18)
> Create Dimension Styles (see Chapter 29)
> Create or *Insert* a standard title block and border

A popular practice is to create a template drawing for <u>each sheet size</u> used for plotting. *Limits* are set to the actual sheet size, and other settings and variables (*LTSCALE, DIMSCALE,* text sizes) are set to the actual desired size for the finished full-size plot. If you then want to create a new drawing and plot on a certain sheet size, start a new drawing using the appropriate template.

For laboratories or offices that use only one sheet size, create template drawings for <u>each plot scale</u> with the appropriate *Limits, Snap, Grid, Layers, LTSCALE, DIMSCALE,* etc.

You can also create templates for different plot scales <u>and</u> different sheet sizes. Make sure you assign appropriate and descriptive names to the template drawings.

Previous AutoCAD Releases

In previous releases of AutoCAD, any drawing file (*.DWG) could be used as a "prototype" drawing. This scheme is similar to the template concept in that a selected prototype drawing is used as a template, but because all prototypes have a .DWG file extension, there could be some confusion as to which files were to be used only as prototypes. The template (*.DWT) scheme in Release 14 provides a more secure method of file management and complies with similar template schemes in other software packages such as word processors and spreadsheets.

CHAPTER EXERCISES

For the first three exercises, you will set up several drawings that will be used in other chapters for creating geometry. Follow the typical steps for setting up a drawing given in this chapter.

1. ***Start from Scratch***
   You are to create a drawing of a wrench. The drawing will be printed full size on an A size sheet (not on an A4 metric sheet), and the dimensions are in millimeters. Start a *New* drawing, select *Start from Scratch,* and select the <u>*English*</u> default settings. Follow these steps:

   A. Set *Units* to *Decimal* and *Precision* to **0.00**.
   B. Set *Limits* to a metric sheet size, **297** x **210;** then *Zoom All* (scale factor is approximately 25).
   C. Set *Snap* to **1**.
   D. Set *Grid* to **10**.
   E. Set *LTSCALE* to **12.5** (.5 times the scale factor of 25).
   F. *Load* the *Center Linetype*.
   G. Create the following *Layers* and assign the *Colors* and *Linetypes* as shown:

WRENCH	*continuous*	*red*
CONSTR	*continuous*	*white*
CENTER	*center*	*green*
TITLE	*continuous*	*yellow*
DIM	*continuous*	*cyan*

   H. *Save* the drawing and assign the name **WRENCH** for use in a later chapter exercise.

2. ***Quick Setup Wizard***
   A drawing of a mechanical part is to be made and plotted full size (1"=1") on an 11" x 8.5" sheet. Use the *Quick Setup Wizard* to assist with the setup.

   A. Begin a *New* drawing. When the *Startup* or *Create new Drawing* dialog box appears, select *Use a Wizard*. Select *Quick Setup*.
   B. In the *Quick Setup* dialog box, select *Decimal* in the *Step 1: Units* tab.
   C. Press *Next* or select the *Step 2: Area* tab. Enter **11** and **8.5** in the two edit boxes so 11 appears in the *Sample Area* image tile along the bottom (X direction) and 8.5 appears along the right side (in the Y direction).
   D. Select the *Done* button.
   E. Check to ensure the *Limits* are set to 11,8.5 and *Grid* is set to .5.
   F. Change the *Snap* to **.250**.
   G. Set *LTSCALE* to **.5**.
   H. *Load* the *Center* and *Hidden Linetypes*.
   I. Create the following *Layers* and assign the given *Colors* and *Linetypes:*

VISIBLE	*continuous*	*red*
CONSTR	*continuous*	*white*
CENTER	*center*	*green*
HIDDEN	*hidden*	*yellow*
TITLE	*continuous*	*white*
DIM	*continuous*	*cyan*

   J. Save the drawing as **BARGUIDE** for use in another chapter exercise.

3. *Quick Setup Wizard*

A floor plan of an apartment has been requested. Dimensions are in feet and inches. The drawing will be plotted at 1/2"=1' scale on an 24" x 18" sheet.

A. Begin a *New* drawing and select the *Quick Setup Wizard*.
B. Select *Architectural* in the *Step 1: Units* tab.
C. In the *Step 2: Area* tab, set *Width* and *Length* to **48'** x **36'** (576" x 432"), respectively (use the apostrophe symbol for feet). Select *Done*.
D. Reset the *Snap* value to **1** (inch).
E. Reset the *Grid* value to **12** (inches).
F. Reset the *LTSCALE* to **12**.
G. Create the following *Layers* and assign the *Colors*:

FLOORPLN	*red*
CONSTR	*white*
TEXT	*green*
TITLE	*yellow*
DIM	*cyan*

H. *Save* the drawing and assign the name **APARTMENT** to be used later.

4. *Quick Setup Wizard*

A drawing is to be created and is to be printed in 1/4"=1" scale on an A size (11" x 8.5") sheet. The *Quick Setup Wizard* can be used to automate the setup. (See Table of *Quick Setup Wizard* Settings, Plot Scale for A Size Sheet column.)

A. Begin a *New* drawing. When the *Startup* or *Create new Drawing* dialog box appears, select **Use a Wizard**. Select *Quick Setup*.
B. In the *Quick Setup* dialog box, select *Decimal* in the *Step 1: Units* tab.
C. Press *Next* or select the *Step 2: Area* tab. Enter **44** and **34** in the two edit boxes so 44 appears in the *Sample Area* image tile along the bottom (X direction) and 34 appears along the right side (in the Y direction).
D. Select the *Done* button.
E. Check to ensure the *Limits* are set to 44,34 and *Snap* and *Grid* are set to 1.
F. Save the drawing as **CH13EX4.**

5. *Advanced Setup Wizard*

In this exercise you will set up a drawing using a standard ANSI title block for printing on an A size (11" x 8.5") sheet for a mechanical engineering application.

A. Begin a *New* drawing. In the *Create New Drawing* dialog box, select **Use a Wizard,** then select **Advanced Setup**.
B. When the *Advanced Setup* dialog box appears, accept all of the default *Units* settings. Proceed to *Step 5: Area*.
C. In the *Area* tab, set the *Width* and *Length*, respectively, to **11** and **8.5**.
D. In the next tab, *Step 6: Title Block*, select the *ANSI A (in)* option.
E. In the *Step 7: Layout* tab, answer *No* to the *Do you want to use advanced paper space layout capabilities?* option.
F. Use *SaveAs* and assign the name **A_ANSI**.

6. *Advanced Setup Wizard*
    Use the *Advanced Setup Wizard* to begin a drawing for an architectural application. The drawing is set up with a title block in Paper Space for a D size sheet and will be plotted in 1/4"=1' scale.

    A. Begin a *New* drawing. In the *Create New Drawing* dialog box, select **Use a Wizard**, then select **Advanced Setup**.
    B. When the *Advanced Setup* dialog box appears, select **Architectural** in the *Step 1: Units* tab. Select a *Precision* of **1/8"**.
    C. Proceed to *Step 5: Area*. Enter *Width* and *Length* values of **144'** and **96'**, respectively. Make sure you use the ' (apostrophe) symbol or else inches are assumed.
    D. In *Step 6: Title Block*, select the *Arch/Eng (in)* title block option.
    E. Finally, in *Step 7: Layout,* select *Yes* to the option for using Paper Space. Also select **Work on my drawing while viewing the layout** to work in Model Space and see the title block in Paper Space. Select *Done*.
    F. Reset *Snap* to **2'** (feet) and *Grid* to **4'** (feet).
    G. Create the following *Layers* and assign the *Colors*:

FLOORPLN	*red*
HVAC	*cyan*
PLUMB	*yellow*
ELEC	*green*
TEXT	*blue*
DIM	*magenta*

    H. Save the drawing in your working directory as **D_ARCH**.

7. **Create Template Drawings**
    In this exercise, you will create several "generic" template drawings that can be used at a later time. Creating the templates now will save you time later when you begin new drawings.

    A. Create a template drawing for use with decimal dimensions and using standard paper A size format. Begin a *New* drawing and select **Start from Scratch**. Select the **English** default settings. Include the following setups:

        1. Set *Units* to *Decimal* and *Precision* to **0.00**.
        2. Set *Limits* to **11** x **8.5**.
        3. Set *Snap* to **.25**.
        4. Set *Grid* to **1**.
        5. Set *LTSCALE* to **.5**.
        6. Create the following *Layers*, assign the *Linetypes* as shown, and assign your choice of *Colors*. Create any other layers you think you may need or any assigned by your instructor.

OBJECT	*continuous*
CONSTR	*continuous*
TEXT	*continuous*
TITLE	*continuous*
VPORTS	*continuous*
DIM	*continuous*
HIDDEN	*hidden*
CENTER	*center*
DASHED	*dashed*

7. Use *SaveAs* and name the template drawing **ASHEET**. Make sure you select ***Drawing Template File (*.dwt)*** from the *Save as Type:* drop-down list in the *Save Drawing As* dialog box. Also, from the ***Save In:*** drop-down list on top, select <u>your working directory</u> as the location to save the template. In the *Template Description* dialog box, enter "**A-size Engineering Sheet 11 × 8.5.**"

B. Using the drawing in the previous exercise, create a template for a standard engineering B size sheet. First, use the *New* command and select the ***Use a Template*** option from the *Create New Drawing* dialog box. Select ***More files...*** from the list in center of the dialog box. When the *Select Template* dialog box appears, select the **ASHEET.DWT** from your working directory (not from the default folder with the AutoCAD supplied templates). When the drawing opens, set *Limits* to **17 x 11**. All other settings and layers are OK as they are. Use *SaveAs* and save the new drawing as a template (**DWT**) drawing in your working directory. Assign the name **BSHEET**. Add the appropriate description in the *Template Description* dialog box.

C. In this exercise, create a template for a standard engineering C size sheet. Use the *New* command and use the **ASHEET** template drawing (use the same methods explained in problem 7. B to use the template). Set *Limits* to **22 x 17**. Set *Snap* to **.5**. Keep all the other settings and layers as they are. Use *SaveAs,* assign the name **CSHEET**, and save the template (**DWT**) file in your working directory. Add the appropriate description in the *Template Description* dialog box.

D. Use the **ASHEET** template drawing (by the methods explained previously), only in this exercise create a template for a standard engineering D size sheet. Set *Limits* to **34 x 22**. All other settings and layers do not need to be changed. Assign the name **DSHEET** when you save it as a template drawing in your working directory. Add the appropriate description in the *Template Description* dialog box.

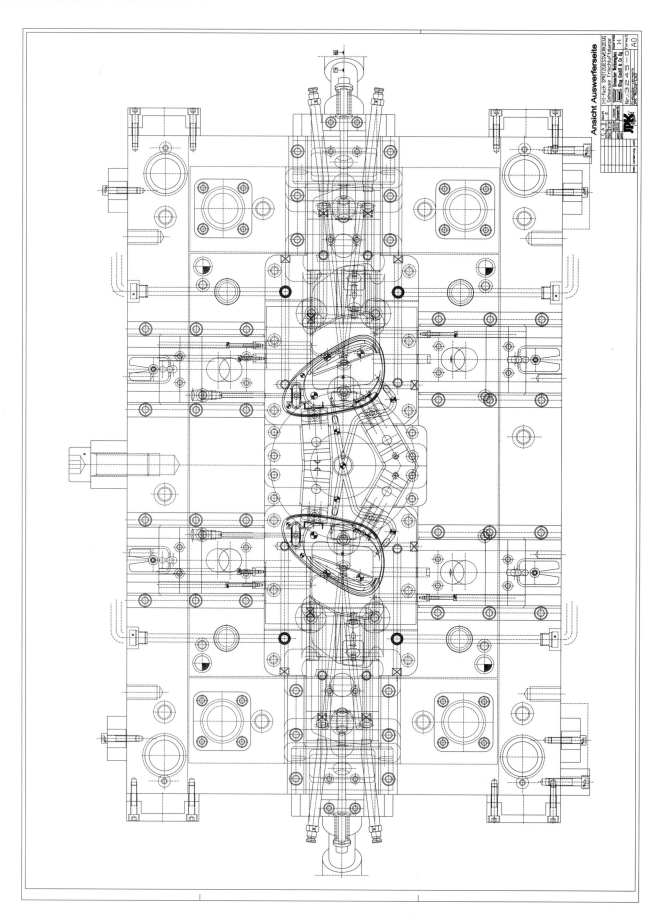

ZKL47_2.DWG Courtesy, AutoCAD Inc. (Release 14 Sample Drawing)

14

PRINTING AND PLOTTING

Chapter Objectives

After completing this chapter you should:

1. know the typical steps for printing or plotting;

2. be able to invoke and use the *Print/Plot Configuration* dialog box;

3. be able to select from available plotting devices and set the paper size and orientation;

4. be able to specify what area of the drawing you want to print or plot;

5. be able to preview the print/plot before creating a plotted drawing;

6. be able to specify a scale for printing or plotting a drawing;

7. know how to set up a drawing for plotting to a standard scale on a standard size sheet;

8. be able to use the tables of *Limits* Settings to determine *Limits,* scale, and paper size settings;

9. know how to configure plot and print devices in AutoCAD.

CONCEPTS

Plotting and printing are accomplished from within AutoCAD by invoking the *Plot* command. Using the *Plot* or *Print* command invokes the *Print/Plot Configuration* dialog box (Fig. 14-1). You have complete control of plotting and printing using the dialog box. In AutoCAD, the term "plotting" can refer to plotting on a pen plotter and/or printing with a printer.

Figure 14-1

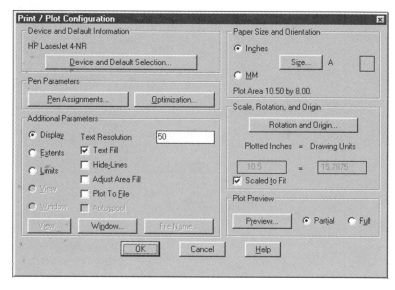

TYPICAL STEPS TO PLOTTING

Assuming the CAD system has been properly configured so the peripheral devices (plotters and/or printers) are functioning, the typical basic steps to printing and plotting using the *Print/Plot Configuration* dialog box are listed below:

1. Use *Save* to ensure the drawing has been saved in its most recent form before plotting (just in case some problem arises while plotting).

2. Make sure the plotter or printer is turned on, has paper and pens loaded, and is ready to accept the plot information from the computer.

3. Invoke the *Print/Plot Configuration* dialog box.

4. Check the upper-left corner of the dialog box to ensure that the intended device has been selected. If not, select the *Device and Default Selection* tile and make the desired choice.

5. Check the upper-right corner of the dialog box to ensure the desired paper size has been selected. If not, use the *Size...* tile to do so.

6. Only when necessary, change other options such as *Rotation and Origin* and *Pen Assignments*.

7. Determine and select which area of the drawing to plot: *Display, Extents, Limits, Window*, or *View*.

8. Enter the desired scale for the print or plot. If no standard scale is needed, toggle *Scaled to Fit* (so the check mark appears in the box).

9. Always *Preview* the plot to ensure the drawing will be printed or plotted as you expect. Select either a *Full* or *Partial* preview. If the preview does not display the plot as you intended, make the appropriate changes. Otherwise, needless time and media could be wasted.

10. If everything is OK, selecting the *OK* tile causes the drawing to be sent to the plotter or printer.

11. For additional plots or prints, you can use the *Preview* command to preview and plot the drawing based on the parameters previously set in the *Print/Plot Configuration* dialog box.

USING THE *PLOT* COMMAND

PLOT

Pull-down Menu	COMMAND (TYPE)	ALIAS (TYPE)	Short-cut	Screen (side) Menu	Tablet Menu
File *Print*	*PLOT or PRINT*	*...*	*Ctrl+P*	*FILE* *Plot*	*W,25*

Invoking *Plot* or *Print* normally invokes the *Print/Plot Configuration* dialog box (Fig. 14-1). (The *Print/Plot Configuration* dialog box can be suppressed by setting the *CMDDIA* system variable to **0**. This action displays a text screen instead and allows changing parameters by keyboard entry or plot script.)

Print/Plot Configuration **Dialog Box**

The *Print/Plot Configuration* dialog box allows you to change plotting parameters such as scale, paper size, pen assignments, rotation, and origin of the drawing on the sheet. Changes made to plotting options are saved (in the ACAD14.CFG file) so they do not have to be re-entered the next time.

Device and Default Selection
Many devices (printers or plotters) can be configured for use with AutoCAD, and all configurations are saved for your selection. For example, you can have both an A size and a D size plotter as well as a laser printer, any one of which could be used to plot the current file.

When you start AutoCAD for the first time, it automatically configures itself to use the Windows system printer. Additional devices for use specifically for AutoCAD can be configured within AutoCAD by using the *Printers* tab of the *Preferences* dialog box. You can also type *Config* to directly open the *Printers* tab of the *Preferences* dialog box. See Configuring Plotters and Printers near the end of this chapter.

Selecting the *Device and Default Selection* tile from the *Print/Plot Configuration* dialog box invokes the *Device and Default Selection* dialog box (Fig. 14-2). Highlighted near the top is the currently selected device along with the other configured choices.

Figure 14-2

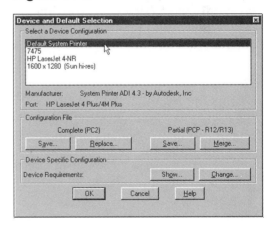

Since multiple devices can be used and there may also be several typical sets of parameters frequently used for each device (scale, pen assignments, etc.), multiple settings of plotting parameters can be saved and retrieved by using the *Save...* and *Replace...* or *Save...* and *Merge...* tiles. The settings can be saved as a .PCP or .PC2 (plotter configuration file) and easily retrieved instead of having to make all of the selections each time you want to change parameters.

A .PCP (partial plot configuration) file keeps the settings you make in the *Pen Parameters* section of the dialog box (see below) but the information is device independent, so a .PCP file can be used for multiple print/plot devices. A .PC2 (complete plot configuration) file is a device-complete file. The file keeps the *Pen Parameter* information but also keeps the name and all other driver and configuration specifications for a <u>particular device</u>, so when a .PC2 file is loaded, it automatically <u>installs and configures the device and makes it the current plotter or printer</u>. The .PC2 files are useful for batch plotting because different printers and plotters can be controlled while AutoCAD is unattended. See Chapter 47, Bonus Features, for information on batch plotting.

Pen Parameters

Selecting the *Pen Assignments* tile produces the dialog box enabling you to specify the pens used for plotting based on screen colors (Fig. 14-3). The <u>screen color drives the plotting pen and/or printer line widths and colors</u>. Select one or several color numbers from the list (color 5 is selected in the example figure). The *Pen, Ltype, Speed,* and *Width* parameters can then be set on the right side of the dialog box.

Figure 14-3

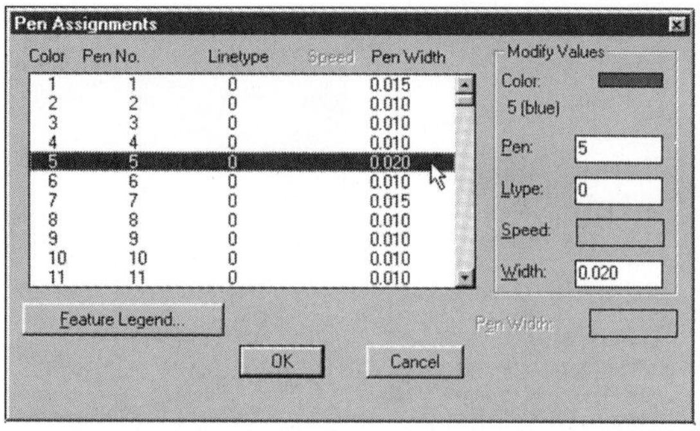

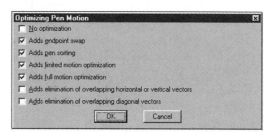

The parameters vary depending on the printer or plotter configured. Generally, multipen plotters and many printer drivers list the number of "pens" available. The *Linetype* column refers to <u>plotter or printer</u> linetypes, which are usually set to 0 (continuous), so AutoCAD's supplied drawing linetypes can be used instead (select *Feature Legend* to see what linetypes are available for your device). Pen plotters allow speed adjustment of pen motion (you may want to decrease *Speed* if the pens appear to skip).

Many printer drivers provide the capability to vary line width using the *Width* option (see the *Pen Width* column in Figure 14-3). This feature allows you to create excellent quality prints by varying the *Pen Widths* for different colors. Therefore, to achieve a print with multiple line widths, simply draw the geometry in the appropriate color (or on a layer with the appropriate color assigned) to determine the line width in the print. You should experiment with possibilities for your printing or plotting device.

The *Optimization* tile produces the dialog box (Fig. 14-4) for specifying up to seven levels of pen motion optimization (default levels are automatically set by most device drivers). Optimization is important for enabling pen plotters to draw lines on a sheet in the most efficient order.

Figure 14-4

Additional Parameters **(What to Plot)**
You specify what part of the drawing you want to plot by selecting the desired button from the lower-left corner of the dialog box (Fig. 14-1). The choices are listed below.

Display
This option plots the current display on the screen. If using viewports, it plots the current viewport display.

Extents
Plotting *Extents* is similar to *Zoom Extents*. This option plots the entire drawing (all objects), disregarding the *Limits*. Use the *Zoom Extents* command in the drawing to make sure the extents have been updated before using this plot option.

Limits
This selection plots the area defined by the *Limits* command (unless plotting a 3D object from other than the plan view).

View
With this option you can plot a view previously saved using the *View* command. Creating *Views* of specific areas of a drawing is very helpful when you plan to make repetitive plots of the same area or plots of multiple areas of one drawing.

Window
Window allows you to plot any portion of the drawing. You must specify the window by **PICK**ing or supplying coordinates for the lower-left and upper-right corners.

Text Resolution

This option allows you to control the resolution, or accuracy, for rounded sections of TrueType fonts such as the curve of the letter "o." The range is from 1 to 100. A low resolution of 5, for example, prints curved text with several very short, straight line segments, whereas a value of 100 produces smooth, rounded curves.

Text Fill

An AutoCAD drawing always displays TrueType fonts in solid filled format. You can, however, control if text is filled or not for the plot or print with this option. *Text Fill* off (no check mark) speeds plots of drawings with large text objects for pen plotters. This option sets the *TEXTFILL* system variable.

Hide Lines

This option removes hidden lines (edges obscured from view) if you are plotting a 3D surface or solid model.

Adjust Area Fill

Using this option forces AutoCAD to adjust for solid-filled areas (*Plines, Solids,* solid *Hatch* patterns) by pulling the pen inside the filled area by one-half the pen width. Use this option for critically accurate pen plots that have solid fills.

Plot to File

Choosing this option writes the plot to a file instead of making a plot. This action generally creates a .PLT file type. The format of the file (for example, PCL or HP/GL language) depends on the device brand and model that is configured. The plot file can be printed or plotted later <u>without</u> AutoCAD, assuming the correct interpreter for the device is available. Selecting this option enables the *File Name* tile for specifying the name and location for the plot file.

Paper Size and Orientation

Inches or *MM* should be checked to correspond to the units used in the drawing. Assuming the drawing was created using the correct units, a scale for plotting can be calculated without inch to millimeter conversion.

The *Size...* tile invokes the dialog box shown in Figure 14-5. You can select from the list of available paper sizes handled by the configured device. User sizes within the maximum plot area can also be defined.

Figure 14-5 —————

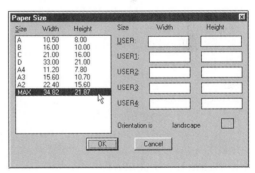

Scale, Rotation, and Origin

Selecting the *Rotation and Origin...* tile invokes the dialog box shown in Figure 14-6. The drawing can be rotated on the paper in 90 degree intervals. The origin specified indicates the location on the paper from point 0,0 (the plotter or printer's home position) from which the plot will be made. Home position for plotters is the <u>lower-left</u> corner (land-scape orientation) and for printers it is the <u>upper-left</u> corner of the paper (portrait orientation).

Figure 14-6 —————

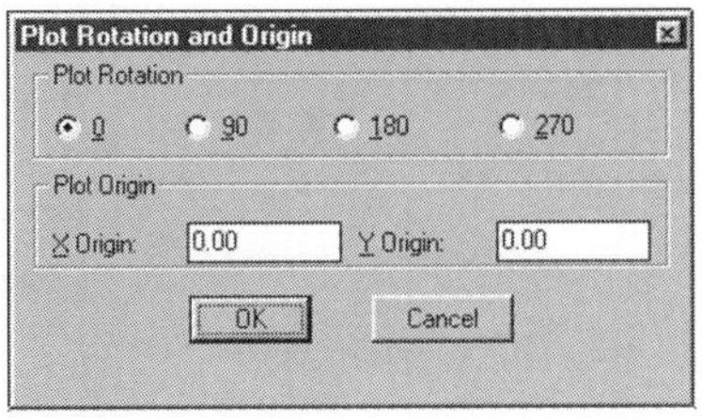

Scaled to Fit **or Plot to Scale**

Near the lower-right corner of the *Print/Plot Configuration* dialog box is the
area where you designate the plot scale (Fig. 14-7). You have two choices:
scale the drawing to automatically fit on the paper or calculate and indicate a
specific scale for the drawing to be plotted. By toggling the *Scaled to Fit* box,
the drawing is automatically sized to fit on the sheet, based on the specified
area to plot (*Display, Limits, Extents*, etc.). If you want to plot the drawing to
a specific scale, <u>remove the check</u> from the *Scaled to Fit* box and enter the
desired ratio in the boxes under *Plotted Inches* (or *Plotted MM*) and *Drawing
Units*. Decimals or fractions can be entered. For example, to prepare for a plot of one-half size
(1/2"=1"), the following ratios can be entered to achieve a plot at the desired scale: 1=2, 1/2=1, or .5=1.
For guidelines on plotting a drawing to scale, see Plotting to Scale in this chapter.

Figure 14-7

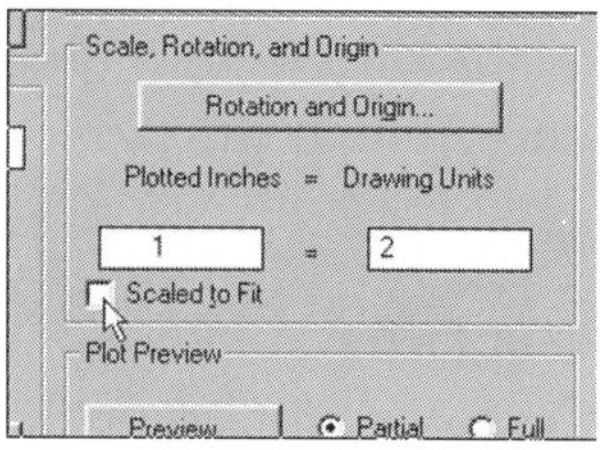

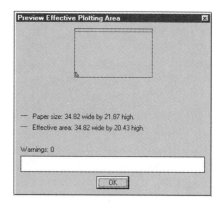

Plot Preview

Selecting the *Partial* tile (Fig. 14-1) displays the effective plotting area
as shown in Figure 14-8. Use the *Partial* option of *Plot Preview* to get
a quick check showing how the drawing will fit on the sheet. Two
rectangles and two sets of dimensions are displayed: the "Paper
size" is given and displayed in red and the "Effective area" is given
and displayed in blue. The "Effective area" is based on the selection
made in the *Additional Parameters* area of the plot dialog, that is,
Display, Extents, Limits, etc.

Figure 14-8

Selecting the *Full* option of *Plot Preview* displays the complete drawing as it will be plotted on the sheet
(Fig. 14-9). This function is particularly helpful to ensure the drawing will plot as you expect. When you
invoke a *Full Preview* in Release 14, the Drawing Editor displays a simulated sheet of paper with the
completed print or plot. The *Real Time Zoom* feature is on by default so you can check specific areas of
the drawing before
making the plot.
Right-clicking produces
the menu shown in
Figure 14-9, allowing you
to use *Real Time Pan* and
other *Zoom* options.
Changing the display
during a *Full Preview* <u>does
not change</u> the area of the
drawing to be plotted
or printed.

Figure 14-9

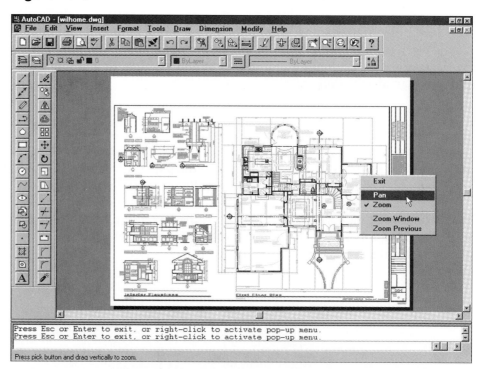

PREVIEW

Pull-down Menu	COMMAND (TYPE)	ALIAS (TYPE)	Short-cut	Screen (side) Menu	Tablet Menu
File Print Preview	PREVIEW	PRE	...	...	X,24

The *Preview* command accomplishes the same action as selecting a *Full Preview* in the *Print/Plot Configuration* dialog box. The advantage to using this command is being able to see a full print preview directly without having to invoke the *Print/Plot Configuration* dialog box first (see Figure 14-9). All functions of this preview (right-click for the *Real time Pan* and *Zoom* menu, etc.) operate the same as a full preview from the *Print/Plot Configuration* dialog box with the exception of one additional option—*Plot*.

You can print or plot during *Preview* directly from the right-click menu (Fig. 14-10). Using the *Plot* option creates a print or plot based on your settings in the *Print/Plot Configuration* dialog box. If you use this feature, ensure you first set the plotting parameters in the dialog box as you want.

Figure 14-10 ·

PLOTTING TO SCALE

When you create a <u>manual</u> drawing, a scale is determined before you can begin drawing. The scale is determined by the proportion between the size of the object on the paper and the actual size of the object. You then complete the drawing in that scale so that the actual object is proportionally reduced or enlarged to fit on the paper.

With a <u>CAD</u> drawing you are not restricted to a sheet of paper while drawing, so the geometry is created full size. Set *Limits* to provide an appropriate amount of drawing space; then the geometry can be drawn using the <u>actual dimensions</u> of the object. The resulting drawing on the CAD system is a virtual full-size dimensional replica of the actual object. Not until the CAD drawing is <u>plotted</u> on a fixed size sheet of paper, however, is it <u>scaled</u> to fit on the sheet.

Plotting an AutoCAD drawing to scale usually involves considering the intended sheet size and plot scale when specifying *Limits* during the initial drawing setup. The scale for plotting is based on the proportion of the paper size to the *Limits* (the reciprocal of the drawing scale factor).

For example, if you want to plot on an 11″ x 8.5″ sheet, you can set the *Limits* to 22 x 17 (2 x the sheet size). The value of **2** is then the drawing scale factor, and the plot scale (to enter in the *Print/Plot Configuration* dialog box) is **1/2** or **1=2** (the reciprocal of the drawing scale factor). If, in another case, the calculated drawing scale factor is **4**, then **1/4** or **1=4** would be the plot scale entered to achieve a drawing plotted at 1/4 actual size.

In order to calculate *Limits* and drawing scale factors correctly, you must know the standard paper sizes.

Standard Paper Sizes

Size	Engineering (")	Architectural (")
A	8.5 x 11	9 x 12
B	11 x 17	12 x 18
C	17 x 22	18 x 24
D	22 x 34	24 x 36
E	34 x 44	36 x 48

Size	Metric (mm)
A4	210 x 297
A3	297 x 420
A2	420 x 594
A1	594 x 841
A0	841 x 1189

Guidelines for Plotting to Scale

Plotting to scale preferably begins with the appropriate initial drawing setup. To correctly set up a drawing for printing or plotting to scale, <u>any two</u> of the following three variables must be known. The third variable can be determined from the other two.

Drawing *Limits*
Print or plot sheet (paper) size
Print or plot scale

Your choice for *Limits* determines the drawing scale factor, which in turn is used to determine the plot scale or vice versa. Even though *Limits* can be changed and plot scale calculated at <u>any time</u> in the drawing process, it is usually done in the first few steps. Following are suggested steps for setting up a drawing for plotting to scale.

1. Set *Units* (*Decimal, Architectural, Engineering,* etc.) and *precision* to be used in the drawing.

2. Set *Limits* to a size, allowing geometry creation full size. *Limits* should be <u>set to the sheet size</u> used for plotting <u>times a factor</u>, if necessary, that provides enough area for drawing. This factor (proportion of the new *Limits* to the sheet size) becomes the <u>drawing scale factor</u>.

$$DSF = \frac{Limits}{Sheet\ size}$$

You must also use a scale factor value that will yield a standard drawing scale ($1/2''=1''$, $1/8''=1'$, 1:50, etc.), instead of a scale that is not a standard ($1/3''=1''$, $3/5''=1'$, 1:23, etc.). See the Tables of *Limits* Settings for standard drawing scales.

3. The drawing scale factor is used as the scale factor (at least a starting point) for changing all size-related variables (*LTSCALE, DIMSCALE,* Hatch Pattern Scale, etc.). The <u>reciprocal</u> of the drawing scale factor is the plot scale to enter in the *Print/Plot Configuration* dialog box.

$$Plot\ scale = \frac{1}{DSF}$$

4. If the drawing is to be created using millimeter dimensions, set *Units* to *Decimal* and use metric values for the sheet size. In this way, the reciprocal of the drawing scale factor is the plot scale, the same as feet and inch drawings. However, multiply the drawing's scale factor by 25.4 (25.4mm = 1″) to determine the factor for changing all size-related variables (*LTSCALE*, *DIM-SCALE*, etc.).

$$\text{DSF(mm)} = \frac{\text{Limits}}{\text{Sheet size}} \times 25.4$$

5. If drawing a border on the sheet, its maximum size cannot exceed the *Plot Area*. The *Plot Area* is given near the upper-right corner of the *Print/Plot Configuration* dialog box, but the values can be different for different devices. Since plotters or printers do not draw all the way to the edge of the paper, the border should not be drawn outside of the *Plot Area*. Generally, approximately 1/4″ to 1/2″ (6mm to 12mm) offset from each edge of the paper is required. To determine the distance in from the edge of the *Limits* after new *Limits* have been set, multiply 1/2 times the drawing scale factor. More precisely, calculate the distance based on the *Plot Area* listed for the current device and multiply by the drawing scale factor.

The **Tables of *Limits* Settings** (starting on the following page) help to illustrate this point. Sheet sizes are listed in the first column. The next series of columns from left to right indicate incremental changes in *Limits* and proportional changes in scale factor.

Simplifying the process of plotting to scale and calculating *Limits* and drawing scale factor can be accomplished by preparing template drawings. One method is to create a template for each sheet size that is used in the lab or office. In this way, the CAD operator begins the session by selecting the template drawing representing the sheet size and then multiples those *Limits* by some factor to achieve the desired *Limits* and drawing scale factor. Another method is to create a separate template drawing for each scale that you expect to use for plotting. In this way a template drawing can be selected with the final *Limits*, drawing scale factor, and plot scale already specified or calculated. This method requires creating a separate template for each expected scale and sheet size. (See Chapter 13 for help creating template drawings.)

Plotting to Scale Using Paper Space

If you are creating the drawing geometry in Model Space but plan to set up your title block and border in Paper Space, it is recommended that you calculate the drawing scale factor (DSF) and set Model Space *Limits* by these same methods. Although setting Model Space *Limits* is not critical in this case, calculating the DSF and *Limits* settings can be of assistance for plotting to scale.

To plot to scale using Paper Space, follow these steps:

1. Set Model Space *Limits* and create the geometry in Model Space as you would otherwise. Use the DSF to determine values for text size, dimension variables, hatch scales, etc.
2. Enable Paper Space (set TILEMODE to 0).
3. If you are using one of AutoCAD's template drawings or another template, the title block and border may already exist. If not, create, *Insert,* or *Xref* the title block and border.
4. Next, set Paper Space *Limits* to the plot or print sheet size.
5. In Model Space, use *Zoom* and enter an *XP* factor. The XP factor is the same as the plot scale (1/DSF).
6. Plot while Paper Space is enabled. Enter 1=1 as the plot scale in the *Print/Plot Configuration* dialog box.

See Chapters 19 and 33 for a complete explanation of using Paper Space.

TABLES OF *LIMITS* SETTINGS

Rather than making calculations of *Limits,* drawing scale factor, and print/plot scale for each drawing, the tables of limits settings on the following pages can be used to make calculating easier. There are five tables, one for each of the following applications:

<u>Table</u>

Mechanical Engineering
Architectural
Metric (using ISO standard metric sheets)
Metric (using U.S. standard engineering sheets)
Civil Engineering

To use the tables correctly, you must know <u>any two</u> of the following three variables. The third variable can be determined from the other two.

Approximate drawing *Limits*
Desired print or plot <u>paper size</u>
Desired print or plot <u>scale</u>

1. If you know the *Scale* and **paper size**:
   Assuming you know the scale you want to use, look along the top row to find the desired <u>scale</u> that you eventually want to print or plot. Find the desired <u>paper size</u> by looking down the left column. The intersection of the row and column yields the *Limits* settings to use to achieve the desired plot scale.

2. If you know the approximate *Limits* and **paper size**:
   Calculate how much space (minimum *Limits*) you require to create the drawing actual size. Look down the left column of the appropriate table to find the <u>paper size</u> you want to use for plotting. Look along that row to find the next larger *Limits* settings than your required area. Use these values to set the drawing *Limits*. The <u>scale</u> to use for the plot is located on top of that column.

3. If you know the *Scale* and approximate *Limits*:
   Calculate how much space (minimum *Limits*) you need to create the drawing objects actual size. Look along the top row of the appropriate table to find the desired <u>scale</u> that you eventually want to plot or print. Look down that column to find the next larger *Limits* than your required minimum area. Set the drawing *Limits* to these values. Look to the left end of that row to give the <u>paper size</u> you need to print or plot the *Limits* in the desired scale.

NOTE: Common scales are given in the tables of limits settings. If you need to create a print or plot in a scale that is not listed in the tables, you may be able to use the tables as a guide by finding the nearest table and standard scale to match your needs, then calculating <u>proportional</u> settings.

MECHANICAL TABLE OF *LIMITS* SETTINGS
(X axis x Y axis)

Paper Size (Inches)	Drawing Scale Factor 1 Scale 1" = 1" Plot 1 = 1	1.33 3/4" = 1" 3 = 4	2 1/2" = 1" 1 = 2	2.67 3/8" = 1" 3 = 8	4 1/4" = 1" 1 = 4	5.33 3/6" = 1" 3 = 16	8 1/8" = 1" 1 = 8
A 11 x 8.5 In.	11.0 x 8.5	14.7 x 11.3	22.0 x 17.0	29.3 x 22.7	44.0 x 34.0	58.7 x 45.3	88.0 x 68.0
B 17 x 11 In.	17.0 x 11.0	22.7 x 14.7	34.0 x 22.0	45.3 x 29.3	68.0 x 44.0	90.7 x 58.7	136.0 x 88.0
C 22 x 17 In.	22.0 x 17.0	29.3 x 22.7	44.0 x 34.0	58.7 x 45.3	88.0 x 68.0	117.0 x 90.7	176.0 x 136.0
D 34 x 22 In.	34.0 x 22.0	45.3 x 29.3	68.0 x 44.0	90.7 x 58.7	136.0 x 88.0	181.3 x 117.3	272.0 x 176.0
E 44 x 34 In.	44.0 x 34.0	58.7 x 45.3	88.0 x 68.0	117.3 x 90.7	176.0 x 136.0	235.7 x 181.3	352.0 x 272.0

ARCHITECTURAL TABLE OF *LIMITS* SETTINGS
(X axis x Y axis)

Paper Size (Inches)	Drawing Scale Factor	12 — 1"=1' — 1=12	16 — 3/4"=1' — 1=16	24 — 1/2"=1' — 1=24	32 — 3/8"=1' — 1=32	48 — 1/4"=1' — 1=48	64 — 3/16"=1' — 1=64	96 — 1/8"=1' — 1=96
A 12 x 9	Ft	12 x 9	16 x 12	24 x 18	32 x 24	48 x 36	64 x 48	96 x 72
	In.	144 x 108	192 x 144	288 x 216	384 x 288	576 x 432	768 x 576	1152 x 864
B 18 x 12	Ft	18 x 12	24 x 16	36 x 24	48 x 32	72 x 48	96 x 64	144 x 96
	In.	216 x 144	288 x 192	432 x 288	576 x 384	864 x 576	1152 x 768	1728 x 1152
C 24 x 18	Ft	24 x 18	32 x 24	48 x 36	64 x 48	96 x 72	128 x 96	192 x 144
	In.	288 x 216	384 x 288	576 x 432	768 x 576	1152 x 864	1536 x 1152	2304 x 1728
D 36 x 24	Ft	36 x 24	48 x 32	72 x 48	96 x 64	144 x 96	192 x 128	288 x 192
	In.	432 x 288	576 x 384	864 x 576	1152 x 768	1728 x 1152	2304 x 1536	3456 x 2304
E 48 x 36	Ft	48 x 36	64 x 48	96 x 72	128 x 96	192 x 144	256 x 192	384 x 288
	In.	576 x 432	768 x 576	1152 x 864	1536 x 1152	2304 x 1728	3072 x 2304	4608 x 3456

METRIC TABLE OF *LIMITS* SETTINGS
FOR METRIC SHEET SIZES
(X axis x Y axis)

Paper Size (mm)	Drawing Scale Factor 25.4 / Scale 1:1 / Plot 1=1	50.8 / 1:2 / 1=2	127 / 1:5 / 1=5	254 / 1:10 / 1=10	508 / 1:20 / 1=20	1270 / 1:50 / 1=50	
A4 297 x 210 mm	297 x 210	594 x 420	1485 x 1050	2970 x 2100	5940 x 4200	14,850 x 10,500	29,700 x 21,000
m	.297 x .210	.594 x .420	1.485 x 1.050	2.97 x 2.10	5.94 x 4.20	14.85 x 10.50	29.70 x 21.00
A3 420 x 297 mm	420 x 297	840 x 594	2100 x 1485	4200 x 2970	8400 x 5940	21,000 x 14,850	42,000 x 29,700
m	.420 x .297	.840 x .594	2.100 x 1.485	4.20 x 2.97	8.40 x 5.94	21.00 x 14.85	42.00 x 29.70
A2 594 x 420 mm	594 x 420	1188 x 840	2970 x 2100	5940 x 4200	11,880 x 8400	29,700 x 21,000	59,400 x 42,000
m	.594 x .420	1.188 x .840	2.97 x 2.10	5.94 x 4.20	11.88 x 8.40	29.70 x 21.00	59.40 x 42.00
A1 841 x 594 mm	841 x 594	1682 x 1188	4205 x 2970	8410 x 5940	16,820 x 11,880	42,050 x 29,700	84,100 x 59,400
m	.841 x .594	1.682 x 1.188	4.205 x 2.970	8.41 x 5.94	16.82 x 11.88	42.05 x 29.70	84.10 x 59.40
A0 1189 x 841 mm	1189 x 841	2378 x 1682	5945 x 4205	11,890 x 8410	23,780 x 16,820	59,450 x 42,050	118,900 x 84,100
m	1.189 x .841	2.378 x 1.682	5.945 x 4.205	11.89 x 8.41	23.78 x 16.82	59.45 x 42.05	118.90 x 84.10

METRIC TABLE OF *LIMITS* SETTINGS
FOR ENGINEERING (8.5 x 11 Format) SHEET SIZES
(X axis x Y axis)

Paper Size (mm)	Drawing Scale Factor 25.4 / Scale 1:1 / Plot 1 = 1	50.8 / 1:2 / 1 = 2	127 / 1:5 / 1 = 5	254 / 1:10 / 1 = 10	508 / 1:20 / 1 = 20	1270 / 1:50 / 1 = 50	2540 / 1:100 / 1 = 100
A 279.4 x 215.9 mm	279.4 x 215.9	558.8 x 431.8	1397 x 1079.5	2794 x 2159	5588 x 4318	13,970 x 10,795	27,940 x 21,590
m	0.2794 x 0.2159	0.5588 x 0.4318	1.397 x 1.0795	2.794 x 2.159	5.588 x 4.318	13.97 x 10.795	27.94 x 21.59
B 431.8 x 279.4 mm	431.8 x 279.4	863.6 x 558.8	2159 x 1397	4318 x 2794	8636 x 5588	21,590 x 13,970	43,180 x 27,940
m	0.4318 x 0.2794	0.8636 x 0.5588	2.159 x 1.397	4.318 x 2.794	8.636 x 5.588	21.59 x 13.97	43.18 x 27.94
C 558.8 x 431.8 mm	558.8 x 431.8	1117.6 x 863.6	2794 x 2159	5588 x 4318	11,176 x 8636	27,940 x 21,590	55,880 x 43,180
m	0.5588 x 0.4318	1.1176 x 0.8636	2.794 x 2.159	5.588 x 4.318	11.176 x 86.36	27.94 x 21.59	55.88 x 43.18
D 863.6 x 558.8 mm	863.6 x 558.8	1727.2 x 1117.6	4318 x 2794	8636 x 5588	17,272 x 11,176	43,180 x 27,940	86,360 x 55,880
m	0.8636 x 0.5588	1.7272 x 1.1176	4.318 x 2.794	8.636 x 5.588	17.272 x 11.176	43.18 x 27.94	86.36 x 55.88
E 1117.6 x 863.6 mm	1117.6 x 863.6	2235.2 x 1727.2	5588 x 4318	11,176 x 8636	22,352 x 17,272	55,880 x 43,180	111,760 x 86,360
m	1.1176 x 0.8636	2.2352 x 1.7272	5.588 x 4.318	11.176 x 8.636	22.352 x 17.272	55.88 x 43.18	111.76 x 86.36

CIVIL TABLE OF *LIMITS* SETTINGS
(For Engineering Units)
(X axis x Y axis)

Paper Size (Inches)	Drawing Scale Factor 120 Scale 1" = 10' Plot 1 = 120	240 1" = 20' 1 = 240	360 1" = 30' 1 = 360	480 1" = 40' 1 = 480	600 1" = 50' 1 = 600
A. 11 x 8.5 In.	1320 x 1020	2640 x 2040	3960 x 3060	5280 x 4080	6600 x 5100
Ft	110 x 85	220 x 170	330 x 255	440 x 340	550 x 425
B. 17 x 11 In.	2040 x 1320	4080 x 2640	6120 x 3960	8160 x 5280	10,200 x 6600
Ft	170 x 110	340 x 220	510 x 330	680 x 440	850 x 550
C. 22 x 17 In.	2640 x 2040	5280 x 4080	7920 x 6120	10,560 x 8160	13,200 x 10,200
Ft	220 x 170	440 x 340	660 x 510	880 x 680	1100 x 850
D. 34 x 22 In.	4080 x 2640	8160 x 5280	12,240 x 7920	16,320 x 10,560	20,400 x 13,200
Ft	340 x 220	680 x 440	1020 x 660	1360 x 880	1700 x 1100
E. 44 x 34 In.	5280 x 4080	10,560 x 8160	15,840 x 12,240	21,120 x 16,320	26,400 x 20,400
Ft	440 x 340	880 x 680	1320 x 1020	1760 x 1360	2200 x 1700

Examples for Plotting to Scale

Following are several hypothetical examples of drawings that can be created using the Guidelines for Plotting to Scale. As you read the examples, try to follow the logic and check the Tables of Limits Settings.

A. A one-view drawing of a mechanical part that is 40″ in length is to be drawn requiring an area of approximately 40″ x 30″, and the drawing is to be plotted on an 8.5″ x 11″ sheet. The template drawing *Limits* are preset to 0,0 and 11,8.5. (AutoCAD's default *Limits* are set to 0,0 and 12,9, which represents an uncut sheet size. Template *Limits* of 11 x 8.5 are more practical in this case.) The expected plot scale is 1/4″=1″. The following steps are used to calculate the new *Limits*.

   1. *Units* are set to *decimal*. Each unit represents 1.00 inch.
   2. Multiplying the *Limits* of 11 x 8.5 by a factor of **4**, the new *Limits* should be set to **44 x 34**, allowing adequate space for the drawing. The drawing scale factor is **4**.
   3. All size-related variable default values (*LTSCALE, DIMSCALE*, etc.) are multiplied by **4**. The plot scale is entered in the *Print/Plot Configuration* dialog box (*Plotted inches=Drawing units*) as **1=4, 1/4=1** or **.25=1** to achieve a plotted drawing of 1/4″=1″.

B. A floor plan of a residence will occupy 60′ x 40′. The drawing is to be plotted on a D size architectural sheet (24″ x 36″). No prepared template drawing exists, so the standard AutoCAD default drawing (12 x 9) *Limits* are used. The expected plot scale is 1/2″=1′.

   1. *Units* are set to *Architectural*. Each unit represents 1″.
   2. The floor plan size is converted to inches (60′ x 40′=720″ x 480″). The sheet size of 36″ x 24″ (or 3′ x 2′) is multiplied by **24** to arrive at *Limits* of **864″ x 576″** (72′ x 48′), allowing adequate area for the floor plan. The *Limits* are changed to those values.
   3. All default values of size-related variables (*LTSCALE, DIMSCALE*, etc.) are multiplied by **24,** the drawing scale factor. The plot scale is entered in the *Print/Plot Configuration* dialog box (*Plotted inches=Drawing units*) as **1=24, 1/2=12** (12 units=1′), or **1/24=1** to achieve a drawing of 1/2″=1′ scale.

C. A roadway cloverleaf is to be laid out to fit in an acquired plot of land measuring 1500′ x 1000′. The drawing will be plotted on D size engineering sheet (34″ x 22″). A template drawing with *Limits* set equal to the sheet size is used. The expected plot scale is 1″=50′.

   1. *Units* are set to *Engineering* (feet and decimal inches). Each unit represents 1.00″.
   2. The sheet size of 34″ x 22″ (or 2.833′ x 1.833′) is multiplied by **600** to arrive at *Limits* of **20400″ x 13200″** (1700′ x 1100′), allowing enough drawing area for the site. The *Limits* are changed to **1700′ x 1100′**.
   3. All default values of size-related variables (*LTSCALE, DIMSCALE*, etc.) are multiplied by **600**, the drawing scale factor. The plot scale is entered in the *Print/Plot Configuration* dialog box (*Plotted inches=Drawing units*) as **1=600** to achieve a drawing of 1″=50′ scale.

   NOTE: Many civil engineering firms use one unit in AutoCAD to represent one foot. This simplifies the problem of using decimal feet (10 parts/foot rather than 12 parts/foot); however, problems occur if architectural layouts are inserted or otherwise combined with civil work.

D. Three views of a small machine part are to be drawn and dimensioned in millimeters. An area of 480mm x 360mm is needed for the views and dimensions. The part is to be plotted on an A4 size sheet. The expected plot scale is 1:2.

   1. *Units* are set to *decimal*. Each unit represents 1.00 millimeter.

2. Setting the *Limits* exactly to the sheet size (297 x 210) would not allow enough area for the drawing. The sheet size is multiplied by a factor of **2** to yield *Limits* of **594 x 420**, providing the necessary 480 x 360 area.

3. The plot scale entered in the *Print/Plot Configuration* dialog box (*Plotted inches=Drawing units*) is **1=2** or **1/2=1** to make a finished plot of 1:2 scale. Since this drawing is metric, the drawing scale factor for changing all size-related variable default values (*LTSCALE, DIM-SCALE*, etc.) is multiplied by 25.4, or 2 x 25.4=approximately **50**.

CONFIGURING PLOTTERS AND PRINTERS

Figure 14-11

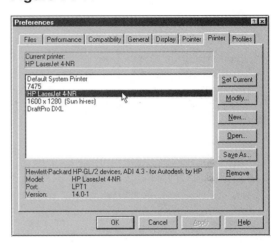

The *Printer* tab of the *Preferences* dialog box allows you to specify which devices you want to use in AutoCAD (Fig. 14-11). Access this dialog box by selecting *Preferences...* from the *Tools* pull-down menu or typing *Preferences*. You can also type *Config* to go directly to the *Printer* tab.

The central area of the tab lists the configured devices. When you begin AutoCAD Release 14 for the first time, the *Windows System Printer* is automatically configured (if you have a previously installed Windows printer). To make one device (from the list of configured devices) the current printer/plotter, highlight it and select the *Set Current* button. Setting a device current can also be accomplished from the *Print/Plot Configuration* dialog box. To configure other plotters or printers, select the *New...* tile.

Figure 14-12

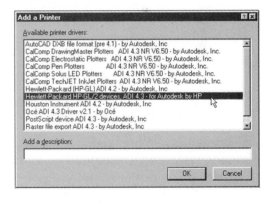

New
Use this option to configure new print or plot devices. The list of available device drivers (manufacturers/device types) appears in the *Add a Printer* dialog box (Fig. 14-12). Selecting a manufacturer/device type, in turn, displays a text screen providing several similar devices to choose from and several questions to answer during the configuration process. You can also add your own description for the device. The device you select and the description then appear in the list in the *Printer* tab of the *Preferences* dialog box.

Modify
If you would like to change the description, change the settings for the particular printer or plotter, or configure a new printer or plotter from the same manufacturer/type list, use this option.

Open
For any single device, you can save the pen settings and optimization (see *Pen Parameters* under the *Plot* command) as well as the device-specific information. This option allows you to select a previously saved .PC2 file. A .PC2 file can be saved from this dialog box or from the *Device and Default Selection* dialog box (from the *Print/Plot Configuration* dialog box). Opening a .PC2 file loads all the information needed to "install" and configure a device as well as making it the current plotter/printer.

SaveAs
Use this option to save a .PC2 file for the selected device. (See *Device and Default Selection* earlier in this chapter.)

Remove

Selecting this button removes the selected printer or plotter from the list of configured devices. A warning message appears before deleting the device driver. If a .PC2 file was saved for the device before using *Remove*, the device can automatically be "reinstalled" and configured by opening the .PC2 file.

Using a Configured Raster File Converter as a Plotter

Using the *Add a Printer* dialog box (Fig. 14-12), you can select *Raster file export* to configure a driver for converting AutoCAD drawings to a variety of raster file formats. See Chapter 32, Raster Images and Other File Formats, for more information on this subject.

CHAPTER EXERCISES

1. *Open* the **GASKETA** drawing that you created in Chapter 9 Exercises. What are the drawing *Limits*?

    A. *Plot* the drawing *Extents* on an 11″ x 8.5″ sheet. Select the *Scale to Fit* box.
    B. Next, *Plot* the drawing *Limits* on the same size sheet. *Scale to Fit*.
    C. Now, *Plot* the drawing *Limits* as before but plot the drawing at **1=1**. Measure the drawing and compare the accuracy with the dimensions given in the exercise in Chapter 9.
    D. Compare the three plots. What are the differences and why did they occur? (When you finish, there is no need to *Save* the changes.)

2. This exercise requires facilities for plotting an engineering C size sheet. *Open* drawing **CH12EX3**. Check to ensure the *Limits* are set at **22 x 17**. *Plot* the *Limits* at **1=1**. Measure the plot for accuracy by comparing with the dimensions given for the exercise in Chapter 12. To refresh your memory, this exercise involved adjusting the *LTSCALE* factor. Does the *LTSCALE* in the drawing yield hidden line dashes of 1/8″ (for the long lines) on the plot? If not, make the *LTSCALE* adjustment and plot again.

3. For this exercise, you will use a previously created template drawing to set up a drawing for plotting to scale. (The drawing will be used in later chapters for creating geometry.) If you know the final plot scale and sheet size, you can plan for the plot during the drawing setup. The *Limits* and other scale-related factors for plotting can be set at this phase. Follow the steps below.

    A drawing of a hammer is needed. The overall dimensions of the hammer are 12 1/2″ x about 5″. The drawing will be plotted on an 11″ x 8.5″ sheet. Begin a *New* drawing and use the template **ASHEET** you created in Chapter 13. Use *Save* and assign the name **HAMMER.DWG**. Follow these steps.

    A. Set *Units* to *Decimal* and *Precision* to .000.
    B. Set *Snap* to .125.
    C. Set *Grid* to .5.
    D. Since the geometry will not fit within *Limits* of 11 x 8.5, plotting at full size (1=1) will not be possible. In order to make a plot to a standard scale and show the geometry at the largest possible size on an 11″ x 8.5″ sheet, consult the Mechanical Table of *Limits* Settings to determine *Limits* settings for the drawing and a scale for plotting. Find the values for setting *Limits* for a plot at **3/4=1** scale. Set *Limits* appropriately; then *Zoom All.*
    E. An *LTSCALE* of .5 (previously set) creates hidden lines with the 1/8″ standard dashes when plotted at 1=1. Therefore, multiply the *LTSCALE* of .5 times the scale factor indicated in the table.
    F. *Save* the drawing.

4. Create a new template for architectural applications to plot at 1/8"=1' scale on a D size sheet. Begin a *New* drawing using the template **DSHEET** and use *Save* to create a new template (DWT) named **D-8-AR** in your working folder (not where the AutoCAD-supplied templates are stored). Set *Units* to *Architectural*. Use the Architectural Table of *Limits* Settings to determine and set the new *Limits* for a **D** size sheet to plot at **1/8"=1'**. Multiply the existing *LTSCALE* (.5) times the scale factor shown in the table. Turn *Snap* and *Grid* off. *Save* as a template (DWT) drawing in your working folder.

5. Create a new template for metric applications to plot at 1:2 scale (1/2 size) on an A size sheet. Begin a *New* drawing using the template **ASHEET**, then use *Save* to create a new template drawing named **A-METRIC-2** in your working folder (not where the AutoCAD-supplied templates are stored). Use the Metric Table of *Limit* Settings to determine and set the new *Limits* for an **A** size sheet to plot at **1:2**. Multiply the existing *LTSCALE*, *Snap*, and *Grid* times the scale factor shown. *Save* as a template (DWT) drawing in your working folder.

6. Create a new template for civil engineering applications to plot at 1"=20' scale on a C size sheet. Begin a *New* drawing using the template **CSHEET** and use *Save* to create a new template (DWT) named **C-CIVIL-20** in your working folder (not where the AutoCAD-supplied templates are stored). Set *Units* to *Engineering*. Use the Civil Table of *Limits* Settings to determine and set the new *Limits* for a **C** size sheet to plot at **1"=20'**. Multiply the existing *LTSCALE* (.5) times the scale factor shown. Turn *Snap* and *Grid* off. *Save* the drawing as a template (DWT) drawing in your working folder.

7. *Open* the **MANFCELL** drawing from Chapter 10. Check to make sure that the *Limits* settings are at **0,0** and **40',30'**. Make two plots of this drawing on your plotter, according to the instructions.

   A. Make one plot of the drawing using a standard architectural scale. The scale you use is your choice and should be based on the sheet sizes available as well as the existing geometry size. *Plot* the drawing *Limits*. You may have to alter the *Limits* in order to plot the *Limits* to a standard scale. Use the Architectural Table of *Limits* Settings for guidance.

   B. Make one plot of the drawing using a standard civil engineering scale. Again, the choice of scale is yours and should be based on your available sheet sizes as well as the existing geometry. *Plot* the drawing *Limits*. You may have to alter the *Limits* for this plot also in order to plot the *Limits* to a standard scale. Use the Civil Table of *Limits* Settings for guidance.

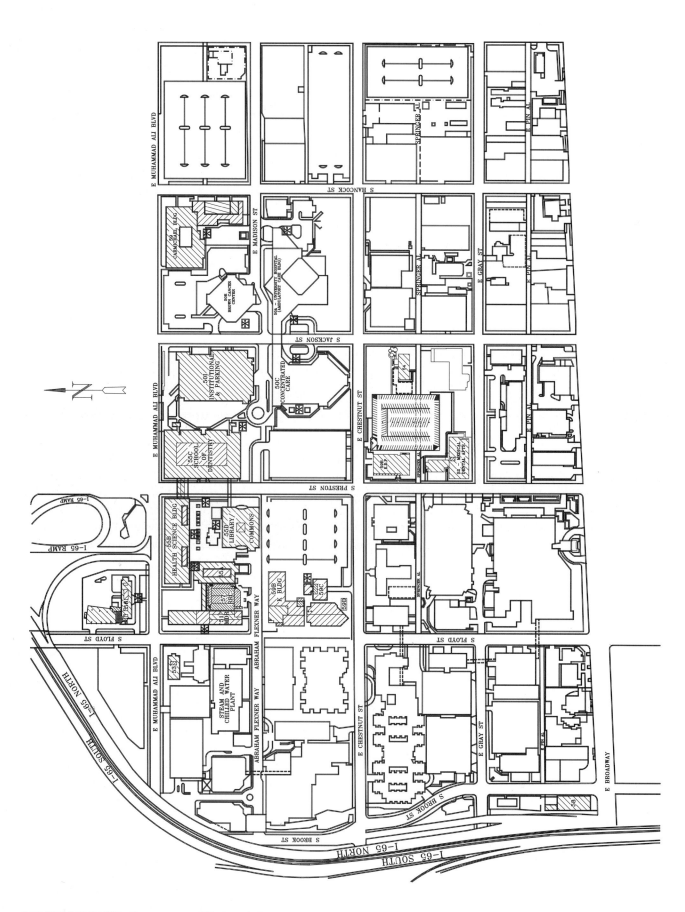

HSCMAP.DWG Courtesy, Michael Anderson

15

DRAW
COMMANDS II

Chapter Objectives

After completing this chapter you should:

1. be able to create construction lines using the *Xline* and *Ray* commands;
2. be able to create *Polygons* by the *Circumscribe*, *Inscribe*, and *Edge* methods;
3. be able to create *Rectangles* by PICKing two corners;
4. be able to use *Donut* to create circles with width;
5. be able to create *Spline* curves passing exactly through the selected points;
6. be able to create *Ellipses* using the *Axis End* method, the *Center* method, and the *Arc* method;
7. be able to use *Divide* to add points at equal parts of an object and *Measure* to add points at specified segment lengths along an object;
8. be able to draw multiple parallel lines with *Mline* and to create multiline styles with *Mstyle*;
9. be able to use the *Sketch* command to create "freehand" sketch lines;
10. be able to use the *Solid* command to create a filled 2D shape with three or four straight edges;
11. be able to create a *Boundary* by PICKing inside a closed area;
12. know that objects forming a closed shape can be combined into one *Region* object.

CONCEPTS

Remember that *Draw* commands create AutoCAD objects. The draw commands addressed in this chapter create more complex objects than those discussed in Chapter 8, Draw Commands I. The draw commands covered previously (*Line, Circle, Arc, Point,* and *Pline*) create simple objects composed of one object. The shapes created by the commands covered in this chapter are more complex. Most of the objects <u>appear</u> to be composed of several simple objects, but each shape is actually treated by AutoCAD as <u>one object</u>. Only the *Divide, Measure* and *Sketch* commands create multiple objects. The following commands are explained in this chapter.

Xline, Ray, Polygon, Rectangle, Donut, Spline, Ellipse, Divide, Measure, Mline, Sketch, Solid, Boundary, and *Region.*

These draw commands can be accessed using any of the command entry methods, including the menus and icon buttons, as illustrated in Chapter 8. Only the icon buttons that appear in the default *Draw* toolbar are given in this chapter by each command, although tools (icons) are available for the other draw commands if you want to customize your own toolbar (see Chapter 45 for information on customizing toolbars).

COMMANDS

XLINE

Pull-down Menu	COMMAND (TYPE)	ALIAS (TYPE)	Short-cut	Screen (side) Menu	Tablet Menu
Draw *Construction Line*	*XLINE*	*XL*	...	DRAW 1 *Xline*	*L,10*

When you draft with pencil and paper, light "construction" lines are used to lay out a drawing. These construction lines are not intended to be part of the finished object lines but are helpful for preliminary layout such as locating intersections, center points, and projecting between views.

There are two types of construction lines—*Xline* and *Ray*. An *Xline* is a line with infinite length, therefore having no endpoints. A *Ray* has one "anchored" endpoint and the other end extends to infinity. Even though these lines extend to infinity, they do not affect the drawing *Limits* or *Extents* or change the display or plot area in any way.

An *Xline* has no endpoints (*Endpoint Osnap* cannot be used) but does have a <u>root</u>, which is the theoretical <u>midpoint</u> (*Midpoint Osnap* can be used). If *Trim* or *Break* is used with *Xlines* or *Rays* such that two endpoints are created, the construction lines become *Line* objects. *Xlines* and *Rays* are drawn on the current layer and assume the current linetype and color (object-specific or *BYLAYER*).

There are many ways that you can create *Xlines* as shown by the options appearing at the command prompt. All options of *Xline* automatically repeat so you can easily draw multiple lines.

    Command: **xline**
    Hor/Ver/Ang/Bisect/Offset/<From point:>

From point

The default option only requires that you specify two points to construct the *Xline* (Fig. 15-1). The first point ("From point") becomes the root and anchors the line for the next point specification. The second point, or "Through point," can be PICKed at any location and can pass through any point (*Osnaps* can be used). If horizontal or vertical *Xlines* are needed, *ORTHO* can be used in conjunction with the "From point:" option.

```
Command: xline
Hor/Ver/Ang/Bisect/Offset/<From point:> PICK
Through point:
```

Figure 15-1

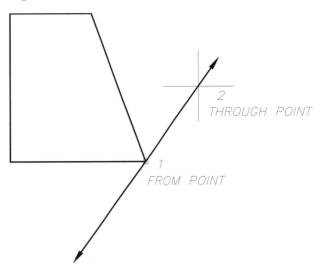

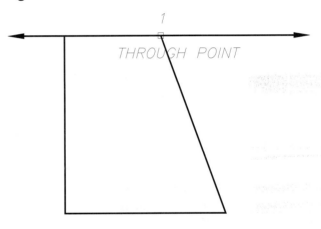

Hor

This option creates a horizontal construction line. Type the letter "H" at the "From point:" prompt. You only specify one point, the "Through point" or root (Fig. 15-2).

Figure 15-2

Ver

Ver creates a vertical construction line. You only specify one point, the "Through point" (root) (Fig. 15-3).

Figure 15-3

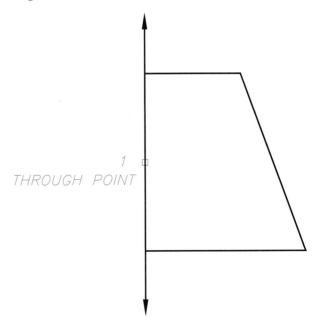

Ang

The *Ang* option provides two ways to specify the desired angle. You can (1) *Enter angle* or (2) select a *Reference* line (*Line, Xline, Ray,* or *Pline*) as the starting angle, then specify an angle from the selected line (in a counter-clockwise direction) for the *Xline* to be drawn (Fig. 15-4).

```
Command: XLINE
Hor/Ver/Ang/Bisect/Offset/<From point>: a
Reference/<Enter angle (0)>:
```

Figure 15-4

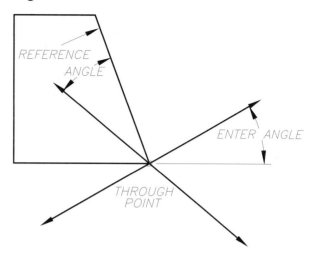

Bisect

This option draws the *Xline* at an angle between two selected points. First, select the angle vertex, then two points to define the angle (Fig. 15-5).

```
Command: XLINE
Hor/Ver/Ang/Bisect/Offset/<From point>: b
Angle vertex point: PICK
Angle start point: PICK
Angle end point: PICK
Angle end point: Enter
Command:
```

Figure 15-5

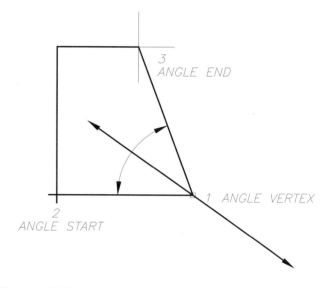

Offset

Offset creates an *Xline* parallel to another line. This option operates similarly to the *Offset* command. You can (1) specify a *Distance* from the selected line or (2) PICK a point to create the *Xline Through*. With the *Distance* option, enter the distance value, select a line (*Line, Xline, Ray,* or *Pline*), and specify on which side to create the offset *Xline*.

```
Command: XLINE
Hor/Ver/Ang/Bisect/Offset/<From point>: o
Offset distance or Through <1.0000>:
(value) or PICK
Select a line object: PICK
Side to offset?
Select a line object: Enter
Command:
```

Figure 15-6

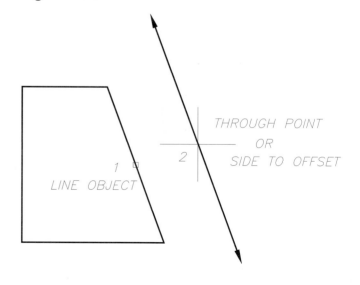

Using the *Through* option, select a line (*Line, Xline, Ray,* or *Pline*); then specify a point for the *Xline* to pass through. In each case, the anchor point of the *Xline* is the "root." (See *Offset,* Chapter 10.)

> Command: **XLINE**
> Hor/Ver/Ang/Bisect/Offset/<From point>: **o**
> Offset distance or Through <1.0000>: **T**
> Select a line object: **PICK**
> Through point: **PICK**
> Select a line object: **Enter**
> Command:

RAY

	COMMAND (TYPE)	ALIAS (TYPE)	Short-cut	Screen (side) Menu	Tablet Menu
Pull-down Menu					
Draw *Ray*	*RAY*	...	...	*DRAW 1* *Ray*	*K,10*

A *Ray* is also a construction line (see *Xline*), but it extends to infinity in only <u>one direction</u> and has one "anchored" endpoint. Like an *Xline*, a *Ray* extends past the drawing area but does not affect the drawing *Limits* or *Extents*. The construction process for a *Ray* is simpler than for an *Xline*, only requiring you to establish a "From point" (endpoint) and a "Through point" (Fig. 15-7). Multiple *Rays* can be created in one command.

> Command: **ray**
> From point: **PICK** or (**coordinates**)
> Through point: **PICK** or (**coordinates**)
> Through point: **PICK** or (**coordinates**)
> Through point: **Enter** to complete the command
> Command:

Figure 15-7 ————

THROUGH POINT

2

1

FROM POINT

Rays are especially helpful for construction of geometry about a central reference point or for construction of angular features. In each case, the geometry is usually constructed in only one direction from the center or vertex. If horizontal or vertical *Rays* are needed, just toggle on *ORTHO*. To draw a *Ray* at a specific angle, use relative polar coordinates at the "through point:" prompt. *Endpoint* and other appropriate *Osnaps* can be used with *Rays*. A *Ray* has one *Endpoint* but no *Midpoint*.

POLYGON

	COMMAND (TYPE)	ALIAS (TYPE)	Short-cut	Screen (side) Menu	Tablet Menu
Pull-down Menu					
Draw *Polygon*	*POLYGON*	*POL*	...	*DRAW 1* *Polygon*	*P,10*

The *Polygon* command creates a regular polygon (all angles are equal and all sides have equal length). A *Polygon* object appears to be several individual objects but, like a *Pline*, is actually <u>one</u> object. In fact, AutoCAD uses *Pline* to create a *Polygon*. There are two basic options for creating *Polygons*: you can specify an *Edge* (length of one side) or specify the size of an imaginary circle for the *Polygon* to *Inscribe* or *Circumscribe*.

Inscribe/Circumscribe

The command sequence for this default method follows:

    Command: **polygon**
    Number of sides: (Enter a value for the number of sides.)
    Edge/<Center of polygon>: **PICK** or (**coordinates**) to specify the center
    Inscribed in circle/Circumscribed about circle: **I** or **C**
    Radius of circle: **PICK** or (**value**) or (**coordinates**)

The orientation of the *Polygon* and the imaginary circle are shown in Figure 15-8. Note that the *Inscribed* option allows control of one-half of the distance <u>across the corners</u>, and the *circumscribed* option allows control of one-half of the distance <u>across the flats</u>.

Using *ORTHO ON* with specification of the *radius of circle* forces the *Polygon* to a 90 degree orientation.

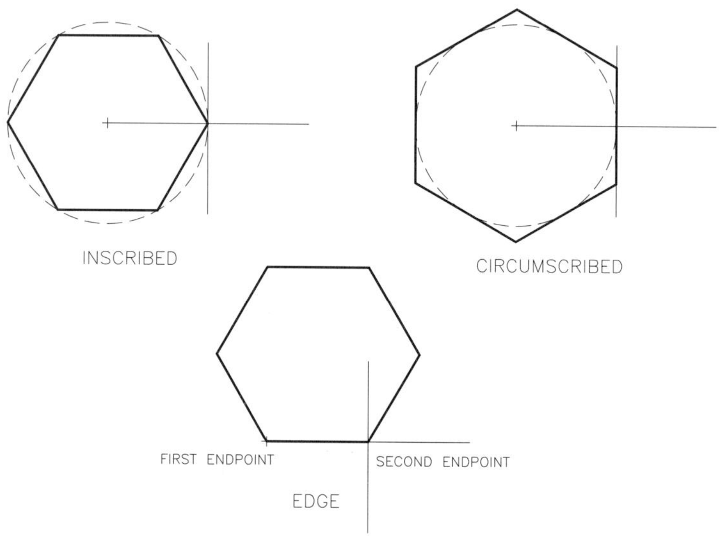

Edge

The *Edge* option only requires you to indicate the number of sides desired and to specify the two endpoints of one edge (Fig. 15-8).

    Command: **polygon**
    Number of sides: (**value**) (Enter a value
    for the number of sides.)
    Edge/<Center of polygon>: **E** (Invokes the edge option.)
    First endpoint of edge: **PICK** or (**coordinates**) (Interactively select or enter coordinates in any format.)
    Second endpoint of edge: **PICK** or (**coordinates**) (Interactively select or enter coordinates in any format.)

Figure 15-8

INSCRIBED   CIRCUMSCRIBED

FIRST ENDPOINT   SECOND ENDPOINT

EDGE

Because *Polygons* are created as *Plines*, *Pedit* can be used to change the line width or edit the shape in some way (see *Pedit*, Chapter 16). *Polygons* can also be *Exploded* into individual objects similar to the way other *Plines* can be broken down into component objects (see *Explode*, Chapter 16).

RECTANG

Pull-down Menu	COMMAND (TYPE)	ALIAS (TYPE)	Short-cut	Screen (side) Menu	Tablet Menu
Draw *Rectangle*	*RECTANG*	*REC*	...	*DRAW 1* *Rectang*	*Q,10*

The *Rectang* command only requires the specification of two diagonal corners for construction of a rectangle, identical to making a selection window (Fig. 15-9). The corners can be PICKed or coordinates can be entered. The rectangle can be any proportion,

Figure 15-9

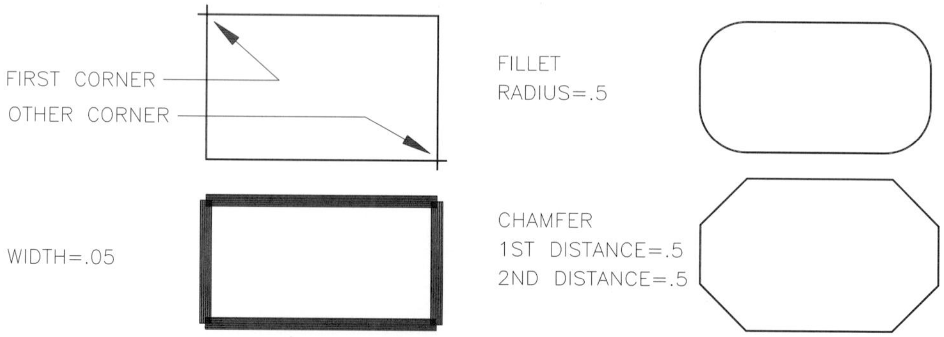

FIRST CORNER
OTHER CORNER

WIDTH=.05

FILLET
RADIUS=.5

CHAMFER
1ST DISTANCE=.5
2ND DISTANCE=.5

but the sides are always horizontal and vertical. The completed rectangle is <u>one AutoCAD object</u>, not four separate objects.

> Command: **rectangle**
> Chamfer/Elevation/Fillet/Thickness/Width/<First corner>: **PICK** or (**coordinates**)
> Other corner: **PICK** or (**coordinates**)
> Command:

Rectangle uses the *Pline* command to construct the shape; therefore, several options of the *Rectangle* command affect the shape as if it were a *Pline* (see *Pline* and *Pedit*). For example, a *Rectangle* can have *width*:

> Chamfer/Elevation/Fillet/Thickness/Width/<First corner>: **w**
> Width for rectangles <0.0000>: (**value**)

The *Fillet* and *Chamfer* options allow you to specify values for the fillet radius or chamfer distances:

> Chamfer/Elevation/Fillet/Thickness/Width/<First corner>: **f**
> Fillet radius for rectangles <0.0000>: (**value**)

> Chamfer/Elevation/Fillet/Thickness/Width/<First corner>: **c**
> First chamfer distance for rectangles <0.0000>: (**value**)
> Second chamfer distance for rectangles <0.0000>: (**value**)

The *Thickness* and *Elevation* options are 3D properties. See Chapter 40, Surface Modeling, for information on *Thickness* and *Elevation*.

DONUT

Pull-down Menu	COMMAND (TYPE)	ALIAS (TYPE)	Short-cut	Screen (side) Menu	Tablet Menu
Draw *Donut*	*DONUT*	*DO*	...	DRAW 1 *Donut*	K,9

A *Donut* is a circle with width (Fig. 15-10). Invoking the command allows changing the inside and outside diameters and creating multiple *Donuts*:

> Command: **donut**
> Inside diameter <current>:
> (**value**) or **Enter**
> Outside diameter <current>:
> (**value**) or **Enter**
> Center of doughnut: **PICK** or
> (**coordinates**)
> Center of doughnut: **PICK** or
> (**coordinates**) or **Enter**

Donuts are actually solid filled circular *Plines* with width. The solid fill for *Donuts*, *Plines*, and other "solid" objects can be turned off with the *Fill* command or *FILLMODE* system variable.

Figure 15-10

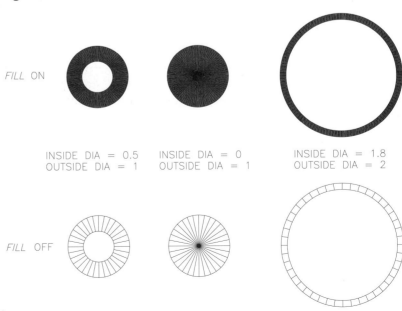

FILL ON

INSIDE DIA = 0.5
OUTSIDE DIA = 1

INSIDE DIA = 0
OUTSIDE DIA = 1

INSIDE DIA = 1.8
OUTSIDE DIA = 2

FILL OFF

SPLINE

Pull-down Menu	COMMAND (TYPE)	ALIAS (TYPE)	Short-cut	Screen (side) Menu	Tablet Menu
Draw Spline	SPLINE	SPL	...	DRAW 1 Spline	L,9

The *Spline* command creates a NURBS (non-uniform rational bezier spline) curve. The non-uniform feature allows irregular spacing of selected points to achieve sharp corners, for example. A *Spline* can also be used to create regular (rational) shapes such as arcs, circles, and ellipses. Irregular shapes can be combined with regular curves, all in one spline curve definition.

Spline is the newer and more functional version of a *Spline*-fit *Pline* (see *Pedit*, Chapter 16). The main difference between a *Spline*-fit *Pline* and a *Spline* is that a *Spline* curve passes through the points selected, while the points selected for construction of a *Spline*-fit *Pline* only have a "pull" on the curve. Therefore, *Splines* are more suited to accurate design because the curve passes exactly through the points used to define the curve (data points) (Fig. 15-11).

Figure 15-11 ————————

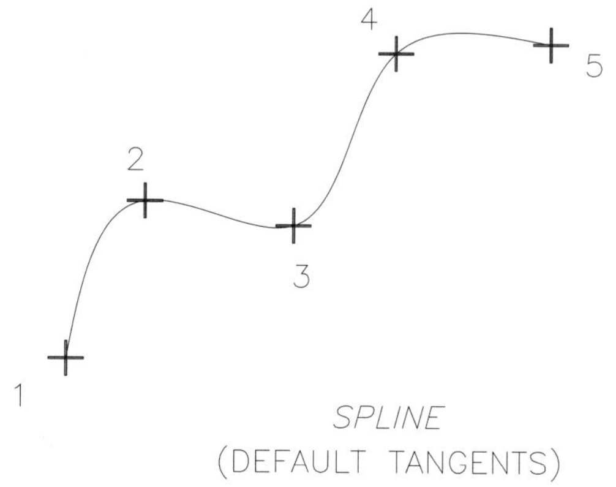

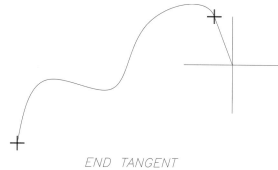

SPLINE
(DEFAULT TANGENTS)

The construction process involves specifying points that the curve will pass through and determining tangent directions for the two ends (for non-closed *Splines*) (Fig. 15-12).

Figure 15-12 ————————

END TANGENT

The *Close* option allows creation of closed *Splines* (these can be regular curves if the selected points are symmetrically arranged) (Fig. 15-13).

```
Command: spline
Object/<Enter first point>:PICK or (coordinates)
Enter point: PICK or (coordinates)
Close/Fit Tolerance/<Enter point>: PICK or (coordinates)
Close/Fit Tolerance/<Enter point>: PICK or (coordinates)
Close/Fit Tolerance/<Enter point>: Enter
Enter start tangent: PICK or Enter (Select direction for
tangent or Enter for default tangent.)
Enter end tangent: PICK or Enter (Select direction for
tangent.)
Command:
```

Figure 15-13 ————————

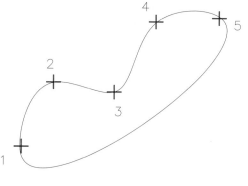

CLOSED SPLINE

The *Object* option allows you to convert *Spline*-fit *Plines* into NURBS *Splines*. Only *Spline*-fit *Plines* can be converted (see *Pedit*, Chapter 16).

Command: **spline**
Object/<Enter first point>: **o**
Select objects to convert to splines.
Select Objects: **PICK**
Select Objects: **Enter**
Command:

A *Fit Tolerance* applied to the *Spline* "loosens" the fit of the curve. A tolerance of 0 (default) causes the *Spline* to pass exactly through the data points. Entering a positive value allows the curve to fall away from the points to form a smoother curve (Fig. 15-14).

> Close/Fit Tolerance/<Enter point>: **F**
> Enter Fit tolerance <0.0000>: (Enter a positive value.)

Figure 15-14

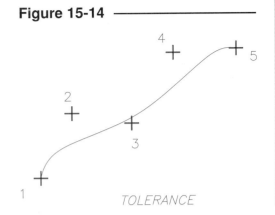

TOLERANCE

ELLIPSE

Pull-down Menu	COMMAND (TYPE)	ALIAS (TYPE)	Short-cut	Screen (side) Menu	Tablet Menu
Draw *Ellipse*	*ELLIPSE*	*EL*	...	*DRAW 1* *Ellipse*	*M,9*

An *Ellipse* is one object. AutoCAD *Ellipses* are (by default) NURBS curves (see *Spline*). There are three methods of creating *Ellipses* in AutoCAD: (1) specify one <u>axis</u> and the <u>end</u> of the second, (2) specify the <u>center</u> and the ends of each axis, and (3) create an elliptical <u>arc</u>. Each option also permits supplying a rotation angle rather than the second axis length.

> Command: **ellipse**
> Arc/Center/<Axis endpoint 1>:**PICK** or (**coordinates**) (This is the first endpoint of either the major or minor axis.)
> Axis endpoint 2: **PICK** or (**coordinates**) (Select a point for the other endpoint of the first axis.)
> <Other axis distance>/Rotation: **PICK** or (**coordinates**) (This distance is measured perpendicularly from the established axis.)

Axis End
This default option requires PICKing three points as indicated in the command sequence above (Fig. 15-15).

Figure 15-15

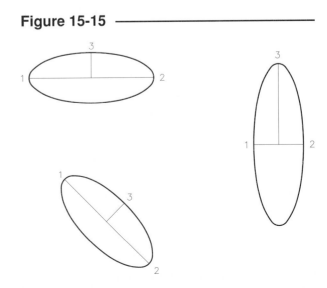

Rotation

If the *Rotation* option is used with the *Axis End* method, the following syntax is used:

> <Other axis distance>/Rotation: **R**
> Rotation around major axis: **PICK** or (**value**)

Figure 15-16

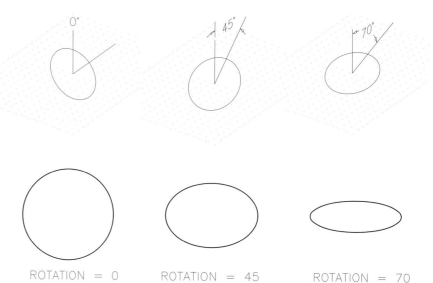

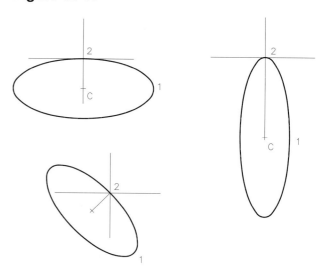

ROTATION = 0   ROTATION = 45   ROTATION = 70

The specified angle is the number of degrees the shape is rotated <u>from the circular position</u> (Fig. 15-16).

Center

With many practical applications, the center point of the ellipse is known, and therefore the *Center* option should be used (Fig. 15-17):

> Command: *ellipse*
> Arc/Center/<Axis endpoint 1>: **C**
> Center of ellipse: **PICK** or (**coordinates**)
> Axis endpoint: **PICK** or (**coordinates**)
> <Other axis distance>/Rotation: **PICK** or (**coordinates**) (This distance is measured perpendicularly from the established axis.)

The *Rotation* option appears and can be invoked after specifying the *Center* and first *Axis endpoint*.

Figure 15-17

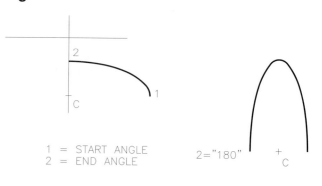

Arc

Use this option to construct an elliptical arc (partial ellipse). The procedure is identical to the *Center* option with the addition of specifying the start- and endpoints for the arc (Fig. 15-18):

> Command: *ellipse*
> Arc/Center/<Axis endpoint 1>: **a**
> <Axis endpoint 1>/Center: **PICK** or (**coordinates**)
> Axis endpoint 2: **PICK** or (**coordinates**)
> <Other axis distance>/Rotation: **PICK** or (**coordinates**)
> Parameter/<start angle>: **PICK** or (**angular value**)
> Parameter/Included/<end angle>: **PICK** or (**angular value**)
> Command:

Figure 15-18

1 = START ANGLE
2 = END ANGLE

2="180"

The *Parameter* option allows you to specify the start point and endpoint for the elliptical arc. The parameters are based on the parametric vector equation: p(u)=c+a*cos(u)+b*sin(u), where c is the center of the ellipse and a and b are the major and minor axes.

DIVIDE

Pull-down Menu	COMMAND (TYPE)	ALIAS (TYPE)	Short-cut	Screen (side) Menu	Tablet Menu
Draw *Point >* *Divide*	*DIVIDE*	*DIV*	...	DRAW 2 *Divide*	V,13

The *Divide* and *Measure* commands add *Point* objects to existing objects. Both commands are found in the *Draw* pull-down menu under *Point* because they create *Point* objects.

The *Divide* command finds equal intervals along an object such as a *Line, Pline, Spline,* or *Arc* and adds a *Point* object at each interval. The object being divided is <u>not</u> actually broken into parts—it remains as <u>one</u> object. *Point* objects are automatically added to display the "divisions."

The point objects that are added to the object can be used for subsequent construction by allowing you to *OSNAP* to equally spaced intervals (*Node*s).

The command sequence for the *Divide* command is as follows:

> Command: *divide*
> Select object to divide: **PICK** (Only one object can be selected.)
> <Number of segments>/Block: (**value**)
> (Enter a value for the number of segments.)
> Command:

Figure 15-19

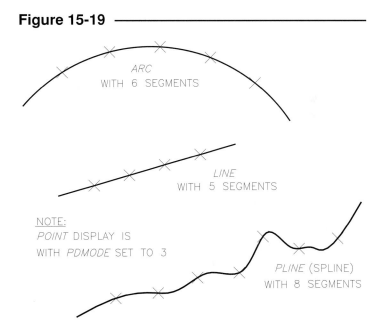

Point objects are added to divide the object selected into the desired number of parts. Therefore, there is <u>one less</u> *Point* added than the number of segments specified.

After using the *Divide* command, the *Point* objects may not be visible unless the point style is changed with the *Point Style...* dialog box (*Format* pull-down menu) or by changing the *PDMODE* variable by command line format (see Chapter 8). A *Regen* must be invoked before the new point style will be displayed. Figure 15-19 shows *Points* displayed using a *PDMODE* of 3.

You can request that *Block*s be inserted rather than *Point* objects along equal divisions of the selected object. Figure 15-20 displays a generic rectangular-shaped block inserted with *Divide*, both aligned and not aligned with a *Line, Arc,* and *Pline.* In order to insert a *Block* using the *Divide* command, the name of an <u>existing</u> *Block* must be given. (See Chapter 21, Blocks.)

Figure 15-20

ARC
BLOCKS ALIGNED

ARC
BLOCKS NOT ALIGNED

LINE
BLOCKS ALIGNED

LINE
BLOCKS NOT ALIGNED

PLINE (SPLINE)
BLOCKS ALIGNED

PLINE (SPLINE)
BLOCKS NOT ALIGNED

MEASURE

Pull-down Menu	COMMAND (TYPE)	ALIAS (TYPE)	Short-cut	Screen (side) Menu	Tablet Menu
Draw Point > Measure	*MEASURE*	*ME*	…	DRAW 2 *Measure*	*V,12*

The *Measure* command is similar to the *Divide* command in that *Point* objects (or *Blocks*) are inserted along the selected object. The *Measure* command, however, allows you to designate the <u>length</u> of segments rather than the <u>number</u> of segments as with the *Divide* command.

    Command: **measure**
    Select object to measure: **PICK** (Only one object can be selected.)
    <Segments length>/Block: (**value**) (Enter a value for length of one segment.)
    Command:

Point objects are added to the selected object at the designated intervals (lengths). One *Point* is added for each interval <u>beginning at the end nearest</u> the end used for object selection (Fig. 15-21). The intervals are of equal length except possibly the last segment, which is whatever length is remaining.

Figure 15-21

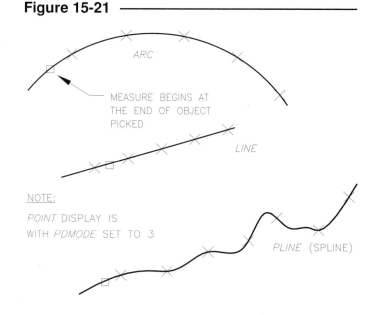

ARC

MEASURE BEGINS AT THE END OF OBJECT PICKED

LINE

NOTE:

POINT DISPLAY IS WITH *PDMODE* SET TO 3

PLINE (SPLINE)

You can request that *Block*s be inserted rather than *Point* objects at the designated intervals of the selected object. Inserting *Block*s with *Measure* requires that an <u>existing</u> *Block* be used.

MLINE

Pull-down Menu	COMMAND (TYPE)	ALIAS (TYPE)	Short-cut	Screen (side) Menu	Tablet Menu
Draw *Multiline*	*MLINE*	*ML*	...	DRAW 1 *Mline*	*M,10*

Mline is short for multiline. A multiline is a set of <u>parallel lines</u> that behave as <u>one object</u>. The individual lines in the set are called line <u>elements</u> and are defined to your specifications. The set can contain up to 16 individual lines, each having its own offset, linetype, or color.

Mlines are useful for creating drawings composed of many parallel lines. An architectural floor plan, for example, contains many parallel lines representing walls. Instead of constructing the parallel lines individually (symbolizing the wall thickness), only <u>one</u> *Mline* need be drawn to depict the wall. The thickness and other details of the wall are dictated by the definition of the *Mline*.

<u>*Mline*</u> is the command to <u>draw</u> multilines. To <u>define</u> the individual line elements of a logical set (offset, linetype, color), use the <u>*Mlstyle*</u> command. An *Mlstyle* (multiline style) is a particular combination of defined line elements. Each *Mlstyle* can be saved in an external file with a .MLN file type. In this way, multilines can be used in different drawings without having to redefine the set. See *Mlstyle* next.

The *Mline* command operates similar to the *Line* command, asking you to specify points:

```
Command: mline
Justification = Top, Scale = 1.00, Style = STANDARD
Justification/Scale/STyle/<From point>: PICK or (coordinates)
<To point>: PICK or (coordinates)
Undo/<To point>: PICK or (coordinates)
Close/Undo/<To point>: PICK or (coordinates)
Close/Undo/<To point>: Enter
Command:
```

Mline provides the following three options for drawing multilines.

Justification
Justification determines how the multiline is drawn between the points you specify. The choices are *Top*, *Zero*, or *Bottom* (Fig. 15-22). *Top* aligns the top of the set of line elements along the points you PICK as you draw. A *Zero* setting uses the center of the set of line elements as the PICK point, and *Bottom* aligns the bottom line element between PICK points.

NOTE: The *Justification* methods are defined for drawing from <u>left to right</u> (positive X direction). In other words, using the *Top* justification, the top of the multiline (as it is defined in the *Mlstyle* dialog box) is between PICK points when you draw from left to right. PICKing points from right to left draws the *Mline* upside down.

Figure 15-22 —————

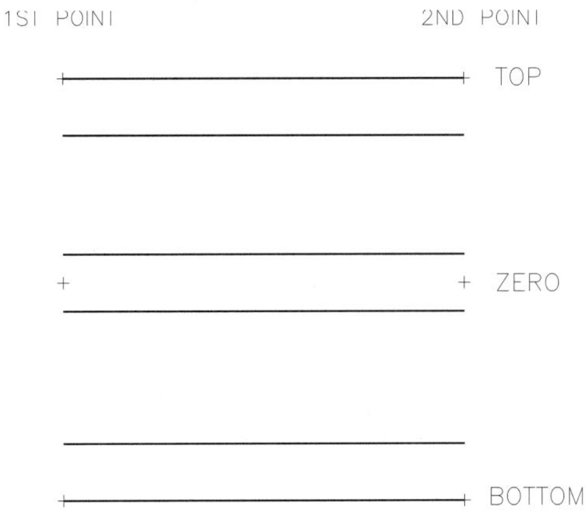

1ST POINT   2ND POINT

├————————————————┤ TOP

———————————————————

———————————————————

+   + ZERO
———————————————————

———————————————————

├————————————————┤ BOTTOM

Scale

The *Scale* option controls the overall <u>width</u> of the multiline. The value you specify for the scale factor increases or decreases the defined multiline width proportionately. In Figure 15-23, a scale factor of 2.0 provides a multiple line pattern that is twice as wide as the *Mlstyle* definition; a scale of 3.0 provides a pattern three times the width.

Figure 15-23

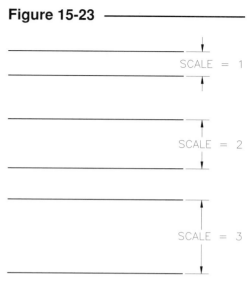

You can also enter negative scale values or a value of 0. A negative scale value flips the order of the multiple line pattern (Fig. 15-24). A value of 0 collapses the multiline into a single line. The utility of the *Scale* option allows you to vary a multiline set proportionally without the need for a complete new multiline definition.

Figure 15-24

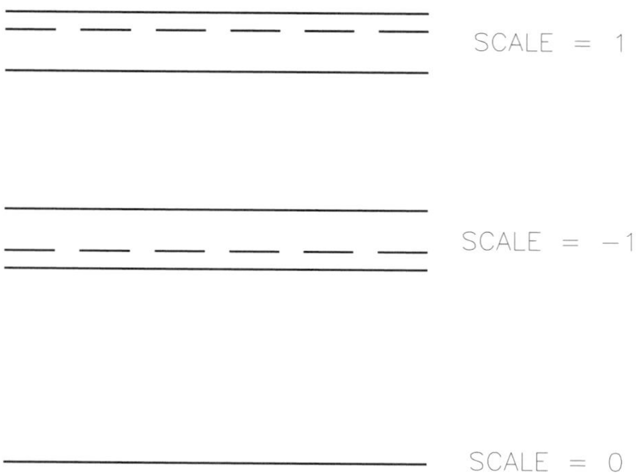

SCALE = 1

SCALE = -1

SCALE = 0

Style

The *Style* option enables you to load different multiline styles provided they have been previously created. Some examples of multiline styles are shown in Figure 15-25. Only one multiline *Style* called STANDARD is provided by AutoCAD. The STANDARD multiline style has two parallel lines defined at one unit width.

Using the *Style* option of the *Mline* command provides only the command line format for setting a current style. Instead, you can use the *Mlstyle* dialog box to select a current multiline style by a more visible method.

Figure 15-25

VARIOUS MULTILINE STYLES

STANDARD (DEFAULT)

SAMPLE 1

SAMPLE 2

MLSTYLE

Pull-down Menu	COMMAND (TYPE)	ALIAS (TYPE)	Short-cut	Screen (side) Menu	Tablet Menu
Format *Multiline Style...*	*MLSTYLE*	...	...	*DRAW 1* *Mline* *Mlstyle:*	*V,5*

The *Mline* command draws multilines, whereas the *Mlstyle* controls the configuration of the multilines that are drawn. Use the *Mlstyle* command to <u>create new</u> multiline definitions or to <u>load, select, and edit</u> previously created multiline styles. *Mlstyle* is <u>not a draw command</u> and is therefore found not in the *Draw* menus but in the *Format* pull-down and screen menus.

Using the *Mlstyle* command by any method invokes the *Multiline Styles* dialog box (Fig. 15-26). The areas of the three main dialog boxes are described next. Creating a new multiline style requires several steps that should follow a specific sequence. The Steps for Creating New Multiline Styles and Steps for Editing Multiline Styles are listed after the sections describing the *Multiline Styles* dialog box and nested dialog boxes.

The *Multiline Styles* Dialog Box

Element Properties... and *Multiline Properties...*
These two tiles produce dialog boxes for defining new or changing existing multiline styles. When creating a new style, use these boxes first to define the line elements and their properties before *Naming, Adding,* and *Saving* the style. Features of these two dialog boxes are described later.

Current
This drop-down list provides you with the names of the currently loaded multiline styles. The image tile in the center of the dialog box provides a representation of the selected *Current* multiline style definition. Therefore, you can select names in the drop-down list to browse through the loaded multiline styles. The *Style* option of the *Mline* command serves the same purpose but operates only in command line format and requires you to know the specific *Mlstyle* name and definition.

Figure 15-26

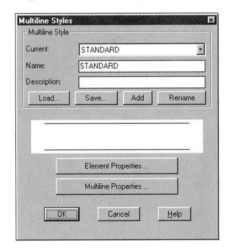

Name
When you create new multiline styles, use the *Name* edit box to enter the name of the style you have created. The *Mlstyle* should be *Named* <u>after</u> assigning *Element Properties* and *Multiline Properties.* Short names are desirable since a *Description* of the multiline configuration can be entered in the edit box just below *Name.*

Description
The *Description* edit box is an optional feature that allows you to provide a description for a new or existing multiline style definition. The description can contain up to 256 characters.

Load...
The *Load* tile invokes the *Load Multiline Styles* dialog box so that you can select an existing multiline definition from an external file (Fig. 15-27). The default multiline definition file is ACAD.MLN, which contains the STANDARD multiline style. This file can be appended or a new .MLN file can be created. An .MLN file can contain <u>more than one</u> multiline style.

Figure 15-27

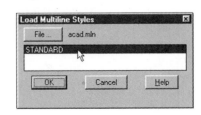

Save...

Selecting the *Save* button invokes the typical save file dialog box (titled *Save Multiline File*) so that you can save a multiline definition to an external file. Two or more multiline styles can be saved to one external file. Use any descriptive file name and an .MLN file extension is automatically appended. Saving a multiline style to an external file allows you to use the multiline definition in other drawings without having to recreate the style.

Add

Once you have defined or edited a new style using the *Element Properties* and *Multiline Properties* dialog boxes, use the *Add* tile. The *Add* tile adds the new multiline definition to the current multiline style drop-down list. Several multiline styles can be created in this manner.

NOTE: New and edited styles require a new *Name* before using *Add*. You must choose the *OK* button in order to save new multiline definitions to the drawing file.

Rename

The *Rename* tile allows you to rename the current multiline style. First, select the style to be renamed so it appears in both the *Current* and *Name* boxes. Next, change the name in the *Name* box and select *Rename*. The new name should appear in both boxes.

The *Element Properties* Dialog Box

The *Element Properties...* button (in the *Multiline Styles* dialog box) invokes the *Element Properties* dialog box (Fig. 15-28). This dialog box provides controls for you to define the offset (spacing), linetype, and color properties of individual line elements of a new or existing multiline style.

The *Elements:* list displays the <u>current</u> settings for the multiline style. Each line entry in a style has an offset, color, and linetype associated with it. A multiline definition can have as many as 16 individual line elements defining it. Select (highlight) any element in the list before changing its properties with the options below.

Figure 15-28

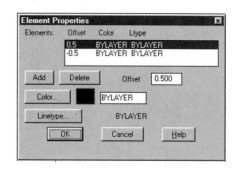

Add

New line elements are added to the list by selecting the *Add* button; then the *Offset, Color,* and *Linetype* properties can be assigned.

Delete

The current line element (the highlighted entry in the *Elements:* list) is deleted by selecting the *Delete* button.

Color

Selecting this tile produces the standard *Select Color* dialog box. *BYLAYER* and *BYBLOCK* settings are valid. Any color selected is assigned to the highlighted line element in the list.

Linetype

Use this button to assign a linetype to the highlighted line element. The standard *Select Linetype* dialog box appears. Linetypes can be *Loaded* from this device.

Offset

Enter a value in the *Offset* edit box for each line element. The offset value is the distance measured perpendicular to the theoretical zero (center) position along the axis of the *Mline*.

The *Multiline Properties* Dialog Box

The *Multiline Properties* dialog box (Fig. 15-29) is also nested and invoked through the *Multiline Styles* dialog box. The options here enable you to control end capping, fill, and joint display properties of the multiline style. This device controls features of the <u>multiline style as a whole</u>, whereas the *Element properties* dialog treats the <u>individual</u> line element properties.

When multiline properties are assigned in this dialog box, the changes are reflected in the image tile upon returning to the *Multiline Styles* dialog box. You can cycle back and forth between these two boxes to "see" the changes as you make them. Figure 15-30 illustrates some examples of the options described below.

Figure 15-29 ──────

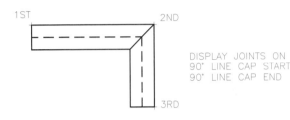

Display Joints

When several segments of an *Mline* are drawn, AutoCAD automatically miters the "joints" so the segments connect with sharp, even corners. The "joint" is the bisector of the angle created by two adjoining segments. The *Display Joints* checkbox toggles on and off this miter line running across all the line elements.

Line

The *Line* checkboxes create <u>straight</u> line caps across either or both of the *Start* or *End* segments of the multiline. The *Line* caps are drawn across all line elements.

Outer Arcs

This option creates an arc connecting and tangent to the two outermost line elements. You can control both the *Start* and *End* segments.

Figure 15-30 ──────

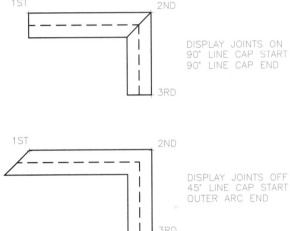

Inner Arcs

The inner arc option creates an arc between interior line segments. This option works with four or more line elements. In the case of an uneven number of line elements (five or more), the middle line is not capped.

Angle

The angle option controls the angle of start and end lines or arcs. Valid values range from 10 to 170 degrees.

Fill

The *Fill* toggle turns off or on the background fill of a multiline. The "X" must appear in the box before the *Color* tile is enabled. Selecting *Color* produces the standard *Select Color* dialog box. The image tile does not display the multiline fill, but fill appears in the drawing.

Steps for Creating New Multiline Styles

1. Use the *Mlstyle* command to invoke the *Multiline Styles* dialog box.

2. Use the *Element Properties* dialog box to define the line elements' offset, linetype, and color properties.

3. Define the multiline end capping, fill, and joint display properties using the *Multiline Properties* dialog box.

4. Return to the *Multiline Styles* dialog box and assign a *Name* to the new style.

5. Use the *Add* button to include the new style to the list of potential current styles and make it the current style.

6. Select *OK* to exit the *Multiline Styles* dialog box and begin drawing *Mlines* with the new style.

7. If you want to use the new style with other drawings, use *Save* in the *Multiline Styles* dialog box to save the style to an external .MLN file.

Steps for Editing Existing Multiline Styles

1. Use the *Mlstyle* command to invoke the *Multiline Styles* dialog box.

2. *Load* the desired multiline style from an external file or select it from the *Current* list. The style name must appear in the *Name* box.

3. If you want to edit the line elements' offset, linetype, and color properties, use the *Element Properties* dialog box.

4. If you want to edit the multiline end capping, fill, and joint display properties, use the *Multiline Properties* dialog box.

5. You must assign a <u>new *Name*</u> to the changed style in the *Multiline Styles* dialog box by editing the old name. You cannot use the same name. An "Invalid name" message appears if you attempt to *Add* using the old name.

6. Then you must *Add* the new (edited) style to the list of styles. This action also makes it the current style.

7. Select *OK* to exit the *Multiline Styles* dialog box and begin drawing *Mlines* with the new style.

8. If you want to use the new (edited) style with other drawings, use *Save* in the *Multiline Styles* dialog box to save the style to an external .MLN file.

Multiline Styles Example

Because the process of creating and editing multiline styles is somewhat involved, here is an example that you can follow to create a new multiline style. The specifications for two multilines are given in Figure 15-31. Follow these steps to create SAMPLE1.

1. Invoke the *Multiline Styles* dialog box by any menu or icon or by typing *Mlstyle*.

2. Select the *Element Properties...* tile to invoke the *Element Properties* dialog box.

Figure 15-31 ———————————————

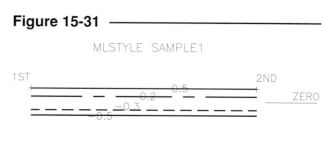

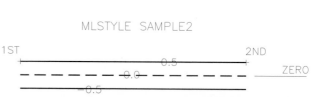

3. Select the *Add* button twice. Four line entries should appear in the *Elements:* list box. The two new entries have a 0.0 offset.

4. Highlight one of the new entries and change its offset by entering 0.2 in the *Offset* edit box. Next, change the linetype by selecting the *Linetype...* button and choosing the *Center* linetype. (If the *Center* linetype is not listed, you should *Load* it.)

5. Highlight the other new entry and change its *Offset* to -0.3. Next, change its *Linetype* to *Hidden*.

6. Select the *OK* button to return to the *Multiline Styles* dialog box. At this point, enter the name SAMPLE1 in the *Name* edit box. Choose the *Add* button to save the new line definition and make it current. Last, choose the *OK* button to save the multiline style to the drawing database.

7. Use *Mline* to draw several segments and to test your new multiline style.

8. Use *Mlstyle* again and enter *Multiline Properties* to create *Line* end caps and a *Color Fill* to the style. Don't forget to change the name (to SAMPLE1-FILL, for example) and *Add* it to the list.

Now try to create SAMPLE2 multiline style on your own (see Fig. 15-31).

Editing Multilines

Because a multiline is treated as one object and its complex configuration is defined by the multiline style, traditional editing commands such as *Trim* and *Extend*, cannot be used for typical editing functions. Instead, a special set of editing functions are designed for *Mlines*. The *Mledit* (multiline edit) command provides the editing functions for existing multilines and is discussed in Chapter 16. Your usefulness for *Mlines* is limited until you can edit them with *Mledit*.

SKETCH

Pull-down Menu	COMMAND (TYPE)	ALIAS (TYPE)	Short-cut	Screen (side) Menu	Tablet Menu
...	*SKETCH*	...	...	...	...

The *Sketch* command is unlike other draw commands. *Sketch* quickly creates many short line segments (individual objects) by following the motions of the cursor. *Sketch* is used to give the appearance of a freehand "sketched" line, as used in Figure 15-32 for the tree and bushes. You do not specify individual endpoints, but rather draw a "freehand line" by placing the "pen" down, moving the cursor, and then picking up the "pen." This action creates a large number of short *Line* segments. *Sketch* line segments are temporary (displayed in a different color) until *Recorded* or until *eXiting Sketch*.

Figure 15-32

CAUTION: *Sketch* can increase the drawing file size greatly due to the relatively large number of line segment endpoints required to define the *Sketch* "line."

Command: *sketch*
Record increment <0.1000>: (**value**) or **Enter** (Enter a value to specify the segment increment length or press Enter to accept the default increment.)
Sketch. Pen eXit Quit Record Erase Connect. (**letter**) (Enter "p" or press button #1 to put the pen down and begin drawing or enter another letter for another option. After drawing a sketch line, enter "p" or press button #1 again to pick up the pen.)

It is important to specify an *increment* length for the short line segments that are created. This *increment* controls the "resolution" of the *Sketch* line (Fig. 15-33).

Too large of an *increment* makes the straight line segments apparent, while too small of an *increment* unnecessarily increases file size. The default *increment* is 0.1 (appropriate for default *Limits* of 12 x 9) and should be changed proportionally with a change in *Limits*. Generally, multiply the default *increment* of 0.1 times the drawing scale factor.

Another important rule to consider whenever using *Sketch* is to <u>turn *SNAP Off* and *ORTHO Off*</u>, unless a "stair-step" effect is desired (Fig. 15-33).

Figure 15-33

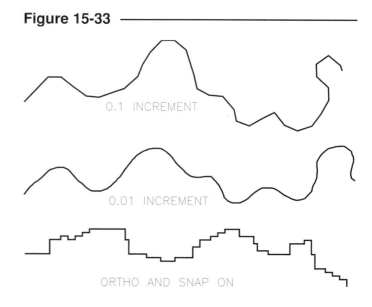

Sketch Options

Options of the *Sketch* command can be activated either by entering the corresponding letter(s) shown in upper case at the *Sketch* command prompt or by pressing the desired mouse or puck button. For example, putting the "pen" up or down can be done by entering *P* at the command line or by pressing button #1. The options of *Sketch* are as follows.

Pen	Lifts and lowers the pen. Position the cursor at the desired location to begin the line. Lower the pen and draw. Raise the pen when finished with the line.
Record	Records all temporary lines sketched so far without changing the pen position. After recording, the lines cannot be *Erased* with the *Erase* option of *Sketch* (although the normal *Erase* command can be used).
eXit	Records all temporary lines entered and returns to the Command: prompt.
Quit	Discards all temporary lines and returns to the Command: prompt.
Erase	Allows selective erasing of temporary lines (before recording). To erase, move backwards from last sketch line toward the first. Press "p" to indicate the end of an erased area. This method works easily for relatively straight sections. To erase complex sketch lines, *eXit* and use the normal *Erase* command with window, crossing, or pickbox object selection.
Connect	Allows connection to the end of the <u>last</u> temporary sketch line (before recording). Move the cursor to the last sketch line and the pen is automatically lowered.
(period)	Draws a straight line (using *Line*) from the last sketched line to the cursor. After adding the straight lines, the pen returns to the up position.

Several options are illustrated in Figure 15-34. The *Pen* option lifts the pen up and down. A *period* (.) causes a straight line segment to be drawn from the last segment to the cursor location. *Erase* is accomplished by entering *E* and making a reverse motion.

As an alternative to the *erase* option of *Sketch*, the *Erase* command can be used to erase all or part of the *Sketch* lines. Using a Window or Crossing Window is suggested to make selection of all the objects easier.

Figure 15-34

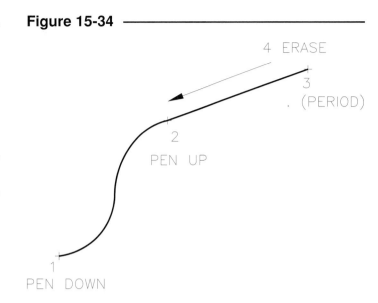

The *SKPOLY* Variable

The *SKPOLY* system variable controls whether AutoCAD creates connected *Sketch* line segments as one *Pline* (one object) or as multiple *Line* segments (separate objects). *SKPOLY* affects <u>newly created</u> *Sketch* lines only (Fig. 15-35).

SKPOLY=0   This setting generates *Sketch* segments as individual *Line* objects. This is the <u>default</u> setting.

SKPOLY=1   This setting generates connected *Sketch* segments as <u>one *Pline* object</u>.

Figure 15-35

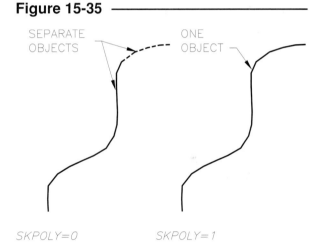

Using editing commands with *Sketch* lines can normally be tedious (when *SKPOLY* is set to the default value of 0). In this case, editing *Sketch* lines usually requires *Zooming* in since the line segments are relatively small. However, changing *SKPOLY* to 1 simplifies operations such as using *Erase* or *Trim*. For example, *Sketch* lines are sometimes used to draw break lines (Fig. 15-36) or to represent broken sections of mechanical parts, in which case use of *Trim* is helpful. If you expect to use *Trim* or other editing commands with *Sketch* lines, change *SKPOLY* to 1 before creating the *Sketch* lines.

Figure 15-36

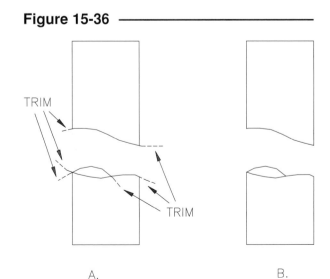

SOLID

Pull-down Menu	COMMAND (TYPE)	ALIAS (TYPE)	Short-cut	Screen (side) Menu	Tablet Menu
Draw Surfaces > 2D Solid	*SOLID*	*SO*	...	DRAW 2 SURFACES Solid:	L,8

The *Solid* command creates a 2D shape that is <u>filled solid</u> with the current color (object or *BYLAYER*). The shape is <u>not</u> a 3D solid. Each 2D solid has either <u>three or four straight edges</u>. The construction process is relatively simple, but the order of PICKing corners is critical. In order to construct a square or rectangle, draw a bow-tie configuration (Fig. 15-37). Several *Solids* can be connected to produce more complex shapes.

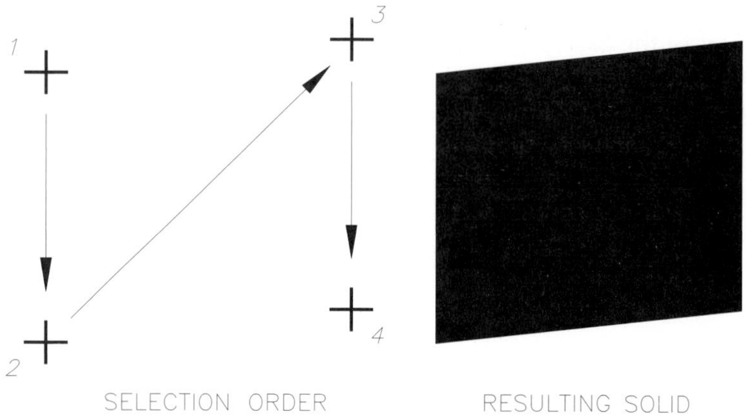

A *Solid* object is useful when you need shapes filled with solid color; for example, three-sided *Solids* can be used for large arrowheads. Keep in mind that solid filled areas may cause problems for some output devices such as plotters with felt pens and thin paper. As an alternative to using *Solid*, enclosed areas can be filled with hatch patterns or solid color (see *Bhatch*, Chapter 26).

Figure 15-37 ——————————————

SELECTION ORDER   RESULTING SOLID

BOUNDARY

Pull-down Menu	COMMAND (TYPE)	ALIAS (TYPE)	Short-cut	Screen (side) Menu	Tablet Menu
Draw Boundary...	*BOUNDARY or -BOUNDARY*	*BO or -BO*	...	DRAW 2 Boundary	Q,9

Boundary finds and draws a boundary from a group of connected or overlapping shapes forming an enclosed area. The shapes can be any AutoCAD objects and can be in any configuration, as long as they form a totally enclosed area. *Boundary* creates either a *Polyline* or *Region* object forming the boundary shape (see *Region* next). The resulting *Boundary* does not affect the existing geometry in any way. *Boundary* finds and includes internal closed areas (called "islands") such as circles (holes) and includes them as part of the *Boundary*.

To create a *Boundary*, select an internal point in any enclosed area (Fig. 15-38). *Boundary* finds the boundary (complete enclosed shape) surrounding the internal point selected. You can use the resulting *Boundary* to construct other geometry or use the generated *Boundary* to determine the *Area* for the shape (such as a room in a floor plan). The same technique of selecting an "internal point" is also used to determine boundaries for sectioning using *Bhatch* (Chapter 26).

Figure 15-38 ——————————————

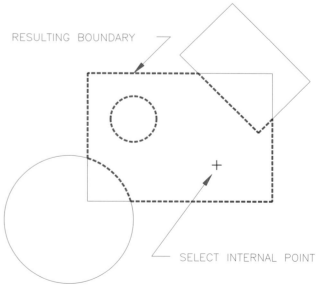

RESULTING BOUNDARY

SELECT INTERNAL POINT

When *Tracking* is turned on, the "rubberband" line extends only in vertical or horizontal directions, similar to the effect of turning on *ORTHO*. Each *Tracking* point specifies only one direction—horizontal or vertical. To indicate which direction you want to track (horizontal or vertical), move the cursor to the first point of tracking, then in either a vertical or horizontal direction. For example, Figure 15-43 illustrates the possible directions for *Tracking* from the *Endpoint* of an existing *Line*.

Figure 15-43

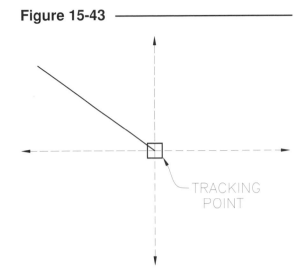

TRACKING POINT

Tracking is invoked like any other *OSNAP*, but other *OSNAPs* can be used <u>after activating tracking</u>. For example, examine this illustration of creating a *Circle* of 2 units radius whose center aligns orthogonally with the *Midpoint* of an adjacent *Line* (Fig. 15-44):

Figure 15-44

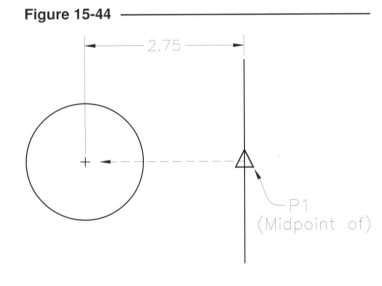

2.75

P1
(Midpoint of)

Command: **CIRCLE**
3P/2P/TTR/<Center point>: **tracking**
(select the *Tracking* icon or menu option)
First tracking point: **mid of** (select the *Midpoint* icon or menu option, then) **PICK** (see P1)
Next point (Press ENTER to end tracking): (move cursor with *ORTHO* on in the desired direction and enter a value) **2.75**
Next point (Press ENTER to end tracking): **Enter**
Diameter/<Radius> <1.5000>: **1**
Command:

A classic example of using *Tracking* effectively is drawing a *Circle* with its center located at the center of an existing rectangle (Figs. 15-45 and 15-46). The coordinate for the <u>center of the rectangle</u> is aligned orthogonally with the midpoint of a horizontal (bottom or top) line and also with the midpoint of a vertical (side) line.

Figure 15-45

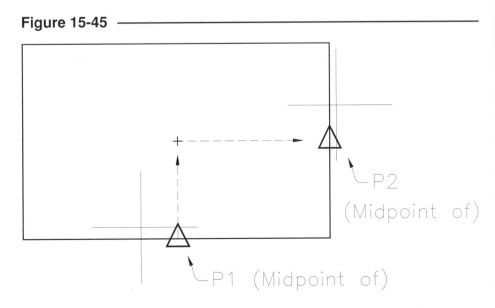

P2
(Midpoint of)

P1 (Midpoint of)

The command syntax for the operation in Figures 15-45 and 15-46 is as follows:

Figure 15-46

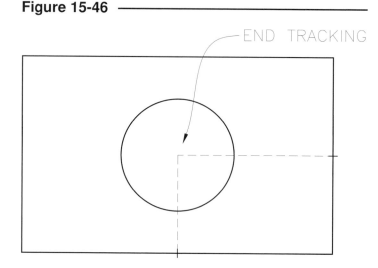

Command: *CIRCLE*
3P/2P/TTR/<Center point>: *tracking*
First tracking point: *mid of* PICK (P1, Fig. 15-45, then move vertically)
Next point (Press ENTER to end tracking):
mid of PICK (P2, Fig. 15-45)
Next point (Press ENTER to end tracking):
Enter (to end *Tracking*)
Diameter/<Radius>: **1.5** (Fig. 15-46)
Command:

NOTE: When *Tracking* two points, <u>it is critical that you first move in the direction you intend from the tracking point.</u> If you are not paying attention to this detail, you can easily become confused.

Another example is given in Figure 15-47. Two *Lines* exist. Two other *Lines* are drawn to complete a four-sided shape such that the new corner aligns orthogonally with the endpoints of the two existing *Lines*:

Figure 15-47

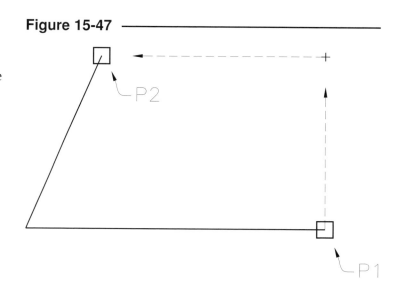

Command: *LINE*
From point: *End of* PICK (P1)
To point: *tracking*
First tracking point: *End of* PICK (P1, then move vertically)
Next point (Press ENTER to end tracking): *End of* PICK (P2) (creates vertical *Line*)
Next point (Press ENTER to end tracking): **Enter**
To point: PICK (P2) (creates horizontal *Line*)
To point: **Enter**
Command:

In summary, *Tracking* can be used for a variety of applications whenever you need to specify points that align orthogonally with other geometry.

POINT FILTERS

Pull-down Menu	COMMAND (TYPE)	ALIAS (TYPE)	Short-cut	Screen (side) Menu	Tablet Menu	Cursor Menu (button 3 or Shift+button 2)
...	*(during command)* .X, .Y, .XY, etc.	...	...	ASSIST .x, .y, .xy, etc.	...	Point Filters >

Point filters provide a feature similar to *Tracking*. Point filters are also used to specify points that align vertically or horizontally with other points. More specifically, point filters find (or "filter") the X, Y, or Z component (or combination) of the selected object's coordinate and use the value for the respective coordinate of the current object point specification. The following examples are valid point filters:

.X
.Y
.Z
.XY
.XZ
.YZ

Figure 15-48 ————

Point filters have been used in AutoCAD for many releases, whereas *Tracking* is new with Release 14. *Tracking* is somewhat easier for most applications because you do not have to specify an X or Y direction; it is assumed by the direction you move the cursor. However, point filters can be used in ways that are not possible with *Tracking,* especially for 3D work. (See Chapters 34 and 37 for information on using 3D point filters.)

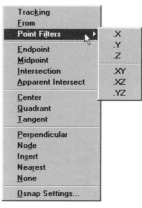

Point filters can be selected from the *OSNAP* cursor menu (Fig 15-48). You can use a point filter whenever you are prompted for a point, such as the *Line* "From point:" or "To point:" prompts or the *Circle* "3P/2P/TTR/ <Center point:>" prompt. Specify a point filter by picking from the menu or entering a period (.) followed by the letter X, Y, or Z, or a combination of two letters.

For example, assume you want to draw a *Line* (Fig. 15-49, highlighted line above) and begin the line at the same X value of an existing line (lower line) so the two lines' endpoints align vertically. When prompted for the "From point:," use an .X point filter to locate and "filter" the existing line's endpoint X ordinate. For accuracy, *OSNAPs* should be used <u>after</u> entering the point filter. The command syntax would read like this:

Figure 15-49 ————————————————————

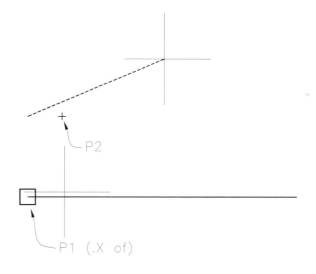

Command: **line**
From point: **.X** (type or select .X)
of **end** (enter OSNAP mode)
of **PICK** (P1)

After entering the point filter, AutoCAD responds with "of." Next, use the *Endpoint OSNAP* mode and PICK the existing line's endpoint (Fig 15-49, P1). AutoCAD filters the X component of the selected point (P1) and passes the value to the *Line* command. AutoCAD responds with:

  (need YZ): **PICK** (PICK to define the Y and Z components)

You must still provide AutoCAD with the Y and Z values. Another point can be PICKed as shown in the figure (P2). To specify the "From point:" of the *Line,* the Y and Z values from the second point (P2) are used, but the X component is the same as that of the first line (P1).

The same strategy can be used to desig-nate the other endpoint of the line. For the "To point:," enter an .X point filter at the prompt and use *Endpoint* to select the other end of the existing line (Fig. 15-50, P3).

Figure 15-50

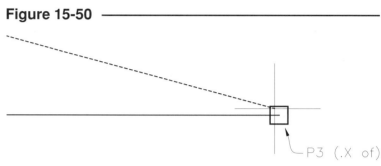

P3 (.X of)

PICK another point (Fig. 15-51, P4) to specify the Y and Z coordinates. The resulting line has the same X ordinates for both of its endpoints as the existing line. The command sequence for the procedure in Figures 15-50 and 15-51 is as follows:

Figure 15-51

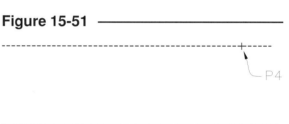

P4

  To point: **.X** (type or select .X)
  of **end** (enter *OSNAP* mode and press Enter)
  of **PICK** (P3)
  (need YZ): **PICK** (P4)
  To point: **Enter**
  Command:

It is possible to specify a point by filtering one existing object's X ordinate and another object's Y ordi-nate. For example you can filter the X component at the *Midpoint* of one existing line and the Y compo-nent at the *Midpoint* of another existing line, similar to the operation shown in Figure 15-45 for construct-ing a *Circle* at the center of a rectangle.

For 2D drawings such as those in the previous examples, only .X and .Y point filters need to be used. The .X and .Y filters can be specified in <u>either order</u>. Other examples of practical 2D applications for point filters are given in Chapter 24, Using Point Filters for Alignment.

Many applications in 3D drawings require X, Y and Z coordinate specification. In addition to the .X, .Y, and .Z filters, the .XY, .XZ, and .YZ filters are useful. Chapters 34 and 37 discuss these concepts.

CHAPTER EXERCISES

1. *Polygon*

   Open the **PLINE1** drawing you created in Chapter 8 Exercises. Use *Polygon* to construct the polygon located at the *Center* of the existing arc, as shown in Figure 15-52. When finished, *SaveAs* **POLYGON1**.

   Figure 15-52 —————————————————

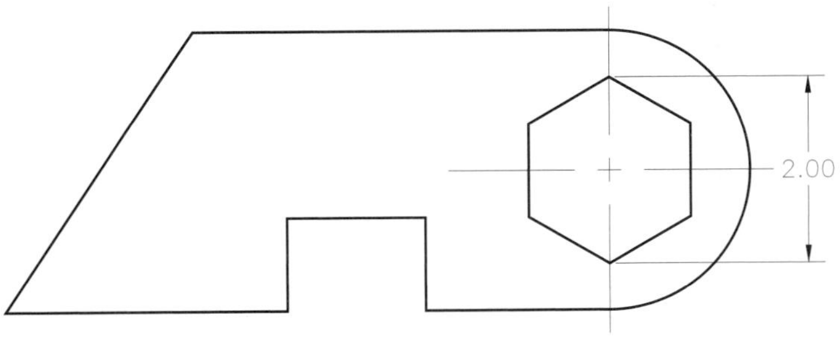

   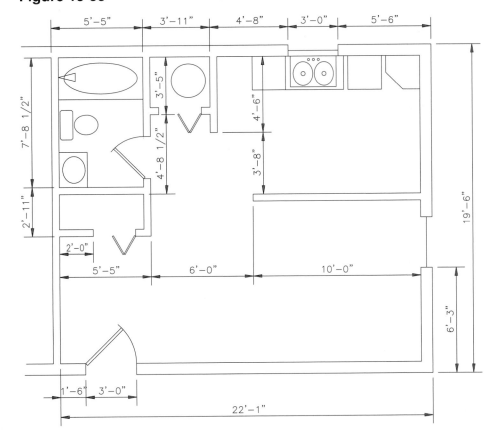

2. *Xline, Ellipse, Rectangle*

   Figure 15-53 ———————————————

   Open the **APARTMENT** drawing that you created in Chapter 13. Draw the floor plan of the efficiency apartment on layer FLOORPLAN as shown in Figure 15-53. Use *SaveAs* to assign the name **EFF-APT**. Use *Xline* and *Line* to construct the floor plan with 8" width for the exterior walls and 5" for interior walls. Use the *Ellipse* and *Rectangle* commands to design and construct the kitchen sink, tub, wash basin, and toilet. *Save* the drawing but do not exit AutoCAD. Continue to Exercise 3.

3. *Polygon, Sketch*

   Figure 15-54 —————————————————

   Create a plant for the efficiency apartment as shown in Figure 15-54. Locate the plant near the entry. The plant is in a hexagonal pot (use *Polygon*) measuring 18" across the corners. Use *Sketch* to create 2 leaves as shown in figure A. Create a *Polar Array* to develop the other leaves similar to figure B. *Save* the EFF-APT drawing.

   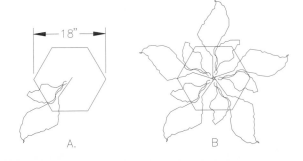

4. *Donut*

Figure 15-55 ————————

Open **SCHEMATIC1** drawing you created in Chapter 10 Exercises. Use the *Point Style* dialog box (*Format* pull-down menu) to change the style of the *Point* objects to dots instead of small circles. Type *Regen* to display the new point style. Turn on the *Node Running Osnap* mode. Use the *Donut* command and set the *Inside diameter* and the *Outside diameter* to an appropriate value for your drawing such that the *Inside diameter* is ½ the value of the *Outside diameter*. Create a *Donut* at each existing *Node*. *SaveAs* **SCHEMATIC1B.** Your finished drawing should look like that in Figure 15-55.

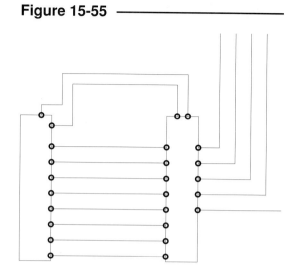

5. *Ellipse, Polygon*

Figure 15-56 ————————

Open the drawing that you set up in Chapter 13 named **WRENCH.** Complete the construction of the wrench shown in Figure 15-56. Center the drawing within the *Limits.* Use *Line, Circle, Arc, Ellipse,* and *Polygon* to create the shape. For the "break lines," create one connected series of jagged *Lines,* then *Copy* for the matching set. Utilize *Trim,*

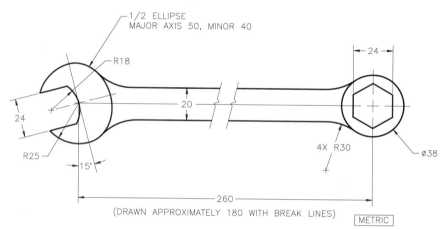

Rotate, and other edit commands where necessary. HINT: Draw the wrench head in an orthogonal position; then rotate the entire shape 15 degrees. Notice the head is composed of ½ of a *Circle* (R25) and ½ of an *Ellipse.* *Save* the drawing when completed.

Figure 15-57 ————————

6. *Xline, Spline*

Create the handle shown in Figure 15-57. Construct multiple *Horizontal* and *Vertical Xlines* at the given distances on a separate layer (shown as dashed lines in Figure 15-57). Construct two *Splines* that make up the handle sides by PICKing the points indicated at the *Xline* intersections. Note the *End Tangent* for one *Spline.* Connect the *Splines* at each end with horizontal *Lines.* *Freeze* the construction (*Xline*) layer. *Save* the drawing as **HANDLE.**

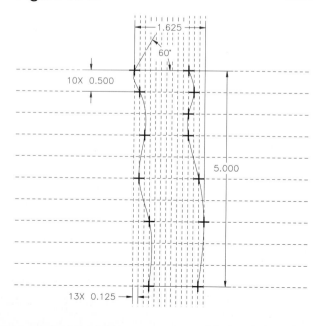

7. *Divide, Measure*

Figure 15-58

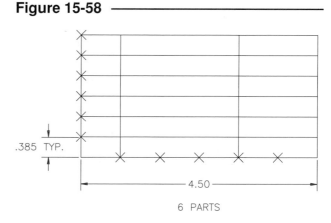

Use the **ASHEET** drawing as a template and assign the name **BILLMATL**. Create the table in Figure 15-58 to be used as a bill of materials. Draw the bottom *Line* (as dimensioned) and a vertical *Line*. Use *Divide* along the bottom *Line* and *Measure* along the vertical *Line* to locate *Points* as desired. Create *Offsets Through* the *Points* using *Node OSNAP*. (*ORTHO* and *Trim* may be of help.)

8. *Mline, Mlstyle, Sketch*

Figure 15-59

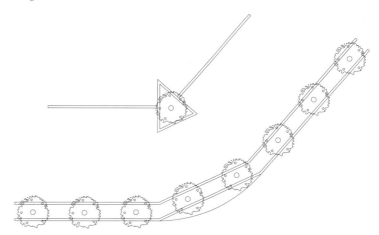

Using the **RET-WALL** drawing that you created in Chapter 8, add the concrete walls, island, and curbs (Fig. 15-59). Use *Mlstyle* to create a multiline style consisting of 2 line elements with *Offsets* of **1** and **-1**. Assign *Start* and *End Caps* at **90** degrees. *Name* the new style **WALL** and *Add* it to the list to make it current. Next, use *Mline* with the *Bottom Justification* and draw one *Mline* along the existing retaining wall centerline. Use *Endpoint* to snap to the ends of the existing 2 *Lines* and *Arc*. Construct a second *Mline* above that using the following coordinates:

    60,123
    192,123
    274,161
    372.7,274

Next, construct the *Closed* triangular island with an *Mline* using *Top Justification*. The coordinates are:

    192,192
    222.58,206.26
    192,233

For the parking curbs, draw two more *Mlines*; each *Mline* snaps to the *Midpoint* of the island and is **100'** long. The curb on the right has an angle of **49** degrees (use polar coordinate entry). Finally, create trees similar to the ones shown using *Sketch*. Remember to set the *SKPOLY* variable. Create one tree and *Copy* it to the other locations. Spacing between trees is approximately **48'**. *Save* as **RET-WAL2**.

9. *Boundary*

Open the **GASKETA** drawing that you created in Chapter 9 Exercises. Use the *Boundary* command to create a boundary of the <u>outside shape only</u> (no islands). Use *Move Last* to displace the new shape to the right of the existing gasket. Use *SaveAs* and rename the drawing to **GASKET-AREA**. Keep this drawing to determine the *Area* of the shape after reading Chapter 17.

10. *Region*

Create a gear, using *Regions*. Begin a *New* drawing, **Start from Scratch** with the **English** defaults settings. Use *Save* and assign the name **GEAR-REGION**.

A. Set *Limits* at **8** x **6** and *Zoom All*. Create a *Circle* of **1** unit *diameter* with the center at **4,3**. Create a second concentric *Circle* with a *radius* of **1.557**. Create a closed *Pline* (to represent one gear tooth) by entering the following coordinates:

From point:	**5.571,2.9191**
To point:	**@.17<160**
To point:	**@.0228<94**
To point:	**@.0228<86**
To point:	**@.17<20**
To point:	**c**

The gear at this stage should look like that in Figure 15-60.

B. Use the *Region* command to convert the three shapes (two *Circles* and one "tooth") to three *Regions*.

C. *Array* the small *Region* (tooth) in a *Polar* array about the center of the gear. There are **40** items that are rotated as they are copied. This action should create all the teeth of the gear (Fig. 15-61).

D. *Save* the drawing. The gear will be completed in Chapter 16 Exercises by using the *Subtract* Boolean operation.

11. *Tracking*

A. In this exercise, you will create a table base using *Tracking* to locate points for construction of holes and other geometry. Begin a *New* drawing and use the default **ACAD.DWT** template. *Save* the drawing and assign the name **TABLE-BASE**. Set the drawing limits to **48** x **32**. Create 2 new *Layers*, named **LEGS** and **TABLE**; then set the **LEGS** layer as *Current*.

B. Use a *Pline* and create the three line segments representing the first leg as shown in Figure 15-62. Begin at the indicated location. (Do not create the dimensions in your drawing.)

Figure 15-60

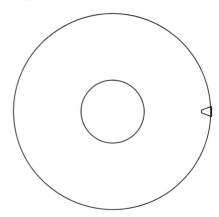

Figure 15-61

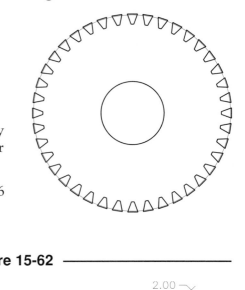

Figure 15-62

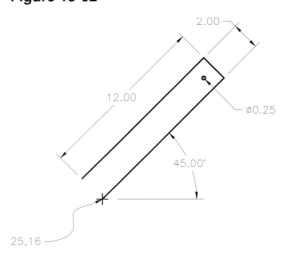

C. Place the first drill hole at the end of the leg using *Circle* with the *Center, Radius* option. Use *Tracking* to indicate the center of the hole (*Circle*). Track vertically and horizontally from the indicated corners of the leg in Figure 15-63. Use *Endpoint OSNAP* to snap to the indicated corners. (See Figure 15-62 for hole dimension.)

Figure 15-63

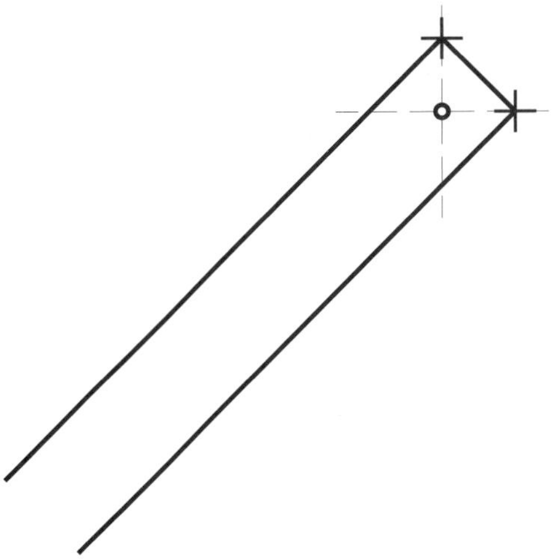

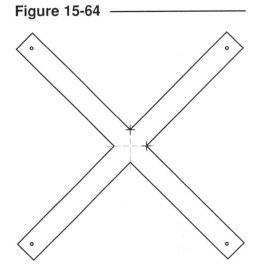

D. Create a *Polar Array* of **4** legs by using *Tracking* in a similar fashion to indicate the center point of the array (in the center of the "X" formed by the four legs). This time use *Tracking* with the *Endpoints* on the bottom of the leg to identify the center point of the *Array* as shown in Figure 15-64.

Figure 15-64

E. Use *Pline* to create a 24″ x 24″ square table top on the right side of the drawing as shown in Figure 15-65. Use *Tracking* with the *Move* command to move the square's center point to the center point of the array of the 4 legs (HINT: using *Tracking, OSNAP* to the square's *Midpoints*).

Figure 15-65

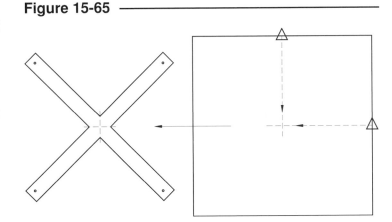

F. Create a smaller square (*Pline*) in the center of the table which will act as a support plate for the legs. Use *Tracking* on the *Midpoint* of the legs to get the square as indicated in Figure 15-66. Use *Trim* to edit the legs and the support plate to look like Figure 15-67.

Figure 15-66 ————————————

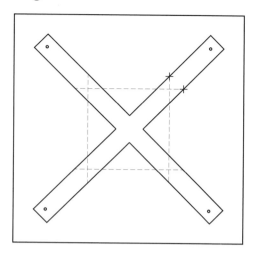

G. Use *Circle* with the *Center, Radius* option to create a 4" diameter circle to represent the welded support for the center post for this table. Finally, add the other 4 drill holes (.25" diameter) on the legs at the intersection of the vertical and horizontal lines representing the edges of the base. Use *Tracking* or *Intersection OSNAP* for this step. The table base should appear as shown in Figure 15-67. *Save* the drawing.

Figure 15-67 ————————————

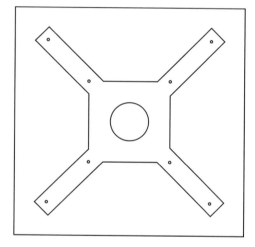

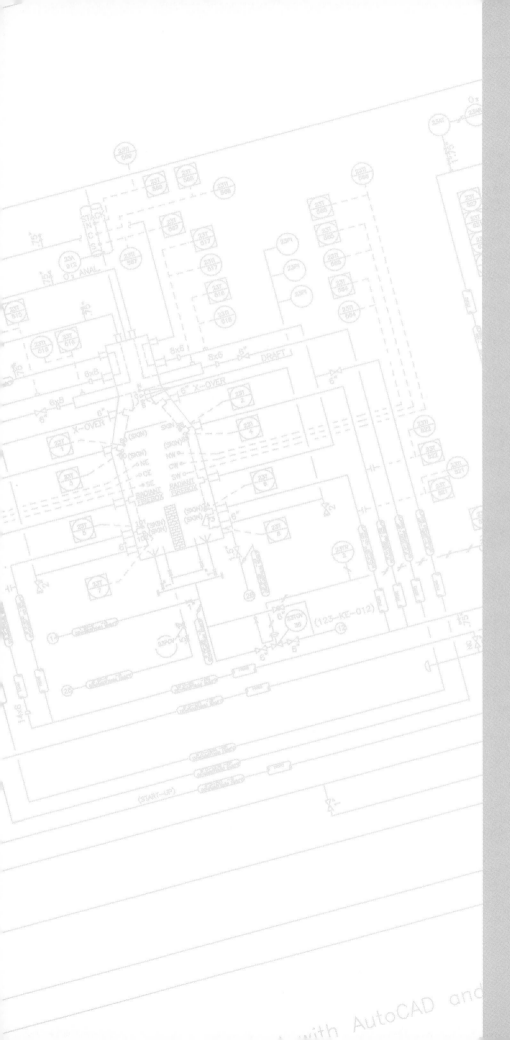

16

MODIFY COMMANDS III

Chapter Objectives

After completing this chapter you should:

1. be able to use *Ddchprop* and *Chprop* to change an object's layer, color, and linetype properties;

2. know how to use *Change* to change points or properties;

3. be able to use *Ddmodify* to modify any type of object;

4. be able to use *Matchprop* and the *Object Properties* toolbar to change properties of an object;

5. be able to *Explode* a *Pline* into its component objects and know that *Explode* can be used with *Blocks* and other composite shapes;

6. be able to *Align* objects with other objects;

7. be able to use all the *Pedit* options to modify *Plines* and to convert *Lines* and *Arcs* to *Plines*;

8. be able to modify *Splines* with *Splinedit*;

9. be able to edit existing *Mlines* using *Mledit*;

10. know that composite *Regions* can be created with *Union*, *Subtract*, and *Intersect*.

CONCEPTS

This chapter examines commands that are similar to, but generally more advanced and powerful than, those discussed in Chapters 9 and 10, Modify Commands I and II. None of the commands in this chapter creates new duplicate objects from existing objects but instead modifies the properties of the objects or convert objects from one type to another. Several commands are used to modify specific types of objects such as *Pedit* (modifies *Plines*), *Splinedit* (modifies *Splines*), *Mledit* (modifies *Mlines*), and *Union, Subtract,* and *Intersect* (modify *Regions*). Only one command in this chapter, *Align*, does not modify object properties but combines *Move* and *Rotate* into one operation. Several of the commands and features discussed in this chapter were mentioned in Chapter 12 (*Ddchprop, Ddmodify, MatchProp,* and *Object Properties* toolbar) but are explained completely here.

Only the commands in this chapter that modify general object properties have icon buttons that appear in the AutoCAD Drawing Editor by default. For example, you can access *Ddmodify, Ddchprop,* and *Matchprop* from the *Object Properties* toolbar and *Explode* by using its icon from the *Modify* toolbar.

Other commands that modify specific objects such as *Pedit, Mledit, Splinedit,* and the Boolean operators (*Union, Subtract,* and *Intersect*) appear in the *Modify II* toolbar. Activate the *Modify II* toolbar by selecting *Toolbars…* from the *View* pull-down menu (Fig. 16-1).

Figure 16-1

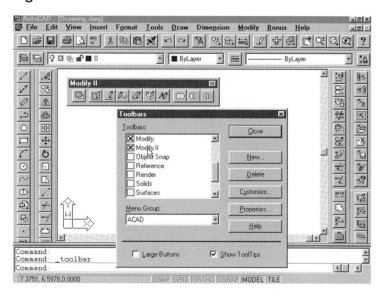

COMMANDS

You can use several methods to change properties of an object or of several objects. If you want to change an object's layer, color, or linetype only, the quickest method is by using the *Object Properties* toolbar (see *Object Properties* toolbar). If you want to change many properties of one or more objects (including layer, color, linetype, linetype scale, text style, dimension style, or hatch style) to match the properties of another existing object, use *Matchprop*. If you want to change any property (including coordinate data) for one object, use *Ddmodify*. You can use *Ddchprop* to change basic properties such as layer, linetype, and color for many objects at one time.

CHPROP and
DDCHPROP

	Pull-down Menu	COMMAND (TYPE)	ALIAS (TYPE)	Short-cut	Screen (side) Menu	Tablet Menu
	Modify Properties...	*CHPROP or DDCHPROP*	*CH*	...	*MODIFY1 Modify*	*Y,14*

Typing *Chprop* (Change Properties) produces the command line format of this command. Typing *Ddchprop*, selecting the icon, or selecting *Properties* from the *Modify* pull-down menu invokes the *Change Properties* dialog box (Fig. 16-2).

Ddchprop and *Chprop* allow you to change properties of several objects with one command. If you use the icon button or pull-down menu, you must select more than one object (in response to the "Select Objects:" prompt) if you want to use the *Change Properties* dialog box. If only one object is selected, the *Modify* dialog box appears instead (see *Ddmodify*).

Figure 16-2

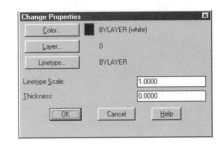

Typing *Chprop* produces the command line format:

    Command: **chprop**
    Select Objects: **PICK**
    Select Objects: **Enter**
    Change what property (Color/LAyer/LType/ltScale/Thickness) ?

Layer
By changing an object's *Layer*, the object is effectively moved to the designated layer. In doing so, if the object's *Color* and *Linetype* are set to *BYLAYER*, the object assumes the color and linetype of the layer to which it is moved.

Color
It may be desirable in some cases to assign explicit *Color* to an object or objects independent of the layer on which they are drawn. The *Change Properties* command allows changing the color of an existing object from one object color to another or from *BYLAYER* assignment to an object color. An object drawn with an object-specific color can also be changed to *BYLAYER* with this option.

Linetype
An object can assume the *Linetype* assigned *BYLAYER* or can be assigned an object-specific *Linetype*. The *Linetype* option of *Change Properties* is used to change an object's *Linetype* to that of its layer (*BYLAYER*) or to any object-specific linetype that has been loaded into the current drawing. An object drawn with an object-specific linetype can also be changed to *BYLAYER* with this option.

Linetype Scale
An object's individual linetype scale can be changed with this option but not the global linetype scale (*LTSCALE*). This is the recommended method to alter an individual object's linetype scale. First, draw all the objects in the current global *LTSCALE*. *Ddchprop* or *Chprop* could then be used with this option to retroactively adjust the linetype scale of specific objects. The result would be similar to setting the *CELTSCALE* before drawing the specific objects. Using *Ddchprop* or *Chprop* to adjust a specific object's linetype scale does not reset the current *LTSCALE* or *CELTSCALE* variables.

Thickness

An object's *Thickness* can be changed by this option. *Thickness* is a three-dimensional quality (Z dimension) assigned to a two-dimensional object (see Chapter 40, Surface Modeling).

CHANGE

Pull-down Menu	COMMAND (TYPE)	ALIAS (TYPE)	Short-cut	Screen (side) Menu	Tablet Menu
...	CHANGE	-CH	...	...	...

The *Change* command allows changing three options: *Points, Properties,* or *Text.*

Point

This option allows changing the endpoint of an object or endpoints of several objects to one new position:

```
Command: change
Select Objects: PICK
Select Objects: Enter
Properties/<Change point>: PICK (Select a point to establish as new endpoint of all objects.
OSNAPs can be used.)
```

The endpoint(s) of the selected object(s) <u>nearest</u> the new point selected at the "Properties/<Change point>:" prompt is changed to the new point (Fig. 16-3).

Figure 16-3

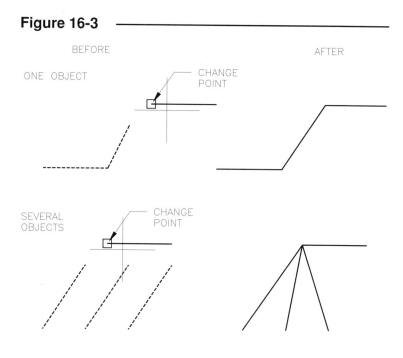

Properties

These options are discussed with the previous command. Selecting *Change* with the *Properties* option offers the same possibilities as using the *Ddchprop* or *Chprop* command. The *Elevation* property (a 3D property) of an object can also be changed with *Change* but not with *Ddchprop* or *Chprop* (see Chapter 40 for information on *Elevation*).

```
Properties/<Change point>: p
Change what property
(Color/Elev/LAyer/LType/ltScale/
Thickness) ?
```

Text

Although the word "*Text*" does <u>not</u> appear as an option, the *Change* command recognizes text created with *Text* or *Dtext* if selected. <u>*Change* does not change *Mtext*</u>. (See Chapter 18 for information on *Text, Dtext,* and *Mtext*.)

You can change the following characteristics of *Text* or *Dtext* objects:

Text insertion point
Text style
Text height
Text rotation angle
Textual content

To change text, use the following command syntax:

Command: *change*
Select Objects: **PICK** (Select one or several lines of existing text.)
Select Objects: **Enter**
Properties/<Change point>: **Enter**
Enter text insertion point: **PICK** or **Enter** (PICK for a new insertion point or press Enter for no change.)
Text Style: ROMANS (Indicates the style of the selected text.)
New style or RETURN for no change: (**text style name**) or **Enter**
New height <0.2000>: (**value**) or **Enter** (Enter a value for new height or Enter if no change.)
New rotation angle: (**value**) or **Enter** (Enter value for new angle or Enter for no change.)
New text <FILLETS AND ROUNDS .125>: (**new text**) or **Enter** (Enter the new <u>complete line</u> of text or press Enter for no change.)

DDMODIFY

Pull-down Menu	COMMAND (TYPE)	ALIAS (TYPE)	Short-cut	Screen (side) Menu	Tablet Menu
Modify Properties...	*DDMODIFY*	*MO*	...	*MODIFY1 Modify*	*Y,14*

Ddmodify allows you to <u>change any property of one object</u>. If using the icon or menus, <u>PICK only one object to invoke the *Modify (object type)* dialog box</u>. If more than one object is selected, the *Change Properties* dialog box appears instead.

The *Ddmodify* command always invokes a dialog box. The power of this feature is apparent because the configuration of the dialog box that appears is <u>specific</u> to the <u>type</u> of object that you select. For example, if you select a *Line,* a dialog box appears, allowing changes to any properties that a *Line* possesses; or, if you select a *Circle* or some *Text,* a dialog box appears specific to *Circle* or *Text* properties.

Figure 16-4 displays the dialog box after selecting a *Line.* Notice the dialog box title. Any aspect of the *Line* can be modified.

The *Properties* section always appears at the top of the *Modify (object type)* dialog box. This section has the same capabilities as the *Change Properties* dialog box. <u>All properties that an object possesses</u>, including position in the drawing, can be modified with the respective *Modify* dialog box.

Figure 16-4 ───────────────

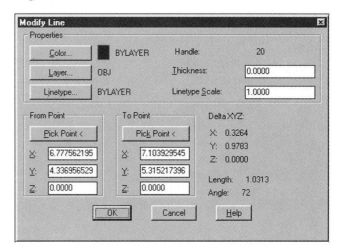

Figure 16-5 displays the dialog box after a *Pline* selection. Properties or individual vertices can be changed. Features of the *Pline* that can be modified here are similar to the capabilities of the *Pedit* command (see *Pedit,* this chapter).

Figure 16-5

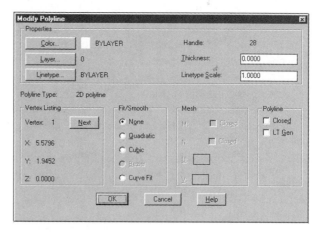

Each *Modify (object type)* dialog box offers a different set of changeable properties and features, depending on the object selected. Notice that the *Modify Circle* dialog box (Fig. 16-6) allows changing the radius and coordinates of the center as well as the typical properties (layer, linetype, etc.). Additional information such as diameter, circumference, and area is given.

Figure 16-6

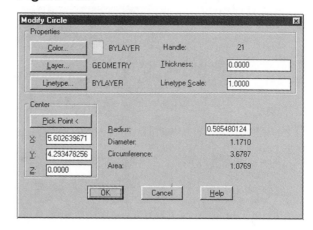

In some cases, the *Modify (object type)* dialog box enables editing by invoking another dialog box. If an *Mtext* object is selected, you can select the *Full editor...* tile (Fig. 16-7), which produces the same dialog box that was originally used to create the text.

Figure 16-7

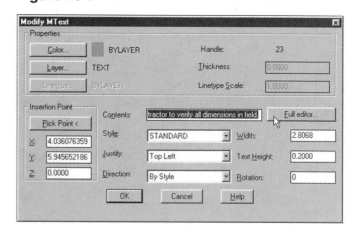

Selecting the *Full editor...* tile produces the *Multiline Text Editor* (Fig. 16-8), enabling you to change any properties that text possesses. (See Chapter 18 for information on *Mtext* and other text objects.)

Figure 16-8

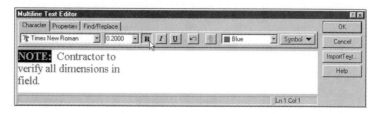

If hatch lines (created with the *Bhatch* command) are selected, the *Modify Hatch* dialog appears. Selecting the *Hatch Edit...* tile (Fig. 16-9) invokes the *Hatchedit* dialog box, providing access to parameters that were used to create the hatched area. (See Chapter 26 for information on hatching.)

Figure 16-9

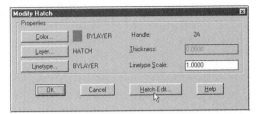

MATCHPROP

Pull-down Menu	COMMAND (TYPE)	ALIAS (TYPE)	Short-cut	Screen (side) Menu	Tablet Menu
Modify *Match Properties*	*MATCHPROP or* *PAINTER*	*MA*	...	*MODIFY1* *Matchprp*	Y,15

Matchprop is explained briefly in Chapter 12 but is explained again in this chapter with the full details of the *Special Properties*.

Matchprop is used to "paint" the properties of one object to another. Simply invoke the command, select the object that has the desired properties ("Source Object"), then select the object you want to "paint" the properties to ("Destination Object"). The command prompt is as follows:

```
Command: matchprop
Select Source Object: PICK
Current active settings = color layer ltype ltscale thickness text dim hatch
Settings/<Select Destination Object(s)>: PICK
Settings/<Select Destination Object(s)>: Enter
Command:
```

Figure 16-10

You can select Several "Destination Objects." The "Destination Object(s)" assume all of the "Current active settings" of the "Source Object."

The *Property Settings* dialog box can be used to set which of the *Basic Properties* and *Special Properties* are to be painted to the "Destination Objects" (Fig. 16-10). At the "Settings/<Select Destination Object(s)>:" prompt, type *S* to display the dialog box. You can control the following *Basic Properties*. Only the checked properties are "painted" to the destination objects.

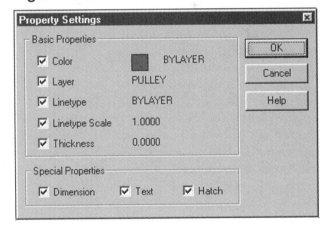

Color	This option paints the object-specific or *BYLAYER* color.
Layer	Move selected objects to the Source Object layer with this option checked.
Linetype	Paint the object-specific or *BYLAYER* linetype of the "Source Object" to the "Destination Object."
Linetype Scale	This option changes the individual object's linetype scale (*CELTSCALE*) not global linetype scale (*LTSCALE*).
Thickness	Thickness is a 3-dimensional quality (see Chapter 40).

R14

The *Special Properties* section allows you to specify features of dimensions, text, and hatch patterns to match, as explained:

Dimension   This setting paints the *Dimension Style.* A *Dimension Style* defines the appearance of a dimension such as text style, size of arrows and text, tolerances if used, and many other features. (See Chapter 29.)

Text   This setting paints the "Source Object's" text *Style.* The text *Style* defines the text font and many other parameters that affect the appearance of the text. (See Chapter 18.)

Hatch   Checking this box paints the hatch properties of the Source Object to the "Destination Object(s)." The properties can include the hatch *Pattern, Angle, Scale,* and other characteristics. (See Chapter 26.)

Matchprop is a simple and powerful command to use, especially if you want to convert dimensions, text or hatch patterns to look like other objects in the drawing. This method works only when you have existing objects in the drawing you want to "match." If you want to convert only layer, linetype, and color properties without changing the other properties or if you do not have existing objects to "match," you can use the *Object Properties* toolbar.

Object Properties **Toolbar**

The three drop-down lists in the Object Properties toolbar (when not "dropped down") generally show the <u>current</u> layer, color, and linetype. However, if an object or set of objects is selected, the information in these boxes changes to display the current <u>object's</u> settings. You can change an object's settings by picking an object (when no commands are in use), then "dropping down" any of the three lists and selecting a different layer, linetype, or color.

Make sure you select (highlight) an object when no commands are in use. Use the pickbox (that appears on the cursor), Window, or Crossing Window to select the desired object or set of objects. The entries in the three boxes (Layer Control, Color Control, and Linetype Control) then change to display the settings for the <u>selected object or objects</u>. If several objects are selected that have different properties, the boxes display no information (go "blank"). Next, use any of the drop-down lists to make another selection (Fig. 16-11). The highlighted object's properties are changed to those selected in the lists. Press the Escape key to complete the process.

Figure 16-11

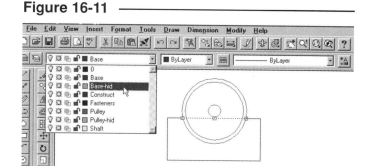

Remember that in most cases color and linetype settings are assigned *BYLAYER.* In this type of drawing scheme, to change the linetype or color properties of an object, you would change only the object's <u>layer</u> (see Figure 16-11). If you are using this type of drawing scheme, refrain from using the Color Control and Linetype Control drop-down lists for changing properties.

This method for changing object properties is about as quick and easy as *Matchprop;* however, only the layer, linetype, or color can be changed with the *Object Properties* toolbar. The *Object Properties* toolbar method works well if you do not have other existing objects to "match" or if you want to change <u>only</u> layer, linetype, and color <u>without</u> matching text, dimension, and hatch styles.

EXPLODE

Pull-down Menu	COMMAND (TYPE)	ALIAS (TYPE)	Short-cut	Screen (side) Menu	Tablet Menu
Modify *Explode*	*EXPLODE*	*X*	...	*MODIFY2* *Explode*	*Y,22*

Many graphical shapes can be created in AutoCAD that are made of several elements but are treated as one object, such as *Plines, Polygons, Blocks, Hatch* patterns, and dimensions. The *Explode* command provides you with a means of breaking down or "exploding" the complex shape from one object into its many component segments (Fig. 16-12). Generally, *Explode* is used to allow subsequent editing of one or more of the component objects of a *Pline, Polygon,* or *Block,* etc., which would otherwise be impossible while the complex shape is considered one object.

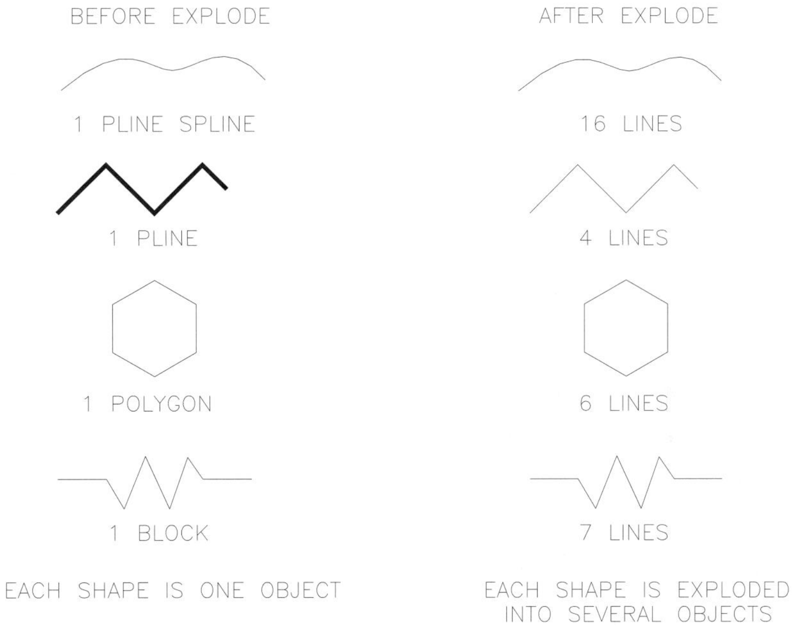

The *Explode* command has no options and is simple to use. You only need to select the objects to *Explode*.

Figure 16-12 ————————————

```
        BEFORE EXPLODE                      AFTER EXPLODE

     1 PLINE SPLINE                        16 LINES

        1 PLINE                              4 LINES

       1 POLYGON                            6 LINES

        1 BLOCK                              7 LINES

EACH SHAPE IS ONE OBJECT           EACH SHAPE IS EXPLODED
                                   INTO SEVERAL OBJECTS
```

    Command: **explode**
    Select Objects: **PICK** (Select one or more *Plines, Blocks,* etc.)
    Select Objects: **Enter** (Indicates selection of objects is complete.)
    Command:

When *Plines, Polygons, Blocks,* or hatch patterns are *Exploded,* they are transformed into *Line, Arc,* and *Circle* objects. Beware, *Plines* having *width* lose their width information when *Exploded* since *Line, Arc,* and *Circle* objects cannot have width. *Exploding* objects can have other consequences such as losing "associativity" of dimensions and hatch objects and increasing file sizes by *Exploding Blocks.*

ALIGN

Pull-down Menu	COMMAND (TYPE)	ALIAS (TYPE)	Short-cut	Screen (side) Menu	Tablet Menu
Modify *3D Operations>* *Align*	*ALIGN*	*AL*	...	*MODIFY2* *Align*	*X,14*

Align provides a means of aligning one shape (a simple object, group of objects, *Pline, Boundary, Region, Block,* or a 3D object) with another shape. *Align* provides a complex motion, usually a combined translation (like *Move*) and rotation (like *Rotate*), in one command.

The alignment is accomplished by connecting source points (on the shape to be moved) to destination points (on the stationary shape). You should use *OSNAP* modes to select the source and destination points to assure accurate alignment. Either a 2D or 3D alignment can be accomplished with this command. The command syntax for alignment in a <u>2D alignment</u> is as follows:

```
Command: align
Select objects: PICK
Select objects: Enter
1st source point: PICK (with OSNAP)
1st destination point: PICK (with OSNAP)
2nd source point: PICK (with OSNAP)
2nd destination point: PICK (with OSNAP)
3rd source point: Enter
<2D> or 3D transformation: Enter
```

This command performs a translation and a rotation in one motion if needed to align the points as designated (Fig. 16-13).

First, the 1st source point is connected to (actually touches) the 1st destination point (causing a translation). Next, the vector defined by the 1st and 2nd source points is aligned with the vector defined by the 1st and 2nd destination points (causing rotation). If no 3rd destination point is given (needed only for a 3D alignment), a 2D alignment is assumed and performed on the basis of the two sets of points.

Figure 16-13

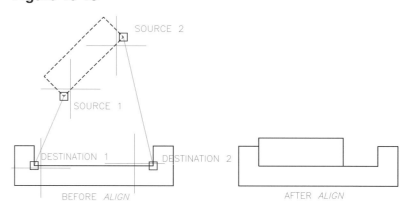

PEDIT

	COMMAND (TYPE)	ALIAS (TYPE)	Short-cut	Screen (side) Menu	Tablet Menu
Pull-down Menu					
Modify *Object >* *Polyline*	*PEDIT*	*PE*	...	*MODIFY1* *Pedit*	*Y,17*

This command provides numerous options for editing *Polylines* (*Plines*). As an alternative, *Ddmodify* can be used to change many of the *Pline*'s features in dialog box form (see Fig. 16-5).

The list of options below emphasizes the great flexibility possible with *Polylines*. The first step after invoking *Pedit* is to select the *Pline* to edit.

```
Command: pedit
Select polyline: PICK (Select the polyline for subsequent editing.)
Close or Open/Join/Width/Edit vertex/Fit/Spline/Decurve/ Ltype gen/Undo/eXit <X>: (option)
(Select the desired option from the screen menu or enter the capitalized letter for the desired option.)
```

Close

Close connects the last segment with the first segment of an existing "open" *Pline*, resulting in a "closed" *Pline* (Fig. 16-14). A closed *Pline* is one continuous object having no specific start or endpoint, as opposed to one closed by PICKing points. A *Closed Pline* reacts differently to the *Spline* option and to some commands such as *Fillet, Pline* option (see *Fillet*, Chapter 10).

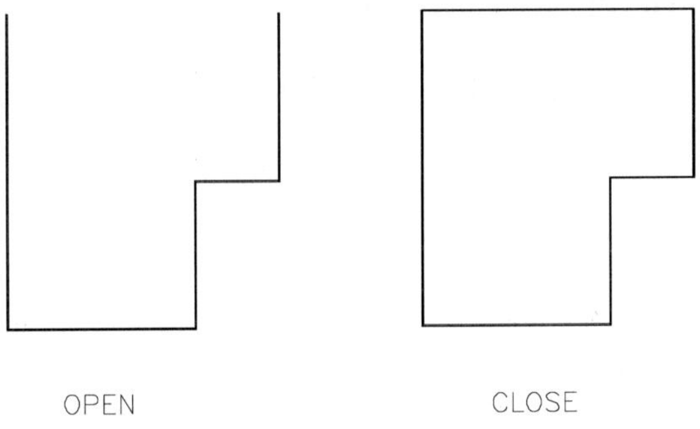

Open

Open removes the closing segment if the *Close* option was used previously (Fig. 16-14).

Figure 16-14

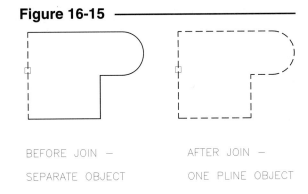

OPEN   CLOSE

Join

This option *Joins*, or connects, any *Plines, Lines,* or *Arcs* that have <u>exact</u> matching endpoints and adds them to the selected *Pline* (Fig. 16-15). Previously closed *Plines* cannot be *Joined*.

Figure 16-15

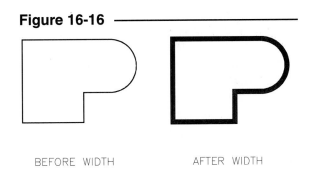

BEFORE JOIN –   AFTER JOIN –

SEPARATE OBJECT   ONE PLINE OBJECT

Width

Width allows specification of a uniform width for *Pline* segments (Fig. 16-16). Non-uniform width can be specified with the *Edit vertex* option.

Edit vertex

This option is covered in the next section.

Figure 16-16

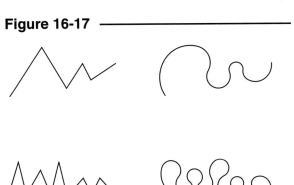

BEFORE WIDTH   AFTER WIDTH

Fit

This option converts the *Pline* from straight line segments to arcs. The curve consists of two arcs for each pair of vertices. The resulting curve can be radical if the original *Pline* consists of sharp angles. The resulting curve passes <u>through all</u> vertices (Fig. 16-17).

Figure 16-17

BEFORE FIT   AFTER FIT

Spline

This option converts the *Pline* to a B-spline (Bezier spline). The *Pline* vertices act as "control points" affecting the shape of the curve. The resulting curve passes through <u>only the end</u> vertices. A *Spline*-fit *Pline* is <u>not the same as a spline curve created with the *Spline*</u> command. This option produces a less versatile version of the newer *Spline* object.

Decurve

Decurve removes the *Spline* or *Fit* curve and returns the *Pline* to its original straight line segments state (Fig. 16-18).

When you use the *Spline* option of *Pedit,* the amount of "pull" can be affected by setting the *SPLINETYPE* system variable to either 5 or 6 <u>before</u> using the *Spline* option. *SPLINETYPE* applies either a quadratic (5=more pull) or cubic (6=less pull) B-spline function (Fig. 16-19).

The *SPLINESEGS* system variable controls the number of line segments created when the *Spline* option is used. The variable should be set <u>before</u> using the option to any value (8=default): the higher the value, the more line segments. The actual number of segments in the resulting curve depends on the original number of *Pline* vertices and the value of the *SPLINETYPE* variable (Fig. 16-20).

Changing the *SPLFRAME* variable to 1 causes the *Pline* frame (the original straight segments) to be displayed for *Splined* or *Fit Plines. Regen* must be used after changing the variable to display the original *Pline* "frame" (Fig. 16-21).

Figure 16-18

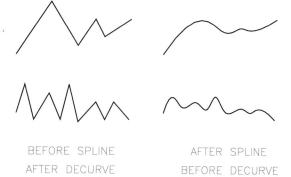

BEFORE SPLINE   AFTER SPLINE
AFTER DECURVE   BEFORE DECURVE

Figure 16-19

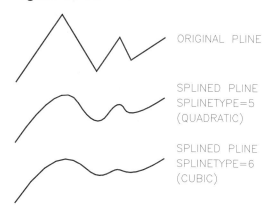

ORIGINAL PLINE

SPLINED PLINE
SPLINETYPE=5
(QUADRATIC)

SPLINED PLINE
SPLINETYPE=6
(CUBIC)

Figure 16-20

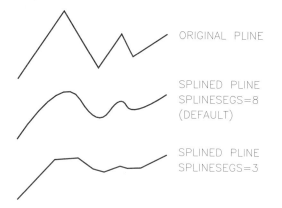

ORIGINAL PLINE

SPLINED PLINE
SPLINESEGS=8
(DEFAULT)

SPLINED PLINE
SPLINESEGS=3

Figure 16-21

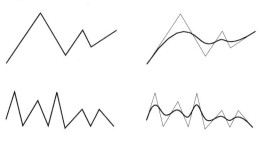

SPLFRAME = 0   SPLFRAME = 1

Ltype gen

This setting controls the generation of non-continuous linetypes for *Plines*. If *Off,* non-continuous
linetype dashes start and stop at each vertex, as if
the *Pline* segments were individual *Line* segments.
For dashed linetypes, each line segment begins and
ends with a full dashed segment (Fig. 16-22). If
On, linetypes are drawn in a consistent pattern,
disregarding vertices. In this case, it is possible for
a vertex to have a space rather than a dash. Using
the *Ltype gen* option <u>retroactively</u> changes *Plines*
that have already been drawn. *Ltype gen* affects
objects composed of *Plines* such as *Polygons,*
Rectangles, and *Boundaries.*

Figure 16-22

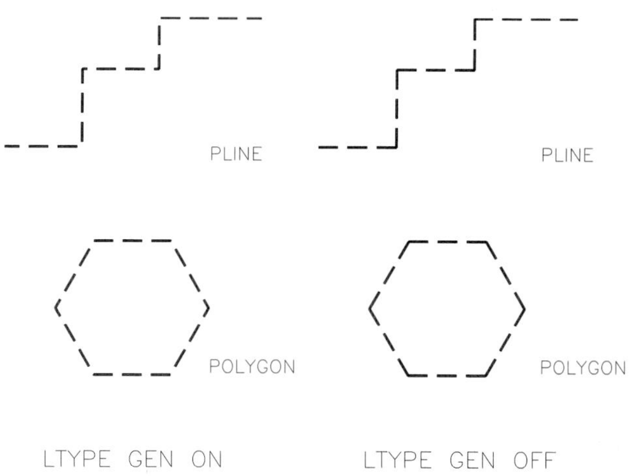

PLINE PLINE

POLYGON POLYGON

LTYPE GEN ON   LTYPE GEN OFF

Similarly, the *PLINEGEN* system variable controls
how <u>new</u> non-continuous linetypes are drawn for *Plines.* A setting of 1 creates a consistent linetype
pattern, disregarding vertices (like *Ltype gen On*). A *PLINEGEN* setting of 0 creates linetypes stopping
and starting at each vertex (like *Ltype gen Off*). However, *PLINEGEN* is <u>not retroactive</u>—it affects only
<u>new</u> *Plines.*

Undo

Undo reverses the most recent *Pedit* operation.

eXit

This option exits the *Pedit* options, keeps the changes, and returns to the Command: prompt.

Vertex Editing

Upon selecting the *Edit Vertex* option from the *Pedit* options list, the group of suboptions is displayed on
the screen menu and command line:

Command: **pedit**
Select polyline: **PICK** (Select the polyline for subsequent editing.)
Close or Open/Join/Width/Edit vertex/Fit/Spline/Decurve/Ltype gen/Undo/eXit <X>: **E** (Invokes the Edit
vertex suboptions.)
Next/Previous/Break/Insert/Move/Regen/Straighten/Tangent/ Width/eXit <N>:

Next

AutoCAD places an **X** marker at the first endpoint of the
Pline. The *Next* and *Previous* options allow you to
sequence the marker to the desired vertex (Fig. 16-23).

Previous

See *Next* above.

Figure 16-23

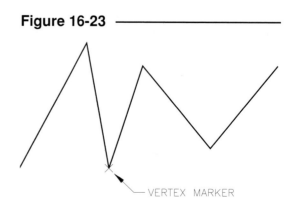

VERTEX MARKER

PREVIOUS AND NEXT OPTIONS WILL MOVE THE
MARKER TO THE DESIRED VERTEX

Break

This selection causes a break between the marked vertex and another vertex you then select using the *Next* or *Previous* option (Fig. 16-24).

Next/Previous/Go/eXit <N>:

Selecting *Go* causes the break. An endpoint vertex cannot be selected.

Figure 16-24

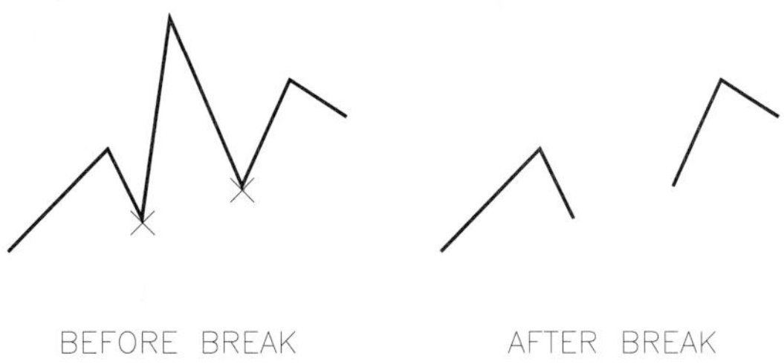

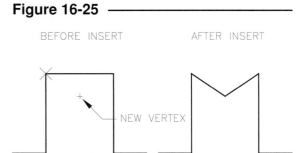

BEFORE BREAK   AFTER BREAK

Insert

Insert allows you to insert a new vertex at any location <u>after</u> the vertex that is marked with the **X** (Fig. 16-25). Place the marker before the intended new vertex, use *Insert*, then PICK the new vertex location.

Figure 16-25

BEFORE INSERT   AFTER INSERT

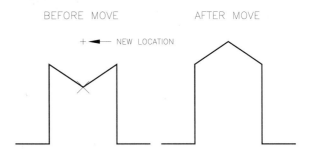

NEW VERTEX

Move

You are prompted to indicate a new location to *Move* the marked vertex (Fig. 16-26).

Regen

Regen should be used after the *Width* option to display the new changes.

Figure 16-26

BEFORE MOVE   AFTER MOVE

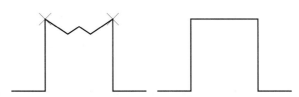

+ ◄— NEW LOCATION

Straighten

You can *Straighten* the *Pline* segments between the current marker and the other marker that you then place by one of these options:

Next/Previous/Go/eXit <N>:

Selecting *Go* causes the straightening to occur (Fig. 16-27).

Figure 16-27

BEFORE STRAIGHTEN   AFTER STRAIGHTEN

Tangent

Tangent allows you to specify the direction of tangency of the current vertex for use with curve *Fitting*.

Width

This option allows changing the *Width* of the *Pline* segment immediately following the marker, thus achieving a specific width for one segment of the *Pline* (Fig. 16-28). *Width* can be specified with different starting and ending values. *Regen* must then be used to display the changes in width after using this option.

Figure 16-28

BEFORE WIDTH   AFTER WIDTH

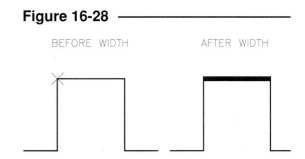

eXit
This option exits from vertex editing, saves changes, and returns to the main *Pedit* prompt.

Grips
Plines can also be edited easily using *Grips* (see Chapter 23). A *Grip* appears on each vertex of the *Pline*. Editing *Plines* with *Grips* is sometimes easier than using *Pedit* because *Grips* are more direct and less dependent on the command interface.

Converting *Lines* and *Arcs* to *Plines*

A very important and productive feature of *Pedit* is the ability to convert *Lines* and *Arcs* to *Plines* and closed *Pline* shapes. Potential uses of this option are converting a series of connected *Lines* and *Arcs* to a closed *Pline* for subsequent use with *Offset* or for inquiry of the area (*Area* command) or length (*List* command) of a single shape. The only requirement for conversion of *Lines* and *Arcs* to *Plines* is that the selected objects must have <u>exact</u> matching endpoints.

To accomplish the conversion of objects to *Plines,* simply select a *Line* or *Arc* object and request to turn it into one:

> Command: **pedit**
> Select polyline: **PICK** (Select one *Line* or *Arc*.)
> Object selected is not a Pline
> Do you want to turn it into one? <Y> **Enter** (Instructs AutoCAD to convert the selected object into a *Pline*. Only <u>one</u> object can be selected.)
> Close or Open/Join/Width/Edit vertex/Fit/Spline/Decurve/Ltype gen/Undo/eXit <X>: **J**
> Select Objects: **PICK** (Select the object or group of objects to join. Multiple objects can be selected.)
> Select Objects: **Enter** (number) segments added to polyline

The resulting conversion is a closed *Polyline* shape.

SPLINEDIT

	COMMAND (TYPE)	ALIAS (TYPE)	Short-cut	Screen (side) Menu	Tablet Menu
Pull-down Menu					
Modify *Object >* *Splinedit*	SPLINEDIT	SPE	...	MODIFY1 *Splinedit*	Y,18

Splinedit is an extremely powerful command for changing the configuration of existing *Splines*. You can use multiple methods to change *Splines*. All of the *Splinedit* methods fall under <u>two sets of options</u>.

The two groups of options that AutoCAD uses to edit *Splines* are based on two sets of points: <u>data points</u> and <u>control points</u>. Data points are the points that were specified when the *Spline* was created—the points that the *Spline* actually <u>passes through</u> (Fig. 16-29).

Figure 16-29

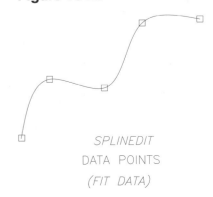

SPLINEDIT
DATA POINTS
(FIT DATA)

Control points are other points <u>outside of the path</u> of the *Spline* that only have a "<u>pull</u>" effect on the curve (Fig. 16-30).

Figure 16-30

Editing *Spline* <u>Data Points</u>

The command prompt displays several levels of options. The top level of options uses the control points method for editing. Select *Fit Data* to use data points for editing. The <u>*Fit Data*</u> methods are <u>recommended</u> for most applications. Since the curve passes directly through the data points, these options offer direct control of the curve path.

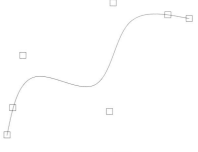

SPLINEDIT
CONTROL POINTS

    Command: *splinedit*
    Select spline: **PICK**
    Fit Data/Close/Move Vertex/Refine/rEverse/Undo/eXit <X>: *f*
    Add/Close/Delete/Move/Purge/Tangents/toLerance/eXit <X>: (**option**)

Add

You can add points to the *Spline*. The *Spline* curve changes to pass through the new points. First, PICK an existing point on the curve. That point and the next one in sequence (in the order of creation) become highlighted. The new point will change the curve between those two highlighted data points (Fig. 16-31).

Figure 16-31

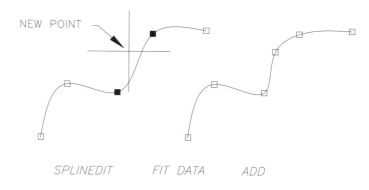

NEW POINT

SPLINEDIT   FIT DATA   ADD

    Select point: **PICK** (Select an existing point
    on the curve before the intended new point.)
    Enter new point: **PICK** (PICK a new point
    location between the two marked points.)

Close/Open

The *Close* option appears only if the existing curve is open, and the *Open* prompt appears only if the curve is closed. Selecting either option automatically forces the opposite change. *Close* causes the two ends to become tangent, forming one smooth curve (Fig. 16-32). This <u>tangent continuity</u> is characteristic of *Closed Splines* only. *Splines* that have matching endpoints do not have tangent continuity unless the *Close* option of *Spline* or *Splinedit* is used.

Figure 16-32

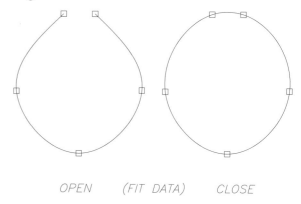

OPEN   (FIT DATA)   CLOSE

Move

You can move any data point to a new location with this option (Fig. 16-33). The beginning endpoint (in the order of creation) becomes highlighted. Type *N* for next or *S* to select the desired data point to move; then PICK the new location.

Figure 16-33

    Next/Previous/Select Point/eXit/<Enter new location> <N>:

Purge

Purge <u>deletes all data points</u> and renders the *Fit Data* set of options unusable. You are returned to the control point options (top level). To reinstate the points, use *Undo*.

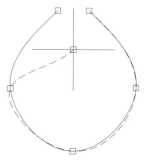

SPLINEDIT   FIT DATA   MOVE

Tangents
You can change the directions for the start and endpoint tangents with this option. This action gives the same control that exists with the "Enter start tangent" and "Enter end tangent" prompts of the *Spline* command used when the curves were created (see Fig. 15-12, *Spline, End Tangent*).

toLerance
Use *toLerance* to specify a value, or tolerance, for the curve to "fall" away from the data points. Specifying a tolerance causes the curve to smooth out, or fall, from the data points. The higher the value, the more the curve "loosens." The *toLerance* option of *Splinedit* is identical to the *Fit Tolerance* option available with *Spline* (see Fig. 15-14, *Spline, Tolerance*).

Editing *Spline* <u>Control Points</u>

Use the top level of command options (except *Fit Data*) to edit the *Spline's* control points. These options are similar to those used for editing the data points; however, the results are different since the curve does not pass through the control points.

```
Command: splinedit
Select spline: PICK
Fit Data/Close/Move Vertex/Refine/rEverse/Undo/eXit <X>:
```

Fit Data
Discussed previously.

Close/Open
These options operate similar to the *Fit Data* equivalents; however, the resulting curve falls away from the control points (Fig. 16-34; see also Fig. 16-32, *Fit Data, Close*).

Figure 16-34

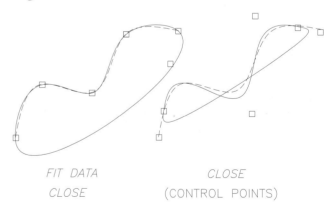

FIT DATA
CLOSE

CLOSE
(CONTROL POINTS)

Move Vertex
Move Vertex allows you to move the location of any control points. This is the control points' equivalent to the *Move* option of *Fit Data* (see Fig. 16-33, *Fit Data, Move*). The method of selecting points (*Next/Previous/Select point/eXit/*) is the same as that used for other options.

Refine
Selecting the *Refine* option reveals another level of options.

```
Add control point/Elevate Order/Weight/eXit <X>:
```

Add control points is the control points' equivalent to *Fit Data Add* (see Fig. 16-31). At the "Select a point on the Spline" prompt, simply PICK a point near the desired location for the new point to appear. Once the *Refine* option has been used, the *Fit Data* options are no longer available.

Elevate order allows you to <u>increase the number of control points</u> uniformly along the length of the *Spline*. Enter a value from *n* to 26, where *n* is the current number of points + one. Once a *Spline* is elevated, it cannot be reduced.

Weight is an option that you use to assign a value to the <u>amount of "pull"</u> that a <u>specific control point</u> has on the *Spline* curve. The higher the value, the more "pull," and the closer the curve moves toward the control point. The typical method of selecting points (*Next/Previous/Select point/eXit/*) is used.

rEverse

The *rEverse* option reverses the direction of the *Spline*. The first endpoint (when created) then becomes the last endpoint. Reversing the direction may be helpful for selection during the *Move* option.

Grips

Splines can also be edited easily using *Grips* (see Chapter 23). The *Grip* points that appear on the *Spline* are identical to the *Fit Data* points. Editing with *Grips* is a bit more direct and less dependent on the command interface.

MLEDIT

	COMMAND (TYPE)	ALIAS (TYPE)	Short-cut	Screen (side) Menu	Tablet Menu
Pull-down Menu					
Modify *Object >* *Multiline*	*MLEDIT or* *-MLEDIT*	...	...	*MODIFY1* *Mledit*	*Y,19*

The *Mledit* command provides tools for editing multilines created with the *Mline* command. Somewhat surprisingly, the typical line editing commands such as *Trim, Extend, Break, Fillet,* and *Chamfer* <u>cannot</u> be used with *Mlines* unless the *Mlines* are *Exploded* (see Other Editing Possibilities for *Mlines* at the end of this section). Instead, the *Mledit* command is required for *Mline* modification and contains several editing functions. These editing functions give you control of multiline intersections.

Invoking the *Mledit* command by any method produces the *Multiline Edit Tools* dialog box. The dialog box presents 12 image tiles that you PICK to activate the desired function. <u>Single-clicking</u> on an image tile displays the option name in the lower-left corner of the dialog box (Fig. 16-35). <u>Double-clicking</u> on an image tile activates that option. The dialog box then disappears, allowing you to select the desired multiline segments for the editing action.

Figure 16-35

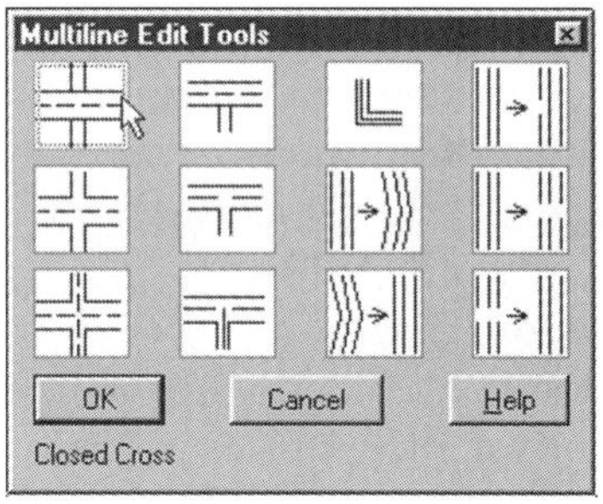

> Command: **mledit** (The *Multiline Edit Tools* dialog box appears. Select desired option.)
> Select mline: **PICK**
> Select mline(or Undo): **PICK** or **Enter**
> Command:

The *Multiline Edit Tools* dialog box is organized in columns as follows:

Intersection—Cross	Intersection—Tee	Corner, Vertices	Lines
Closed Cross	Closed Tee	Corner Joint	Cut Single
Open Cross	Open Tee	Vertex Add	Cut All
Merged Cross	Merged Tee	Vertex Delete	Weld

The options are described and illustrated next.

Closed Cross
Use this option to trim one of two intersecting *Mlines*. The <u>first</u> *Mline* PICKed is <u>trimmed</u> to the outer edges of the second. The <u>second</u> *Mline* is "<u>closed</u>." All line elements in the first multiline are trimmed (Fig. 16-36).

Figure 16-36

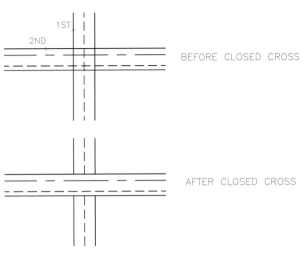

Open Cross
This option <u>trims both</u> intersecting *Mlines*. Both *Mlines* are "open." <u>All</u> line elements of the <u>first</u> *Mline* PICKed are trimmed to the outer edges of the second. Only the outer line elements of the second multiline are trimmed, while the inner lines continue through the intersection (Fig. 16-37).

Figure 16-37

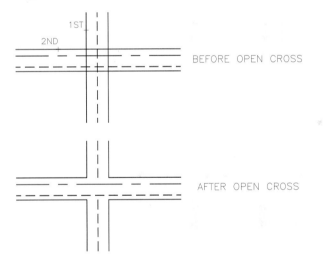

Merged Cross
With this option, the <u>outer</u> line elements of <u>both</u> intersecting *Mlines* are trimmed and the <u>inner line elements merge</u>. The inner line elements merge at the second *Mline's* next set (Fig. 16-38). A full merge (both inner lines continue through) occurs only if the second *Mline* PICKed has no or only one inner line element.

Figure 16-38

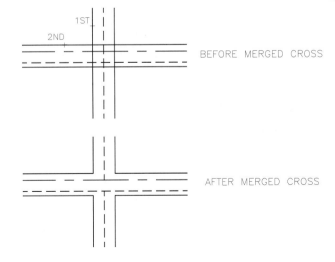

Closed Tee

As indicated by the image tile, a "T" intersection is created rather than a "crossing" intersection. The <u>first</u> *Mline* PICKed is <u>trimmed</u> to the <u>nearest</u> (to the PICK point) outer edge of the second. Only the side of the first *Mline* nearest the PICK point remains. The second *Mline* is not affected (Fig. 16-39).

Figure 16-39

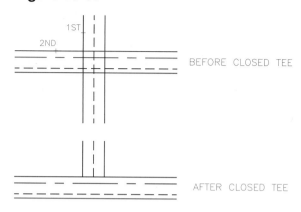

Open Tee

With this "T" intersection, both outer edges of two intersecting *Mlines* are "open." All line elements of the <u>first</u> *Mline* PICKed are <u>trimmed</u> at the outer edges of the second. Only the outer line element of the second is trimmed (Fig. 16-40).

Figure 16-40

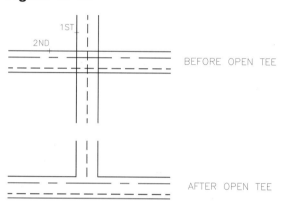

Merged Tee

This "T" option allows the <u>inner line elements</u> of both intersecting *Mlines* to *merge*. The merge occurs at the second *Mline's* first inner element. Only the side of the first line nearest the PICK point remains (Fig. 16-41).

Figure 16-41

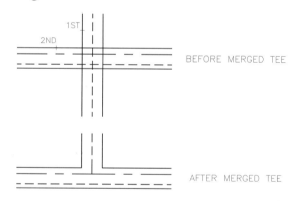

Corner Joint

This option <u>trims both</u> *Mlines* to create a <u>corner</u>. Only the PICKed sides of the *Mlines* remain and the extending portions of both (if any) are trimmed. All of the inner line elements merge (Fig. 16-42).

Figure 16-42

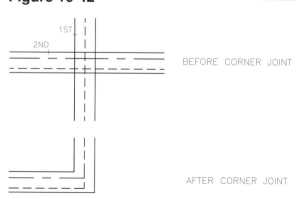

Add Vertex

If you want to add a new corner (vertex) to an existing *Mline,* use this feature. A new vertex is added <u>where you PICK</u> the *Mline.* However, it is not readily apparent that a new vertex exists, nor does the *Mline* visibly change in any way. You must <u>further edit</u> the *Mline* with *Stretch* or Grips (see Chapter 23) to create a new "corner" (Fig. 16-43).

Figure 16-43 ———————

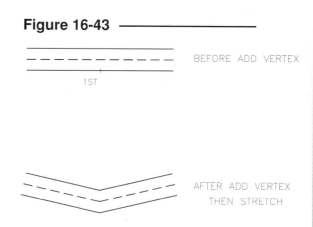

Delete Vertex

Use this feature to remove a corner (vertex) of an *Mline.* The vertex <u>nearest</u> the location you PICK is deleted. The resulting *Mline* contains only one straight segment between the adjacent two vertices. Unlike *Add Vertex,* the *Mline* immediately changes and no further editing is needed to see the effect (Fig. 16-44).

Figure 16-44 ———————

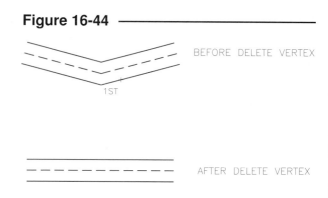

Cut Single

The *Cut* options are used to break (cut) a space in one *Mline. Cut Single* <u>breaks any line element</u> that is selected. Similar to *Break 2Points,* the break occurs <u>between the two PICK points</u> (Fig. 16-45). The break points can be on either side of a vertex.

Figure 16-45 ———————

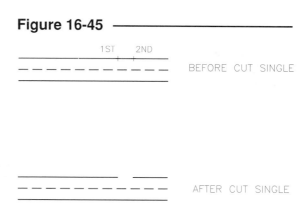

Cut All

This *Cut* option <u>breaks all line elements</u> of the selected *Mline.* Any line elements can be selected. The line elements are cut at the PICK points in a direction perpendicular to the axis of the *Mline* (Fig. 16-46). NOTE: Although the *Mline* appears to be cut into two separate *Mlines,* it <u>remains one single object</u>. Using *Stretch* or Grips to "move" the *Mline* causes the cut to close again.

Figure 16-46 ———————

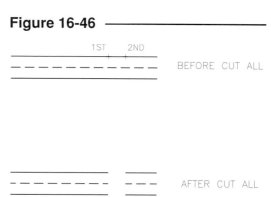

Weld

This option reverses the action of a *Cut*. The *Mline* is restored to its original continuous configuration. PICK on both sides of the "break" (Fig. 16-47).

Figure 16-47

Like many other dialog box-based commands, *Mledit* can also be used in command line format by using the hyphen (-) symbol as a prefix to the command. Key in "-MLEDIT" to force the command line interface. The following prompt appears:

    Command: **-mledit**
    Mline editing option AV/DV/CC/OC/MC/CT/OT/MT/CJ/CS/CA/WA:

Entering the acronym activates the related option.

AV	Add Vertex
DV	Delete Vertex
CC	Closed Cross
OC	Open Cross
MC	Merged Cross
CT	Closed Tee
OT	Open Tee
MT	Merged Tee
CJ	Corner Joint
CS	Cut Single
CA	Cut All
WA	Weld All

NOTE: *Mledit* operates only with co-planar *Mlines*

Other Editing Possibilities for *Mlines*

Although it appears that *Mledit* contains the necessary tools to handle all editing possibilities for *Mlines*, that is not necessarily the case. You may experience situations where *Mledit* cannot handle your desired editing request and you have to resort to traditional editing commands. Only in the case that *Mledit* cannot perform as you want, *Explode* the *Mline*, which converts each line element to an individual object. That action allows you to use Grips, *Trim*, *Extend*, *Break*, *Fillet*, or *Chamfer* on the individual line elements. Once an *Mline* is *Exploded*, it cannot be converted back to an *Mline* nor can *Mledit* be used with it.

Boolean Commands

Region combines one or several objects forming a closed shape into one object, a *Region*. The appearance of the object(s) does not change after the conversion, even though the resulting shape is one object (see *Region*, Chapter 15).

Although the *Region* appears to be no different than a closed *Pline*, it is more powerful because several *Regions* can be combined to form complex shapes (called "composite *Regions*") using the three Boolean operations explained next. As an example, a set of *Regions* (converted *Circles*) can be combined to form the sprocket with only one *Subtract* operation, as shown in Chapter 15, Figure 15-41.

The Boolean operators, *Union*, *Subtract*, and *Intersect*, can be used with *Regions* as well as solids. Any number of these commands can be used with *Regions* to form complex geometry. To illustrate each of the Boolean commands, consider the shapes shown in Figure 16-48. The *Circle* and the closed *Pline* are <u>first</u> converted to *Regions*; then *Union*, *Subtract*, or *Intersection* can be used.

Figure 16-48

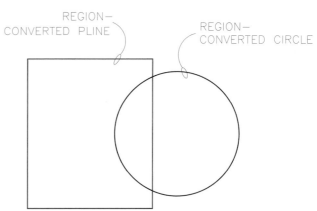

REGION—CONVERTED PLINE

REGION—CONVERTED CIRCLE

UNION

Pull-down Menu	COMMAND (TYPE)	ALIAS (TYPE)	Short-cut	Screen (side) Menu	Tablet Menu
Modify *Boolean >* *Union*	UNION	UNI	...	MODIFY2 *Union*	X,15

Union combines <u>two or more</u> *Regions* (or solids) into <u>one</u> *Region* (or solid). The resulting composite *Region* has the encompassing perimeter and area of the original *Regions*. Invoking *Union* causes AutoCAD to prompt you to select objects. You can select only existing *Regions* (or solids).

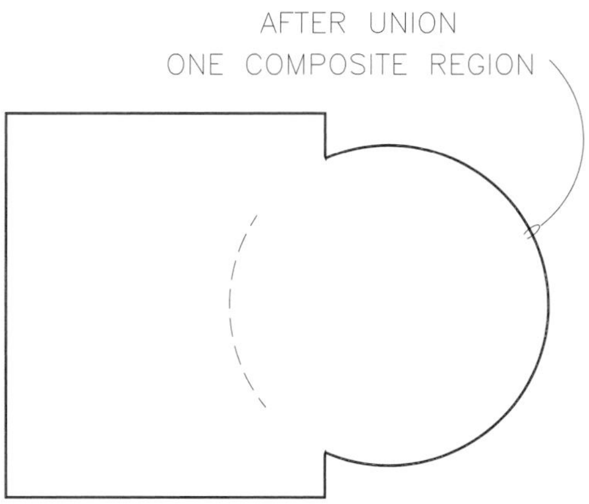

Command: **union**
Select Objects: **PICK** (*Region*)
Select Objects: **PICK** (*Region*)
Select Objects: **Enter**
Command:

The selected *Regions* are combined into one composite *Region* (Fig. 16-49). Any number of Boolean operations can be performed on the *Region(s)*.

Figure 16-49

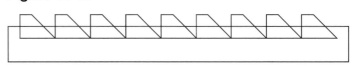

AFTER UNION
ONE COMPOSITE REGION

Several *Regions* can be selected in response to the "Select Objects:" prompt. For example, a composite *Region* such as that in Figure 16-50 can be created with one *Union*.

Figure 16-50

BEFORE— 10 REGIONS

AFTER UNION— ONE REGION

Two or more *Regions* can be *Unioned* even if they do not overlap. They are simply combined into one object although they still appear as two.

SUBTRACT

Pull-down Menu	COMMAND (TYPE)	ALIAS (TYPE)	Short-cut	Screen (side) Menu	Tablet Menu
Modify Boolean > Subtract	SUBTRACT	SU	...	MODIFY2 Subtract	X,16

Subtract enables you to remove one *Region* (or set of *Regions*) from another. The *Regions* must be created before using *Subtract* (or another Boolean operator). *Subtract* also works with solids (as do the other Boolean operations).

There are two steps to *Subtract*. First, you are prompted to select the *Region* or set of *Regions* to "subtract from" (those that you wish to <u>keep</u>), then to select the *Regions* "to subtract" (those you want to <u>remove</u>). The resulting shape is one composite *Region* comprising the perimeter of the first set minus the second (sometimes called "difference").

```
Command: subtract
Select solids and regions to subtract from...
Select Objects: PICK
Select Objects: Enter
Select solids and regions to subtract...
Select Objects: PICK
Select Objects: Enter
Command:
```

Figure 16-51

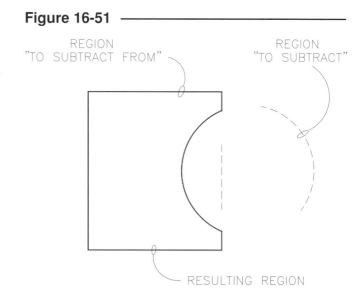

Consider the two shapes previously shown in Figure 16-48. The resulting *Region* shown in Figure 16-51 is the result of *Subtracting* the circular *Region* from the rectangular one.

Keep in mind that <u>multiple</u> *Regions* can be selected as the set to keep or as the set to remove. For example, the sprocket illustrated previously (Fig. 15-41) was created by subtracting several circular *Regions* in one operation. Another example is the removal of material to create holes or slots in sheet metal (discussed in Chapter 40, Surface Modeling).

INTERSECT

Pull-down Menu	COMMAND (TYPE)	ALIAS (TYPE)	Short-cut	Screen (side) Menu	Tablet Menu
Modify Boolean > Intersect	INTERSECT	IN	...	MODIFY2 Intrsect	X,17

Intersect is the Boolean operator that finds the common area from two or more *Regions*.

Consider the rectangular and circular *Regions* previously shown (Fig. 16-48). Using the *Intersect* command and selecting both shapes results in a *Region* comprising only that area that is shared by both shapes (Fig. 16-52):

Figure 16-52

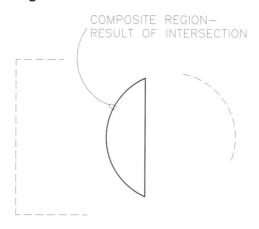

Command: **intersect**
Select Objects: **PICK**
Select Objects: **PICK**
Select Objects: **Enter**
Command:

If more than two *Regions* are selected, the resulting *Intersection* is composed of only the common area from all shapes (Fig 16-53). If all of the shapes selected do not overlap, a null *Region* is created (all shapes disappear because no area is common to all).

Intersect is a powerful operation when used with solid modeling. Techniques for saving time and file space using Boolean operations are discussed in Chapter 38, Solid Modeling Construction.

Figure 16-53

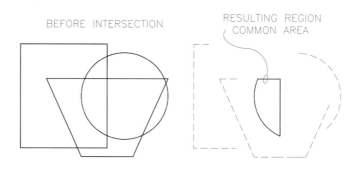

CHAPTER EXERCISES

1. *Ddchprop* or *Chprop*

   Open the **PIVOTARM CH9** drawing that you worked on in Chapter 9 Exercises. *Load* the *Hidden* and *Center Linetypes*. Make two *New Layers* named **HID** and **CEN** and assign the matching linetypes and yellow and green colors, respectively. Check the *Limits* of the drawing; then calculate and set an appropriate *Ltscale* (usually .5 times the drawing scale factor). Use *Chprop* or *Ddchprop* to change the *Lines* representing the holes in the front view to the **HID** layer as shown in Figure 16-54. Use *SaveAs* and name the drawing **PIVOTARM CH16**.

Figure 16-54

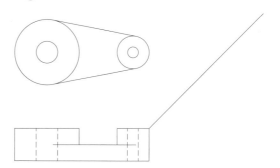

2. *Change*

   Open **CH8EX3**. *Erase* the *Arc* at the top of the object. *Erase* the *Points* with a window. Invoke the *Change* command. When prompted to *Select objects*, **PICK** all of the inclined *Lines* near the top. When prompted to <Change point>:, enter coordinate **6,8**. The object should appear as that in Figure 16-55. Use *SaveAs* and assign the name **CH16EX2**.

Figure 16-55

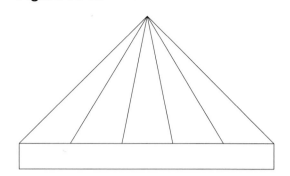

3. *Modify Circle*

   A design change is required for the bolt holes in **GASKETA** (from Chapter 9 Exercises). *Open* **GASKETA** and invoke the *Modify* dialog box (*Ddmodify*). Change each of the four bolt holes to **.375** diameter (Fig. 16-56). *Save* the drawing.

Figure 16-56

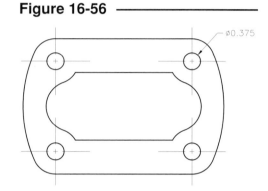

4. *Align*

   Open the **PLATES** drawing you created in Chapter 10. The three plates are to be stamped at one time on a single sheet of stock measuring 15″ x 12″. Place the three plates together to achieve optimum nesting on the sheet stock.

   A. Use *Align* to move the plate in the center (with 9 holes). Select the *1st source* and *destination points* (1S, 1D) and *2nd source* and *destination points* (2S, 2D) as shown in Figure 16-57.

Figure 16-57

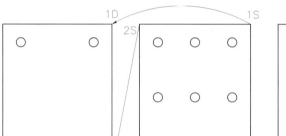
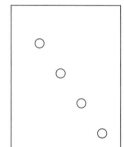

   B. After the first alignment is complete, use *Align* to move the plate on the right (with 4 diagonal holes). The *source* and *destination points* are indicated in Figure 16-58.

Figure 16-58

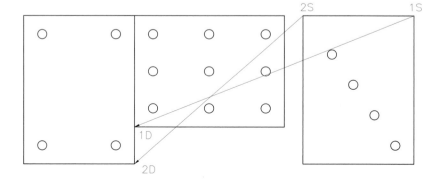

C. Finally, draw the sheet stock outline (15" x 12") using *Line* as shown in Figure 16-59. The plates are ready for production. Use *SaveAs* and assign the name **PLATENEST**.

Figure 16-59 ——————

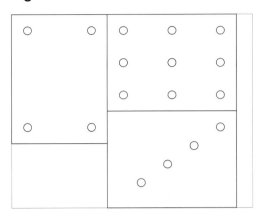

5. *Explode*

Figure 16-60 ——————

Open the **POLYGON1** drawing that you completed in Chapter 15 Exercises. You can quickly create the five-sided shape shown (continuous lines) in Figure 16-60 by *Exploding* the *Polygon*. First, *Explode* the *Polygon* and *Erase* the two *Lines* (highlighted). Draw the bottom *Line* from the two open *Endpoints*. Do not exit the drawing.

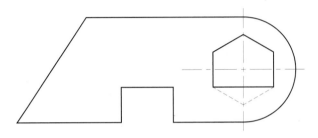

6. *Pedit*

Figure 16-61 ——————

Use *Pedit* with the *Edit vertex* options to alter the shape as shown in Figure 16-61. For the bottom notch, use *Straighten*. For the top notch, use *Insert*. Use *SaveAs* and change the name to **PEDIT1.**

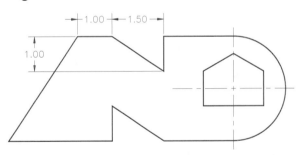

7. *Pline, Pedit*

Figure 16-62 ——————

A. Create a line graph as shown in Figure 16-62 to illustrate the low temperatures for a week. The temperatures are as follows:

X axis	Y axis
Sunday	20
Monday	14
Tuesday	18
Wednesday	26
Thursday	34
Friday	38
Saturday	27

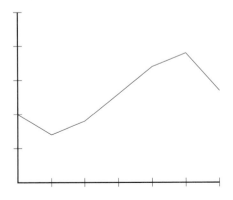

Use equal intervals along each axis. Use a *Pline* for the graph line. *Save* the drawing as **TEMP-A**. (You will label the graph at a later time.)

B. Use *Pedit* to change the *Pline* to a *Spline*. Note that the graph line is no longer 100% accurate because it does not pass through the original vertices (see Fig. 16-63). Use the *SPLFRAME* variable to display the original "frame" (*Regen* must be used after).

Use the **Modify Object** dialog box and try the **Cubic** and **Quadratic** options. Which option causes the vertices to have more pull? Find the most accurate option. Set *SPL-FRAME* to **0** and *SaveAs* **TEMP-B**.

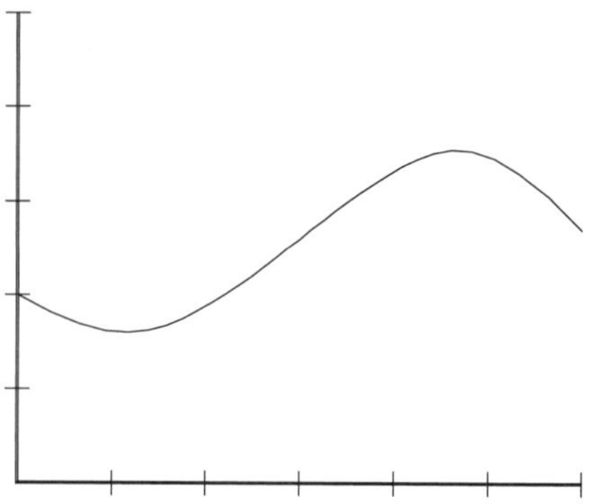

C. Use *Pedit* to change the curve from *Pline* to *Fit Curve*. Does the graph line pass through the vertices? *Saveas* **TEMP-C**.

D. *Open* drawing **TEMP-A**. *Erase* the *Splined Pline* and construct the graph using a *Spline* instead (Fig. 16-64, see Exercise 7A for data). *SaveAs* **TEMP-D**. Compare the *Spline* with the variations of *Plines*. Which of the four drawings (A, B, C, or D) is smoothest? Which is the most accurate?

E. It was learned that there was a mistake in reporting the temperatures for that week. Thursday's low must be changed to 38 degrees and Friday's to 34 degrees. *Open* **TEMP-D** (if not already open) and use *Splinedit* to correct the mistake. Use the *Move* option of *Fit Data* so that the exact data points can be altered as shown in Figure 16-65. *SaveAs* **TEMP-E**.

Figure 16-63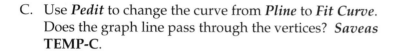

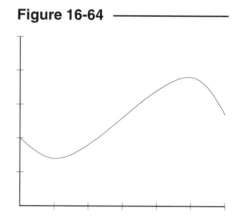

Figure 16-64

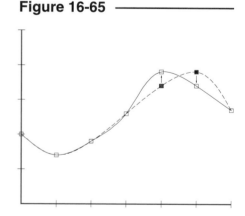

Figure 16-65

8. **Converting** *Lines, Circles,* **Figure 16-66**
   and *Arcs* **to** *Plines*

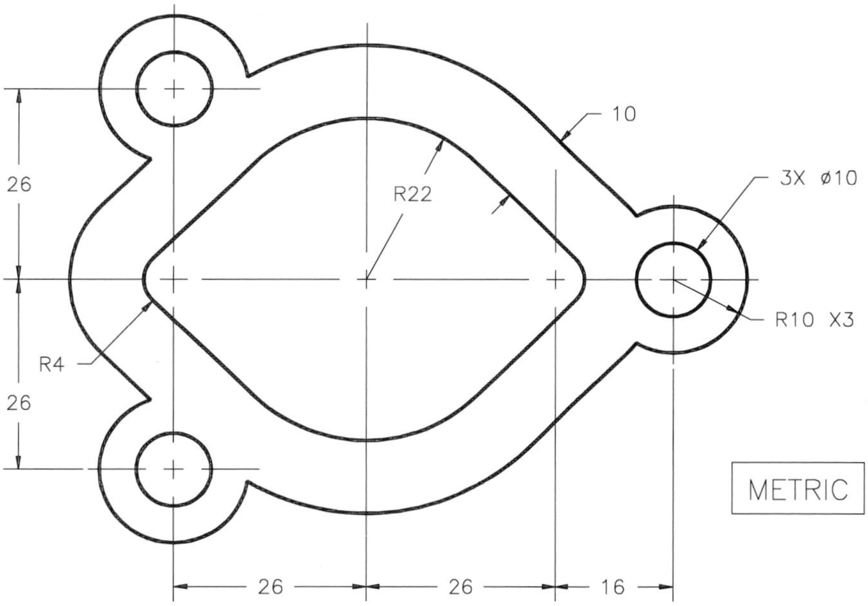

   A. Begin a *New* drawing
      and use **A-METRIC** as
      a template (that you
      created in Chapter 6
      Exercises). Use *Save*
      and assign the name
      GASKETC. Change
      the *Limits* for plotting
      on an A sheet at 2:1
      (refer to the Metric
      Table of *Limits* Settings
      and set *Limits* to
      1/2 x *Limits* specified
      for 1:1 scale). Change
      the *LTSCALE* to **6**.
      Begin but do not com-
      plete the drawing of
      the gasket shown in
      Figure 16-66. First, draw
      only the <u>inside</u> shape using *Lines* and *Circles* (with *Trim*) or *Arcs*. Then convert the *Lines* and
      Arcs to one closed *Pline* using **Pedit**. Finally, locate and draw the 3 bolt holes. *Save* the
      drawing.

   B. *Offset* the existing inside shape to create the outside shape. Use *Offset* to create concentric
      circles around the bolt holes. Use *Trim* to complete the gasket. *Save* the drawing and create a
      plot at 2:1 scale.

9. *Splinedit*   **Figure 16-67** —   **Figure 16-68** —

   Open the **HANDLE** drawing that you created in
   Chapter 15 Exercises. During the testing and analysis
   process, it was discovered that the shape of the handle
   should have a more ergonomic design. The finger side
   (left) should be flatter to accommodate varying sizes of
   hands, and the thumb side (right) should have more of a
   protrusion on top to prevent slippage.

   First, add more control points uniformly along the
   length of the left side with *Splinedit, Refine*. *Elevate* the
   Order from 4 to **6**, then use *Move Vertex* to align the
   control points as shown in Figure 16-67.

   On the right side of the handle, *Add* two points under
   the *Fit Data* option to create the protrusion shown in
   Figure 16-68. You may have to *Reverse* the direction of
   the *Spline* to add the new points between the two highlighted ones. *SaveAs* **HANDLE2**.

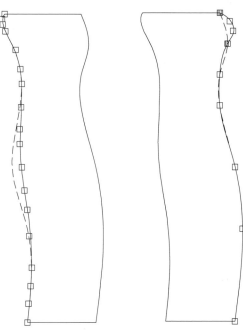

10. *Mline, Mledit, Stretch*

Draw the floor plan of the storage room shown in Figure 16-69. Use *Mline* objects for the walls and *Plines* for the windows and doors. Use *Mledit* to treat the intersections as shown. When your drawing is complete according to the given specifications, use *Stretch* to center the large window along the top wall. Save the drawing as **STORE ROOM**.

Figure 16-69

11. **Hammer Drawing**

Open the **HAMMER** drawing that you set up in Chapter 14 Exercises. Create the hammer shown in Figure 16-70. Make use of *Offset* for determining the center points for *Arc* and *Circle* radii (use the CONSTR layer for construction lines). Use *Fillet* wherever possible. When you finish, *Save* the drawing and create a plot to scale.

Figure 16-70

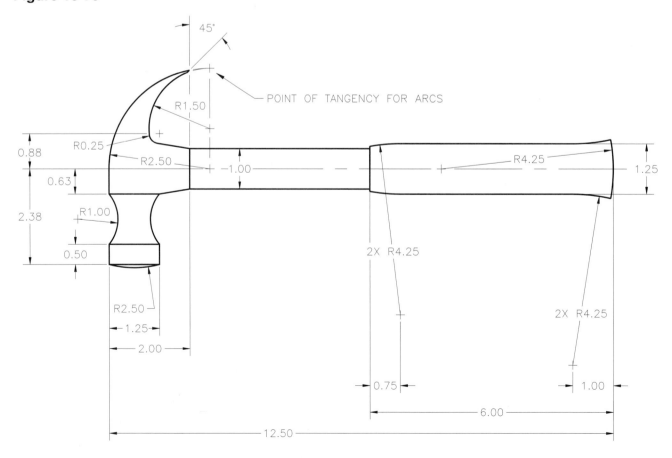

12. Gear Drawing

Complete the drawing of the gear you began in Chapter 15 Exercises called **GEAR-REGION**. If you remember, three shapes were created (two *Circles* and one "tooth") and each was converted to a *Region*. Finally, the "tooth" was *Arrayed* to create the total of 40 teeth.

A. To complete the gear, *Subtract* the small circular *Region* and all of the teeth from the large circular *Region*. First, use **Subtract**. At the "Select solids and regions to subtract from…" prompt, PICK the large circular *Region*. At the "Select solids and regions to subtract…" prompt, use a window to select <u>everything</u> (the large circular *Region* is automatically filtered out). The resulting gear should resemble Figure 16-71. *Save* the drawing as **GEAR-REGION 2**.

Figure 16-71 ————————————

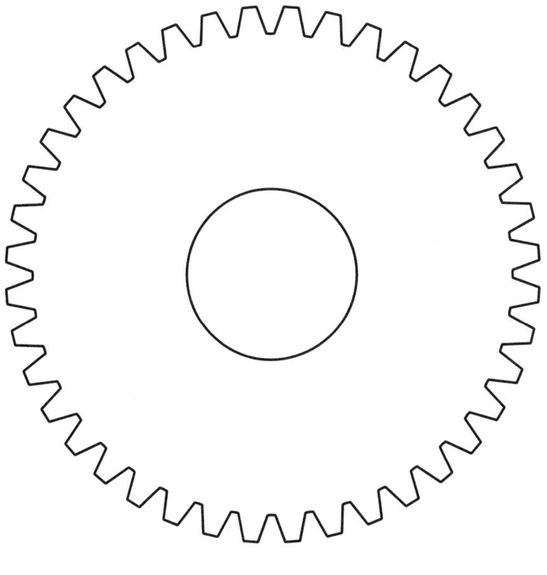

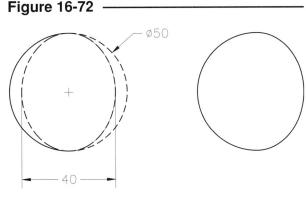

B. Consider the steps involved if you were to create the gear (as an alternative) by using *Trim* to remove 40 small sections of the large *Circle* and all unwanted parts of the teeth. *Regions* are clearly easier in this case.

13. Wrench Drawing

Create the same wrench you created in Chapter 15 again, only this time use region modeling. Refer to Chapter 15, Exercise 5 for dimensions (exclude the break lines). Begin a *New* drawing and use the **A-METRIC** template. Use *SaveAs* and assign the name **WRENCH-REG**.

Figure 16-72 ————————————

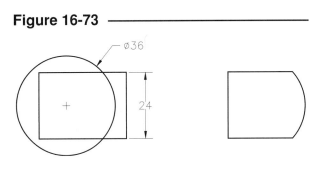

A. Set *Limits* to **372,288** to prepare the drawing for plotting at 3:4 (the drawing scale factor is 33.87). Set the **GRID** to **10**.

B. Draw a *Circle* and an *Ellipse* as shown on the left in Figure 16-72. The center of each shape is located at **60,150**. *Trim* half of each shape as shown (highlighted). Use *Region* to convert the two remaining halves into a *Region* as shown on the right of Figure 16-72.

C. Next, create a *Circle* with the center at **110,150** and a diameter as shown in Figure 16-73. Then draw a closed *Pline* in a rectangular shape as shown. The height of the rectangle must be drawn as specified; however, the width of the rectangle can be drawn <u>approximately</u> as shown on the left. Convert each shape to a *Region*; then use *Intersect* to create the region as shown on the right side of the figure.

Figure 16-73 ————————————

D. *Move* the rectangular-shaped region **68** units to the left to overlap the first region as shown in Figure 16-74. Use *Subtract* to create the composite region on the right representing the head of the wrench.

E. Complete the construction of the wrench in a manner similar to that used in the previous steps. Refer to Chapter 15 Exercises for dimensions of the wrench. Complete the wrench as one *Region*. *Save* the drawing.

Figure 16-74 ——————————————

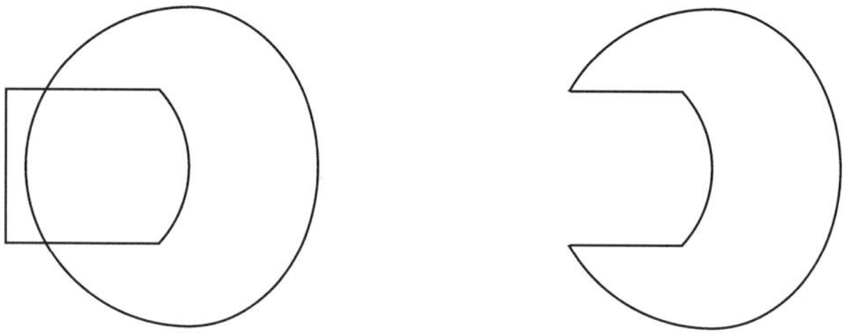

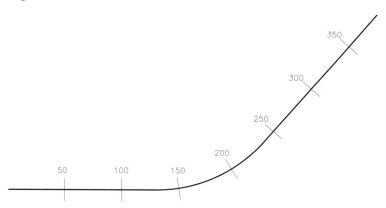

14. **Retaining Wall**

Figure 16-75 ——————————————

Open the **RET-WALL** drawing that you created in Chapter 8 Exercises. Annotate the wall with 50 unit stations as shown in Figure 16-75.

HINT: Convert the centerline of the retaining wall to a *Pline*; then use the *Measure* command to place *Point* objects at the 50 unit stations. *Offset* the wall on both sides to provide a construction aid in the creation of the perpendicular tick marks. Use *Node* and *Perpendicular* Osnaps. *Save* the drawing as **RET-WAL3**.

EXAMPLE 1.

17

INQUIRY COMMANDS

Chapter Objectives

After completing this chapter you should:

1. be able to list the *Status* of a drawing;

2. be able to *List* the AutoCAD database information about an object;

3. know how to list the entire database of all objects with *Dblist*;

4. be able to calculate the *Area* of a closed shape with and without "islands";

5. be able to find the *Distance* between two points;

6. be able to report the coordinate value of a selected point using the *ID* command;

7. know how to list the *Time* spent on a drawing or in the current drawing session;

8. be able to use *Setvar* to change system variable settings or list current settings.

CONCEPTS

AutoCAD provides several commands that allow you to find out information about the current drawing status and specific objects in a drawing. These commands as a group are known as "*Inquiry* commands" and are grouped together in the menu systems. The *Inquiry* commands are located on the Standard toolbar in a flyout fashion (Fig. 17-1). You can also use the *Tools* pull-down menu to access the *Inquiry* commands (Fig. 17-2).

Figure 17-1

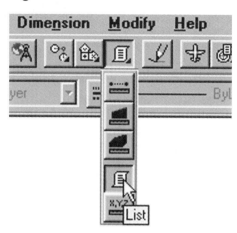

Figure 17-2

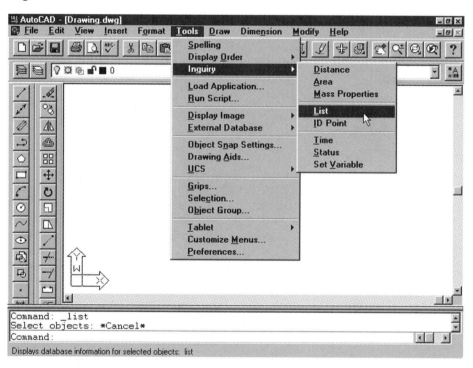

Using *Inquiry* commands, you can find out such information as the amount of time spent in the current drawing, the distance between two points, the area of a closed shape, the database listing of properties for specific objects (coordinates of endpoints, lengths, angles, etc.), and current settings for system variables as well as other information. The *Inquiry* commands are:

Status, List, Dblist, Area, Distance, ID, Time, and *Setvar*

COMMANDS

STATUS

Pull-down Menu	COMMAND (TYPE)	ALIAS (TYPE)	Short-cut	Screen (side) Menu	Tablet Menu
Tools *Inquiry >* *Status*	*STATUS*	...	...	*TOOLS 1* *Status*	...

The *Status* command gives many pieces of information related to the current drawing. Typing or **PICK**ing the command from the icon or one of the menus causes a text screen to appear similar to that shown in Figure 17-3. The information items are:

Figure 17-3

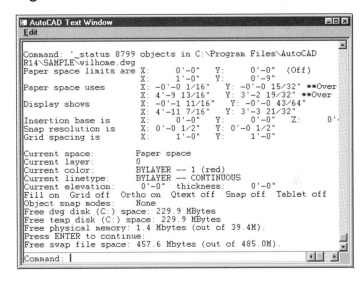

Total number of objects in the current drawing
Paper space limits: values set by the *Limits* command in Paper Space
Paper space uses: area used by the objects (drawing extents) in Paper Space
Model space limits: values set by the *Limits* command in Model Space
Model space uses: area used by the objects (drawing extents) in Model Space
Display shows: current display or windowed area
Insertion basepoint: point specified by the *Base* command or default (0,0)
Snap resolution: value specified by the *Snap* command
Grid spacing: value specified by the *Grid* command
Current space: Paper Space or Model Space
Current layer: name
Current color: current color assignment
Current linetype: current linetype assignment
Current elevation, thickness: 3D properties – current height above the XY plane and Z dimension
On or off status: *FILL, GRID, ORTHO, QTEXT, SNAP, TABLET*
Object Snap Modes: current *Running OSNAP* modes
Free dwg disk: space on the current drawing hard disk drive
Free temp disk: space on the current temporary files hard disk drive
Free physical memory: amount of free RAM (total RAM)
Free swap file space: amount of free swap file space (total allocated swap file)

LIST

Pull-down Menu	COMMAND (TYPE)	ALIAS (TYPE)	Short-cut	Screen (side) Menu	Tablet Menu
Tools Inquiry > List	LIST	LS or LI	...	TOOLS 1 List	U,8

The *List* command displays the database list of information in text window format for one or more specified objects. The information displayed depends on the <u>type</u> of object selected. Invoking the *List* command causes a prompt for you to select objects. AutoCAD then displays the list for the selected objects (see Figs. 17-4 and 17-5).

A *List* of a *Line* and an *Arc* is given in Figure 17-4.

For a *Line,* coordinates for the endpoints, line length and angle, current layer, and other information are given.

For an *Arc,* the center coordinate, radius, start and end angles, and length are given.

The *List* for a *Pline* is shown in Figure 17-5. The location of each vertex is given, as well as the length and perimeter of the entire *Pline*.

Note that in Release 14, *Plines* are created and listed as *Lwpolylines,* or "lightweight polylines." In previous releases, complete data for each *Pline* vertex (starting width, ending width, color, etc.) was stored along with the coordinate values of the vertex, then repeated for each vertex. In Release 14, the data common to all vertices is stored only once and only the coordinate data is stored for each vertex. Because this data structure saves file space, the new *Plines* are known as light weight *Plines*.

Figure 17-4

Figure 17-5

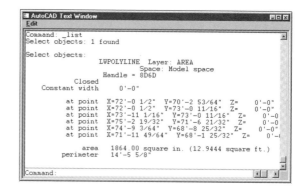

DBLIST

Pull-down Menu	COMMAND (TYPE)	ALIAS (TYPE)	Short-cut	Screen (side) Menu	Tablet Menu
...	DBLIST	...	...	...	...

The *Dblist* command is similar to the *List* command in that it displays the database listing of objects; however, *Dblist* gives information for <u>every</u> object in the current drawing! This command is generally used when you desire to send the list to a printer or when only a few objects are in the drawing. If you use this command in a complex drawing, be prepared to page through many screens of information. Press Escape to cancel *Dblist* and return to the Command: prompt. Press F2 to close the text window and return to the Drawing Editor.

AREA

	Pull-down Menu	COMMAND (TYPE)	ALIAS (TYPE)	Short-cut	Screen (side) Menu	Tablet Menu
	Tools *Inquiry >* *Area*	*AREA*	*AA*	...	*TOOLS 1* *Area*	*T,7*

The *Area* command is helpful for many applications. With this command AutoCAD calculates the area and the perimeter of any enclosed shape in a matter of milliseconds. You specify the area (shape) to consider for calculation by PICKing the *Object* (if it is a closed *Pline, Polygon, Circle, Boundary, Region* or other closed object) or by PICKing points (corners of the outline) to define the shape. The options are given below.

First Point

The command sequence for specifying the area by PICKing points is shown below. This method should be used only for shapes with <u>straight</u> sides. An example of the *Point* method (PICKing points to define the area) is shown in Figure 17-6.

Figure 17-6

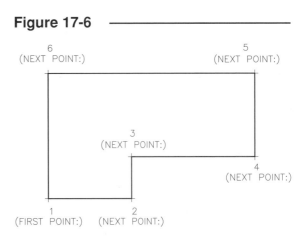

```
Command: area
<First point>/Object/Add/Subtract: PICK (Locate the first point to designate the shape.)
Next point: PICK (Select the second point to define the shape.)
Next point: PICK (Select the third point.)
Next point: PICK (Continue selecting points until all corners have been selected to completely define
the shape.)
Next point: Enter
Area = nn.nnn  perimeter = nn.nnn
Command:
```

Object

If the shape for which you want to find the area and perimeter is a *Circle, Polygon, Ellipse, Boundary, Region*, or closed *Pline*, the *Object* option of the *Area* command can be used. Select the shape with one PICK (since all of these shapes are considered as one object by AutoCAD).

The ability to find the area of a closed *Pline, Region*, or *Boundary* is extremely helpful. Remember that <u>any</u> closed shape, even if it includes *Arcs* and other curves, can be converted to a closed *Pline* with the *Pedit* command (as long as there are no gaps or overlaps) or can be used with the *Boundary* command. This method provides you with the ability to easily calculate the area of any shape, curved or straight. In short, convert the shape to a closed *Pline, Region*, or *Boundary* and find the *Area* with the *Object* option.

Add, Subtract

Add and *Subtract* provide you with the means to find the area of a closed shape that has islands, or negative spaces. For example, you may be required to find the surface area of a sheet of material that has several punched holes. In this case, the area of the holes is subtracted from the area defined by the perimeter shape. The *Add* and *Subtract* options are used specifically for that purpose. The following command sequence displays the process of calculating an area and subtracting the area occupied by the holes.

> Command: **area**
> <First point>/Object/Add/Subtract: **a** (Defining the perimeter shape is prefaced by the *Add* option.)
> <First point>/Object/Subtract: **o** (Assuming the perimeter shape is a *Boundary* or closed object.)
> (Add mode) Select object: **PICK** (Select the closed object.)
>
> Area = nn.nnn, Perimeter = nn.nnn
> Total area = nn.nnn
>
> (Add mode) Select object: **Enter** (Indicates completion of the *Add* mode.)
> <First point>/Object/Subtract: **s** (Change to the *Subtract* mode.)
> <First point>/Object/Add: **o** (Assuming the holes are *Circle* objects.)
> (Subtract mode) Select object: **PICK** (Select the first *Circle* to subtract.)
>
> Area = nn.nnn, Perimeter = nn.nnn
> Total area = nn.nnn
>
> (Subtract mode) Select object: **PICK** (Select the second *Circle* to subtract.)
>
> Area = nn.nnn, Perimeter = nn.nnn
> Total area = nn.nnn
>
> (Subtract mode) Select object: **Enter** (Completion of *Subtract* mode.)
> <First point>/Object/Add: **Enter** (Completion of *Area* command.)
> Command:

Make sure that you press Enter between the *Add* and *Subtract* modes.

An example of the last command sequence used to find the area of a shape minus the holes is shown in Figure 17-7. Notice that the object selected in the first step is a closed *Pline* shape, including an *Arc*.

Figure 17-7 ————————————

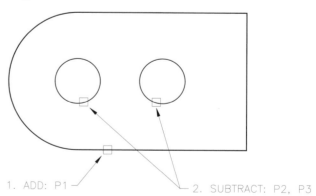

1. ADD: P1   2. SUBTRACT: P2, P3

DISTANCE

Pull-down Menu	COMMAND (TYPE)	ALIAS (TYPE)	Short-cut	Screen (side) Menu	Tablet Menu
Tools Inquiry > Distance	DIST	DI	...	*TOOLS 1* Dist	T,8

The *Distance* command reports the distance between any two points you specify. *OSNAPs* can be used to "snap" to the existing points. This command is helpful in many engineering or architectural applications, such as finding the clearance between two mechanical parts, finding the distance between columns in a building, or finding the size of an opening in a part or doorway. The command is easy to use.

```
Command: dist
First point: PICK (Use OSNAPs if needed.)
Second point: PICK
Distance = nn.nnn, Angle in XY Plane = nn, Angle from XY Plane = nn
Delta X = nn.nnn, Delta Y = nn.nnn, Delta Z = nn.nnn
```

AutoCAD reports the absolute and relative distances as well as the angle of the line between the points.

ID

Pull-down Menu	COMMAND (TYPE)	ALIAS (TYPE)	Short-cut	Screen (side) Menu	Tablet Menu
Tools Inquiry > ID Point	ID	...	...	*TOOLS 1* ID	U,9

The *ID* command reports the coordinate value of any point you select with the cursor. If you require the location associated with a specific object, an *OSNAP* mode (*Endpoint, Midpoint, Center,* etc.) can be used.

```
Command: ID
Point: PICK or (coordinate) (Select a point or enter a coordinate. OSNAPs can be used.)
X = nn.nnn, Y = nn.nnn, Z = nn.nnn
Command:
```

NOTE: *ID* also sets AutoCAD's "last point." The last point can be referenced in commands by using the @ (at) symbol with relative rectangular or relative polar coordinates. *ID* creates a "blip" at the location of the coordinates you enter or PICK.

TIME

Pull-down Menu	COMMAND (TYPE)	ALIAS (TYPE)	Short-cut	Screen (side) Menu	Tablet Menu
Tools Inquiry > Time	TIME	...	...	TOOLS 1 Time	...

This command is useful for keeping track of the time spent in the current drawing session or total time spent on a particular drawing. Knowing how much time is spent on a drawing can be useful in an office situation for bidding or billing jobs. The *Time* command reports the information shown in Figure 17-8.

The *Total editing time* is automatically kept, starting from when the drawing was first created until the current time. Plotting and printing time is not included in this total, nor is the time spent in a session when changes are discarded.

Figure 17-8

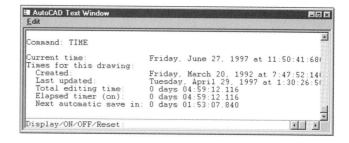

Display
The *Display* option causes *Time* to repeat the display with the updated times.

ON/OFF/Reset
The *Elapsed timer* is a separate compilation of time controlled by the user. The *Elapsed timer* can be turned *ON* or *OFF* or can be *Reset*.

Time also reports when the next automatic save will be made. The time interval of the Automatic Save feature is controlled by the *SAVETIME* system variable. To set the interval between automatic saves, type *SAVETIME* at the command line and specify a value for time (in minutes). The default interval is 120 minutes (see also *SAVETIME*, Chapter 2).

SETVAR

Pull-down Menu	COMMAND (TYPE)	ALIAS (TYPE)	Short-cut	Screen (side) Menu	Tablet Menu
Tools Inquiry > Set Variable	SETVAR	SET	...	TOOLS 1 Setvar	U,10

The settings (values or on/off status) that you make for many commands, such as *Limits, Grid, Snap, Running Osnaps, Fillet* values, *Pline* width, etc. are saved in system variables. In AutoCAD Release 14 there are 286 system variables. The variables store the settings that are used to create and edit the drawing. *Setvar* ("set variable") gives you access to the system variables. (See Appendix A for a complete list of the system variables, including an explanation, default setting, and possible settings for each.)

Setvar allows you to perform two functions: (1) change the setting for any system variable and (2) display the current setting for one or all system variables. To change a setting for a system variable using *Setvar*, just use the command and enter the name of the variable. For example, the following syntax lists values for the *GRID* setting and current *Fillet* radius:

Command: **setvar**
Variable name or ?: **gridunit**
New value for GRIDUNIT <0.5000,0.5000>:

Command: **setvar**
Variable name or ?: **filletrad**
New value for FILLETRAD <0.5000>:

With recent releases of AutoCAD, <u>*Setvar* is not needed to set system variables</u>. You can enter the variable name directly at the Command: prompt without using *Setvar* first.

Command: **filletrad**
New value for FILLETRAD <0.5000>:

To list the <u>current settings for all system variables</u>, use *Setvar* with the *?* (question mark) option. The complete list of system variables is given in a text window with the current setting for each variable (Fig. 17-9).

You can also use *Help* to list the system variables with a short explanation for each. In the *Help Topics* dialog box, select the *Contents* tab, select *Command Reference*, then *System Variables*.

Figure 17-9

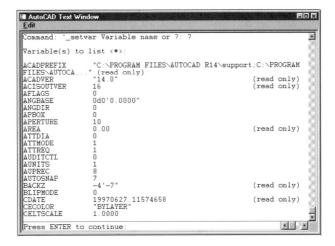

MASSPROP

The *Massprop* command is used to give mass and volumetric properties for AutoCAD solids. See Chapter 39, Advanced Solids Features, for use of *Massprop*.

CHAPTER EXERCISES

1. *List*

   A. *Open* the **PLATENEST** drawing from Chapter 16 Exercises. Assume that a laser will be used to cut the plates and holes from the stock, and you must program the coordinates. Use the *List* command to give information on the *Line*s and *Circle*s for the one plate with four holes in a diagonal orientation. Determine and write down the coordinates for the 4 corners of the plate and the centers of the 4 holes.

B. *Open* the **EFF-APT** drawing from Chapter 15 Exercises. Use *List* to determine the area of the inside of the tub. If the tub were filled with 10" of water, what would be the volume of water in the tub?

2. *Area*

A. *Open* the **EFF-APT** drawing. The entry room is to be carpeted at a cost of $12.50 per square yard. Use the *Area* command (with the PICK points option) to determine the cost for carpeting the room, not including the closet.

B. *Open* the **PLATENEST** drawing from Chapter 16 Exercises. Use *SaveAs* to create a file named **PLATE-AREA**. Using the *Area* command, calculate the wasted material (the two pieces of stock remaining after the 3 plates have been cut or stamped). HINT: Use *Boundary* to create objects from the waste areas for determining the *Area*.

C. Create a *Boundary* (with islands) of the plate with four holes arranged diagonally. *Move* the new boundary objects **10** units to the right. Using the *Add* and *Subtract* options of *Area*, calculate the surface area for painting 100 pieces, both sides. (Remember to press Enter between the *Add* and *Subtract* operations.) *Save* the drawing.

3. *Dist*

A. *Open* the **EFF-APT** drawing. Use the *Dist* command to determine the best location for installing a wall-mounted telephone in the apartment. Where should the telephone be located in order to provide the most equal access from all corners of the apartment? What is the farthest distance that you would have to walk to answer the phone?

B. Using the *Dist* command, determine what length of pipe would be required to connect the kitchen sink drain (use the center of the far sink) to the tub drain (assume the drain is at the far end of the tub). Calculate only the direct distance (under the floor).

4. *ID*

Open the **PLATENEST** drawing once again. You have now been assigned to program the laser to cut the plate with 4 holes in the corners. Use the *ID* command (with *OSNAPs*) to determine the coordinates for the 4 corners and the hole centers.

5. *Time*

Using the *Time* command, what is the total amount of editing time you spent with the **PLATENEST** drawing? How much time have you spent in this session? How much time until the next automatic save?

6. *Setvar*

Using the **PLATENEST** drawing again, use *Setvar* to list system variables. What are the current settings for *Fillet* radius (**FILLETRAD**), the last point used (**LASTPOINT**), *Linetype Scale* (**LTSCALE**), automatic file save interval (**SAVETIME**), and text *Style* (**TEXTSTYLE**).

18

CREATING AND EDITING TEXT

Chapter Objectives

After completing this chapter you should:

1. be able to create lines of text in a drawing using *Text* and *Dtext*;

2. be able to create and format <u>paragraph</u> text using *Mtext*;

3. be able to *Justify* text using each of the methods;

4. be able to <u>create</u> text styles with the *Style* <u>command</u>;

5. know how to import external text into AutoCAD;

6. know that *Ddedit* can be used to edit the content of existing text and *Ddmodify* can be used to modify any property of existing text;

7. be able use *Spell* to check spelling and be able to *Find* and *Replace* text in a drawing;

8. know how features such as *Qtext* (quick text), *TEXTFILL*, and font mapping can be used to control the appearance of text in a drawing or print/plot.

CONCEPTS

The *Dtext, Mtext,* and *Text* commands provide you with a means of creating text in an AutoCAD drawing. "Text" in CAD drawings usually refers to sentences, words, or notes created from alphabetical or numerical characters that appear in the drawing. The numeric values that are part of specific dimensions are generally not considered "text," since dimensional values are a component of the dimension created automatically with the use of dimensioning commands.

Text in technical drawings is typically in the form of notes concerning information or descriptions of the objects contained in the drawing. For example, an architectural drawing might have written descriptions of rooms or spaces, special instructions for construction, or notes concerning materials or furnishings (Fig. 18-1). An engineering drawing may contain, in addition to the dimensions, manufacturing notes, bill of materials, schedules, or tables (Fig. 18-2). Technical illustrations may contain part numbers or assembly notes. Title blocks also contain text.

Figure 18-1

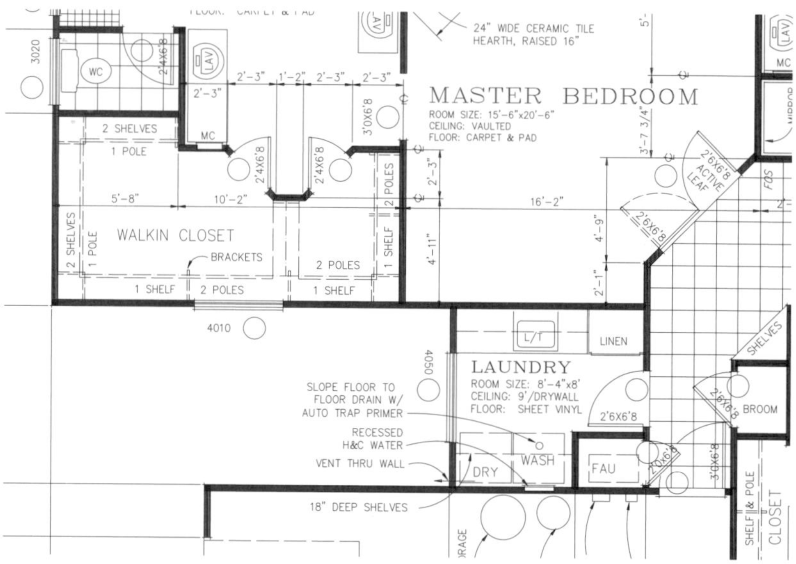

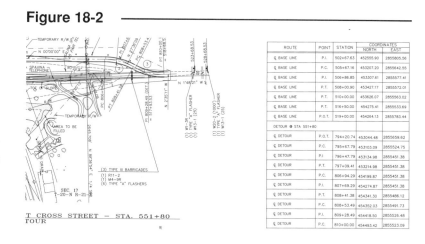

A line of text or paragraph of text in an AutoCAD drawing is treated as an object, just like a *Line* or a *Circle*. Each text object can be *Erased, Moved, Rotated,* or otherwise edited as any other graphical object. The letters themselves can be changed individually with special text editing commands. A spell checker is available by using the *Spell* command. Since text is treated as a graphical element, the use of many lines of text in a drawing can slow regeneration time and increase plotting time significantly.

Figure 18-2

ROUTE	POINT	STATION	COORDINATES NORTH	COORDINATES EAST
℄ BASE LINE	P.I.	502+67.63	452555.90	2855805.56
℄ BASE LINE	P.C.	505+67.16	453207.20	2855642.55
℄ BASE LINE	P.I.	506+86.85	453307.61	2855577.41
℄ BASE LINE	P.T.	508+00.90	453427.17	2855572.01
℄ BASE LINE	P.T.	510+00.00	453626.07	2855563.02
℄ BASE LINE	P.T.	516+50.00	454275.41	2855533.69
℄ BASE LINE	P.O.T.	519+00.00	454264.13	2855783.44
DETOUR @ STA 551+80				
℄ DETOUR	P.O.T.	794+20.74	453044.48	2855659.62
℄ DETOUR	P.C.	795+67.79	453103.09	2855524.75
℄ DETOUR	P.I	796+47.79	453134.98	2855451.38
℄ DETOUR	P.T.	797+09.41	453214.98	2855451.38
℄ DETOUR	P.C.	806+94.29	454199.87	2855451.38
℄ DETOUR	P.I.	807+69.29	454274.87	2855451.38
℄ DETOUR	P.T.	808+41.38	454341.30	2855486.12
℄ DETOUR	P.C.	808+53.49	454352.03	2855491.73
℄ DETOUR	P.I.	809+28.49	454418.50	2855526.48
℄ DETOUR	P.C.	810+00.00	454493.42	2855523.09

The *Dtext, Mtext,* and *Text* commands perform basically the same function; they create text in a drawing. *Mtext* is the newest and most sophisticated method of text entry. With *Mtext* (multiline text) you can create a paragraph of text that "wraps" within a text boundary (rectangle) that you specify. An *Mtext* paragraph is treated as one AutoCAD object. *Dtext* (dynamic text) displays each character in the drawing as it is typed and allows entry of multiple lines of text. *Text* allows only one line of text and requires an **Enter** or **Return** to display each line of text.

Many options for text justification are available. *Justification* is the method of aligning multiple lines of text. For example, if text is right justified, the right ends of the lines of text are aligned.

The form or shape of the individual letters is determined in AutoCAD by the text *Style*. Creating a *Style* begins with selecting a Windows standard TrueType or AutoCAD-supplied font file. Font files supplied with AutoCAD have file extensions of .TTF (TrueType) or .SHX (AutoCAD compiled shape files). The AutoCAD .SHX fonts are located in the C:\Program Files\AutoCAD R14\Fonts directory (by the default installation). The TrueType fonts are installed with the other Windows fonts in the C:\Windows\Fonts directory. Additional fonts can be purchased or may already be on your computer (supplied with Windows, word processors, or other software).

After a font for the *Style* is selected, other parameters (such as width and obliquing angle) can be specified to customize the *style* to your needs. Only one *style*, called *Standard*, has been created as part of the traditional default template drawing (ACAD.DWT) and uses the TXT.SHX font file. Other template drawings may have two or more created *styles*.

If any other style of text is desired, it must be created with the *Style* command. When a new *style* is created, it becomes the current one used by the *Dtext*, *Mtext*, or *Text* command. If several *styles* have been created in a drawing, a particular one can be recalled or made current by using the *style* <u>option</u> of the *Dtext*, *Mtext*, or *Text* commands.

In summary, the *Style* <u>command</u> allows you to design <u>new</u> styles with your choice of options, such as fonts, width factor, and obliquing angle, whereas the *style* <u>option</u> of *Dtext*, *Mtext*, and *Text* allows you to select from <u>existing</u> styles in the drawing that you previously created.

Commands related to creating or editing text in an AutoCAD drawing include:

Text	Places one line of text in a drawing, but the text is not visible until after pressing Enter.
Dtext	Places individual lines of text in a drawing and allows you to see each letter as it is typed.
Mtext	Places text in paragraph form (with word wrap) within a text boundary and allows many methods of formatting the appearance of the text.
Style	Creates text styles for use with any of the text creation commands. You can select from font files, specify other parameters to design the appearance of the letters, and assign a name for each style.
Spell	Checks the spelling of existing text in a drawing.
Ddedit	Invokes a dialog box for editing text. If you select *Text* or *Dtext* for editing, you can change the text (characters) only; if you select *Mtext* objects, you can edit individual characters and change the appearance of individual characters or the entire paragraph(s).
Ddmodify	A broad editing command for any type of text object (*Text*, *Dtext*, or *Mtext*), allowing revision of many aspects of text such as height, rotation angle, insertion point, style, or the text (characters).
Qtext	Short for quick-text, temporarily displays a line of text as a box instead of individual characters in order to speed up regeneration time and plotting time.

TEXT CREATION COMMANDS

The commands for creating text are formally named *Dtext*, *Mtext*, and *Text* (these are the commands used for typing). The *Draw* pull-down and screen (side) menus provide access to the two commonly used text commands, *Multiline Text...* (*Mtext*) and *Single-Line Text* (*Dtext*) (Fig. 18-3). Only the *Mtext* command has an icon button (by default) near the bottom of the *Draw* toolbar (Fig 18-4). The *Text* command must be typed at the command line.

Figure 18-3 ———— **Figure 18-4**

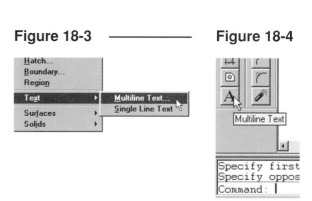

DTEXT

	COMMAND (TYPE)	ALIAS (TYPE)	Short-cut	Screen (side) Menu	Tablet Menu
Pull-down Menu					
Draw *Text >* *Single Line Text...*	*DTEXT*	*DT*	...	DRAW 2 *Dtext*	*K,8*

Dtext (dynamic text) lets you insert text into an AutoCAD drawing. *Dtext* displays each character in the drawing as it is typed. You can enter multiple lines of text without exiting the *Dtext* command. The lines of text do not "wrap." The options are presented below:

> Command: **dtext**
> Justify/Style/<Start point>:

Start Point

The *Start point* for a line of text is the <u>left end</u> of the baseline for the text (Fig. 18-5). *Height* is the distance from the baseline to the top of upper case letters. Additional lines of text are automatically spaced below and left justified. The *rotation angle* is the angle of the baseline (Fig. 18-6).

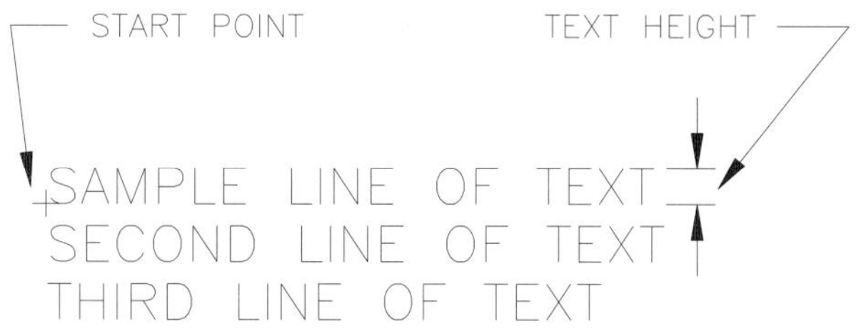

The command sequence for this option is:

> Command: **dtext**
> Justify/Style/<Start point>: **PICK** or (**coordinates**)
> Height <0.20>: **Enter** or (**value**)
> Rotation Angle <0>: **Enter** or (**value**)
> Text: (Type the desired line of text and press **Enter**.)
> Text: (Type another line of text and press **Enter**.)
> Text: **Enter**
> Command:

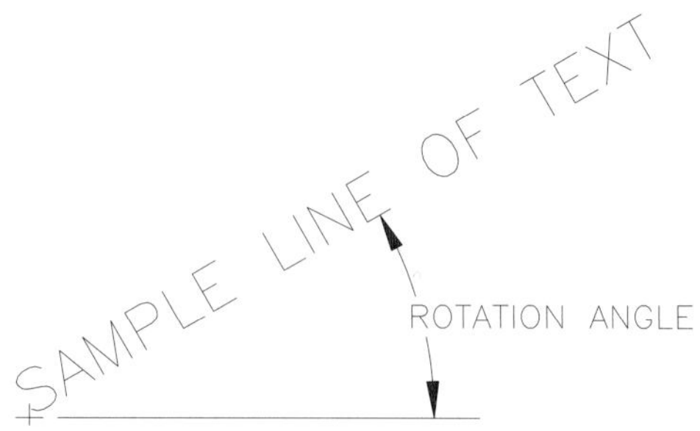

NOTE: When the "Text:" prompt appears, you can also PICK a new location for the next line of text anywhere in the drawing.

Figure 18-5

Figure 18-6

Justify

If you want to use one of the justification methods, invoking this option displays the choices at the prompt:

> Command: **dtext**
> Justify/Style/<Start point>: **J** (Invokes the justification options.)
> Align/Fit/Center/Middle/Right/TL/TC/TR/ML/MC/MR/BL/BC/BR: (**choice**) (Type capital letters.)

After specifying a justification option, you can enter the desired text in response to the "Text:" prompt. The text is not justified until <u>after</u> you press Enter.

Align

Aligns the line of text between the two points specified (P1, P2). The text height is adjusted automatically (Fig. 18-7).

Fit

Fits (compresses or extends) the line of text between the two points specified (P1, P2). The text height does not change (Fig. 18-7).

Figure 18-7

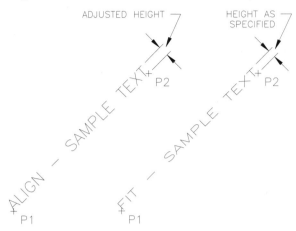

Center

Centers the baseline of the first line of text at the specified point. Additional lines of text are centered below the first (Fig 18-8).

Middle

Centers the first line of text both vertically and horizontally about the specified point. Additional lines of text are centered below it (Fig. 18-8).

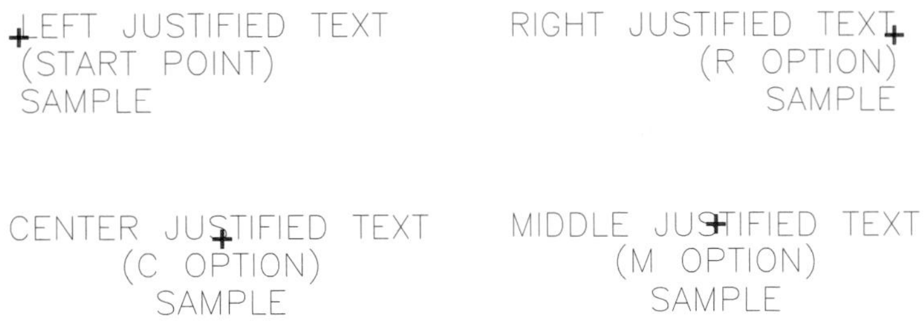

Right

Creates text that is right justified from the specified point (Fig. 18-8).

Figure 18-8

LEFT JUSTIFIED TEXT
(START POINT)
SAMPLE

RIGHT JUSTIFIED TEXT
(R OPTION)
SAMPLE

CENTER JUSTIFIED TEXT
(C OPTION)
SAMPLE

MIDDLE JUSTIFIED TEXT
(M OPTION)
SAMPLE

TL

Top Left. Places the text in the drawing so the top line (of the first line of text) is at the point specified and additional lines of text are left justified below the point. The top line is defined by the upper case and tall lower case letters (Fig. 18-9).

Figure 18-9

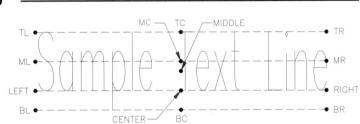

TC

Top Center. Places the text so the top line of text is at the point specified and the line(s) of text are centered below the point (Fig. 18-9).

TR

Top Right. Places the text so the top right corner of the text is at the point specified and additional lines of text are right justified below that point (Fig. 18-9).

ML

Middle Left. Places text so it is left justified and the middle line of the first line of text aligns with the point specified. The middle line is half way between the top line and the baseline, not considering the bottom (extender) line (Fig. 18-9).

MC

Middle Center. Centers the first line of text both vertically and horizontally about the midpoint of the middle line. Additional lines of text are centered below that point (Fig. 18-9).

MR

Middle Right. Justifies the first line of text at the right end of the middle line. Additional lines of text are right justified (Fig. 18-9).

BL

Bottom Left. Attaches the bottom (extender) line of the first line of text to the specified point. The bottom line is determined by the lowest point of lower case extended letters such as y, p, q, j, and g. If only upper-case letters are used, the letters appear to be located above the specified point. Additional lines of text are left justified (Fig. 18-9).

BC

Bottom Center. Centers the first line of text horizontally about the bottom (extender) line (Fig. 18-9).

BR

Bottom Right. Aligns the bottom (extender) line of the first line of text at the specified point. Additional lines of text are right justified (Fig. 18-9).

NOTE: Because there is a separate baseline and bottom (extender) line, the *MC* and *Middle* points do not coincide and the *BL, BC, BR* and *Left, Center, Right* options differ. Also, because of this feature, when all upper case letters are used, they ride above the bottom line. This can be helpful for placing text in a table because selecting a horizontal *Line* object for text alignment with *BL, BC,* or *BR* options automatically spaces the text visibly above the *Line.*

Style   (option of *Dtext, Mtext,* or *Text*)

The *style* option of the *Dtext, Mtext,* or *Text* command allows you to select from the existing text styles which have been previously created as part of the current drawing. The style selected from the list becomes the current style and is used when placing text with *Dtext, Mtext,* or *Text.*

Since only one text *style, Standard,* is available in the traditional (English) template drawing (ACAD.DWT) and the metric template drawing (ACADISO.DWT), other styles must be created before the *style* option of *Text* is of any use. Various text styles are created with the *Style* command (this topic is discussed later).

Use the *Style* option of *Dtext, Mtext,* or *Text* to list existing styles for the drawing. An example listing is shown below:

```
Command: dtext
Justify/Style/<Start point>: s
Style name (or ?) <STANDARD>: ?
Text style(s) to list <*>: Enter
Text styles:

Style name: ARCH-ITALIC   Font typeface: CityBlueprint
  Height: 0.0000  Width factor: 1.0000  Obliquing angle: 15
  Generation: Normal

Style name: ARCHITECT1   Font typeface: CityBlueprint
  Height: 0.0000  Width factor: 1.0000  Obliquing angle: 0
  Generation: Normal
```

Style name: COURIER   Font typeface: Courier New
  Height: 0.0000 Width factor: 1.0000 Obliquing angle: 15
  Generation: Normal

Style name: HELVETICA1   Font typeface: Arial
  Height: 0.0000 Width factor: 1.0000 Obliquing angle: 15
  Generation: Normal

Style name: ROMANSIMPLEX   Font files: romans.shx
  Height: 0.0000 Width factor: 1.0000 Obliquing angle: 0
  Generation: Normal

Style name: STANDARD   Font files: ISOCP
  Height: 0.0000 Width factor: 1.0000 Obliquing angle: 0
  Generation: Normal

Current text style: STANDARD
Justify/Style/<Start point>:

TEXT

Pull-down Menu	COMMAND (TYPE)	ALIAS (TYPE)	Short-cut	Screen (side) Menu	Tablet Menu
...	*TEXT*	...	...	...	...

Text is essentially the same as *Dtext* except that the text is not dynamically displayed one letter at a time as you type, but rather appears in the drawing only after pressing Enter. The other difference is that *Dtext* repeatedly displays the "Text:" prompt to allow entering multiple lines of text, whereas *Text* allows only one line. Otherwise, all the options and capabilities of *Text* are identical to *Dtext*.

    Command: **text**
    Justify/Style/<Start point>: **PICK**
    Height <0.2000>: **Enter** or (**value**)
    Rotation angle <0>: **Enter** or (**value**)
    Text: **Sample line of text**. (The line of text appears in the drawing after pressing Enter.)
    Command:

If you want to type another line of text below the previous line with the *Text* command, use *Text* again, but press Enter at the first prompt. The *Text* command then responds with the "Text:" prompt, at which time you can enter the next line of text. The new line is automatically spaced below and uses the same height, justification, and other options as the previous line.

MTEXT

Pull-down Menu	COMMAND (TYPE)	ALIAS (TYPE)	Short-cut	Screen (side) Menu	Tablet Menu
Draw *Text >* *Multiline Text...*	*MTEXT or* *-MTEXT*	*T, -T or* *MT*	...	*DRAW 2* *Mtext*	*J,8*

Multiline Text (*Mtext*) has more editing options than other text commands. You can apply underlining, color, bold, italic, font, and height changes to individual characters or words within a paragraph or multiple paragraphs of text.

Mtext allows you to create paragraph text defined by a text boundary. The <u>text boundary</u> is a reference rectangle that specifies the paragraph width. The *Mtext* object that you create can be a line, one paragraph, or several paragraphs. AutoCAD references *Mtext* (created with one use of the *Mtext* command) as one object, regardless of the amount of text supplied. Like *Text* and *Dtext*, several justification methods are possible.

> Command: **mtext**
> Current text style: STANDARD. Text height: 0.2000
> Specify first corner: **PICK**
> Specify opposite corner or [Height/Justify/Rotation/Style/Width]: **PICK** or **(option)**

You can PICK two corners to invoke the *Multiline Text Editor,* or enter the first letter of one of these options: *Height, Justify, Rotation, Style,* or *Width.* <u>All of the options</u> can also be accessed <u>within the Multiline Text Editor</u>.

Using the default option the *Mtext* command, you supply a "first corner" and "opposite corner" to define the diagonal corners of the text boundary (like a window). Although this boundary confines the text on two or three sides, one or two arrows indicate the direction text flows if it "spills" out of the boundary (Fig 18-10). (See Text Flow and *Justification.*)

Figure 18-10

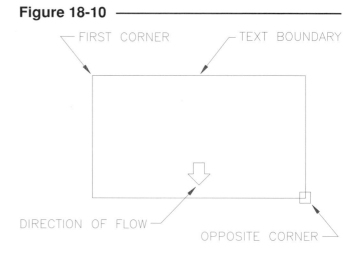

After you PICK the two points defining the text boundary, the *Multiline Text Editor* appears ready for you to enter the text (Fig. 18-11). Enter the desired text. The text wraps based on the width you defined for the text boundary. You can right-click for a menu allowing you to *Cut, Copy,* and *Paste* selected text. Select the *OK* button to have the text entered into the drawing within the text boundary.

Figure 18-11

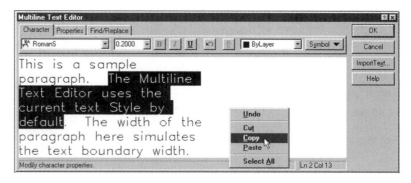

There are three tabs in the *Multiline Text Editor: Character* tab, *Properties* tab , and *Find and Replace* tab. Using the options in these tabs is <u>interactive</u>—text in the editor immediately reflects the changes made for most options in these tabs. There is also a button to *Import Text.*

Import Text
See Importing External Text into AutoCAD.

Find/Replace Tab
See Editing Text later in this chapter.

Properties **Tab**

Use the properties tab to specify the format of the <u>entire paragraph</u>. Although this is the second tab, it is recommended that you <u>format the entire paragraph(s) here before editing individual characters</u> using the *Character* tab (Fig. 18-12). The following options are available.

Figure 18-12

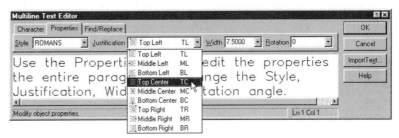

Style
Choose from a drop-down list of existing text styles (see *Style* command).

Justification
This property determines how the paragraph is located and direction of flow with respect to the text boundary (see Text Flow and *Justification*).

Width
Previous paragraph widths used are displayed in this drop-down list. You can enter a new value in the edit box to change the width of the existing text boundary. If a *Width* of 0 is entered or "no wrap" is selected, the lines of text will not "wrap" within the text boundary.

Rotation
The entire paragraph can be rotated to any angle. Changes made here are <u>not</u> reflected in the text appearing in the editor but only in the drawing itself. (See *Rotation*.)

You can type *MTPROP* at the command line to directly access the *Properties* tab of the *Multiline Text Editor*.

Character **Tab**

After formatting the entire paragraph, use the *Character* tab to alter individual characters in the paragraph(s) (Fig. 18-13). Using the options in this tab, first select (highlight) the desired characters or words, then set the desired options. The following options are available.

Figure 18-13

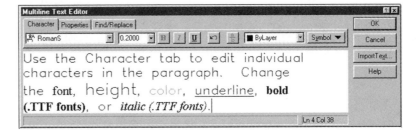

Font
Choose from any font in the drop-down list. Your selection here <u>overrides the text *Style*</u> used for the entire paragraph(s). Even though you can change the font for the entire *Mtext* object (paragraph), it is recommended to set the paragraph to the desired *Style* (in the *Properties* tab), rather than changing all characters to a different font here. See following NOTE.

Height
Select from the list or enter a new value for the height of selected words or letters. Your selection overrides the text *Height* used globally for the paragraph.

Bold, Italic, Underline
Select (highlight) the desired letters or words then PICK the desired button. Only authentic TrueType fonts (not the AutoCAD-supplied .SHX equivalents) can be bolded or italicized.

Stack/Unstack

If creating a stacked fraction, use a / (slash) between the numerator and denominator. If creating stacked text, place a ^ (caret) before the bottom text. Highlight the fraction or text, then use this option to stack or unstack the fraction or text.

Text Color

Select individual text, then use this drop-down list to select a color for the selected text. This selection overrides the layer color.

Symbol

Common symbols (plus/minus, diameter, degrees) can be inserted. Selecting *Other…* produces a character map to select symbols from (See *Other Symbols* below).

Other Symbols

The steps for inserting symbols from the *Character Map* dialog box (Fig. 18-14) are as follows:

Figure 18-14

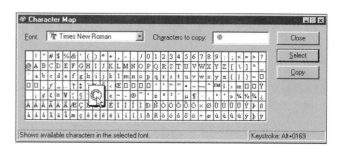

1.   Highlight the symbol.
2.   Double-click or PICK *Select* so the item appears in the *Characters to copy:* edit box.
3.   Select the *Copy* button to copy the item(s) to the Windows Clipboard.
4.   *Close* the dialog box.
5.   In the *Multiline Text Editor,* move the cursor to the desired location to insert the symbol.
6.   Finally, right-click and select *Paste* from the menu.

NOTE: The font, color, and height options in the *Character* tab override the properties of the entire paragraph. For example, changing the font in the *Character* tab overrides the text style's font used for the paragraph so it is possible to have one font used for the paragraph and others for individual characters within the paragraph. This is analogous to object-specific color and linetype assignment in that you can have a layer containing objects with different linetypes and colors than the linetype and color assigned to the layer. To avoid confusion, it is recommended to create text *Styles* to be used globally for the paragraphs, layer color to determine color for the paragraphs, and text *Height* for global paragraph formatting. Then if needed, use the *Character* tab to change fonts, colors, or height for <u>selected</u> text rather than for the entire paragraph.

Text Flow and *Justification*

When you PICK two corners to define the text boundary, you determine the width of the paragraph and the direction that the text flows. Unlike using *Mtext* in Release 13, you can draw the text boundary in any direction from the "first corner" to the "other corner." Text does not always fit within the boundary but is confined on two or three sides and "spills" out of the boundary (up, down, or both) in the direction of the arrow(s) displayed when you draw the boundary (see Figure 18-10).

Justification is the method of aligning the text with respect to the text boundary. The *Justification* options are *TL, TC, TR, ML, MC, MR, BL, BC,* and *BR (Top Left, Top Centered, Top Right, Middle Left, Bottom Left,* etc.) The text paragraph (*Mtext* object) is effectively "attached" to the boundary based on the *Justification* option selected. *Justification* can be specified by two methods: in command line format before specifying the text boundary or in the *Properties* tab of the *Multiline Text Editor.*

The *Justificaiton* methods are illustrated in Figure 18-15. The illustration shows the relationship among the *Justification* option, the text boundary, the direction of flow, and the resulting text paragraph.

Figure 18-15

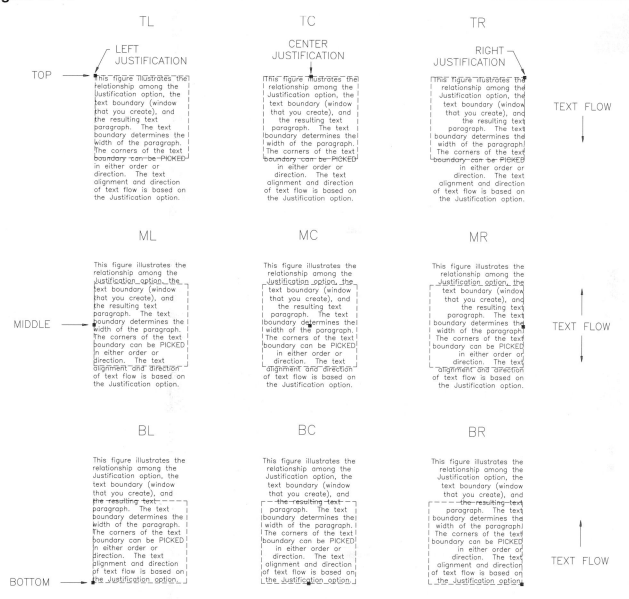

If you want to adjust the text boundary after creating the *Mtext* object, you can use *Ddedit* or *Ddmodify* (see Editing Text) or use Grips (Chapter 23) to stretch the text boundary. If you activate the Grips, four grips appear at the text boundary corners and one grip appears at the defined *Justification* point (Fig. 18-16).

Figure 18-16

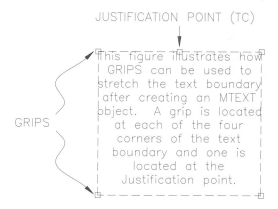

Rotation

The *Rotation* option specifies the rotation angle of the entire *Mtext* object (paragraph) including the text boundary. *Rotation* can be specified by two methods: in command line format before specifying the text boundary or in the *Properties* tab of the *Multiline Text Editor.* It is recommended to use the command line method to specify a rotation angle so you can see the rotated text boundary as you specify the corners (Fig. 18-17). Using this method you can enter a value, or the angle can be specified by PICKing two points. If you use the *Rotation* option within the *Properties* tab of the *Multiline Text Editor,* the text boundary is rotated after the fact.

Figure 18-17

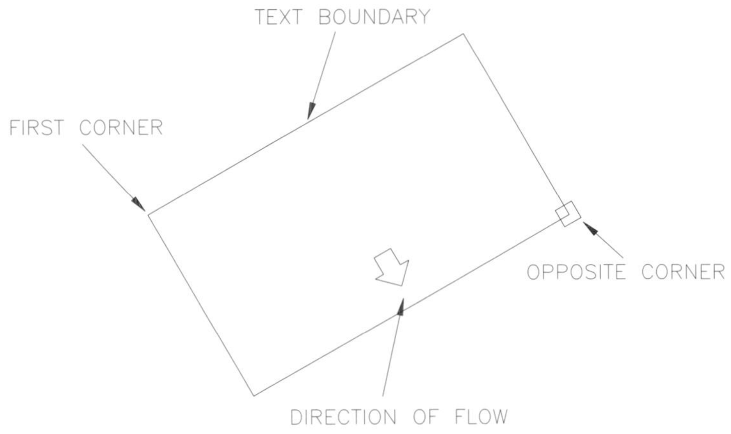

Calculating Text Height for Scaled Drawings

To achieve a specific height in a drawing intended to be plotted to scale, multiply the desired text height for the plot by the drawing scale factor. (See Chapter 14, Plotting.) For example, if the drawing scale factor is 48 and the desired text height on the plotted drawing is 1/8", enter **6** (1/8 x 48) in response to the "Height:" prompt of the *Dtext, Mtext,* or *Text* command.

If you know the plot scale (for example, ¼"= 1') but not the drawing scale factor (DSF), calculate the reciprocal of the plot scale to determine the DSF, then multiply the intended text height for the plot by the DSF, and enter the value in response to the "Height:" prompt. For example, if the plot scale is ¼"=1', then the DSF = 48 (reciprocal of 1/48).

If *Limits* have already been set and you do not know the drawing scale factor, use the following steps to calculate a text height to enter in response to the "Height:" prompt to achieve a specific plotted text height.

1.   Determine the sheet size to be used for plotting (for example, 36" x 24").
2.   Decide on the text height for the finished plot (for example, .125").
3.   Check the *Limits* of the current drawing (for example, 144' x 96' or 1728" x 1152").
4.   Divide the *Limits* by the sheet size to determine the drawing scale factor (1728"/36" = 48).
5.   Multiply the desired text height by the drawing scale factor (.125 x 48 = 6).

TEXT STYLES, FONTS, EXTERNAL TEXT, AND EXTERNAL EDITORS

STYLE

Pull-down Menu	COMMAND (TYPE)	ALIAS (TYPE)	Short-cut	Screen (side) Menu	Tablet Menu
Format Text Style...	STYLE or -STYLE	ST	...	DRAW 2 Dtext Style:	U,2

Text styles can be created by using the *STYLE* command. A text *Style* is created by selecting a font file as a foundation and then specifying several other parameters to define the configuration of the letters.

Using the *Style* command invokes the *Text Style* dialog box (Fig 18-18). All options for creating and modifying text styles are accessible from this device. Selecting the font file is the initial step in creating a style. The font file selected then becomes a foundation for "designing" the new style based on your choices for the other parameters (*Effects*). Recommended steps for creating text *Styles* are as follows:

Figure 18-18 ─────────────

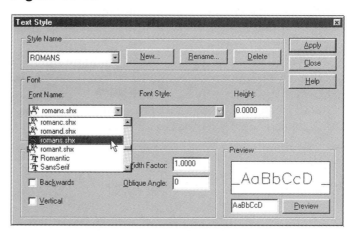

1. Use the *New* button to create a new text *Style*. This button opens the *New Text Style* dialog box (Fig. 18-19). By default, AutoCAD automatically assigns the name Style*n*, where *n* is a number that starts at 1. Enter a descriptive name into the edit box. Style names can be up to 31 characters long and can contain letters, numbers, and the special characters dollar sign ($), underscore (_), and hyphen (-).

2. Select a font to use for the style from the *Font Name* drop-down list (see Fig. 18-18). TrueType equivalents of the traditional AutoCAD fonts files (.SHX) and authentic TrueType fonts are available.

Figure 18-19 ─────────────

3. Select parameters in the *Effects* section of the dialog box (Fig. 18-20, lower-left). Changes to the *Upside down, Backwards, Width Factor,* or *Oblique Angle* are displayed immediately in the *Preview* tile.

4. You can enter specific words or characters in the edit box in the lower-right, then press *Preview* to view the characters in the new style.

5. Select *Apply* to save the changes you made in the *Effects* section to the new style.

6. Create other new *Styles* using the same procedure listed in steps 1 through 5.

7. Select *Close*. The new (or last created) text style that is created automatically becomes the current style inserted when *Dtext, Mtext, Text,* or dimensioning is used.

Figure 18-20 ─────────────

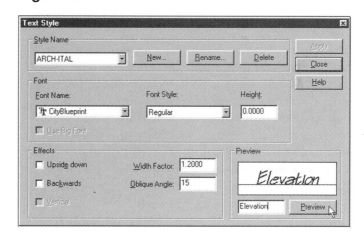

You can modify existing styles using this dialog box by selecting the existing *Style* from the list, making the changes, then selecting *Apply* to save the changes to the existing *Style*.

Rename
Select an existing *Style* from the drop-down list, then select *Rename*. Enter a new name in the edit box.

Delete
Select an existing *Style* from the drop-down list, then select *Delete*. You cannot delete the current *Style* or *Styles* that have been used for creating text in the drawing.

Alternately, you can enter *-Style* (use the apostrophe prefix) at the command prompt to display the command line format of *Style* as shown below:

```
Command: -style
Text style name (or ?) <STANDARD>: name (Enter new Style name)
New style. Specify full font name or font filename <txt>: name (Enter desired font file)
Height <0.0000>: Enter or (value)
Width factor <1.0000>: Enter or (value)
Obliquing angle <0>: Enter or (value)
Backwards? <N> Enter or Y
Upside-down? <N> Enter or Y
Vertical? <N>  Enter or Y
(name) is now the current text style.
Command:
```

The options that appear in both the *Text Style* dialog box and in the command line format are described in detail here.

Height <0.000>

The height should be 0.000 if you want to be prompted again for height each time the *Dtext, Mtext,* or *Text* command is used. In this way, the height is variable for the style each time you create text in the drawing. If you want the height to be constant, enter a value other than 0. Then, *Dtext, Mtext,* or *Text* will not prompt you for a height since it has already been specified. A specific height assignment with the *Style* command also overrides the *DIMTXT* setting (see Chapter 29).

Width factor <1.000>

A *width factor* of 1 keeps the characters proportioned normally. A value less than 1 compresses the width of the text (horizontal dimension) proportionally; a value of greater than 1 extends the text proportionally.

Obliquing angle <0>

An angle of 0 keeps the font file as vertical characters. Entering an angle of 15, for example, would slant the text forward from the existing position, or entering a negative angle would cause a back-slant on a vertically oriented font (Fig. 18-21).

Figure 18-21

Backwards? <Y/N>

Backwards characters can be helpful for special applications, such as those in the printing industry.

Upside-down? <Y/N>

Each letter is created upside-down in the order as typed (Fig. 18-22). This is different than entering a rotation angle of 180 in the *Dtext* or *Text* commands. (Turn this book 180 degrees to read the figure.)

Figure 18-22

Vertical? <Y/N>

Vertical letters are shown in Figure 18-23. The normal rotation angle for vertical text when using *Dtext*, *Mtext*, or *Text* is 270. Only .SHX fonts can be used for this option. *Vertical* text does not display in the *Preview* image tile.

Figure 18-23 —

Since specification of a font file is an initial step in creating a style, it seems logical that different styles could be created using <u>one</u> font file but changing the other parameters. It is possible, and in many cases desirable, to do so. For example, a common practice is to create styles that reference the same font file but have different obliquing angles (often used for lettering on isometric

Figure 18-24 ——————————————————————————

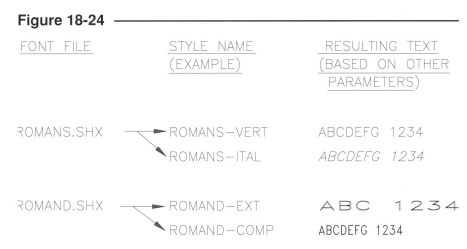

planes). This can be done by using the *Style* command to assign the same font file for each style, but assign different parameters (obliquing angle, width factor, etc.) and a unique name to each style. The relationship between fonts, styles, and resulting text is shown in Figure 18-24.

.SHX Fonts

The .SHX fonts are fonts created especially for AutoCAD drawings by Autodesk. They are generally smaller files and are efficient to use for AutoCAD drawings where file size and regeneration time is critical. These fonts, however, are composed of single line segments and are not solid-filled such as most TrueType fonts (use *Zoom* to closely examine and compare .SHX vs. TrueType fonts).

Because the *Multiline Text Editor* is a Windows-compliant dialog box, it can only display fonts that are recognized by Windows. Since AutoCAD .SHX fonts are not recognized by Windows, AutoCAD supplies a TrueType equivalent when you select an .SHX or any other non-TrueType font for display in the *Multiline Text Editor* only. When you then use a text creation command, AutoCAD creates the text in the drawing with the true .SHX font.

All AutoCAD .SHX shape fonts are Unicode fonts in Release 14. A Unicode font can contain 65,535 characters with shapes for many languages (see *Big Fonts*). Unicode fonts contain many more characters than are shown on your keyboard in your system; therefore, to use a character not directly available from the keyboard, you can enter the escape sequence ***U+nnnn*** where *nnnn* represents the Unicode hexadecimal value for the character (see Special Text Characters).

Big Fonts

Big Fonts are used for characters not found in the *English* language. Text files for some alphabets contain thousands of non-ASCII characters. For these applications, AutoCAD provides a special type of shape definition known as a *Big Font* file that contains many other characters.

When you check *Use Big Font* (grayed out in the *Text Style* dialog box, Fig. 18-20) the *Font Style* box changes to a *Big Font Name* box. You can select *Use Big Font* only when using .SHX fonts. Set a *Style* to use both regular and *Big Font* files for other than English languages. Otherwise, for most applications, select only the regular .SHX font files.

TrueType Fonts

AutoCAD uses the Windows operating system directly to display TrueType text on the screen for most applications. However, when text in AutoCAD has been transformed (mirrored, upside-down, backward, oblique, has a width factor not equal to 1, or is in an orientation that is not co-planar with the screen), AutoCAD must draw the TrueType text. Text that has been transformed might appear slightly more bold in some circumstances, especially at lower resolutions. This difference is only in the screen display of the font and does not affect the plotted drawing.

For TrueType fonts only, the value specified for text *Height* might not represent the actual height of upper case letters. The height specified represents the height of a capital letter plus an "ascent" area reserved for accent marks and other marks used in non-English languages. TrueType fonts also have a "descent" area for portions of characters with extenders (such as the characters y, j, p, g, and q).

Proxy Fonts

For third-party or custom .SHX fonts that have no TrueType equivalent, AutoCAD supplies up to eight different TrueType fonts called proxy fonts. Proxy fonts appear different in the *Multiline Text Editor* from the font they represent to indicate that they are substitutions for the fonts used in the drawing.

Importing External Text into AutoCAD

You can import ASCII text files created in other text editors or word processors into an AutoCAD drawing. There are three methods: use the *Import Text* button in *Multiline Text Editor,* use *Copy* and *Paste* (using the Windows Clipboard), or "drag and drop" a file icon from the Windows Explorer.

Importing text in ASCII or RTF files can save you drawing time. For example, you can create a text file of standard notes that you include in many drawings or use information prepared for a report or other document. Instead of entering (typing) this information in the drawing, you can import the file. The imported text becomes an AutoCAD text object, which you can edit as if you created it in AutoCAD. Imported text retains its original formatting properties from the text editor in which it was created.

Importing with the *Multiline Text Editor*

To import text into the *Multiline Text Editor,* first use the *Mtext* command. PICK two points to define the text boundary as usual. When the *Multiline Text Editor* appears, select the *Import Text* button. The *Open* dialog appears for you to locate and select the desired file. The file to import must be in an ASCII or RTF format. (An ASCII text file often has a .TXT file extension.) The selected text appears in the editor and can be manipulated just as any text you entered.

You can also *Copy* text (to the Windows Clipboard) from any Windows document and *Paste* it into AutoCAD. The best method to use with *Copy* and *Paste* is to use the *Multiline Text Editor*. Open the editor, then use the Alt+Tab key combination to switch to the other document. Highlight the desired text and select *Copy* from the *Edit* menu or right-click menu. Switch back to the *Multiline Text Editor*, right-click for the pop-up menu, and select *Paste*. The imported text can be edited like any other AutoCAD *Mtext* object.

Using Drag and Drop to Import Text Files

You can also use the "drag-and-drop" feature to insert ASCII or RTF text into a drawing. Open the Windows Explorer (file manager) and size the window so AutoCAD is also visible on the screen. "Drag" the file name or icon and "drop" it into an AutoCAD drawing. The pasted text uses the formats and fonts defined by the current AutoCAD text style. The width of the lines or paragraph is determined by line breaks and carriage returns in the original document. Imported text files are limited to a maximum of 16 KB. You can only drag files with a file extension of .TXT or .RTF into an AutoCAD drawing.

Using an External Text Editor for *Multiline Text*

You can specify a different text editor to use instead of the "internal" *Multiline Text Editor*. For example, you may prefer the Windows Notepad or other editor, although these editors do not offer the text formatting capabilities available in the *Multiline Text Editor*.

To specify an external text editor, you can use the *Files* tab of the *Preferences* dialog box from the *Tools* pull-down menu (Fig. 18-25) or use the *MTEXTED* system variable. Enter "internal" to use the internal AutoCAD *Multiline Text Editor*.

If you use an external text editor (other than the AutoCAD internal editor), you must enter the for-matting codes as you enter the text using the text editor. Text features that can be controlled are under-line, overline, height, color, spacing, stacked text, and other options. The following table lists the codes and the resulting text:

Figure 18-25

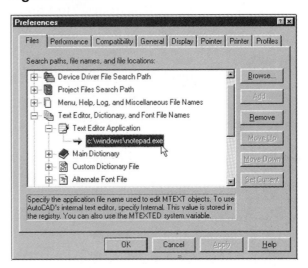

Format Codes for Creating *Multiline* Text in External Text Editors

Format code	Purpose	Type this. . .	To produce this
\O. . .\o	Turns overline on and off	You have \Omany\o choices	You have $\overline{\text{many}}$ choices
\L. . .\l	Turns underline on and off	You have \Lmany\l choices	You have <u>many</u> choices
\~	Inserts a nonbreaking space	Keep these\~words together	Keep these words together
\\	Inserts a backslash	slash\\backslash	slash\ backslash
\{. . .\}	Inserts an opening and closing brace	The \{bracketed\} word	The {bracketed} word
\C*value;*	Changes to the specified color	Change \C2;these colors	Change these colors
\F*filename;*	Changes to the specified font file	Change \Farial;these fonts	Change **these fonts**
\H*value;*	Changes to the specified text height	Change \H2;these sizes	Change **these sizes**
\S. . .^. . .	Stacks the subsequent text at the \ or ^ symbol	1.000\S+0.010^0.000;	$1.000^{+0.010}_{-0.000}$
\T*value;*	Adjusts the space between characters, from .75 to 4 times	\T2;TRACKING	T R A C K I N G
\Q*angle;*	Changes obliquing angle	\Q20;OBLIQUE	*OBLIQUE*
\W*value;*	Change width factor to produce wide text	\W2;Wide	**WIDE**
\A	Sets the alignment value to 0 (bottom), 1 (center), or 2 (top)	\A1;Center\S1/2	Center $\frac{1}{2}$
\P	Ends paragraph	First paragraph\PSecond paragraph	First paragraph Second paragraph

For example, you may enter the following text in the Windows Notepad editor using the *Mtext* command.

```
\C1;This sample line of text is red in color.\P
\C7;\H.25;This text has a height of 0.25.\P
\H.2;\Fromanc;This line is an example of romanc.shx.\P
\Ftxt;This line contains stacked fractions such as \S1/2.
```

Figure 18-26 displays the text as it might appear in the AutoCAD drawing editor.

Figure 18-26 ———————————————————————

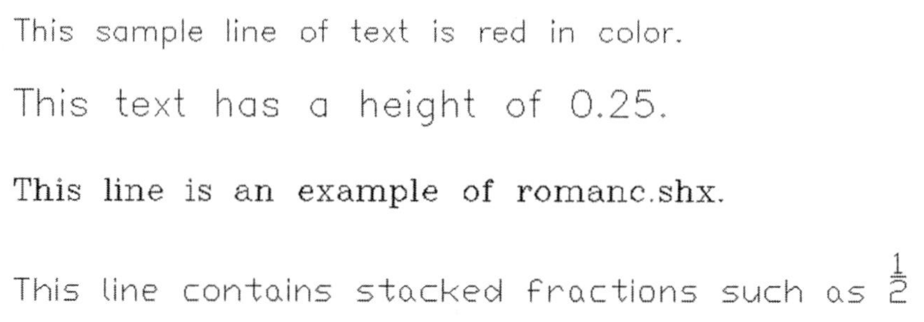

This sample line of text is red in color.

This text has a height of 0.25.

This line is an example of romanc.shx.

This line contains stacked fractions such as $\frac{1}{2}$

Special Text Characters

Special characters that are often used in drawings can be entered easily in the internal *Multiline Text Editor* by selecting the *Symbol* button (see Fig. 18-13). In an external editor or with the *Text* and *Dtext* commands, you can code the text by using the "%%" symbols or by entering the Unicode values for high-ASCII (characters above the 128 range). The %% symbols can be used with *Text* and *Dtext* commands, but the Unicode values must be used for creating *Mtext* with an external text editor. The following codes are typed in response to the "Text:" prompt in the *Dtext* or *Text* command or entered in the external text editor.

External Editor	*Text or Dtext*	Result	Description
\U+2205	%%c	ø	diameter (metric)
\U+00b0	%%d	°	degrees
	%%o	‾	overscored text
	%%u	___	underscored text
\U+00b1	%%p	±	plus or minus
	%%nnn	varies	ASCII text character number
\U+nnnn		varies	Unicode text hexadecimal value

For example, entering "**Chamfer 45%%d**" at the "Text:" prompt of the *Dtext* command draws the following text: **Chamfer 45°**.

EDITING TEXT

SPELL

	Pull-down Menu	**COMMAND (TYPE)**	**ALIAS (TYPE)**	**Short-cut**	**Screen (side) Menu**	**Tablet Menu**
	Tools *Spelling*	*SPELL*	*SP*	...	*TOOLS 1* *Spell*	*T,10*

AutoCAD has an internal spell checker that can be used to spell check and correct existing text in a drawing after using *Dtext*, *Mtext*, or *Text*. The *Check Spelling* dialog box (Fig. 18-27) has options to *Ignore* the current word or *Change* to the suggested word. The *Ignore All* and *Change All* options treat every occurrence of the highlighted word.

AutoCAD matches the words in the drawing to the words in the current dictionary. If the speller indicates that a word is misspelled but it is a proper name or an acronym you use often, it can be added to a custom dictionary. Choose *Add* if you want to leave a word unchanged but add it to the current custom dictionary.

Figure 18-27

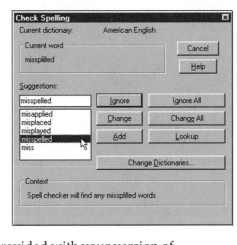

Selecting the *Change Dictionaries...* tile produces the dialog shown in Figure 18-28. You can select from other *Main Dictionaries* that are provided with your version of AutoCAD. The current main dictionary can also be changed in the *Files* tab of the *Preferences* dialog box (see Fig. 18-25), and the name is stored in the *DCTMAIN* system variable.

The Release 14 default custom dictionary is SAMPLE.CUS (see Fig. 18-28). If you use the *Add* function (in the *Check Spelling* dialog box), the selected word is added to the current custom dictionary. You can also create a custom dictionary "on the fly" by entering any name in the *Custom dictionary* edit box. A file extension of .CUS should be used with the name (although other extensions will work). A custom dictionary name must be specified before you can add words. Words can be added by entering the desired word in the *Custom dictionary words* edit box or by using the *Add* tile in the *Check Spelling* dialog box. Custom dictionaries can be changed during a spell check. The custom dictionary can also be changed in the *Files* tab of the *Preferences* dialog box (see Figure 18-25), and its name is stored in the *DCTCUST* system variable.

Figure 18-28

DDEDIT

Pull-down Menu	COMMAND (TYPE)	ALIAS (TYPE)	Short-cut	Screen (side) Menu	Tablet Menu
Modify Object > Text...	*DDEDIT*	*ED*	...	*Modify1 Ddedit*	*Y,21*

Ddedit invokes a dialog box for editing existing text in a drawing. You can edit <u>individual characters</u> or the entire line or paragraph. If the selected text was created by the *Text* or *Dtext* command, the *Edit Text* dialog appears displaying one line of text (Fig. 18-29). If *Mtext* created the selected text, the *Multiline Text Editor* box appears (or current text editor).

Figure 18-29

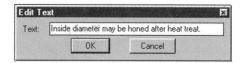

DDMODIFY

Pull-down Menu	COMMAND (TYPE)	ALIAS (TYPE)	Short-cut	Screen (side) Menu	Tablet Menu
Modify Properties...	*DDMODIFY*	*MO*	...	*MODIFY1 Modify*	*Y,14*

Invoking this command prompts you to select objects and causes the *Modify Text* or *Modify Mtext* dialog box to appear. As described in Chapter 16, the dialog box that appears is <u>specific</u> to the type of object that is PICKed— kind of a "smart" dialog box. If a line of text (*Dtext* or *Text*) is PICKed in response to the "Select objects:" prompt, the dialog box shown in Figure 18-30 appears.

Modify Text allows <u>any</u> kind of editing to text, including an *Upside down, Backward, Width factor,* and *Obliquing angle* change.

Figure 18-30

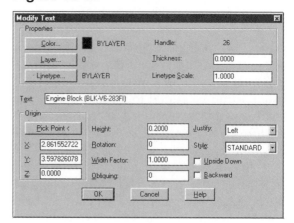

If a paragraph of text (created with *Mtext*) is selected, a similar *Modify Mtext* dialog box appears (Fig. 18-31). You can change the properties of the text such as *Style, Justification, Width*, or *Text Height*. Choose the *Full editor...* button to edit the text content in the *Multiline Text Editor* (or current text editor).

Figure 18-31

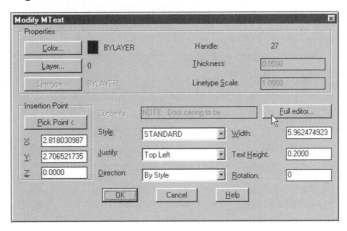

Find/Replace

The *Multiline Text Editor* includes a feature for finding and replacing text on a word-by-word basis. The *Find/Replace* tool replaces text based on content and case (upper, lower) but does not change character formatting or text properties. *Find/Replace* operates only for *Mtext* objects.

To *Find* Text

To *Find* text, follow these steps:

Figure 18-32

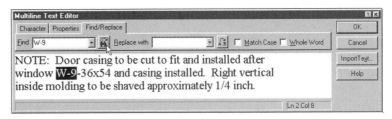

1. Use *Properties* from the *Modify* menu (or otherwise invoke *Ddmodify*) and select an *Mtext* object.
2. In the *Modify Mtext* dialog box (see Fig. 18-31), select *Full editor*....
3. When the *Multiline Text Editor* appears, select the *Find/Replace* tab.
4. Enter the word to find in the *Find* edit box (Fig. 18-32).
5. Check *Match Case* to locate only text that matches the case (upper and lower) of the *Find* box entry.
6. If you want AutoCAD to find the text entry as an entire word (not as part of another word), check the *Whole Word* box.
7. Choose the *Find* button.

To *Replace* Text

To *Replace* text, follow these steps:

1. Use *Properties* from the *Modify* menu (or otherwise invoke *Ddmodify*) and select an *Mtext* object.
2. In the *Modify Mtext* dialog box (see Fig. 18-31), select *Full editor. . ..*
3. When the *Multiline Text Editor* appears, select the *Find/Replace* tab.
4. Enter the word to be replaced (old word) in the *Find* edit box (Fig. 18-32).
5. Enter the new word (to replace the old new word), including the desired case, in the *Replace* box.
6. Choose the *Find* button.

If you want to replace a word with the same word but with the a change of a case (upper or lower) for one or more letters, AutoCAD makes the replacement correctly without using the *Match Case* option. The *Match Case* and *Whole Word* options are needed for the *Find* function only.

QTEXT

Pull-down Menu	COMMAND (TYPE)	ALIAS (TYPE)	Short-cut	Screen (side) Menu	Tablet Menu
...	QTEXT	...	...	...	...

Qtext (quick text) allows you to display a line of text <u>as a box</u> in order to speed up drawing and plotting times. Because text objects are treated as graphical elements, a drawing with much text can be relatively slower to regenerate and take considerably more time plotting than the same drawing with little or no text.

When *Qtext* is turned *ON* and the drawing is regenerated, each text line is displayed as a rectangular box (Fig. 18-33). Each box displayed represents one line of text and is approximately equal in size to the associated line of text.

For drawings with considerable amounts of text, *Qtext ON* noticeably reduces regeneration time.

Figure 18-33

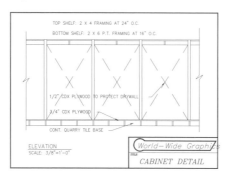

QTEXT OFF   QTEXT ON

For check plots (plots made during the drawing or design process used for checking progress), the drawing can be plotted with *Qtext ON*, requiring considerably less plotting time. *Qtext* is then turned *OFF* and the drawing must be *Regenerated* to make the final plot.

When *Qtext* is turned *ON*, the text remains in a readable state until a *Regen* is invoked or caused. When *Qtext* is turned *OFF*, the drawing must be regenerated to read the text again.

TEXTFILL

Pull-down Menu	COMMAND (TYPE)	ALIAS (TYPE)	Short-cut	Screen (side) Menu	Tablet Menu
File Print Text Fill	TEXTFILL	...	...	FILE Plot Text Fill	...

The *TEXTFILL* variable controls the display of TrueType fonts for <u>printing and plotting only</u>. *TEXTFILL* does not control the display of fonts for the screen—fonts always appear filled in the Drawing Editor. If *TEXTFILL* is set to 1 (on), these fonts print and plot with solid-filled characters (Fig. 18-34). If *TEXTFILL* is set to 0 (off), the fonts print and plot as outlined text (Fig. 18-35). The variable controls text display globally and retroactively.

Figure 18-34

TEXTFILL ON

Figure 18-35

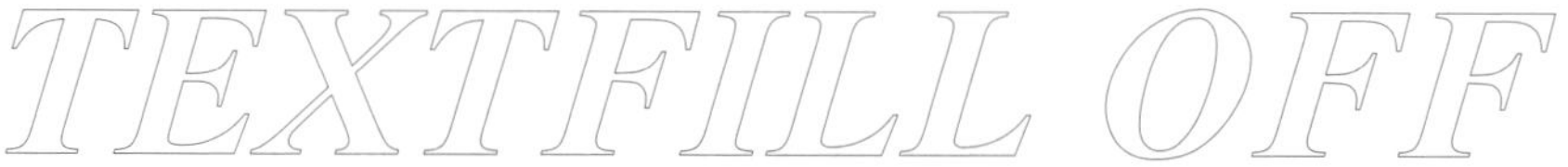

TEXTFILL OFF

You can control this feature most easily in the *Additional Parameters* section of the *Plot Configuration* dialog box (Fig. 18-36). *Text Fill* is on (checked) by default. See also Text Fill, Chapter 14.

You can also control the *Text Resolution* for printing and plotting. See Text Resolution, Chapter 14.

Figure 18-36

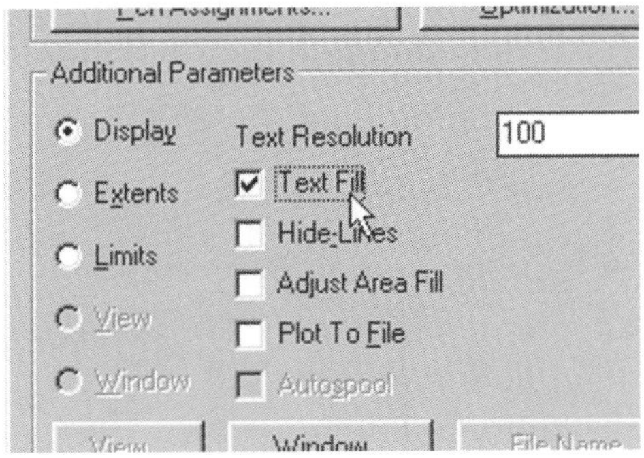

Substituting Fonts

Alternate Fonts

Fonts in your drawing are specified by the font file referenced by the current text *Style* and by individual font formats specified for sections of *Mtext*. Text font files are not stored in the drawing; instead, text objects reference the font files located on the current computer system. Occasionally, a drawing is loaded that references a font file not available on your system. This occurrence is more common with the ability to create text styles using TrueType fonts.

By default, AutoCAD substitutes the SIMPLEX.SHX for unreferenced fonts. In other words, when AutoCAD loads a drawing that contains a font not on your system, the SIMPLEX.SHX font is automatically substituted. You can use the *FONTALT* system variable or the *Files* tab of the *Preferences* dialog box (Fig. 18-37) to specify other font files that you want to use as an alternate font whenever a drawing with unreferenced fonts is loaded. Typically, .SHX fonts are used as alternates.

Figure 18-37

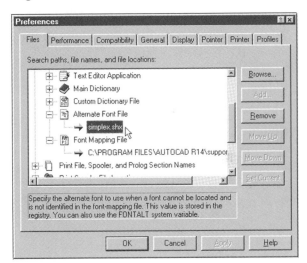

Font Mapping

You can also set up a <u>font mapping table</u> listing several fonts and substitutes to be used whenever a text object is encountered that references one of the fonts. The current table name is stored in the *FONTMAP* system variable (the default is none, defined by a period [.]). A font mapping table is a plain ASCII file with a .FMP file extension and contains one font mapping per line. Each line contains the base font followed by a semicolon and the substitute font.

You can use font mapping to simplify problems that may occur when exchanging drawings with clients. You can also use a font mapping table to substitute fast-drawing .SHX files while drawing and for test plots, then use another table to substitute back the more complex fonts for the final draft. For example, you might want to substitute the ROMANS.SHX font for the Times TrueType font and the TXT.SHX font for the Arial TrueType font. (The second font in each line is the new font that you want to appear in the drawing.)

```
times;romans.shx
arial;txt.shx
```

AutoCAD provides a sample font mapping table called ACAD.FMP located in the C:\Program Files\ AutoCAD R14\Sample directory. The table has 16 substitutions which are used to map the Release 13 PostScript fonts to the newer Release 14 TrueType fonts. The ACAD.FMP contents are shown here.

```
cibt;CITYB__.TTF
cobt;COUNB__.TTF
eur;EURR____.TTF
euro;EURRO__.TTF
par;PANROMAN.TTF
rom;ROMANTIC.TTF
romb;ROMAB__.TTF
romi;ROMAI__.TTF
sas;SANSS__.TTF
sasb;SANSSB__.TTF
sasbo;SANSSBO_.TTF
saso;SANSSO__.TTF
suf;SUPEF__.TTF
te;TECHNIC_.TTF
teb;TECHB__.TTF
tel;TECHL__.TTF
```

You can edit this table with an ASCII text editor or create a new font mapping table. You can specify a new font mapping file in the *Files* tab of the *Preferences* dialog box (Fig. 18-37) or use the *FONTMAP* system variable.

A font mapping table <u>forces</u> the listed substitutions, while an alternate font substitutes the specified font for <u>any</u> font but <u>only</u> when an unreferenced font is encountered.

When a drawing is opened and AutoCAD begins to locate the fonts referenced by text *Styles* in the drawing, AutoCAD follows the following priority:

1. AutoCAD uses the *FONTMAP* value(s) if defined.
2. If not defined, uses the font defined in the text *Style*.
3. If not found, Windows substitutes a similar font (if .TTF) or uses the *FONTALT* specified (if .SHX).
4. If an .SHX is not found, it prompts you for a new font.

Text Attributes

When you want text to be entered into a drawing and associated with some geometry, such as a label for a symbol or number for a part, *Block Attributes* can be used. *Attributes* are text objects attached to *Blocks*. When the *Blocks* are *Inserted* into a drawing, the text *Attributes* are also inserted; however, the content of the text can be entered at the time of the insertion. See Chapter 21, Blocks, and Chapter 22, Block Attributes.

CHAPTER EXERCISES

1. *Dtext*

   Open the **EFF-APT** drawing. Make *Layer* **TEXT** *Current* and use *Dtext* to label the three rooms: **KITCHEN, LIVING ROOM,** and **BATH**. Use the *Standard style* and the *Start point* justification option. When prompted for the *Height:*, enter a value to yield letters of 3/16″ on a 1/4″=1′ plot (3/16 x the drawing scale factor = text height). *Save* the drawing.

2. *Style, Dtext, Ddedit*

Open the drawing of the temperature graph you created as **TEMP-D**. Use *Style* to create two styles named **ROMANS** and **ROMANC** based on the *romans.shx* and *romanc.shx* font files (accept all defaults). Use *Dtext* with the *Center* justification option to label the days of the week and the temperatures (and degree symbols) with the **ROMANS** style as shown in Figure 18-38. Use a *Height* of **1.6**. Label the axes as shown using the **ROMANC** style. Use *Ddedit* for editing any mistakes. Use *SaveAs* and name the drawing **TEMPGRPH**.

Figure 18-38

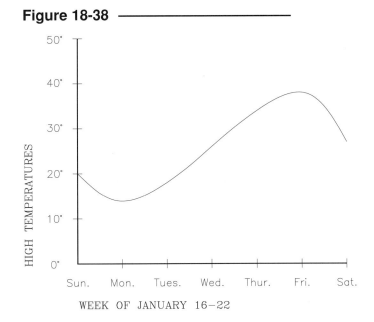

3. *Style, Dtext*

Open the **BILLMATL** drawing created in the Chapter 15 Exercises. Use *Style* to create a new style using the *romans.shx* font. Use whatever justification methods you need to align the text information (not the titles) as shown in Figure 18-39. Next, type the *Style* command to create a new style that you name as **ROMANS-ITAL**. Use the *romans.shx* font file and specify a **15** degree *obliquing angle*. Use this style for the **NO.**, **PART NAME**, and **MATERIAL**. *SaveAs* **BILLMAT2**.

Figure 18-39

NO.	PART NAME	MATERIAL
1	Base	Cast Iron
2	Centering Screw	N/A
3	Slide Bracket	Mild Steel
4	Swivel Plate	Mild Steel
5	Top Plate	Cast Iron

4. *Edit Text*

Open the **EFF-APT** drawing. Create a new style named **ARCH1** using the *CityBlueprint* (.TTF) font file. Next, invoke the *Ddmodify* command. Use this dialog box to modify the text style of each of the existing room names to the new style as shown in Figure 18-40. *SaveAs* **EFF-APT2**.

Figure 18-40

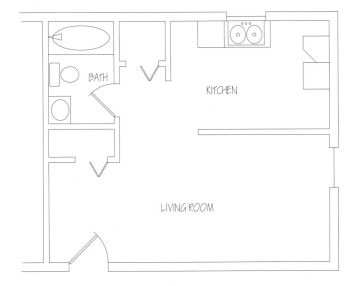

5. *Dtext, Mtext*

Open the **CBRACKET** drawing from Chapter 10 Exercises. Using *romans.shx* font, use *Dtext* to place the part name and METRIC annotation (Fig 18-41). Use a *Height* of **5** and **4**, respectively, and the *Center Justification* option. For the notes, use *Mtext* to create the boundary as shown. Use the default *Justify* method (*TL*)and a *Height* of **3**. Use *Ddedit* or *Ddmodify* if necessary. *SaveAs* **CBRACTXT**.

Figure 18-41 ———————————————

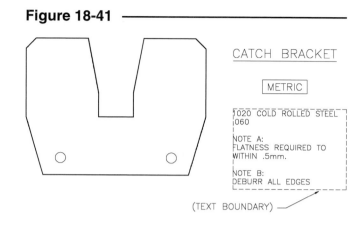

6. *Style*

Create two new styles for each of your template drawings: **ASHEET**, **BSHEET**, and **CSHEET**. Use the *romans.shx* style with the default options for engineering applications or *CityBlueprint* (.TTF) for architectural applications. Next, design a style of your choosing to use for larger text as in title blocks or large notes.

Figure 18-42 ———————————————

7. *Import Text, Ddedit, Ddmodify*

Use a text editor such as Windows Wordpad or Notepad to create a text file containing words similar to "Temperatures were recorded at Sanderson Field by the National Weather Service." Then *Open* the **TEMPGRPH** drawing and use the *Import Text...* option of *Mtext* to bring the text into the drawing as a note in the graph as shown in Figure 18-42. Use *Ddedit* to edit the text if desired or use *Ddmodify* to change the text style or height. *Save* the drawing.

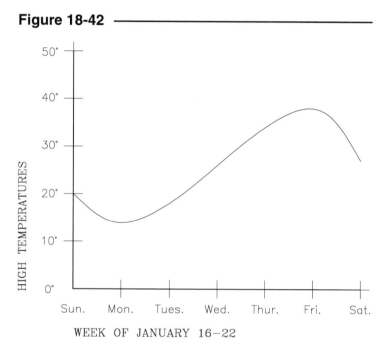

8. *Create a Title Block*

A. Begin a *New* drawing and assign the name **TBLOCK**. Create the title block as shown in Figure 18-43 or design your own, allowing space for eight text entries. The dimensions are set for an A size sheet. Draw on *Layer* **0**. Use a *Pline* with **.02** *width* for the boundary and *Lines* for the interior divisions. (No *Lines* are needed on the right side and bottom because the title block will fit against the border lines.)

Figure 18-43 ———————————————

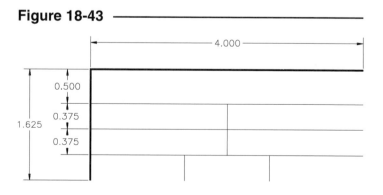

B. Create two text *Styles* using *romans.shx* and *romanc.shx* font files. Insert text similar to that shown in Figure 18-44. Examples of the fields to create are:

> Company or School Name
> Part Name or Project Title
> Scale
> Designer Name
> Checker or Instructor Name
> Completion Date
> Check or Grade Date
> Project or Part Number

Figure 18-44 —————————

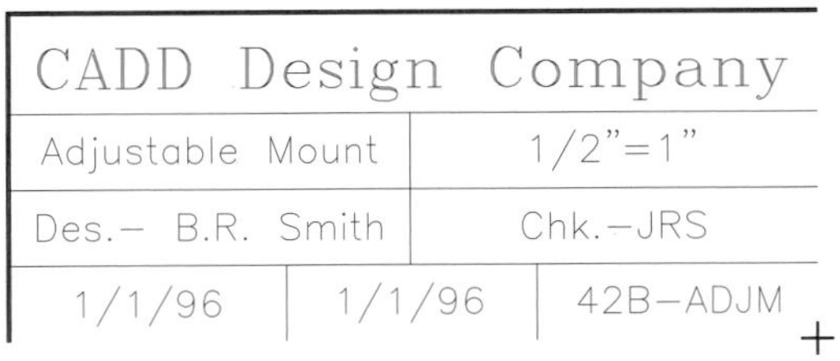

CADD Design Company		
Adjustable Mount	1/2"=1"	
Des.— B.R. Smith	Chk.—JRS	
1/1/96	1/1/96	42B—ADJM

Choose text items relevant to your school or office. *Save* the drawing.

9. *Mtext*

Open the **STORE ROOM** drawing that you created in Chapter 16 Exercises. Use *Mtext* to create text paragraphs giving specifications as shown in Figure 18-45. Format the text below as shown in the figure. Use *CityBlueprint* (.TTF) as the base font file and specify a base *Height* of **3.5"**. Use a *Color* of your choice and *CountryBlueprint* (.TTF) font to emphasize the first line of the Room paragraph. All paragraphs use the *TC Justify* methods except the Contractor Notes paragraph, which is *TL*. Save the drawing as **STORE ROOM2**.

> Room: <u>STORAGE ROOM</u>
> 11'-2" x 10'-2"
> Cedar Lined - 2 Walls

> Doors: 2 - 2268 DOORS
> Fire Type A
> Andermax

> Windows: 2 - 2640 CASEMENT WINDOWS
> Triple Pane Argon Filled
> Andermax

Notes: <u>Contractor Notes:</u>

> Contractor to verify all dimensions in field. Fill door and window roughouts after door and window placement.

Figure 18-45 ————————————————

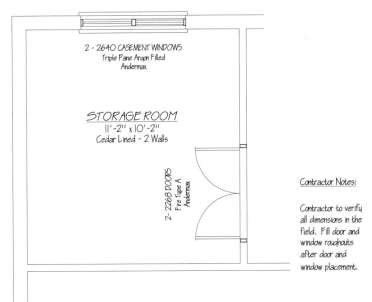

PLANPROF.DWG Courtesy, AutoCAD Inc. (Release 13 Sample Drawing)

19

TILED VIEWPORTS AND PAPER SPACE VIEWPORTS

Chapter Objectives

After completing this chapter you should:

1. know the difference between tiled viewports and paper space viewports and the purpose of the *TILEMODE* variable;

2. be able to create, *Save*, and *Restore* tiled viewport configurations using the *Vports* command and *Tiled Viewport Layout* dialog box;

3. be able to use the options of *Mview* to create viewports in paper space;

4. be able to use the *XP* option of *Zoom* to scale model space geometry to paper space units.

CONCEPTS

This chapter gives an introduction to the two types of viewports in AutoCAD (tiled viewports and paper space viewports) and the system variable (*TILEMODE*) that controls which type of viewport can be enabled. Tiled viewports are simple to understand and easy to use during construction of geometry. Paper space (or floating) viewports are more complex and are used mainly for arranging several views on a sheet for printing or plotting. Only an introduction to tiled and paper space viewports is given in this chapter. Advanced features of paper space viewports are discussed in Chapter 33 and using paper space viewports for 3D applications is discussed in Chapter 42. Using tiled viewports for constructing 3D models is discussed in Chapter 35.

Tiled viewports divide the screen into several areas that fit together with no space between, like tiles. Tiled viewports are created by the *Vports* command. Each tiled viewport can show a different area of the drawing by using display commands—especially helpful for 3D models. *Vports* divides the screen into multiple sections, allowing you to view different parts of a drawing on one screen. Tiled viewports can make construction and editing of complex drawings more efficient than repeatedly using display commands to view detailed areas. Tiled viewports were introduced with AutoCAD Release 10.

Paper space (floating) viewports are used for plotting several views of a drawing, or several drawings, on one sheet. Paper space viewports are used to set up a sheet of paper for plotting. The *Mview* command is used to create viewports in paper space. *Mview* allows several viewports to exist on the screen, but the viewports can be any size and rectangular proportion, unlike tiled viewports. Paper space viewports are generally used when the drawing is complete and you are ready to prepare a plot. Paper space viewports were introduced with AutoCAD Release 11.

AutoCAD allows you to work on your drawing in two separate spaces—model space and paper space. The space that you have been drawing in up to this time is called model space. The Drawing Editor begins a drawing in model space by default. The model geometry objects (which represent the subject of the drawing) are almost always drawn in model space. Paper space is normally used to set up the configuration of the plotting sheet.

Another fundamental difference between tiled viewports and paper space viewports is the space in which these viewports exist. Tiled viewports exist only when model space is active; therefore, they are sometimes referred to as "model space viewports" or "tiled model space." Paper space viewports do not exist in the same space as the model geometry. In order to create or use paper space viewports, paper space must first be activated. Release 14 of AutoCAD refers to viewports created in paper space as "floating viewports."

Although both tiled viewports and paper space viewports can be used in one drawing, they cannot be used at the same time. Therefore, the *TILEMODE* variable controls which type of viewport can be enabled at a specific point in time. Moreover, *TILEMODE* effectively controls which space is active—model space or paper space.

Commands related to creating and activating tiled and floating viewports are located in the *View* pulldown menu (Fig. 19-1). Note that only one of the two options in the menu (*Tiled Viewports* or *Floating Viewports*) is available at any one time, based on the setting of *TILEMODE* or whether *Model Space (Tiled)* or *Paper Space* is checked in the menu.

Figure 19-1 ——————

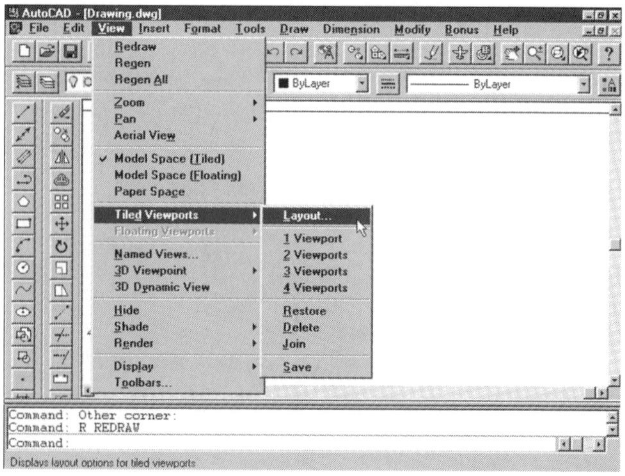

TILEMODE

Pull-down Menu	COMMAND (TYPE)	ALIAS (TYPE)	Short-cut	Screen (side) Menu	Tablet Menu
View *Model Space (Tiled)*	*TILEMODE*	*TM or TI*	...	*VIEW 1* *Tilemode*	*L,3*

When paper space viewports were introduced with Release 11, a new system variable called *TILEMODE* was also introduced. Since only <u>one</u> type of viewport can be used at a time, the *TILEMODE* system variable is used to enable one or the other type of viewports. More importantly, *TILEMODE* is a toggle that switches between using only model space or using paper space.

When *TILEMODE* is set to 1 (*TILEMODE* is *On*), only tiled viewports in model space can be used. The viewports themselves must be created with the *Vports* command. The variable can be typed, or you can select *Model Space (Tiled)* from the *View* pull-down menu. *TILEMODE* can also be toggled by double-clicking the word "TILE" on the Status Line. *TILEMODE On* is the default setting for the AutoCAD template drawing ACAD.DWT; so normally, model space is active and only tiled viewports can be used.

If *TILEMODE* is set to 0 (*TILEMODE* is *Off*), paper space is enabled and paper space-style viewports are available. Viewports in paper space must be created with the *Mview* command. Change *TILEMODE* to *Off* by typing the variable, selecting *Paper Space* from the *View* pull-down menu, or double-click the word "TILE" on the Status Line. Therefore, setting *TILEMODE* to 0 disables display of model space and tiled viewports and automatically enables paper space.

Whether or not viewports of any kind have been created with either the *Vports* or the *Mview* commands, the setting of *TILEMODE* displays the following:

<u>TILEMODE setting</u>	<u>Resulting display</u>
1	model space only—tiled (*Vports*) viewports can be used
0	paper space—paper space (*Mview*) viewports can be used

Switching the *TILEMODE* setting does not affect any objects you have drawn. The setting may affect the visibility of objects in the display.

This chapter discusses both tiled viewports and paper space viewports. Tiled viewports are discussed first because they are easier to use and understand.

TILED VIEWPORTS

All controls for tiled viewports are included in the *Vports* command. Tiled viewports are useful for displaying several views of one or more 2D drawings simultaneously on the screen. Using tiled viewports for displaying several views of a 3D drawing is discussed in Chapter 35.

VPORTS

Pull-down Menu	COMMAND (TYPE)	ALIAS (TYPE)	Short-cut	Screen (side) Menu	Tablet Menu
View *Tiled Viewports >*	*VPORTS*	...	...	*VIEW 1* *Vports*	*M,3*

The *Vports* command allows <u>tiled</u> viewports to be created on the screen. *Vports* divides the screen into several areas. *Vports* (tiled viewports) are available only when the *TILEMODE* variable is set to **1**. Tiled *Vports* affect <u>only the screen display</u>. The viewport configuration <u>cannot be plotted</u>. If the *Plot* command is used, only the <u>current</u> viewport display is plotted.

Figure 19-2 displays the AutoCAD drawing editor after the *Vports* command was used to divide the screen into tiled viewports. Tiled viewports always fit together like tiles with no space between. The shape and location of the viewports are not flexible as with paper space viewports.

Figure 19-2

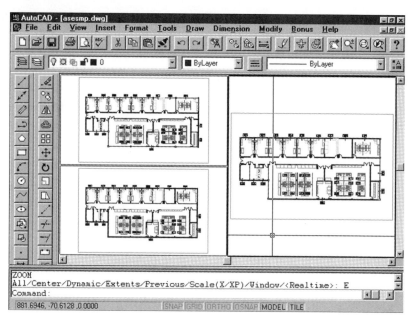

The *View* pull-down menu (Fig. 19-1) can be used to display the *Tiled Viewport Layout* dialog box (Fig. 19-3). This option in the pull-down menu is enabled only if *TILEMODE*=1. The dialog box allows you to PICK the image tile representing the layout that you want.

Figure 19-3

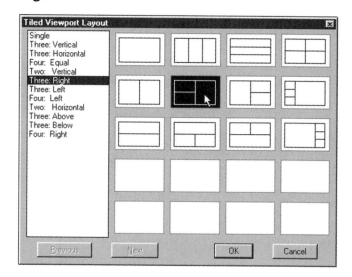

After you select the desired layout, the previous display appears in <u>each</u> of the viewports. For example, if a full view of the office layout was displayed when you use the *Vports* command or the related dialog box, the resulting display in <u>each</u> viewport would be the same full view of the office (Fig. 19-2). It is up to you then to use viewing commands (*Zoom, Pan,* etc.) in the active viewport to specify what areas of the drawing you want to see in each viewport. There is no automatic viewpoint configuration option with *Vports*.

A popular arrangement of view-points for construction and editing of 2D drawings is a combination of an overall view and one or two *Zoomed* views (Fig. 19-4). You cannot draw or project from one viewport to another. Keep in mind that there is only <u>one model</u> (drawing) but several views of it on the screen. Notice that the active or current viewport displays the cursor, while moving the pointing device to another viewport displays only the pointer (small arrow) in that view-port.

A viewport is made active by PICKing in it. Any display com-mands (*Zoom, Vpoint, Redraw*, etc.) and drawing aids (*SNAP, GRID, ORTHO*) used affect only the <u>current viewport</u>. Draw and edit commands that affect the model are potentially apparent in all viewports (for every display of the affected part of the model). *Redrawall* and *Regenall* can be used to redraw and regenerate all viewports.

Figure 19-4

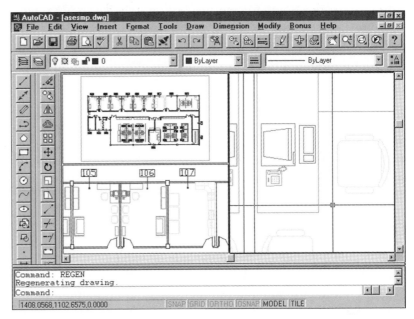

You can begin a drawing command in one viewport and finish in another. In other words, you can toggle viewports within a command. For example, you can use the *Line* command to PICK the "From point:" in one viewport, then make another viewport current to PICK the "To point:".

Using the command line format, the syntax for *Vports* is as follows:

        Command: **vports**
        Save/Restore/Delete/Join/SIngle/?/2/<3>/4: **Enter** or (**option**)
        Horizontal/Vertical/Above/Below/Left/<Right>: **Enter** or (**option**)
        Regenerating drawing.
        Command:

Save
Allows you to assign a name and save the current viewport configuration. A viewport configuration is the particular arrangement of viewports and the display settings (*Zoom, Pan*, etc.) in each viewport. Up to 31 characters can be used when assigning a name. The configuration can be *Restored* at a later time.

Restore
Redisplays a previously *Saved* viewport configuration. AutoCAD prompts for the assigned name.

Delete
Deletes a named viewport configuration. AutoCAD prompts for the assigned name.

Join

This option allows you to combine (join) two adjacent viewports. The viewports to join must share a common edge the full length of each viewport. For example, if four equal viewports were displayed, two adjacent viewports could be joined to produce a total of three viewports. You must select a *dominant* viewport. The *dominant* viewport determines the display to be used for the new viewport.

SIngle

Changes back to a single screen display using the current viewport's display.

?

Displays the identification numbers and screen positions of named (saved) and active viewport configurations. The screen positions are relative to the lower-left corner of the screen (0,0) and the upper-right (1,1).

2, 3, 4

Use these options to create 2, 3, or 4 viewports. You can choose the configuration. The possibilities are illustrated if you use the *Tiled Viewport Layout* dialog box.

The *Vports* command can be used most effectively when constructing and editing drawings, whereas, paper space viewports are generally used when the model is complete and ready to prepare a plot. (See Chapter 35 for more information on using tiled viewports with 3D models.)

PAPER SPACE (FLOATING) VIEWPORTS

The two spaces that AutoCAD provides are model space and paper space. When you start AutoCAD and begin a drawing, model space is active by default. Objects that represent the subject of the drawing (model geometry) are normally drawn in model space. Dimensioning is also performed in model space because it is associative—directly associated to the model geometry. The model geometry is usually completed before using paper space.

Paper space represents the <u>paper that you plot or print on</u>. When you enter paper space for the first time, you see a blank "sheet." In order to see any model geometry, viewports in paper space must be created with *Mview* (like cutting rectangular holes) so you can "see" into model space. Any number or size of rectangular-shaped viewports can be created in paper space. Since there is only <u>one model space</u> in a drawing, you see the same model space geometry in each viewport. You can, however, control which

layers are *Frozen* and *Thawed* and the scale of the geometry displayed <u>in each viewport</u>. Since paper space represents the actual paper used for plotting, plot from paper space at a scale of 1=1.

Consider this brief example to explain the basics of using paper space. In order to keep this example simple, only one viewport is created to set up a drawing for plotting.

First, the part geometry is created in model space as usual (Fig. 19-5). Associative dimensions are also created in model space. This step is the same method that you would have normally used to create a drawing.

Figure 19-5

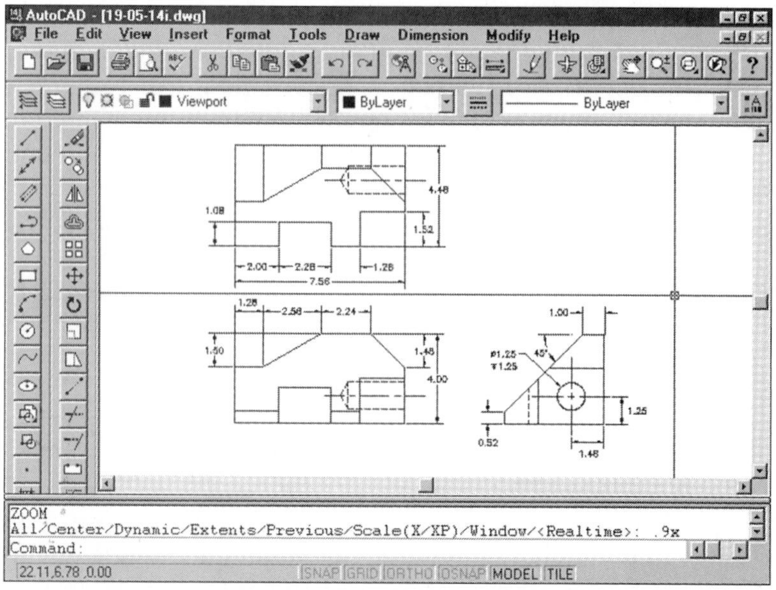

When the part geometry is complete, enable paper space. To do this, change the setting of the *TILEMODE* variable to 0. (See the previous *TILEMODE* command table for several possible methods of toggling the *TILEMODE* variable.)

By setting the *TILEMODE* variable to 0, you are automatically switched to paper space. When you enable paper space for the first time in a drawing, a "blank sheet" appears. The *Limits* command is used to set the paper space *Limits* to the sheet size intended for plotting. Objects such as a title block and border are created in paper space (Fig. 19-6). Normally, only objects that are annotations for the drawing (tile blocks, tables, border, company logo, etc.) are drawn in paper space.

Figure 19-6 ──────────────

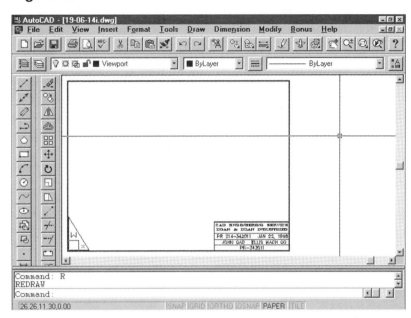

A viewport must be created in the "paper" with the *Mview* command in order for you to "look" into model space. All of the model geometry in the drawing initially appears in the viewport. At this point, there is no specific scale relation between paper space units and the size of the geometry in the viewport (Fig. 19-7).

You can control the scale of model space to paper space and what part of model space geometry you see in a viewport. Two methods are used to control the geometry that is visible in a particular viewport:

Figure 19-7 ──────────────

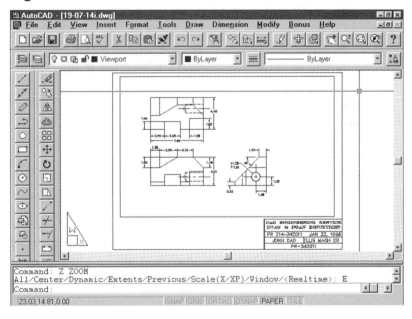

1. **Display commands**
   The *Zoom, Pan, View, Vpoint* and other display commands allow you to specify the part of model space you want to see in a viewport. There is even a *Zoom XP* option used to scale the geometry in model space units "times paper space" units. *Zoom XP* is the method used to define the scale of model geometry for the final plot.

2. **Viewport-specific layer visibility control**
   You can control what layers are visible in specific viewports. This function is often used for displaying different model space geometry in separate viewports. You can control which layers appear in which viewports by using the *Vplayer* command (command line format) or using the icons in the *Layer/Linetype Properties* dialog box or the *Layer Control* drop-down list (*Freeze/Thaw in current viewport* and *Freeze/Thaw in new viewport*).

For example, the *Zoom* command could be used in model space to display a detail of the model geometry in a viewport. In this case, the model geometry is scaled to 3/4 size by using *Zoom 3/4XP*.

While in a drawing session with paper space enabled, the *Mspace* (Model Space) and *Pspace* (Paper Space) commands allow you to switch between paper space (outside a viewport) and floating model space (inside a viewport) so you can draw or edit in either space. Commands that are used will affect the objects or display of the current space. An object <u>cannot be in both spaces</u>. You can, however, draw in paper space and <u>OSNAP to objects in model space</u>. When drawing is completed, activate paper space and plot at 1=1 since paper space *Limits* are set to the <u>actual paper size</u>.

When typing the *Pspace* and *Mspace* commands or selecting *Paper Space* or *Model Space (Floating)* from the *View* menu, the cursor control switches between paper space (outside the viewport) and model space (inside the viewport). The "crosshairs" can move completely across the screen when in paper space, but only within the viewport when using *Mspace* or *Model Space (Floating)*. The figures in this chapter show the cursor at 100% screen size to indicate this feature more clearly (see Figures 19-7 and 19-8). (Cursor size is variable using the *Pointer* tab of the *Preferences* dialog box.)

Figure 19-8

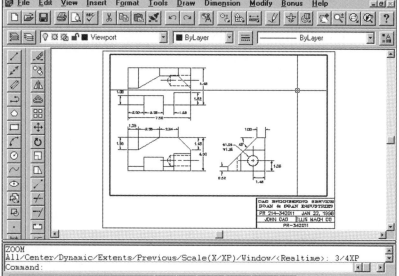

Reasons for Using Paper Space

Paper space is intended to be an aid to plotting. Since paper space represents a sheet of paper, it provides you with a means of preparing specific plotting sheet configurations.

In the previous example, paper space was used with the typical 2D drawing by creating the model geometry as usual but using paper space for the border, title block, tables, or other annotations. If you are creating a typical 2D drawing, you don't have to use paper space and probably shouldn't use paper space. The typical methods for creating drawings and plots all in model space are still valid and should be used except for special cases.

Paper space should be used if you want to prepare plots with <u>multiple views</u> or <u>multiple drawings</u> as listed here:

1. Use paper space to plot several views of the same drawing on one sheet, possibly in different scales, such as a main view and a detail. (Alternately, model space only can be used to accomplish the same result by using the *DIMLFAC* variable for dimensioning, for example, an enlarged detail.)

2. Use paper space to plot several drawings on one sheet. This action is accomplished by using *Xref* to bring several drawings into model space and by using layer visibility control to display individual drawings in separate viewports.

3. Use paper space to plot different views (e.g., top, front, side) of a 3D model on one sheet.

This chapter should supply you with the information to create viewports to display a single drawing. For information on using paper space for the applications listed here, see Chapter 33, Advanced Paper Space Viewports, and Chapter 42, Creating 2D Drawings from 3D Models.

Guidelines for Using Paper Space

Although there are other alternatives, the typical steps using paper space to set up a drawing for plotting are listed here:

1. Create the part geometry in model space. Associative dimensions are also created in model space.

2. Switch to paper space by changing *TILEMODE* to 0 by any method.

3. Set up paper space using the following steps.

   A. Set *Limits* in paper space equal to the sheet size used for plotting. (Paper space has its own *Limits, Snap,* and *Grid* settings.)
   B. *Zoom All* in paper space.
   C. Set the desired *Snap* and *Grid* for paper space.
   D. Make a layer named BORDER or TITLE and *Set* that layer as current.
   E. Draw, *Insert,* or *Xref* a border and a title block.

4. Make viewports in paper space.

   A. Make a layer named VIEWPORT (or other descriptive name) and set it as the *Current* layer (it can be turned *Off* later if you do not want the viewport objects to appear in the plot).
   B. Use the *Mview* command to make viewports. Each viewport contains a view of model space geometry.

5. Control the display of model space graphics in each viewport. Complete these steps for each viewport.

   A. Use the *Mspace* (*MS* or *Floating Model Space*) command to "go into" model space (in the viewport). Move the cursor and PICK to activate the desired viewport.
   B. Use *Zoom XP* to scale the display of the model units to paper space units. This action dictates the plot scale for the model space graphics. The *Zoom XP* factor is the same as the plot scale factor that would otherwise be used (reciprocal of the "drawing scale factor"). For example: *Zoom .5XP* or *Zoom 1/2XP* would scale the model geometry at 1/2 times paper space units.
   C. Control the layer visibility for each viewport by using the *Vplayer* command or the icons in the *Layer/Linetype Properties* dialog box or the *Layer Control* drop-down list (*Freeze/Thaw in current viewport* and *Freeze/Thaw in new viewports*).

6. Plot from paper space at a scale of 1=1.

   A. Use the *Pspace* (*PS*) command to switch to paper space.
   B. If desired, turn *Off* the VIEWPORT layer so the viewport borders do not plot.
   C. Plot the drawing from paper space at a plot scale of 1=1 since paper space is set to the actual paper size. The paper space geometry will plot full size, and the resulting model space geometry will be plotted to the same <u>scale</u> as the *Zoom XP* factor.

PAPER SPACE COMMANDS AND VARIABLES

MVIEW

Pull-down Menu	COMMAND (TYPE)	ALIAS (TYPE)	Short-cut	Screen (side) Menu	Tablet Menu
View *Floating Viewports >*	MVIEW	MV	...	*VIEW 1* *Mview*	M,4

Mview provides several options for creating paper space viewport objects (sometimes called "floating viewports"). *Mview* must be used in paper space (*TILEMODE*=0), so if you have switched to model space, invoking *Mview* (by typing or screen menu) automatically displays the reminder "** Command not allowed unless TILEMODE is set to 0 **." The *Floating Viewports* option in the *View* pull-down menu is enabled only when *TILEMODE*=0. There is no limit to the number of paper space viewports that can be created in a drawing. The command syntax to create one viewport (the default option) is as follows:

```
Command: mview
ON/OFF/Hideplot/Fit/2/3/4/Restore/<First Point>: PICK or (coordinates)
Other corner: PICK or (coordinates)
Regenerating drawing.
Command:
```

This action creates one rectangular viewport between the diagonal corners specified. The other options are described here briefly:

ON
On turns on the display of model space geometry in the selected viewport. Select the desired viewport to turn on.

OFF
Turn off the display of model space geometry in the selected viewport with this option. Select the desired viewport to turn off.

Hideplot
Hideplot causes hidden line removal in the selected viewport during a <u>plot</u>. This is used for 3D surface or solid models. Select the desired viewport.

Fit
Fit creates a new viewport and fits it to the size of the current display. If you are *Zoomed* in paper space, the resulting viewport is the size of the *Zoomed* area. The new viewport becomes the current viewport.

Restore
Use this option to create new viewports in the same relative pattern as a named tiled viewport configuration (see Tiled Viewports).

2/3/4
These options create a number of viewports within a rectangular area that you specify. The 2 option allows you to arrange 2 viewports either *Vertically* or *Horizontally*. The 4 option automatically divides the specified area into 4 equal viewports. Figure 19-9 displays the possible configurations using the *3* option. Using this option yields the following prompt:

Horizontal/Vertical/Above/Below/Left/<Right>:

Figure 19-9 ——————————

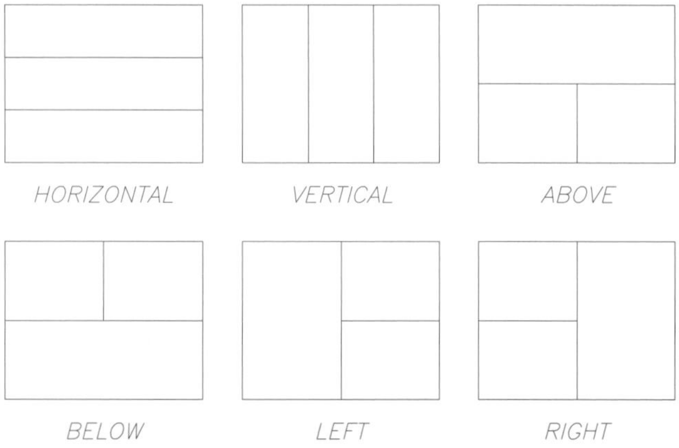

HORIZONTAL   VERTICAL   ABOVE

BELOW   LEFT   RIGHT

After making the desired selection, AutoCAD prompts: "Fit/<First Point>:". The *Fit* option automatically fills the current display with the specified number of viewport configurations (similar to the *Fit* option described earlier).

Paper space viewports created with *Mview* are treated by AutoCAD as <u>objects</u>. Like other objects, *Mview* viewports can be affected by most editing commands. For example, you could use *Mview* to create one viewport, then use *Copy* or *Array* to create other viewports. You could edit the size of viewports with *Stretch* or *Scale*. Additionally, you can use *Move* (Fig. 19-10) to relocate the position of viewports. Delete a viewport using *Erase*. You must be in paper space to PICK the viewport objects (borders). The viewports created by *Mview* or other editing methods are always rectangular and have a vertical and horizontal orientation (they cannot be *Rotated*).

Figure 19-10

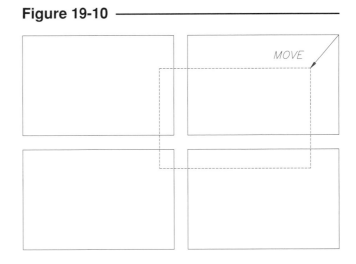

Floating viewports can also overlap (Fig. 19-10). This feature makes it possible for geometry appearing in different viewports to occupy the same area on the screen or on a plot.

NOTE: Avoid creating one viewport completely within another's border because visibility and selection problems may result.

Model Space, Paper Space, and Floating Model Space

Previously, it was stated that there are two spaces: paper space and model space. The *TILE-MODE* variable controls which of these two spaces is active. However, if paper space is enabled (*TILEMODE*=0) and viewports in paper space have been created, there are three possible spaces: model space, paper space, and floating model space. <u>Floating model space is model space inside a paper space viewport.</u> The *Pspace* and *Mspace* commands control whether the cursor is outside or inside a viewport (when *TILEMODE*=0) (Fig. 19-11).

Figure 19-11

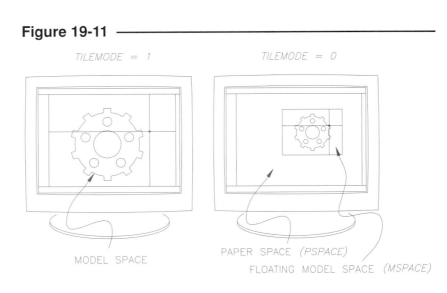

If you prefer typing, the correct combination of *TILEMODE*, *Pspace*, and *Mspace* must be used to work in one of the three spaces (normal model space, paper space, and floating model space). If you use the *View* pull-down menu, selecting one of the three options (*Model Space Tiled*, *Paper Space*, or *Model Space Floating*) automatically toggles the correct *TILEMODE* setting and invokes the *Pspace* or *Mspace* command.

PSPACE

Pull-down Menu	COMMAND (TYPE)	ALIAS (TYPE)	Short-cut	Screen (side) Menu	Tablet Menu
View *Paper Space*	PSPACE	PS	...	VIEW 1 *Pspace*	L,5

The *Pspace* command switches from model space (inside a viewport) to paper space (outside a viewport). The cursor is displayed in paper space and can move <u>across the entire screen</u> when in paper space (see Fig. 19-7), while the cursor appears inside the current viewport only if the *Mspace* command is used (see Fig. 19-8). If you type this command or select it from the screen menu, paper space must be enabled by setting *TILEMODE* to 0. If *TILEMODE* is set to 1, the following message appears:

Command: **pspace**
** Command not allowed unless TILEMODE is set to 0 **
Command:

Selecting *Pspace* from the pull-down menu also automatically sets *TILEMODE* to 0 so paper space is enabled. Therefore, it is possible to use the menu selection to go from *TILEMODE*=1 directly to paper space.

MSPACE

Pull-down Menu	COMMAND (TYPE)	ALIAS (TYPE)	Short-cut	Screen (side) Menu	Tablet Menu
View *Model Space (Floating)*	*MSPACE*	*MS*	...	VIEW 1	L,4

The *Mspace* command switches from paper space to model space <u>inside a viewport</u>. If floating viewports exist in the drawing when you use the command, you are switched to the last active viewport. The cursor appears only <u>in</u> that viewport. The current viewport displays a heavy border. You can switch to another viewport (make another current) by PICKing in it.

If you select *Model Space (Floating)* from the pull-down menu, *TILEMODE* is automatically set to 0. If no viewports exist, the *Mview* command is automatically invoked. If you type *MSpace* or use the screen menu, *TILEMODE* must previoulsy be set to 0, there must be at least one paper space viewport, and it must be *On* (not turned *Off* by the *Mview* command) or AutoCAD issues a message and cancels the command.

ZOOM XP Factors

Paper space objects such as title block and border should correspond to the paper on a 1=1 scale. Paper space is intended to represent the plotting sheet. *Limits* in <u>paper space</u> should be set to the exact paper size, and the finished drawing is plotted <u>from paper space</u> to a scale of 1=1. The model space geometry, however, should be true scale in real-world units, and *Limits* in <u>model space</u> are generally set to accommodate that geometry.

When model space geometry appears in a viewport in paper space, the size of the <u>displayed</u> geometry can be controlled so that it appears and plots in the correct scale. This action is accomplished with the *XP* option of the *Zoom* command. *XP* means "times paper space." Thus, model space geometry is *Zoomed* to some factor "times paper space."

Since paper space is set to the actual size of the paper, the *Zoom XP* factor that should be used for a viewport is <u>equivalent to the plot scale</u> that would otherwise be used for plotting that geometry in model space. The *Zoom XP* factor is the reciprocal of the "drawing scale factor" (Chapter 13). *Zoom XP* <u>only</u> while you are "in" the desired model space viewport. Fractions or decimals are accepted.

For example, if the model space geometry would normally be plotted at 1/2"=1" or 1:2, the *Zoom* factor would be .5*XP* or 1/2*XP*. If a drawing would normally be plotted at 1/4"=1', the *Zoom* factor would be 1/48*XP*. Other examples are given below:

1:5	*Zoom* .2*XP* or 1/5 XP
1:10	*Zoom* .1*XP* or 1/10XP
1:20	*Zoom* .05*XP* or 1/20XP
1/2"=1"	*Zoom* 1/2*XP*
3/8"=1"	*Zoom* 3/8*XP*
1/4"=1"	*Zoom* 1/4*XP*
1/8"=1"	*Zoom* 1/8*XP*
3"=1'	*Zoom* 1/4*XP*
1"=1'	*Zoom* 1/12*XP*
3/4'=1'	*Zoom* 1/16*XP*
1/2'=1'	*Zoom* 1/24*XP*
3/8"=1'	*Zoom* 1/32*XP*
1/4"=1'	*Zoom* 1/48*XP*
1/8"=1'	*Zoom* 1/96*XP*

Refer to the Tables of Limits Settings, Chapter 14, for other plot scale factors.

Using the *Advanced Setup Wizard* and Template Drawings

The *Advanced Setup Wizard* available in the *Create New Drawing* dialog box can be used to create one paper space viewport when you begin a drawing. Alternately, several template drawings are available already set up with one paper space viewport and a title block and border.

Advanced Setup Wizard

When you use the *New* command, the *Create New Drawing* dialog box appears with an option for the *Advanced Setup Wizard* (see Chapters 6 and 13 for introductory information). In the steps for setting up a drawing using this utility, *Step 7: Layout* provides you with the option to use paper space and how to begin working in the drawing.

In *Step 7: Layout,* there are two decisions to make (Fig. 19-12). First, decide if you want to use paper space or not. The question, "Do you want to use advanced paper space capabilities?" means, "Do you want to use paper space?" (No special advanced capabilities are activated here—only the choice to use paper space or not.) Selecting *No* (not the default in this tab) allows you to work in model space (*TILEMODE*=1). Select *Yes* if you want to use paper space. Selecting the *Yes* option, however, creates only one paper space viewport. The viewport is automatically created on layer VIEWPORT. If you have selected a title block (*Step 6*), the title block is inserted into paper space. If you select *Yes*, the three options that follow are enabled.

Figure 19-12

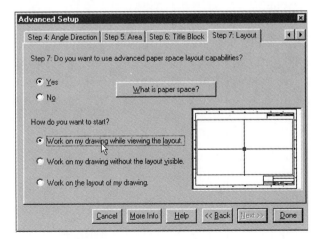

Next, make a decision on how you want to begin working. Below is a translation of the three choices for *How do you want to start?*

Option	Meaning
Work on my drawing while viewing the layout.	Tilemode=0, model space current
Work on my drawing without the layout visible.	Tilemode=1, model space current
Work on the layout of my drawing.	Tilemode=0, paper space current

Keep in mind that using the *Advanced Setup Wizard* automatically changes a number of system variables based on your input to *Step 5: Area*. The system variables may or (most likely) may not be the settings you need for printing or plotting a drawing to a specific scale, so some adjustments may be necessary. See Chapter 13, Advanced Drawing Setup, for more information on the *Setup Wizards*.

Template Drawings Table

Most of the AutoCAD-supplied template drawings contain one paper space viewport and a title block inserted in paper space. The size of the title block and *Limits* values are preset in the template drawings, not created on the fly as with the *Setup Wizards*.

To use a template drawing, select the *Use a Template* option in the *Create New Drawing* dialog box. Select the desired template (DWT) drawing from the list including all templates in the TEMPLATE folder (usually in the C:\Program Files\AutoCAD R14\Template subdirectory). The list of template drawings supplied with AutoCAD Release 14 are described in the following table. Note that all but two of the templates are set up to use paper space.

DWT File	Sheet Size Intended	Limits (mod. Sp.)	Units, Precn.	Title Block	Paper Space	Layers	Line-types	Dimen. Styles
Acad	—	12, 9	.0000	no	no	Layer 0	no [1]	Standard
Acadiso	—	420, 297	.0000	no	no	Layer 0	no [1]	yes [9]
Ansi_a	11"x 8.5"	11, 8.5	.0000	yes	yes [1]	yes [2]	no [1]	Standard
Ansi_b	17"x 11"	17, 11	.0000	yes	yes [1]	yes [2]	no [1]	Standard
Ansi_c	22"x 17"	22, 17	.0000	yes	yes [1]	yes [2]	no [1]	Standard
Ansi_d	34"x 22"	34, 22	.0000	yes	yes [1]	yes [2]	no [1]	Standard
Ansi_e	44"x 34"	44, 34	.0000	yes	yes [1]	yes [3]	no [1]	Standard
Ansi_v	11"x 8.5" vert	8.5, 11	.0000	yes	yes [1]	yes [2]	no [1]	Standard
Archeng	36" x 24"	36, 24	.0000	yes	yes [1]	yes [3]	no [1]	Standard
Din_a0	1189 x 841 mm	1189, 841	.0000	yes	yes [1]	yes [4]	no [1]	yes [8]
Din_a1	841 x 594 mm	841, 594	.0000	yes	yes [1]	yes [4]	no [1]	yes [8]
Din_a2	594 x 420 mm	594, 420	.0000	yes	yes [1]	yes [4]	no [1]	yes [8]
Din_a3	420 x 297 mm	420, 297	.0000	yes	yes [1]	yes [5]	no [1]	yes [8]
Din_a4	210 x 297 vert.	210, 297	.0000	yes	yes [1]	yes [4]	no [1]	yes [8]
Gs24x36	36" x 24"	34.5, 23	.0000	yes	yes [1]	yes [2]	no [1]	Standard
Iso_a0	1188 x 840 mm	1189, 841	.0000	yes	yes [1]	yes [6]	no [1]	yes [9]
Iso_a1	840 x 594 mm	841, 594	.0000	yes	yes [1]	yes [6]	no [1]	yes [9]
Iso_a2	594 x 420 mm	594, 420	.0000	yes	yes [1]	yes [6]	no [1]	yes [9]
Iso_a3	420 x 297 mm	420, 297	.0000	yes	yes [1]	yes [6]	no [1]	yes [9]
Iso_a4	210 x 297 vert.	210, 297	.0000	yes	yes [1]	yes [6]	no [1]	yes [9]
Jis_a0	1189 x 841 mm	1189, 841	.0000	yes	yes [1]	yes [7]	no [1]	yes [10]

(continued)

(*continued*)

DWT File	Sheet Size Intended	Limits (mod. Sp.)	Units, Precn.	Title Block	Paper Space	Layers	Line-types	Dimen. Styles
Jis_a1	841 x 594 mm	841, 594	.0000	yes	yes [1]	yes [7]	no [1]	yes [10]
Jis_a2	594 x 420 mm	594, 420	.0000	yes	yes [1]	yes [7]	no [1]	yes [10]
Jis_a3	420 x 297 mm	420, 297	.0000	yes	yes [1]	yes [7]	no [1]	yes [10]
Jis_a4l	297 x 210 mm	297, 210	.0000	yes	yes [1]	yes [7]	no [1]	yes [10]
Jis_a4r	210 x 297 vert.	210, 297	.0000	yes	yes [1]	yes [7]	no [1]	yes [10]

[1] Paper Space is enabled (TILEMODE=0) and title block is in Paper Space.
[2] Layers: 0, TB, and Title_block
[3] Layers: 0 and Title_block
[4] Layers: 0, Rahmen025, Rahmen035, Rahmen05, Rahmen07
[5] Layers: 0, Rahmen025, Rahmen05, Rahmen07
[6] Layers: 0, Frame025, Frame035, Frame050, Frame070, Frame100, Frame140, Frame200, Tb
[7] Layers: 0, Defpoints, Object
[8] Dimension Styles: DIN
[9] Dimension Styles: ISO-25, ISO-35, ISO-4, ISO-5, ISO-7, ISO1, ISO1-4, ISO2, Standard
[10] Dimension Styles: JIS

For more information on standard plotting scales and intended sheet sizes for the *Limits* settings in these templates, see the Tables of Limits Settings in Chapter 14.

R14

Advanced Applications of Paper Space Viewports

In this chapter, the basic capabilities of paper space are discussed. These basic capabilities include creating one or more viewports for a single drawing and scaling the geometry in the viewport(s) to print or plot to scale. However, setting up a drawing with only one viewport offers very little advantage over using model space exclusively and is generally more difficult.

Paper Space can be used in ways that are not possible with model space exclusively; that is, setting up multiple views of one drawing at different scales, displaying multiple drawings (*Xrefs*) in one drawing, and displaying several views of a 3D model. These are the best applications for using model space. These topics are discussed in Chapter 33, Advanced Paper Space Viewports, and Chapter 42, Creating 2D Drawings from 3D Models.

CHAPTER EXERCISES

1. *Vports* **(Tiled Viewports)**

   Open the **ASESMP** drawing. This drawing is available if you used the *Full Installation* option when AutoCAD was installed on your computer or network. The drawing is usually located in the C:\Program Files\AutoCAD R14\Sample directory (assuming the default installation location).

A. Invoke ***Tiled Viewports*** by the pull-down menu; then select ***Layout....*** When the *Tiled Viewport Layout* dialog box appears, choose the ***Three: Right*** option (or image tile just left of center). The resulting viewport configuration should appear as Figure 19-2.

B. Using ***Zoom*** and ***Pan*** in the individual viewports, produce a display with an overall view on the top and two detailed views as shown in Figure 19-4.

C. Type the ***Vports*** command. Use the ***Save*** option to save the viewport configuration as **3R**. Click in the bottom left viewport to make it the current viewport. Use ***Vports*** again and change the display to a ***Single*** screen. Now use ***SaveAs*** to save the drawing as **ASESMP-2** and <u>reset the path to your working directory</u>.

D. ***Open* ASESMP-2**. Use the *?* option of ***Vports***; then ***Restore* 3R** to ensure the viewport configuration was saved as expected. Invoke a ***Single*** viewport again. Finally, use ***Vports*** or the ***Tiled Viewport Layout*** dialog box to create another viewport configuration of your choosing. ***Save*** the viewport configuration and assign an appropriate name. ***Save*** the **ASESMP-2** drawing.

2. ***TILEMODE, Mview***

In this exercise, you will use an existing drawing, enable paper space, draw a border and title block, create one viewport, and print or plot the drawing to scale.

A. ***Open*** the **CBRACKET** drawing you worked on in Chapter 10 Exercises. If you have already created a title block and border, ***Erase*** them.

B. Change ***TILEMODE*** to **0** by any method (type, double-click the word ***TILE*** on the Status Line, or select ***Paper Space*** from the ***View*** pull-down menu). You should be presented with a "blank sheet" and the paper space icon (drafting triangle) should appear.

C. Set ***Limits*** (in paper space) to an A size sheet, **11 x 8.5**. Set ***Snap*** to **.5**. Set layer **TITLE** *Current*. Draw a title block and border (in paper space). HINT: Use *Pline* with a *width* of .02, and draw the border with a .5 unit margin within the edge of the *Limits* (border size of 10 x 7.5). You can create a title block similar to the TBLOCK you created in Chapter 18 (Fig. 18-44) with a space for your school or company name, your name, part name, date, and scale. Enter a scale of "**1:1 METRIC**" in the title block. Save the drawing as **CBRACKET-PS**.

Figure 19-13 ——————————————

CADD Design Company	
Catch Bracket	1:1 METRIC
Des.– B.R. Smith	Ch– JAL
1/1/98	1/1/98	CBRACKET

D. Create layer **VPORTS** and make it ***Current***. Use ***Mview*** to create one viewport. Pick diagonal corners for the viewport at **1,2** and **10,7.5**. The saddle drawing should appear in the viewport at no particular scale, similar to Figure 19-13.

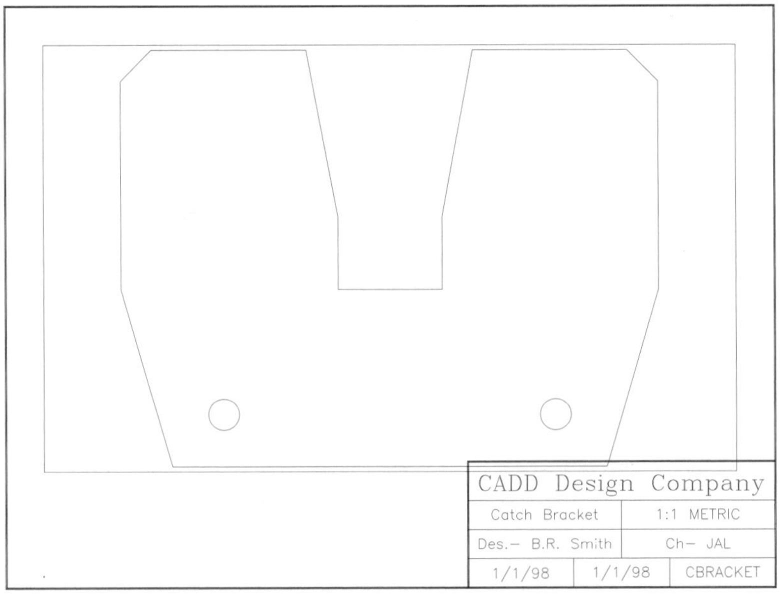

E. Next, use the *Mspace* command to bring the cursor into model space. Use *Zoom* with a *.03937XP* factor to scale the model space geometry to paper space. (The conversion from inches to milimeters is 25.4 and from milimeters to inches is .03937. Model space units are milimeters and paper space units are inches, therefore enter a *Zoom XP* factor of .03937). You may have to *Pan* (in model space) to center the drawing within the viewport.

F. Finally, use the *Pspace* command to activate paper space. Make layer 0 current and *Freeze* layer **VPORTS**. Your completed drawing should look like that in Figure 19-14. *Save* the drawing.

Figure 19-14 ───────────────────────

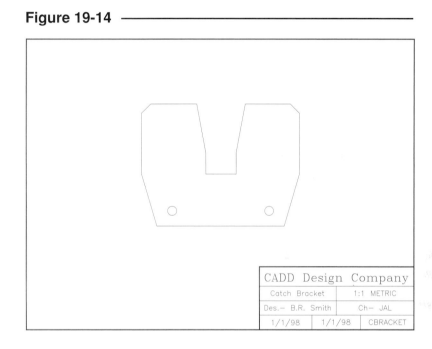

CADD Design Company	
Catch Bracket	1:1 METRIC
Des.– B.R. Smith	Ch– JAL
1/1/98 1/1/98	CBRACKET

3. *TILEMODE, Mview*

This exercise gives you more experience using an existing drawing, creating one paper space view-port, and printing or plotting to scale.

A. *Open* the **EFF-APT2** drawing you worked with last in Chapter 18 Exercises. If you have already created a title block and border, *Erase* them.

B. Change *TILEMODE* to **0** by any method. You should be presented with a "blank sheet" and the paper space icon (drafting triangle) should appear.

C. Set *Limits* in paper space to a C size sheet, **24 x 18**, and *Zoom All*. Set *Snap* to **.25**. Make layer **TITLE** the *Current* layer. Draw a title block and border (in paper space). HINT: Use *Pline* with a *width* of .03, and draw the border with a .75 unit margin within the edge of the *Limits* (border size of 22.5 x 16.5). You can create a title block similar to the TBLOCK you created in Chapter 18 (Fig. 18-44) with a space for your school or company name, your name, part name, date, and scale. Enter a scale of **1/2″=1′** in the title block. *Save* the drawing as **EFF-APT-PS**.

D. Create a new layer named **VIEWPORT** and make it *Current*. Use *Mview* to create one viewport. Pick diagonal corners for the viewport at **1,1** and **23,17**. The apartment drawing should appear in the viewport at no particular scale.

E. Use the *Mspace* command to bring the cursor into model space. Use *Zoom* with a *1/24XP* factor. (The proportion of 1/2″=1′ is the same as 1/2″=12″, or 1=1/24). You may have to *Pan* (in model space) to center the drawing within the viewport.

F. Next, use the *Pspace* command to place your cursor in paper space. *Freeze* layer **VIEWPORT**. Your completed drawing should like that in Figure 19-15. *Save* the drawing.

Figure 19-15

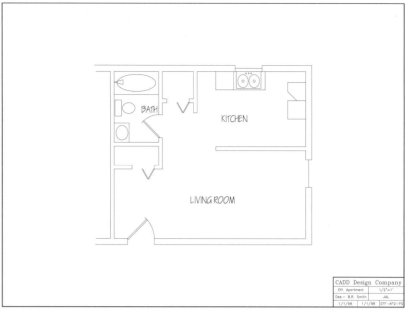

4. **Using a Template Drawing**

This exercise gives you experience using a template drawing with paper space already set up with a title block and border. Use the template drawing to draw the Hammer (Fig. 19-16).

Figure 19-16

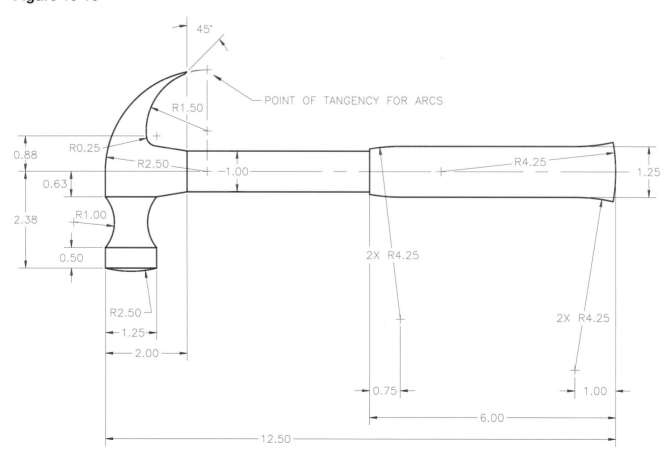

A. Begin a *New* drawing. In the *Create New Drawing* dialog box, select *Use a Template*. Select the *Ansi_b.dwt* template. This template loads with a tile block and border in paper space and has model space active. The drawing has *Limits* in paper space and in model space set up at 17 x 11—correct for printing 1"=1" on a B size engineering sheet.

B. Create the following layers and assign linetypes. Assign colors of your choice.

    GEOMETRY   *Continuous*
    CENTER   *Center*

    Notice the *Title_block* layer already exists.

C. Change **TILEMODE** to **1**. Changing *TILEMODE* allows you to create geometry in model space without having to see the tile block during construction. Draw the Hammer according to the dimensions given. *Save* the drawing as **HAMMER-PS**.

D. When you are finished with the drawing, change **TILEMODE** to **0**. Use *Mspace* to ensure model space is active. Use *Zoom Extents* to center the geometry, then *Zoom 1XP* to correctly size model space units to paper space units.

F. Use the *Pspace* command to activate paper space. Your completed drawing should look like that in Figure 19-17. In paper space, complete the title block with appropriate information, including the drawing scale (1"=1") and the part name. *Save* the drawing. Make a plot or print on a B size sheet at a scale of 1=1.

Figure 19-17

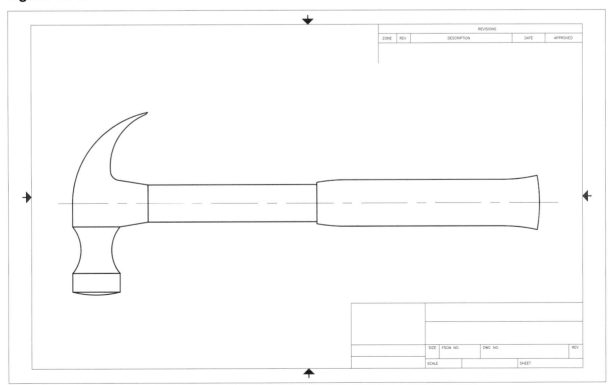

5. *Advanced Setup Wizard*

This exercise gives you practice using the *Advanced Setup Wizard* to set up a drawing to work in paper space and model space. You will construct the Wedge Block in Figure 19-18.

A. Begin a *New* drawing. In the *Create New Drawing* dialog box, select **Use a Wizard**, then select **Advanced Setup**. In the *Advanced Setup* dialog box that appears, take the following steps:

Figure 19-18 —————————————————

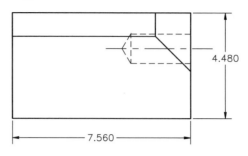

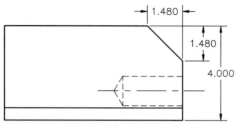

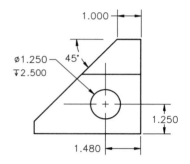

Step 1: Units	use **Decimal** with **.000** *Precision*
Step 2: Angle	accept default setting
Step 3: Angle Measure	accept default setting
Step 4: Angle Direction	accept default setting
Step 5: Area	enter **18** for **Width** and **12** for **Length**
Step 6: Title Block	select the *Ansi_b* title block
Step 7: Layout	answer **Yes** to **"Do you want to use paper space?"** and select **Work on my drawing while viewing the layout.**

B. When the drawing appears, the title block is visible in paper space and model space is active (Fig. 19-19). The *Advanced Setup Wizard* sets many system variables based on your input to *Area*, but often these settings are not appropriate. Change the *Snap* to **.125** and *Grid* to **.5**. Change *LTSCALE* to **.5**. *Save* the drawing as **WEDGEBK**.

Figure 19-19 —————————————————

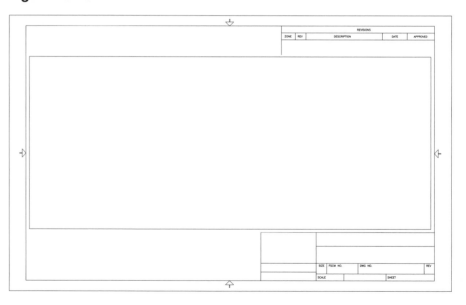

C. Use the *Mspace* command or select *Model Space (Floating)* from the *View* pull-down menu to activate model space and begin drawing. Alternately, you can change *TILEMODE* to **1** while you draw. Create layers for hidden lines, centerlines, and object lines. Create the drawing geometry as shown in Figure 19-18.

D. When you are finished with the views, ensure **TILEMODE** is set to **0** (title block is visible) and make model space active (**Mspace**). To scale the model space geometry to paper space, first **Zoom Extents**, then use **Zoom** with a **.5XP** factor. (The drawing is scaled to .5=1, or 1/2"=1" when plotted from paper space at 1=1.) You may also need to use **Pan** to move the geometry about within the viewport so it does not overlap with the title block.

E. Next, use **Pspace**. **Freeze** layer **Viewport**. Your completed drawing should look like that in Figure 19-20. Fill in the title block with appropriate information. **Save** the drawing. From paper space, print or plot the drawing on a B size sheet and enter a scale of **1=1**.

Figure 19-20

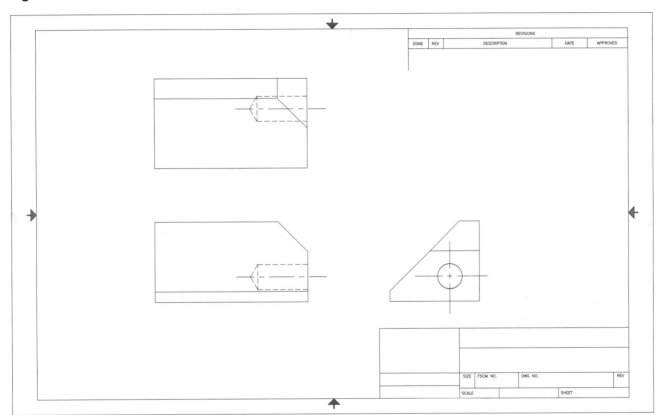

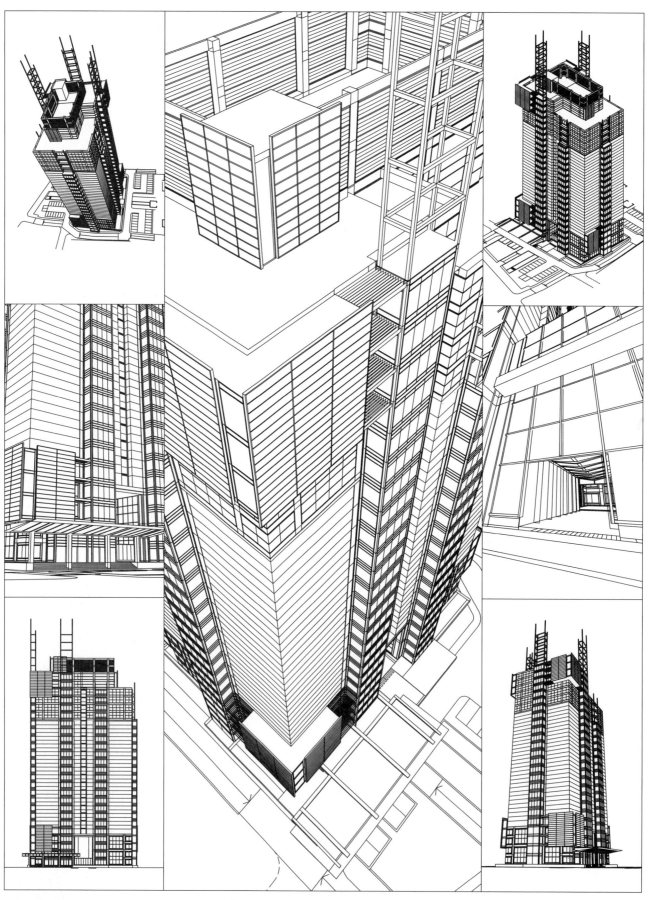

TOWERS.DWG Courtesy Autodesk, Inc. (Release 14 Sample Drawing)

20

ADVANCED SELECTION SETS

Chapter Objectives

After completing this chapter you should:

1. be able to enable or disable Noun/Verb order of editing using the *Object Selection Settings* dialog box or the *PICKFIRST* system variable;

2. be able to control whether objects are added to or replace the selection set by using the *Object Selection Settings* dialog box or *PICKADD* variable;

3. know that the pickbox *Auto* window/crossing window feature can be controlled using the *Object Selection Settings* dialog box or *PICKAUTO* variable;

4. be able to set the preferred window dragging style with the *Object Selection Settings* dialog box or *PICKDRAG* variable;

5. be able to use *Object Selection Filters* in a complex drawing to find a selection set for future use with an editing command;

6. be able to create named selection sets (Object Groups) with the *Group* command and use options in the *Object Grouping* dialog box for appropriate applications.

CONCEPTS

When you "Select Objects:", you determine which objects in the drawing are affected by the subsequent editing action by specifying a <u>selection set</u>. You can select the set of objects in several ways. The fundamental methods of object selection (such as PICKing with the pickbox, a *Window*, a *Crossing Window*, etc.) are explained in Chapter 4, Selection Sets. This chapter deals with advanced methods of specifying selection sets and the variables that control your preferences for how selection methods operate. Specifically, this chapter explains:

> Selection Set Variables
> Object Filters
> Object Groups

SELECTION SET VARIABLES

Four variables allow you to customize the way you select objects. Keep in mind that selecting objects occurs <u>only for editing</u> commands; therefore, the variables discussed in this chapter affect how you select objects when you <u>edit</u> AutoCAD objects. The variable names and the related action are briefly explained here:

Variable Name	Default Setting	Related Action
PICKFIRST	1	Enables and disables Noun/Verb command syntax. Noun/Verb means *PICK* objects (noun) *FIRST* and then use the edit command (verb).
PICKADD	1	Controls whether objects are *ADD*ed to the selection set when *PICK*ed or replace the selection set when picked. Also controls whether the SHIFT key + #1 button combination removes or adds selected objects to the selection set.
PICKAUTO	1	Enables or disables the *PICK*box *AUTO*matic window/crossing window feature for object selection.
PICKDRAG	0	Enables or disables single *PICK* window *DRAG*ging. When *PICKDRAG* is set to 1, you start the window by pressing the *PICK* button, then draw the window by holding the button down and *DRAG*ging to specify the diagonal corner, and close the window by releasing the button. In other words, windowing is done with one PICK and one release rather than with two PICKs.

Like many system variables, the selection set variables listed above hold an integer value of either 1 or 0. The 1 designates a setting of *ON* and 0 designates a setting of *OFF*. System variables that hold an integer value of either 1 or 0 "toggle" a feature *ON* or *OFF*.

Changing the Settings

Settings for the selection set variables can be changed in either of two ways:

1. You can type the variable name at the Command: prompt (just like an AutoCAD command) to change the setting.

2. The *Object Selection Settings* dialog box (Fig. 20-1) can be PICKed from the *Tools* pull-down menu or called by typing *Ddselect*. The four variables listed above can be changed in this dialog box, but the syntax in the dialog box is <u>not</u> the same as the variable name. A check in the checkbox by each choice does <u>not</u> necessarily mean a setting of *ON* for the related variable.

When you change any of these four variables, the setting is recorded in the system registry, rather than in the current drawing file as with most variable settings. In this way, the change (which is generally the personal preference of the operator) is established at the workstation, not the drawing file.

DDSELECT

	Pull-down Menu	COMMAND (TYPE)	ALIAS (TYPE)	Short-cut	Screen (side) Menu	Tablet Menu
	Tools *Selection...*	*DDSELECT*	*SE*	...	TOOLS 2 *Ddselect*	*X,9*

The *Object Selection Settings* dialog box (Fig. 20-1) displays the default settings for AutoCAD Release 14. Notice the syntax used in the dialog box does not reflect the variables' names. To avoid confusion, it is suggested that you initially use either the variable names to change the settings in command line format or the dialog box to change the settings, but not both.

Figure 20-1 ———

PICKFIRST

NOTE: If the *GRIPS* variable is set to 1, object grips are enabled. Object grips do not hinder your ability to use *PICKFIRST*, but they may distract your attention from the current topic. Setting the *GRIPS* variable to 0 disables object grips. You can type *GRIPS* at the command prompt to change the setting. See Chapter 23 for a full discussion of object grips.

Changing the setting of *PICKFIRST* is easily accomplished in command line mode (typing). However, the *Object Selection Settings* dialog box can also be used to toggle the setting (Fig. 20-1). The choice is titled *Noun/Verb Selection*, and a check appearing in the box means that *PICKFIRST* is set to 1 (*ON*).

The *PICKFIRST* system variable enables or disables the ability to select objects <u>before</u> using a command. If *PICKFIRST* is set to 1, or *ON*, you can select objects at the Command: prompt <u>before</u> a command is used. *PICKFIRST* means that you PICK the objects FIRST and then invoke the desired command. *PICK-FIRST* set to 1 makes the small pickbox appear at the cursor when you are not using a command (at the open Command: prompt). *PICKFIRST* set to 1 enables you to select objects with the pickbox, AUto window, or Crossing Window methods when a command is not in use.

This order of editing is called "Noun/Verb"; the <u>objects are the nouns</u> and the <u>command is the verb</u>. Noun/Verb editing is preferred by some users because you can decide what objects need to be changed, then decide how (what command) you want to change them. Noun/Verb editing allows AutoCAD to operate like some other CAD systems; that is, the objects are PICKed FIRST and then the command is chosen.

The command syntax for Noun/Verb editing is given next using the *Move* command as an example:

> Command: **PICK** (Use the cursor pickbox or auto window/crossing window to select objects.)
> Command: **PICK** (Continue selecting desired objects.)
> Command: **Move** (Enter the desired command and AutoCAD responds with the number of objects selected.) 2 found
> Base point or displacement: **PICK** or **(coordinates)**
> Second point of displacement: **PICK** or **(coordinates)**
> Command:

Notice that as soon as the edit command is invoked, AutoCAD reports the number of objects found and uses these as the selection set to act on. You do <u>not</u> get a chance to "Select objects:" within the command. The selection set PICKed <u>immediately</u> before the command is used for the editing action. The command then passes through the "Select objects:" step to the next prompt in the sequence. All editing commands operate the same as with Verb/Noun syntax order with the exception that the "Select objects:" step is bypassed.

<u>Only</u> the pickbox, auto window, and crossing window can be used for object selection with Noun/Verb editing. The other object selection methods (*ALL, Last, Previous, Fence, Window Polygon,* and *Crossing Polygon*) are only available when you are presented with the "Select objects:" prompt.

NOTE: To disable the cursor pickbox, *PICKFIRST* <u>and</u> *GRIPS* must be OFF.

PICKADD

The *PICKADD* variable controls whether objects are ADDed to the selection set when they are PICKed or whether selected objects replace the last selection set. This variable is *ON* (set to 1) by default. Most AutoCAD operators work in this mode.

Until you reached this section, it is probable that all PICKing you did was with *PICKADD* set to 1. In other words, every time you selected an object it was added to the selection set. In this way, the selection set is <u>cumulative</u>; that is, each object PICKed is added to the current set. This mode also allows you to use multiple selection methods to build the set. You can PICK with the pickbox, then with a window, then with any other method to continue selecting objects. The "Select objects:" process can only be ended by pressing Enter.

With the default option (when *PICKADD* is set to 1 or *ON*), the Shift+#1 key combination allows you to <u>deselect</u>, or remove, objects from the current selection set. This has the same result as using the *Remove* option. Deselecting is helpful if you accidentally select objects or if it is easier in some situations to select *ALL* and then deselect (Shift+#1) a few objects.

When the *PICKADD* variable is set to 0 (*OFF*), objects that you select <u>replace</u> the last selection set. Let's say you select five objects and they become highlighted. If you then select two other objects with a window, they would become highlighted and the other five objects automatically become deselected and unhighlighted. The two new objects would replace the last five to define the new selection set. Figure 20-2 illustrates a similar scenario.

Figure 20-2

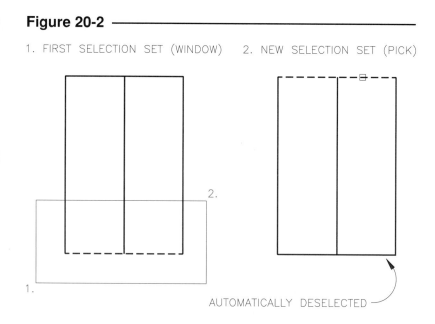

1. FIRST SELECTION SET (WINDOW) 2. NEW SELECTION SET (PICK)

AUTOMATICALLY DESELECTED

The *PICKADD* variable also controls whether the Shift+#1 button combination removes from or adds to the selection set. When *PICKADD* is set to 1 (*ON*), the Shift+#1 combination deselects (removes) objects from the selection set. When *PICKADD* is set to 0 (*OFF*), the Shift+#1 combination toggles objects in or out of the selection set, depending on the object's current state. In other words (when *PICKADD* is *OFF*), if the object is included in the set (highlighted), Shift+#1 deselects (unhighlights) it, or if the object is not in the selection set, Shift+#1 adds it.

PICKADD can easily be turned *On* (1) or *Off* (0) by toggling the Ctrl+K key sequence. You can also change the variable by typing in *PICKADD* at the Command: prompt. Changing the *PICKADD* variable in the *Object Selection Settings* dialog box is accomplished by making the desired choice in the checkbox. Changing the setting in this way is confusing because a check by *Use Shift to Add* means that *PICKADD* is *OFF*! Normally, a check means the related variable is *ON*. To avoid confusion, use only one method (typing, Ctrl+K, or dialog box) to change the setting until you are familiar with it.

It may occur to you that you cannot imagine a practical application for using *PICKADD OFF* and that it makes perfect sense to operate AutoCAD with *PICKADD ON*. This is true for most applications; however, if you use *GRIPS* often, setting *PICKADD* to *OFF* simplifies the process of changing **warm** *Grips* to **cold**. *Grips* are discussed in Chapter 23.

PICKAUTO

The *PICKAUTO* variable controls automatic windowing when the "Select objects:" prompt appears. Automatic windowing (*Implied Windowing*) is the feature that starts the first corner of a window or crossing window when you PICK in an open area. Once the first corner is established, if the cursor is moved to the right, a window is created, and if the cursor is moved to the left, a crossing window is started. See Chapter 4 for details of this feature.

When *PICKAUTO* is set to 1 (or *Implied Windowing* is checked in the dialog box, Fig. 20-1), the automatic window/crossing is available whenever you select objects. When *PICKAUTO* is set to 0 (or no check appears by *Implied Windowing*), the automatic windowing feature is disabled. However, the *PICKAUTO* variable is overridden if *GRIPS* are *ON* or if the *PICKFIRST* variable is *ON*. In either of these cases, the cursor pickbox and auto windowing are enabled so that objects can be selected at the open Command: prompt.

Auto windowing is a helpful feature and can be used to increase your drawing efficiency. The default setting of *PICKAUTO* is the typical setting for AutoCAD users.

PICKDRAG

Figure 20-3

PICKDRAG=1 (ON)　　　*PICKDRAG= 0 (OFF)*

This variable controls the method of drawing a selection window. *PICKDRAG* set to 1 or *ON* allows you to draw the window or crossing window by clicking at the first corner, holding down the button, dragging to the other corner, and releasing the button at the other corner (Fig. 20-3). With *PICK-DRAG ON* you can specify diagonal corners of the window with one press and one release rather than with two clicks. Many GUIs (graphical user interfaces) of other software use this method of mouse control.

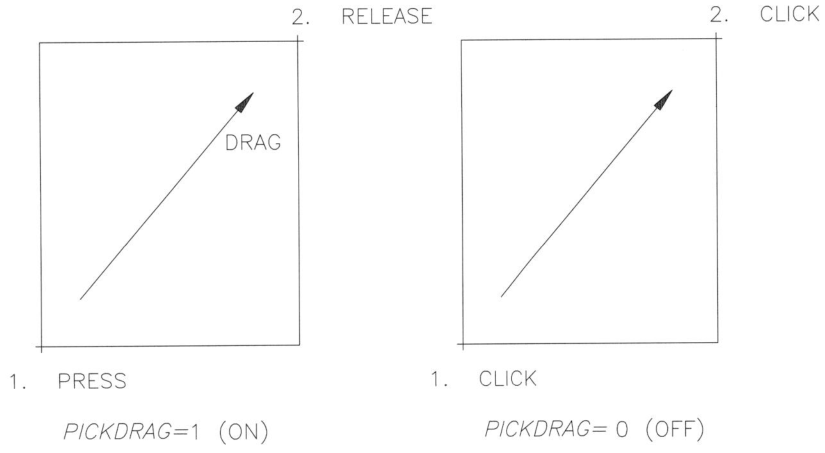

The default setting of *PICKDRAG* is 0 or *OFF*. This method allows you to create a window by clicking at one corner and again at the other corner. Releases of AutoCAD previous to 12 use this method of window creation exclusively. Most AutoCAD users are accustomed to this style, which accounts for the default setting of *PICKDRAG* to 0 (*OFF*).

OBJECT SELECTION FILTERS

Large and complex AutoCAD drawings require advanced methods of specifying a selection set. For example, consider working with a drawing of a manufacturing plant layout and having to select all of the doors on all floors, or having to select all metal washers of less than 1/2" diameter in a complex mechanical assembly drawing. Rather than spend several minutes PICKing objects, it may be more efficient to use the object filters dialog box that AutoCAD provides.

The object selection filters feature allows you to specify particular criteria, and AutoCAD will search (filter) the drawing to find objects that match that criteria. You can specify the criteria based on any object type or property. Once AutoCAD finds the objects that match the criteria, that selection set may be used in any editing command requiring the selection of objects. For example, at the "Select objects:" prompt, you can invoke the *Object Selection Filters* dialog box and create a criteria or select a previously created named filter (criteria) from the list. AutoCAD uses the specified criteria to find the desired object set to use with the command. Multiple selection sets can be named and saved.

FILTER

Pull-down Menu	COMMAND (TYPE)	ALIAS (TYPE)	Short-cut	Screen (side) Menu	Tablet Menu
...	FILTER	FI	...	ASSIST Filters	...

Using the *Filter* command by any method produces the *Object Selection Filters* dialog box. Figure 20-4 displays the dialog box when it first appears. The *Select Filter* cluster near the upper right is where you specify the search criteria. After you designate the filters and values, use *Add to List* to cause your choices to appear in the large area at the top of the dialog box. All filters appearing at the top of the box are applied to the drawing when *Apply* is selected.

Figure 20-4

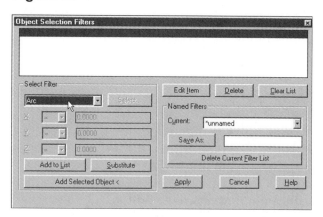

Select Filter (Drop-down List)

Activating the drop-down list reveals the possible selection criteria (filters) that you can use, such as *Arc* or *Layer*. You can select one or more of the selection filters shown in the following list. You can also specify a set of values that applies to each filter, for example, *Arcs* with a *radius* of less than 1.00" or *Layers* that begin with "AR."

Arc	Attribute Tag	Body
Arc Center	Block Rotation	Circle Center
Arc Radius	Block Position	Circle
Attribute	Block	Circle Radius
Attribute Position	Block Name	Color

Dimension	Linetype Scale	Text
Dimension Style	Multiline Style	Text Height
Elevation	Multiline	Text Position
Ellipse	Normal Vector	Text Value
Ellipse Center	Point Position	Text Rotation
Hatch	Point	Text Style Name
Hatch Pattern Name	Polyline	Thickness
Image	Ray	Tolerance
Image Position	Region	Trace
Layer	Shape Position	3d face
Leader	Shape Name	Viewport
Line	Shape	Viewport Center
Line Start	Solid	Xdata
Line End	Solid Body	Xline
Linetype	Spline	

The list also provides a series of typical database grouping operators which include AND, OR, XOR, and NOT. For example, you may want to select all *Text* in a drawing but *NOT* the *Text* in a specific text style.

Select

If you select a filter that has multiple values, such as *Layer, Block Name,* and *Text Style Name,* the *Select* button is activated and enables you to select a value from the list (Fig. 20-5). If the *Select* button is grayed out, no existing values are available.

X

If you select a filter that requires alpha-numeric input (numbers or text strings), the X field is used. For example, if you selected *Arc Radius* as a filter, you would select the operator (=, <, >, etc.) from the first drop-down list and key in a numeric value in the edit box (Fig. 20-6). Other examples requiring numeric input are *Circle Radius, Text Rotation,* and *Text Height*.

If you select a filter that requires a text string, the string is entered in the X field edit box. Examples are *Attribute Tag* and *Text Value*.

The X field is also used (in conjunction with the Y and Z fields) for filters that require coordinate input in an *X,Y,Z* format. Example filters include *Arc Center, Block Position, Line Start*.

Y,Z

These fields are used for coordinate entry of the Y and Z coordinates when needed as one of the criteria for the filter (see X above). Additionally, you may use standard database relational operators such as equal, less than, or greater than, to more effectively specify the desired value.

Figure 20-5

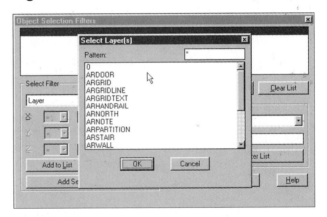

Figure 20-6

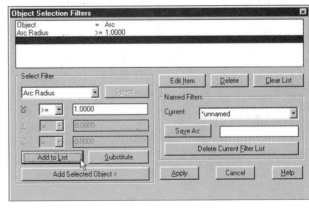

Add to List

Use the *Add to List* button after selecting a filter to force the new selection to appear in the list at the top of the dialog box.

Substitute

After choosing a value from the *Select* listing or entering values for X, Y, and Z, *Substitute* replaces the value of the <u>highlighted</u> item in the list (at the top of the dialog box) with the new choice or value.

Add Selected Object

Use this button to select an object from the drawing that you want to include in the filter list.

Edit Item

Once an item is added to the list, you must use *Edit Item* to change it. First, highlight the item in the filter list (on top); then select *Edit Item*. The filter and values then appear in the *Select Filter* cluster ready for you to specify new values in the X,Y,Z fields or you can select a new filter by using the *Select* button.

Delete

This button simply deletes the highlighted item in the filter list.

Clear List

Use *Clear List* to delete all filters in the existing filter list.

Named Filters

The *Named Filters* cluster contains features that enable you to save the current filter configuration to a file for future use with the current drawing. This eliminates the need to rebuild a selection filter used previously.

Current

By default, the current filter list of criteria is **unnamed*, in much the same fashion as the default dimension style. Once a filter list has been saved by name, it appears as a selectable object filter configuration in the drop-down list of named filters.

Save As

Once object filter(s) have been specified and added to the list (on top), you can save the list by name. First, enter a name for the list; then PICK the *Save As* button. The list is automatically set as the current object filter configuration and can be recalled in the future for use in the drawing.

Delete Current Filter List

Deletes the named filter configuration displayed in the *Current* field.

Apply

The *Apply* button applies the current filter list to the current AutoCAD drawing. Using *Apply* causes the dialog box to disappear and the "Select objects:" prompt to appear. <u>You must specify a selection set (by any method) to be tested against the filter criteria</u>. If you want the filters to apply to the entire drawing (except *Locked* and *Frozen* layers), enter **ALL**. If you only need to search a smaller area of the drawing, PICK objects or use a window or other selection method.

The application of the current filters can be accomplished in two ways:

1. Within a command (transparently)
    At the "Select objects:" prompt, you can invoke *Filter* transparently by selecting from a menu or

typing "*'filter*" (prefaced by an apostrophe). Then set the desired filters, select *Apply*, and specify the selection set to test against the criteria. The resulting (filtered) selection set is used for the current command.

2. At the Command: prompt
From the Command: prompt (when no command is in use), type or select *Filter*, specify the filter configuration, and then use *Apply*. You are prompted to "Select objects:". Specify *All* or use a selection method to specify the area of the drawing for the filter to be applied against. Press Enter to complete the operation. The filtered selection set is stored in the selection set <u>buffer</u> and can be recalled by using the *Previous* option in response to the next "Select objects:" prompt.

OBJECT GROUPS

Figure 20-7

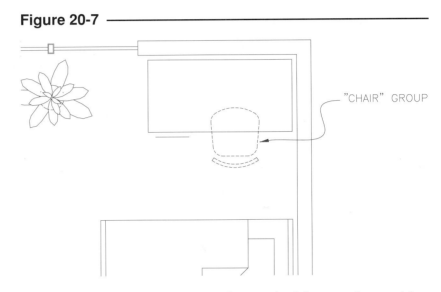

"CHAIR" GROUP

Often, a group of objects that are related in some way in a drawing may require an editing action. For example, all objects representing a chair in a floor plan may need to be selected for *Copying* or changing color or changing to another layer (Fig. 20-7). Or you may have to make several manipulations of all the fasteners in a mechanical drawing or all of the data points in a civil engineering drawing. In complex drawings containing a large number of objects, the process of building such a selection set (with the pickbox, window, or other methods) may take considerable effort. One disadvantage of the traditional selection process (without *Groups*) is that selection sets could not be saved and recalled for use at a later time. AutoCAD introduced the *Group* command to identify and organize named selection sets.

A <u>Group</u> is a <u>set of objects</u> that has an <u>assigned name</u> and description. The *Group* command allows you to determine the objects you want to include in the group and to assign a name and description. Usually the objects chosen to participate in a group relate to each other in some manner. Objects can be members of more than one group.

Once a group is created, it can be assigned a *Selectable* status. If a group is *Selectable*, the entire group can be automatically selected (highlighted) at any "Select Objects:" prompt by PICKing any individual member, or by typing the word "group" or the letter "G," then giving the group name. For example, if all the objects in a mechanical drawing representing fasteners were assigned to a group named "fasten," the group could be selected during the *Move* command as follows:

```
Command: move
Select objects: g
Enter group name: fasten
18 found
Select objects: Enter
Base point or displacement:
        etc.
```

If the group is assigned a <u>non-selectable</u> status, the group <u>cannot</u> be selected as a whole by any method. Non-selectablility prevents accidentally *Copying* the entire group, for example. If a group is assigned a <u>selectable</u> status, the group can be manipulated as a whole <u>or</u> each member can be edited individually. The Ctrl+A key sequence (or *PICKSTYLE* variable) controls whether PICKing highlights the individual member or the entire group. In other words, you can PICK individual members <u>or</u> an entire selectable group "on the fly" by toggling Ctrl+A.

A group can be thought of as a set of objects that has a level of distinction beyond that of a typical selection set (specified "on the fly" by the pickbox, window, or other method), but not as formal as a *Block*. A *Block* is a group of objects that is combined into <u>one object</u> (using the *Block* command), but individual entities <u>cannot</u> be edited separately. (Blocks are discussed in Chapter 21.) Using groups for effective drawing organization is similar to good layer control; however, with groups you have the versatility to allow members of a group to reside on different layers.

To summarize, using groups involves two basic activities: (1) use the *Group* command to define the objects and assign a name and (2) at any later time, locate (highlight) the group for editing action at the "Select Objects:" prompt by PICKing a member or typing "G" and giving its name.

GROUP

Pull-down Menu	COMMAND (TYPE)	ALIAS (TYPE)	Short-cut	Screen (side) Menu	Tablet Menu
Tools Object Group...	GROUP or -GROUP	G or -G	...	TOOLS 2 Group	X,8

Invoking the *Group* command by any method produces the *Object Grouping* dialog box (Fig. 20-8). (The hyphen can be entered as a prefix, such as "-group," to present the command line equivalent.) The main purposes of this interface are to <u>create</u> groups and <u>change</u> the properties of existing groups.

The focus of the dialog box is the *Group Name* list (on top), which gives the list of existing groups and indicates each group's selectable status. If a group is <u>selectable</u>, the entire group can be highlighted by typing the word "group" or letter "G" at the "Select Objects:" prompt, then entering the group name. Alternately, you can PICK one member to highlight the entire group, or use Ctrl+A to PICK only one member. A group that is <u>not selectable</u> cannot be selected as a whole.

Typically, your first activity with the *Object Grouping* dialog box is to create a new group.

Figure 20-8 ——————

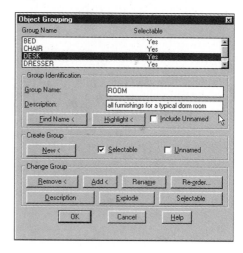

Creating a Named Group

1. Enter the desired name in the *Group Name:* edit box. This must be done as the <u>first</u> step.

2. Enter a unique description for the group in the *Description* edit box. Use up to 64 characters to describe the relationship or characteristic features of the group.

3. Determine if the group is to be selectable or not and indicate so in the *Selectable* check box.

4. Use the *New* button to PICK the desired members (objects) to be included in the group. The dialog box is temporarily hidden until all desired objects are selected. Once the selection set is chosen (press Enter), the dialog box reappears with the new group added to the list.

All of the options and details of the *Group Identification* and *Create Group* clusters are explained here:

Group Name:

The *Group Name:* edit box (just below the list) is used to enter the name of a new group that you want to create or to display the name of an existing group selected from the list above that requires changing. When creating a new group, enter the desired name before choosing the set of objects (using the *New* tile).

Description

Including a *Description* for new groups is optional, but this feature is useful in organizing various selection sets in a drawing. A higher level of distinction is added to the groups if a description is included. The description may be up to 64 characters. If a group name is selected from the list, both the group name and description appear in the *Group Identification* cluster.

Find Name

Figure 20-9 ─────

Use this option to find the group name for any object in the drawing. (The name currently appearing in the *Group Name:* edit box does not have an effect on this option.) Using the *Find Name* tile temporarily removes the *Object Grouping* dialog box and prompts you to "Pick a member of a group." After doing so, AutoCAD responds with the names of the group(s) of which the selected object is a member (Fig. 20-9). If an object is PICKed that is not assigned to a group, AutoCAD responds with "Not a group member." This option can be used immediately upon entering the *Object Grouping* dialog box, with no other action required.

Highlight

Choosing this tile highlights all members of the group appearing in the *Group Name:* box. The dialog box temporarily disappears so the drawing can be viewed. Use this to verify which objects are members of the current group.

Include Unnamed

Figure 20-10 ──────────

When a group is copied (with *Copy*, *Mirror*, etc.), the objects in the new group have a collective relationship like any other group; however, the group is considered an "unnamed group." Toggling the *Include Unnamed* checkbox forces any unnamed groups to be included in the list above (Fig. 20-10). In this way, you can select and manipulate copied groups like any other group but without having to go through the process to create the group from scratch. The copied group is given a default name such as *An*, where n is the number of the group created since the beginning of the drawing. You can also create an unnamed group using the *Unnamed* toggle and the *New* tile. An unnamed group can be *Renamed*.

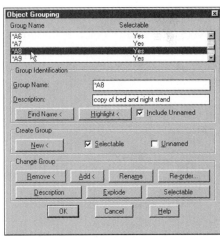

New

The *New* tile in the *Create Group* cluster is used to select the objects that you want to assign as group members. You are returned to the drawing and prompted to "Select objects:". (See Creating a Named Group.)

Selectable

Use this toggle to specify the selectable status for the new group to create. If the "X" appears, the new group is created as a *Selectable* group. A selectable group can be selected as a whole by <u>two methods</u>. When prompted to "Select objects:" during an editing command, you can (1) PICK any individual member of a selectable group or (2) enter "G" and the group name.

When using the PICK method, the Ctrl+A key sequence (or changing the *PICKSTYLE* variable) controls whether an individual member is selected or the entire group. When *PICKSTYLE* is set to 1 or 3, PICKing any member of a <u>selectable</u> group highlights (selects) the entire group. If *PICKSTYLE* is set to 0 or 2, PICKing any member of a selectable group highlights <u>only</u> that member. The Ctrl+A key sequence automatically changes the *PICKSTYLE* variable setting to <Group on> or <Group off>. Therefore, any member, or the entire group, of a selectable group can be PICKed by toggling Ctrl+A. *PICKSTYLE* has no effect on non-selectable groups.

The *PICKSTYLE* variable can also be changed from the *Object Selection Settings* dialog box (*Ddselect*). Toggling the "X" in the checkbox sets *PICK-STYLE* to <Group on> (1 or 3) (Fig. 20-11). (*PICKSTYLE* also affects the selection of associative hatch objects. See Chapter 26 and Apendix A.) The *PICKSTYLE* variable is saved with the drawing file.

Figure 20-11

For drawings that contain objects that are members of multiple groups, normally the *Selectable* status of the groups (or all but one group) should be toggled off to avoid activating all groups when a member is selected. Groups or group members that are on *Frozen* layers cannot be selected.

Unnamed
If you want to create a group with a default name (*A*n, where *n* is the number of the group created since the beginning of the drawing), use this option. If the "X" appears in the checkbox, you cannot enter a name in the *Group Name* edit box. Use the *New* tile to select the objects to include in the unnamed group. If you keep unnamed groups, toggle *Include Unnamed* on to include these groups in the list. An unnamed group can be *Renamed*.

Typically, you should name groups that you create. When you make a copy of an existing group, assign a new name (with *Rename*) if changes are made to the members. However, if you keep an unnamed group that is an identical copy, a good strategy is to keep the default (*A*n) name as an identifier and add a description like "copy of fasten group." It is suggested that unnamed groups be <u>non-selectable</u> to avoid accidentally selecting the entire group by PICKing one object for a *Copy* or other operation. Remember, you can always change the selectable status when you want to access the entire unnamed group.

Once a group exists, it can be changed in a number of ways. The options for changing a group appear in the *Change Group* cluster of the *Object Grouping* dialog box.

Changing Properties of a Group

1. Select the group to change from the list on top of the *Object Grouping* dialog box (Fig. 20-10). Once a group is selected, the name and description appear in the related edit boxes. A group must be selected before the *Change Group* cluster options are enabled.

2. Select the desired option from the choices in the *Change Group* area.

Remove
The *Remove* tile allows you to <u>remove any member</u> of the specified group. The dialog box disappears temporarily and you are prompted to "Remove objects." If all members of a group are removed, the object group still remains defined in the list box.

Add

Use this option to add new members to the group appearing in the *Group Name:* edit box. The "Select objects:" prompt appears.

Rename

First, select the desired group to rename from the list. Next, type over the existing name in the *Group Name:* edit box and enter the desired new name. Then choose the *Rename* button.

Re-order

See Re-ordering Group Members at the end of this section.

Description

You can change the description for the specified object group with this option. First, select the desired group from the list. Next, edit the contents of the *Description* edit box. Then, select the *Description* tile.

Explode

Use this option to remove a group definition from the drawing. The *Explode* option breaks the current group into its original objects. The group definition is removed from the object group list box; however, the individual group members are not affected in any other way.

Selectable

This option changes the selectable status for a group. Highlight a group name from the list; then choose the *Selectable* tile. The *Yes* or *No* indicator in the Selectable column of the list toggles as you click the tile. See *Selectable* in the discussion of the *Create Group* cluster earlier.

Re-ordering Group Members

When you select objects to be included in new groups, AutoCAD assigns a number to each member. The number corresponds to the sequence that you select the objects. For special applications, the number associated with individual members is critical. For example, a group may be formed to generate a complete tool path where each member of the group represents one motion of the sequence.

The *Re-order* button (in the *Object Grouping* dialog box) invokes the *Order Group* dialog box (Fig. 20-12). This dialog box allows you to change the order of individual group members. To reverse the order, simply select *Reverse Order*. To re-order members of a group to another sequence, use these steps:

Figure 20-12 ——————

1. In the *Order Group* dialog box, select the group to re-order from the list on top.

2. Select the *Highlight* option to display each group member one at a time. In the small dialog box that appears, select *Next* repeatedly to go completely through the current sequence of group members. The sequence begins with 0, so the first object has position 0. Then select OK to return to the *Order Group* dialog box.

3. In the *Number of Objects* edit box, enter the number of objects that you want to reposition.

4. In the *Remove from Position* edit box, enter the current position of the object to re-order.

5. Enter the new position in the *Replace at Position* edit box.

6. Select *Re-order*.

7. Select *Highlight* again to verify the new order.

Operation of this dialog box is difficult and is not explained well in the documentation. It is possible to accidentally remove or duplicate members from the group using this device. A more straightforward and stable alternative for re-ordering is to *Explode* the group and make a new group composed of the same members, but PICK the members <u>in the desired order</u>.

Groups Examples

Imagine that you have the task to lay out the facilities for a college dormitory. Using the floor plan provided by the architect, you begin to go through the process of drawing furniture items to be included in a typical room, such as a desk, chair, and accessory furnishings. The drawing may look like that in Figure 20-13.

Figure 20-13 ――――――――――

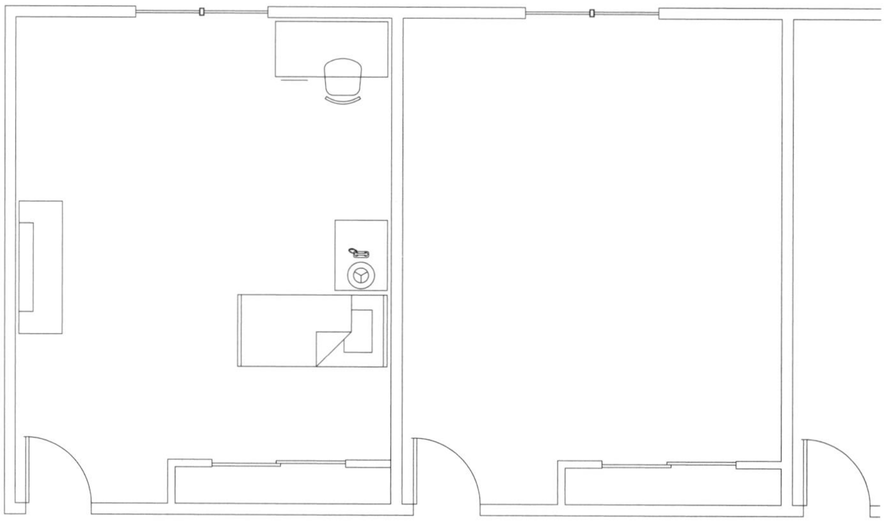

Since each room is to house two students, all objects representing each furniture item must be *Copied*, *Mirrored*, or otherwise manipulated several times to complete the layout for the entire dorm. To *Copy* the chair, for example, by the traditional method (without creating groups), each of the 12 objects (*Lines* and *Arcs*) comprising the chair must be selected <u>each time</u> the chair is manipulated. Instead, combining the individual objects into a group named "CHAIR" would make selection of these items for each operation much easier.

To make the CHAIR group, the *Group* command is used to activate the *Object Grouping* dialog box. The CHAIR group is created by selecting all *Lines* and *Arcs* as members (see previous Figure 20-7). The group is made *Selectable* as shown in Figure 20-14.

Now, any time you want to *Copy* the chair, at the "Select objects:" prompt enter the letter "G" and "CHAIR" to cause the entire group to be highlighted. Alternately, the CHAIR group can be selected by PICKing any *Line* or *Arc* on the chair since the group is selectable.

Figure 20-14 ――――――――――

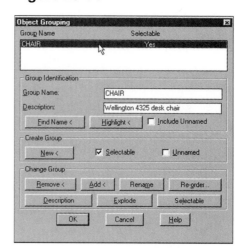

The same procedure is used to combine several objects into a group for each furniture item. The complete suite of groups may appear as shown in the dialog box shown in Figure 20-15. Toggling on *Include Unnamed* forces the unnamed group to appear in the list. The unnamed group was automatically created when the CHAIR group was copied.

Using the named groups, completing the layout is simplified. Each group (CHAIR, DESK, etc.) is selected and *Copied* to complete the room layout.

The next step in the drawing is to lay out several other rooms. Assuming that other rooms will have the same arrangement of furniture, another group could be made of the entire room layout. (Objects can be members of more than one group.) A group called ROOM is created in the same manner as before, and all objects that are part of furniture items are selected to be members of the group. If the ROOM group is made selectable, all items can be highlighted with one PICK and copied to the other rooms (Fig. 20-16).

After submitting the layout to the client for review, a request is made for you to present alternative room layouts. Going back to the drawing, you discover that each time you select an individual furniture item to manipulate, the entire ROOM group is highlighted. To remedy the situation, the ROOM group is changed to non-selectable with the *GROUP* command. Now the individual furniture items can be selected since the original groups (CHAIR, DESK, BED, etc.) are still *Selectable* (Fig. 20-17).

Figure 20-15

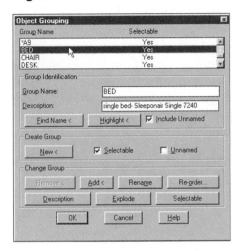

Figure 20-16

Figure 20-17

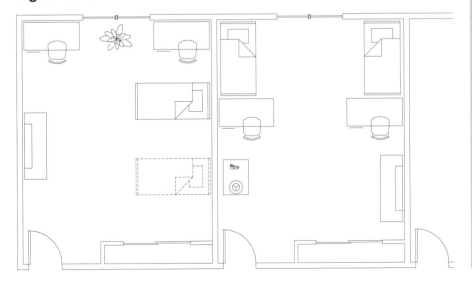

R13

Another, more typical application is the use of *Groups* in conjunction with *Blocks*. A block is a set of objects combined into one object with the *Block* command. A block always behaves as <u>one object</u>. In our dorm room layout, each furniture item could be *Inserted* as a previously defined block. The entire set of blocks (bed, chair, desk, etc.) could be combined into a group called ROOM. In this case, the ROOM group would be set to *Selectable*. This would allow you to select <u>all</u> furniture items as a group by PICKing only one. However, if you wanted to select only <u>one</u> furniture item (Block), use the <u>Ctrl+A</u> key sequence to toggle "<Group off>." With "Group off," individual members of any group (blocks in this case) can be PICKed without having to change the *Selectable* status. See Chapter 21 for more information on *Blocks*.

CHAPTER EXERCISES

1. *PICKFIRST*

   Check to ensure that **PICKFIRST** is set to **1** (*ON*). Also make sure *GRIPS* are set to 0 (*OFF*) by typing **GRIPS** at the command prompt or invoking the **Object Selection Settings** dialog box. For each of the following editing commands you use, make sure you PICK the objects FIRST, then invoke the editing command.

   A. *Open* the **PLATES** drawing that you created as an exercise in Chapter 10 (not the PLATNEST drawing). *Erase* 3 of the 4 holes from the plate on the left, leaving only the hole at the lower-left corner (**PICK** the *Circles*; then invoke *Erase*). Create a *Rectangular Array* with **7** rows and **5** columns and **1** unit between each hole (PICK the *Circle* before invoking *Array*). The new plate should have 35 holes as shown in Figure 20-18, plate A.

   Figure 20-18

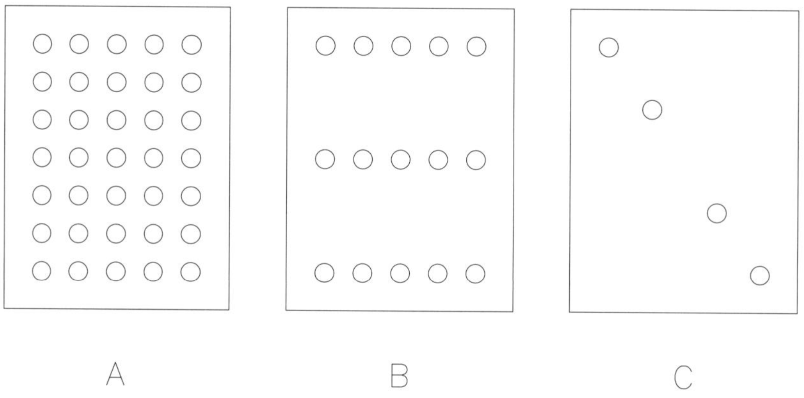

   A   B   C

   B. For the center plate, **PICK** the 3 holes on the <u>vertical</u> center; then invoke *Copy*. Use the *Multiple* option to create the additional 2 sets of 3 holes on each side of the center column. Your new plate should look like that in Figure 20-18, plate B. Use *SaveAs* to assign a new name, **PLATES2**.

   C. For the last plate (on the right), **PICK** the two top holes and *Move* them upward. Use the *Center* of the top hole as the "Base point or displacement" and specify a "Second point of displacement" as **1** unit from the top and left edges. Compare your results to Figure 20-18, plate C. *Save* the drawing (as **PLATES2**).

   Remember that you can leave the *PICKFIRST* setting *ON* always. This means that you can use both Noun/Verb or Verb/Noun editing at any time. You always have the choice of whether you want to PICK objects first or use the command first.

2. *PICKADD*

Change *PICKADD* to **0** (*OFF*). Remember that if you want to PICK objects and add them to the existing highlighted selection set, hold down the **SHIFT** key; then **PICK**.

A. *Open* the **ARRAY1** drawing. (The settings for *GRIPS* and *PICKFIRST* have not changed by *Opening* a new drawing.) *Erase* the holes (highlighted *Circles* shown in Figure 20-19) leaving only 4 holes. Use either Noun/Verb or Verb/Noun editing. *Saveas* **FLANGE1**.

B. *Open* drawing **ARRAY2**. Select all the holes and *Rotate* them **180** degrees to achieve the arrangement shown in Figure 20-20. Any selection method may be used. Try several selection methods to see how *PICKADD* reacts. (The *Fence* option can be used to select all holes without having to use the **Shift** key.) *SaveAs* **FLANGE2**. Change the *PICKADD* setting back to **1** (*ON*).

3. *PICKDRAG*

Open the **CH12EX3** drawing. Set *PICKDRAG* to **1** (*ON*). Use *Stretch* to change the front and side views to achieve the new base thickness as indicated in Figure 20-21. (Remember to hold down the PICK button when dragging the mouse or puck.) When finished, *SaveAs* **HOLDRCUP**. Change the *PICKDRAG* setting back to **0** (*OFF*).

Figure 20-19 ———————————

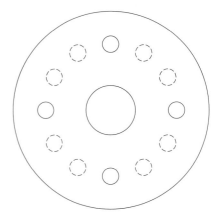

Figure 20-20 ———————————

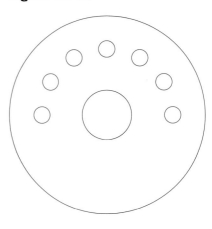

Figure 20-21 ———————————

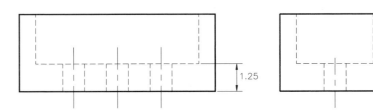

4. **Object Selection Filters**

This exercise involves using object selection filters to change the properties of objects in the drawing. Open the **ASESMP.DWG** from the C:\Program Files\AutoCAD R14\ Sample directory (assuming you have the full installation of AutoCAD).

A. Notice that all the chairs in the drawing are cyan color. Using *List*, verify that any chair is on the CHAIRS layer. We want to move all of the chairs in the drawing to the FURNITURE layer.

B. At the Command: prompt, invoke the *Object Selection Filters* dialog box by any method. In the *Select Filter* area, open the drop-down list. Scroll down and choose *Layer*, or type "L" and then choose *Layer*.

C. After the drop-down list closes, choose *Select...* to display the layers available for selection in the current drawing. Choose **CHAIRS**; then choose *OK* and the *Select Layer(s)* dialog box closes. Select the *Add to List* tile to make "Layer=CHAIRS" appear in the filter criteria list at the top of the dialog box (Fig. 20-22).

Figure 20-22 —————————

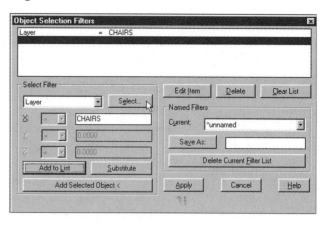

D. Choose *Apply* to apply the filtering criteria to the drawing. When the "Select objects:" prompt appears, type *ALL*; then press **Enter**. From all the objects in the drawing, AutoCAD will filter out those objects that <u>do not</u> meet the filtering criteria and select only those objects that do. When the "Select objects:" prompt returns, press **Enter** to complete the selection process. The new selection set (all of the chairs) is stored in the selection set buffer.

E. Invoke the *Change Properties* command by any method. At the "Select objects:" prompt, type *P* for *Previous* and press **Enter**. AutoCAD uses the 40 chairs stored previously in the buffer as the selection set. Press **Enter** a second time to complete the selection process and the *Change Properties* dialog box should appear. From the dialog box, select the *Layer* button. Choose **FUR-NITURE**, then *OK* to close the *Select Layers* dialog box and *OK* to close the *Change Properties* dialog box. All of the chairs in the drawing should change to the color of the FURNITURE layer (green). Use *List* to verify the layer of any chair. Do not save your changes and do not close the drawing.

5. **Object Selection Filters**

Object selection filters enable you to globally select all instances of an object or block for some editing action. In this exercise, we will use the *Erase* command and invoke the *Object Selection Filters* <u>transparently</u> to find and erase all instances of the room tag *Block*.

A. Using the ASESMP drawing again (from exercise 4), invoke the *Erase* command.

B. At the "Select objects:" prompt, invoke the *Object Selection Filters* dialog box by typing *'FILTER* or selecting from a menu or icon.

C. Choose *Clear List* to start with a new filter criteria list (this action removes "Layer=CHAIRS").

D. In the *Select Filter* area, open the drop-down list and select *Block* from the list. After the drop-down list closes and displays "Block", choose *Add to List*. This will add "Object = Block" to the filter criteria.

E. From the drop-down list, choose *Block Name*; then PICK the *Select...* tile to produce a list of blocks in the drawing. Select **RMTAG** and **RMTAG2** (hold down the **Shift** key to select both items); then choose *OK* for the *Select Layer(s)* dialog box to close. "RMTAG, RMTAG2" appears in the X= field but not in the filter criteria listing. PICK *Add to List* and you see "Block Name=RMTAG, RMTAG2" in the list at the top of the box (Fig. 20-23).

F. PICK *Apply* to apply the filtering criteria to the drawing. At the "Select objects:" prompt, type *ALL* or make a crossing window around the entire drawing. From all the objects in the drawing, AutoCAD filters out those objects that are NOT the blocks desired and selects only RMTAG and RMTAG2. When the "Select objects:" prompt returns, press **Enter** to complete the selection process. The selection set is used for the *Erase* command, and all of the room tag blocks are erased from the drawing.

Figure 20-23

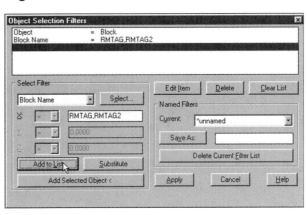

G. Do not save your changes to the ASESMP drawing.

6. **Object Groups**

A. *Open* the **EFF-APT2** drawing that you last worked on in Chapter 18 exercises. Create a *Group* for each of the three bathroom fixtures. *Name* one group **WASHBASIN**, give it a *Description*, and include the *Ellipse* and the surrounding *Rectangle* as its members. Make two more groups named **WCLOSET** and **BATHTUB** and select the appropriate members. Next, combine the three fixture groups into one *Group* named **BATHROOM** and give it a description. Then, draw a bed of your own design and combine its members into a *Group* named **SINGLEBED**. Make all five groups *Selectable*. Use *SaveAs* to save and rename the drawing to **2BR-APT**.

Figure 20-24

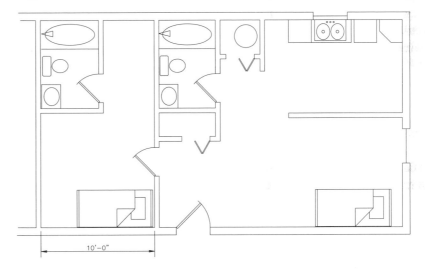

10'-0"

B. A small bedroom and bathroom are to be added to the apartment, but a 10' maximum interior span is allowed. One possible design is shown in Figure 20-24. Other more efficient designs are possible. Draw the new walls for the addition, but design the bathroom and bedroom door locations to your personal specifications. Use *Copy* or *Mirror* where appropriate. Make the necessary *Trims* and other edits to complete the floor plan.

C. Next, *Copy* or *Mirror* the bathroom fixture groups to the new bathroom. If your design is similar to that in the figure, the BATHROOM group can be copied as a whole. If you need to *Copy* only individual fixtures, take the appropriate action so you don't select the entire BATHROOM group. *Copy* or *Mirror* the **SINGLEBED**. Once completed, activate the *Object Grouping* dialog box and toggle on *Include Unnamed*. Can you explain what happened? Assign a *Description* to each of the unnamed groups and *Save* the drawing.

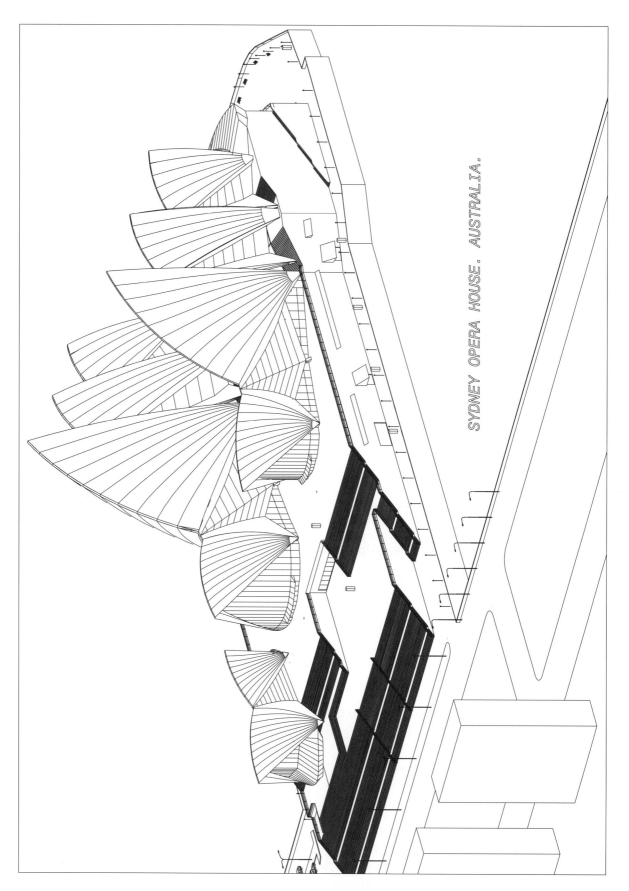

SYDNEY OPERA HOUSE. AUSTRALIA.

OPERA.DWG Courtesy of Autodesk, Inc. (Release 14 Sample Drawing)

21

BLOCKS

Chapter Objectives

After completing this chapter you should:

1. understand the concept of creating and inserting symbols in AutoCAD drawings;

2. be able to use the *Block* command to transform a group of objects into one object that is stored in the current drawing's block definition table;

3. be able to use the *Insert* and *Minsert* commands to bring *Blocks* into drawings;

4. know that *color* and *linetype* of *Blocks* are based on conditions when the *Block* is made;

5. be able to convert *Blocks* to individual objects with *Explode*;

6. be able to use *Wblock* to prepare .DWG files for insertion into other drawings;

7. be able to redefine and globally change previously inserted *Blocks*;

8. be able to define an insertion base point for .DWG files using *Base*.

CONCEPTS

A *Block* is a <u>group</u> of objects that are combined into <u>one</u> object with the *Block* command. The typical application for *Blocks* is in the use of symbols. Many drawings contain symbols, such as doors and windows for architectural drawings, capacitors and resistors for electrical schematics, or pumps and valves for piping and instrumentation drawings. In AutoCAD, symbols are created first by constructing the desired geometry with objects like *Line*, *Arc*, and *Circle*, then transforming the set of objects comprising the symbol into a *Block*. A description of the objects comprising the *Block* is then stored in the drawing's "block definition table." The *Blocks* can then each be *Insert*ed into a drawing many times and treated as a single object. Text can be attached to *Blocks* (called *Attributes*) and the text can be modified for each *Block* when inserted.

Figure 21-1 compares a shape composed of a set of objects and the same shape after it has been made into a *Block* and *Insert*ed back into the drawing. Notice that the original set of objects is selected (highlighted) individually for editing, whereas, the *Block* is only one object.

Figure 21-1

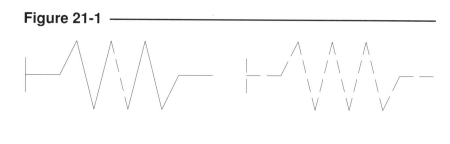

10 OBJECTS   1 OBJECT

Since an inserted *Block* is one object, it uses less file space than a set of objects that is copied with *Copy*. The *Copy* command creates a duplicate set of objects, so that if the original symbol were created with 10 objects, 3 copies would yield a total of 40 objects. If instead the original set of 10 were made into a *Block* and then *Insert*ed 3 times, the total objects would be 13 (the original 10 + 3).

Upon *Insert*ing a *Block*, its scale can be changed and rotational orientation specified without having to use the *Scale* or *Rotate* commands (Fig. 21-2). If a design change is desired in the *Blocks* that have already been *Insert*ed, the original *Block* can be redefined and the previously inserted *Blocks* are automatically updated. *Blocks* can be made to have explicit *Linetype* and *Color*, regardless of the layer they are inserted onto, or they can be made to assume the *Color* and *Linetype* of the layer onto which they are *Insert*ed.

Figure 21-2

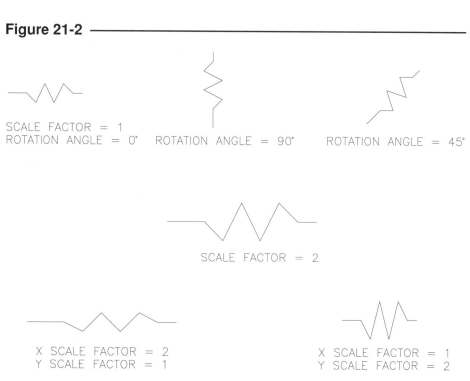

SCALE FACTOR = 1
ROTATION ANGLE = 0°   ROTATION ANGLE = 90°   ROTATION ANGLE = 45°

SCALE FACTOR = 2

X SCALE FACTOR = 2   X SCALE FACTOR = 1
Y SCALE FACTOR = 1   Y SCALE FACTOR = 2

Blocks can be <u>nested</u>; that is, one *Block* can reference another *Block*. Practically, this means that the definition of *Block* "C" can contain *Block* "A" so that when *Block* "C" is inserted, *Block* "A" is also inserted as part of *Block* "C" (Fig. 21-3).

Figure 21-3 ————————————————————————————

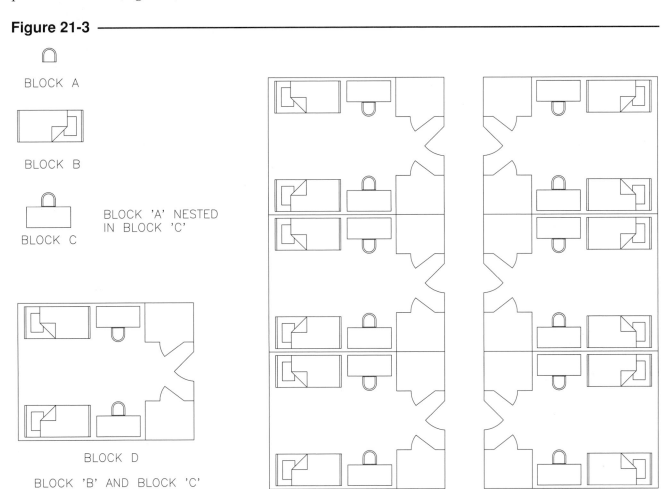

BLOCK A

BLOCK B

BLOCK C

BLOCK 'A' NESTED
IN BLOCK 'C'

BLOCK D

BLOCK 'B' AND BLOCK 'C'
NESTED IN BLOCK 'D'

Blocks created within the current drawing can be copied to disk as complete and separate drawing files (.DWG file) by using the *Wblock* command (Write Block). This action allows you to *Insert* the *Blocks* into other drawings. Specifically, when you use the *Insert* command, AutoCAD first searches for the supplied *Block* name in the current drawing's block definition table. If the designated *Block* is not located there, AutoCAD searches the directories for a .DWG file with the designated name.

Commands related to using *Blocks* are:

Bmake	Invokes a dialog box for creating *Blocks*
Block	Creates a *Block* from individual objects
Insert	Inserts a *Block* into a drawing
Ddinsert	Invokes a dialog box for inserting a *Block*
Minsert	Permits a multiple insert in a rectangular pattern
Explode	Breaks a *Block* into its original set of multiple objects
Wblock	Writes an existing *Block* or a set of objects to a file on disk
Base	Allows specification of an insertion base point
Purge	Deletes uninserted *Blocks* from the block definition table
Rename	Allows renaming *Blocks*

COMMANDS

BLOCK and
BMAKE

	Pull-down Menu	COMMAND (TYPE)	ALIAS (TYPE)	Short-cut	Screen (side) Menu	Tablet Menu
	Draw *Block >* *Make...*	*BLOCK* or *BMAKE*	*B or -B*	...	*DRAW2* *Bmake*	*N,9*

Selecting the icon button, using the pull-down or screen menu, or typing *Bmake* produces the *Block Definition* dialog box shown in Figure 21-4. This dialog box provides the same functions as using the *Block* command (command line equivalent).

To make a *Block*, first create the *Lines, Circles, Arcs,* or other objects comprising the shapes to be combined into the *Block*. Next, use the *Bmake* or *Block* command to transform the objects into one object—a *Block*.

Figure 21-4 ———————

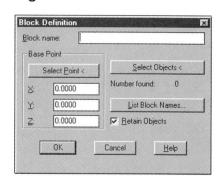

If you are using the *Block Definition* dialog box, enter the desired *Block* name in the *Block name* box. Then use the *Select Objects* tile to return to the drawing temporarily to select the objects you wish to comprise the *Block*. After selection of objects, the dialog box reappears. Use the *Select Point* button in the *Base Point* section of the dialog box if you want to use a point other than the default 0,0 as the "insertion point" when the *Block* is later inserted (usually select a point in the center or corner of the set of objects). When you select *OK*, the objects comprising the *Block* disappear and the new *Block* is defined and stored in the drawing's block definition table awaiting future insertions. A check appearing in the *Retain Objects* checkbox forces AutoCAD to retain the original "template" objects (similar to using *Oops* after the *Block* command), even though the definition of the *Block* remains in the table.

The *List Block Names* tile can be chosen to list all existing *Blocks* in the current drawing (Fig. 21-5).

Figure 21-5 ———————

If you prefer to type, use *Block* to produce the command line equivalent of *Bmake*. The command syntax is as follows:

    Command: **Block**
    Block name (or ?): (**name**) (Enter a descriptive name for the *Block* up
    to 31 characters.)
    Insertion base point: **PICK** or (**coordinates**) (Select a point to be
    used later for insertion.)
    Select objects: **PICK**
    Select objects: **PICK** (Continue selecting all desired objects.)
    Select objects: **Enter**

The *Block* then <u>disappears</u> as it is stored in the current drawing's "block definition table." The *Oops* command can be used to restore the original set of "template" objects (they reappear), but the definition of the *Block* remains in the table. Using the *?* option of the *Block* command lists the *Blocks* stored in the block definition table.

Block Color and *Linetype* Settings

The <u>color</u> and <u>linetype</u> of an inserted *Block* are determined by one of the following settings when the *Block* is created:

1. When a *Block* is inserted, it is drawn on its original layer with its original *Color* and *Linetype* (when the objects were <u>created</u>), regardless of the layer or *color* and *linetype* settings that are current when the *Block* is inserted (unless conditions 2 or 3 exist).

2. If a *Block* is created on <u>Layer 0</u> (Layer 0 is current when the original objects comprising the *Block* are created), then the *Block* assumes the *color* and *linetype* of any layer that is current when it is inserted (Fig. 21-6).

Figure 21-6

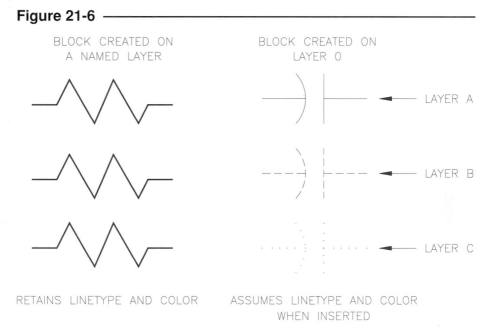

3. If the *Block* is created with the special *BYBLOCK linetype* and *color* setting, the *Block* is inserted with the *Color* and *Linetype* settings that are <u>current during insertion</u>, whether the *BYLAYER* or explicit object *Color* and *Linetype* settings are current.

	COMMAND (TYPE)	ALIAS (TYPE)	Short-cut	Screen (side) Menu	Tablet Menu
Pull-down Menu					
Insert Block...	*INSERT or DDINSERT*	*I or -I*	...	*INSERT Ddinsert*	*T,5*

INSERT and DDINSERT

Once the *Block* has been created, it is inserted back into the drawing at the desired location(s) with the *Insert* command. *Insert* also allows the *Blocks* to be scaled or rotated upon insertion. The command syntax is given here:

Command: **insert**
Block name (or ?): **name** (Type the name of an existing block or .DWG file to insert.)
Insertion point: **PICK** or (**coordinates**) (Give the desired location of the *Block*.)
X scale factor <1>/Corner/XYZ: **PICK** or (**value**) (Specifies the size of the *Block* in the X direction.)
Y scale factor (default=X): (**value**) or **Enter** (Specifies the size in the Y direction.)
Rotation angle: **PICK** or (**value**) (Enter an angle for *Block* rotation.)

Selecting the "*X scale factor*" with the cursor specifies both X and Y factors. The rotation angle can be forced to 90 degree increments by turning *ORTHO* (F8) *On*.

The *Insert* dialog box can be invoked by using the pull-down menu, icon buttons, or tablet menu or by typing *Ddinsert* (Fig. 21-7). Selecting the *Block* tile causes another box to pop up, listing the *Blocks* previously defined in the drawing's block definition table (Fig. 21-8). The desired *Block* is selected from the list. Selecting the *File* tile causes a box to pop up, allowing selection of any drawing (.DWG files) from any accessible drive and directory for insertion. The *Insert* dialog box provides explicit value entry of insertion point coordinates, scale, and rotation angle. *Explode* can also be toggled, which would insert the *Block* as multiple objects (see *INSERT* with *).

Figure 21-7 ──────────

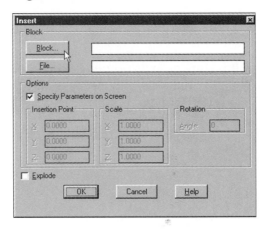

Figure 21-8 ──────────

Insert **Presets**

Sometimes it is desirable to see the *Block* in the intended scale factor or rotation angle before you choose the insertion point. *Insert* presets allow you to specify a rotation angle or scale factor <u>before</u> you dynamically drag the *Block* to PICK the insertion point. (Normally, you have to select the insertion point before the prompts for scale factor and rotation angle appear.) Presets can be used with the *Insert* command or *Insert* dialog box. To do this, type in one of the following characters <u>at the "Insertion point:" prompt</u>:

R	rotation angle
S	scale factor (uniform)
X	X scale factor only
Y	Y scale factor only
Z	Z scale factor only

For example, to insert a *Block* at a 45 degree angle using the *Insert* dialog box, the command syntax reads as follows:

Command: **ddinsert** (select *Block* name from the list)
Insertion point: **R** (specifies *Rotation* preset)
Rotation angle: **45** (rotates the *Block* to 45 degrees during dynamic insertion)
Insertion point: X scale factor <1> / Corner / XYZ: **Enter** or (**value**)
Y scale factor (default=X): **Enter** or (**value**)
Command:

This action allows you to see the *Block* at the prescribed rotation angle as you dynamically drag it to PICK the insertion point.

MINSERT

	COMMAND (TYPE)	ALIAS (TYPE)	Short-cut	Screen (side) Menu	Tablet Menu
Pull-down Menu					
...	*MINSERT*	...	...	...	...

This command allows a <u>multiple insert</u> in a rectangular pattern (Fig. 21-9). *Minsert* is actually a combination of the *Insert* and the *Array Rectangular* commands. The *Blocks* inserted with *Minsert* are associated (the group is treated as one object) and cannot be edited independently (unless *Exploded*).

Examining the command syntax yields the similarity to a *Rectangular Array*.

Figure 21-9

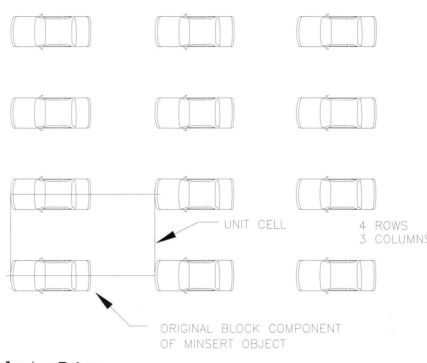

UNIT CELL

4 ROWS
3 COLUMNS

ORIGINAL BLOCK COMPONENT
OF MINSERT OBJECT

Command: **Minsert**
Block name (or ?): **name**
Insertion point: **PICK** or
  (**coordinates**)
X scale factor
  <1>/Corner/XYZ:
  (**value**) or **PICK**
Y scale factor (default=X): (**value**) or **Enter**
Rotation angle: (**value**) or **PICK**
Number of rows (---): (**value**)
Number of columns (||||): (**value**)
Unit cell or distance between rows: (**value**) or **PICK** (Value specifies Y distance from *Block* corner
  to *Block* corner; PICK allows drawing a unit cell rectangle.)
Distance between columns: (**value**) or **PICK** (Specifies X distance between *Block* corners.)
Command:

EXPLODE

Pull-down Menu	COMMAND (TYPE)	ALIAS (TYPE)	Short-cut	Screen (side) Menu	Tablet Menu
Modify *Explode*	*Explode*	X	...	MODIFY2 *Explode*	Y,22

Explode breaks a <u>previously</u> inserted *Block* back into its original set of objects (Fig. 21-10), which allows you to edit individual objects comprising the shape. *Blocks* that have been inserted with differing X and Y scale factors or *Blocks* that have been *Minsert*ed can be *Explode*d in Release 13 and 14. There are no options for this command.

Command: **explode**
Select objects: **PICK**
Select objects: **Enter**
Command:

Inserting with an * (asterisk) symbol accomplishes the same goal as using *Insert* normally, then *Explode*.

Figure 21-10

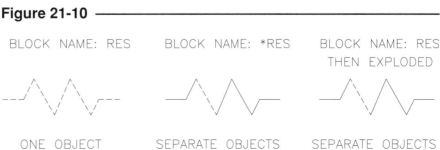

BLOCK NAME: RES

BLOCK NAME: *RES

BLOCK NAME: RES
THEN EXPLODED

ONE OBJECT

SEPARATE OBJECTS

SEPARATE OBJECTS

INSERT with *

Using the *Insert* command with the asterisk (*) allows you to insert a *Block,* not as one object, but as the original set of objects comprising the *Block.* In this way, you can edit individual objects in the *Block,* otherwise impossible if the *Block* is only one object (Fig. 21-10).

The normal *Insert* command is used; however, when the desired <u>Block name</u> is entered, it is prefaced by the asterisk (*) symbol:

> Command: **insert**
> Block name (or ?): ***** (**name**) (Type the * symbol, then the name of an existing block or .DWG file to insert.)
> Command:

This action accomplishes the same goal as using *Insert;* then *Explode.*

WBLOCK

Pull-down Menu	COMMAND (TYPE)	ALIAS (TYPE)	Short-cut	Screen (side) Menu	Tablet Menu
File *Export...*	WBLOCK	W	...	*FILE* *Export*	W,24

The *Wblock* command writes a *Block* out to disk as a separate and complete drawing (.DWG) file. The *Block* used for writing to disk can exist in the current drawing's *Block* definition table or can be created by the *Wblock* command. Remember that the *Insert* command inserts *Blocks* (from the current drawing's block definition table) or finds and accepts .DWG files and treats them as *Blocks* upon insertion.

If you are using an existing *Block,* a copy of the *Block* is essentially transformed by the *Wblock* command to create a complete AutoCAD drawing (.DWG) file. The original block definition remains in the current drawing's block definition table. In this way, *Blocks* that were originally intended for insertion into the current drawing can be inserted into other drawings.

If you want to transform a set of objects to be used as a *Block* in other drawings but not in the current one, you can use *Wblock* to transform (a copy of) the objects in the current drawing into a separate .DWG file. This action does not create a *Block* in the current drawing.

As an alternative, if you want to create symbols specifically to be inserted into other drawings, each symbol could be created initially as a separate .DWG file. Figure 21-11 illustrates the relationship among a *Block,* the current drawing, and a *WBlock.*

To create *Wblocks* (.DWG files) <u>from existing Blocks,</u> follow this command syntax:

> Command: **Wblock**
> (At this point, the *Create Drawing File* dialog box appears, prompting you to supply a name for the .DWG file to be created. Typically, a new descriptive name would be typed in the edit box rather than selecting from the existing names.)
> Block name: (**name**) (Enter the name of the desired existing *Block.* If the file name given in the previous step is the same as the existing *Block* name, an "=" symbol can be entered at this prompt.)
> Command:

A copy of the existing *Block* is then created in the current or selected directory as a *Wblock* (.DWG file).

To create a *Wblock* (.DWG file) to be used as a *Block* in other drawings <u>but not in the current drawing,</u> follow the same steps as before, but when prompted for the "Block name:" press Enter or select *blank* from the screen menu. The next steps are like the *Block* command prompts:

Figure 21-11

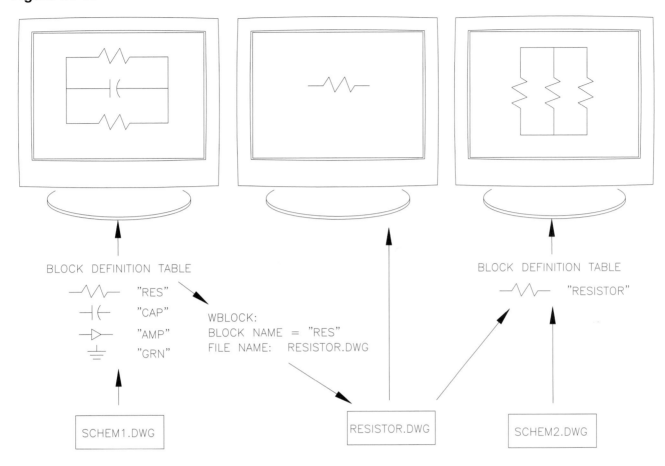

Command: **Wblock**
(The *Create Drawing File* dialog box appears, prompting you to supply a name for the .DWG file to be created)
Block name: (**Enter**) or (**blank**)
Insertion base point: **PICK** or (**coordinates**) (Pick a point to be used later for insertion.)
Select objects: **PICK**
Select objects: **Enter** (Press Enter to complete selection.)
Command:

Figure 21-12

An alternative method for creating a *Wblock* is using the *Export Data* dialog box accessed from the *File* pull-down menu. Make sure you select *Drawing (*.DWG)* as the type of file to export (bottom left of the box, Fig. 21-12). You can create a *Wblock* (.DWG file) from an existing *Block* or from objects that you select from the screen. After specifying the .DWG name you want to create and the dialog box disappears, you are presented with the same command syntax as above.

When *Wblocks* are *Inserted*, the *Color* and *Linetype* settings of the *Wblock* are determined by the settings current when the original objects comprising the *Wblock* were created. The three possible settings are the same as those for *Blocks* (see the *Block* command, *Color* and *Linetype* Settings).

When a *Wblock* is *Inserted*, its parent (original) layer is also inserted into the current drawing. *Freezing* either the parent layer or the layer that was current during the insertion causes the *Wblock* to be frozen.

Redefining *Blocks*

If you want to change the configuration of a *Block*, even after it has been inserted, it can be accomplished by redefining the *Block*. In doing so, all of the previous *Block* insertions are automatically and globally updated (Fig. 21-13). AutoCAD stores two fundamental pieces of information for each *Block* insertion—the insertion point and the *Block* name. The actual block definition is stored in the block definition table. Redefining the *Block* involves changing that definition.

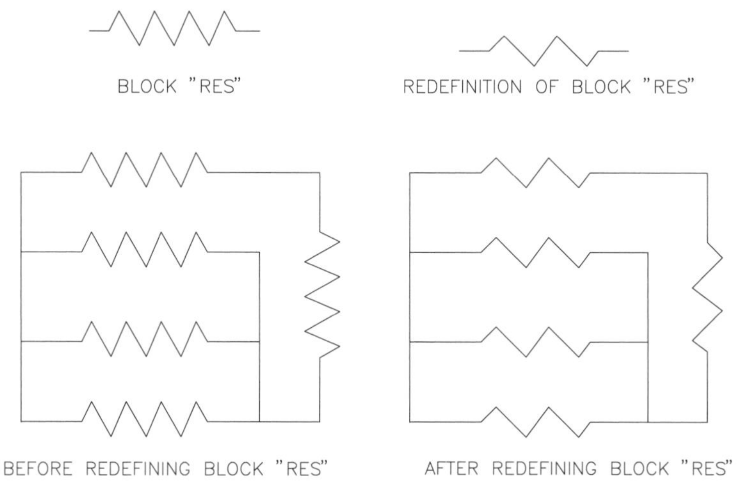

To redefine a *Block*, use the *Block* command. First, draw the new geometry or change the original "template" set of objects. (The change cannot be made

Figure 21-13 ─────────────

BLOCK "RES"   REDEFINITION OF BLOCK "RES"

BEFORE REDEFINING BLOCK "RES"   AFTER REDEFINING BLOCK "RES"

using an inserted *Block* unless it is *Exploded* because a *Block* cannot reference itself.) Next, use the *Block* command and select the new or changed geometry. The old *Block* is redefined with the new geometry as long as the original Block name is used.

```
Command: block
Block name (or ?): name (Enter the original Block name.)
Block (name) already exists.
Redefine it? <N>: Yes (Answering Y or yes causes the redefinition.)
Insertion base point: PICK or (coordinates) (Select a point to be used later for insertion.)
Select objects: PICK
Select objects: PICK (Continue selecting all desired objects.)
Select objects: Enter
Command:
```

The *Block* is redefined and all insertions of the original *Block* display the new geometry.

The *Block* command can also be used to redefine *Wblocks* that have been inserted. In this case, enter the *Wblock* name (.DWG filename) at the "Block name:" prompt to redefine (actually replace) a previously inserted *Wblock*.

BASE

Pull-down Menu	COMMAND (TYPE)	ALIAS (TYPE)	Short-cut	Screen (side) Menu	Tablet Menu
Draw *Block >* *Base*	*BASE*	...	...	DRAW2 *Base*	...

The *Base* command allows you to specify an "insertion base point" (see the *Block* command) in the current drawing for subsequent insertions. If the *Insert* command is used to bring a .DWG file into another drawing, the insertion base point of the .DWG is 0,0 by default. The *Base* command permits you to specify another location as the insertion base point. The *Base* command is used in the symbol drawing, that is, used in the drawing <u>to be inserted</u>. For example, while creating

Figure 21-14

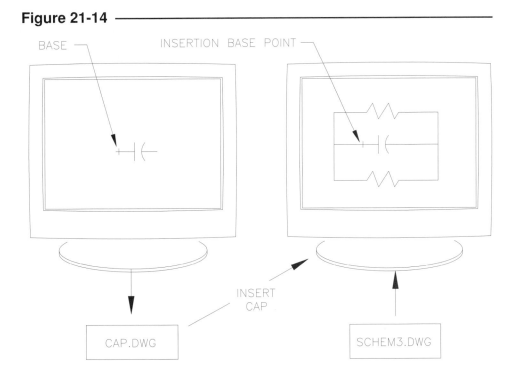

separate symbol drawings (.DWGs) for subsequent insertion into other drawings, the *Base* command is used to specify an appropriate point on the symbol geometry for the *Insert* command to use as a "handle" other than point 0,0 (see Fig. 21-14).

Blocks Search Path

In AutoCAD Release 14 you can use the *Preferences* dialog box to specify the search path used when the *Insert* command attempts to locate .DWG files. As you remember, when *Insert* is used and a *Block* name is entered, AutoCAD searches the drawing's block definition table and, if no *Block* by the specified name is found, searches the specified path for a .DWG file by the name. (If you use the *Insert* dialog box, you can select *File…* button to locate a .DWG file.) If the paths of the intended files are specified in the

"Support File Search Path," you can type *Insert* and the name of the file, and AutoCAD will locate and insert the file. Otherwise, you have to specify the name of the file prefaced by the path each time you *Insert* a new file.

The *Files* tab of the *Preferences* dialog box allows you to list and modify the path that AutoCAD searches for drawings to *Insert* (Fig. 21-15). The search path is also used for finding font files, menus, plug-ins, linetypes, and hatch patterns.

Suppose all of your symbols were stored in a folder called D:\Dwg\Symbols. Access the *Preferences* dialog box by typing *Preferences* or selecting from the *Tools* pull-down menu. In the *Files* tab, open the *Support File Search Path* section.

Figure 21-15

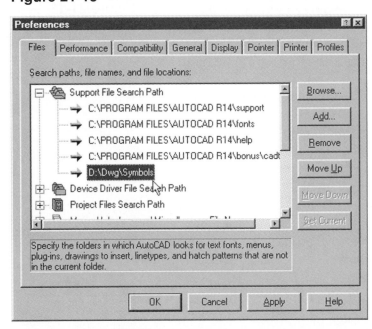

Select *Add..* or *Browse...* to enter the location of your symbols as shown in the figure. Use *Move Up* or *Move Down* to specify an order (priority) for searching. Then each time you type *"Insert"*, you could enter the name of a file without the path and AutoCAD can locate it.

PURGE

Pull-down Menu	COMMAND (TYPE)	ALIAS (TYPE)	Short-cut	Screen (side) Menu	Tablet Menu
File Drawing Utilities > Purge >	PURGE	PU	...	FILE Purge	X,25

Purge allows you to selectively delete a *Block* that is not referenced in the drawing. In other words, if the drawing has a *Block* defined but not appearing in the drawing, it can be deleted with *Purge*. In fact, *Purge* can selectively delete <u>any named object</u> that is not referenced in the drawing. Examples of unreferenced named objects are:

Blocks that have been defined but not *Inserted*;
Layers that exist without objects residing on the layers;
Dimstyles that have been defined, yet no dimensions are created in the style (see Chapter 29);
Linetypes that were loaded but not used;
Shapes that were loaded but not used;
Text Styles that have been defined, yet no text has been created in the *Style*;
Mlstyles (multiline styles) that have been defined, yet no *Mlines* have been drawn in the style.

Since these named objects occupy a small amount of space in the drawing, using *Purge* to delete unused named objects can reduce the file size. This may be helpful when drawings are created using template drawings that contain many unused named objects or when other drawings are *Inserted* that contain many unused *Layers* or *Blocks*, etc.

Purge is especially useful for getting rid of unused *Blocks* because unused Blocks can occupy a huge amount of file space compared to other named objects (*Blocks* are the only geometry-based named objects). Although some other named objects can be deleted by other methods (Windows 95/NT dialog boxes with right-click options), *Purge* is the only method for deleting *Blocks*.

Purge can be invoked by the methods shown in the command table. Using the *File* pull-down menu as shown in Figure 21-16 reveals all of the *Purge* options. The command options (when typing *Purge*) are listed next.

Figure 21-16

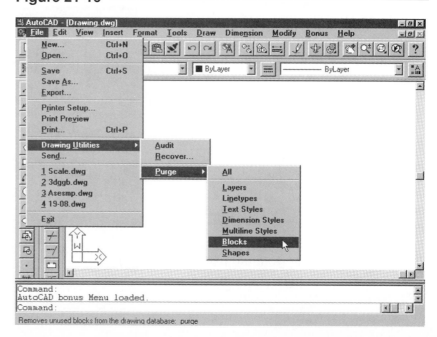

Command: **purge**
Purge unused Blocks/Dimstyles/LAyers/LTypes/SHapes/STyles/Mlinestyles/All: (**option**)
Names to purge <*>: **Enter** or (**name**)
Verify each name to be purged? <Y>

You can select one type of object to be *Purged* or select *All* to have all named objects listed. AutoCAD lists all named objects matching the type selected. For example, if you specified *Blocks* to purge, AutoCAD responds with a list of <u>unused</u> *Blocks* that may look something like this:

Purge block DESK? <N>
Purge block DESK2? <N>
Purge block CHAIR? <N>
Purge block TABLE? <N>

Answering with a "Y" causes the deletion of the *Blocks*. The other named object options operate similarly. Purge can be used <u>anytime</u> in a drawing session in Release 14 or Release 13.

RENAME and *DDRENAME*

Pull-down Menu	COMMAND (TYPE)	ALIAS (TYPE)	Short-cut	Screen (side) Menu	Tablet Menu
Format *Rename...*	*RENAME or* *DDRENAME*	*REN or* *-REN*	...	*FORMAT* *Ddrename*	*V,1*

These utility commands allow you to rename a *Block* or <u>any named object</u> that is part of the current drawing. *Rename* and the related dialog box accessed by the *Ddrename* command allow you to rename the named objects listed here:

Blocks, Dimension Styles, Layers, Linetypes, Text Styles, User Coordinate Systems, Views, and *Viewport* configurations.

The *Rename* command can be used to rename objects one at a time. For example, the command sequence might be as follows:

Command: **rename**
Block/Dimstyle/LAyer/LType/Style/Ucs/VIew/VPort: **B** (use the *Block* option)
Old block name: **DESK**
New block name: **DESK-30**
Command:

Wildcard characters are not allowed with the *Rename* command but are allowed with the dialog box.

Optionally, if you prefer to use dialog boxes, you may type *Ddrename* or select *Rename...* from the *Format* pull-down menu to access the dialog box shown in Figure 21-17.

You may select from the *Named Objects* list to display the related objects existing in the drawing. Then select or type the old name so it appears in the *Old Name:* edit box. Specify the new name in the *Rename To:* edit box. <u>You must then PICK *the Rename To:* tile</u> and the new names will appear in the list. Then select the *OK* tile to confirm.

The *Rename* dialog box can be used with wildcard characters (see Chapter 44) to specify a list of named objects for renaming.

Figure 21-17

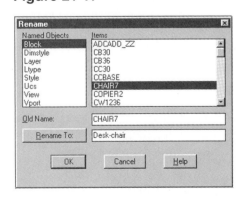

For example, if you desired to rename all the "DIM-*" layers so that the letters "DIM" were replaced with only the letter "D", the following sequence would be used.

1. In the *Old Name* edit box, enter "DIM-*" and press Enter. All of the layers beginning with "DIM-" are highlighted.

2. In the *Rename To:* edit box, enter "D-*." Next, the *Rename To:* tile must be PICKed. Finally, PICK the *OK* tile to confirm the change.

CHAPTER EXERCISES

1. **Block, Insert**
    In the next several exercises, you will create an office floor plan, then create pieces of furniture as *Blocks* and *Insert* them into the office. All of the block-related commands are used.

    A. Start a *New* drawing. Select *Start from Scratch* and use the *English* defaults. Use *Save* and assign the name **OFF-ATT**. Set up the drawing as follows:

1. *Units*	Architectural	**1/2" Precision**
2. *Limits*	48' x 36'	(1/4"=1' scale on an A size sheet), drawing scale factor = 48
3. *Snap*	3	
4. *Grid*	12	
5. *Layers*	FLOORPLAN	continuous	colors of your choice, all different

	FURNITURE	continuous
	ELEC-HDWR	continuous
	ELEC-LINES	hidden
	DIM-FLOOR	continuous
	DIM-ELEC	continuous
	TEXT	continuous
	TITLE	continuous

        6. *Text Style* CityBlueprint   **CityBlueprint (TrueType font)**
        7. *Ltscale* 24

    B. Create the floor plan shown in Figure 21-18. Center the geometry in the *Limits*. Draw on layer **FLOORPLAN**. Use any method you want for construction (e.g., *Line, Pline, Xline, Mline, Offset*).

    C. Create the furniture shown in Figure 21-19. Draw on layer **FURNITURE**. Locate the pieces anywhere for now. Do not make each piece a *Block*. *Save* the drawing as **OFF-ATT**.

        Now make each piece a *Block*. Use the **name** as indicated and the **insertion base point** as shown by the "blip." Next, use the *Block* command again but with the **?** option to list the block definition table. Use *SaveAs* and rename the drawing **OFFICE**.

    D. Use *Insert* to insert the furniture into the drawing, as shown in Figure 21-20. You may use your own arrangement for the furniture, but *Insert* the same number of each piece as shown. *Save* the drawing.

Figure 21-18

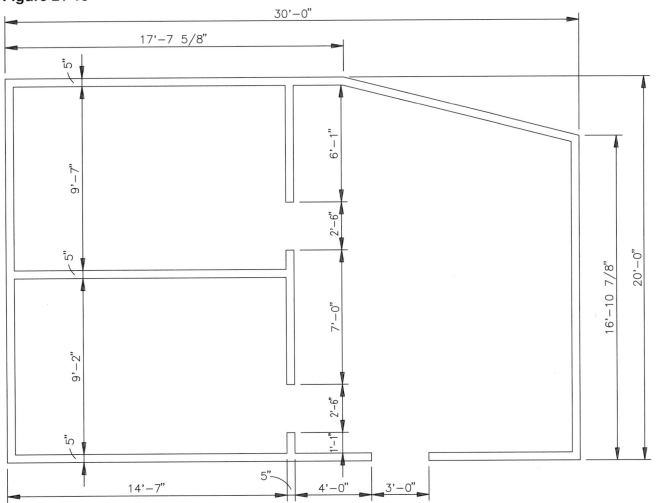

Figure 21-19

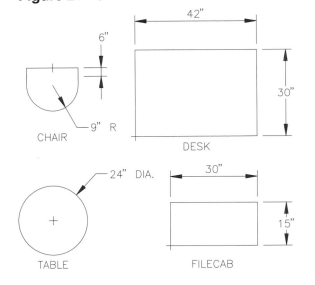

Figure 21-20

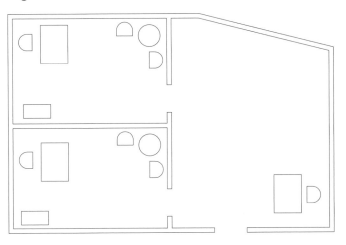

2. **Creating a .DWG file for** *Insertion, Base*

Figure 21-21

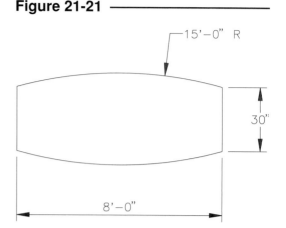

Begin a *New* drawing. Assign the name **CONFTABL**.
Create the table as shown in Figure 21-21 on *Layer* **0**.
Since this drawing is intended for insertion into the
OFFICE drawing, use the *Base* command to assign an
insertion base point at the lower-left corner of the table.

When you are finished, *Save* the drawing.

3. *Insert, Explode, Divide*

Figure 21-22

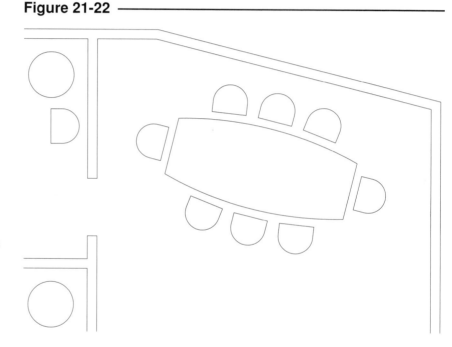

A. *Open* the **OFFICE**
drawing. Ensure that
layer **FURNITURE** is
current. Use *Insert* to
bring the **CONFTABL**
drawing in as a *Block* in
the placement shown in
Figure 21-22.

Notice that the CONF-
TABL assumes the line-
type and color of the
current layer, since it was
created on layer **0**.

B. *Explode* the CONFTABL.
The *Exploded*
CONFTABL
returns to *Layer* **0**, so
use *Chprop* to change it
back to *Layer* **FURNITURE**. Then use the *Divide* command (with the *Block* option) to insert
the **CHAIR** block as shown in Figure 21-22. Also *Insert* a **CHAIR** at each end of the table.
Save the drawing.

Figure 21-23

4. *Wblock, BYBLOCK setting*

A. *Open* the **EFF-APT2** drawing you worked on in Chapter 18
Exercises. Use the *Wblock* command to transform the plant
into a .DWG file (Fig. 21-23). Use the name **PLANT** and
specify the *Insertion base point* at the center. Do not save the
EFF-APT2 drawing.

B. *Open* the **OFFICE** drawing and *Insert* the **PLANT** into one of
the three rooms. The plant probably appears in a different
color than the current layer. Why? Check the *Layer* listing to
see if any new layers came in with the PLANT block. *Erase*
the PLANT block.

PLANT

C. *Open* the **PLANT** drawing. Change the *Color* and *Linetype* setting of the plant objects to *BYBLOCK*. *Save* the drawing.

D. *Open* the **OFFICE** drawing again and *Insert* the **PLANT** onto the **FURNITURE** layer. It should appear now in the current layer's *color* and *linetype*. *Insert* a **PLANT** into each of the 3 rooms. *Save* the drawing.

5. **Redefining a** *Block*

Figure 21-24 ────────────────

After a successful meeting, the client accepts the proposed office design with one small change. The client requests a slightly larger chair than that specified. *Explode* one of the **CHAIR** blocks. Use the *Scale* command to increase the size <u>slightly</u> or otherwise redesign the chair in some way. Use the *Block* command to redefine the **CHAIR** block. All previous insertions of the CHAIR should reflect the design change. *Save* the drawing. Your design should look similar to that shown in Figure 21-24. *Plot* to a standard scale based on your plotting capabilities.

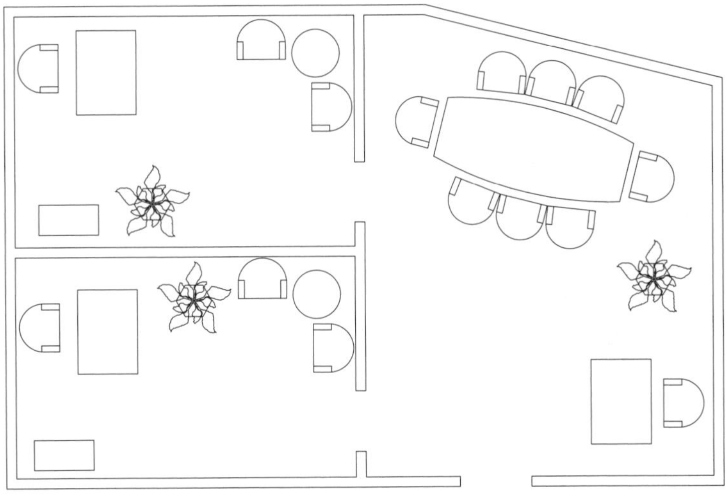

6. *Rename*

Open the **OFFICE** drawing. Enter *Ddrename* (or select from the *Format* pull-down menu) to produce the *Rename* dialog box. Rename the *Blocks* as follows:

New Name	Old Name
DSK	DESK
CHR	CHAIR
TBL	TABLE
FLC	FILECAB

Next, use the *Rename* dialog box again to rename the *Layers* as indicated below:

New Name	Old Name
FLOORPLN-DIM	DIM-FLOOR
ELEC-DIM	DIM-ELEC

Use the *Rename* dialog box again to change the *Text Style* name as follows:

New Name	Old Name
ARCH-FONT	CITYBLUEPRINT

Use *SaveAs* to save and rename the drawing to **OFFICE-REN**.

7. *Purge*

Using the Windows Explorer, check the file size of **OFFICE-REN.DWG.** Now use *Purge*. Answer *Yes* to *Purge* any unreferenced named objects. *Exit AutoCAD* and *Save Changes*. Check the file size again. Is the file size slightly smaller?

8. Create the process flow diagram shown in Figure 21-25. Create symbols (*Blocks*) for each of the valves and gates. Use the names indicated (for the *Blocks*) and include the text in your drawing. *Save* the drawing as **PFD**.

Figure 21-25

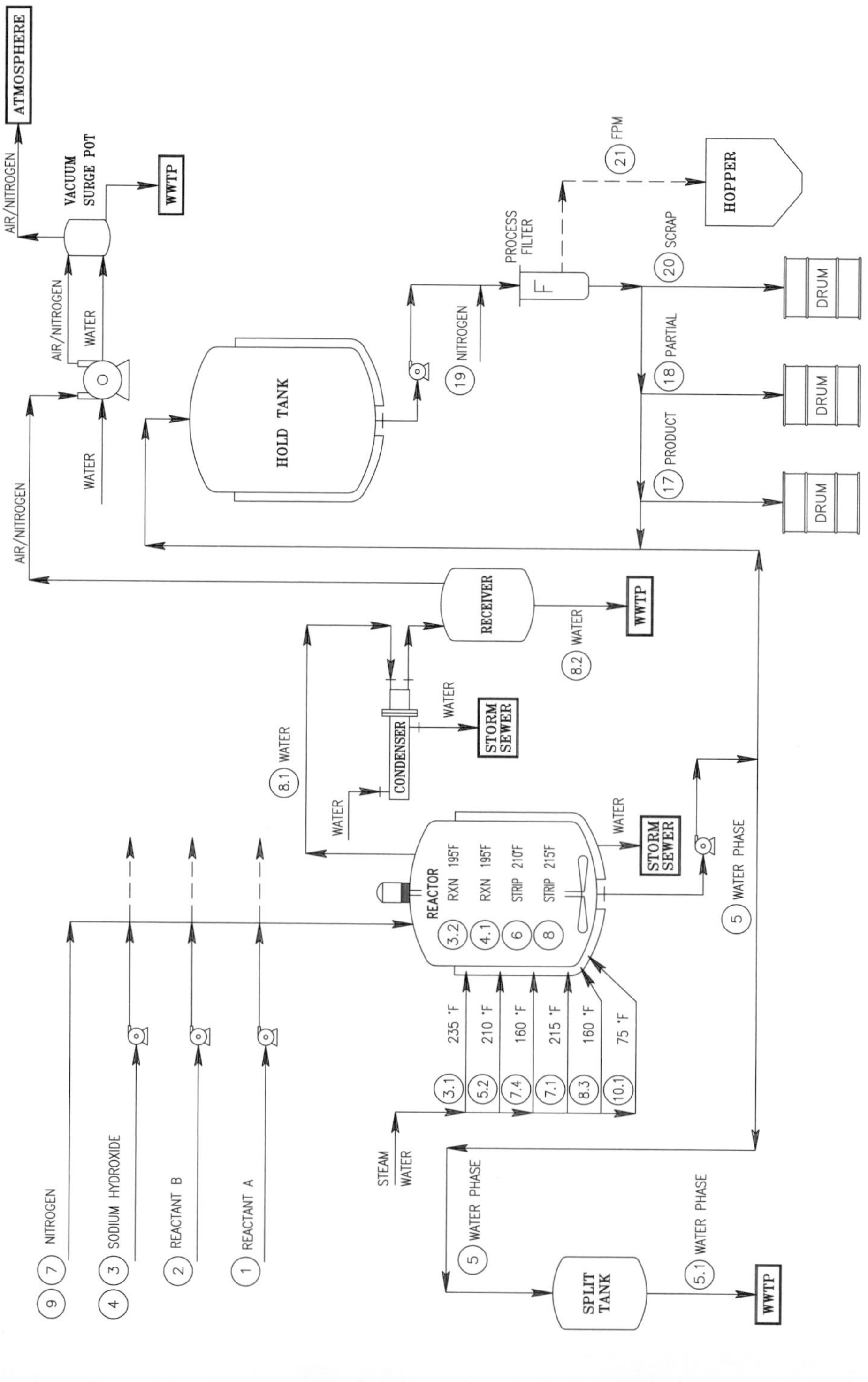

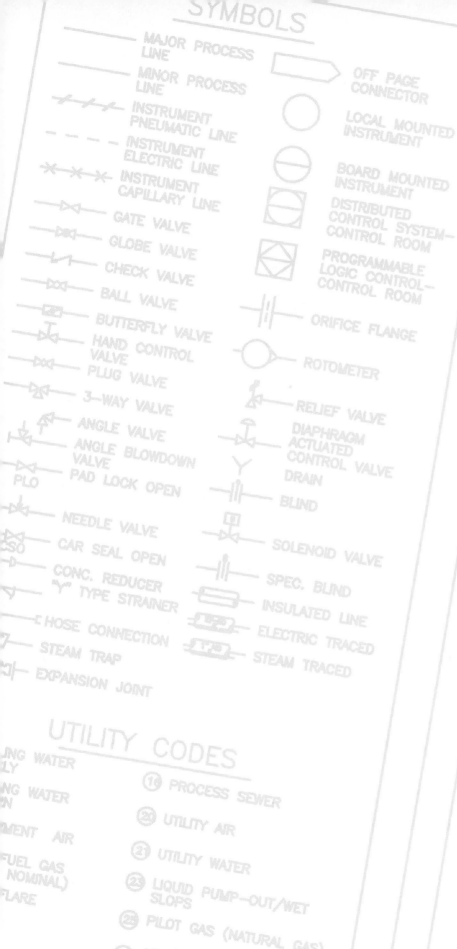

22

BLOCK ATTRIBUTES

Chapter Objectives

After completing this chapter you should:

1. be able to define block attributes using the *Attdef* command and the *Attribute Definition* dialog box;

2. be able to specify *Invisible, Constant, Verify,* and *Preset* attribute modes;

3. be able to control the display of attributes in a drawing with *Attdisp;*

4. know how to edit existing attributes with the *Edit Attributes* dialog box;

5. be able to globally or individually edit attributes using the *Attedit* command;

6. be able to redefine attributes for existing blocks using *Attredef;*

7. be able to create an extract file containing a drawing's attributes in an external text file;

8. know how to create a template file in a text editor for specifying the structure of an extract file.

CONCEPTS

A block <u>attribute</u> is a line of text (numbers or letters) associated with a block. An attribute can be thought of as a label or description for a block. A block can have multiple attributes. The attributes are included with the drawing objects when the block is defined (using the *Block* command). When the block is *Inserted*, its attributes are also *Inserted*.

Since *Blocks* are typically used as symbols in a drawing, attributes are text strings that label or describe each symbol. For example, you can have a series of symbols such as transistors, resistors, and capacitors prepared for creating electrical schematics. The associated attributes can give the related description of each block, such as ohms or wattage values, model number, part number, and cost. If your symbols are doors, windows, and fixtures for architectural applications, attached attributes can include size, cost, and manufacturer. A mechanical engineer can have a series of blocks representing fasteners with attached attributes to specify fastener type, major diameter, pitch, length, etc. Attributes could even be used to automate the process of entering text into a title block, assuming the title block was *Inserted* as a block or separate .DWG file.

Attributes can add another level of significance to an AutoCAD drawing. Not only can attributes automate the process of placing the text attached to a block, but the inserted attribute information can be <u>extracted</u> from a drawing to form a bill of materials or used for cost analysis, for example. Extracted text can be imported to a database or spreadsheet program for further processing and analysis.

Attributes are created with the *Attdef* (attribute define) command. *Attdef* operates similarly to *Text* or *Dtext*, prompting you for text height, placement, and justification. During attribute definition, parameters can be adjusted that determine how the attributes will appear when they are inserted.

This chapter discusses defining attributes, inserting attributed blocks, displaying and editing attributes in a drawing, and extracting attributes from drawings.

CREATING ATTRIBUTES

Steps for Creating and Inserting Block Attributes

1. Create the objects that will comprise the block. Do not use the *Block* command yet; only draw the geometry.

2. Use the *Attdef* or *Ddattdef* command to create and place the desired text strings associated with the geometry comprising the proposed block.

3. Use the *Block* command to convert the drawing geometry (objects) and the attributes (text) into a named block. When prompted to "Select objects:" for the block, select both the drawing objects and the text (attributes). The block and attributes disappear but are defined in the drawing's block definition table.

4. Use *Insert* to insert the attributed block into the drawing. Edit the text attributes as necessary (if parameters were set as such) during the insertion process.

ATTDEF and
DDATTDEF

Pull-down Menu	COMMAND (TYPE)	ALIAS (TYPE)	Short-cut	Screen (side) Menu	Tablet Menu
Draw *Block >* *Define Attributes...*	ATTDEF or DDATTDEF	AT or -AT	...	DRAW 2 Ddattdef	...

Attdef (attribute define) allows you to define attributes for future combination with a block. If you intend to associate the text attributes with drawing objects (*Line, Arc, Circle,* etc.) for the block, it is usually preferred to draw the objects before using *Attdef*. In this way, the text can be located in reference to the drawing objects.

Attdef allows you to create the text and provides justification, height, and rotation options similar to the *Text* and *Dtext* commands. You can also define parameters, called Attribute Modes, specifying how the text will be inserted—*Invisible, Constant, Verify,* or *Preset*.

The *Attdef* command differs, depending on the method used to invoke it. If you select from any menus or the icon or type *Ddattdef*, the *Attribute Definition* dialog box appears (Figure 22-1). If you type *Attdef*, the command line format is used. The command line format is presented here first:

Figure 22-1

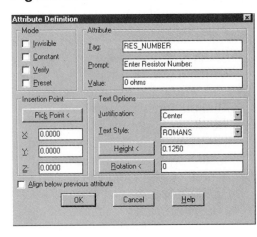

```
Command: attdef
Attribute modes—Invisible:N Constant:N Verify:N Preset:N
Enter (ICVP) to change, RETURN when done: Enter or (letter)
Attribute tag: Enter tag, no spaces.
Attribute prompt: Enter desired prompt, spaces allowed.
Default attribute value: Enter desired default text.
Justify/Style/<Start point>: PICK or (option)
Height <0.2000>: Enter, PICK or (value)
Rotation angle <0>: Enter, PICK or (value)
Command:
```

The first option, "attribute modes," is described later. For this example, the default attribute modes have been accepted.

The *Attribute Definition* dialog box has identical options that are found in the command line format, including justification options. The same entries have been made here to define the first attribute example shown in Figure 22-2.

Figure 22-2

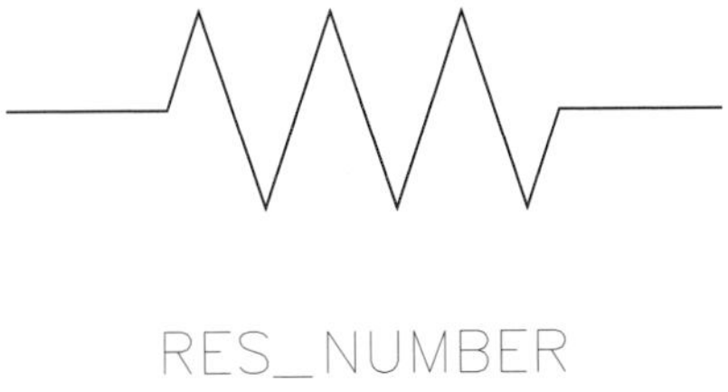

RES_NUMBER

Attribute Tag

This is the descriptor for the <u>type of text</u> to be entered, such as MODEL_NO, PART_NO, or NAME. Spaces cannot be used in the Attribute Tag.

Attribute Prompt

The prompt is what words (<u>prompt) you want to appear</u> when the block is *Inserted* and when the actual text (values) must be entered. Spaces and punctuation can be included.

Default Attribute Value
This is the <u>default text</u> that appears with the block when it is *Inserted*. Supply the typical or expected words or numbers.

Justify
Any of the typical text justification methods can be selected using this option. (See Chapter 18, Inserting and Editing Text, if you need help with justification.)

Style
This option allows you to select from existing text *Styles* in the drawing. Otherwise, the attribute is drawn in the current *Style*.

As an example, assume that you are using the *Attdef* or *Ddattdef* command to make attributes attached to a resistor symbol to be *Blocked* (Fig. 22-2). The scenario may appear as follows or as presented in the *Attribute Definition* dialog box in Figure 22-1:

```
Command: attdef
Attribute modes—Invisible:N Constant:N Verify:N Preset:N
Enter (ICVP) to change, RETURN when done: Enter
Attribute tag: RES_NUMBER
Attribute prompt: Enter Resistor Number:
Default attribute value: 0
Justify/Style/<Start point>: j
Align/Fit/Center/Middle/Right/TL/TC/TR/ML/MC/MR/BL/BC/BR: c
Center point: PICK
Height <0.2000>: .125
Rotation angle <0>: Enter
```

The attribute would then appear at the selected location, as shown in Figure 22-2. Keep in mind that the *Lines* comprising the resistor were created before creating attributes. Also remember that the resistor and the attribute are not yet *Blocked*. The *Attribute Tag* <u>only</u> is displayed until the *Block* command is used.

Two other attributes could be created and positioned beneath the first one by pressing **Enter** when prompted for the "<Start point>:" (Command line format) or selecting the *"Align below previous attribute"* checkbox (dialog box format). The command syntax may read like this:

```
Command: attdef
Attribute modes—Invisible:N Constant:N Verify:N Preset:N
Enter (ICVP) to change, RETURN when done: Enter
Attribute tag: RESISTANCE
Attribute prompt: Enter Resistance:
Default attribute value: 0 ohms
Justify/Style/<Start point>: Enter
Command:

Command: attdef
Attribute modes—Invisible:N Constant:N Verify:N Preset:N
Enter (ICVP) to change, RETURN when done: Enter
Attribute tag: PART_NO
Attribute prompt: Enter Part Number:
Default attribute value: 0-0000
Justify/Style/<Start point>: Enter
Command:
```

The resulting unblocked symbol and three attributes appear as shown in Figure 22-3. Only the *Attribute Tags* are displayed at this point. The *Block* command has not yet been used.

Figure 22-3

RES_NUMBER
RESISTANCE
PART_NO

The *Attributes Modes* define the appearance or the action required when you *Insert* the attributes. The options are as follows.

Invisible

An *Invisible* attribute is not displayed in the drawing after insertion. This option can be used to prevent unnecessary information from cluttering the drawing and slowing regeneration time. The *Attdisp* command can be used later to make these attributes visible.

Constant

This option gives the attribute a fixed value for all insertions of the block. In other words, the attribute always has the same text and cannot be changed. You are not prompted for the value upon insertion.

Verify

When attributes are entered in command line format, this option forces you to verify that the attribute is correct when the block is inserted. This option has no effect when attributes are entered in dialog box format (when *ATTDIA*=1).

Preset

This option prevents you from having to enter a value for the attribute during insertion if you are using the command line format. In this case, you are not prompted for the attribute value during insertion. Instead, the default value is used for the block, although it can be changed later with the *Attedit* command.

The *Attribute Modes* are toggles (Yes or No). In command line format, the options are toggled by entering the appropriate letter, which reverses its current position (Y or N). For example, to create an *Invisible* attribute, type **"I"** at the prompt:

    Command: **attdef**
    Attribute modes—Invisible:N Constant:N Verify:N Preset:N
    Enter (ICVP) to change, RETURN when done: **I**
    Attribute modes—Invisible:Y Constant:N Verify:N Preset:N
    Enter (ICVP) to change, RETURN when done: **Enter**

Note that the second prompt (fourth line) reflects the new position of the previous request.

Alternately, attribute modes could be defined using the *Attribute Definition* dialog box (Fig. 22-4). The *Mode* cluster of checkboxes (upper-left corner) makes setting attribute modes simpler than in command line format.

Figure 22-4

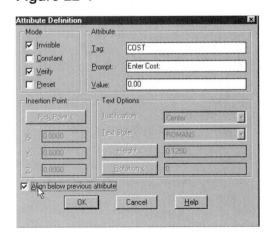

Using our previous example, two additional attributes can be created with specific *Invisible* and *Verify* attribute modes. Either the dialog box or command line format can be used. The following command syntax shows the creation of an *Invisible* attribute:

    Command: **attdef**
    Attribute modes—Invisible:N Constant:N
    Verify:N Preset:N
    Enter (ICVP) to change, RETURN when done: **I**

Attribute modes—Invisible:Y Constant:N Verify:N Preset:N
Enter (ICVP) to change, RETURN when done: **Enter**
Attribute tag: **MANUFACTURER**
Attribute prompt: **Enter Manufacturer:**
Default attribute value: **Enter**
Justify/Style/<Start point>: **Enter**
Height <0.1250>: **Enter**
Rotation angle <0>: **Enter**
Command:

The first attribute has been defined. Next, a second attribute is defined having *Invisible* and *Verify* parameters. (Figure 22-4 also shows the correct entries for creation of this attribute.)

Command: **attdef**
Attribute modes—Invisible:Y Constant:N Verify:N Preset:N
Enter (ICVP) to change, RETURN when done: **v**
Attribute modes—Invisible:Y Constant:N Verify:Y Preset:N
Enter (ICVP) to change, RETURN when done: **Enter**
Attribute tag: **COST**
Attribute prompt: **Enter Cost:**
Default attribute value: **0.00**
Justify/Style/<Start point>: **Enter**
Command:

Tags for *Invisible* attributes appear as any other attribute tags. The *Values*, however, are invisible when the block is inserted.

Using the *Block* Command to Create an Attributed Block

Next, the *Block* command is used to transform the drawing objects and the text (attributes) into an attributed block. When you are prompted to "Select objects:" in the *Block* command, select the attributes in the order you desire their prompts to appear on insertion. In other words, the first attribute selected during the *Block* command (RES_NUMBER) is the first attribute prompted for editing during the *Insert* command. PICK the attributes one at a time in order. Then PICK the drawing objects (a crossing window can be used). Figure 22-5 shows this sequence.

Figure 22-5

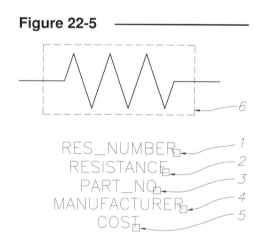

If the attributes are selected with a window, they are inserted in reverse order of creation, so you would be prompted for COST first, MANUFACTURER second, and so on:

Command: **block**
Block name (or ?): **res**
Insertion base point: **PICK**
Select objects: **PICK** (1)
Select objects: **PICK** (2)
Select objects: **PICK** (3)
Select objects: **PICK** (4)
Select objects: **PICK** (5)

Select objects: **PICK** (objects)
Select objects: **Enter**
Command:

As with other blocks, this one disappears after using the *Block* command. Use the *Oops* command to redisplay the original objects if you expect to make further changes. The RES block definition remains in the block definition table.

Inserting Attributed Blocks

Use the *Insert* command to bring attributed blocks into your drawing as you would to bring any block into your drawing. After the normal *Insert* command prompts, the attribute prompts (that you defined with *Attdef* or *Ddatte*) appear. In this step, enter the desired attribute *values* for each block insertion. The prompts appear in dialog box fashion (*Enter Attributes* dialog, Fig. 22-7) or in command line format, depending on the setting of the *ATTDIA* variable. (The command line format is the default setting, *ATTDIA*=0. See Appendix A, System Variables.)

The command line format follows:

Command: *insert*
Block name (or ?): **res**
Insertion point: **PICK**
X scale factor <1> / Corner / XYZ: **Enter**
Y scale factor (default=X): **Enter**
Rotation angle <0>: **Enter**
Enter attribute values
Enter Resistor Number: <0>: **R1**
Enter Resistance: <0 ohms>: **4.7K**
Enter Part Number: <0-0000>: **R-4746**
Enter Manufacturer: **Electro Supply Co.**
Enter Cost <0.00>: **.37**
Verify Attribute values
Enter Cost <.37> : **Enter**
Command:

Figure 22-6

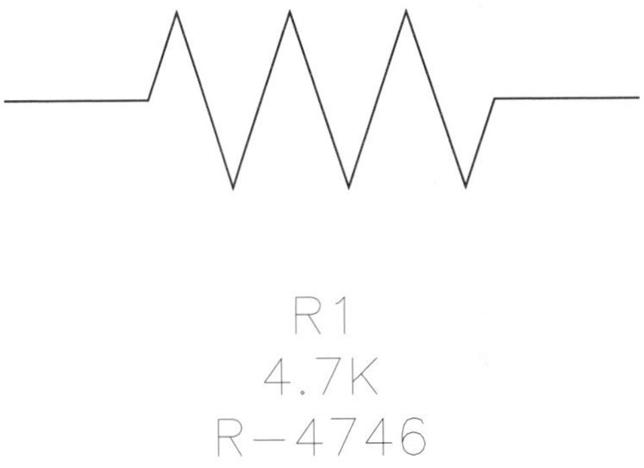

R1
4.7K
R—4746

Notice the prompts appearing on insertion are those that you specified as the "Attribute prompt:" with the *Attdef* command or *Ddattdef* dialog box. The order of the prompts matches the order of selection during the *Block* command. Also note the repeated prompt (*Verify*) for the COST attribute.

Entering the attribute values as indicated above yields the block insertion shown in Figure 22-6. Notice the absence of the last two attribute values (since they are *Invisible*).

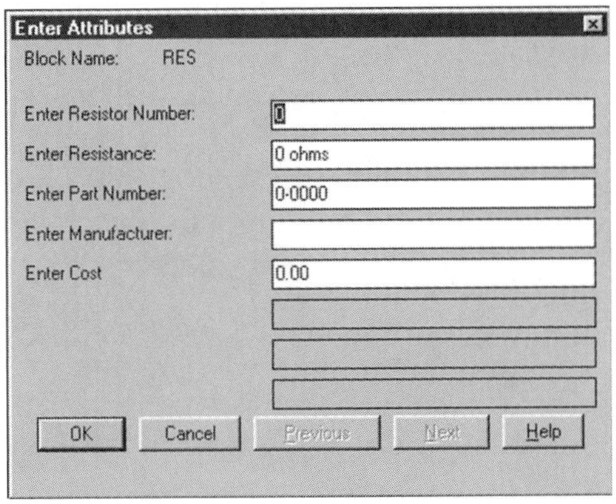

The *ATTDIA* Variable

If the *ATTDIA* system variable is set to a value of 1, the *Enter Attributes* dialog box (Fig. 22-7) is invoked automatically by the *Insert* command for entering attribute values. The dialog box, in that case, would have appeared rather than the command line format for entering attributes. The *Verify* and *Preset* Attribute Mode options have no effect on the dialog box format of attribute value entry. The dialog box can hold multiple screens of 10 attributes in each, all connected to 1 block.

Figure 22-7

The *ATTREQ* Variable

When *Inserting* blocks with attributes, you can force the attribute requests to be suppressed. In other words, you can disable the prompts asking for attribute values when the blocks are *Inserted*. To do this, use the *ATTREQ* (attribute request) system variable. A setting of **1** (the default) turns attribute requesting on, and a value of **0** disables the attribute value prompts.

The attribute prompts can be disabled when you want to *Insert* several blocks but do not want to enter the attribute values right away. The attribute values for each block can be entered at a later time using the *Attedit* or *Ddatte* commands. These attribute editing commands are discussed on the following pages.

DISPLAYING AND EDITING ATTRIBUTES

ATTDISP

Pull-down Menu	COMMAND (TYPE)	ALIAS (TYPE)	Short-cut	Screen (side) Menu	Tablet Menu
View *Display >* *Attribute Display*	*ATTDISP*	...	...	VIEW 2 *Attdisp*	L,1

The *Attdisp* (attribute display) command allows you to control the visibility state of all inserted attributes contained in the drawing:

    Command: **attdisp**
    Normal/ON/OFF <current value>: (option)
    Command:

Attdisp has three positions. Changing the state forces a regeneration

Figure 22-8 ────────────────────────

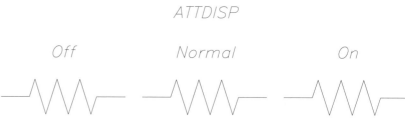

On
All attributes (normal and *Invisible*) in the drawing are displayed.

Off
No attributes in the drawing are displayed.

Normal
Normal attributes are displayed; *Invisible* attributes are not displayed.

Figure 22-8 illustrates the RES block example with each of the three *Attdisp* options. The last two attributes were defined with *Invisible* modes, but are displayed with the *On* state of *Attdisp*. (Since *Attdisp* affects attributes globally, it is not possible to display the three options at one time in a drawing.)

When attributes are defined using the *Invisible* mode of the *Attdef* command, the attribute values are normally not displayed with the block after *Insertion*. However, the *Attdisp On* option allows you to override the *Invisible* mode. Turning all attributes *Off* with *Attdisp* can be useful for many applications when it is desirable to display or plot only the drawing geometry. The *Attdisp* command changes the *ATTMODE* system variable.

DDATTE

Pull-down Menu	COMMAND (TYPE)	ALIAS (TYPE)	Short-cut	Screen (side) Menu	Tablet Menu
Modify Object > Attribute > Single...	*DDATTE*	*ATE*	...	*MODIFY1 Ddatte*	*Y,20*

The *Ddatte* (Dialog Attribute Edit) command invokes the *Edit Attributes* dialog box (Fig. 22-9). This dialog box allows you to edit attributes of <u>existing</u> blocks in the drawing. The configuration and operation of the box are identical to the *Enter Attributes* dialog box (Fig. 22-7).

When the *Ddatte* command is entered, AutoCAD requests that you select a block. Any existing attributed block can be selected for attribute editing.

> Command: **ddatte**
> Select block: **PICK**

Figure 22-9 ⎯⎯⎯⎯⎯⎯⎯

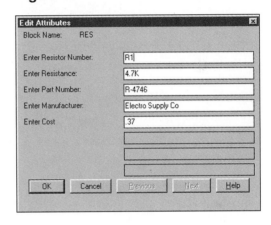

At this point, the dialog box appears. After the desired changes have been made, selecting the *OK* tile updates the selected block with the indicated changes.

MULTIPLE DDATTE

Using *Ddatte* with the *Multiple* modifier causes AutoCAD to repeat the *Ddatte* command requesting multiple block selections over and over until you cancel the command with Escape. You must type *Ddatte* (or any other command) for it to operate with the *Multiple* modifier. (See Chapter 44 for other uses of the Multiple modifier.)

> Command: **MULTIPLE**
> Multiple command: **DDATTE**
> Select block: **PICK**

This form of the command can be used to query blocks in an existing drawing. For example, instead of using the *Attdisp* command to display the *Invisible* attributes, *Multiple Ddatte* can be used on selected blocks to display the *Invisible* attributes in dialog box form. Another example is the use of *Attdisp* to turn visibility of <u>all</u> attributes *Off* in a drawing, thus simplifying the drawing appearance and speeding regenerations. The *Multiple Ddatte* command could then be used to query the blocks to display the attributes in dialog box form.

ATTEDIT

	Pull-down Menu	COMMAND (TYPE)	ALIAS (TYPE)	Short-cut	Screen (side) Menu	Tablet Menu
	Modify Object > Attribute > Global	*ATTEDIT*	*-ATE*	...	*MODIFY1 Attedit*	...

Attedit (Attribute Edit) is similar to *Ddatte* in that you can edit selected existing attributes. *Attedit*, however, is more powerful in that it allows many options for editing attributes, including editing any properties of selected attributes or global editing of values for all attributes in a drawing:

Command: **attedit**
Edit attributes one at a time? <Y>

The following prompts depend on your response to the first prompt:

N (No)

Indicates that you want global editing. (In other words, all attributes are selected.) Attributes can then be further selected by block name, tag, or value. <u>Only attribute values can be edited</u> using this global mode.

Y (Yes)

Allows you to PICK each attribute you want to edit. You can further filter (restrict) the selected set by block name, tag, and value. You can then <u>edit any property or placement</u> of the attribute.

Next, the following prompts appear, which allow you to filter, or further restrict, the set of attributes for editing. You can specify the selection set to include only specific block names, tags, or values:

Block name specification <*>: Enter desired block name(s)
Attribute tag specification <*>: Enter desired tag(s)
Attribute value specification <*>: Enter desired value(s)

During this portion of the selection set specification, sets of names, tags, or values can be entered only to filter the set you choose. Commas and wildcard characters can be used (see Chapter 44 for information on using wildcards). Pressing Enter accepts the default option of all (*) selected.

Attribute values are <u>case sensitive</u> (upper- and lowercase letters must be specified exactly). If you have entered null values (for example, if *ATTREQ* was set to 0 during block insertion), you can select all null values to be included in the selection set by using the backslash symbol (\).

The subsequent prompts depend on your previous choice of global or individual editing.

No (Global Editing—Editing Values Only)

Global edit of attribute values.
Edit only attributes visible on screen? <Y> **Enter** or (N)

An *N* response indicates all attributes in the selection set to this point can be edited. If your choice is *No*, you are reminded that the drawing will be regenerated after the command, and the next prompt is skipped. If you select *Y*, this prompt appears:

Select Attributes: **PICK**

Beware! The attributes <u>do not highlight as you select</u> them, but highlight when you press Enter to complete the selection. Finally, this prompt appears:

String to change:
New string:

At this point, you can enter any string, and AutoCAD searches for it in the selected attributes and replaces all occurrences of the string with the new string. No change is made if the string is not found. The string can be any number of characters and can be embedded within an attribute text. Remember that only the attribute values can be edited with the global mode.

Yes (Individual Editing—Editing Any Property)

After filtering for the block name, attribute tag, and value, AutoCAD prompts:

> Select Attributes: **PICK** (The attributes <u>do not highlight</u> during selection.)
> Value/Position/Height/Angle/Style/Layer/Color/Next <N>:

A marker appears at one of the attributes. Use the *Next* option to move the marker to the attribute you wish to edit. You can then select any other option. After editing the marked attribute, the change is displayed immediately and you can position the marker at another attribute to edit.

The *Value* option invokes this prompt, allowing you to change a specific string:

> Change or Replace? <R>: *c*
> String to change:
> New string:

The *Replace* option responds to this prompt, providing for an entire new value:

> New Attribute value:

You can also change the attribute's *Position, Height, Angle, Style, Layer,* or *Color* by selecting the appropriate option. The prompts are specific and straightforward for these options. There is a surprising amount of flexibility with these options.

Global and Individual Editing Example

Assume that the RES block and a CAP block were inserted several times to create the partial schematic shown in Figure 22-10. The *Attdisp* command is set to *On* to display all the attributes, including the last two for each block, which are normally *Invisible*.

Figure 22-10 ———————————————————————————————————————

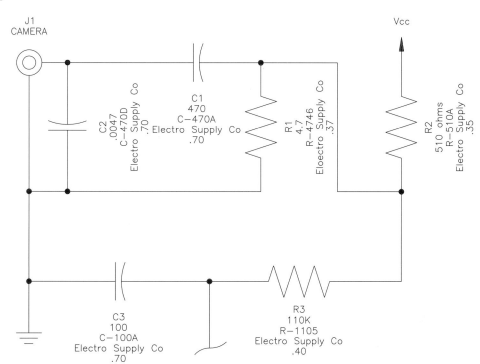

Attedit is used to edit the attributes of RES and CAP. If you wish to change the <u>value</u> of one or several attributes, the global editing mode of *Attedit* would generally be used, since <u>values only</u> can be edited in global mode. For this example, assuming that the supplier changed, the MANUFACTURER attribute for all blocks in the drawing is edited. The command syntax is as follows:

    Command: **attedit**
    Edit attributes one at a time? <Y> **N**
    Global edit of attribute values.
    Edit only attributes visible on screen? <Y> **Enter**
    Block name specification <*>: **Enter**
    Attribute tag specification <*>: **MANUFACTURER**
    Attribute value specification <*>: **Enter**
    Select Attributes: **PICK** (Select specifically the MANUFACTURER attributes.)
    6 attributes selected.
    String to change: **Electro Supply Co**
    New string: **Sparks R Us**
    Command:

All instances of the "Electro Supply Co" redisplay with the new string. Note that resistor R1 did not change because the string did not match (due to misspelling). *Ddatte* can be used to edit the value of that attribute individually.

Figure 22-11

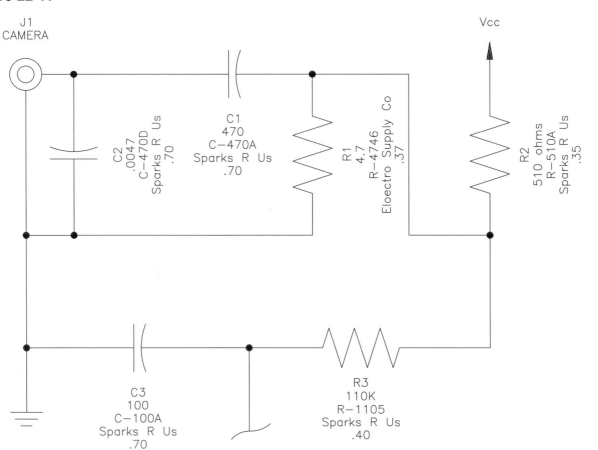

The individual editing mode of *Attedit* can be used to change the <u>height</u> of text for the RES_NUMBER and CAP_NUMBER attribute for both the RES and CAP blocks. The command syntax is as follows:

```
Command: attedit
Edit attributes one at a time? <Y> Enter
Block name specification <*>: Enter
Attribute tag specification <*>: Enter
Attribute value specification <*>: Enter
Select Attributes:  PICK
6 attributes selected.
Value/Position/Height/Angle/Style/Layer/Color/Next <N>: H
New height <0.1250>: .2
Value/Position/Height/Angle/Style/Layer/Color/Next <N>: Enter
Value/Position/Height/Angle/Style/Layer/Color/Next <N>: H
New height <0.1250>: .2
        etc.
```

The result of the new attribute text height for CAP and RES blocks is shown in Figure 22-12. *Attdisp* has been changed to *Normal* (*Invisible* attributes do not display).

Figure 22-12 ———

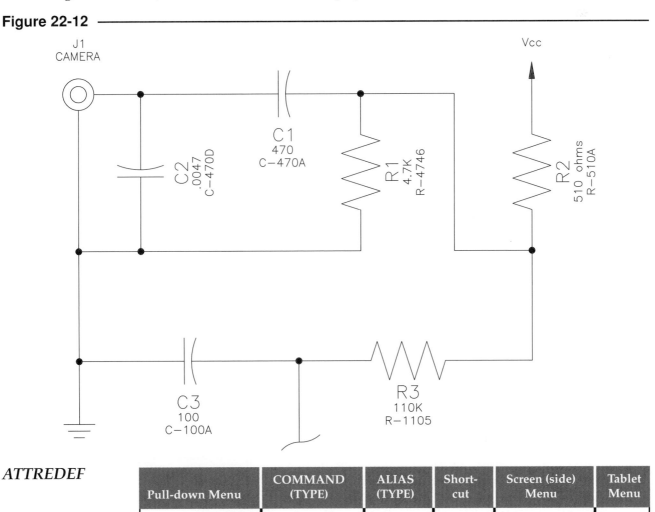

ATTREDEF

Pull-down Menu	COMMAND (TYPE)	ALIAS (TYPE)	Short-cut	Screen (side) Menu	Tablet Menu
...	*ATTREDEF*	...	...	...	...

The *Attredef* command (Attribute redefine) enables you to redefine an attributed block. *Attredef* operates similarly to the method of redefining a block discussed previously in Chapter 21; that is, the old block definition (only with attributes in this case) is replaced with a newer one. To use the *Attredef* feature, the new objects (to be *Blocked*) with attributes defined must be in place <u>before</u> invoking the command.

Creating the new objects and attributes is easily accomplished by *Inserting* the old block, *Exploding* it, then making the desired changes to the attributes and objects. Figure 22-13 illustrates a new RES block definition (in the lower-right corner) before using *Attredef*.

Figure 22-13 ————————————

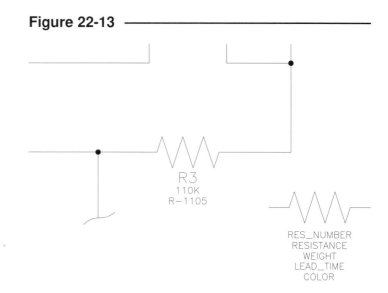

Once the new geometry and attributes have been created, invoke *Attredef*. The command operates only in command line format as follows:

    Command: **attredef**
    Name of Block you wish to redefine: **RES**
    Select objects for new Block...
    Select objects: **PICK** attributes in order
    Select objects: **PICK** attributes in order
        (etc.)
    Select objects: **PICK** objects (geometry)
    Select objects: **Enter**
    Insertion base point of new Block: **PICK**
    Verify attribute values
    Verify attribute values
        (etc.)
    Command:

The block definition is updated, and all instances of the block automatically and globally change to display the new changes. All attributes that were changed reflect the default values. *Ddatte* can be used later to make the desired corrections to the individual attributes.

Figure 22-14 displays one of the RES blocks after redefining. Notice that the forth attribute is invisible as it was before it was *Exploded*. For this attribute, the mode was not changed. The new last attribute, however, was changed to a normal attribute mode and therefore appears after using *Attredef*

Figure 22-14 ————————————

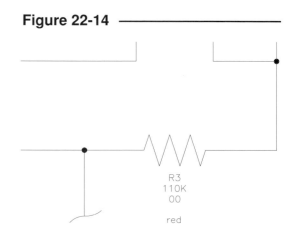

Other Attribute Editing Methods

Keep in mind that if you want to specify new attributes after *Exploding* the old block, there are three alternative methods: you can use *Ddedit*, *Ddmodify*, or *Attdef*. *Ddedit*, the dialog box for editing text, recognizes the text as attributes but allows you to change only the attribute *Tags*. Figure 22-15 illustrates using the *Ddedit* dialog box for changing the RES_NUMBER tag.

Figure 22-15 ————————————

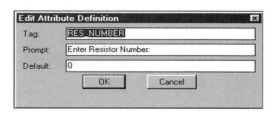

Optionally, the *Ddmodify* dialog box can be invoked. Not only does *Ddmodify* recognize the text as attributes, but the *Tag, Prompt,* default *Value,* and attribute modes can be modified. Figure 22-16 displays this dialog box for the same operation. Notice all of the possible changes that can be made.

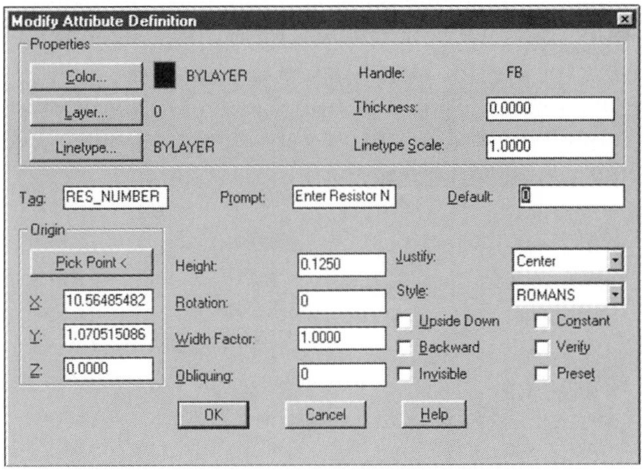

Instead of using *Ddedit* or *Ddmodify* to change <u>existing</u> attributes, new attributes can be created "from scratch" with *Attdef.* Remember that if you want to change the attributes when redefining a block, one of these three operations must take place before using *Attredef.*

Figure 22-16 ————————————

Grip Editing Attribute Location

Grips can be used to "move" the location of the attributes after an attributed *Block* has been inserted. To do this (with Grips on), PICK the *Block* so grips appear at each attribute and at the block insertion point. Next, PICK the attribute you want to move so its grip becomes "hot." The attribute can be moved to any location and still retain its association with the block. (See Chapter 23, Fig. 23-15.)

EXTRACTING ATTRIBUTES

The *Attext* (attribute extract) command and related *Ddattext* (attribute extract dialog box) allow you to extract a list of the existing blocks and text attributes from the current drawing. In other words, a list of all, or a subset, of the blocks and attributes that are in a drawing can be written to a separate file (extracted) in text form. The resulting text file can be used as a report indicating a variety of information, such as the name, number, location, layer, and scale factor of the blocks, as well as the block attributes. If desired, the extract file can then be imported to a database or spreadsheet package for further analysis or calculations.

Steps for Creating Extract Files

1. In AutoCAD, *Insert* all desired blocks and related attributes into the drawing. *Save* the drawing.

2. With a word processor or text editor, create a <u>template</u> file. The template file specifies what information should be included in the <u>extract</u> file and how the extract file should be structured. (A template file is not required if you specify a *DXF* format for the extract file.)

3. In AutoCAD, use the *Attext* or *Ddattext* command to specify the name of the template file to be used and to create the extract file.

4. Examine the extract file in the text editor or word processor or import the extract file into a database or spreadsheet package.

Creating the Template File (Step 2)

The template file is created in a text editor (such as MS-DOS Editor, Norton Editor®, or Windows Wordpad®) or word processor (Microsoft Word® or Corel WordPerfect®) in non-document mode. The template file must be in straight ASCII form (no internal word processing codes). The template file <u>must</u> be given a .TXT file extension.

Each <u>line</u> of the template file specifies a <u>column</u> of information, or field, to be written in the extract file, including the name of the column, its width (number of characters or numerals), and the numerical precision. Each block that matches a line in the template file creates an column in the extract file.

The possible fields that AutoCAD allows you to specify and the format that you must use for each are shown below. Use any of the fields you want when creating the template file. The first two columns below are included in the template file (<u>not the comments in the third column</u>):

BL:LEVEL	Nwww000	(*Block* nesting level)
BL:NAME	Cwww000	(*Block* name)
BL:X	Nwwwddd	(X coordinate of *Block* insertion point)
BL:Y	Nwwwddd	(Y coordinate of *Block* insertion point)
BL:Z	Nwwwddd	(Z coordinate of *Block* insertion point)
BL:NUMBER	Nwwwddd	(*Block* counter; same for all members of *MINSERT*)
BL:HANDLE	Cwww000	(*Block* handle; same for all members of *MINSERT*)
BL:LAYER	Cwww000	(*Block* insertion layer name)
BL:ORIENT	Nwwwddd	(*Block* rotation angle)
BL:XSCALE	Nwwwddd	(X scale factor of *Block*)
BL:YSCALE	Nwwwddd	(Y scale factor of *Block*)
BL:ZSCALE	Nwwwddd	(Z scale factor of *Block*)
BL:XEXTRUDE	Nwwwddd	(X component of *Block's* extrusion direction)
BL:YEXTRUDE	Nwwwddd	(Y component of *Block's* extrusion direction)
BL:ZEXTRUDE	Nwwwddd	(Z component of *Block's* extrusion direction)
BL:SPACE	Cwww000	(Space between fields)
Attribute Tag	Cwww000	(Attribute tag, character)
Attribute Tag	Nwwwddd	(Attribute tag, numeric)

All items in the first column (in `Courier` font) must be spelled exactly as shown, if used. Use <u>spaces</u> rather than tabs or indents to separate the second column entries. In the second column, the first letter must be **N** or **C**, representing a numerical or character field. The three "w"s indicate digits that specify the field width (number of characters or numbers). The last three digits (**0** or **d**) specify the number of decimal places you desire for the field (**0** indicates that no decimals can be specified for that field).

The template file can specify any or all of the possible fields in any order, but should be listed in the order you wish the extract file to display. Each template file must include <u>at least one</u> attribute field. If a block contains any of the specified attributes, it is listed in the extract file; otherwise it is skipped. If a block contains some, but not all, of the specified attributes, the blank fields are filled with spaces or zeros.

A sample template file for the schematic drawing example in Figure 22-12 is shown here:

BL:NAME	C004000
BL:X	N005002
BL:Y	N005002
BL:SPACE	C002000
RES_NUMBER	C004000
RESISTANCE	C009000
CAP_NUMBER	C004000
CAPACITANCE	C006000
PART_NO	C007000
MANUFACTURER	C012000
COST	N005002

Creating the Extract File (Step 3)

ATTEXT and DDATEXT

Pull-down Menu	COMMAND (TYPE)	ALIAS (TYPE)	Short-cut	Screen (side) Menu	Tablet Menu
...	*ATTEXT or DDATTEXT*	...	...	...	...

Once you have created a template file, you can use the *Attext* command (for command line format) or invoke the *Attribute Extraction* dialog box by the *Ddattext* command. The command line format syntax is as follows:

Command: **attext**
CDF, SDF, or DXF Attribute extract (or Objects)? <C>: **Enter** or (**option**)

Instead of using the *Attext* command, you may prefer to use the *Ddattext* command, which invokes the *Attribute Extraction* dialog box (Fig. 22-17). This dialog box serves the same functions as the *Attext* command.

You can use the *Select Objects* option (or the *Objects* option in command line format) if you want to PICK only certain block attributes to be included in the extract file.

The *File Format* specifies the structure of the extract file. You can specify either a *SDF*, *CDF*, or *DXF* format for the extract file.

Figure 22-17 ────────

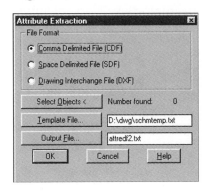

SDF
SDF (Space Delimited File) uses spaces to separate the fields. The *SDF* format is more readable because it appears in columnar format. Numerical fields are right justified, whereas character fields are left justified. Therefore, it may be necessary to include a space field (BL:SPACE or BL:DUMMY) after a numeric field that would otherwise be followed immediately by a character field. This method is used in the example template file shown previously.

CDF
CDF (Comma Delimited File) uses a character that you specify to separate the fields of the extract file. The default character for *CDF* is a comma (,). Some database packages require a *CDF* format for files to be imported.

DXF
The *DXF* format is a variation of the standard AutoCAD *DXF* (Data Interchange File format); however, it contains only block and attribute information. This format contains more information than the *SDF* and *CDF* files and is generally harder to interpret. *DXF* is a standard format and therefore does <u>not</u> request a template file. (Selecting *Export* from the *File* pull-down menu, then *DXX Extract*, allows you to create a *DXF* format extract file only. The *Attext* or *Ddattext* commands must be typed if you want an *SDF* or *CDF* format extract file.)

After specifying the desired format, you must specify the *Template File* (name of previously created template file) and *Output File* (name of the extract file to create). The default extract file name is the same as the drawing name but with the .TXT extension.

PICKing the *Template File* tile invokes the *Template File* dialog box (Fig. 22-18). PICKing the *Output File* tile invokes the *Output File* dialog box (Fig. 22-19). If you use the command line format (*Attext*), the dialog boxes automatically appear (when *FILEDIA*=1).

The *Template File* dialog box requests the name of the file you wish to use as a template. This file must have been previously created. The template file must have a .TXT file extension.

Enter the desired name for the extract file in the edit box of the *Extract File* dialog box. AutoCAD uses the <u>drawing</u> name as the default extract file name and appends the .TXT file extension.

If the *FILEDIA* variable is set to 0 (file dialog boxes turned off), the dialog boxes do not appear, and requests for template and extract file names appear in command line format as follows:

> Template file <default>: Enter name of template file with .TXT extension.
> Extract filename <drawing name>: Enter desired name of extract file.

Figure 22-18

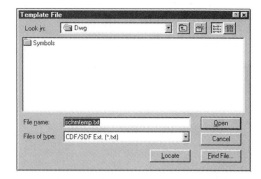

Figure 22-19

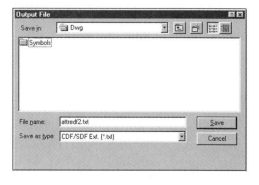

Finally, AutoCAD then creates the extract file based on the specified parameters, displaying a message similar to the following:

> 6 records in extract file.
> Command:

If you receive an error message, check the format of your template file. An error can occur if, for example, you have a BL:NAME field with a width of 10 characters, but a block in the drawing has a name 12 characters long.

Below is the sample extract file created using the electrical schematic drawing and the previous example template file.

Sample Extract File

```
RES   7.25 5.75  R1   4.7                    R-4746 Sparks R Us  0.37
RES  10.00 5.75  R2   510 ohms               R-510A Sparks R Us  0.35
CAP   6.00 7.00             C1   470         C-470A Sparks R Us  0.70
CAP   3.75 5.75             C2   .0047       C-470D Sparks R Us  0.70
CAP   4.50 2.75             C3   100         C-100A Sparks R Us  0.70
RES   8.00 2.75  R3   110K                   R-1105 Sparks R Us  0.40
```

This extract file lists the block name, X and Y location of the block resistor number, resistance value, capacitor number, capacitance value, manufacturer, and cost. Compare the resulting extract file above with the matching template file shown next. Remember that each <u>line</u> in the template file creates a <u>column</u> in the extract file.

Sample (Matching) Template File

```
BL:NAME          C004000
BL:X             N005002
BL:Y             N005002
BL:SPACE         C002000
RES_NUMBER       C004000
RESISTANCE       C009000
CAP_NUMBER       C004000
CAPACITANCE      C006000
PART_NO          C007000
MANUFACTURER     C012000
COST             N005002
```

The extract file below was created from the template file shown above using the *CDF* option:

```
'RES', 7.25, 5.75,'','R1','4.7K','','','R-4746','Sparks R Us', 0.37
'RES',10.00, 5.75,'','R2','510 ohms','','','R-510A','Sparks R Us', 0.35
'CAP', 6.00, 7.00,'','','','C1','470','C-470A','Sparks R Us', 0.70
'CAP', 3.75, 5.75,'','','','C2','.0047','C-470D','Sparks R Us', 0.70
'CAP', 4.50, 2.75,'','','','C3','100','C-100A','Sparks R Us', 0.70
'RES', 8.00, 2.75,'','R3','110K','','','R-1105','Sparks R Us', 0.40
```

CHAPTER EXERCISES

1. **Create an Attributed Title Block**

   A. Open the **TBLOCK** drawing that you created in Chapter 18, Exercise 8. Because we need to create block attributes, *Erase* the existing text so only the lines remain. (Alternately, you can begin a *New* drawing and follow the instructions for Chapter 18, Exercise 8A. Do not complete 8B.)

   B. *SaveAs* **TBLOCKAT**. Check the drawing to ensure that 2 text *Styles* exist using *romans* and *romanc* font files (if not, create the two *Styles*). Next, define attributes similar to those described below using *Attdef* or *Ddattdef*. Substitute your personalized *Tags*, *Prompts*, and *Values* where appropriate or as assigned (such as your name in place of B. R. Smith):

Tag	Prompt	Value	Mode
COMPANY		CADD Design Co.	const
PROJ_TITLE	Enter Project Title	PROJ-	
SCALE	Enter Scale	1"=1"	
DES_NAME		Des.-B.R.Smith	const
CHK_NAME	Enter Checker Name	Ch.-	
COMP_DATE	Enter Completion Date	1/1/98	
CHK_DATE	Enter Check Date	1/1/98	
PROJ_NO	Enter Project Number	PROJ	verif

The completed attributes should appear similar to those shown in Figure 22-20. *Save* the drawing.

C. Use the *Base* command to assign the *Insertion base point*. **PICK** the lower-right corner (shown in Figure 22-20 by the "blip") as the *Insertion base point*.

Since the entire TBLOCKAT drawing (.DWG file) will be inserted as a block, you do not have to use the *Block* command to define the attributed block. *Save* the drawing (as **TBLOCKAT**).

D. Begin a *New* drawing and use the **ASHEET** drawing as a *template*. Set the **TITLE** layer *current* and draw a border with a *Pline* of .02 *width*. Set the *ATTDIA* variable to **1** (*On*). Use the *Insert* or *Ddinsert* command to bring the **TBLOCKAT** drawing in as a block. The *Enter Attributes* dialog box should appear. Enter the attributes to complete the title block similar to that shown in Figure 22-21. Do not *Save* the drawing.

Figure 22-20 ——————————

COMPANY		
PROJ_TITLE		SCALE
DES_NAME		CHK_NAME
COMP_DATE	CHK_DATE	PROJ_NO

+

Figure 22-21 ——————————

CADD Design Company		
Adjustable Mount		1/2"=1"
Des.− B.R. Smith		Chk.−JRS
1/1/98	1/1/98	42B−ADJM

2. **Create Attributed Blocks for the OFF-ATT Drawing**

A. *Open* the **OFF-ATT** drawing (not OFFICE-REN) that you prepared in Chapter 21 Exercises. The drawing should have the office floor plan completed and the geometry drawn for the furniture blocks. The furniture has not yet been transformed to *Blocks*. You are ready to create attributes for the furniture, as shown in Figure 22-22.

Include the information on the next page for the proposed blocks. Use the same *Tag*, *Prompt*, and *Mode* for each proposed block. The *Values* are different, depending on the furniture item.

Figure 22-22 ——————————

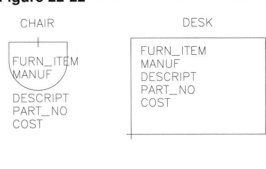

CHAIR

FURN_ITEM
MANUF

DESCRIPT
PART_NO
COST

DESK

FURN_ITEM
MANUF
DESCRIPT
PART_NO
COST

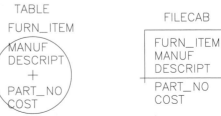

TABLE

FURN_ITEM

MANUF
DESCRIPT
+
PART_NO
COST

FILECAB

FURN_ITEM
MANUF
DESCRIPT

PART_NO
COST

Tag	Prompt	Mode
FURN_ITEM	Enter furniture item	
MANUF	Enter manufacturer	
DESCRIPT	Enter model or size	
PART_NO	Enter part number	Invisible
COST	Enter dealer cost	Invisible

Values

CHAIR	DESK	TABLE	FILE CABINET
WELLTON	KERSEY	KERSEY	STEELMAN
SWIVEL	50" x 30"	ROUND 2'	28 x 3
C-143	KER-29	KER-13	3-28-L
79.99	249.95	42.50	129.99

B. Next, use the *Block* command to make each attributed block Use the block *names* shown (Figure 22-22). Select the attributes in the order you want them to appear upon insertion; then select the geometry. Use the *Insertion base points* as shown. *Save* the drawing.

C. Set the *ATTDIA* variable to **0** (*Off*). *Insert* the blocks into the office and accept the default values. Create an arrangement similar to that shown in Figure 22-23.

Figure 22-23

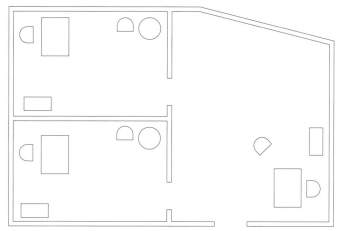

D. Use the *ATTDISP* variable and change the setting to *Off*. Do the attributes disappear?

E. A price change has been reported for all KERSEY furniture items. Your buyer has negotiated an additional 10% off the current dealer cost. Type in the *Multiple Ddatte* command and select each block one at a time to view the attributes. Make the COST changes as necessary. Use *SaveAs* and assign the new name **OFF-ATT2**.

F. Change the *ATTDISP* setting back to *Normal*. It would be helpful to increase the *Height* of the FURN_ITEM attribute (CHAIR, TABLE, etc.). Use the *Attedit* command and answer *Yes* to "Edit attributes one at a time?" Enter **FURN_ITEM** in response to "Attribute tag specification" and accept the default for the other two prompts. Select all the FURN_ITEM attributes (no highlighting occurs at this time). Increase the *Height* to **3"** for each. If everything looks correct, *Save* the drawing.

3. **Extract Attributes**

    A. A cost analysis of the office design is required to check against the $3,400.00 budget allocated for furnishings. This can be accomplished by extracting the attributes into a report. Use a text editor or word processor (in ASCII or non-document mode) to generate the template file below. (Ensure that you have no extra spaces or lines.) Name the file **OFF-TMPL.TXT**.

```
BL:NAME     C008000
BL:X        N008002
BL:Y        N008002
BL:SPACE    C002000
FUR_ITEM    C014000
MANUF       C012000
COST        N006002
```

    B. Use the *Ddattext* command. Select *SDF* format. Specify **OFF-EXT.TXT** for the extract file name. The resulting extract file should appear similar to that below:

```
DESK      165.00  348.00  DESK          KERSEY    224.95
DESK      165.00  228.00  DESK          KERSEY    224.95
DESK      405.00  174.00  DESK          KERSEY    224.95
CHAIR     156.00  327.00  CHAIR         WELLTON    79.99
CHAIR     153.00  204.00  CHAIR         WELLTON    79.99
CHAIR     249.00  336.00  CHAIR         WELLTON    79.99
CHAIR     249.00  216.00  CHAIR         WELLTON    79.99
CHAIR     447.00  153.00  CHAIR         WELLTON    79.99
CHAIR     387.00  189.00  CHAIR         WELLTON    79.99
TABLE     276.00  204.00  TABLE         KERSEY     38.25
TABLE     276.00  327.00  TABLE         KERSEY     38.25
FILECAB   450.00  225.00  FILE CABINET  STEELMAN  129.99
FILECAB   135.00  132.00  FILE CABINET  STEELMAN  129.99
FILECAB   135.00  249.00  FILE CABINET  STEELMAN  129.99
```

    C. Total the COST column to acquire the total furnishings cost so far for the job. If you purchase the $880 conference table and 8 more chairs, how much money will be left in the budget to purchase plants and wall decorations?

4. Use *Attredef* to redefine the CHAIR block in the OFF-ATT drawing. Make a *Copy* of one chair and *Explode* it. Change the geometry in some way (add arms to the chair, for example). Use *Ddedit* or *Ddmodify* to change the **COST** attribute. Increase the default cost value by 15%. Then use *Attredef* to define the new block and to redefine all the existing CHAIR blocks. Use *SaveAs* and assign the name **OFF-ATT3**. Finally, create a new extract file and calculate the total cost for the office (the original template file can be used again).

5. Create the electrical schematic illustrated in Figure 22-24. Use *Blocks* for the symbols. The text associated with the symbols should be created as attributes. Use the following block names and tags:

Block Names: RES
                 CAP
                 GRD
                 AMP

Attribute Tags: PART_NUM
                 MANUF_NUM
                 RESISTANCE
                 CAPACITANCE

(Only the RES and CAP blocks have values for resistance or capacitance.) When you are finished with the drawing, create an extract file reporting information for all attribute tags (X and Y coordinate data is not needed). Save the drawing as **SCHEM2**.

Figure 22-24 ───

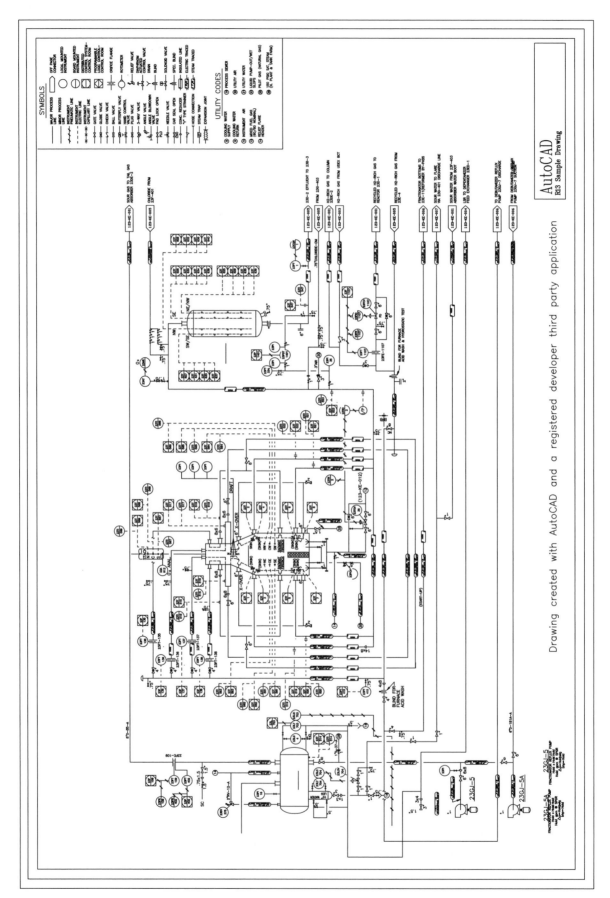

PNID.DWG Courtesy of Autodesk, Inc. (Release 13 Sample Drawing)

23

GRIP EDITING

Chapter Objectives

After completing this chapter you should:

1. be able to use the *GRIPS* variable to enable or disable object grips;

2. be able to activate the grips on any object;

3. be able to make an object's grips warm, hot, or cold;

4. be able to use each of the grip editing options, namely, STRETCH, MOVE, ROTATE, SCALE, AND MIRROR;

5. be able to use the Copy and Base suboptions;

6. be able to use the auxiliary grid that is automatically created when the Copy suboption is used;

7. be able to change grip variable settings using the *Grips* dialog box or the command line format.

CONCEPTS

Grips provide an alternative method of editing AutoCAD objects. The object grips are available for use by setting the *GRIPS* variable to **1**. Object grips are small squares appearing on selected objects at endpoints, midpoints, or centers, etc. The object grips are activated (made visible) by **PICK**ing objects with the cursor pickbox only <u>when no commands are in use</u> (at the open Command: prompt). Grips are like small, magnetic *OSNAPs* (*Endpoint, Midpoint, Center, Quadrant*, etc.) that can be used for snapping one object to another, for example. If the cursor is moved within the small square, it is automatically "snapped" to the grip. Grips can replace the use of *OSNAP* for many applications. The grip option allows you to STRETCH, MOVE, ROTATE, SCALE, MIRROR, or COPY objects without invoking the normal editing commands or *OSNAPs*.

As an example, the endpoint of a *Line* could be "snapped" to the endpoint of an *Arc* (shown in Fig. 23-1) by the following steps:

Figure 23-1

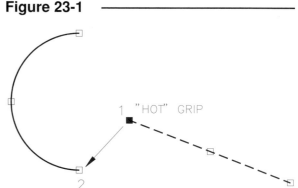

1. Activate the grips by selecting both objects. Selection is done when no commands are in use (during the open Command: prompt).
2. Select the grip at the endpoint of the *Line* (1). The grip turns **hot** (red).
3. The ** STRETCH ** option appears in place of the Command: prompt.
4. STRETCH the *Line* to the endpoint grip on the *Arc* (2). **PICK** when the cursor "snaps" to the grip.
5. The *Line* and the *Arc* should then have connecting endpoints. The Command: prompt reappears. Press Escape twice to cancel (deactivate) the grips.

GRIPS FEATURES

GRIPS and *DDGRIPS*

Pull-down Menu	COMMAND (TYPE)	ALIAS (TYPE)	Short-cut	Screen (side) Menu	Tablet Menu
Tools *Grips...*	*GRIPS or* *DDGRIPS*	*GR*	*...*	TOOLS 2 *Ddgrips*	*X,10*

Grips are enabled or disabled by changing the setting of the system variable, *GRIPS*. A setting of 1 enables or turns *ON* GRIPS and a setting of **0** disables or turns *OFF GRIPS*. This variable can be typed at the Command: prompt, or the *Grips* dialog box can be invoked from the *Options* pull-down menu, or by typing *Ddgrips* (Fig. 23-2). Using the dialog box, toggle *Enable Grips* to turn *GRIPS ON*. The default setting in AutoCAD Release 14 for the *GRIPS* variable is **1** (*ON*). (See Grip Settings near the end of the chapter for explanations of the other options in the *Grips* dialog box.)

Figure 23-2

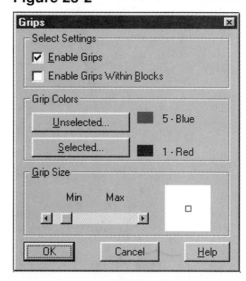

The *GRIPS* variable is saved in the system registry rather than in the current drawing as with most other system variables. Variables saved in the system registry are effective for any drawing session on that <u>particular computer</u>, no matter which drawing is current. The reasoning is that grip-related variables (and selection set-related variables) are a matter of personal preference and therefore should remain constant for a particular CAD station.

When *GRIPS* have been enabled, a small pickbox (3 pixels is the default size) appears at the center of the cursor crosshairs. (The pickbox also appears if the *PICKFIRST* system variable is set to **1**.) This pickbox operates in the same manner as the pickbox appearing during the "Select objects:" prompt. Only the pickbox, *AUTO window*, or *AUTO crossing window* methods can be used for selecting objects to activate the grips. (These three options are the only options available for Noun/Verb object selection as well.)

Activating Grips on Objects

Figure 23-3

The grips on objects are activated by selecting desired objects with the cursor pickbox, window, or crossing window. This action is done when no commands are in use (at the open Command: prompt). When an object has been selected, two things happen: the grips appear and the object is highlighted. The grips are the small blue (default color) boxes appearing at the endpoints, midpoint, center, quadrants, vertices, insertion point, or other locations depending on the object type (Fig. 23-3). Highlighting indicates that the object is included in the selection set.

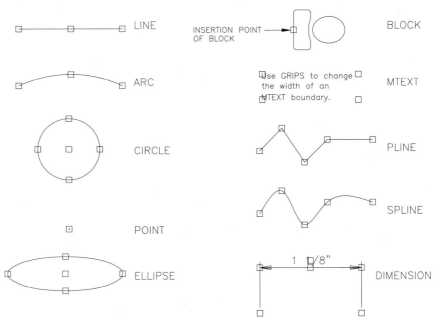

Warm, Hot, and Cold Grips

Grips can have three states: **warm, hot,** or **cold** (Fig. 23-4).

Figure 23-4

A grip is always **warm** first. When an object is selected, it is **warm**—its grips are displayed in blue (default color) and the object is highlighted. The grips can then be made **hot** or **cold**. **Cold** grips are also blue, but the object becomes <u>unhighlighted</u>.

Cold grips are created by <u>deselecting</u> a highlighted object that has **warm** grips. In other words, removing the **warm** grip object from the selection set with SHIFT+#1, or using the *Remove* mode, makes its grips **cold**. A **cold** grip can be used as a point to snap <u>to</u>. A **cold** grip's object is not in the selection set and therefore is not affected by the MOVE, ROTATE, or other editing action. Pressing Escape <u>once</u> changes all **warm** grips to **cold** and clears the selection set. Pressing Escape again deactivates all grips. (In short, press Escape <u>twice</u> to cancel grips entirely.)

A **hot** grip is red (by default) and its object is almost always highlighted. Any grip can be changed to **hot** by selecting the grip itself. A <u>hot grip is the default base point</u> used for the editing action such as MOVE, ROTATE, SCALE, or MIRROR, or is the stretch point for STRETCH. When a **hot** grip exists, a new series of prompts appear in place of the Command: prompt that displays the various grip editing options. The grip editing options are also available from a right-click cursor menu (Fig. 23-5). <u>A grip must be changed to **hot** before the editing options appear</u>. A grip can be transformed from **cold** to **hot,** but the object is not highlighted (included in the selection set). Two or more grips can be made **hot** simultaneously by pressing SHIFT while selecting <u>each</u> grip.

Figure 23-5

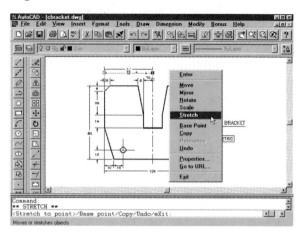

If you have made a *Grip* **hot** and want to deactivate it, possibly to make another *Grip* **hot** instead, press Escape once. This returns the object to a **warm** state. In effect, this is an undo only for the **hot** Grip. Pressing Escape twice makes the *Grips* **cold**, and pressing Escape again cancels all *Grips*.

The three states of grips can be summarized as follows:

Cold	blue grips unhighlighted object	The object is not part of the selection set, but cold grips can be used to "snap" to or used as an alternate base point.
Warm	blue grips highlighted object	The object is included in selection set and is affected by the editing action.
Hot	red grips highlighted object	The base point or control point for the editing action depending on which option is used. The object is included in selection set.

Pressing Escape demotes only the highest grips one level:

Grip State	Press Escape once	Press Escape twice	Press Escape three times
only **cold**	grips are deactivated		
warm, **cold**	**warm** demoted to **cold**	grips are deactivated	
hot, warm, cold	**hot** demoted to **warm**	**warm** demoted to **cold**	grips are deactivated

Grip Editing Options

When a **hot grip** has been activated, the grip editing options are available. The Command: prompt is replaced by the STRETCH, MOVE, ROTATE, SCALE, or MIRROR grip editing options. You can sequentially cycle through the options by pressing the Space bar or Enter. The editing options are displayed in Figures 23-6 through 23-10.

Alternately, you can <u>right-click when a grip is **hot**</u> to activate the grip menu (see Fig. 23-5). This menu has the same options available in command line format with the addition of *Reference, Properties…*, and *Go to URL….* The options are described in the following figures.

** STRETCH **
<Stretch to point>/Base point/Copy/Undo/eXit:

Figure 23-6

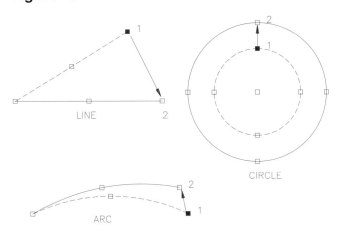

** MOVE **
<Move to point>/Base point/Copy/Undo/eXit:

Figure 23-7

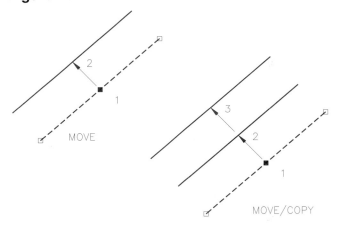

** ROTATE **
<Rotate angle>/Base point/Copy/Undo/Reference/eXit:

Figure 23-8

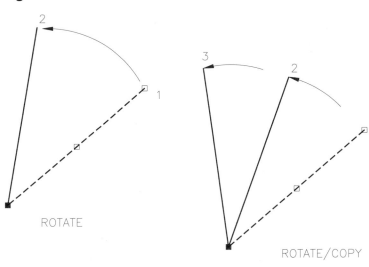

** SCALE **
<Scale factor>/Base point/Copy/Undo/Reference/eXit:

Figure 23-9

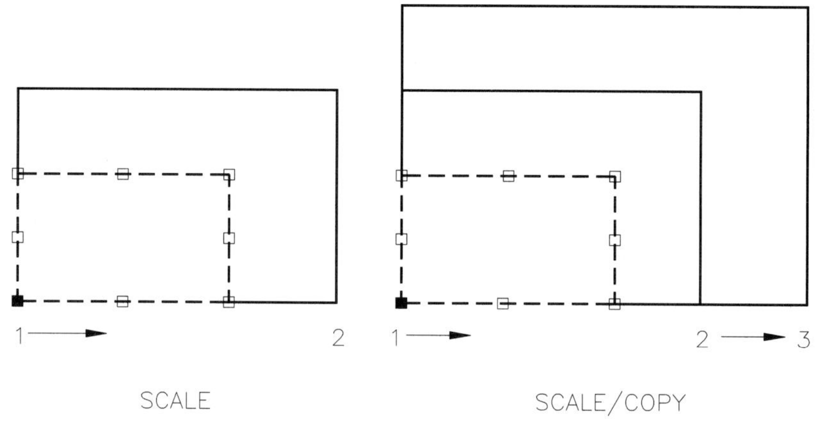

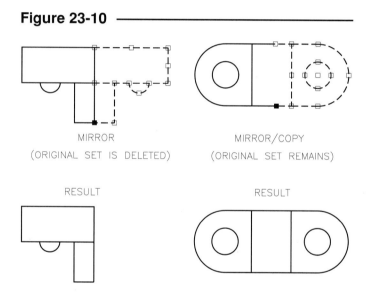

SCALE   SCALE/COPY

** MIRROR **
<Second point>/Base point/
Copy/Undo/eXit:

Figure 23-10

MIRROR
(ORIGINAL SET IS DELETED)

MIRROR/COPY
(ORIGINAL SET REMAINS)

RESULT   RESULT

The *Grip* options are easy to understand and use. Each option operates like the full AutoCAD command by the same name. Generally, the editing option used (except for STRETCH) affects all highlighted objects. The **hot** grip is the base point for each operation. The suboptions, Base and Copy, are explained next.

NOTE: The STRETCH option differs from other options in that STRETCH affects <u>only</u> the object that is attached to the **hot** grip, rather than affecting all highlighted (**warm**) objects.

Base
The Base suboption appears with all of the main grip editing options (STRETCH, MOVE, etc.). Base allows using any other grip as the base point instead of the **hot** grip. Type the letter *B* or select from the right-click cursor menu to invoke this suboption.

Copy
Copy is a suboption of every main choice. Activating this suboption by typing the letter *C* or selecting from the right-click cursor menu invokes a Multiple copy mode, such that whatever set of objects is STRETCHed, MOVEd, ROTATEd, etc., becomes the first of an unlimited number of copies (see the previous five figures). The Multiple mode remains active until exiting back to the Command: prompt.

Undo

The Undo option, invoked by typing the letter **U** or selecting from the right-click cursor menu will undo the last Copy or the last Base point selection. Undo functions <u>only</u> after a Base or Copy operation.

Reference

This option operates similarly to the reference option of the *Scale* and *Rotate* commands. Use *Reference* to enter or PICK a new reference length (SCALE) or angle (ROTATE). (See *Scale* and *Rotate*, Chapter 9.) With grips, *Reference* is only enabled when the SCALE or ROTATE options are active.

Properties... **(Right-Click Menu Only)**

Selecting this option from the right-click grip menu (see Fig. 23-5) activates the *Modify (object)* dialog box. The object with the **hot** grip is the subject of the dialog box. Any property of the selected object can be changed with this dialog box. If more than one hot grip exists, the *Change Properties* dialog box appears. (See *Ddmodify* and *Ddchprop*, Chapter 16.)

Go to URL... **(Right-Click Menu Only)**

This option connects to the Internet URL (uniform resource locator) if one is attached to the object with the **hot** grip. A URL is an Internet address. Using the Internet tools, you can attach a URL to an AutoCAD object (see Chapter 43, Internet Tools).

Figure 23-11

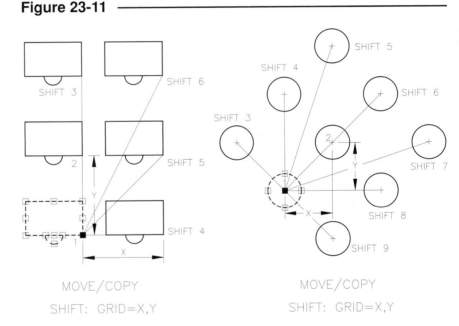

MOVE/COPY
SHIFT: GRID=X,Y

MOVE/COPY
SHIFT: GRID=X,Y

Auxiliary Grid

An auxiliary grid is <u>automatically</u> established on creating the first Copy (Fig. 23-11). The grid is activated by pressing Shift while placing the subsequent copies. The subsequent copies are then "snapped" to the grid in the same manner that *SNAP* functions. The spacing of this auxiliary grid is determined by the location of the first Copy, that is, the X and Y intervals between the base point and the second point.

For example, a "polar array" can be simulated with grips by using ROTATE with the Copy suboption (Fig. 23-12).

The "array" can be constructed by making one Copy, then using the auxiliary grid to achieve equal angular spacing. The steps for creating a "polar array" are as follows:

1. Select the object(s) to array.

2. Select a grip on the set of objects to use as the center of the array. Cycle to the ROTATE option by pressing Enter or selecting from the right-click cursor menu. Next, invoke the Copy suboption.

Figure 23-12

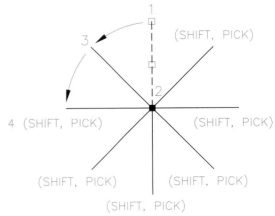

ROTATE/COPY
SHIFT — ANGLES EQUAL

3. Make the first copy at any desired location.

4. After making the first copy, activate the auxiliary angular grid by holding down SHIFT while making the other copies.

5. Cancel the grips or select another command from the menus.

Editing Dimensions

One of the most effective applications of grips is as a dimension editor. Because grips exist at the dimension's extension line origins, arrowhead endpoints, and dimensional value, a dimension can be changed in many ways and still retain its associativity (Fig. 23-13). See Chapter 28, Dimensioning, for further information about dimensions, associativity, and editing dimensions with *Grips*.

Figure 23-13

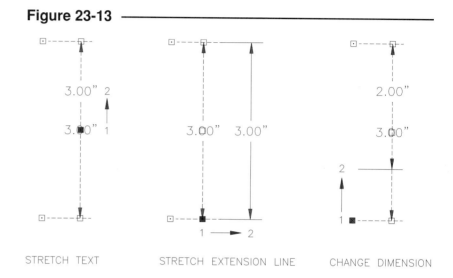

STRETCH TEXT   STRETCH EXTENSION LINE   CHANGE DIMENSION

GUIDELINES FOR USING GRIPS

Although there are many ways to use grips based on the existing objects and the desired application, a general set of guidelines for using grips is given here:

1. Create the objects to edit.

2. Select **warm** and **cold** grips first. This is usually accomplished by selecting <u>all</u> grips first (**warm** state), then <u>deselecting</u> objects to establish the desired **cold** grips.

3. Select the desired **hot** grip(s). The *Grip* options should appear in place of the Command: line.

4. Press Space or Enter to cycle to the desired editing option (STRETCH, MOVE, ROTATE, SCALE, MIRROR) or select from the right-click cursor menu.

5. Select the desired suboptions, if any. If the *Copy* suboption is needed or the Base point needs to be re-specified, do so at this time. *Base* or *Copy* can be selected in either order.

6. Make the desired STRETCH, MOVE, ROTATE, SCALE, or MIRROR.

7. Cancel the grips by pressing Escape twice or selecting a command from a menu.

Figure 23-14

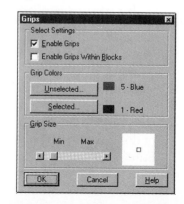

GRIPS SETTINGS

Several settings are available in the *Grips* dialog box (Fig. 23-14) that control the way grips appear or operate. The settings can also be changed by typing in the related variable name at the command prompt.

Enable Grips

If an "X" appears in the checkbox, *Grips* are enabled for the workstation. A check sets the *GRIPS* variable to 1 (on). Removing the check disables grips and sets *GRIPS*=0 (off). The default setting is on.

Enable Grips Within Blocks

When no check appears in this box, only one grip at the block's insertion point is visible (Fig. 23-15). This allows you to work with a block as one object. The related variable is *GRIPBLOCK*, with a setting of 0=off. This is the default setting (disabled).

When the box is checked, *GRIP-BLOCK* is set to 1. All grips on all objects contained in the block are visible and functional (Fig. 23-15). This allows you to use any grip on any object within the block. This does <u>not</u> mean that the block is <u>Exploded</u>—the block retains its one-object integrity and individual entities in the block cannot be edited independently. This feature permits you to use the grips only on each of the block's components.

Figure 23-15 ——————————————————————

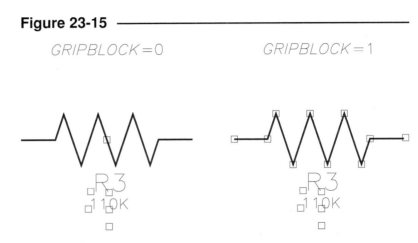

Notice in Figure 23-15 how the insertion point is not accessible when *GRIPBLOCK* is set to 1. Notice also that there are two grips on each of the *Normal* attributes and only one on the *Invisible* attribute and that these grips do not change with the two *GRIPBLOCK* settings. Making one of these grips hot allows you to "move" the attribute to any location and the attribute still retains its association with the block.

Grip Colors Unselected

This setting enables you to change the color of **warm** and **cold** grips. PICKing the *Unselected...* tile produces the *Select Color* dialog box (identical to that used with other color settings). The default setting is blue (ACI number 5). Alternately, the *GRIPCOLOR* variable can be typed and any ACI number from 0 to 255 can be entered.

Grip Colors Selected

The color of **hot** grips can be specified with this option. The *Selected...* tile produces the *Select Color* dialog box. You can also type *GRIPHOT* and enter any ACI number to make the change. The default color is red (1).

Grip Size

Use the slider bar to interactively increase or decrease the size of the grip "box." The *GRIPSIZE* variable could alternately be used. The default size is 3 pixels square (*GRIPSIZE*=3).

NOTE: All *Grip*-related variable settings are saved in the system registry rather than in the current drawing file. This generally means that changes in these variables remain with the computer, not the drawing being used.

Point Specification Precedence

When you select a point with the input device, AutoCAD uses a point specification precedence to determine which location on the drawing to find. The hierarchy is listed here:

1. Object Snap (*OSNAP*).
2. Explicit coordinate entry (absolute, relative, or polar coordinates).

3. *ORTHO.*
4. Point filters (XY filters).
5. *Grips* auxiliary grid (rectangular and circular).
6. *Grips* (on objects).
7. *SNAP* (F9 grid snap).
8. Digitizing a point location.

Practically, this means that *OSNAP* has priority over any other point selection mode. As far as *Grips* are concerned, *ORTHO* overrides *Grips*, so <u>turn off *ORTHO*</u> if you want to snap to *Grips*. Although *Grips* override *SNAP* (F9), it is suggested that <u>*SNAP* be turned *Off*</u> while using *Grips* to simplify PICKing.

More Grips

Because *Grips* have such a wide range of editing potential and because different AutoCAD objects react to *Grips* in different ways, discussion of grip editing is integrated into other chapters. For example, dimensions and surface models have special editing capabilities for *Grips*; therefore, those topics are covered in the related chapters.

CHAPTER EXERCISES

1. ***Open*** the **TEMP-D** drawing. Use the grips on the existing *Spline* to generate a new graph displaying the temperatures for the following week. **PICK** the *Spline* to activate **warm** grips. Use the **STRETCH** option to stretch the first data point grip to the value indicated below for Sunday's temperature. **Cancel** the grips; then repeat the steps for each data point on the graph. ***Save*** the drawing as **TEMP-F**.

Figure 23-16 ————————

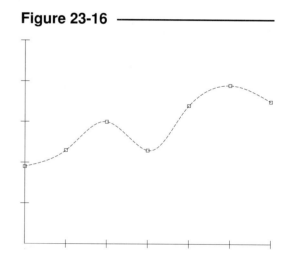

X axis	Y axis
Sunday	29
Monday	32
Tuesday	40
Wednesday	33
Thursday	44
Friday	49
Saturday	45

2. ***Open*** **CH16EX2** drawing. Activate **grips** on the *Line* to the far right. Make the top grip **hot**. Use the **STRETCH** option to stretch the top grip to the right to create a vertical *Line*. **Cancel** the grips.

    Next, activate the **grips** for all the *Lines* (including the vertical one on the right); then make the vertical *Line* grips **cold** as shown in Figure 23-17. **STRETCH** the top of all inclined *Lines* to the top of the vertical *Line* by making the common top grips **hot**, then stretching to the cold **grip** of the vertical *Line*. ***Save*** the drawing as **CH23EX2**.

Figure 23-17 ————————

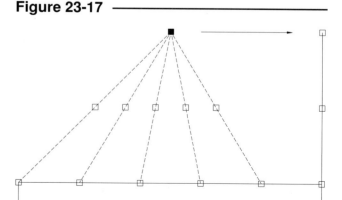

3. This exercise involves all the options of grip editing to create a Space Plate. Begin a *New* drawing or use the **ASHEET** *template* and use *SaveAs* to assign the name **SPACEPLT**.

Figure 23-18

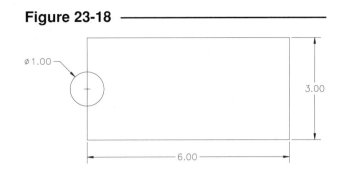

A. Set the *Snap* value to **.125**.
Draw the geometry shown in Figure 23-18 using the *Line* and *Circle* commands.

B. Activate the **grips** on the *Circle*. Make the center grip **hot**. Cycle to the **MOVE** option. Enter *C* for the **Copy** option. You should then get the prompt for **MOVE (multiple)**. Make two copies as shown in Figure 23-19. The new *Circles* should be spaced evenly, with the one at the far right in the center of the rectangle. If the spacing is off, use **grips** with the **STRETCH** option to make the correction.

Figure 23-19

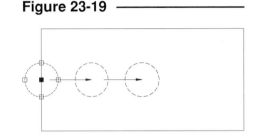

C. Activate the **grips** on the number 2 *Circle* (from the left) to make them **warm**. Also activate the **grips** on the bottom horizontal *Line*, but make them **cold**. Make the center *Circle* grip **hot** and cycle to the **MIRROR** option. Enter *C* for the **Copy** option. (You'll see the **MIRROR (multiple)** prompt.) Then enter *B* to specify a new base point as indicated in Figure 23-20. Turn *On* **ORTHO** and specify the mirror axis as shown to create the new *Circle* (shown in Fig. 23-20 in hidden linetype).

Figure 23-20

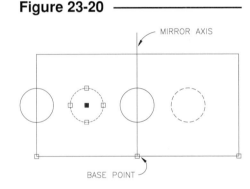

D. Use *Trim* to trim away the outer half of the *Circle* and the interior portion of the vertical *Line* on the left side of the Space Plate. Activate the **grips** on the two vertical *Lines* and the new *Arc*. Make the common grip **hot** (on the *Arc* and *Line* as shown in Fig. 23-21) and **STRETCH** it downward .5 units. (Note how you can affect multiple objects by selecting a common grip.) **Stretch** the upper end of the *Arc* upward .5 units.

Figure 23-21

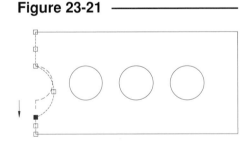

E. *Erase* the *Line* on the right side of the Space Plate. Use the same method that you used in step C to **MIRROR** the *Lines* and *Arc* to the right side of the plate (as shown in Fig. 23-22).

(REMINDER: Make sure that you make the grips on the bottom *Line* **cold**. After you select the **hot** grip, use the **Copy** option <u>and</u> the **Base** point option. Use *ORTHO*.)

Figure 23-22

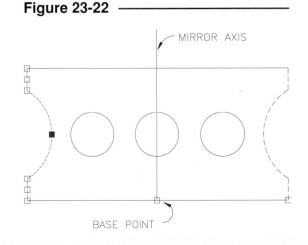

F. In this step, you will **STRETCH** the top edge upward one unit and the bottom edge downward one unit by selecting <u>multiple</u> **hot** grips.

Select the desired horizontal <u>and</u> attached vertical *Lines*. Hold down Shift while selecting <u>each</u> of the endpoint grips, as shown in Figure 23-23. Although they appear **hot**, you must select one of the two again to activate the **STRETCH** option.

Figure 23-23

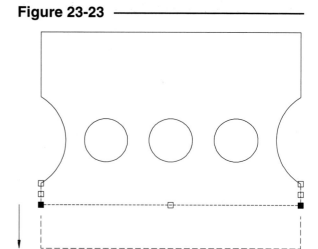

G. In this step, two more *Circles* are created by the **ROTATE** option (see Fig. 23-24). Select the three existing *Circles* to make the grips **warm**. Deselect the center *Circle* so that it will not be copied. PICK the center grip to make it **hot**. Cycle to the **ROTATE** option. Enter *C* for the **Copy** option. Make sure *ORTHO* is *On* and create the new *Circles*.

Figure 23-24

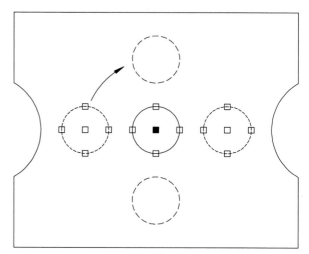

H. Select the center *Circle* and make the center grip **hot** (Fig. 23-25). Cycle to the **SCALE** option. The scale factor is **1.5**. Since the *Circle* is a 1 unit diameter, it can be interactively scaled (watch the *Coords* display), or you can enter the value. The drawing is complete. *Save* the drawing (as **SPACEPLT**).

Figure 23-25

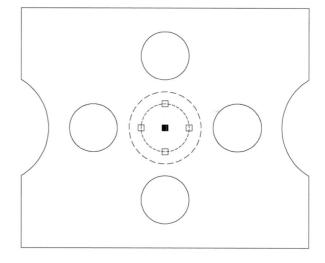

4. *Open* the **STORE ROOM2** drawing from Chapter 18 Exercises. Use the grips to **STRETCH**, **MOVE**, or **ROTATE** each text paragraph to achieve the results shown in Figure 23-26. *SaveAs* **STORE ROOM3**.

Figure 23-26 ────────────

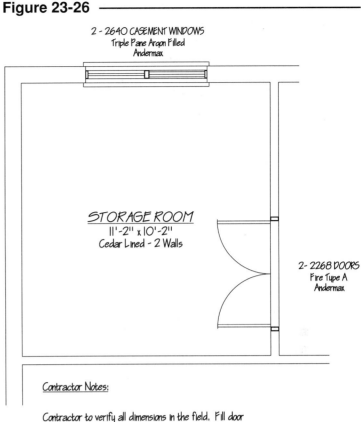

2 - 2640 CASEMENT WINDOWS
Triple Pane Argon Filled
Andermax

STORAGE ROOM
11'-2" x 10'-2"
Cedar Lined - 2 Walls

2- 2268 DOORS
Fire Type A
Andermax

Contractor Notes:

Contractor to verify all dimensions in the field. Fill door and window roughouts after door and window placement.

5. *Open* the **T-PLATE** drawing that you created in Chapter 10. An order has arrived for a modified version of the part. The new plate requires two new holes along the top, a 1″ increase in the height, and a .5″ increase from the vertical center to the hole on the left (Fig. 23-27). Use grips to **STRETCH** and **Copy** the necessary components of the existing part. Use *SaveAs* to rename the part to **TPLATEB**.

Figure 23-27 ────────────

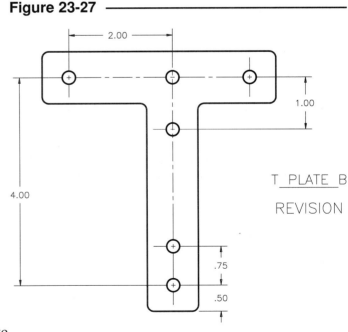

2.00

1.00

4.00

T_PLATE_B

REVISION

.75

.50

6. *Open* the **OFFICE** drawing that you worked on in Chapter 21 Exercises. Use grips to **STRETCH**, **MOVE**, **ROTATE**, **SCALE**, **MIRROR**, and **Copy** the furniture *Blocks*. Experiment with each option. Change the *GRIPBLOCK* variable to enable all the grips on the *Blocks* for some of the editing. *Save* any changes that you like.

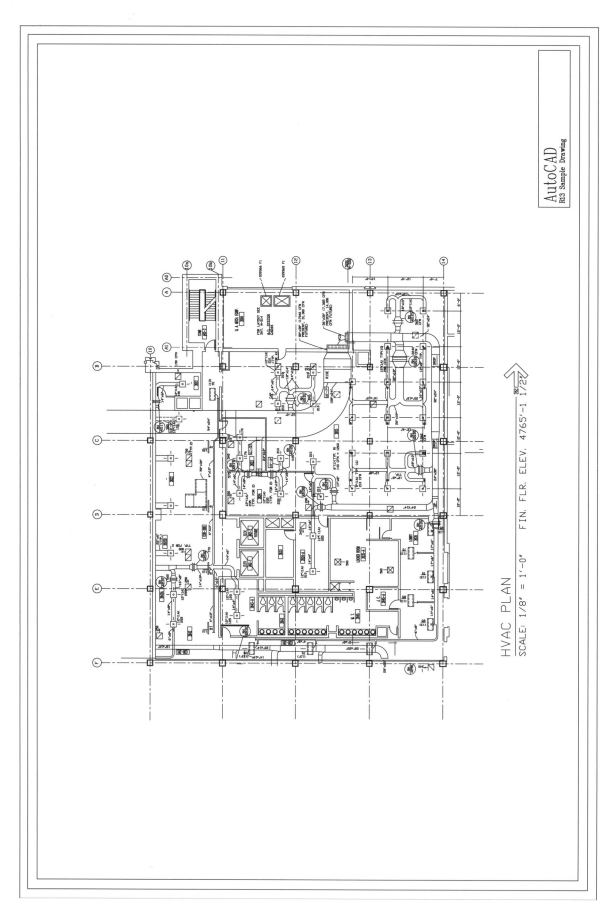

24

MULTIVIEW DRAWING

Chapter Objectives

After completing this chapter you should:

1. be able to draw projection lines using *ORTHO* and *SNAP*;

2. be able to use *Xline* and *Ray* to create construction lines;

3. know how to use *Offset* for construction of views;

4. be able to use point filters and *Tracking* for alignment of lines and views;

5. be able to use construction layers for managing construction lines and notes;

6. be able to use linetypes and layers to draw and manage ANSI standard linetypes;

7. know how to create fillets, rounds, and runouts;

8. know the typical guidelines for creating a three-view multiview drawing.

CONCEPTS

Multiview drawings are used to represent 3D objects on 2D media. The standards and conventions related to multiview drawings have been developed over years of using and optimizing a system of representing real objects on paper. Now that our technology has developed to a point that we can create 3D models, some of the methods we use to generate multiview drawings have changed, but the standards and conventions have been retained so that we can continue to have a universally understood method of communication.

This chapter illustrates methods of creating 2D multiview drawings with AutoCAD (without a 3D model) while complying with industry standards. (Creating 2D drawings from 3D models is addressed in Chapter 42.) Many techniques can be used to construct multiview drawings with AutoCAD because of its versatility. The methods shown in this chapter are the more common methods because they are derived from traditional manual techniques. Other methods are possible.

PROJECTION AND ALIGNMENT OF VIEWS

Projection theory and the conventions of multiview drawing dictate that the views be aligned with each other and oriented in a particular relationship. AutoCAD has particular features, such as *SNAP*, *ORTHO*, construction lines (*Xline, Ray*), Object Snap, *Tracking*, and point filters, that can be used effectively for facilitating projection and alignment of views.

Using *ORTHO* and *OSNAP* to Draw Projection Lines

ORTHO (F8) can be used effectively in concert with *OSNAP* to draw projection lines during construction of multiview drawings. For example, drawing a *Line* interactively with *ORTHO ON* forces the *Line* to be drawn in either a horizontal or vertical direction.

Figure 24-1 simulates this feature while drawing projection *Lines* from the top view over to a 45 degree miter line (intended for transfer of dimensions to the side view). The "From point:" of the *Line* originated from the *Endpoint* of the *Line* on the top view.

Figure 24-1

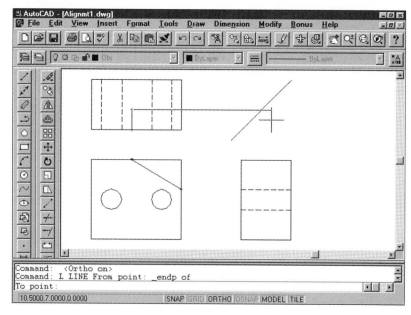

Figure 24-2 illustrates the next step. The vertical projection *Line* is drawn from the *Intersection* of the 45 degree line and the last projection line. *ORTHO* forces the *Line* to the correct vertical alignment with the side view.

Remember that any draw <u>or</u> edit command that requires PICKing is a candidate for *ORTHO* and/or *OSNAP*.

NOTE: *OSNAP* overrides *ORTHO*. If *ORTHO* is *ON* and you are using an *OSNAP* mode to PICK the "To point:" of a *Line*, the *OSNAP* mode has priority; and, therefore, the construction may not result in an orthogonal *Line*.

Figure 24-2

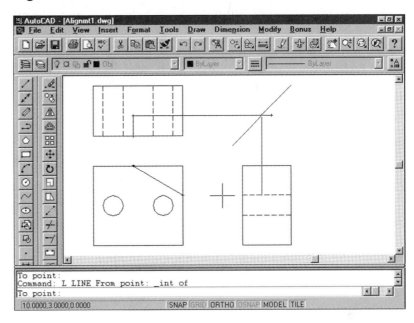

Using *Xline* and *Ray* for Construction Lines

Another strategy for constructing multiview drawings is to make use of the AutoCAD construction line commands *Xline* and *Ray*.

Xlines can be created to "box in" the views and ensure proper alignment. *Ray* is suited for creating the 45 degree miter line for projection between the top and side views. The advantage to using this method is that horizontal and vertical *Xlines* can be created quickly.

These lines can be *Trimmed* to become part of the finished view, or other lines could be drawn "on top of" the construction lines to create the final lines of the views. In either case, <u>construction lines should be drawn on a separate layer</u> so that the layer can be frozen before plotting. If you intend to *Trim* the construction lines so that they become part of the final geometry, draw them originally on the view layers.

Figure 24-3

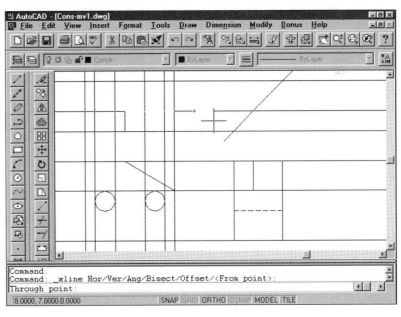

Using *Offset* for Construction of Views

An alternative to using the traditional miter line method for construction of a view by projection, the *Offset* command can be used to transfer distances from one view and to construct another. The *Distance* option of *Offset* provides this alternative.

For example, assume that the side view was completed and you need to construct a top view (Fig. 24-4). First, create a horizontal line as the inner edge of the top view (shown highlighted) by *Offset* or other method. To create the outer edge of the top view (shown in phantom linetype), use *Offset* and PICK points (1) and (2) to specify the *distance*. Select the existing line (3) as the "*Object to Offset:*", then PICK the "*Side to offset?*" at the current cursor position.

Figure 24-4

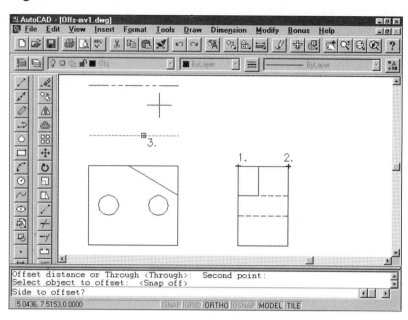

Realignment of Views Using *ORTHO* and *SNAP*

Another application of *ORTHO* and *OSNAP* is the use of *Move* to change the location of an entire view.

For example, assume that the views of the multiview in Figure 24-5 are complete and ready for dimensioning; however, there is not enough room between the front and side views. You can invoke the *Move* command, select the entire view with a *window* or other option, and "slide" the entire view outward. *ORTHO* assures the proper alignment. Make sure *SNAP* is *ON* during the *Move* (if *SNAP* was used in the original construction). This action assures that the final position of the view is on a *SNAP* point and that the objects retain their orientation with respect to the *SNAP*.

An alternative to *Mov*ing the view interactively is use of coordinate specification (absolute, relative rectangular, or relative polar or direct distance entry). (See Chapter 9, Fig. 9-5, for an example of moving a view with direct distance entry.)

Figure 24-5

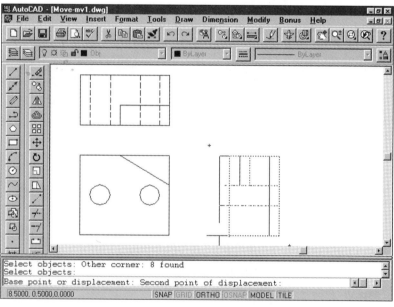

Figure 24-6

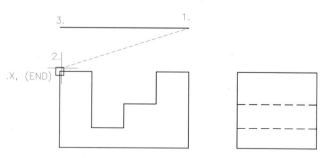

Using Point Filters for Alignment

If you want to draw a *Line* so that it ends or begins in alignment with another object or view, X or Y point filters can be used. For example, Figure 24-6 shows the construction of a *Line* in the top view. It is desired to end the *Line* in alignment with the X position directly above the left side of the front view. An *.X* point filter is used to accomplish this action. *ORTHO is On*. The following command syntax is used for the construction:

Figure 24-7

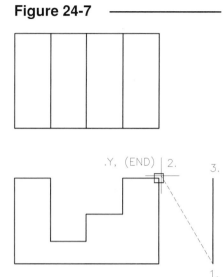

Command: **line**
From point: **PICK** (1.)
To point: **.x** (Type ".X"; then press **Enter**.)
of Endpoint of PICK (2.) (AutoCAD uses only the X value of the point selected.)
(need YZ): **PICK** (anywhere near 3. AutoCAD uses the Y and Z points selected.)
To point: **Enter**
Command:

Another possibility is to use a .Y point filter for construction in alignment with the Y value (height) of existing objects (Fig. 24-7). In this case, enter a **.Y** in response to the "To point:" prompt and press Enter. The point you PICK (2) supplies only the Y value. AutoCAD responds with "(need XZ)". PICK again near (3) to supply the X value. Make sure *ORTHO* is *On*.

Using *Tracking* for Alignment

Tracking can also be used for drawing or moving objects into alignment with other objects in much the same way as point filters are used. *Tracking*, a new Release 14 feature, is a method of specifying points that align orthogonally to other points. *Tracking* is used during a command similar to the way *OSNAPs* are used to locate points on objects. For example, when a command prompts you to PICK a point or enter coordinate values, you can use *Tracking* to find a point orthogonal to a *Midpoint, Endpoint,* or other point. See *Tracking* in Chapter 15.

As an example, you can use *Tracking* to move a misplaced view into alignment with another view (Fig. 24-8). In this example, *Tracking* is used in conjunction with *OSNAPs* during the *Move* command to align the top view orthogonally with the front view. The command sequence is as follows:

Figure 24-8

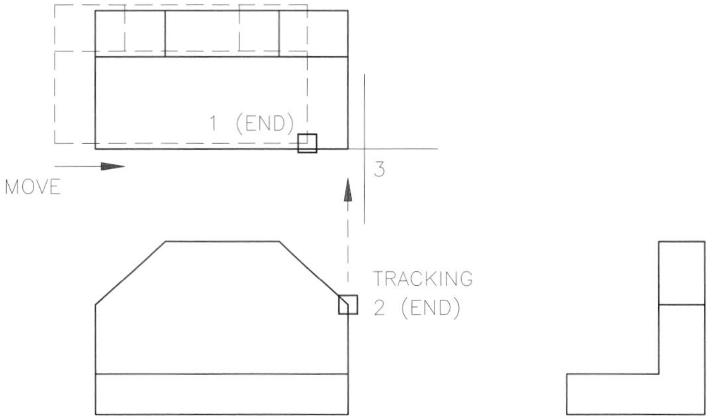

Command: **move**
Select objects: **PICK** (top view objects)
Select objects: **PICK** (top view objects)
Select objects: **Enter**
Base point or displacement: **endp of**
PICK (P1)
Second point of displacement: **tracking**
First tracking point: **endp of PICK** (P2)
Next point (Press ENTER to end tracking): **PICK** (P3)
Next point (Press ENTER to end tracking): **Enter**
Command:

Using *Tracking* before specifying the second point (P2) ensures that the next point will be aligned with the *Endpoint* of the front view. The next point (P3) can be PICKed anywhere above (P2) and it will remain in vertical alignment with (P2).

USING CONSTRUCTION LAYERS

The use of layers for isolating construction lines can make drawing and editing faster and easier. The creation of multiview drawings can involve construction lines, reference points, or notes that are not intended for the final plot. Rather than *Erasing* these construction lines, points, or notes before plotting, they can be created on a separate layer and turned *Off* or made *Frozen* before running the final plot. If design changes are required, as they often are, the construction layers can be turned *On*, rather than having to recreate the construction.

There are two strategies for creating construction objects on separate layers:

1. Use Layer 0 for construction lines, reference points, and notes. This method can be used for fast, simple drawings.

2. Create a new layer for construction lines, reference points, and notes. Use this method for more complex drawings or drawings involving use of *Blocks* on Layer 0.

For example, consider the drawing during construction in Figure 24-9. A separate layer has been created for the construction lines, notes, and reference points.

In Figure 24-10, the same drawing is shown ready for making the final plot. Notice that the construction Layer has been *Frozen*.

If you are plotting the *Limits*, and the construction layer has objects outside the *Limits*, the construction layer should be *Frozen*, rather than being turned *Off*, unless only *Xlines* and *Rays* exist on the layer.

Figure 24-9 ───────────────

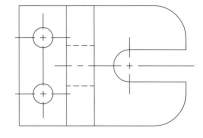

Figure 24-10 ───────────────

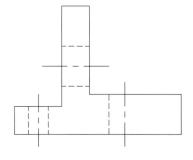

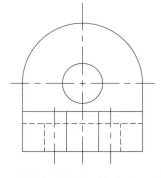

USING LINETYPES

Different types of lines are used to represent different features of a multiview drawing. Linetypes in AutoCAD are accessed by the *Linetype* command or through the *Linetype* tab of the *Layer & Linetype Properties* dialog box. *Linetypes* can be changed retroactively by the *Change Properties* or the *Modify* dialog boxes. *Linetypes* can be assigned to individual objects specifically or to layers (*BYLAYER*). See Chapter 12 for a full discussion on this topic.

AutoCAD complies with the ANSI and ISO standards for linetypes. The principal AutoCAD linetypes used in multiview drawings and the associated names are shown in Figure 24-11.

Figure 24-11

CONTINUOUS

HIDDEN

CENTER

PHANTOM

DASHED

Many other linetypes are provided in AutoCAD. Refer to Chapter 12 for the full list and illustration of the linetypes.

Other standard lines are created by AutoCAD automatically. For example, dimension lines can be automatically drawn when using dimensioning commands (Chapter 28), and section lines can be automatically drawn when using the *Hatch* command (Chapter 26).

AutoCAD linetypes do not have a specified thickness, but this can be accomplished by either creating *Plines* and specifying *Width* or by plotting with specific width pens. (See Chapters 15 and 16 for information on *Plines* and Chapter 14 for information on plotting.)

Drawing Hidden and Center Lines

Figure 24-12 illustrates a typical application of AutoCAD *Hidden* and *Center* linetypes. Notice that the horizontal center line in the front view does not automatically locate the short dashes correctly, and the hidden lines in the right side view incorrectly intersect the center vertical line.

Although AutoCAD supplies ANSI standard linetypes, the application of those linetypes does not always follow ANSI standards. For example, you do <u>not</u> have control over the

Figure 24-12

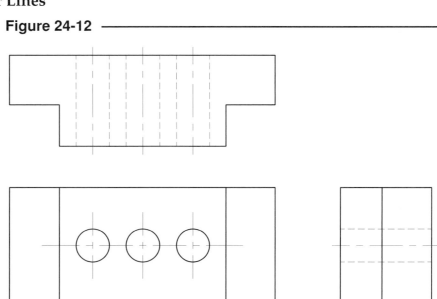

placement of the individual dashes of center lines and hidden lines. (You have control of only the endpoints of the lines and the *Ltscale*.) Therefore, the short dashes of center lines may not cross exactly at the circle centers, or the dashes of hidden lines may not always intersect as desired.

You do, however, have control of the endpoints of the lines. Draw lines with the *Center* linetype such that the endpoints are symmetric about the circle or group of circles. This action assures that the short dash occurs at the center of the circle (if an odd number of dashes are generated). Figure 24-13 illustrates correct and incorrect technique.

Figure 24-13 ——————————————

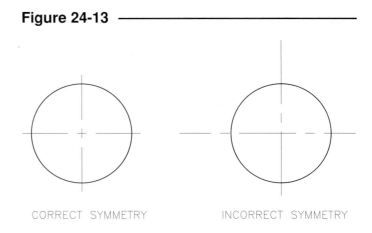

CORRECT SYMMETRY   INCORRECT SYMMETRY

You can also control the relative size of non-continuous linetypes with the *Ltscale* variable. In some cases, the variable can be adjusted to achieve the desired results.

For example, Figure 24-14 demonstrates the use of *Ltscale* to adjust the center line dashes to the correct spacing. Remember that *Ltscale* adjusts linetypes globally (all linetypes across the drawing).

Figure 24-14 ——————————————

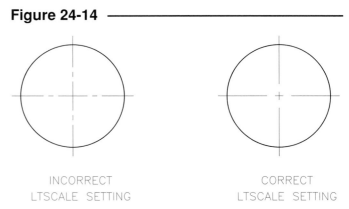

INCORRECT   CORRECT
LTSCALE SETTING   LTSCALE SETTING

When the *Ltscale* has been optimally adjusted for the drawing globally, use the *Ddmodify* dialog box to adjust the linetype scale of individual objects. In this way, the drawing lines can originally be created to the global linetype scale without regard to the *Celtscale*. The finished drawing linetype scale can be adjusted with *Ltscale* globally; then only those objects that need further adjusting can be fine-tuned retroactively with *Ddmodify* or *Ddchprop*.

The *Center* command (a dimensioning command) can be used to draw center lines automatically with correct symmetry and spacing (see Chapter 28, Dimensioning).

Managing Linetypes and Colors

There are two strategies for assigning linetypes and colors: *BYLAYER* and object-specific assignment. In either case, thoughtful layer utilization for linetypes will make your drawings more flexible and efficient.

BYLAYER Linetypes and Colors

The *BYLAYER* linetype and color settings are recommended when you want the most control over linetype visibility and plotting. This is accomplished by creating layers with the *Layer Control* dialog box and using *Set Color* and *Set Ltype* for each layer. After you assign *Linetypes* to specific layers, you simply set the layer (with the desired linetype) as the *Current* layer and draw on that layer in order to draw objects in a specific linetype.

A template drawing for creating typical multiview drawings could be set up with the layer and linetype assignments similar to that shown in Figure 24-15. Notice the layer names, associated colors, and assigned *Ltypes*.

Using this strategy (*BYLAYER Linetype* and *Color* assignment) gives you flexibility. You can control the linetype visibility by controlling the layer visibility. You can control the associated plotting pen for each linetype (in the *Plot Control* dialog box) if each linetype is assigned a specific color. You can also <u>retroactively</u> change the linetype and color of an existing object by changing the object's *Layer* property with the *Change Properties* or the *Modify* dialog box. Objects changed to a different layer assume the *color* and *linetype* of the new layer.

Figure 24-15

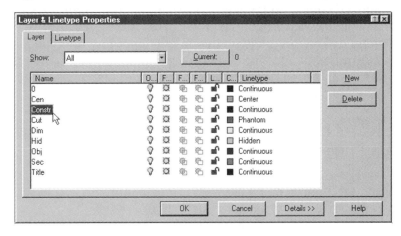

Another strategy for multiview drawings involving several parts, such as an assembly, is to create layers for each linetype <u>specific to each part</u>, as shown in Figure 24-16. With this strategy, each part has the complete set of linetypes, but only one color per part in order to distinguish the part from others in the display.

Related layer <u>groups</u> can be selected within the dialog box using the *Filters* dialog box. If the *Layer* command is used instead, wildcards can be typed for layer selection. For example, entering "?????-HID" selects all of the layers with hidden lines, or entering "MOUNT*" would select all of the layers associated with the "MOUNT" part.

Figure 24-16

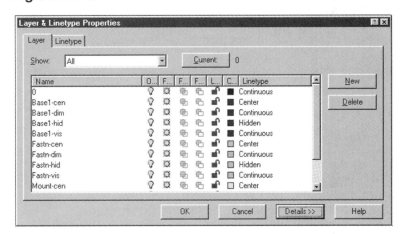

Object *Linetypes and Colors*
Although this method can be complex, object-specific linetype and color assignment can also be managed easier by skillful utilization of layers. One method is to create one layer for each part or each group of related geometry (Fig. 24-17). The *BYLAYER* color and linetype settings should be left to the defaults. Then object-specific linetype settings can be assigned using the *Linetype* command or through the *Linetype* tab of the *Layer & Linetype Properties* dialog box, and

Figure 24-17

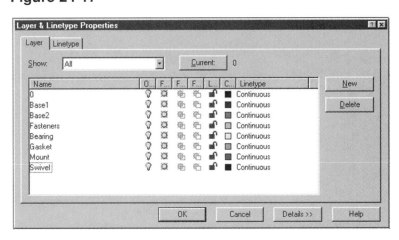

color settings can be assigned using the *Color* command or *Ddcolor* dialog box. (Colors provide the means for controlling plotted line weights). For assemblies, all lines related to one part would be drawn on one layer. Remember that you can draw everything with one linetype and color setting, then use *Ddchprop* to retroactively set the desired color and linetype for each set of objects. Visibility of parts can be controlled by *Freezing* or *Thawing* part layers. You cannot isolate and control visibility of linetypes or colors by this method.

CREATING FILLETS, ROUNDS, AND RUNOUTS

Many mechanical metal or plastic parts manufactured from a molding process have slightly rounded corners. The otherwise sharp corners are rounded because of limitations in the molding process or for safety. A convex corner is called a <u>round</u> and a concave corner is called a <u>fillet</u>. These fillets and rounds are created easily in AutoCAD by using the *Fillet* command.

The example in Figure 24-18 shows a multiview drawing of a part with sharp corners before the fillets and rounds are drawn.

Figure 24-18 —————————

The corners are rounded using the *Fillet* command. First, use *Fillet* to specify the *Radius*. Once the *Radius* is specified, just select the desired lines to *Fillet* near the end to round.

If the *Fillet* is in the middle portion of a *Line* instead of the end, *Extend* can be used to reconnect the part of the *Line* automatically trimmed by *Fillet*, or *Fillet* can be used in the *Notrim* mode.

The finish marks (V-shaped symbols) indicate machined surfaces. Finished surfaces have sharp corners.

Figure 24-19 —————————

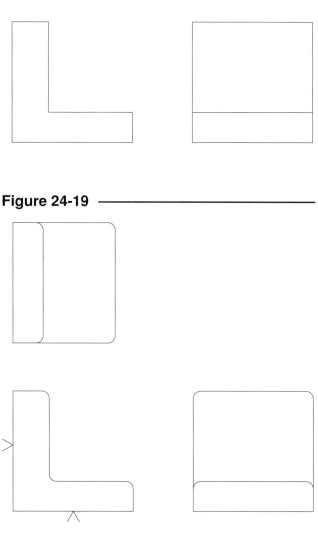

A <u>runout is a visual representation</u> of a complex fillet or round. For example, when two filleted edges intersect at less than a 90 degree angle, a runout should be drawn as shown in the top view of the multiview drawing (Fig. 24-20).

Figure 24-20

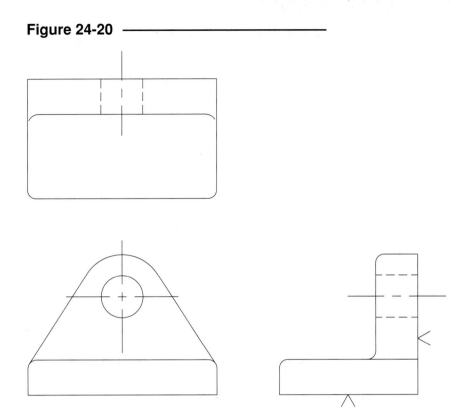

A close-up of the runouts is shown in Figure 24-21. No AutoCAD command is provided for this specific function. The *3point* option of the *Arc* command can be used to create the runouts, although other options can be used. Alternately, the *Circle TTR* option can be used with *Trim* to achieve the desired effect. As a general rule, use the same radius or slightly larger than that given for the fillets and rounds, but draw it less than 90 degrees.

Figure 24-21

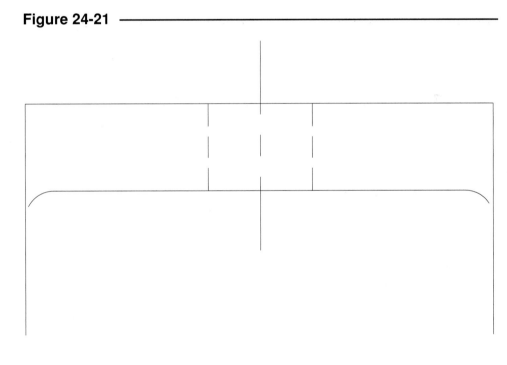

GUIDELINES FOR CREATING A TYPICAL THREE-VIEW DRAWING

Following are some guidelines for creating the three-view drawing in Figure 24-22. This object is used only as an example. The steps or particular construction procedure may vary, depending on the specific object drawn. Dimensions are shown in the figure so you can create the multiview drawing as an exercise.

Figure 24-22

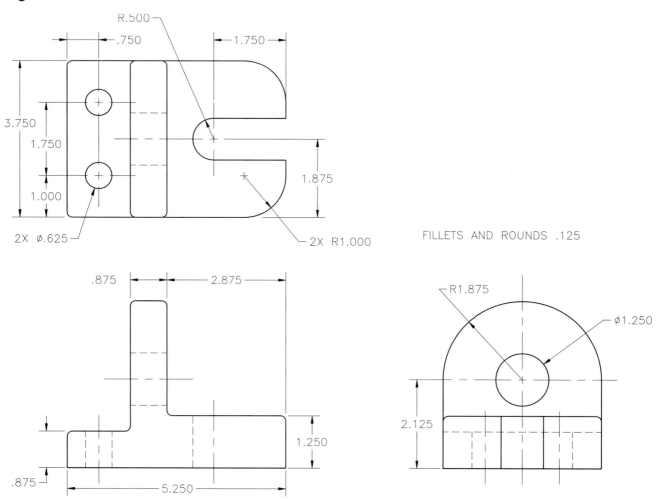

1. Drawing Setup

   Units are set to *Decimal* with 3 places of *Precision*. *Limits* of .22 x 17 are set to allow enough drawing space for both views. The finished drawing can be plotted on a B size sheet at a scale of 1"=1" or on an A size sheet at a scale of 1/2"=1". *Snap* is set to an increment of .125. *Grid* is set to an increment of .5. *Ltscale* is not changed from the default of 1. Layers are created (OBJ, HID, CEN, DIM, BORDER, and CONSTR) with appropriate *Ltypes* and *Colors* assigned.

2. An outline of each view is "blocked in" by drawing the appropriate *Lines* and *Circles* on the OBJ layer similar to that shown in Figure 24-23. Ensure that *SNAP* is *ON*. *ORTHO* should be turned *ON* when appropriate. Use the cursor to ensure that the views align horizontally and vertically. Note that the top edge of the front view was determined by projecting from the *Circle* in the right side view.

Figure 24-23 ─────────────

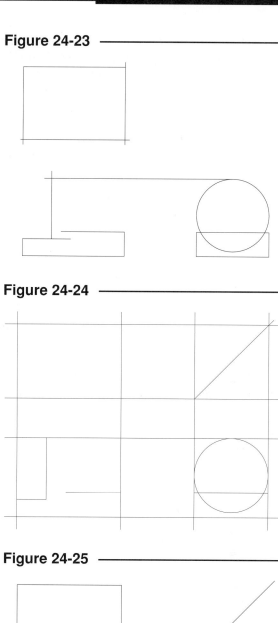

Another method for construction of a multiview drawing is shown in Figure 24-24. This method uses the *Xline* and *Ray* commands to create construction and projection lines. Here all the views are blocked in and some of the object lines have been formed. The construction lines should be kept on a separate layer, except in the case where *Xlines* and *Rays* can be trimmed and converted to the object lines. (The following illustrations do not display this method because of the difficulty in seeing which are object and construction lines.)

Figure 24-24 ─────────────

3. This drawing requires some projection between the top and side views (Fig. 24-25). The CONSTR layer is set as *Current*. Two *Lines* are drawn from the inside edges of the two views (using *OSNAP* and *ORTHO* for alignment). A 45 degree miter line is constructed for the projection lines to "make the turn." A *Ray* is suited for this purpose.

Figure 24-25 ─────────────

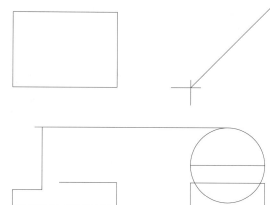

4. Details are added to the front and top views
(Fig. 24-26). The projection line from the side view to
the front view (previous figure) is *Trimmed*. A *Circle*
representing the hole is drawn in the side view and
projected up and over to the top view and to the front
view. The object lines are drawn on layer OBJ and
some projection lines are drawn on layer CONSTR.
The horizontal projection lines from the 45 degree
miter line are drawn on layer HID awaiting *Trimming*.
Alternately, those two projection lines could be drawn
on layer CONSTR and changed to the appropriate
layer with *Change Properties* after *Trimming*.

Figure 24-26

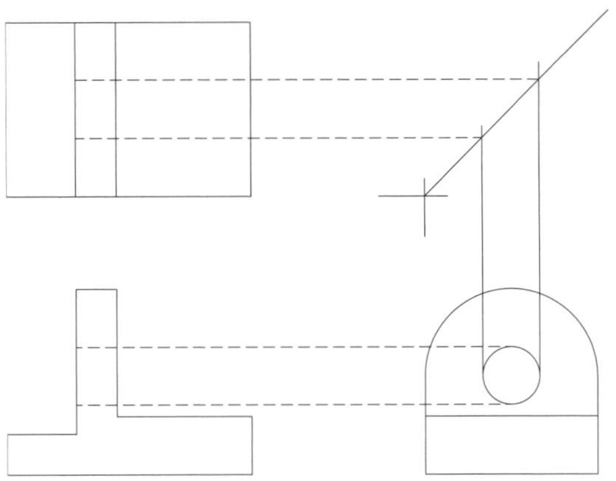

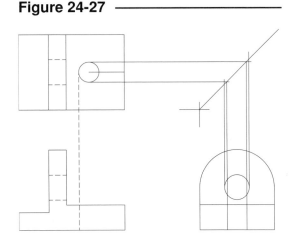

5. The hidden lines used for projection to the top view
and front view (previous figure) are *Trimmed*. The slot
is created in the top view with a *Circle* and projected to
the side and front views. It is usually faster and easier
to draw round object features in their circular view
first, then project to the other views. Make sure you
use the correct layers (OBJ, CONSTR, HID) for the
appropriate features. If you do not, *Ddmodify* can be
used retroactively.

Figure 24-27

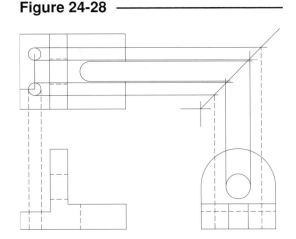

6. The lines shown in the previous figure as projection
lines or construction lines for the slot are *Trimmed*
or *Erased*. The holes in the top view are drawn on
layer OBJ and projected to the other views. The
projection lines and hidden lines are drawn on
their respective layers.

Figure 24-28

7. *Trim* the appropriate hidden lines. *Freeze* layer CONSTR. On layer OBJ, use *Fillet* to create the rounded corners in the top view. Draw the correct center lines for the holes on layer CEN. The value for *Ltscale* should be adjusted to achieve the optimum center line spacing. *Ddmodify* can be used to adjust individual object linetype scale.

Figure 24-29 ————————————————

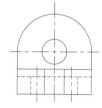
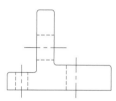
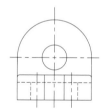

8. Fillets and rounds are added using the *Fillet* command. The runouts are created by drawing a *3point Arc* and *Trimming* or *Extending* the *Line* ends as necessary. Use *Zoom* for this detail work.

Figure 24-30 ————————————————

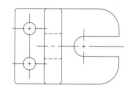

9. Add a border and a title block using *Pline*. Include the part name, company, draftsperson, scale, date, and drawing file name in the title block. The drawing is ready for dimensioning and man-ufacturing notes.

Figure 24-31 ————————————————

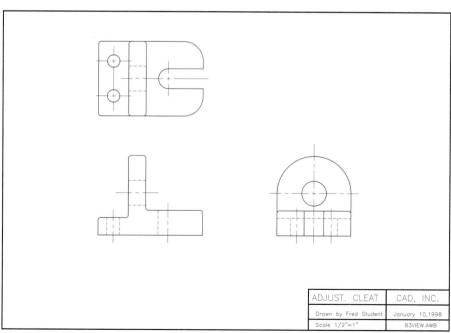

ADJUST. CLEAT	CAD, INC.
Drawn by Fred Student	January 10,1998
Scale 1/2"=1"	83VIEW.AWB

CHAPTER EXERCISES

1. *Open* the **PIVOTARM CH16** drawing.

    A. Create the right side view. Use *OSNAP* and *ORTHO* to create *Lines* or *Rays* to the miter line and down to the right side view as shown in Figure 24-32. *Offset* may be used effectively for this purpose instead. Use *Extend, Offset,* or *Ray* to create the projection lines from the front view to the right side view.

Figure 24-32 ─────────────

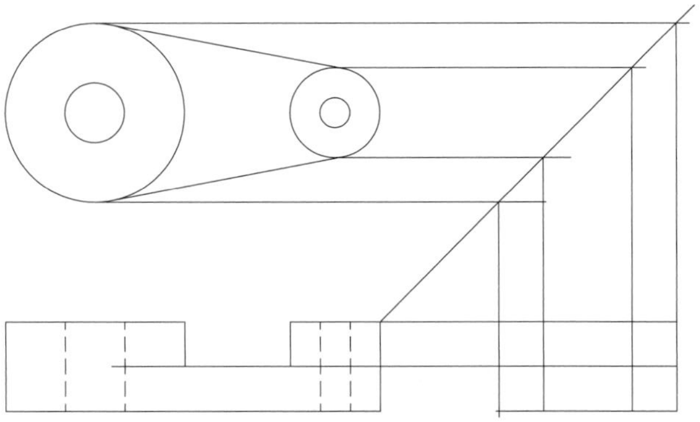

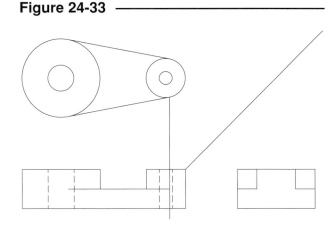

    B. *Trim* or *Erase* the unwanted projection lines, as shown in Figure 24-33. Draw a *Line* or *Ray* from the *Endpoint* of the diagonal *Line* in the top view down to the front to supply the boundary edge for *Trimming* the horizontal *Line* in the front view as shown.

Figure 24-33 ─────────────

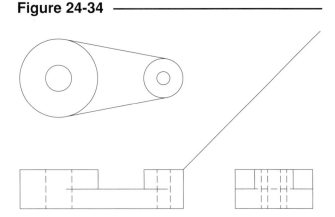

    C. Next, create the hidden lines for the holes by the same fashion as before. Use previously created *Layers* to achieve the desired *Linetypes*. Complete the side view by adding the horizontal hidden *Line* in the center of the view.

Figure 24-34 ─────────────

D. Another hole for a set screw must be added to the small end of the Pivot Arm. Construct a *Circle* of **4**mm diameter with its center located **8**mm from the top edge in the side view as shown in Figure 24-35. Project the set screw hole to the other views.

Figure 24-35

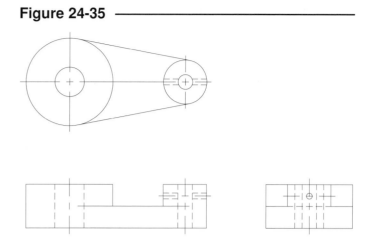

E. Make new layers **CONSTR, OBJ,** and **TITLE** and change objects to the appropriate layers with *Change Properties*. *Freeze* layer **CONSTR**. Add centerlines on the **CEN** layer as shown in Figure 24-35. Change the *Ltscale* to **18**. To complete the PIVOTARM drawing, draw a *Pline* border (*width* **.02** x scale factor) and *Insert* the **TBLOCKAT** drawing that you created in Chapter 22 Exercises. *SaveAs* **PIVOTARM CH24.**

For exercises 2 through 5, construct and plot the multiview drawings as instructed. Use an appropriate *template* drawing for each exercise unless instructed otherwise. Use conventional practices for *layers* and *linetypes*. Draw a *Pline* border with the correct *width* and *Insert* your **TBLOCK** or **TBLOCKAT** drawing.

2. Make a two-view multiview drawing of the Clip. *Plot* the drawing full size (**1=1**). Use the **ASHEET** template drawing to achieve the desired plot scale. *Save* the drawing as **CLIP**.

Figure 24-36

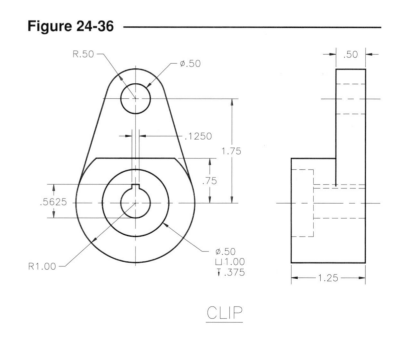

CLIP

3. Make a three-view multiview drawing of the Bar Guide and *Plot* it **1=1**. Use the **BAR-GUIDE** drawing you set up in Chapter 13 Exercises. Note that a partial *Ellipse* will appear in one of the views.

Figure 24-37

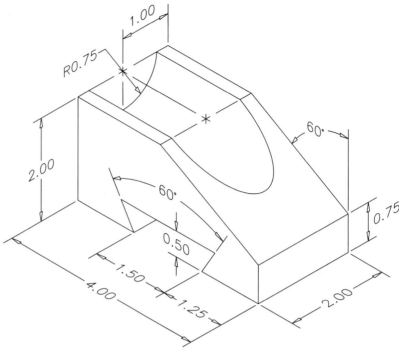

4. Construct a multi-view drawing of the Saddle. Three views are needed. The channel along the bottom of the part intersects with the saddle on top to create a slotted hole visible in the top view. *Plot* the drawing at **1=1**. *Save* as **SADDLE**.

Figure 24-38

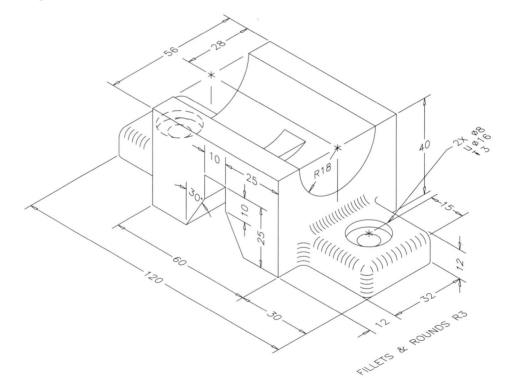

5. Draw a multiview of the Adjustable Mount shown in Figure 24-39. Determine an appropriate template drawing to use and scale for plotting based on your plotter capabilities. *Plot* the drawing to an accepted scale. *Save* the drawing as **ADJMOUNT**.

Figure 24-39

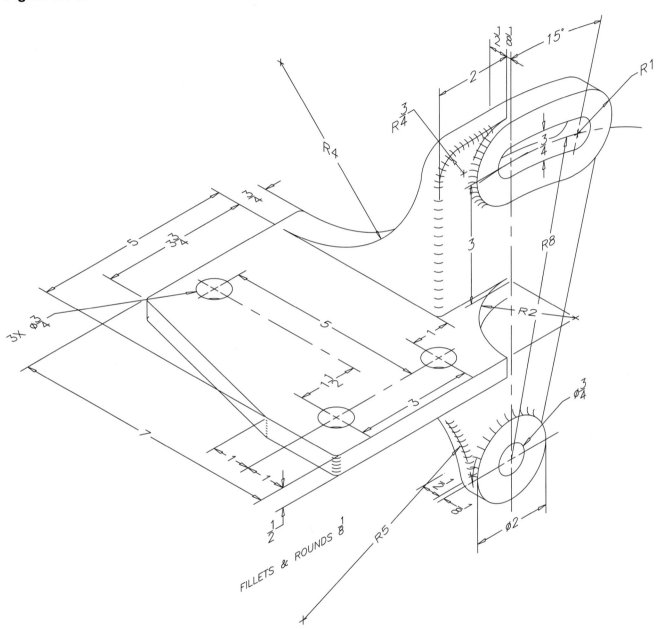

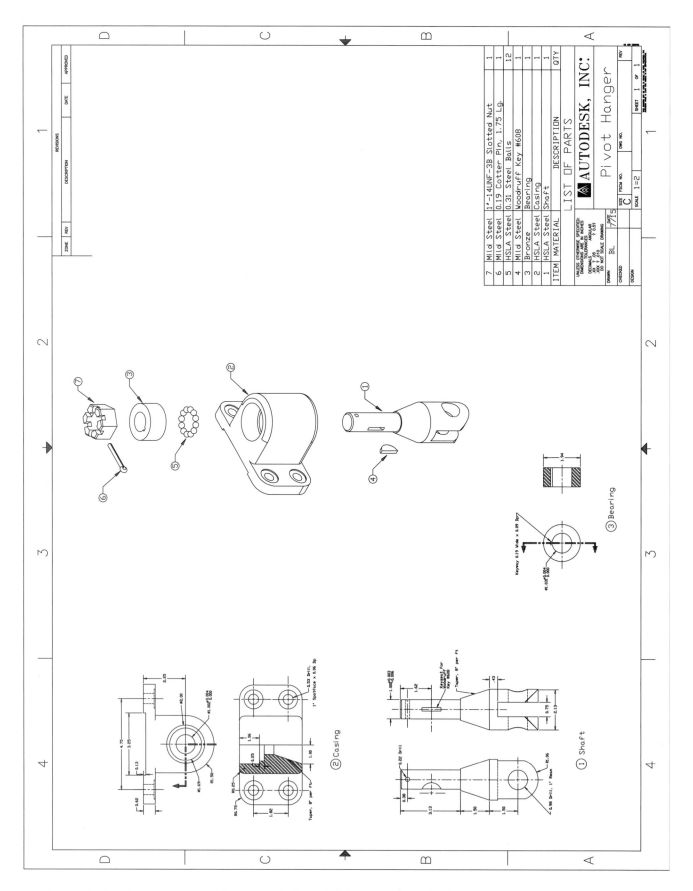

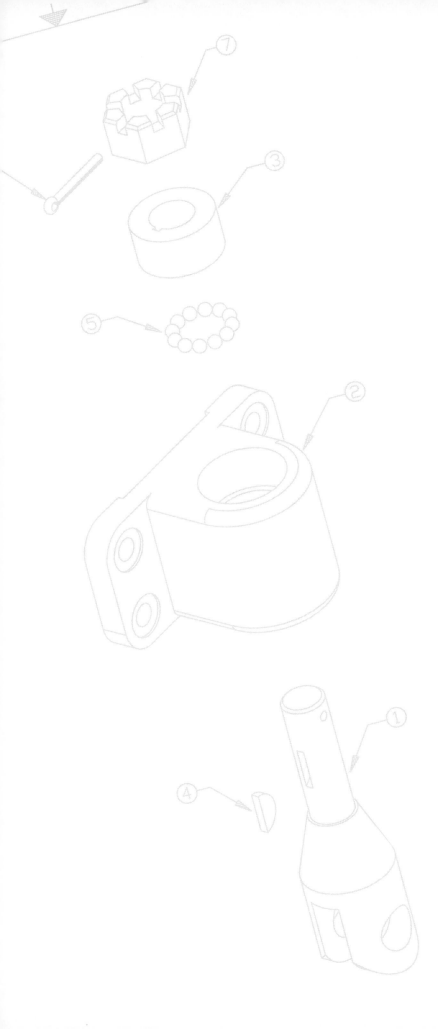

25

PICTORIAL DRAWINGS

Chapter Objectives

After completing this chapter you should be able to:

1. activate the *Isometric Style* of *Snap* for creating isometric drawings;

2. draw on the three isometric planes by toggling *Isoplane* using Ctrl+E;

3. create isometric ellipses with the *Isocircle* option of *Ellipse*;

4. construct an isometric drawing in AutoCAD;

5. create Oblique Cavalier and Cabinet drawings in AutoCAD.

CONCEPTS

Isometric drawings and oblique drawings are pictorial drawings. Pictorial drawings show three principal faces of the object in one view. A pictorial drawing is a drawing of a 3D object as if you were positioned to see (typically) some of the front, some of the top, and some of the side of the object. All three dimensions of the object (width, height, and depth) are visible in a pictorial drawing.

Multiview drawings differ from pictorial drawings because a multiview only shows two dimensions in each view, so two or more views are needed to see all three dimensions of the object. A pictorial drawing shows all dimensions in the one view. Pictorial drawings depict the object similar to the way you are accustomed to viewing objects in everyday life, that is, seeing all three dimensions. Figure 25-1 and Figure 25-2 show the same object in multiview and in pictorial representation, respectively. Notice that multiview drawings use hidden lines to indicate features that are normally obstructed from view, whereas <u>hidden lines are normally omitted</u> in isometric drawings (unless certain hidden features must be indicated for a particular function or purpose).

Types of Pictorial Drawings

Pictorial drawings are classified as follows:

1. Axonometric drawings
    a. Isometric drawings
    b. Dimetric drawings
    c. Trimetric drawings

2. Oblique drawings

Axonometric drawings are characterized by how the angle of the edges or axes (axon-) are measured (-metric) with respect to each other.

Isometric drawings are drawn so that each of the axes have equal angular measurement. ("Isometric" means equal measurement.) The isometric axes are always drawn at 120 degree increments (Fig. 25-3). All rectilinear lines on the object (representing horizontal and vertical edges—not inclined or oblique) are drawn on the isometric axes.

A 3D object seen "in isometric" is thought of as being oriented so that each of three perpendicular faces (such as the top, front, and side) are seen equally. In other words, the angles formed between the line of sight and each of the principal faces are equal.

Figure 25-1

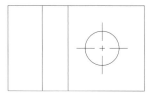
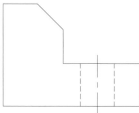

Figure 25-2

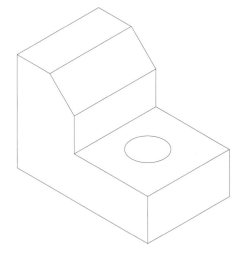

Figure 25-3

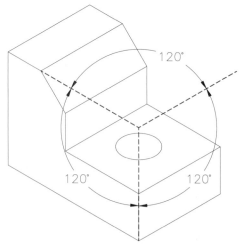
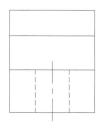

Dimetric drawings are constructed so that the angle between any two of the three axes is equal. There are many possibilities for dimetric axes. A common orientation for dimetric drawings is shown in Figure 25-4. For 3D objects seen from a dimetric viewpoint, the angles formed between the line of sight and each of two principal faces are equal.

Figure 25-4

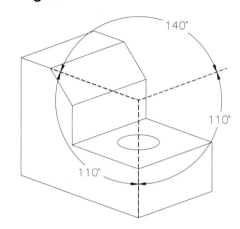

Trimetric drawings have three unequal angles between the axes. Numerous possibilities exist. A common orientation for trimetric drawings is shown in Figure 25-5.

Oblique drawings are characterized by a vertical axis and horizontal axis for the two dimensions of the front face and a third (receding) axis of either 30, 45, or 60 degrees (Fig. 25-6). Oblique drawings depict the true size and shape of the front face, but add the depth to what would otherwise be a typical 2D view. This technique simplifies construction of drawings for objects that have contours in the profile view (front face) but relatively few features along the depth. Viewing a 3D object from an oblique viewpoint is not possible.

Figure 25-5

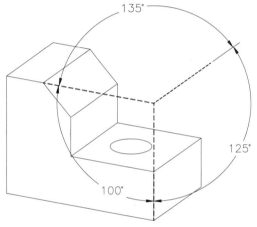

This chapter will explain the construction of isometric and oblique drawings in AutoCAD.

Pictorial Drawings Are 2D Drawings

Isometric, dimetric, trimetric, and oblique drawings are <u>2D drawings</u>, whether created with AutoCAD or otherwise. Pictorial drawing was invented before the existence of CAD and therefore was intended to simulate a 3D object on a 2D plane (the plane of the paper). If AutoCAD is used to create the pictorial, the geometry lies on a 2D plane—the XY plane. All coordinates defining objects have X and Y values with a Z value of 0. When the *Isometric* style of *Snap* is activated, an isometrically structured *SNAP* and *GRID* appear on the XY plane.

Figure 25-6

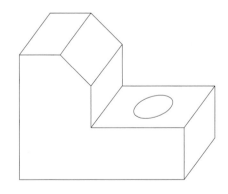

Figure 25-7 illustrates the 2D nature of an isometric drawing created in AutoCAD. Isometric lines are created on the XY plane. The *Isometric SNAP* and *GRID* are also on the 2D plane. (The *Vpoint* command was used to give other than a *Plan* view of the drawing in this figure.)

Although pictorial drawings are based on the theory of projecting 3D objects onto 2D planes, it is physically possible to achieve an axonometric (isometric, dimetric, or trimetric) viewpoint of a 3D object using a 3D CAD system. In AutoCAD, the *Vpoint* command is used to specify the observer's position in 3D space with respect to a 3D model. Chapter 35 discusses the specific commands and values needed to attain axonometric viewpoints of a 3D model.

Figure 25-7

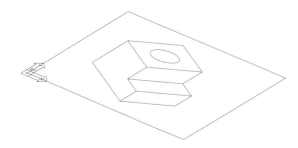

ISOMETRIC DRAWING IN AutoCAD

AutoCAD provides the capability to construct isometric drawings. An isometric *SNAP* and *GRID* are available, as well as a utility for creation of isometrically correct ellipses. Isometric lines are created with the *Line* command. There are no special options of *Line* for isometric drawing, but isometric *SNAP* and *GRID* can be used to force *Lines* to an isometric orientation. Begin creating an isometric drawing in AutoCAD by activating the *Isometric Style* option of the *Snap* command. This action can be done using any of the options listed in the following command table.

SNAP

Pull-down Menu	COMMAND (TYPE)	ALIAS (TYPE)	Short-cut	Screen (side) Menu	Tablet Menu
Tools *Drawing Aids...*	*SNAP*	*SN*	*F9 or* *Ctrl+B*	*TOOLS 2* *Ddrmodes*	*W,10*

```
Command: snap
Snap spacing or ON/OFF/Aspect/Rotate/Style <1.0000>: s
Standard/Isometric <S>: I
Vertical spacing <1.0000>: Enter
Command:
```

Alternately, toggling the indicated checkbox in the lower-right corner of the *Drawing Aids* dialog box activates the *Isometric SNAP* and *GRID* (Fig. 25-8).

Figure 25-8 —————

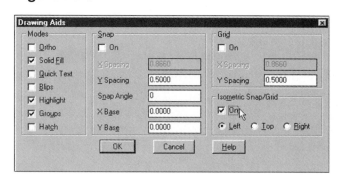

Figure 25-9 illustrates the effect of setting the *Isometric SNAP* and *GRID*. Notice the new position of the cursor.

Using <u>Ctrl+E</u> (pressing the Ctrl key and the letter "E" simultaneously) toggles the cursor to one of three possible *Isoplanes* (AutoCAD's term for the three faces of the isometric pictorial). If *ORTHO* is *ON*, only <u>isometric</u> lines are drawn; that is, you can only draw *Lines* aligned with the isometric axes. *Lines* can be drawn on only two axes for each isoplane. Ctrl+E allows drawing on the two axes aligned with another face of the object. *ORTHO* is *OFF* in order to draw inclined or oblique lines (not on the isometric axis). The functions of *GRID* (F7) and *SNAP* (F9) remain unchanged.

Figure 25-9 —————

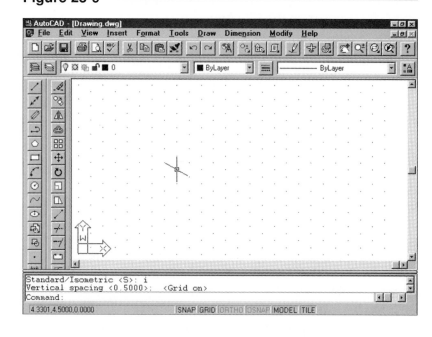

With *SNAP ON*, toggle *Coords* (F6) several times and examine the readout as you move the cursor. The <u>absolute coordinate format is of no</u> <u>particular assistance</u> while drawing in isometric because of the configuration of the *GRID*. The relative polar format, however, is very helpful. Use relative polar format for *Coords* while drawing in isometric (see Fig. 25-10).

The effects of changing the *Isoplane* are shown in the following figures. Press Ctrl+E to change *Isoplane*.

With *ORTHO ON*, drawing a *Line* is limited to the two axes of the current *Isoplane*. Only one side of a cube, for example, can be drawn on the current *Isoplane*. Watch *Coords* (in a polar format) to give the length of the current *Line* as you draw (lower-left corner of the screen).

Figure 25-10

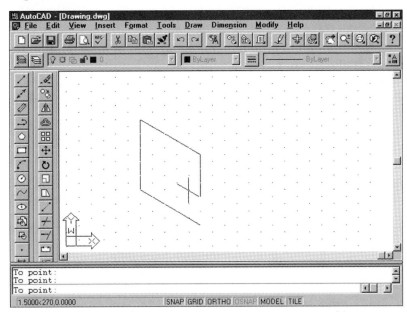

Toggling Ctrl+E switches the cursor and the effect of *ORTHO* to another *Isoplane*. One other side of a cube can be constructed on this *Isoplane* (Fig. 25-11).

Toggling Ctrl+E again forces the cursor and the effect of *ORTHO* to the third *Isoplane*. The third side of the cube can be constructed similarly.

Figure 25-11

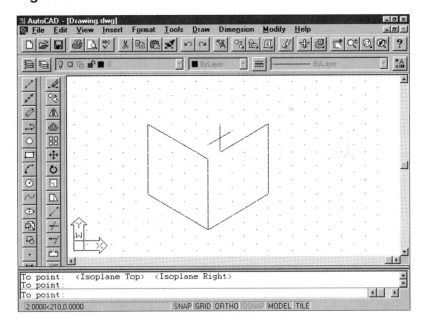

Isometric Ellipses

Isometric ellipses are easily drawn in AutoCAD by using the *Isocircle* option of the *Ellipse* command. This option appears only when the isometric *SNAP* is *ON*.

	Pull-down Menu	COMMAND (TYPE)	ALIAS (TYPE)	Short-cut	Screen (side) Menu	Tablet Menu
ELLIPSE	*Draw* *Ellipse*	*ELLIPSE*	*EL*	...	DRAW 1 *Ellipse*	M,9

Although the *Isocircle* option does not appear in the pull-down or digitizing tablet menus, it can be invoked as an option of the *Ellipse* command. You must type "I" to use the *Isocircle* option. The command syntax is as follows:

Command: **ellipse**
Arc/Center/Isocircle/<Axis endpoint 1>: **I**
Center of circle: **PICK** or (**coordinates**)
<Circle radius>/Diameter: **PICK** or (**coordinates**)
Command:

After selecting the center point of the *Isocircle*, the isometrically correct ellipse appears on the screen on the current *Isoplane*. Use Ctrl+E to toggle the ellipse to the correct orientation. When defining the radius interactively, use *ORTHO* to force the rubberband line to an isometric axis (Fig. 25-12).

Since isometric angles are equal, all isometric ellipses have the same proportion (major to minor axis). The only differences in isometric ellipses are the size and the orientation (*Isoplane*).

Figure 25-12 ────────────────

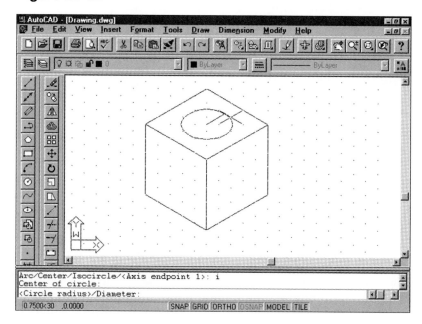

Figure 25-13 shows three ellipses correctly oriented on their respective faces. Use Ctrl+E to toggle the correct *Isoplane* orientation: *Isoplane Top*, *Isoplane Left*, or *Isoplane Right*.

When defining the radius or diameter of an ellipse, it should always be measured in an isometric direction. In other words, an isometric ellipse is always measured on the two isometric axes (or center lines) parallel with the plane of the ellipse.

If you define the radius or diameter interactively, use *ORTHO ON*. If you enter a value, AutoCAD automatically applies the value to the correct isometric axes.

Figure 25-13 ────────────────

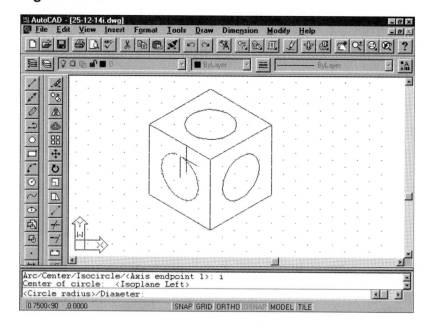

Creating an Isometric Drawing

In this exercise, the object in Figure 25-14 is drawn in isometric.

The initial steps to create an isometric drawing begin with the typical setup (see Chapter 6, Drawing Setup):

1. Set the desired *Units*.

2. Set appropriate *Limits*.

3. Set the *Isometric Style* of *Snap* and specify an appropriate value for spacing.

Figure 25-14 ————————

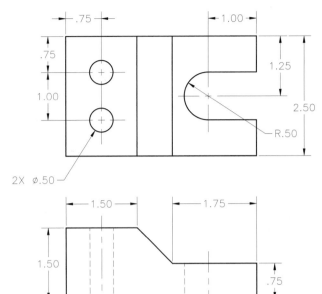

4. The next step involves creating an isometric framework of the desired object. In other words, draw an isometric box equal to the overall dimensions of the object. Using the dimensions given in Figure 25-14, create the encompassing isometric box with the *Line* command (Fig. 25-15).

   Use *ORTHO* to force isometric *Lines*. Watch the *Coords* display (in a relative polar format) to give the current lengths as you draw or use direct distance entry.

Figure 25-15 ————————

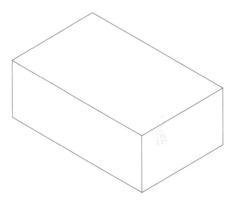

5. Add the lines defining the lower surface. Define the needed edge of the upper isometric surface as shown.

Figure 25-16 ————————

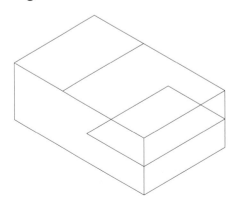

6. The <u>inclined</u> edges of the inclined surface can be drawn (with *Line*) only when *ORTHO* is *OFF*. <u>Inclined</u> lines in isometric cannot be drawn by transferring the lengths of the lines, but only by defining the <u>ends</u> of the inclined lines on <u>isometric</u> lines, then connecting the endpoints. Next, *Trim* or *Erase* the necessary *Lines*.

Figure 25-17

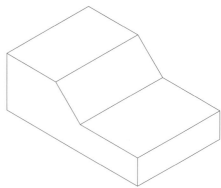

7. Draw the slot by constructing an *Ellipse* with the *Isocircle* option. Draw the two *Lines* connecting the circle to the right edge. *Trim* the unwanted part of the *Ellipse* (highlighted) using the *Lines* as cutting edges.

Figure 25-18

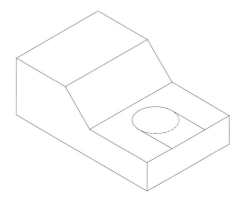

8. *Copy* the far *Line* and the *Ellipse* down to the bottom surface. Add two vertical *Lines* at the end of the slot.

Figure 25-19

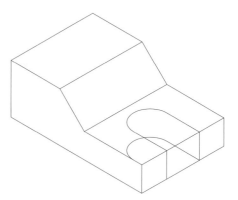

9. Use *Trim* to remove the part of the *Ellipse* that would normally be hidden from view. *Trim* the *Lines* along the right edge at the opening of the slot.

Figure 25-20

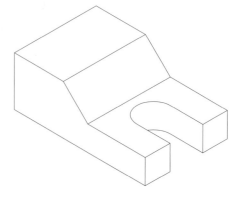

10. Add the two holes on the top with *Ellipse, Isocircle* option. Use *ORTHO ON* when defining the radius. *Copy* can also be used to create the second *Ellipse* from the first.

Figure 25-21

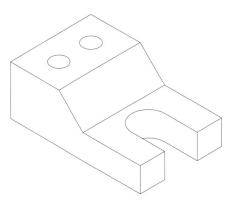

Dimensioning Isometric Drawings in AutoCAD

Refer to Chapter 28, Dimensioning, for details on how to dimension isometric drawings.

OBLIQUE DRAWING IN AutoCAD

Figure 25-22

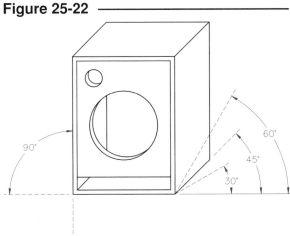

Oblique drawings are characterized by having two axes at a 90 degree orientation. Typically, you should locate the <u>front face</u> of the object along these two axes. Since the object's characteristic shape is seen in the front view, an oblique drawing allows you to create all shapes parallel to the front face true size and shape as you would in a multiview drawing. Circles on or parallel to the front face can be drawn as circles. The third axis, the receding axis, can be drawn at a choice of angles, 30, 45, or 60 degrees, depending on whether you want to show more of the top or the side of the object.

Figure 25-22 illustrates the axes orientation of an oblique drawing, including the choice of angles for the receding axis.

Another option allowed with oblique drawings is the measurement used for the receding axis. Using the full depth of the object along the receding axis is called <u>Cavalier</u> oblique drawing. This method depicts the object (a cube with a hole in this case) as having an elongated depth (Fig. 25-23).

Figure 25-23

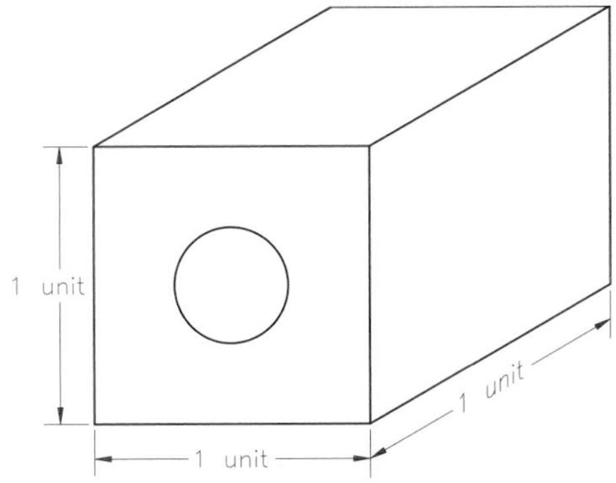

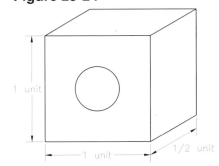

Using 1/2 or 3/4 of the true depth along the receding axis gives a more realistic pictorial representation of the object. This is called a <u>Cabinet</u> oblique (Fig. 25-24).

No functions or commands in AutoCAD are intended specifically for oblique drawing. However, *SNAP* and *GRID* can be aligned with the receding axis using the *Rotate* option of the *Snap* command to simplify the construction of edges of the object along that axis. The steps for creating a typical oblique drawing are given next.

Figure 25-24

The object in Figure 25-25 is used for the example. From the dimensions given in the multiview, create a cabinet oblique with the receding axis at 45 degrees.

Figure 25-25 —————————

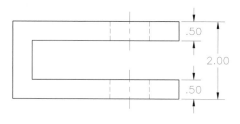

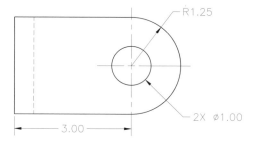

1. Create the characteristic shape of the front face of the object as shown in the front view.

Figure 25-26 —————————

2. Use *Copy* with the *Multiple* option to copy the front face back on the receding axis. Polar coordinates can be entered as the "second point of displacement." For example, the first *Copy* should be located at **@.25<45** relative to the "Base point." (Remember to calculate 1/2 of the actual depth.) As an alternative, a *Line* can be drawn at 45 degrees and *Divided* to place *Points* at .25 increments. *Multiple Copies* of the front face can be *OSNAP*ed with the *Node* option.

Figure 25-27 —————————

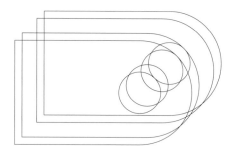

3. Draw the *Line* representing the edge on the upper-left of the object along the receding axis. Use *Endpoint OSNAP* to connect the *Lines*. Make a *Copy* of the *Line* or draw another *Line* .5 units to the right. Drop a vertical *Line* from the *Intersection* as shown.

Figure 25-28 —————————

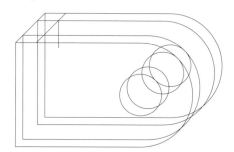

4. Use *Trim* and *Erase* to remove the unwanted parts of the *Lines* and *Circles* (those edges that are normally obscured).

Figure 25-29 ——————

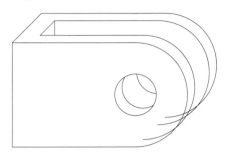

5. *Zoom* with a *window* to the lower right corner of the drawing. Draw a *Line Tangent* to the edges of the arcs to define the limiting elements along the receding axis. *Trim* the unwanted segments of the arcs.

Figure 25-30 ——————

The resulting cabinet oblique drawing should appear like that in Figure 25-31.

Figure 25-31 ——————

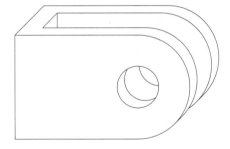

CHAPTER EXERCISES

Isometric Drawing

For exercises 1, 2, and 3 create isometric drawings as instructed. To begin, use an appropriate template drawing and draw a *Pline* border and insert the **TBLOCKAT**.

1. Create an isometric drawing of the cylinder shown in Figure 25-32. *Save* the drawing as **CYLINDER** and *Plot* so the drawing is *Scaled to Fit* on an A size sheet.

Figure 25-32

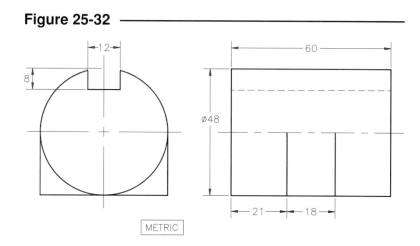

2. Make an isometric drawing of the Corner Brace shown in Figure 25-33. *Save* the drawing as **CRNBRACE**. *Plot* at **1=1** scale on an A size sheet.

Figure 25-33

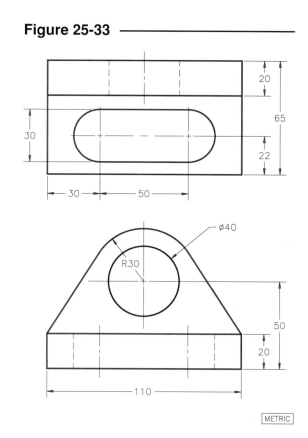

3. Draw the Support Bracket (Fig. 25-34) in isometric. The drawing can be *Plotted* at **1=1** scale on an A size sheet. *Save* the drawing and assign the name **SBRACKET**.

Figure 25-34 —————————————

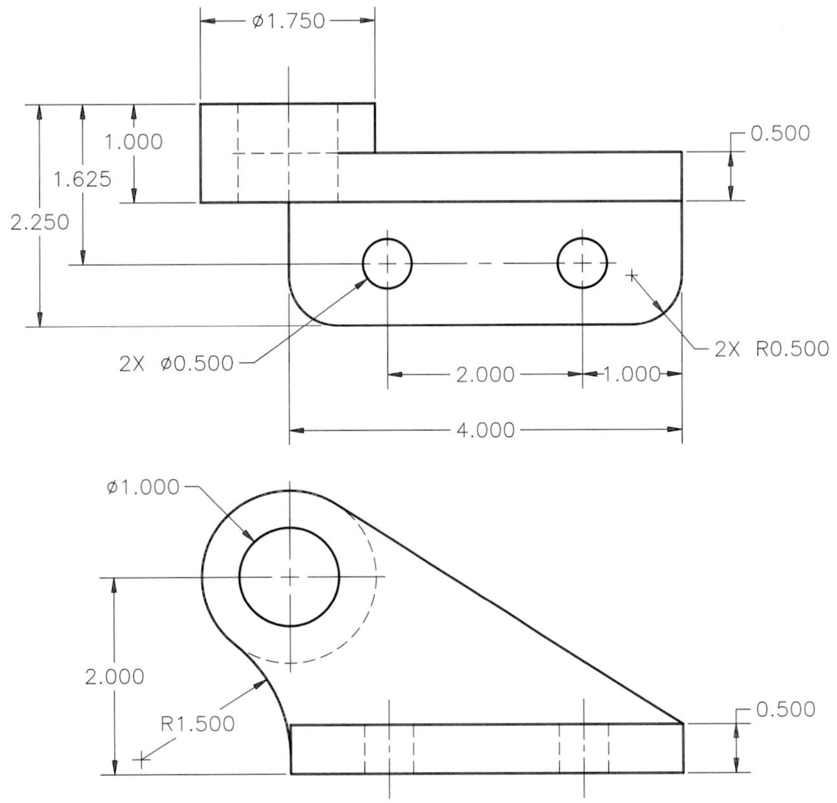

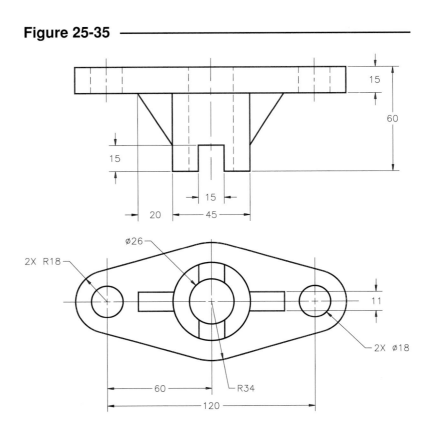

Oblique Drawing

For exercises 4 and 5, create oblique drawings as instructed. To begin, use an appropriate template drawing and draw a *Pline* border and *Insert* the **TBLOCKAT**.

4. Make an oblique cabinet projection of the Bearing shown in Figure 25-35. Construct all dimensions on the receding axis 1/2 of the actual length. Select the optimum angle for the receding axis to be able to view the 15 x 15 slot. *Plot* at **1=1** scale on an A size sheet. *Save* the drawing as **BEARING**.

Figure 25-35 —————————————

5. Construct a cavalier oblique drawing of the Pulley showing the circular view true size and shape. The illustration in Figure 25-36 gives only the side view. All vertical dimensions in the figure are diameters. *Save* the drawing as **PULLEY** and make a *Plot* on an A size sheet at **1=1**.

Figure 25-36

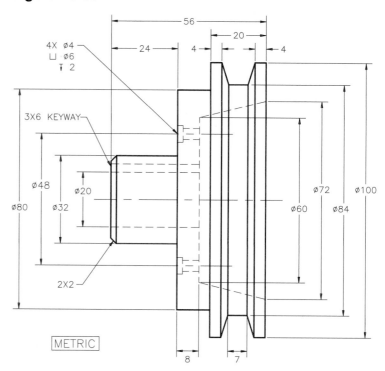

26

SECTION VIEWS

Chapter Objectives

After completing this chapter you should:

1. be able to use the *Bhatch* command to select associative hatch patterns;

2. be able to specify a *Scale* and *Angle* for hatch lines;

3. know how to define a boundary for hatching using the *Pick Points* and *Select Objects* methods;

4. be able to Preview the hatch and make necessary adjustments, then *Apply* the hatch pattern;

5. be able to use *Hatch* to create non-associative hatch lines and discard the boundary;

6. know how to use *Hatchedit* to modify parameters of existing hatch patterns in the drawing;

7. be able to edit hatched areas using *Grips*, selection options, *Convert*, and *Draworder*.

CONCEPTS

A section view is a view of the interior of an object after it has been imaginarily cut open to reveal the object's inner details. A section view is only one of two or more views of a multiview drawing describing the object. For example, a multiview drawing of a machine part may contain three views, one of which is a section view.

Hatch lines (also known as section lines) are drawn in the section view to indicate the solid material that has been cut through. Each combination of section lines is called a hatch pattern, and each pattern is used to represent a specific material. In full and half section views, hidden lines are omitted since the inside of the object is visible.

ANSI (American National Standards Institute) and ISO (International Organization of Standardization) have published standard configurations for section lines, and AutoCAD Release 14 supports those standards. However, ANSI no longer specifies section line pattern standards.

A cutting plane line is drawn in an adjacent view to the section view to indicate the plane that imaginarily cuts through the object. Arrows on each end of the cutting plane line indicate the line of sight for the section view. ANSI dictates that a thick dashed or phantom line be used as the standard cutting plane line.

This chapter discusses the AutoCAD methods used to draw hatch lines for section views and related cutting plane lines. The *Bhatch* (boundary hatch) command allows you to select an enclosed area and select the hatch pattern and the parameters for the appearance of the hatch pattern; then AutoCAD automatically draws the hatch (section) lines. Existing hatch lines in the drawing can be modified using *Hatchedit*. Cutting plane lines are created in AutoCAD by using a dashed linetype. The line itself should be created with the *Pline* command to achieve line width.

DEFINING HATCH PATTERNS AND HATCH BOUNDARIES

A hatch pattern is composed of many lines that have a particular linetype, spacing, and angle. Many standard hatch patterns are provided by AutoCAD for your selection. Rather than having to draw each section line individually, you are required only to specify the area to be hatched and AutoCAD fills the designated area with the selected hatch pattern. An AutoCAD hatch pattern is inserted as one object. For example, you can *Erase* the inserted hatch pattern by selecting only one line in the pattern, and the entire pattern in the area is *Erased*.

Figure 26-1

In a typical section view (Fig. 26-1), the hatch pattern completely fills the area representing the material that has been cut through. With the *Bhatch* command you can define the boundary of an area to be hatched simply by pointing inside of an enclosed area.

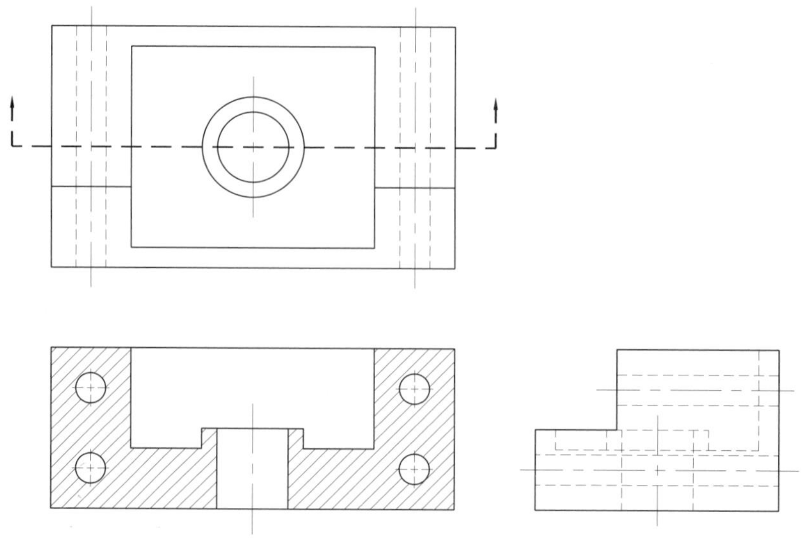

Both the *Hatch* and the *Bhatch* commands fill a specified area with a selected hatch pattern. *Hatch* requires that you select <u>each object</u> defining the boundary, whereas the *Bhatch* command finds the boundary automatically when you point inside it. Additionally, *Hatch* operates in command line format, whereas *Bhatch* operates in dialog box fashion. For these reasons, *Bhatch* is superior to *Hatch* and is recommended in most cases for drawing section views.

Hatch patterns created with *Bhatch* are <u>associative</u>. Associative hatch patterns are associated to the boundary geometry such that when the shape of the boundary changes (by *Stretch, Scale, Rotate, Move, Ddmodify, Grips,* etc.), the hatch pattern automatically reforms itself to conform to the new shape (Fig. 26-2). For example, if a design change required a larger diameter for a hole, *Ddmodify* could be used to change the diameter of the hole, and the surrounding section lines would automatically adapt to the new diameter.

Figure 26-2

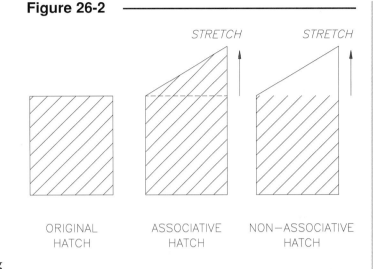

Once the hatch patterns have been drawn, any feature of the existing hatch pattern (created with *Bhatch*) can be changed retroactively using *Hatchedit*. The *Hatchedit* dialog box gives access to the same options that were used to create the hatch (in the *Boundary Hatch* dialog box). Changing the scale, angle, or pattern of any existing section view in the drawing is a simple process.

Steps for Creating a Section View Using the *Bhatch* Command

1. Create the view that contains the area to be hatched using typical draw commands such as *Line, Arc, Circle,* or *Pline.* If you intend to have text or dimensions inside the area to be hatched, add them before hatching.

2. Invoke the *Bhatch* command. The *Boundary Hatch* dialog box appears (Fig. 26-3).

3. Specify the *Pattern Type* to use. Select the desired pattern by clicking on the image tile until the desired pattern appears, select from the *Pattern* drop-down list or select the *Pattern* tile to allow you to select from the *Hatch pattern palette* image tiles.

4. Specify the *Scale* and *Angle* in the dialog box.

5. Define the area to be hatched by PICKing an internal point (AutoCAD automatically traces the boundary) or by individually selecting the objects.

6. *Preview* the hatch to make sure everything is as expected. Adjust hatching parameters as necessary and *Preview* again.

7. *Apply* the hatch. The hatch pattern is automatically drawn and becomes an associated object in the drawing.

8. If other areas are to be hatched, additional internal points or objects can be selected to define the new area for hatching. The parameters previously used are again applied to the new hatch area by default. You may want to *Inherit Properties* from a previously applied hatch.

9. For mechanical drawings, draw a cutting plane line in a view adjacent to the section view. The *Pline* command with a *Dashed* or *Phantom* linetype is used. Arrows at the ends of the cutting plane line indicate the line of sight for the section view.

10. If any aspect of the hatch lines needs to be edited at a later time, *Hatchedit* can be used to change those properties. If the hatch boundary is changed by *Stretch, Rotate, Scale, Move, Ddmodify,* etc., the hatched area will conform to the new boundary.

The *Bhatch* and *Hatchedit* commands with all dialog boxes and options are explained in detail on this and the following pages.

BHATCH

Pull-down Menu	COMMAND (TYPE)	ALIAS (TYPE)	Short-cut	Screen (side) Menu	Tablet Menu
Draw *Hatch*	*BHATCH*	*BH or H*	...	DRAW 2 *Bhatch*	*P,9*

Bhatch allows you to create hatch lines for a section view (or for other purposes) by simply PICKing inside a closed boundary. A closed boundary refers to an area completely enclosed by objects. *Bhatch* locates the closed boundary automatically by creating a <u>temporary *Pline*</u> that follows the outline of the hatch area, fills the area with hatch lines, and then deletes the boundary (default option) after hatching is completed. *Bhatch* ignores all objects or parts of objects that are not part of the boundary.

Any method of invoking *Bhatch* yields the *Boundary Hatch* dialog box (Fig. 26-3). Typically, the first step in the *Boundary Hatch* dialog box is the selection of a hatch pattern.

Figure 26-3

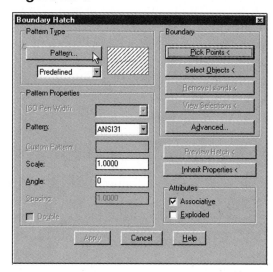

Boundary Hatch Dialog Box Options

Pattern Type
This option allows you to specify the type of the hatch pattern: *Predefined, User-defined,* or *Custom*. Use *Predefined* for standard hatch pattern styles that AutoCAD provides (in the ACAD.PAT file).

Predefined
There are three ways to select from *Predefined* patterns: (1) select the *Pattern...* tile to produce the *Hatch pattern palette* dialog box displaying hatch pattern names and image tiles (Fig. 26-4), (2) select the *Pattern:* drop-down list to PICK the pattern name, or (3) select the single image tile (in the *Boundary Hatch* dialog box) to sequence through the possible choices.

Figure 26-4

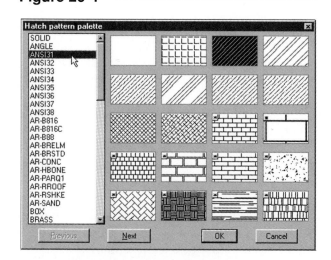

R14

The *Hatch pattern palette* dialog box (Fig. 26-4) allows you to select a predefined pattern by its image tile or by its name. There are four screens of image tiles. Use the *Next* and *Previous* buttons to cycle through the four screens.

User-defined

To define a simple hatch pattern "on the fly," select the *User-defined* tile. This causes the *Pattern* and *Scale* options to be disabled and the *Angle, Spacing,* and *Double* options to be enabled. Creating a *User-defined* pattern is easy. Specify the *Angle* of the lines, the *Spacing* between lines, and optionally create *Double* (perpendicular) lines. All *User-defined* patterns have continuous lines.

Custom

Custom patterns are previously created user-defined patterns stored in other than the ACAD.PAT file. Custom patterns can contain continuous, dashed and dotted line combinations. See the AutoCAD Customization Guide for information on creating and saving custom hatch patterns.

The list of AutoCAD predefined patterns is shown below. The patterns are defined and stored in the ACAD.PAT file.

R14

Pattern Type	Description
SOLID	- Solid color, current color setting
ANGLE	- Angle steel
ANSI31	- ANSI Iron, Brick, Stone masonry
ANSI32	- ANSI Steel
ANSI33	- ANSI Bronze, Brass, Copper
ANSI34	- ANSI Plastic, Rubber
ANSI35	- ANSI Fire brick, Refractory material
ANSI36	- ANSI Marble, Slate, Glass
ANSI37	- ANSI Lead, Zinc, Magnesium, Sound/Heat/Elec Insulation
ANSI38	- ANSI Aluminum
AR-B816	- 8x16 Block elevation stretcher bond
AR-B816C	- 8x16 Block elevation stretcher bond with mortar joints
AR-B88	- 8x8 Block elevation stretcher bond
AR-BRELM	- Standard brick elevation english bond with mortar joints
AR-BRSTD	- Standard brick elevation stretcher bond
AR-CONC	- Random dot and stone pattern
AR-HBONE	- Standard brick herringbone pattern @ 45 degrees
AR-PARQ1	- 2x12 Parquet flooring: pattern of 12 x 12
AR-RROOF	- Roof shingle texture
AR-RSHKE	- Roof wood shake texture
AR-SAND	- Random dot pattern
BOX	- Box steel
BRASS	- Brass material
BRICK	- Brick or masonry-type surface
BRSTONE	- Brick and stone
CLAY	- Clay material
CORK	- Cork material

CROSS	- A series of crosses
DASH	- Dashed lines
DOLMIT	- Geological rock layering
DOTS	- A series of dots
EARTH	- Earth or ground (subterranean)
ESCHER	- Escher pattern
FLEX	- Flexible material
GRASS	- Grass area
GRATE	- Grated area
HEX	- Hexagons
HONEY	- Honeycomb pattern
HOUND	- Houndstooth check
INSUL	- Insulation material
LINE	- Parallel horizontal lines
MUDST	- Mud and sand
NET	- Horizontal/vertical grid
NET3	- Network pattern 0-60-120
PLAST	- Plastic material
PLASTI	- Plastic material
SACNCR	- Concrete
SQUARE	- Small aligned squares
STARS	- Star of David
STEEL	- Steel material
SWAMP	- Swampy area
TRANS	- Heat transfer material
TRIANG	- Equilateral triangles
ZIGZAG	- Staircase effect
ISO02W100	- dashed line
ISO03W100	- dashed space line
ISO04W100	- long dashed dotted line
ISO05W100	- long dashed double-dotted line
ISO06W100	- long dashed triple-dotted line
ISO07W100	- dotted line
ISO08W100	- long dashed short dashed line
ISO09W100	- long dashed double-short-dashed line
ISO10W100	- dashed dotted line
ISO11W100	- double-dashed dotted line
ISO12W100	- dashed double-dotted line
ISO13W100	- double-dashed double-dotted line
ISO14W100	- dashed triple-dotted line
ISO15W100	- double-dashed triple-dotted line

Figure 26-5 displays each of the AutoCAD hatch patterns defined in the ACAD.PAT file. Note that the patterns are not shown to scale.

Figure 26-5

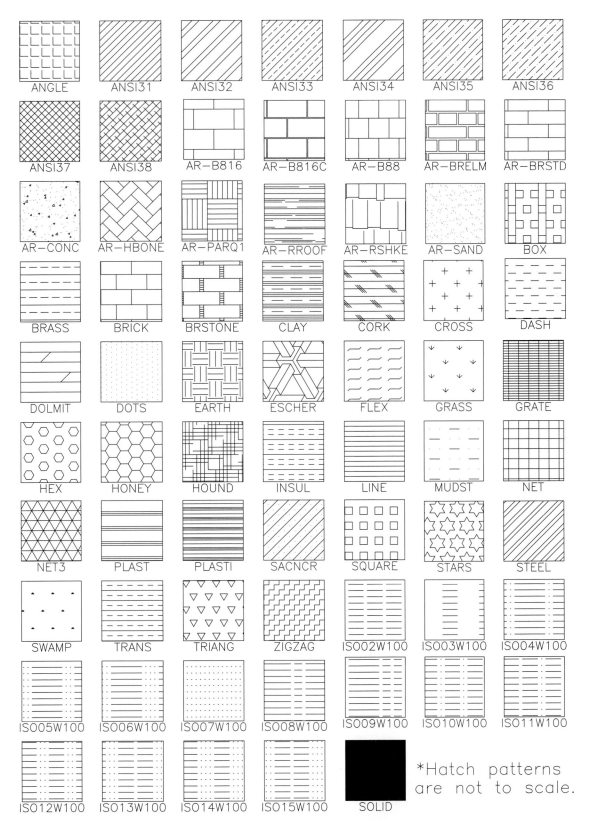

Since hatch patterns have their own linetype, the *Continuous* linetype or a layer with the *Continuous* line-type <u>should be current</u> when hatching. After selecting a pattern, specify the desired *Scale* and *Angle* of the pattern or ISO pen width.

Pattern Properties
This cluster (Fig. 26-3) is used to specify the type of hatch pattern to be drawn and the parameters that govern its appearance.

ISO Pen Width
You must select an ISO hatch pattern for this tile to be enabled. Selecting an *ISO Pen Width* from the drop-down list automatically sets the scale and enters the value in the *Scale* edit box. See *Scale*.

Pattern...
Selecting the *Pattern* drop-down list displays the name of each predefined pattern. Making a selection sets that as the current pattern and causes the pattern to display in the window above. Using the up and down arrow keys causes the name and display (in the image tile) to sequence.

Scale
The value entered in this edit box is a scale factor that is applied to the existing selected pattern. Normally, this scale factor should be changed proportionally with changes in the drawing *Limits*. Like many other scale factors (*LTSCALE, DIMSCALE*), AutoCAD defaults are set to a value of 1, which is appropriate for the default *Limits* of 12 x 9. If you have calculated the drawing scale factor, it can be applied here (see Chapter 13). In the example, *Limits* are set to 22 x 17, so a *Scale* of 1.8 is used. The *Scale* value is stored in the *HPSCALE* system variable.

ISO hatch patterns are intended for use with metric drawings; therefore, the scale (spacing between hatch lines) is much greater than for inch drawings. Because *Limits* values for metric sheet sizes are greater than for inch-based drawings (25.4 times greater than comparable inch drawings), the ISO hatch pattern scales are automatically compensated. If you want to use an ISO pattern with inch-based draw-ings, calculate a hatch pattern scale based on the drawing scale factor and multiply by .039 (1/25.4).

Angle
The *Angle* specification determines the angle (slant) of the hatch lines. The default angle of 0 represents whatever angle is displayed in the pattern's image tile. Any value entered deflects the existing pattern by the specified value (in degrees). The value entered in this box is held in the *HPANG* system variable.

Spacing
This option is enabled only if *User-defined* pattern is specified. Enter a value for the distance between lines.

Double
Only for a *User-defined* pattern, check this box to have a second set of lines drawn at 90 degrees to the original set.

Boundary—Selecting the Hatch Area

Once the hatch pattern and options have been selected, you must indicate to AutoCAD what area(s) should be hatched. Either the *Pick Points* method, *Select Objects* method, or a combination of both can be used to accomplish this.

Pick Points <
This tile should be selected if you want AutoCAD to automatically determine the boundaries for hatch-ing. You only need to select a point <u>inside</u> the area you want to hatch. The point selected must be inside a <u>completely closed shape</u>. When the *Pick Points* tile is selected, AutoCAD gives the following prompts:

Select internal point: **PICK**
Selecting everything...
Selecting everything visible...
Analyzing the selected data...
Analyzing the internal islands...
Select internal point: **PICK** another area or **Enter**

When an internal point is PICKed, AutoCAD traces and highlights the boundary (Fig. 26-6). The interior area is then analyzed for islands to be included in the hatch boundary. Multiple boundaries can be designated by selecting multiple internal points. Type *U* to undo the last one, if necessary.

Figure 26-6

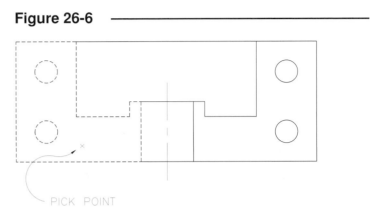

PICK POINT

The location of the point selected is usually not critical. However, the point must be PICKed inside the expected boundary. If there are any gaps in the area, a complete boundary cannot be formed and a boundary error message appears (Fig. 26-7).

Figure 26-7

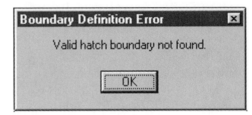

Select Objects<
Alternately, you can designate the boundary with the *Select Objects* method. Using the *Select Objects* method, you specify the boundary objects rather than let AutoCAD locate a boundary. With the *Select Objects* method, no temporary *Pline* boundary is created as with the *Pick Points* method. Therefore, the selected objects must form a closed shape with no gaps or overlaps. If gaps exist (Fig. 26-8) or if the objects extend past the desired hatch area (Fig. 26-9), AutoCAD cannot interpret the intended hatch area correctly, and problems will occur.

Figure 26-8

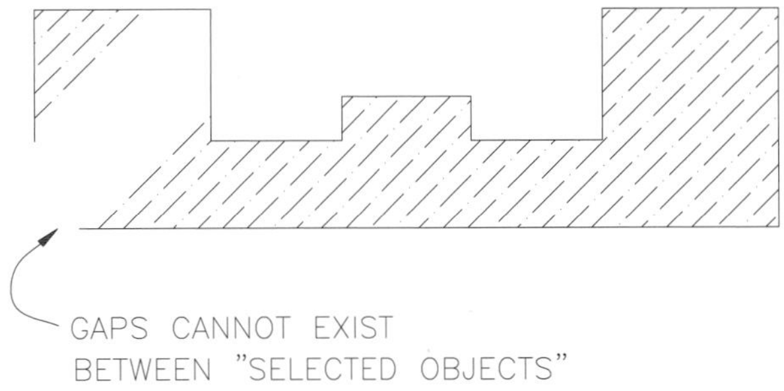

GAPS CANNOT EXIST
BETWEEN "SELECTED OBJECTS"

The *Select Objects* option can be used after the *Pick Points* method to select specific objects for *Bhatch* to consider before drawing the hatch pattern lines. For example, if you had created text objects or dimensions within the boundary found by the *Pick Points* method, you must then use the *Select Objects* option to select the text and dimensions. Using this procedure, the hatch lines are automatically "trimmed" around the text and dimensions (see Fig. 26-18). If you do not use *Select Objects* to indicate these items, *Bhatch* does not recognize the text and dimensions and draws the hatch lines right over them.

Figure 26-9

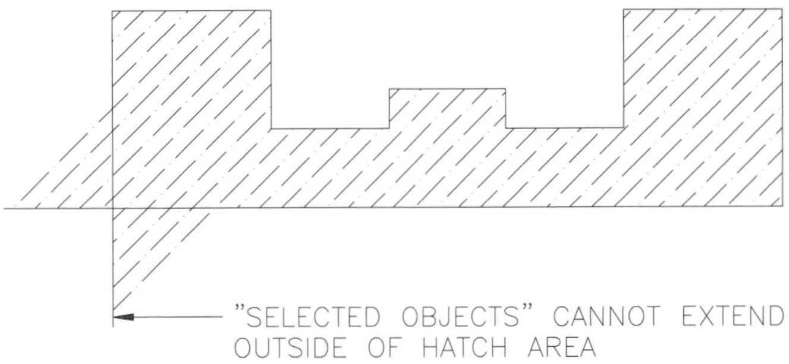

"SELECTED OBJECTS" CANNOT EXTEND
OUTSIDE OF HATCH AREA

R13

Remove Islands <

PICKing this tile allows you to select specific islands to remove from those AutoCAD has found within the outer boundary. If hatch lines have been drawn, be careful to *Zoom* in close enough to select the desired boundary and not the hatch pattern.

View Selections <

Clicking the *View Selections* tile causes AutoCAD to highlight all selected boundaries. This can be used as a check to ensure the desired areas are selected.

Advanced...

This tile invokes the *Advanced Options* dialog box (see *Advanced Options* Dialog Box).

Preview Hatch <

You should always use the *Preview Hatch* option after specifying the hatch parameters and selecting boundaries, but before you *Apply* the hatch. This option allows you to temporarily look at the hatch pattern in your drawing with the current settings applied and allows you to adjust the settings, if necessary, before using *Apply*. After viewing the drawing, pressing the *Continue* tile redisplays the *Boundary Hatch* dialog box, allowing you to make adjustments.

Inherit Properties <

This option allows you to select a hatch pattern from one existing in the drawing. The dialog box disappears and the "Select hatch objects:" prompt appears. PICKing a previously drawn hatch pattern resets all the parameters in the *Boundary Hatch* dialog box for subsequent application to another hatch area.

Associative

This checkbox toggles (on or off) the associative property for newly applied hatch patterns. Associative hatch patterns automatically update by conforming to the new boundary shape when the boundary is changed (see Fig. 26-2). A non-associative hatch pattern is static even when the boundary changes.

Exploded

Bhatch normally draws a hatch pattern as one object, similar to a *Block*. Checking this box causes the hatch patterns to be pre-exploded when inserted. (This is similar to *Inserting* a *Block* with an asterisk [*] prefixing the *Block* name.) Exploded hatch patterns allow you to edit the hatch lines individually. Exploded hatch patterns are <u>non-associative</u>.

Apply

This is typically the <u>last step</u> to the hatching process with *Bhatch*. Selecting *Apply* causes AutoCAD to create the hatch using all the parameters specified in the *Boundary Hatch* dialog box and return to the Command: prompt for further drawing and editing.

Advanced Options Dialog Box

Figure 26-10 ─────────────

Define Boundary Set

By default, AutoCAD examines <u>all objects on the screen</u> when determining the boundaries by the *Pick Points* method. (The *From Everything on Screen* box is checked by default when you begin the *Bhatch* command.) For complex drawings, examining all objects can take some time. In that case, you may want to specify a smaller boundary set for AutoCAD to consider. Clicking the *Make New Boundary Set* tile clears the dialog boxes and permits you to select objects or select a window to define the new set.

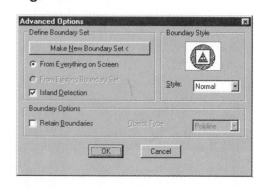

Style

This section allows you to specify how the hatch pattern is drawn when the area inside the defined boundary contains text or closed areas (islands). Select the text from the list or PICK the icons displayed in the window. These options are applicable <u>only</u> when interior objects (islands) have been included in the selection set (considered for hatching). Otherwise, if only the outer shape is included in the selection set, the results are identical to the *Ignore* option.

Figure 26-11

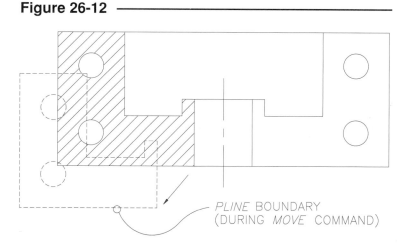

Normal
This should be used for most applications of *Bhatch*. Text or closed shapes within the outer border are considered in such cases. Hatching will begin at the outer boundary and move inward alternating between applying and not applying the pattern as interior shapes or text are encountered (Fig. 26-11).

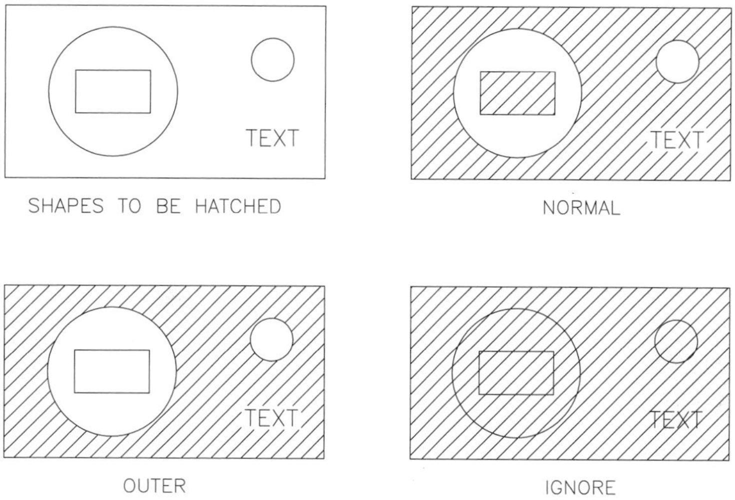

Outer
This option causes AutoCAD to hatch only the outer closed shape. Hatching is turned off for all interior closed shapes (Fig. 26-11).

Ignore
Ignore draws the hatch pattern from the outer boundary inward ignoring any interior shapes. The resulting hatch pattern is drawn through the interior shapes (Fig. 26-11).

Island Detection

Removing the check in this box causes AutoCAD to trace the outer boundary only and ignore any internal closed areas (islands) when the hatch pattern is applied.

Retain Boundaries

When AutoCAD uses the *Pick Points* method to locate a boundary for hatching, a temporary *Pline* is created for hatching, then discarded after the hatching process. Checking this box forces AutoCAD to <u>keep</u> the boundary. When the box is checked, *Bhatch* creates two objects—the hatch pattern and the boundary object. (You can specify whether you want to create a *Pline* or a *Region* boundary in the *Object Type* drop-down list.) Using this option and erasing the hatch pattern accomplishes the same results as using the *Boundary* command.

Figure 26-12

After using *Bhatch*, the *Pline* or *Region* can be used for other purposes. To test this function, complete a *Bhatch* with the *Retain Boundaries* box checked; then use *Move* (make sure you select the boundary) to reveal the new boundary object (Fig. 26-12).

Object Type

Bhatch creates *Polyline* or *Region* boundaries. This option is enabled only when the *Retain Boundaries* box is checked (see Fig. 26-10).

HATCH

Pull-down Menu	COMMAND (TYPE)	ALIAS (TYPE)	Short-cut	Screen (side) Menu	Tablet Menu
...	*HATCH*	*-H*	...	...	...

Hatch must be typed at the keyboard since it is not available from the menus. *Hatch* operates in a command line format with many of the options available in the *Bhatch* dialog box. However, *Hatch* creates a <u>non-associative</u> hatch pattern and <u>does not use the *Pick points*</u> method ("select internal point:") to create a boundary. Generally, *Bhatch* would be used instead of *Hatch* except for special cases.

You can use *Hatch* to find a boundary from existing objects. The objects must form a complete closed shape, just as with *Bhatch*. The shape can be composed of <u>one</u> closed object such as a *Pline, Polygon, Spline,* or *Circle* or composed of <u>several</u> objects forming a closed area such as *Lines* and *Arcs*. For example, the shape in Figure 26-13 can be hatched correctly with *Hatch,* whether it is composed of one object (*Pline, Region,* etc.) or several objects (*Lines, Arc,* etc.).

Figure 26-13 ────────────

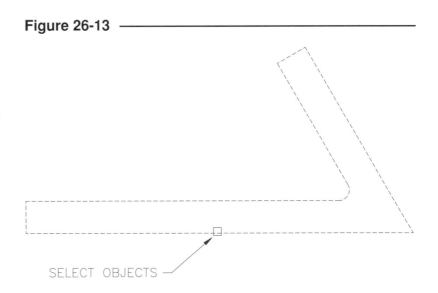

SELECT OBJECTS

If you select several objects to define the boundary, the objects must comprise <u>only the boundary shape</u> and not extend past the desired boundary. If objects extend past the desired boundary (Fig. 26-14) or if there are gaps, the resulting hatch will be <u>incorrect</u> as shown in Figures 26-8 and 26-9.

Figure 26-14 ────────────

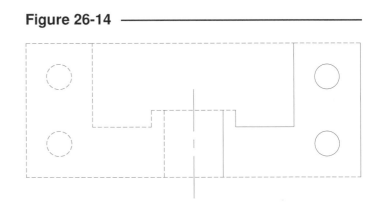

```
Command: hatch
Enter pattern name or [?/Solid/User defined]<ANSI31>: (style or pattern name)
Scale for pattern <1.0000>: (value) or Enter
Angle for pattern <0>: (value) or Enter
Select hatch boundaries or RETURN for direct hatch option
Select objects: PICK
Select objects: Enter
Command:
```

Direct Hatch

With the *Direct Hatch* method of *Hatch*, you specify points to define the boundary, not objects. The significance of this method is that you can create a hatch pattern without using existing objects as the boundary. In addition, you can select whether you want to retain or discard the boundary after the pattern is applied (Fig. 26-15).

Figure 26-15

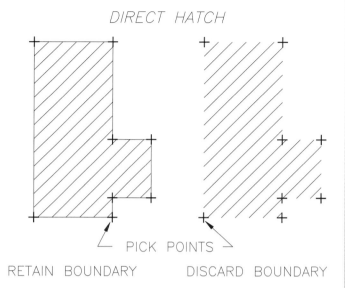

DIRECT HATCH

PICK POINTS

RETAIN BOUNDARY   DISCARD BOUNDARY

```
Command: hatch
Enter pattern name or [?/Solid/User defined] <ANSI31>: (pattern name, style) or Enter
Scale for pattern <1.0000>: (value) or Enter
Angle for pattern <0>: (value) or Enter
Select hatch boundaries or RETURN for direct hatch option
Select objects: Enter
Retain polyline? <N> Y or Enter
From point: PICK or (coordinates)
Arc/Close/Length/Undo/<Next point>: PICK or (coordinates)
Arc/Close/Length/Undo/<Next point>: PICK or (coordinates)
Arc/Close/Length/Undo/<Next point>: PICK or (coordinates)
Arc/Close/Length/Undo/<Next point>: c
From point or RETURN to apply hatch: Enter
Command:
```

For special cases when you do not want a hatch area boundary to appear in the drawing (Fig. 26-16), you can use the *Direct Hatch* method to specify a hatch area, then discard the boundary *Pline*.

Figure 26-16

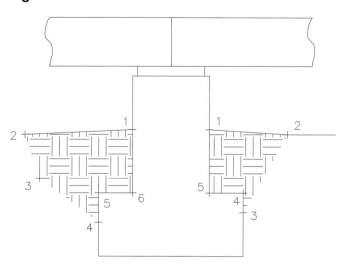

EDITING HATCH PATTERNS AND BOUNDARIES

HATCHEDIT

	COMMAND (TYPE)	ALIAS (TYPE)	Short-cut	Screen (side) Menu	Tablet Menu
Pull-down Menu					
Modify Object > Hatchedit	*HATCHEDIT*	*HE*	...	*MODIFY1 Hatchedt*	*Y,16*

Figure 26-17

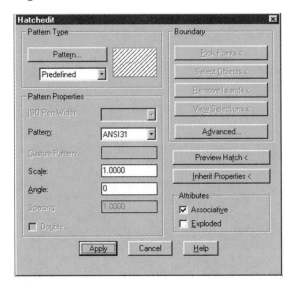

Hatchedit allows you to modify an <u>existing associative</u> hatch pattern in the drawing. This feature of AutoCAD makes the hatching process more flexible because you can hatch several areas to quickly create a "rough" drawing, then retroactively fine-tune the hatching parameters when the drawing nears completion with *Hatchedit*.

Invoking the *Hatchedit* command by any method produces the *Hatchedit* dialog box (Fig. 26-17). The dialog box provides options for changing the *Pattern, Scale, Angle,* and *Style* properties of the existing hatch. You can also change the hatch to non-associative and *Exploded*.

Apparent in Figure 26-17, the *Hatchedit* dialog box is essentially the same as the *Boundary Hatch* dialog box with some of the options disabled. For an explanation of the options that are available with *Hatchedit*, see *Boundary Hatch* earlier in this chapter.

Exploded or Associative Hatches?

Exploded hatches allow you to edit individual hatch lines. For example, you may want to *Trim* an area out of a hatch pattern to place a construction note, etc. To explode a hatch pattern, use either the *Hatchedit* dialog box or the *Explode* command.

Although exploded hatches give you the ability to edit the hatch lines individually, the exploded hatch loses its associativity and therefore loses its ability to be edited with *Hatchedit* or to update when the related boundary is changed.

It is generally preferred to retain hatch associativity. The associative feature usually gives you more flexibility in editing hatch patterns. If dimensions or text objects are <u>created before you use *Bhatch*</u> and you use the *Select Objects* button to indicate those objects when you specify the boundary, the hatch pattern is automatically drawn correctly around those objects as if you had *Trimmed* the hatch pattern around the objects (Fig. 26-18).

Figure 26-18

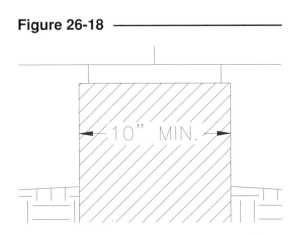

Using Grips with Hatch Patterns

Grips can be used effectively to edit hatch pattern boundaries of associative hatches. Do this by selecting the boundary with the pickbox or automatic window/crossing window (as you would normally activate an object's grips). Associative hatches retain the association with the boundary after the boundary has been changed using grips (see Fig. 26-2, earlier this chapter).

Beware, when selecting the hatch boundary, ensure you select only the boundary and not the hatch object. If you edit the hatch object with grips, the associative feature is lost. This can happen if you select the hatch object grip (Fig. 26-19). In this case, AutoCAD issues the following prompt:

Figure 26-19

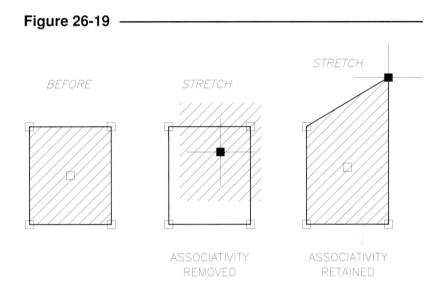

BEFORE

STRETCH

STRETCH

ASSOCIATIVITY
REMOVED

ASSOCIATIVITY
RETAINED

Command: (select grip on hatch object)
STRETCH
<Stretch to point>/Base point/Copy/Undo/eXit: **PICK**
Hatch boundary associativity removed.
Command:

If you activate grips on both the hatch object and the boundary and make a boundary grip hot, the boundary and related hatch retain associativity (Fig. 26-19).

Object Selection Features of Hatch Patterns

When you select a hatch object for editing (PICK only one line of an entire hatch pattern), you actually select the entire hatch object. By default, when you PICK any element of the hatch pattern, the hatch object can be *Moved, Copied*, grip edited, etc. However, when you edit the hatch object but not the boundary associativity is removed (see Using Grips with Hatches).

In Release 14, you can easily control whether just the hatch object or the hatch object and boundary together are selected when you PICK the hatch object. Do this by selecting *Associative Hatch* in the *Object Selection Settings* dialog box (Fig. 26-20). Type *Ddselect* or choose *Selection...* from the *Tools* pull-down menu to produce the dialog box. Your choice for the *Associative Hatch* option sets the *PICKSTYLE* variable (see Appendix A).

Figure 26-20

With *Associative Hatch* selected, PICKing the hatch object results in highlighting both the hatch object and boundary; therefore, you always edit the hatch object and the boundary together (Fig. 26-21). When *Associative Hatch* is not selected, PICKing the hatch object results in only the hatch object selection; therefore, you run the risk of editing the hatch object independent of the boundary and losing associativity.

Figure 26-21

OBJECT SELECTION WITH ASSOCIATIVE HATCH CHECKED   OBJECT SELECTION WITH ASSOCIATIVE HATCH NOT CHECKED

Lightweight Hatches

Previous releases of AutoCAD stored data such as width, color, linetype, and pattern (as well as coordinate data) for every *Pline* vertex and *Hatch* object element. To reduce the file size and speed up drawing time, Release 14 stores *Plines* and *Hatch* objects in a more optimized format called "lightweight" *Plines* and *Hatch* objects. Release 14 *Plines* and *Hatch* objects are automatically "lightweight."

Hatch objects in previous releases of AutoCAD were listed as unnamed *Blocks*. Using the *List* command reveals the difference between the old "Block Reference" and the new "Hatch" object.

When you work with Release 13 and earlier drawings, you can reduce the file size by converting *Hatch* objects to "lightweight" hatches. (*Plines* are automatically converted upon opening the old drawing. See Lightweight *Plines*, Chapter 8.) There are two methods for converting pre-Release 14 *Hatch* objects:

1. Use *Hatchedit*
2. Use *Convert*

To convert *Hatch* objects one at a time, invoke the *Hatchedit* dialog box, Then, simply select the hatch object and choose *Apply*. The conversion is complete. To convert all *Hatch* objects in the drawing at one time, use *Convert*.

CONVERT

Pull-down Menu	COMMAND (TYPE)	ALIAS (TYPE)	Short-cut	Screen (side) Menu	Tablet Menu
...	*CONVERT*	...	...	...	...

Convert can be used to convert pre-Release 14 *Plines* and *Hatch* objects to the new, more optimized "lightweight" data structure (see previous discussion and Lightweight *Plines*, Chapter 8). *Convert* can make the conversion of these objects one at a time or can convert all *Plines* and *Hatch* patterns globally. The command is simple to use. The following syntax shown here is for the conversion of all *Hatch* objects in the drawing:

```
Command: Convert
Hatch/Polyline/<All>: h
Select/<All>: Enter
8 hatch objects converted.
Command:
```

Solid *Hatch* Fills

In Release 14, a *Solid* hatch pattern is available. This feature is welcomed by professionals who create hatch objects such as walls for architectural drawings and require the solid filled areas. There are many applications for this feature.

You can use the *Fill* command (or *FILLMODE* system variable) to control the visibility of the solid filled areas. *Fill* also controls the display of solid filled TrueType fonts (see Chapter 18). If you want to display solid filled *Hatch* objects as solid, set *Fill* to *On* (Fig. 26-22). Setting *Fill* to *Off* causes AutoCAD <u>not</u> to display the solid filled areas. *Regen* must be used after *Fill* to display the new visibility state. This solid fill control can be helpful for saving toner or ink during test prints or speeding up plots when much solid fill is used in a drawing.

Figure 26-22 ————————————————————————

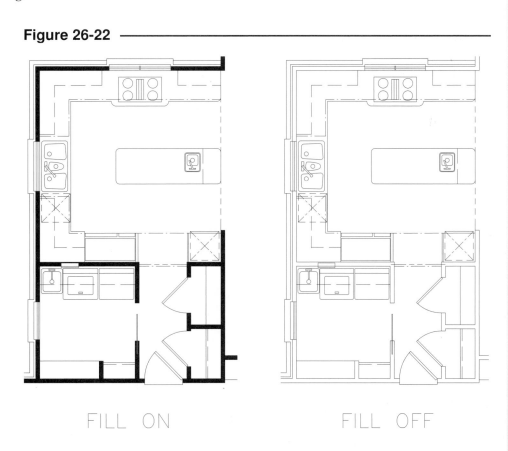

FILL ON   FILL OFF

DRAWORDER

Pull-down Menu	COMMAND (TYPE)	ALIAS (TYPE)	Short-cut	Screen (side) Menu	Tablet Menu
Tools *Display Order >*	*DRAWORDER*	*DR*	...	*TOOLS 1* *Drawordr*	*T,9*

New possibilities exist in Release 14 such as using solid hatch patterns and raster images in combination with filled text, other hatch patterns and solid images, etc. With these solid areas and images, some control of which objects are in "front" and "back or "above" and "under" must be provided. This control is provided by the *Draworder* command.

For example, if a company logo were created using filled *Dtext*, a *Sand* hatch pattern, and a *Solid* hatched circle, the object that was created last would appear "in front" (Fig. 26-23, top logo). The *Draworder* command is used to control which objects appear in front and back or above and under. This capability is necessary for creating prints and plots in black and white and in color.

Figure 26-23

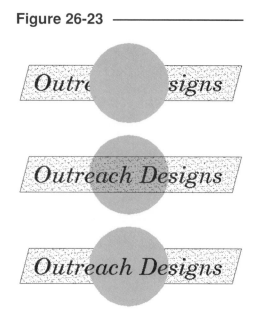

The *Draworder* command is simple to use. Simply select objects and indicate *Front*, *Back*, *Above*, or *Under*.

```
Command: draworder
Select objects: PICK
Select objects: Enter
Above object/Under object/Front/<Back>: (option)
Regenerating drawing.
Command:
```

In Figure 26-23, three (of four) possibilities are shown for the three objects—*Dtext*, *Solid* hatch and boundary, and *Sand* hatch and boundary.

DRAWING CUTTING PLANE LINES

Most section views (full, half, and offset sections) require a cutting plane line to indicate the plane on which the object is cut. The cutting plane line is drawn in a view <u>adjacent</u> to the section view because the line indicates the plane of the cut from its edge view. (In the section view, the cutting plane is perpendicular to the line of sight, therefore, not visible as a line.)

Standards provide two optional line types for cutting plane lines. In AutoCAD, the two linetypes are *Dashed* and *Phantom*, as shown in Figure 26-24. Arrows at the ends of the cutting plane line indicate the <u>line-of-sight</u> for the section view.

Figure 26-24

Cutting plane lines should be drawn or plotted in a <u>heavy</u> line weight. This is accomplished in AutoCAD by using the *Pline* command to draw the cutting plane line having width assigned using the *Width* option. For plotting full size, a width of .02 for small plots (8.5 x 11, or 17 x 22) to .03 for large plots is recommended. Keep in mind that appearance of the *Pline Width* is affected by the drawing plot scale; and, therefore, *Width* should be increased or decreased by that proportion. For example, if you changed the *Limits*, multiply .02 times the drawing scale factor. (See Chapter 13 for information on drawing scale factor.)

For the example section view, a cutting plane line is created in the top view to indicate the plane on which the object is cut and the line of sight for the section view (Fig. 26-25). (The cutting plane could be drawn in the side view, but the top would be much clearer.) First, a *New* layer named CUT is created and the *Dashed linetype* is assigned to the layer. The layer is *Set* as the *Current* layer. Next, construct a *Pline* with the desired *Width* to represent the cutting plane line.

Figure 26-25

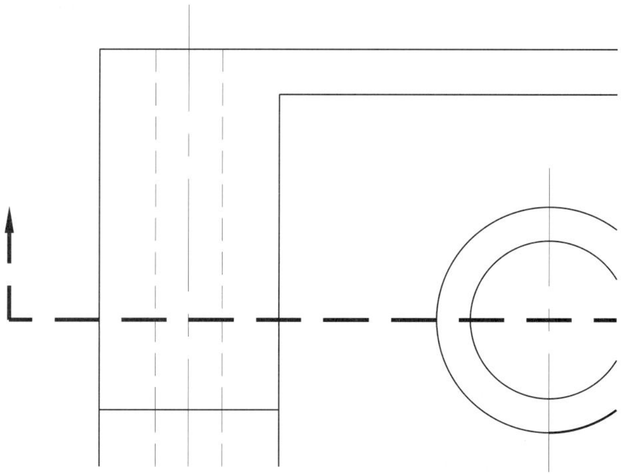

The resulting cutting plane line appears as shown in Figure 26-26 but without the arrow heads. The horizontal center line for the hole (top view) was *Erased* before creating the cutting plane line.

The last step is to add arrowheads to the ends of the cutting plane line. Arrowheads can be drawn by three methods: (1) use the *Solid* command to create a three-sided solid area; (2) use a tapered *Pline* with beginning width of 0 and ending width of .08, for example; or (3) use the *Leader* command to create the arrowhead (see Chapter 28, Dimensioning, for details on *Leader*). The arrowhead you create can be *Scaled*, *Rotated*, and *Copied* if needed to achieve the desired size, orientation, and locations.

Figure 26-26

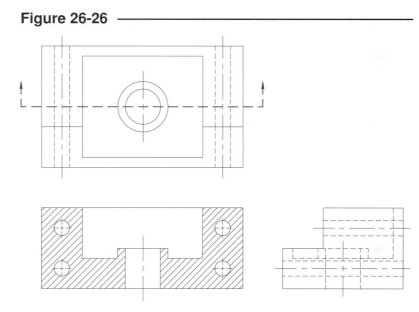

The resulting multiview drawing with section view is complete and ready for dimensioning, drawing a border, and inserting a title block (Fig. 26-26).

CHAPTER EXERCISES

For the following exercises, create the section views as instructed. Use an appropriate template drawing for each unless instructed otherwise. Include a border and title block for each drawing.

1. **Open** the **SADDLE** drawing that you created in Chapter 24. Convert the front view to a full section. **Save** the drawing as **SADL-SEC**. **Plot** the drawing at **1=1** scale.

2. Make a multiview drawing of the Bearing shown in Figure 26-27. Convert the front view to a full section view. Add the necessary cutting plane line in the top view. **Save** the drawing as **BEAR-SEC**. Make a **Plot** at full size.

Figure 26-27

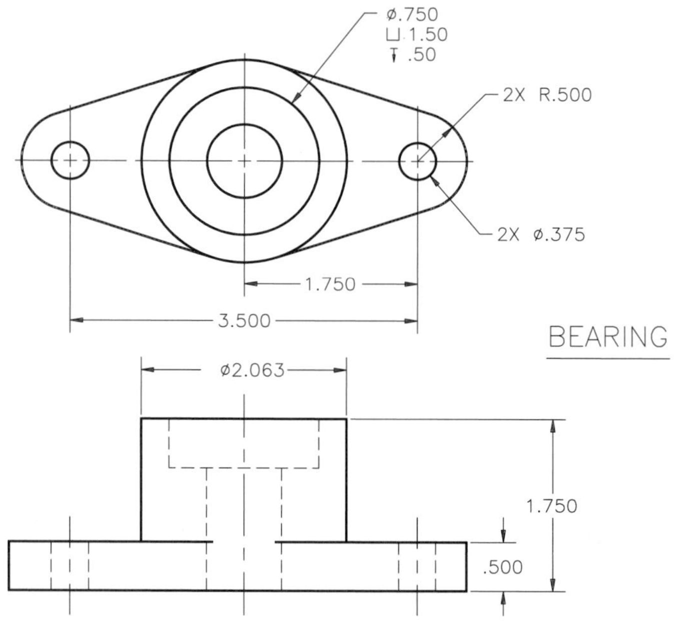

3. Create a multiview drawing of the Clip shown in Figure 26-28. Include a side view as a full section view. You can use the **CLIP** drawing you created in Chapter 24 and convert the side view to a section view. Add the necessary cutting plane line in the front view. **Plot** the finished drawing at **1=1** scale and **SaveAs CLIP-SEC**.

Figure 26-28

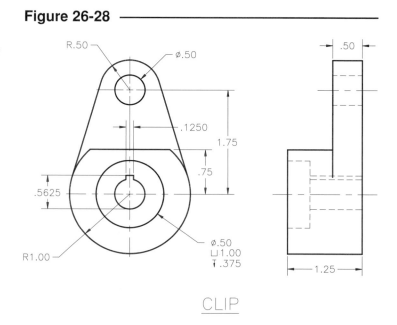

4. Create a multiview drawing, including two full sections of the Stop Block as shown in Figure 26-29. Section B–B' should replace the side view shown in the figure. *Save* the drawing as **SPBK-SEC**. *Plot* the drawing at **1=2** scale.

Figure 26-29

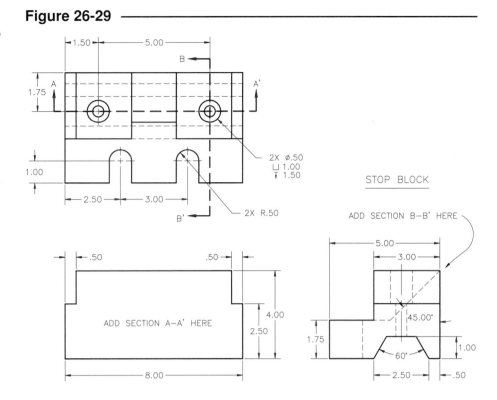

STOP BLOCK

5. Make a multiview drawing, including a half section of the Pulley. All vertical dimensions are diameters. Two views (including the half section) are sufficient to describe the part. Add the necessary cutting plane line. *Save* the drawing as **PUL-SEC** and make a *Plot* at **1:1** scale.

Figure 26-30

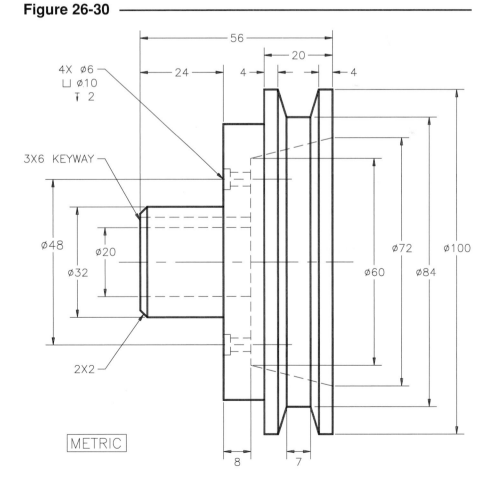

6. Draw the Grade Beam foundation detail in Figure 26-31. Do not include the dimensions in your drawing. Use the *Bhatch* command to hatch the concrete slab with **AR-CONC** hatch pattern. Use *Sketch* to draw the grade line and *Hatch* with the **EARTH** hatch pattern. *Save* the drawing as **GRADBEAM**.

Figure 26-31 ─────────────────────────────────

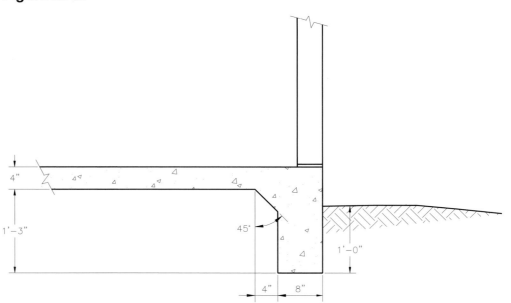

27

AUXILIARY VIEWS

Chapter Objectives

After completing this chapter you should:

1. be able to use the *Rotate* option of *Snap* to change the angle of the *SNAP*, *GRID*, and *ORTHO*;

2. know how to use the *Offset* command to create parallel line copies;

3. be able to use *Xline* and *Ray* to create construction lines for auxiliary views.

CONCEPTS

AutoCAD provides no features explicitly for the creation of auxiliary views in a 2D drawing. No new commands are discussed in this chapter. However, three particular features that have been discussed earlier can assist you in construction of auxiliary views. Those features are the *SNAP* rotation, the *Offset* command, and the *Xline* and *Ray* commands.

An auxiliary view is a supplementary view among a series of multiviews. The auxiliary view is drawn in addition to the typical views that are mutually perpendicular (top, front, side). An auxiliary view is one that is normal (the line-of-sight is perpendicular) to an inclined surface of the object. Therefore, the auxiliary view is constructed by projecting in a 90 degree direction from the edge view of an inclined surface in order to show the true size and shape of the inclined surface. The edge view of the inclined surface could be at any angle (depending on the object), so lines are typically drawn parallel and perpendicular relative to that edge view. Hence, the *SNAP* rotation feature, the *Offset* command, and the *Xline* and *Ray* commands can provide assistance in this task.

Figure 27-1

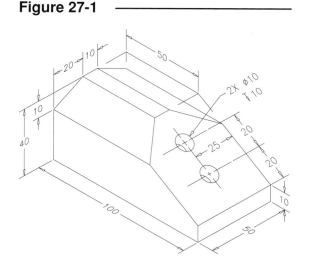

An example mechanical part used for the application of these AutoCAD features related to auxiliary view construction is shown in Figure 27-1. As you can see, there is an inclined surface that contains two drilled holes. To describe this object adequately, an auxiliary view should be created to show the true size and shape of the inclined surface.

This chapter explains the construction of a partial auxiliary view for the example object in Figure 27-1.

CONSTRUCTING AN AUXILIARY VIEW

Setting Up the Principal Views

Figure 27-2

To begin this drawing, the typical steps are followed for drawing setup (Chapter 13). Because the dimensions are in millimeters, *Limits* should be set accordingly. For example, to provide enough space to draw the views full size and to plot full size on an A sheet, *Limits* of 279 x 216 are specified.

In preparation for the auxiliary view, the principal views are "blocked in," as shown in Figure 27-2. The purpose of this step is to ensure that the desired views fit and are optimally spaced within the allocated

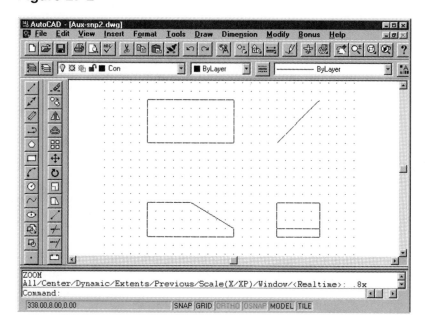

Limits. If there is too little or too much room, adjustments can be made to the *Limits*. Notice that space has been allotted between the views for a partial auxiliary view to be projected from the front view.

Before additional construction on the principal views is undertaken, initial steps in the construction of the partial auxiliary view should be performed. The projection of the auxiliary view requires drawing lines perpendicular and parallel to the inclined surface. One or more of the three alternatives (explained next) can be used.

Using *Snap Rotate*

One possibility to construct an auxiliary view is to use the *Snap* command with the *Rotate* option. This action permits you to rotate the *SNAP* to any angle about a specified base point. The *GRID* automatically follows the *SNAP*. Turning *ORTHO ON* forces *Lines* to be drawn orthogonally with respect to the rotated *SNAP* and *GRID*.

Figure 27-3 displays the *SNAP* and *GRID* after rotation. The command syntax is given below.

In this and the following figures, the cursor size is changed from the default 5% of screen size to 100% of screen size to help illustrate the orientation of the *SNAP, GRID,* and cursor when *SNAP* is *Rotated*. You can change the cursor size in the *Pointer* tab of the *Preferences* dialog box.

Figure 27-3

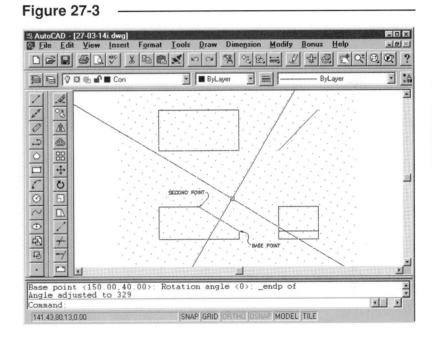

```
Command: snap
Snap spacing or ON/OFF/Aspect/Rotate/Style <current>: r
Base point <0.00,0.00>: PICK or (coordinates) (Starts a rubberband line.)
Rotation angle <0>: PICK or (value) (PICK to specify the second point to define the angle.
See Fig. 27-3.)
Command:
```

PICK (or specify coordinates for) the endpoint of the *Line* representing the inclined surface as the "Base point." At the "Rotation angle:" prompt, a value can be entered or another point (the other end of the inclined *Line*) can be PICKed. Use *OSNAP* when PICKing the *Endpoints*. If you want to enter a value but don't know what angle to rotate to, use *List* to display the angle of the inclined *Line*. The *GRID* and *SNAP* should align with the inclined plane as shown in Figure 27-3.

After rotating the *SNAP* and *GRID*, the partial auxiliary view can be "blocked in," as displayed in Figure 27-4. Begin by projecting *Lines* up from and perpendicular to the inclined surface. (Make sure *ORTHO* is *ON*.) Next, two *Lines* representing the depth of the view should be constructed parallel to the inclined surface and perpendicular to the previous two projection lines. The depth dimension of the object in the auxiliary view is equal to the depth dimension in the top or right view. *Trim* as necessary.

Locate the centers of the holes in the auxiliary view and construct two *Circles*. It is generally preferred to construct circular shapes in the view

Figure 27-4

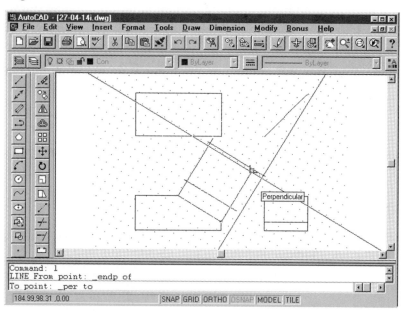

in which they appear as circles, then project to the other views. That is particularly true for this type of auxiliary since the other views contain ellipses. The centers can be located by projection from the front view or by *Offsetting Lines* from the view outline.

Next, project lines from the *Circles* and their centers back to the inclined surface. Use of a hidden line layer can be helpful here. While the *SNAP* and *GRID* are rotated, construct the *Lines* representing the bottom of the holes in the front view. (Alternately, *Offset* could be used to copy the inclined edge down to the hole bottoms; then *Trim* the unwanted portions of the *Lines*.)

Figure 27-5

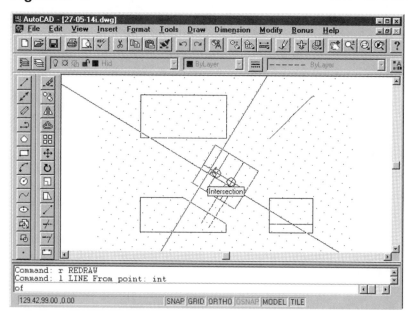

Rotating *SNAP* Back to the Original Position

Before details can be added to the other views, the *SNAP* and *GRID* should be rotated back to the original position. It is very important to rotate back using the <u>same basepoint</u>. Fortunately, AutoCAD remembers the original basepoint so you can accept the default for the prompt. Next, enter a value of **0** when rotating back to the original position. (When using the *Snap Rotate* option, the value entered for the angle of rotation is absolute, not relative to the current position. For example, if the *Snap* was rotated to 45 degrees, rotate back to 0 degrees, not -45.)

> Command: **snap**
> Snap spacing or ON/OFF/Aspect/Rotate/Style <2.00>: **r**
> Base point <150.00,40.00>: **Enter** (AutoCAD remembers the previous basepoint)
> Rotation angle <329>: **0**
> Command:

Construction of multiview drawings with auxiliaries typically involves repeated rotation of the *SNAP* and *GRID* to the angle of the inclined surface and back again as needed.

With the *SNAP* and *GRID* in the original position, details can be added to the other views as shown in Figure 27-6. Since the two circles appear as ellipses in the top and right side views, project lines from the circles' centers and limiting elements on the inclined surface. Locate centers for the two *Ellipses* to be drawn in the top and right side views.

Figure 27-6

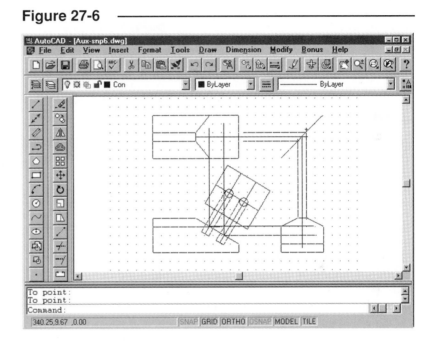

Use the *Ellipse* command to construct the ellipses in the top and right side views. Using the *Center* option of *Ellipse*, specify the center by PICKing with the *Intersection OSNAP* mode. *OSNAP* to the appropriate construction line *Intersection* for the first axis endpoint. For the second axis endpoint (as shown), use the actual circle diameter, since that dimension is not foreshortened.

Figure 27-7

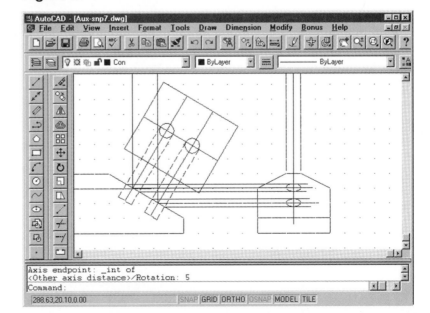

The remaining steps for completing the drawing involve finalizing the perimeter shape of the partial auxiliary view and *Copying* the *Ellipses* to the bottom of the hole positions. The *SNAP* and *GRID* should be rotated back to project the new edges found in the front view (Fig. 27-8).

At this point, the multiview drawing with auxiliary view is ready for centerlines, dimensioning, and construction or insertion of a border and title block.

Figure 27-8

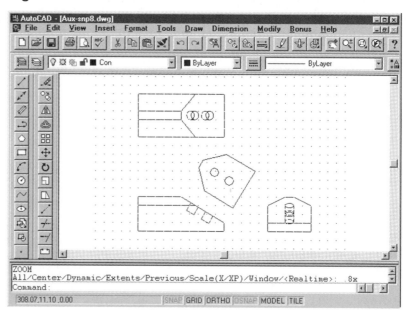

Using the *Offset* Command

Another possibility, and an alternative to the *SNAP* rotation, is to use *Offset* to make parallel *Lines*. This command can be particularly useful for construction of the "blocked in" partial auxiliary view because it is not necessary to rotate the *SNAP* and *GRID*.

OFFSET

Pull-down Menu	COMMAND (TYPE)	ALIAS (TYPE)	Short-cut	Screen (side) Menu	Tablet Menu
Modify *Offset*	*OFFSET*	*O*	...	*MODIFY1* *Offset*	*V,17*

Invoke the *Offset* command and specify a distance. The first distance is arbitrary. Specify an appropriate value between the front view inclined plane and the nearest edge of the auxiliary view (20 for the example). *Offset* the new *Line* at a distance of 50 (for the example) or PICK two points (equal to the depth of the view).

Note that the *Offset* lines have lengths equal to the original and therefore require no additional editing (Fig. 27-9).

Figure 27-9

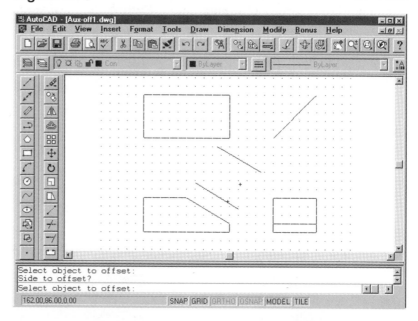

Next, two *Lines* would be drawn between *Endpoints* of the existing offset lines to complete the rectangle. *Offset* could be used again to construct additional lines to facilitate the construction of the two circles in the partial auxiliary view (Fig 27-10).

From this point forward, the construction process would be similar to the example given previously (Figs. 27-5 through 27-8). Even though *Offset* does not require that the *SNAP* be *Rotated*, the complete construction of the auxiliary view could be simplified by using the rotated *SNAP* and *GRID* in conjunction with *Offset*.

Figure 27-10

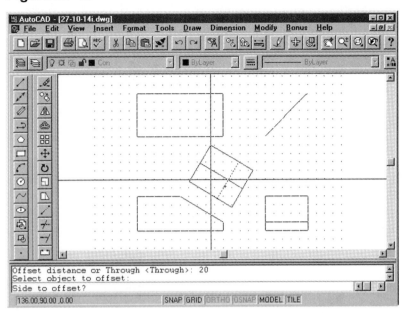

Using the *Xline* and *Ray* Commands

As a third alternative for construction of auxiliary views, the *Xline* and *Ray* commands could be used to create construction lines.

XLINE

Pull-down Menu	COMMAND (TYPE)	ALIAS (TYPE)	Short-cut	Screen (side) Menu	Tablet Menu
Draw Construction Line	XLINE	XL	...	DRAW 1 Xline	L,10

RAY

Pull-down Menu	COMMAND (TYPE)	ALIAS (TYPE)	Short-cut	Screen (side) Menu	Tablet Menu
Draw Ray	RAY	...	...	DRAW 1 Ray	K,10

The *Xline* command offers several options shown below:

    Command: **Xline**
    Hor/Ver/Ang/Bisect/Offset/<From point:>

The *Ang* option can be used to create a construction line at a specified angle. In this case, the angle specified would be that of the inclined plane or perpendicular to the inclined plane. The *Offset* option works well for drawing construction lines parallel to the inclined plane, especially in the case where the angle of the plane is not known.

Figure 27-11 illustrates the use of *Xline Offset* to create construction lines for the partial auxiliary view. The *Offset* option operates similarly to the *Offset* command described previously. Remember that an *Xline* extends to infinity but can be *Trimmed*, in which case it is converted to *Ray* (*Trim* once) or to a *Line* (*Trim* twice). See Chapter 15 for more information on the *Xline* command.

Figure 27-11

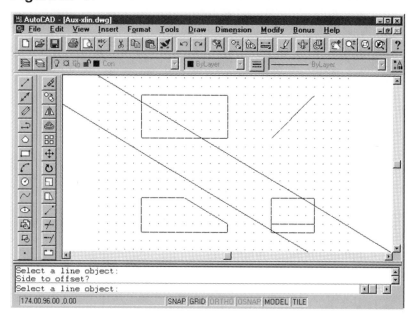

The *Ray* command also creates construction lines; however, the *Ray* has one anchored point and the other end extends to infinity.

```
Command: Ray
From point: PICK or (coordinates)
To point: PICK or (coordinates)
```

Rays are helpful for auxiliary view construction when you want to create projection lines perpendicular to the inclined plane. In Figure 27-12, two *Rays* are constructed from the *Endpoints* of the inclined plane and *Perpendicular* to the existing *Xlines*. Using *Xlines* and *Rays* in conjunction is an <u>excellent method</u> for "blocking in" the view.

There are two strategies for creating drawings using *Xlines* and *Rays*. First, these construction lines can be created on a separate layer and set up as a framework for the object lines. The object lines would then be drawn "on top of" the construction lines using *Osnaps*, but would be drawn on the object layer. The con-

Figure 27-12

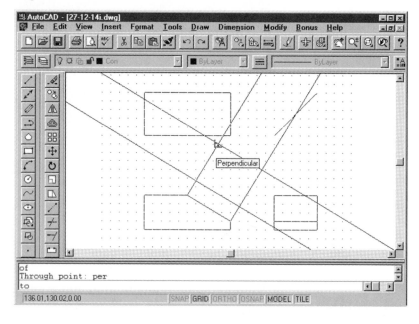

struction layer would be *Frozen* for plotting. The other strategy is to create the construction lines on the object layer. Through a series of *Trims* and other modifications, the *Xlines* and *Rays* are transformed to the finished object lines.

Now that you are aware of several methods for constructing auxiliary views, use any one method or a combination of methods for your drawings. No matter which methods are used, the final lines that are needed for the finished auxiliary view should be the same. It is up to you to use the methods that are the most appropriate for the particular application or are the easiest and quickest for you personally.

Constructing Full Auxiliary Views

The construction of a full auxiliary view begins with the partial view. After initial construction of the partial view, begin the construction of the full auxiliary by projecting the other edges and features of the object (other than the inclined plane) to the existing auxiliary view.

The procedure for constructing full auxiliary views in AutoCAD is essentially the same as that for partial auxiliary views. Use of the *Offset, Xline,* and *Ray* commands and *SNAP* and *GRID* rotation should be used as illustrated for the partial auxiliary view example. Because a full auxiliary view is projected at the same angle as a partial, the same rotation angle and basepoint would be used for the *SNAP*.

CHAPTER EXERCISES

For the following exercises, create the multiview drawing, including the partial or full auxiliary view as indicated. Use the appropriate template drawing based on the given dimensions and indicated plot scale.

1. Make a multiview drawing with a partial auxiliary view of the example used in this chapter. Refer to Figure 27-1 for dimensions. *Save* the drawing as **CH27EX1**. *Plot* on an A size sheet at **1=1** scale.

2. Recreate the views given in Figure 27-13 and add a partial auxiliary view. *Save* the drawing as **CH27EX2** and *Plot* on an A size sheet at **2=1** scale.

Figure 27-13

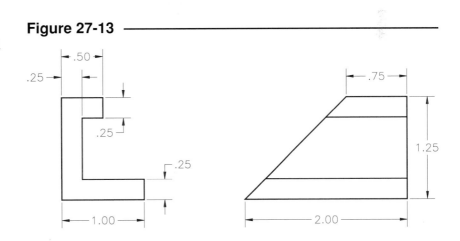

3. Recreate the views shown in Figure 27-14 and add a partial auxiliary view. *Save* the drawing as **CH27EX3**. Make a *Plot* on an A size sheet at **1=1** scale.

Figure 27-14

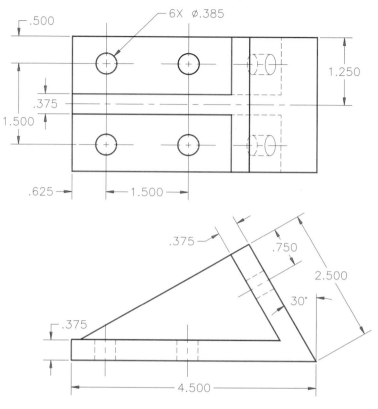

4. Make a multiview drawing of the given views in Figure 27-15. Add a full auxiliary view. *Save* the drawing as **CH27EX4**. *Plot* the drawing at an appropriate scale on an A size sheet.

Figure 27-15

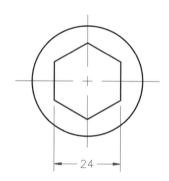

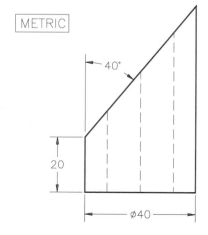

5. Draw the front, top, and a partial auxiliary view of the holder. Make a *Plot* full size. Save the drawing as **HOLDER**.

Figure 27-16 ———————————————————————————————————————

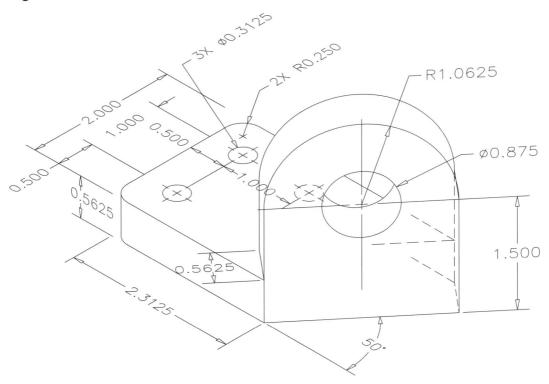

6. Draw three principal views and a full auxiliary view of the V-block shown in Figure 27-17. *Save* the drawing as **VBLOCK**. *Plot* to an accepted scale.

Figure 27-17 ——————————————————————————————

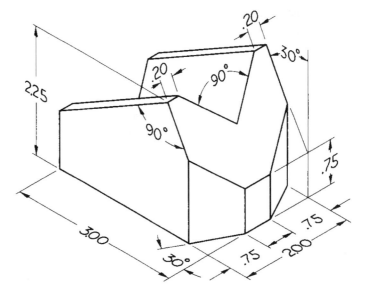

7. Draw two principal views and a partial auxiliary of the angle brace. *Save* as **ANGLBRAC**. Make a plot to an accepted scale on an A or B size sheet.

Figure 27-18

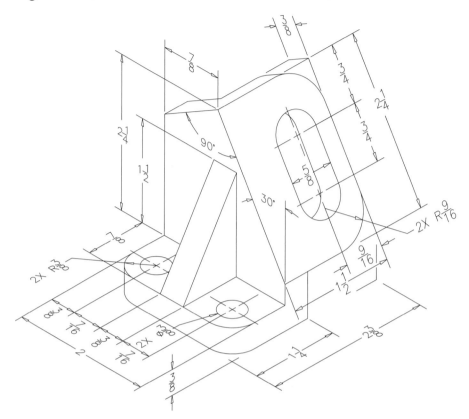

28

DIMENSIONING

Chapter Objectives

After completing this chapter you should:

1. be able to create linear dimensions with *Dimlinear*;

2. be able to append *Dimcontinue* and *Dimbaseline* dimensions to existing dimensions;

3. be able to create *Angular, Diameter,* and *Radius* dimensions;

4. know how to affix notes to drawings with *Leaders*;

5. know that *Dimordinate* can be used to specify Xdatum and Ydatum dimensions;

6. be able to create and apply geometric dimensioning and tolerancing symbols using the *tolerance* command and dialog boxes;

7. know the possible methods for editing associative dimensions and dimensioning text;

8. know how to dimension isometric drawings.

CONCEPTS

As you know, drawings created with CAD systems should be constructed with the same dimensions and units as the real-world objects they represent. The importance of this practice is evident when you begin applying dimensions to the drawing geometry in AutoCAD. The features of the object that you specify for dimensioning are automatically measured, and those values are used for the dimensioning text. If the geometry has been drawn accurately, the dimensions will be created correctly.

The main components of a dimension are:

Figure 28-1

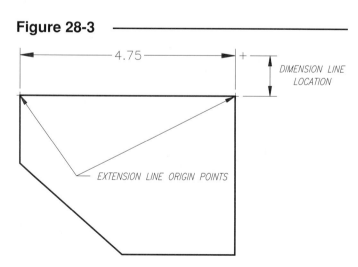

1. Dimension line
2. Extension lines
3. Dimension text (usually a numeric value)
4. Arrowheads or tick marks

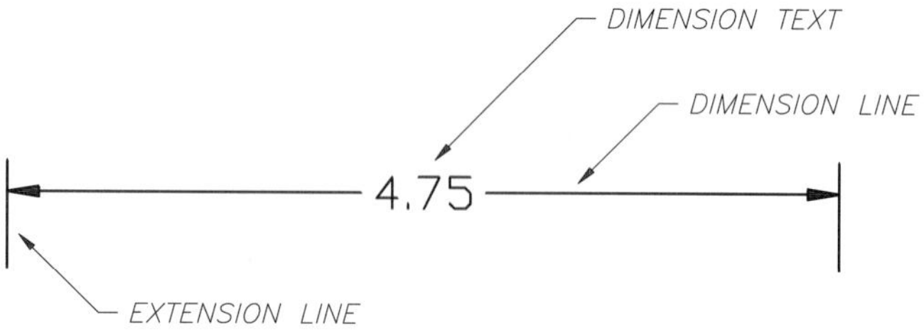

AutoCAD dimensioning is <u>semi-automatic</u>. When you invoke a command to create a linear dimension, AutoCAD only requires that you PICK an object or specify the extension line origins (where you want the extension lines to begin) and PICK the location of the dimension line (distance from the object). AutoCAD then measures the feature and draws the extension lines, dimension line, arrowheads, and dimension text.

For linear dimensioning commands, there are <u>two ways</u> to specify placement for a dimension in AutoCAD: you can PICK the <u>object</u> to be dimensioned or you can PICK the two <u>extension line origins</u>. The simplest method is to select the object because it requires only one PICK (Fig. 28-2):

Figure 28-2

> Command: *dimlinear*
> First extension line origin or press ENTER to select: **Enter**
> Select object to dimension: **PICK**

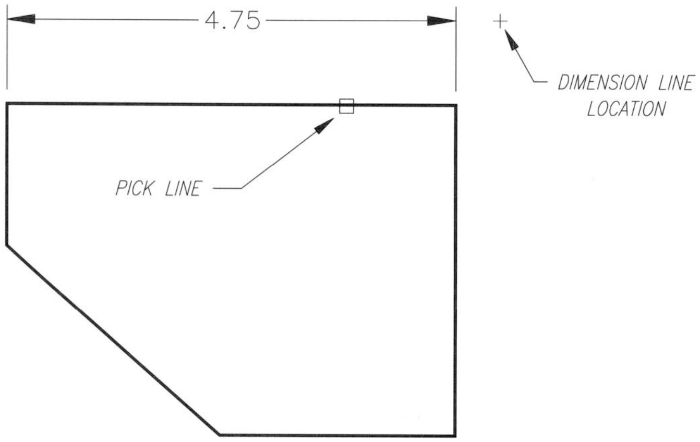

The other method is to PICK the extension line origins (Fig. 28-3). *Osnaps* should be used to PICK the object (endpoints in this case) so that the dimension is <u>associated</u> with the object.

Figure 28-3

> Command: *dimlinear*
> First extension line origin or press ENTER to select: **PICK**
> Second extension line origin: **PICK**

Once the dimension is attached to the object, you specify how far you want the dimension to be placed from the object (called the "dimension line location").

Dimensioning in AutoCAD is <u>associative</u> (by default). Because the extension line origins are "associated" with the geometry, the dimension text automatically updates if the geometry is *Stretched, Rotated, Scaled,* or likewise edited using grips.

Because dimensioning is semi-automatic, <u>dimensioning variables</u> are used to control the way dimensions are created. Dimensioning variables can be used to control features such as text or arrow size, direction of the leader arrow for radial or diametrical dimensions, format of the text, and many other possible options. Groups of variable settings can be named and saved as <u>Dimension Styles</u>. Dimensioning variables and dimension styles are discussed in Chapter 29.

Dimensioning commands can be invoked by the typical methods. The *Dimension* pull-down menu contains the dimension creation and editing commands (Fig. 28-4). A *Dimension* toolbar can also be activated by using the *Toolbars* dialog box (Fig. 28-5).

Figure 28-4 ——————————————————————————— **Figure 28-5** ———————

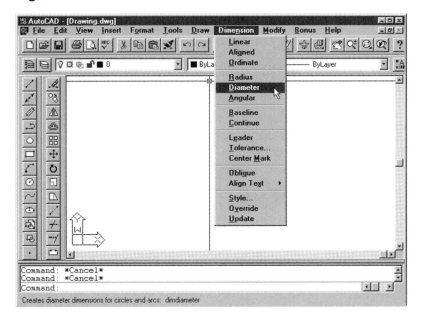

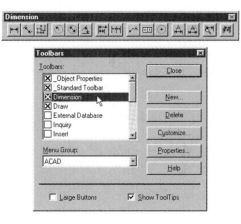

DIMENSION DRAWING COMMANDS

DIMLINEAR

Pull-down Menu	COMMAND (TYPE)	ALIAS (TYPE)	Short-cut	Screen (side) Menu	Tablet Menu
Dimension Linear	DIMLINEAR	DIMLIN or DLI	...	DIMNSION Linear	W,5

Dimlinear creates a <u>horizontal, vertical, or rotated</u> dimension. If the object selected is a horizontal line (or the extension line origins are horizontally oriented), the resulting dimension is a horizontal dimension. This situation is displayed in the previous illustrations (Figs. 28-2 and 28-3), or if the selected object or extension line origins are vertically oriented, the resulting dimension is vertical (Fig. 28-6).

Figure 28-6 ——————————————

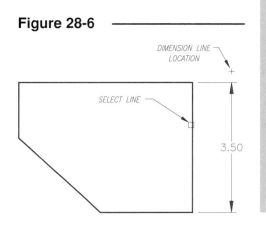

When you dimension an inclined object (or if the selected extension line origins are diagonally oriented), a vertical <u>or</u> horizontal dimension can be made, depending on where you drag the dimension line in relation to the object. If the dimension line location is more to the side, a vertical dimension is created (Fig. 28-7), or if you drag farther up or down, a horizontal dimension results (Fig 28-8).

If you select the extension line origins, it is very important to PICK the <u>object's endpoints</u> if the dimensions are to be truly associative (associated with the geometry). *OSNAP* should be used to find the object's *Endpoint*, *Intersection*, etc., unless that part of the geometry is located at a *SNAP* point.

Figure 28-7

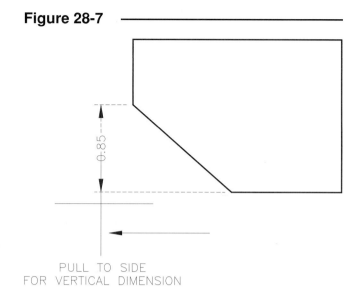

PULL TO SIDE
FOR VERTICAL DIMENSION

Command: ***dimlinear***
First extension line origin or press ENTER to select: ***endp of* PICK**
Second extension line origin: ***endp of* PICK**
Dimension line location (Mtext/Text/Angle/Horizontal/Vertical/Rotated): **PICK** (where you want the dimension line placed)
Command:

Figure 28-8

When you pick the location for the dimension line, AutoCAD automatically measures the object and inserts the correct numerical value. The other options are explained next.

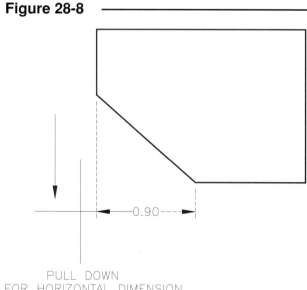

PULL DOWN
FOR HORIZONTAL DIMENSION

Rotated
If you want the dimension line to be drawn at an angle instead of vertical or horizontal, use this option. Selecting an inclined line, as in the previous two illustrations, would normally create a horizontal or vertical dimension. The *Rotated* option allows you to enter an <u>angular value</u> for the dimension line to be drawn. For example, selecting the diagonal line and specifying the appropriate angle would create the dimension shown in Figure 28-9. This object, however, could be more easily dimensioned with the *Dimaligned* command (see *Dimaligned*).

Figure 28-9

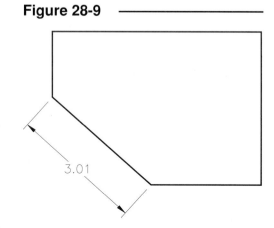

A *Rotated* dimension should be used when the geometry has "steps" or any time the desired dimension line angle is different than the dimensioned feature (when you need extension lines of different lengths). Figure 28-10 illustrates the result of using a *Rotated* dimension to give the correct dimension line angle and extension line origins for the given object. In this case, the extension line origins were explicitly PICKed. The feature of *Rotated* that makes it unique is that you specify the angle that the dimension line will be drawn.

Figure 28-10

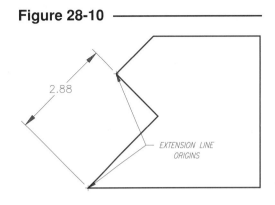

Text
Using the *Text* option allows you to enter any value or characters in place of the AutoCAD-measured text. The measured value is given as a reference at the command prompt.

> Command: **DIMLINEAR**
> First extension line origin or press ENTER to select:
> **Enter**
> Select object to dimension: **PICK**
> Dimension line location (Mtext/Text/Angle/Horizontal/
> Vertical/Rotated): **T**
> Dimension text <0.25>: (**text** or **Value**)

Figure 28-11

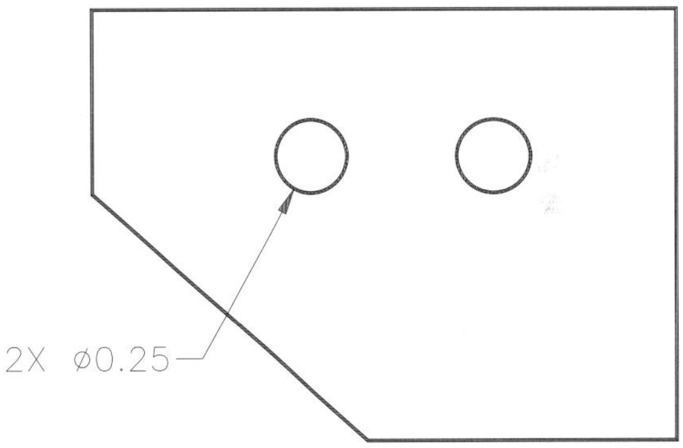

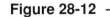

Entering a value or text at the prompt (above) causes AutoCAD to display that value or text instead of the AutoCAD-measured value. If you want to keep the AutoCAD-measured value but add a prefix or suffix, use the < and > (less-than, greater-than) symbols to represent the actual (AutoCAD) value:

> Dimension text <0.25>: **2X <>**
> Dimension line location (Mtext/Text/Angle/Horizontal/Vertical/Rotated): **PICK**
> Dimension text = 0.25
> Command:

The "2X <>" response produces the dimension text shown in Figure 28-11.

NOTE: Changing the AutoCAD-measured value should be discouraged. If the geometry is drawn accurately, the dimensional value is correct. If you specify other dimension text, the text value is <u>not</u> updated in the event of *Stretching, Rotating,* or otherwise editing the associative dimension.

Mtext
This option allows you to change the existing or insert additional text to the AutoCAD-supplied numerical value. The text is entered by the *Multiline Text Editor.* The less-than and greater-than symbols (< >) represent the AutoCAD-supplied dimensional value. Place text or numbers inside the symbols if you

Figure 28-12

want to override the correct measurement (not advised) or place text outside the symbols if you want to add annotation to the numerical value. For example, entering "2X " before the symbols (Fig. 28-12) creates a dimensional value as shown in Figure 28-11.

Keep in mind—the power of using the *Multiline Text Editor* is the availability of all the text creation and editing features in the dialog box. For example, you can specify fonts, text height, bold, italic, underline, stacked text or fractions, and layer from the *Character* tab. In addition, you can use the *Symbol* drop-down box to insert a degree, plus/minus, or diameter symbol with the text value. The *Properties* tab gives access to text style, justification, and rotation angle options.

Angle

This creates text drawn at the angle you specify. Use this for special cases when the text must be drawn to a specific angle other than horizontal. (It is also possible to make the text automatically align with the angle of the dimension line using the *Dimension Styles* dialog box. See Chapter 29.)

Horizontal

Use the *Horizontal* option when you want to force a horizontal dimension for an inclined line and the desired placement of the dimension line would otherwise cause a vertical dimension.

Vertical

This option forces a *Vertical* dimension for any case.

DIMALIGNED

Pull-down Menu	COMMAND (TYPE)	ALIAS (TYPE)	Short-cut	Screen (side) Menu	Tablet Menu
Dimension Aligned	DIMALIGNED	DIMALI or DAL	...	DIMNSION Aligned	W,4

An *Aligned* dimension is aligned with (at the same angle as) the selected object or the extension line origins. For example, when *Aligned* is used to dimension the angled object shown in Figure 28-13, the resulting dimension aligns with the *Line*. This holds true for either option—PICKing the object or the extension line origins. If a *Circle* is PICKed, the dimension line is aligned with the selected point on the *Circle* and its center. The command syntax for the *Dimaligned* command accepting the defaults is:

Figure 28-13 ———————

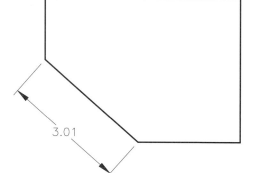

```
Command: dimaligned
First extension line origin or press ENTER to select: PICK
Second extension line origin: PICK
Dimension line location (Mtext/Text/Angle): PICK
Command:
```

The three options (*Mtext/Text/Angle*) operate similar to those for *Dimlinear*.

Mtext

The *Mtext* option calls the *Multiline Text Editor*. You can alter the AutoCAD-supplied text value or modify other visible features of the dimension text such as fonts, text height, bold, italic, underline, stacked text or fractions, layer, symbols, text style, justification, and rotation angle (see *Dimlinear, Mtext*).

Text

You can change the AutoCAD-supplied numerical value or add other annotation to the value in command line format (see *Dimlinear, Text*).

Angle

Enter a value for the angle that the text will be drawn.

The typical application for *Dimaligned* is for dimensioning an angled but <u>straight</u> feature of an object, as shown in Figure 28-13. *Dimaligned* should not be used to dimension an object feature that contains "steps," as shown in Figure 28-10. *Dimaligned* always draws <u>extension lines of equal length</u>.

DIMBASELINE

	Pull-down Menu	COMMAND (TYPE)	ALIAS (TYPE)	Short-cut	Screen (side) Menu	Tablet Menu
	Dimension Baseline	DIMBASELINE	DIMBASE or DBA	...	DIMNSION *Baseline*	W,2

Dimbaseline allows you to create a dimension that uses an extension line origin from a previously created dimension. Successive *Dimbaseline* dimensions can be used to create the style of dimensioning shown in Figure 28-14.

A baseline dimension must be connected to an existing dimension. If *Dimbaseline* is invoked immediately after another dimensioning command, you are required only to specify the second extension line origin since AutoCAD knows to use the <u>previous</u> dimension's <u>first</u> extension line origin:

Figure 28-14 ——————

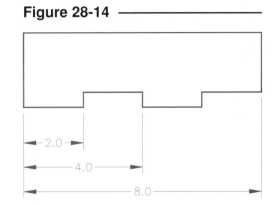

    Command: **Dimbaseline**
    Specify a second extension line origin or (Undo/<Select>): **PICK**
    Dimension text = nn.nnnn
    Specify a second extension line origin or (Undo/<Select>): **Enter**
    Command:

The <u>previous dimension's first extension line</u> is used also for the baseline dimension (Fig. 28-15). Therefore, you specify only the second extension line origin. Note that you are not required to specify the dimension line location. AutoCAD spaces the new dimension line automatically, based on the setting of the dimension line increment variable (Chapter 29).

Figure 28-15 ——————

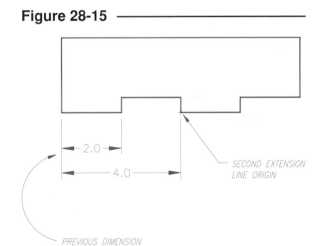

If you wish to create a *Dimbaseline* dimension using a dimension other than the one just created, use the "Select" option (Fig. 28-16):

    Command: **Dimbaseline**
    Specify a second extension line origin or (Undo/<Select>): **S**
    Select base dimension: **PICK** (existing extension line)
    Specify a second extension line origin or (Undo/<Select>): **PICK**
    Dimension text = nn.nnnn
    Specify a second extension line origin or (Undo/<Select>): **Enter**
    Command:

Figure 28-16 ——————

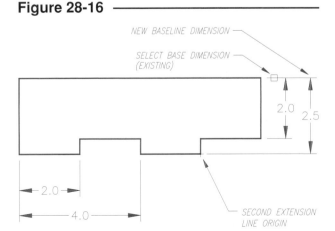

The extension line selected as the base dimension becomes the first extension line for the new *Dimbaseline* dimension.

The *Undo* option can be used to undo the last baseline dimension created in the current command sequence.

Dimbaseline can be used with rotated, aligned, angular, and ordinate dimensions.

DIMCONTINUE

	Pull-down Menu	COMMAND (TYPE)	ALIAS (TYPE)	Short-cut	Screen (side) Menu	Tablet Menu
	Dimension Continue	DIMCONTINUE	DIMCONT or DCO	...	DIMNSION Continue	W,1

Dimcontinue dimensions continue in a line from a previously created dimension. *Dimcontinue* dimension lines are attached to, and drawn the same distance from, the object as an existing dimension.

Dimcontinue is similar to *Dimbaseline* except that an existing dimension's <u>second</u> extension line is used to begin the new dimension. In other words, the new dimension is connected to the <u>second</u> extension line, rather than to the <u>first</u>, as with a *Dimbaseline* dimension (Fig. 28-17).

Figure 28-17

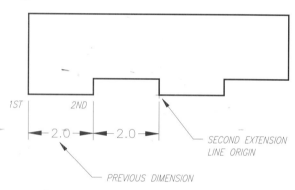

The command syntax is as follows:

    Command: **Dimcontinue**
    Specify a second extension line origin or
    (Undo/<Select>): **PICK**
    Dimension text = *nn.nnnn*
    Specify a second extension line origin or (Undo/<Select>): **Enter**
    Command:

Assuming a dimension was just drawn, *Dimcontinue* could be used to place the next dimension, as shown in Figure 28-17.

Figure 28-18

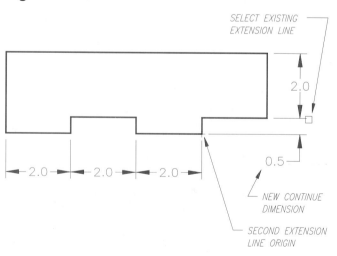

If you want to create continued dimension and attach it to an extension line <u>other</u> than the previous dimension's second extension line, you can use the *Select* option to pick an extension line of any other dimension. Then, use *Dimcontinue* to create a continued dimension from the selected extension line (Fig. 28-18).

The *Undo* option can be used to undo the last continued dimension created in the current command sequence. *Dimcontinue* can be used with rotated, aligned, angular, and ordinate dimensions.

DIMDIAMETER

	COMMAND (TYPE)	ALIAS (TYPE)	Short-cut	Screen (side) Menu	Tablet Menu
Pull-down Menu					
Dimension Diameter	*DIMDIAMETER*	*DIMDIA or DDI*	...	*DIMNSION Diameter*	X,4

The *Dimdiameter* command creates a diametrical dimension by selecting any *Circle*. Diametrical dimensions should be used for full 360 degree *Circles* and can be used for *Arcs* of more than 180 degrees.

> Command: **Dimdiameter**
> Select arc or circle: **PICK**
> Dimension text = *nn.nnnn*
> Dimension line location (Mtext/Text/Angle): **PICK**
> Command:

You can PICK the circle at any location. AutoCAD allows you to adjust the position of the dimension line to any angle or length (Fig. 28-19). Dimension lines for diametrical or radial dimensions should be drawn to a regular angle, such as 30, 45, or 60 degrees, never vertical or horizontal.

A typical diametrical dimension appears as the example in Figure 28-19. According to ANSI standards, a diameter dimension line and arrow should point inward (toward the center) for holes and small circles where the dimension line and text do not fit within the circle. Use the default variable settings for *Dimdiameter* dimensions such as this.

For dimensioning large circles, ANSI standards suggest an alternate method for diameter dimensions where sufficient room exists for text and arrows inside the circle (Fig. 28-20). To create this style of dimension in Release 14, the variables that control these features should be set to *Arrows Only Fit* and *User Defined* in the *Format* Dimension Style dialog box (see Chapter 29).

Notice that the *Diameter* command creates center marks at the *Circle's* center. Center marks can also be drawn by the *Dimcenter* command (discussed later). AutoCAD uses the center and the point selected on the *Circle* to maintain its associativity.

Mtext/Text

The *Mtext* or *Text* options can be used to modify or add annotation to the default value. The *Mtext* option summons the *Multiline Text Editor* and the *Text* option uses command line format. Both options operate similar to the other dimensioning commands (see *Dimlinear, Mtext,* and *Text*).

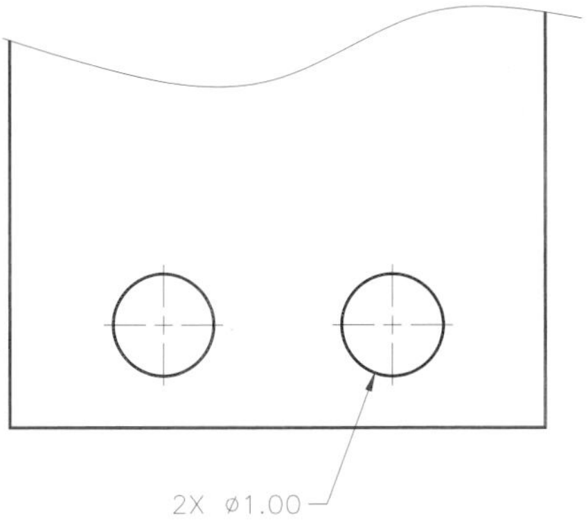

Notice that with diameter dimensions AutoCAD automatically creates the Ø (phi) symbol before the dimensional value. This is the latest ANSI standard for representing diameters. If you prefer to use a prefix before or suffix after the dimension value, it can be accomplished with the *Mtext* or *Text* option. Remember that the < > symbols represent the AutoCAD-measured value so the additional text should be inserted on <u>either side</u> of the symbols. Inserting a prefix by this method <u>does not override</u> the Ø (phi) symbol (Fig. 28-21).

Figure 28-19

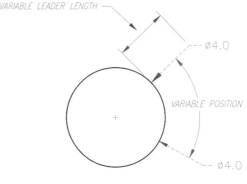

Figure 28-20

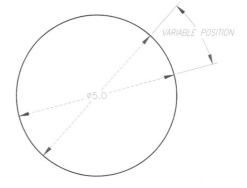

Figure 28-21

A prefix or suffix can alternately be added to the measured value by using the *Dimension Styles* dialog box and entering text or values in the *Prefix* or *Suffix* edit boxes. Using the dialog box method, however, <u>overrides</u> the Ø symbol (see Chapter 29).

Angle
With this option, you can specify an angle (other than the default) for the text to be drawn by entering a value.

DIMRADIUS

Pull-down Menu	COMMAND (TYPE)	ALIAS (TYPE)	Short-cut	Screen (side) Menu	Tablet Menu
Dimension Radius	*DIMRADIUS*	*DIMRAD or DRA*	...	*DIMNSION Radius*	*X,5*

Dimradius is used to create a dimension for an arc of anything less than half of a circle. ANSI standards dictate that a *Radius* dimension line should point outward (from the arc's center), unless there is insufficient room, in which case the line can be drawn on the outside pointing inward, as with a leader. The text can be located inside an arc (if sufficient room exists) or is forced outside of small *Arcs* on a leader.

```
Command: Dimradius
Select arc or circle: PICK
Dimension text = nn.nnnn
Dimension line location (Mtext/Text/Angle): PICK
Command:
```

Figure 28-22 ─────────────

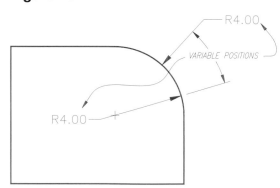

Assuming the defaults, a *Dimradius* dimension can appear on either side of an arc, as shown in Figure 28-22. Placement of the dimension line is variable. Dimension lines for arcs and circles should be positioned at a regular angle such as 30, 45, or 60 degrees, never vertical or horizontal.

When the radius dimension is dragged outside of the arc, a center mark is automatically created (Fig. 28-22). When the radius dimension is dragged inside the arc, no center mark is created. (Center marks can be created using the *Center* command discussed next.)

Figure 28-23 ─────────────

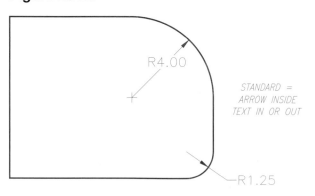

According to ANSI standards, the dimension line and arrow should point <u>outward</u> from the center for radial dimensions (Fig. 28-23). The text can be placed inside or outside of the *Arc*, depending on how much room exists. To create a *Dimradius* dimension to comply with this standard, dimension variables must be changed from the defaults. Using the *Dimension Styles* dialog box, then the *Format* dialog box, check *User Defined* and set *Fit* to *Arrows Only* to achieve *Dimradius* dimensions as shown in Figure 28-23 (see Chapter 29 for more information on these variables).

Figure 28-24 ─────────────

For very small radii, such as that shown in Figure 28-24, there is insufficient room for the text and arrow to fit inside the *Arc*. In this case, AutoCAD automatically forces the text outside with the leader pointing inward toward the center. <u>No changes</u> have to be made to the default settings for this to occur.

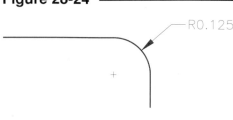

Mtext/Text

These options can be used to modify or add annotation to the default value. AutoCAD automatically inserts the letter "R" before the numerical text whenever a *Dimradius* dimension is created. This is the correct notation for radius dimensions. The *Mtext* option calls the *Multiline Text Editor*, and the *Text* option uses the command line format for entering text. Remember that AutoCAD uses the < > symbols to represent the AutoCAD-supplied value. Entering text inside the < > symbols overrides the measured value. Entering text before the symbols adds a prefix <u>without overriding the "R" designation</u>. Alternately, text can be added by using the *Prefix* and *Suffix* options of the *Dimension Styles* dialog box series; however, a *Prefix* entered in the edit box <u>replaces</u> the letter "R." (See Chapter 29.)

Angle

With this option, you can specify an angle (other than the default) for the <u>text</u> to be drawn by entering a value.

DIMCENTER

	COMMAND (TYPE)	ALIAS (TYPE)	Short-cut	Screen (side) Menu	Tablet Menu
Pull-down Menu					
Dimension Center Mark	*DIMCENTER*	*DCE*	...	*DIMNSION Center*	*X,2*

The *Dimcenter* command draws a center mark on any selected *Arc* or *Circle*. As shown earlier, the *Dimdiameter* command and the *Dimradius* command sometimes create the center marks automatically.

The command requires you only to select the desired *Circle* or *Arc* to acquire the center marks:

Command: **dimcenter**
Select arc or circle: **PICK**
Command:

No matter if the center mark is created by the *Center* command or by the *Diameter* or *Radius* commands, the center mark can be either a small cross or complete center lines extending past the *Circle* or *Arc* (Fig 28-25). The type of center mark drawn is controlled by the *Mark* or *Line* setting in the *Geometry* dialog box from the *Dimension Style* series. It is suggested that a new *Dimension Style* be created with these settings just for drawing center marks. (See Chapter 29.)

When dimensioning, short center marks should be used for *Arcs* of less than 180 degrees, and full center lines should be drawn for *Circles* and for *Arcs* of 180 degrees or more (Fig. 28-26).

NOTE: Since the center marks created with the *Dimcenter* command are <u>not</u> associative, they may be *Trimmed*, *Erased*, or otherwise edited, as shown in Figure 28-27. The center marks created with the *Dimradius* or *Dimdiameter* commands <u>are</u> associative and cannot be edited.

The lines comprising the center marks created with *Dimcenter* can be *Erased* or otherwise edited. In the case of a 180 degree *Arc*, two center mark lines can be shortened using *Break,* and one line can be *Erased* to achieve center lines as shown in Figure 28-27.

Figure 28-25

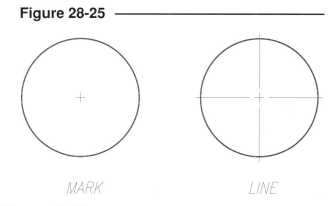

MARK　　　　　　　LINE

Figure 28-26

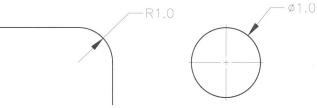

Figure 28-27

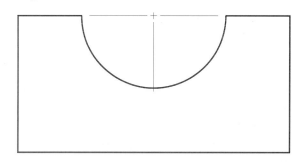

DIMANGULAR

Pull-down Menu	COMMAND (TYPE)	ALIAS (TYPE)	Short-cut	Screen (side) Menu	Tablet Menu
Dimension Angular	DIMANGULAR	*DIMANG or DAN*	...	*DIMNSION Angular*	*X,3*

The *Dimangular* command provides many possible methods of creating an angular dimension.

A typical angular dimension is created between two *Lines* that form an angle (of other than 90 degrees). The dimension line for an angular dimension is radiused with its center at the vertex of the angle (Figure 28-28). A *Dimangular* dimension automatically adds the degree symbol (°) to the dimension text. The dimension text format is controlled by the current settings for *Units* in the *Dimension Style* dialog box.

Figure 28-28 ————————————

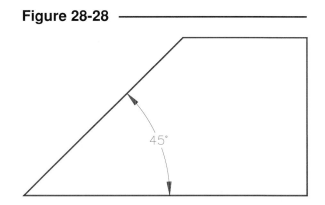

AutoCAD automates the process of creating this type of dimension by offering options within the command syntax. The default options create a dimension, as shown here:

```
Command: Dimangular
Select arc, circle, line, or press ENTER: PICK
Second line: PICK
Dimension arc line location (Mtext/Text/Angle): PICK the desired location of the dimension line
Dimension text = nn.nnnn
Command:
```

Dimangular dimensioning offers some very useful and easy-to-use options for placing the desired dimension line and text location.

At the "Dimension arc line location (Mtext/Text/Angle):" prompt, you can move the cursor around the vertex to dynamically display possible placements available for the dimension. The dimension can be placed in any of four positions as well as any distance from the vertex (Fig. 28-29). Extension lines are automatically created as needed.

The *Dimangular* command offers other options, including dimensioning angles for *Arcs*, *Circles*, or allowing selection of any three points.

Figure 28-29 ————————————

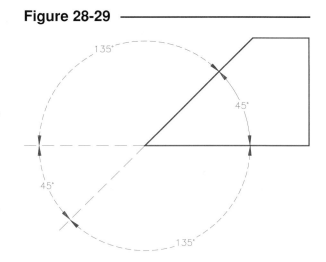

If you select an *Arc* in response to the "Select arc, circle, line, or press ENTER:" prompt, AutoCAD uses the *Arc's* center as the vertex and the *Arc's* endpoints to generate the extension lines. You can select either angle of the *Arc* to dimension (Fig. 28-30).

Figure 28-30

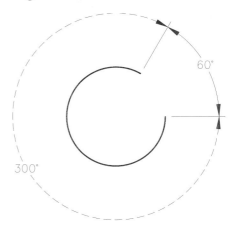

If you select a *Circle*, AutoCAD uses the PICK point as the first extension line origin. The second extension line origin does not have to be on the *Circle*, as shown in Figure 28-31.

If you press **Enter** in response to the "Select arc, circle, line, or press ENTER:" prompt, AutoCAD responds with the following:

    Angle vertex: **PICK**
    First angle endpoint: **PICK**
    Second angle endpoint: **PICK**

This option allows you to apply an *Angular* dimension to a variety of shapes.

Figure 28-31

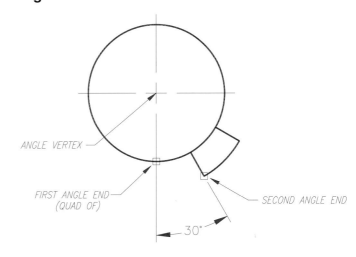

LEADER

Pull-down Menu	COMMAND (TYPE)	ALIAS (TYPE)	Short-cut	Screen (side) Menu	Tablet Menu
Dimension *Leader*	*LEADER*	**LEAD** **or LE**	...	*DIMNSION* *Leader*	**R,7**

The *Leader* command allows you to create an associative leader similar to that created with the *Diameter* command. The *Leader* command is intended to give dimensional notes such as the manufacturing or construction specifications shown here:

    Command: *leader*
    From point: **PICK**
    To point: **PICK**
    To point (Format/Annotation/Undo)<Annotation>:
    CASE HARDEN
    <Mtext>: **Enter**
    Command:

Figure 28-32

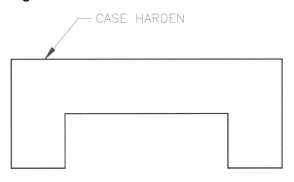

At the "From point:" prompt, select the desired location for the arrow. You should use *SNAP* or an *OSNAP* option (such as *Nearest*, in this case) to ensure the arrow touches the desired object.

Figure 28-33 —————————

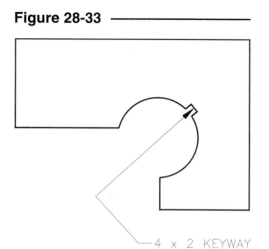

A short horizontal line segment called the "hook line" is automatically added to the last line segment drawn if the leader line is 15 degrees or more from horizontal. Note that command syntax for the *Leader* in Figure 28-32 indicates only one line segment was PICKed. A *Leader* can have as many segments as you desire (Fig. 28-33).

If you do not enter text at the "Annotation:" prompt, another series of options are available:

```
Command: leader
From point: PICK
To point: PICK
To point (Format/Annotation/Undo)<Annotation>: Enter or (option) or (text)
Annotation (or press ENTER for options): Enter or (text)
Tolerance/Copy/Block/None/<Mtext>: Enter or (option)
Command:
```

Format
This option produces another list of choices:

Spline/STraight/Arrow/None/<Exit>:

Spline/STraight
You can draw either a *Spline* or straight version of the leader line with these options. The resulting *Spline* leader line has the characteristics of a normal *Spline*. An example of a *Splined* leader is shown in Figure 28-34.

Figure 28-34 —————————

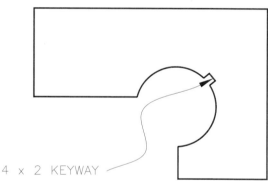

Arrow/None
This option draws the leader line with or without an arrowhead at the start point.

Annotation
This option prompts for text to insert at the end of the *Leader* line.

Mtext
The *Multiline Text Editor* appears with this option. Text can be entered into paragraph form using the *Mtext* text boundary mechanism (see *Mtext*, Chapter 18).

Tolerance
This option produces a feature control frame using the *Geometric Tolerances* dialog boxes (see *Tolerance*).

Copy
You can copy existing *Text, Dtext,* or *Mtext* objects from the drawing to be placed at the end of the *Leader* line. The copied object is associated with the *Leader* line.

Block
An existing *Block* of your selection can be placed at the end of the *Leader* line. The same prompts as the *Insert* command are used.

None
Using this option draws the *Leader* line with no annotation.

Undo
This option undoes the last vertex point of the *Leader* line.

Since a *Leader* is associative, it is affected by the current dimension style settings. You can control the *Leader's* arrowhead type, scale, color, etc., with the related dimensioning variables (see Chapter 29).

DIMORDINATE

Pull-down Menu	COMMAND (TYPE)	ALIAS (TYPE)	Short-cut	Screen (side) Menu	Tablet Menu
Dimension Ordinate	*DIMORDINATE*	*DIMORD or DOR*	...	*DIMNSION Ordinate*	*W,3*

Ordinate dimensioning is a specialized method of dimensioning used in the manufacturing of flat components such as those in the sheet metal industry. Because the thickness (depth) of the parts is uniform, only the width and height dimensions are specified as Xdatum and Ydatum dimensions.

Dimordinate dimensions give an Xdatum or a Ydatum distance between object "features" and a reference point on the geometry treated as the origin, usually the lower-left corner of the part. This method of dimensioning is relatively simple to create and easy to understand. Each dimension is composed only of one leader line and the aligned numerical value.

Figure 28-35 ───────────────

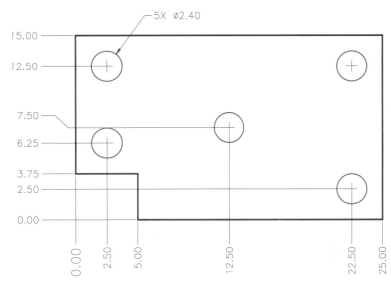

To create ordinate dimensions in AutoCAD, the *UCS* command should be used first to establish a new 0,0 point. *UCS*, which stands for user coordinate system, allows you to establish a new coordinate system with the origin and the orientation of the axes anywhere in 3D space (see Chapters 34 and 36 for complete details). In this case, we need to change only the location of the origin and leave the orientation of the axes as is. Type **UCS** and use the **Origin** option to PICK a new origin as shown (Fig. 28-36).

Figure 28-36 ───────────────

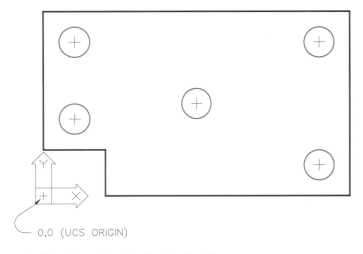

When you create a *Dimordinate* dimension, AutoCAD only requires you to (1) PICK the object "Feature" and then (2) specify the "Leader endpoint:". The dimension text is automatically aligned with the leader line.

It is not necessary in most cases to indicate whether you are creating an Xdatum or a Ydatum. Using the default option of *Dimordinate*, AutoCAD makes the determination based on the direction of the leader you specify (step 2). If the leader is <u>perpendicular</u> (or almost perpendicular) to the X axis, an Xdatum is created. If the leader is (almost) <u>perpendicular</u> to the Y axis, a Ydatum is created. The command syntax for a *Dimordinate* dimension is this:

        Command: **Dimordinate**
        Select feature: **PICK**
        Leader endpoint
        (Xdatum/Ydatum/Mtext//Text): **PICK**
        Dimension text = *nn.nnnn*
        Command:

A *Dimordinate* dimension is created in Figure 28-37 by PICKing the object feature and the leader endpoint. That's all there is to it. The dimension is a Ydatum; yet AutoCAD automatically makes that determination, since the leader is perpendicular to the Y axis. It is a good practice to turn *ORTHO ON* in order to ensure the leader lines are drawn horizontally or vertically.

An Xdatum *Dimordinate* dimension is created in the same manner. Just PICK the object feature and the other end of the leader line (Fig. 28-38). The leader is perpendicular to the X axis; therefore, an Xdatum dimension is created.

The leader line does not have to be purely horizontal or vertical. In some cases, where the dimension text is crowded, it is desirable to place the end of the leader so that sufficient room is provided for the dimension text. In other words, draw the leader line at an angle. (*ORTHO* must be turned *Off* to specify an offset leader line.) AutoCAD automatically creates an offset in the leader as shown in the 7.50 Ydatum dimension in Figure 28-39. As long as the leader is more perpendicular to the X axis, an X datum is drawn and vice versa.

Figure 28-37 ───────────────

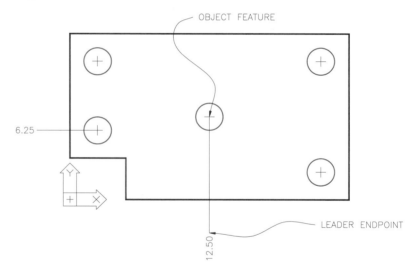

Figure 28-38 ───────────────

Figure 28-39 ───────────────

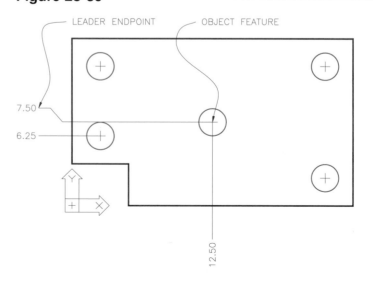

Xdatum/Ydatum

The *Xdatum* and *Ydatum* options are used to specify one of these dimensions explicitly. This is necessary in case the leader line that you specify is more perpendicular to the <u>other</u> axis that you want to measure along. The command syntax would be as follows:

```
Command: Dimordinate
Select feature: PICK
Leader endpoint (Xdatum/Ydatum/Mtext/Text): X
Leader endpoint: PICK
Dimension text = nn.nnnn
Command:
```

Mtext/Text

The *Mtext* and *Text* options operate similar to the same options of other dimensioning commands. The *Mtext* option summons the *Multiline Text Editor* and the *Text* option uses command line format. Any text placed inside the < > characters overrides the AutoCAD-measured dimensional value. Any additional text entered on either side of the < > symbols is treated as a prefix or suffix to the measured value. All other options in the *Mtext* dialog box are usable for ordinate dimensions.

TOLERANCE

Pull-down Menu	COMMAND (TYPE)	ALIAS (TYPE)	Short-cut	Screen (side) Menu	Tablet Menu
Dimension Tolerance...	TOLERANCE	TOL	...	DIMNSION *Toleranc*	X,1

Geometric dimensioning and tolerancing (GDT) has become an essential component of detail drawings of manufactured parts. Standard symbols are used to define the geometric aspects of the parts and how the parts function in relation to other parts of an assembly. The *Tolerance* command in AutoCAD produces a series of dialog boxes for you to create the symbols, values, and feature control frames needed to dimension a drawing with GDT. Invoking the *Tolerance* command produces the *Symbol* dialog box (Fig. 28-42) and the *Geometric Tolerance* dialog box. These dialog boxes can also be accessed by the *Leader* command. Unlike general dimensioning in AutoCAD, geometric dimensioning is <u>not associative</u>; however, you can use *Ddedit* to edit the components of an existing feature control frame.

Some common examples of geometric dimensioning may be a control of the flatness of a surface, the parallelism of one surface relative to another, or the positioning of a group of holes to the outside surfaces of the part. Most of the conditions are controls that cannot be stated with the other AutoCAD dimensioning commands.

This book does not provide instruction on how to use geometric dimensioning, rather how to use AutoCAD to apply the symbology. The ASME Y14.5M - 1994 Dimensioning and Tolerancing standard is the authority on the topic in the United States. The symbol application presented in AutoCAD Release 14 is actually based on the 1982 release of the Y14.5 standard. Some of the symbols in the 1994 version are modified from the 1982 standard.

The dimensioning symbols are placed in a feature control frame (FCF). This frame is composed of a minimum of two sections and a maximum of three different sections (Fig. 28-40).

Figure 28-40

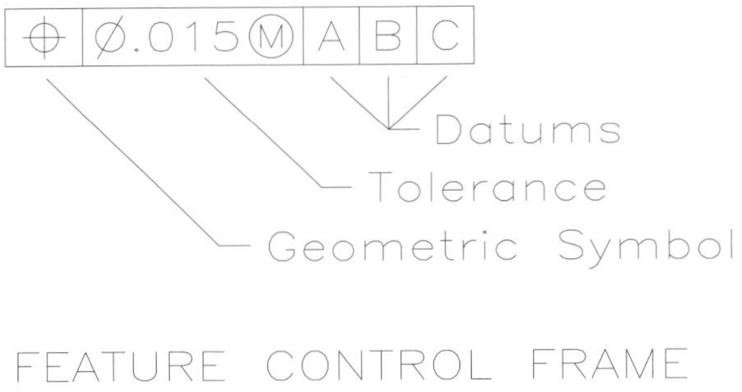

⊕ | Ø.015Ⓜ | A | B | C

Datums
Tolerance
Geometric Symbol

FEATURE CONTROL FRAME

The first section of the FCF houses one of 14 possible geometric characteristic symbols (Fig. 28-41).

The second section of the FCF contains the tolerance information. The third section includes any datum references.

In AutoCAD, the *Tolerance* command allows you to specify the values and symbols needed for a feature control frame. The lines comprising the frame itself are automatically generated. Invoking *Tolerance* by any method produces the *Symbol* dialog box (Fig. 28-42).

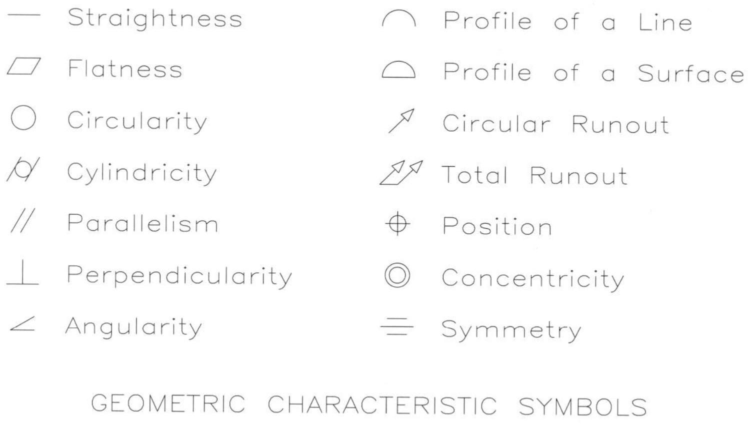

Here you select the geometric characteristic symbol to use for the new feature control frame. If you want to specify a *Datum Identifier* only and do not need to select a symbol, click *OK* to bypass this step.

Once the symbol has been selected, the *Geometric Tolerance* dialog box (Fig. 28-43) appears for you to specify the tolerance values and datum. The areas of the *Geometric Tolerance* dialog box are explained here.

Sym

This image tile displays the symbol previously selected from the *Symbol* dialog box. If you want to change the symbol, click on the image tile to return to the *Symbol* dialog box for another selection.

Tolerance 1

This area is used to specify the first tolerance value in the feature control frame. This value specifies the amount of allowable deviation for the geometric feature. The three sections in this cluster are:

> *Dia*
> If a cylindrical tolerance zone is specified, a diameter symbol can be placed before the value by clicking in this area.

> *Value*
> Enter the tolerance value in this edit box.

> *MC*
> A material condition symbol can be placed after the value by choosing this tile. The *Material Condition* dialog box appears for your selection (Fig. 28-44):

> > *M*   Maximum material condition
> > *L*   Least material condition
> > *S*   Regardless of feature size

Once you select a material condition or cancel, the *Geometric Tolerance* dialog box reappears.

Figure 28-41

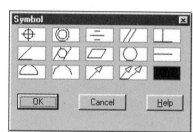

GEOMETRIC CHARACTERISTIC SYMBOLS

Figure 28-42

Figure 28-43

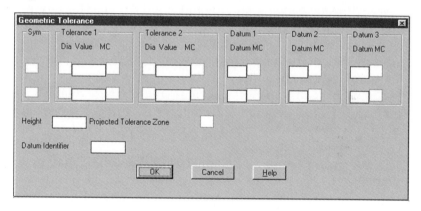

Figure 28-44

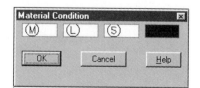

Tolerance 2
Using this section creates a second tolerance area in the feature control frame. GDT standards, however, require only one tolerance area. The options in this section are identical to *Tolerance 1*.

Datum 1
This section allows you to specify the primary datum reference for the current feature control frame. A datum is the exact (theoretical) geometric feature that the current feature references to establish its tolerance zone.

> *Datum*
> Enter the primary datum letter (A, B, C, etc.) for the FCF in this edit box.
>
> *MC*
> Choosing the *MC* tile allows you to specify a material condition modifier for the datum using the *Material Condition* dialog box (Fig. 28-44).

Datum 2
This section is used if you need to create a second datum reference in the feature control frame. The options are identical to those for *Datum 1*.

Datum 3
Use this section if you need to specify a third datum reference for the current feature control frame.

Height
The edit box allows entry of a value for a projected tolerance zone in the feature control frame. A projected tolerance zone specifies a permissible cylindrical tolerance that is projected above the surface of the part. The axis of the control feature must remain within the stated tolerance.

Projected Tolerance Zone
Click this section to insert a projected tolerance zone symbol (a circled "P") after the value.

Datum Identifier
This option creates a datum feature symbol. The edit box allows entry of the value (letter) indicating the datum. A hyphen should be included on each side of the datum letter, such as "-A-".

After using the dialog boxes and specifying the necessary values and symbols, AutoCAD prompts for you to PICK a location on the drawing for the feature control symbol or datum identifier:

    Enter tolerance location: **PICK**

The feature control symbol or datum identifier is drawn as specified.

Editing Feature Control Frames

The AutoCAD-generated feature control frames are not associative and can be *Erased*, *Moved*, or otherwise edited without consequence to the related geometry objects. Feature control frames created with *Tolerance* are treated as one object; therefore, editing an individual component of a feature control frame is not possible unless *Ddedit* is used.

Ddedit is normally used to edit text, but if an existing feature control frame is selected in response to the "<Select an annotation object>/undo:" prompt, the *Geometric Tolerance* dialog box appears. The edit boxes in this dialog box contain the values and symbols from the selected FCF. Making the appropriate changes updates the selected FCF. (See Chapter 18 for detailed information on *Ddedit*.)

Basic Dimensions

Basic dimensions are often required in drawings using GDT. Basic dimensions in AutoCAD are created by using the *Dimension Styles* dialog box (or by entering a negative value in the *DIMGAP* variable). The procedure is explained briefly in the following example (application 5) and discussed further in Chapter 29, Dimension Styles and Dimension Variables.

GDT-Related Dimension Variables

Some aspects of how AutoCAD draws the GDT symbols can be controlled using dimension variables and dimension styles (discussed fully in Chapter 29). These variables are:

DIMCLRE	Controls the color of the FCF
DIMCLRT	Controls the color of the tolerance text
DIMGAP	Controls the gap between the FCF and the text, and controls the existance of a basic dimension box
DIMTXT	Controls the size of the tolerance text
DIMTSTSTY	Controls the style of the tolerance text

Geometric Dimensioning and Tolerancing Example

Six different geometric dimensioning and tolerancing examples are shown on the SPACER drawing in Figure 28-45. Each example is indicated on the drawing and in the following text by a number, 1 through 6. The purpose of the examples is to explain how to use the geometric dimensioning features of AutoCAD. In each example, any method shown in the *Tolerance* command table can be used to invoke the command and dialog boxes.

Figure 28-45

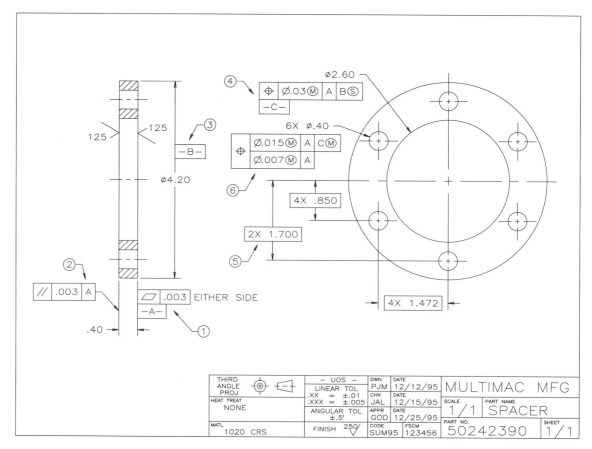

1. Flatness Application

The first specification applied is flatness. In addition to one of the surfaces being controlled for flatness, it is also identified as a datum surface. Both conditions are applied in the same application:

> Command: **tolerance** (The *Symbol* dialog box appears.)
> PICK the *flatness* symbol, then the **OK** button. (The *Geometric Tolerance* dialog box appears, Fig. 28-46.)
> PICK the *Tolerance 1 Value* box and type ".003" as a tolerance.
> PICK the *Datum Identifier* box and type "-A-".
> PICK the **OK** button. (The *Geometric Tolerance* dialog box disappears.)
> Enter tolerance location: **PICK** (PICK a point on the right extension line where you want the middle of the left vertical line of the flatness part of the FCF to be attached, Fig. 28-47.)
> Command:

Because either side of the SPACER can be chosen for the flatness specification, the note "EITHER SIDE" is entered next to the FCF in Figure 28-45.

Figure 28-46

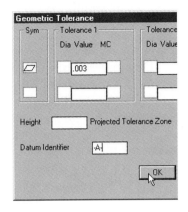

Figure 28-47

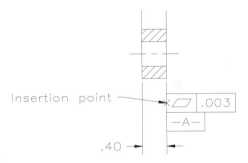

2. Parallelism Application

The second application is a parallelism specification to the opposite side of the part. This control is applied using a *Leader*:

> Command: **leader**
> From point: **PICK** (PICK a point on the left extension line, Fig. 28-45.)
> To point: **PICK** (PICK a location on the middle of the right vertical line of the FCF.)
> To point (Format/Annotation/Undo)<Annotation>: **Enter**
> Annotation (or RETURN for options): **Enter**
> Tolerance/Copy/Block/None/<Mtext>: **T** (The *Symbol* dialog box appears.)
> PICK the *parallelism* symbol, then **OK** to activate the *Geometric Tolerance* dialog box.
> PICK the *Tolerance 1 Value* box and type ".003" as a tolerance.
> PICK the *Datum 1 Datum* box and type "A".
> PICK the **OK** button. (The *Geometric Tolerance* dialog box clears and the parallelism FCF is placed to the left of the leader.)
> Command:

If the leader had projected to the right of the controlled surface, the FCF would be placed on the right of the leader.

3. Datum Application

The third application is the B datum on the 4.20 diameter. This datum is used later in the position specification.

> Command: **tolerance** (The *Symbol* dialog box appears.)
> PICK the **OK** button. (The *Geometric Tolerance* dialog box appears.)
> PICK the *Datum Identifier* box and type "-B-". (The *Geometric Tolerance* dialog box clears.)
> Enter tolerance location: **PICK** (PICK a point on the dimension line of the 4.20 diameter.)
> Command:

4. Position Application

The fourth application is a position specification of the 2.60 diameter hole. It is identified as a C datum because it will be used in the position specification of the six mounting holes. This specification creates a relationship between inside and outside diameters and a perpendicularity requirement to datum A.

> Command: **tolerance** (The *Symbol* dialog box appears.)
> PICK the **position** symbol, then the **OK** button to produce the *Geometric Tolerance* dialog box.
> PICK the *Tolerance 1* **Dia** box. (This places a diameter symbol in front of the tolerance.)
> PICK the *Tolerance 1* **Value** box and type ".03" as a tolerance.
> PICK the *Tolerance 1* **MC** box. (The *Material Condition* dialog box appears.)
> PICK the circled **M** and the **OK** button.
> PICK the *Datum 1* **Datum** box and type "**A**".
> PICK the *Datum 2* **Datum** box and type "**B**".
> PICK the *Datum 2* **MC** box to produce the *Material Condition* dialog box.
> PICK the circled **S** and then the **OK** button.
> PICK the **Datum Identifier** box and type "**-C-**".
> PICK the **OK** button to return to the drawing.
> Enter tolerance location: **PICK** (PICK a point that places the FCF below the 2.60 diameter [Fig. 28-45].)
> Command:

If you need to move the FCF with *Move* or grips, select any part of the FCF or its contents since it is treated as one object.

5. Basic Dimension Application

The fifth application concerns basic dimensions. Position uses basic (theoretically exact) location dimensions to locate holes because the tolerances are stated in the FCF. Before applying these dimensions, the *Dimension Style* must be changed. (See Chapter 29 for a complete discussion of Dimension Styles.)

> Command: **ddim** (Or use any method to invoke the *Dimension Styles* dialog box.)
> PICK the **Annotation** button. (The *Annotation* dialog box appears.)
> PICK the *Tolerance* **Methods** drop-down list.
> PICK **Basic**, then the **OK** button.
> PICK the **OK** button in the *Dimension Styles* dialog box.
> Command: **dimlinear** (Use the *dimlinear* command to apply each of the three linear basic dimensions shown in Fig. 28-45.)

6. Composite Position Application

The sixth application is the composite position specification. A composite specification consists of two separate lines but only one geometric symbol. Achieving this result requires entering data on the top and bottom lines in the *Geometric Tolerance* dialog box.

> Command: **tolerance**
> PICK the **position** symbol, then the **OK** button in the *Symbol* dialog box.
> PICK the *Tolerance 1* **Dia** box in the *Geometric Tolerance* dialog box.
> PICK the *Tolerance 1* **Value** box and enter ".015" as a tolerance.
> PICK the *Tolerance 1* **MC** box. (The *Material Condition* dialog box appears.)
> PICK the circled **M** and then **OK**.
> PICK the *Datum 1* **Datum** box and type "**A**".
> PICK the *Datum 2* **Datum** box and type "**C**".
> PICK the *Datum 2* **MC** box. (The *Material Condition* dialog box appears.)
> PICK the circled **M** and then the **OK** button.

PICK the **Sym** box under the position symbol to produce the *Symbol* dialog box.
PICK the **position** symbol, then the **OK** button.
PICK the *Tolerance 1* **Dia** box on the second line.
PICK the *Tolerance 1* **Value** box on the second line and type ".007" as a tolerance.
PICK the *Tolerance 1* **MC** box on the second line. (The *Material Condition* dialog box appears.)
PICK the circled **M** and the **OK** button.
PICK the *Datum 1* **Datum** box and type "**A**".
PICK the **OK** button. (The *Geometric Tolerance* dialog box clears.)
Enter tolerance location: **PICK** (PICK a point that places the FCF below the .40 diameter, Fig. 28-45.)
Command:

Datum Targets

The COVER drawing (Fig. 28-48) uses datum targets to locate the part in 3D space. AutoCAD Release 14 provides no commands to apply datum targets. The best way to apply these symbols is to use *Blocks* and attributes. The target circle diameter is 3.5 times the letter height. The dividing line is always drawn horizontally through the center of the circle.

Figure 28-48

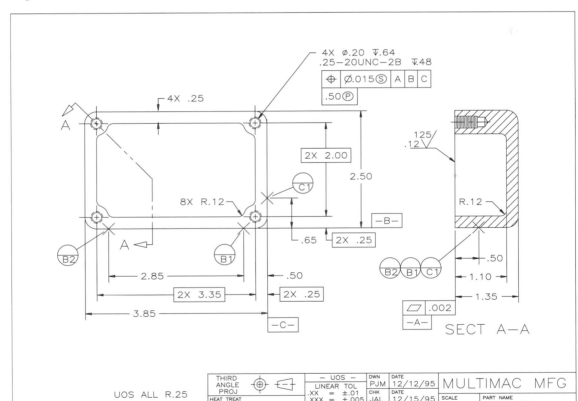

Projected Tolerance Zone Application

The COVER drawing also uses a projected tolerance with the position specification. A projected tolerance zone is used primarily with internally threaded holes and dowel holes. In this case, the concern is not the position and orientation of the threaded hole, rather the position and orientation of the fastener inserted into the hole, specifically the shank of the fastener; therefore, the tolerance zone projects above the surface of the part and not within the part. This condition is especially important when the hole allowance is small.

Command: **tolerance**
PICK the **position** tile and **OK** in the *Symbol* dialog box.
PICK the **Height** box in the *Geometric Tolerance* dialog box and enter ".**50**".
PICK the **Projected Tolerance Zone** box. A circled *P* appears.
PICK the **OK** button in the *Geometric Tolerance* box.
Enter tolerance location: **PICK** (PICK a point that places the FCF below the .20 diameter, Fig. 28-48.)
Command:

This section has presented only the mechanics of GDT symbol application in AutoCAD Release 14. Geometric dimensioning and tolerancing can be a complicated and detailed subject to learn. Many different combination possibilities may appear in a feature control frame. However, for any one feature, there are very few possibilities. Knowing which of the possibilities is best comes from a fundamental knowledge of GDT. Knowledge of GDT increases your understanding of design, tooling, manufacturing, and inspection concepts and processes.

DIM

In AutoCAD releases previous to Release 13, dimensioning commands had to be entered at the "Dim:" prompt and could not be entered at the "Command:" prompt. The "Dim:" prompt puts AutoCAD in the dimensioning mode so earlier release dimensioning commands can be used. This convention can also be used in Release 14. The *DIM* command must be typed in Release 14 to produce the "Dim:" prompt. Enter *E* or *Exit* or press Escape to exit the dimensioning mode and return to the "Command:" prompt. The syntax is as follows:

Command: **dim**
Dim:

Release 12 and previous dimensioning command names do not have the "dim" prefix; therefore, using the "Dim:" prompt is required for them. Almost all of the Release 12 commands have the same name as the Release 14 version but without the "dim" prefix. For example, *Dimdiameter* is "*Diameter*" in Release 12. In Release 14, you can enter these older dimensioning command names (without the "dim" prefix) at the "Dim:" prompt; for example,

Command: **dim**
Dim: **diameter**

With the exception of a few dimensioning commands and options, the Release 12 dimensioning commands are the same as the Release 14 versions without the "dim" prefix. However, several other older commands can be used in Release 14 at the "Dim:" prompt.

UNDO or *U*
Erases the last dimension objects or dimension variable setting.

UPDATE
Performs the same action as *Dimstyle, Apply.*

STYLE
Changes the current <u>text</u> style.

HORIZONTAL
Performs the same action as *Dimlinear, Horizontal.*

VERTICAL

Performs the same action as *Dimlinear, Vertical.*

ROTATED

Performs the same action as *Dimlinear, Rotate.*

Another use for the "Dim:" prompt is for the *Viewport* option of *DIMLFAC*, which is not available by any other method (see *DIMLFAC*, Chapter 33).

EDITING DIMENSIONS

Associative Dimensions

AutoCAD creates associative dimensions by default. Associative dimensions contain "definition points" that define points in the dimensions such as extension line origins, placement of dimension line, the selected points of *Circles* and *Arcs*, and the centers of those *Circles* and *Arcs*. If the geometry is modified by certain editing commands (more specifically, if the definition points are changed), the dimension components, including the numerical value, automatically update. When you create the first associative dimension, AutoCAD automatically creates a new layer called DEFPOINTS that contains all the definition points. The layer should be kept in a *Frozen* state. If you alter the points on this layer, existing dimensions lose their associativity and you eliminate the possibility of automatic editing.

All of the dimensioning commands are associative by default, and the resulting dimensions would be affected by the editing commands listed below. The *DIMASO* variable can be changed to make new dimensions <u>unassociative</u> if you desire (see Chapter 29). The commands that affect associative dimensions are *Extend, Mirror, Rotate, Scale, Stretch, Trim* (linear dimensions only), *Array* (if rotated in a *Polar* array), and grip editing options.

These commands cause the changed dimension to adjust to the changed angle or length. For example, if you use *Stretch* to change some geometry with associative dimensions, the numerical values and extension lines automatically update as the related geometry changes (Fig. 28-49). Remember to use a <u>Crossing Window</u> for selection with *Stretch* and to include the extension line <u>origin</u> (definition point) in the selection set.

Figure 28-49

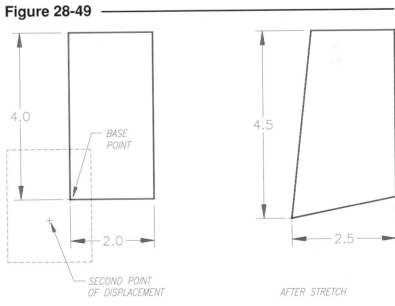

Another feature of associative dimensions is the ability to use *Trim* and *Extend*. For example, you can *Trim* an object and at the same time *Trim* its associated dimension (Fig. 28-50). In this case select the geometry to *Trim* as you would normally, then <u>also select the dimension to *Trim*</u> as shown.

Figure 28-50

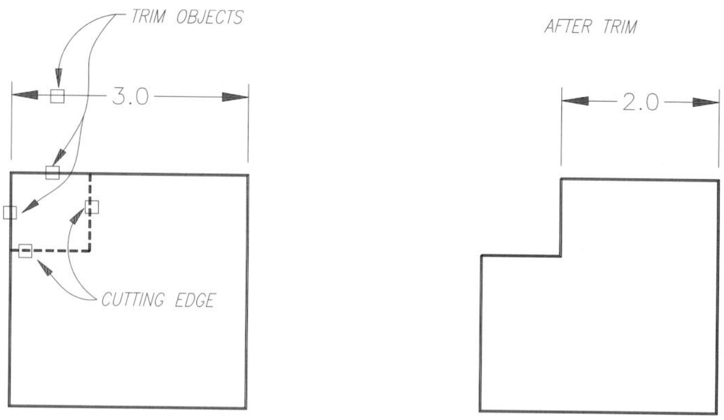

Grip Editing Dimensions

Grips can be used effectively for editing dimensions. Any of the grip options (STRETCH, MOVE, ROTATE, SCALE, and MIRROR) is applicable. Depending on the type of dimension (linear, radial, angular, etc.), grips appear at several locations on the dimension when you activate the grips by selecting the dimension at the Command: prompt. Associative dimensions offer the most powerful editing possibilities, although non-associative dimension components can also be edited with grips. There are many ways in which grips can be used to alter the measured value and configuration of dimensions.

Figure 28-51 shows the grips for each type of associative dimension. Linear dimensions (horizontal, vertical, and rotated) and aligned and angular dimensions have grips at each extension line origin, the dimension line position, and a grip at the text. A diameter dimension has grips defining two points on the diameter as well as one defining the leader length. The radius dimension has center and radius grips as well as a leader grip.

Figure 28-51

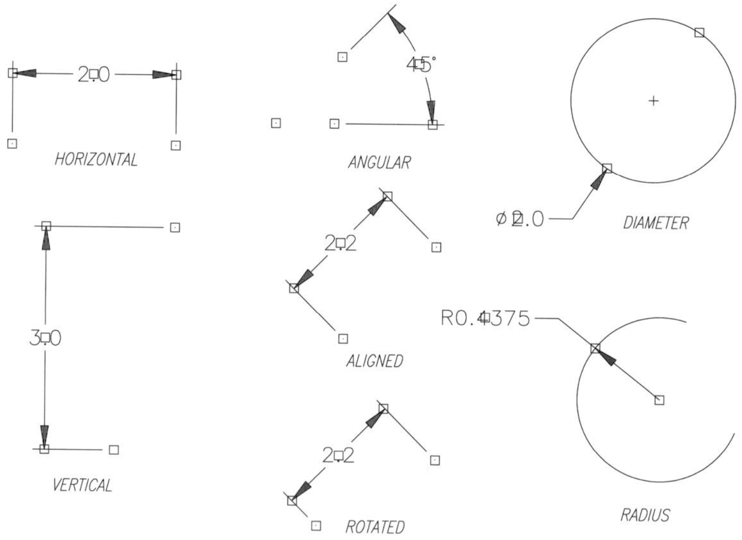

With dimension grips, a wide variety of editing options are possible. Any of the grips can be PICKed to make them **hot** grips. All grip options are valid methods for editing dimensions.

For example (Fig. 28-52), a horizontal dimension value can be increased by stretching an extension line origin grip in a horizontal direction. A vertical direction movement changes the length of the extension line. The dimension line placement is changed by stretching its grips. The dimension text can be stretched to any position by manipulating its grip.

An angular dimension can be increased by stretching the extension line origin grip. The numerical value automatically updates.

Figure 28-52

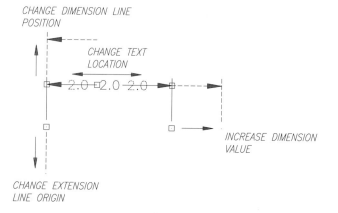

Stretching a rotated dimension's extension line origin allows changing the length of the dimension as well as the length of the extension line. An aligned dimension's extension line origin grip also allows you to change the aligned <u>angle</u> of the dimension.

Figure 28-53

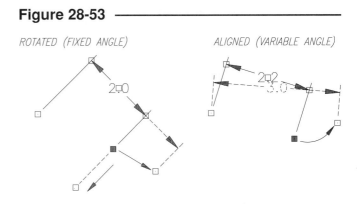

ROTATED (FIXED ANGLE)   ALIGNED (VARIABLE ANGLE)

Rotating the <u>center</u> grip of a radius dimension (with the ROTATE option) allows you to reposition the location of the dimension around the *Arc*. Note that the text remains in its original horizontal orientation.

You can move the dimension text independent of the dimension line with grips for dimensions that have a *DIMFIT* variable setting of 5. Use an existing dimension, and change the *DIMFIT* setting to 5 (or use the *Dimension Style* dialog box, then the *Format* dialog box, and set *Fit* to *No leader*). The dimension text can be moved to any location without losing its associativity. See *Fit: (DIMFIT)*, Chapter 29.

Figure 28-54

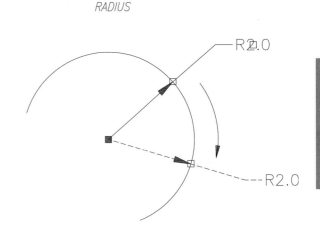

RADIUS

Many other possibilities exist for editing dimensions using grips. Experiment on your own to discover some of the possibilities that are not shown here.

Exploding Associative Dimensions

Associative dimensions are treated as one object. If *Erase* is used with an associative dimension, the entire dimension (dimension line, extension lines, arrows, and text) is selected and *Erased*.

Explode can be used to break an associative dimension into its component parts. The individual components can then be edited. For example, after *Exploding*, an extension line can be *Erased* or text can be *Moved*.

There are two main drawbacks to *Exploding* associative dimensions. First, the associative property is lost. Editing commands or *Grips* cannot be used to change the entire dimension but affect only the component objects. More important, Dimension Styles cannot be used with unassociative dimensions. Second, the dimension is *Exploded* onto Layer 0 and loses its *color* and *linetype* properties. This can include additional work to reestablish the desired layer, *color*, and *linetype*.

Dimension Editing Commands

Several commands are provided to facilitate easy editing of existing dimensions in a drawing. Most of these commands are intended to allow variations in the appearance of dimension <u>text</u>. These editing commands operate <u>only</u> with <u>associative</u> dimensions.

DIMTEDIT

	COMMAND (TYPE)	ALIAS (TYPE)	Short-cut	Screen (side) Menu	Tablet Menu
Pull-down Menu					
Dimension Align Text >	*DIMTEDIT*	*DIMTED*	...	**DIMNSION** *Dimtedit*	*Y,2*

Dimtedit (text edit) allows you to change the position or orientation of the text for a single associative dimension. To move the position of text, this command syntax is used:

> Command: **Dimtedit**
> Select dimension: **PICK**
> Enter text location (Left/Right/Home/Angle): **PICK**

Figure 28-55 ————

At the "Enter text location" prompt, drag the text to the desired location. The selected text and dimension line can be changed to any position, while the text and dimension line retain their associativity.

Angle
The **Angle** option works with any *Horizontal, Vertical, Aligned, Rotated, Radius,* or *Diameter* dimensions. You are prompted for the new <u>text</u> angle (Fig. 28-56):

> Command: **dimtedit**
> Select dimension: **PICK**
> Enter text location (Left/Right/Home/Angle): **a**
> Enter text angle: **45**
> Command:

Figure 28-56 ————

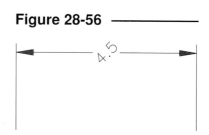

Home
The text can be restored to its original (default) rotation angle with the **Home** option. The text retains its right/left position.

Right/Left
The **Right** and **Left** options automatically justify the dimension text at the extreme right and left ends, respectively, of the dimension line. The arrow and a short section of the dimension line, however, remain between the text and closest extension line (Fig. 28-57).

Center
The *Center* option brings the text to the center of the dimension line and sets the rotation angle to 0 (if previously assigned; Fig. 28-57).

Figure 28-57 ————

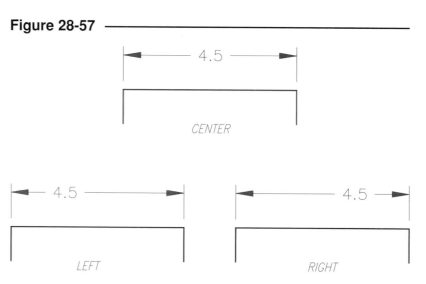

DIMEDIT

Pull-down Menu	COMMAND (TYPE)	ALIAS (TYPE)	Short-cut	Screen (side) Menu	Tablet Menu
Dimension Oblique	DIMEDIT	DIMED or DED	...	DIMNSION Dimedit	Y,1

The *Dimedit* command allows you to change the angle of the extension lines to an obliquing angle and provides several ways to edit dimension text. Two of the text editing options (*Home, Rotate*) duplicate those of the *Dimtedit* command. Another feature (*New*) allows you to change the text value and annotation.

> Command: **dimedit**
> Dimension Edit (Home/New/Rotate/Oblique) <Home>:

Home
This option moves the dimension text back to its original angular orientation angle without changing the left/right position. *Home* is a duplicate of the *Dimtedit* option of the same name.

New
You can change the text value and annotation of existing text. The same mechanism appears when you create the dimension and use the *Mtext* option—that is, the *Multiline Text Editor* appears. Remember that AutoCAD draws the dimension value in place of the < > characters. Add a prefix before these characters or a suffix after them. Entering text to replace the < > characters (erase them) causes your new text to appear instead of the AutoCAD-supplied value.

Placing text or numbers <u>inside</u> the symbols overrides the correct measurement and creates static, non-associative text. The original AutoCAD-measured value <u>can be restored</u>, however, using the *New* option. Simply invoke *Dimedit* and *New* but do not enter any value in the *Multiline Text Editor*. After selecting the desired dimension, the AutoCAD-measured value is restored.

Rotate
AutoCAD prompts for an angle to rotate the text. Enter an absolute angle (relative to angle 0). This option is identical to the *Angle* option of *Dimtedit*.

Oblique
This option is unique to *Dimedit*. Entering an angle at the prompt affects the extension lines. Enter an absolute angle:

> Enter obliquing angle (press ENTER for none):

Normally, the extension lines are perpendicular to the dimension lines. In some cases, it is desirable to set an obliquing angle for the extension lines, such as when dimensions are crowded and hard to read (Fig. 28-58) or for dimensioning isometric drawings (see Dimensioning Isometric Drawings).

Figure 28-58 —————

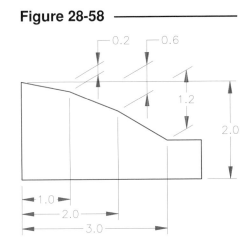

DDMODIFY

	Pull-down Menu	COMMAND (TYPE)	ALIAS (TYPE)	Short-cut	Screen (side) Menu	Tablet Menu
	Modify Properties...	DDMODIFY	MO	...	*MODIFY1 Modify*	Y,14

The *Ddmodify* command (discussed in Chapter 16) can be used effectively for a wide range of dimension editing purposes. Selecting a dimension at the "Select objects:" prompt produces the *Modify Dimension* dialog box (Fig. 28-59). This dialog box provides access to the *Mtext, Geometry, Format,* and *Annotation* dialog boxes.

Use the *Contents:* edit box to edit the dimension text. Overwriting or deleting the < > symbols results in creating new, non-associative text (not recommended). Add a prefix before or suffix after the < > symbols without affecting the AutoCAD-measured value or associativity feature.

Figure 28-59

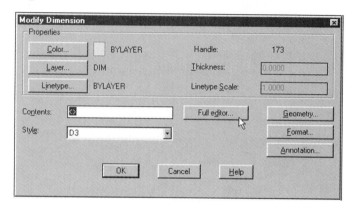

You can also edit text and all aspects of the text (height, font, style, bold, italic, etc.) by selecting the *Full editor...* button. This action produces the *Multiline Text Editor*.

If you want to change other aspects of the dimension object, such as the appearance of the dimension lines, extension lines, or how the text value is displayed, select the *Geometry, Format,* or *Annotation* tiles. This action produces a series of dialog boxes that change <u>dimension variables</u>. Keep in mind that when you use this interface for changing dimension variables, the change creates a <u>dimension style override</u> and applies the change to the selected dimension object. For information on dimension variables, dimension styles, and dimension style overrides, read Chapter 29, Dimension Styles and Dimension Variables.

Customizing Dimensioning Text

As discussed earlier, you can specify dimensioning text other than what AutoCAD measures and supplies as the default text, using the ***Text*** option of the individual dimension creation commands. In addition, the text can be modified at a later time using the ***Ddmodify*** command or the ***New*** option of ***Dimedit***, both of which provide access to the *Multiline Text Editor*.

The less-than and greater-than symbols (< >) represent the AutoCAD-supplied dimensional value. Placing text or numbers <u>inside</u> the symbols overrides the correct measurement (not advised) and creates static text. Text placed inside the < > symbols is not associative and is <u>not</u> updated in the event of *Stretching, Rotating,* or otherwise editing the associative dimension. The original text can only be retrieved using the *New* option of *Dimedit*.

Text placed outside of the < > symbols, however, acts only as a prefix/suffix to the measured value. In the case of a diameter or radius dimension, AutoCAD automatically inserts a diameter symbol (Ø) or radius designator (R) before the dimension value. Inserting a prefix with the *Multiline Text Editor* <u>does not override</u> the AutoCAD symbols; however, entering a *Prefix* or *Suffix* using the *Dimension Styles* dialog box <u>overrides</u> the symbols. This feature is important in correct ANSI standard dimensioning for entering the number of times a feature occurs. For example, one of two holes having the same diameter would be dimensioned "2X Ø1.00" (see Fig. 28-21).

You can also enter special characters by using the Unicode values or "%%" symbols. The following codes can be entered <u>outside</u> the < > symbols for *Dimedit New*, *Mtext*, or the *Text* option of dimensioning commands:

Enter:	Enter:	Result	Description
\U+2205	%%c	ø	diameter (metric)
\U+00b0	%%d	°	degrees
	%%o	‾‾‾	overscored text
	%%u	___	underscored text
\U+00b1	%%p	±	plus or minus
	%%*nnn*	varies	ASCII text character number
\U+*nnnn*		varies	Unicode text hexadecimal value

DIMENSIONING ISOMETRIC DRAWINGS

Dimensioning isometric drawings in AutoCAD is accomplished using *Dimaligned* dimensions and then adjusting the angle of the extension lines with the *Oblique* option of *Dimedit*. The technique follows two basic steps:

1. Use *Dimaligned* or the *Vertical* option of *Dimlinear* to place a dimension along one edge of the isometric face. Isometric dimensions should be drawn on the isometric axes lines (vertical or at a 30° rotation from horizontal; Fig. 28-60).

Figure 28-60 ————————————————

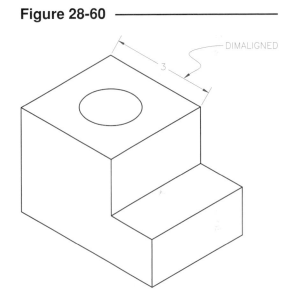

2. Use the *Oblique* dimensioning option of *Dimedit* and select the dimension just created. When prompted to "Enter obliquing angle," enter the desired value (**30** in this case) or PICK two points designating the desired angle. The extension lines should change to the designated angle. In AutoCAD, the possible obliquing angles for isometric dimensions are **30**, **150**, **210**, or **330**.

Figure 28-61 ————————————————

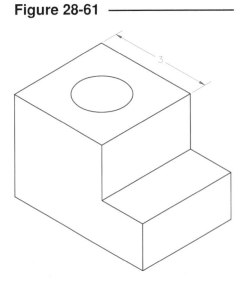

Place isometric dimensions so that they align with the face of the particular feature. Dimensioning the object in the previous illustrations would continue as follows.

Create a *Dimlinear, vertical* dimension along a vertical edge.

Figure 28-62

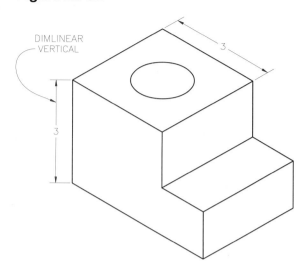

Use *Dimedit, Oblique* to force the dimension to an isometric axis orientation. Enter a value of **150** or PICK two points in response to the "Enter obliquing angle:" prompt.

Figure 28-63

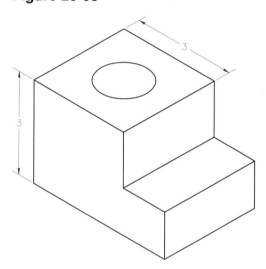

For isometric dimensioning, the extension line origin points <u>must</u> be aligned with the isometric axes. If not, the dimension is not properly oriented. This is important when dimensioning an isometric ellipse (such as this case) or an inclined or oblique edge. Construct centerlines for the ellipse on the isometric axes. Next, construct an *Aligned* dimension and *OSNAP* the extension line origins to the centerlines. Finally, use *Dimedit Oblique* to reorient the angle of the extension lines.

Figure 28-64

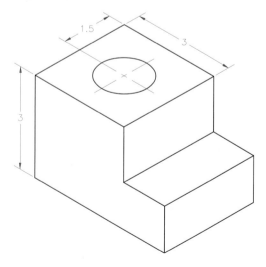

Using the same technique, other appropriate *Dimlinear vertical* or *Aligned* dimensions are placed and reoriented with *Oblique*. Use a *Leader* to dimension a diameter of an *Isocircle*, since *Dimdiameter* cannot be used for an ellipse.

Figure 28-65 ─────────────

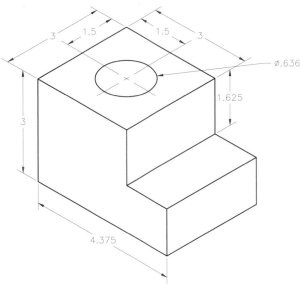

The dimensioning text can be treated two ways. (1) For quick and simple isometric dimensioning, <u>unidirectional</u> dimensioning (all values read from the bottom) is preferred since there is no automatic method for drawing the numerical values in an isometric plane (Fig. 28-65). (2) As a better alternative, text *Styles* could be created with the correct obliquing angles (30 and -30 degrees). The text must also be rotated to the correct angle using the *Rotate* option of the dimension commands or by using *Dimedit Rotate* (Fig. 28-66). Optionally, *Dimension Styles* could be created with the correct variables set for text style and rotation angle for dimensioning on each isoplane.

Figure 28-66 ─────────────

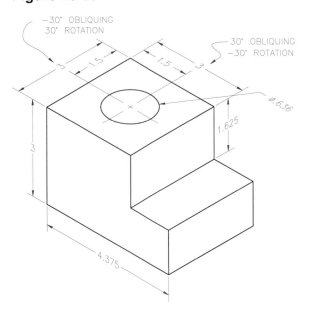

CHAPTER EXERCISES

Only four exercises are offered in this chapter to give you a start with the dimensioning commands. Many other dimensioning exercises are given at the end of Chapter 29, Dimension Styles and Dimension Variables. Since most dimensioning practices in AutoCAD require the use of dimensioning variables and dimension styles, that information should be discussed before you can begin dimensioning effectively.

The units that AutoCAD uses for the dimensioning values are based on the *Units* and *Precision* settings in the *Annotation* dialog box accessed through the *Dimension Styles* dialog box. You can dimension these drawings using the default settings, or if you are adventurous, you can change the *Units* and *Precision* settings for each drawing to match the dimensions shown in the figures. (Type *DDIM*, select *Annotation...*, then *Units*.)

Other dimensioning features may appear different than those in the figures because of the variables set in the AutoCAD default STANDARD dimension style. For example, the default settings for diameter and radius dimensions may draw the text and dimension lines differently than you desire. After reading Chapter 29, those features in your exercises can be changed retroactively by changing the dimension style.

1. *Open* the **PLATES** drawing that you created in Chapter 10 Exercises. *Erase* the plate on the right. Use *Move* to move the remaining two plates apart, allowing 5 units between. Create a *New* layer called **DIM** and make it *Current*. Create a *Text Style* using *Romans* font. Dimension the two plates, as shown in Figure 28-67. *Save* the drawing as **PLATES-D**.

Figure 28-67 ───

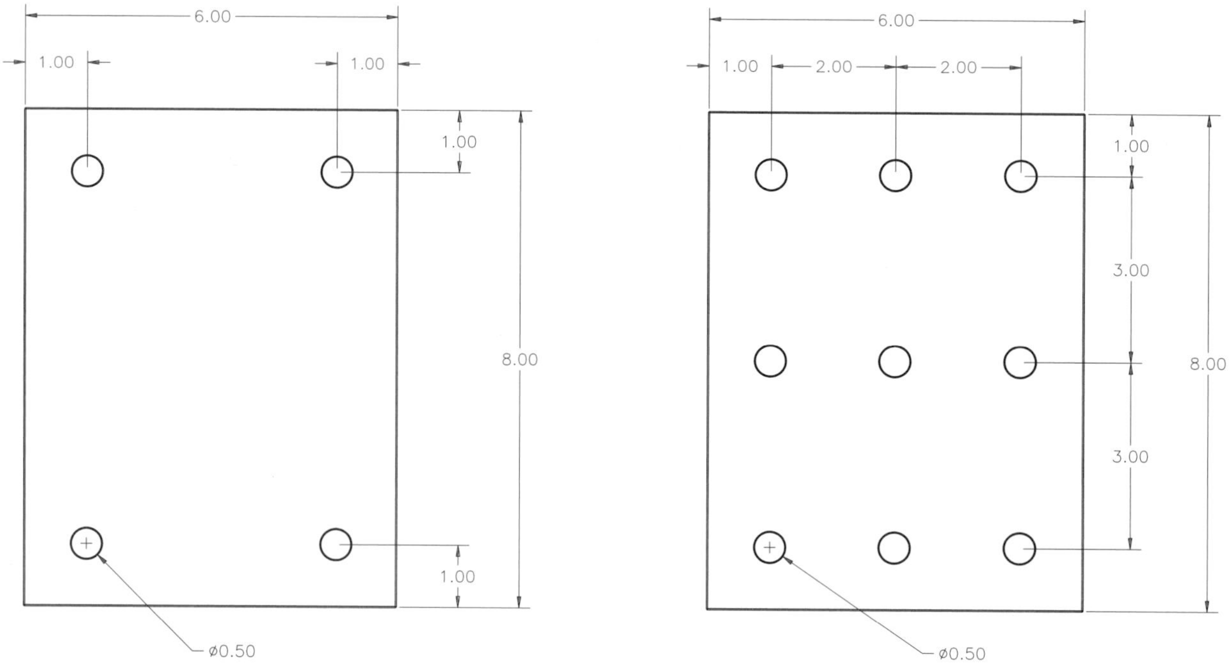

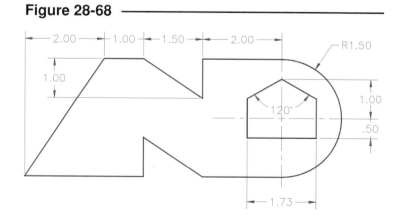

2. *Open* the **PEDIT1** drawing. Create a *New* layer called **DIM** and make it *Current*. Create a *Text Style* using *Romans* font. Dimension the part as shown in Figure 28-68. *Save* the drawing as **PEDIT1-DIM**. Draw a *Pline* border of *.02 width* and *Insert* **TBLOCK** or **TBLOCKAT**. *Plot* on an A size sheet using *Scale to Fit*.

Figure 28-68 ──────────────────────────────────

3. *Open* the **GASKETA** drawing that you created in Chapter 9 Exercises. Create a *New* layer called **DIM** and make it *Current*. Create a *Text Style* using *Romans* font. Dimension the part as shown in Figure 28-69. *Save* the drawing as **GASKETD**.

    Draw a *Pline* border with .02 *width* and *Insert* **TBLOCK** or **TBLOCKAT** with an **8/11** scale factor. *Plot* the drawing *Scaled to Fit* on an A size sheet.

Figure 28-69

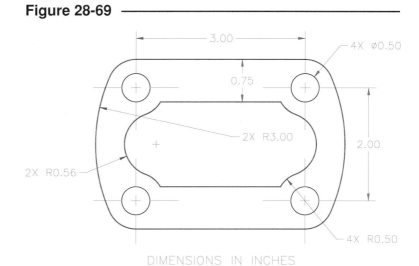

4. *Open* the **BARGUIDE** multiview drawing that you created in Chapter 24 Exercises. Create the dimensions on the **DIM** layer. Keep in mind that you have more possibilities for placement of dimensions than are shown in Figure 28-70.

Figure 28-70

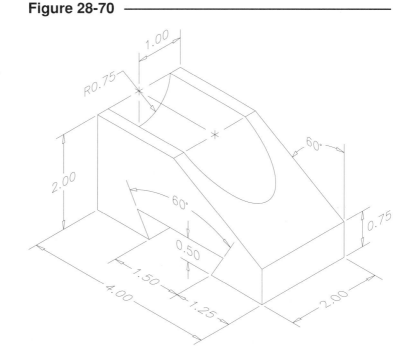

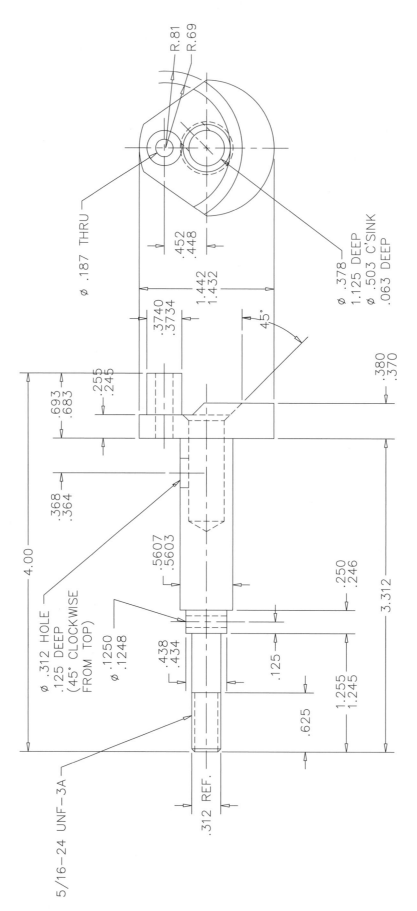

29

DIMENSION STYLES AND DIMENSION VARIABLES

Chapter Objectives

After completing this chapter you should:

1. be able to *Save* and restore dimension styles with the *Dimension Styles* dialog box or in command line format with the *Dimstyle* command;

2. know how to create dimension style families and specify variables for each child;

3. know how to create and apply dimension style overrides by either command line or dialog box method;

4. be able to control dimension variable settings using the *Geometry, Format,* and *Annotation* dialog boxes;

5. be able to use this chapter as a reference for specifying settings using formal dimension variable names in command line format;

6. be able to modify dimensions using *Update, Dimoverride,* and *Matchprop;*

7. know the guidelines for dimensioning.

CONCEPTS

Dimension Variables

Since a large part of AutoCAD's dimensioning capabilities are automatic, some method must be provided for you to control the way dimensions are drawn. A set of 59 <u>dimension variables</u> allows you to affect the way dimensions are drawn by controlling sizes, distances, appearance of extension and dimension lines, and dimension text formats.

Figure 29-1

An example of a dimension variable is *DIMSCALE* (if typed) or *Overall Scaling* (if selected from a dialog box, Fig. 29-1). This variable controls the overall size of the dimension features (such as text, arrowheads, and gaps). Changing the value from 1.00 to 1.50, for example, makes all of the size-related features of the drawn dimension 1.5 times as large as the default size of 1.00 (Fig. 29-2). Other examples of features controlled by dimension variables are arrowhead type, orientation of the text, text style, units and precision, suppression of extension lines, fit of text and arrows (inside or outside of extension lines), and direction of leaders for radii and diameters (pointing in or out). The dimension variable changes that you make affect the <u>appearance</u> of the dimensions that you create.

Figure 29-2

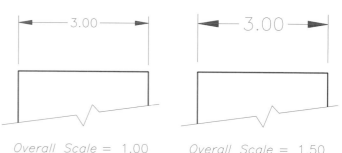

Overall Scale = 1.00
(DEFAULT)

Overall Scale = 1.50

There are two basic ways to control dimensioning variables:

1. Use the *Dimension Styles* dialog box series (Fig. 29-3).

2. Type the dimension variable name in command line format.

The dialog boxes employ "user-friendly" terminology and selection, while the command line format uses the formal dimension variable names.

Changes to dimension variables are usually made <u>before</u> you create the affected dimensions. Dimension variable changes are <u>not automatically retroactive</u>, in contrast to *LTSCALE*, for example, which can be continually modified to adjust the spacing of existing non-continuous lines. Changes to dimensioning variables affect only <u>newly created</u> dimensions unless those changes are *Saved* to an existing *Dimension Style* that was in effect when previous dimensions were created. Generally, dimensioning variables should be set <u>before</u> creating the desired dimensions although it is possible to modify dimensions and *Dimension Styles* retroactively.

Dimension Styles

Figure 29-3

All associative dimensions are part of a <u>dimension style</u>. The default (and the only supplied) dimension style is named STANDARD (Fig. 29-3). Logically, this dimension style has all of the default dimension variable settings for creating a dimension with the typical size and appearance. Similar to layers, you can create, name, and specify settings for any number of dimension styles. Each dimension style contains the dimension variable settings that you select. <u>A dimension style is a group of dimension variable settings that has been saved under a name you assign</u>. When you select a dimension style from the *Current:* list, AutoCAD remembers and resets that <u>particular combination of dimension variable settings</u>.

Imagine dimensioning a complex drawing <u>without</u> having dimension styles (Fig. 29-4). In order to create a dimension using limit dimensioning (as shown in the diameter dimension), for example, you would change the desired variable settings, then "draw" the dimension. To draw another dimension without extension lines (as shown in the interior slot), you would have to reset the previous variables and make changes to other variables in order to place the special dimensions as you prefer. This same process would be repeated each time you want to create a new type of dimension. If you needed to add another dimension with the limits, you would have to reset the same variables as before.

Figure 29-4

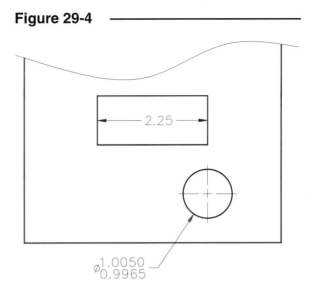

To simplify this process, you can create and save a dimension style for each particular "style" of dimension. Each time you want to draw a particular style of dimension, select the style name from the list and begin drawing. A dimension style could contain all the default settings plus only one or two dimension variable changes or a large number of variable changes.

NOTE: When you make dimension variable changes that you want to keep, make sure you save them to an existing or new style name (by PICKing the *Save* tile).

Another advantage of using dimension styles is that <u>existing</u> dimensions in the drawing can be globally modified by making a change and saving it to the dimension style(s). Knowing that all dimensions are part of a dimension style, making a change in the variable settings using the *Geometry...*, *Format...*, *or Annotation...*dialog boxes and *Saving* that change results in an <u>automatic update</u> for all existing dimensions drawn with that style.

Dimension Style Families

In the previous chapter, you learned about the various <u>classifications</u> of dimensions, such as linear, angular, radial, diameter, ordinate, and leader. You may want the appearance of each of these classifications of dimensions to vary, for example, all radius dimensions to appear with the dimension line pointing out and all diameter dimensions to appear with the dimension line inside pointing in. It is logical to assume that a new dimension style would have to be created for each variation. However, a new dimension style name is not necessary for each classification of dimension. AutoCAD provides <u>dimension style families</u> for the purpose of providing variations within each named dimension style.

The dimension style <u>names</u> that you create are assigned to a dimension style <u>family</u>. Each dimension style family has a <u>parent</u> and can have six <u>children</u>. The children are <u>linear, angular, diameter, radial, ordinate, and leader</u>. If the *Parent* button is selected in the *Dimension Styles* dialog box (Fig. 29-3) when a variable change is made, the variable setting generally affects the entire family. If any one of the buttons for the children (*Linear, Angular, Radial*, etc.) is selected when saving a variable, the variable affects only that child (classification of dimension) (Fig. 29-6). When a dimension is drawn, AutoCAD knows what classification of dimension is created and applies the selected variables to that child.

AutoCAD refers to these children as $0, $2, $3, $4, $6, and $7 as suffixes appended to the parent name (see following table). AutoCAD automatically applies the appropriate suffix code to the dimension style name when a child is created. Although the codes do not appear in the *Dimension Styles* dialog box, you can display the code by using the *List* command and selecting an existing child dimension.

Child Type	Suffix Code
linear	$0
angular	$2
diameter	$3
radial	$4
ordinate	$6
leader	$7

For example, you may want to set the DETAILS dimension style to draw the arrows and dimension lines inside the arc but drag the dimension text outside of the arc for *Radial* dimensions (as shown in Fig. 29-5). First, click on the *Radial* button. Second, make the necessary variable changes in the *Format* dialog box (explained later, see *Radius* and *Diameter* Variable Settings). Third, save the changes to the dimension style, assigning a name (DETAILS) and selecting the *Save* button in the *Dimension Styles* dialog box (as shown in Fig. 29-6). When a *Radius* dimension is drawn, the dimension line, text, and arrow should appear as shown in Figure 29-5. Using *List* to display information about the dimension reports the dimension style name as DETAILS$4.

Figure 29-5

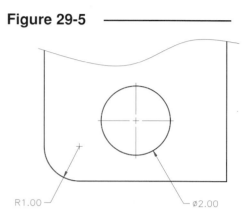

R1.00   ⌀2.00

In summary, a dimension style family is simply a set of dimension variables related by name. Variations within the family are allowed, in that each child (classification of dimension) can possess its own subset of variables. Therefore, each child inherits all the variables of the parent in addition to any others that may be assigned individually.

NOTE: Once a child has been created (special variables have been set for one type of dimension), then variables are changed and saved for the parent; the changes are automatically updated for the existing dimensions (of that style) in the drawing but are <u>not</u> updated for existing or new children (of that style). For example, let's assume you set variables and saved them to a dimension style family (*Parent*), then set other variables and saved them for a child of that style *(Radial,* for example). Later you changed the *Overall Scale* and *Saved* it for the *Parent.* All the dimensions of that style in the drawing automatically update to the new setting except the radial dimensions. To minimize this problem: (1) assure you make all familywide settings before creating children, and (2) when new settings are made to the style (*Parent*) after drawing dimensions, make the same changes and *Save* them for each child or use *Dimoverride* or *Update* to apply new variable settings to existing child dimensions in the drawing. Alternately, you could create a separate dimension style for radial dimensions, etc. instead of creating a radial child. (See Modifying Existing Dimensions, later in this chapter.)

Dimension Style Overrides

When you make the variable change and <u>do not *Save*</u> it to the current dimension style, AutoCAD remembers the setting nevertheless. The change is appended to the <u>current</u> style as a <u>dimension style override</u>. This is particularly helpful if you need a special feature on one or two dimensions but do not want to update the dimension style (and all existing dimensions of the style) with the new change.

A dimension style override is created by <u>changing a variable setting but not saving it</u> to the current dimension style. You can create an override by using the *Dimension Styles* dialog box series to set a variable and exiting without using *Save* or by changing a variable setting by command line format. (If you type the name of a dimension variable and change a value at the command prompt "on the fly," the current dimension style is used as a base style.) When an override is made, a new style name is created with a "+" placed as a prefix and the new style is set current. Each new dimension created with the new modified style applies the overridden variables. If you list one of these dimensions (using *List*), the new style name and the overridden variables are displayed. Dimension style overrides <u>can be applied only to the *Parent* style</u> and cannot be created for an individual child.

DIMENSION STYLES

DDIM

Pull-down Menu	COMMAND (TYPE)	ALIAS (TYPE)	Short-cut	Screen (side) Menu	Tablet Menu
Dimension Style...	*DDIM*	*D*	...	*DIMNSION Ddim*	*Y,5*

The *Ddim* command produces the *Dimension Styles* dialog box (Fig. 29-6). This is the primary interface for creating new dimension styles and making existing dimension styles current. This dialog box also provides access to the three boxes that enable you to change dimension variables. The radio buttons are used to select a classification of dimension (child) for applying variable changes. (Please read the previous sections of this chapter for an introduction to these concepts.) The available options are explained next.

Figure 29-6 ——————————

Current:

Select the desired <u>existing</u> style from the drop-down list to set it as the current style. The dimension style appearing in the box is the current style that is used for drawing subsequent dimensions. Additionally, the current style is the template used for any operations within the *Dimension Style* dialog box series, such as (1) creating a new style, (2) making modifications to the current style by changing dimension variables, then using *Save*, or (3) renaming the style. A name appearing with a "+" (plus) symbol before the name, "+STANDARD," for example, means that a variable has been changed but not yet *Saved* to the dimension style (see Dimension Style Overrides).

Name:

Enter a name in the box to be used for creating a new style or renaming an existing style. The name appearing in the *Current:* box is the template for the new style or renamed style.

Save

Select this tile after creating a new style or making the desired variable changes to the current style. Changing a variable setting and exiting the dialog box without saving creates a dimension style override.

Rename

Select this option to replace the old style name (appearing in the *Current:* box) with the new name (appearing in the *Name:* box). Enter the desired new name in the *Name:* box, then select *Rename*.

Geometry...

This tile produces the *Geometry* dialog box (explained in detail in the following section). The *Geometry* box enables you to make dimension variable changes that affect the appearance of <u>arrowheads</u>, <u>dimension lines</u>, and <u>extension lines</u>.

Format...

This tile produces the *Format* dialog box (explained in detail in the following section). You can make variable changes related to <u>location and orientation of dimension text</u> in the *Format* box.

Annotation...

The *Annotation* dialog box is the interface for dimension variables controlling the <u>content of the dimension text</u>.

Family

Use this section to specify the classification of dimension that you want to affect using the variable controls in the *Geometry, Format,* and *Annotation* boxes. Select the desired family member <u>first</u>; then make the desired adjustments. The family members are:

Parent

Select this button to make familywide dimension variable changes.

Linear, Angular, Radial, Diameter, Ordinate, and Leader

Select the desired classification (child) for subsequent dimension variable changes.

Creating a New Dimension Style Family Using the *Dimension Styles* Dialog Box

1. Select the *Parent* button in the *Dimension Styles* dialog box (see Fig. 29-6).

2. Select the desired dimension style name from the *Current:* list to use as a template. Whatever style appears in this edit box is used as a template. The new style will contain all of the variable settings of the template plus any changes you make and *Save*. Initially, the STANDARD style is used as a template.

3. Type over or modify the name in the *Name:* edit box to assign the new name.

4. Change the desired dimension variables from the *Geometry..., Format...,* or *Annotation...* dialog boxes (discussed in detail later). Familywide variables such as *Overall Scale* are set at this stage. Select the *OK* tile in each of these boxes to keep your changes.

5. After returning to the *Dimension Styles* dialog box, PICK *Save* to assign the variable changes to the new dimension style. The new style becomes current. Then select the *OK* tile. *Cancel* cancels the creation of the new dimension style.

NOTE: Do not change the STANDARD style or you lose your ability to easily restore all of the default settings. Generally, make dimension styles with new names using STANDARD as a <u>template</u>.

Setting Dimension Variables for a Child Using the *Dimension Styles* Dialog Box

1. Select the desired <u>existing</u> dimension style family name from the *Current:* list (see Fig. 29-6).

2. Select the desired classification of dimension (child) such as *Linear, Angular,* or *Radial.*

3. Change the desired dimension variables from the *Geometry..., Format...,* or *Annotation...* dialog boxes (discussed in detail later). Select the *OK* tile in each of these boxes to keep your changes.

4. After returning to the *Dimension Styles* dialog box, PICK *Save* to assign the variable changes to the dimension style family. Then select the *OK* tile.

Restoring a Dimension Style Using the *Dimension Styles* Dialog Box

1. Select the desired style name from the *Current:* list. The style should then appear in the *Name:* edit box. Select the *OK* tile.

2. The selected dimension style becomes the current style and remains so until another is made current. Any dimensions created while the style is current contain the features dictated by the style's variable settings.

Modifying a Dimension Style and Updating Existing Dimensions

1. Select the desired style name from the *Current:* list in the *Dimension Styles* dialog box.

2. Change the desired dimension variables from the *Geometry..., Format...,* or *Annotation...* dialog boxes. Select the *OK* tile in each of these boxes to keep your changes. When you select the *Save* tile in the main dialog box, the current dimension style is updated.

3. As you exit the dialog box, you will notice some activity in the Drawing Editor. Most dimensions originally created with that style are <u>automatically</u> updated with the new variables.

4. Dimensions that were created as children (special variables set and saved for that type of dimension) do not automatically update when a variable change is made and *Saved* for the style (*Parent*). To update existing children, either (1) select the child (*Radial, Diameter,* etc.), make the same variable changes and select *Save,* or (2) use *Dimoverride* or *Update* to apply new variable settings to existing child dimensions in the drawing.

(See Modifying Existing Dimensions later in this chapter.)

DIMSTYLE

Pull-down Menu	COMMAND (TYPE)	ALIAS (TYPE)	Short-cut	Screen (side) Menu	Tablet Menu
Dimension *Update*	*DIMSTYLE*	*DIMSTY* *or DST*	...	*DIMNSION* *Dimstyle*	*Y,3*

The *Dimstyle* command is the command line format equivalent to the *Dimension Style* dialog box. Operations that are performed by the dialog box can also be accomplished with the *Dimstyle* command. The *Apply* option offers one other important feature that is <u>not</u> available in the dialog box:

```
Command: dimstyle
dimension style: STANDARD
Dimension Style Edit (Save/Restore/STatus/Variables/Apply/?) <Restore>:
```

Save
Use this option to <u>create a new</u> dimension style. The new style assumes all of the current dimension variable settings comprised of those from the current style plus any other variables changed by command line format (overrides). The new dimension style becomes the current style.

```
Dimension Style Edit (Save/Restore/STatus/Variables/Apply/?) <Restore>: s
?/Name for new dimension style:
```

Restore
This option prompts for an existing dimension style name to be restored. The restored style becomes the current style. You can use the "press ENTER to select:" option to PICK a dimension object that references the style you want to restore.

STatus

Status gives the <u>current settings</u> for all dimension variables. The displayed list comprises the settings of the current dimension style and any overrides. All 59 dimension variables are listed:

Dimension Style Edit (Save/Restore/STatus/Variables/Apply/?) <Restore>: **st**

DIMALT Off	Alternate units selected
DIMALTD 2	Alternate unit decimal places
DIMALTF 25.4000	Alternate unit scale factor
DIMALTTD 2	Alternate tolerance decimal places
etc.	

Variables

You can use this option to <u>list</u> the dimension variable settings for <u>any existing dimension style</u>. You cannot modify the settings.

Dimension Style Edit (Save/Restore/STatus/Variables/Apply/?) <Restore>: **v**
?/Enter dimension style name or press ENTER to select dimension:

Enter the name of any style or "press ENTER to select:" to PICK a dimension object that references the desired style. A list of all variables and settings appears.

~stylename

This variation can be entered in response to the *Variables* option. Entering a dimension style name preceded by the tilde symbol (~) displays the <u>differences between the current style and the (~) named style</u>.

For example, assume that you wanted to display the <u>differences</u> between STANDARD and the current dimension style (LIMITS-1, for example). Use *Variables* with the ~ symbol. (The ~ symbol is used as a wildcard to mean "all but.")

Command: **dimstyle**
dimension style: LIMITS-1
Dimension Style Edit (Save/Restore/STatus/Variables/Apply/?) <Restore>: **v**
?/Enter dimension style name or press ENTER to select dimension: **~standard**
Differences between STANDARD and current settings:

	STANDARD	Current Setting
DIMLIM	Off	On
DIMTFAC	1.0000	0.7000
DIMTM	0.0000	0.1000
DIMTP	0.0000	0.0500

?/Enter dimension style name or press ENTER to select dimension: *Cancel*
Command:

This option is very useful for keeping track of your dimension styles and their variable settings.

Apply

Use *Apply* to <u>update dimension objects</u> that you PICK with the current variable settings. The current variable settings can contain those of the current dimension style plus any overrides. The selected object loses its reference to its original style and is "adopted" by the applied dimension style family.

Dimension Style Edit (Save/Restore/STatus/Variables/Apply/?) <Restore>: **a**
Select objects:

This is a useful tool for <u>changing an existing dimension object from one style to another</u>. Simply make the desired dimension style current; then use *Apply* and select the dimension objects to change to that style.

Using the *Apply* option of *Dimstyle* is identical to using the *Update* command. The *Update* command can be invoked by using the *Dimension Update* button, using *Update* from the *Dimension* pull-down menu, or typing *Dim*: (Enter), then *Update*.

DIMENSION VARIABLES

Now that you understand how to create and use dimension styles and dimension style families, let's explore the dimension variables. <u>Two methods</u> can be used to set dimension variables: the *Dimension Styles* dialog box series and command line format. First, we will examine the *Geometry, Format,* and *Annotation* dialog boxes (accessible through the *Dimension Styles* dialog box). These dialog boxes contain an interface for setting the dimension variables. Dimension variables can alternately be set by typing the formal name of the variable and changing the desired value (discussed after this section on dialog boxes). However, the dialog boxes offer a more understandable terminology, buttons, check-boxes, lists, and several image tiles that automatically change the variables when you click on the image.

Changing Dimension Variables Using the Dialog Boxes

In this section, the three main dialog boxes for changing variables are explained. The headings for the paragraphs below include the <u>option</u> title in the dialog box and the related <u>dimension variable name</u> in parentheses.

The following figures usually display the AutoCAD default setting for the variable and an example of changing the setting to another value. These AutoCAD default settings (in the STANDARD dimension style) are for the ACAD.DWT template drawing, which is used for *Start from Scratch, English Default Settings*. If you use the *Start from Scratch, Metric Default Settings; Quick Setup Wizard, Advanced Setup Wizard*; or other template drawings, dimension variables may be automatically reset or other dimension styles may exist. See Using *Setup Wizards* and Template Drawings later in this chapter.

Geometry...

Choosing the *Geometry...* button in the *Dimension Styles* dialog box (Fig. 29-6) opens the *Geometry* dialog box (Fig. 29-7). Use this box to change variables that control the <u>appearance of arrowheads, dimension lines, and extension lines</u>. The *Geometry* dialog box contains five specific areas, which are explained in the following sections:

Overall Scale (DIMSCALE)
The *Overall Scale* value globally affects the scale of <u>all size-related features</u> of dimension objects, such as arrowheads, text height, extension line gaps (from the object), extensions (past dimension lines), etc. All other size-related values (variables) appearing in the dialog box series are <u>multiplied by the Overall Scale</u>. Therefore, to keep all features proportional, <u>change this one setting rather than each of the others individually</u>.

Figure 29-7

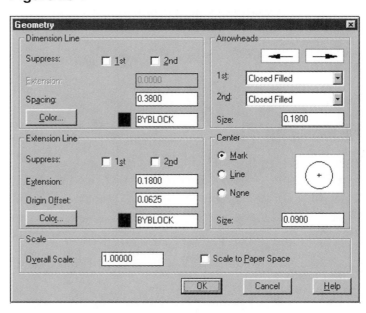

Although this area is located at the bottom of the box, it is probably the most important option. Because the *Overall Scale* should be set as a familywide variable, enter this value as the <u>first step</u> in creating a dimension style. Select *Parent* family member before making this setting.

Figure 29-8

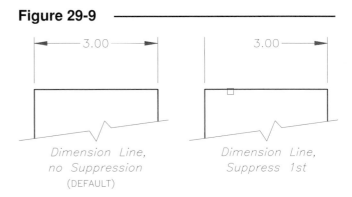

Changes in this variable should be based on the *Limits* and plot scale. You can use the drawing scale factor to determine this value. (See Drawing Scale Factor, Chapter 13.) The *Overall Scale* value is stored in the *DIMSCALE* variable.

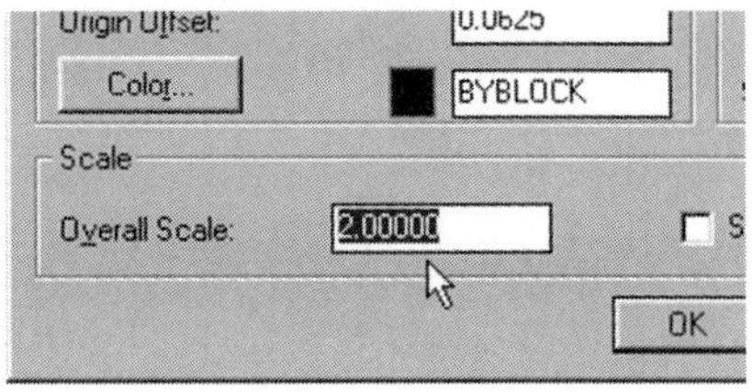

Scale to Paper Space
Checking this box forces dimensions to appear in the same size for all viewports (Paper Space viewports created with *Mview*). Toggling this on sets *DIMSCALE* to 0. See Chapter 32 for detailed information on this subject.

Dimension Line
Suppress 1st, 2nd (DIMSD1, DIMSD2)
This area allows you to suppress (not draw) the *1st* or *2nd* dimension line or both (Fig. 29-9). The <u>first</u> dimension line would be on the "First extension line origin" side or nearest the end of object PICKed in response to "Select object to dimension." These toggles change the *DIMSD1* and *DIMSD2* dimension variables.

Figure 29-9

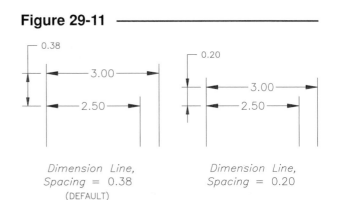

Dimension Line,
no Suppression
(DEFAULT)

Dimension Line,
Suppress 1st

Extension (DIMDLE)
The *Extension* edit box is disabled unless the *Oblique* or *Architectural Tick* arrowhead type is selected (in the *Arrowheads* section). The *Extension* value controls the length of dimension line that extends past the dimension line for *Oblique* or *Architectural Tick* "arrows" (Fig. 29-10).

Figure 29-10

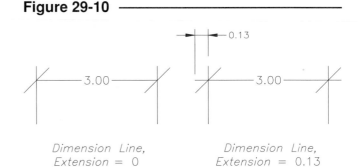

Dimension Line,
Extension = 0

Dimension Line,
Extension = 0.13

Spacing (DIMDLI)
The *Spacing* edit box reflects the value that AutoCAD uses in baseline dimensioning to "stack" the dimension line above or below the previous one (Fig. 29-11). This value is held in the *DIMDLI* variable.

Figure 29-11

Dimension Line,
Spacing = 0.38
(DEFAULT)

Dimension Line,
Spacing = 0.20

Color (DIMCLRD)

The *Color* button allows you to choose the color for the dimension line (Fig. 29-12). Assigning a specific color to the dimension lines, extension lines, and dimension text enables you to print or plot these features with different line widths or colors (pen assignments are made in the *Print/Plot Configuration* dialog box). This feature corresponds to the *DIMCLRD* variable.

Figure 29-12

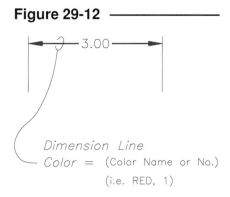

Extension Line

Suppress 1st, 2nd (DIMSE1, DIMSE2)

This area is similar to the *Dimension Line* area and can be easily confused with it, so be careful! This area allows you to suppress the *1st* or *2nd* extension line or both (Fig. 29-13). These options correspond to the *DIMSE1* and *DIMSE2* dimension variables.

Figure 29-13

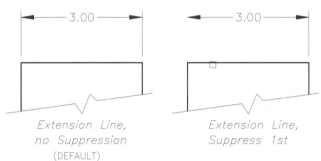

Extension (DIMEXE)

The *Extension* edit box reflects the value that AutoCAD uses to set the distance for the extension line to extend beyond the dimension line (Fig. 29-14). This value is held in the *DIMEXE* variable. Generally, this value does not require changing since it is automatically multiplied by the *Overall Scale*.

Figure 29-14

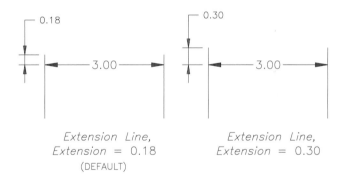

Origin Offset (DIMEXO)

The *Origin Offset* value specifies the distance between the origin points and the extension lines (Fig. 29-15). This offset distance allows you to PICK the object corners, yet the extension lines maintain the required gap from the object. This value rarely requires input since it is affected by *Overall Scale*. The value is stored in the *DIMEXO* variable.

Figure 29-15

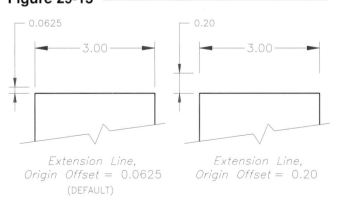

Color (*DIMCLRE*)

The *Color* button allows you to choose the color for the extension lines (Fig. 29-16). This feature corresponds to the *DIMCLRE* variable.

Figure 29-16

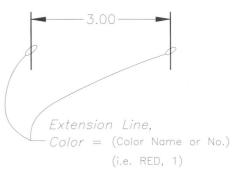

Arrowheads

1st, 2nd (*DIMBLK, DIMBLK1, DIMBLK2*)

This area contains two drop-down lists of various arrowhead types, including dots and ticks. Each list corresponds to the *1st* or *2nd* arrowhead created in the drawn dimension. The image tiles display each arrowhead type selected. Click in the first image tile to change both arrowheads, or click in each to change them individually (Fig. 29-17). The variables affected are *DIMBLK, DIMBLK1, DIMBLK2*, and *DIMSAH* (see Dimension Variables Table for details).

Figure 29-17

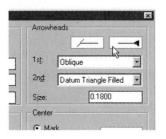

Size (*DIMASZ*)

The size of the arrow can be specified in the *Size* edit box (Fig. 29-18). The *DIMASZ* variable holds the value. Remember that *Size* is multiplied by *Overall Scale*.

Figure 29-18

Center

Mark, Line, None (*DIMCEN*)

The radio buttons determine how center marks are drawn when the dimension commands *Dimcenter, Dimdiameter*, or *Dimradius* are used (Fig. 29-19). The image tile displays the *Mark, Line,* or *None* radio button feature specified. This area actually controls the value of <u>one</u> dimension variable, *DIMCEN* (by using a 0, positive, or negative value; see Fig. 29-20).

Figure 29-19

Size (*DIMCEN*)

The *Size* edit box controls the size of the short dashes and extensions past the arc or circle. The value is stored in the *DIMCEN* variable (Fig. 29-20). Only positive values can be (and need to be) entered in the <u>dialog box</u>.

Figure 29-20

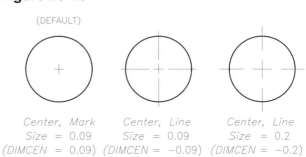

Format...

The *Format* dialog box (Fig. 29-21) provides control of the <u>location and orientation of the text with respect to the dimension line</u>.

Figure 29-21

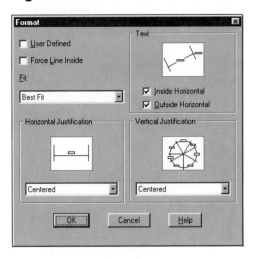

User Defined (DIMUPT)

Checking the *User Defined* box enables you to move the dimension text value lengthwise along the dimension line when you select the location for the dimension line. The default position (not checked) always keeps the text in the center of the dimension line. In Release 14, *Diameter* and *Radius* dimension text can automatically be dragged assuming the default variable settings. If you want to draw *Radius* dimensions with the arrow and dimension line inside the arc (the correct standard for most applications), you must enable *User Defined* to allow dragging the text outside of the arc, as on a leader (see *Radius* and *Diameter* Variable Settings). The value for this switch is stored in the *DIMUPT* variable.

Figure 29-22

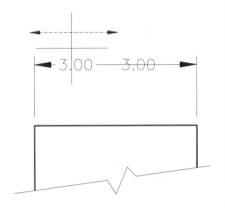

User Defined CHECKED

Force Line Inside (DIMTOFL)

The *Force Line Inside* toggle sets the value of *DIMTOFL* on or off. When this variable is on, a dimension line is drawn between the extension lines when the arrows are drawn outside the extension lines (Fig. 29-23).

Figure 29-23

Force Line Inside
NOT CHECKED
(DEFAULT)

Force Line Inside
CHECKED

Fit (DIMFIT)

For *Linear, Aligned, Angular, Baseline,* and *Continue* dimensions, the *Fit* drop-down list determines how text and arrows are treated <u>only if there is insufficient room</u> between extension lines for text, arrows, and extension lines. In most cases where space permits all components to fit inside, the *Fit* setting has no effect on the placement. If there is absolutely not enough room for text or arrows, all options behave similarly—text and arrows are placed outside. The six *Fit* options set the value of the dimension variable *DIMFIT* to 0 through 5.

For *Diameter* and *Radius* dimensions, the text and arrows are normally dragged outside of the arc or circle, but the *Fit* option can be used to force dimension line and arrows inside (see *Radius* and *Diameter* Variable Settings).

The *Text and Arrows* option <u>keeps the text and arrows together</u> always. If space does not permit both features to fit between the extension lines, it places the text and arrows outside the extension lines (Fig. 29-24). (*DIMFIT* = 0.)

Figure 29-24 ————————————

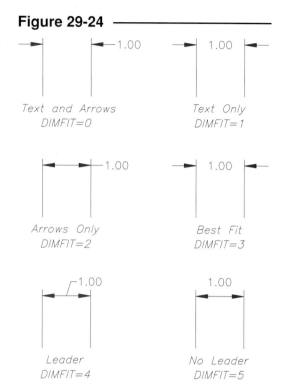

The *Text Only* option keeps the <u>text between the extension lines</u> and the arrows on the outside. If the text absolutely cannot fit, it is placed outside the extension lines (Fig. 29-24). (*DIMFIT* = 1.)

The *Arrows Only* option places the text on the outside and the <u>arrows on the inside</u> unless the arrows cannot fit, in which case they are placed on the outside as well (Fig. 29-24). (*DIMFIT* = 2.)

The *Best Fit* option behaves similarly to *Text Only*, except that if the text cannot fit, the text is placed outside the extension lines and the arrows are placed inside. However, if the arrows cannot fit either, both text and arrows are placed outside the extension lines. This is the default setting for the STANDARD style. (*DIMFIT* = 3.)

The *Leader* option enables the creation of leaders. If text cannot fit between the extension lines, then leaders to the text are created automatically (Fig. 29-24). (*DIMFIT* = 4.)

A setting of *No Leader* (*DIMFIT*=5) causes the dimension text to appear above the dimension line, similar to the *Leader* option (*DIMFIT*=4) when there is insufficient room for the dimensioning components (Fig. 29-24). A setting of 5 (or any other *DIMFIT* setting) has <u>no effect</u> on the <u>appearance</u> of the dimension when there is enough room for the dimensioning components.

There is an <u>important benefit</u> to creating a dimension with a *Fit* setting of *No Leader*; that is, the text can be edited with *Dimtedit* or with grips as if it were "detached" from the dimension line. In Figure 29-25, a dimension was created with *Fit* setting of *No Leader*. Although there is no apparent difference between this dimension and one created with another *Fit* setting (when sufficient room exists), the dimension text can be edited independent of the other dimension components. Using **grips** or the **Dimtedit** command (but <u>not</u> *Dimedit*), the text component can be moved independently to any location and the dimension retains its associativity. Figure 29-25 illustrates the use of grips to edit the dimension text.

Figure 29-25 ————————————

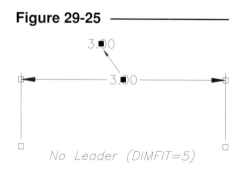

No Leader (DIMFIT=5)

Radius and *Diameter* Variable Settings

For creating mechanical drawing dimensions according to ANSI standards, the default settings in AutoCAD Release 14 are correct for creating *Diameter* dimensions but not for *Radius* dimensions. The following variable settings are recommended for creating *Radius* and *Diameter* dimensions for mechanical applications.

For *Diameter* dimensions, the default settings produce ANSI-compliant dimensions that suit most applications—that is, text and arrows are on the outside of the circle or arc pointing inward toward the center. For situations where large circles are dimensioned, *Fit* (*DIMFIT*) and *User Defined* (*DIMUPT*) can be changed to force the dimension inside the circle (Fig. 29-26).

Figure 29-26

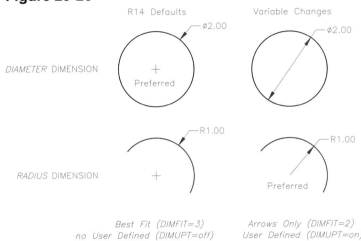

For *Radius* dimensions the default settings produce incorrect dimensioning practices. Normally (when space permits) you want the dimension line and arrow to be inside the arc, while the text can be inside or outside. To produce ANSI-compliant *Radius* dimensions, set *Fit* (*DIMFIT*) and *User Defined* (*DIMUPT*) as shown in Figure 29-26. Radius dimensions can be outside the arc in cases where there is insufficient room inside.

Text–Inside/Outside Horizontal (DIMTIH, DIMTOH)
The *Text* section in the upper-right of the dialog box sets the <u>orientation of the dimension text to horizontal or aligned</u> (with the dimension line). The setting can be selected from the drop-down list or by <u>clicking on the image tile</u> (Fig. 29-27). The results of the changes are obvious by the configuration in the image tile. Your choice sets the value for variables *DIMTIH* and *DIMTOH*.

Figure 29-27

Figure 29-28

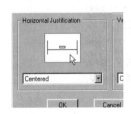

Horizontal Justification (DIMJUST)
The *Horizontal Justification* section determines the <u>horizontal location of the text with respect to the dimension line</u>. Change the setting with the drop-down list or by clicking the image tile. The choice sets the value for *DIMJUST*.

Vertical Justification (DIMTAD, DIMTIH)
The *Vertical Justification* section operates similarly to the *Horizontal Justification* feature. Changes can be made through selection from the drop-down list or by clicking on the image tile. The possible orientation options are displayed in the image tile. The selections modify the values of the *DIMTAD* (four possible settings) and *DIMTIH* variables together; therefore, eight different images are possible. Experiment with these settings and examine the changes to the image tile.

Figure 29-29

Figure 29-30

Annotation...

The *Annotation* dialog box (Fig. 29-30) controls the <u>format of the annotation (text) of a dimension</u>. You can vary the AutoCAD-supplied numerical value that is drawn in the dimension in several ways. Features such as the text style and height, prefix and/or suffix, alternate units (for inch and metric notation), and several variations of tolerance and limit dimensions are possible. These features are displayed in the dialog box in four separate areas.

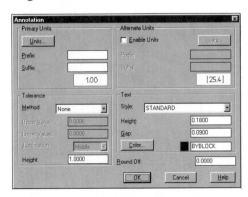

NOTE: The image tile in the *Primary Units* section of the *Annotation* dialog box (just below the *Suffix*: edit box) <u>does not</u> display the units and precision selected. It displays only the options selected from the *Tolerance* section below and should therefore be located in the *Tolerance* section.

Primary Units **(DIMUNIT)**

The *Primary Units* section (top left) controls the display of units, precision, and prefix and/or suffix. Selecting the *Units* tile summons the *Primary Units* dialog (Fig. 29-31).

You can select the *Unit* type from the drop-down list (Fig. 29-32). These are the same unit types available with the *Units* and *Ddunits* command (*Decimal, Scientific, Engineering, Architectural,* and *Fractional*) with the addition of *Architectural (stacked)* and *Fractional (stacked)* and *Windows Desktop.* The *Windows Desktop* option displays AutoCAD units based on the settings made for units display in Windows 95/NT Control Panel (settings for decimal separator and number grouping symbols). Remember that your selection affects the <u>units drawn in dimension objects, not the global drawing units</u>. Examples of the resulting dimension objects for the two fractional and two architectural unit choices are shown in Figure 29-33. The choice for *Primary Units* is stored in the *DIMUNIT* variable. This drop-down list is disabled for an *Angular* family member.

Figure 29-31 ———

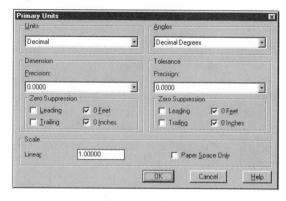

Figure 29-32 ———

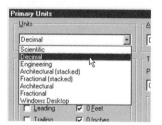

Figure 29-33 ———

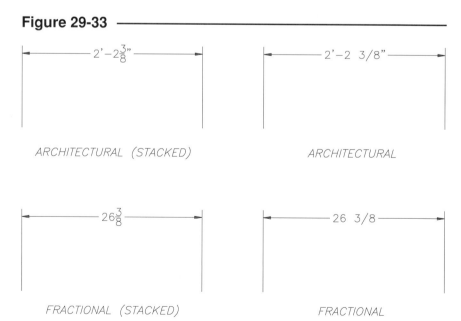

Dimension, Precision (DIMDEC)

The *Precision* drop-down list in the *Dimension* section (Fig. 29-34) specifies the number of places for decimal dimensions or denominator for fractional dimensions. This setting does not alter the drawing units precision. This value is stored in the *DIMDEC* variable.

Figure 29-34

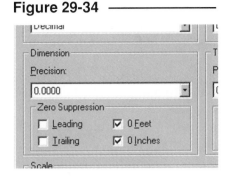

Dimension, Zero Suppression (DIMZIN)

The *Zero Suppression* section controls how zeros are drawn in a dimension when they occur. A check in one of these boxes means that zeros are <u>not drawn for that case</u>. This sets the value for *DIMZIN*.

Angles (DIMAUNIT)

The *Angles* section (Fig. 29-31) sets the unit type for angular dimensions from a drop-down list, including *Decimal degrees, Deg/Min/Sec, Grads, Radians,* and *Surveyor* (units). This list is enabled only for *Parent* and *Angular* family members. The variable used for the angular units is *DIMAUNIT*.

DIMADEC (Decimal Places for Angular Units)

DIMADEC is the only new dimensioning variable for Release 14. This variable sets the number of places of precision (decimal places) for angular dimension text. *DIMADEC* can be set only for *Parent* and *Angular* family members.

To make the desired setting for angular dimensions, set the *DIMADEC* variable for the entire family (although only *Angular* dimensions are affected) by typing *DIMADEC* <u>at the command prompt.</u> For example, to display 2 places of precision for angular dimensions, change *DIMADEC* from -1 to 2. Then use *Dimstyle, Save* (in command line format) to save the new setting to the current dimension style. A positive value sets the precision to that number of places for the entire family, although only angular dimensions are affected. A setting of -1 (default) dictates that the angular units are controlled by the *DIMDEC* variable (*Primary Units Precision*) setting.

There is no method for changing this variable for the *Parent* using the *Dimension Style* dialog boxes. You can, however, select the *Angular* member in the *Family* list (in the *Dimension Style* dialog box) and make the desired change in the *Precision* drop-down list. This action creates an *Angular* child and sets the *DIMADEC* variable for that child. Creating an *Angular* child to set this variable adds unnecessary complexity since a *Parent* setting affects only angular dimensions; therefore, it is recommended that you use the command line format to set the variable for the entire dimension style.

Tolerance (DIMTDEC, DIMTZIN)

Use the *Tolerance* cluster area of the *Primary Units* dialog box (Fig. 29-35) to set values when drawing tolerance dimensions (see *Tolerance*). The *Precision* and *Zero Suppression* operate identically to the *Dimension* section (see *Dimension, Precision* and *Dimension, Zero Suppression*). The affected variables are *DIMTDEC* and *DIMTZIN*.

Figure 29-35

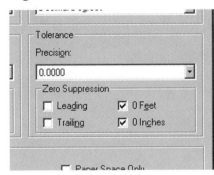

Scale, Linear (DIMLFAC)

The value appearing in the *Linear Scale* section of the *Primary Units* dialog box is a <u>multiplier</u> for the AutoCAD-measured numerical value. The default is 1. Entering a 2 would cause AutoCAD to draw a value two times the actual measured value (Fig. 29-36). This feature is used when a drawing is created in some scale other than the actual size, such as in an enlarged detail created in the same drawing as the full view. For example, if you wanted to make a detail (enlarged view) of a portion of a part, create the detail two times the actual size, enter .5 in the *Linear Scale* edit box, dimension the detail, and AutoCAD will display the actual part measurements rather than the actual measured value of the enlarged AutoCAD objects. This value is stored in the *DIMLFAC* variable.

Figure 29-36

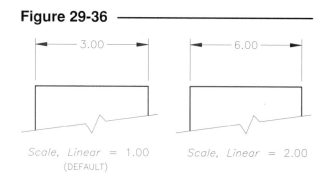

Scale, Linear = 1.00 (DEFAULT)   *Scale, Linear* = 2.00

Alternate Units (DIMALTx)

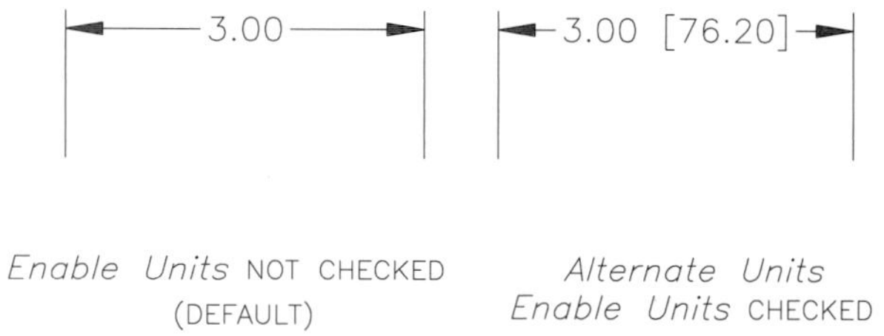

You can specify that AutoCAD draws <u>alternate units</u> (a <u>second set</u> of numerical values) when placing a dimension (Fig. 29-37). In the *Annotation* dialog box (Fig. 29-30), the *Enable Units* toggle can be checked to enable the *Units...* button of the *Alternate Units* cluster. The *Alternate Units* dialog box features are nearly the same as *Primary Units* with the exception of the *Linear Scale*. The value appearing in this field is a multiplier for the AutoCAD-measured value. Note that *Linear Scale* in the *Aternate Units* dialog box is set to 25.4, the inch-to-milimeter conversion factor (Fig. 29-38). In many industries, both inch units and metric units are required for all dimensions.

Figure 29-37

Enable Units NOT CHECKED (DEFAULT)   *Alternate Units Enable Units* CHECKED

Figure 29-38

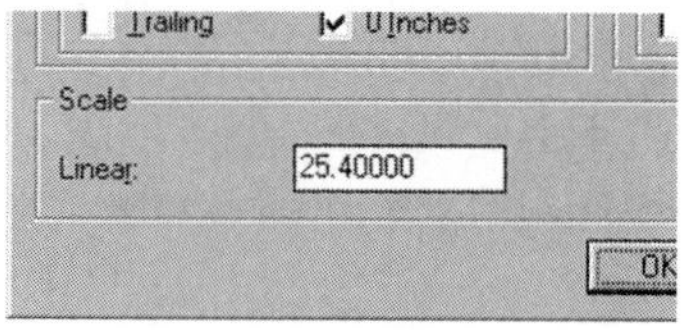

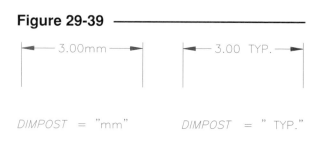

Because the alternate unit options parallel the primary unit options, a parallel set of variables is used to store the related values. The variable names are *DIMALTU* for alternate units, *DIMALTD* for alternate units decimal places, *DIMALTF* for alternate linear scale, *DIMALTZ* for alternate units zero suppression, *DIMALTTZ* for alternate units tolerance zero suppression, and *DIMALTTD* for alternate units tolerance decimal places. *DIMALT* stores the on/off toggle for enabling alternate units.

Prefix/Suffix (DIMPOST, DIMAPOST)

In both the *Primary Units* and *Alternate Units* sections of the *Annotation* dialog box (Fig. 29-30), the *Prefix* and *Suffix* edit box holds any text that you want to add to the AutoCAD-supplied dimensional value. For example, entering the string "mm" or a "TYP." in the *Suffix* edit box would produce text as shown in Figure 29-39. (In such a case, don't forget the space between the numerical value and the suffix.) The string is stored in the variable *DIMPOST* and in *DIMAPOST* for alternate units prefix or suffix.

Figure 29-39

DIMPOST = "mm"   *DIMPOST* = " TYP."

If you use the *Prefix* box to enter letters or values, any AutoCAD-supplied symbols (for radius and diameter dimensions) are overridden (not drawn). For example, if you want to specify that a specific hole appears twice, you should indicate by designating a "2X" before the diameter dimension. However, doing so by this method overrides the phi (Ø) symbol that AutoCAD inserts before the value. Instead, use the *Mtext/Text* options within the dimensioning command or use *Dimedit* to add a prefix to an existing dimension having an AutoCAD-supplied symbol (see Fig. 28-21 and related text, and *Dimedit*).

Tolerance
Method (**DIMTOL, DIMLIM, DIMGAP**)

Figure 29-40

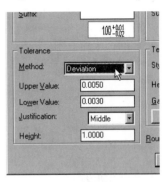

The *Tolerance Method* displays a drop-down list with five types: *None*, *Symmetrical*, *Deviation*, *Limits*, and *Basic*. Each format is displayed in the image tile <u>above</u> this section (actually in the *Primary Units* section) (Fig. 29-40). You can click in the image tile above to cycle through the formats. The four possibilities (other than *None*) are illustrated in Figure 29-41. The *Symmetrical* and *Deviation* methods create plus/minus dimensions and turn on the *DIMTOL* variable. The *Limits* method creates limit dimensions and turns the *DIMLIM* variable on. The *Basic* method creates a basic dimension by drawing a box around the dimensional value, which is accomplished by changing the *DIMGAP* to a negative value.

Upper Value/Lower Value (**DIMTP, DIMTM**)

An *Upper Value* and *Lower Value* can be entered in the edit boxes for the *Deviation* and *Limits* method types. In these cases, the *Upper Value* (*DIMTP*—DIM Tolerance Plus) is added to the measured dimension and the *Lower Value* (*DIMTP*—DIM Tolerance Minus) is subtracted (Fig. 29-41). An *Upper Value* only is needed for *Symmetrical* and is applied as both the plus and minus value.

Figure 29-41

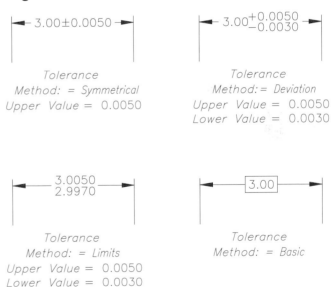

Height (**DIMTFAC**)

The height of the <u>tolerance text</u> is controlled with the *Height* edit box value. The entered value is a <u>proportion</u> of the primary dimension value height. For example, a value of .75 would draw the tolerance text at 75% of the primary text height (Fig. 29-42). The *Tolerance Height* value is <u>not reflected in the image tile above</u>. The setting affects the *Symmetrical*, *Deviation*, and *Limits* tolerance methods. The value is stored in the *DIMTFAC* (Text FACtor) variable.

Figure 29-42

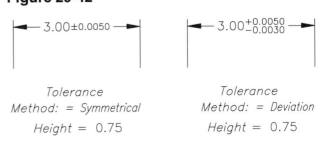

Justification (DIMTOLJ)

Justification places the tolerance dimension values with either a top, middle, or bottom alignment with the primary dimension value. The result is displayed in the image tile. The *Justification* feature is enabled only for *Symmetrical* and *Deviation* dimension methods (Fig. 29-43).

Figure 29-43

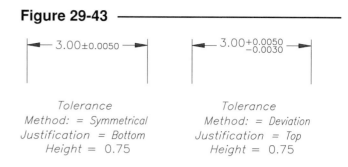

Text

Style (DIMTXSTY)

The *Text* cluster controls the text *Style, Height,* and *Color* of the dimension text. The text styles are chosen from a drop-down list of existing styles. The advantage is different dimension styles can have different text styles. The text style used for dimensions remains constant (as defined by the dimension style) and does not change when other text styles in the drawing are made current (as defined by *Style, Dtext, Text,* or *Mtext*). The *DIMTXSTY* variable holds the text style name for the dimension style.

Figure 29-44

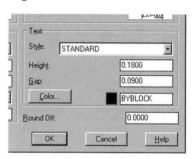

Height (DIMTXT)

This value specifies the primary text height; however, this value is multiplied by the *Overall Scale (DIMSCALE)* to determine the actual drawn text height. Change *Height* if you want to increase or decrease only the text height (Fig. 29-45). The value is stored in *DIMTXT*. Change instead the *Overall Scale* to change all size-related features (arrows, gaps, text, etc.) proportionally.

Figure 29-45

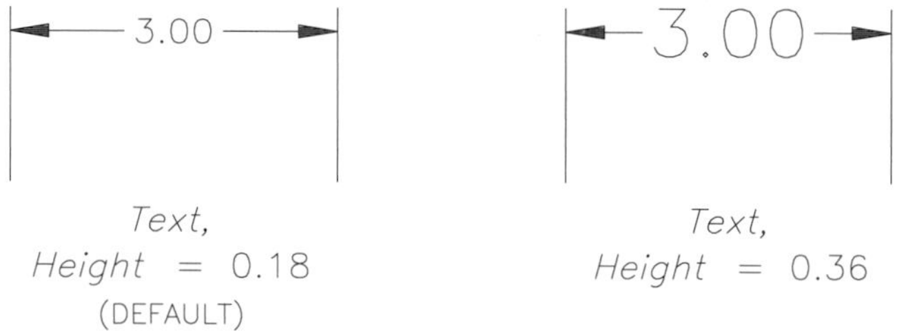

Gap (DIMGAP)

The *Gap* edit box sets the distance between the dimension text and its dimension line (Fig. 29-46). The "gap" is actually an invisible box around the text. Changing the *DIMGAP* variable to a negative value makes the box visible and is used for creating a *Basic* dimension (see *Tolerance, Method*).

Figure 29-46

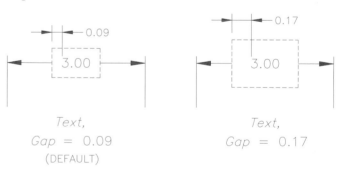

Color (DIMCLRT)

Select this edit box to activate the standard *Select Color* dialog box. The color choice is assigned to the dimension text only (Fig. 29-47). This is useful for printing or plotting the dimension text in a darker color or wider pen.

Figure 29-47

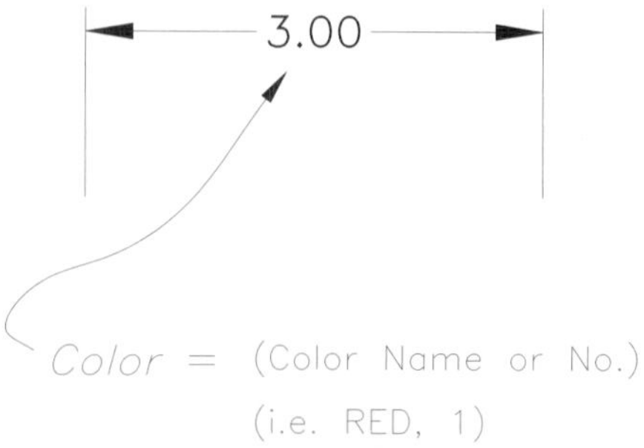

Round Off (DIMRND)

The *Round Off* edit box rounds the primary dimension value to the decimal place (or fraction) specified (Fig. 29-48). Normally, AutoCAD values are kept to 14 significant places but are rounded to the place dictated by the dimension *Primary Units, Precision* (*DIMDEC*). Use this feature to round up or down appropriately to the specified decimal or fraction.

Figure 29-48

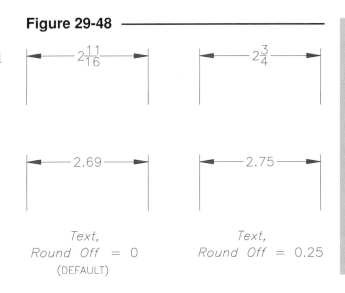

Text,
Round Off = 0
(DEFAULT)

Text,
Round Off = 0.25

R13

Changing Dimension Variables Using the Command Line Format

(VARIABLE NAME)	Pull-down Menu	COMMAND (TYPE)	ALIAS (TYPE)	Short-cut	Screen (side) Menu	Tablet Menu
	Tools *Inquiry >* *Set Variable*	*(VARIABLE NAME)*	...	...	TOOLS 1 *Setvar*	U,10

As an alternative to setting dimension variables through the *Geometry, Format,* and *Annotation* dialog boxes, you can type the dimensioning variable name at the Command: prompt. There is a noticeable difference in the two methods—the dialog boxes use <u>different nomenclature</u> than the formal dimensioning variable names used in command line format; that is, the dialog boxes use descriptive terms that, if selected, make the appropriate change to the dimensioning variable. The formal dimensioning variable names accessed by command line format, however, all begin with the letters *DIM*, are limited to eight characters, and are accessible only by typing.

Another important but subtle difference in the two methods is the act of saving dimension variable settings to a dimension style. Remember that all drawn dimensions are part of a dimension style, whether it is STANDARD or some user-created style. The act of saving variable changes made through the *Dimension Styles* dialog boxes is accomplished by selecting the *Save* tile upon exiting. When you change dimension variables by the command line format, the changes become <u>overrides</u> until you use the *Save* option of the *Dimstyle* command or *Dimension Styles* dialog box. When a variable change is made, it becomes an <u>override that is applied to the current dimension style</u> and affects only the newly drawn dimensions. Variable changes must be *Saved* to become a permanent part of the style and to retroactively affect all dimensions created with that style.

In order to access and change a variable's setting by name, simply type the variable name at the Command: prompt. For dimension variables that require distances, you can enter the distance (in any format accepted by the current *Units* settings) or you can designate by (PICKing) two points:

For example, to change the value of the *DIMSCALE* to **.5**, this command syntax is used:

```
Command: dimscale
New value for dimscale <1.0000>: .5
Command:
```

Dimension Variables Table

This table is a summary of the dimensioning variables. Each variable has a brief description, the type of variable it is, and its <u>default setting</u> (taken from the *AutoCAD Command Reference*, Appendix A, System Variables, p. 669). Remember that the dimension variables and the current settings can be listed using the *STatus* and *Variables* options of the *Dimstyle* command.

Variable	Characteristics	Description
DIMADEC	Type: Integer Saved in: Drawing Initial value: -1	Controls the number of places of precision displayed for angular dimension text. The default value (-1) uses the *DIMDEC* system variable setting to generate places of precision. When *DIMADEC* is set to a value from 0 to 8, angular dimensions display precisions that differ from their linear dimension values: -1 Angular dimension is drawn using the number of decimal places corresponding to the *DIMDEC* setting. 0-8 Angular dimension is drawn using the number of decimal places corresponding to the *DIMADEC* setting.
DIMALT	Type: Switch Saved in: Drawing Initial value: Off	When turned on, enables alternate units dimensioning. See also *DIMALTD, DIMALTF, DIMALTZ (DIMALTTZ, DIMALTTD)*, and *DIMAPOST*.
DIMALTD	Type: Integer Saved in: Drawing Initial value: 2	Controls alternate units decimal places. If *DIMALT* is enabled, *DIMALTD* governs the number of decimal places displayed in the alternate measurement.
DIMALTF	Type: Real Saved in: Drawing Initial value: 25.4000	Controls alternate units scale factor. If *DIMALT* is enabled, *DIMALTF* multiplies linear dimensions by a factor to produce a value in an alternate system of measurement.
DIMALTTD	Type: Integer Saved in: Drawing Initial value: 2	Sets the number of decimal places for the tolerance values of an alternate units dimension. *DIMALTTD* sets this value when entered on the command line or set in the Alternate Units section of the *Annotation* dialog box.
DIMALTTZ	Type: Integer Saved in: Drawing Initial value: 0	Toggles suppression of zeros for tolerance values. *DIMALTTZ* sets this value when entered on the command line or set in the Alternate Units section of the *Annotation* dialog box.
DIMALTU	Type: Integer Saved in: Drawing Initial value: 2	Sets the units format for alternate units of all dimension style family members except angular: 1 Scientific 2 Decimal 3 Engineering 4 Architectural 5 Fractional *DIMALTU* sets this value when entered on the command line or set in the Alternate Units section of the *Annotation* dialog box.

R14

R13

Variable	Characteristics	Description
DIMALTZ	Type: Integer Saved in: Drawing Initial value: 0	Toggles suppression of zeros for alternate unit dimension values: 0 Turns off suppression of zeros. 1 Turns on suppression of zeros. *DIMALTZ* sets this value when entered on the command line or set in the Alternate Units section of the *Annotation* dialog box.
DIMAPOST	Type: String Saved in: Drawing Initial value: ""	Specifies a text prefix or suffix (or both) to the alternate dimension measurement for all types of dimensions except angular. For instance, if the current *Units* mode is Architectural, *DIMALT* is enabled, *DIMALTF* is 25.4, *DIMALTD* is 2, and *DIMAPOST* is set to "mm," a distance of 10 units would be edited as 10"[254.00mm]. To disable an established prefix or suffix (or both), set it to a single period (.).
DIMASO	Type: Switch Saved in: Drawing Initial value: On	Controls the creation of associative dimension objects. Off No association between the dimension and points on the object. The lines, arcs, arrowheads, and text of a dimension are drawn as separate objects. On Creates an association between the dimension and definition points located on a feature of the object, such as an intersection of two lines. If the feature is moved, so must the definition point be. The elements are formed into a single object associated with the geometry used to define it. The *DIMASO* value is not stored in a dimension style.
DIMASZ	Type: Real Saved in: Drawing Initial value: 0.1800	Controls the size of dimension line and leader line arrowheads. Also controls the size of hook lines. Multiples of the arrowhead size determine whether dimension lines and text are to fit between the extension lines. Also used to scale arrowhead blocks if set to *DIMBLK*. *DIMASZ* has no effect when *DIMTSZ* is other than zero.
DIMAUNIT	Type: Integer Saved in: Drawing Initial value: 0	Sets the angle format for angular dimension: 0 Decimal degrees 1 Degrees/minutes/seconds 2 Gradians 3 Radians 4 Surveyor's units *DIMAUNIT* sets this value when entered on the command line or set from the Primary Units section of the *Annotation* dialog box.
DIMBLK	Type: String Saved in: Drawing Initial value: ""	Sets the name of a block to be drawn instead of the normal arrowhead at the ends of the dimension line or leader line. To disable an established block name, set it to a single period (.).

Variable	Characteristics	Description
DIMBLK1	Type: String Saved in: Drawing Initial value: ""	If *DIMSAH* is on, *DIMBLK1* specifies user-defined arrowhead blocks for the first end of the dimension line. This variable contains the name of a previously defined block. To disable an established block name, set it to a single period (.).
DIMBLK2	Type: String Saved in: Drawing Initial value: ""	If *DIMSAH* is on, *DIMBLK2* specifies user-defined arrowhead blocks for the second end of the dimension line. This variable contains the name of a previously defined block. To disable an established block name, set it to a single period (.).
DIMCEN	Type: Real Saved in: Drawing Initial value: 0.0900	Controls drawing of circle or arc center marks and centerlines by the *DIMCENTER, DIMDIAMETER,* and *DIMRADIUS* dimensioning commands. 0 No center marks or lines are drawn. <0 Centerlines are drawn. >0 Center marks are drawn. The absolute value specifies the size of the mark portion of the centerline. *DIMRADIUS* and *DIMDIAMETER* draw the center mark or line only if the dimension line is placed outside the circle or arc.
DIMCLRD	Type: Integer Saved in: Drawing Initial value: 0	Assigns colors to dimension lines, arrowheads, and dimension leader lines. Also controls the color of leader lines created with the *LEADER* command. The color can be any valid color number or special color label *BYBLOCK* or *BYLAYER*. Using the *SETVAR* command, supply the color number. Integer equivalents for *BYBLOCK* and *BYLAYER* are 0 and 256, respectively. From the Command: prompt, set the color values by entering *DIMCLRD* and then a standard color name or *BYBLOCK* or *BYLAYER*.
DIMCLRE	Type: Integer Saved in: Drawing Initial value: 0	Assigns colors to dimension extension lines. The color can be any valid color number or the special color label *BYBLOCK* or *BYLAYER*. See *DIMCLRD*.
DIMCLRT	Type: Integer Saved in: Drawing Initial value: 0	Assigns colors to dimension text. The color can be any valid color number or the special color label *BYBLOCK* or *BYLAYER*. See *DIMCLRD*.
DIMDEC	Type: Integer Saved in: Drawing Initial value: 4	Sets the number of decimal places for the tolerance values of a primary units dimension. *DIMDEC* stores this value when entered on the commend line or set in the Primary Units section of the *Annotation* dialog box.
DIMDLE	Type: Real Saved in: Drawing Initial value: 0.0000	Extends the dimension line beyond the extension line when oblique strokes are drawn instead of arrowheads.
DIMDLI	Type: Real Saved in: Drawing Initial value: 0.3800	Controls the dimension line spacing for baseline dimensions. Each baseline dimension is offset by this amount, if necessary, to avoid drawing over the previous dimension.

R13

Variable	Characteristics	Description
DIMEXE	Type: Real Saved in: Drawing Initial value: 0.1800	Determines how far to extend the extension line beyond the dimension line.
DIMEXO	Type: Real Saved in: Drawing Initial value: 0.0625	Determines how far extension lines are offset from origin points. If you point directly at the corners of an object to be dimensioned, the extension lines stop just short of the object.
DIMFIT	Type: Integer Saved in: Drawing Initial value: 3	Controls the placement of text and arrowheads inside or outside extension lines based on the available space between the extension lines: 0 Places text and arrowheads between the extension lines if space is available. Otherwise places both text and arrowheads outside extension lines. 1 If space is available, places text and arrowheads between the extension lines. When enough space is available for text, places text between the extension lines and arrowheads outside them. When not enough space is available for text, places both text and arrowheads outside extension lines. 2 If space is available, places text and arrowheads between the extension lines. When space is available for the text only, AutoCAD places the text between the extension lines and the arrowheads outside. When space is available for the arrowheads only, AutoCAD places them between the extension lines and the text outside. When no space is available for either text or arrowheads, AutoCAD places them both outside the extension lines. 3 Places whatever best fits between the extension lines. 4 Creates leader lines when there is not enough space for text between extension lines. Horizontal justification controls whether the text is drawn to the right or the left of the leader. For more information, see *DIMJUST*. 5 Places the dimension text above the dimension line when there is not enough room between extension lines for arrows and text. Dimension text with this setting can be moved with *Dimtedit* or grips to any location as if it were detached but still retains its associativity.
DIMGAP	Type: Real Saved in: Drawing Initial value: 0.0900	Sets the distance around the dimension text when you break the dimension line to accommodate dimension text. Also sets the gap between annotation and a hook line created with the *LEADER* command. A negative *DIMGAP* value creates basic dimensioning-dimension text with a box around its full extents.

R13

R14

R13

Variable	Characteristics	Description
		AutoCAD also used *DIMGAP* as the minimum length for pieces of the dimension line. When calculating the default position for the dimension text, it positions the text inside the extension lines only if doing so breaks the dimension lines into two segments at least as long as *DIMGAP*. Text placed above or below the dimension line is moved inside if there is room for the arrowheads, dimension text, and a margin between them at least as large as *DIMGAP*: 2 * (*DIMASZ* + *DIMGAP*).
DIMJUST	Type: Integer Saved in: Drawing Initial value: 0	Controls horizontal dimension text position: 0 Center justifies the text between the extension lines. 1 Positions the text next to the first extension line. 2 Positions the text next to the second extension line. 3 Positions the text above and aligned with the first extension line. 4 Positions the text above and aligned with the second extension line.
DIMLFAC	Type: Real Saved in: Drawing Initial value: 1.0000	Sets a global scale factor for linear dimensioning measurements. All linear distances measured by dimensioning (including radii, diameters, and coordinates) are multiplied by the *DIMLFAC* setting before being converted to dimension text. *DIMLFAC* has no effect on angular dimensions, and it is not applied to the values held in *DIMTM, DIMTP,* or *DIMRND*. If you are creating a dimension in Paper Space and *DIMLFAC* is nonzero, AutoCAD multiplies the distance measured by the absolute value of *DIMLFAC*. In model space, negative values are ignored, and the value 1.0 is used instead. AutoCAD computes a value for *DIMLFAC* if you try to change *DIMLFAC* from the Dim prompt while in Paper Space and you select the Viewport option. Dim: **dimlfac** Current value <1.0000> New value (Viewport): **v** Select viewport to set scale: AutoCAD calculates the scaling of model space to Paper Space and assigns the negative of this value to *DIMLFAC*.
DIMLIM	Type: Switch Saved in: Drawing Initial value: Off	When turned on, generates dimension limits as the default text. Setting *DIMLIM* on forces *DIMTOL* to be off.
DIMPOST	Type: String Saved in: Drawing Initial value: ""	Specifies a text prefix or suffix (or both) to the dimension measurement. For example, to establish a suffix for millimeters, set *DIMPOST* to mm; a distance of 19.2 units would be displayed as 19.2mm. If tolerances are enabled, the suffix is applied to the tolerances as well as to the main dimension. To separate *DIMPOST* values into prefix and suffix parts of the dimension text, use the <> mechanism; this allows AutoCAD to use the *DIMPOST* values as text. Use this mechanism for angular dimension.

Variable	Characteristics	Description
DIMRND	Type: Real Saved in: Drawing Initial value: 0.0000	Rounds all dimensioning distances to the specified value. For instance, if *DIMRND* is set to 0.25, all distances round to the nearest 0.25 unit. If you set *DIMRND* to 1.0, all distances round to the nearest integer. Note that the number of digits edited after the decimal point depends on the precision set by *DIMDEC*. *DIMRND* does not apply to angular dimensions.
DIMSAH	Type: Switch Saved in: Drawing Initial value: Off	Controls use of user-defined arrowhead blocks at the ends of the dimension line: Off Normal arrowheads or user-defined arrowhead blocks set by *DIMBLK* are used. On User-defined arrowhead blocks are used. *DIMBLK1* and *DIMBLK2* specify different user-defined arrowhead blocks for each end of the dimension line.
DIMSCALE	Type: Real Saved in: Drawing Initial value: 1.0000	Sets the overall scale factor applied to dimensioning variables that specify sizes, distances, or offsets. It is not applied to tolerances or to measured lengths, coordinates, or angles. Also affects the scale of leader objects created with the *LEADER* command. 0.0 AutoCAD computes a reasonable default value based on the scaling between the current Model Space viewport and Paper Space. If you are in Paper Space, or in Model Space and not using the Paper Space feature, the scale factor is 1.0. >0 AutoCAD computes a scale factor that leads text sizes, arrowhead sizes, and other scaled distances to plot at their face values.
DIMSD1	Type: Switch Saved in: Drawing Initial value: Off	When turned on, suppresses drawing of the first dimension line.
DIMSD2	Type: Switch Saved in: Drawing Initial value: Off	When turned on, suppresses drawing of the second dimension line.
DIMSE1	Type: Switch Saved in: Drawing Initial value: Off	When turned on, suppresses drawing of the first extension line.
DIMSE2	Type: Switch Saved in: Drawing Initial value: Off	When turned on, suppresses drawing of the second extension line.
DIMSHO	Type: Switch Saved in: Drawing Initial value: On	When turned on, controls redefinition of dimension objects while dragging. Associative dimensions recompute dynamically as they are dragged. Radius or diameter leader length input uses dynamic dragging and ignores *DIMSHO*. On some computers, dynamic dragging can be very slow, so you can set *DIMSHO* to off to drag the original image instead. The *DIMSHO* value is not stored in a dimension style.

Variable	Characteristics	Description
DIMSOXD	Type: Switch Saved in: Drawing Initial value: Off	When turned on, suppresses drawing of dimension lines outside the extension lines. If the dimension lines would be outside the extension lines and DIMTIX is on, setting DIMSOXD to on suppresses the dimension line. If DIMTIX is off, DIMSOXD has no effect.
DIMSTYLE	(Read-only) Type: String Saved in: Drawing	Sets the current dimension style by name. To change the dimension style, use DDIM or the DIMSTYLE dimensioning command.
DIMTAD	Type: Integer Saved in: Drawing Initial value: 0	Controls vertical position of text in relation to the dimension line: 0 Centers the dimension text between the extension lines. 1 Places the dimension text above the dimension line except when the dimension line is not horizontal and text inside the extension lines is forced horizontal (DIMTIH = 1). The distance from the dimension line to the baseline of the lowest line of text is the current DIMGAP value. 2 Places the dimension text on the side of the dimension line farthest away from the defining points. 3 Places the dimension text to conform to a JIS representation.
DIMTDEC	Type: Integer Saved in: Drawing Initial value: 4	Sets the number of decimal places for the tolerance values for a primary units dimension.
DIMTFAC	Type: Real Saved in: Drawing Initial value: 1.0000	Specifies a scale factor for text height of tolerance values relative to the dimension text height as set by DIMTXT. $$DIMTFAC = \frac{\text{Tolerance height}}{\text{Text height}}$$ For example, if DIMTFAC is set to 1.0, the text height of tolerances is the same as the dimension text. If DIMTFAC is set to 0.75, the text height of tolerances is three-quarters the size of dimension text. Use DIMTFAC for plus and minus tolerance strings when DIMTOL is on and DIMTM is not equal to DIMTP or when DIMLIM is on.
DIMTIH	Type: Switch Saved in: Drawing Initial value: On	Controls the position of dimension text inside the extension lines for all dimension types except ordinate dimensions: Off Aligns text with the dimension line. On Draws text horizontally.
DIMTIX	Type: Switch Saved in: Drawing Initial value: Off	Draw text between extension lines: Off The result varies with the type of dimension. For linear and angular dimensions, AutoCAD places text inside the extension lines if there is sufficient room. For radius and diameter dimensions, setting DIMTIX off forces the text outside the circle or arc.

Variable	Characteristics	Description
		On Draws dimension text between the extension lines even if AutocAD would ordinarily place it outside those lines.
DIMTM	Type: Real Saved in: Drawing Initial value: 0.0000	When *DIMTOL* or *DIMLIM* is on, sets the minimum (or lower) tolerance limit for dimension text. AutoCAD accepts signed values for *DIMTM*. If *DIMTOL* is on and *DIMTP* and *DIMTM* are set to the same value, AutoCAD draws a ± symbol followed by the tolerance value. If *DIMTM* and *DIMTP* values differ, the upper tolerance is drawn above the lower and a plus sign is added to the *DIMTP* value if it is positive. For *DIMTM*, AutoCAD uses the negative of the value you enter (adding a minus sign if you specify a positive number and a plus sign if you specify a negative number). No sign is added to a value of zero.
DIMTOFL	Type: Switch Saved in: Drawing Initial value: Off	When turned on, draws a dimension line between the extension lines, even when the text is placed outside the extension lines. For radius and diameter dimensions (while *DIMTIX* is off), draws a dimension line and arrowheads inside the circle or arc and places the text and leader outside.
DIMTOH	Type: Switch Saved in: Drawing Initial value: On	When turned on, controls the position of dimension text outside the extension lines: 0 Aligns text with the dimension line. 1 Draw text horizontally.
DIMTOL	Type: Switch Saved in: Drawing Initial value: Off	When turned on, appends dimension tolerances to dimension text. Setting *DIMTOL* on forces *DIMLIM* off.
DIMTOLJ	Type: Integer Saved in: Drawing Initial value: 1	Sets the vertical justification for tolerance values relative to the nominal dimension text: 0 Bottom 1 Middle 2 Top
DIMTP	Type: Real Saved in: Drawing Initial value: 0.0000	When *DIMTOL* or *DIMLIM* is on, sets the maximum (or upper) tolerance limit for dimension text. AutoCAD accepts signed values for *DIMTP*. If *DIMTOL* is on and *DIMTP* and *DIMTM* are set to the same value, AutoCAD draws a ± symbol followed by the tolerance value. If *DIMTM* and *DIMTP* values differ, the upper tolerance is drawn above the lower and a plus sign is added to the *DIMTP* value if it is positive.
DIMTSZ	Type: Real Saved in: Drawing Initial value: 0.0000	Specifies the size of oblique strokes drawn instead of arrowheads for linear, radius, and diameter dimension: 0 Draws arrows. >0 Draws oblique strokes instead of arrows. Size of oblique strokes is determined by this value multiplied by the *DIMSCALE* value. Also determines if dimension lines and text fit between extension lines.

R13

Variable	Characteristics	Description
DIMTVP	Type: Real Saved in: Drawing Initial value: 0.0000	Adjusts the vertical position of dimension text above or below the dimension line. AutoCAD uses the *DIMTVP* value when *DIMTAD* is off. The magnitude of the vertical offset of text is the product of the text height and *DIMTVP*. Setting *DIMTVP* to 1.0 is equivalent to setting *DIMTAD* to on. AutoCAD splits the dimension line to accommodate the text only if the absolute value of *DIMTVP* is less than 0.7.
DIMTXSTY	Type: String Saved in: Drawing Initial value: "STANDARD"	Specifies the text style of the dimension.
DIMTXT	Type: Real Saved in: Drawing Initial value: 0.1800	Specifies the height of dimension text, unless the current text style has a fixed height.
DIMTZIN	Type: Integer Saved in: Drawing Initial value: 0	Toggles suppression of zeros for tolerance values. *DIMZIN* stores this value when entered on the command line or set in the Primary Units section of the *Annotation* dialog box.
DIMUNIT	Type: Integer Saved in: Drawing Initial value: 2	Sets the units format for all dimension style family members except angular: 1 Scientific 2 Decimal 3 Engineering 4 Architectural (stacked) 5 Fractional (stacked) 6 Architectural 7 Fractional 8 Windows Desktop (decimal format using Control Panel settings for decimal separator and number grouping symbols)
DIMUPT	Type: Switch Saved in: Drawing Initial value: Off	Controls cursor functionality for User Positioned Text: 0 Cursor controls only the dimension line location. 1 Cursor controls the text position as well as the dimension line location.
DIMZIN	Type: Integer Saved in: Drawing Initial value: 0	Controls the suppression of the inches portion of a feet-and-inches dimension when the distance is an integral number of feet, or the feet portion when the distance is less than one foot: 0 Suppresses zero feet and precisely zero inches 1 Includes zero feet and precisely zero inches 2 Includes zero feet and suppresses zero inches 3 Includes zero inches and suppresses zero feet *DIMZIN* stores this value when entered on the command line or set in the Primary Units section of the *Annotation* dialog box.

R13

R13

R14

R13

Using *Setup Wizards* and Template Drawings

The previous dimension variable table gives default settings for the ACAD.DWT template drawing. This drawing is basically the same that is traditionally used when you begin a drawing session in previous releases of AutoCAD. In Release 14, the ACAD.DWT drawing is the template or basis used when you begin AutoCAD with the *Startup* dialog box toggled off, choose *Start from Scratch, English Default Settings* in the *Create New Drawing* dialog box, select the ACAD.DWT as a template drawing, or choose all the defaults in the *Setup Wizards*. In other cases, one or more dimension variables are automatically changed. In some template drawings, pre-made dimension styles are available.

Start from Scratch, Metric Default Settings or **ACADISO.DWT**

When you choose *Start from Scratch, Metric Default Settings*, AutoCAD uses the ACADISO.DWT template drawing. The *DIMSCALE* retains a setting of 1, but many other dimension variables are changed to accommodate metric sizes (English default setting x 25.4). The following settings are made to the following dimension variables (that differ from the ACAD.DWT setting):

DIMASZ	2.5000
DIMCEN	2.5000
DIMDLI	3.7500
DIMEXE	1.2500
DIMEXO	0.6250
DIMGAP	0.6250
DIMTXT	2.5000

Setup Wizards

When you use a *Setup Wizard*, many system variables are changed according to the values you enter for *Limits* (*Step 2: Units* in *Quick Setup* or *Step 5: Units* in *Advanced Setup*). *DIMSCALE is the only dimensioning variable changed* by the *Setup Wizards*. The affected system variables are changed from the default settings based on the following formula:

$$N=D(Y/9)$$

where N is the new system variable setting,
D is the system variable default value,
Y is the Area (*Limits*) Y value that you enter, and
9 is the default *Area* (*Limits*) Y value.

For example, entering *Limits* values of 24 x 18 would set *DIMSCALE* to 2, or using the formula above, $2=1(18/9)$. Beware, settings made by the *Setup Wizards* are correct for printing or plotting only on an A size sheet and are incorrect for any other sheet size. See Chapters 6, 13, and 14 for discussion on Drawing Scale Factor and Plotting to Scale.

Template Drawings

Template drawings may or may not include dimensioning variable changes or dimension styles, depending on the selected template. In general, the ANSI templates (Ansi_a through Ansi_e) use the same default settings (no changes) for all dimension variables, including *DIMSCALE*, as the ACAD.DWT template. The *Din, ISO,* and *JIS* templates include one or more *Dimension Styles* with appropriate variable settings. See Table of AutoCAD-Supplied Template Drawing Settings, Chapter 13, for a more complete description.

R14

MODIFYING EXISTING DIMENSIONS

Even with the best planning, a full understanding of dimension variables, and the correct use of *Dimension Styles*, it is probable that changes will have to be made to existing dimensions in the drawing due to design changes, new plot scales, or new industry/company standards. There are several ways that changes can be made to existing dimensions while retaining associativity and membership to a dimension style. The possible methods are discussed in this section.

Modifying a Dimension Style

Existing dimensions in a drawing can be modified by making one or more variable changes to the dimension style family or child, then *Saving* those changes to the dimension style. This process can be accomplished using either the *Dimension Style* dialog box series or command line format. When you use the *Save* option (in the *Dimension Style* dialog box or in the *Dimstyle* command), the existing dimensions in the drawing <u>automatically update</u> to display the new variable settings. For the individual steps for this procedure, see Modifying a Dimension Style and Updating Existing Dimensions in the Dimension Styles section earlier in this chapter.

This method is generally preferred in cases where all dimensions in a style need updating. One draw-back of this method is that children created previously (special settings for *Radial, Diameter*, etc.) in the dimension style being modified <u>do not automatically update.</u> The solution is to make the same variable changes for the child (select *Radial, Diameter*, etc. in the *Dimension Styles* dialog box, then make variable changes and *Save*) or use *Dimoverride* or *Update* (discussed later).

Creating Dimension Style Overrides and Using *Update*

You can modify existing dimensions in a drawing without making permanent changes to the dimension style by creating a dimension style override, then using *Update* to apply the new setting to an existing dimension. This method is preferred if you wish <u>to modify one or two dimensions</u> without modifying all dimensions referencing (created in) the style.

To do this, use either the command line format or *Dimension Styles* dialog box series to set the new variable. In the *Dimension Styles* dialog box, make the desired variable settings in the *Geometry, Format,* or *Annotation* dialog boxes, but <u>do not use *Save*—use *OK* instead.</u> This creates an override to the current style. In command line format, simply enter the formal dimension variable name and make the change to create an override to the current style. Next, use *Update* to apply the current style settings plus the override settings to existing dimensions that you PICK. The overrides remain in effect for the current style unless the variables are reset to the original values. You can <u>clear overrides</u> for a dimension style by selecting the original style name (without the "+" prefix) in the *Dimension Style* dialog box to make it current, then answer *No* to "save changes to current style?"

UPDATE

	COMMAND (TYPE)	ALIAS (TYPE)	Short-cut	Screen (side) Menu	Tablet Menu
Pull-down Menu					
Dimension *Update*	*DIM* *UPDATE*	*DIM* *UP*	...	*DIMNSION* *Update*	...

Update can be used to update existing dimensions in the drawing to the current settings. The current settings are determined by the current *Dimension Style* and any dimension variable overrides that are in effect (see previous explanation, Creating Dimension Style Overrides and Using *Update*). This is an excellent method of modifying one or more existing dimensions without making permanent changes to the dimension style that the selected dimensions reference. *Update* has the same effect as using *Dimstyle, Apply*.

Update is actually a Release 12 command. In Release 12, dimensioning commands could only be entered at the Dim: prompt. For example, to create a linear dimension, you had to type *Dim* and press Enter, then type *Linear*. In Release 13, all dimensioning commands were upgraded to top-level commands with the *Dim-* prefix added, so you can type *Dimlinear* at the command prompt, for example. *Update*, although very useful, was never upgraded. In Release 14 the command is given prominence by making available an *Update* button and an *Update* option in the *Dimension* pull-down menu. However, if you prefer to type, you must first type *Dim*, press Enter, and then enter *Update*. (See *DIM* in Chapter 28.) The command syntax is as follows:

```
Command: Dim
Dim: Update
Select objects: PICK (select a dimension object to update)
Select objects: Enter
Dim: press Esc or type Exit
Command:
```

For example, if you wanted to change the *DIMSCALE* of several existing dimensions to a value of 2, change the variable by typing *DIMSCALE* or change *Overall Scale* in the *Geometry* dialog box (do not *Save*). Next use *Update* and select the desired dimension. That dimension is updated to the new setting. The command syntax is the following:

```
Command: Dimscale
New value for DIMSCALE <1.0000>: 2
Command: Dim
Dim: Update
Select objects: PICK (select a dimension object to update)
Select objects: PICK (select a dimension object to update)
Select objects: Enter
Dim: Exit
Command:
```

Beware, after *Updating* the selected dimensions, the overrides for that style remain in effect. You should then reset the variable to its original value unless you want to keep the override for creating other new dimensions. You can clear overrides for a dimension style by selecting the original style name (without the "+" prefix) in the *Dimension Style* dialog box to make it current, then answer *No* to "save changes to current style?"

DIMOVERRIDE

Pull-down Menu	COMMAND (TYPE)	ALIAS (TYPE)	Short-cut	Screen (side) Menu	Tablet Menu
Dimension Override	DIMOVERRIDE	DIMOVER or DOV	...	...	Y,4

Dimoverride grants you a great deal of control to edit existing dimensions. The abilities enabled by *Dimoverride* are unique in that no other dimensioning commands or options allow you to retroactively modify dimensions in such a way.

Dimoverride enables you to make variable changes to dimensions that exist in your drawing without changing the dimension style that the dimension references (was created under). For example, using *Dimoverride*, you can make a variable change and select existing dimension objects to apply the change. The existing dimension does not lose its reference to the parent dimension style nor is the dimension style changed in any way. In effect, you can override the dimension styles for selected dimension objects. There are two steps: set the desired variable and select dimension objects to alter.

```
Command: dimoverride
Dimension variable to override (or Clear to remove overrides): (variable name)
Current value <current  value> New value: (value)
Dimension variable to override: (variable name) or Enter
Select objects: PICK
Select objects: PICK or Enter
Command:
```

The *Dimoverride* feature differs from creating overrides in two ways: <u>overrides</u> (variable changes that are made but not *Saved*) (1) are <u>appended</u> to the *Parent* dimension style and (2) are applied to <u>newly created</u> dimensions only and are not retroactive. In contrast, *Dimoverride* affects only the selected existing dimensions and does not change the parent dimension style.

Dimoverride is useful as a "backdoor" approach to dimensioning. Once dimensions have been created, you may want to make a few modifications, but you do not want the changes to affect dimension styles (resulting in an update to all existing dimensions that reference the dimension styles). *Dimoverride* offers that capability. You can even make one variable change to affect all dimensions globally without having to change all the dimension styles. For example, you may be required to make a test plot of the drawing in a different scale than originally intended, necessitating a new global *Dimscale*. Use *Dimoverride* to make the change for the plot:

```
Command: dimoverride
Dimension variable to override (or Clear to remove overrides): dimscale
Current value <1.0000> New value: .5
Dimension variable to override: Enter
Select objects: (window entire drawing) Other corner: 128 found
Select objects: Enter
Command:
```

This action results in having all the existing dimensions reflect the new *DIMSCALE*. No other dimension variables or any dimension styles are affected. Only the selected objects contain the overrides. *Dimoverride* does not append changes (overrides) to the original dimension styles, so no action must be taken to clear the overrides from the styles. After making the plot, *Dimoverride* can be used with the *Clear* option to change the dimensions (by object selection) back to their original appearance. The *Clear* option is used to clear overrides from <u>dimension objects, not from dimension styles</u>.

Clear
The *Clear* option removes the overrides from the <u>selected dimension objects</u>. It does not remove overrides from the current dimension style:

```
Command: dimoverride
Dimension variable to override (or Clear to remove overrides): c
Select objects: PICK
Select objects: Enter
Command:
```

The dimension then displays the variable settings as specified by the dimension style it references without any overrides (as if the dimension were originally created without the overrides). Using *Clear* does not remove the overrides that are appended to the dimension style so that if another dimension is drawn, the overrides apply.

Dimoverride is a very powerful and useful tool because it provides capabilities that are not available by any other method. You should experiment with it now so you can use it later to simplify otherwise difficult dimension editing situations.

Other Methods for Modifying Existing Dimensions

MATCHPROP

Matchprop can be used to "convert" an existing dimension to the style (including overrides) of another dimension in the drawing. For example, if you have two linear dimensions, one has *Oblique* arrows and *Romans* text font (*Dimension Style* = "Oblique") and one is a typical linear dimension (*Dimension Style* = "Standard") as in Fig. 29-49, before. You can convert the typical dimension to the "Oblique" style by using *Matchprop*, selecting the "Oblique" dimension as the "source object" (to match), then selecting the typical dimension as the "destination object" (to convert). The typical dimension then references the "Oblique" dimension style and changes appearance accordingly (Fig. 29-49, after). Note that *Matchprop* does not alter the dimension text value.

Figure 29-49

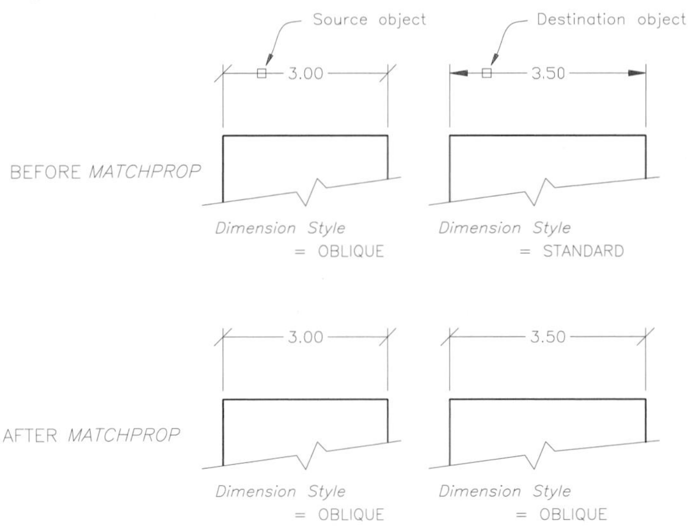

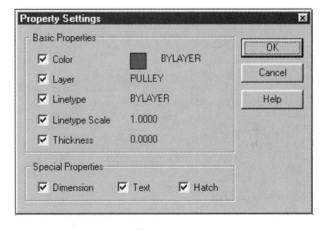

Using the *Settings* option of the *Matchprop* command, you can display the *Property Settings* dialog box (Fig. 29-50). The *Dimension* box under *Special Properties* must be checked to "convert" existing dimensions as illustrated above.

Using *Matchprop* is a fast and easy method for modifying dimensions from one style to another. However, this method is applicable only if you have existing dimensions in the drawing with the desired appearance that you want others to match.

Figure 29-50

DDMODIFY

Remember that *Ddmodify* can be used to edit existing dimensions. Using *Ddmodify* and selecting one dimension invokes the *Modify Dimension* dialog box. Here you can modify any aspect of the dimension, including text, and access is given to the *Geometry, Format,* and *Annotation* dialog boxes. Any changes to dimension variables through *Ddmodify* result in <u>overrides to the dimension object only</u> but do not affect the dimension style. *Ddmodify* has essentially the same result as *Dimoverride*, except only one dimension object at a time can be modified. *Ddmodify* can also be used to modify the dimension text value. *Ddmodify* with dimensions is discussed in Chapter 28.

DIMTEDIT, DIMEDIT

Dimension text of existing dimensions can be modified using either *Dimtedit* or *Dimedit*. These commands allow you to change the text position with respect to the dimension line and the angle of the text. *Dimtedit* can be used with the *New* option to restate the original AutoCAD-measured text value if needed. See Chapter 28 for a full explanation of these commands.

Grips

Grips can be used to effectively alter dimension length, position, text location and more. Keep in mind the possibilities of moving the dimension text with grips for dimensions with *DIMFIT* of 5. See Chapter 28 for a full discussion of the possibilities of Grip editing dimensions.

GUIDELINES FOR DIMENSIONING IN AutoCAD

Listed in this section are some guidelines to use for dimensioning a drawing using dimensioning variables and dimension styles. Although there are other strategies for dimensioning, two strategies are offered here as a framework so you can develop an organized approach to dimensioning.

In almost every case, dimensioning is one of the last steps in creating a drawing, since the geometry must exist in order to dimension it. You may need to review the steps for drawing setup, including the concept of drawing scale factor (Chapter 13).

Strategy 1. Dimensioning a Single Drawing

This method assumes that the fundamental steps have been taken to set up the drawing and create the geometry. Assume this has been accomplished:

> Drawing setup is completed: *Units, Limits, Snap, Grid, Ltscale, Layers*, border, and title block.
> The drawing geometry (objects comprising the subject of the drawing) has been created.

Now you are ready to dimension the drawing subject (of the multiview drawing, pictorial drawing, floor plan, or whatever type of drawing).

1. Create a *Layer* (named DIM, or similar) for dimensioning if one has not already been created. Set *Continuous* linetype and appropriate color. Make it the *current* layer.

2. Set the *Overall Scale (DIMSCALE)* based on drawing *Limits* and expected plotting size.

   For plotted dimension text of 3/16":

   Multiply *Overall Scale (DIMSCALE)* times the drawing scale factor. The default *Overall Scale* is set to 1, which creates dimensioning text of approximately 3/16" (default *Text Height:* or *DIMTXT* =.18) when plotted full size. All other size-related dimensioning variables' defaults are set appropriately.

   For plotted dimension text of 1/8":

   Multiply *Overall Scale* times the drawing scale factor, times .7. Since the *Overall Scale* times the scale factor produces dimensioning text of .18, then .18 x .7 = .126 or approximately 1/8". (See Optional Method for Fixed Dimension Text Height.) (*Overall Scale [DIMSCALE]* is one variable that usually remains constant throughout the drawing and therefore should generally have the same value for every dimension style created.)

3. Make the other dimension variable changes you expect you will need to create most dimensions in the drawing with the appearance you desire. When you have the basic dimension variables set as you want, *Save* this new dimension style family (select *Parent*) to the TEMPLATE (or other descriptive name) style. (Remember you must select *Save* and exit the dialog box to save the new dimension style.) If you need to make special settings for types of dimensions (*Linear, Diameter,* etc.), create "children" at this stage and *Save* the changes to the style. This style is the fundamental style to use for creating most dimensions and should be used as a template for creating other dimension styles. If you need to reset or list the original (default) settings, the STANDARD style can be restored.

4. Create all the relatively simple dimensions first. These are dimensions that are easy and fast and require no other dimension variable changes. Begin with linear dimensions; then progress to the other types of dimensions.

5. Create the special dimensions next. These are dimensions that require variable changes. Create appropriate dimension styles by changing the necessary variables, then *Save* each set of variables (relating to a particular style of dimension) to an appropriate dimension style name. Specify dimension variables for the classification of dimension (children) in each dimension style when appropriate. The dimension styles can be created "on the fly" or as a group before dimensioning. Use TEMPLATE as your base dimension style when appropriate.

   For the few dimensions that require variable settings unique to that style, you can create dimension variable overrides. Change the desired variable(s), but do not *Save* to the style so the other dimensions in the style are not affected.

6. When all of the dimensions are in place, make the final adjustments. Several methods can be used:

   a. If modifications need to be made <u>familywide,</u> change the appropriate variables and *Save* the changes to the dimension style. This action automatically updates existing dimensions that reference that style.

   b. To modify the appearance of selected dimensions, you can create dimension style overrides, then use *Update* or *Dimstyle, Apply* to update the selected dimensions to the new settings. Keep in mind that the dimension style overrides are still in effect and are applied to new dimensions. Clear the overrides by setting the original style current.

   c. Alternately, use *Ddmodify* to change variable settings for selected dimensions. These changes are made as overrides to the selected object only and do not affect the dimension style. This action has the same result as using *Dimoverride* but for only one dimension.

   d. To change one dimension to adopt the appearance of another, use *Matchprop*. Select the dimension to match first, then the dimension(s) to convert. *Dimstyle, Apply* can be used for this same purpose.

   e. Use *Dimoverride* to make changes to selected dimensions or to all dimensions <u>globally</u> by windowing the entire drawing. *Dimoverride* has the advantage of allowing you to assign the variables to change and selecting the objects to change all in one command. What is more, the changes are applied as overrides only to the selected dimensions but <u>do not alter the original dimension styles</u> in any way.

   f. If you want to change only the dimension text value, use *Ddmodify* or *Dimedit, New*. *Dimedit, New* can also be used to reapply the AutoCAD-measured value if text was previously changed. The location of the text can be changed with *Dimtedit* or Grips.

   g. Grips can be used effectively to change the location of the dimension text or to move the dimension line closer or further from the object. Other adjustments are possible, such as rotating a *Radius* dimension text around the arc.

Strategy 2. Creating Dimension Styles as Part of Template Drawings

1. Begin a *New* drawing or *Open* an existing *Template*. Assign a descriptive name.

2. Create a DIM *Layer* for dimensioning with *continuous* linetype and appropriate color (if one has not already been created).

3. Set the *Overall Scale* accounting for the drawing *Limits* and expected plotting size. Use the guidelines given in Strategy 1, step 2. Make any other dimension variable changes needed for general dimensioning or required for industry or company standards.

4. Next, *Save* a dimension style named "TEMPLATE." This should be used as a template when you create most new dimension styles. The *Overall Scale* is already set appropriately for new dimension styles in the drawing.

5. Create the appropriate dimension styles for expected drawing geometry. Use TEMPLATE dimension style as a base style when appropriate.

6. *Save* and *Exit* the newly created template drawing.

7. Use this template in the future for creating new drawings. Restore the desired dimension styles to create the appropriate dimensions.

Using a template drawing with prepared dimension styles is a preferred alternative to repeatedly creating the same dimension styles for each new drawing.

Optional Method for Fixed Dimension Text Height in Template Drawings

To summarize Strategy 1, step 2., the default *Overall Scale* (*DIMSCALE* =1) times the default *Text Height* (*DIMTXT* =.18) produces dimensioning text of approximately 3/16" when plotted to 1=1. To create 1/8" text, multiply *Overall Scale* times .7 (.18 x .7 = .126). As an alternative to this method, try the following.

For 1/8" dimensions, for example, multiply the initial values of the size-related variables by .7; namely

Text Height	(*DIMTXT*)	.18 x .7 = .126
Arrow Size	(*DIMASZ*)	.18 x .7 = .126
Extension Line Extension	(*DIMEXE*)	.18 x .7 = .126
Dimension Line Spacing	(*DIMDLI*)	.38 x .7 = .266
Text Gap	(*DIMGAP*)	.09 x .7 = .063

Save these settings in your template drawing(s). When you are ready to dimension, simply multiply *Overall Scale* (1) times the drawing scale factor.

Although this method may seem complex, the drawing setup for individual drawings is simplified. For example, assume your template drawing contained preset *Limits* to the paper size, say 11 x 8.5. In addition, the previously mentioned dimension variables were set to produce 1/8" (or whatever) dimensions when plotted full size. Other variables such as *LTSCALE* could be appropriately set. Then, when you wish to plot a drawing to 1=1, everything is preset. If you wish to plot to 1=2, simply multiply all the size-related variables (*Limits*, *Overall Scale*, *Ltscale*, etc.) times 2!

CHAPTER EXERCISES

For each of the following exercises, use the existing drawings, as instructed. Create dimensions on the DIM (or other appropriate) layer. Follow the Guidelines for Dimensioning given in the chapter, including setting an appropriate *Overall Scale* based on the drawing scale factor. Use dimension variables and create and use dimension styles when needed.

1. **Dimensioning a Multiview**

    Figure 29-51 ——————————————————

    Open the **SADDLE** drawing that you created in Chapter 24 Exercises. Set the appropriate dimensional *Units* and *Precision*. Add the dimensions as shown. Because the illustration in Figure 29-51 is in isometric, placement of the dimensions can be improved for your multiview. Use optimum placement for the dimensions. *Save* the drawing as **SADDL-DM** and make a *Plot* to scale.

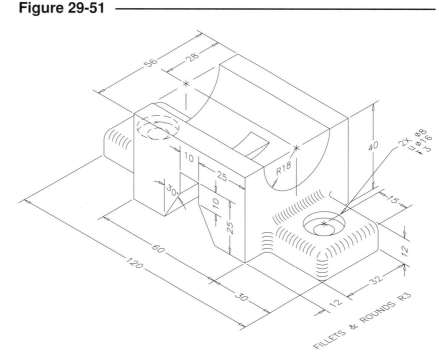

2. **Architectural Dimensioning**

    Open the **OFFICE** drawing that you completed in Chapter 21. Dimension the floor plan as shown in Figure 21-18. Add *Text* to name the rooms. *Save* the drawing as **OFF-DIM** and make a *Plot* to an accepted scale and sheet size based on your plotter capabilities.

3. **Dimensioning an Auxiliary**

    Figure 29-52 ——————————————————

    Open the **ANGLBRAC** drawing that you created in Chapter 27. Dimension as shown in Figure 29-52, but <u>convert the dimensions to *Decimal*</u> with **Precision** of **.000.** Use the Guidelines for Dimensioning. Dimension the slot width as a *Limit* dimension— **.6248/.6255**. *Save* the drawing as **ANGL-DIM** and *Plot* to an accepted scale.

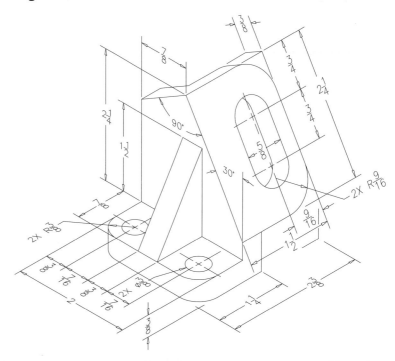

4. **Isometric Dimensioning**

Figure 29-53

Dimension the support bracket that you created as an isometric drawing named **SBRACKET** in the Chapter 25 Exercises. All of the dimensions shown in Figure 29-53 should appear on your drawing in the optimum placement. *Save* the drawing as **SBRCK-DM** and *Plot* to **1=1** on and A size sheet.

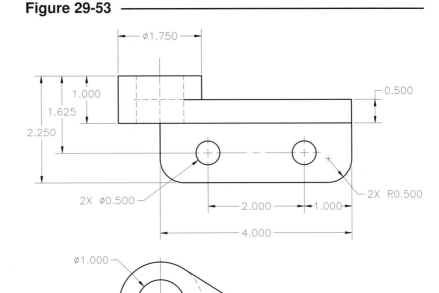

5. **Dimensioning a Multiview**

Figure 29-54

Open the **ADJMOUNT** drawing that you completed in Chapter 24. Add the dimensions shown in Figure 29-54 but <u>convert the dimensions to *Decimal*</u> with **Precision** of **.000** and check **Trailing** under **Zero Suppression**. Set appropriate dimensional **Units** and **Precision**. Calculate and set an appropriate *DIMSCALE*. Use the Guidelines for Dimensioning given in this chapter. Save the drawing as **ADJM-DIM** and make a *Plot* to an accepted scale.

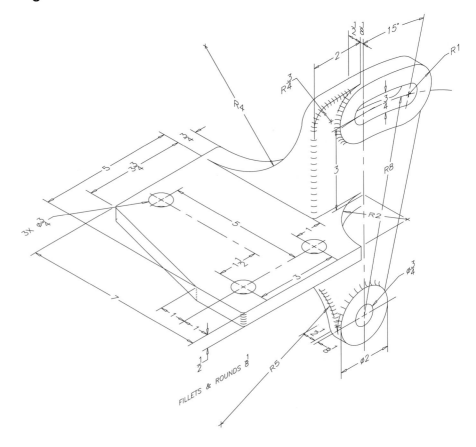

6. **Geometric Dimensioning and Tolerancing**

    This exercise involves Flatness, Profile of a Surface, and Position applications. *Open* the dimensioned **ANGL-DIM** drawing you worked on in Exercise 3 and add the following geometric dimensions:

    A. **Flatness** specification of **.002** to the bottom surface.

    B. **Profile of a Surface** specification of **.005** to the angled surface relative to the bottom surface and the right side surface of the 1/4 dimension. The 30° angle must be *Basic*.

    C. **Position** specification of **.01** for the two mounting holes relative to the bottom surface and two other surfaces that are perpendicular to the bottom. Remember, the location dimensions must be *Basic* dimensions.

    Use *SaveAs* to save and name the drawing **ANGL-TOL**.

7. **Geometric Dimensioning and Tolerancing**

    This exercise involves a cylindricity and runout application. Draw the idler shown in Figure 29-55. Apply a **Cylindricity** specification of **.002** to the small diameter and a **Circular Runout** specification of **.007** to the large diameter relative to the small diameter as shown. The P730 neck has a **.07** radius and **30°** angle. *Save* the drawing as **IDLERDIM**.

Figure 29-55 ───

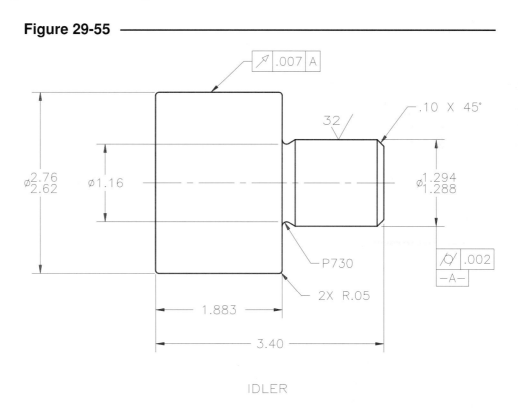

IDLER

FLOOR PLAN   (LIVING AREA: 3,257 SQUARE FEET)

SCALE: 1/4" = 1'-0"

30

XREFERENCES

Chapter Objectives

After completing this chapter you should:

1. know the differences and similarities between an *Xrefed* drawing and an *Inserted* drawing;

2. be able to *Attach* an externally referenced drawing to the current drawing;

3. understand that an *Overlay* drawing cannot be nested;

4. be able to *Reload*, *Unload*, or *Detach* an externally referenced drawing;

5. be able to *Bind* an entire *Xrefed* drawing and *Xbind* individual named objects;

6. be able to *Xclip* an externally referenced drawing so only a portion of the *Xref* appears;

7. be able to control demand loading with *INDEXCTL* and *XLOADCTL*;

8. know several techniques for managing *Xrefs* using the *External Reference* dialog box, using project names, setting *Xref Paths*, keeping log files, and using backup procedures.

CONCEPTS

AutoCAD's Xreference (external reference) feature enables you to view another drawing from within the current drawing. The externally referenced drawing (*Xref*) is visible in the same screen as the current drawing; however, you cannot draw in or edit the *Xref* drawing.

Similar to the *Insert* command's ability to bring another drawing into the current drawing as a *Block*, the *Xref* command can bring any drawing (or *WBlock*) into the current drawing but not as a permanent part of the current drawing. Comparisons between an *Inserted* drawing and an *Xrefed* drawing are outlined below:

INSERT	XREF
Any drawing can be *Inserted* as a *Block*.	Any drawing can be *Xrefed*.
The drawing comes in as a *Block*.	The drawing comes in as an *Xref*.
The *Block* drawing is a permanent part of the current drawing.	The *Xref* is not permanent, only "attached" or "overlayed."
The current drawing file size increases approximately equal to the *Block* drawing size.	The current drawing increases only by a small amount (enough to store information about loading the *Xref*).
The *Inserted Block* drawing is static. It never changes.	Each time the current drawing is opened, it loads the most current version of the *Xref* drawing.
If the original drawing that was used as the *Block* is changed, the *Inserted Block* does not change because it is not linked to the original drawing.	If the original *Xref* drawing is changed, the changes are automatically reflected when the current drawing is *Opened*.
Objects of the *Inserted* drawing cannot be changed in the current drawing unless *Exploded*.	The *Xref* cannot be changed in the current drawing. Changes are typically made in the original drawing.
The current drawing can contain multiple *Block*s.	The current drawing can contain multiple *Xref*s.
A *Block* drawing cannot be converted to an *Xref*.	An *Xref* can be converted to a *Block*. The *Bind* option makes it a *Block*—a permanent part of the current drawing.
A *Block* is *Inserted* on the current layer.	An *Xref* drawing's layers are also *Xrefed* and visibility of its layers can be controlled independently.
Only the drawing geometry of a *Block* is *Inserted*.	Any "named objects" of an *Xref* can be referenced in the current drawing if *Xbind* is used.
*Block*s can be nested.	An *Attached Xref* can be nested.
Any *Block* that is referenced by another *Block* is "nested."	An *Overlay Xref* cannot be nested.
A "circular" reference is not allowed with *Block*s (X references Y and Y references X) because nested *Block*s are also referenced.	Circular references are automatically detected (when the drawing you are *Xrefing* contains a nested *Xref* of the current drawing). You have the option of continuing or not continuing.

INSERT	XREF
When a *Block* is inserted, you can control the properties of the block's layers.	When a drawing is *Attached* or *Overlayed*, you can control the properties of the *Xref's* layers.
When a *Block* is inserted, all objects originally comprising the *Block* are visible.	You can *Xclip* the *Xref* so that only portions of the entire *Xref* drawing are visible.

As you can see, an *Xrefed* drawing has similarities and differences to an *Inserted* drawing. The following figures illustrate both the differences and similarities between an *Xrefed* drawing and an *Inserted* one.

The relationship between the current (parent) drawing and a drawing that has been *Inserted* and one that has been *Xreferenced* is illustrated here (Fig. 30-1). The *Inserted* drawing becomes a permanent part of the current drawing. No link exists between the parent drawing and the original *Inserted* drawing. In contrast, the *Xrefed* drawing is <u>not</u> a permanent part of the parent drawing but is a dynamic one-way link between the two drawings. In this way, when the original *Xrefed* drawing is edited, the changes are reflected in the parent drawing (if a *Reload* is invoked or when the parent drawing is *Opened* next). The file size of the parent drawing for the *Inserted* case is the sum of both the drawings whereas the file size of the parent drawing for the *Xref* case is only the original size plus the link information.

Figure 30-1

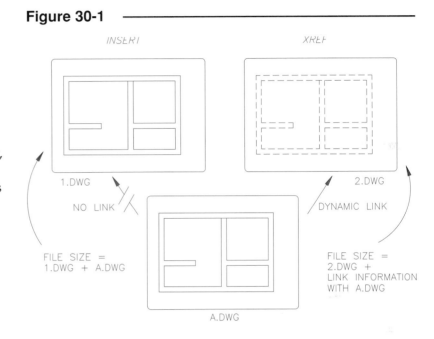

Some of the <u>similarities</u> between *Xreferenced* drawings and *Inserted* drawings are shown in the following figures.

Any number of drawings can be *Xrefed* or *Inserted* into the current drawing (Fig. 30-2). For example, component parts can be *Xrefed* to compile an assembly drawing.

Figure 30-2

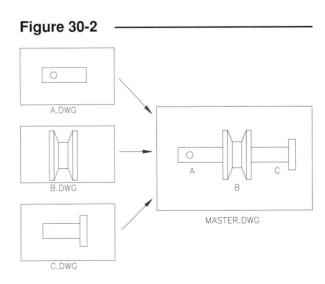

The current drawing can contain nested *Xrefs*. This feature is also similar to *Blocks*. For example, the OFFICE drawing can *Xref* a TABLE drawing and the TABLE drawing can *Xref* a CHAIR drawing. Therefore, the CHAIR drawing is considered a "nested" *Xref* in the OFFICE drawing (Fig. 30-3).

One of the main features and purposes of an *Xref* drawing is that it <u>cannot be edited</u>. Even though an *Xrefed* drawing's named objects (layers, linetypes, text styles, etc.) are attached, you cannot draw or edit on any of its layers. You do, however, have complete control over the <u>visibility</u> of the *Xref* drawing's layers. The *State* of the layers (*On*, *Off*, *Freeze*, and *Thaw*) of the *Xref* drawing can be controlled like any layers in the parent drawing.

Figure 30-3

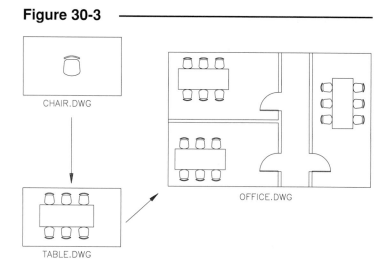

The named objects (*Layers*, *Text Styles*, *Blocks*, *Views*, etc.) that may be part of an *Xref* drawing become <u>dependent</u> objects when they are *Xrefed* to the parent drawing. These dependent objects cannot be renamed, changed, or used in the parent drawing. They can, however, be converted individually (permanently attached) to the parent drawing with the *Xbind* command.

Xref drawings have many applications. *Xrefs* are particularly useful in a networked office or laboratory environment. For example, several people may be working on one project, each person constructing individual components of a project. The components may be mechanical parts of an assembly; or electrical, plumbing, and HVAC layouts for a construction project; or several areas of a plant layout. In any case, each person can *Xref* another person's drawing as an external reference without fear of the original drawing being edited.

As an example of the usefulness of *Xrefs*, a project coordinator could *Open* a new drawing, *Xref* all components of an assembly, analyze the relationships among components, and plot the compilation of the components (Fig. 30-2). The master drawing may not even contain any objects other than *Xrefs*, yet each time it is *Opened*, it would contain the most up-to-date component drawings.

In another application, an entire team can access (*Xref*, *Attach*) the same master layout, such as a floor plan, assembly drawing, or topographic map. Figure 30-4 represents a mechanical design team working on an automobile assembly and all accessing the master body drawing. Each team member "sees" the master drawing, but cannot change it. If any changes are made to the original master drawing, all team members see the updates whenever they *Reload* the *Xref* or *Open* their drawing.

Figure 30-4

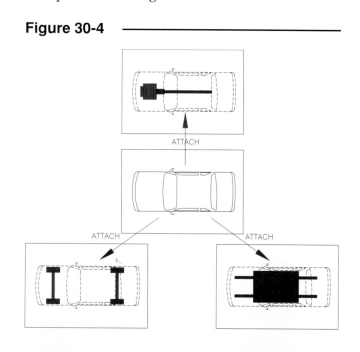

An enhancement in Release 14 is the capability to clip portions of the *Xref* drawing that you do not want to see. Once you attach an *Xref*, you can use the *Xclip* command to draw a rectangular or polygonal (any shaped) boundary to determine what area of the *Xref* to view—all areas outside the boundary become invisible. For example, assume a site plane is *Xrefed* in order to calculate drainage around a house (Fig. 30-5). *Xclip* can be used to display only a portion of the Xref drawing needed for the calculations (Fig. 30-6).

Figure 30-5

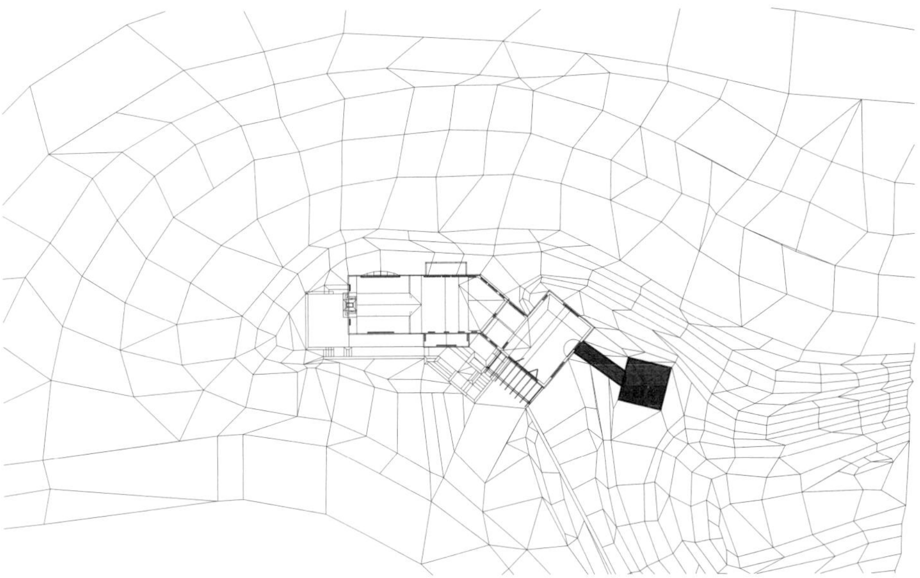

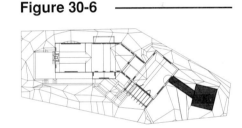

Figure 30-6

Internal features in Release 14 known as "demand loading," "spatial indexing," and "layer indexing" allow AutoCAD to <u>load only the portion of the drawing that appears in the clip boundary</u> into the current drawing. Demand loading and spatial and layer indexing can save a considerable amount of time and memory when Xreferencing large drawings such as maps or large floor plans, because only the small portion of the drawing that is needed (in the clip boundary) is actually loaded.

The simplicity of using Xreferences is that one command, *Xref*, controls almost all of the external referencing features. Several options permit you to create, sever, or change the link between the parent drawing and the *Xref* drawing.

An *Xref* Example

Assume that you are an interior designer in an architectural firm. The project drawings are stored on a local network so all team members have access to all drawings related to a particular project. Your job is to design the interior layout comprised of chairs, tables, desks, and file cabinets for an office complex. You use a prototype drawing named INTERIOR.DWG for all of your office interiors. It is a blank drawing that contains only block definitions (not yet *Inserted*) of each chair, desk, etc., as shown in Figure 30-7. The *Block* definitions are CHAIR, CONCHAIR, DESK, TABLE, FILECAB, and CONTABLE.

Figure 30-7

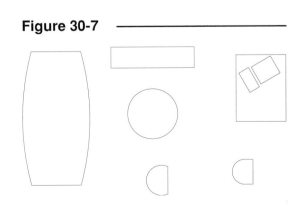

An architect on the team has almost completed the floorplan for the project. You use the *Attach* option of the *Xref* command to "see" the architect's floor plan drawing, OFFICEX.DWG. The *Xrefed* drawing is visible; however, it cannot be altered. The visibility of the individual dependent layers can be controlled with the *Layer* command or *Layer Control* dialog box. For example, layers showing the HVAC, electrical layout, and other details are *Frozen* to yield only the floor plan layer needed for the interior layout (Fig. 30-8).

Figure 30-8

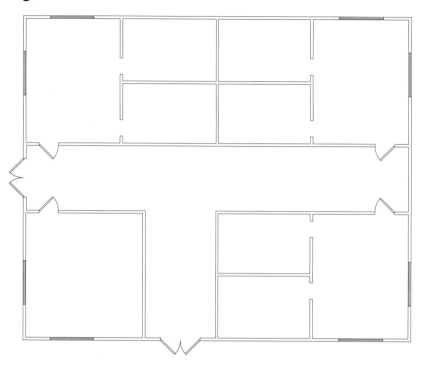

Next, you draw objects in the current drawing or, in this case, *Insert* the office furniture *Block*s that are defined in the current drawing (INTERIOR.DWG). The resulting drawing would display a complete layout of the office floor plan plus the *Inserted* furniture—as if it were one drawing (see Figure 30-9). You *Save* your drawing and go home for the evening.

Figure 30-9

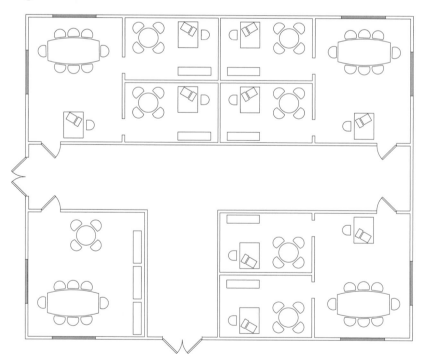

That evening, the architect makes a change to the floor plan. He works on the original OFFICEX drawing that you *Xrefed*. The individual office entry doors in the halls are moved to be more centrally located. The new layout appears in Figure 30-10.

Figure 30-10

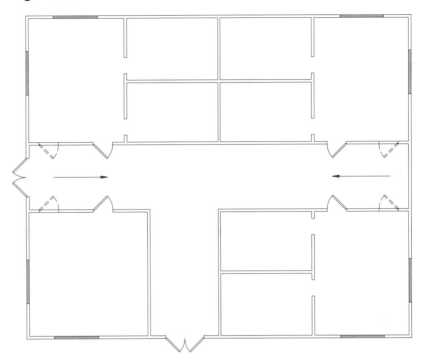

The next day, you *Open* the INTE-RIOR drawing. The automatic loading of the OFFICEX drawing displays the latest version of the office floor plan with the design changes. The appropriate changes to the interior drawing must be made, such as relocating the furniture by the hall doors (Fig. 30-11). When all parts of the office complex are completed, the INTE-RIOR (parent) drawing could be plotted to show the office floor plan and the interior layout. The resulting plot would include objects in the parent drawing and the visible *Xref* drawing layers.

Figure 30-11

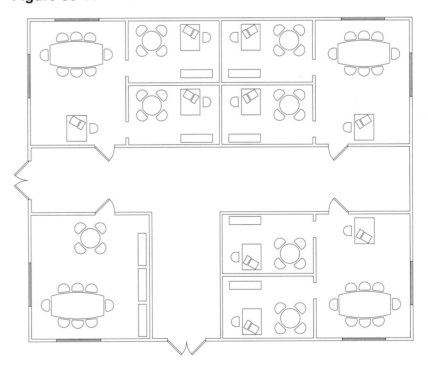

AutoCAD's ability to externally reference drawings makes team projects much more flexible and efficient. If the *Xref* capability did not exist, the OFFICEX drawing would have to be *Inserted* as a *Block*, and design changes made later to the OFFICEX drawing would not be apparent. The original OFFICEX *Block* would have to be *Erased*, *Purged*, and the new drawing *Inserted*.

The new *Xclip* feature in Release 14 gives you the capability to isolate only a portion of an *Xrefed* drawing. For example, assume you wanted to discuss a particular room with the architect, so you open a *New* drawing and *Xref* the INTERIOR drawing. The OFFICEX drawing also is *Xrefed* because it is nested (previously *Xrefed* into the INTERIOR drawing) so the new drawing appears the same as the original INTERIOR drawing (Fig. 30-11). To isolate only the room in discussion, *Xclip* is used to draw a rectangular boundary around the desired area (Fig. 30-12).

Figure 30-12 ————————————

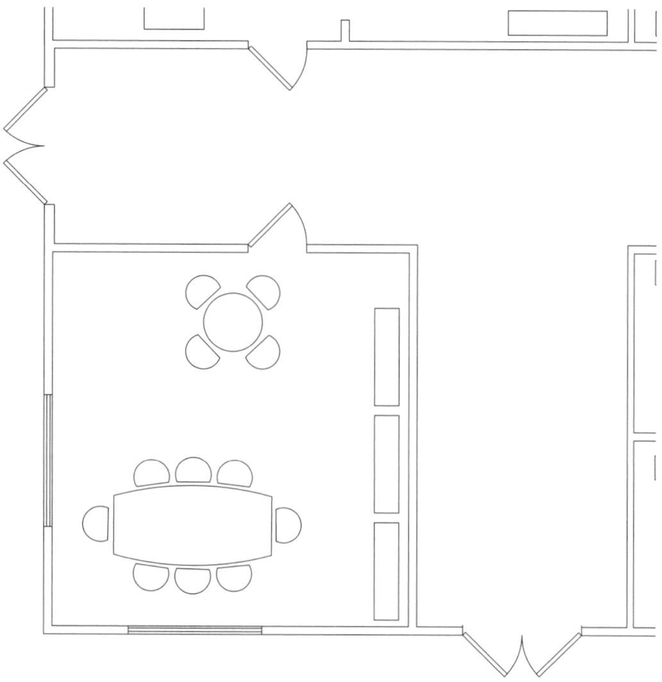

Often in the termination of a project, the final set of drawings is sent to the client. Because the *Xref* capability makes it feasible to work on a design as a set of individual component drawings, it is necessary to send the complete set of drawings to the customer only or to combine the set into one drawing to submit. *Bind* can be used to transform all the *Xref* drawings into *Blocks* in the current drawing. *Bind* brings an *Xrefed* drawing into the current drawing as if it were *Inserted*, so it becomes a permanent part of the parent drawing (as a *Block*). This method combines the set of drawings into one final drawing.

Another technique can be used to send the entire set of drawings but alleviates the problem of having to duplicate the *Xref* path and/or directory structure for the customer. In this case a *PROJECTNAME* is saved in the parent drawing. The customer then sets up the same project name on his or her system, but the directory structure (for locating the *Xrefs*) can be different than on the original system (where the drawing set was created). The parent drawing looks for its *Xrefs* only by *PROJECTNAME*, whose drive and directory mappings can be different on different systems.

XREF AND RELATED COMMANDS AND VARIABLES

The *Xref* command and related commands can be invoked through the *Reference* toolbar. Activate this toolbar using the *Toolbars* dialog box (Fig. 30-13). If you prefer the pull-down menus, the *Insert* and *Modify* pull-downs contain *Xref* and related commands.

Figure 30-13 ————————————

XREF and -XREF	Pull-down Menu	COMMAND (TYPE)	ALIAS (TYPE)	Short-cut	Screen (side) Menu	Tablet Menu
	Insert External Reference...	*XREF or* -XREF	*XR or* -XR	...	INSERT *Xref*	*T,4*

The *Xref* command invoked by any method produces the *External Reference* dialog box (Fig. 30-14). This one dialog box provides you with a means to create and manage *Xrefs*. The central area of the dialog box (discussed later) lists drawings that are currently *Attached* or *Overlayed* (see *List View* and *Tree View*). The *Xref* options are activated by the buttons on the right side of the dialog box.

Figure 30-14

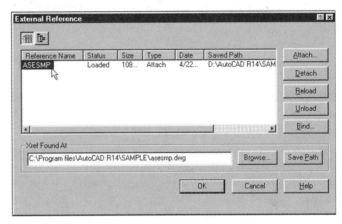

If you prefer to type, you can use the -*Xref* command (notice the hyphen prefix). Using -*Xref* displays the following prompt:

Command: **-xref**
?/Bind/Detach/Path/Unload/Reload/Overlay/<Attach>:

Whether you use *Xref* or -*Xref*, the *Attach, Overlay, Detach, Reload, Unload, Bind,* and *Path* options are the same.

Attach
This (the default) option attaches one drawing to another by making the *Xref* drawing visible in the "parent" drawing. Even though the *Xref* drawing is visible, it cannot be edited from within the parent drawing. *Attach* creates a one-way link between two drawings (Fig. 30-15).

Figure 30-15

Selecting the *Attach* button from the *External References* dialog box produces the *Attach Xref* dialog box (Fig. 30-16). Here you specify the *Reference Type, Attachment* (Attach) or *Overlay*. Ensure the *Attachment* button is pressed. (See *Overlay* next.)

Figure 30-16

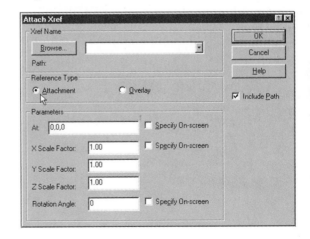

Enter the name (and path if needed) of the drawing to attach in the *Xref Name* edit box. Alternately, you can select *Browse...*, which produces the *Select file to attach* dialog box (Fig. 30-17). (Typing the *-Xref* command and using the *Attach* option produces the *Select file to attach* dialog box directly—the *Attach Xref* dialog box does not appear.) Once the desired file is specified, the *Attach Xref* dialog box appears again for further action.

Figure 30-17

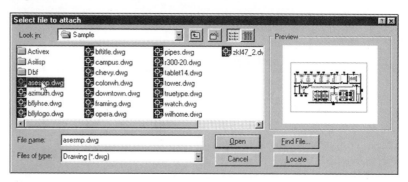

Notice the similarity of the edit boxes (Fig. 30-16) to the *Insert* command. The "X scale factor:", "Y scale factor:", and "Rotation angle:" prompts operate identically to the *Insert* command. Similar to the action of *Insert*, the basepoint of the *Xref* drawing is its 0,0 point unless a different basepoint has been previously defined with the *Base* command.

This action causes the specified drawing to appear in the current drawing similar to the way a drawing appears when it is *Inserted* as a *Block*. Each time the current drawing is *Opened*, the *Attached* drawing is loaded as an *Xref*. More than one reference drawing can be *Attached* to the current drawing. A single reference drawing can be *Attached* to any number of insertion points in the current drawing. The dependent objects in the reference drawing (*Layers*, *Blocks*, *Text styles*, etc.) are assigned new names, "external drawing | old name" (see Dependent Objects and Names).

If you *Attach* a drawing that has a long file name (greater than 8 characters) or contains spaces or unsupported characters, the *Substitute Block Name* dialog box appears (Fig. 30-18). Use the name that AutoCAD assigns or assign your own name of 8 characters or less. The new name appears in the *External Reference* dialog box list (see Fig. 30-24). Consider the new name an internal alias that AutoCAD uses to reference the actual specified drawing.

Figure 30-18

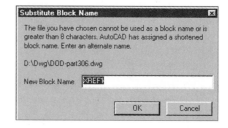

Overlay
An *Xref Overlay* is similar to an *Attached Xref* with one main difference—an *Overlay* cannot be nested. In other words, if you *Attach* a drawing that (itself) has an *Overlay*, the *Overlay* does not appear in your drawing. On the other hand, if the first drawing is *Attached*, it appears when the parent drawing is *Attached* to another drawing (Fig. 30-19).

Figure 30-19

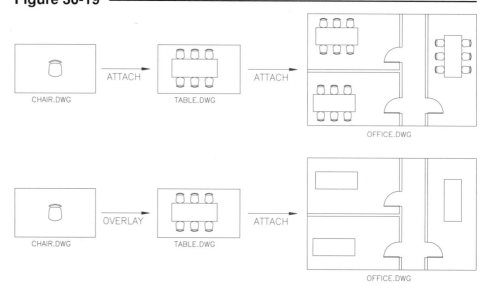

To produce an *Overlay* using the dialog boxes, you must first select the *Attach* button from the *External References* dialog box to produce the *Attach Xref* dialog box, then select *Overlay* in the *Reference Type* section (Fig. 30-16). Ensure the *Overlay* button is pressed. If you type the *-Xref* command, simply use the *Overlay* option and the *Select file to overlay* dialog box appears.

The *Overlay* option enables "circular" *Xrefs* to occur by preventing unwanted nested *Xrefs*. This is helpful in a networking environment where many drawings *Xref* other drawings. For example, assume drawing B has drawing A as an *Overlay*. As you work on drawing C, you *Xref* and view only drawing B without drawing B's overlays—namely drawing A. This occurs because drawing A is an *Overlay* to B not *Attached*. If all drawings are *Overlays*, no nesting occurs (Fig. 30-20).

Figure 30-20

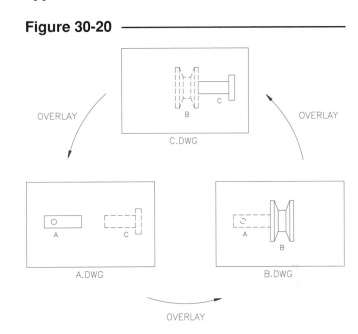

The overlay concept is particularly useful in a concurrent engineering project, where all team members must access each other's work simultaneously. In the case of a design team working on an automobile assembly, each team member can *Attach* the master body drawing and also *Overlay* the individual subassembly drawings (Fig. 30-21). *Overlay* enables each member to *Xref* all drawings without bringing in each drawing's nested Xreferences.

One limitation of *Overlay* is that (*Block*) attributes are not visible in an *Overlay* *Xref*. An *Attached Xref* displays attribute information, but the information cannot be changed in the parent drawing.

Figure 30-21

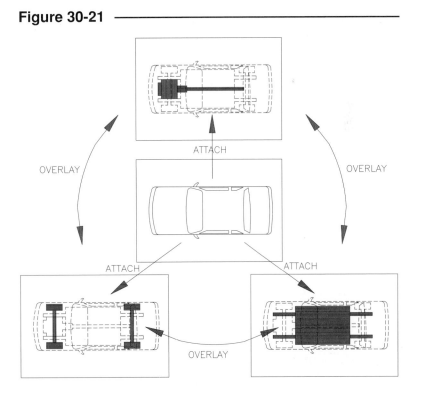

There are times when you cannot foresee the potential use of a drawing and therefore do not *Overlay* rather than *Attach*. For example, in a large design office it is quite possible to *Xref* a drawing on the network that includes your current drawing as an *Attached Xref*, thus causing a circular reference. Release 14 has a provision that allows you to continue with the *Xref* of the first drawing. If this occurs, an *AutoCAD Alert* dialog box appears giving notification of the circular reference and presents the option to continue or not (Fig. 30-22). Answering *Yes* causes AutoCAD to load the first *Xref* (selected drawing), but drops the nested (circular) reference. Answering *No* cancels the *Attach*.

Figure 30-22

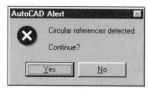

Detach

Detach severs the link between the parent drawing and the *Xref*. Using this option causes the *Xrefed* drawing to "drop" from view immediately. When the parent drawing is *Opened*, the previously *Attached* drawing is <u>no longer</u> loaded as an *Xref*.

Path

You can specify the drive and directory location of the external reference using the *Path* option. This is necessary if an *Xrefed* drawing has been relocated to another drive or directory on the computer drive or network drives. For example, if the ASESMP drawing (that your parent drawing referenced) was relocated to another drive or directory, it could not be loaded when the parent drawing was opened and would be listed as "not found" in the *External Reference* dialog box (Fig. 30-23). Use the *Path* option to specify the new location. Enter the path in the edit box, or use the *Browse...* tile to invoke the *Select new path* dialog box (essentially the same interface as the

Figure 30-23

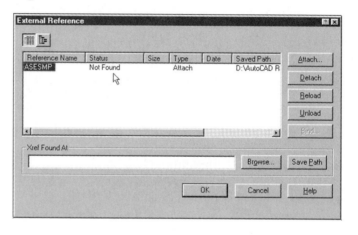

Select file to attach dialog box, Fig. 30-17). Using the *Path* option forces a reload. Using the *Path* option of the typed version of the command, *-Xref*, the path must also be typed.

When you archive (back up) drawings, remember also to back up the drives and directories where the *Xref* drawings are located (or use the *Bind* option when the project is finished). One possible file management technique is to specify a directory especially for *Xrefs* (see *PROJECTNAME*). In a networking environment, access rights to *Xref* directories must be granted.

Reload

This option forces a *Reload* of the external drawing at any time while the current drawing is in the Drawing Editor. This ensures that the current drawing contains the <u>most recent version</u> of the external drawing. In a networking environment, you may not be able to *Attach*, *Overlay*, *Bind*, or *Reload* a drawing based on whether the referenced drawing is currently being edited (open) by another person or by the setting of the *XLOADCTL* system variable in the referenced drawing (see *XLOADCTL*).

Unload

The *Unload* option can be used to increase computing speed in a drawing. In cases when you need to view (load) drawings only on an as-need basis, *Unload* them. This causes the current (parent) drawing to open more quickly, speeds regenerations, and uses less memory. The unloaded *Xref* is not visible, but the link information remains in the current drawing. To view the *Xref* again, use *Reload*. This method is preferred over using *Detach* when you expect to use the *Xref* again.

Bind

The *Bind* option <u>converts</u> the *Xref* drawing to a *Block* in the current drawing, then terminates the external reference partnership. The original *Xrefed* drawing is not affected. The names of the dependent objects in the external drawing are changed to avoid possible conflicts in the case that the parent and *Xref* drawing have dependent objects with the same name (see Dependent Objects and Names). If you want to only bind selected named objects, use the *Xbind* command.

When the *Bind* button is selected (in the *External Reference* dialog box), the *Bind Xrefs* dialog box appears for you to select the *Bind Type* (see Fig. 30-28). This option determines how the <u>names of dependent objects</u> (such as layers and blocks) appear in the drawing after they become a permanent part of the parent drawing (see Dependent Objects and Names).

List View and *Tree View*

The central area of the *External Reference* dialog box is used to list and give the status of drawings that have been *Xrefed*. Use the two buttons in the upper-left corner of the dialog box (Fig. 30-24) to display the information in a list that can be ordered alphabetically by column (*List View*) or in a hierarchical tree structure (*Tree View*).

Figure 30-24

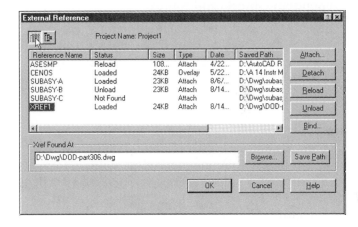

List View

This option (default) displays the *Xrefs* alphabetically by name. You can sort the list by column in forward or reverse order (*Reference Name, Size, Type, Date,* etc.) by selecting the column heading and clicking once or twice. You can resize the column width by dragging the column separator left or right. The *Reference Name* can also be edited (up to 31 characters, no spaces). The *Type* can be *Attach* or *Overlay*. The *Status* column displays the state of each reference. The following states are possible:

List View	Description
Loaded	XREF was found when drawing was opened or reloaded
Unload	XREF will be unloaded when the dialog box is closed
Unloaded	XREF is currently unloaded
Reload	XREF will be reloaded when the dialog box is closed
Not found	XREF was not found when drawing was opened or reloaded
Unresolved	XREF was found but could not be read by AutoCAD
Unreferenced	Nested XREF is attached to an XREF that is no longer attached (the path to its parent XREF may have been changed to reference a different file)
Orphaned	Nested XREF is attached to an XREF that is unloaded

Tree View

The *Tree View* shows the *Xrefs* in a hierarchical tree structure. This option is particularly useful if you have <u>nested</u> *Xrefs* (Fig. 30-25). Note that the icons also indicate the type and status of each referenced drawing (*Overlay, Attach, Reload, Unload, Not found*, etc.).

Figure 30-25

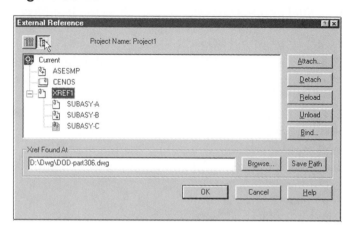

Dependent Objects and Names

The named objects that have been created as part of a drawing become dependent objects when that drawing is *Xrefed*. The named objects in a drawing can be any of the following:

> *Blocks*
> *Layers*
> *Linetypes*
> *Dimension Styles*
> *Text Styles*

These dependent objects <u>cannot</u> be renamed or changed in the parent drawing. Dependent text or dimension styles cannot be used in the parent drawing. You cannot draw on dependent layers; you can only control the visibility of the dependent layers. You <u>can</u>, however, *Bind* the entire *Xref* and the dependent named objects, or you can bind individual dependent objects with the *Xbind* command. *Xbind* converts an individual dependent object into a permanent part of the parent drawing. (See *Xbind*.)

When an *Xref* contains named objects, the names of the dependent objects are changed (in the parent drawing). A new naming scheme is necessary because conflicts would occur if the *Xref* drawing and the parent drawing both contained *Layers* or *Blocks*, etc., with the same names, as would happen if both drawings used the same template drawing. When a drawing is *Attached*, its original object names are prefixed by the drawing name and separated by a pipe (|) symbol.

For example, if the ASESMP drawing is *Xrefed* and it contains a layer named "kitchen," it is renamed (in the parent drawing) to Asesmp | kitchen (Fig. 30-26).

Figure 30-26

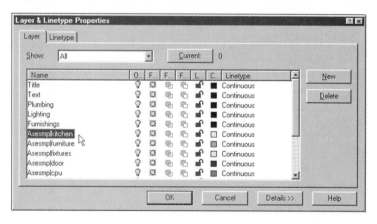

Because the dependent objects cannot be used in the parent drawing
(you cannot insert a dependent block nor can you draw on a dependent
layer), these objects cannot be renamed or deleted either. If you attempt
to delete or rename a dependent layer a warning appears (Fig. 30-27).
(Keep in mind that dependent blocks, layers, etc. can be used if *Bind* or
Xbind is used to make them a permanent part of the parent drawing).

Figure 30-27

Likewise, if the *Xref* drawing contains a block named VENDING, it is renamed in the parent drawing to
ASESMP | VENDING. Using the *Block* command to reveal a listing of blocks in the current drawing
yields the following display. Note the tally below of "User Blocks," "External References," and
"Dependent Blocks":

```
Block name (or ?): ?
Block(s) to list <*>:

Defined blocks.
 ASESMP                 Xref: resolved
 XREF1                  Xref: resolved
 SUBASY-A               Xref: resolved
 SUBASY-B               Xref: resolved
 SUBASY-C               Xref: not found
 CENOS                  Xref: resolved
 ASESMPIVENDING         Xdep: ASESMP
 ASESMPIRANGE           Xdep: ASESMP
 ASESMPIDISHWASH        Xdep: ASESMP
 ASESMPIREF1            Xdep: ASESMP
 ASESMPISINKDBL1        Xdep: ASESMP
 ASESMPIDESK4           Xdep: ASESMP
 ASESMPIDESK3           Xdep: ASESMP
 ASESMPIDESK2           Xdep: ASESMP
 ASESMPISOFA2           Xdep: ASESMP
 ASESMPITABLE1          Xdep: ASESMP
 ASESMPITABLE2          Xdep: ASESMP
 FIX-FLOR-60W-30
 FIX-23A
 FIX-A14-100W
```

User Blocks	External References	Dependent Blocks	Unnamed Blocks
3	6	11	0

Command:

The same naming scheme operates with all dependent objects—*Dimension Styles, Text Styles, Views,* etc.

Dependent Object Names with *Bind* and *Xbind*
If you want to *Bind* the *Xref* drawing, the link is severed and the *Xref* drawing becomes a *Block* in the
parent drawing. Or, if you want to bind only <u>individual named objects</u>, the *Xbind* command can be used
to bring in specific *Blocks* or *Layer* names. The names are converted from the previous dependent
naming scheme (using the *Xref* drawing name as a prefix with a | [pipe] separator) to another name
depending on the command and option you use.

If you use *Xbind* or use *Bind* with the default *Bind* option, the names of dependent objects keep the *Xref* drawing name, but the | (pipe) symbol is changed to a $#$ separator, where # represents a random number. For example, assume you use the *Bind* button (in the *External References* dialog box) with the ASESMP drawing from the previous example. The *Bind Xrefs* dialog box appears (Fig. 30-28). The default (*Bind*) option is the traditional (pre-Release 14) naming scheme. That is, the *Xref* drawing name prefix stays with the object name.

Figure 30-28

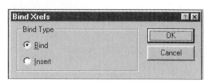

Using the *Bind* option, the resulting layer names would appear in the *Layer & Linetype Properties* dialog box listing (Fig. 30-29). Notice the dependent layer previously named "Asesmp|kitchen" is changed to "Asesmp0kitchen." These layers <u>can be renamed</u> at this point because they are now a permanent part of the current drawing. Using the *Xbind* command to "import" selected named objects results in the same layering scheme illustrated here.

Figure 30-29

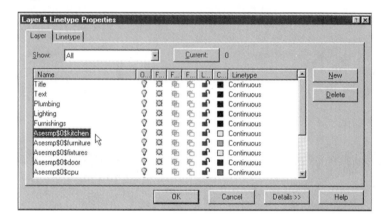

On the other hand, if you select the *Insert* option in the *Bind Type* dialog box, the *Xref* drawing name prefix is dropped from the object names when they are "imported." Therefore, the named objects assume their <u>original name</u>, assigned when they were created in the original drawing. Using the previous *Xref* (ASESMP) example, layer names would appear as shown in the *Layer & Linetype Properties* dialog box as "Kitchen," "Furniture," "Doors," etc. (Fig. 30-30).

Figure 30-30

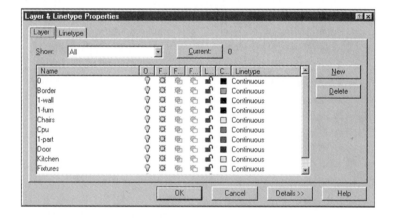

This method (*Bind, Insert*) is preferred if you want to *Bind* an *Xrefed* drawing that was created from the same template drawing as the current (parent) drawing. In this way, the same layers, dimension styles, text styles, etc. are not duplicated when you *Bind* the *Xref*. However, this method should <u>not be used</u> when <u>two different</u> *Blocks* exist with the same name or two layers exist with different linetype and color settings that should be maintained. When a dependent *Block* or layer having the same name as that in the parent drawing is "imported" with *Bind, Insert*, the *Block* definition, and the layer properties of the parent drawing take precedence. Therefore, it is possible to accidentally redefine a *Block* or lose linetype and color properties of a layer with this scheme. There is no *Insert* option for *Xbind*.

XATTACH

Pull-down Menu	COMMAND (TYPE)	ALIAS (TYPE)	Short-cut	Screen (side) Menu	Tablet Menu
Insert *External Reference* *Attach...*	*XATTACH*	*XA*	...	*INSERT* *Xref* *Attach...*	...

Xattach is a new Release 14 command. It is a separate command from *Xref* but accomplishes the same action as *Xref, Attach*. If you use *Xref*, the *External References* dialog box appears, then you must select the *Attach...* tile to invoke the *Attach Xref* dialog box. However, you can use *Xattach* to invoke the *Attach Xref* dialog box directly (Fig. 30-31). Therefore, use the *Xattach* command when you want to *Attach* or *Overlay* an *Xref* but do not want to view or manage the list of current Xrefs.

Figure 30-31

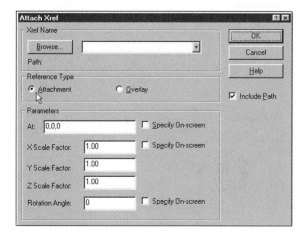

XBIND and -XBIND

Pull-down Menu	COMMAND (TYPE)	ALIAS (TYPE)	Short-cut	Screen (side) Menu	Tablet Menu
Modify *Object>* *External Reference>*	*XBIND or* *-XBIND*	*XB or* *-XB*	...	*MODIFY1* *Xbind*	*X,19*

The *Xbind* command is a separate command; it is not an option of the *Xref* command as are most other Xreference controls. *Xbind* is similar to *Bind*, except that *Xbind* binds an individual dependent object, whereas *Bind* converts the entire *Xref* drawing to a *Block*. Any of the listed dependent named objects (see Dependent Objects and Names) that exist in a drawing can be converted to a permanent and usable part of the current drawing with *Xbind*. In effect, *Xbind* makes a copy of the named object since the original named object is not removed from the dependent drawing.

The *Xbind* command produces the *Xbind* dialog box (Fig. 30-32). All *Attached* and *Overlayed Xrefs* are listed. Expand any *Xref* to reveal the dependent object types (*Block, Dimstyle*, etc.) that can be imported. Expand each dependent object type to list the named objects of the type for that particular drawing. Select the desired object, then pick *Add ->* to add the object to the current (parent) drawing. The following object types are valid.

Figure 30-32

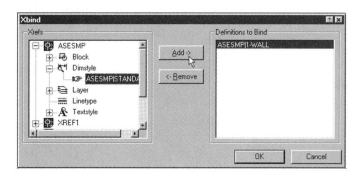

Block

Select this option to *Xbind* a *Block* from the *Xrefed* drawing. The *Block* can then be inserted into the current drawing and has no link to the original *Xref* drawing.

Dimstyle

This option allows you to make a dependent Dimension Style a permanent part of the current drawing. Dimensions can be created in the current drawing that reference the new Dimension Style.

Layer

Any dependent layer can be brought into the current drawing with this option. The geometry that resides on that layer does <u>not</u> become part of the current drawing, only the layer name and its properties. *Xbinding* a layer also *Xbinds* the layer's color and linetype.

Linetype

A linetype from an *Xrefed* drawing can be bound to the current drawing using this option. This action has the same effect as loading the linetype.

Textstyle

If a text style exists in an *Xrefed* drawing that you want to use in the current drawing, use this option to bring it into the current drawing. New text can be created or existing text can be modified to reference the new text style.

If you use (type) the *-Xbind* command, the command line version of the command is invoked and individual dependent objects must be typed:

```
Command: -xbind
Block/Dimstyle/Layer/Ltype/Style: b
Dependent Block name(s):
```

You must know and type the exact name of the dependent object name, including the dependent drawing name prefix and pipe character; therefore, this method is cumbersome compared to using the *Xbind* dialog box.

VISRETAIN

Pull-down Menu	COMMAND (TYPE)	ALIAS (TYPE)	Short-cut	Screen (side) Menu	Tablet Menu
Format *Layer...* *Details>>* *Retain changes to xref-dependent layers*	*VISRETAIN*	...	...	*FORMAT* *Layer* *Details>>* *Retain changes to...*	...

This variable controls the dependent (*Xref* drawing) layer visibility, color, and linetype settings in the parent drawing. Remember that you cannot draw or edit on a dependent layer; however, you can change the layer's settings (*Freeze/Thaw, On/Off, Color, Linetype*, etc.) with the *-Layer* command, *Layer & Linetype Properties* dialog box, or *Layer Control* drop-down box. The setting of *VISRETAIN* determines if the settings that have been made are retained when the *Xref* is reloaded.

R14

When *VISRETAIN* is set to 1 in the parent drawing, all the layer settings (*Freeze/Thaw*, *On/Off*, *Color*, *Linetype*, etc.) are retained for the Xreferenced dependent layers. When you *Save* the drawing, the settings for dependent layers are saved with the current (parent) drawing, regardless of whether the settings have changed in the *Xref* drawing itself. In other words, if *VISRETAIN*=1 in the parent drawing, the layer settings for dependent layers are saved. If *VISRETAIN*=0, then dependent layer properties settings are determined by the settings in the *Xref* drawing.

VISRETAIN also determines whether or not changes to the path of nested *Xrefs* are saved. You can change the path of an *Xref* or of a nested *Xref* (location AutoCAD searches for *Xrefs*) with the *Path* option of *-Xref* or the *Save Path* button in the *External Reference* dialog box.

In Release 14, you can set *VISRETAIN* in the *Layer* tab of the *Layer & Linetype Properties* dialog box. To set *VISRETAIN* to 1, select the checkbox next to *Retain changes to xref-dependent layers* (Fig. 30-33, lower left). No check appearing in the box means that *VISRETAIN* is set to 0. (You must open the *Details>>* section of the dialog box to use this option.)

Figure 30-33

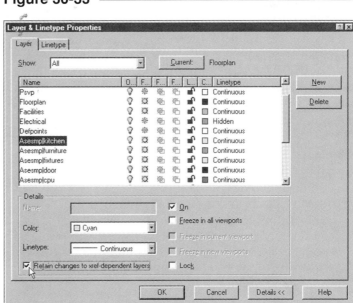

	COMMAND	ALIAS	Short-	Screen (side)	Tablet
Pull-down Menu	**(TYPE)**	**(TYPE)**	**cut**	**Menu**	**Menu**
Modify *Object>* *Clip*	*XCLIP*	*XC*	...	*MODIFY1* *Xclip*	*X,18*

XCLIP

When working with *Xrefs* you may want to display only a specific part of the attached drawing. For example, you may *Xref* an entire manufacturing plant layout (floor plan) in order to redesign an office space, or you may need to *Xref* a large map of a city just to design a new interchange at one location. In these cases, loading and displaying the entire *Xref* would occupy most of the system resources just for display, even though only a small area is needed.

R14

Using the *Xclip* command, you can visually clip an <u>*Xref* drawing or a</u> <u>*Block*</u> to display only the portion inside the area defined by the clip boundary (Fig. 30-34). You can define the boundary as a rectangle or as a many-sided polygon of virtually any shape. All geometry inside the boundary is displayed normally. The portions of the *Xrefed* drawing (or *Block*) that fall outside the boundary are not only restricted from the display, but a feature called "demand loading" can be used to prevent the unwanted portions of the *Xref* from loading so system resources are not used unnecessarily (see Demand Loading, *INDEXCTL* and

Figure 30-34

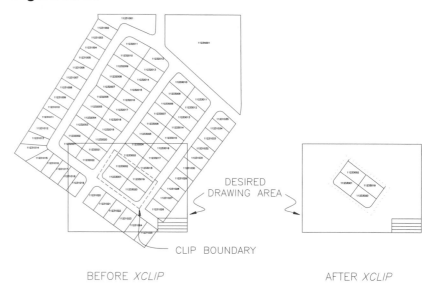

BEFORE *XCLIP*

DESIRED DRAWING AREA

CLIP BOUNDARY

AFTER *XCLIP*

XLOADCTL). Clipping does not edit or change the *Xref* in any way, it only prevents portions from being displayed:

```
Command: xclip
Select objects: PICK (the Xref object)
Select objects: Enter
ON/OFF/Clipdepth/Delete/generate Polyline/<New boundary>: Enter
Specify clipping boundary:
Select polyline/Polygonal/<Rectangular>: Enter
First corner: PICK
Other corner: PICK
Command:
```

The options of *Xclip* are as follows.

New Boundary

Rectangular
You define a rectangular boundary by PICKing diagonal corners. The clipping boundary is on the XY plane of the current UCS.

Polygonal
Select points as the vertices of a polygon. The polygon can have any shape and number of sides.

Select Polyline
Select an existing polyline (*Pline*) to use for the boundary. The existing polyline must consist of straight segments but can be open.

ON/OFF
ON displays only the *Xref* geometry inside the clipping boundary. *OFF* displays the entire *Xref*, ignoring the clipping boundary.

Clipdepth
Use this option to set front and back clipping planes on the *Xref* object. This option is useful for 3D geometry when you want to clip portions of the object behind a back clipping plane and in front of a front clipping plane. You must specify a clip distance from and parallel to the clipping boundary.

Delete
You can delete an existing boundary for the selected *Xref* or *Block* with this option. *Erase* cannot be used to delete a clipping boundary.

Generate Polyline
Automatically draws a *Pline* along an existing clipping boundary. The new *Pline* assumes the current layer, linetype, and color settings. This option is helpful if you want to change the clipping boundary using *Pedit* with the new *Pline*, then use the *Polyline* option to establish the new boundary.

XCLIP

Pull-down Menu	COMMAND (TYPE)	ALIAS (TYPE)	Short-cut	Screen (side) Menu	Tablet Menu
Modify *Object>* *External Reference>* *Frame*	*XCLIPFRAME*	...	...	...	...

XCLIPFRAME is a system variable that can be used to control the display of the clipping boundary created by the *Xclip* command. For example, in some cases it may be desirable to display the boundary, while in other cases no boundary may be preferred (Fig. 30-35). The options are 0 (turns clipping boundary off) and 1 (turns clipping boundary on).

Figure 30-35

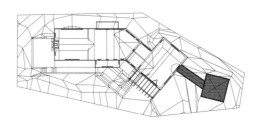

XCLIPFRAME = 1

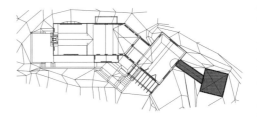

XCLIPFRAME = 0

Demand Loading

When *Xref* clipping is used to display only a selected area of an *Xrefed* drawing or block, it logically follows that the area outside the clipping boundary is of little or no interest in the current drawing. In order to conserve system resources and save time loading and regenerating these unwanted drawing areas, AutoCAD provides a feature called "demand loading."

Demand loading allows AutoCAD to load only the information (about the *Xref* drawing) that is needed to display the geometry in the clipping boundary. To take maximum advantage of demand loading, the demand loading feature must be enabled in the parent drawing and the *Xref* drawings must be saved with layer and spatial indexes enabled. When a clipping boundary is formed, AutoCAD uses the <u>layer</u>

index and spatial index in the *Xrefed* drawing to categorize which objects (of the *Xref* or *Block*) are fully within, partially within, and not within the clipping boundary and which layers are visible and not visible:

Spatial Index	Updated in the *Xref* drawing when a clipping boundary is formed to display a portion of the *Xref's* objects
Layer Index	Updated in the *Xref* drawing when layer visibility is changed (*Freeze*, *Thaw*, *On*, *Off*)

Using the layer index and spatial index, AutoCAD determines which portions of the *Xrefed* drawing file to load. This process provides performance advantages; however, certain demand loading controls must be used. The controls for demand loading are the *INDEXCTL* and *XLOADCTL* system variables.

INDEXCTL

Pull-down Menu	COMMAND (TYPE)	ALIAS (TYPE)	Short-cut	Screen (side) Menu	Tablet Menu
File *Saveas...* *Options...* *Index type*	*INDEXCTL*	...	...	*FILE* *Saveas* *Options...* *Index type*	...

When a drawing is used as an *Xref*, and particularly a clipped *Xref*, demand loading can increase performance in the parent drawing. In order for this to occur, spatial indexing and/or layer indexing must be enabled in the *Xref* drawing(s). *INDEXCTL* controls how a drawing is loaded <u>when it is used as an *Xref*</u>. In other words, <u>*INDEXCTL* is set in the *Xref* drawing</u> (the drawing to be referenced), <u>not in the parent drawing</u> (the current drawing when using the *Xref* command).

Considering this in another way, assume you *Open* a drawing (parent drawing) and *Xref* two other drawings. Next, you create a clipping boundary around each *Xref*, so only a portion of each *Xref* is visible. Then you *Save* the parent drawing. In order to take advantage of demand loading and realize a performance increase when you *Open* the parent drawing again, layer and spatial indexing must be enabled in <u>each of the two *Xrefed* drawings</u> for demand loading to occur. *Open* each (original) *Xref* drawing, and set *INDEXCTL* to 1, 2, or 3, then *Save*.

The *INDEXCTL* variable can be set by command line format or can be set by using *SaveAs*, then selecting the *Options...* button in the *SaveAs* dialog box. The *Export Options* dialog appears (Fig. 30-36) with four options for *Index Type*.

Figure 30-36

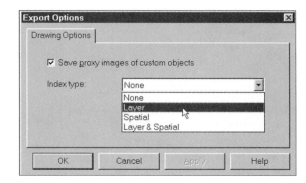

The *INDEXCTL* system variable is set to 0 by default, in which no spatial or layer indexes are created. The settings for *INDEXCTL* are as follows:

Setting	*Export Options* Setting	Effect
0	*None*	no indexes created
1	*Layer*	layer indexes created
2	*Spatial*	spatial indexes created
3	*Layer & Spatial*	both layer and spatial indexes created

To take full advantage of demand loading, set *INDEXCTL* to 3 in the drawing you anticipate using as an *Xrefed* drawing.

NOTE: <u>Only drawings in Release 14 format</u> can be *Saved* with an *INDEXCTL* setting. Drawings created and saved in previous releases are loaded in their entirety, so no benefit is obtained from demand loading.

XLOADCTL

Pull-down Menu	COMMAND (TYPE)	ALIAS (TYPE)	Short-cut	Screen (side) Menu	Tablet Menu
Tools *Preferences...* *Performance* *External reference file* *demand load:*	*XLOADCTL*	...	...	*TOOLS2* *Prefernc* *Performance* *External* *reference...*	...

The *XLOADCTL* variable set in the parent drawing (1) turns on or off demand loading for the current drawing and (2) controls what access other users have to *Xrefed* drawings.

Demand loading is dynamic. If the clipping boundary is changed, the spatial index in the *Xrefed* drawing is automatically updated and if an *Xref* layer is frozen or thawed, the layer index is updated. This unique feature allows information from the *Xrefed* drawing to be loaded "on demand." However, there is one drawback to demand loading—that is, when demand loading is enabled, AutoCAD places a lock on all *Xrefed* drawings so that it can read and modify the information it needs (indexes) on demand. Therefore, other users can *Open* the *Xrefed* drawings but cannot *Save* changes to them. Alternately, you can use the *Copy* option of demand loading, which creates a temporary copy of the *Xrefed* drawing(s) for use by the parent drawing and leaves the original *Xref* drawing available for use by others to edit and *Save*.

R14

The options of *XLOADCTL* are as follows.

Setting	*Preferences* Setting	Effect
0	*Disabled*	demand loading is disabled *Xref* drawing is loaded in its entirety others can access, edit, and save the *Xref* file
1	*Enabled*	demand loading is enabled locks original *Xref* drawing for use by parent drawing only
2	*Enabled with copy*	demand loading is enabled creates a temporary copy of the *Xref* for use by the parent drawing others can access, edit, and save original *Xref* file

The default setting for *XLOADCTL* is 1. This setting is good for stand-alone users. If you are in a net-working environment, a setting of 1 ensures others cannot alter the *Xref*, while a setting of 2 allows other users involved in a particular project (related to the *Xrefed* drawing) to continue working on the original *Xref* files.

XLOADCTL can also be set in the *Performance* tab of the *Preferences* dialog box (Fig. 30-37, lower left). The three options (*Disabled*, *Enabled*, and *Enabled with copy*) are explained in the previous table.

Figure 30-37

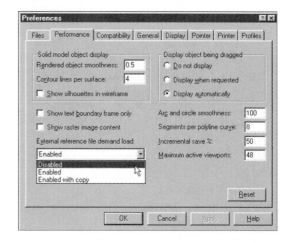

XLOADPATH

	Pull-down Menu	COMMAND (TYPE)	ALIAS (TYPE)	Short-cut	Screen (side) Menu	Tablet Menu
	Tools *Preferences...* *Files* *Temporary external reference file location*	*XLOADPATH*	...	...	*TOOLS2* *Prefernc* *Files* *Temporary external refs..*	...

This system variable is used when *XLOADCTL* is set to 2 (*Enabled with copy*). Normally, when AutoCAD creates the copy of *Xref* drawings the temporary copy is placed in the directory specified as *Temporary Drawing File Location* in the *Files* tab of the *Preferences* dialog box. You can use the *XLOADPATH* variable directly (in command line format) or the *Preferences* dialog box to specify another location for temporary files.

In the *Files* tab of the *Preferences* dialog box (Fig. 30-38), expand *Temporary Drawing File Location*, highlight the previously specified location (the default is C:\Windows\Temp\), then select the *Browse...* button to specify a new location for temporary files.

Figure 30-38

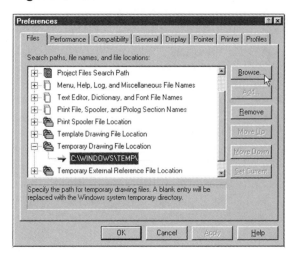

As a summary for setting demand loading and related controls, Figure 30-39 illustrates the appropriate drawing files (parent drawing and related *Xref* drawings) for setting the *XLOAD-CTL*, *XLOADPATH*, and *INDEXCTL* variables.

Figure 30-39

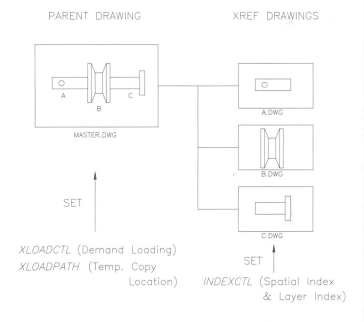

PROJECTNAME

Pull-down Menu	COMMAND (TYPE)	ALIAS (TYPE)	Short-cut	Screen (side) Menu	Tablet Menu
Tools Preferences... Files Project files...	PROJECTNAME	...	...	TOOLS2 Prefernc Files Project files..	...

Project names make it easier for AutoCAD users to exchange drawings that contain *Xrefs*. The *PRO-JECTNAME* variable allows you to specify a project name that AutoCAD uses to point to an alternate search path when trying to locate *Xrefs*.

With previous releases of AutoCAD, if you load a drawing containing several *Xrefs* given to you by a colleague from another company, AutoCAD attempts to locate the *Xrefs* by searching the hard-coded paths. (The path of an *Xref* is "hard coded" when the drawing is originally *Attached* or *Overlayed* or if the *Path* option is used in the *External Reference* dialog box). The *Xrefs* are not found unless you have an identical directory structure on your system or network, and the *Xrefs* have been copied into the directories accordingly. In Release 14, if an *Xref* cannot be found because the drawing is not located in the hard-coded path, AutoCAD then uses the drawing's *PROJECTNAME* to search for the *Xref* in the location specified by your system registry.

There are two components to this scheme: (1) the *PROJECTNAME* variable is saved in the parent drawing file, and (2) the directory location for that project name is saved in your system registry. This scheme makes it possible for collaborating designers to exchange drawings containing *Xrefs* when the designers have different directory structures. The drawing that is exchanged contains the project name, but the location (path) for that project name can be different for each system (saved in the system registries). Each designer's system contains the same project name but has a different drive and/or directory mapping. In other words, if an *Xref* cannot be found in the hard-coded path, AutoCAD then searches for the *Xref* by using the *PROJECTNAME* variable in the drawing file, but the <u>location</u> for the project name is saved in your system registry.

The value for the drawing's *PROJECTNAME* variable (*Project1*, for example) can be set using the *PROJECTNAME* variable in command line format or using the *Files* tab of the *Preferences* dialog box. The *PROJECTNAME* variable is saved in the <u>parent drawing file</u>. However, you must use the *Files* tab of the *Preferences* dialog box (but <u>not</u> the *PROJECTNAME* variable in command line format) to specify the directory mapping (location), which is then stored in <u>your system registry</u>.

To assign a project name for the drawing file, use the *PROJECTNAME* variable at the command line or use the *Files* tab of the *Preferences* dialog box (Fig. 30-40). Using the dialog box, follow these steps.

1. Double-click *Project Files Search Path*, then select *Add*. A project folder named *Projectx* appears, where x is the next available number.
2. Enter a new name or press Enter to accept *Projectx*.
3. Select *OK*.

This action assigns a project name to the <u>drawing file</u>.

Figure 30-40

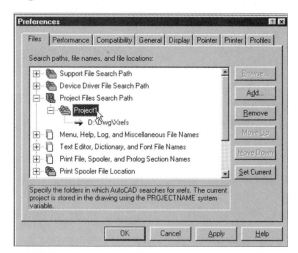

To assign a search path to be saved in your system registry, you <u>must</u> use the *Files* tab of the *Preferences* dialog box. You <u>cannot</u> use the *PROJECTNAME* variable or any other method of command line entry to add, remove, or modify the directory location for a project name in the system registry. Follow these steps:

1. Double-click or expand *Project Files Search Path*.
2. Select an existing or make a new project name (*Project1*, for example).
3. Highlight the desired project name, then select *Add*.
4. Select *Browse*. In the *Choose directory* dialog box, select a path.
5. Select *OK*.

XREFCTL

Pull-down Menu	COMMAND (TYPE)	ALIAS (TYPE)	Short-cut	Screen (side) Menu	Tablet Menu
...	XREFCTL	...	...	...	...

AutoCAD has a tracking mechanism for the *Xref* activity in drawings. An external ASCII log file can be kept for all drawings that contain information on external references. The *XREFCTL* variable controls the creation of the *Xref* log files.

If *XREFCTL* is set to 1, a log file registers each *Attach*, *Overlay*, *Bind*, *Detach*, *Reload*, and *Unload* of each external reference for every drawing. These files have the same name as the related drawing with a file extension ".XLG." The log files are placed in the same directory as the related drawing. If a drawing is *Opened* again and the *Xref* command is used, the *Xref* activity (*Attach*, *Overlay*, *Bind*, etc.) is appended to the existing log file. The .XLG files have no <u>direct</u> connection to the drawing, so they can be deleted if desired without consequence to the related drawing file. If *XREFCTL* is set to 0 (default setting), no log is created or appended. The *XREFCTL* variable is saved in the system registry, so once it is set, it affects all drawings.

An example .XLG file is shown here. This file lists the activity from an earlier example in this chapter when the OFFICEX drawing was *Xrefed* to the INTERIOR drawing.

Loading D:\Dwg\Interior.dwg into Drawing Editor.

===============================

Drawing: D:\Dwg\Interior.dwg
Date/Time: 08/24/97 08:55:16
Operation: Attach Xref

===============================

Attach Xref OFFICEX: D:\Dwg\Officex.dwg

Searching in ACAD search path

Update block symbol table:
 Appending symbol: OFFICEX|DUPOUT
 Appending symbol: OFFICEX|L1-TBLCK
 Appending symbol: OFFICEX|DESK
 Appending symbol: OFFICEX|CHAIR
 Appending symbol: OFFICEX|CONFCHAIR
 Appending symbol: OFFICEX|TABLE
 Appending symbol: OFFICEX|CONFTABLE
 Appending symbol: OFFICEX|FILECAB
 Appending symbol: OFFICEX|DOOR
 Appending symbol: OFFICEX|WINDOW
Block update complete.

Update layer symbol table:
 Appending symbol: OFFICEX|FLOORPLAN
 Appending symbol: OFFICEX|FACILITIES

Appending symbol: OFFICEX|ELECTRICAL
Appending symbol: OFFICEX|TEXT
Duplicate DEFPOINTS layer entry ignored.
Appending symbol: OFFICEX|PSVP
Appending symbol: OFFICEX|TBLCK
Layer update complete.

Update linetype symbol table:
Appending symbol: OFFICEX|BORDER
Appending symbol: OFFICEX|BORDER2
Appending symbol: OFFICEX|BORDERX2
Appending symbol: OFFICEX|CENTER
Appending symbol: OFFICEX|CENTER2
Appending symbol: OFFICEX|CENTERX2
Appending symbol: OFFICEX|DASHDOT
Appending symbol: OFFICEX|DASHDOT2
Appending symbol: OFFICEX|DASHDOTX2
Appending symbol: OFFICEX|DASHED
Appending symbol: OFFICEX|DASHED2
Appending symbol: OFFICEX|DASHEDX2
Appending symbol: OFFICEX|DIVIDE
Appending symbol: OFFICEX|DIVIDE2
Appending symbol: OFFICEX|DIVIDEX2
Appending symbol: OFFICEX|DOT
Appending symbol: OFFICEX|DOT2
Appending symbol: OFFICEX|DOTX2
Appending symbol: OFFICEX|HIDDEN
Appending symbol: OFFICEX|HIDDEN2
Appending symbol: OFFICEX|HIDDENX2
Appending symbol: OFFICEX|PHANTOM
Appending symbol: OFFICEX|PHANTOM2
Appending symbol: OFFICEX|PHANTOMX2
Linetype update complete.

Update text style symbol table:
Appending symbol: OFFICEX|STANDARD
Appending symbol: OFFICEX|ROMANS
Appending symbol: OFFICEX|COMPLEX
Appending symbol: OFFICEX|SIMPLEX
Text style update complete.

Update registered application symbol table:
Registered application update complete.

Update dimension style symbol table:
Appending symbol: OFFICEX|ST1
Appending symbol: OFFICEX|ST2
Appending symbol: OFFICEX|AR
Dimension style update complete.
OFFICEX loaded.

Managing *Xref* Drawings

Because of the external reference feature, the contents of a drawing can be stored in multiple drawing files and directories. In other words, if a drawing has *Xrefs*, the drawing is actually composed of several drawings, each possibly located in a different directory. This means that special procedures to handle drawings linked in external reference partnerships should be considered when drawings are to be backed up or sent to clients. Four possible solutions to consider are listed here:

1. Create a project name for the drawing with the *PROJECTNAME* variable. The project name is stored in the drawing file. Notify the client so the client can create the same project name on his or her system (with the *Files* tab of the *Preferences* dialog box) to store the drawings and *Xrefs* (see *PRO-JECTNAME*).

2. Modify the current drawing's path to the external reference drawing so both are stored in the same directory; then archive them together.

3. Archive the directory structure of the external reference drawing along with the drawing which references it.

4. Make the external reference drawing a permanent part of the current drawing with the *Bind* option of the *Xref* command prior to archiving. This option is preferred when a finished drawing set is sent to a client.

CHAPTER EXERCISES

1. *Xref Attach*

    Use the **SLOTPLATE2** drawing that you modified last (with *Stretch*) in Chapter 9 Exercises. The slot plate is to be manufactured by a stamping process. Your job is to nest as many pieces as possible within the largest stock sheet size of 30" x 20" that the press will handle.

    A. *Open* the **SLOTPLATE2** drawing and use the *Base* command to specify a basepoint at the plate's lower-left corner. *Save* the drawing.

    B. Begin a *New* drawing using the **DSHEET** prototype and assign the name **SLOTNEST**. Draw the boundary of the stock sheet (30" x 20"). *Xref Attach* the **SLOTPLATE2** into the sheet drawing multiple times as shown in Figure 30-41 to determine the optimum nesting pattern for the slot plate and to minimize wasted material. The slot plate can be rotated to any angle but cannot be scaled. Can you fit 12 pieces within the sheet stock? *Save* the drawing.

Figure 30-41

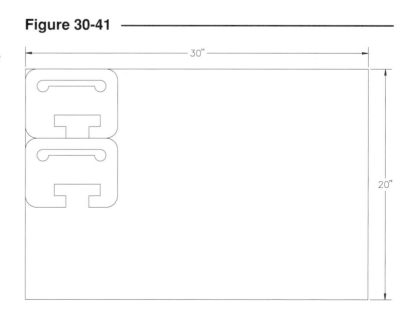

C. In the final test of the piece, it was determined that a symmetrical orientation of the "T" slot in the bottom would allow for a simplified assembly. *Open* the **SLOTPLATE2** drawing and use *Stretch* to center the "T" slot about the vertical axis as shown in Figure 30-42. *Save* the new change.

Figure 30-42 ─────────────

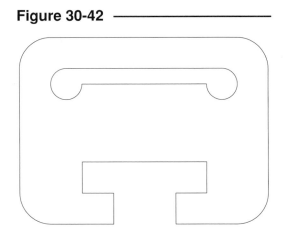

D. *Open* the **SLOTNEST** drawing. Is the design change reflected in the nested pieces?

2. *Attach, Detach, Xbind, VISRETAIN*

As the interior design consultant for an architectural firm, you are required to place furniture in an efficiency apartment. You can use some of the furniture drawings (*Blocks*) that you specified for a previous job, the office drawing, and *Insert* them into the apartment along with some new *Blocks* that you create.

A. Use the **CSHEET** as a *template* and set *Architectural Units* and set *Limits* of **48′** x **36′**. (Scale factor is 24 and plot scale is 1/2″=1′ or 1=24). Assign the name **INTERIOR**. *Xref* the **OFFICE** drawing that you created in Chapter 21 Exercises. Examine the layers (using the *Layer Control* drop-down list) and list (*?*) the *Blocks*. *Freeze* the **TEXT** layer. Use *Xbind* to bring the **CHAIR**, **DESK**, and **TABLE** *Blocks* and the **FURNITURE** *Layer* into the INTERIOR drawing. *Detach* the **OFFICE** drawing. Then *Rename* the new *Blocks* and the new *Layer* to the original names (without the **OFFICE0** prefix). In the INTERIOR drawing, create a new *Block* called **BED**. Use measurements for a queen size (60″ x 80″). *Save* the INTERIOR drawing.

B. *Xref* the **EFF-APT2** drawing (from Chapter 18 Exercises) into the INTERIOR.DWG. Change the *Layer* visibility to turn *Off* the **EFF-APT2 | TEXT** layer. *Insert* the furniture *Blocks* into the apartment on *Layer* **FURNITURE**. Lay out the apartment as you choose; however, your design should include at least one insertion of each of the furniture *Blocks* as shown in Figure 30-43. When you are finished with the design, *Save* the INTERIOR drawing.

Figure 30-43 ─────────────

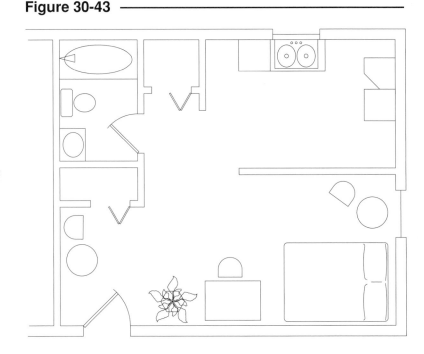

C. Assuming you are now the architect, a design change is requested by the client. In order to meet the fire code, the entry doorway must be moved farther from the end of the hall within the row of several apartments. *Open* the **EFF-APT2** drawing and use *Stretch* to relocate the entrance **8'** to the right as shown in Figure 30-44. *Save* the EFFAPT2 drawing.

Figure 30-44

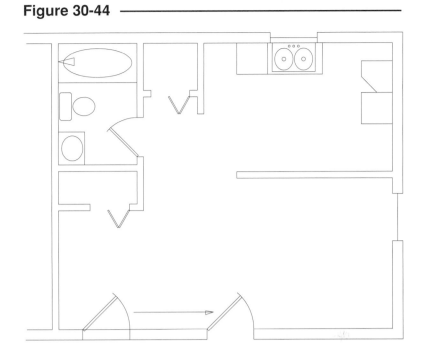

D. Next (as the interior design consultant), *Open* the **INTERIOR** drawing. Notice that the design change made by the architect is reflected in your INTERIOR drawing. Does the text appear in the drawing? *Freeze* the **TEXT** layer again, but this time change the *VISRETAIN* variable to **1**. Make the necessary alterations to the interior layout to accommodate the new entry location. *Save* the INTERIOR drawing.

3. *Xref Attach*

As project manager for a mechanical engineering team, you are responsible for coordinating an assembly composed of 5 parts. The parts are being designed by several people on the team. You are to create a new drawing and *Xref* each part, check for correct construction and assembly of the parts, and make the final plot.

A. The first step is to create two new parts (as <u>separate</u> drawings) named **SLEEVE** and **SHAFT**, as shown in Figure 30-45. Use appropriate drawing setup and layering techniques. In each case, draw the two views on <u>separate layers</u> (so that appearance of each view can be controlled by layer visibility). Use the *Base* command to specify an insertion point at the right end of the rectangular views at the centerline.

Figure 30-45

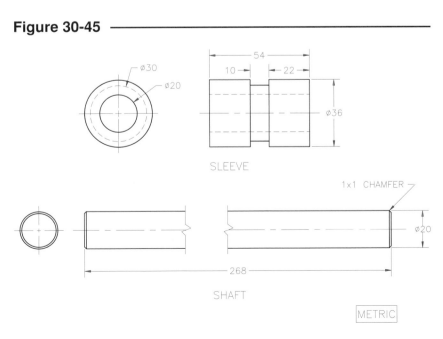

B. You will eventually *Xref* the drawings **SHAFT**, **SLEEVE**, **PUL-SEC** (from Chapter 26 Exercises), **SADDLE** (from Chapter 24 Exercises), and **BOLT** (from Chapter 10 Exercises) to achieve the assembly as shown in Figure 30-46. Before creating the assembly, *Open* each of the drawings and use *Chprop* to move the desired view (with all related geometry, no dimensions) to a new layer. You may also want to set a *Base* point for each drawing.

Figure 30-46

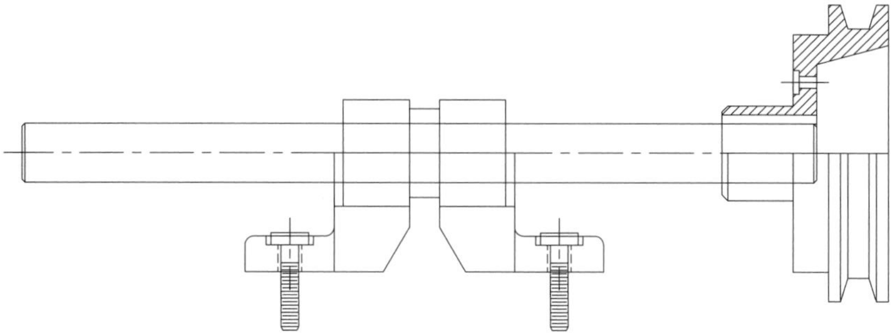

C. Finally, begin a *New* drawing (or use a *Template*) and assign the name **ASSY**. Set appropriate *Limits*. Make a *Layer* called **Xref** and set it *Current*. *Xref* the drawings to complete the assembly. The **BOLT** drawing must be scaled to acquire a diameter of **6mm**. Use the *Layer Control* drop-down list to achieve the desired visibility for the *Xrefs*. Set *VISRETAIN* to **1**. Make a *Plot* to scale. *Save* the **ASSY** drawing.

4. *Xref Overlay*

A. Assume you are now the project coordinator/checker for the assembly. (If it is possible in your lab or office, exchange the SHAFT drawings [by diskette or by network] with another person so that you can perform this step for each other.) Begin a *New* drawing and make a *Layer* called **CHECK**. Next, *Xref Overlay* the **SHAFT** drawing (of your partner). On the CHECK layer, insert some *text* by any method and a dimension similar to that shown in Figure 30-47. *Save* the CHECK drawing. (When completed, exchange the CHECK and SHAFT drawings back.)

Figure 30-47

3 X 3 NECK
(ADD FEATURE)

18

B. *Open* the **SHAFT** drawing. In order to see the notes from the checker, *Xref Overlay* the **CHECK** drawing. The instructions for the change in design should appear. (Notice that the *Overlay* option allows a circular *Xref* since SHAFT references CHECK and CHECK references SHAFT.) Make the changes by adding the 3 x 3 NECK to the geometry (Fig. 30-48). *Save* the **SHAFT** drawing (with the *Overlay*).

Figure 30-48

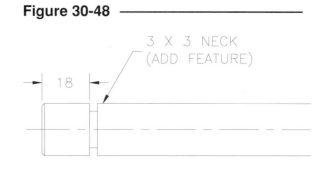

3 X 3 NECK
(ADD FEATURE)

18

C. Now *Open* the **ASSY** drawing. The most recent *Xrefs* are automatically loaded, so the design change should appear (Fig. 30-49). Why don't the checker notes (from CHECK drawing) appear? Finally, *Save* the drawing and make another plot.

Figure 30-49

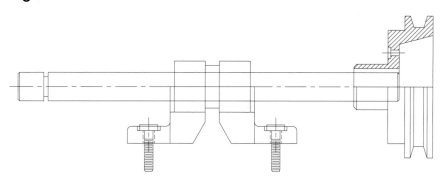

5. *XCLIP, XLOADCTL, INDEXCTL*

A. *Open* the **ASESMP** drawing from the sample drawings provided with AutoCAD (if AutoCAD was installed with the "Full" installation and accepting all the other defaults, ASESMP.DWG is located in the C:\Program Files\AutoCAD R14\Sample directory). Use *SaveAs* to change the name to **SMP-OFF** and change the location of the drawing (using *SaveAs*) to your working directory.

B. Begin a *New* drawing. Use *Save* and assign the name **SMP-CLIP**. Next, use *Xref* and *Attach* the **SMP-OFF** drawing. Use an *Insertion point* of **0,0**, accept the defaults for scale factors and rotation angle, then *Zoom All*. The entire **SMP-OFF** drawing should appear.

C. Create a clipping boundary using the *Xclip* command. Make a *Rectangular* boundary around room 101 (upper-left corner office). *Zoom* in to display only the displayed office, 101. Now, use the *Layer* command or *Layer Control* drop-down list to *Freeze* the following layers:

Figure 30-50

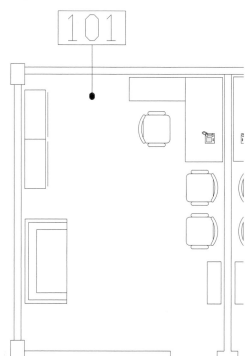

    Smp-off l border
    Smp-off l cpu
    Smp-off l door
    Smp-off l kitchen
    Smp-off l fixtures

Your drawing should look like Figure 30-50.

D. Use *XLOADCTL* or use the *Performance* tab of the *Preferences* dialog box to set the variable to 0 or *Disabled*. *Save* the drawing (SMP-CLIP).

E. Now *Open* **SMP-CLIP** again and use your stopwatch or count seconds to time how long it takes to load the drawing and *Xref* drawing. At this point, demand loading is disabled and layer and spatial indexing are not being utilized, so the drawing and *Xref* loading time should be at its maximum for your system.

F.   Next, *Open* the **SMP-OFF** drawing. Use the *INDEXCTL* variable or the *SaveAs* command with the *Options…* button to set indexing on for both layer and spatial indexes. *Save* the SMP-OFF drawing.

G.   Open the **SMP-CLIP** drawing again. Use *XLOADCTL* or use the *Performance* tab of the *Preferences* dialog box to set the variable to **1** or *Enabled*. This action turns on demand loading, and you should now take advantage of the spatial and layer indexing. *Save* the drawing.

H.   *Open* the **SMP-CLIP** drawing again and time the loading process. You should experience a slight increase in loading time now that demand loading is enabled. This drawing is relatively small, so a performance increase may not be noticeable if you are using a fast system; however, when using large (file size) *Xrefs* or multiple *Xrefs*, this feature can offer a big performance increase.

31

OBJECT LINKING AND EMBEDDING (OLE)

Chapter Objectives

After completing this chapter you should:

1. be able to use AutoCAD's Object Linking and Embedding (OLE) feature to copy information from a document (in a software program such as AutoCAD, Microsoft Word, or Microsoft Excel) and temporarily store it in the Clipboard;

2. be able to cut information from a document and temporarily store it in the Clipboard;

3. be able to paste information from the Clipboard into a drawing;

4. be able to embed information from another software program into an AutoCAD drawing;

5. be able to link information from another software program into an AutoCAD drawing;

6. be able to manage OLE information in a drawing.

CONCEPTS

The term "graphics communication tool" is often used to describe AutoCAD's many advanced graphic capabilities. For many years people have been communicating design and conceptual information using AutoCAD's 2D and 3D graphic features. The Microsoft Windows/NT operating platforms have now enabled AutoCAD users to extend their graphic communication abilities within, and beyond, the Drawing Editor.

You can now easily communicate 2D and 3D AutoCAD designs with other Windows applications and benefit from the Windows Object Linking and Embedding (OLE) feature. The standard OLE features of <u>linking</u> and <u>embedding</u> operate just as smoothly and easily from within AutoCAD as they have for some years in other Windows software such as Microsoft® (MS) Word, MS Excel, and others.

AutoCAD's support of the OLE features allows you to include data and information created in an application, such as MS Word, within your AutoCAD drawing. You can also take AutoCAD drawing information and use it within MS Word documents, MS Excel spreadsheets, and so on. This feature works best if all the applications using or sharing the common information support the OLE feature. But it is not absolutely necessary for a software application to support the OLE feature for that application to display AutoCAD drawing information by way of the copy and paste functions available within MS Windows.

AutoCAD's OLE features utilize the standard MS Windows *Copy, Cut,* and *Paste* commands common to all Windows programs. OLE, however, extends these Windows commands so information copied from one application (such as a MS Word document) and pasted into another application (like an AutoCAD drawing) can be *linked* back to the original source document, also known as the <u>server</u>. The source document is a separate document; it is only displayed within the <u>client</u> document (the AutoCAD drawing in this example).

Object Linking and Embedding enables you to create a bill of materials within, for example, MS Word, which is designed specifically for word processing, then paste it into an AutoCAD drawing. First, create and format the bill of materials within MS Word (Fig. 31-1). Next, *Copy* the formatted text onto the Clipboard and then *Paste* it into the AutoCAD drawing. In AutoCAD (the client), it will have the same appearance that it has in the MS Word (the source) document.

Figure 31-1

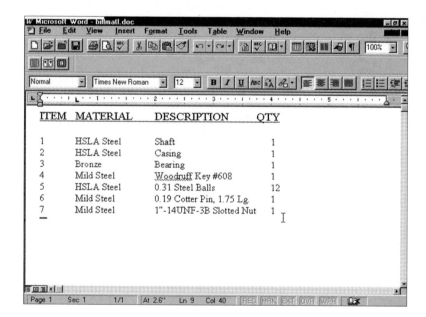

If the bill of materials is pasted into the AutoCAD drawing as a <u>linked</u> object, then the text document that appears in AutoCAD is simply a picture of the original source document (Fig. 31-2). If the source document is changed within MS Word (its original software application), then you can cause that source to update automatically or "manually" in AutoCAD, depending on the update options you set within the OLE *Links* dialog box. While the drawing file is open, you can also double-click on the linked object's picture in AutoCAD to cause the source application to be opened (in this case, MS Word) with the document loaded and ready for editing. Upon saving and closing the bill of materials document, the

Figure 31-2

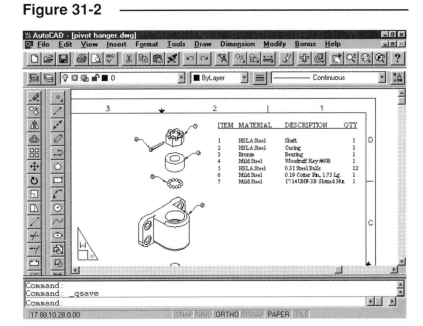

"picture" of the bill of materials in the AutoCAD drawing is updated automatically. You can also select *Update* from the *File* pull-down menu of the source application program to update the linked OLE object dynamically while editing it.

There are two strategies for pasting information from other applications into your AutoCAD drawing:

1. <u>Embed the object when pasting</u>
    Embedding is similar to using the AutoCAD *Insert* command. When embedded, the object becomes a permanent part of the drawing. If the source application that created the object supports OLE, you can double-click on the embedded object and the source application is started so you can edit the contents of the object for the insertion of the (embedded) objects in AutoCAD only. The original source document remains unchanged.

2. <u>Link the object when pasting</u>
    Linking is similar to the AutoCAD *Xref* command. When the object is linked, it is not part of the drawing; it is only displayed in the drawing. When you double-click on the linked object, the source application is started with the original document loaded and ready for editing. Any changes made to the original source document are reflected in the drawing automatically or upon a manual update, depending on your update setting in the OLE *Links* dialog box. Similar to an *Xref,* when the original document is edited, the updated version appears in the client application.

Four basic AutoCAD commands are used to copy, cut, link, or embed objects:

OLE Feature	AutoCAD Command	OLE Result
Copy	*Copyclip*	Embedded or linked
Cut	*Cutclip*	Embedded
Paste	*Pasteclip*	Embedded
Paste Link	*Pastespec*	Embedded or linked

The main difference between linked and embedded is how the information is stored—with the client (application that "displays" the document) or server (source application that created the document).

The Windows Clipboard

MS Windows manages the OLE features. The Windows Clipboard is a key mechanism used throughout the process of linking and embedding information in an AutoCAD drawing. The Clipboard is a Windows feature which stores information temporarily in memory so that it can be utilized (or pasted) into another application. This enables you to create the information only once and then share it with other documents where it is needed.

For example, assume the bill of materials mentioned earlier was created, spell checked, and formatted in MS Word, then selected and copied to the common memory area called the Clipboard. Then the information was pasted (similar to the AutoCAD *Insert* command) into an AutoCAD drawing with the same content and appearance it had in the original document. The same information could then be pasted (inserted) into a spreadsheet with the identical content and appearance it has in the AutoCAD drawing and in the original MS Word document that created it.

The Clipboard retains only the most recent information that was cut or copied from an application currently running in the Windows environment. When another piece of information (e.g., graphical objects or text) is cut or copied, it replaces the last group of information stored within the Clipboard's memory area.

The Clipboard's information can be saved to a file (normally with a .CLP extension) if you activate the Windows Clipboard utility program and manually save the information.

The options encountered when performing a paste vary depending upon the AutoCAD paste command being used (*Pasteclip, Pastespec*) and the type of object information being pasted from the Clipboard. The following table lists some common object types and options that are available when pasting objects into an AutoCAD drawing:

Object Type:	Paste Objects As:				
	AutoCAD Block	AutoCAD Text	Bitmap	Picture	Image
AutoCAD Vector Objects (*Lines, Arcs, Circles, Blocks, Plines*, etc.)	✔				
Other Vector Objects (primarily Windows metafile)	✔				
AutoCAD Text Objects (*Dtext, Mtext, Text*)	✔	✔		✔	✔
Formatted Text Objects (such as MS Word text or MS Excel spreadsheet)	✔	✔	✔	✔	✔

To summarize, first create the document (to be copied) in its source application (MS Word, for example). While the document is open, use *Copy* or similar OLE command in that application to copy the document contents (called an OLE object) to the Clipboard. Next, open the client application (AutoCAD, for example) and use *Paste, Paste Special,* or a similar OLE command to paste the OLE object. The OLE object appears in the client application just as it did in its original format (in the source application). Later, you can double-click on the OLE object to open the original (source) application and modify the object. If the object is linked, the original source document is updated; if the object is embedded, only the OLE object is changed, but the source document (file) is not changed.

OLE RELATED COMMANDS

AutoCAD OLE related commands covered in this chapter are:

Copyclip   Copies objects to the Clipboard
Copylink   Copies all objects in the current display to the Clipboard for linking to other OLE
               applications
Cutclip   Cuts (*Erases*) objects from the drawing and copies them to the Clipboard
Pasteclip   Inserts data from the Clipboard
Pastespec   Inserts data from the Clipboard and controls the format of the data
Insertobj   Inserts a linked or embedded object
Olelinks   Updates, changes, and cancels existing OLE links

Most OLE related commands are available from the *Edit* pull-down menu (Fig. 31-3). The *Insertobj* command is accessible from the *Insert* pull-down menu.

Figure 31-3

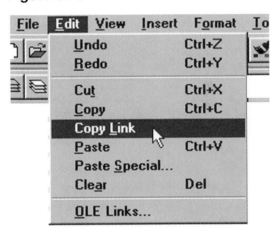

COPYCLIP	**Pull-down Menu**	**COMMAND (TYPE)**	**ALIAS (TYPE)**	**Short-cut**	**Screen (side) Menu**	**Tablet Menu**
	Edit *Copy*	*COPYCLIP*	...	*Ctrl+C*	*EDIT* *Copyclip*	*T,14*

You can use the *Copyclip* command to select the objects from the current drawing to be placed on the Clipboard. *Copyclip* prompts you to "Select objects:." You can use any of the normal AutoCAD selection options (like *Window, Crossing, Fence, Group,* etc.). You can also pre-select the objects (known as Noun-Verb selection) you want placed on the Clipboard, and then issue the *Copyclip* command to avoid the "Select Objects:" prompt:

    Command: **copyclip**
    Select objects: **PICK** (select objects to be copied from the drawing and placed on the Clipboard)
    Select objects: **Enter**
    Command:

AutoCAD objects that are selected with *Copyclip* remain in the drawing. Copies of the objects are temporarily stored on the Clipboard.

COPYLINK

Pull-down Menu	COMMAND (TYPE)	ALIAS (TYPE)	Short-cut	Screen (side) Menu	Tablet Menu
Edit Copy Link	COPYLINK	...	...	EDIT Copylink	...

The *Copylink* command is used to copy the entire contents of the current display in the Drawing Editor. You are not prompted to select individual objects. Whatever is currently displayed in the Drawing Editor is copied to the Clipboard. This information can then be pasted into another application as linked or embedded objects.

```
Command: copylink
Command:
```

No change is made to the objects in the Drawing Editor. Copies of the objects are temporarily stored in the Clipboard.

CUTCLIP

Pull-down Menu	COMMAND (TYPE)	ALIAS (TYPE)	Short-cut	Screen (side) Menu	Tablet Menu
Edit Cut	CUTCLIP	...	Ctrl+X	EDIT Cutclip	T,13

The *Cutclip* command places selected information (objects) onto the Clipboard. However, unlike *Copyclip,* it <u>deletes</u> (*Erases*) the selected objects from the drawing. If desired, you can use the AutoCAD *Oops* command to "unerase" the objects deleted from the drawing by *Cutclip.* If you use *Oops,* you must use it before you delete any other objects from the drawing since *Oops* brings back only the last set of *Erased* objects.

```
Command: cutclip
Select objects: PICK (select objects to be removed from the drawing and placed on the Clipboard)
Select objects: Enter
Command:
```

This command is typically used only when you do not need the selected items in the drawing any longer but wish to use them in another application. In this case, *Cutclip* allows you to clear (*Erase*) them out of the current drawing and then paste them into another application. For example, you may want to open a second AutoCAD session, then cut objects from one Drawing Editor and paste them into the second AutoCAD session. This action helps to reduce the current drawing file size and also eliminates the need to draw the items again for another drawing. Since the objects temporarily remain on the Clipboard, the deleted items can be pasted into another drawing or document where they are needed.

PASTECLIP

Pull-down Menu	COMMAND (TYPE)	ALIAS (TYPE)	Short-cut	Screen (side) Menu	Tablet Menu
Edit Paste	PASTECLIP	...	Ctrl+V	EDIT Pasteclp	U,13

You can copy information from another application and paste it into your drawing with AutoCAD's *Pasteclip* command. The pasted object is <u>embedded</u> within the AutoCAD drawing. This concept is very similar to the AutoCAD *Insert* command. The embedded information becomes a permanent part of the drawing like most other objects.

First, you must use *Copy* or similar OLE command from another application (MS Word, MS Excel, AutoCAD, etc.) to bring objects onto the Clipboard. Next, use the *Pasteclip* command in AutoCAD:

    Command: **pasteclip**
    Command:

If the embedded object's source application supports OLE, the object is displayed in the original application's format. The embedded object can be edited using the original application that created it by double-clicking on the object. This <u>is not true,</u> however, if the source of the object is another AutoCAD drawing.

Although AutoCAD supports OLE, embedded AutoCAD objects cannot be edited in this manner (by double-clicking to start the original application) because the pasted objects become *Block* references in the drawing. Embedded AutoCAD objects must be edited like any other internally referenced block. You can *Explode* them or update the *Block* reference (see Chapter 21, Blocks).

Embedded objects are <u>copies</u> of the original information. They have <u>no connection,</u> or link, to the original source document in which they were created. Therefore, editing the original objects using the source application does not update the embedded (copied) object. Alternately, if you use the *Pastespec* command (discussed next), you have the choice of embedding or linking the objects.

PASTESPEC

Pull-down Menu	COMMAND (TYPE)	ALIAS (TYPE)	Short-cut	Screen (side) Menu	Tablet Menu
Edit Paste special...	PASTESPEC	...	...	EDIT Pastespe	...

The *Pastespec* command gives you the choice of embedding or linking objects that are pasted into the drawing. *Pastespec* produces the *Paste Special* dialog box (Fig. 31-4). Your selection of *Paste* or *Paste Link* determines if the objects from the Clipboard are <u>embedded</u> or <u>linked</u>, respectively.

If the information copied from an application that supports OLE is <u>linked</u> within your AutoCAD drawing, a "picture" of the original document is displayed that links (or references) the original source document that was used to create it.

Figure 31-4

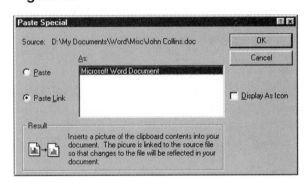

To use *Pastespec,* first you must use *Copy* or similar OLE command from another application (MS Word, MS Excel, AutoCAD, etc.) to bring objects onto the Clipboard. Next, use the *Pastespec* command in AutoCAD:

    Command: **pastespec** (the *Paste Special* dialog box appears)
    Command:

Displayed near the top of the *Paste Special* dialog box is the *Source:* application which created the object that is currently contained on the Clipboard (see previous figure, 31-4). The two radio buttons on the left allow you to embed (*Paste*) or link (*Paste Link*) the objects.

The list of (Paste) *As:* options varies depending on the type of information contained on the Clipboard and the current *Paste/Paste Link* radio button you have selected. If the current information on the Clipboard is from an AutoCAD drawing, the *Paste Link* option is <u>not</u> available (grayed out) (Fig. 31-5). All of the options in this dialog box are described in detail next.

Figure 31-5

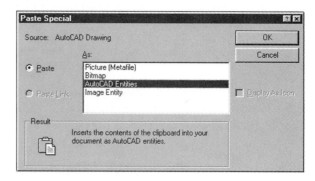

Paste

When the *Paste* radio button is selected, the object becomes <u>embedded</u> within the drawing. You can select the type of object to be inserted in the drawing by making a selection from the (Paste) *As:* list (see Fig. 31-5). The list of object types varies depending on the source of the objects on the Clipboard.

Paste Link

Selecting the *Paste Link* button causes AutoCAD to display a "picture" of the object in the drawing. The information displayed is not actually copied into the drawing; it is <u>linked</u> back to the source document from which it was originally copied. This option is available only when the information contained on the Clipboard was created with an application that supports OLE. When *Paste Link* is selected, only one choice appears in the (Paste) *As:* list—the source application (see Fig. 31-4).

Result

The *Result* box at the bottom of the dialog box describes the current object on the Clipboard. It also indicates how the pasted objects are managed in the drawing based on your selections within the dialog box (*Paste/Paste Link* and *Paste As:* list). This is a very helpful feature that dynamically explains your options as they are selected.

Display As Icon

You can display the source application's icon instead of displaying the actual information pasted from the Clipboard by choosing *Display As Icon* (Fig. 31-6). Since only the application's icon is visible, you must double-click on it to view the linked or embedded information. This icon is a <u>shortcut</u> that points to the location of the information pasted from the Clipboard. This can be useful in a situation where the linked information is relevant to the project depicted in your drawing, but it is not necessary to display that information directly in the drawing.

Figure 31-6

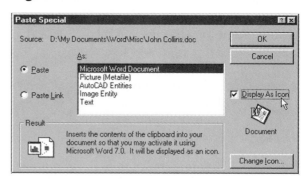

For example, you may be displaying a *Block* of a pump in the drawing. Specific text information such as the pump's specifications, manufacturer's model number, cost, and ordering information could be included within a Word document which is pasted into your drawing and displayed as an icon. The information is important for reference, but it is not necessary to display the information directly in your drawing.

If the information represented by the icon is embedded, double-clicking on the icon activates the application used to create it and displays the contents of the information that was pasted into the drawing in its original format. Changes made to the information as a result of this action are changed for the embedded information only and <u>are not reflected</u> in the source (original) document.

If the *Icon* information is linked, double-clicking on the icon activates the application used to create it and in turn displays <u>the original file</u> that contains the information. The information referenced by the pasted icon becomes highlighted in the source document. Once the source document is saved and closed, the updated version is linked to the icon displayed within your drawing. In this case, the source document itself has actually been changed.

Change Icon
Whenever you have selected the *Display As Icon* checkbox, a preview of the application's default icon is displayed directly beneath the checkbox area. A new button also appears labeled *Change Icon*. This allows you to change the icon if you so desire.

INSERTOBJ

Pull-down Menu	COMMAND (TYPE)	ALIAS (TYPE)	Short-cut	Screen (side) Menu	Tablet Menu
Insert *OLE Object...*	*INSERTOBJ*	...	...	*INSERT* *Insertob*	*T,1*

Insertobj also allows you to paste specific information into your drawing as either a <u>linked</u> or <u>embedded</u> object. It functions in much the same way as the *Pastespec* command. The <u>key difference</u> between *Insertobj* and *Pastespec* is that *Insertobj* allows you to paste a <u>file</u> without having the application or document currently open and a particular portion of it selected. Because you are not copying objects from the Clipboard, you do not have to use *Copy* or similar OLE command before using *Insertobj*. *Insertobj* produces the *Insert Object* dialog box (Fig. 31-7).

Figure 31-7

```
Command: insertobj (produces the Insert Object
dialog box)
Command:
```

Create New
The options in this command afford you some flexibility. If the file doesn't exist, you can create a new file by selecting *Create New*, then the application you want to use to create it. The applications available to you (registered on your system) are listed in the *Object Type* listing. The selected application is opened when you select *OK* so that you can create the new document. When you are finished creating the new document and you close the application, the new document (object) is <u>embedded</u> into your drawing.

Create from File
When the *Create from File* radio button is selected (Fig. 31-8), you can select the *Browse* button to look for files available in your available drives and folders. After selecting a specific file, you can paste it into your drawing as an <u>embedded</u> or <u>linked</u> object. If the object is to be linked, the display of the linked object may be limited to an icon based on the file type and the OLE capabilities of the source application.

Figure 31-8

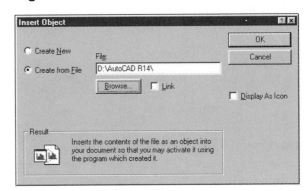

MANAGING OLE OBJECTS

OLELINKS

Pull-down Menu	COMMAND (TYPE)	ALIAS (TYPE)	Short-cut	Screen (side) Menu	Tablet Menu
Edit *OLE Links...*	*OLELINKS*	...	...	*EDIT* *OLElinks*	...

When you have *Linked* an object within an AutoCAD drawing, it can be automatically or manually updated whenever changes are made to the original (source) document. The *Olelinks* command activates the *Links* dialog box with options for the maintenance of object links. If no objects are currently *Linked* within the drawing, the *OLE Links...* option in the *Edit* menu is grayed out and typing *Olelinks* invokes no response from AutoCAD.

> Command: **olelinks** (activates the *Links* dialog box if links exist in the current drawing)
> Command:

The following are features included in the *Links* dialog box (Fig. 31-9).

Figure 31-9

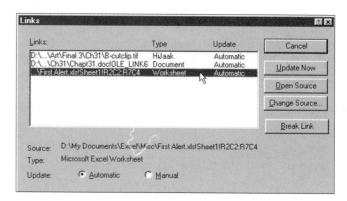

Links, *Type*, and *Update* Columns
The list box in the main section of the dialog provides a listing of all linked objects in the drawing. The *Links* column displays the source information (file name and path). The *Type* is the source application type. The *Update* column lists the current status of the update option.

Source
Found in the lower-left corner of the dialog, this description specifies the drive, path, application, and additional information such as object type or selection area, depending on the source application.

Type
Also listed in the lower-left corner of the dialog is the *Type* of link for the highlighted item. This is the application-specific name (source application) of the currently selected linked object. The application-specific name is sometimes followed by an OLE link number assigned to the object by AutoCAD.

Update
Two *Update* radio buttons are located at the bottom of the dialog. The *Automatic* option causes updating of the linked object whenever its source document is modified while the drawing is open. Setting the *Update* function to *Manual* prevents the link from updating when revisions are made to the source. You can use *Update Now* or use *Olelinks* to update the link when *Manual* is selected.

Update Now
This button forces the links selected from the list to be updated. When you close the dialog box, listed data reflects the newly updated status of the source object.

Open Source
This option opens the currently selected link with the source application so it can be modified. The linked information is highlighted when the source application is opened.

Break Link
Use this option to sever the link of the object to its source file and automatically convert it to a "Static OLE object" or Windows metafile object. It can no longer be updated if the source is changed. The object can no longer be edited with the source application or with AutoCAD commands.

For example, if a spreadsheet displaying a bill of materials for the drawing is created in MS Excel and then linked inside the drawing, it can be modified at any time with MS Excel simply by double-clicking on the spreadsheet picture displayed on the drawing. If, however, the link is broken, the spreadsheet picture is converted to a metafile picture object and can no longer be edited using MS Excel. Static OLE objects cannot be edited or removed with AutoCAD commands. To remove a static OLE object, right-click on the object and select *Cut* from the cursor menu that appears.

MOUSE-ACTIVATED OLE EDITING FEATURES

You can access two additional editing features for embedded and linked objects by locating the mouse pointer on an OLE object and selecting it.

Resizing and Repositioning an OLE Object

By clicking once on an OLE object you can activate the object for editing its size or location. After selecting the linked or embedded object once, you will see "handles" around the boundary of the object (Fig. 31-10).

Figure 31-10

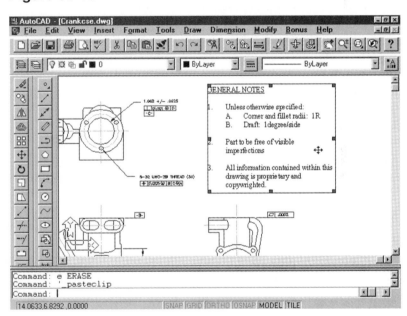

Selecting a midpoint handle allows you to stretch the object in one direction at a time, vertically or horizontally. A two-arrow cursor appears when the pointer moves near a midpoint or corner handle. Selecting a corner handle scales the object proportionally, allowing you to maintain its aspect (width-to-height) ratio. The latter is generally the preferred method since stretching the object in only one direction changes the proportions of the text or image, often making it appear distorted or unreadable.

You can restore a resized or distorted OLE object to its original size and proportion if needed. Do this by right-clicking on the OLE object, then choose *Cut,* and then *Paste* it back into the drawing again. This action restores the original size and aspect ratio (see Right-Click Editing Menu).

You can also reposition an OLE object in the drawing by clicking once and moving it to a new location. When you left-click on the object once, simply place the cursor anywhere inside the boundary of the object so a four-arrow cursor appears (see Fig. 31-10). Press the left mouse to move (reposition) the OLE object in the drawing, much like you would use *Real Time Pan* to reposition an entire drawing.

Right-Click Editing Menu

Cut, Copy, Clear, Undo

When you place the cursor inside the boundary of an OLE object and right-click on the mouse, the OLE cursor menu appears (Fig. 31-11). You can select the standard OLE *Cut* and *Copy* commands through this menu to cut or copy the OLE object to the Clipboard. You can also *Clear* the OLE object, which deletes (*Erases*) it from the drawing. *Undo* undoes the last action taken on the object.

Figure 31-11

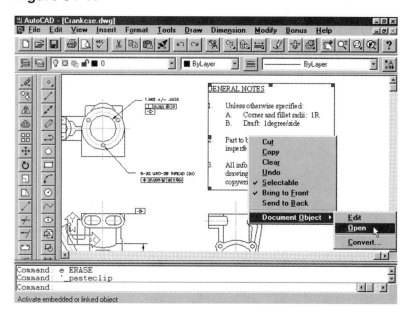

Selectable

An OLE pasted object is *Selectable* by default. When the object is *Selectable* (a check appears in the cursor menu before the word *Selectable*), you can PICK it (place the cursor on the object and left-click) so its size/position handles appear. An <u>OLE object cannot be included in an AutoCAD selection set</u> nor can it be edited by typical AutoCAD commands, even when it is *Selectable*; however, it <u>can</u> be *Erased*. You can remove the check to make the OLE object unselectable. When an OLE object is not selectable, you cannot resize or reposition it by left-clicking it. You can right-click on it to produce the OLE cursor menu to make it *Selectable* again.

Bring to Front, Send to Back

The drawing order in which the OLE object is displayed relative to other objects in the drawing can be controlled through the cursor menu by using *Bring to Front* or *Send to Back*. These two options accomplish the same action as using the *Draworder* command options (see Chapters 26 and 32 for details on the *Draworder* command).

(Object Type) Object>

The option description (*Object Type*) that appears at the bottom of the cursor menu differs based on the source application of the selected OLE object. Selecting this option produces a cascading menu so you can *Edit, Open,* or *Convert...* the OLE object. Selecting *Edit* or *Open* starts the source application so the document can be edited in the application in which it was created. The *Convert...* option produces the *Convert* dialog box (Fig. 31-12) which allows you to *Convert to:* or *Activate as:*. The options available in this dialog differ based on the selected OLE object type. Figure 31-12 displays options for converting a Microsoft Word document object.

Figure 31-12

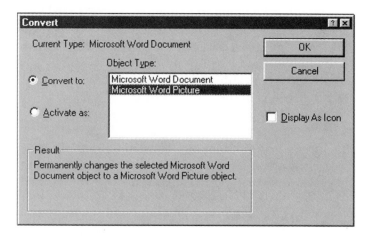

CHAPTER EXERCISES

1. *Copyclip, Pasteclip*

   In this exercise, you will create a block reference and a triangle in one drawing, copy them to the Windows Clipboard, and then paste them into a another drawing.

   A. Begin a *New* drawing and select *Start from Scratch, English* default settings. Draw a *Circle* of radius **2**. Draw four smaller *Circles* inside the first circle with radii of **.25**, as shown in Figure 31-13. Next, make a *Block* containing all five circles and name the *Block* **C5**. Make sure *Retain Objects* is <u>not</u> checked if you use the *Block Definition* dialog box. Now *Insert Block* **C5** back into your drawing. Draw a triangle next to the **C5** *Block* using the *Line* command. Your drawing should appear as that shown in Figure 31-13.

   Figure 31-13 ——————

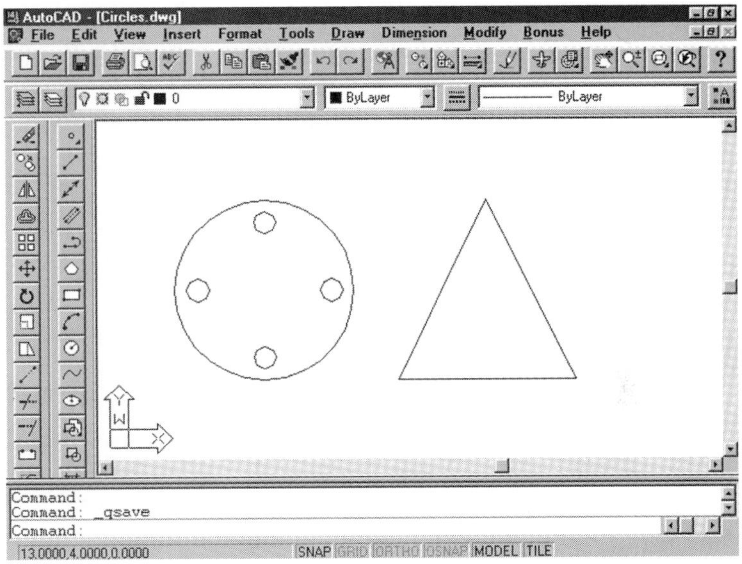

   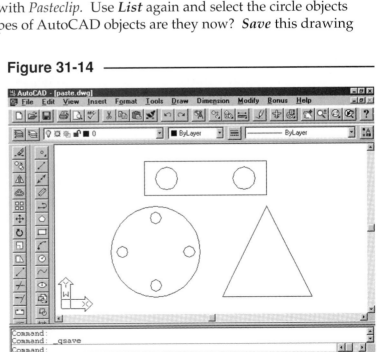

   B. Assume you need to use these same objects in another drawing. Use *Copyclip* to copy all of the objects to the Windows Clipboard. A copy of the block **C5** and the three lines that make up the triangle reside on the Clipboard. *Save* the current drawing as **CIRCLES**.

   C. Begin a *New* drawing. Use the *Pasteclip* command to paste the information from the Clipboard into the new drawing. Use *List*, select the object you just inserted with *Pasteclip*, and verify that it is a *Block* reference with an AutoCAD-generated name.

   D. *Explode* the *Block* that was inserted with *Pasteclip*. Use *List* again and select the circle objects and the triangle segments. What types of AutoCAD objects are they now? *Save* this drawing as **PASTE**.

   E. *Open* the **CIRCLES** drawing and draw a *Rectangle* and two more *Circles* as shown in Figure 31-14. Then use *Cutclip* and select the *Rectangle* and two new *Circles*, then press **Enter** to finish the *Cutclip* command. These objects should now be deleted from the drawing and placed on the Clipboard. Next, *Open* the **PASTE** drawing and issue the *Pasteclip* command. What happens? *Save* the PASTE drawing.

   Figure 31-14 ——————

2. *Copy, Pasteclip*

For this exercise, you will write a short paragraph of General Notes to be used in an AutoCAD drawing. Since the notes are text objects, it is easier to use a word processor to write and format the notes, *Copy* them to the Clipboard, and then use *Pasteclip* in AutoCAD to insert the notes into your drawing.

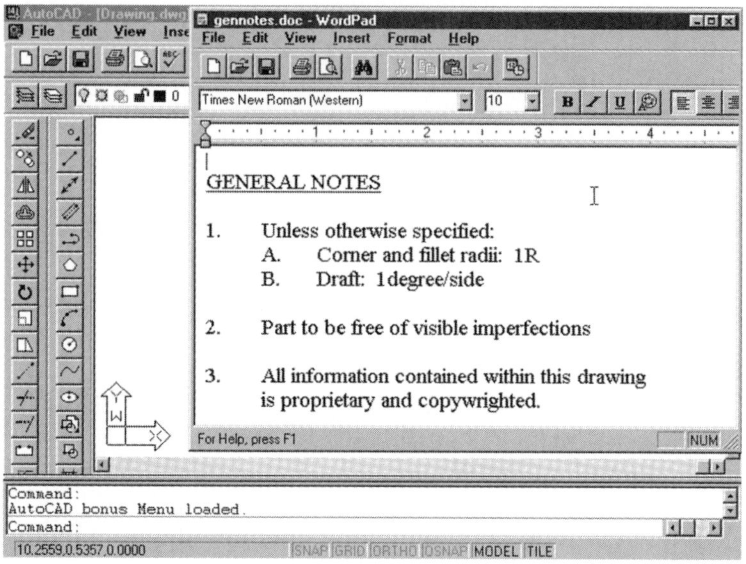

A. Begin a *New* drawing (***Start from Scratch, English*** default settings). Also, while AutoCAD is running, start **WordPad** and compose a short paragraph of General Notes similar to the note shown in Figure 31-15 (the word "copyrighted" is intentionally misspelled). After typing the notes, <u>highlight the entire block of text</u> and the select *Copy* from the *Edit* pull-down menu. This action copies the WordPad document onto the Clipboard.

Figure 31-15

B. Now return to the new drawing you started in step A (you can use the **Ctrl+Tab** key sequence to switch to AutoCAD). In AutoCAD, select *Pasteclip*. The information from the Clipboard (the General Notes copied previously from WordPad) is pasted into the top left portion of the drawing area. You can see the "handles" (blue boxes) that allow you to adjust the size of the pasted object. Select one of the <u>corners</u> of the General Notes border and drag it in or out to resize the note. Notice that the text within the pasted image changes proportionally as you resize the image from one of the corner resize buttons. Now select a resize button at one of the <u>midpoints</u> of the image. What happens to the text within the image?

C. The notes from the WordPad document are still stored within the current memory of the Clipboard. Select the *Mtext* command. When the *Multiline Text Editor* appears, do not enter in any text, but type a Ctrl+V (hold down the Control key and press the "v" key). Select the *OK* button in the *Multiline Text Editor.* What happened? Save this drawing as **PASTETXT**.

D. *Open* the **CIRCLES** drawing. Create a *New Layer,* assign the **red** color to the layer. Use *Ddchprop* to "move" the triangle objects (*Lines*) to the new layer. Next, use *Bhatch* to fill the triangle with the *Solid* pattern. The triangle color should be red.

E. Use *Pasteclip* in this drawing to paste the note contained on the Clipboard. Use the resize buttons to make the text block about the same size as the triangle. Now, place the mouse pointer on the pasted object (the General Notes) and press the **PICK** button once (this action selects the pasted note). Press and hold down the **PICK** button, and move the pasted object so it is on top of the triangle. With the pointer on the pasted note, **right-click** to produce the Cursor menu, then select *Send to Back.* Can you read the notes? If not, produce the cursor menu again and select *Bring to Front. Save* the CIRCLES drawing.

3. **Double-clicking on an Embedded Object**

   Assume that you printed the **PASTETXT** drawing and noticed the misspelled word. Follow this procedure to change the spelling of the word "copywrighted" to "copyrighted."

   A. *Open* the **PASTETXT** drawing. Point to the pasted (embedded) note object and then double-click on it using the mouse. What happens?

   B. Now correct the misspelled word in **WordPad**. Then select *Exit* from the *File* pull-down menu in WordPad. Next, view the notes in AutoCAD. What happened to the notes? *Save* the drawing.

4. *Copylink, Paste*

   If you want to copy the entire contents of the display to the Clipboard and then paste the objects into another application, you can accomplish this using *Copylink.*

   A. *Open* the **CIRCLES** drawing you created in Exercise 1. Select *Copy Link* from the *Edit* pull-down menu. The contents of the current display are copied onto the Clipboard.

   B. Next, open **WordPad**. Type in a line or two of text. Then select *Paste* from the *Edit* pull-down menu in WordPad. This action pastes the contents of the Clipboard (AutoCAD objects in this case) into the WordPad document. You can also choose *Paste Special* instead of *Paste* from the WordPad *Edit* pull-down to allow you to choose whether to embed or link the contents of the Clipboard into this document. (The WordPad text is not an AutoCAD object and therefore does not appear in WordPad.)

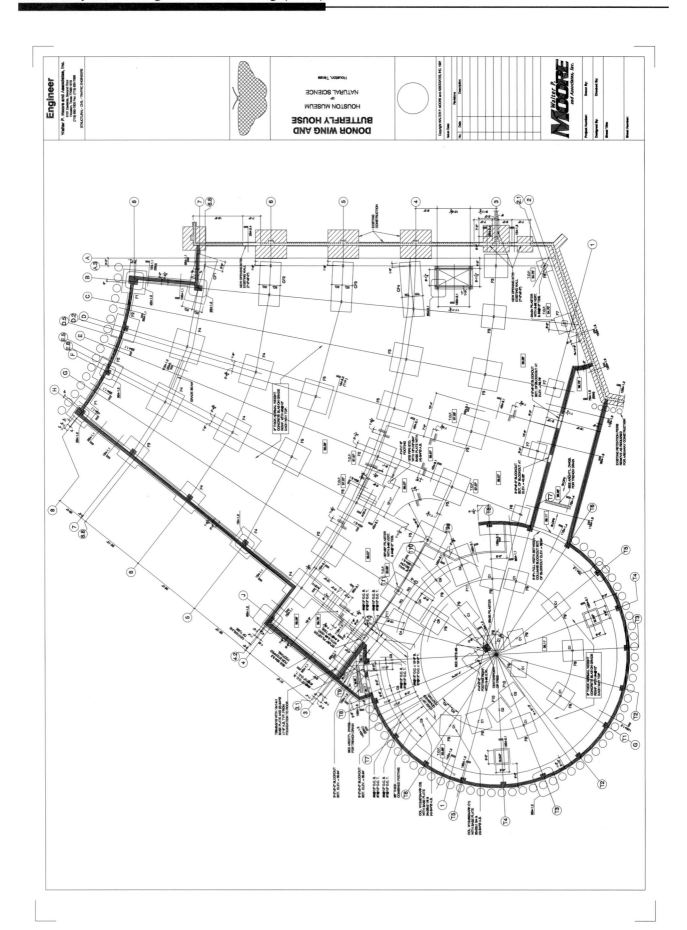

BFLYHSE.DWG Courtesy, AutoCAD, Inc. (Release 14 Sample Drawing)

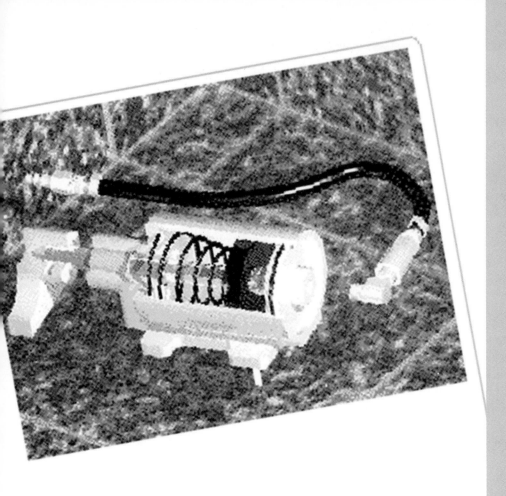

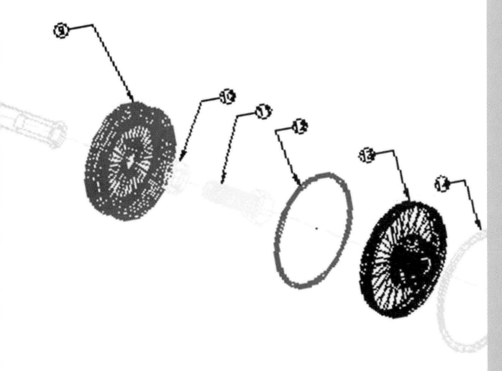

32

ATTACHING RASTER IMAGES AND IMPORTING VECTOR FILES

Chapter Objectives

After completing this chapter you should:

1. know the difference between raster files and vector files;

2. be able to *Attach* raster images within your drawing;

3. be able to use *Imageadjust* and *Imagequality* to control the appearance of an attached image;

4. be able to use *Imageclip* to create a clip boundary for an attached image;

5. know that certain images can be controlled for *Transparency*;

6. be able to *Import* vector file formats into your drawing and convert the contents into AutoCAD objects;

7. know the individual commands for importing and exporting specific file types (*Bmpout*, *3dsin*, *Psin*, etc.);

8. be able to *Export* a variety of vector file formats from your drawing with this one command.

CONCEPTS

The previous chapter discussed how to utilize textual and graphic data from other applications within your AutoCAD drawings using AutoCAD's Object Linking and Embedding (OLE) feature. This chapter discusses how to utilize graphic information in the form of raster file and vector file formats within your drawings.

This chapter discusses two different but related topics: (1) attaching raster files and (2) importing and exporting vector file formats. When a raster image is attached to your drawing, the image that appears is actually linked to the original raster file similar to the way text or spreadsheet information in an AutoCAD drawing is linked to the original file when using OLE. When other types of graphic file formats are imported into your drawings, such as vector files, AutoCAD converts the contents of the file into AutoCAD objects and permanently adds them to the drawing.

Raster Files versus Vector Files

Raster images are "pictures" composed of many dots of various shades of gray to produce black and white images or many dots of various colors to produce color images. For example, when you look closely at a color or black and white picture in a newspaper, you can actually see the small dots that make up the image. The quality, or resolution, of the image depends upon the process used to create the image. Some software can produce images with many dots per inch (dpi) yielding a highquality (resolution) image that looks very distinct and clear. Images that have lower resolution have fewer dots per inch and produce a lower quality image. The number of dots used to create the image determines the resolution, or quality, of the picture. Figure 32-1 displays an enlarged view of a raster image to reveal the dots composing the lines and arcs.

Figure 32-1

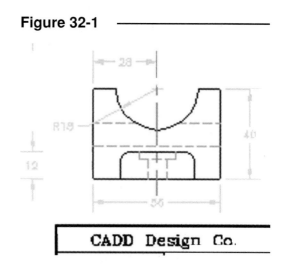

A few of the applications that create raster images frequently attached within AutoCAD drawings are computer graphics or desktop publishing programs (used for creating company logos, for example), document management software, and mapping and geographic information systems (which produce aerial photographs or satellite photographs).

In contrast, CAD programs such as AutoCAD, which are the primary graphics creation and storage media for the engineering, architecture, design, and construction industries, generate vector data. A vector file defines a line, for example, as a vector between two endpoints. Therefore, lines in vector drawings do not enlarge when zoomed in like the dots in a raster image do (Fig. 32-2). The thickness of the displayed line on a computer monitor (for a vector drawing) is always one pixel, the size of which is determined by the display device. Therefore, vector drawings are used when great detail is needed. Raster images are useful for insertion into a CAD (vector) drawing and can be helpful in the development of a CAD drawing but cannot be used for drawings that require detailed informa-

Figure 32-2

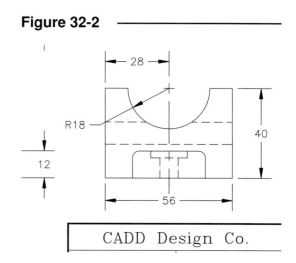

tion because of the static configuration of the dots. AutoCAD stores its vector data in the .DWG format but can also create other forms of vector files such as .DXF or .DWF discussed in this chapter. (See Chapter 43, Internet Tools, for more information on vector and raster files.)

Attaching Raster Images and Importing Vector Files

Two techniques for utilizing raster file and vector file formats within AutoCAD drawings discussed in this chapter are:

1. attaching a raster image, which links the image to the original file, and

2. importing a vector file, which converts the file contents into AutoCAD objects.

Because AutoCAD can store graphic information only in a vector format, it treats an attached raster image differently because it is <u>not</u> vector information (made up of lines, arcs, and circles). An attached raster image in AutoCAD is actually a picture of the original raster file that was selected when attaching it to your drawing. Attached raster images are displayed within a <u>frame</u> in your drawing and are *linked* to (or reference) the original file. On the other hand, if an image is imported into AutoCAD, it must be converted to AutoCAD objects for AutoCAD to store the geometry as vector data.

Exporting Raster and Vector Files

Because AutoCAD's primary function is to create and manipulate vector data (AutoCAD objects), it specializes in handling vector file formats, especially the .DWG format. Although AutoCAD can create raster files such as .BMP, .TIF, and .TGA, these formats have little usefulness for storing drawing information. Only images generated by AutoCAD's rendering capabilities are well suited for saving in a raster format (see Chapter 41, Rendering). Therefore, AutoCAD's file exportation capabilities are centered around vector formats such as .DWG, .DWF, .DXF, .SAT, .WMF, and .EPS. These file formats are also discussed in this chapter.

Related Commands

This chapter discusses the following AutoCAD commands related to using raster images and vector file formats within your drawing:

Image	Inserts various types of raster images into your drawing
Imageattach	Attaches, or links, a new raster image object and definition within your drawing
Imageframe	Controls whether the image frame is displayed on screen or hidden from view
Imageadjust	Allows you to control the brightness, contrast, and fade values of a raster image
Imageclip	Enables you to create new clipping boundaries for raster image objects
Imagequality	Allows you to control the display quality of raster images
Import	Imports a variety of vector file formats into AutoCAD
3dsin, 3dsout	Imports and exports 3D Studio files
Acisin, Acisout	Imports and exports ACIS solid models
Dxfin, Dxfout	Imports and exports AutoCAD drawings in an exchange format
Psin, Psout	Imports and exports Encapsulated PostScript files
Psdrag, Psquality	Adjusts manipulation and quality of PostScript images
Psfill	Allows filling a closed shape with a PostScript pattern
Export	Exports a variety of vector file formats from AutoCAD

A few other commands, which are covered in more detail in previous or later chapters, are also discussed in this chapter in the context of attaching, importing, and manipulating raster images and vector files.

ATTACHING RASTER IMAGES

The procedure for attaching a raster image in AutoCAD is very similar to using the *Xref* command (see Chapter 30, Xreferences). A raster image can be *Attached, Detached, Reloaded,* and the *Path* modified, just as you might do with an *Xref*. When you attach an image, it is displayed within a frame in the drawing. You can view the original image, but the image (and all the dots that define it) does not become a permanent part of the drawing. Like an *Xref,* you see only the picture of the original file within its frame. Since raster images normally have large file sizes, this method of "attaching" can save a tremendous amount of space within your drawing.

A few example applications of attaching a raster image to an AutoCAD drawing are listed here:

A. You want to display a digital (raster) picture of the true, or intended, object. For example, you can take a digital photograph of a house and attach it within your construction drawing of an addition to the house.
B. You want to use images of company or client logos within a title block.
C. You need to compare and verify site plan information, such as overlaying an aerial photograph to verify the contours shown on your AutoCAD drawing.
D. You need to compare or reconstruct an old "paper" drawing by scanning it (creating a raster image) and attaching it in an AutoCAD drawing.

You can edit the appearance of the attached image but not the individual parts (or dots) of the image within AutoCAD. You can adjust the quality, color, contrast, brightness, and transparency of an attached image with AutoCAD's various image editing commands and system variables. Attached raster images can also be edited with typical AutoCAD editing commands such as *Copy, Move, Rotate, Scale,* and *Trim*. Attached images can be edited with Grips as well. The special image attaching and editing commands are discussed next.

IMAGE and -*IMAGE*

Pull-down Menu	COMMAND (TYPE)	ALIAS (TYPE)	Short-cut	Screen (side) Menu	Tablet Menu
Insert Raster Image...	*IMAGE*	*IM or* -*IM*	...	*INSERT Image*	*T,3*

Use the *Image* command to control the attachment of raster images within your drawing. There is no practical limit to the number or size of images you can attach within your drawing. You can have more than one image displayed within any viewport.

Although you can attach a variety of file types (BMP, TIF, JPG, GIF, TGA, etc.), AutoCAD determines how to handle the image (for display and editing) based upon the image file contents not the image file type. The following raster file types are accepted by AutoCAD's *Image* and *Imageattach* commands:

Image File Type	File Extensions	Description
BMP	*.BMP, *.RLE, *.DIB	Windows device-independent bitmap format
CALS-1	*.RST, *.GP4, *.MIL, *.CAL, *.CG4	Mil R-Raster-1
FLIC	*.FLC, *.FLI	Autodesk Animator FLIC
GIF	*.GIF	Compuserve Graphic Interchange Format
JPEG	*.JPG	JEPG raster
PCX	*.PCX	PC Paintbrush Exchange
PICT	*.PCT	Macintosh PICT1, PICT2
PNG	*.PNG	Portable Network Graphics
TARGA	*.TGA	Truevision
TIFF	*.TIF	Tagged Image File Format

The *Image* command invokes the *Image* dialog box (Fig. 32-3). In this dialog box you can *Attach, Detach, Reload, Unload,* locate a file using *Browse,* and save a new path using *Save Path.* The operation of and options within this dialog box are similar to those in the *External Reference* dialog box (see Chapter 30).

Figure 32-3

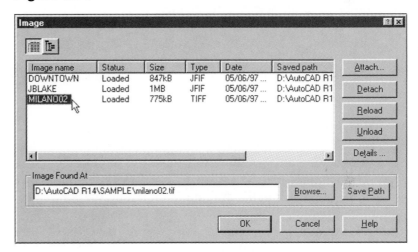

Tree View, List View
The two buttons located in the top left corner of this dialog box allow you to control the appearance of the list of currently attached raster images in the center of the dialog box. The list can be displayed in a *List View* or *Tree View.*

List View is the default setting for the list when you open the *Image* dialog box. When *List View* is active, column headings appear at the top of the list for *Image name, Status, Size, Type, Date,* and *Saved Path* for each image currently attached to the drawing.

When *Tree View* is selected, AutoCAD displays the list of attached images in a tree structure. This format allows you to view the list of images currently attached to the drawing in a hierarchical structure.

Attach

When you first activate the *Image* dialog box the only available button is *Attach*. Use this button to attach an image to the drawing in the same way you attach an *Xref*. Selecting the *Attach* button invokes the *Attach Image File* dialog box (Fig. 32-4). Alternately, this dialog box can be opened directly using the *Imageattach* command.

From the *Attach Image File* dialog box you can select from a variety of raster image file types (listed in the previous table). The features and functionality of the *Attach Image File* dialog box are the same

Figure 32-4

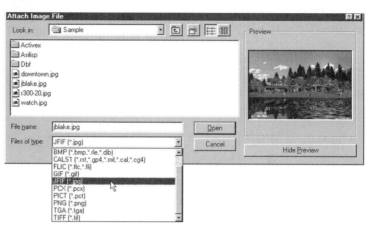

as those discussed in Chapter 30 for attaching *Xref* files. However, the *Attach Image File* dialog box has the additional option of *Hide Preview/Show Preview* that allows you to turn on and off the *Preview* display area. If the preview is turned off, the *Preview* display box does not display the thumbnail picture of the file currently selected from the file list. Use this feature to increase the speed at which you can select files from the dialog box when you know the image files by name. Turn the *Preview* display on again to view the thumbnail of the selected image file.

When an image file has been selected for attachment (select *Open* in the *Attach Image File* dialog box), the *Attach Image* dialog box appears for you to specify the attachment parameters (Fig. 32-5). Specifying the *At:* (or insertion) point and *Rotation angle* is similar to inserting a *Block* or attaching an *Xref*. Use *Scale factor* to match the image geometry scale to the scale of the geometry in the drawing.

The *Scale factor* is based on the ratio of image units to the current AutoCAD units. You can enter a value in the edit box if you know the scale of the geometry contained within the image file, otherwise use the default setting, *Specify on screen*. Select the *Details* button to display image size and resolution information (Fig. 32-5, bottom).

Figure 32-5

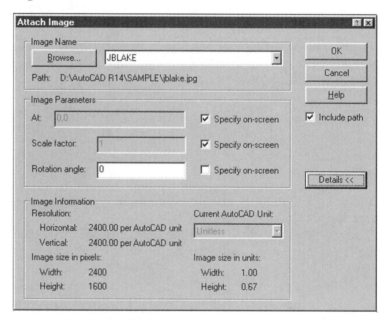

If you want to attach a second image in a drawing, selecting the *Attach* button produces the *Attach Image* dialog box directly. You must then select *Browse* to produce the *Attach Image File* dialog box to select from folders and files. When the same image is attached more than once in a drawing, only one reference is displayed in the list in the *Attach Image File* dialog box, but a different insertion point, scale, clipping boundary, etc. can be specified for each instance of the image in your drawing.

Detach

You can detach image files appearing in the list by highlighting the desired file(s), then selecting *Detach*. This action deletes all references to the image within the drawing (breaks the link to the image file), but does not delete the file from the storage device (local or network drive). This is essentially the same as detaching an *Xref* from a drawing.

Reload

You can *Unload* an image to remove its display while you work on the drawing, but the attachment information is not deleted from the drawing. You can then *Reload* the image picture by selecting the *Reload* button. *Reload* causes AutoCAD to reread the file referenced in the *Saved Path* and redisplay it within the image frame. This is also useful if the referenced image file has been changed and you want the changes to be displayed during the current drawing session.

Unload

This feature unloads the raster image picture from the display without deleting the attachment information from your drawing. Only the image's frame remains visible in the drawing. Unloading an image can dramatically decrease the amount of time AutoCAD needs to redisplay, or regenerate, the drawing since the image is not resident in current memory. Use this feature when you do not need to see the image in order to work on the drawing.

Unloading a drawing is not the same as turning off the display of the image. You can turn off the display of an image within the drawing using *Ddmodify* (see *Ddmodify*, this chapter).

Details...

You can quickly access specific information about an image by selecting an image from the list and selecting the *Details* button (in the *Image* dialog box). This activates the *Image File Details* dialog box that lists details of the currently selected image as shown in Figure 32-06.

Figure 32-6

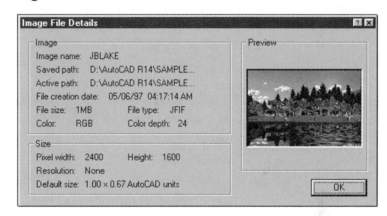

Image Found At

A text box at the bottom of the *Image* dialog box displays the current location of the highlighted image file in the list. You can choose another image file to attach to the selected *Image name* (and keep the same name from the list) by selecting the *Browse* button to the right of this text box.

You may want to do this, for example, if you have an older version of an aerial photograph currently attached and you want to temporarily display a newer photo of the same aerial view. You can *Browse*, find and select the newer raster image, then exit the *Image* dialog box to verify that it is a more appropriate image to display in your drawing.

If you prefer the newer file, you can then return to the *Image* dialog box and save the current path for the attached image by selecting the attached image from the list, then selecting *Save Path*. In this case, the *Image name* originally stored by AutoCAD does not change, but the path and file name change.

Browse

Selecting this button takes you to the *Attach Image File* dialog box, enabling you to search folders and locate raster files to attach.

Save Path

Select this function when you have used *Browse* to specify a new image file to display in your drawing and you want to make it a permanent change. If you do not use *Save Path* prior to exiting the drawing, any changes made to the original path or file name specified for the named image are lost.

IMAGEATTACH

Pull-down Menu	COMMAND (TYPE)	ALIAS (TYPE)	Short-cut	Screen (side) Menu	Tablet Menu
Insert Raster Image... Attach...	IMAGEATTACH	IAT	...	INSERT Image Attach...	...

Imageattach allows you to attach a raster image directly without having to use the *Image* command first. When you use *Imageattach,* either the *Attach Image File* (see Fig. 32-4) or *Attach Image* dialog box (see Fig. 32-5) appears, depending on whether you already have images in the drawing. The capabilities and operation of these dialog boxes are the same as discussed previously regarding the *Image* command (see *Image*).

IMAGEFRAME

Pull-down Menu	COMMAND (TYPE)	ALIAS (TYPE)	Short-cut	Screen (side) Menu	Tablet Menu
Modify Object> Image> Frame	IMAGEFRAME	...	...	MODIFY1 Imagefrm	...

The *Imageframe* command allows you to control whether the image frames are displayed on the screen or hidden from view. When you attach an image within your drawing it is displayed inside a "frame" that defines the image's outermost boundary. The *Imageframe* options are *Off* and *On.* Setting *Imageframe* to *Off* hides the frames so they are no longer visible in the AutoCAD Drawing Editor, but the images remain visible:

        Command: **imageframe**
        ON/OFF <ON>: (option)
        Command:

The *Imageframe* setting is global; that is, it affects all images in the drawing. All image frames are *Off* or *On.*

Images can be modified by using the general editing commands (*Move, Copy, Scale, Rotate,* Grips, etc.) or by using special image editing commands (discussed next). The general editing commands require you to select the image's frame; therefore, turning the image frame *Off* makes the image unselectable for editing with general editing commands. You must select the image's frame whenever AutoCAD prompts you to "Select objects:" to modify. The image becomes selectable again once you turn the display of its frame *On.*

The special image adjusting commands (*Imageadjust, Imageclip,* and *Imagequality*) allow you to select images whether the image frames are *On* or *Off.*

IMAGEADJUST

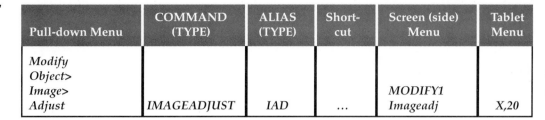

Pull-down Menu	COMMAND (TYPE)	ALIAS (TYPE)	Short-cut	Screen (side) Menu	Tablet Menu
Modify Object> Image> Adjust	IMAGEADJUST	IAD	...	MODIFY1 *Imageadj*	X,20

You can adjust an image's *Brightness, Contrast,* and *Fade* values by using *Imageadjust.* The *Imageadjust* command produces the *Image Adjust* dialog box (Fig. 32-7). The *Image Adjust* dialog box displays a thumbnail preview of the selected image and allows you to change the following settings.

Figure 32-7 ———————

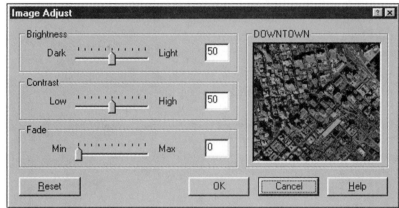

Brightness
This slider and edit box control the brightness of the image. The edit box values range from 0 to 100. The higher the value, the brighter the image becomes. *Brightness* indirectly affects the range of *Contrast* for the image.

Contrast
Use this slider or edit box to control the contrast of the image. *Contrast* is the ratio of light to dark (for black and white images) or pure hue to no color (for color images). The higher the value, the more each pixel is forced to its primary or secondary color. *Contrast* indirectly affects *Fade.* The DONTOWN.JPG image is displayed in Figure 32-8 with *Contrast* set to 90 and *Brightness* set to 70.

Figure 32-8 ———————

Fade
The *Fade* effect of the image controls how much the image blends with the background color. The higher the value, the more the image is blended with the current background color. A value of 100 allows you to blend the image completely into the background. If you change the screen background color, the image will fade to the new color. For plotting, white is used for the background fade color.

Reset
Select this button to reset the values for *Brightness, Contrast,* and *Fade* back to the default settings of 50, 50, and 0, respectively.

R14

IMAGECLIP

Pull-down Menu	COMMAND (TYPE)	ALIAS (TYPE)	Short-cut	Screen (side) Menu	Tablet Menu
Modify Object> Image Clip	*IMAGECLIP*	*ICL*	...	*MODIFY1 Imageclp*	*X,22*

You can clip an image with *Imageclip* in the same manner that you clip *Xrefs* with *Xclip*. When specifying the area to clip, you can use a rectangle or polygon to define the clip boundary. Everything outside the clip boundary is hidden from view. Figure 32-9 displays the DOWNTOWN.JPG image with a polygonal clip boundary.

Specifying the clip boundary rectangle or polygon is similar to using AutoCAD's *Rectangle* command or the *Window Polygon* option when selecting objects. *Rectangle* is the default boundary.

Figure 32-9 ─────────────

```
Command: imageclip
Select image to clip: PICK (select an
image frame)
ON/OFF/Delete/<New boundary>: Enter
Polygonal/<Rectangular>: Enter
First point: PICK
Select second corner: PICK
Command:
```

Once a clipping boundary has been specified for an image you can turn the boundary *Off*. This action makes the entire image visible again (Fig. 32-10). You can also turn the boundary back *On*, which displays the image with its original clip boundary (see Fig. 32-9). Alternately, you can *Delete* the clip boundary so the entire image is visible.

If you use *Imageclip* and select an image that already has a boundary, the following prompt appears:

```
Command: imageclip
ON/OFF/Delete/<New boundary>: Enter
Delete old boundary? No/<Yes>:
```

Figure 32-10 ─────────────

IMAGEQUALITY

Pull-down Menu	COMMAND (TYPE)	ALIAS (TYPE)	Short-cut	Screen (side) Menu	Tablet Menu
Modify Object> Image> Quality	*IMAGEQUALITY*	...	...	*MODIFY1 Imagequa*	...

This command allows you to adjust the quality setting that affects display performance. High-quality images take longer to display. When you change the *Imagequality* setting, the display updates immediately without causing a *Regen*. The images attached to your drawing are always plotted using a high-quality display, regardless of the *Imagequality* setting. The setting you choose for *Imagequality* affects all images in the drawing globally.

```
Command: imagequality
High/Draft <High>: (option)
Command:
```

TRANSPARENCY

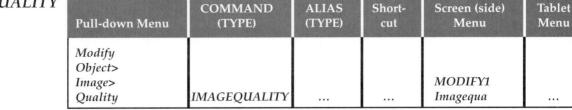

Pull-down Menu	COMMAND (TYPE)	ALIAS (TYPE)	Short-cut	Screen (side) Menu	Tablet Menu
Modify Object> Image> Transparency	*TRANSPARENCY*	...	...	*MODIFY1 Transpar*	...

Transparency allows you to control whether the <u>background</u> pixels in an image are transparent or opaque. Many image file formats allow images with transparent pixels. When setting image transparency to *On*, AutoCAD recognizes transparent pixels so that graphics on the screen (AutoCAD objects or another image) show through those pixels. Transparency can be set for both bitonal and non-bitonal (Alpha RGB or gray-scale) images. When you attach an image to your drawing, the image's transparency setting is *Off* by default:

```
Command: transparency
ON/OFF <OFF>: (option)
Command:
```

This unique feature allows you to put one image on top of another, then set *Transparency* for the top image *On*. (Use *Draworder* to specify which image is "on top.") The transparency feature makes the "top" image appear in the same scene as the "background" image (Fig. 32-11, right image).

Figure 32-11

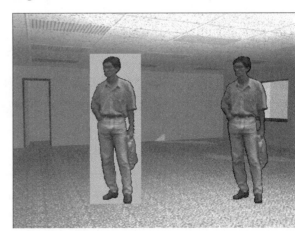

DDMODIFY

Pull-down Menu	COMMAND (TYPE)	ALIAS (TYPE)	Short-cut	Screen (side) Menu	Tablet Menu
Modify Properties	DDMODIFY	MO	...	MODIFY1 Modify	Y,14

Using *Ddmodify* with an image produces the *Modify Image* dialog box (Fig. 32-12). This dialog box gives you the ability to modify an image's properties as you might with other image-related commands and in ways not possible with the image-related commands. (*Imageframe* must be *On* to select the image.) For example, you can remove the check for *Show Image* to turn off the display of the image. This feature is not possible by other means.

The *Modify Image* dialog box also allows you to turn on and off the display of images not perpendicular to the viewpoint (*Show Non-Ortho Image*), turn the clipping boundary on and off (*Show Clipped Image*), and turn *Transparency* on and off. The *Adjust Image* button produces the *Image Adjust* dialog box (see Fig. 32-7).

Figure 32-12

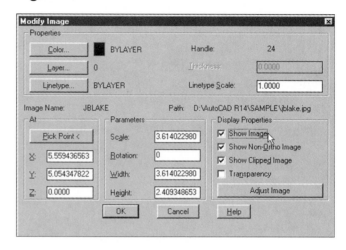

DRAWORDER

Pull-down Menu	COMMAND (TYPE)	ALIAS (TYPE)	Short-cut	Screen (side) Menu	Tablet Menu
Tools Display Order>	DRAWORDER	DR	...	TOOLS1 Drawordr	...

The *Draworder* command is very useful when utilizing raster images in your drawing. Often you will attach the raster image so that it appears "on top" of another image or other AutoCAD objects. The *Draworder* command allows you to select an attached image and then specify the drawing order placement of the raster image as *Above, Under, Front,* or *Back*:

```
Command: draworder
Select objects: (select image frame)
Select objects: Enter
Above object/Under object/Front/<Back>: (option)
Command:
```

For additional information about *Draworder,* see Chapter 26, Hatching and Creating Section Views.

SAVEIMG

Pull-down Menu	COMMAND (TYPE)	ALIAS (TYPE)	Short-cut	Screen (side) Menu	Tablet Menu
Tools Display Image> Save...	SAVEIMG	...	...	TOOLS1 Saveimg	...

The *Saveimg* (save image) command allows you to save the current drawing or rendered image as either a .BMP, .TGA, or .TIF (raster) format. The saved image can later be viewed by using the *Replay* command or can be attached by using *Imageattach*. See Chapter 41, Rendering, for more information on this command.

BMPOUT

Pull-down Menu	COMMAND (TYPE)	ALIAS (TYPE)	Short-cut	Screen (side) Menu	Tablet Menu
...	*BMPOUT*	...	...	...	...

The *Bmpout* command enables you to export selected objects to a Windows device-independent (.BMP) bitmap (raster) file format. AutoCAD creates a compressed .BMP file that uses less disk space <u>but might not be readable</u> by some applications. See Chapter 41, Rendering, for information on saving .BMP images generated by AutoCAD.

IMPORTING AND EXPORTING VECTOR FILES

Importing vector file formats into your drawing converts the graphic information, or objects, within the selected file into AutoCAD objects and inserts them onto your drawing. The converted information becomes a permanent part of your drawing. Depending on the file being imported, this can dramatically increase the file size of your AutoCAD drawing.

Since importing vector files converts the contents into AutoCAD objects, you can edit them using typical AutoCAD editing methods. An imported object is initially defined in your drawing as a *Block* reference. The first step in editing them with AutoCAD is to *Explode* the block. You will find that AutoCAD often converts the information into *Lines*, *Arcs*, *Circles*, *Plines*, *Solids* (2D), or solids (3D).

IMPORT

Pull-down Menu	COMMAND (TYPE)	ALIAS (TYPE)	Short-cut	Screen (side) Menu	Tablet Menu
...	*IMPORT*	*IMP*	...	...	T,2

The *Import* command is used for importing vector information into your drawing and produces the *Import File* dialog box (Fig. 32-13). The *Import File* dialog box has the same layout and basic functionality as the *Attach Image File* dialog box. The available file types you can import to your drawing are listed in the following table:

Figure 32-13

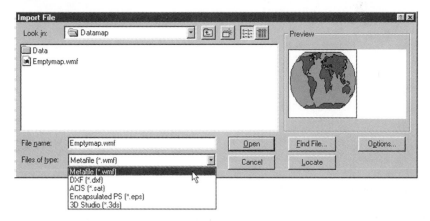

Vector File Type	File Extensions	Description
Metafile	*.WMF	Windows Metafile Format
DXF	*.DXF	Drawing Interchange Format
ACIS	*.SAT	ACIS Technology solid model
EPS	*.EPS	Encapsulated PostScript
3D Studio	*.3DS	Autodesk 3D Studio file

There is an additional feature in the *Import* dialog box that is not in the *Attach Image File* dialog box. The *Options* button (below the *Preview*) produces the *Import Options* dialog box (Fig. 32-14) but is active only when the *Metafile (WMF)* type is specified in the *Files of type* list box. The *Wide Lines* option lets you control whether the imported .WMF file maintains relative line widths or imports line widths of 0. The *Wire Frame (No Fills)* option controls whether the metafile is converted into a wireframe (lines only) or keeps solid-filled objects.

Figure 32-14

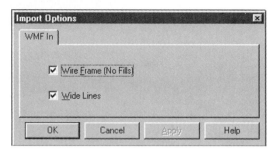

When the metafile is imported, it is converted to a *Block* reference. Use *Explode* to break the *Block* into component objects such as *Lines, Arcs,* and *Plines.*

3DSIN

Pull-down Menu	COMMAND (TYPE)	ALIAS (TYPE)	Short-cut	Screen (side) Menu	Tablet Menu
Insert *3D Studio...*	*3DSIN*	...	...	*INSERT* *3DSin*	...

3D Studio Max® is a 3D modeling, rendering, and animation product that is produced by Kinetix®, a division of Autodesk. 3D Studio Max creates models, complete with materials, lights, cameras, etc., that can be stored in .3DS file format.

The *3dsin* command in AutoCAD enables you to read in geometry and rendering data from 3D Studio (.3DS) files. *3dsin* produces the *3D Studio File Import* dialog box (Fig. 32-15). AutoCAD is capable of importing 3D Studio information such as meshes, materials, mapping, lights, and cameras. 3D Studio procedural materials and smoothing groups <u>are not</u> imported.

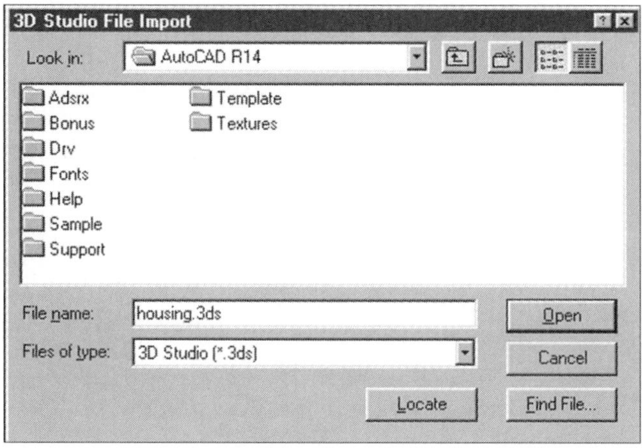

Once you have selected a .3DS file to import and select the *Open* button, the *3D Studio File Import Options* dialog box is displayed, enabling you to specify the objects to import, the layers to place them on, and how to handle objects constructed of multiple materials (Fig. 32-16).

Figure 32-15

Figure 32-16

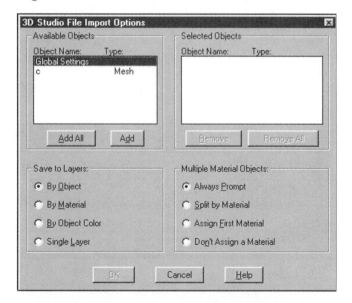

3DSOUT

Pull-down Menu	COMMAND (TYPE)	ALIAS (TYPE)	Short-cut	Screen (side) Menu	Tablet Menu
File *Export...*	*3DSOUT*	...	...	*FILE* *Export*	...

You can convert your AutoCAD geometry and rendering data into 3D Studio data by using the *3dsout* command. *3dsout* first opens the *3D Studio Output File* dialog box (similar to the *3D Studio File Import* dialog box, Fig. 32-15). Once you give the output file name and select *Save*, the *3D Studio File Export Options* dialog box appears, giving you the opportunity to specify the division method, output, mode, smoothing angle, and welding threshold for the export process (Fig. 32-17).

Figure 32-17 ─────────

ACISIN

Pull-down Menu	COMMAND (TYPE)	ALIAS (TYPE)	Short-cut	Screen (side) Menu	Tablet Menu
Insert *ACIS Solid...*	*ACISIN*	...	...	*INSERT* *ACISin*	...

AutoCAD uses the ACIS® solid modeling engine to construct solids. This modeler is used also by other CAD programs. The *Acisin* and *Acisout* commands allow you to exchange solid model geometry with other programs that use the ACIS modeling engine. ACIS solid model geometry can be stored as a .SAT (ASCII) file format.

In AutoCAD, use the *Acisin* command to produce the *Select ASIC File* dialog box for importing ACIS (.SAT) files into AutoCAD. The *Select ACIS File* dialog box is the same as the *3D Studio File Import* dialog box (see Fig. 32-15) except for the *Files of type* section.

ACISOUT

Pull-down Menu	COMMAND (TYPE)	ALIAS (TYPE)	Short-cut	Screen (side) Menu	Tablet Menu
File *Export...*	*ACISOUT*	...	...	*FILE* *Export*	...

You can use the *Acisout* command to export AutoCAD solid or *Region* data from your drawing into an ACIS format. The ACIS file format allows AutoCAD to store solid model objects as an ASCII (.SAT) file type. The *Create ACIS File* dialog box that appears is the same as the *3D Studio File Import* dialog box except for the *Files of type* section (see Fig. 32-15). *Acisout* ignores all selected objects that are not solids or *Regions*.

STLOUT

Pull-down Menu	COMMAND (TYPE)	ALIAS (TYPE)	Short-cut	Screen (side) Menu	Tablet Menu
*File Export... *.stl*	*STLOUT*	...	...	*FILE Export *.stl*	...

You can create a file to be used with Stereo Lithography Apparatus for creating prototypes from AutoCAD solid models. The *Stlout* command creates an ASCII format STL file defining the 3D geometry. For more information, see Chapter 39, Advanced Solids Features.

DWFOUT

Pull-down Menu	COMMAND (TYPE)	ALIAS (TYPE)	Short-cut	Screen (side) Menu	Tablet Menu
File Export...	*DWFOUT*	...	...	*FILE Export*	...

Your drawing can be viewed on the Internet by exporting it with the *Dwfout* command. *Dwfout* converts your drawing into a web format (.DWF) file (in vector format). Only model space data in the current viewport is exported, however. Paper space information is not supported and cannot be exported. The *DWF Export Options* dialog box becomes available so you can set the level of decimal precision and use file compression if necessary. See Chapter 43, Internet Tools, for more details.

DXFIN

Pull-down Menu	COMMAND (TYPE)	ALIAS (TYPE)	Short-cut	Screen (side) Menu	Tablet Menu
...	*DXFIN*	...	...	...	...

Autodesk initiated an interchangeable CAD drawing (vector file) standard in the early 1980s. This standard, called the Drawing Interchange File Format (.DXF), was intended to be utilized by AutoCAD and other CAD programs for exchanging vector drawing data in a simple ASCII format. Because the ASCII format is used, DXF files can be very large. The DXF format has become one of the most widely used formats (along with IGES format) for exchanging drawing data.

Importing vector information from other CAD-related applications is accomplished by using the *Dxfin* command. This command produces the *Select DXF File* dialog box (not shown), similar in format to the other file import dialog boxes. <u>The *Dxfin* command must be used in a new drawing to function properly</u>. You can also *Audit* the information (have AutoCAD evaluate the integrity of the data) being imported during the *Dxfin* process by checking *Audit after each DXFIN or DXBIN* in the *General* tab of the *Preferences* dialog box (Fig. 32-18).

Figure 32-18

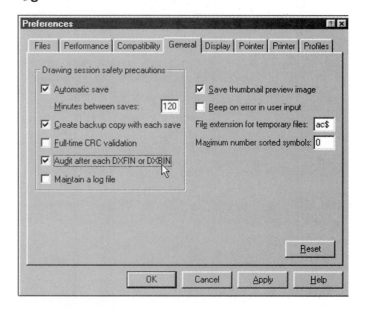

DXFOUT

Pull-down Menu	COMMAND (TYPE)	ALIAS (TYPE)	Short-cut	Screen (side) Menu	Tablet Menu
File *Export...*	*DXFOUT*	...	...	*FILE* *Export*	...

Use the *Dxfout* command to export your drawing into a Drawing Interchange Format (.DXF) file. *Dxfout* produces the *Create DXF File* dialog box (not shown, similar to the other file export dialog boxes). You can choose to export the drawing as either an AutoCAD Release 13/AutoCAD LT for Windows 95 (*AutoCAD R13/LT95 DXF*) or an AutoCAD Release 12/AutoCAD LT Version 2 (*AutoCAD R12/LT2 DXF*) Drawing Interchange Formats.

In the *Create DXF File* dialog box you can also specify options for the *Dxfout* process by selecting the *Options* button which produces the *Export Options* dialog box (Fig. 32-19). Here you can specify whether you want the interchange file to be created in ASCII format or binary format as well as the degree of accuracy used (in decimal places) for defining the geometry. Drawing Interchange Files, whether created as ASCII or binary format, are saved as a .DXF file type.

Figure 32-19

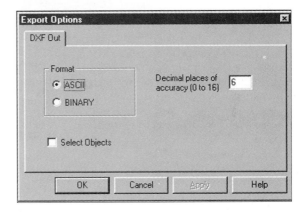

Because an ASCII file is composed of recognizable alphanumeric characters, the ASCII format .DXF files can be quite large, so the binary format is sometimes used. The advantage to the ASCII file is that it is readable in a text editor or word processor (see Fig. 1-4), whereas the binary format is not.

DXBIN

Pull-down Menu	COMMAND (TYPE)	ALIAS (TYPE)	Short-cut	Screen (side) Menu	Tablet Menu
Insert *Drawing Exchange...*	*DXBIN*	...	...	*INSERT* *DXBin*	...

The *Dxbin* command allows you to import Drawing Exchange Binary (.DXB) files into AutoCAD. These .DXB files are formatted with a specially coded binary format and are produced by programs such as AutoShade. *Dxbin* produces the *Select DXB File* dialog box (not shown).

PSIN

Pull-down Menu	COMMAND (TYPE)	ALIAS (TYPE)	Short-cut	Screen (side) Menu	Tablet Menu
Insert *Encapsulated* *Postscript*	*PSIN*	...	...	*INSERT* *PSin*	...

You can use *Psin* to insert a PostScript (.EPS) file into your drawing. PostScript files are often used to describe images that were originally pixel based, but the PostScript format is technically a vector-based format. When a PostScript file is imported into AutoCAD, it is treated similarly to other vector-based files in that the image is initially stored as a *Block* reference and, when *Exploded,* is converted to AutoCAD objects such as a *Solid* (2D solid) or *Pline*.

AutoCAD inserts the PostScript image into your drawing as an anonymous *Block* representing the size and shape of the image. When importing a PostScript image you are asked to specify the "Insertion point:" and "Scale factor:" for the image. Depending on the setting of *Psdrag*, discussed next, AutoCAD displays the bounding box of the PostScript image and its file name or the actual image rather than a bounding box.

> Command: **psin** (opens the *Select PostScript File* dialog box; select a .EPS file, then select *Open*)
> Insertion point<0,0,0>: **PICK** or (**coordinates**)
> Scale factor: **PICK** or (**value**)
> Command:

PSDRAG

Pull-down Menu	COMMAND (TYPE)	ALIAS (TYPE)	Short-cut	Screen (side) Menu	Tablet Menu
...	PSDRAG	...	...	...	...

You can control the appearance of a PostScript image when it is dragged into position with the *Psdrag* setting. *Psdrag* can be set to *0* (off) which displays only the image's "bounding box," or to *1* (on) which allows the rendered PostScript image to be displayed while the image is being dragged into position with the cursor:

> Command: **psdrag**
> PSIN drag mode <1>: (0 for *Off* or 1 for *On*)
> Command:

Generally *Psdrag* is set to 0, which allows you to drag the image into position faster since the image contents do not have to be displayed. After you PICK or assign an "Insertion point" and "Scale factor" (while using *Psin*), the image is fully displayed no matter what the *Psdrag* setting. *Psdrag* affects the image's appearance only while dragging. *Psdrag* can be affected by the *Psquality* setting (see next).

PSQUALITY

Pull-down Menu	COMMAND (TYPE)	ALIAS (TYPE)	Short-cut	Screen (side) Menu	Tablet Menu
...	PSQUALITY	...	...	...	...

Psquality is a system variable that controls the rendering quality of a PostScript image when it is imported. *Psquality* can also control if the image is drawn as filled objects or as outlines:

> Command: **psquality**
> New value for PSQUALITY <75>: (can be specified as an integer <0, 0, or >0)

Possible settings for *Psquality* are as follows:

0   Turns off the image contents and displays the image (during and <u>after</u> insertion) as a bounding box. The file name and path are printed in the center of the bounding box.
<0   Sets the number of pixels per AutoCAD drawing unit for the PostScript resolution.
>0   Sets the number of pixels per drawing unit but uses the absolute value.

When *Psqaulity* is set to 0, it overrides the *Psdrag* setting.

PSFILL

	COMMAND (TYPE)	ALIAS (TYPE)	Short-cut	Screen (side) Menu	Tablet Menu
Pull-down Menu					
...	*PSFILL*	...	...	...	...

Psfill is used to fill a two-dimensional *Pline* outline with a PostScript fill pattern when creating an image in AutoCAD for exportation to a PostScript file with *Psout*. When you use *Psfill* to fill an outline with a pattern, AutoCAD does not display the PostScript fill patterns on the screen nor does it plot the pattern. *Psfill* is intended for use with *Psout*.

```
Command: psfill
Select polyline: PICK
PostScript fill pattern (. = none) <.>/?: ?
Grayscale RGBcolor Allogo Lineargray Radialgray Square Waffle Zigzag Stars Brick Specks
PostScript fill pattern (. = none) <.>/?: (type in desired pattern name for printed output)
Scale <1.0000>: (value)
ForeGroundGray <100>: (value)
BackGroundGray <0>: (value)
Command:
```

The PostScript fill pattern definition can contain arguments or parameters that control how the fill pattern appears. The PostScript patterns "Grayscale, RGBcolor, Allogo, Lineargray," etc. are patterns defined in the ACAD.PSF file. For more information, see PostScript Support in the AutoCAD *User's Guide*.

PSOUT

	COMMAND (TYPE)	ALIAS (TYPE)	Short-cut	Screen (side) Menu	Tablet Menu
Pull-down Menu					
File *Export...*	*PSOUT*	...	...	*FILE* *Export*	...

You can export parts or all of your drawing as a PostScript (.EPS) file using *Psout*. PostScript files generated by *Psout* can contain a PostScript rendering of an AutoCAD model, for example. AutoCAD normally exports *Arcs*, *Circles*, *Plines*, and filled *Regions* as PostScript primitives instead of vectors. One exception is when the AutoCAD objects cannot be represented in PostScript, such as extruded objects, in which case they are output as vectors. In this case, the information is exported as wireframe images just as AutoCAD displays them in your drawing. *HIDE* and *SHADE* have no effect on the output created when you use *Psout*.

Psout produces the *Create PostScript File* dialog box (not shown). This dialog box is the same as the *3D Studio File Import* dialog box (see Fig. 32-15) except for the *Files of type* section and the additional *Options* button, which allow you to set PostScript specific output options via the *Export Options* dialog box (Fig. 32-20).

Figure 32-20

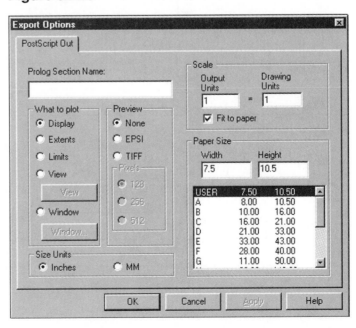

In the *Export Options* dialog box (Fig. 32-20) you can specify what part of the drawing should be written to the .EPS file by selecting *Display, Extents, Limits, View,* or *Window* in the *What to plot* section. EPS files can contain a preview which speeds and simplifies the selection and display of a PostScript image in other software programs. The preview can be saved in EPSI or TIFF format, along with the size in pixels for the preview image. The *Prolog Section Name* specifies a name for the prolog section to be read from the ACAD.PSF file. Because a PostScript file can contain all information necessary for printing to a specific size paper, that information can also be specified in this dialog box. For more information, see Using PostScript Files in the AutoCAD *User's Guide*.

EXPORT

Pull-down Menu	COMMAND (TYPE)	ALIAS (TYPE)	Short-cut	Screen (side) Menu	Tablet Menu
File Export...	*EXPORT*	*EXP*	...	*FILE Export*	...

AutoCAD can *export* your entire drawing or portions of it to many vector file formats other than .DWG. This can be accomplished by two methods: (1) use the *Export* command and select the desired file type to export, or (2) use the specific export command as discussed earlier, such as *Dxfout, 3dsout,* or *Psout.* (.BMP raster files and .DXX data extract files can also be created using *Export.*)

Use the *Export* command to produce the *Export Data* dialog box (Fig. 32-21). Here you can choose the file format you want AutoCAD to create during the export process. The available file formats that AutoCAD can export are listed in the following table along with the equivalent specific AutoCAD commands. The *Options* button allows you to set export options when you select files of type .DWF, .DXF, or .EPS (as described previously).

Figure 32-21

File Extensions	AutoCAD Equivalent Command	Description
*.3DS	*3DSOUT*	3D Studio format
*.BMP	*BMPOUT*	Windows device-independent bitmap format
*.DWG	*WBLOCK*	AutoCAD drawing file format
*.DWF	*DWFOUT*	AutoCAD drawing web file format
*.DXF	*DXFOUT*	AutoCAD Release 14, AutoCAD R13/LT95, or AutoCADR12/LT2 formats
*.DXX	*ATTEXT*	Attribute extract DXF format
*.EPS	*PSOUT*	Encapsulated PostScript format
*.SAT	*ACISOUT*	ACIS solid object format
*.STL	*STLOUT*	Solid object stereolithography format
*.WMF	*WMFOUT*	Windows Metafile format

CHAPTER EXERCISES

1. *Image Attach*

   You can use the *Image* command to enhance the appearance of your drawing by attaching a raster image such as a company logo, for example.

   A. *Open* the **SADDL-DM** drawing from Chapter 29 Exercises. *Zoom* in to the title block area in the lower-right corner of the drawing. You will attach a raster image in the title block as a company logo.

   B. Use the *Image* command, and in the *Image* dialog box choose the *Attach...* button. Locate a raster image file called **ADESK2.TGA** in the **AutoCAD R14/Textures** folder. Select this file and *Open* it. In the *Attach Image* dialog box choose *OK*. PICK the insertion point as shown in Figure 32-22 and scale the image by dragging it so that the image frame fits within the first text area of the title block.

   Figure 32-22

   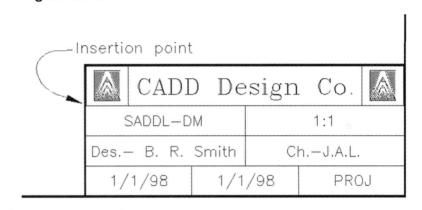

   C. Select the attached image and *Copy* it to the other side of this title area. Compare your results to Figure 32-22.

   D. Use *Imageadjust* and select one of the image's frames. Set *Brightness* to **75** and *Contrast* to **100**. The logo should now appear as a pure black and white image. Do the same for the other image. *Zoom All* and save this drawing as **SADDL-DM2**.

2. *Image Attach, Imageclip*

   It is often useful to display only a portion of an attached image in your drawing. You can accomplish this with *Imageclip*.

   A. Begin a *New* drawing and select *Start from Scratch, English* default settings. Use the *Image* command and select *Attach...*. Locate the **WATCH.JPG** file from the **AutoCAD R14/Sample folder, then select *Open*.** In the *Attach Image* dialog box, ensure *Include path* is checked. Ensure *Specify on screen* is checked and attach the watch image at an *Insertion point* of **2,2**, a *Scale factor* of **5**, and *Rotation angle* of **0**.

   B. Now use the *Image* command again to attach the watch a second time to your drawing at the point **8,2**. Use the same *Scale factor* and *Rotation angle* as the first attachment, but this time specify these parameters in the *Attach Image* dialog box. (You will notice that once you have an image attached to your drawing you receive the *Attach Image* dialog automatically when you select the *Attach* button.)

   C. Issue the *Image* command again and notice that although you attached the WATCH.JPG image twice it is referenced only once in the list of attached images. Select *Cancel* in the *Image* dialog box.

D. Now use the *Imageclip* command to create the two clipping boundaries as shown in Figure 32-23.

Figure 32-23

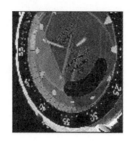 

First attachment of WATCH.JPG with a rectangular clip boundary turned On

Second attachment of WATCH.JPG with a polygonal clip boundary turned on

E. After you have created the clip boundaries, use the *Imageclip* command again to turn the boundaries *OFF* and then back *ON* again. You will find that the clip boundary is saved even though you turn it *OFF* to display the entire image. Save this drawing as **CLIPPING**.

3. *Draworder*

Raster images are also useful for reference purposes during drawing construction. You can utilize aerial photographs, for example, in order to construct your drawing in relation to existing structures. When using images as reference information, you often need to reorder the display of objects on your drawing.

A. Begin a *New* drawing and select *Start from Scratch, English* default settings. Use the *Image* command and select *Attach...*. Locate and *Open* the **DOWNTOWN.JPG** file from the **AutoCAD R14/Sample** folder. Attach the downtown image to your drawing at **2,2**, with a *Scale factor* of **7**, and *Rotation angle* of **0**.

B. Use *Imageadjust*, select the image, and set the *Brightness* to **75**.

C. Next, *Zoom* in to the lower-right portion of the downtown image, being careful that you can still see the image frame on the screen. Using the *Pline* command, draw a closed object as shown in Figure 32-24 (make sure you use *close* to finish the last segment of the shape). Use the *Hatch* command to hatch the object with a *Solid* pattern as shown.

Figure 32-24

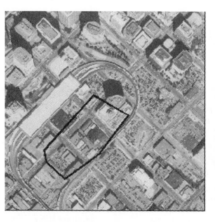 

Draw an enlcosed boundary using the Pline command.

Hatch the enlcosed boundary using a solid hatch patterr..

D. Notice that the solid hatch pattern is displayed on top of the downtown image. Now use *Draworder*, select the solid hatch pattern, and send it to the *Back*. It is no longer visible on screen since it is now in back of the downtown image (you may have to *Regen*). You can reverse the display of these two objects by using *Draworder* again, and selecting the downtown image (by selecting its frame), and sending it to the *Back* of the display order. Save this drawing as **DWGORDER**.

4. *Image Found At*

You might encounter situations when you already have an image attached to your drawing and you need to redisplay a more current or totally different version of the image on your drawing.

A. *Open* the drawing called **CLIPPING**. Use the *Image* command and select **WATCH** from the list of attached images. Choose the *Browse* button in the *Image Found At* section, and *Open* the **DOWNTOWN.JPG** image file. Then select *OK* to close the dialog box.

B. The WATCH image name, attached at two different locations, now displays the contents of the DOWNTOWN file. You did not use the *Save Path* option in the *Image* dialog box, so this is not a permanent change of the image picture. Use *Open* (do not save changes), then select the same CLIPPING drawing to *Open*. Notice that the WATCH.JPG file is still the image that is displayed for the WATCH attachment name.

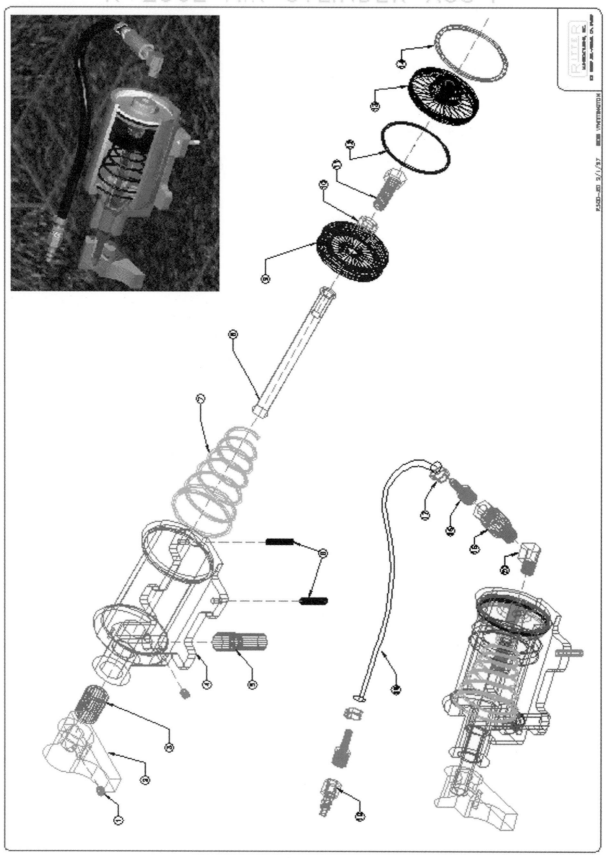

R300-20.DWG Courtesy, AutoCAD, Inc. (Release 14 Sample Drawing)

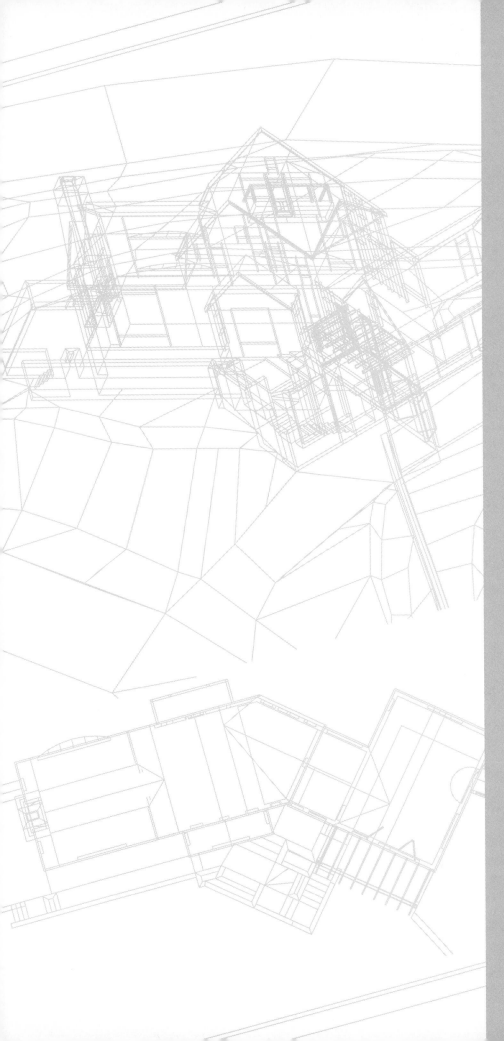

33

ADVANCED PAPER SPACE VIEWPORTS

Chapter Objectives

After completing this chapter you should:

1. be able to control viewport-specific layer visibility using the *Vplayer* command, the *Layer* command, or the *Layer Control* drop-down list;

2. be able to dimension in viewports using the *Scale to Paper Space* (*DIMSCALE*=0) option;

3. be able to use paper space to display and plot or print multiple viewports of one drawing;

4. be able to use paper space to display and plot or print viewports of several *Xrefed* drawings;

5. know that paper space viewports can be used to display different views of a 3D model.

CONCEPTS

You have already learned the basic concepts, commands and procedures for creating a paper space viewport, zooming the model geometry to size (*Zoom XP*), and plotting to scale at 1=1 from paper space. These topics were covered in Chapter 19, Tiled Viewports and Paper Space Viewports. The commands, variables, and concepts that were discussed in Chapter 19 are listed here:

> Guidelines for Using Paper Space
> *TILEMODE*
> *Mview*
> *Pspace*
> *Mspace*
> *Zoom XP*

If you are not familiar with these commands and concepts, please review Chapter 19.

This chapter deals with advanced capabilities of paper space viewports. The ideas included in this chapter are listed below.

> Reasons for using multiple viewports
> Controlling layer visibility in specific viewports
> Dimensioning in paper space viewports
> Controlling linetype scaling in viewports
> Creating multiple views of one drawing using paper space viewports
> Displaying several *Xrefed* drawings, one in each of multiple viewports
> Using *Mvsetup* for 2D drawings

Reasons for Using Paper Space

Paper Space is intended to be an aid for plotting and printing. Because paper space represents a sheet of paper, it provides you with a means of preparing drawing views to be plotted or printed in a specific configuration.

In Chapter 19, an example is given using only <u>one</u> paper space viewport. The model space geometry, including associative dimensions, appears in the viewport and annotations such as notes, bill of materials, title block, and border appear in paper space. The model geometry (units) is scaled to paper space units using *Zoom XP* so that you can then plot at 1=1 from paper space and model geometry is scaled correctly.

Using <u>one</u> paper space viewport to display and plot model geometry is unnecessary in most cases because it is simpler to create the title block, border, and annotations in model space, and then calculate the appropriate plot scale based on the drawing scale factor. There are a few cases, however, where one only paper space viewport may be used such as in companies that standardize using paper space for all template drawings.

Paper space viewports are most useful for the following situations. These cases are drawing applications that require multiple views or multiple drawings on one plotted or printed sheet.

1. Use paper space to plot or print several views of the same drawing on one sheet, possibly at different scales, such as a main view and one or several detail views.

2. Use paper space to plot several drawings on one sheet. This is accomplished by creating multiple viewports, then using *Xref* to bring several drawings into model space and using layer visibility

to control which drawings are visible in which viewports. This method has been popular in the past (before Release 14); however, *Xref* clipping (*Xclip*) has made the same results possible without using paper space.

3. Use paper space to plot different views of a 3D model (e.g., top, front, side). There are also other more automated methods that can be used to create views of a 3D model, depending on the type of model (wireframe, surface, or solid). Using paper space and other methods for creating these views is discussed in Chapter 42, Creating 2D Drawings from 3D Models.

In order to use paper space for the applications listed above, you should be familiar with the commands discussed next in addition to those commands covered in Chapter 19 (*TILEMODE, Mview, Mspace, Pspace,* and *Zoom XP*).

ADVANCED PAPER SPACE COMMANDS

LAYER

	COMMAND (TYPE)	**ALIAS (TYPE)**	**Short-cut**	**Screen (side) Menu**	**Tablet Menu**
Pull-down Menu					
Format Layer...	*LAYER or -LAYER*	*LA or -LA*	...	*FORMAT Layer*	*U,5*

The *Layer* command produces the *Layer* tab of the *Layer & Linetype Properties* dialog box. This dialog box and the *Layer Control* drop-down list can be used to control which layers are visible in which viewports. Two small icon buttons in the central area of the *Layer & Linetype Properties* dialog box and one icon that appears in the *Layer Control* drop-down list are used for viewport-specific layer visibility.

Figure 33-1 displays the *Layer & Linetype Properties* dialog box with the *Freeze in Current Viewport* and *Freeze in New Viewports* columns expanded (so the complete headings can be read). These icons are grayed out when *TILEMODE* is set to 1. If paper space is enabled (*TILEMODE* is set to 0) and a viewport is active (the cursor appears in a viewport), use these icons to freeze layers <u>in the current viewport</u> or to freeze layers <u>in new viewports</u> (the selected layer will be frozen for any new viewports that are created).

Figure 33-1

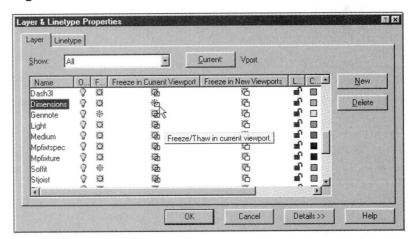

Freeze in Current Viewport

For example, assume you have two paper space viewports created and each viewport displays a different part of the same model space geometry (Fig. 33-2). Notice that in the left viewport all layers are visible compared to the right viewport, in which the dimension layer is not visible. To freeze the "Dimensions" layer in the right viewport only, first make the right viewport active, then use *Layer* to produce the *Layer & Linetype Properties* dialog box and select the *Freeze in*

Figure 33-2

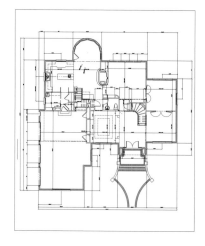 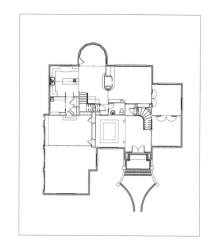

Current Viewport icon to change this layer to a frozen state (snowflake; see Fig. 33-1). If the left viewport were made active and then the *Layer & Linetype Properties* dialog box invoked, all layers would appear thawed (sunshine) for the current (left) viewport.

When a viewport is active, the *Freeze in Current Viewport* icons that appear give the visibility state of layers <u>specific to the viewport that is current</u> when the list is displayed. In other words, the icons (*Freeze in Current Viewport*) may display different information depending on which viewport is current when the list is viewed. Changing a *Freeze/Thaw* icon affects the layer's state globally (for all viewports in the drawing), while changing a *Freeze in Current Viewport* icon affects the layer's state only <u>for the current viewport</u>.

Paper space itself is considered a viewport when dealing with layer visibility. That is, if you are in paper space (*Pspace* is active) and you PICK the *Freeze in Current Viewport* icon, the highlighted layer is frozen in paper space only (the current viewport).

Freeze in New Viewports

This icon prevents the selected layer from appearing in any viewports that are created from that time on. Any layers selected to *Freeze in New Viewports* do not appear in subsequently created viewports.

This feature is helpful when you need several viewports, each to display a separate drawing. In other words, you want to *Xref* several drawings, but only want one drawing to appear in each viewport. Assume you created one viewport and *Xrefed* a drawing so it (all its layers) appeared in the viewport. If you were to create a second viewport, the same drawing would normally appear in the second viewport (Fig. 33-3). If you were to *Xref* a second drawing into model space, it would also normally appear in both viewports.

Figure 33-3

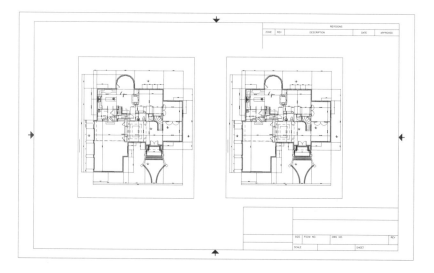

To avoid this problem, after creating one viewport and *Xrefing* the first drawing but before creating the second viewport, use *Freeze in New Viewports* for all layers of the *Xref* drawing (Fig. 33-4, all "Xref1 |*" layers). Next, create the second viewport. The original *Xref* drawing does not appear in the second viewport (right) since all of its layers are frozen for that viewport and all other new viewports (Fig. 33-5). (The *Freeze in New Viewports* icon displays the same information for a specific layer no matter which viewport is active when the list is displayed.)

Figure 33-4

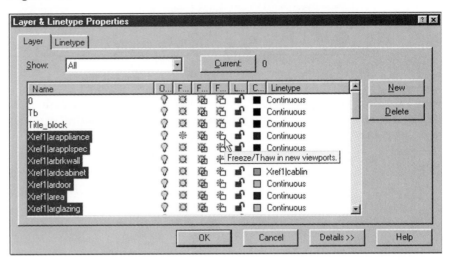

Figure 33-5

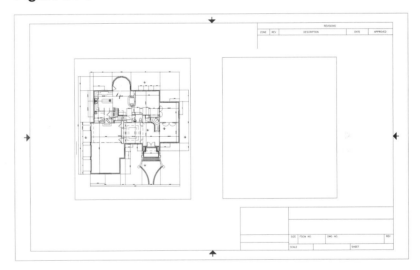

Layer Control Drop-down List

The *Layer Control* drop-down list includes the icon (for each layer) to *Freeze in Current Viewport* but does not include the icon to *Freeze in New Viewports* (Fig. 33-6). This column of icons (*Freeze in Current Viewport*) serves the identical function as the same column in the *Layer & Linetype Properties* dialog box, so either could be used to freeze layers for the current viewport (see *Layer, Freeze in Current Viewport* in the previous section).

Figure 33-6

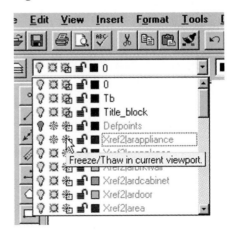

VPLAYER

Pull-down Menu	COMMAND (TYPE)	ALIAS (TYPE)	Short-cut	Screen (side) Menu	Tablet Menu
...	*VPLAYER*	...	...	...	...

The *Vplayer* (Viewport Layer) command provides options for control of the layers you want to see in each viewport (viewport-specific layer visibility control). With *Vplayer* you can specify the state (*Freeze/Thaw* and *Off/On*) of the drawing layers in any existing or new viewport. You do <u>not</u> have to make the desired viewport current before using *Vplayer*.

The *Layer* command allows you to *Freeze/Thaw* layers <u>globally</u>, whereas, *Vplayer* only allows you to *Freeze/Thaw* layers that are <u>viewport-specific</u>. The *TILEMODE* variable must be set to 0 for the *Vplayer* command to operate, and a layer's global (*Layer*) settings must be *On* and *Thawed* to be affected by the *Vplayer* settings.

For example, you may have two paper space viewports set up, such as in Figure 33-7. In order to control the visibility of the DIM (dimensioning) layer for a specific viewport, the command syntax shown below would be used. In this case, the DIM layer is *Frozen* in the right viewport.

Notice in the command sequence below that if floating model space is active, AutoCAD automatically switches to paper space to allow you to make the selection (viewport objects can only be PICKed from paper space):

Figure 33-7

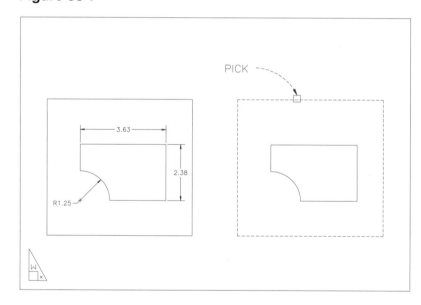

```
Command: vplayer
?/Freeze/Thaw/Reset/Newfrz/Vpvisdflt: freeze
Layer(s) to Freeze: dim
All/Select/<Current>: S (Use the Select option.)
Switching to Paper space.
Select objects: PICK (Select the desired viewport object. See Fig. 33-7.)
Select objects: Enter
Switching to Model space.
?/Freeze/Thaw/Reset/Newfrz/Vpvisdflt: Enter
Regenerating drawing.
Command:
```

The *Vplayer* **Current** option automatically selects the current viewport. The **All** option automatically selects all viewports. Note that the *Select* option switches to paper space so that the viewport objects (borders) can be selected. The *Freeze* or *Thaw* does not take place until the command is completed and the drawing is automatically regenerated. All of the options are explained briefly here.

?

This option <u>lists the frozen layers</u> in the current viewport or a selected viewport. AutoCAD automatically switches to paper space if you are in model space to allow you to select the viewport object.

Freeze

This option controls the visibility of specified layers in the <u>current or selected viewports</u>. You are prompted for layer(s) to *Freeze*, then to select which viewports for the *Freeze* to affect.

Thaw

This option turns the visibility control back to the global settings of *On/Off/Freeze/Thaw* in the <u>Layer</u> command. The procedure for specification of layers and viewports is identical to the *Freeze* option procedure.

Reset

Reset returns the layer visibility status to the <u>default</u> if one was specified with *Vpvisdflt*, which is described next. You are prompted for "Layer(s) to Reset" and viewports to select by "All/Select/<Current>:".

Newfrz

This option <u>creates new layers</u>. The new layers are frozen in all viewports. This is a shortcut when you want to create a new layer that is visible only in the current viewport. Use *Newfrz* to create the new layer, then use *Vplayer Thaw* to make it visible only in that (or other selected) viewport(s). You are prompted for "New viewport frozen layer name(s):". You may enter one name or several separated by commas.

Vpvisdflt

This option allows you to set up <u>layer visibility</u> for <u>new</u> viewports. This is handy when you want to create new viewports but do not want any of the <u>existing</u> layers to be visible in the new viewports. This option is particularly helpful for *Xrefs*. For example, suppose you created a viewport and *Xrefed* a drawing "into" the viewport. Before creating new viewports and *Xrefing* other drawings "into" them, use *Vpvisdflt* to set the default layer visibility to *Off* for the <u>existing</u> *Xrefed* layers in new viewports. (See the following application example.) You are prompted for a list of layers and whether they should be *Thawed* or *Frozen*.

Dimensioning in Floating Viewports Using *Scale to Paper Space* (*DIMSCALE*)

The option in the *Dimension Styles, Geometry* dialog box (*DIMSCALE* variable) has a special meaning when you dimension in paper space. As you know, this variable allows you to control the size of the dimensioning features (arrows, text, etc.). *Scale to Paper Space* (*DIMSCALE*) can be set to ensure that dimensions appearing in <u>different</u> viewports of different *Zoom XP* factors have the <u>same</u> text sizes, arrow sizes, etc. (Fig. 33-8, lower-right).

Figure 33-8

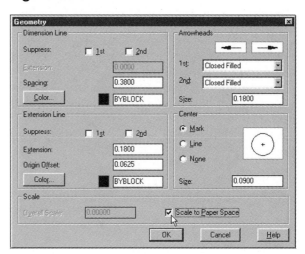

If the *Scale to Paper Space* option is checked (or *DIMSCALE* is set to 0), AutoCAD computes a value based on the scaling between the <u>current viewport</u> and paper space. In other words, a *DIMSCALE* of 0 leads to dimension text sizes, arrow sizes, and other scaled distances to plot at their original face values by counteracting the effects of the viewport's *Zoom* factor. If you are in paper space, AutoCAD uses a scale factor of 1.0.

For example, assume that you have completed the model geometry for a part with associative dimensions in model space as you would normally. Next, enable paper space and create two viewports to display two views, as shown in Figure 33-9. *Zoom XP* factors have been applied to model geometry in each viewport as follows:

Left	*Zoom 1XP*
Right	*Zoom 2XP*

Notice how the dimension sizes are displayed according to the *Zoom XP* values.

In order to force the dimensions to appear the same size in each viewport, switch to paper space and check *Scale to Paper Space* (or set *DIMSCALE* to 0). This forces the viewport-specific *DIMSCALE* to be divided by the *Zoom XP* factor.

Next, switch to floating model space, activate the desired viewport, and use the *Dimstyle* command with the *Apply* option and select the three existing dimensions, as shown in Figure 33-10. Notice that dimensions appearing in both viewports can only be displayed in one size (there is only one model space geometry).

An alternative is to create dimensions on separate layers for each viewport (or use *Ddchprop* to change layers for existing dimensions). In this way you can have viewport-specific layer visibility for dimensions and avoid the problem of having dimensions appearing in both viewports (Fig. 33-11).

Figure 33-9

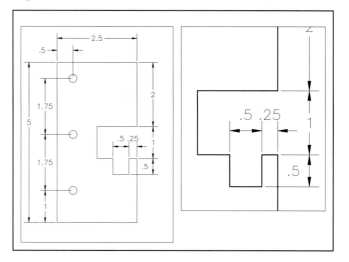

Figure 33-10

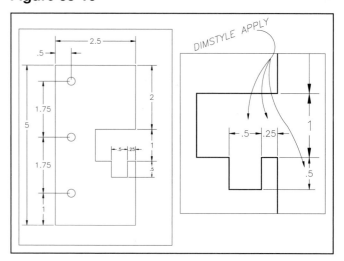

Figure 33-11

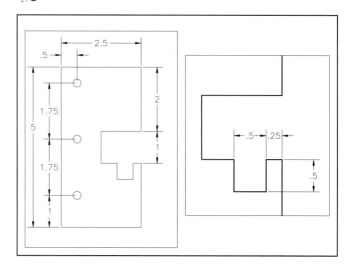

Dimensioning in Paper Space Using *DIMLFAC*

Even though part geometry should be dimensioned in model space (so the dimensions are associated with the model), you can place dimensions in paper space for particular situations. The *DIMLFAC* variable has special meaning when placing <u>dimensions in paper space</u> related to geometry in model space. If you set *DIMLFAC* to a negative value equal to the *Zoom XP* factor of the desired viewport, AutoCAD generates a measured length (dimension text value) relative to the model space units instead of paper space units. This is done automatically if you use the *Viewport* option of *DIMLFAC*, as described below. If you are creating a dimension in model space, negative values are ignored and the value of 1.0 is used instead.

AutoCAD computes a value of *DIMLFAC* for you if you change *DIMLFAC* from the "Dim:" prompt while in paper space and you select the *Viewport* option. You <u>must be in paper space</u> and invoke *DIMLFAC* from the Dim: prompt (not the Command: prompt) to enable the *Viewport* option:

> Command: **dim**
> Dim: **dimlfac**
> Current value <1.0000> New value (Viewport): **v**
> Select viewport to set scale: **PICK** the desired viewport object.
> Dimlfac set to -*nnn.nn*
> Dim:

AutoCAD calculates the scaling of model space to paper space (*Zoom XP* factor) and assigns the negative of this value to *DIMLFAC*. Setting the *DIMLFAC* variable with the *Viewport* option causes the measured value to be calculated relative to the model space geometry rather than indicating the actual paper space measurement.

For example, suppose an additional dimension is required for the detailed view and you need to create the dimension in paper space since there is insufficient room in the viewport. First, switch to paper space; then create the associative dimension in paper space (see Fig. 33-12 above the right viewport). *OSNAP* can be used to snap to the model space geometry.

When dimensioning in paper space using the <u>default</u> *DIMLFAC* setting, the dimensional value as measured by AutoCAD appears in <u>paper space units</u> as in Figure 33-12. Obviously, this is an <u>incorrect</u> value for the model space geometry.

Figure 33-12 —————————

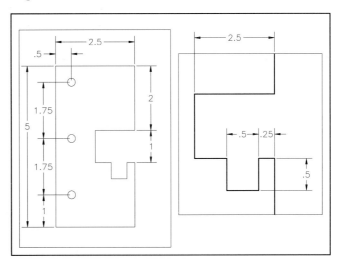

DIMLFAC should be set for paper space using the *Viewport* option (at the Dim: prompt). Select the viewport object (border). Create the dimension again, or use *Dimstyle Apply* to update an existing dimension to the new *DIMLFAC* setting. The resulting dimension is displayed relative to model space units as shown in Figure 33-13.

The VIEWPORT layer can be *Frozen* with the *Layer* dialog box to prepare the drawing for plotting, as shown in Figure 33-13.

Figure 33-13

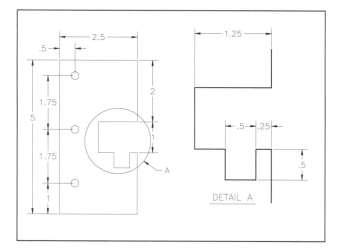

PSLTSCALE

	Pull-down Menu	COMMAND (TYPE)	ALIAS (TYPE)	Short-cut	Screen (side) Menu	Tablet Menu
	Format *Linetype... Details>> Use paper space units...*	*PSLTSCALE*	...	...	*FORMAT Linetype Details>> Use p.s. units*	...

LTSCALE controls linetype scaling <u>globally</u> for the drawing, whereas the *PSLTSCALE* variable controls the linetype scaling for non-continuous lines only in <u>paper space</u>. A *PSLTSCALE* setting of 1 (the default setting) displays all non-continuous lines relative to paper space units so that all lines in both spaces and in different viewports appear the same regardless of viewport *Zoom* values. *LTSCALE* can still be used to change linetype scaling globally regardless of the *PSLTSCALE* setting.

PSLTSCALE can also be set using the *Linetype* tab of the *Layer & Linetype Properties* dialog box (Fig. 33-14, lower-left). A check appearing in the *Use paper space units for scaling* checkbox sets *PSLTSCALE* to 1. This check normally appears since the default setting is 1.

Figure 33-14

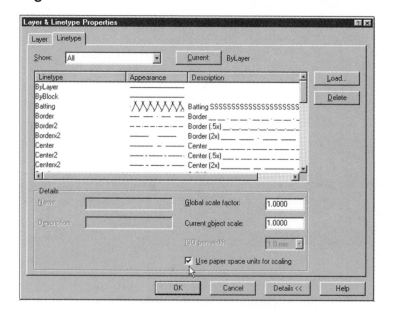

If *PSLTSCALE* is set to 0, linetype spacing is controlled by the *LTSCALE* variable and based on the drawing units in the space the objects are created in (paper or model). In other words, if *PSLTSCALE* is 0, the linetypes appear different lengths if the units in paper and model spaces are scaled differently or if different viewports are scaled differently as in Figure 33-15.

If *PSLTSCALE* is changed, a *Regenall* must be used to display the effects of the new setting.

Figure 33-15

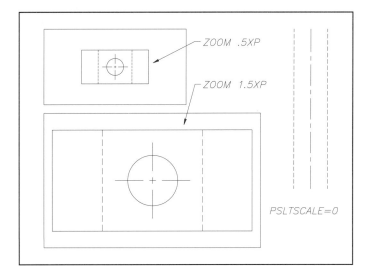

If *PSLTSCALE* is set to 1 (and *TILEMODE* is 0), the linetype scales for both spaces are <u>scaled to paper space units</u> so that all dashed lines <u>appear equal</u>. The model geometry in viewports can have different *Zoom XP* factors, yet the linetypes display the same (as shown in Fig. 33-16).

Figure 33-16

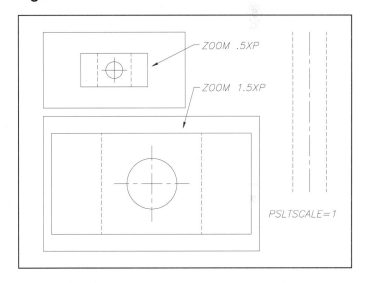

EXAMPLES OF USING MULTIPLE PAPER SPACE VIEWPORTS

The reasons for using multiple paper space viewports are listed near the beginning of the chapter. Following are examples of the first two applications listed: using paper space viewports to display two views of one drawing and using paper space to display several *Xrefed* drawings on one sheet. Applications of paper space for 3D drawings are discussed in Chapter 42.

1. Viewports for Different Views of One Drawing

In this case, a full view of the part is displayed in one viewport and a detail of the part is displayed in a different scale in a second viewport.

1. Create the part geometry in model space.

Begin AutoCAD and create the part geometry as you normally would (model space is enabled by default when you enter the Drawing Editor). Associative dimensions are placed on the part in model space (Fig. 33-17).

Figure 33-17

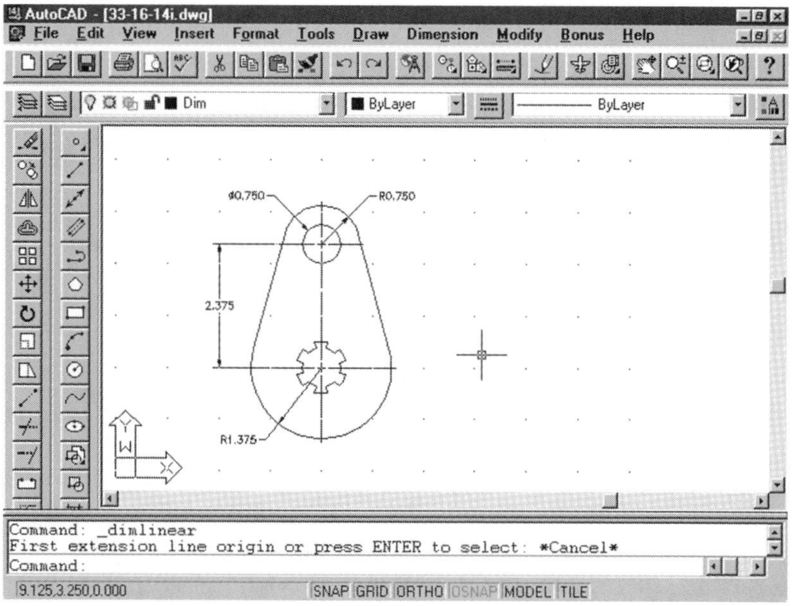

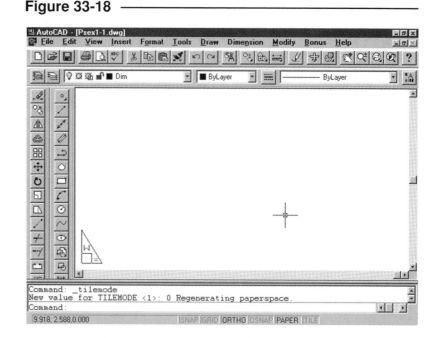

2. Switch to paper space by changing *TILEMODE* to 0.

When *TILEMODE* is changed from 1 to 0, you are automatically switched from model space to paper space. As a reminder that paper space is enabled, the coordinate system icon (in the lower-left corner of the Drawing Editor) changes to a drafting triangle (Fig. 33-18). Also, the word "TILE" at the bottom of the screen is grayed out.

Figure 33-18

3. Set up paper space to mimic the plot sheet.

Paper space can have *Limits*, *Snap*, and *Grid* settings independent of those settings in model space. Set paper space <u>*Limits* equal to the paper size</u>. *Snap* and *Grid* spacing can be set to an appropriate value.

Make a new layer called BORDER or TITLE for the title block and border. Draw, *Insert*, or *Xref* the title block and border on the new layer (Fig. 33-19).

Figure 33-19

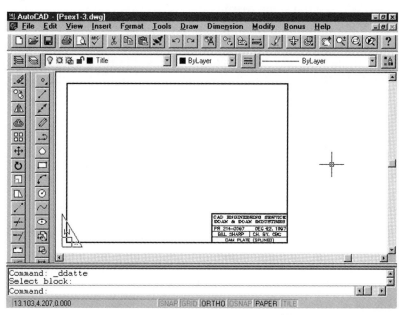

4. Make viewports in paper space.

Make a layer named VIEWPORT and set it as the *Current* layer. Next, draw viewport objects with the *Mview* command <u>while in paper space</u> in order to see part geometry created in model space. Paper space viewports are windows into model space, hence the term "floating model space" (Fig. 33-20).

Figure 33-20

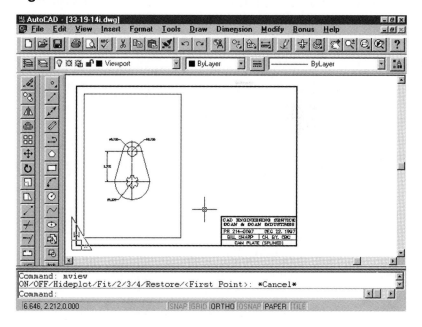

Paper space viewports created with *Mview* are actually <u>objects</u> drawn on the current layer. Because they are objects, they are affected by *Move*, *Copy*, *Array*, *Scale*, and *Erase* and can overlap one another. Subsequent viewports can be created by *Copying* or *Arraying* existing viewports or by using the *Mview* command. You can use *Mview* to make 2, 3, or 4 viewports or to *Fit* multiple viewports into a window (see *Mview*, Chapter 19).

When several paper space viewports exist, you see the same display of model space in each floating viewport by default. Remember that there is only one model space, so you will see it in each viewport (see Figure 33-21). You can, however, control the part of model space you wish to see in each viewport by using display or layer commands.

A second viewport is created by the *Mview* command or other commands (like *Copy* and *Stretch*, for example). The new viewport is also a window into model space (Fig. 33-21). By default, the display is the same geometry as in the first viewport.

Figure 33-21 ————————————

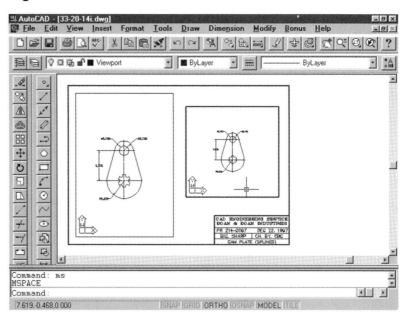

You can switch between drawing in paper space and model space by using *Pspace* (*PS*) and *Mspace* (*MS*) commands. Alternately, you can double-click on the word "PAPER" or "MODEL" at the bottom of the screen to switch between paper space and floating model space (model space inside a viewport). If you are in floating model space, the "crosshairs" cursor is visible only within the viewport (as shown in Fig. 33-21) and the "arrow" pointer appears when you move outside the viewport. If paper space is active, the cursor can move entirely across the screen but cannot move "into" a viewport. The current viewport is highlighted by a thick border. You can make a different viewport current by PICKing a point inside it.

5. Control the display of model space graphics in each viewport with display commands and layer visibility controls.

Use the *Mspace* (*MS*) command to activate model space. Move the cursor and PICK to activate the desired viewport. Use *Zoom XP* to scale the display of the model units to paper space units. In this case, *Zoom 1XP* would be used to scale the model geometry in the left viewport to 1 times paper space units. The geometry in the right viewport is scaled to 2 times paper space units by using *Zoom 2XP* (Fig. 33-22).

When switched to floating model space, display commands that are invoked affect the display of the current viewport only. Draw and edit commands, however, affect the model geometry, so these effects are potentially visible in all viewports.

Figure 33-22 ————————————

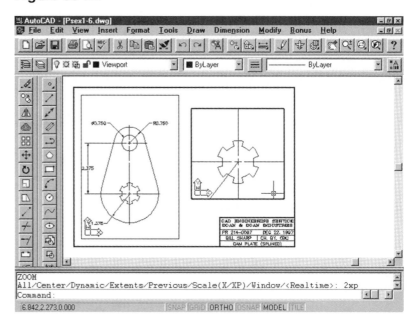

Next, set the viewport-specific layer visibility. Use the *Vplayer* command or the *Freeze in Current Viewport* icons in the *Layer & Linetype Properties* dialog box or *Layer Control* drop-down list to control which layers are visible in which viewports. The desired layer(s) can be *Frozen* or *Thawed* for each viewport. For example, the dimension layer could be frozen for the right viewport only to yield a display as shown in Figure 33-23.

Figure 33-23

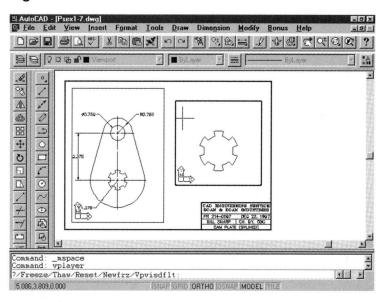

6.   Make the final adjustments in paper space and plot at a scale of 1=1.

Use the *Pspace* (*PS*) command to switch to paper space. Annotations can be added to the drawing in paper space (such as the "Detail A" notes; Fig. 33-24).

Because the viewports are on a separate layer, you can control the display of the rectangular viewport borders (objects) by controlling the layer's visibility. Turning the VIEWPORT layer (or whatever name is assigned) *Off* prevents the viewport objects from displaying. This is useful when the drawing is ready for plotting.

Plot the drawing from paper space at a plot scale of 1=1. Since paper space is set to the actual paper size, the paper space geometry plots full size, and the model space geometry plots at the scale determined by the *Zoom XP* factor.

Figure 33-24

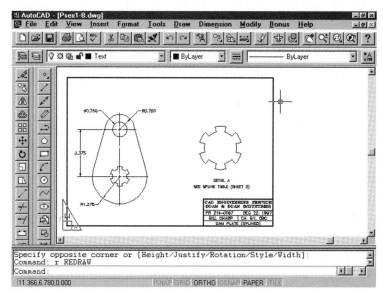

2.   Viewports for *Xrefed* Drawings

In the following example, multiple viewports are created. The drawing also contains several *Xrefs*, each *Xref* appearing in a separate viewport. Viewport-specific layer visibility control is important for this application to ensure that <u>only one *Xref*</u> is displayed in <u>each viewport</u>. This example describes a typical procedure. Slightly different applications may require different procedures.

First, a *New* drawing is begun and *Units* are set to *Architectural*. Paper space is enabled immediately by setting *TILEMODE* to 0, then setting *Limits* in paper space equal to a D size sheet (36 x 24). After doing a *Zoom All*, the XREF, VIEWPORT, and TITLE layers are created. Next, a *Pline* border is drawn and a title block is drawn, *Inserted*, or *Xrefed* on layer TITLE.

Then the VIEWPORT layer is set as the current layer, and *Mview* is used to create a viewport. The drawing, at this point, should appear similar to that in Figure 33-25.

The *MS* (*Mspace*) command is used to activate model space so the *Xrefed* drawings will come into model space. The current layer is then set to XREF. The *Xref* command with the *Attach* option is used to bring in the first drawing—the DOOR.DWG. Accepting all the defaults, the drawing may not appear until *Zoom Extents* is used to make the geometry visible in the viewport.

In order to make the drawing appear in the viewport and plot to the correct scale, a *Zoom XP* factor should be used. The original door drawing was created full size and would normally be plotted on a D size sheet at a scale of 3/4"=1'. Since paper space is set to the actual paper size, the drawing is scaled proportionally by using a *Zoom* factor of *1/16XP* as before. *Pan* is used to center the door in the viewport. The resulting drawing would be similar to Figure 33-26.

Before creating another viewport, *PS* is used to activate paper space and the current layer is set to VIEWPORT. A second viewport is created with the *Mview* or *Copy* command. The current layer is set again to XREF and model space is activated with *MS*. Another detail drawing, WALL, is *Attached* with *Xref*. The intended plot scale for the WALL drawing is 3"=1', accomplished by using *Zoom 1/4XP*. The drawing, at this point, appears similar to Figure 33-27.

Figure 33-25 ———————————

Figure 33-26 ———————————

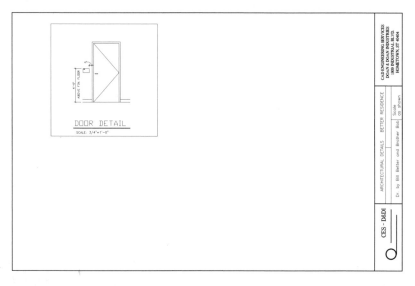

Figure 33-27 ———————————

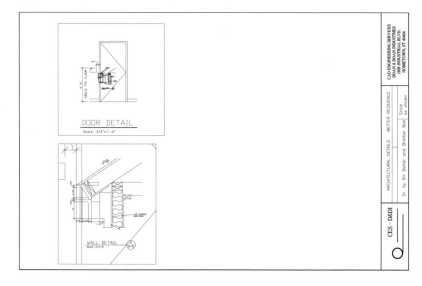

Note that <u>both drawings appear in each viewport</u> (Fig. 33-27). They have each been *Xrefed* into model space, and there is only one model space (this is apparent if *TILEMODE* is changed to 1). Viewport specific layer control must be exercised to display only one *Xref* per viewport. Using the *Layer* tab of the *Layer & Linetype Properties* dialog box, the visibility of dependent layers can be controlled for each viewport. With the bottom viewport active (WALL detail), all the DOOR drawing layers are selected and frozen for the <u>current</u> viewport, and <u>new</u> viewports, as shown in Figure 33-28. Confirming the dialog box (PICK *OK*) regenerates the new display.

Figure 33-28 ───────────

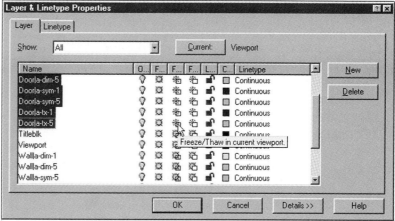

Next, visibility for the top viewport must be specified. The top viewport is made active and the dialog box invoked again. Remember that the *Layer* tab of the *Layer & Linetype Properties* dialog box displays settings specific to the <u>active</u> viewport. All of the WALL drawing-dependent layers are set to *Freeze in Current Viewport* and *Freeze in New Viewports*. The correct layer visibility appears like that in Figure 33-29. The dependent WALL and DOOR layers do not appear in any new viewports.

Figure 33-29 ───────────

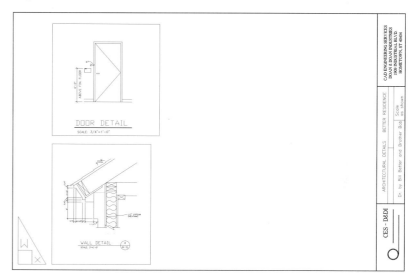

Alternatively, the *Vplayer* command could be used to accomplish the viewport-specific layer visibility. (See the alternate method at the end of this section.)

Next, paper space is activated (with *PS*) and the current layer is set to VIEWPORT. One more viewport is created with *Mview*. The WALL* and DOOR* layers do not appear in the new viewport because of the previous action to *Freeze in New Viewports*. Model space is activated and the right viewport is made current. Layer XREF is set current. *Xref* is used to *Attach* the SECTION detail drawing. It is *Zoomed* to *1/16XP* to achieve a scale of 3/4"=1'.

Now, even though the DOOR and WALL layers appear only in the appropriate viewports, all of the SECTION-dependent layers appear in all viewports. The *Vplayer* command can be used to control layer visibility as an alternative to the *Layer Control* dialog box. *Vplayer* can be used to set a default visibility setting for specific layers with the *Freeze* option. *Vplayer* can be invoked from paper space or model space. The asterisk (*) is used to specify all SECTION layers, as shown in the following command sequence:

```
Command: vplayer
?/Freeze/Thaw/Reset/Newfrz/Vpvisdflt: f
Layer(s) to Freeze: SECTION*
All/Select/<Current>: s  (use the Select option)
Select objects: PICK  (top viewport object)
Select objects: PICK  (bottom viewport object)
Select objects: Enter
?/Freeze/Thaw/Reset/Newfrz/Vpvisdflt: Enter
Regenerating drawing.
Command:
```

Lastly, the VIEWPORT layer is *Frozen* so the viewport borders do not appear. The resulting drawing, as shown in Figure 33-30, is ready for plotting.

Figure 33-30 ───────────

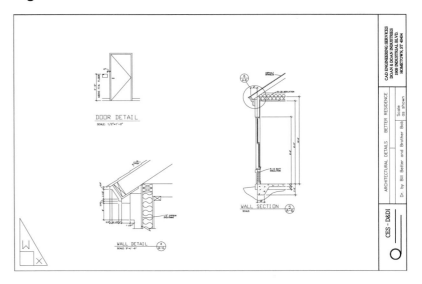

Alternate Layer Visibility Control Method

As an alternative method of layer control for the previous example, the *Vpvisdflt* option of *Vplayer* can be used to set the default visibility settings for new viewports. The procedure is as follows:

1. Create the first viewport and *Xref* the DOOR drawing as before.

2. Invoke the *Vplayer* command with the *Vpvisdflt* option as follows:

```
Command: vplayer
?/Freeze/Thaw/Reset/Newfrz/Vpvisdflt: v
Layer name(s) to change default viewport visibility: DOOR*
Change default viewport visibility to Frozen/<Thawed>: f
?/Freeze/Thaw/Reset/Newfrz/Vpvisdflt: Enter
Command:
```

The DOOR layers are frozen for any <u>new</u> viewports that are created.

3. Create the second viewport and *Xref* the WALL drawing as before.

4. Invoke *Vplayer* and use the *Vpvisdflt* option to freeze the WALL* layers for <u>new</u> viewports.

5. Invoke *Vplayer* and use the *Freeze* option to freeze the WALL* layers for the first viewport.

6. Continue this sequence for the last viewport.

Use *Xref* with or without Paper Space?

If you want to create one drawing that displays several *Xrefed* drawings, is it better to use paper space viewports (each viewport displays one drawing) or to use *Xref* without paper space (and *Xclip* if necessary)?

In this chapter the procedure is explained for creating paper space viewports to control the display of several *Xrefs*. In previous versions of AutoCAD, this method was the only possibility to achieve such results. In Release 14 it is possible to use *Xref* and display several drawings in model space (*TILEMODE=1*) and still control layer visibility (these procedures are discussed in Chapter 30, Xreferences). Demand loading, the *Xclip* feature, spatial indexing, and layer indexing make this possible.

In general, it is simpler to use *Xref* <u>without</u> paper space viewports to display several drawings in one (parent) drawing. Spatial indexing, layer indexing, and demand loading should be enabled to maximize the performance benefits. If you *Xref* several drawings, the layer visibility for each *Xref* can be controlled, just as if you created paper space viewports.

If you want to create several displays of one drawing (a main view and several details, for example) and control the display of each, paper space viewports should be used in the original drawing rather than using *Xref* and *Xclip* to display several views of one drawing. Using paper space viewports, layer visibility can be controlled <u>for each viewport,</u> so it is possible to control the geometry <u>and the layers</u> that are visible in each viewport (see Fig. 33-23). Using *Xref* and *Xclip* to create separate views of one drawing provides control of the geometry in each clip boundary, but not control of layer visibility—all clip boundaries must display the same layers for any one drawing.

USING *MVSETUP* AND PAPER SPACE FOR 2D DRAWINGS

Mvsetup is an AutoLISP program supplied with AutoCAD that enables you to automatically create, scale, and align paper space viewports and insert title blocks for 2D or 3D model geometry. Several ANSI and ISO title blocks and borders are available for insertion.

Mvsetup can be used when *TILEMODE* is *On* or *Off*. If you choose to use *Mvsetup* when *TILEMODE* is *On*, AutoCAD draws a "bounding box" border and ends the command. However, using *Mvsetup* in conjunction with paper space (with *TILEMODE Off*) displays the power of this routine. This section describes the use of *Mvsetup* with paper space for 2D drawings. A popular application for *Mvsetup* is displaying 3D objects in a typical multiview-type viewport configuration. *Mvsetup* is discussed for that purpose in Chapter 42.

MVSETUP

Pull-down Menu	COMMAND (TYPE)	ALIAS (TYPE)	Short-cut	Screen (side) Menu	Tablet Menu
...	*MVSETUP*	...	...	...	...

There are two applications for *Mvsetup*: for model space setup (*TILEMODE=1*) and for paper space setup (*TILEMODE=0*). One application of *Mvsetup* with paper space (*TILEMODE* is 0) is for the setup of paper space viewports and insertion of title blocks for 2D drawings.

Mvsetup can be used with *TILEMODE* at either setting (1 or 0). If paper space has not been enabled (*TILEMODE* = 1), *Mvsetup* asks if you want to enable it. For *TILEMODE* set to 0, the command syntax is listed here. The options are explained here:

    Command: **mvsetup**
    Initializing...
    Align/Create/Scale viewports/Options/Title block/Undo:

Create
Using this option displays the following prompt:

    Delete objects/Undo/<Create viewports>:

Create viewports

    Available Mview viewport layout options:

    0.   None
    1.   Single
    2.   Std. Engineering
    3.   Array of Viewports

1. Single
This option prompts for a first and other point to define the viewport.

2. Std. Engineering
This option is for 3D drawings and is discussed in Chapter 42.

3. Array of Viewports
Selecting this option prompts for a bounding box and a number of viewports along the X and Y axes. You can also specify a distance between viewports. The resulting viewports are of equal size.

Delete objects
Use *Delete objects* to delete existing <u>viewports</u>.

Undo
Undoes the previous *Mvsetup* operation (without exiting *Mvsetup*).

Options
Options is used for <u>title block and border insertion options</u>. Using this displays the following prompt:

    Choose option to set—Layer/LImits/Units/Xref:

Layer
This selection allows you to specify a layer on which to insert a title block and border. The layer can be existing or you can name a new one.

Limits
If desired, this option resets the *Limits* after inserting the title block and border to the drawing extents defined by the border.

Units
You may specify inch or millimeter units for use in paper space.

Xref
This option is used to specify if the title block and border are to be *Attached* as an *Xref* or *Inserted* as a Block.

Title block
Choosing this option displays the prompt: Delete objects/Origin/Undo/<Insert title block>:

Delete Objects
This option allows you to delete (erase) objects from paper space without having to leave the *Mvsetup* routine. It operates like the *Erase* command.

Origin
You can relocate the plotter origin point for the sheet with this option. You are prompted to PICK a point.

Undo
This option undoes the last *Mvsetup* operation (without exiting *Mvsetup*).

Insert title block
This option provides the following options and title blocks with borders for insertion.

Available title block options:

0:	None	
1:	ISO A4 Size(mm)	
2:	ISO A3 Size(mm)	
3:	ISO A2 Size(mm)	
4:	ISO A1 Size(mm)	
5:	ISO A0 Size(mm)	
6:	ANSI-V Size(in)	(Vertical A size)
7:	ANSI-A Size(in)	
8:	ANSI-B Size(in)	
9:	ANSI-C Size(in)	
10:	ANSI-D Size(in)	
11:	ANSI-E Size(in)	
12:	Arch/Engineering (24 x 36in)	
13:	Generic D size Sheet (24 x 36in)	

Add/Delete/Redisplay/<Number of entry to load>:

Entering a number causes the appropriate size border and title block to be created. Custom borders can be created and inserted instead of the AutoCAD-supplied ones by using the *Add* option. The *Add* option allows you to add title block options to the list shown above. *Delete* will delete options from the list.

Typical Steps for Using *Mvsetup* (with Paper Space) for 2D Drawings

The typical steps for using *Mvsetup* for this application are as follows:

1. Create the 2D drawing in model space.

2. Load *Mvsetup* and use *Options* to set the preferences for title block insertion and to specify the units to be used.

3.   Use *Title block* to *Insert* or *Xref* one of many AutoCAD-supplied or user-supplied borders and title blocks.

4.   Use the *Create* option to make one or an array of paper space viewports. (The *"Std. Engineering"* option is intended for 3D drawings.)

5.   Use *Scale viewports* to set the scale factor (*Zoom XP* factor) for the geometry displayed in the viewports.

Using *Mvsetup* to Create a Single Viewport in Paper Space

Following is a simple example of using *Mvsetup* to create a paper space viewport and insert a title and border for a completed 2D drawing. Assume that you have just completed the drawing shown in Figure 33-31. The geometry and dimensions are created in model space.

Normally, the next step is to set *TILEMODE* to 0, which enables paper space. Invoke *Mvsetup*. From the first prompt, select *Options*. Use *Layer* to define an existing layer or create a new layer (named TITLE or similar descriptive name) for title block and border insertion.

Figure 33-31 ————————————————————

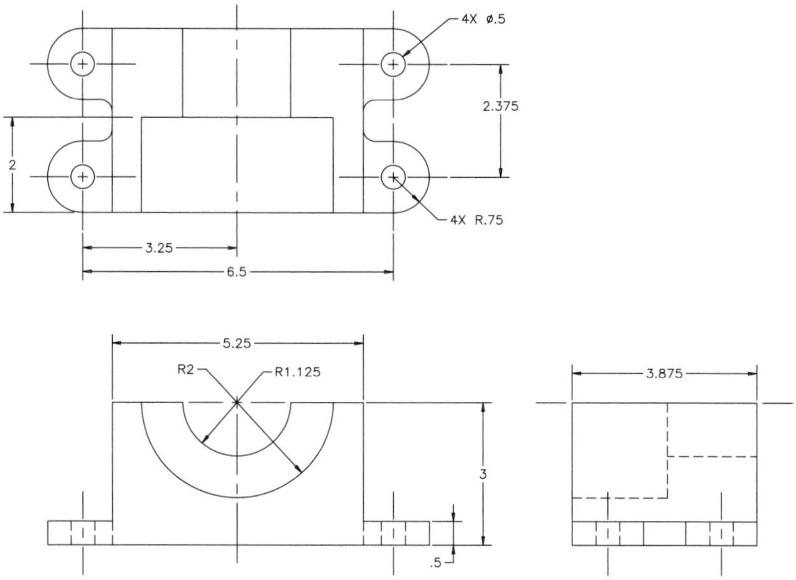

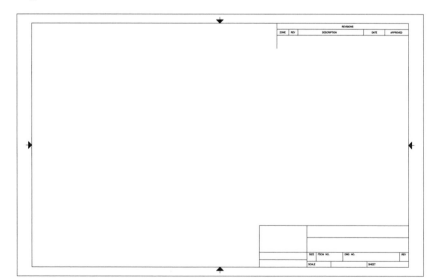

The *Limits* option can be used to automatically set paper space *Limits* equal to the paper space extents after the border and title block are inserted (the drawing *Extents* is equal to the border size). Also use the *Xref* option to specify whether the title block and border are to be *Inserted* or *Xrefed*. Next, use the *Title block* option and then *Insert title block*. Since this drawing is intended to be plotted on a B size sheet, for this example, option *8. ANSI-B size* is selected. The resulting title block appears in paper space (Fig. 33-32).

Figure 33-32 ————————————————————

Next, layer VIEWPORT (or other descriptive name) must be created and made current for the viewport objects. To make the viewports, the *Create* option of the *Mvsetup* command is invoked; then *Create viewports* is selected. From the list of available viewport layouts, *1. Single* is designated. PICK the "First point" and the "Other corner" to define the viewport size. By default, the model geometry is displayed in the viewport at maximum size (AutoCAD causes a *Zoom Extents*; Fig. 33-33).

Figure 33-33

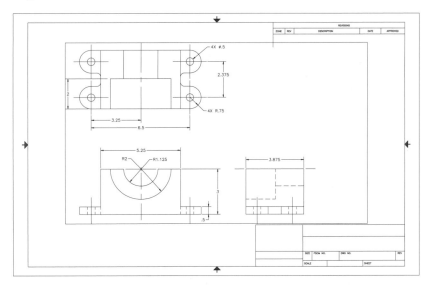

In order to display the model geometry to scale, the *Scale viewports* option from the main prompt is selected. AutoCAD requires you to enter the ratio of "paper space units" to "model space units." This ratio is the *Zoom XP* factor; for example, 1 paper space unit to 2 model space units equals *Zoom 1/2XP*.

The resulting drawing should appear as shown in Figure 33-34. Enter the desired text into the title block. If desired, *Freeze* the VIEW-PORT layer. The drawing is ready for plotting.

Figure 33-34

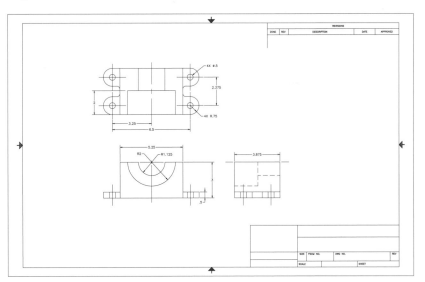

The *Mvsetup* command is an AutoLISP program that can be used to automate the process of setting up paper space viewports. *Mvsetup* can save time and effort compared to the previous methods discussed in which the steps are not as automated.

CHAPTER EXERCISES

1. *TILEMODE, Mview*

   This first exercise will serve as a review of the basic procedures and commands for creating one paper space viewport and using *Xref* to display and print an existing drawing.

   A. *Open* the **ASSY** drawing that you created in the Chapter 30 Exercises (make sure all of the *Xrefed* drawings are also located in your working directory). Use *Xref* with the *Bind* option to bind all of the *Xrefed* drawings. Use *SaveAs* to save and name the drawing as **ASSY-PS**.

   B. Set *TILEMODE* to **0**. Set *Limits* in paper space to the paper size that you intend to plot on. Set paper space *Snap* and *Grid* to an appropriate size (**.250″** or **6mm**). Make a *New Layer* named **TITLE** and set it as the *Current* layer. Draw a border and draw, *Insert*, or *Xref* a title block (in paper space).

   C. Make a *New Layer* called **VIEWPORT** or **VPORT** and set it as *Current*. Next, use *Mview* to create one viewport. Make the viewport as large as possible while staying within the confines of the title block and border. Type *MS* to activate model space in the viewport, and use *Zoom XP* to scale the model geometry times paper space (remember that the geometry is in millimeter units). Enter the scale in the title block (scale=XP factor).

   D. Turn *Off* the **VIEWPORT** layer and *Save* the drawing (as ASSY-PS). *Plot* the drawing from paper space at **1=1**. Your drawing should appear as Figure 33-35.

   Figure 33-35

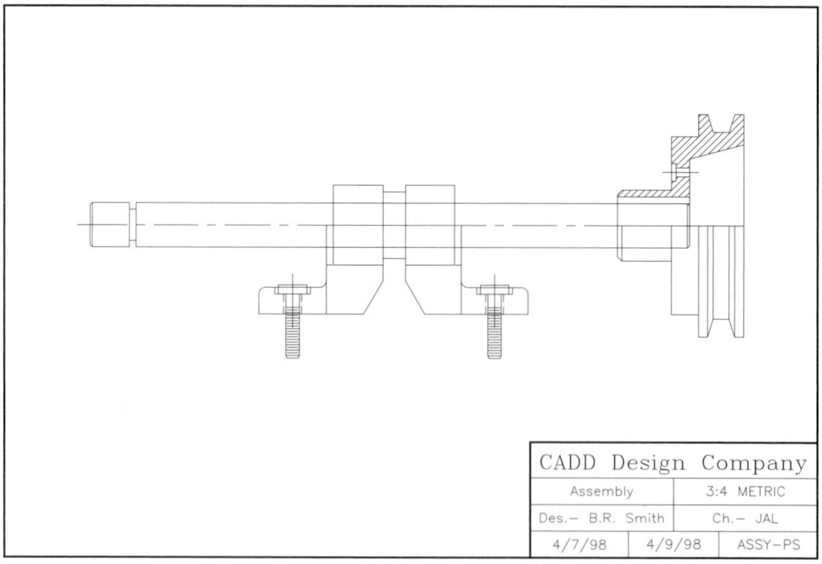

2. *Mvsetup*

   In this exercise, using an existing drawing you will enable paper space, then use *Mvsetup* to insert a border and title block, to create one viewport, and to scale model space geometry to paper space units.

   A. *Open* the **SADDL-DM** drawing from Chapter 29 Exercises. If you have already created a title block and border, *Erase* them.

   B. Change *TILEMODE* to 1 by any method (type, double-click the word *TILE* on the Status line, or select *Paper Space* from the *View* pull-down menu). You should be presented with a "blank sheet," and the paper space icon (drafting triangle) should appear.

C. Invoke *Mvsetup*. Next, select *Options*, then *Layer*. Enter **Title** to cause AutoCAD to insert the title block on layer **TITLE**. Next, use the *Limits* option, and answer *Yes* to cause AutoCAD to set Limits to the title block and border size. Finally, use the *Title block* option, then *Insert Title Block*, and select 8: *ANSI-B Size (in)* option to insert (in paper space) a title block and border for a B size sheet (17 x 11). Exit *Mvsetup* and set *Snap* to **.5**. The drawing at this point should appear like that in Figure 33-36.

Figure 33-36 —————————————

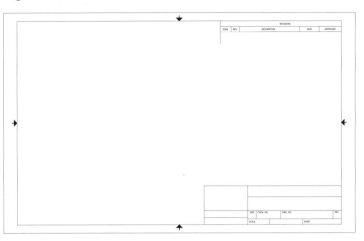

D. Create a new layer named **VIEW-PORT** and make it *Current*. Use *Mvsetup* again to create one viewport. Use the *Create* option, then enter **1** to select the "1: Single" option. Enter **2,1.5** at the "First point:" prompt and **14,9** at the "Other point:" prompt. The saddle drawing should appear in the viewport at no particular scale, similar to Figure 33-37.

Figure 33-37 —————————————

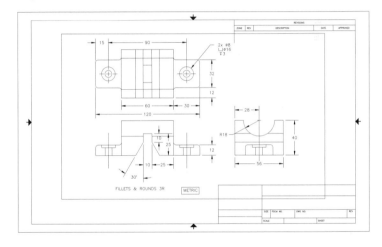

E. Next, use the *Scale Viewports* option of *Mvsetup* to scale the geometry. Select the viewport (border) to scale when prompted. To scale the model space geometry to paper space, enter 1 at the "Number of paper space units. <1.0>:" prompt and enter **25.4** at the "Number of model space units. <1.0>:" prompt. (The conversion from inches to millimeters is 25.4.) Press Enter to end the *Mvsetup* command. The size of the model space geometry changes only slightly but is scaled exactly to plot to scale. Use *Pspace* to activate paper space. Make layer **TITLE** current and *Freeze* layer **VIEWPORT**. Your completed drawing should look like that in Figure 33-38. Finally, *Save* the drawing as **SADDLE-PS** and make a plot at 1=1 on an ANSI B size sheet.

Figure 33-38 —————————————

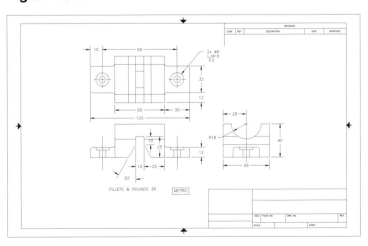

3. *TILEMODE, Mview, DIMSCALE*

In this exercise, you will use paper space with 3 viewports to show an overall view of a drawing and two details in different scales. *DIMSCALE* (*Overall Scaling*) will be set to 0 (*Scale to Paper Space*) to produce dimension features appearing equal in size.

A. *Open* the **EFF-APT2** drawing that you worked on in Chapter 30 Exercises. *Erase* the plant. Create a text *Style* using *Roman Simplex* font. Create dimensions for the interior of each room on the **DIM** layer. *Zoom* in, and on another *New Layer* named **DIM2** give the dimensions for the wash basin in the Bath and the counter for the Kitchen sink.

B. Change *TILEMODE* to **0**. Set *Limits* to the paper size for plotting. Set appropriate *Snap* and *Grid* increments. Set *Layer* **TITLE** *Current*. Draw, *Insert*, or *Xref* a border and title block.

C. Make a *New Layer* named **VIEWPORT** and set it as the *Current* layer. Use *Mview* to make one viewport on the right, occupying approximately 2/3 of the page. Use *Mview* again to make two smaller viewports in the remaining space on the left.

D. Activate the large viewport (in *MS*) and *Zoom XP* to achieve the largest possible display of the apartment to an <u>accepted scale</u>. In the top-left viewport, use display controls (including *Zoom XP*) to produce a scaled detail of the Bath. Produce a <u>scaled</u> detail of the kitchen sink in the lower-left viewport.

E. Use *Layer* to *Freeze* layer **DIM2** in the large viewport and *Freeze* layer **DIM** in the two small viewports. Change to *PS*. Set *DIMSCALE* to **0** (*Overall Scaling* checkbox will be set to *Scale to Paper Space*). Return to *MS* and use *Dimstyle Apply* to update the dimensions in the small viewports so that the dimensions appear the same size in all viewports.

F. Return to paper space and turn *Off Layer* **VIEWPORT**. Use *Text* (in paper space) to label the detail views and give the scale for each. *Save* the drawing as **EFF-APT3** and *Plot* from paper space at **1=1**. The final plot should look similar to Figure 33-39.

Figure 33-39 ────────────────────

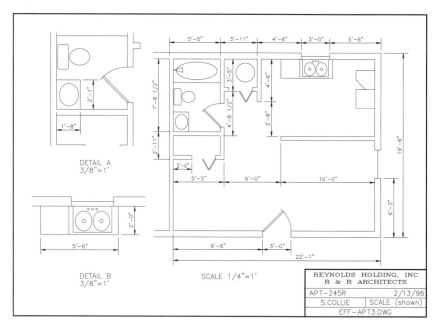

4. *TILEMODE, Mview, Xref,* **Viewport-Specific Layer Visibility**

In this exercise, you will create three paper space viewports and display three *Xrefed* drawings, one in each viewport. Viewport-specific layer visibility control will be exercised to produce the desired display.

A. Begin a *New* drawing using a *Template* to correspond to the sheet size you intend to plot on. Create *New Layers* named **TITLE, XREF,** and **VIEWPORT**. Set *TILEMODE* to **0** and prepare paper space with *Limits, Snap, Grid,* a border, and title block. On the **VIEWPORT** layer, use *Mview* to create a viewport occupying about 1/2 of the page on the right side.

B. Make the **XREF** layer *Current.* Change to *MS.* In the viewport, *Xref* the **GASKETD** drawing from Chapter 28 Exercises. Use *Zoom* with an *XP* value to scale it to paper space. Use the *Layer Control* dialog box to *Freeze* the **GASKETD | DIM** layer and to *Freeze* all **GASKETD*** layers for *New VP.*

C. Option 1.   Change to *PS* and create two more viewports on *Layer* **VIEWPORT** equal in size to the first. Set *Layer* **XREF** *Current* and *Xref* the **GASKETB** and the **GASKETC** drawings. Change to *MS* and use the *Layer Control* dialog box or *Vplayer* to constrain only one gasket to appear in each viewport.

   Option 2.   Change to *PS* and create one more viewport on *Layer* **VIEWPORT** equal in size to the first. Set *Layer* **XREF** *Current* and *Xref* only the **GASKETB** drawing. Change to *MS* and use the *Layer Control* dialog box or *Vplayer* to set the **GASKETB** visibility for the existing and new viewports. Repeat these steps for **GASKETC.**

D. Use *Zoom XP* values to produce a scaled view of each new gasket. Change to *PS* and *Freeze* layer **VIEWPORT.** Use *Text* to label the gaskets. *Plot* the drawing to scale. Your plot should look similar to Figure 33-40. *Save* the drawing as **GASKETS.**

Figure 33-40

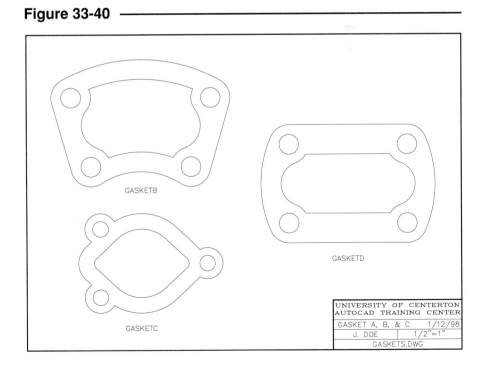

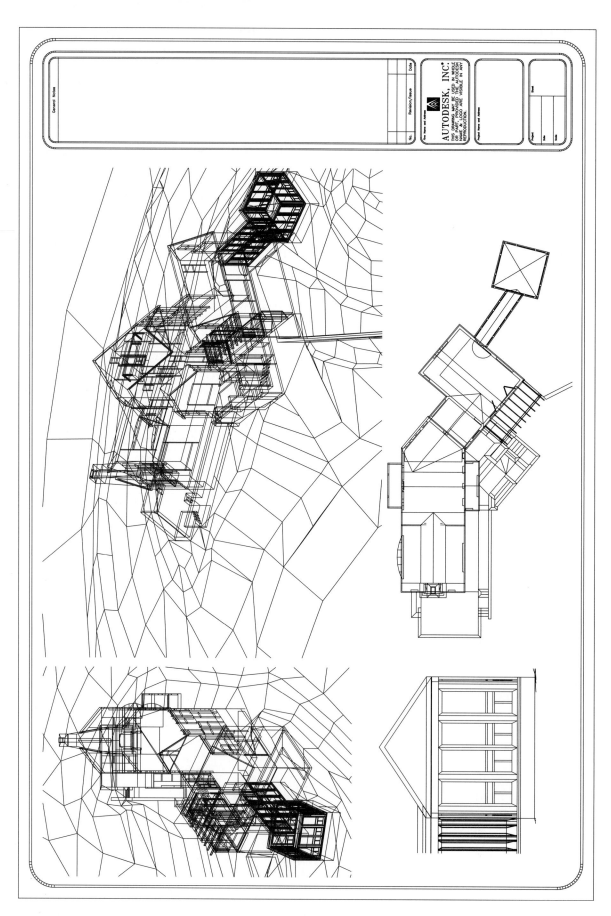

SITE-3D.DWG Courtesy of Autodesk, Inc. (Release 13 Sample Drawing)

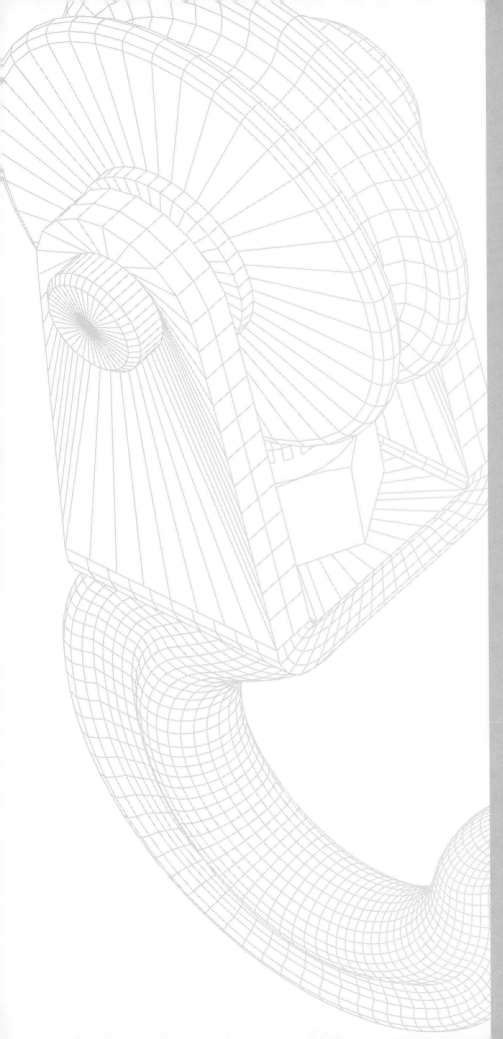

34

3D MODELING BASICS

Chapter Objectives

After completing this chapter you should:

1. know the characteristics of wireframe, surface, and solid models;

2. know the six formats for 3D coordinate entry;

3. understand the orientation of the World Coordinate System (WCS);

4. be able to use the right-hand rule for orientation of the X, Y, and Z axes and for determining positive and negative rotation about an axis;

5. be able to control the appearance and positioning of the Coordinate System icon with the *Ucsicon* command.

CONCEPTS

Three basic types of 3D (three-dimensional) models created by CAD systems are used to represent actual objects. They are:

1.   Wireframe models
2.   Surface models
3.   Solid models

These three types of 3D models range from a simple description to a very complete description of an actual object. The different types of models require different construction techniques, although many concepts of 3D modeling are the same for creating any type of model on any type of CAD system.

Wireframe Models

"Wireframe" is a good descriptor of this type of modeling. A wireframe model of a cube is like a model constructed of 12 coat-hanger wires. Each wire represents an <u>edge</u> of the actual object. The <u>surfaces</u> of the object are <u>not</u> defined; only the boundaries of surfaces are represented by edges. No wires exist where edges do not exist. The model is see-through since it has no surfaces to obscure the back edges. A wireframe model has complete dimensional information but contains no volume. Examples of wireframe models are shown in Figures 34-1 and 34-2.

Wireframe models are relatively easy and quick to construct; however, they are not very useful for visualization purposes because of their "transparency." For example, does Figure 34-1 display the cube as if you are looking towards a top-front edge or looking toward a bottom-back edge? Wireframe models tend to have an optical illusion effect, allowing you to visualize the object from two opposite directions unless another visual clue such as perspective is given.

With AutoCAD a wireframe model is constructed by creating 2D objects in 3D space. The *Line, Circle, Arc,* and other 2D *Draw* and *Edit* commands are used to create the "wires," but 3D coordinates must be specified. The cube in Figure 34-1 was created with 12 *Line* segments. AutoCAD provides all the necessary tools to easily construct, edit, and view wireframe models.

Figure 34-1 ——————

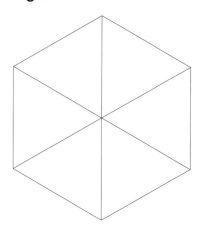

Figure 34-2 ——————

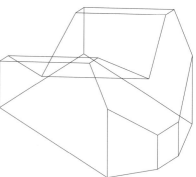

A wireframe model offers many advantages over a 2D engineering drawing. Wireframe models are useful in industry for providing computerized replicas of actual objects. A wireframe model is dimensionally complete and accurate for all three dimensions. Visualization of a wireframe is generally better than a 2D drawing because the model can be viewed from any position or a perspective can be easily attained. The 3D database can be used to test and analyze the object <u>three dimensionally</u>. The sheet metal industry, for example, uses wireframe models to calculate flat patterns complete with bending allowances. A wireframe model can also be used as a foundation for construction of a surface model. Because a wireframe describes edges but not surfaces, wireframe modeling is appropriate to describe objects with planar or single-curved surfaces, but not compound curved surfaces.

Surface Models

Surface models provide a better description of an object than a wireframe, principally because the surfaces as well as the edges are defined. A surface model of a cube is like a cardboard box—all the surfaces and edges are defined, but there is nothing inside. Therefore, a surface model has volume but no mass. A surface model provides an excellent visual representation of an actual 3D object because the front surfaces obscure the back surfaces and edges from view. Figure 34-3 shows a surface model of a cube, and Figure 34-4 displays a surface model of a somewhat more complex shape. Notice that a surface model leaves no question as to which side of the object you are viewing.

Figure 34-3

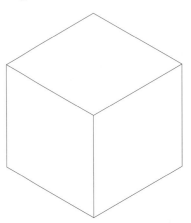

Surface models require a relatively tedious construction process. Each surface must be constructed individually. Each surface must be created in, or moved to, the correct orientation with respect to the other surfaces of the object. In AutoCAD, a surface is constructed by defining its edges. Often, wireframe models are used as a framework to build and attach surfaces. The complexity of the construction process of a surface is related to the number and shapes of its edges. AutoCAD Release 14 is not a complete surface modeler. The tools provided allow construction of simple planar and single curved surfaces, but there are few capabilities for construction of double-curved or other complex surfaces. No NURBS (Non-Uniform Rational B-Splines) surfacing capabilities exist in Release 14, although these capabilities are available in other Autodesk products such as AutoSurf™ and Mechanical Desktop™.

Figure 34-4

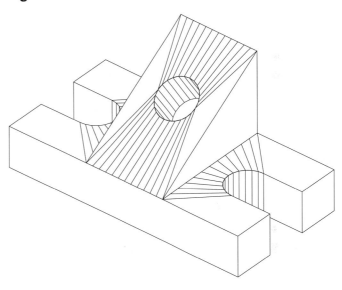

Most CAD systems, including AutoCAD, can display surface and solid models in both wireframe and "hidden" representation. Figure 34-4 shows the object "hidden" (with the *Hide* command). Figure 34-5 displays the same object in "wireframe" (the default) representation. Surface modeling systems use wireframe representation during the construction and editing process to speed computing time. The user activates the "hidden" representation after completing the model to enhance visibility. A reasonable amount of computing time is required to calculate the surface model "visibility." That is, the process required to determine which surfaces would obscure other surfaces for every position of the model would require noticeable computing time, so wireframe display is used during the model construction process.

Figure 34-5

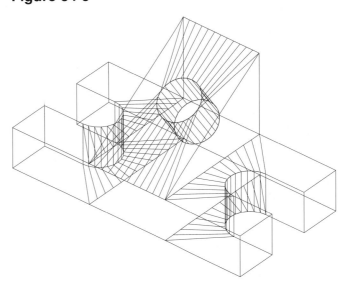

Solid Models

Solid modeling is the most complete and descriptive type of 3D modeling. A solid model is a complete computerized replica of the actual object. A solid model contains the complete surface and edge definition, as well as description of the interior features of the object. If a solid model is cut in half (sectioned), the interior features become visible. Since a solid model is "solid," it can be assigned material characteristics and is considered to have mass. Because solid models have volume and mass, most solid modeling systems include capabilities to automatically calculate volumetric and mass properties.

Solid model construction techniques are generally much simpler (and much more fun) than those of surface models. AutoCAD's solid modeler, called ACIS, is a <u>hybrid</u> modeler. That is, ACIS is a combination CSG (Constructive Solid Geometry) and B-Rep (Boundary Representation) modeler. CSG is characterized by its simple and straightforward construction techniques of combining primitive shapes (boxes, cylinders, wedges, etc.) utilizing Boolean operations (*Union*, *Subtract*, and *Intersect*, etc.). Boundary Representation modeling defines a model in terms of its edges and surfaces (boundaries) and determines the solid model based on which side of the surfaces the model lies. The user interface and construction techniques used in ACIS (primitive shapes combined by Boolean operations) are CSG-based, whereas the B-Rep capabilities are invoked automatically to display models in mesh representation and are transparent to the user. Figure 34-6 displays a solid model constructed of simple primitive shapes combined by Boolean operations.

Figure 34-6 ─────────────

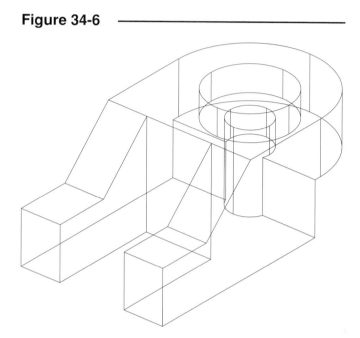

The CSG modeling techniques offer you the advantage of complete and relatively simple editing. CSG construction typically begins by specifying dimensions for simple primitive shapes such as boxes or cylinders, then combines the primitives using Boolean operations to create a "composite" solid. Other primitives and/or composite solids can be combined by the same process. Several repetitions of this process can be continued until the desired solid model is finally achieved. CSG construction techniques are discussed in detail in Chapter 38.

Figure 34-7 ─────────────

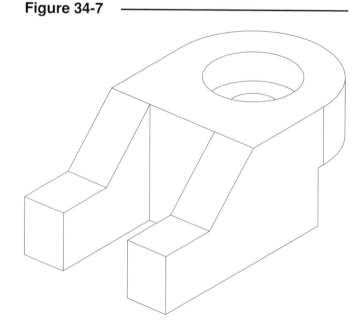

Solid models, like surface models, are capable of either wireframe or "hidden" display. Generally, wireframe display is used during construction (Fig. 34-6), and hidden representation is used to display the finished model (Fig. 34-7).

3D COORDINATE ENTRY

When creating a model in three-dimensional drawing space, the concept of the X and Y coordinate system, which is used for two-dimensional drawing, must be expanded to include the third dimension, Z, which is measured from the origin in a direction perpendicular to the plane defined by X and Y. Remember that two-dimensional CAD systems use X and Y coordinate values to define and store the location of drawing elements such as *Lines* and *Circles*. Likewise, a three-dimensional CAD system keeps a database of X, Y, and Z coordinate values to define locations and sizes of two- and three-dimensional elements. For example,

Figure 34-8

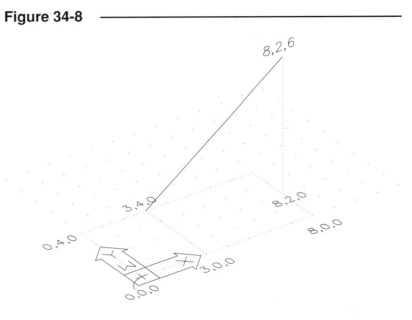

a *Line* is a two-dimensional object, yet the location of its endpoints in three-dimensional space must be specified and stored in the database using X, Y, and Z coordinates (Fig. 34-8). The X, Y, and Z coordinates are always defined in that order, delineated by commas. The AutoCAD Coordinate Display (*Coords*), however, displays only X and Y values, even though a three-dimensional model can be displayed in the Drawing Editor.

The icon that appears in the lower-left corner of the AutoCAD Drawing Editor is the Coordinate System icon (sometimes called the UCS icon) (Fig. 34-9). The Coordinate System icon displays the directions for the X and the Y axes and the orientation of the XY plane and can be made to locate itself at the origin, 0,0. The X coordinate values increase going to the right along the X axis, and the Y values increase going upward along the Y axis.

Figure 34-9

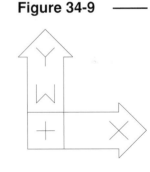

The Z axis is not indicated by the Coordinate System icon, but is assumed to be in a direction <u>perpendicular</u> to the XY plane. In other words, in the default orientation (when you begin a new drawing), the XY plane is <u>parallel</u> with the screen and the Z axis is <u>perpendicular</u> to, and out of, the screen (Fig. 34-10).

Figure 34-10

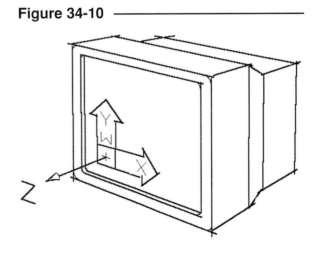

3D Coordinate Entry Formats

Because construction in three dimensions requires the definition of X, Y, <u>and</u> Z values, the methods of coordinate entry used for 2D construction must be expanded to include the Z value. The five methods of command entry used for 2D construction are valid for 3D coordinates with the addition of a Z value specification. Relative polar coordinate specification (@dist<angle) is expanded to form two other coordinate entry methods available explicitly for 3D coordinate entry. The six methods of coordinate entry for 3D construction follow:

1. **Interactive coordinates**   **PICK**   Use the cursor to select points on the screen. *OSNAP* or point filters must be used to select a point in 3D space; otherwise, points selected are <u>on the XY plane</u>.

2. **Absolute coordinates**   **X,Y,Z**   Enter explicit X, Y, and Z values relative to point 0,0.

3. **Relative rectangular coordinates**   **@X,Y,Z**   Enter explicit X, Y, and Z values relative to the last point.

4. **Cylindrical coordinates (relative)**   **@dist<angle,Z**   Enter a distance value, an angle in the XY plane value, and a Z value, all relative to the last point.

5. **Spherical coordinates (relative)**   **@dist<angle<angle**   Enter a distance value, an angle <u>in</u> the XY plane value, and an angle <u>from</u> the XY plane value, all relative to the last point.

6. **Direct distance entry**   **dist,direction**   Enter (type) a value, and move the cursor in the desired direction. To draw in 3D effectively, *ORTHO* must be *On*.

Cylindrical and spherical coordinates can be given <u>without</u> the @ symbol, in which case the location specified is relative to point 0,0,0 (the origin). This method is useful if you are creating geometry centered around the origin. Otherwise, the @ symbol is used to establish points in space relative to the last point.

Examples of each of the six 3D coordinate entry methods are illustrated in the following section. In the illustrations, the orientation of the observer has been changed from the default plan view in order to enable the visibility of the three dimensions.

Interactive Coordinate Specification

Figure 34-11 illustrates using the <u>interactive</u> method to PICK a location in 3D space. *OSNAP* <u>must</u> be used in order to PICK in 3D space. Any point PICKed with the input device without *OSNAP* will result in a location <u>on the XY plane</u>. In this example, the *Endpoint OSNAP* mode is used to establish the "To point:" of a second *Line* by snapping to the end of an existing vertical *Line* at 8,2,6:

```
Command: Line
From point: 3,4,0
To point: Endpoint of PICK
```

Figure 34-11

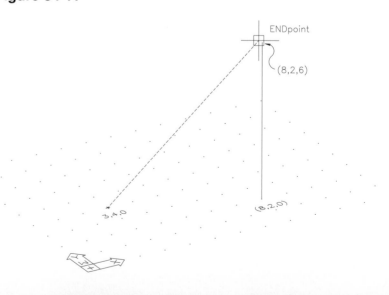

ENDpoint

(8,2,6)

3,4,0

(8,2,0)

Absolute Coordinates

Figure 34-12 illustrates the <u>absolute</u> coordinate entry to draw the *Line*. The endpoints of the *Line* are given as explicit X,Y,Z coordinates:

```
Command: Line
From point: 3,4,0
To point: 8,2,6
Command:
```

Figure 34-12

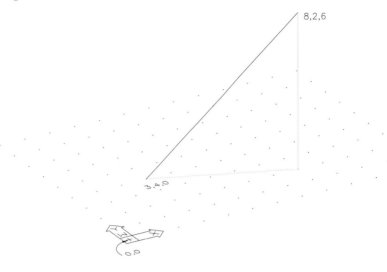

Relative Rectangular Coordinates

Relative rectangular coordinate entry is displayed in Figure 34-13. The "From point:" of the *Line* is given in absolute coordinates, and the "To point:" end of the *Line* is given as X,Y,Z values <u>relative</u> to the last point:

```
Command: Line
From point: 3,4,0
To point: @5,-2,6
Command:
```

Figure 34-13

Cylindrical Coordinates (Relative)

Cylindrical and spherical coordinates are an extension of polar coordinates with a provision for the third dimension. Relative cylindrical coordinates give the distance <u>in</u> the XY plane, angle <u>in</u> the XY plane, and Z dimension and can be relative to the last point by prefixing the @ symbol. The *Line* in Figure 34-14 is drawn with absolute and relative cylindrical coordinates. (The *Line* established is approximately the same *Line* as in the previous figures.)

```
Command: Line
From point: 3,4,0
To point: @5<-22,6
Command:
```

Figure 34-14

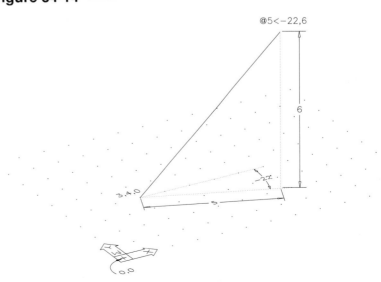

Spherical Coordinates (Relative)
Spherical coordinates are also an extension of polar coordinates with a provision for specifying the third dimension in angular format. Spherical coordinates specify a distance, an angle <u>in</u> the XY plane, and an angle <u>from</u> the XY plane and can be relative to the last point by prefixing the @ symbol. The distance specified is a <u>3D distance</u>, not a distance in the XY plane. Figure 34-15 illustrates the creation of approximately the same line as in the previous figures using absolute and relative spherical coordinates.

Figure 34-15

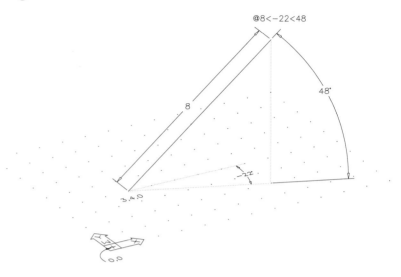

```
Command: Line
From point: 3,4,0
To point: @8<-22<48
Command:
```

Direct Distance Entry
Direct distance entry operates as it does in 2D, except <u>ORTHO must be On</u> in order to draw effectively by this method in 3D space. When *ORTHO* is *On*, objects drawn in 3D space are limited to a <u>plane parallel with the current XY plane.</u> For example, to draw a square in 3D space with sides of 3 units (Fig. 34-16), begin by using *Line* and specifying the "From point:" at the *Endpoint* of the existing diagonal *Line* shown (point 8,2,6). To draw the first 3 unit segment, turn on *ORTHO*, move the cursor in the desired direction, and enter "3." The *ORTHO* feature "locks" the *Line* to a plane parallel to the current XY plane.
Next, move the cursor in the desired (90 degree) direction and enter "3," and so on. The command syntax follows:

Figure 34-16

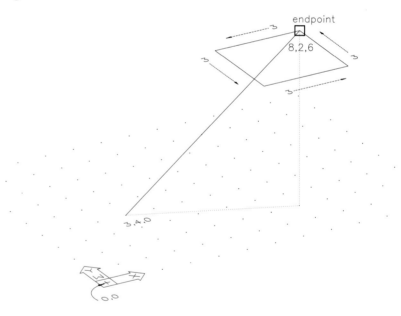

```
Command: LINE
From point: end of
To point: <Ortho on>  3
To point: 3
To point: 3
To point: 3
To point: Enter
Command:
```

Point Filters

Point filters are used to filter X and/or Y and/or Z coordinate values from a location PICKed with the pointing device. Point filtering makes it possible to build an X,Y,Z coordinate specification from a combination of point(s) selected on the screen and point(s) entered at the keyboard. A .XY (read "point XY") filter would extract, or filter, the X and Y coordinate value from the location PICKed and then prompt you to enter a Z value. Valid point filters are listed below. (See also Point Filters, Chapter 15.)

.X   Filters (finds) the X component of the location PICKed with the pointing device.
.Y   Filters the Y component of the location PICKed.
.Z   Filters the Z component of the location PICKed.
.XY   Filters the X and Y components of the location PICKed.
.XZ   Filters the X and Z components of the location PICKed.
.YZ   Filters the Y and Z components of the location PICKed.

The .XY filter is the most commonly used point filter for 3D construction and editing. Because 3D construction often begins on the XY plane of the current coordinate system, elements in Z space are easily constructed by selecting existing points on the XY plane using .XY filters and then entering the Z component of the desired 3D coordinate specification by keyboard. For example, in order to draw a line in Z space two units above an existing line on the XY plane, the .XY filter can be used in combination with *Endpoint OSNAP* to supply the XY component for the new line. See Figure 34-17 for an illustration of the following command sequence:

Figure 34-17

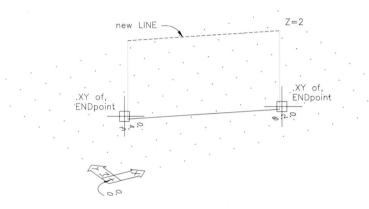

```
Command: Line
From point: .XY of
END of PICK (Select one Endpoint of the existing line on the XY plane.)
(need Z) 2
To point: .XY of
END of PICK (Select the other Endpoint of the existing line.)
(need Z) 2
Command:
```

Typing **.XY** and pressing **Enter** cause AutoCAD to respond with "of" similar to the way "of" appears after typing an *OSNAP* mode. When a location is specified using an .XY filter, AutoCAD responds with "(need Z)" and, likewise, when other filters are used, AutoCAD prompts for the missing component(s).

Tracking

Tracking is only useful for drawing in 3D when it is used to draw <u>in the current XY plane</u>. To construct geometry in the current XY plan, *Tracking* operates exactly as it does in 2D. It may be helpful to create User Coordinate Systems (UCSs) in order to facilitate drawing on a 2D plane in 3D space (see Coordinate Systems in this chapter).

R14

COORDINATE SYSTEMS

In AutoCAD two kinds of coordinate systems can exist, the <u>World Coordinate System</u> (WCS) and one or more <u>User Coordinate Systems</u> (UCS). The World Coordinate System <u>always</u> exists in any drawing and cannot be deleted. The user can also create and save multiple User Coordinate Systems to make construction of a particular 3D geometry easier. <u>Only one coordinate system can be active</u> at any one time, either the WCS or one of the user-created UCSs.

The World Coordinate System (WCS) and WCS Icon

The World Coordinate System (WCS) is the default coordinate system in AutoCAD for defining the position of drawing objects in 2D or 3D space. The WCS is always available and cannot be erased or removed but is deactivated temporarily when utilizing another coordinate system created by the user (UCS). The icon that appears (by default) at the lower-left corner of the Drawing Editor (Fig. 34-9) indicates the orientation of the WCS. The coordinate system icon indicates only X and Y directions, so Z is assumed to be perpendicular to the XY plane. The icon, whose appearance (*ON, OFF*) is controlled by the *Ucsicon* command, appears for the WCS and for any UCS. However, the letter "W" appears in the icon slightly above the origin <u>only</u> when the <u>WCS</u> is active.

Figure 34-18 ————————————

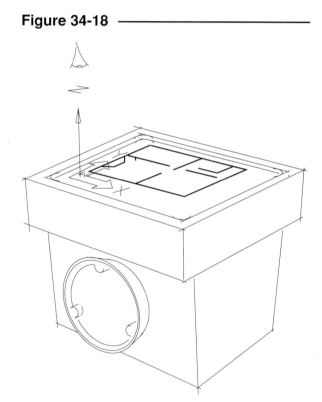

The orientation of the WCS with respect to Earth may be different among CAD systems. In AutoCAD the WCS has an architectural orientation such that the XY plane is a horizontal plane with respect to Earth, making Z the height dimension. A 2D drawing (X and Y coordinates only) is thought of as being viewed from above, sometimes called a <u>plan</u> view. Therefore, in a 3D AutoCAD drawing, X is the width dimension, Y is the depth dimension, and Z is height. This default orientation is like viewing a floor <u>plan</u>—from above (Fig. 34-18).

Some CAD systems that have a mechanical engineering orientation align their World Coordinate Systems such that the XY plane is a vertical plane intended for drawing a front view. In other words, some mechanical engineering CAD systems define X as the width dimension, Y as height, and Z as depth.

User Coordinate Systems (UCS) and Icons

There are no User Coordinate Systems that exist as part of the AutoCAD default template drawing (ACAD.DWT) as it comes "out of the box." UCSs are created to suit the 3D model when and where they are needed.

Creating geometry is relatively simple when having to deal only with X and Y coordinates, such as in creating a 2D drawing, or when creating simple 3D geometry with uniform Z dimensions. However,

3D models containing complex shapes on planes not parallel with the XY plane are good candidates for UCSs (Fig. 34-19).

Figure 34-19

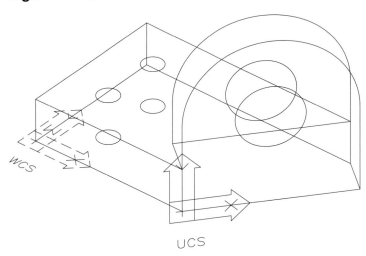

A User Coordinate System is thought of as a construction plane created to simplify creation of geometry on a specific plane or surface of the object. The user creates the UCS, aligning its XY plane with a surface of the object, such as along an inclined plane, with the UCS origin typically at a corner or center of the surface.

The user can then create geometry aligned with that plane by defining only X and Y coordinate values of the current UCS (Fig. 34-20). The *SNAP* and *GRID* automatically align with the current coordinate system, providing *SNAP* points and enhancing visualization of the construction plane. Practically speaking, it is easier in some cases to specify only X and Y coordinates with respect to a specific plane on the object rather than calculating X, Y, and Z values with respect to the World Coordinate System.

Figure 34-20

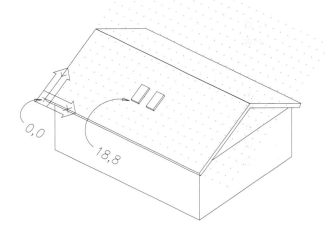

User Coordinate Systems can be created by any of several options of the *UCS* command. Once a UCS has been created, it becomes the current coordinate system. Only one coordinate system can be active; therefore, it is suggested that UCSs be saved (by using the *Save* option of *UCS*) for possible future geometry creation or editing.

When a UCS is created, the icon at the lower-left corner of the screen can be made to automatically align itself with the UCS along with *SNAP* and *GRID*. The letter "W," however, appears on the icon only when the WCS is active (when the WCS is the current coordinate system). When creating 3D geometry, it is recommended that the *ORigin* option of the *Ucsicon* command be used to place the icon always at the origin of the current UCS, rather than in the lower-left corner of the screen. Since the origin of the UCS is typically specified as a corner or center of the construction plane, aligning the Coordinate System icon with the current origin aids your visualization of the UCS orientation. The Coordinate Display (*Coords*) in the Status line always displays the X, Y and Z values of the current coordinate system, whether it is WCS or UCS.

THE RIGHT-HAND RULE

AutoCAD complies with the right-hand rule for defining the orientation of the X, Y, and Z axes. The right-hand rule states that if your right hand is held partially open, the thumb, first, and middle fingers define positive X, Y, and Z directions, respectively, and positive rotation about any axis is like screwing in a light bulb.

More precisely, if the thumb and first two fingers are held out to be mutually perpendicular, the thumb points in the positive X direction, the first finger points in the positive Y direction, and the middle finger points in the positive Z direction (Fig. 34-21).

In this position, looking toward your hand from the tip of your middle finger is like the default AutoCAD viewing orientation—positive X is to the right, positive Y is up, and positive Z is toward you.

Figure 34-21 ————————————

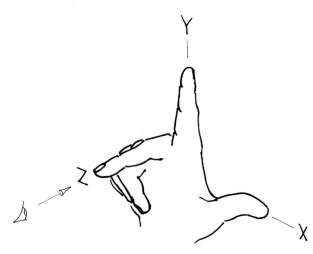

Positive rotation about any axis is <u>counterclockwise looking down the axis toward the origin</u>. For example, when you view your right hand, positive rotation about the X axis is as if you look down your thumb toward the hand (origin) and twist your hand counterclockwise (Fig. 34-22).

Figure 34-22 ————————————

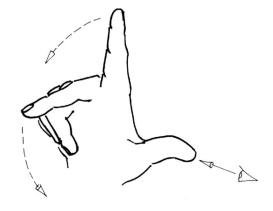

Figure 34-23 shows the Coordinate System icon in a +90 degree rotation about the X axis (like the previous figure). This orientation is typical for setting up a <u>front</u> view UCS for drawing on a plane parallel with the front surface of an object.

Figure 34-23 ————————

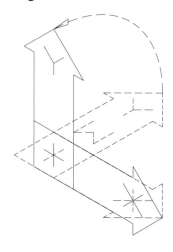

Positive rotation about the Y axis would be as if looking down your first finger toward the hand (origin) and twisting counterclockwise (Fig. 34-24).

Figure 34-24 ————————

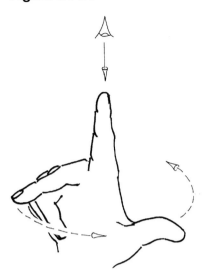

Figure 34-25 illustrates the Coordinate System icon with a +90 degree rotation about the Y axis (like the previous figure). To avoid confusion in relating this figure to Figure 34-24, consider that the icon is oriented on the XY plane as a horizontal plane in its original (highlighted) position, whereas Figure 34-24 shows the hand in an upright position. (Compare Figures 34-18 and 34-10.)

Figure 34-25 ——

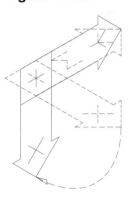

Positive rotation about the Z axis would be as if looking toward your hand from the end of your middle finger and twisting counterclockwise (Fig. 34-26).

Figure 34-26 ————————

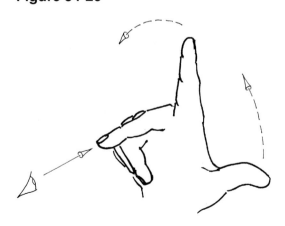

Figure 34-27 shows the Coordinate System icon with a +90 degree rotation about the Z axis. Again notice the orientation of the icon is horizontal, whereas the hand (Fig. 34-26) is upright.

Figure 34-27 ————————

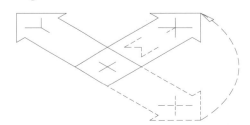

UCS ICON CONTROL

UCSICON

Pull-down Menu	COMMAND (TYPE)	ALIAS (TYPE)	Short-cut	Screen (side) Menu	Tablet Menu
View Display > UCS Icon >	UCSICON	...	...	VIEW 2 UCSicon	L,2

The *Ucsicon* command controls the appearance and positioning of the Coordinate System icon. In order to aid your visualization of the current UCS or the WCS, it is <u>highly</u> recommended that the Coordinate System icon be turned *ON* and positioned at the *ORigin*.

    Command: **ucsicon**
    ON/OFF/All/Noorigin/ORigin<ON>: **on**
    Command:

This option causes the Coordinate System icon to appear (Fig. 34-28).

Figure 34-28

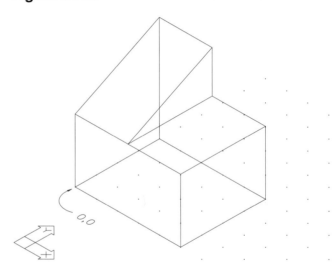

The *Ucsicon* command must be invoked again to use the *ORigin* option:

    Command: **ucsicon**
    ON/OFF/All/Noorigin/ORigin<ON>: **or**
    Command:

This setting causes the icon to move to the origin of the current coordinate system (Fig. 34-29).

The icon does not always appear at the origin after using some viewing commands like *Vpoint*. This is because *Vpoint* causes a *Zoom Extents*, forcing the geometry against the border of the graphics screen (or viewport) area, which prevents the icon from aligning with the origin. In order to cause the icon to appear at the origin, try using *Zoom* with a **.9X** magnification factor. This action usually brings the geometry slightly in from the border and allows the icon to align itself with the origin.

Figure 34-29

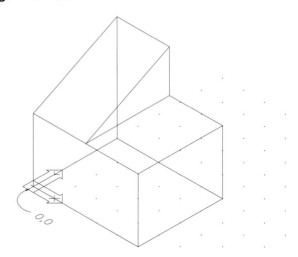

Options of the *Ucsicon* command are:

ON   Turns the Coordinate System icon on.

OFF   Turns the Coordinate System icon off.

All   Causes the *Ucsicon* settings to be effective for all viewports.

Noorigin Causes the icon to appear always in the lower-left corner of the screen, not at the origin.

ORigin   Forces the placement and orientation of the icon to align with the origin of the current coordinate system.

Now that you know the basics of 3D modeling, you will develop your skills in viewing and displaying 3D models (Chapter 35, 3D Viewing and Display). It is imperative that you are able to view objects from different viewpoints in 3D space before you learn to construct them.

CHAPTER EXERCISES

1. What are the three types of 3D models?

2. What characterizes each of the three types of 3D models?

3. What kind of modeling techniques does AutoCAD's solid modeling system use?

4. What are the five formats for 3D coordinate specification?

5. Examine the 3D geometry shown in Figure 34-30. Specify the designated coordinates in the specified formats below.

    A. Give the coordinate of corner D in absolute format.

    B. Give the coordinate of corner F in absolute format.

    C. Give the coordinate of corner H in absolute format.

    D. Give the coordinate of corner J in absolute format.

    E. What are the coordinates of the line that define edge F-I?

    F. What are the coordinates of the line that define edge J-I?

Figure 34-30 ——————————————————————

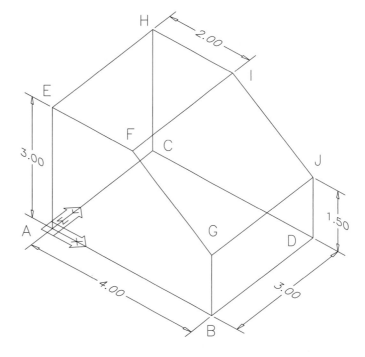

G. Assume point E is the "last point." Give the coordinates of point I in relative rectangular format.

H. Point I is now the last point. Give the coordinates of point J in relative rectangular format.

I. Point J is now the last point. Give the coordinates of point A in relative rectangular format.

J. What are the coordinates of corner G in cylindrical format (from the origin)?

K. If E is the "last point," what are the coordinates of corner C in relative cylindrical format?

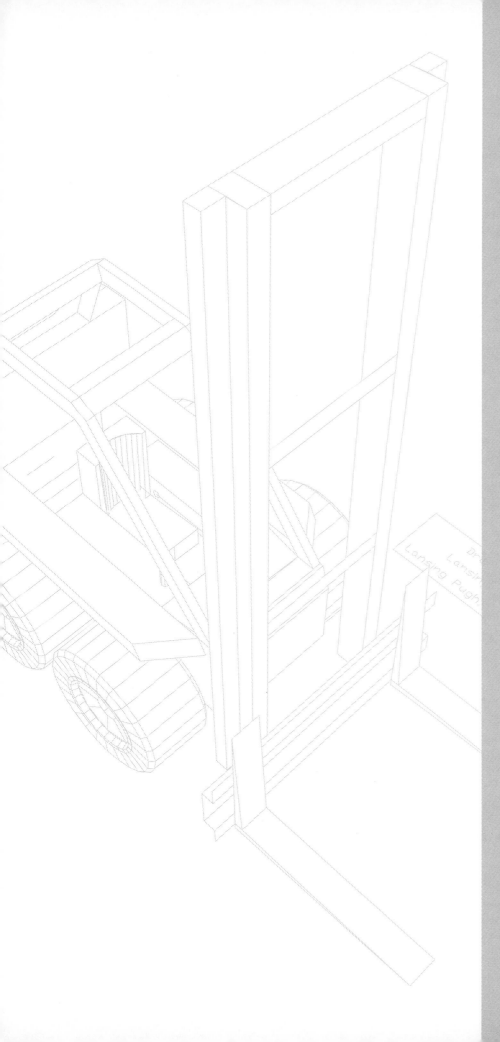

35

3D VIEWING AND DISPLAY

Chapter Objectives

After completing this chapter you should:

1. be able to recognize a 3D model in wireframe, perspective, *Hide*, *Shade*, and *Render* display representations;
2. be able to use *3D Viewpoint* to quickly attain the desired view;
3. be able to use *Vpoint* to view a 3D model from various viewpoints;
4. be able to use the *Vector*, *Tripod*, and *Rotate* options of *Vpoint*;
5. be able to use *Dview* to specify a *TArget* and *CAmera* point and to generate a perspective with the *Distance* option;
6. be able to use the *PAn*, *Zoom*, *TWist*, *CLip*, and *Hide* options of *Dview*;
7. be able to create several configurations of tiled viewports with the *Vports* command;
8. know how to suppress "hidden lines" of a solid or surface model with the *Hide* command;
9. know how to *Shade* a solid model using four *SHADEDGE* options.

AutoCAD'S 3D VIEWING AND DISPLAY CAPABILITIES

Viewing Commands

These commands allow you to change the direction from which you view a 3D model or otherwise affect the view of the 3D model:

Vpoint   *Vpoint* allows you to change your viewpoint of a 3D model. The object remains stationary while the viewpoint of the observer changes. Three options are provided: *Vector, Tripod,* and *Rotate.* Many useful *3D Viewpoint* options are available in Release 14.

Plan   The *Plan* command automatically gives the observer a plan (top) view of the object. The plan view can be with respect to the WCS (World Coordinate System) or an existing UCS (User Coordinate System).

Dview   This command allows you to dynamically (interactively) rotate the viewpoint of the observer about 3D objects. *Dview* also allows generation of perspective views.

Zoom   Although *Zoom* operates in 3D just as in 2D, the *Zoom Previous* option restores the previous 3D viewpoint.

View   The *View* command can be used with 3D viewing to *Save* and *Restore* 3D viewpoints.

Vports   The *Vports* command creates tiled viewports. *Vports* does not actually change the viewpoint of the model but divides the screen into several viewports and allows you to display several viewpoints or sizes of the model on one screen.

Display Commands

In most CAD systems, <u>surface and solid models are shown in a wireframe display</u> during the construction and editing process. A "hidden" display may hinder your ability to select needed edges of the object during the construction process and would require additional computing time. Once the model is completed and the desired viewpoint has been attained, these commands can change the appearance of a 3D surface or solid model from the default wireframe representation by displaying the surfaces:

Hide   This command removes normally "hidden" edges and surfaces from a solid or surface model, making it appear as an opaque rather than transparent object.

Shade   The *Shade* command fills the surfaces with the object's color and calculates light reflection by applying gradient shading (variable gray values) to the surfaces.

Render   *Render* allows you to create and place lights in 3D space, adjust the light intensity, and assign materials (color and reflective qualities) to the surfaces. This is the most sophisticated of the visualization capabilities offered in AutoCAD.

When using the viewing commands *Vpoint, Dview,* and *Plan,* it is important to imagine that the <u>observer moves about the object</u> rather than imagining that the object rotates. The object and the coordinate system (WCS and icon) always remain <u>stationary</u> and always keep the same orientation with respect to Earth. Since the observer moves and not the geometry, the objects' coordinate values retain their integrity, whereas if the geometry rotated within the coordinate system, all coordinate values of the objects would change as the object rotated. The *Vpoint, Dview,* and *Plan* commands change only the viewpoint of the <u>observer</u>.

Figure 35-1 ——————————————

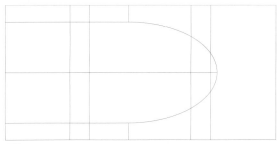

Plan

Figure 35-2 ——————————————

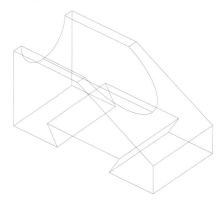

Vpoint (wireframe representation)

Figure 35-3 ——————————————

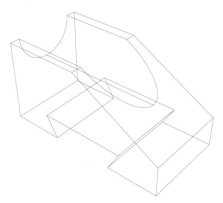

Dview, Distance (perspective)

Figure 35-4 ——————————————

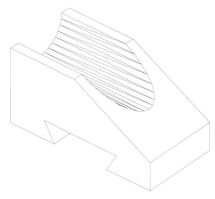

Hide (surface or solid model)

Figure 35-5 ——————————————

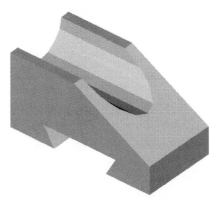

Shade (surface or solid model)

Figure 35-6 ——————————————

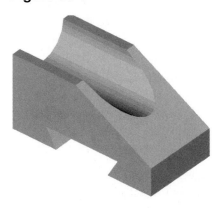

Render (surface or solid model)

As the 3D viewing commands are discussed, it would be helpful if you opened a 3D drawing to practice using the viewing commands. The FORKLIFT drawing is an excellent 3D model to practice using the 3D viewing commands. The FORKLIFT.DWG is not copied to your computer during installation but is located on the AutoCAD Release 14 CD-ROM. It is located in the Acad/Bonus/Sample/History directory along with many other interesting and useful sample drawings. When you open the drawing, remember to set the *Ucsicon* to the *Origin*.

3D VIEWING COMMANDS

3D VIEWPOINT

Pull-down Menu	COMMAND (TYPE)	ALIAS (TYPE)	Short-cut	Screen (side) Menu	Tablet Menu
View *3D Viewpoint >*	...	...	...	*VIEW 1* *Vpoint*	*O,3–Q,5*

AutoCAD Release 13 introduced *3D Viewpoint*, which really speeds up the process of attaining common views of 3D objects. All of these options actually use the *Vpoint* command and automatically enter in coordinate values. The other options of *Vpoint* can be used for other particular viewing angles (see *Vpoint* next). These options can be selected from the *View* pull-down menu (Fig. 35-7).

Figure 35-7

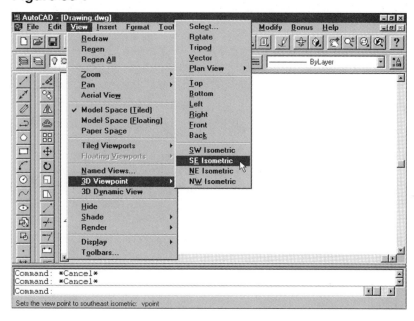

The *3D Viewpoint* options are accessible from the Standard toolbar (Fig. 35-8). If you prefer a separate toolbar, one can be invoked from the toolbar list and made to float or dock.

Figure 35-8

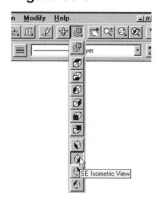

Each of the *3D Viewpoint* options is described next. The FORKLIFT drawing is displayed for several viewpoints (the *Hide* command was used in these figures for clarity). It is important to remember that for each view, imagine you are viewing the object from the indicated position in space. The object does not rotate.

Top

The object is viewed from the top (Fig. 35-9). Selecting this option shows the XY plane from above (the default orientation when you begin a *New* drawing). Notice the position of the WCS icon. This orientation should be used periodically during construction of a 3D model to check for proper alignment of parts.

Figure 35-9

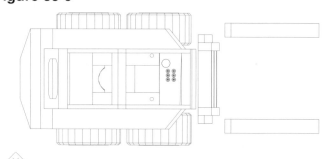

Bottom

This option displays the object as if you are looking up at it from the bottom. AutoCAD uses the *Vpoint* command and automatically enters in coordinates to show this view. The coordinate system icon appears backwards when viewed from the bottom.

Left

This is looking at the object from the left side. Again the *Vpoint* command is used by AutoCAD to give this view.

Right

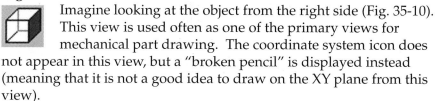

Imagine looking at the object from the right side (Fig. 35-10). This view is used often as one of the primary views for mechanical part drawing. The coordinate system icon does not appear in this view, but a "broken pencil" is displayed instead (meaning that it is not a good idea to draw on the XY plane from this view).

Figure 35-10

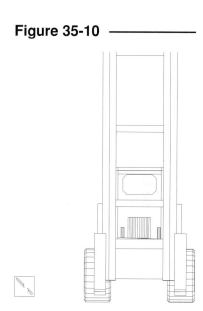

Front

Selecting this option displays the object from the front view. This is a common view that can be used often during the construction process. The front view is usually the "profile" view for mechanical parts (Fig. 35-11).

Figure 35-11

Back

Back displays the object as if the observer is behind the object. Remember that the object does not rotate; the observer moves.

SW Isometric

Isometric views are used more than the "orthographic" views (front, top, right, etc.) for constructing a 3D model. Isometric viewpoints give more information about the model because you can see three dimensions instead of only two dimensions (as in the previously discussed views).

SE Isometric

The southeast isometric is generally the first choice for displaying 3D geometry. If the object is constructed with its base on the XY plane so that X is length, Y is depth, and Z equals height, this orientation shows the front, top, and right sides of the object. Try to use this viewpoint as your principal mode of viewing during construction. Note the orientation of the WCS icon (Fig. 35-12).

NE Isometric

The northeast isometric shows the right side, top, and back (if the object is oriented in the manner described earlier).

NW Isometric

This viewpoint allows the observer to look at the left side, top, and back of the 3D object (Fig. 35-13).

NOTE: *3D Viewpoint* always displays the object (or orients the observer) with respect to the World Coordinate System. For example, the *Top* option always shows the plan view of the WCS XY plane. Even if another coordinate system (UCS) is active, AutoCAD temporarily switches back to the WCS to attain the selected viewpoint.

Figure 35-12

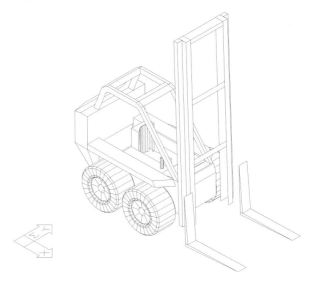

Figure 35-13

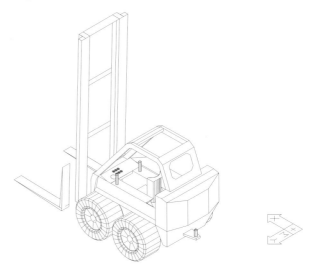

VPOINT

Pull-down Menu	COMMAND (TYPE)	ALIAS (TYPE)	Short-cut	Screen (side) Menu	Tablet Menu
View *3D Viewpoint >*	*VPOINT*	*-VP*	...	*VIEW 1* *Vpoint*	*N,4*

The *3D Viewpoint* options actually use the *Vpoint* command to generate the typical views of a 3D object. In some cases, it is necessary to display a 3D object from a direction other than the typical *Vpoints* attained through the presets. For example, an object with 45 degree angled planes or with regular proportions would present a visually confusing description if a perfect isometric view were given (some lines in the view would overlap and angles would align). Instead, a viewpoint slightly off of a pure isometric would show the angled surfaces and regular proportions more clearly. In cases such as these, one of the three options of the *Vpoint* command should be used: *Rotate*, *Vector*, or *Tripod* (or *Axes*).

The *Vpoint* command displays a parallel projection rather than a perspective projection. In a parallel projection, all visual rays from the observer to the object are parallel as if the observer were (theoretically) an infinite distance from the object. The resulting display shows all parallel edges of the object as parallel in the display. This projection differs from a perspective projection, where parts of the object that are farther from the observer appear smaller and parallel edges converge to a point.

Rotate

The name of this option is somewhat misleading because the object is not rotated. The *Rotate* option prompts for two angles in the WCS (by default) which specify a vector indicating the direction of viewing. The two angles are (1) the angle in the XY plane and (2) the angle from the XY plane. The observer is positioned along the vector looking toward the origin.

```
Command: Vpoint
Rotate <View point><0,0,1>: R
Enter angle in XY plane from X axis <270>: 315
Enter angle from XY plane <90>: 35
Command:
```

The first angle is the angle <u>in</u> the XY plane at which the observer is positioned looking toward the origin. This angle is just like specifying an angle in 2D. The second angle is the angle that the observer is positioned <u>up or down from</u> the XY plane. The two angles are given with respect to the WCS (Fig. 35-14).

Angles of **315** and **35** specified in response to the *Rotate* option display an almost <u>perfect isometric</u> viewing angle. An isometric drawing often displays some of the top, front, and right sides of the object. For some regularly proportioned objects, perfect isometric viewing angles can cause visualization difficulties, while a slightly different angle can display the object more clearly. Figure 35-15 and Figure 35-16 display a cube from an almost perfect isometric viewing angle (**315** and **35**) and from a slightly different viewing angle (**310** and **40**), respectively.

Figure 35-14

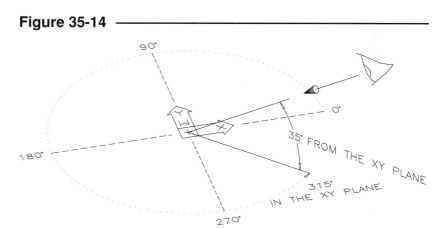

Figure 35-15 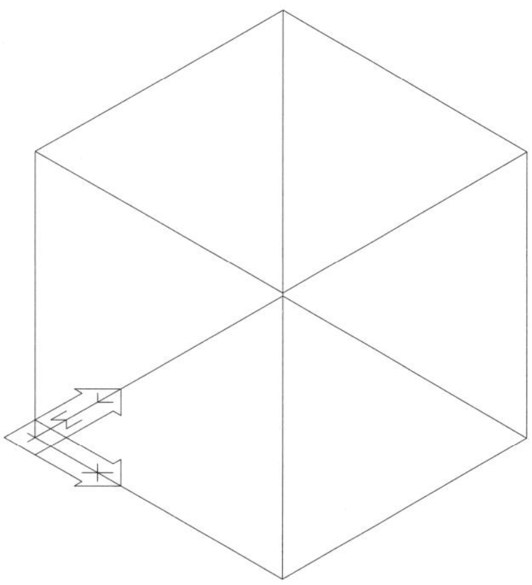 ────────── **Figure 35-16** ──────────

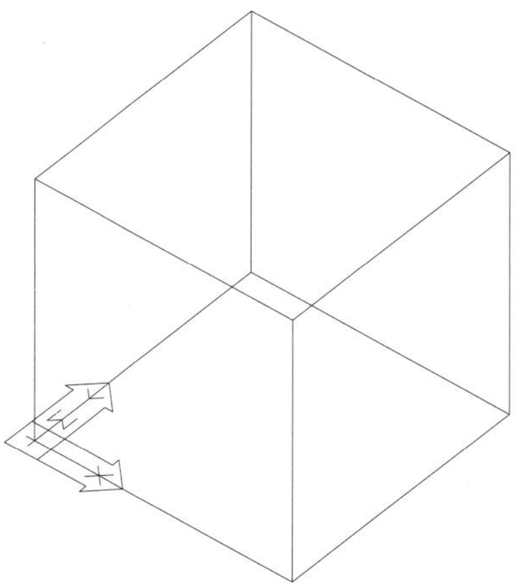

The *Vpoint Rotate* option can be used to display 3D objects from a front, top, or side view. *Rotate* angles for common views are:

Top	270, 90
Front	270, 0
Right side	0,0
Southeast isometric	315, 35.27
Southwest isometric	225, 35.27

Vector

Another option of the *Vpoint* command is to enter X,Y,Z coordinate values. The coordinate values indicate the position in 3D space at which the observer is located. The coordinate values do not specify an absolute position but rather specify a <u>vector</u> passing through the coordinate position and the origin. In other words, the observer is located at <u>any point along the vector looking toward the origin</u>.

Because the *Vpoint* command generates a parallel projection and since parallel projection does not consider a distance (which is considered only in perspective projection), the magnitude of the coordinate <u>values</u> is of no importance, only the relationship among the values. Values of 1,-1,1 would generate the same display as 2,-2,2.

The command syntax for specifying a perfect isometric viewing angle is as follows:

```
Command: Vpoint
Rotate <View point><0,0,1>: 1,-1,1
Command:
```

Coordinates of 1,-1,1 generate a display from a perfect isometric viewing angle.

Figure 35-17 illustrates positioning the observer in space, using coordinates of 1,-1,1. Using *Vpoint* coordinates of 1,-1,1 generates an isometric display similar to *Rotate* angles of 315, 35, or *SE Isometric*.

Other typical views of an object can be easily achieved by entering coordinates at the *Vpoint* command. Consider the following coordinate values and the resulting displays:

Figure 35-17

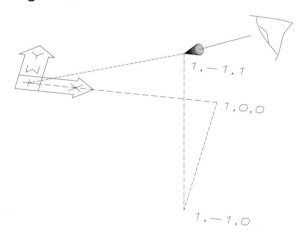

Coordinates	Display
0,0,1	Top
0,-1,0	Front
1,0,0	Right side
1,-1,1	Southeast isometric
-1,-1,1	Southwest isometric

Tripod or Axes

This option displays a three-pole axes system which rotates dynamically on the screen as the cursor is moved within a "globe." The axes represent the X, Y, and Z axes of the WCS. When you PICK the desired viewing position, the geometry is displayed in that orientation. Because the viewing direction is specified by PICKing a point, it is difficult to specify an exact viewpoint.

The *Tripod* option does not appear on the command line as one of the possible *Vpoint* methods. This option must be invoked by pressing Enter at the "Rotate<Viewpoint> <(coordinates)>:" prompt or by selecting *Axes* or *Tripod* from one of the menus. When the *Tripod* method is invoked, the current drawing temporarily disappears and a three-pole axes system appears at the center of the screen. The axes are dynamically rotated by moving the cursor (a small cross) in a small "globe" at the upper-right of the screen (Fig. 35-18).

Figure 35-18

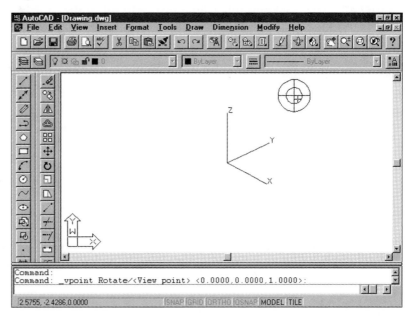

The three-pole axes indicate the orientation of the X, Y, and Z axes for the new *Vpoint*. The center of the "globe" represents the North Pole, so moving the cursor to that location generates a plan, or top, view. The small circle of the globe represents the Equator, so moving the cursor to any location on the Equator generates an elevation view (front, side, back, etc.). The outside circle represents the South Pole, so locating the cursor there shows a bottom view. When you PICK, the axes disappear and the current drawing is displayed from the new viewpoint. The command format for using the *Tripod* method is as follows:

Command: **Vpoint**
Rotate <Viewpoint><(current coordinates)>: **Enter** (axes appear)
PICK (Select desired cursor location.)
Regenerating drawing. (Current drawing appears showing the new viewpoint.)

Using the *Tripod* method of *Vpoint* is quick and easy. Because no exact *Vpoint* can be given, it is difficult to achieve the exact *Vpoint* twice. Therefore, if you are working with a complex drawing that requires a <u>specific</u> viewpoint of a 3D model, use another option of *Vpoint* or save the desired *Vpoint* as a named *View*.

Using *Zoom .9X* with *Vpoint*

In the FORKLIFT drawing (Figs. 35-9, 35-12, and 35-13), the Coordinate System icon was previously turned *ON* and forced to appear at the *ORigin*. This action was accomplished by the *Ucsicon* command (see *Ucsicon*, Chapter 34). The icon does not <u>always</u> appear at the origin after using the *Vpoint* command. This is because *Vpoint* causes a *Zoom Extents*, forcing the geometry near the border of the graphics screen area. This sometimes prevents the icon from aligning with the origin since the origin can also be against the border of the graphics screen area. In order to cause the icon to appear at the origin, try using *Zoom* with a .9X magnification factor. This action usually brings the geometry slightly away from the border and allows the icon to "pop" into place at the origin.

Isometric, Dimetric, and Trimetric Viewpoints

Assuming your 3D model is at least partially complete, an <u>axonometric-type</u> of viewpoint can be achieved. This is done by using the *Vpoint* command or the *3D Viewpoint* options.

In Chapter 25, Pictorial Drawings, the three types of axonometric drawings (isometric, dimetric, and trimetric) are discussed and illustrated. Keep in mind that a true axonometric drawing is a <u>2D</u> drawing made to show three sides of an object. This section discusses viewing <u>3D</u> drawings (wireframe, surface, or solid models) from viewpoints that <u>simulate</u> isometric, dimetric, or trimetric views.

Isometric *Viewpoints*

An isometric view is characterized by having equal angles between all three axes. To achieve an isometric-type of a viewpoint, use *3D Viewpoint,* as discussed earlier. Alternately, the *Vpoint* command can be used with either the *Rotate* or *Vector* option. Those settings are as follows:

> *Rotate*
> Any combination of **45, 135, 225,** or **315** entered as the "angle in the XY plane" and either **35.27** or **-35.27** as the "angle from the XY plane" yields an isometric-type *Vpoint*.

> *Vector*
> In response to "Viewpoint <0,0,1>:" enter any combination of positive or negative **1**s (ones) as coordinate entry to yield some type of an isometric *Vpoint*.

Dimetric *Viewpoints*

Figure 35-19

A dimetric view is characterized by having equal angles between two of the three axes. A dimetric-type of a viewpoint can be attained most easily by using the *Rotate* option of *Vpoint*. Although there are many possible viewing angles technically considered dimetric, the correct settings for a typical dimetric viewing angle (Fig. 35-19) are as follows:

> *Rotate*
> In response to the prompts "Enter angle in XY plane from X axis" and "Enter angle from XY plane", enter values of **315** and **15.54**, respectively.

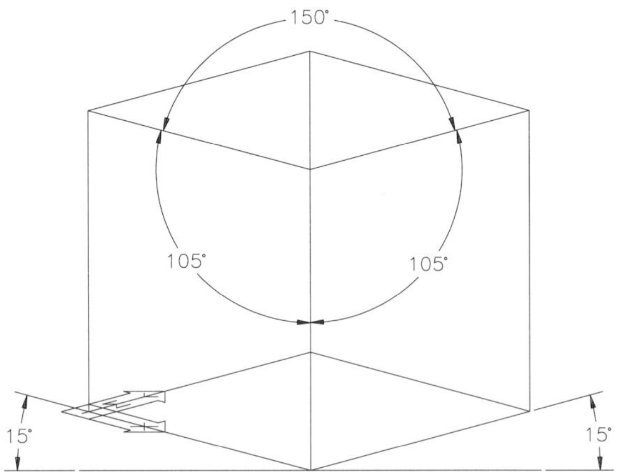

A wireframe model of a cube in dimetric appears in Figure 35-19. This typical dimetric-type orientation has two axes at 15 degrees from horizontal.

You can enter **45, 135, 225,** or **315** as the "angle in the XY plane from the X axis." This assures that two of the three angles between axes are equal. Any value can be entered as the "angle from the XY plane." Technically, this creates a dimetric view.

Trimetric *Viewpoints*

Figure 35-20

The definition of a trimetric drawing requires that the angles between the axes be <u>unequal</u>. Therefore, <u>any</u> viewpoint other that an isometric or dimetric falls into this category. However, a typical combination of angles can be approximated by entering the following values with the *Rotate* option of the *Vpoint* command:

> Angle in the XY plane from the X axis: **294.55**
> Angle from the XY plane: **22.7**

These entries result in the viewpoint shown in Figure 35-20. The resulting angles between two of the axes and horizontal are 10 and 40 degrees.

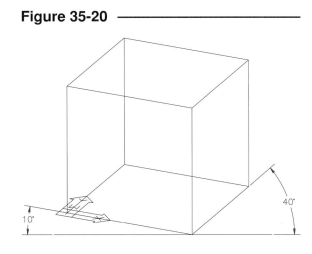

DDVPOINT

Pull-down Menu	COMMAND (TYPE)	ALIAS (TYPE)	Short-cut	Screen (side) Menu	Tablet Menu
View *3D Viewpoint >* *Select...*	*DDVPOINT*	*VP*	...	*VIEW 1* *Ddvpoint*	*N,5*

The *Ddvpoint* command produces the *Viewpoint Presets* dialog box (Fig. 35-21). This tool is an interface for attaining viewpoints that could otherwise be attained with the *3D Viewpoint* or the *Vpoint* command.

Ddvpoint serves the same function as the *Rotate* option of *Vpoint*. You can specify angles *From: X Axis* and *From: XY Plane*. Angular values can be entered in the edit boxes, or you can PICK anywhere in the image tiles to specify the angles. PICKing in the enclosed boxes results in a regular angle (e.g., 45, 90, or 10, 30) while PICKing near the sundial-like "hands" results in irregular angles (see pointer in Fig. 35-21). The *Set to Plan View* tile produces a plan view.

Figure 35-21 ──────────────

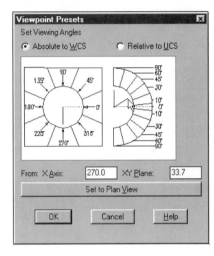

The *3D Viewpoint* options are newer and faster for producing standard views and are suggested for most cases. *Ddvpoint*, however, offers one option not available by other means. The *Relative to UCS* radio button calculates the viewing angles with respect to the current UCS rather than the WCS. Normally, the viewing angles should be absolute to WCS, but certain situations may require this alternative. Viewing angles relative to the current UCS can produce some surprising viewpoints if you are not completely secure with the model and observer orientation in 3D space.

NOTE: When you use this tool, ensure that the *Absolute to WCS* radio button is checked unless you are sure the specified angles should be applied relative to the current UCS.

PLAN

Pull-down Menu	COMMAND (TYPE)	ALIAS (TYPE)	Short-cut	Screen (side) Menu	Tablet Menu
View *3D Viewpoint >* *Plan View >*	*PLAN*	...	...	*VIEW 1* *Plan*	*N,3*

This command is useful to quickly display a plan view of any UCS:

    Command: **plan**
    <Current UCS>/Ucs/World: (**letter**) or **Enter**

Responding by pressing Enter causes AutoCAD to display the plan (top) view of the current UCS. Typing *W* causes the display to show the plan view of the World Coordinate System. The *World* option does <u>not</u> cause the WCS to become the active coordinate system but only displays its plan view. Invoking the *UCS* option displays the following prompt:

    ?/Name of UCS:

The *?* displays a list of existing UCSs. Entering the name of an existing UCS displays a plan view of that UCS.

ZOOM

The *Zoom* command can be used effectively with 3D models just as with 2D drawings. However, two options of *Zoom* are particularly applicable to 3D work: *Previous* and *Scale*.

Previous

This option of *Zoom* restores the previous display. When you are using *3D Viewpoint, Vpoint, Dview,* and *Plan* for viewing a 3D model, *Zoom Previous* will display the previous viewpoint.

Scale

When a *Vpoint* is specified, AutoCAD automatically causes a *Zoom Extents* that leaves the 3D model nearly against the borders of the screen. If you are using the *Ucsicon* with the *ORigin* option, invoking a *Zoom .9X* or *.8X* usually allows room for the icon to "pop" back to the origin position.

VIEW

The *View* command can also be used effectively for 3D modeling. When a desirable viewpoint is achieved by the *Vpoint* or *Dview* commands, the viewpoint can be saved with the *Save* option of *View*. *Restoring* the *View* is often easier than using *Vpoint* or *Dview* again.

DVIEW

Pull-down Menu	COMMAND (TYPE)	ALIAS (TYPE)	Short-cut	Screen (side) Menu	Tablet Menu
View *3D Dynamic View*	*DVIEW*	*DV*	...	VIEW 1 *Dview*	R,5

Dview (Dynamic View) has many capabilities for displaying a 3D model. It can be used like *Vpoint* to specify different viewpoints of an object. *Dview* can be used to change the *TArget* (point to look <u>at</u>) and the *CAmera* (point to view <u>from</u>) as well as to *Zoom, PAn,* and establish clipping planes. *Dview* is commonly used to generate <u>perspective projections</u> of the current drawing.

Because *Dview* dynamically displays rotation or other movement of objects on the screen, you are first given the chance to select objects for this action. If your 3D drawing is large and complex, select only the "framework" of the set of objects; otherwise, select all objects for dynamic action. If too many objects are selected, the dynamic motion is slow and difficult to see.

As an alternative, if a null selection set is used (**Enter** is pressed in response to the "Select objects:" prompt), the DVIEWBLOCK house appears, which is used temporarily for dynamic movement (Fig. 35-22).

Figure 35-22 ⎯⎯⎯⎯⎯⎯⎯⎯⎯⎯⎯⎯⎯

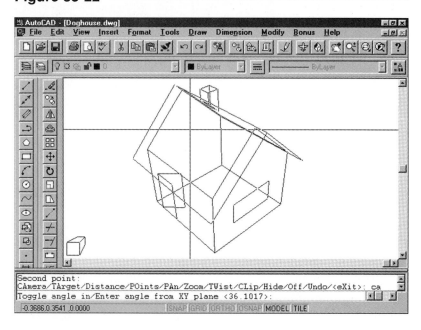

When the dynamic action of the command is completed, the entire 3D object reappears on the screen in the new position. The command syntax for *Dview* is:

> Command: **dview**
> Select objects: **PICK** or **Enter** (Select objects or press Enter for the DVIEWBLOCK house.)
> Select objects: **Enter** (Completes the selection process.)
> CAmera/TArget/Distance/POints/PAn/Zoom/TWist/CLip/Hide/Off/ Undo/<eXit>: (**option**)

The *Dview* options are given here in the best order for understanding related concepts.

eXit

This option should be used to save the resulting *Dview* display after using any of the other options to generate the desired view or perspective of the 3D model. *Dview* must be exited before any other AutoCAD commands can be used. Using the *Dview eXit* option saves the current display and returns to the Command: prompt.

CAmera

CAmera allows you to specify a point representing the <u>observer's location</u>, like the *Vpoint* command, only *Dview CAmera* is <u>dynamic</u>. Although it appears that the object is being dynamically rotated, imagine that the <u>observer</u> is changing position. When you PICK the desired position, AutoCAD records a coordinate value representing the *CAmera* position. (The *CAmera* position can be specified by entering coordinate values when using the *POints* option rather than the *CAmera* option.)

When the *CAmera* option is selected, the selected objects appear to rotate with movement of the cursor. If the cursor is positioned at the center of the screen, the objects are viewed from the positive X axis on the XY plane (like *Vpoint Rotate* angles 0, 0). Moving the cursor to the upper-left displays the objects from a viewing angle similar to a SE Isometric viewpoint (or enter angles of 35, -45) (Fig. 35-23). Vertical cursor movement allows you to view from above or below, and horizontal cursor movement allows you to view from side to side. Since *Dview* displays a wireframe image, it may be difficult to see which "side" of the object you are viewing. The Coordinate System icon reappears only after PICKing a *CAmera* position.

Figure 35-23

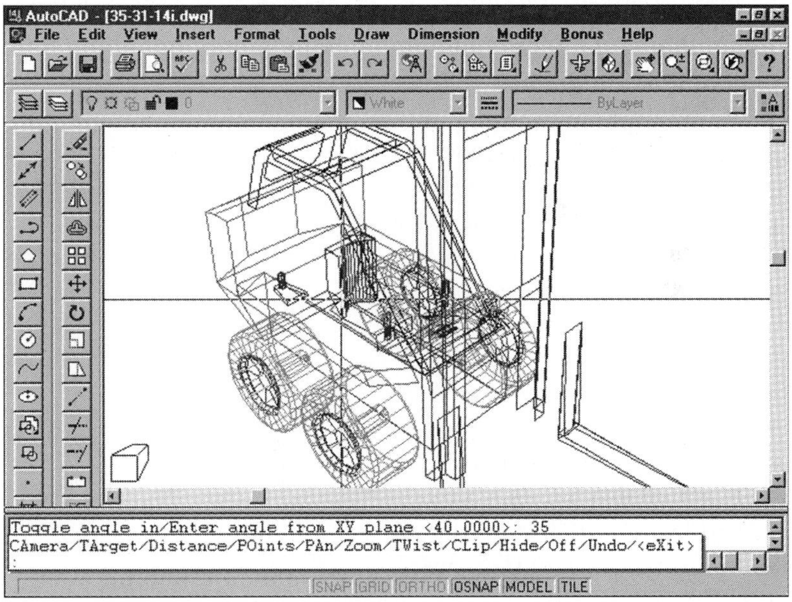

Rather than PICKing a *CAmera* position interactively, you can specify angles in the XY plane and from the XY plane. This method is like the *Vpoint* command, except the angles are given in the <u>reverse order</u>; first specify the angle <u>from</u> the XY plane, then the angle <u>in</u> the XY plane.

> Command: **dview**
> Select objects: **PICK** or **Enter** (Select objects or press Enter for the DVIEWBLOCK house.)
> Select objects: **Enter** (Completes the selection process.)
> CAmera/TArget/Distance/POints/PAn/Zoom/TWist/CLip/Hide/Off/ Undo/<eXit>: **ca**
> Toggle angle in/Enter angle from XY plane<0.00>: (**value**) or **PICK**
> Toggle angle from/Enter angle in XY plane from XY axis<0.00>: (**value**)

TArget

The *TArget* is the point that the *CAmera* (observer position) is focused on (looking toward). For comparison, when using the *Vpoint* command, the observer always looks toward the <u>origin</u>. A default *TArget* is automatically calculated by AutoCAD approximately at the <u>center of the selected geometry</u>. The *TArget* option allows you to specify any position to focus on. This is accomplished with the *TArget* option either interactively or by specifying angles.

Selecting a *TArget* position can also be accomplished by specifying angles <u>from</u> the XY plane and <u>in</u> the XY plane. The current *TArget* angles are always <u>opposite</u> the current *CAmera* angles. Another method for specifying the *TArget* is the *POints* option (see *POints*).

If you use the interactive method, use caution. It is not reasonable to interactively PICK a point in 3D space on a 2D screen. Therefore, <u>use *OSNAPs*</u> whenever PICKing a the *TArget* position. The *TArget* option is useful for viewing specific areas of a large 3D model such as an architectural plan. For example, the *CAmera* (observer) can be stationed in the middle of a room, and a separate display of each end of the room can be generated by *OSNAPing* to two *TArget*s, one at each end of the room (Fig. 35-24 and 35-25).

Figure 35-24 ─────────────

Figure 35-25 ─────────────

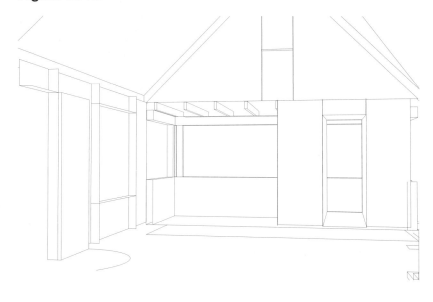

POints

Another method for specifying the *TArget* and/or the *CAmera* is the *POints* option. This option allows specification of the *CAmera* and/or *TArget* points <u>either</u> by specifying coordinate values or by interactive selection. Either use *OSNAPs* to interactively PICK a *CAmera/TArget* or enter coordinate values to designate an <u>exact</u> *CAmera/TArget* position (assuming you are familiar with the coordinate geometry of the 3D model). The *POints* option can also be used to list the current *TArget* coordinates established automatically by AutoCAD (unless you previously specified a *TArget*):

```
CAmera/TArget/Distance/POints/PAn/Zoom/TWist/CLip/Hide/Off/ Undo/<eXit>: po
Enter target point<(current coordinates)>: (coordinates) or (PICK)
Enter camera point<(current coordinates)>: (coordinates) or (PICK)
```

Distance

The distance option of *Dview* generates <u>perspective</u> projections. *Distance* takes into account the distance that the *CAmera* is from the *TArget*, whereas all other *Dview* options (and *Vpoint* options) use a vector between the *CAmera* and *TArget* to generate <u>parallel</u> projections.

Distance is used to adjust the <u>amount</u> of perspective, not the <u>size</u> of the display. Disregard the size of the display as you use *Distance*. The size of the display can be adjusted <u>later</u> with the *Zoom* option of *Dview*.

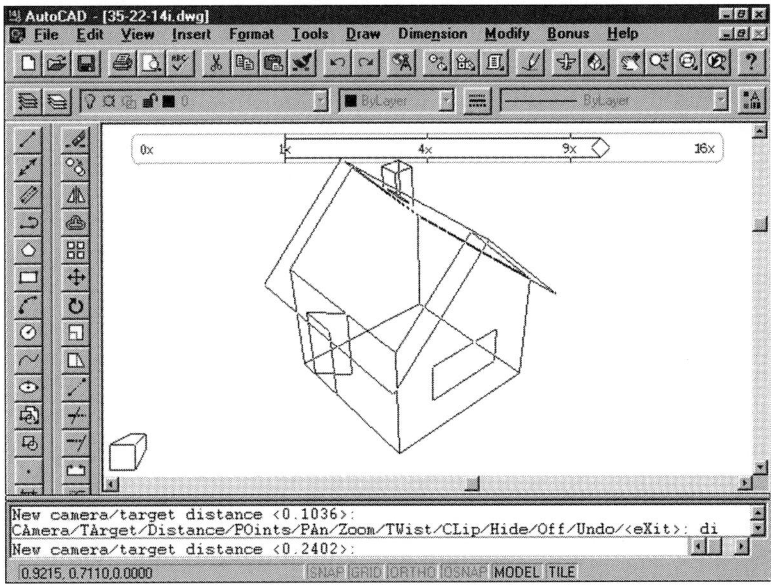

Distance uses a slide bar along the top of the screen to allow you to interactively adjust the amount of perspective (distance) for the display. The default value (1X on the slide bar) represents the previously specified distance between the *CAmera* and *TArget* points (Fig. 35-26).

A default distance value of 1 is used unless you previously specified the points in response to the *Vpoint* command or entered values in response to the *POints* option of *Dview*. Often the initial (default) distance (1) is too small to give the slider bar any usefulness, especially for metric, civil, or architectural applications (since the default of 1 represents one millimeter or inch). As a starting

Figure 35-26

point, try entering a value of about 3 to 4 times the longest dimension of the model. A value of 220 is a good value to begin with for the FORKLIFT drawing.

Zoom

The *Zoom* option of *Dview* changes the size of the 3D objects with respect to the screen display. *Zoom* <u>does not affect the amount of perspective</u>. The *Zoom* option should be used after the *Distance* option to adjust the size of the geometry so that it is visible and appropriately sized on the screen. *Distance* often makes the geometry too large or too small so that it must be adjusted with *Zoom*.

Zoom allows you to use the slider bar for interactive adjustment or to enter a camera lens size at the keyboard. The "lens length" represents that of a 35mm camera, with 50mm being the default length (zoom factor of 1X). The *Coords* display (on the Status line) dynamically lists the lens length as the slider bar is moved:

    CAmera/TArget/Distance/POints/PAn/Zoom/TWist/CLip/Hide/Off/ Undo/<eXit>: **z**
    Adjust lens length<50.000mm>: **PICK** or (**value**) (Use slider bar or enter a value.)
    CAmera/TArget/Distance/POints/PAn/Zoom/TWist/CLip/Hide/Off/ Undo/<eXit>:

PAn

The *PAn* option of *Dview* operates similarly to the full *PAN* command. *Dview PAn*, however, is <u>dynamic</u> and fully 3D. As you use *Dview PAn*, you can watch the 3D model as it is moved and see the 3D features of the object adjust in real time as your viewpoint of the object changes. *Dview PAn* allows panning both vertically and horizontally. The command syntax is similar to the full *PAN* command:

    CAmera/TArget/Distance/POints/PAn/Zoom/TWist/CLip/Hide/Off/ Undo/<eXit>: *pa*
    Displacement base point: **PICK** or (**coordinates**) (Usually, *PAn* is used interactively.)
    Second point of displacement: move and **PICK**
    CAmera/TArget/Distance/POints/PAn/Zoom/TWist/CLip/Hide/Off/ Undo/<eXit>:

PAn is generally used to center the geometry on the screen after *Distance* and *Zoom* are used. The three *Dview* options, *Distance, Zoom,* and *PAn,* can be used as a related group of commands to generate perspective projections and to size and center the 3D model appropriately on the screen.

Off

Once a perspective projection of a drawing has been generated (*Dview Distance* used), draw and edit commands <u>cannot</u> be used. Perspective drawings in AutoCAD are intended for display and output only. The drawing <u>can</u> be plotted and rendered, and slides can be made while perspective is in effect. When the perspective mode is active, a small icon representing a perspective box is displayed in the lower-left corner of the screen (see Fig. 35-22).

The *Off* option of *Dview* turns off <u>perspective</u> mode only and allows further editing and drawing with the existing geometry. When *Off* is used, a parallel projection of the geometry is generated, but all other changes to the display of the geometry made with *Dview* remain in effect. Figure 35-27 shows the FORKLIFT drawing in parallel projection (*Off* used).

Figure 35-27 —————————————————

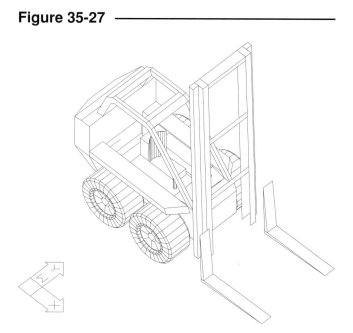

Figure 35-28 shows the FORKLIFT in perspective mode (*Distance* used). There is <u>no</u> *On* option of *Dview*; therefore, in order to change back to perspective mode, you must use the *Distance* option again. Fortunately, AutoCAD remembers the previous *Distance* setting, so accepting the default value generates the previous perspective.

Figure 35-28 —————————————————

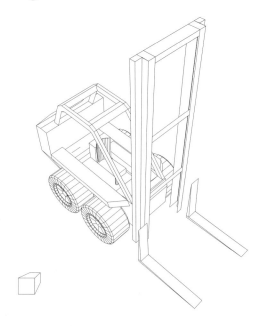

TWist

TWist is used to rotate the 3D geometry on the screen. The 3D model twists (rotates) about the line of sight (between *CAmera* point and *TArget* point). *TWist* operates interactively so you can see the position of the 3D model as you move the cursor. As an alternative, an angle can be entered at the keyboard. The *TWist* angle is measured counterclockwise starting at the 0 degree angle in X positive direction. *TWist* is typically used to tilt the 3D model for a particular effect. The *Coords* display (on the Status line) lists the current twist angle interactively. Figure 35-29 displays the DVIEWBLOCK house during a *TWist*.

Figure 35-29

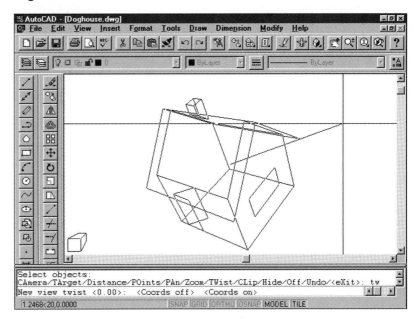

CAmera/TArget/Distance/POints/PAn/Zoom/TWist/CLip/Hide/Off/ Undo/<eXit>: **tw**
New view twist<0.000>: move and **PICK** or (**value**)
CAmera/TArget/Distance/POints/PAn/Zoom/TWist/CLip/Hide/Off/ Undo/<eXit>:

CLip

In many applications, it is desirable to display only a part of a 3D model. For example, an architect may want to display a view of a room in a house but not include the roof in the display.

The *CLip* option of *Dview* is used to establish a *Front* or *Back* clipping plane in a 3D model. Any geometry <u>behind</u> a *Back* clipping plane or any geometry in <u>front</u> of a *Front* clipping plane is <u>not</u> displayed. In the architectural example, a *Front* clipping plane could be established just below the roof of the model. Everything in front of (between the *CAmera* and) the clipping plane would not be included in the resulting display. A view of the room's interior would then be visible from above without being obscured by the roof. Likewise, a *Back* clipping plane could be established in a model to eliminate portions of the model that can distract the viewer's attention from the focus of the display.

The *Front* and *Back* clipping planes are like invisible walls <u>perpendicular to the line of sight</u> between the *CAmera* and the *TArget*. Like the other options of *Dview*, *CLip* is dynamic and uses slider bars, so you can move clipping planes dynamically forward and backward. As an alternative, you can enter a distance from the *TArget* (for *Back* clipping planes) or from the *CAmera* (for *Front* clipping planes). Clipping planes can be used in a parallel or perspective projection. The command syntax for creating a *Front* clipping plane is as follows:

CAmera/TArget/Distance/POints/PAn/Zoom/TWist/CLip/Hide/Off/ Undo/<eXit>: **cl**
Front/Back/<Off>: **f**
Eye/On/Off<Distance from target><1.0000>: move and **PICK** or (**value**)

Establishing a clipping plane distance automatically turns it
On. *Off* can be used to disable an established clipping plane.
Figure 35-30 displays the FORKLIFT drawing in hidden repre-
sentation with a front clipping plane established.

Figure 35-30 ————————

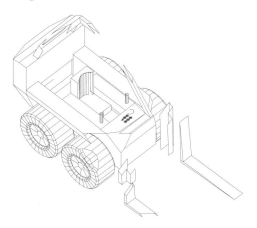

Hide

The *Hide* option of *Dview* removes hidden lines from a surface or solid model for the current screen
display. *Hide* causes all edges and surfaces that would normally be obscured by other surfaces or edges
to be removed so that the model appears solid rather than "see through." There are no options of *Hide*.
Hide remains active only for the current display. If another option of *Dview* is used or if *Dview* is exited,
hidden lines reappear. Figures 35-31 and 35-32 display the FORKLIFT drawing before and after *Hide* has
been performed.

Figure 35-31 ————————————————

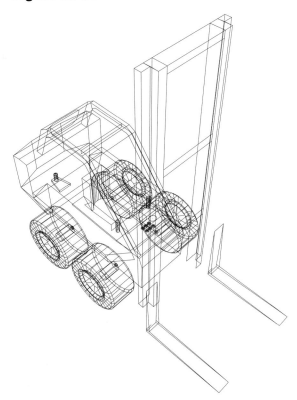

Figure 35-32 ————————————————

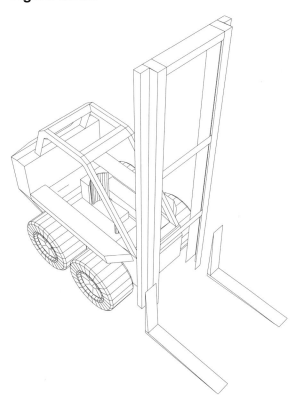

Undo

Undo reverses the effect of the last *Dview* option. Multiple *Undo*s can be performed to step backward through the *Dview* operations.

Saving *Dview* Images

Many *Dview* options such as *Distance*, *Hide*, and *Clip* are intended to be used for display of a 3D model, not necessarily to enhance drawing and editing capabilities. As a matter of fact, you cannot draw in any of these modes. When these *Dview* options are used to generate the desired view of the object, you can plot the display or "save" the image by using *Mslide* or *Render* and *Saveimg* (see Chapters 41 and 44).

VPORTS

Pull-down Menu	COMMAND (TYPE)	ALIAS (TYPE)	Short-cut	Screen (side) Menu	Tablet Menu
View *Tiled Viewports >*	*VPORTS*	...	...	*VIEW 1* *Vports*	*M,3*

The *Vports* command allows <u>tiled</u> viewports to be created on the screen. *Vports* allows the screen to be divided into several areas. Tiled viewports are different than paper space viewports. (See Chapter 19, Tiled and Paper Space Viewports.)

Vports (tiled viewports) are available only when the *TILEMODE* variable is set to **1**. Tiled *Vports* affect <u>only the screen display</u>. The viewport configuration <u>cannot be plotted</u>. If the *Plot* command is used, only the <u>current</u> viewport display is plotted. Figure 35-33 displays the AutoCAD drawing editor after the *Vports* command was used to divide the screen into tiled viewports.

Figure 35-33

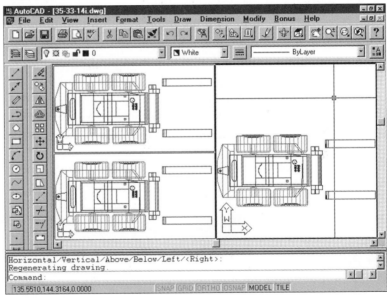

After you select the desired layout, the previous display appears in <u>each</u> of the viewports. For example, if a top view of the 3D model is displayed when you use the *Vports* command or the related dialog box, the resulting display in each of the viewports would be the top view (see Fig. 35-33). It is up to you then to use viewing commands (*Vpoint, Dview, Plan, Zoom,* etc.) to specify what viewpoints or areas of the model you want to see in each viewport. There is no automatic viewpoint configuration option with *Vports*.

A popular arrangement of viewpoints for construction and editing of 3D models is a combination of principal views (top, front, side, etc.) and an isometric or other pictorial type of viewpoint (Fig. 35-34). You can use *Zoom* with magnification factors to size the model in each viewport and use *PAn* to "align" the views. There is, however, <u>no automatic method</u> of alignment of views to achieve a true orthogonal projection. You cannot draw or project from one viewport to another.

A viewport is made active by PICKing a point in it. Any <u>display</u> commands (*Zoom, Vpoint, Redraw*, etc.) and drawing aids (*SNAP, GRID, ORTHO*) used affect only the <u>current viewport</u>. <u>Draw and edit</u> commands that affect the model are potentially apparent in *all* <u>viewports</u> (for every display of the affected part of the model). *Redrawall* and *Regenall* can be used to redraw and regenerate all viewports.

The *Vports* command can be used most effectively when constructing and editing 3D geometry, whereas paper space viewports are generally used when the model is complete and ready to prepare a plot. *Vports* is typically used to display a different *Vpoint*

Figure 35-34

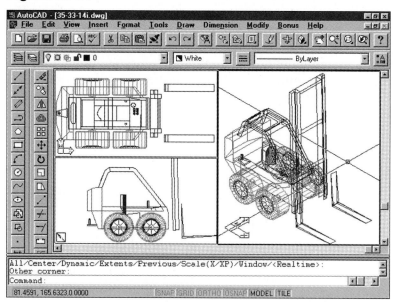

of the 3D model in each viewport, as in Figure 35-34. In this way, you can create 3D geometry and view the construction from different *Vpoints* in order to enhance your visibility in all three dimensions.

3D DISPLAY COMMANDS

Commands that are used for changing the appearance of a surface or solid model in AutoCAD are *Hide, Shade*, and *Render*. By default, surface and solid models are shown in wireframe representation in order to speed computing time during construction. As you know, wireframe representation can somewhat hinder your visualization of a model because it presents the model as transparent. To display surfaces and solid models as opaque and to remove the normally obscured edges, *Hide, Shade*, or *Render* can be used.

Wireframe models are not affected by these commands since they do not contain surfaces. Wireframe models can only be displayed in wireframe representation.

HIDE

Pull-down Menu	COMMAND (TYPE)	ALIAS (TYPE)	Short-cut	Screen (side) Menu	Tablet Menu
View *Hide*	*HIDE*	*HI*	...	VIEW 2 *Hide*	*M,2*

The *Hide* command suppresses the hidden lines (lines that would normally be obscured from view by opaque surfaces) for the current display or viewport:

    Command: **hide**

There are no options for the command. The current display may go blank for a period of time depending on the complexity of the model(s); then the new display with hidden lines removed temporarily appears.

The hidden line display is maintained only for the <u>current display</u>. Once a regeneration occurs, the model is displayed in wireframe representation again. You <u>cannot *Plot*</u> the display generated by *Hide*. Instead, use the *Hide Lines* option of the *Print/Plot Configuration* dialog box (or *Hideplot* option of *Mview* for paper space viewports).

Figure 35-35 displays the FORKLIFT model in the default wireframe representation, and Figure 35-36 illustrates the use of *Hide* with the same drawing.

Figure 35-35 ——————————————————   **Figure 35-36** ——————————————————

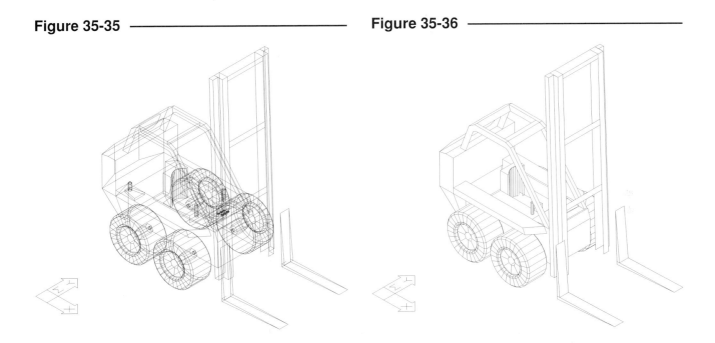

SHADE

Pull-down Menu	COMMAND (TYPE)	ALIAS (TYPE)	Short-cut	Screen (side) Menu	Tablet Menu
View *Shade >*	*SHADE*	*SHA*	...	*VIEW 2* *Shade*	N,2

The *Shade* command allows you to shade a surface or solid model in a full-screen display or in the current viewport. A shaded image is present only for the duration of the current display (until a *Regen*) and <u>cannot be plotted</u>. The image can be saved using the *Mslide* command (see Chapter 44). You can think of the shaded display as being projected in front of the current drawing. You cannot draw or edit the drawing while a shaded image is displayed. If you want to view or edit the original drawing again, use *Regen*.

Shade automatically positions one light in space and sets a default intensity. The light is located at the camera position. Several options allow you to fill the model surfaces and/or edges with the object color.

Two system variables control how the shade appears: *SHADEDGE* and *SHADEDIF*. *SHADEDGE* controls how the surfaces and edges are shaded. The only lighting control is a percentage of diffuse and ambient light adjusted by *SHADEDIF*. Any changes made to the variables are displayed on the <u>next</u> use of the *Shade* command.

If you select *Shade* from the *View* pull-down menu, the *SHADEDGE* options appear on a cascading menu (Fig. 35-37) and prevent you from having to change the variable setting by other means. By contrast, if you <u>type</u> the *Shade* command, the *SHADEDGE* variable must be previously set. The four possible settings are given next (with the integer values used when typing).

Figure 35-37 ─────────

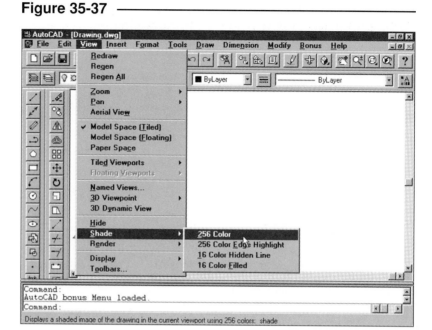

256 Color, (0)

The object surfaces are shaded in gradient values of the object color (Fig. 35-38). The object's visible edges are not highlighted. This option requires 256 color display capabilities of the monitor and video card. (Keep in mind that these figures have been converted to gray scale.)

Figure 35-38 ─────────

Figure 35-39

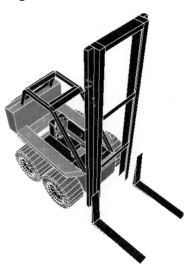

256 Color Edge Highlight, (1)
This option shades the surfaces in gradient values of the object color and highlights the visible edges in the background color (Fig. 35-39). This option also requires a 256 color display.

Figure 35-40

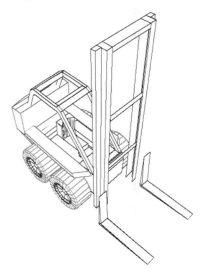

16 Color Hidden Line, (2)
This option simulates hidden line removal. The resulting display is similar to using the *Hide* command. All surfaces are painted in the background color, and the visible edges display the object color (Fig. 35-40). This display works well for monitors with only 16 colors or for monochrome displays.

Figure 35-41

16 Color Filled, (3)
The surfaces are filled solid with the object's color <u>without</u> gradient shading. The visible edges are drawn in the background color (Fig. 35-41). This display also works well for display capabilities of 16 colors and for monochrome displays.

SHADEDIF

When you use 256 color shading, *SHADEDIF* can be used to control the contrast of those values. Two kinds of lights are used with *Shade:* ambient and diffuse. The *SHADEDIF* variable controls the amount of diffuse (reflected) lighting versus the ambient (overall) light. The default setting of 70 specifies 70% diffuse and 30% ambient light. The <u>higher the *SHADEDIF*</u> setting, the greater the reflected light and the <u>greater the contrast</u> of surfaces on the object. The lower the *SHADEDIF* setting, the lower the contrast and the more all surfaces are lighted equally.

AutoCAD has sophisticated capabilities for generating a photo-realistic surface or solid object. With these capabilities, you can place lights in 3D space, assign light intensity and color, attach materials, and assign backgrounds in the model. *Render* is the command that causes the current rendering parameters to be applied and generates the rendered display. Although much time and effort can be spent adjusting parameters, *Render* can be used anytime to generate a display using the default parameters. (See Chapter 41, Rendering.)

CHAPTER EXERCISES

1. In this exercise, you use an AutoCAD sample drawing to practice with *3D Viewpoint*. **Open** the **CAMPUS** drawing from the Sample directory. (Assuming AutoCAD is installed using the default location, find the sample drawings in C:/Program Files/AutoCAD R14/Sample.)

    A. Use any method to generate a *Top* view. Examine the view and notice the orientation of the coordinate system icon.

    B. Generate a *SE Isometric* viewpoint to orient yourself. Examine the icon and find north (Y axis). Consider how you (the observer) are positioned as if viewing <u>from</u> the southeast.

    C. Produce a *Front* view. This is a view looking north. Notice the new icon. Remember that you cannot see the XY plane, so it is not a good idea to draw on that plane from this viewpoint.

    D. Generate a *Right* view. Produce a *Top* view again.

    E. Next, view the CAMPUS from a *SE Isometric* viewpoint. Your view should appear like that in Figure 35-42. Notice the position of the WCS icon.

    Figure 35-42

    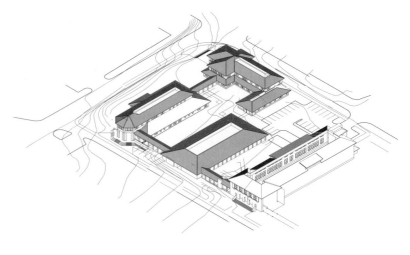

    F. Finally, generate a *NW Isometric* and *NE Isometric*. In each case, examine the WCS icon to orient your viewing direction.

    G. Experiment and practice more with *3D Viewpoint* if you desire. <u>Do not *Save*</u> the drawing.

2. The FORKLIFT sample drawing is an excellent 3D model for practice using the 3D viewing commands because it is a relatively small file (fast to generate) and is easily recognizable from any view (front, back, top, etc.). The FORKLIFT.DWG is not copied to your computer during installation but is located on the AutoCAD Release 14 CD-ROM. It is located in the Acad/Bonus/Sample/History directory along with many other interesting and useful sample drawings.

   A. *Open* the **FORKLIFT** drawing. Notice the *Ucsicon* is *On*. Force the **Ucsicon** to the **Origin** (the cross appears at the intersection of the X and Y arrows only when the icon is at the origin).

   B. Generate a *SE Isometric* viewpoint so you can see the top, front, and side of the forklift. *Zoom* with a *Window* and read the name of the creator of the drawing (Lansing Pugh, Architect). Use *Zoom Previous* to achieve the previous display.

   C. Next, generate a *Top* view, then a *Front* view. Often the front view is the profile view, not necessarily the front of the object. For mechanical applications, the front view should show the most about the object. Find which view (of the *3D Viewpoint* options) displays the actual front of the forklift.

   D. Generate a *SW Isometric* view. *Zoom* (out) until the icon orients itself at the origin. Next, generate a *NW Isometric*. Is the icon at the origin?

   E. Use *Vpoint* with the *Tripod* option to view the forklift looking from the southeast direction and from slightly above. HINT: PICK the lower-right (southeast) quadrant inside the small circle.

   F. Use *Vpoint* with the *Tripod* option again to view the forklift from the southwest and slightly above. View the forklift again from the back side (PICK on the "equator" at the top of the small circle).

   G. <u>Do not *Save*</u> the drawing.

3. Begin a *New* drawing to practice with *Dview*. Use *Start from Scratch, English* default settings. Immediately invoke the *Dview* command. When prompted to "Select objects:", press **Enter**. The DVIEWBLOCK house should appear as you view from above.

   A. Use the *Zoom* option of *Dview* to make the house smaller by moving the slider bar or entering a scale factor of **.5**. Now use the *CAmera* option. Place the cursor slightly to the upper-left of center of the screen and PICK. Use the *Hide* option of *Dview* to remove hidden edges.

   B. Next, invoke the *Distance* option. Since the default distance is so small, you can enter a value of **35** or use *Distance* with the slider bar to the maximum position twice. Note the perspective generation. To compare to a parallel projection, use the *Off* option. To turn the perspective on again, use the *Distance* option and accept the default value.

   C. Now use the *CLip* option and set up a *Front* clipping plane. Slide the plane into the house slowly until the roof begins to disappear, then **PICK**. Turn the front clipping plane *Off*; then use a *Back* clipping plane in the same manner. Finally, turn the back clipping plane *Off*.

   D. Use the *TWist* option. *TWist* the house around approximately 90 degrees. Use *TWist* again to bring the house to its normal position.

   E. Experiment more if you like. Do <u>not</u> *Save* the drawing.

4. *Open* the **FORKLIFT** drawing. Generate a display similar to that in Figure 35-32. Use *Dview* with the *CAmera, Distance, Zoom, PAn,* and *Hide* options. If you want to save the drawing, make sure you use *SaveAs* to rename the drawing to **FORKLIFT2** and to save it <u>in your working directory</u>.

5. *Open* the **CAMPUS** drawing again.

    A. Generate a *Top* view. Use *Vports* or the *Tiled Viewports Layout* dialog box to generate *3* viewports with the large vertical viewport on the *right*. The CAMPUS should appear in a top view in all viewports.

    Figure 35-43 ────────────────

    B. Use the *Vpoint* command or *3D Viewpoint* to create a specific *Vpoint* for each viewport, as given below and shown in Figure 35-43:

Upper-left	*Top* view
Lower-left	*Front* view
Right	*SE Isometric*

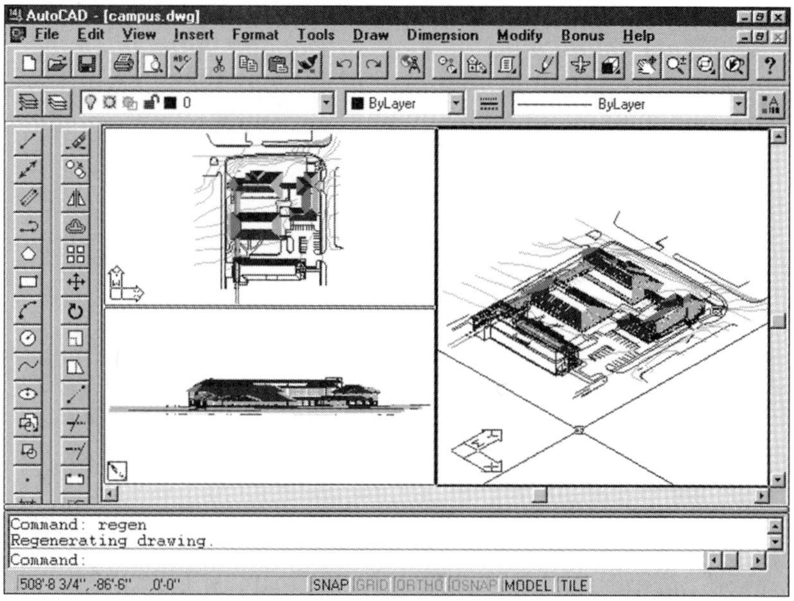

6. *Open* the **FORKLIFT** drawing again.

    A. Use the *Hide* command to remove hidden edges.

    B. *Shade* the forklift with each of the *SHADEDGE* options:

    256 Color
    256 Color Edge Highlight
    16 Color Hidden Line
    16 Color Filled

    Figure 35-44 ────────────────

    C. Next, use *Vports* or the *Tiled Viewports Layout* dialog box to generate *2 Vertical* viewports. *Zoom Extents* in each viewport. In the left viewport, use *Hide*, and in the right, *Shade* the forklift with the most realistic option. The drawing should look similar to that in Figure 35-44.

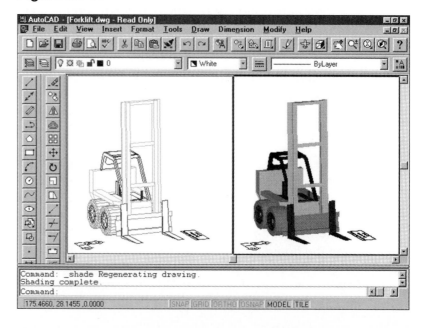

7. In this exercise, you will create a simple solid model that you can use for practicing creation of User Coordinate Systems in Chapter 36. You will also get an introduction to some of the solid modeling construction techniques discussed in Chapter 38.

    A. Begin a *New* drawing and name it *SOLID1*. Turn *On* the *Ucsicon* and force it to appear at the *ORigin*.

    B. Type *Box* to create a solid box. When the prompts appear, use **0,0,0** as the "Corner of box." Next, use the *Length* option and give dimensions for the Length, Width, and Height as **5, 4,** and **3**. A rectangle should appear. Change your viewpoint to *SE Isometric* to view the box. *Zoom* with a magnification factor of *.6X*.

Figure 35-45

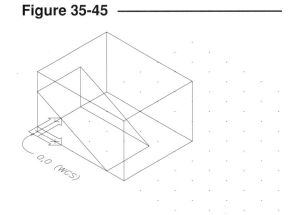

    C. Type the *Wedge* command. When the prompts appear, use **0,0,0** as the "Corner of wedge." Again use the *Length* option and give the dimensions of **4, 2.5,** and **2** for *Length, Width,* and *Height*. Your solid model should appear as that in Figure 35-45.

    D. Use *Rotate* and select only the wedge. Use **0,0** as the "Base point" and enter **-90** as the "Rotation angle." The model should appear as Figure 35-46.

Figure 35-46

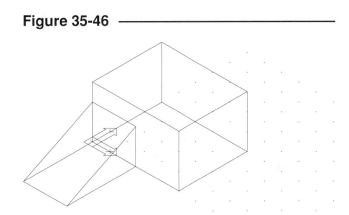

    E. Now use *Move* and select the wedge for moving. Use **0,0** as the "Base point" and enter **0,4,3** as the "Second point of displacement."

    F. Finally, type *Union*. When prompted to "Select objects:", PICK both the wedge and the box. The finished solid model should look like Figure 35-47. *Save* the drawing.

Figure 35-47

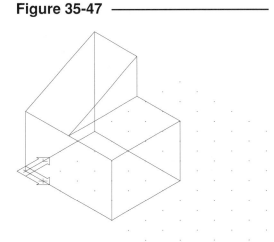

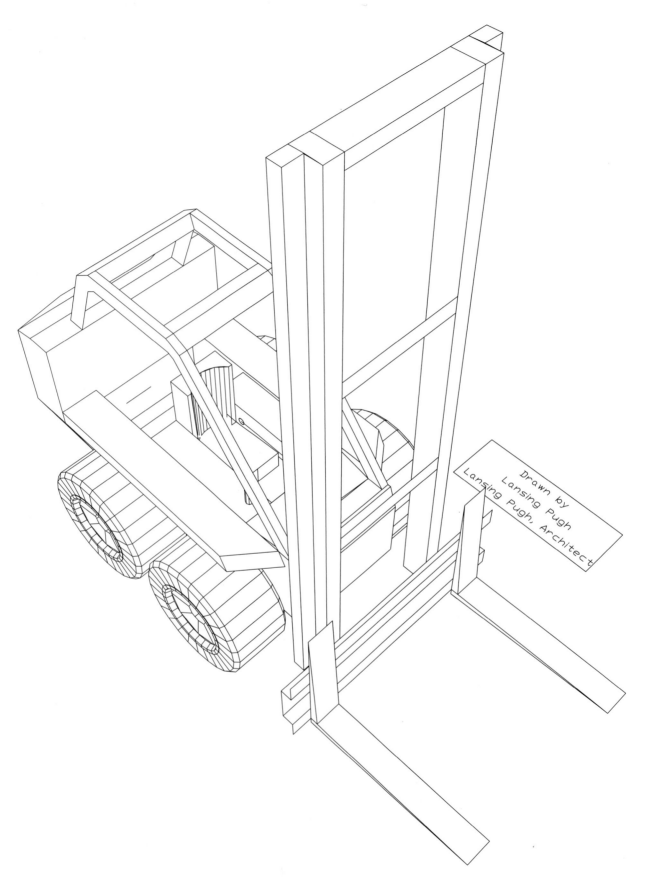

Drawn by
Lansing Pugh
Lansing Pugh, Architect

FORKLIFT.DWG Courtesy of Autodesk, Inc. (Release 14 Sample Drawing)

36

USER COORDINATE SYSTEMS

Chapter Objectives

After completing this chapter you should:

1. know how to create, *Save*, *Restore*, and *Delete* User Coordinate Systems;

2. be able to create UCSs by the *Origin*, *Zaxis*, *3point*, *Object*, *View*, *X*, *Y*, or *Z* methods;

3. know how to create UCSs using the *UCS Orientation* dialog box;

4. be able to use the *UCSFOLLOW* variable to automatically display a plan view of new UCSs.

CONCEPTS

No User Coordinate Systems exist as part of the AutoCAD default template drawing (ACAD.DWT) as it comes "out of the box." UCSs are created to simplify construction of the 3D model when and where they are needed.

When you create a multiview or other 2D drawing, creating geometry is relatively simple since you only have to deal with X and Y coordinates. However, when you create 3D models, you usually have to consider the Z coordinates, which makes the construction process more complex. Constructing some geometries in 3D can be very difficult, especially when the objects have shapes on planes not parallel with, or perpendicular to, the XY plane or not aligned with the WCS (World Coordinate System).

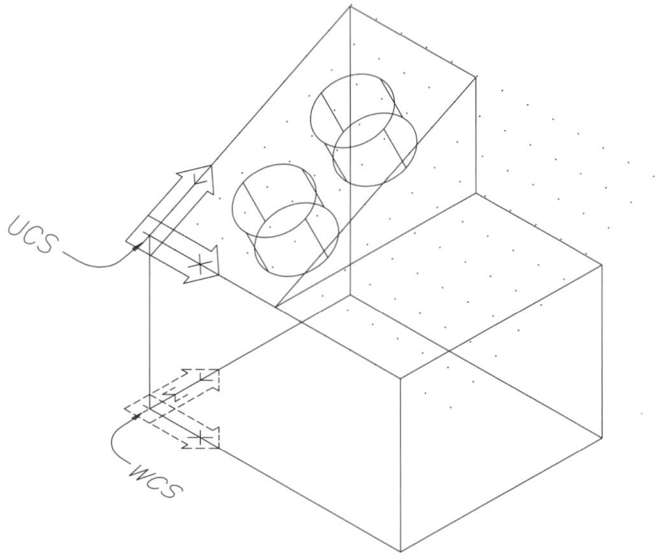

UCSs are created when needed to simplify the construction process of a 3D object. For example, imagine specifying the coordinates for the centers of the cylindrical shapes in Figure 36-1, using only world coordinates. Instead, if a UCS were created on the face of the object containing the cylinders (the inclined plane), the construction process would be much simpler. To draw on the new UCS, only the X and Y coordinates (of the UCS) would be needed to specify the centers, since anything drawn on that plane has a Z value of 0. The Coordinate Display (*Coords*) in the Status line always displays the X, Y, and Z values of the <u>current</u> coordinate system, whether it is the WCS or a UCS.

Figure 36-1

Generally, you would create a UCS aligning its XY plane with a surface of the object, such as along an inclined plane, with the UCS origin typically at a corner or center of the surface. Any coordinates that you specify (in any format) are assumed to be user coordinates (coordinates that lie in or align with the current UCS). Even when you PICK points interactively, they fall on the XY plane of the current UCS. The *SNAP* and *GRID* automatically align with the current coordinate system XY plane, providing *SNAP* points and enhancing your visualization of the construction plane.

UCS COMMANDS AND VARIABLES

User Coordinate Systems are created by using any of several options of the *UCS* command. Once a UCS has been created, it becomes the current coordinate system. <u>Only one coordinate system can be active</u>. If you switch among several UCSs and the WCS, you should save the UCSs when you create them (by using the *Save* option of *UCS*) and restore them at a later time (using the *Restore* option).

When creating 3D models, it is recommended that the <u>*ORigin* option of the *Ucsicon* command be used</u> to place the icon always at the origin of the current UCS, rather than in the lower-left corner of the screen. Since the origin of the UCS is typically specified as a corner or center of the construction plane, aligning the Coordinate System icon with the current origin aids the user's visualization of the UCS orientation.

UCS

	Pull-down Menu	COMMAND (TYPE)	ALIAS (TYPE)	Short-cut	Screen (side) Menu	Tablet Menu
	Tools *UCS >*	*UCS*	...	...	*TOOLS 2* *UCS*	*W,7*

The *UCS* command allows you to create, save, restore, and delete UCSs. There are many possibilities for creating UCSs to align with any geometry:

Command: **ucs**
Origin/ZAxis/3point/OBject/View/X/Y/Z/Previous/Save/Restore /Delete/?<World>: (**letter**) (Enter capitalized letters of desired option.)

The *UCS* command options are available through the *Tools* pull-down menu.

Keep in mind, <u>UCSs have no association with views or viewpoints.</u> Creating or restoring a UCS <u>does not change the view</u> (unless the *UCS-FOLLOW* variable is *On*), and <u>changing a viewpoint never changes the UCS.</u>

Figure 36-2

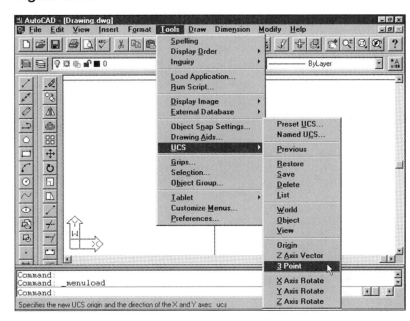

The *UCS* icon group is available on the Standard toolbar (Fig. 36-3). If you prefer, a separate *UCS* toolbar can be brought on the screen and made to float or dock.

When you create UCSs, AutoCAD prompts for points. These points can be entered as coordinate values at the keyboard, or points on existing geometry can be PICKed. *OSNAP*s should be used to PICK points in 3D space. Understanding the right-hand rule is imperative when creating UCSs.

(If you completed the SOLID1 drawing from Chapter 35, Exercise 7, *Open* the drawing and try the *UCS* options as you read along through this chapter.)

Figure 36-3

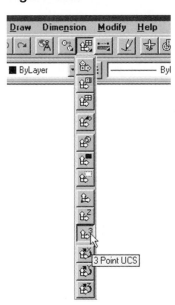

Origin

This option defines a new UCS by specifying a new X,Y,Z location for the origin. The <u>orientation</u> of the UCS (direction of X,Y,Z axes) remains the same as in its previous position; only the location of the origin changes. AutoCAD prompts:

Origin point<0,0,0>:

Coordinates may be specified in any format or PICKed (use *OSNAPs* in 3D space).

Figure 36-4

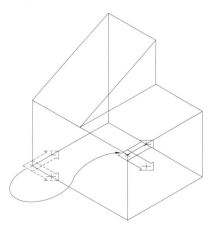

ZAxis

You can define a new UCS by specifying an <u>origin</u> and a direction for the <u>Z axis</u>. Only the two points are needed. The X <u>or</u> Y axis generally remains parallel with the current UCS XY plane, depending on how the Z axis is tilted. AutoCAD prompts:

Origin point <0,0,0>:
Point on positive portion of the Z axis <default>:

Figure 36-5

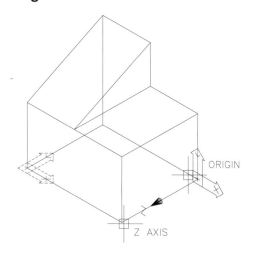

3point

The new UCS is defined by (1) the <u>origin</u>, (2) a point on the <u>X axis</u> (positive direction), and (3) a point on the <u>Y axis</u> (positive direction) or <u>XY plane</u> (positive Y). This is the most universal of all the UCS options (works for most cases). It helps if you have geometry established that can be PICKed with *OSNAP* to establish the points. The prompts are:

Origin point <0,0,0>:
Point on positive portion of the X axis <default>:
Point on positive-Y portion of the UCS XY plane <default>:

Figure 36-6

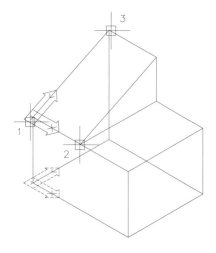

Object

This option creates a new UCS aligned with the selected object. You need to PICK only one point to designate the object. The orientation of the new UCS is based on the type of object selected and the XY plane that was current when the object was created. This option is intended for use primarily with <u>wireframe</u> or <u>surface</u> model objects.

AutoCAD prompts:

Figure 36-7

> Select object to align UCS:

The following list gives the orientation of the UCS using the *Object* option for each type of object.

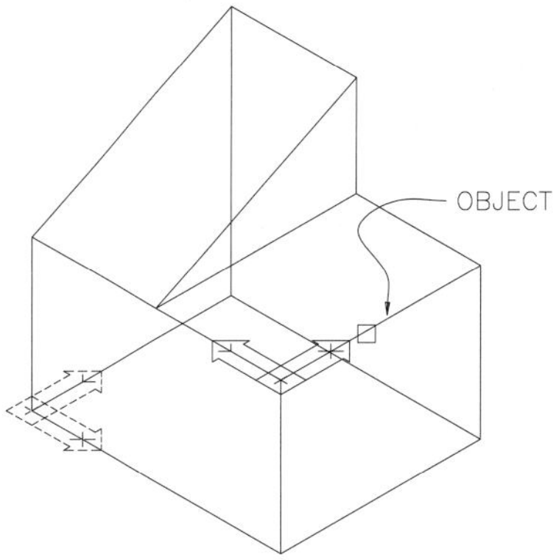

Object	Orientation of New UCS
Line	The end nearest the point PICKed becomes the new UCS origin. The new X axis aligns with the *Line*. The XY plane keeps the same orientation as the previous UCS.
Circle	The center becomes the new UCS origin with the X axis passing through the PICK point.
Arc	The center becomes the new UCS origin. The X axis passes through the endpoint of the *Arc* that is closest to the pick point.
Point	The new UCS origin is at the *Point*. The X axis is derived by an arbitrary but consistent "arbitrary axis algorithm." (See the *AutoCAD Customization Guide*.)
Pline (2D)	The *Pline* start point is the new UCS origin with the X axis extending from the start point to the next vertex.
2D Solid	The first point of the *Solid* determines the new UCS origin. The new X axis lies along the line between the first two points.
Dimension	The new UCS origin is the middle point of the dimension text. The direction of the new X axis is parallel to the X axis of the UCS that was current when the dimension was drawn.
3Dface	The new UCS origin is the first point of the *3Dface*, the X axis aligns with the first two points, and the Y positive side is on that of the first and fourth points. (See Chapter 40, Surface Modeling, *3Dface*.)
Text, Insertion, or *Attribute*	The new UCS origin is the insertion point of the object, while the new X axis is defined by the rotation of the object around its extrusion direction. Thus, the object you PICK to establish a new UCS will have a rotation angle of 0 in the new UCS.

View

This *UCS* option creates a UCS parallel with the screen (perpendicular to the viewing angle). The UCS <u>origin</u> remains unchanged. This option is handy if you wish to use the current viewpoint and include a border, title, or other annotation. There are no options or prompts.

Figure 36-8

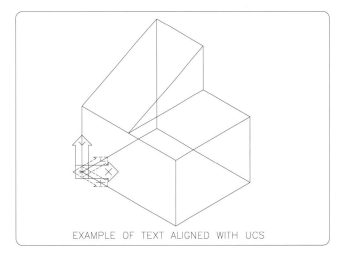

EXAMPLE OF TEXT ALIGNED WITH UCS

X   Y   Z

Each of these options rotates the UCS about the indicated axis according to the right-hand rule. The command prompt is:

Figure 36-9

Rotation angle about *n* axis:

The angle can be entered by PICKing two points or by entering a value. This option can be repeated or combined with other options to achieve the desired location of the UCS. It is imperative that the right-hand rule is followed when rotating the UCS about an axis (see Chapter 34).

X, 90° ROTATION

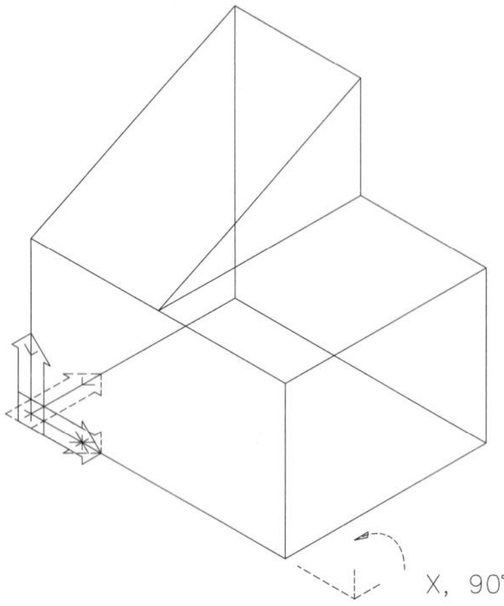

Previous

Use this option to restore the previous UCS. AutoCAD remembers the ten previous UCSs used. *Previous* can be used repeatedly to step back through the UCSs.

World

Using this option makes the WCS (World Coordinate System) the current coordinate system.

Save

Invoking this option prompts for a name for saving the current UCS. Up to 31 characters can be used in the name. Entering a name causes AutoCAD to save the current UCS. The *?* option and the *Dducs* command list all previously saved UCSs.

Restore

Any *Save*d UCS can be restored with this option. The *Restored* UCS becomes the <u>current</u> UCS. The ? option and the *Dducs* command list the previously saved UCSs.

Delete

You can remove a *Save*d UCS with this option. Entering the name of an existing UCS causes AutoCAD to delete it.

?

This option lists the origin and X, Y, and Z axes for any saved UCS you specify.

It is important to remember that changing to another UCS does <u>not</u> change the display (unless *UCSFOL-LOW* is activated). Only <u>one</u> coordinate system (WCS or a UCS) can be current at a time. If you are using viewports, several *Viewpoints* can be displayed, but only <u>one</u> UCS can be current and it appears in all viewports.

DDUCSP

	COMMAND (TYPE)	ALIAS (TYPE)	Short-cut	Screen (side) Menu	Tablet Menu
Pull-down Menu					
Tools *UCS >* *Preset UCS...*	*DDUCSP*	*UCP*	...	*TOOLS 2* *Dducsp*	*W,9*

If you prefer to do things a bit more automatically, the *Dducsp* command (UCS presets) lets you set up new UCSs simply by selecting an image tile in the *UCS Orientation* dialog box (Fig. 36-10). These options result in the same functions accomplished by the *UCS* command.

Figure 36-10 ——————————————

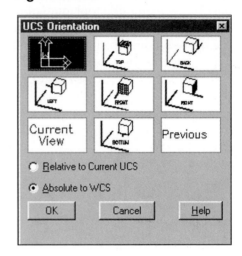

The images displayed in this dialog box are somewhat misleading. The selections <u>do not necessarily reposition the UCS on the face of the object</u> indicated in the images but <u>rotate the coordinate system to the correct orientation</u>.

For example, if the current coordinate system were the WCS, selecting the *Right* option would rotate the UCS to the correct orientation <u>keeping the same origin</u>, but would not move the UCS to the rightmost face of the object (Fig. 36-11).

Figure 36-11 ————————————————————————————

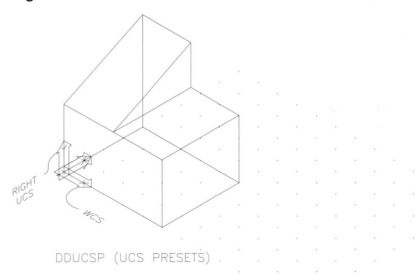

DDUCSP (UCS PRESETS)

NOTE: Always check the settings of the two buttons near the bottom of the dialog box. <u>Generally, UCSs are established *Absolute to the WCS.*</u> Be careful when creating a UCS *Relative to the Current UCS*—you may get some unexpected results.

DDUCS

Pull-down Menu	COMMAND (TYPE)	ALIAS (TYPE)	Short-cut	Screen (side) Menu	Tablet Menu
Tools UCS > Named UCS...	DDUCS	UC	...	TOOLS 2 Dducs	W,8

The *Dducs* command invokes the *UCS Control* dialog box (Fig. 36-12). By using this tool, named UCSs (those that have been previously *Saved*) can be made the current UCS by highlighting the desired name from the list and checking the *Current* tile. This performs the same action as *UCS, Restore*. Named UCSs can also be *Renamed*, *Deleted*, or *Listed* from this dialog box.

Figure 36-12 ————

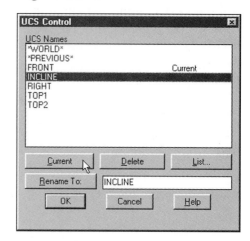

UCSFOLLOW

Pull-down Menu	COMMAND (TYPE)	ALIAS (TYPE)	Short-cut	Screen (side) Menu	Tablet Menu
...	UCSFOLLOW	...	...	...	...

The *UCSFOLLOW* system variable, if set to 1 (*On*), causes the <u>plan view</u> of a UCS to be displayed automatically <u>when a UCS is made current</u>. The default setting for *UCSFOLLOW* is 0 (*Off*). *UCSFOLLOW* can be set separately for each viewport.

For example, consider the case shown in Figure 36-13. Two tiled viewports (*Vports*) are being used, and the *UCS-FOLLOW* variable is turned *On* for the <u>left</u> viewport only. Notice that a UCS created on the inclined surface is current. The left viewport shows the plan view of this UCS automatically, while the right viewport shows the model in the same orientation no matter what coordinate system is current. If the WCS were made active, the right viewport would keep the same viewpoint, while the left would automatically show a plan view when the change was made.

Figure 36-13 ————

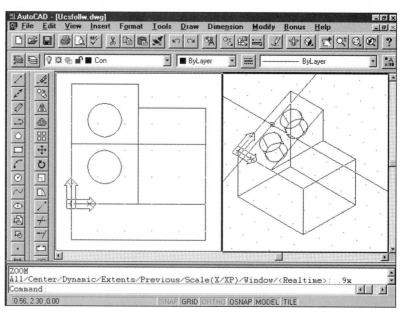

NOTE: When *UCSFOLLOW* is set to 1, the display does not change <u>until</u> a new UCS is created or restored.

UCSFOLLOW is the only variable or command that directly links a change in UCS to an automatic change in viewpoint. Normally, changing to a different current UCS does not change the display, except in this case (when *UCSFOLLOW* is set to 1). On the other hand, making a change in viewpoint <u>never</u> causes a change in UCS.

3D Basics and Wireframe, Surface, and Solid Modeling

Now that you understand the basics of 3D modeling, you are ready to progress to constructing and editing wireframe, surface, and solid models. Almost all the topics discussed in Chapters 34, 35, and 36 are applicable to <u>all types</u> of 3D modeling.

In the progression from an elementary to a complete description of an actual object, the three types of models range from wireframe to surface to solid. The next several chapters, however, present the modeling types in a different order: wireframe models, solid models, and surface models. This is a logical order from simple to complex with respect to command functions and construction and editing complexity. That is to say, wireframe construction and editing is the simplest, and surface construction and editing is the most tedious and complex.

CHAPTER EXERCISES

1. Using the model in Figure 36-14, assume a UCS was created by rotating about the X axis 90 degrees from the existing orientation (World Coordinate System).

    A. What are the absolute coordinates of corner F?

    B. What are the absolute coordinates of corner H?

    C. What are the absolute coordinates of corner J?

2. Using the model in Figure 36-14, assume a UCS was created by rotating about the X axis 90 degrees from the existing orientation (WCS) and then rotating 90 degrees about the (new) Y.

    A. What are the absolute coordinates of corner F?

    B. What are the absolute coordinates of corner H?

    C. What are the absolute coordinates of corner J?

Figure 36-14

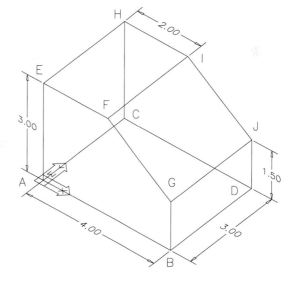

3. ***Open*** the **SOLID1** drawing that you created
    in Chapter 35, Exercise 7. View the object
    from a *SE Isometric Vpoint*. Make sure that
    the *UCSicon* is *On* and set to the *ORigin*.

    A. Create a *UCS* with a vertical XY plane
        on the front surface of the model as
        shown in Figure 36-15. *Save* the UCS as
        FRONT.

    B. Change the coordinate system back to
        the *World*. Now, create a *UCS* with the
        XY plane on the top horizontal surface
        and with the orientation as shown in
        Figure 36-15 as TOP1. *Save* the UCS as
        TOP1.

    C. Change the coordinate system back to
        the *World*. Next, create the *UCS* shown
        in the figure as TOP2. *Save* the UCS
        under the name **TOP2**.

    D. Activate the *WCS* again. Create and *Save* the **RIGHT** UCS.

    E. Use *SaveAs* and change the drawing name to **SOLIDUCS**.

Figure 36-15 ─────────────────

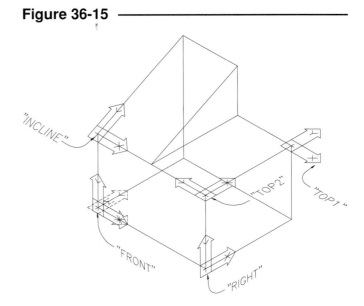

4. Using the same drawing, make the *WCS* the active coordinate system. In this exercise, you will
    create a model and screen display like that in Figure 36-13.

    A. Create **2 *Vertical*** tiled viewports. Use *Zoom* in each viewport if needed to display the model
        at an appropriate size. Make sure the *UCSicon* appears at the origin. Activate the <u>left</u> view-
        port. Set *UCSFOLLOW On* in that viewport only. (The change in display will not occur until
        the next change in UCS setting.)

    B. Use the right viewport to create the **INCLINE UCS,** as shown in Figure 36-15. *Regenall* to
        display the new viewpoint.

    C. In either viewport create two cylinders as follows: Type *cylinder*; specify **1.25,1** as the *center*,
        .5 as the *radius*, and **-1** (negative 1) as the *height*. The new cylinder should appear in both
        viewports.

    D. Use *Copy* to create the second cylinder. Specify the existing center as the "Basepoint, of dis-
        placement." To specify the basepoint, you can enter coordinates of **1.25,1** or **PICK** the center
        with *Osnap*. Make the copy **1.5** units above the original (in a Y direction). Again, you can
        PICK interactively or use relative coordinates to specify the second point of displacement.

    E. Use *SaveAs* and assign the name **SOLID2**.

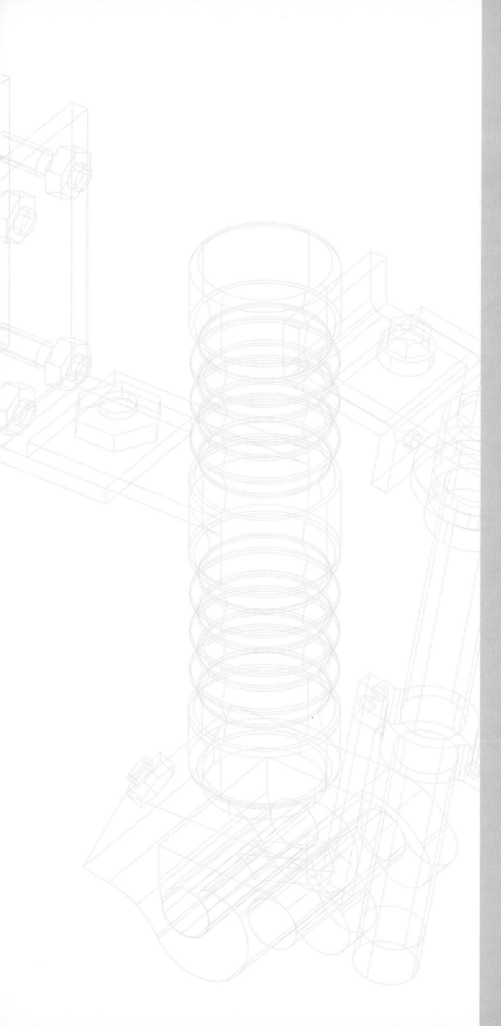

37

WIREFRAME MODELING

Chapter Objectives

After completing this chapter you should:

1. understand how common 2D draw and edit commands are used to create 3D wireframe models;

2. gain experience in specification of 3D coordinates;

3. gain experience with 3D viewing commands;

4. be able to apply point filters to 3D model construction;

5. gain fundamental experience with creating, *Saving*, and *Restoring* User Coordinate Systems;

6. be able to create, *Save*, and *Restore* tiled viewport configurations;

7. be able to manipulate the Coordinate System icon.

CONCEPTS

Wireframe models are created in AutoCAD by using common draw commands that create 2D objects. When draw commands are used with X ,Y, and Z coordinate specification, geometry is created in 3D space. A true wireframe model is simply a combination of 2D elements in 3D space.

A *Line*, for example, can be drawn in 3D space by specifying 3D coordinates in response to the "From point:" and "To point:" prompts (Fig. 37-1):

> Command: **line**
> From point: **3,2,0**
> To point: **8,2,6**
> Command:

Figure 37-1

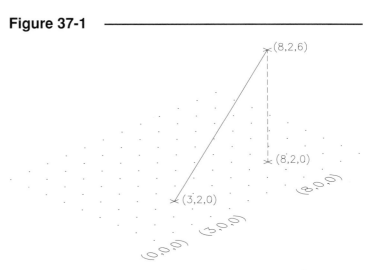

The same procedure of entering 3D coordinate values can be applied to other draw commands such as *Arc, Circle,* etc. These 2D objects are combined in 3D space to create wireframe models. No draw commands are used <u>specifically</u> for creating 3D wireframe geometry (with the exception of *3Dpoly*). The use of 3D viewing and display, User Coordinate Systems (UCSs), and 3D coordinate entry are the principal features of AutoCAD, along with the usual 2D object creation commands, and they allow you to create wireframe models. Specification of 3D coordinate values (Chapter 34), 3D display and viewing (Chapter 35), and the use of UCSs (Chapter 36) aid in creating wireframe models as well as surface and solid models. The *3Dpoly* command is discussed at the end of this chapter.

WIREFRAME MODELING TUTORIAL

Because you are experienced in 2D geometry creation, it is efficient to apply your experience, combined with the concepts already given in the previous three chapters, to create a wireframe model in order to learn fundamental modeling techniques. Using the tutorial in this chapter, you create the wireframe model of the stop plate shown in Figure 37-2 in a step-by-step fashion. The Wireframe Modeling Tutorial employs the following fundamental modeling techniques:

> 3D viewing commands
> 3D coordinate entry
> Point filters
> *Vports*
> User Coordinate Systems
> *Ucsicon* manipulation

Figure 37-2

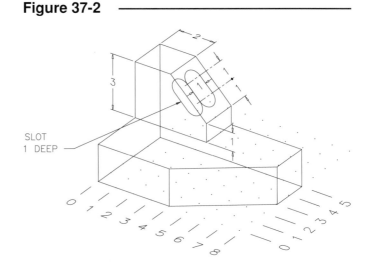

It is suggested that you follow along and work on your computer as you read through the pages of this tutorial. The Wireframe Modeling Tutorial has 22 steps. The steps give written instructions with illustrations.

Additional hints, such as explicit coordinate values and menus that can be used to access the commands, are given on pages immediately following the tutorial (titled Hints). Try to complete the wireframe modeling tutorial without looking at the hints given on the following pages. Refer to Hints only if you need assistance.

1. Set up your drawing.

    A. Start AutoCAD. Use ***Start from Scratch, English*** default settings. Use ***Save*** and name the file **WIREFRAME**.

    B. Examine the figure on the previous page showing the completed wireframe model. Notice that all of the dimensions are to the nearest unit. For this exercise, assume generic units.

    C. Keep the default ***Decimal Units***. Set ***Units Precision* .00** (two places to the right of the decimal).

    D. Keep the default *Limits* (assuming they are already set to 0,0 and 12,9 or close to those values). It appears from Figure 37-2 that the drawing space needed is a minimum of 8 units by 5 units on the XY plane.

    E. Turn on the *SNAP*.

    F. Turn on the *GRID*.

    G. Make a layer named **MODEL**. Set it as the ***Current*** layer. Set a ***Color,*** if desired.

2. Draw the base of the part on the XY plane.

    A. Use ***Line***. Begin at the origin (0,0,0) and draw the shape of the base of the stop plate. (It is a good practice to begin construction of any 3D model at 0,0,0.) Use *SNAP* as you draw and watch the *Coords* display. See Figure 37-3.

    B. Change your ***Vpoint*** with the ***Rotate*** option. Specify angles of **305, 30**. Next, ***Zoom .9X***.

    C. Turn the ***Ucsicon ON*** and force it to the ***ORigin***. See Figure 37-4 to see if you have the correct *Vpoint*.

Figure 37-3 ——————————————————

0,0

3. Begin drawing the lines bounding the top "surface" of the base by using absolute coordinates and .XY filters.

    A. Draw the first two *Lines*, as shown in Figure 37-4, by entering <u>absolute</u> coordinates. You cannot PICK points interactively (without *OSNAP*) since it results in selections only <u>on the XY plane</u> of the current coordinate system. Do not exit the *Line* command when you finish.

Figure 37-4 ——————————————————

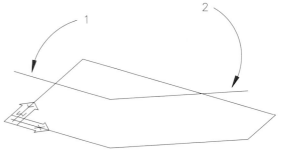

B. Draw *Lines* 3 and 4 using **.XY** point filters (Fig. 37-5). Do not exit the *Line* command.

C. Complete the shape defining the top surface (*Line* 5) by using the **Close** option. The completed shape should look like Figure 37-5.

Figure 37-5

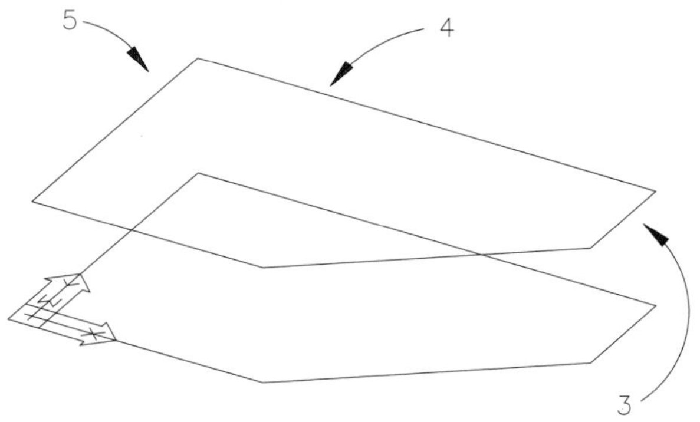

4. There is an easier way to construct the top "surface"; that is, use *Copy* to create the top "surface" from the bottom "surface."

A. *Erase* the 5 lines defining the top shape.

B. Use **Copy** and select the 5 lines defining the bottom shape. You can PICK **0,0,0** as the "Base Point." The "Second point of displacement" can be specified by entering absolute coordinates or relative coordinates or by using an .XY filter. Compare your results with Figure 37-5.

5. Check your work and *Save* it to disk.

A. Use **Plan** to see if all of the *Lines* defining the top shape appear to be aligned with those on the bottom. The correct results at this point should look like Figure 37-3 since the top shape is directly above the bottom.

B. Use **Zoom Previous** to display the previous view (same as *Vpoint, Rotate*, 305,30.)

C. If you have errors, correct them. You can use *Undo* or *Erase* the incorrect *Lines* and redraw them. See Hints for further assistance. You may have to use *OSNAP* or enter absolute coordinates to begin a new *Line* at an existing *Endpoint*.

D. *Save* your drawing.

6. Draw the vertical *Lines* between the base and top "surfaces" using the following methods.

Figure 37-6

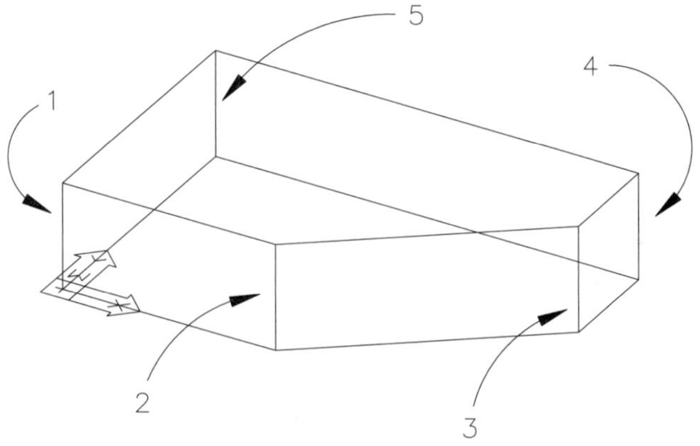

A. Enter absolute coordinates (X, Y, and Z values) to draw the first two vertical *Lines* shown in Figure 37-6 (1, 2).

B. Use another method (interactive coordinate entry) to draw the remaining three *Lines* (3, 4, 5). You have learned that if you PICK a point, with or without *SNAP*, the selected point is on the <u>XY plane</u> of the current coordinate system. *OSNAP*, however, allows you to PICK points in <u>3D space</u>. Begin each of the three *Lines* ("From point:") by using *SNAP* and picking the appropriate point on the XY plane. The second point of each *Line* ("To point:") can only be specified by using *OSNAP* (use **Endpoint**).

C. Once again, there is an easier method. *Erase* four lines and <u>leave one</u>. You can use *Copy* with the *Multiple* option to copy the one vertical *Line* around to the other positions. Make sure you select the *Multiple* option. In response to the "Base point or displacement:" prompt, select the base of the remaining *Line* (on the XY plane) with *SNAP ON*. In response to "second point of displacement:" make the multiple copies.

D. Use a *Plan* view to check your work. *Zoom Previous*. Correct your work, if necessary.

7. Draw two *Lines* defining a vertical plane. Often in the process of beginning a 3D model (when geometry has not been created in some positions), it may be necessary to use absolute coordinates. This is one of those cases.

A. Use absolute coordinates to draw the two *Lines* shown in Figure 37-7. Refer to Figure 37-2 for the dimensions.

B. *Zoom Extents* and then *Zoom .9X* so that the coordinate system icon appears at the origin. Your screen should look like Figure 37-7.

Figure 37-7 ——————————

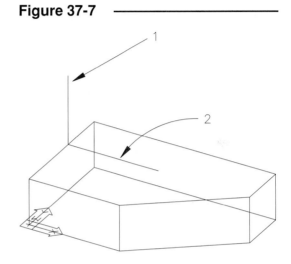

8. Rather than drawing the next objects by defining absolute coordinates, you can establish a vertical construction plane and just PICK points with *SNAP* and *GRID* on the construction plane.

A. This can be done by creating a new UCS (User Coordinate System). Invoke the *UCS* command by typing **UCS** or by selecting it from the menus. Use the *3point* option and select the line *Endpoints* in the order indicated (Fig. 37-8) to define the origin, point on X axis, and point on Y axis.

Figure 37-8 ——————————

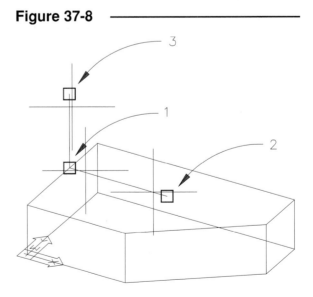

B. The Coordinate System icon should appear in the position shown in Figure 37-9. Notice that the *GRID* and *SNAP* have followed the new UCS. Move the cursor and note the orientation of the cursor. You can now draw on this vertical construction plane simply by PICKing points on the XY plane of the new UCS!

If your UCS does not appear like that in the figure, *UNDO* and try again.

Figure 37-9 ———————————

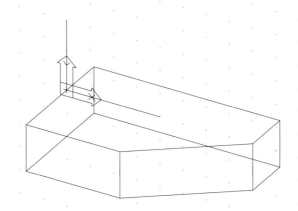

C. Draw the *Lines* to complete the geometry on the XY plane of the UCS. Refer to Figure 37-2 for the dimensions. The stop plate should look like that in Figure 37-10.

Figure 37-10 ———————————

9. Check your work; save the drawing and the newly created UCS.

A. View your drawing from a plan view of the UCS. Invoke *Plan* and accept the default option "*<current UCS>*." If your work appears to be correct, use *Zoom Previous*.
B. Save your new UCS. Invoke the *UCS* command as before. Use the *Save* option and assign the name **FRONT**.

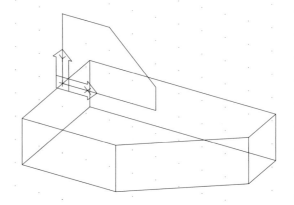

10. Create a second vertical surface behind the first with *Copy* (Fig. 37-11). You can give coordinates relative to the (new) current UCS.

Figure 37-11 ———————————

A. Invoke *Copy*. Select the four lines (highlighted) comprising the vertical plane (on the current UCS). Do <u>not</u> select the bottom line defining the plane. Specify **0,0** as the "Base point:" and give either absolute or relative coordinates (with respect to the current UCS) as the "second point of displacement." Note that the direction to *Copy* is <u>negative</u> Z.

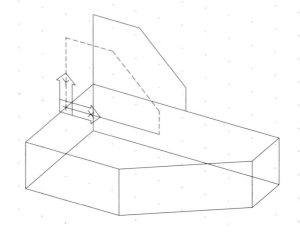

B. Alternately, using *Copy* with *OSNAP* would be valid here since you can PICK (with *OSNAP*) the back corner.

11. Draw the horizontal lines between the two vertical "surfaces" using one of several possible methods.

A. Use the **Line** command to connect the two vertical surfaces with four horizontal edges as shown in Figure 37-12. With *SNAP* on, you can **PICK** points on the XY plane of the UCS. Use *OSNAP* to **PICK** the *Endpoints* of the lines on the back vertical plane.

B. As another possibility, you could draw only one **Line** by the previous method and **Copy** it with the **Multiple** option to the other three locations. In this case, **PICK** the "Base point" on the XY plane (with *SNAP ON*). Specifying the "second point(s) of displacement" is easily done (also at *SNAP* points).

Figure 37-12

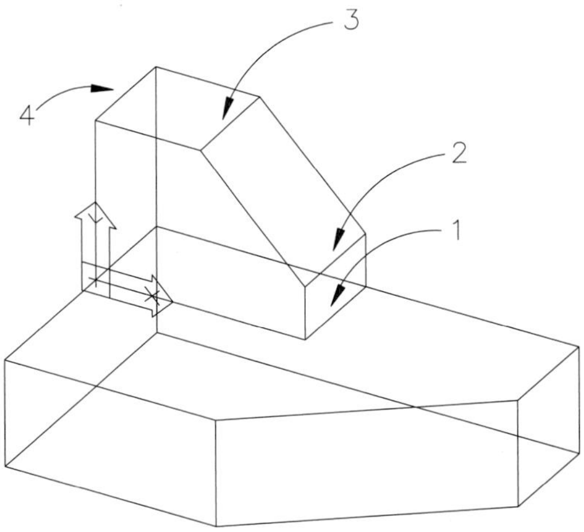

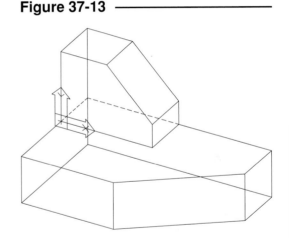

12. A wireframe model uses 2D objects to define the bounding edges of "surfaces." There should be no objects that do not fulfill this purpose. If you examine Figure 37-13, you notice two lines that should be removed (highlighted). These lines do not define edges since they each exist between coplanar "surfaces." Remove these lines with *Trim*.

Trimming 2D objects in 3D space requires special attention. You have two choices for using *Trim* in this case: *Projmode = None* or *View*. You can use the *View* option of *Projmode* (to trim any objects that appear to intersect from the current view) or you must select the vertical *Line* as a cutting edge (because with the *None* option of *Projmode*, you cannot select a cutting edge that is perpendicular to the XY plane of the current coordinate system). In other words, when *Projmode=None*, you can select only the vertical *Line* as the cutting edge, but when *Projmode=View*, either the vertical or horizontal *Line* can be used as the cutting edge (Fig. 37-14).

Figure 37-13

A. Use *Trim* and select either *Line* (highlighted, Fig. 37-14) as the *cutting edge*. Change the **Projmode** to **View** and *Trim* the indicated *Line*s. Use *Trim* again to trim the other indicated horizontal *Line* (previous figure).

B. Change back to the WCS. Invoke **UCS**; make the **World** coordinate system current.

Figure 37-14

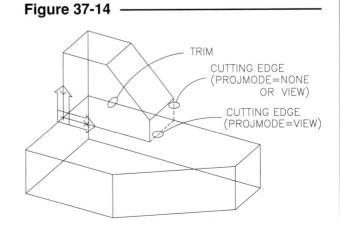

TRIM

CUTTING EDGE
(PROJMODE=NONE
OR VIEW)

CUTTING EDGE
(PROJMODE=VIEW)

13. Check your work and *Save* your drawing.

Figure 37-15 ────────────

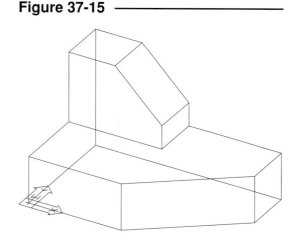

A. Make sure your wireframe model looks like that in Figure 37-15.
B. Use *Save* to secure your drawing.

14. It may be wise to examine your model from other viewpoints. As an additional check of your work and as a review of 3D viewing, follow these steps.

A. Invoke **UCS** and *Restore* the UCS you saved called **FRONT**. Now use *Plan* with the *Current UCS* option. This view represents a front view and should allow you to visualize the alignment of the two vertical planes constructed in the last several steps.
B. Next, use **UCS** and restore the **World** Coordinate System. Then use *Plan* again with the *Current UCS* option. Check your model from this viewing angle.
C. Finally, use **Vpoint Rotate** and enter angles of **305** and **30**. This should return to your original viewpoint.

15. Instead of constantly changing *Vpoints*, it would be convenient to have several *Vpoints* of your model visible on the screen <u>at one time</u>. This can be done with the *Vports* (Viewports) command. In order to enhance visualization of your model for the following constructions, divide your screen into four sections or viewports with *Vports* and then set a different *Vpoint* in each viewport.

A. The first step is to type **Vports** or select from the **View** pull-down menu *Tiled Viewports >, Layout....* Specify **4** (four) viewports. Your screen should appear as that in Figure 37-16. Move your cursor around in the viewports. The cursor appears only in the <u>current</u> viewport. Only <u>one</u> viewport can be current at a time. Any change you make to the model appears in all viewports. You cannot draw <u>between</u> viewports; however, you can begin a draw or edit command in one viewport and finish in another. Experiment by drawing and *Erasing* a **Line**.

B. Now configure the upper-left viewport with a top view. Make the upper-left viewport the current viewport. To produce a top view, use the **Plan** command. Next, use **Zoom** with a **.9X** magnification factor to bring the edges of the model slightly away from the viewport borders.

Figure 37-16 ────────────

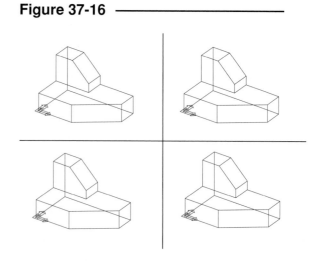

C. Now make the lower-left viewport current. In order to configure this for a front view, use *3D Viewpoint* with the *Front* option. Next, *Zoom .9x* in this viewport. Use the *Ucsicon* command with the *All* option to make the icon appear at the origin for all viewports. Check your results with Figure 37-17.

Figure 37-17 ————————

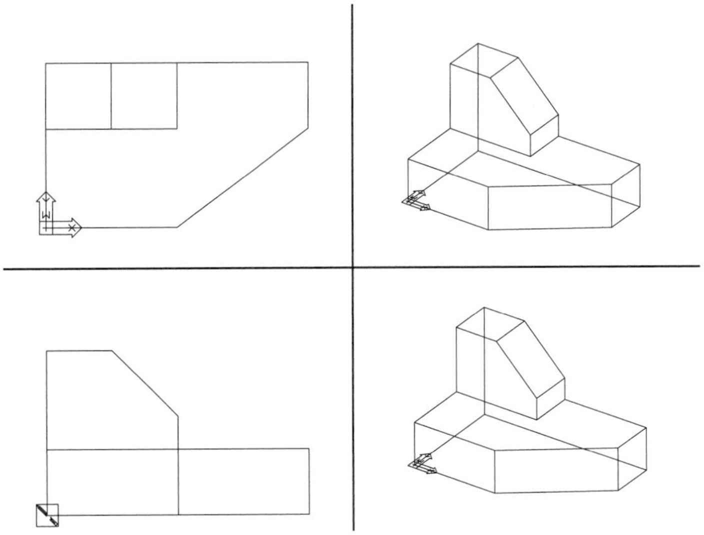

Note that by using viewports you can achieve multiple views of the model; however, only <u>one coordinate system</u> can be current at a time. The Coordinate System icon appears in all viewports, indicating the current coordinate system orientation. The *Ucsicon ORigin* and *All* settings force the icons to appear at the origin in all viewports. Since the XY plane of the coordinate system is not visible in the lower-left viewport, a different symbol appears in place of the Coordinate System icon. This icon, a broken pencil, indicates that it is not a good idea to draw from the viewpoint because you cannot see the XY plane.

16. In the next construction, add the milled slot in the inclined surface of the stop plate. Because the slot is represented by geometry parallel with the inclined surface, a UCS should be created on the surface to facilitate the construction.

Figure 37-18 ————————

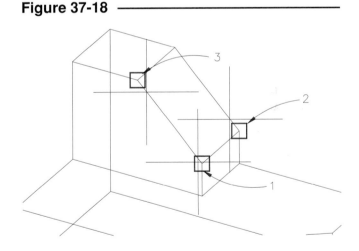

A. Make the lower-right viewport current. Use the *UCS 3point* option. Select the three points indicated in Figure 37-18 for (1) the "Origin," (2) "Point on the positive X axis," and (3) "Point on the positive Y axis portion on the XY plane." Make sure you *OSNAP* with the *Endpoint* option.

B. The icon should move to the new UCS origin. Its orientation should appear as that in Figure 37-19. Notice that the new coordinate system is current in all viewports.

Figure 37-19 ————————

C. Save this new UCS by using the *UCS* command with the *Save* option. Assign the name **AUX** (short for "auxiliary view").

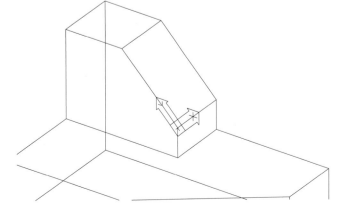

(Remember to check the Hints <u>if</u> you have trouble.)

17. To enhance your visualization of the inclined surface, change your viewpoint for the lower-right viewport.

Figure 37-20

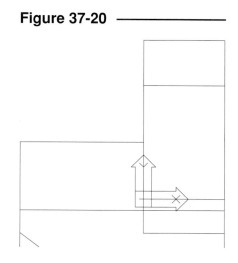

   A. Make sure that the lower-right viewport is current. Use the **Plan** command and accept the default **<Current UCS>**. This action should yield a view normal to the inclined surface—an auxiliary view.

   B. It would also be helpful to **Zoom** in with a Window to only the inclined surface. Do so. Your results (for that viewport) should look like Figure 37-20.

18. The next several steps guide you through the construction of the slot. Refer to Figure 37-2 for the dimensions of the slot.

   A. Turn on the *SNAP* and *GRID* for the viewport.

Figure 37-21

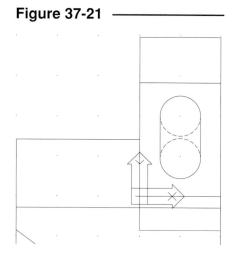

   B. Draw two **Circles** with a **.5** radius. The center of the lower circle is at coordinate **1,1** (of the UCS). The other circle center is one unit above.

   C. Draw two vertical **Lines** connecting the circles. Use **Tangent** or **Quadrant** *OSNAP*.

   D. Use **Trim** to remove the inner halves of the *Circles* (highlighted). Notice that changes made in one viewport are reflected in the other viewports.

19. The slot has a depth of 1 unit. Create the bottom of the slot by *Copy*ing the shape created on the top "surface."

   A. With the current UCS, the direction of the *Copy* is in a <u>negative</u> Z direction. Use **Copy** and select the two *Arcs* and two *Lines* comprising the slot. Use the **Center** of an *Arc* or the origin of the UCS as the "basepoint." The "second point of displacement" can be specified using absolute or relative coordinates.

Figure 37-22

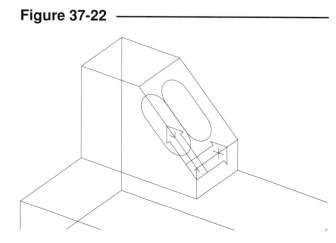

   B. Check to ensure that the copy has been made correctly. Compare your work to that in Figure 37-22. No other lines are needed to define the slot. The "surface" on the inside of the slot has a smooth transition from curved to straight, so no additional edges (represented by lines) exist.

20. The geometry defining the stop plate is complete at this point. No other edges are needed to define the "surfaces." However, you may be called on to make a design change to this part in the future. Before you change your display back to a *Single* viewport and save your work, it is helpful to *Save* the viewport configuration so that it can be recalled in a future session. Unless you save the current viewport configuration, you have to respecify the *Vpoints* for each viewport.

   A. Use the **Vports** command and the **Save** option (not the *SAVE* command). You can specify any name for the viewport configuration. A descriptive name for this configuration is **FTIA** (short for Front, Top, Isometric, and Auxiliary). Using this convention tells you how many viewports (by the number of letters in the name) and the viewpoints of each viewport when searching through a list of named viewports.

   B. When changing back to a single viewport, the <u>current</u> viewport becomes the single viewport. Make the upper-right viewport (isometric-type viewpoint) current. Invoke the **Vports** command and the **SIngle** option. The resulting screen should display only the isometric-type viewpoint (like Fig. 37-23). If needed, the viewport configuration can be recalled in the future by using the **Restore** option of **Vports**.

   Figure 37-23 ————————

   C. Before saving your drawing, you should make the WCS current by using the **UCS** command. Notice that changing display options (like *Vpoint* or *Vports*) has no effect on the UCS. UCS controls are independent from display controls (unless the *UCSFOLLOW* variable is set to 1).

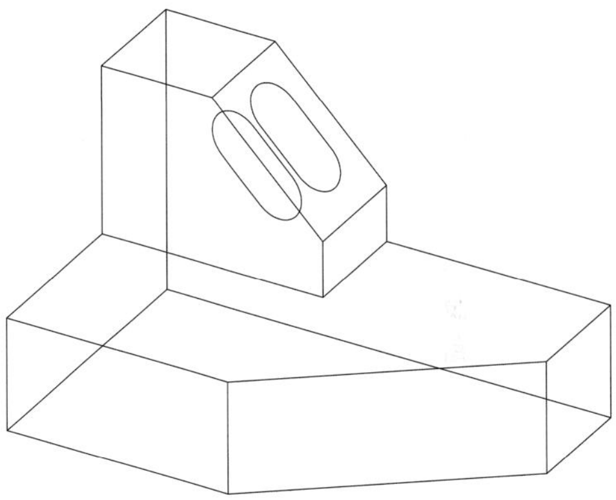

21. *Save* your drawing. Use this drawing for practicing with display commands covered in the previous chapters, 3D Viewing and Display, and User Coordinate Systems. The stop plate is especially good for practicing with *Dview*.

22. When you are finished experimenting with display and view commands, *Exit* AutoCAD and do not save your changes.

Hints

H1.   A. Start AutoCAD by the method you normally use at your computer workstation. Select the *File* pull-down menu, *New....* In the dialog box, select *Start from Scratch*, then *English* default settings. Next, use *Save* and assign the name **WIREFRAME**.
       B. (No help needed.)
       C. *Format* pull-down menu, *Units....*
       D. (No help needed.)
       E. Press **F9**. Make sure the word **SNAP** appears on the Status line.
       F. Press **F7**. The *GRID* should be visible.
       G. *Format* pull-down menu, *Layer...*, select *New*, type **MODEL** in the edit box, then *Current* and *OK*.

H2.   A. *Draw* pull-down menu, *Line*. Make sure *SNAP* is *ON*. The *Coords* display helps you keep track of the line lengths as you draw. You may have to press **F6** until *Coords* displays a cursor tracking or polar display.
       B. Type **Vpoint**. Type **r**. Enter **305** for the first angle and **30** for the second. Type **Z** (for *Zoom*). Enter **.9X**
       C. *View* pull-down, *Display>*, then **UCS icon**, then check *Origin*.

H3.   A.   Line 1   From point: **0,0,2**
                      To point: **4,0,2**
              Line 2   To point: **8,3,2**
                      To point: do not press **Enter** yet

      B.   If you happened to exit the *Line* command, you can use *Line* again and attach to the last end-point by entering an "@" in response to the "From point:" prompt. If that does not work, use *OSNAP* **Endpoint** to locate the end of the last *Line*.

<u>Using point filters:</u> At the "To point:" prompt, type **.XY** and press **Enter**. AutoCAD responds with "of." **PICK** a point on the XY plane <u>directly below</u> the desired location for the *Line* end (Fig 37-24). AutoCAD responds with "(need Z)." Enter a value of **2**. The result-ing *Line* end coordinate has the X and Y

Figure 37-24

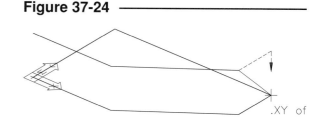

values of the point you PICKed and the Z value of 2. In other words, since you can PICK points only on the XY plane, using the .XY point filter locates only the X and Y coordinates of the selected point and allows you to specify the Z coordinate separately.

      C.   Enter **C** in response to the "To point:" prompt. AutoCAD *closes* the first and last *Line* seg-ments created in one use of the *Line* command. If you used the *Line* command two or more times to complete the shape defining the top "surface," this option does not work. In that case, use absolute coordinates, .XY filters, or *OSNAP Endpoint* to connect to the final end-point.

H4.   A.   Type *E* for *Erase* or use the *Modify* pull-down, *Erase*.
      B.   Type *CP* for *Copy* or use the *Modify* pull-down, *Copy*. Select the 5 *Lines* in response to "Select objects:." PICK or enter **0,0,0** as the "Base point or displacement." Enter either **0,0,2** (absolute coordinates) or **@0,0,2** (relative coordinates) in response to "Second point of dis-placement:."

H5.   A.   Type *Plan* and then the *W* option or select *View* pull-down, *3D Viewpoint >*, *Plan View >*, *World UCS*.
      B.   Type *Z* for *Zoom* and type *P* for *Previous*.
      C.   If you have mistakes, correct them by first *Erasing* the incorrect *Lines*. Begin redrawing at the last correct *Line Endpoint*. You have to use absolute coordinates, .XY filters, or *OSNAP Endpoint* to attach the new *Line* to the old.
      D.   Type *Save* or select *File* pull-down, *Save*.

H6.   A.   Line 1   From point: **0,0,0**
                      To point: **0,0,2**
              Line 2   From point: **4,0,0**
                      To point: **4,0,2**
      B.   Line 3   From point:
                      PICK point a. (with *SNAP* on)
                      To point:
                      Endpoint of, **PICK** point b.

Figure 37-25

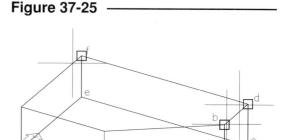

Line 4	From point:
	PICK point c.
	To point:
	Endpoint of, **PICK** point d.
Line 5	From point:
	PICK point e.
	To point:
	Endpoint of, **PICK** point f.

    C. Type *E* for *Erase*. *Erase* the last four vertical lines, leaving the first. Type *CP* for *Copy*. Select the remaining vertical line. At the next prompt, type *M* for *Multiple*. When asked for the "Base Point:," select the bottom of the line (on the XY plane—make sure *SNAP* is On). At the "Second point of displacement:" prompt, pick the other locations for the new lines.

    D. *View* pull-down, *3D Viewpoint >*, *Plan View >*, *World UCS*. Examine your work. Type *Z* for *Zoom*. Type *P* for *Previous*.

H7.   A.

Line 1	From point:	**0,3,2**
	To point:	**0,3,5**
Line 2	From point:	**0,3,2**
	To point:	**4,3,2**

    B. Type *Z*. Type *E* for *Extents*. Press **Enter** to invoke the last command (*Zoom*). Type **.9X**.

H8.   A. Type *UCS*. Type *3* for *3point*. When prompted for the first point ("Origin"), invoke the cursor (*OSNAP*) menu and select *Endpoint*. **PICK** point 1 (Fig. 37-8). When prompted for the second point ("Point on positive portion of the X-axis"), again use *Endpoint* and **PICK** point 2. Repeat for the third point ("Point on positive Y portion of the UCS XY plane").

    B. If you have trouble establishing this UCS, you can type the letter *U* to undo one step. Repeat until the icon is returned to the origin of the WCS orientation. If that does not work for you, type *UCS* and use the *World* option. Repeat the steps given in the previous instructions. Make sure you use *OSNAP* correctly to **PICK** the indicated *Endpoint*s.

    C. When constructing the *Lines* representing the vertical plane, ensure that *SNAP* (F9) and *GRID* (F7) are *ON*. Notice how you can PICK points only on the XY plane of the UCS with the cursor. This makes drawing objects on the plane very easy. Use the *Line* command to draw the remaining edges defining the vertical "surface."

H9.   A. *View* pull-down menu, *3D Viewpoint >*, *Plan View >*, *Current UCS*. Check your work. Then type *Z*, then *P*. Correct your work if necessary.

    B. Type *UCS*. Type *S* for the *Save* option. Enter the name **FRONT**.

H10.   A. Type *CP* or select *Copy* from the *Modify* pull-down menu. Select the indicated four *Lines* (Figure 37-11). For the basepoint, enter or **PICK 0,0**. For the second point of displacement, enter **0,0,-2**.

    B. Alternately, perform the same sequence except, when prompted for the "second point of displacement:", select *Endpoint* from the *OSNAP* menu and **PICK** the corner indicated in Figure 37-26.

Figure 37-26

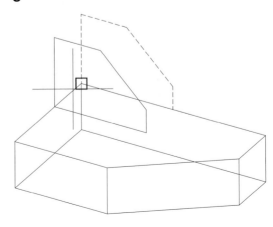

H11.   A. Type **L** for *Line*. At the "From point:" prompt
        PICK point (a) on the UCS XY plane (Fig. 37-27).
        You should be able to *SNAP* to the point. At the
        "To point:" prompt, select the **Endpoint** *OSNAP*
        option and **PICK** point (b). Repeat the steps for the
        other 3 *Lines*.

        B. Alternately, draw only the first *Line* (line 1,
        Fig. 37-12). Type **CP** for *Copy*. Select the first *Line*.
        Type **M** for *Multiple*. At the "Base point:" prompt,
        PICK point (a) (Fig. 37-27). Since this point is on
        the XY plane, PICK (with *SNAP ON*). For the
        second points of displacement, **PICK** the other
        corners of the vertical surface on the XY plane
        (with *SNAP ON*).

Figure 37-27

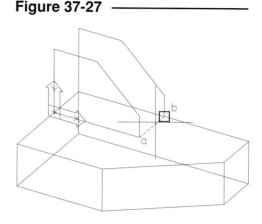

H12.   A. ***Modify*** pull-down, ***Trim***. Select either the vertical or horizontal *Line* (as indicated in
        Fig. 37-14) as the cutting edge. Change ***Projmode*** to ***View***. Complete the trim operation.
        With *Projmode* now set to *View*, ***Trim*** the other *Line* indicated (Fig. 35-13).

        B. Type **UCS**, then **W** for *World*. The WCS icon should appear at the origin (Fig. 37-15).

H13.   A. If your wireframe model is not correct, make the necessary changes before the next step.

        B. ***File*** pull-down, ***Save***.

H14.   A. ***Tools*** pull-down, **UCS >**, ***Named UCS....*** In the *UCS Control* dialog box, select **FRONT**.
        PICK the *Current* tile, then **OK**. Make sure you PICK ***Current***. Then, from the *View* pull-
        down, select **3D Viewpoint >**, **Plan View >**, **Current UCS**. Examine your figure from this
        viewpoint.

        B. Invoke the *UCS Control* dialog box again and set **World** as **Current**. Next, use the *View* pull-
        down menu again to see a **Plan View** of the **World** coordinate system.

        C. Type **Vpoint**, then **R** for the *Rotate* option. Enter **305** for the "angle in the XY plane from the
        X axis." Enter **30** for the "angle from the XY plane."

H15.   A. No help needed for creating viewports. To experiment with creating and erasing *Lines*,
        invoke the **Line** command and **PICK** points in the current viewport (wherever the cursor
        appear). Now, use *Line* again and **PICK** only the "From point:." Move your cursor to
        another viewport and make it current (**PICK**). **PICK** the "To point:" in the new viewport.
        Notice that you can change the current viewport <u>within</u> a command to facilitate using the
        same draw command (or edit command) in multiple viewports. ***Erase*** the experimental
        Lines.

        B. **PICK** the top-left viewport to make it current. Viewing commands that you use affect only
        the current viewport. Select (***View*** pull-down), **3D Viewpoint >**, **Plan View >**, **World UCS**.
        Type **Z** for *Zoom*, then **.9X**.

        C. **PICK** the lower-left viewport. Select **3D Viewpoint**, **Front**. Next, type **Z**, then **.9X**. Type
        Ucsicon, then **A** for the *All* option.

H16.   A.   **PICK** the lower-right viewport. Type *UCS*, then **3**. When prompted for the "Origin," select *Endpoint* from the *OSNAP* menu and **PICK** point 1 (see Fig. 37-18). Use *Endpoint OSNAP* to select the other two points.

         B.   No help needed.

         C.   Select the *Tools* pull-down menu, *UCS, Named UCS....* When the *UCS Control* dialog appears, **PICK** the current UCS "***No Name*.**" In the edit box, type over the *No Name* with the name **AUX**. **PICK** *Rename*. **PICK** *OK*.

H17.   A.   **PICK** the lower-right viewport. Select *View* pull-down, *3D Viewpoint >, Plan View >, Current UCS*.

         B.   Type **Z**. Make a Window around the entire inclined surface.

H18.   A.   Check to see if *SNAP* and *GRID* are *ON*. If not, press **F9** (*SNAP*) or **F7** (*GRID*).

         B.   Select the *Draw* pull-down menu, then *Circle >* with the *Center, Radius* option. For the center, **PICK** point **1,1** (watch the *Coords* display). Enter a value of **.5** for the radius. Repeat this for the second *Circle*, but the center is at **1,2**.

Figure 37-28

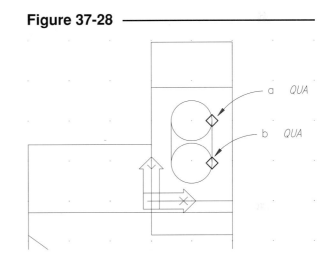

         C.   Type **L** for *Line*. At the "From point:" prompt, select the *Quadrant OSNAP* mode and **PICK** point (a) (Fig. 37-28). At the "To point:" prompt, use *Quadrant* again and **PICK** point (b).

         D.   *Modify* pull-down, *Trim*. Select the two *Lines* as *Cutting edges* and *Trim* the inner halves of the circles, as shown in Figure 37-21.

H19.   A.   Type **CP**. Select the two arcs and two lines comprising the slot. At the "Base point or displacement:" prompt, enter **0,0,0**. At the "second point of displacement:" prompt, enter **0,0,-1**.

         B.   No help needed.

H20.   A.   Type *Vports*, then **S** for *Save*. At the "Name for new viewport configuration prompt," enter **FTIA**.

         B.   **PICK** the upper-right viewport. Type *Vports*. Type **SI** for the *Single* option. Activate *Vports* again. Type **?** to display a list of viewports. The name FTIA should appear. Type *Vports* again. This time type **R** for *Restore*. Give the name **FTIA.** Finally, type **U** to undo the last action.

         C.   Select the *Tools* pull-down menu and then *UCS >, Named UCS....* **PICK** *World* from the list and make it *Current*.

H21.   *File* pull-down, *Save*.

H22.   *File* pull-down, *Exit* AutoCAD.

COMMANDS

The draw and edit commands that operate for 2D drawings are used to create 3D elements by entering X,Y,Z values, or by using UCSs. *Pline* cannot be used with 3D coordinate entry. Instead, the *3Dpoly* command is used especially for creating 3D polylines.

3DPOLY

Pull-down Menu	COMMAND (TYPE)	ALIAS (TYPE)	Short-cut	Screen (side) Menu	Tablet Menu
Draw *3D Polyline*	3DPOLY	3P	...	DRAW 1 3dpoly	O,10

A line created by *3Dpoly* is a *Pline* with 3D coordinates. The *Pline* command only allows 2D coordinate entry, whereas the *3Dpoly* command allows you to create <u>straight</u> polyline segments in 3D space by specifying 3D coordinates. A *3Dpoly* line has the "single object" characteristic of normal *Plines*; that is, several *3Dpoly* line segments created with one *3Dpoly* command are treated as one object by AutoCAD.

Figure 37-29

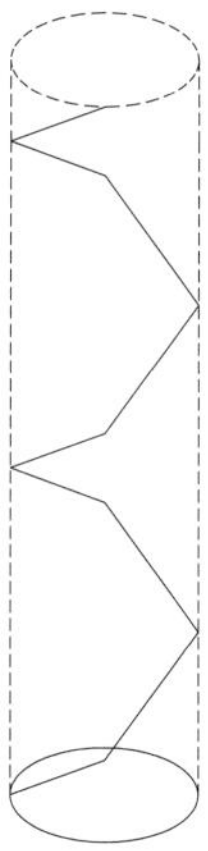

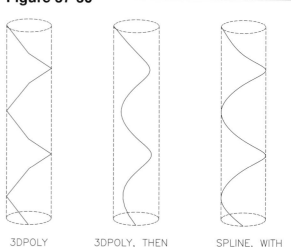

```
Command: 3dpoly
From point: PICK or (X,Y,Z coordinates)
Close/Undo/<Endpoint of line>: PICK or (X,Y,Z coordinates) or
(option)
```

If you PICK points or specify coordinate values, AutoCAD simply connects the points with straight polyline segments in 3D space. For example, *3Dpoly* could be used to draw line segments in a spiral fashion by specifying coordinates or PICKing points (Fig. 37-29).

Close
This option closes the last point to the first point entered in the command sequence.

Undo
This option deletes the last segment and allows you to specify another point for the new segment.

A *3Dpoly* does <u>not</u> have the other features or options of a *Pline* such as *Arc* segments and line *Width*.

Special options of *Pedit* can be used with *3Dpoly* lines to *Close, Edit Vertex, Spine curve,* or *Decurve* the *3Dpoly*. You can also use grips to edit the *3Dpoly*.

Figure 37-30

Although some applications may require a helix, *3Dpoly* and *Pedit* <u>cannot</u> be used in conjunction to do this. The *Spine curve* option of the *Pedit* command converts the *3Dpoly* to a polyline version of a B-Spline, not a true NURBS spline. A *Spline*-fit *Polyline* does not pass through the vertices that were specified when the *Pline* or *3DPoly* was created. The *Spline* command, however, <u>can</u> be used to specify points in 3D to create a true helix (Fig. 37-30).

3DPOLY   3DPOLY, THEN   SPLINE, WITH
                   PEDIT, SPLINE CURVE   3D COORDINATES

Lines and *Arcs* <u>cannot</u> be converted to *3Dpoly* lines using *Pedit*. You can, however, convert *Lines* and *Arcs* to *Plines* if they lie on the XY plane of the current UCS. Spline-fit *Plines* and *3Dpoly* lines can be converted to NURBS splines with the *Object* option of the *Spline* command, but the conversion does not realign the new *Spline* to pass through the original vertices.

CHAPTER EXERCISES

1–4. **Introduction to Wireframes**

Create a wireframe model of each of the objects in Figures 37-31 through 37-34. Use each dimension marker in the figure to equal one unit in AutoCAD. Begin with the lower-left corner of the model located at 0,0,0. Assign the names **WFEX1**, **WFEX2**, **WFEX3**, and **WFEX4**.

Figure 37-31 ————————————

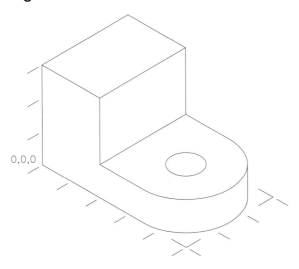

Figure 37-32 ————————————

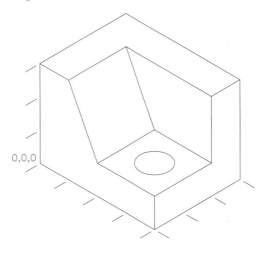

Figure 37-33 ————————————

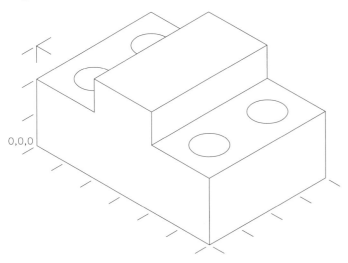

Figure 37-34 ————————————

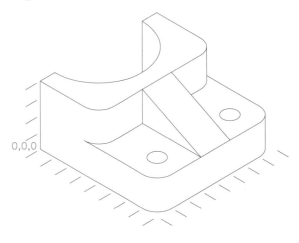

5. Create a wireframe model of the corner brace shown in Figure 37-35. *Save* the drawing as **CBRAC-WF**.

Figure 37-35

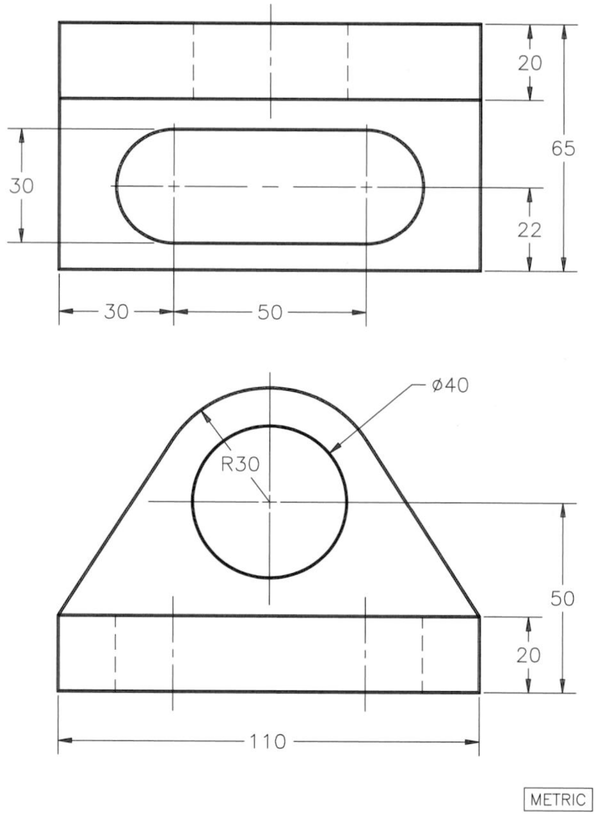

METRIC

6. Make a wireframe model of the V-block (Figure 37-36). *Save* as **VBLCK-WF**.

Figure 37-36

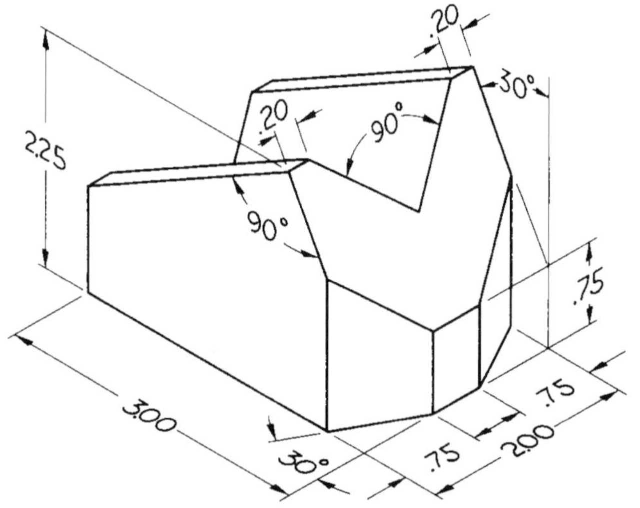

7.   Make a wireframe model of the bar guide (Figure 37-37). *Save* as **BGUID-WF**.

Figure 37-37

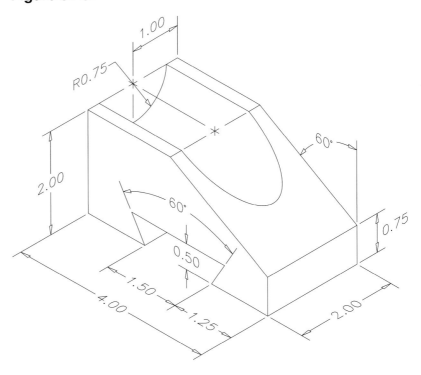

8.   Make a wireframe model of the angle brace (Figure 37-38). *Save* as **ANGLB-WF**.

Figure 37-38

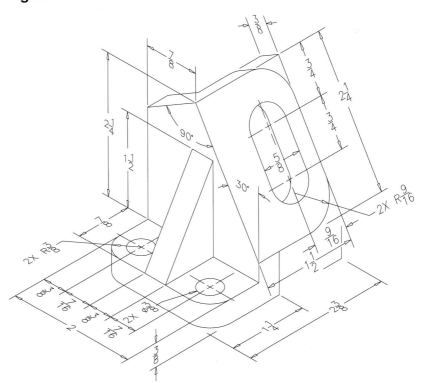

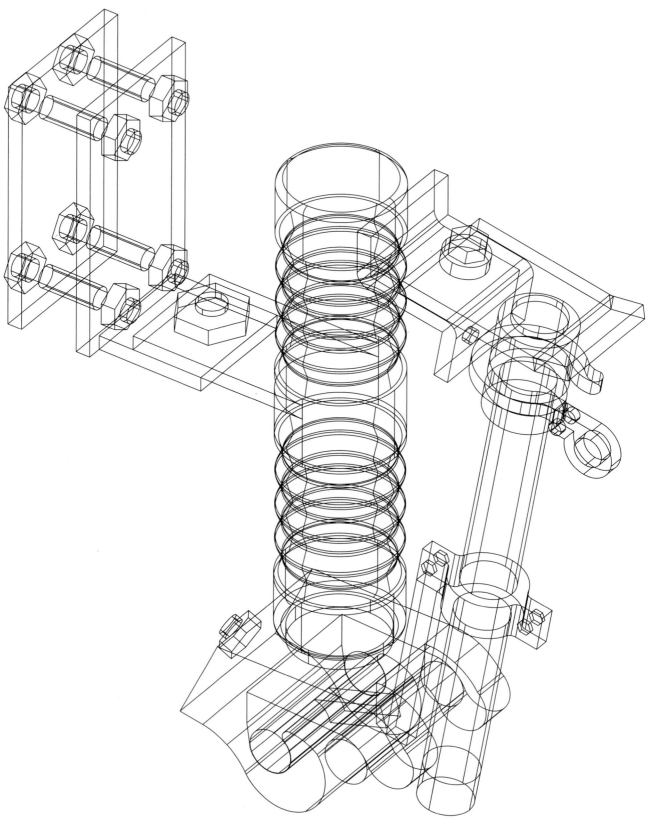

DISTUN-2.DWG Courtesy of University of Louisville, Engineering Graphics

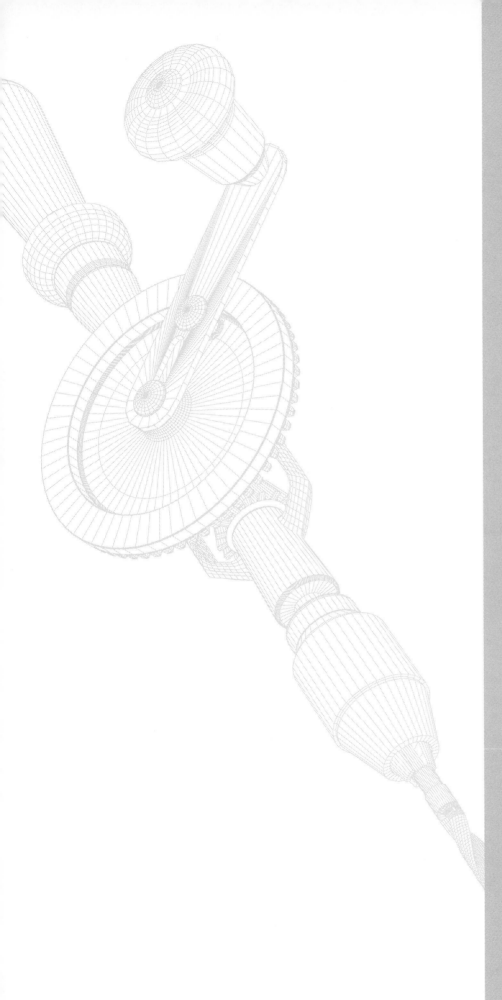

38

SOLID MODELING CONSTRUCTION

Chapter Objectives

After completing this chapter you should:

1. be able to create solid model primitives using the following commands: *Box, Wedge, Cone, Cylinder, Torus,* and *Sphere;*

2. be able to create swept solids from existing 2D shapes using the *Extrude* and *Revolve* commands;

3. be able to *Move* objects in 3D space;

4. know how to assemble a solid to another solid in 3D space using *Align;*

5. be able to rotate a solid about any axis in 3D space with *Rotate3D;*

6. be able to mirror a solid about any axis in 3D space with *Mirror3D;*

7. be able to create rectangular and circular copies with *3Darray;*

8. be able to combine multiple primitives into one composite solid using *Union, Subtract,* and *Intersect;*

9. know how to create beveled edges and rounded corners using *Chamfer* and *Fillet.*

CONCEPTS

The ACIS solid modeler is included in AutoCAD Release 14. With the ACIS modeler, you can create complex 3D parts and assemblies using Boolean operations to combine simple shapes, called *"primitives."* This modeling technique is often referred to as CSG or Constructive Solid Geometry modeling. ACIS also provides a method for analyzing and sectioning the geometry of the models (Chapter 39).

The techniques used with ACIS for construction of many solid models follow three general steps:

1. Construct simple 3D <u>primitive</u> solids, or create 2D shapes and convert them to 3D solids by extruding or revolving.

2. Create the primitives <u>in location</u> relative to the associated primitives or <u>move</u> the primitives into the desired location relative to the associated primitives.

3. Use <u>Boolean operations</u> (such as *Union*, *Subtract*, or *Intersect*) to combine the primitives to form a <u>composite solid</u>.

A solid model is an informationally complete representation of the shape of a physical object. Solid modeling differs from wireframe or surface modeling in two fundamental ways: (1) the information is more complete in a solid model and (2) the method of construction of the model itself is relatively easy to construct and edit. Solid model construction is accomplished by creating regular geometric 3D shapes such as boxes, cones, wedges, cylinders, etc. (primitive solids) and combining them using union, subtract, and intersect (Boolean operations) to form composite solids.

CONSTRUCTIVE SOLID GEOMETRY TECHNIQUES

AutoCAD uses Constructive Solid Geometry (CSG) techniques for construction of solid models. CSG is characterized by solid primitives combined by Boolean operations to form composite solids. The CSG technique is a relatively fast and intuitive way of modeling that imitates the manufacturing process.

Primitives

Solid primitives are the basic building blocks that make up more complex solid models. The ACIS primitives commands are:

BOX	Creates a solid box or cube
CONE	Creates a solid cone with a circular or elliptical base
CYLINDER	Creates a solid cylinder with a circular or elliptical base
EXTRUDE	Creates a solid by extruding (adding a Z dimension to) a closed 2D object (*Pline, Circle, Region*)
REVOLVE	Creates a solid by revolving a shape about an axis
SHPERE	Creates a solid sphere
TORUS	Creates a solid torus
WEDGE	Creates a solid wedge

Primitives can be created by entering the command name or by selecting from the menus or icons.

Boolean Operations

Primitives are combined to create complex solids by using Boolean operations. The ACIS Boolean operators are listed below. An illustration and detailed description are given for each of the commands.

UNION   Unions (joins) selected solids.

SUBTRACT   Subtracts one set of solids from another.

INTERSECT   Creates a solid of intersection (common volume) from the selected solids.

Primitives are created at the desired location or are moved into the desired location before using a Boolean operator. In other words, two or more primitives can occupy the same space (or part of the same space), yet are separate solids. When a Boolean operation is performed, the solids are combined or altered in some way to create one solid. AutoCAD takes care of deleting or adding the necessary geometry and displays the new composite solid complete with the correct configuration and lines of intersection.

Figure 38-1

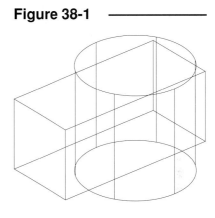

Consider the two solids shown in Figure 38-1. When solids are created, they can occupy the same physical space. A Boolean operation is used to combine the solids into a composite solid and it interprets the resulting utilization of space.

UNION

Union creates a union of the two solids into one composite solid (Fig. 38-2). The lines of intersection between the two shapes are calculated and displayed by AutoCAD. (*Hide* has been used with *DISPSILH*=1 for this figure.)

Figure 38-2

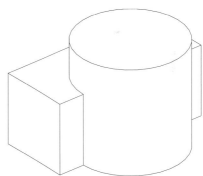

SUBTRACT

Subtract removes one or more solids from another solid. ACIS calculates the resulting composite solid. The term "difference" is sometimes used rather than "subtract." In Figure 38-3, the cylinder has been subtracted from the box.

Figure 38-3

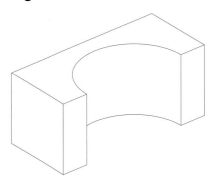

INTERSECT

Intersect calculates the intersection between two or more solids. When *Intersect* is used with *Regions* (2D surfaces), it determines the shared <u>area</u>. Used with solids, as in Figure 38-4, *Intersect* creates a solid composed of the shared <u>volume</u> of the cylinder and the box. In other words, the result of *Intersect* is a solid that has only the volume which is part of both (or all) of the selected solids.

Figure 38-4 —

SOLID PRIMITIVES COMMANDS

This section explains the commands that allow you to create primitives used for construction of composite solid models. The commands allow you to specify the dimensions and the orientation of the solids. Once primitives are created, they are combined with other solids using Boolean operations to form composite solids.

NOTE: If you use a pointing device to PICK points, use *OSNAP* when possible to PICK points in 3D space. If you do not use *OSNAP*, the selected points are located on the current XY construction plane, so the true points may not be obvious. It is recommended that you use *OSNAP* or enter values.

Figure 38-5 ——————

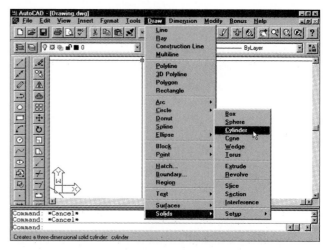

The solid modeling commands are located in groups. The commands for creating solid primitives are located in a group in the *Draw* menus under *Solids* (Fig. 38-5). Commands for moving solids and the Boolean commands are found in the *Modify* menus. You can bring the *Solids* toolbar and the *Modify II* toolbar (which contains the Boolean commands) to the screen by selecting *Toolbars...* from the *View* pull-down menu (Fig. 38-6).

NOTE: When you begin to construct a solid model, <u>create the primitives at location 0,0,0</u> (or some other <u>known point</u>) rather than PICKing points anywhere in space. It is very helpful to know where the primitives are located in 3D space so they can be moved or rotated in the correct orientation with respect to other primitives when they are assembled to composite solids.

Figure 38-6 ——————

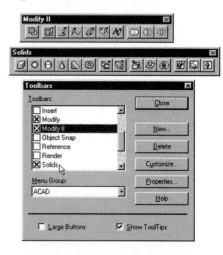

BOX

Pull-down Menu	COMMAND (TYPE)	ALIAS (TYPE)	Short-cut	Screen (side) Menu	Tablet Menu
Draw *Solids >* *Box*	*BOX*	...	...	DRAW 2 SOLIDS Box	J,7

Box creates a solid box primitive to your dimensional specifications. You can specify dimensions of the box by PICKing or by entering values. The box can be defined by (1) giving the corners of the base, then height, (2) by locating the center and height, or (3) by giving each of the three dimensions. The base of the box is oriented parallel to the current XY plane:

Command: **box**
Center/<Corner of box> <0,0,0>: **PICK** or (**coordinates**) or (**letter**) or **Enter**

Options for the *Box* command are listed as follows.

Corner of box
Pressing **Enter** begins the corner of a box at 0,0,0 of the current coordinate system. In this case, the box can be moved into its desired location later. As an alternative, a coordinate position can be entered or **PICK**ed as the starting corner of the box. AutoCAD responds with:

Cube/Length/<Other corner>:

The other corner can be PICKed or specified by coordinates. The *Cube* option requires only one dimension to define the cube.

Length
The *Length* option prompts you for the three dimensions of the box in the order of X, Y, and Z. AutoCAD prompts for *Length* (X dimension), *Width* (Y dimension), and *Height* (Z dimension).

Figure 38-7 shows a box created at 0,0,0 with width, depth, and height dimensions of 5, 4, and 3.

Figure 38-7 ————————————

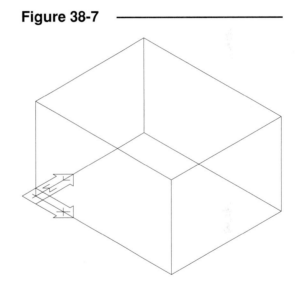

Center
With this option, you first locate the center of the box, then specify the three dimensions of the box. AutoCAD prompts:

Center of box: **PICK** or (**coordinates**)

Specify the center location. AutoCAD responds with:

Cube/Length/<corner of box>:

The resulting solid box is centered about the specified point (0,0,0 for the example; Fig. 38-8).

Figure 38-8 ————————————

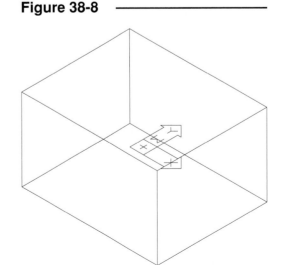

Figure 38-9 illustrates a *Box* created with the *Center* option using *OSNAP* to snap to the *Center* of the top of the cylinder. Note that the center of the box is the <u>volumetric</u> center, not the center of the base.

Figure 38-9

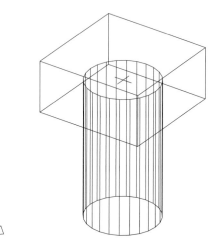

CONE

Pull-down Menu	COMMAND (TYPE)	ALIAS (TYPE)	Short-cut	Screen (side) Menu	Tablet Menu
Draw Solids > Cone	CONE	...	...	DRAW 2 SOLIDS Cone	M,7

Cone creates a right circular or elliptical solid cone ("right" means the axis forms a right angle with the base). You can specify the center location, radius (or diameter), and height. By default, the orientation of the cylinder is determined by the current UCS so that the base lies on the XY plane and height is perpendicular (in a Z direction). Alternately, the orientation can be defined by using the *Apex* option.

Center Point
Using the defaults (center at 0,0,0, PICK a radius, enter a value for height), the cone is generated in the orientation shown in Figure 38-10. The cone here may differ in detail (number of contour lines), depending on the current setting of the *ISOLINES* variable. The default prompts are shown as follows:

Figure 38-10

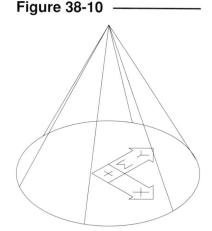

    Command: **cone**
    Elliptical/<Center point> <0,0,0>: **PICK** or (**coordinates**)
    Diameter/<Radius>: **PICK** or (**coordinates**)
    Apex/<Height>: **PICK** or (**value**)
    Command:

Invoking the *Apex* option (after *Center point* of the base and *Radius* or *Diameter* have been specified) displays the following prompt:

Figure 38-11

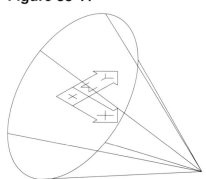

    Apex: **PICK** or (**coordinates**)

Locating a point for the apex defines the height and orientation of the cone (Fig. 38-11). The axis of the cone is aligned with the line between the specified center point and the *Apex* point, and the height is equal to the distance between the two points.

The solid model in Figure 38-12 was created with a *Cone* and a *Cylinder*. The cylinder was created first, then the cone was created using the *Apex* option of *Cone*. The orientation of the *Cone* was generated by PICKing the *Center* of one end of the cylinder for the "Center point" of the base of the cone and the *Center* of the cylinder's other end for the *Apex*. *Union* created the composite model.

Figure 38-12

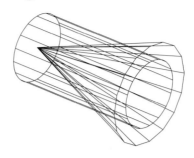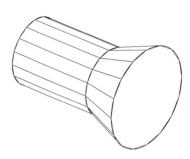

Elliptical

This option draws a cone with an elliptical base (Fig. 38-13). You specify two axis endpoints to define the elliptical base. An elliptical cone can be created using the *Center, Apex,* or *Height* options:

Figure 38-13

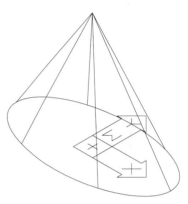

```
Elliptical/<center point> <0,0,0>: e
Center/<Axis endpoint>: c
Center of ellipse <0,0,0>: PICK or (coordinates)
Axis endpoint: PICK or (coordinates)
Other axis distance: PICK or (coordinates)
Apex/<Height>:
```

CYLINDER

Pull-down Menu	COMMAND (TYPE)	ALIAS (TYPE)	Short-cut	Screen (side) Menu	Tablet Menu
Draw *Solids >* *Cylinder*	*CYLINDER*	...	...	DRAW 2 SOLIDS *Cylinder*	L,7

Cylinder creates a cylinder with an elliptical or circular base with a center location, diameter, and height you specify. Default orientation of the cylinder is determined by the current UCS, such that the circular plane is coplanar with the XY plane and height is in a Z direction. However, the orientation can be defined otherwise by the *Center of other end* option.

Center Point

The default options create a cylinder in the orientation shown in Figure 38-14:

Figure 38-14

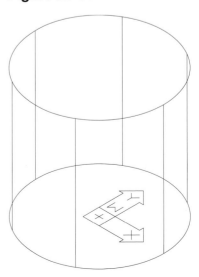

```
Command:  cylinder
Elliptical/<Center point> <0,0,0>: PICK or (coordinates) or
Enter
Diameter/<Radius>: PICK or (coordinates)
Center other end/<Height>: Pick or (coordinates)
```

Elliptical

This option draws a cylinder with an elliptical base (Fig. 38-15). Specify two axis endpoints to define the elliptical base. An elliptical cylinder can be created using the *Center, Center other end*, or *Height* options.

Figure 38-15

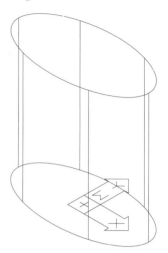

Elliptical/<center point> <0,0,0>: *e*
Center/<Axis endpoint>: *c*
Center of ellipse <0,0,0>: **PICK** or **(coordinates)**
Axis endpoint: **PICK** or **(coordinates)**
Other axis distance: **PICK** or **(coordinates)**
Center other end/<Height>:

The **Center other end** option of *Cylinder* is similar to *Cone Apex* option in that the height and orientation are defined by the *Center point* and the *Center other end*. In Figure 38-16, a hole is created in the *Box* by using *Center other end* option and *OSNAP*ing to the diagonal lines' *Midpoints*, then *Subtract*ing the *Cylinder* from the *Box*.

Figure 38-16

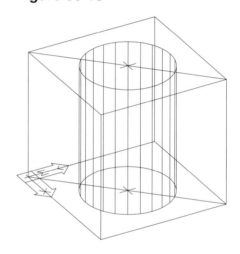

WEDGE

Pull-down Menu	COMMAND (TYPE)	ALIAS (TYPE)	Short-cut	Screen (side) Menu	Tablet Menu
Draw Solids > Wedge	WEDGE	WE	...	DRAW 2 SOLIDS Wedge	N,7

Corner of Wedge

Wedge creates a wedge solid primitive. The base of the wedge is always parallel with the current UCS XY plane, and the <u>slope of the wedge is along the X axis</u>.

Command: *wedge*
Center/<Corner of wedge> <0,0,0>: **PICK** or **(coordinates)**
Cube/Length/<other corner>: **PICK** or **(coordinates)**

Accepting all the defaults, a *Wedge* can be created as shown in Figure 38-17, with the slope along the X axis.

Figure 38-17

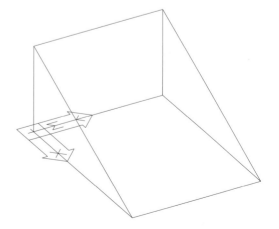

Invoking the **Length** option prompts you for the *Length, Width,* and *Depth.* More precisely, AutoCAD means width (X dimension), depth (Y dimension), and height (Z dimension).

Center

The point you specify as the center is actually in the center of an imaginary box, half of which is occupied by the wedge. Therefore, the center point is actually at the <u>center of the sloping side of the wedge</u> (Fig. 38-18).

Figure 38-18 ─────────

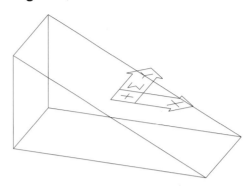

SPHERE

Pull-down Menu	COMMAND (TYPE)	ALIAS (TYPE)	Short-cut	Screen (side) Menu	Tablet Menu
Draw Solids > Sphere	SPHERE	...	...	DRAW 2 SOLIDS Sphere	K,7

Sphere allows you to create a solid sphere by defining its center point and radius or diameter:

Figure 38-19 ─────────

Command: **sphere**
<Center point> <0,0,0>: **PICK** or **(coordinates)**
Diameter/<radius>: **PICK** or **(coordinates)**
Command:

Creating a *Sphere* with the default options would yield a sphere similar to that in Figure 38-19.

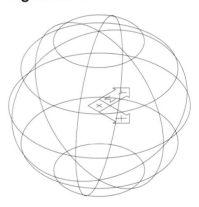

TORUS

Pull-down Menu	COMMAND (TYPE)	ALIAS (TYPE)	Short-cut	Screen (side) Menu	Tablet Menu
Draw Solids > Torus	TORUS	TOR	...	DRAW 2 SOLIDS Tours	O,7

Torus creates a torus (donut shaped) solid primitive using the dimensions you specify. Two dimensions are needed: (1) the radius or diameter of the tube and (2) the radius or diameter from the axis of the torus to the center of the tube. AutoCAD prompts:

Command: **torus**
<Center of torus> <0,0,0>: **PICK** or **(coordinates)** or **Enter**
Diameter/<radius> of torus: **PICK** or **(coordinates)**
Diameter/<radius> of tube: **PICK** or **(coordinates)**
Command:

Figure 38-20 shows a *Torus* created using the default orientation with the axis of the tube aligned with the Z axis of the UCS and the center of the torus at 0,0,0. (*Hide* was used for this display.)

Figure 38-20

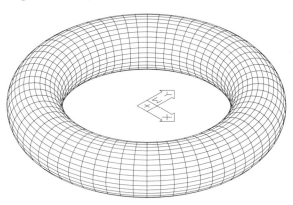

A self-intersecting torus is allowed with the *Torus* command. A self-intersecting torus is created by specifying a torus radius less than the tube radius. Figure 38-21 illustrates a torus with a torus radius of 3 and a tube radius of 4.

Figure 38-21

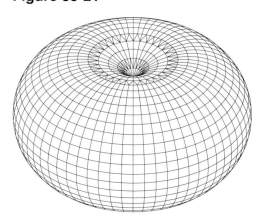

EXTRUDE

	Pull-down Menu	COMMAND (TYPE)	ALIAS (TYPE)	Short-cut	Screen (side) Menu	Tablet Menu
	Draw *Solids >* *Extrude*	EXTRUDE	EXT	...	DRAW 2 SOLIDS *Extrude*	P,7

Extrude (like *Revolve*) is a "sweeping operation." Sweeping operations use existing 2D objects to create a solid.

Extrude means to add a Z (height) dimension to an otherwise 2D shape. This command extrudes an existing closed 2D shape such as a *Circle, Polygon, Ellipse, Pline, Spline,* or *Region*. Only <u>closed 2D shapes</u> can be extruded. The closed 2D shape cannot be self-intersecting (crossing over itself).

Two methods determine the direction of extruding: <u>perpendicular</u> to the shape and along a <u>*Path*</u>. With the default method, the selected 2D shape is extruded <u>perpendicular to the plane</u> of the shape <u>regardless</u> of the current UCS orientation. Using the *Path* method, you can extrude the existing closed 2D shape along any existing path determined by a *Line, Arc, Spline,* or *Pline*.

The versatility of this command lies in the fact that <u>any closed shape</u> that can be created by (or converted to) a *Pline, Spline, Region*, etc., no matter how complex, can be transformed into a solid and can be extruded perpendicularly or along a *Path*. *Extrude* can simplify the creation of many solids that may otherwise take much more time and effort using typical primitives and Boolean operations. Create the closed 2D shape first, then invoke *Extrude*:

Command: **extrude**
Select objects: **PICK**
Path/<Height of Extrusion>: **(value)**
Extrusion taper angle <0>: **(value)** or **Enter**
Command:

Figure 38-22 shows a *Pline* before and after using *Extrude*.

Figure 38-22 ─────────────────

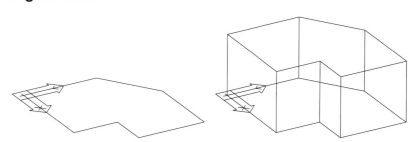

A taper angle can be specified for the extrusion. The resulting solid has sides extruded inward at the specified angle (Fig. 38-23). This is helpful for developing parts for molds that require a slight draft angle to facilitate easy removal of the part from the mold. Entering a negative taper angle results in the sides of the extrusion sloping outward.

Figure 38-23 ─────────────────

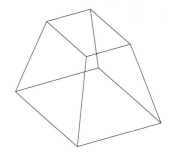

Many complex shapes based on closed 2D *Splines* or *Plines* can be transformed to solids using *Extrude* (Fig. 38-24).

Figure 38-24 ─────────────────

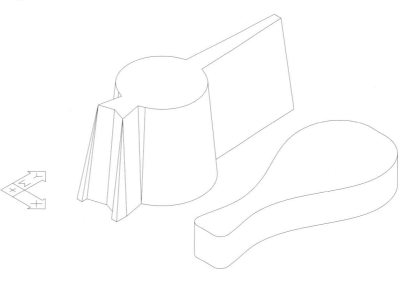

The *Path* option allows you to sweep the 2D shape along an existing line or curve called a "path." The *Path* can be composed of a *Line*, *Arc*, *Ellipse*, *Pline*, or *Spline* (or can be different shapes converted to a *Pline*). This path must lie in a plane. Figure 38-25 shows a closed *Pline* extruded along a *Pline* path and a curved *Spline* path:

Figure 38-25

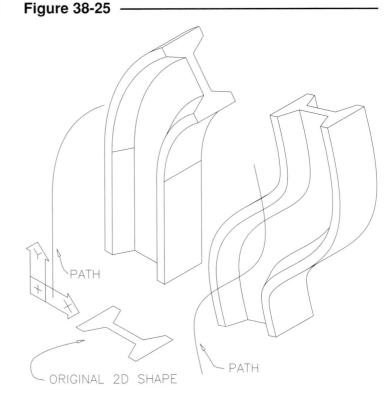

    Command: **extrude**
    Select objects: **PICK**
    Path/<Height of Extrusion>: **path**
    Select path: **PICK**
    Command:

The path cannot lie in the same plane as the 2D shape to be extruded, since the plane of the 2D shape is always extruded perpendicular along the path. If one of the endpoints of the path is not located on the plane of the 2D shape, AutoCAD will automatically move the path to the center of the profile temporarily. Notice how the original 2D shape was extruded perpendicularly to the path (Fig. 38-26).

Figure 38-26

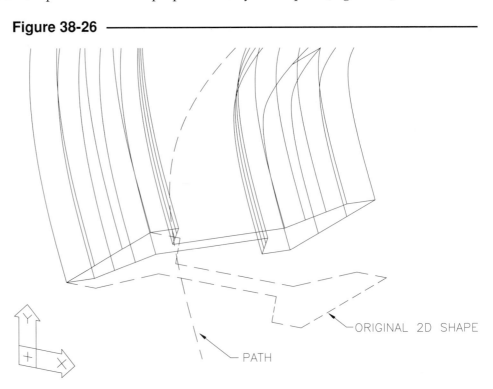

REVOLVE

Pull-down Menu	COMMAND (TYPE)	ALIAS (TYPE)	Short-cut	Screen (side) Menu	Tablet Menu
Draw *Solids >* *Revolve*	*REVOLVE*	*REV*	...	*DRAW 2* *SOLIDS* *Revolve*	Q,7

Revolve creates a swept solid. *Revolve* creates a solid by revolving a 2D shape about a selected axis. The 2D shape to revolve can be a *Pline, Polygon, Circle, Ellipse, Spline,* or a *Region* object. Only one object at a time can be revolved. *Splines* or *Plines* selected for revolving <u>must be closed</u>. The command syntax for *Revolve* (accepting the defaults) is:

        Command: **revolve**
        Select objects: **PICK**
        Select objects: **Enter** (Indicates completion of selection process.)
        Axis of revolution - Object/X/Y/<Start point of axis>: **PICK**
        End point of axis: **PICK**
        Angle of revolution <full circle>: (**value**) or **Enter**
        Command:

Figure 38-27 illustrates a possibility for an existing *Pline* shape and the resulting *revolved* shape generated through a full circle.

Figure 38-27

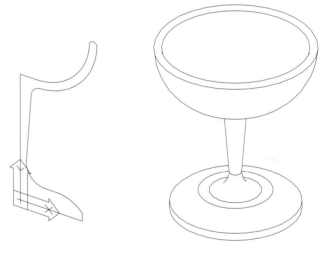

Because *Revolve* acts on an existing object, the 2D shape intended for revolution should be created in or moved to the desired orientation. There are multiple options for selecting an axis of revolution.

Object
A *Line* or single segment *Pline* can be selected for an axis. The positive axis direction is from the closest endpoint PICKed to the farthest.

X or Y
Uses the positive *X* or *Y* axis of the current UCS as the positive axis direction.

Start point of axis
Defines two points in the drawing to use as an axis (length is irrelevant). Select any two points in 3D space. The two points do <u>not</u> have to be coplanar with the 2D shape.

If the *Object* or *Start point* options are used, the selected object or the two indicated points do <u>not</u> have to be coplanar with the 2D shape. The axis of revolution used is <u>always on the plane of the 2D shape</u> to revolve and is aligned with the direction of the object or selected points.

Figure 38-28 demonstrates a possible use of the *Object* option of *Revolve*. In this case, a *Pline* square is revolved about a *Line* object. Note that the *Line* used for the axis is <u>not</u> on the same plane as the *Pline*. *Revolve* uses an axis <u>on the plane</u> of the revolved shape aligned with the direction of the selected axis. In this case, the endpoints of the *Line* are 0,-1,-1 and 0,1,1 so the shape is actually revolved about the Y axis.

Figure 38-28

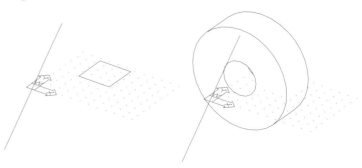

After defining the axis of revolution, *Revolve* requests the number of degrees for the object to be revolved. Any angle can be entered.

Another possibility for revolving a 2D shape is shown in Figure 38-29. The shape is generated through 270 degrees.

Figure 38-29

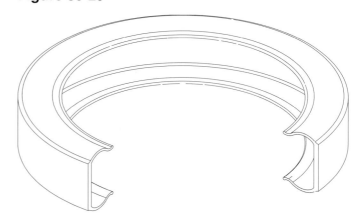

COMMANDS FOR MOVING SOLIDS

When you create the desired primitives, Boolean operations are used to construct the composite solids. However, the primitives must be in the correct position and orientation with respect to each other before Boolean operations can be performed. You can either create the primitives in the desired position during construction (by using UCSs) or move the primitives into position after their creation. Several methods that allow you to move solid primitives are explained in this section.

When constructing 3D geometry, it is critical that the objects are located in space at some <u>known</u> position. Do <u>not</u> create primitives at any convenient place in the drawing; know the position. The location is important when you begin the process of moving primitives to assemble 3D composite solids. Of course, *OSNAPs* can be used, but sometimes it is necessary to use coordinate values in absolute, rectangular, or polar format. A good practice is to <u>create primitives at the final location</u> if possible (use UCSs when needed) or <u>create the primitives at 0,0,0</u>, then <u>move</u> them preceding the Boolean operations.

MOVE

Pull-down Menu	COMMAND (TYPE)	ALIAS (TYPE)	Short-cut	Screen (side) Menu	Tablet Menu
Modify *Move*	*MOVE*	*M*	...	*MODIFY2* *Move*	*V,19*

The *Move* command that you use for moving 2D objects in 2D drawings can also be used to move 3D primitives. Generally, *Move* is used to change the position of an object in one plane (translation), which is typical of 2D drawings. *Move* can also be used to move an ACIS primitive in 3D space <u>if</u> *OSNAPs* or 3D coordinates are used.

Move operates in 3D just as you used it in 2D. Previously, you used *Move* only for repositioning objects in the XY plane, so it was only necessary to PICK or use X and Y coordinates. Using *Move* in 3D space requires entering X, Y, and Z values or using *OSNAPs*.

For example, to create the composite solid used in the figures in Chapter 36, a *Wedge* primitive was *Moved* into position on top of the *Box*. The *Wedge* was created at 0,0,0, then rotated. Figure 38-30 illustrates the movement using absolute coordinates described in the syntax below:

Figure 38-30

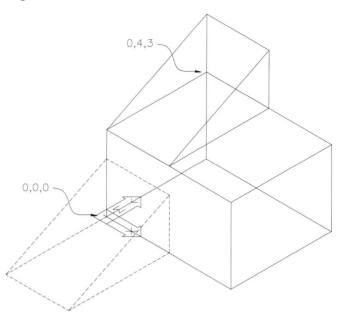

```
Command: move
Select objects: PICK
Select objects: Enter
<Base point of displacement>/Multiple:
0,0,0
Second point of displacement: 0,4,3
Command:
```

Alternately, you can use *OSNAPs* to select geometry in 3D space. Figure 38-31 illustrates the same *Move* operation using *Endpoint OSNAPs* instead of entering coordinates.

Figure 38-31

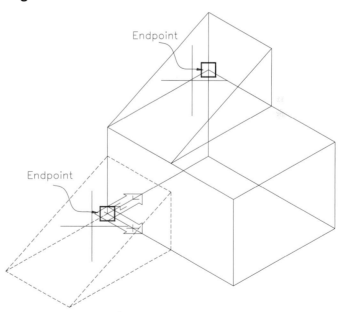

ALIGN

Pull-down Menu	COMMAND (TYPE)	ALIAS (TYPE)	Short-cut	Screen (side) Menu	Tablet Menu
Modify *3D Operations >* *Align*	*ALIGN*	*AL*	...	*MODIFY2* *Align*	*X,14*

Align is an ADS application that is loaded automatically when typing or selecting this command from the menus. *Align* is discussed in Chapter 16 but only in terms of 2D alignment.

Align is, however, a very powerful 3D command because it automatically performs 3D translation and rotation if needed. *Align* is more intuitive and in many situations is easier to use than *Move*. All you have to do is select the points on two 3D objects that you want to align (connect).

Align provides a means of aligning one shape (an object, a group of objects, a block, a region, or a 3D solid) with another shape. The alignment is accomplished by connecting source points (on the shape to be moved) to destination points (on the stationary shape). You can use *OSNAP* modes to select the source and destination points, assuring accurate alignment. Either a 2D or 3D alignment can be accomplished with this command. The command syntax for 3D alignment is as follows:

> Command: **align**
> Select objects: **PICK** (Select object to move, not the stationary object.)
> Select objects: **Enter**
> 1st source point: **PICK** (use *OSNAP*)
> 1st destination point: **PICK** (use *OSNAP*)
> 2nd source point: **PICK** (use *OSNAP*)
> 2nd destination point: **PICK** (use *OSNAP*)
> 3rd source point: **PICK** (use *OSNAP*)
> 3rd destination point: **PICK** (use *OSNAP*)

After the source and destination points have been designated, lines connecting those points temporarily remain until *Align* performs the action (Fig. 38-32).

Align performs a translation (like *Move*) and two rotations (like *Rotate*), each in separate planes to align the points as designated. The motion automatically performed by *Align* is actually done in three steps.

Figure 38-32 ————————————————

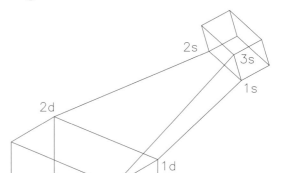

Initially, the first source point is connected to the first destination point (translation). These two points always physically touch.

Next, the vector defined by the first and second source points is aligned with the vector defined by the first and second destination points. The length of the segments between the first and second points on each object is of no consequence because AutoCAD only considers the vector direction. This second motion is a rotation along one axis.

Figure 38-33 ————————————————

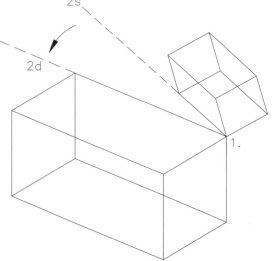

Finally, the third set of points are aligned similarly. This third motion is a rotation along the other axis, completing the alignment.

Figure 38-34

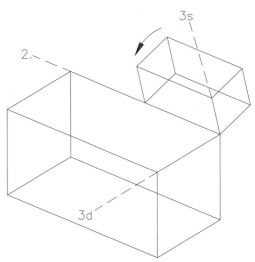

In some cases, such as when cylindrical objects are aligned, only two sets of points have to be specified. For example, if aligning a shaft with a hole (Fig. 38-35), the first set of points (source and destination) specify the attachment of the base of the shaft with the bottom of the hole (use *Center OSNAP*s). The second set of points specify the alignment of the axes of the two cylindrical shapes. A third set of points is <u>not</u> required because the radial alignment between the two objects is not important. When only two sets of points are specified, AutoCAD asks if a 2D or 3D alignment is desired. The command syntax is as follows:

Figure 38-35

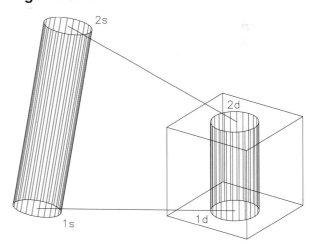

    Command: *align*
    Select objects: **PICK**
    Select objects: **Enter**
    1st source point: cen of **PICK**
    1st destination point: cen of **PICK**
    2nd source point: cen of **PICK**
    2nd destination point: cen of **PICK**
    3rd source point: **Enter**
    <2d> or 3d transformation: **3**
    Command:

ROTATE3D

Pull-down Menu	COMMAND (TYPE)	ALIAS (TYPE)	Short-cut	Screen (side) Menu	Tablet Menu
Modify *3D Operations >* *Rotate 3D*	*ROTATE3D*	...	...	*MODIFY2* *Rotate3D*	*W,22*

Rotate3D is very useful for any type of 3D modeling, particularly with CSG, where primitives must be moved, rotated, or otherwise aligned with other primitives before Boolean operations can be performed.

Rotate3D allows you to rotate a 3D object about any axis in 3D space. Many alternatives are available for defining the desired rotational axis. Following is the command sequence for rotating a 3D object using the default (*2points*) option:

> Command: **rotate3d**
> Select objects: **PICK**
> Select objects: **Enter** (Indicates completion of selection process.)
> Axis by Object/Last/View/Xaxis/Yaxis/Zaxis/<2points>: **PICK** or (**coordinates**) (Select the first point to define the rotational axis.)
> 2nd point on axis: **PICK** or (**coordinates**) (Select the second point to define rotational axis.)
> <Rotation angle>/Reference: **PICK** or (**value**) (Select two points to define the rotation angle or enter a value. If two points are PICKed, the angle between the points in the XY plane of the current UCS determine the angle of rotation.)

The options are explained next.

2points

The power of *Rotate3D* (over *Rotate*) is that any points or objects in <u>3D space</u> can be used to define the axis for rotation. When using the default (*2points* option), remember you can use *OSNAP* to select points on existing 3D objects. Figure 38-36 illustrates the *2points* option used to select two points with *OSNAP* on the solid object selected for rotating.

Figure 38-36

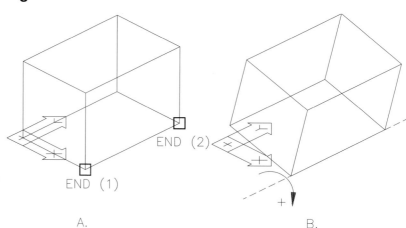

END (2)

END (1)

A.

B.

Object

This option allows you to rotate about a selected 2D object. You can select a *Line, Circle, Arc,* or 2D *Pline* segment. The rotational axis is aligned with the selected *Line* or *Pline* segment. Positive rotation is determined by the right-hand rule and the "arbitrary axis algorithm." When selecting *Arc* or *Circle* objects, the rotational axis is perpendicular to the plane of the *Arc* or *Circle* passing through the center. You <u>cannot</u> select the edge of a solid object with this option.

Last

This option allows you to rotate about the axis used for the last rotation.

Figure 38-37

View

The *View* option allows you to pick a point on the screen and rotates the selected object(s) about an axis perpendicular to the screen and passing through the selected point (Fig. 38-37).

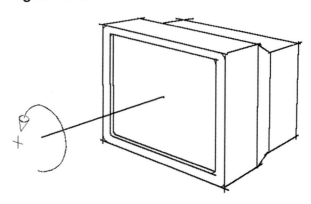

Xaxis

With this option, you can rotate the selected objects about the X axis of the current UCS or any axis <u>parallel to</u> the X axis of the current UCS. You are prompted to pick a point on the X axis. The point selected defines an axis for rotation parallel to the current X axis passing through the selected point. You can use *OSNAP* to select points on existing 3D objects (Fig. 38-38). The current X axis can be used if the point you select is on the X axis.

Figure 38-38

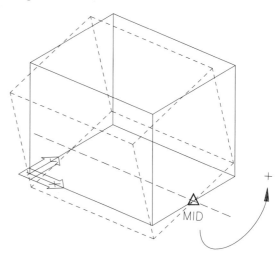

Yaxis

This option allows you to use the Y axis of the current UCS or any axis <u>parallel to</u> the Y axis of the current UCS as the axis of rotation. The point you select defines a rotational axis parallel to the current Y axis passing through the point. *OSNAP* can be used to snap to existing geometry (Fig. 38-39).

Figure 38-39

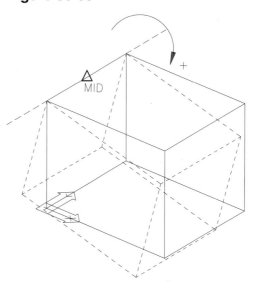

Zaxis

With this option, you can use the Z axis of the current UCS or any axis <u>parallel to</u> the Z axis of the current UCS as the axis of rotation. The point you select defines a rotational axis parallel to the current Z axis passing through the point. Figure 38-40 indicates the use of the *Midpoint OSNAP* to establish a vertical (parallel to Z) rotational axis.

Figure 38-40

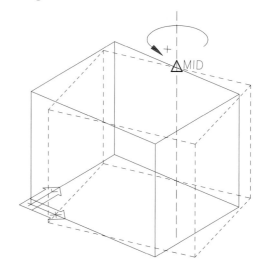

Reference
After you have specified the axis for rotation, you must specify the rotation angle. You are presented with the following prompt:

> <Rotation angle>/Reference: **r** (Indicates the *Reference* option.)
> Reference angle: **PICK** or (**value**) (PICK two points; *OSNAP*s can be used. You can enter a value.)
> New angle: **PICK** or (**value**) (Specify the new angle by either method.)

The angle you specify for the reference (relative) is used instead of angle 0 (absolute) for the starting position. You can enter either a value or PICK two points to specify the angle. You then specify a new angle. The *Reference* angle you select is rotated to the absolute angle position you specify as the "New angle."

Figure 38-41 illustrates how *Endpoint OSNAP*s are used to select a *Reference* angle. The "New angle" is specified as **90**. AutoCAD rotates the reference angle to the 90 degree position.

Figure 38-41 ————————————————————

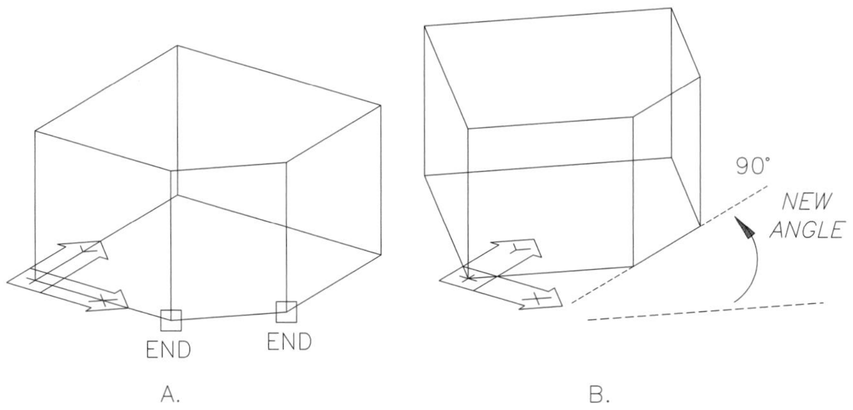

MIRROR3D

Pull-down Menu	COMMAND (TYPE)	ALIAS (TYPE)	Short-cut	Screen (side) Menu	Tablet Menu
Modify *3D Operations >* *Mirror 3D*	MIRROR3D	...	...	*MODIFY2* *Mirror3D*	W,21

Mirror3D operates similar to the 2D version of the command *Mirror* in that mirrored replicas of selected objects are created. With *Mirror* (2D) the selected objects are mirrored about an axis. The axis is defined by a vector lying in the XY plane. With *Mirror3D*, selected objects are mirrored about a <u>plane</u>. *Mirror3D* provides multiple options for specifying the plane to mirror about:

> Command: **mirror3d**
> Select objects: **PICK** (Select one or multiple objects to mirror.)
> Select objects: **Enter** (Indicate completion of the selection process.)
> Plane by Object/Last/Zaxis/View/XY/YZ/ZX/<3points>: **PICK** or (**letters**) (PICK points for the default option or enter a letter(s) to designate option.)

The options are listed and explained next. A phantom icon is shown in the following figures <u>only</u> to aid your visualization of the mirroring plane.

3points

The *3points* option mirrors selected objects about the plane you specify by selecting three points to define the plane. You can PICK points (with or without *OSNAP*) or give coordinates. *Midpoint OSNAP* is used to define the 3 points in Figure 38-42 (A) to achieve the result in (B).

Figure 38-42

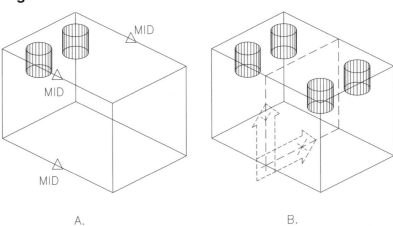

A.

B.

Object

Using this option establishes a mirroring plane with the plane of a 2D object. Selecting an *Arc* or *Circle* automatically mirrors selected objects using the plane in which the *Arc* or *Circle* lies. The plane defined by a *Pline* segment is the XY plane of the *Pline* when the *Pline* segment was created. Using a *Line* object or edge of an ACIS solid is not allowed because neither defines a plane. Figure 38-43 shows a box mirrored about the plane defined by the *Circle* object. Using *Subtract* produces the result shown in (B).

Figure 38-43

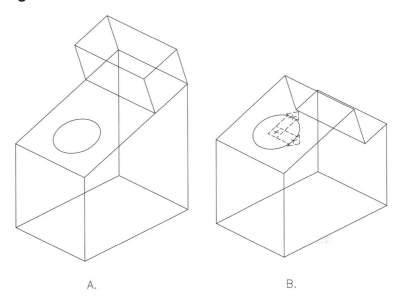

A.

B.

Last

Selecting this option uses the plane that was last used for mirroring.

Zaxis

With this option, the mirror plane is the <u>XY</u> plane perpendicular to a Z vector you specify. The first point you specify on the Z axis establishes the location of the XY plane origin (a point through which the plane passes). The second point establishes the Z axis and the orientation of the XY plane (perpendicular to the Z axis). Figure 38-44 illustrates this concept. Note that this option requires only two PICK points.

Figure 38-44

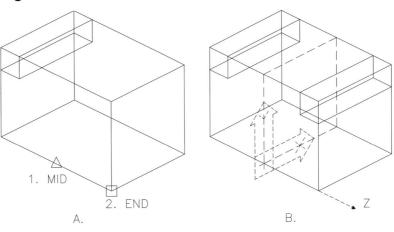

A.

B.

View

The *View* option of *Rotate3D* uses a mirroring plane <u>parallel</u> with the screen and perpendicular to your line of sight based on your current viewpoint. You are required to select a point on the plane. Accepting the default (0,0,0) establishes the mirroring plane passing through the current origin. Any other point can be selected. You must change *Vpoint* to "see" the mirrored objects (Fig. 38-45).

Figure 38-45

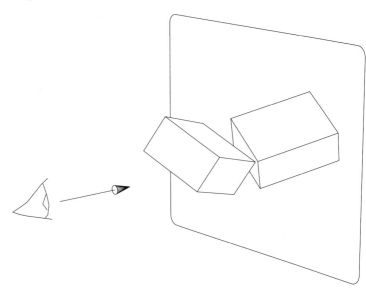

XY

This option situates a mirroring plane parallel with the current XY plane. You can specify a point through which the mirroring plane passes. Figure 38-46 represents a plane established by selecting the *Center* of an existing solid.

Figure 38-46

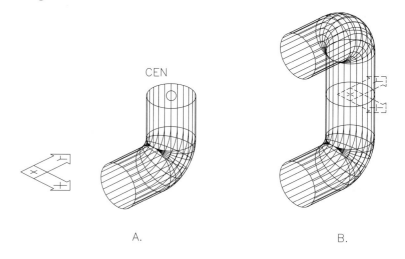

CEN

A.

B.

YZ

Using the *YZ* option constructs a plane to mirror about that is parallel with the current YZ plane. Any point can be selected through which the plane will pass (Fig. 38-47).

Figure 38-47

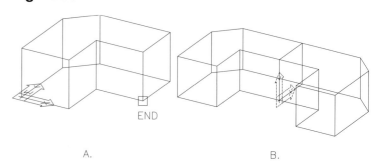

END

A.

B.

ZX

This option uses a plane parallel with the current ZX plane for mirroring. Figure 38-48 shows a point selected on the *Midpoint* of an existing edge to mirror two holes.

Figure 38-48 ─────────────

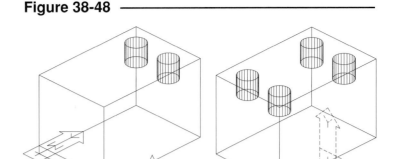

A.   B.

3DARRAY

	COMMAND (TYPE)	ALIAS (TYPE)	Short-cut	Screen (side) Menu	Tablet Menu
Pull-down Menu					
Modify 3D Operations > 3D Array	*3DARRAY*	*3A*	...	*MODIFY2 3Darray*	*W,20*

Rectangular

With this option of *3Darray*, you create a 3D array specifying three dimensions—the number of and distance between rows (along the Y axis), the number/distance of columns (along the X axis), and the number/distance of levels (along the Z axis). Technically, the result is an array in a <u>prism</u> configuration <u>rather than a rectangle</u>.

```
Command: 3darray
Initializing...  3DARRAY loaded.
Select objects: PICK
Select objects: Enter
Rectangular or Polar array (R/P): r
Number of rows (---) <1>: (value)
Number of columns (|||) <1>: (value)
Number of levels (...) <1>: (value)
Distance between rows (---): PICK or (value)
Distance between columns (|||): PICK or (value)
Distance between levels (...): PICK or (value)
Command:
```

The selection set can be one or more objects. The entire set is treated as one object for arraying. All values entered must be positive.

Figures 38-49 and 38-50 illustrate creating a *Rectangular 3Darray* of a cylinder with 3 rows, 4 columns, and 2 levels.

Figure 38-49

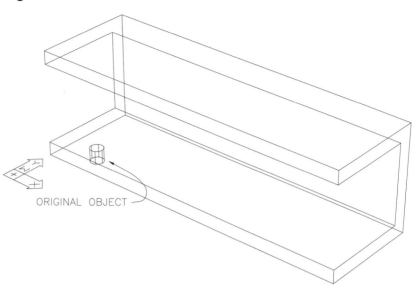

ORIGINAL OBJECT

The cylinders are *Subtracted* from the extrusion to form the finished part.

Figure 38-50

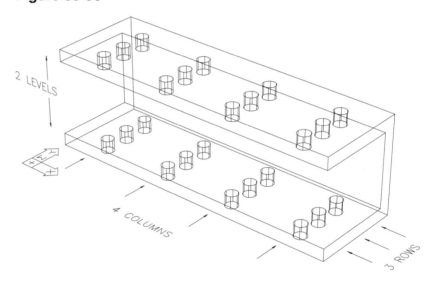

2 LEVELS

4 COLUMNS

3 ROWS

Polar
Similar to a *Polar Array* (2D), this option creates an array of selected objects in a <u>circular</u> fashion. The only difference in the 3D version is that an array is created about an <u>axis of rotation</u> (3D) rather than a point (2D). Specification of an axis of rotation requires two points in 3D space:

```
Command: 3darray
Select objects: PICK
Select objects: Enter
Rectangular or Polar array (R/P): p
Number of items: 8
Angle to fill <360>: Enter or (value)
Rotate objects as they are copied? <Y>: Enter or N
Center point of array: PICK or (coordinates)
Second point on axis of rotation: PICK or (coordinates)
Command:
```

In Figures 38-51 and 38-52, a *3Darray* is created to form a series of holes from a cylinder. The axis of rotation is the center axis of the large cylinder specified by PICKing the *Center* of the top and bottom circles.

Figure 38-51

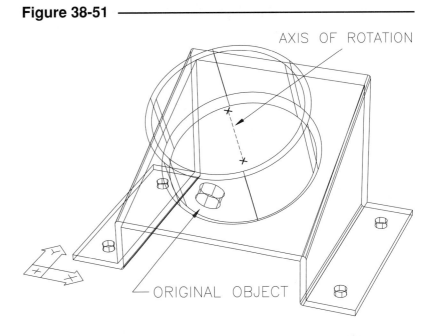

After the eight items are arrayed, the small cylinders are subtracted from the large cylinder to create the holes.

Figure 38-52

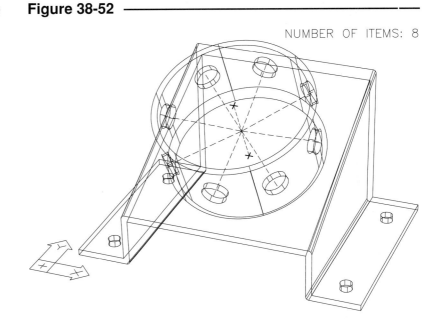

BOOLEAN OPERATION COMMANDS

Once the individual 3D primitives have been created and moved into place, you are ready to put together the parts. The primitives can be "assembled" or combined by Boolean operations to create composite solids. The Boolean operations found in AutoCAD are listed in this section: *Union*, *Subtract*, and *Intersect*.

UNION

Pull-down Menu	COMMAND (TYPE)	ALIAS (TYPE)	Short-cut	Screen (side) Menu	Tablet Menu
Modify Boolean > Union	UNION	UNI	...	MODIFY2 *Union*	X,15

Union joins selected primitives or composite solids to form one composite solid. Usually, the selected solids occupy portions of the same space, yet are separate solids. *Union* creates one solid composed of the total encompassing volume of the selected solids. (You can union solids even if the solids do not overlap.) All lines of intersections (surface boundaries) are calculated and displayed by AutoCAD. Multiple solid objects can be unioned with one *Union* command:

    Command: **solunion**
    Select objects: **PICK** (Select two or more solids.)
    Select objects: **Enter** (Indicate completion of the selection process.)
    Command:

Two solid boxes are combined into one composite solid with *Union* (Fig 38-53). The original two solids (A) occupy the same physical space. The resulting union (B) consists of the total contained volume. The new lines of intersection are automatically calculated and displayed. *Hide* was used to enhance visualization in (B).

Figure 38-53 ────────────────────

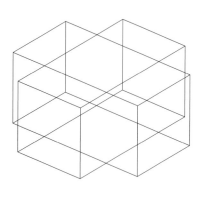

A.

B.

Because the volume occupied by any one of the primitives is included in the resulting composite solid, any redundant volumes are immaterial. The two primitives in Figure 38-54 (A) yield the same enclosed volume as the composite solid (B). (*Hide* has been used with *DISPSILH*=1 for this figure.)

Figure 38-54 ────────────────────

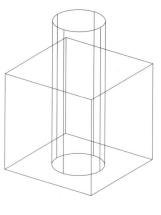
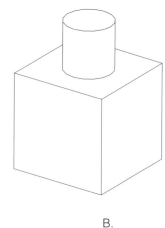

A.

B.

Multiple objects can be selected in response to the *Union* "Select objects:" prompt. It is not necessary, nor is it efficient, to use several successive Boolean operations if one or two can accomplish the same result.

Two primitives that have coincident faces (touching sides) can be joined with *Union*. Several "blocks" can be put together to form a composite solid.

Figure 38-55 illustrates how several primitives having coincident faces (A) can be combined into a composite solid (B). Only one *Union* is required to yield the composite solid.

Figure 38-55

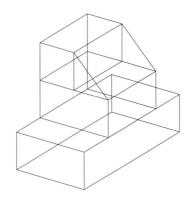
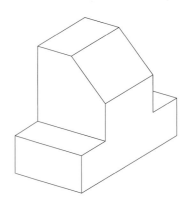

A. B.

SUBTRACT

Pull-down Menu	COMMAND (TYPE)	ALIAS (TYPE)	Short-cut	Screen (side) Menu	Tablet Menu
Modify Boolean > Subtract	SUBTRACT	SU	...	MODIFY2 Subtract	X,16

Subtract takes the difference of one set of solids from another. *Subtract* operates with *Regions* as well as solids. When using solids, *Subtract* subtracts the <u>volume</u> of one set of solids from another set of solids. Either set can contain only one or several solids. *Subtract* requires that you first select the set of solids that will remain (the "source objects"), then select the set you want to subtract from the first:

```
Command: subtract
Select solids and regions to subtract from...
Select objects: PICK
Select objects: Enter
Select solids and regions to subtract...
Select objects: PICK
Select objects: Enter
Command:
```

The entire volume of the solid or set of solids that is subtracted is completely removed, leaving the remaining volume of the source set.

To create a box with a hole, a cylinder is located in the same 3D space as the box (see Figure 38-56). *Subtract* is used to subtract the entire volume of the cylinder from the box. Note that the cylinder can have any height, as long as it is at least equal in height to the box.

Because you can select more than one object for the objects "to subtract from" and the objects "to subtract," many possible construction techniques are possible.

Figure 38-56

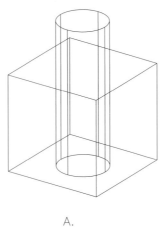
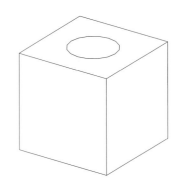

A. B.

If you select multiple solids in response to the select objects "to subtract from" prompt, they are <u>automatically</u> unioned. This is known as an <u>*n*-way Boolean</u> operation. Using *Subtract* in this manner is very efficient and fast.

Figure 38-57 illustrates an *n*-way Boolean. The two boxes (A) are selected in response to the objects "to subtract from" prompt. The cylinder is selected as the objects "to subtract...". *Subtract* joins the source objects (identical to a *Union*) and subtracts the cylinder. The resulting composite solid is shown in (B). (*Hide* has been used with *DISPSILH*=1 for this figure.)

Figure 38-57 ————————————

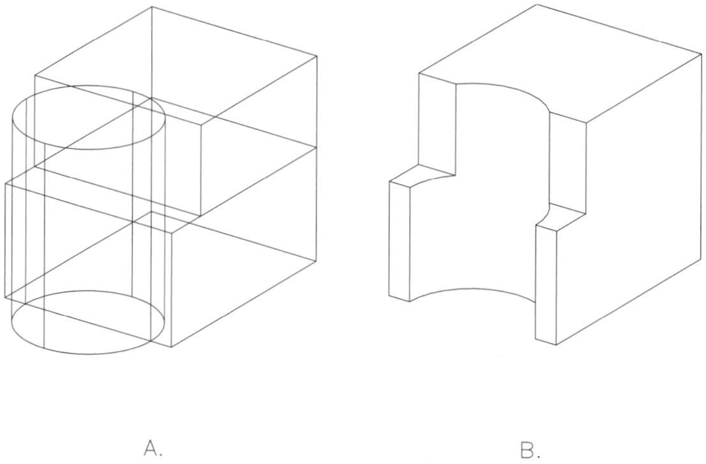

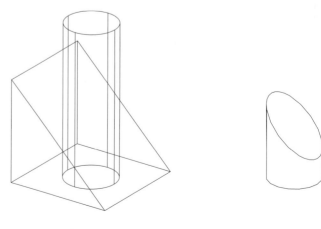

A.   B.

INTERSECT

Pull-down Menu	COMMAND (TYPE)	ALIAS (TYPE)	Short-cut	Screen (side) Menu	Tablet Menu
Modify *Boolean >* *Intersect*	*INTERSECT*	*IN*	...	MODIFY2 *Intrsect*	X,17

Intersect creates composite solids by calculating the intersection of two or more solids. The intersection is the common volume <u>shared</u> by the selected objects. Only the 3D space that is <u>part of all</u> of the selected objects is included in the resulting composite solid. *Intersect* requires only that you select the solids from which the intersection is to be calculated.

    Command: **intersect**
    Select objects: **PICK** (Select all desired solids.)
    Select objects: **Enter** (Indicates completion of the selection process.)
    Command:

An example of *Intersect* is shown in Figure 38-58. The cylinder and the box share common 3D space (A). The result of the *Intersect* is a composite solid that represents that common space (B). (*Hide* has been used with *DISPSILH*=1 for this figure.)

Figure 38-58 ————————————

A.   B.

Intersect can be very effective when used in conjunction with *Extrude*. A technique known as <u>reverse drafting</u> can be used to create composite solids that may otherwise require several primitives and several Boolean operations. Consider the composite solid shown in Figure 38-55 A. Using *Union*, the composite shape requires four primitives.

A more efficient technique than unioning several box primitives is to create two *Pline* shapes on vertical planes (Fig. 38-59). Each *Pline* shape represents the outline of the desired shape from its respective view: in this case, the front and side views. The *Pline* shapes are intended to be extruded to occupy the same space. It is apparent from this illustration why this technique is called reverse drafting.

Figure 38-59

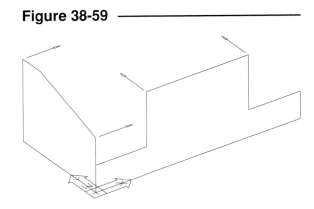

The two *Pline* "views" are extruded with *Extrude* to comprise the total volume of the desired solid (Fig. 38-60 A). Finally, *Intersect* is used to calculate the common volume and create the composite solid (B.).

Figure 38-60

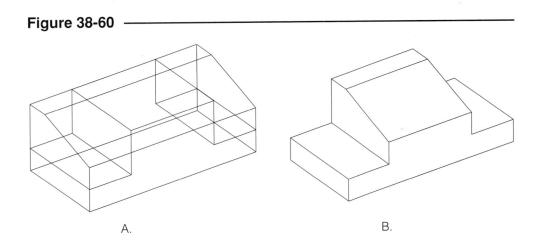

A.   B.

CHAMFER

Pull-down Menu	COMMAND (TYPE)	ALIAS (TYPE)	Short-cut	Screen (side) Menu	Tablet Menu
Modify *Chamfer*	*CHAMFER*	*CHA*	...	*MODIFY2* *Chamfer*	*W,18*

Chamfering is a machining operation that bevels a sharp corner. *Chamfer* chamfers selected edges of an AutoCAD solid as well as 2D objects. Technically, *Chamfer* (used with a solid) is a Boolean operation because it creates a wedge primitive and then adds to or subtracts from the selected solid.

When you select a solid, *Chamfer* recognizes the object as a solid and <u>switches to the solid version of prompts and options</u>. Therefore, all of the 2D options are not available for use with a solid, <u>only the "distances" method</u>. When using *Chamfer*, you must both select the "base surface" and indicate which edge(s) on that surface you wish to chamfer:

> Command: **chamfer**
> (TRIM mode) Current chamfer Length = 0.5000, Angle = 30.0000
> Polyline/Distance/Angle/Trim/Method/<Select first line>: **PICK** (Select solid)
> Select base surface: **PICK** (Select any edge of the desired solid face)
> Next/<OK>: **N** or **Enter**
> Enter base surface distance <1.0000>: **Enter** or **(value)**
> Enter other surface distance <0.5000>: **Enter** or **(value)**
> Loop/<Select edge>: **PICK** (Select edge to be chamfered)
> Loop/<Select edge>: **Enter**
> Command:

When AutoCAD prompts to select the "base surface," only an edge can be selected since the solids are displayed in wireframe. When you select an edge, AutoCAD highlights one of the two surfaces connected to the selected edge. Therefore, you must use the "Next/<OK>:" option to indicate which of the two surfaces you want to chamfer (Fig. 38-61 A). The two distances are applied to the object, as shown in Figure 38-61 B.

Figure 38-61

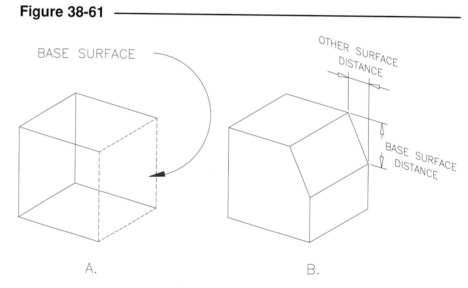

BASE SURFACE

OTHER SURFACE DISTANCE

BASE SURFACE DISTANCE

A.

B.

You can chamfer multiple edges of the selected "base surface" simply by PICKing them at the "<Select Edge>:" prompt (Fig. 38-62). If the base surface is adjacent to cylindrical edges, the bevel follows the curved shape.

Figure 38-62

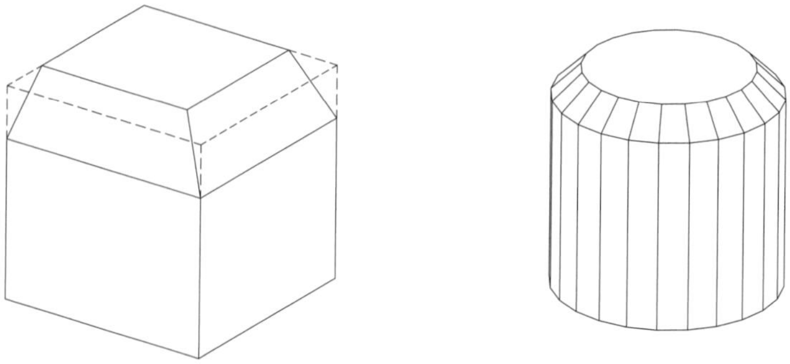

A.

B.

Loop

The *Loop* option chamfers the entire perimeter of the base surface. Simply PICK any edge on the base surface.

Loop/<Select edge>: *1*
Edge/<Select edge loop>: **PICK**
Edge/<Select edge loop>: **Enter**
Command:

Edge

The *Edge* option switches back to the "Select edge" method.

Figure 38-63 ————————

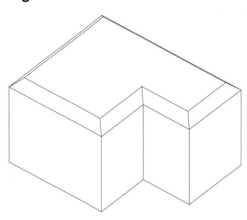

FILLET

Pull-down Menu	COMMAND (TYPE)	ALIAS (TYPE)	Short-cut	Screen (side) Menu	Tablet Menu
Modify *Fillet*	*FILLET*	*F*	...	*MODIFY 2* *Fillet*	*W,19*

Fillet creates fillets (concave corners) or rounds (convex corners) on selected solids, just as with 2D objects. Technically, *Fillet* creates a rounded primitive and automatically performs the Boolean needed to add or subtract it from the selected solids.

When using *Fillet* with a solid, the command <u>switches</u> to a special group of prompts and options for 3D filleting, and the <u>2D options become invalid</u>. After selecting the solid, you must specify the desired radius and then select the edges to fillet. When selecting edges to fillet, the edges must be PICKed individually. Figure 38-64 depicts concave and convex fillets created with *Fillet*. The selected edges are highlighted.

Figure 38-64 ————————

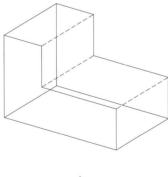

A.

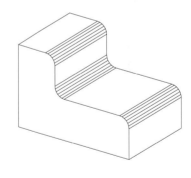

B.

Command: *fillet*
(TRIM mode) Current fillet radius = 0.5000
Polyline/Radius/Trim/<Select first object>: **PICK** (Select solid)
Chain/Radius/<Select edge>: Enter radius: **(value)**
Chain/Radius/<Select edge>: **PICK**
Chain/Radius/<Select edge>: **PICK**
Chain/Radius/<Select edge>: **Enter**
n edges selected for fillet.
Command:

Curved surfaces can be treated with *Fillet,* as shown in Figure 38-65. If you want to fillet intersecting concave or convex edges, *Fillet* handles your request, providing you specify all edges in <u>one</u> use of the command. Figure 38-65 shows the selected edges (highlighted) and the resulting solid. Make sure you select <u>all</u> edges together (in one *Fillet* command).

Figure 38-65

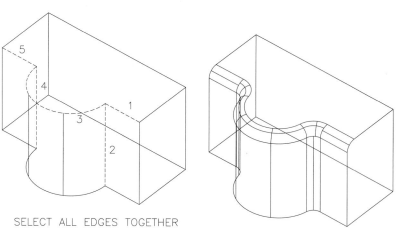

SELECT ALL EDGES TOGETHER

Chain

The *Chain* option allows you to fillet a series of connecting edges. Select the edges to form the chain (Fig. 38-66). If the chain is obvious (only one direct path), you can PICK only the ending edges, and AutoCAD will find the most direct path (series of connected edges):

Figure 38-66

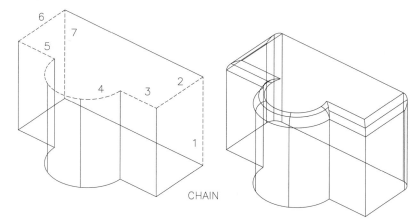

CHAIN

```
Chain/Radius/<Select edge>: c
Edge/Radius/<Select edge
chain>: PICK
```

Edge

This option cycles back to the "<Select edge>:" prompt.

Radius

This method returns to the "Enter radius:" prompt.

DESIGN EFFICIENCY

Now that you know the complete sequence for creating composite solid models, you can work toward improving design efficiency. The typical construction sequence is (1) create primitives, (2) ensure the primitives are in place by using UCSs or any of several move and rotate options, and (3) combine the primitives into a composite solid using Boolean operations. The typical step-by-step, "building-block" strategy, however, may not lead to the most efficient design. In order to minimize computation and construction time, you should <u>minimize the number of Boolean operations</u> and, if possible, the <u>number of primitives</u> you use.

For any composite solid, usually several strategies could be used to construct the geometry. You should plan your designs ahead of time, striving to minimize primitives and Boolean operations.

For example, consider the procedure shown in Figure 38-55. As discussed, it is more efficient to accomplish all unions with one *Union*, rather than each union as a separate step. Even better, create a closed *Pline* shape of the profile; then use *Extrude*. Figure 38-57 is another example of design efficiency based on using an *n*-way Boolean. Multiple solids can be unioned automatically by selecting them at the select objects "to subtract from" prompt of *Subtract*. Also consider the strategy of reverse drafting, as shown in Figure 38-59. Using *Extrude* in concert with *Intersect* can minimize design complexity and time.

In order to create efficient designs and minimize Boolean operations and primitives, keep these strategies in mind:

- Execute as many subtractions, unions, or intersections as possible within one *Subtract*, *Union*, or *Intersect* command.

- Use *n*-way Booleans with *Subtract*. Combine solids (union) automatically by selecting <u>multiple</u> objects "to subtract from," and then select "objects to subtract."

- Make use of *Plines* or regions for complex profile geometry; then *Extrude* the profile shape. This is almost always more efficient for complex curved profile creation than using multiple Boolean operations.

- Make use of reverse drafting by extruding the "view" profiles (*Plines* or *Regions*) with *Extrude*, then finding the common volume with *Intersect*.

CHAPTER EXERCISES

1. What are the typical three steps for creating composite solids?

2. Consider the two solids in Figure 38-67. They are two extruded hexagons that overlap (occupy the same 3D space).

   A. Sketch the resulting composite solid if you performed a *Union* on the two solids.

   B. Sketch the resulting composite solid if you performed an *Intersect* on the two solids.

   C. Sketch the resulting composite solid if you performed a *Subtract* on the two solids.

Figure 38-67

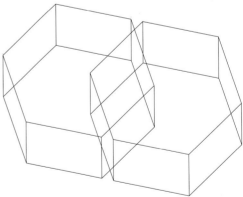

For the following exercises, use a *Template* drawing or begin a *New* drawing. Turn *On* the *Ucsicon* and set it to the *Origin*. Set the *Vpoint* with the *Rotate* option to angles of **310, 30**.

3. Begin a drawing and assign the name **CH38EX3**.

    Figure 38-68

    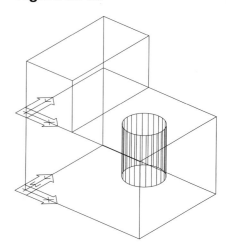

    A. Create a *box* with the lower-left corner at **0,0,0**. The *Lengths* are **5**, **4**, and **3**.

    B. Create a second *box* at a new UCS as shown in Figure 38-68 (use the *ORigin* option). The *box* dimensions are **2** x **4** x **2**.

    C. Create a *Cylinder*. Use the same UCS as in the previous step. The *cylinder Center* is at **3.5,2** (of the *UCS*), the *Diameter* is **1.5**, and the *Height* is **-2**.

    D. *Save* the drawing.

    E. Perform a *Union* to combine the two boxes. Next, use *Subtract* to subtract the cylinder to create a hole. The resulting composite solid should look like that in Figure 38-69. *Save* the drawing.

    Figure 38-69

    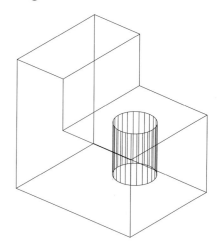

4. Begin a drawing and assign the name **CH38EX4**.

    Figure 38-70

    A. Create a *Wedge* at point **0,0,0** with the *Lengths* of **5**, **4**, **3**.

    B. Create a *3point UCS* option with an orientation indicated in Figure 38-70. Create a *Cone* with the *Center* at **2,3** (of the *UCS*) and a *Diameter* of **2** and a *Height* of **-4**.

    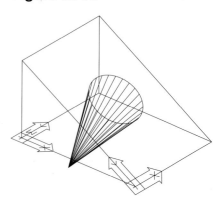

C. *Subtract* the cone from the wedge. The resulting composite solid should resemble Figure 38-71. *Save* the drawing.

Figure 38-71

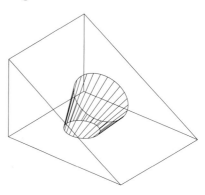

5. Begin a drawing and assign the name **CH38EX5**. Display a *Plan* view.

Figure 38-72

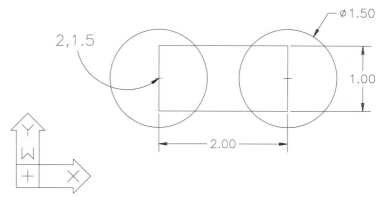

A. Create 2 *Circles* as shown in Figure 38-72, with dimensions and locations as specified. Use *Pline* to construct the rectangular shape. Combine the 3 shapes into a *Region* by using the **Region** and **Union** commands or converting the outside shape into a *Pline* using *Trim* and *Pedit*.

B. Change the display to an isometric-type **Vpoint**. **Extrude** the *Region* or *Pline* with a **Height** of **3** (no *taper angle*).

C. Create a *Box* with the lower-left corner at **0,0**. The **Lengths** of the box are **6, 3, 3**.

D. *Subtract* the extruded shape from the box. Your composite solid should look like that in Figure 38-73. *Save* the drawing.

Figure 38-73

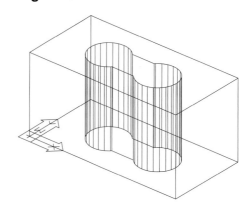

6. Begin a drawing and assign the name **CH38EX6**. Display a *Plan* view.

Figure 38-74

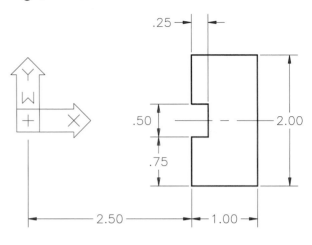

A. Create a closed *Pline* shape symmetrical about the X axis with the locational and dimensional specifications given in Figure 38-74.

B. Change to an isometric-type *Vpoint*. Use *Revolve* to generate a complete circular shape from the closed *Pline*. Revolve about the **Y** axis.

C. Create a *Torus* with the *Center* at **0,0**. The *Radius of torus* is **3** and the *Radius of tube* is **.5**. The two shapes should intersect.

D. Use *Hide* to generate a display like Figure 38-75.

Figure 38-75

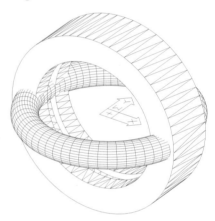

E. Create a *Cylinder* with the *Center* at **0,0,0**, a *Radius* of **3**, and a *Height* of **8**.

F. Use *Rotate3D* to rotate the revolved *Pline* shape **90** degrees about the X axis (the *Pline* shape that was previously converted to a solid—not the torus). Next, move the shape up (positive Z) **6** units with *Move*.

G. Move the torus up **4** units with *Move*.

H. The solid primitives should appear as those in Figure 38-76. (*Hide* has been used for the figure.)

Figure 38-76

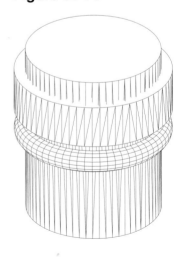

I. Use *Subtract* to subtract both revolved shapes from the cylinder. Use *Hide*. The solid should resemble that in Figure 38-77. (*Hide* has been used with *DISPSILH*=1 for this figure.) *Save* the drawing.

Figure 38-77

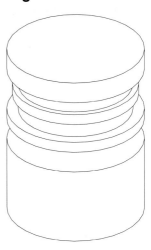

7. Begin a *New* drawing or use a *Template*. Assign the name **FAUCET**.

A. Draw 3 closed *Pline* shapes, as shown in Figure 38-78. Assume symmetry about the longitudinal axis. Use the WCS and create 2 new *UCS*s for the geometry. Use *3point Arcs* for the "front" profile.

Figure 38-78

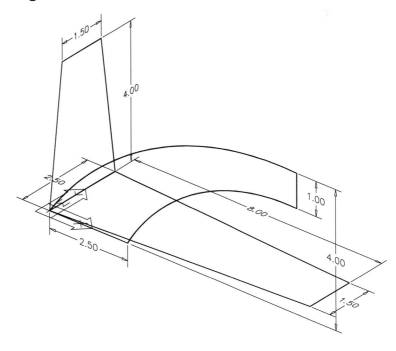

B. *Extrude* each of the 3 profiles into the same space. Make sure you specify the correct positive or negative *Height* value.

Figure 38-79

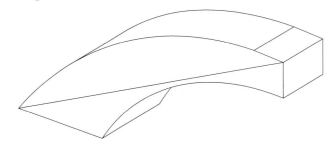

C. Finally, use *Intersect* to create the composite solid of the Faucet. *Save* the drawing.

D. (Optional) Create a nozzle extending down from the small end. Then create a channel for the water to flow (through the inside) and subtract it from the faucet.

8. Construct a solid model of the bar guide in Figure 38-80. Strive for the most efficient design. It is possible to construct this object with one *Extrude* and one *Subtract*. Save the model as **BGUID-SL**.

Figure 38-80 ————————————————

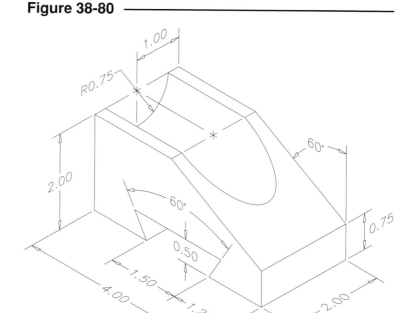

9. Make a solid model of the V-block shown in Figure 38-81. Several strategies could be used for construction of this object. Strive for the most efficient design. Plan your approach by sketching a few possibilities. Save the model as **VBLOK-SL**.

Figure 38-81 ————————————————

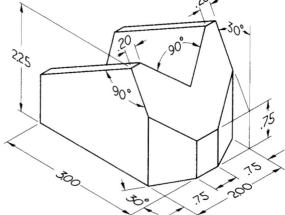

10. Construct a composite solid model of the support bracket using efficient techniques. Save the model as **SUPBK-SL**.

Figure 38-82

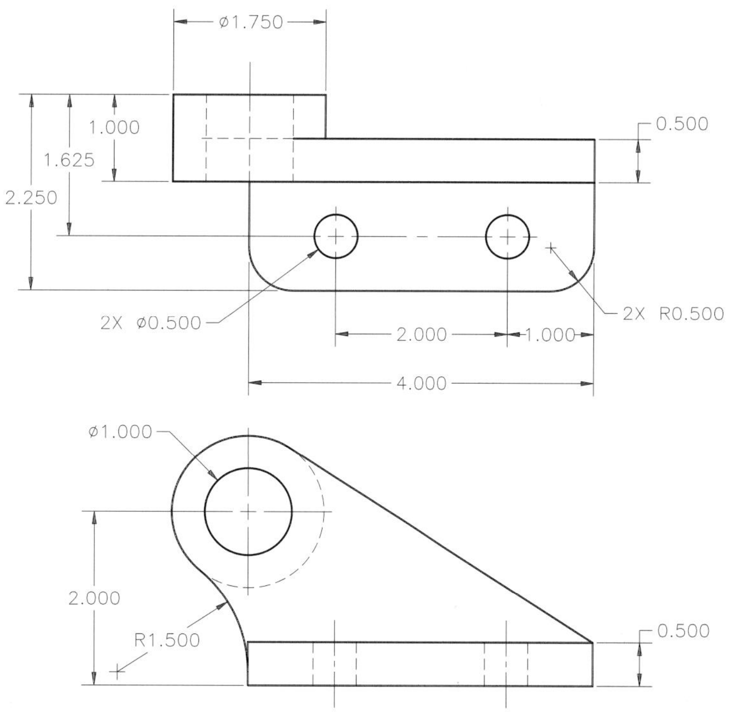

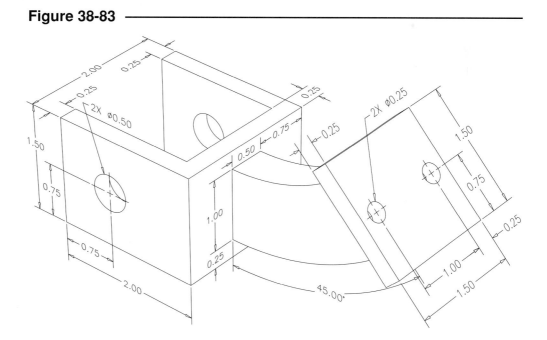

11. Construct the swivel shown in Figure 38-83. The center arm requires *Extruding* the 1.00 x 0.50 rectangular shape along an arc path through 45 degrees. Save the drawing as **SWIVEL**.

Figure 38-83

12. Construct a solid model of the angle brace shown in Figure 38-84. Use efficient design techniques. Save the drawing as **AGLBR-SL**.

Figure 38-84

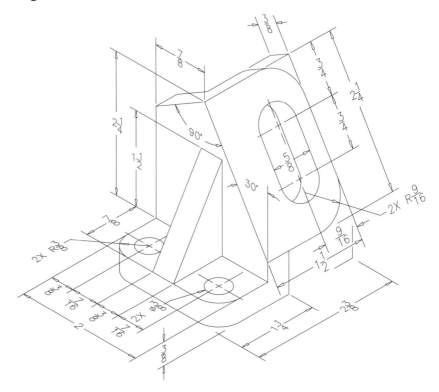

13. Construct a solid model of the saddle shown in Figure 38-85. An efficient design can be utilized by creating *Pline* profiles of the top "view" and the front "view," as shown in Figure 38-86. Use *Extrude* and *Intersect* to produce a composite solid. Additional Boolean operations are required to complete the part. The finished model should look like Figure 38-87 (with *Hide* performed). Save the drawing as **SADL-SL**.

Figure 38-85

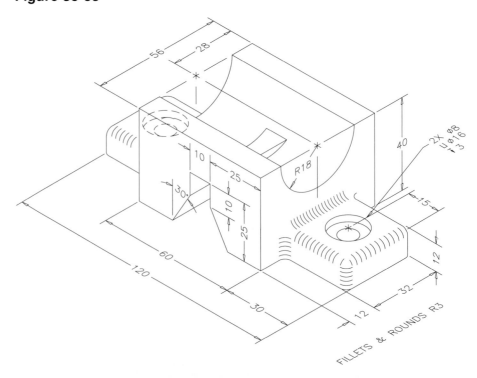

Figure 38-86 ────────────────────────

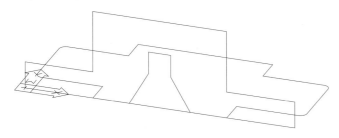

Figure 38-87 ────────────────────────

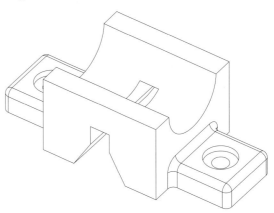

14. Construct a solid model of a bicycle handle bar. Create a centerline (Fig. 38-88) as a *Path* to extrude a *Circle* through. Three mutually perpendicular coordinate systems are required: the **WORLD**, the **SIDE**, and the **FRONT**. The centerline path consists of three separate *Plines*. First, create the 370 length *Pline* with 60 radii arcs on each end on the WCS. Then create the drop portion of the bars using the SIDE *UCS*. Create three *Circles* using the FRONT *UCS*, and extrude one along each *Path*.

Figure 38-88 ────────────────────────

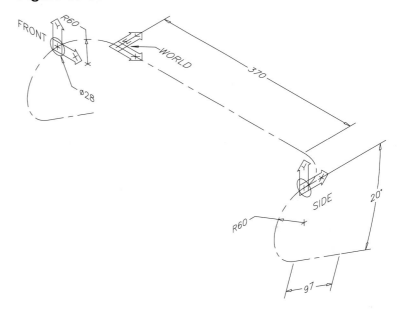

Plot the bar and *Hide Lines* as shown in Figure 38-89. *Save* the drawing as **DROPBAR**.

Figure 38-89 ────────────────────────

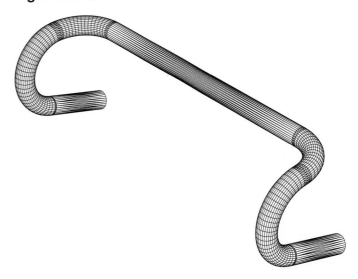

15. Create a solid model of the pulley. All vertical dimensions are diameters. Orientation of primitives is critical in the construction of this model. Try creating the circular shapes on the XY plane (circular axis aligns with Z axis of the WCS). After the construction, use *Rotate3D* to align the circular axis of the composite solid with the Y axis of the WCS. *Save* the drawing as **PULLY-SL**.

Figure 38-90 ──────────────

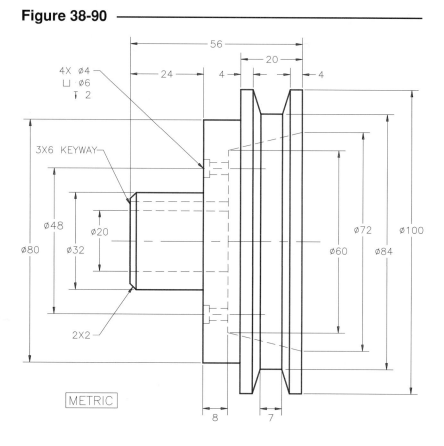

16. Create a composite solid model of the adjustable mount (Fig. 38-91). Use *Move* and other methods to move and align the primitives. Use of efficient design techniques is extremely important with a model of this complexity. Assign the name **ADJMT-SL**.

Figure 38-91 ──────────────

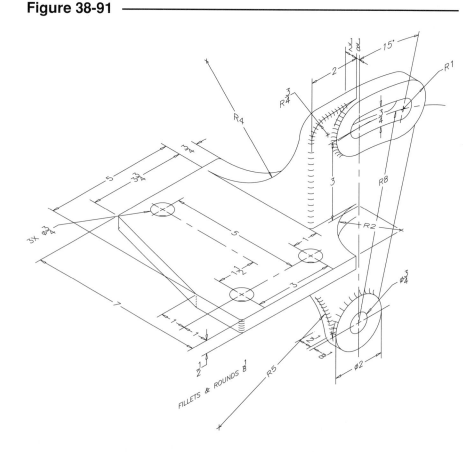

39

ADVANCED SOLIDS FEATURES

Chapter Objectives

After completing this chapter you should be able to:

1. use the *ISOLINES*, *DISPSILH*, and *FACETRES* variables to control the display of tessellation lines, silhouette lines, and mesh density for solid models;

2. calculate mass properties of a solid model using *Massprop*;

3. determine if *Interference* exists between two or more solids and create a solid equal in volume to the interference;

4. create a 2D section view for a solid model using *Section* and *Bhatch*;

5. use *Slice* to cut a solid model at any desired cutting plane and retain one or both halves;

6. convert AME (AutoCAD Release 12) solid models to AutoCAD Release 14 ACIS solid models with *AMECONVERT*;

7. use *STLOUT* to create a file suitable for use with rapid prototyping apparatus.

CONCEPTS

Several topics related to solid modeling capabilities for AutoCAD Release 14 ACIS models are discussed in this chapter. The topics are categorized in the following sections:

> Solid Modeling Display Variables
> Analyzing Solid Models
> Creating Sections from Solids
> Converting Solids

SOLID MODELING DISPLAY VARIABLES

AutoCAD solid models are displayed in wireframe representation by default. Wireframe representation requires less computation time and less complex file structure, so your drawing time can be spent more efficiently. When you use *Hide* or *Shade*, the solid models are automatically meshed before they are displayed with hidden lines removed or as a shaded image. This meshed version of the model is apparent when you use *Hide* on cylindrical or curved surfaces.

Three variables control the display of solids for wireframe, hidden, and meshed representation. The *ISOLINES* variable controls the number of tessellation lines that are used to visually define cylindrical surfaces for wireframes. The *DISPSILH* variable can be toggled on or off to display silhouette lines for wireframe displays. *FACETRES* is the variable that controls the density of the mesh apparent with *Hide*. The variables are accessed by typing the names at the Command: prompt.

ISOLINES

This variable sets the number of tessellation lines that appear on a curved surface when shown in wireframe representation. The default setting for *ISOLINES* is 4 (Fig. 39-1). A solid of extrusion shows fewer tessellation lines (the current *ISOLINES* setting less 4) to speed regeneration time.

A higher setting gives better visualization of the curved surfaces but takes more computing time (Fig. 39-2). After changing the *ISOLINES* setting, *Regen* the drawing to see the new display.

Figure 39-1 ———————— **Figure 39-2** ————————

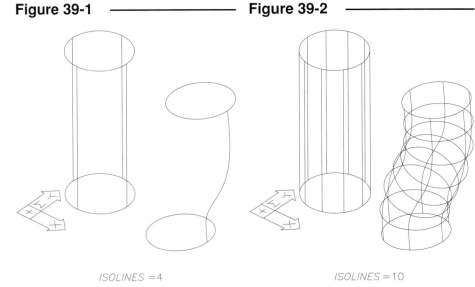

ISOLINES =4   ISOLINES =10

NOTE: In an isometric view attained by *3D Viewpoint*, the 4 lines (like Fig. 39-1) appear to overlap—they align when viewed from any perfect isometric angle.

DISPSILH

This variable can be turned on to display the limiting element contour lines, or silhouette, of curved shapes for a wireframe display (Fig. 39-3). The default setting is 0 (off). Since the silhouette lines are <u>viewpoint dependent</u>, significant computing time is taken to generate the display. You should <u>not</u> leave *DISPSILH* on 1 during construction.

Figure 39-3

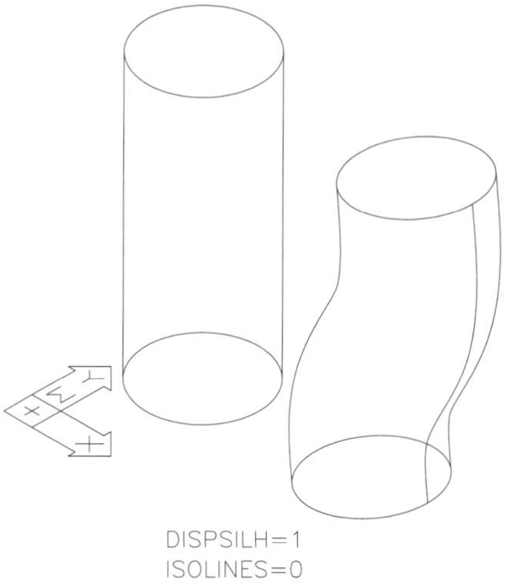

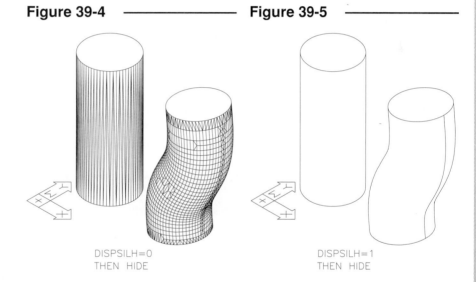

DISPSILH=1
ISOLINES=0

DISPSILH has a special function when used with *Hide*. When *DISPSILH* has the default setting of 0 and a *Hide* is performed, the solids appear opaque but display the mesh lines (Fig. 39-4). If *DISPSILH* is set to 1 before *Hide* is performed, the solids appear opaque but do <u>not</u> display the mesh lines (Fig. 39-5).

Figure 39-4

Figure 39-5

DISPSILH=0
THEN HIDE

DISPSILH=1
THEN HIDE

FACETRES

FACETRES controls the <u>density of the mesh</u> that is automatically created when a <u>*Hide* or a *Shade*</u> is performed. The default setting is .5, as shown in Figure 39-6. Decreasing the value produces a coarser mesh (Fig. 39-7), while increasing the value produces a finer mesh. The higher the value, the more computation time involved to generate the display or plot. The density

Figure 39-6

Figure 39-7

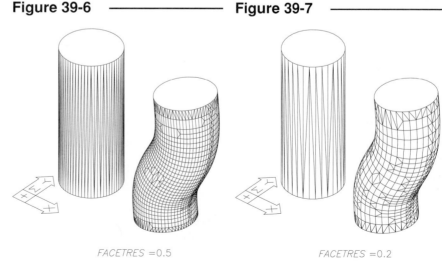

FACETRES =0.5

FACETRES =0.2

of the mesh is actually a factor of both the *FACETRES* setting and the *VIEWRES* setting. Increasing *VIEWRES* also makes the mesh more dense. *FACETRES* can be set to any value between .01 and 10.

The *Preferences* dialog box can be used to change the settings for the *ISOLINES, DISPSILH,* and *FACET-RES* variables (Fig. 39-8). The upper-left corner of the *Performance* tab (*Solid model object display*) has two edit boxes and one checkbox that allow changing these variables as follows.

Rendered object smoothness	*FACETRES*
Contour lines per surface	*ISOLINES*
Show silhouettes in wireframe	*DISPSILH*

Figure 39-8

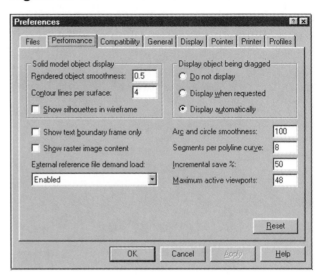

ANALYZING SOLID MODELS

Two commands in AutoCAD allow you to inquire about and analyze the solid geometry. *Massprop* calculates a variety of properties for the selected ACIS solid model. AutoCAD does the calculation and lists the information in screen or text window format. The data can be saved to a file for future exportation to a report document or analysis package. The *Interfere* command finds the interference of two or more solids and highlights the overlapping features so you can make necessary alterations.

MASSPROP

Pull-down Menu	COMMAND (TYPE)	ALIAS (TYPE)	Short-cut	Screen (side) Menu	Tablet Menu
Tools *Inquiry >* *Mass Properties*	*MASSPROP*	...	...	*TOOLS 1* *Massprop*	*U,7*

Since solid models define a complete description of the geometry, they are ideal for mass properties analysis. The *Massprop* command automatically computes a variety of mass properties.

Mass properties are useful for a variety of applications. The data generated by the *Massprop* command can be saved to an .MPR file for future exportation in order to develop bills of material, stress analysis, kinematics studies, and dynamics analysis.

Applying the *Massprop* command to a solid model produces a text screen displaying the following list of calculations (Fig. 39-9):

Figure 39-9

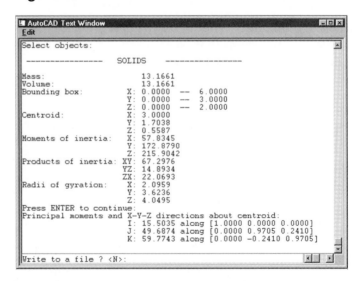

Mass	Mass is the quantity of matter that a solid contains. Mass is determined by density of the material and volume of the solid. Mass is not dependent on gravity and, therefore, different from but proportional to weight. Mass is also considered a measure of a solid's resistance to linear acceleration (overcoming inertia).
Volume	This value specifies the amount of space occupied by the solid.
Bounding Box	These lengths specify the extreme width, depth, and height of the selected solid.
Centroid	The centroid is the geometrical center of the solid. Assuming the solid is composed of material that is homogeneous (uniform density), the centroid is also considered the center of mass and center of gravity. Therefore, the solid can be balanced when supported only at this point.
Moments of Inertia	Moments convey how the mass is distributed around the X, Y, and Z axes of the current coordinate system. These values are a measure of a solid's resistance to <u>angular</u> acceleration (mass is a measure of a solid's resistance to <u>linear</u> acceleration). Moments of inertia are helpful for stress computations.
Products of Inertia	These values specify the solid's resistance to <u>angular</u> acceleration with respect to two axes at a time (XY, YZ, or ZX). Products of inertia are also useful for stress analysis.
Radii of Gyration	If the object were a concentrated solid mass without holes or other features, the radii of gyration represent these theoretical dimensions (radius about each axis) such that the same moments of inertia would be computed.
Principal Moments and X, Y, Z Directions	In structural mechanics, it is sometimes important to determine the orientation of the axes about which the moments of inertia are at a maximum. When the moments of inertia about centroidal axes reach a maximum, the products of inertia become zero. These particular axes are called the principal axes, and the corresponding moments of inertia with respect to these axes are the principal moments (about the centroid).

Notice the "Mass:" is reported as having the same value as "Volume." This is because AutoCAD Release 14 solids cannot have material characteristics assigned to them, so AutoCAD assumes a density value of 1. To calculate the mass of a selected solid in AutoCAD, use a reference guide (such as a machinist's handbook) to find the material density and multiply the value times the reported volume (mass = volume x density).

INTERFERE

Pull-down Menu	COMMAND (TYPE)	ALIAS (TYPE)	Short-cut	Screen (side) Menu	Tablet Menu
Draw Solids > Interference	*INTERFERE*	*INF*	...	DRAW 2 SOLIDS *Interfer*	...

In AutoCAD, unlike real life, it is possible to create two solids that occupy the same physical space. *Interfere* checks solids to determine whether or not they interfere (occupy the same space). If there is interference, *Interfere* reports the overlap and allows you to create a new solid from the interfering volume, if you desire. Normally, you specify two sets of solids for AutoCAD to check against each other:

Command: *Interfere*
Select the first set of solids...
Select objects: **PICK**
Select objects: **Enter**
1 solid selected.
Select the second set of solids...
Select objects: **PICK**
Select objects: **Enter**
1 solid selected.
Comparing 1 solid against 1 solid.
Interfering solids (first set): 1
                (second set): 1
Interfering pairs: 1
Create interference solids? <N>: **y**
Command:

Figure 39-10

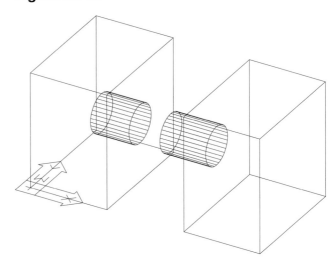

If you answer "yes" to the last prompt, a new solid is created equal to the exact size and volume of the interference. The original solids are not changed in any way. If no interference is found, AutoCAD reports "Solids do not interfere."

For example, consider the two solids shown in Figure 39-10. (The parts are displayed in wireframe representation.) The two shapes fit together as an assembly. The locating pin on the part on the right should fit in the hole in the left part.

Figure 39-11

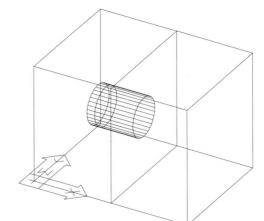

Sliding the parts together until the two vertical faces meet produces the assembly shown in Figure 39-11. There appears to be some inconsistency in the assembly of the hole and the pin. Either the pin extends beyond the hole (interference) or the hole is deeper than necessary (no interference). Using *Interfere*, you can find an overlap and create a solid is created by answering "yes" to "Create interference solids?" Use *Move* with the *Last* selection option to view and analyze the solid of interference.

You can compare <u>more than two</u> solids against each other with *Interfere*. This is accomplished by selecting <u>all desired solids at the first prompt</u> and none at the second:

    Select the first set of solids... **PICK**
    Select the second set of solids... **Enter**

AutoCAD then compares all solids in the first set against each other. If more than one interference is found, AutoCAD highlights intersecting solids, one pair at a time.

CREATING SECTIONS FROM SOLIDS

Two AutoCAD commands are intended to create sections from solid models. *Section* is a drafting feature that creates a 2D "section view." The cross-section is determined by specifying a cutting plane. A cross-section view is automatically created based on the solid geometry that intersects the cutting plane. The original solid is not affected by the action of *Section*. *Slice* actually cuts the solid at the specified cutting plane. *Slice* therefore creates two solids from the original one and offers the possibility to retain both halves or only one. Many options are available for placement of the cutting plane.

SECTION

Pull-down Menu	COMMAND (TYPE)	ALIAS (TYPE)	Short-cut	Screen (side) Menu	Tablet Menu
Draw *Solids >* *Section*	*SECTION*	*SEC*	...	*DRAW 2* *SOLIDS* *Section*	...

Section creates a 2D cross-section of a solid or set of solids. The cross-section created by *Solsect* is considered a traditional 2D section view. The cross-section is defined by a cutting plane, and the resulting section is determined by any solid material that passes through the cutting plane. The cutting plane can be specified by a variety of methods. The options for establishing the cutting plane are listed in the command prompt:

```
Command: section
Select objects: PICK
Select objects: Enter
Sectioning plane by Entity/Last/Zaxis/View/XY/YZ/ZX /<3points>:
```

For example, assume a cross-section is desired for the geometry shown in Figure 39-12. To create the section, you must define the cutting plane.

Figure 39-12 ───────────────────

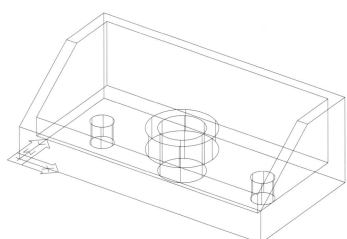

For this case, the *ZX* option is used and the requested point is defined to establish the position of the plane, as shown in Figure 39-13. The cross-section will be created on this plane.

Once the cutting plane is established, a cross-section is automatically created by *Section*. The resulting geometry is a *Region* created on the <u>current</u> layer.

Figure 39-13 ────────────

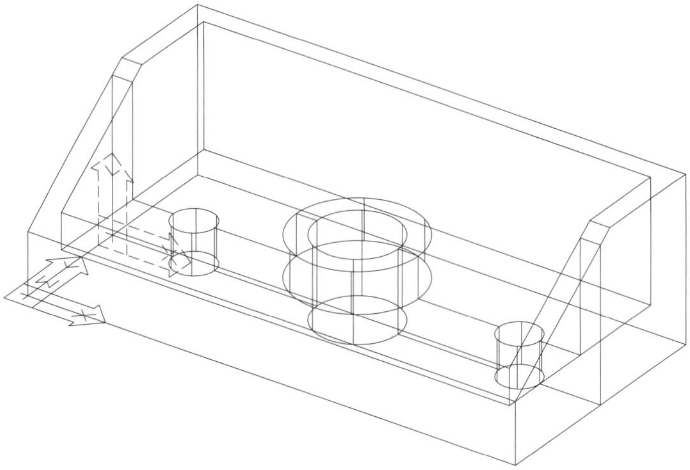

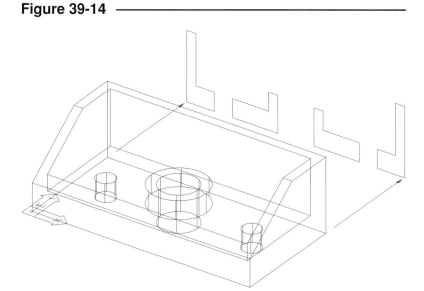

The resulting *Region* can be moved if needed as shown in Figure 39-14. If you want to use the *Region* to create a 2D section complete with hatch lines, several steps are required. First, the *Region* must be *Exploded* into individual *Regions* (4, in this case). Then each *Region* must be *Exploded* again to break the shapes into *Lines* that *Bhatch* can interpret as boundaries. You may need to create a *UCS* on the plane of the shapes before using *Bhatch*. Other *Lines* may be needed to make a complete section view.

Figure 39-14 ────────────

SLICE

	COMMAND (TYPE)	ALIAS (TYPE)	Short-cut	Screen (side) Menu	Tablet Menu
Pull-down Menu					
Draw *Solids >* *Slice*	*SLICE*	*SL*	...	*DRAW 2* *SOLIDS* *Slice*	...

Slice creates a true solid section. *Slice* cuts an ACIS solid or set of solids on a specified cutting plane. The original solid is converted to two solids. You have the option to keep both halves or only the half that you specify. Examine the following command syntax:

```
Command: slice
Select objects: PICK
Select objects: Enter
Slicing plane by Object/Last/Zaxis/View/XY/YZ/ZX/<3points>: (option)
Point on XY plane <0,0,0>: PICK
Both sides/<Point on desired side of the plane>: PICK
Command:
```

Entering *B* at the "Both sides/<Point on desired side of plane>:" prompt retains the solids on both sides of the cutting plane. Otherwise, you can pick a point on either side of the plane to specify which half to keep.

Figure 39-15

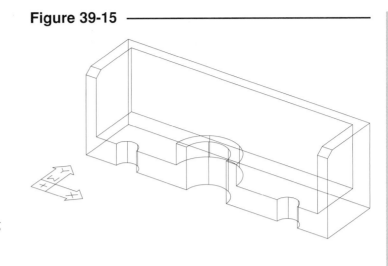

For example, using the solid model shown previously in Figure 39-12, you can use *Slice* to create a new sectioned solid shown here. The *ZX* method is used to define the cutting plane midway through the solid. Next, the new solid to retain was specified by PICKing a point on that geometry. The resulting sectioned solid is shown in Figure 39-15. Note that the solid on the near side of the cutting plane was not retained.

CONVERTING SOLIDS

AutoCAD Release 14 provides several utilities for converting solids to and from other file formats. Older AutoCAD solid models created with AME (Advanced Modeling Extension) can be converted to a Release 14 ACIS model with some success. Other file utilities allow you to import and export .SAT files and to create files for use with stereolithography apparatus.

AMECONVERT

Pull-down Menu	COMMAND (TYPE)	ALIAS (TYPE)	Short-cut	Screen (side) Menu	Tablet Menu
...	*AMECONVERT*	...	...	...	...

Ameconvert is a conversion utility to convert older solid models (created with AME Release 2 or 2.1) to AutoCAD Release 14 ACIS solid models. The command is simple to use; however, the conversion may change the model slightly. If the selected solids to convert are not AME Release 2 or 2.1, AutoCAD ignores the request.

The newer Release 14 solid modeler, ACIS, creates solids of higher accuracy than the older AME modeler. Because of this new accuracy, objects that are converted may change in appearance and form. For example, two shapes that were originally considered sufficiently close and combined by Boolean operations in the old modeler may be interpreted as slightly offset in the ACIS modeler. This can occur for converted features such as fillets, chamfers, and through holes. Occasionally, a solid created in the older AME model is converted into two or more solids in Release 14 ACIS modeler.

Solids that are converted to Release 14 ACIS solids can be edited using Release 14 solids commands. Converted solids that do not convert as expected can usually be edited in Release 14 to produce an equivalent, but more accurate, solid.

STLOUT

Pull-down Menu	COMMAND (TYPE)	ALIAS (TYPE)	Short-cut	Screen (side) Menu	Tablet Menu
File Export... **.stl*	*STLOUT*	...	...	*FILE Export* **.stl*	...

The *Stlout* command converts an ACIS solid model to a file suitable for use with rapid prototyping apparatus. Use *Stlout* if you want to use an ACIS solid to create a prototype part using stereo lithography or sintering technology. This technology reads the CAD data and creates a part by using a laser to solidify microthin layers of plastic or wax polymer or powder. A complex 3D prototype can be created from a CAD drawing in a matter of hours.

Stlout writes an ASCII or binary .STL file from an ACIS solid model. The model must reside entirely within the positive X,Y,Z octant of the WCS (all part geometry coordinates must be positive):

```
Command: stlout
Select a single solid for STL output:
elect objects: PICK
Create a binary STL file ? <Y>: Enter Y or N to create an ASCII file.
Command:
```

AutoCAD displays the *Create .STL File* dialog box for you to designate a name and path for the file to create.

When you design parts with AutoCAD to be generated by stereolithography apparatus, it is a good idea to create the profile view (that which contains the most complex geometry) parallel to the XY plane. In this way, the aliasing (stair-step effect) on angled or curved surfaces, caused by incremental passes of the laser, is minimized.

ACISIN

Pull-down Menu	COMMAND (TYPE)	ALIAS (TYPE)	Short-cut	Screen (side) Menu	Tablet Menu
Insert ACIS Solid...	*ACISIN*	...	...	*INSERT ACISin*	...

The *Acisin* command imports an ACIS solid model stored in a .SAT (ASCII) file format. This utility can be used to import .SAT files describing a solid model created by AutoCAD Release 14 or other CAD systems that create ACIS solid models. The *Select ACIS File* dialog box appears, allowing you to select the desired file. AutoCAD reads the file and builds the model in the current drawing.

ACISOUT

Pull-down Menu	COMMAND (TYPE)	ALIAS (TYPE)	Short-cut	Screen (side) Menu	Tablet Menu
File Export... **.sat*	*ACISOUT*	...	...	*FILE Export* **.sat*	...

This utility is used to export an ACIS solid model to a .SAT (ASCII) file format that can later be read by AutoCAD Release 14 or other CAD systems utilizing the ACIS modeler. The *Create ACIS File* dialog box is used to define the desired name and path for the file to be created.

R13

CHAPTER EXERCISES

1. *Open* the **SADL-SL** drawing that you created in Chapter 38 Exercises. Calculate *Mass Properties* for the saddle. Write the report out to a file named **SADL-SL.MPR**. Use a text editor or the DOS TYPE command to examine the file.

2. *Open* the **SADL-SL** drawing again. Use *Slice* to cut the model in half longitudinally. Use an appropriate method to establish the "slicing plane" in order to achieve the resulting model, as shown in Figure 39-16. Use *SaveAs* and assign the name **SADL-CUT**.

Figure 39-16 ─────────────

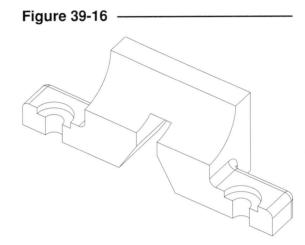

3. *Open* the **PULLY-SL** drawing that you created in Chapter 38 Exercises.

Figure 39-17 ─────────────

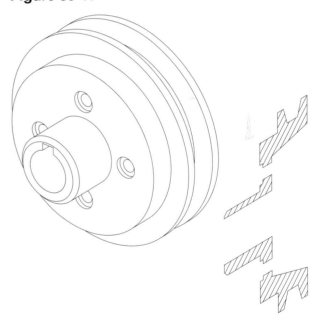

Make a *New Layer* named **SECTION** and set it *Current*. Then use the *Section* command to create a full section "view" of the pulley. Establish a vertical cutting plane through the center of the model. Remove the section view object (*Region*) with the *Move* command, translating **100** units in the **X** direction. The model and the new section view should appear as in Figure 39-17 (*Hide* was performed on the pulley to enhance visualization). Complete the view by establishing a *UCS* at the section view, then adding the *Bhatch,* as shown in Figure 39-17. Finally, create the necessary *Lines* to complete the view. *SaveAs* **PULLY-SC**.

4. *Open* the **SADL-SL** drawing that you created in the Chapter 38 Exercises. Change the *ISOLINES* setting to display **10** tessellation lines. Use the *Hide* command to create a meshed hidden display. Change the *FACETRES* setting to display a coarser mesh and use *Hide* again. Make a plot of the model with the coarse mesh and with hidden lines (check the *Hide Lines* box in the *Plot Configuration* dialog box).

Next, change the *FACETRES* setting to display a fine mesh. Use *Hide* to reveal the change. Make a plot of the model with *Hide Lines* checked to display the fine mesh. Then set the *DISPSILH* variable to **1** and make another plot with lines hidden. What is the difference in the last two plots? *Save* the **SADL-SL** file with the new settings.

Watch

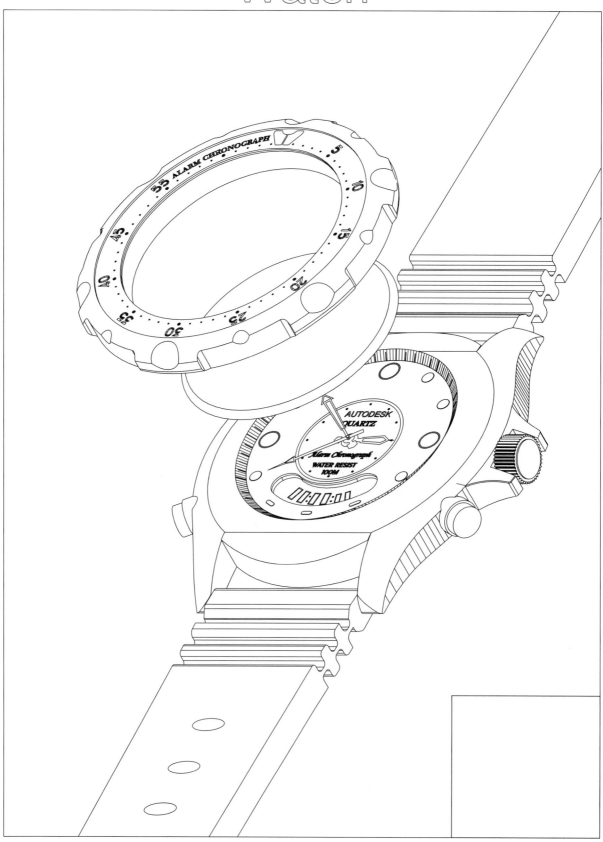

WATCH.DWG Courtesy of Autodesk, Inc. (Release 14 Sample Drawing)

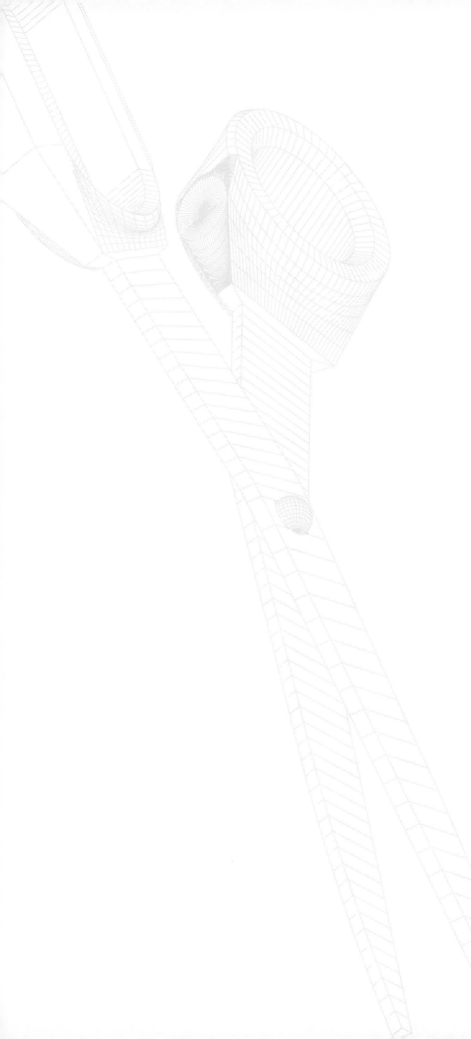

40

SURFACE MODELING

Chapter Objectives

After completing this chapter you should:

1. be able to create planar surfaces bounded by straight edges using *3Dface*;

2. be able to edit *3Dfaces* with grips and create invisible edges between *3Dfaces* using *Edge*;

3. be able to create meshed surfaces bounded by straight sides using *3Dmesh*;

4. be able to create geometrically defined meshed surfaces using *Rulesurf*, *Tabsurf*, *Revsurf*, and *Edgesurf*;

5. be able to edit polygon meshes (*3Dmesh*, *Rulesurf*, *Tabsurf*, *Revsurf*, and *Edgesurf*) using *Pedit*, *Explode*, and Grips;

6. be able to create *Regions* for use with surface models;

7. be able to use surface modeling primitives (*3D Objects*) to aid in construction of complex surface models;

8. be able to use *Thickness* and *Elevation* to create surfaces from 2D draw commands.

CONCEPTS

A surface model is more sophisticated than a wireframe model because it contains the description of surfaces as well as the description of edges. Surface models can contain descriptions of complex curves, whereas surfaces on a wireframe are assumed to be planar or single curved. Surface models are superior to wireframe models in that they provide better visualization cues. The surface information describes how the surfaces appear and gives the model a "solid" look, since surfaces that are nearer to the observer naturally obscure the surfaces and edges that are behind.

Generally, surface models do not describe physical objects as completely and as accurately as solid models. A surface model can be compared to a cardboard box—it has surfaces but is hollow inside. If a surface model is sectioned (cut in half), there is only air inside, whereas a solid model can be sectioned to reveal its interior features. Therefore, surface models have volume, whereas solid models have volume and mass. Curved surfaces are defined by discrete meshes composed of straight edges; therefore, small "gaps" may exist where a curved surface is adjacent to a planar surface.

Surface modeling is similar to wireframe and solid modeling in that 3D coordinate values must be used to create and edit the geometry. However, the construction and editing process of surface modeling is complex and somewhat tedious compared to wireframe and solid modeling. Each surface must be defined by describing the edges that bound the surface. Each surface must be constructed individually in location or moved into location with respect to other surfaces that comprise the 3D model.

Surface modeling is the type of 3D modeling that is best suited for efficiently defining complex curved shapes such as automobile bodies, aircraft fuselages, and ship hulls. Thus, surface modeling is a necessity for 3D modeling in many industries.

The surface modeling capabilities of AutoCAD Release 14 provide solutions for only a small portion of the possible applications. AutoCAD Release 14 does not utilize NURBS (Non-Uniform Rational B-Spline) surfacing techniques, which is the current preferred surface modeling technology. NURBS technology is utilized, however, in Autodesk's AutoSurf™ and Mechanical Desktop™ product line.

The categories of surfaces and commands that AutoCAD Release 14 provides to create surface models are as follows:

Surfaces with straight edges (usually planar)

> *3Dface*   A surface defined by 3 or 4 straight edges

Meshed surfaces (polygon meshes)

> *3Dmesh*   A planar, curved, or complex surface defined by a mesh
> *Pface*   A surface with any number of vertices and faces

Geometrically defined meshed surfaces (polygon meshes)

> *Edgesurf*   A surface defined by "patching" together 4 straight or curved edges
> *Rulesurf*   A surface created between 2 straight or curved edges
> *Revsurf*   A surface revolved from any 2D shape about an axis
> *Tabsurf*   A surface created by sweeping a 2D shape in the direction specified by a vector

Surface model primitives

> *3D Objects...*   A menu of complete 3D primitive surface models (box, cone, wedge, sphere, etc.) is available. These simple 3D shapes can be used as a basis for construction of more complex shapes.

Thickness, Solid

> AutoCAD's early methods for creation of planar "extrusions" of 2D shapes are discussed briefly.

Regions can be used to simplify construction of complex planar surfaces for use with surface models. The region modeler utilizes Boolean operations, which makes creation of surfaces with holes and/or complex outlines relatively easy.

A visualization enhancement for surfaces is provided by the *Hide* command. Normally, during construction, surfaces are displayed by default (like solids) in <u>wireframe</u> representation. *Hide* causes a regeneration of the display showing surfaces as <u>opaque</u>, therefore obscuring other surfaces or objects behind them.

The surface modeling commands can be accessed easily from the *Draw* pull-down menu (Fig. 40-1). A *Surfaces* toolbar can also be invoked (Fig. 40-2) by selecting *Toolbars...* from the *View* pull-down menu.

The surfacing commands are discussed first in this chapter. Then, application of the surfacing commands to create a complete 3D model is discussed.

Figure 40-1

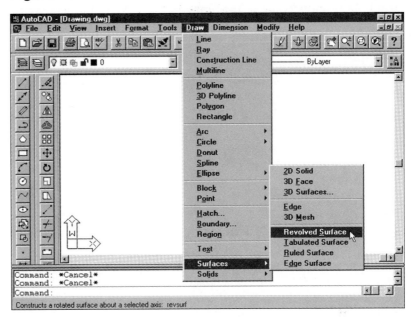

Figure 40-2

SURFACES WITH STRAIGHT EDGES

3DFACE

Pull-down Menu	COMMAND (TYPE)	ALIAS (TYPE)	Short-cut	Screen (side) Menu	Tablet Menu
Draw *Surfaces >* *3D Face*	*3DFACE*	*3F*	...	DRAW 2 SURFACES *3Dface*	*M,8*

3Dface creates a surface bounded by three or four straight edges. The three- or four-sided surface can be connected to other *3Dfaces* within the same *3Dface* command sequence, similar to the way several *Line* segments created in one command sequence are connected. Beware, there is <u>no *Undo*</u> option for *3Dface*. The command sequence is:

    Command: *3dface*
    First Point: **PICK** or (**coordinates**)
    Second point: **PICK** or (**coordinates**)
    Third point: **PICK** or (**coordinates**)
    Fourth point: **PICK** or (**coordinates**)
    Third point: **PICK** or (**coordinates**)
    Fourth point: **PICK** or (**coordinates**)
    Third point: **Enter** (Indicates completion of the *3Dface* sequence.)

After completing four points, AutoCAD connects the next point ("Third point:") to the previous fourth point. The next fourth point is connected to the previous third point, and so on. Figure 40-3 shows the sequence for attaching several *3Dface* segments in one command sequence.

In the example, each edge of the *3Dface* is <u>coplanar</u>, as is the entire sequence of *3Dfaces* (the Z value of every point on each edge is 0). The edges must be <u>straight</u> but not necessarily coplanar. Entering specific coordinate values, using point filters, or using *OSNAP* in 3D space allows you to create geometry with *3Dface* that is not on a plane. The following command sequence creates a *3Dface* as a complex curve by entering a different Z value for the fourth point (Fig. 40-4):

> Command: **3dface**
> First Point: **PICK**
> Second point: **PICK**
> Third point: **PICK**
> Fourth point: **.XY**
> of **PICK**
> (need Z): **2**
> Third point: **Enter**
> Command:

Although it is possible to use *3Dface* to create complex curved surfaces, the surface curve is not visible as it would be if a mesh were used. A *3Dmesh* provides superior visibility and flexibility and is recommended for such a surface.

Figure 40-5 displays a possibility for creating nonplanar geometry in one command sequence. .XY filters can be used to PICK points, or absolute X,Y,Z values can be entered.

Figure 40-3

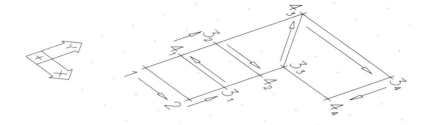

Figure 40-4

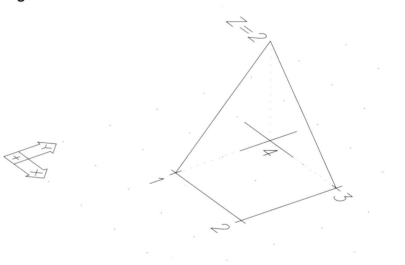

Figure 40-5

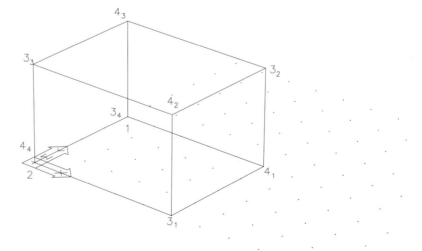

The lines between *3Dface* segments in Figure 40-3 are visible even though the segments are coplanar. AutoCAD provides two methods for making invisible edges. The first method requires that you enter the letter *"I"* immediately before the <u>first</u> of the two points that define an invisible edge. For the shape shown in Figure 40-6, enter the letter *"I"* <u>before</u> points 3_1, 3_2, and 3_3. The second, and much easier, method for defining invisible edges of *3Dfaces* is to use the *Edge* command.

EDGE

Pull-down Menu	COMMAND (TYPE)	ALIAS (TYPE)	Short-cut	Screen (side) Menu	Tablet Menu
Draw *Surfaces >* *Edge*	*EDGE*	...	...	DRAW 2 SURFACES *Edge:*	...

The first method for creating invisible edges described above takes careful planning. The same action shown in Figure 40-6 can be accomplished much more easily and <u>retroactively</u> by this second method. AutoCAD provides the *Edge* command to create invisible edges after construction of the *3Dface*. The command prompt is:

Figure 40-6

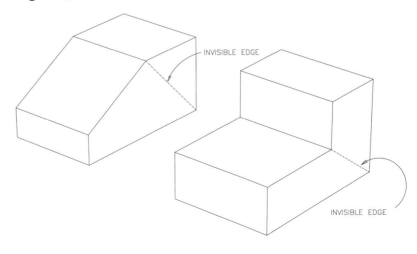

Command: **edge**
Display<select edge>: **PICK**

Selecting edges converts visible edges to invisible edges. <u>Any</u> edges on the *3Dface* can be made invisible by this method. In Release 14, the *Midpoint OSNAP* option is automatically invoked to simplify selection of the edges (Fig. 40-6). The *Display* option causes selected invisible edges to become visible again.

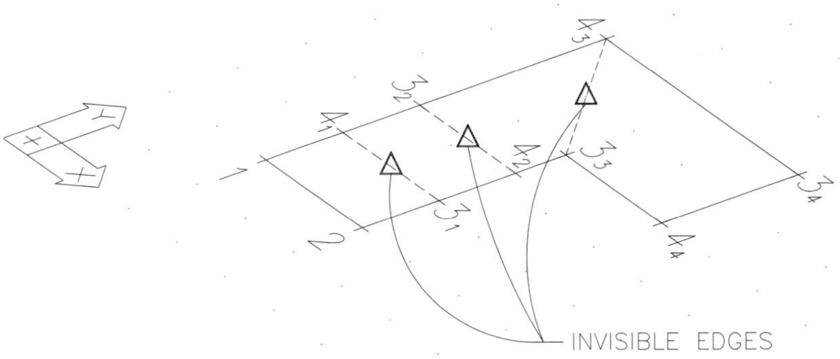

Applications of *3Dface*

3Dface is used to construct 3D surface models by creating individual (planar) faces or connected faces. For relatively simple models, the faces are placed in space as they are constructed. A construction strategy to use for more complex models is to first construct a wireframe model and then attach surfaces (*3Dfaces* or other surfaces) to the wireframe. The wireframe geometry can be constructed on a separate layer; then the layer can be turned off after the surfacing is complete.

Figure 40-7

Figures 40-7 and 40-8 show some applications of *3Dface* for construction of surface models. Remember, *3Dface* can be used <u>only</u> for construction of surfaces with straight edges.

The edges made invisible by *Edge* are displayed in these figures in a hidden linetype. *Hide* has also been used to enhance visualization. You may want to use *Hide* periodically to enhance the visibility of surfaces.

Figure 40-8

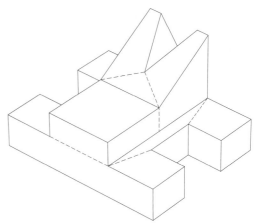

Editing *3Dfaces*

Grips can be used to edit *3Dfaces*. Each *3Dface* has three or four grips (one for each corner). Activating grips (PICKing the *3Dface* at the open Command: prompt), makes the individual *3Dface* surface become highlighted and display its **warm** grips. If you then make one of the three or four grips **hot**, all of the normal Grip editing options are available, that is, STRETCH, MOVE, ROTATE, SCALE, and MIRROR. The single activated *3Dface* surface can then be edited. MOVE, for example, could be used to move the single activated surface to another location.

Figure 40-9

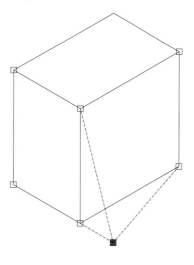

The STRETCH option, however, is generally the most useful Grip editing option, since all of the other options change the entire surface. The STRETCH option allows you to change one (or more) corner(s) of the *3Dface* rather than the entire surface, as with the other options. For example, Figure 40-9 shows six *3Dfaces* connected to form a box. One corner of the box could be relocated by activating the grips on two of the *3Dfaces*. Two grips (common to one intersection) can be made **hot** simultaneously by holding down SHIFT while PICKing the **hot** grip. When the STRETCH option appears at the command line, stretch the common **hot** grips to the new location, as shown (highlighted) in the figure.

Explode or *Pedit* <u>cannot</u> be used to edit *3Dfaces*.

Figure 40-10

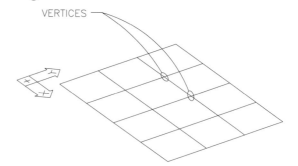

CREATING MESHED SURFACES

Meshes are versatile methods for generating a surface. *Ai_mesh* can be used to create a simple planar surface bounded by <u>four</u> straight edges (Fig. 40-10), or *3Dmesh* can be used to create a complex surface defining an irregular shape bounded by <u>four</u> sides (Fig. 40-11). The surface created with *Ai_mesh* or *3Dmesh* differs from a *3Dface* because the surface is defined by a "mesh." A mesh is a series of vertices (sometimes called "nodes") arranged in rows and columns connected by lines. *Pedit* or grips can be used to edit the resulting mesh (see Editing Polygon Meshes, this chapter).

Figure 40-11

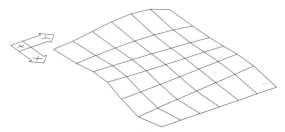

AI_MESH

Pull-down Menu	COMMAND (TYPE)	ALIAS (TYPE)	Short-cut	Screen (side) Menu	Tablet Menu
Draw Surfaces > 3D Surfaces... Mesh	*AI_MESH*	...	...	*DRAW 2 SURFACES 3Dobjec: Mesh*	...

Although no icon button is provided in AutoCAD Release 14 for this command, you can invoke *Ai_mesh* through the *3D Objects* dialog box (see Surface Modeling Primitives) or by typing. *Ai_mesh* can be used to create a simple planar surface bounded by <u>four</u> straight edges (Fig. 40-10). Using this command causes:

Command: **ai_mesh**
First corner: **PICK** or (**coordinates**)
Second corner: **PICK** or (**coordinates**)
Third corner: **PICK** or (**coordinates**)
Fourth corner: **PICK** or (**coordinates**)
Mesh *M* size: (**value**) (Enter a value for the number of vertices in the *M* direction.)
Mesh *N* size: (**value**) (Enter a value for the number of vertices in the *N* direction.)
Command:

Figure 40-12 ——————————————

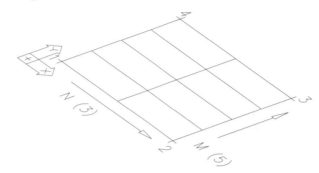

First, specify the four corners of the mesh. The number of vertices along each side is defined by the *M* size and the *N* size. The *M* direction is <u>perpendicular</u> to the first side specified, and the *N* direction is <u>perpendicular</u> to the second side specified (Fig. 40-12).

Figure 40-13 ——————————————

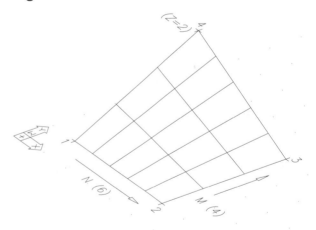

The values specified for the *M* size and *N* size <u>include</u> the vertices along the edges of the surface. The *M* and *N* sizes are the numbers of rows and columns of vertices (density of the mesh).

Ai_mesh can quickly generate a complex curved surface with straight edges by specifying nonplanar coordinates for the corners. The following command sequence is used to generate the surface shown in Figure 40-13:

Command: **ai_mesh**
First corner: **1,1,0**
Second corner: **5,1,0**
Third corner: **5,5,0**
Fourth corner: **1,5,2**
Mesh *M* size: **4**
Mesh *N* size: **6**
Command:

Figure 40-14 ——————————————

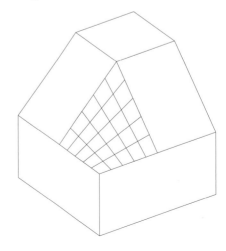

Ai_mesh visually defines the curvature of a surface better than *3Dface* because of the mesh lines.

Figure 40-14 illustrates an application for a nonplanar *Ai_mesh*.

3DMESH

Pull-down Menu	COMMAND (TYPE)	ALIAS (TYPE)	Short-cut	Screen (side) Menu	Tablet Menu
Draw Surfaces > 3D Mesh	*3DMESH*	...	...	DRAW 2 SURFACES 3Dmesh:	...

3Dmesh is a versatile tool for generating a complex surface. It is used to create a complex surface defining an irregular shape bounded by <u>four</u> sides (Fig. 40-11). *Pedit* or grips can be used to edit the *3Dmesh* retroactively.

3Dmesh prompts for coordinate values for the placement of <u>each</u> vertex, allowing you to create a variety of shapes. A numbering scheme is used to label each vertex. The first vertex in the <u>first</u> row in the M direction is labeled *0,0*, the second in the first row is labeled *0,1*, and so on. The first vertex in the <u>second</u> row is labeled *1,0*, the second in that row is *1,1*, and so on.

Since each vertex can be given explicit coordinates, a surface can be created to define any shape. An example is shown in Figure 40-15. This command is useful for developing geographical topographies. The command syntax for this method of defining a *3Dmesh* is as follows:

Command: **3dmesh**
Mesh M size: (**value**) (Enter a value for the number of vertices in the *M* direction.)
Mesh N size: (**value**) (Enter a value for the number of vertices in the *N* direction.)
Vertex (0,0): **PICK** or (**coordinates**)
Vertex (0,1): **PICK** or (**coordinates**)
Vertex (0,2): **PICK** or (**coordinates**)
(continues sequence for all vertices in the row)
Vertex (1,0): **PICK** or (**coordinates**)
Vertex (1,1): **PICK** or (**coordinates**)
Vertex (1,2): **PICK** or (**coordinates**)
(continues sequence for all vertices in the row)
Vertex (2,0): **PICK** or (**coordinates**)
Vertex (2,1): **PICK** or (**coordinates**)
Vertex (2,2): **PICK** or (**coordinates**)
(continues sequence for all vertices in the row)
(sequence continues for all rows)
Command:

Figure 40-15 ——————————————

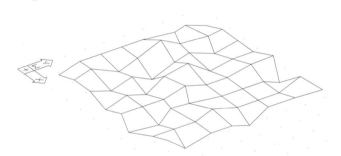

Obviously, using *3Dmesh* is very tedious. For complex shapes such as topographical maps, an AutoLISP program can be written to read coordinate data from an external file to generate the mesh.

PFACE

Pull-down Menu	COMMAND (TYPE)	ALIAS (TYPE)	Short-cut	Screen (side) Menu	Tablet Menu
...	*PFACE*	...	...	...	...

The *Pface* (polyface mesh) command can be used to create a polygon mesh of virtually any topology. *Pface* is the most versatile of any of the 3D mesh commands. It allows creation of any shape, any number of vertices, and even objects that are not physically connected, yet are considered the same object.

The *Pface* command, however, is designed primarily for use by applications that run with AutoCAD and for use by other software developers. *Pedit* <u>cannot</u> be used to edit *Pface*s.

The procedure for creation of geometry with *Pface* is somewhat like the previous procedure for *3Dmesh*. There are two basic steps to *Pface*. AutoCAD first prompts for the coordinate location for each vertex. Any number of vertices can be defined.

> Command: **Pface**
> Vertex 1: (**coordinates**)
> Vertex 2: (**coordinates**)
> (The sequence continues until pressing **Enter**.)

Secondly, AutoCAD prompts for the vertices to be assigned to each face. A face can be attached to any set of vertices.

> Face 1, vertex 1: (**number**) (Enter number of vertex for face to be attached.)
> Face 1, vertex 2: (**number**) (Press **Enter** to complete face.)
> Face 2, vertex 1: (**number**)
> Face 2, vertex 2: (**number**) (Press **Enter** twice to complete command.)
> Command:

Specifying a *Pface* mesh of any size can be tedious. The geometrically defined mesh commands (*Rulesurf, Tabsurf, Revsurf,* and *Edgesurf*) are more convenient to use for most applications.

CREATING GEOMETRICALLY DEFINED MESHES

Geometrically defined meshes are meshed surfaces that are created by "attaching" a surface to existing geometry. In other words, geometrically defined meshed surfaces <u>require existing geometry</u> to define them. *Edgesurf, Rulesurf, Tabsurf,* and *Revsurf* are the commands that create these meshes. The geometry used may differ for the application but always consists of either two or four objects (*Lines, Circles, Arcs,* or *Plines*).

Controlling the Mesh Density

Geometrically defined mesh commands (*Edgesurf, Rulesurf, Revsurf,* and *Tabsurf*) do not prompt the user for the number of vertices in the *M* and *N* direction. Instead, the number of vertices is determined by the settings of the *SURFTAB1* and *SURFTAB2* variables. The number of vertices includes the endpoints of the edge:

> Command: **surftab1**
> New value for SURFTAB1 <6>: (**value**)
> Command:

SURFTAB1 and *SURFTAB2* must be set <u>before</u> using *Edgesurf, Rulesurf, Revsurf,* and *Tabsurf*. These variables are <u>not</u> retroactive for previously created surface meshes.

The individual surfaces (1 by 1 meshes between vertices) created by geometrically defined meshes are composed of a series of <u>straight</u> edges connecting the vertices. These edges do not curve but can only change direction between vertices to approximate curved edges. The higher the settings of *SURFTAB1* and *SURFTAB2*, the more closely these straight edges match the defining curved edges.

Note that the defining edges of geometrically defined mesh <u>do not become part of the surface</u>; they remain separate objects. Therefore, you can construct surface models by utilizing existing wireframe model objects as the defining edges. If the wireframe model is on a separate layer, that layer can be *Frozen* after constructing the surfaces to reveal only the surfaces.

EDGESURF

Pull-down Menu	COMMAND (TYPE)	ALIAS (TYPE)	Short-cut	Screen (side) Menu	Tablet Menu
Draw Surfaces > Edge Surface	*EDGESURF*	...	...	DRAW 2 SURFACES *Edgsurf:*	R,8

An *Edgesurf* is a meshed surface generated between <u>four existing</u> edges. The four edges can be of any shape as long as they have connecting endpoints (no gaps, no overlaps). The four edges can be *Lines*, *Arcs*, or *Plines*. The four edges can be selected in any order. The surface that is generated between the edges is sometimes called a "Coon's surface patch." *Edgesurf* interpolates the edges and generates a smooth transitional mesh, or patch, between the four shapes.

The command syntax is as follows:

```
Command: edgesurf
Select edge 1: PICK
Select edge 2: PICK
Select edge 3: PICK
Select edge 4: PICK
Command:
```

Figure 40-16 ————————————————

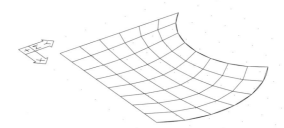

Figure 40-16 displays an *Edgesurf* generated from four planar edges. Settings for *SURFTAB1* and *SURFTAB2* are 6 and 8.

The relation of *SURFTAB1* and *SURFTAB2* for any *Edgesurf* are given here:

SURFTAB1	first edge picked
SURFTAB2	second edge picked

Edges used to define an *Edgesurf* may be <u>nonplanar</u>, resulting in a smooth, nonplanar surface, as displayed in Figure 40-17. Remember, the edges may be any *Pline* shape in 3D space. Use *Hide* to enhance your visibility of nonplanar surfaces.

Figure 40-17 ————————————————

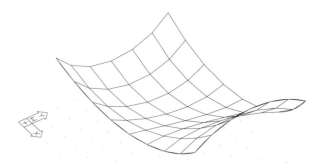

RULESURF

	COMMAND (TYPE)	ALIAS (TYPE)	Short-cut	Screen (side) Menu	Tablet Menu
Pull-down Menu					
Draw Surfaces > Ruled Surface	*RULESURF*	...	...	*DRAW 2 SURFACES Rulsurf:*	*Q,8*

A *Rulesurf* is a polygon meshed surface created between <u>two edges</u>. The edges can be *Lines, Arcs,* or *Plines*. The command syntax only asks for the two "defining curves." The two defining edges must <u>both</u> be open or closed.

Figure 40-18 displays two planar *Rulesurfs*.

Figure 40-18

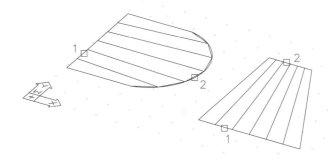

Figure 40-19 displays possibilities for creating *Rulesurfs* between two edges that are nonplanar.

The command syntax for *Rulesurf* is as follows:

    Command: **rulesurf**
    Select first defining curve: **PICK**
    Select second defining curve: **PICK**
    Command:

Figure 40-19

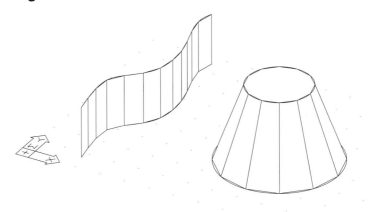

A cylinder can be created with *Rulesurf* by utilizing two *Circles* lying on different planes as the defining curves. A cone can be created by creating a *Rulesurf* between a *Circle* and a *Point* (Fig. 40-20), or a complex shape can be created using two identical closed *Pline* shapes (Fig. 40-20).

Figure 40-20

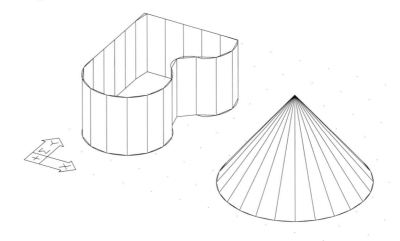

For open defining edges (not *Circles* or closed *Plines*), the *Rulesurf* is generated connecting the two <u>selected</u> ends of the defining edges. Figure 40-21 illustrates the two possibilities for generating a *Rulesurf* between the same two straight nonplanar *Lines* based on the endpoints selected.

Since the curve is stretched between only two edges, the number of vertices along the *defining curve* is determined only by the value previously specified for *SURFTAB1*.

Figure 40-21 ————————————

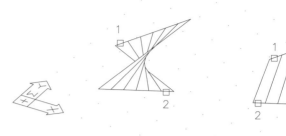

SURFTAB1	defining curve
SURFTAB2	not used

TABSURF

Pull-down Menu	COMMAND (TYPE)	ALIAS (TYPE)	Short-cut	Screen (side) Menu	Tablet Menu
Draw Surfaces > Tabulated Surface	*TABSURF*	...	...	DRAW 2 SURFACES *Tabsurf:*	P,8

A *Tabsurf* is generated by an existing *path curve* and a *direction vector*. The *curve path* (generatrix) is extruded in the direction of, and equal in length to, the *direction vector* (directrix) to create a swept surface. The path curve can be a *Line, Arc, Circle, 2D Polyline,* or *3D Polyline*. The direction vector can be a *Line* or an open 2D *Pline*. If a curved *Pline* is used, the surface is generated in a direction connecting the two end vertices of the *Pline*.

A *Circle* can be swept with *Tabsurf* to generate a cylinder (Fig. 40-22). The *Tabsurf* has only N direction; therefore, the current setting of *SURFTAB1* controls the number of vertices along the *path curve*.

The relation of *SURFTAB1* and *SURFTAB2* for *Tabsurf* are:

SURFTAB1	Path curve
SURFTAB2	not used

The *Tabsurf* is generated in the direction <u>away from</u> the end of the direction vector that is selected. This is evident in both Figures 40-22 and 40-23.

The command sequence is:

    Command: **tabsurf**
    Select path curve: **PICK**
    Select direction vector: **PICK**

Figure 40-22 ————————————

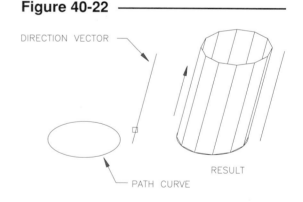

Figure 40-23 ————————————

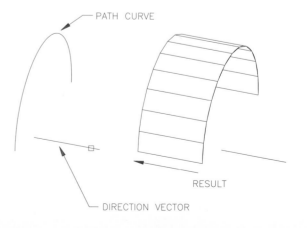

REVSURF

Pull-down Menu	COMMAND (TYPE)	ALIAS (TYPE)	Short-cut	Screen (side) Menu	Tablet Menu
Draw Surfaces > Revolved Surface	*REVSURF*	...	...	DRAW 2 SURFACES *Revsurf:*	O,8

This command creates a surface of revolution by revolving an existing *path curve* around an *axis*. The path curve can be a *Line, Arc, Circle*, 2D or 3D *Pline*. The *axis* can be a *Line* or open *Pline*. If a curved *Pline* is used as the *axis*, only the endpoint vertices are considered for the axis of revolution.

```
Command: revsurf
Select path curve: PICK
Select axis of revolution: PICK
Start angle<0>: PICK or (value)
Included angle (+=ccw, -=cw)<Full circle>: PICK or (value)
Command:
```

The structure of the command allows many variations. The *path curve* can be <u>open or closed</u>. The *axis* can be in any plane. The *start angle* and *included angle* allow for complete (closed) or partial (open) revolutions of the surface in any orientation. Figures 40-24 and 40-25 show possible *Revsurfs* created with *path curves* generated through 360 degrees.

The wine glass was created with an open *path curve* generated through 360 degrees (Fig. 40-24).

The number of vertices in the direction of revolution is controlled by the setting of *SURFTAB2*. The *SURFTAB1* setting controls the number of vertices along the length of the *path curve*.

The relation of *SURFTAB1* and *SURFTAB2* for *Revsurf* is given here:

SURFTAB1	*Axis of revolution*
SURFTAB2	*Path curve*

A torus can be created by using a closed *path curve* generated through 360 degrees (Fig. 40-25). In this case, the surface begins at the *path curve*, is revolved all the way around the axis, and closes on itself.

When generating a curve path through <u>less than</u> 360 degrees, you must specify the included angle. Entering a positive angle specifies a counterclockwise direction of revolution, and a negative angle specification causes a clockwise revolution. As an alternative, you can pick different points on the *axis of revolution*. The end of the line PICKed represents the end nearest the origin for positive rotation using the right-hand rule.

Figure 40-24 ───────

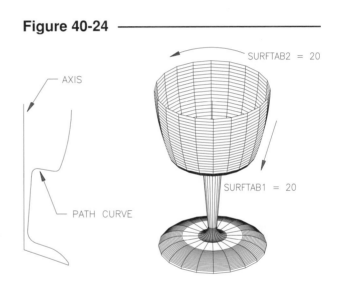

Figure 40-25 ───────

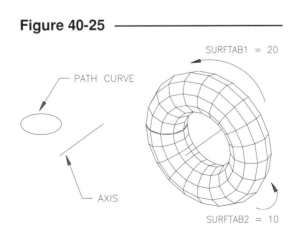

Figure 40-26 displays a *Revsurf* generated through 90 degrees. The end of the *axis of revolution* PICKed specifies the direction of positive rotation.

Figure 40-26 ───────────

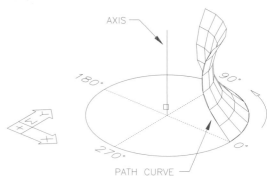

Figure 40-27 displays a *Revsurf* generated in a negative revolution direction, accomplished by specifying a negative angle or by PICKing a point specifying an inverted origin (of the right-hand rule) for rotation.

Figure 40-27 ───────────

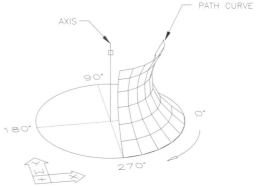

EDITING POLYGON MESHES

There are several ways that existing AutoCAD polygon meshed surfaces (*3Dmesh, Rulesurf, Tabsurf, Revsurf,* and *Edgesurf*) can be changed. The use of *Pedit*, Grips, and *Explode* each provide special capabilities.

PEDIT with Polygon Meshes

One of the options of *Pedit* is *Edit vertex*, which allows editing of individual vertices of a *Pline*. Vertex editing can also be accomplished with *3Dmeshes, Revsurfs, Edgesurfs, Tabsurfs,* and *Rulesurfs*. *Pedit* does not allow editing of *3Dfaces*. See Chapter 16 for information on using *Pedit* with 2D objects.

```
Command: Pedit
Select Polyline: PICK (Select the surface.)
Edit vertex/Smooth surface/Desmooth/Mclose/Nclose/Undo /eXit<X>:
```

Each *Pedit* option for editing Polygon Meshes is explained and illustrated here.

Edit Vertex
Invoking this option causes the display of another set of options at the command prompt, which allows you to locate the vertex that you wish to edit:

```
Vertex(0,0). Next/Previous/Left/Right/Up/Down/Move/ REgen/eXit<N>:
```

Pressing Enter locates the marker (X) at the next vertex. Selecting *Left/Right* or *Up/Down* controls the direction of the movement in the *N* and *M* directions (as specified by *SURFTAB1* and *SURFTAB2* when the surfaces were created; Fig. 40-28).

Figure 40-28

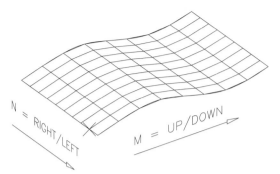

Once the vertex has been located, use the *Move* option to move the vertex to any location in 3D space (Fig. 40-29).

Figure 40-29

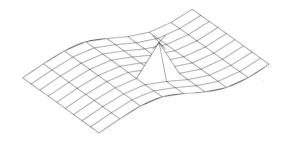

Smooth surface

Invoking this option causes the surface to be smoothed using the current setting of *SURFTYPE* (described later) (Fig. 40-30).

Figure 40-30

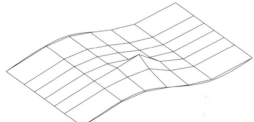

Desmooth

The *Desmooth* option reverses the effect of *Smooth*.

Mclose/Nclose

The *Mclose* and *Nclose* options cause the surface to close in either the *M* or *N* directions. The last and first set of vertices automatically connect. Figure 40-31 displays a half-cylindrical surface closed with *Nclose*.

Figure 40-31

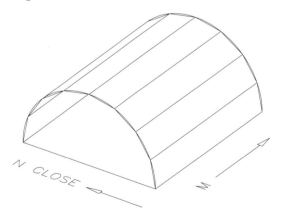

Undo

Undo reverses the last operation with *Pedit*.

eXit

This option is used to keep the editing changes, exit the *Pedit* command, and return to the Command: prompt.

Variables Affecting Polygon Meshes

SURFTYPE

Pull-down Menu	COMMAND (TYPE)	ALIAS (TYPE)	Short-cut	Screen (side) Menu	Tablet Menu
...	*SURFTYPE*	...	...	...	...

This variable affects the degree of smoothing of a surface when using the *Smooth* option of *Pedit*. The command prompt syntax is shown here:

Command: **Surftype**
New value for Surftype<6>: (**value**)

The values allowed are **5**, **6**, and **8**. The options are illustrated in the following figures.

Original polygon mesh (Fig. 40-32)

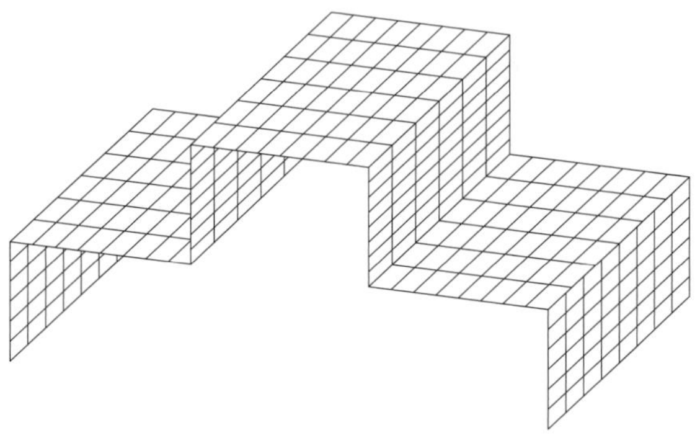

SURFTYPE=5
Quadratic B-spline surface (Fig. 40-33)

SURFTYPE=6
Cubic B-spline surface (Fig. 40-34)

SURFTYPE=8
Bezier surface (Fig. 40-35)

NOTE: Changing *SUFRTYPE* by command line format or using the *Set Spline Fit Variables* dialog box sets the smoothing only for <u>new</u> polygon meshes. If you want to change an <u>existing</u> polygon mesh, use *Ddmodify*, then select *Quadratic*, *Cubic*, or *Bezier*.

Figure 40-32 ————————————————

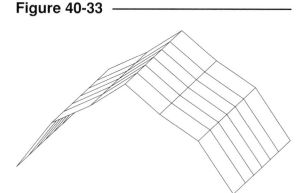

Figure 40-33 ————————————————

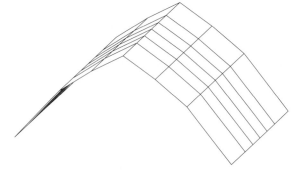

Figure 40-34 ————————————————

Figure 40-35 ————————————————

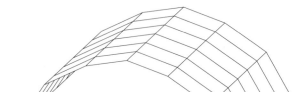

SPLFRAME

	COMMAND (TYPE)	ALIAS (TYPE)	Short-cut	Screen (side) Menu	Tablet Menu
Pull-down Menu					
...	*SPLFRAME*	...	...	...	...

The *SPLFRAME* (spline frame) variable causes the original polygon mesh to be displayed along with the smoothed version of the surface. The values allowed are **0** for off and **1** for on. *SPLFRAME* can also be used with 2D polylines when using the *Spline* option.

SPLFRAME is set to **1** in Figure 40-36 to display the original polygon mesh.

Figure 40-36

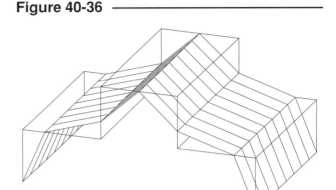

Editing Polygon Meshes with Grips

Grips can be used to edit the individual vertices of a polygon mesh (*3Dmesh*, *Edgesurf*, *Rulesurf*, *Tabsurf*, and *Revsurf*). Normally, the STRETCH option of Grip editing would be used to stretch (move) the location of a single vertex grip. The other Grip editing options (MOVE, SCALE, ROTATE, MIRROR) affect the entire polygon mesh. Selecting the mesh at the open Command: prompt activates all of the grips (one for each vertex) and makes them **warm**. Selecting one of the individual grips again makes that one **hot** and able to be relocated with the STRETCH option.

For example, a *3Dmesh* could be generated to define a surface with a rough texture, such as a chipped stone (Fig. 40-37).

Grip editing STRETCH option could be used to relocate individual vertices of the mesh. The result is a surface with a rough or bumpy surface (Fig. 40-38).

Although the concept is simple, some care should be taken to assure that the individual vertices are moved in the desired plane. Since Grip editing is interactive, any points PICKed are moved to the <u>current XY plane</u>. Careful use of UCSs can ensure that the vertices are STRETCHed in the desired XY plane.

Figure 40-37

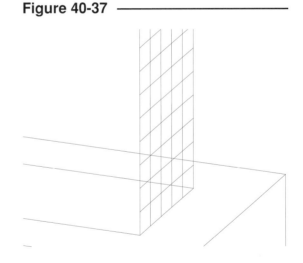

Figure 40-38

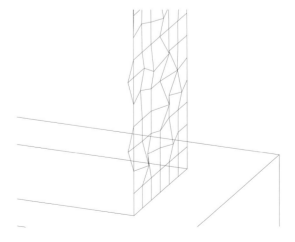

Editing Polygon Meshes with *Explode*

The *Explode* command can be used to allow editing 3Dmeshes, geometrically defined meshes (*Edgesurf, Revsurf*, etc.), 3D primitives (*3D Objects...*), and *Regions*.

Figure 40-39

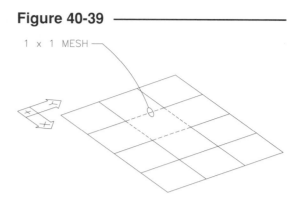

1 x 1 MESH

When used with other AutoCAD objects, *Explode* will "break down" the object group into a lower level of objects; for example, a *Pline* can be broken into its individual segments with *Explode*. Likewise, *Explode* will "break down" a 3D polygon mesh, for example, into individual 1 by 1 meshes (in effect, *3Dfaces*). *Explode* will affect any polygon meshes (*3Dmesh, Edgesurf, Revsurf, Tabsurf, Rulesurf*). *Explode* converts a polygon mesh into its individual 1 by 1 meshes (*3Dfaces*) (Fig. 40-39).

CONSTRUCTION TECHNIQUES FOR POLYGON MESHES AND *3DFACES*

Surfaces are created individually and combined to form a complete surface model. The surfaces can be created and moved into place in order to "assemble" the complete model, or surfaces can be created <u>in place</u> to form the finished model. Because polygon meshes require existing geometry to define the surfaces, the construction technique generally used is:

1.   Construct a wireframe model (or partial wireframe) defining the edges of the surfaces on a separate layer.
2.   Attach polygon mesh and other surfaces to the wireframe using another layer.

The creation of UCSs greatly assists in constructing 3D elements in 3D space during construction of the wireframe and surfaces. The sequence in the creation of a complete surface model is given here as a sample surface modeling strategy.

Surface Model Example 1

Consider the object in Figure 40-40. To create a surface model of this object, first create a partial wireframe.

Figure 40-40

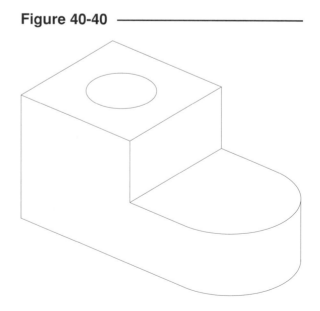

1. Create part of a wireframe model using *Line* and *Circle* elements. Create the wireframe on a separate layer. Begin the geometry at the origin. Use *Vpoint* to enable visualization of three dimensions (Fig. 40-41).

Figure 40-41

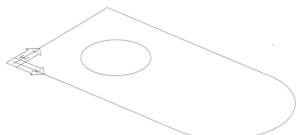

2. *Copy* the usable objects to the top plane by specifying a Z dimension at the "second point of displacement" prompt (Fig. 40-42).

Figure 40-42

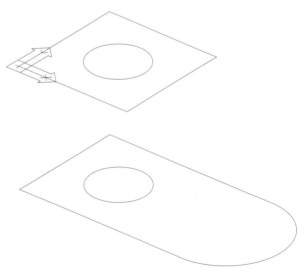

3. Create a UCS on this new plane using the *Origin* option. *OSNAP* the *Origin* to the existing geometry. Draw the connecting *Line* and *Trim* the extensions (Fig. 40-43).

Figure 40-43

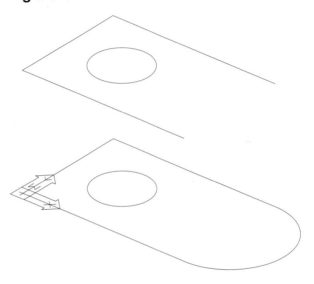

4. *Copy* the *Arc* and two attached *Lines* specifying a Z dimension for the "second point of displacement." *Copy* the *Line* drawn last (on the top plane) specifying a negative Z dimension. *Trim* the extending *Line* ends. A new *UCS* is not necessary on this plane (Fig. 40-44).

    This is not a complete wireframe, but it includes all the edges that are necessary to begin attaching surfaces.

Figure 40-44

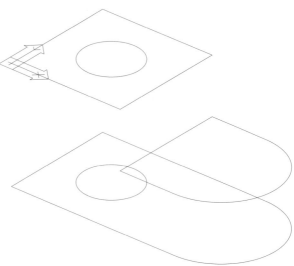

5. On a new layer, a 2-segment *3Dface* is created by PICKing the *Endpoints* of the wireframe objects in the order indicated. Remember that *3Dface* segments connect at the third and fourth points (Fig. 40-45). (*Hide* is used in this figure to provide visibility of the *3Dface*.)

Figure 40-45

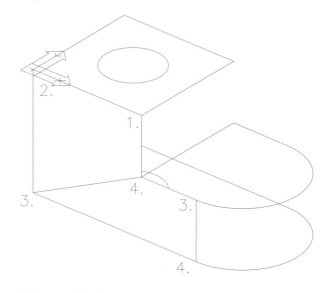

6. The *3Dface* is *Copied* to the opposite side of the model. A vertical *3Dface* is created between the top and the intermediate plane (Fig. 40-46). Another vertical *3Dface* is created on the back (not visible). (*Hide* is used again in this figure. Notice how the *Circle* is treated.)

Figure 40-46

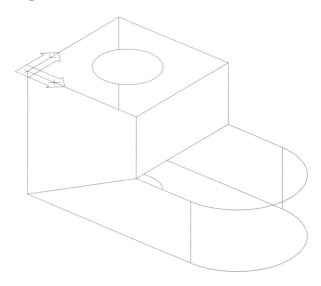

7. *Rulesurf* is used to create the rounded surface shown between the two *Arcs* (Fig. 40-47). The *SURFTAB1* and *SURFTAB2* variables were previously set to 12.

Figure 40-47

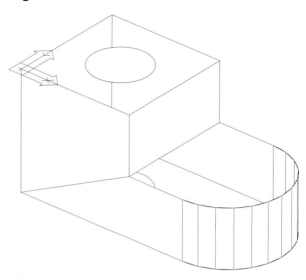

8. Next, a *Rulesurf* is used to create the cylindrical hole surface between the *Circles* on the top and bottom planes. The wireframe layer was *Frozen* before *Hide* was used for this figure (Fig. 40-48).

Figure 40-48

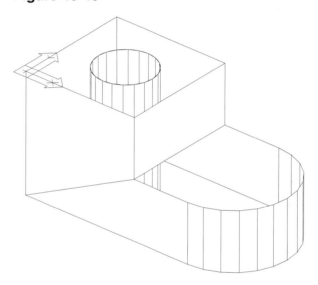

9. The surface connecting the *Arc* and three *Lines* is created by *Edgesurf* (Fig. 40-49).

10. The only remaining surfaces to create are the top and bottom. <u>There is no</u> polygon mesh, *3Dface*, or *3Dmesh* that can be generated to create a simple surface configuration such as that on top—a surface with a hole! It must be accomplished by creating several adjoining surfaces.

Figure 40-49

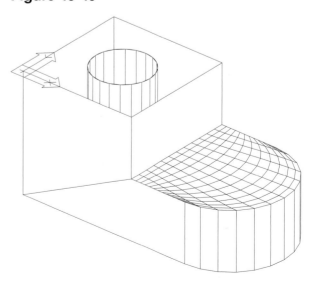

Several methods are possible for construction of a surface with a hole. (The top surface has been isolated here for simplicity.) Probably the simplest method is shown here (Fig. 40-50). First, the *Circle* must be <u>replaced</u> by two 180 degree *Arcs*. Next a *Rulesurf* is created between one *Arc* and one edge as shown.

Figure 40-50 ───────────

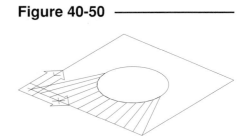

11. The top plane is completed by creating another *Rulesurf* and two triangular *3Dfaces* adjoining the *Rulesurfs* (Fig. 40-51). (Wireframe elements are <u>not</u> required for the *3Dface*.)

Figure 40-51 ───────────

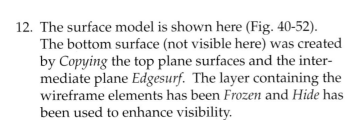

12. The surface model is shown here (Fig. 40-52). The bottom surface (not visible here) was created by *Copying* the top plane surfaces and the intermediate plane *Edgesurf*. The layer containing the wireframe elements has been *Frozen* and *Hide* has been used to enhance visibility.

Figure 40-52 ───────────

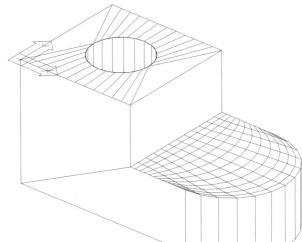

13. The mesh lines shown in Figure 40-53 are not visible when a *Render* is performed. This figure illustrates the completed model after using *Render*. *Shade* also provides options to display or to hide the edge and mesh lines.

Figure 40-53 ───────────

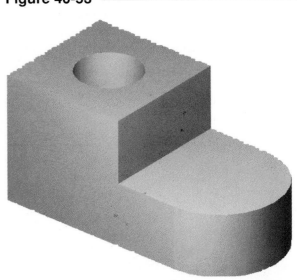

Limitations of Polygon Meshes and *3Dface*

From the previous example of the creation of a relatively simple 3D surface model, you can imagine how construction of some shapes with *3Dface* or polygon meshes can be somewhat involved. The particular characteristics of a surface that cause some difficulty in creation are the surfaces that embody many edges or that have holes, slots, or other "islands." Examine the surfaces shown in Figure 40-54. Although some of these shapes are fairly common, the construction techniques using *3Dfaces* and polygon meshes could be very involved.

3Dface and polygon meshes would <u>not</u> be an efficient construction technique to use in the creation of shapes such as these. Another AutoCAD feature, Region Modeling, can be used to easily create such shapes. Region models can be combined with *3Dfaces*, polygon meshes, and other surfacing techniques to create surface models.

Figure 40-54

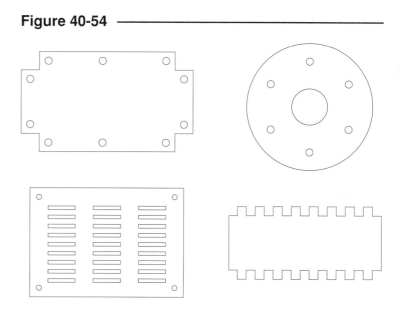

REGION MODELING APPLICATIONS

Region modeling is sometimes defined in the AutoCAD documentation as "2D solid modeling," <u>not</u> "surface modeling." This is because region modeling applies solid modeling techniques, namely, Boolean operations, to 2D closed objects. Practically, however, *Regions* can be considered surfaces.

A region is created by applying the *Region* command to <u>closed</u> 2D objects such as *Plines, Circles, Ellipses,* and *Polygons* (see Chapters 15 and 16). *Regions* are always <u>planar</u>. <u>Composite</u> regions can be constructed by combining multiple *Regions* using Boolean operations. AutoCAD allows you to utilize the Boolean operations *Union, Subtract,* and *Intersect* with *Regions* to create planar surfaces with holes and complex shapes. These surfaces can then be combined with other surfaces (regions, *3Dfaces,* or polygon meshes) on other planes to create complex surface models.

Region modeling solves the problem of creating relatively simple surfaces that are not easily accomplished with the polygon mesh commands such as *3Dmesh, Rulesurf, Revsurf, Tabsurf,* and *Edgesurf.* Region modeling makes creation of surfaces with islands (holes) especially easy.

For example, consider the creation of a rectangular surface with a circular hole. This problem is addressed in the previous Surface Model Example 1 (Fig. 40-51). Using the polygon mesh commands, this relatively simple surface requires four *Lines,* two *Arcs* as a wireframe structure, then two *Rulesurfs* and two *3Dfaces* to complete the surface.

With the Region Modeler, the same surface can be created by using *Subtract* with a circular *Region* (create a *Circle*, then use *Region*) and a rectangular *Region* (draw a closed rectangle with *Line*, *Pline*, or *Rectangle*, then use *Region*; Fig. 40-55).

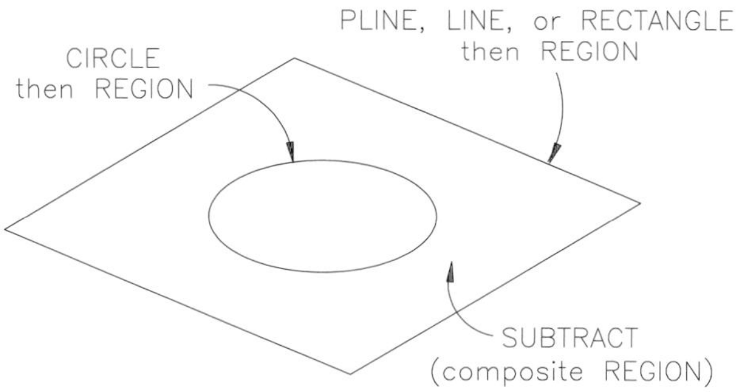

Consider the surfaces shown in Figure 40-54. Imagine the amount of work involved in creating the surfaces using commands such as *3Dface*, *3Dmesh*, *Rulesurf*, *Revsurf*, *Tabsurf*, and *Edgesurf*. Next, consider the work involved to create these shapes as regions. Surfaces such as these are very common for mechanical applications for surface models.

Figure 40-55

CIRCLE then REGION

PLINE, LINE, or RECTANGLE then REGION

SUBTRACT (composite REGION)

SURFACE MODELING PRIMITIVES

3D SURFACES...

Pull-down Menu	COMMAND (TYPE)	ALIAS (TYPE)	Short-cut	Screen (side) Menu	Tablet Menu
Draw Surfaces > 3D Surfaces...	*AI_BOX, AI_WEDGE, etc.*	...	...	*DRAW 2 SURFACES 3Dobjec:*	...

This command allows you to create 3D surface <u>primitives</u>. A primitive is a simple 3D geometric shape such as a *box, cone, dish, dome, mesh, pyramid, sphere, torus,* or *wedge* (the mesh is discussed earlier). These primitives are pre-created shapes composed of *3Dfaces* or polygon meshes. The primitives in the *3D Objects* icon menu are surface models (Fig. 40-56).

Surface primitives can be created by selecting the image tiles in the *3D Objects* dialog box, by selecting from the *SURFACES* screen menu, or by typing the object name prefaced by "AI_"; for example "*AI_BOX*" summons the prompts for constructing a box.

AutoCAD prompts you for the dimensions of the primitive. Depending on the type of primitive chosen, the prompts differ. The prompts for a *box* are given here:

Figure 40-56

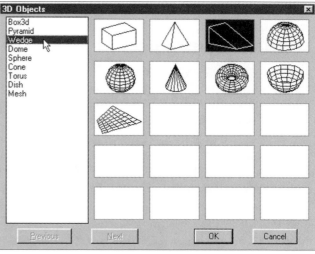

    Command: **ai_box**
    Corner of box: **PICK** or (**value**)
    Length: **PICK** or (**value**)
    Cube<Width>: **PICK** or (**value**)
    Height: **PICK** or (**value**)
    Rotation angle about Z axis: **PICK** or (**value**)
    Command:

These primitives are composed of AutoCAD surfaces. The 3D primitives (*3D Surfaces*) are provided to simplify the process of creating complex surface models. Generally, the primitives have to be edited and combined with other surfaces to build the complex model you need for a particular application. In this case, editing commands (*Pedit, Explode*, Grip editing, etc.) discussed in this chapter can be used along with other surfacing capabilities (*3Dmesh, 3Dface, Revsurf, Edgesurf, Region*, etc.). The editing methods differ depending on the type of surface. *Pedit, Explode*, and Grips can be used with any polygon mesh (*torus, sphere, dome, dish, cone*). *Pedit* <u>cannot</u> be used with *3Dfaces* (*box, wedge, pyramid*); however, *Explode* and Grip editing can be used. See Editing Polygon Meshes and Editing *3Dfaces* in this chapter.

For example, the *Box* can be *Exploded* into its individual *3Dfaces*. Figure 40-57 exhibits the box in its original form. Figure 40-58 illustrates the *box* after it has been *Exploded* and the top *3Dface* removed to reveal the inside. (*Hide* has been performed in these figures to enhance visualization.)

Figure 40-57 ————————————

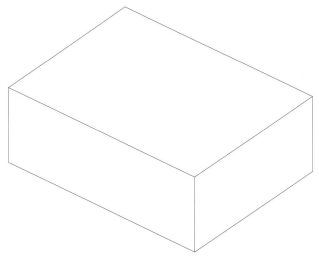

Figure 40-58 ————————————

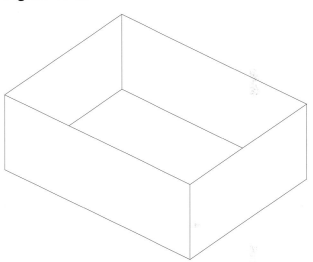

Surface Model Example 2

The following example is given as a sample of surface model construction strategy utilizing *3D Objects* combined with *3Dfaces* and polygon meshes.

In this example, the object in Figure 40-59 is constructed from 3D primitives (*3D Objects*), edited, and combined with other surfaces.

Figure 40-59 ————————————

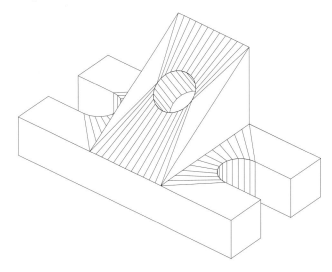

1. The drawing is set up by creating a layer for the surface model and setting the *Ucsicon ON* and at the *Origin*. The construction begins by selecting a *3D Box* from the *3D Objects* dialog box.

Figure 40-60 ⎯⎯⎯⎯⎯⎯⎯⎯⎯⎯⎯

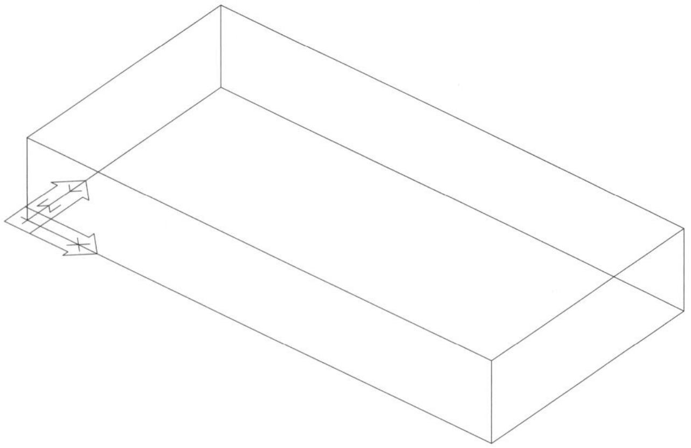

2. The *3D Box* is *Exploded* and the top and side faces are *Erased* (Fig. 40-61).

Figure 40-61 ⎯⎯⎯⎯⎯⎯⎯⎯⎯⎯⎯

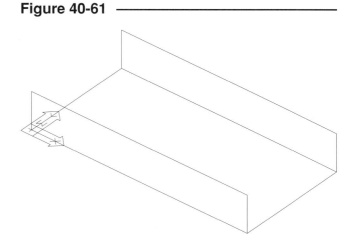

3. A *UCS* is created with the *Origin* option with its location (using *OSNAP*) on the top plane.

Figure 40-62 ⎯⎯⎯⎯⎯⎯⎯⎯⎯⎯⎯

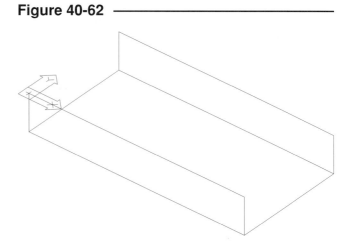

4. A new *Layer* is created for a wireframe. Wireframe elements (*Line, Arc*) are added to prepare for creating *3Dfaces* and polygon meshes (Fig. 40-63).

Figure 40-63

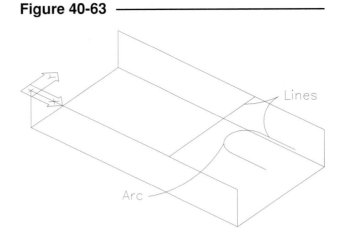

5. *SURFTAB1* and *SURFTAB2* are set to 12. *Rulesurf* and *3Dface* are used to create part of the top surface. Alternately, a composite *Region* could be created instead of the *Rulesurf* and two *3Dfaces* shown. (*Hide* is used in this figure to define the surfaces.)

Figure 40-64

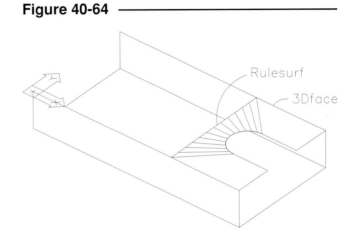

Figure 40-65

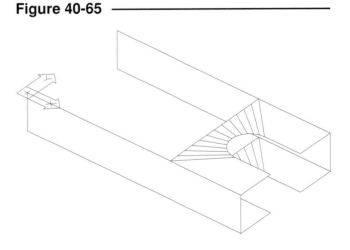

6. The bottom surface (part of the original *3D Box*) is erased. The new surfaces and wireframe elements are *copied* to the bottom plane (Fig. 40-65).

7. A *Rulesurf* and four *3Dfaces* are created, connecting the top and bottom surfaces. Use of layers is critical but somewhat laborious at this point. Only the wireframe elements can be selected as the defining edges for *Rulesurf*; therefore, turning the surface model layer *off* is helpful to allow selection of the wireframe elements. Alternately, press Ctrl while PICKing to cycle the selection of only the wireframe element.

Figure 40-66

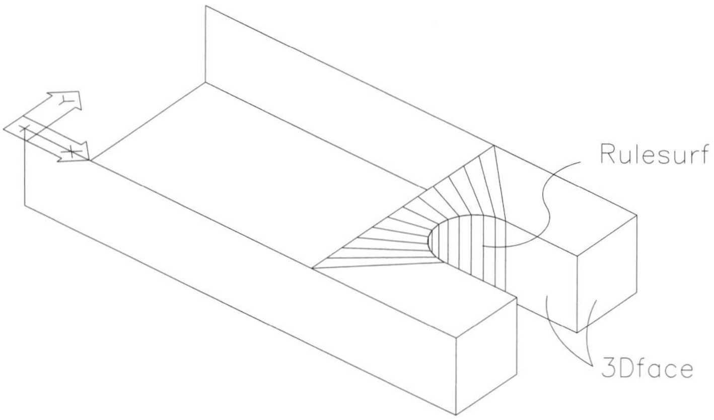

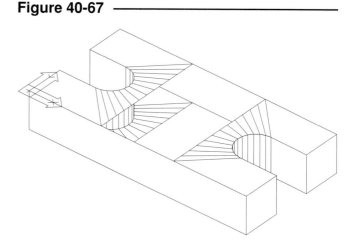

8. Next, the eleven surfaces are mirrored to the other side (Fig. 40-67). *Mirror3D* is used with the *YZ* option and PICKing the *Midpoint* of the long surface across the front as the "point on YZ plane." A *3Dface* is created to fill the opening on the bottom surface.

Figure 40-67

9. The *3D Objects* dialog box is used again to select a *Wedge*. The *Wedge* is placed in the center of the top surface.

Figure 40-68

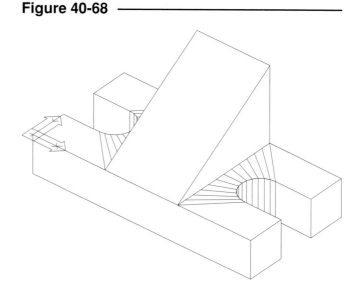

10. *Explode* breaks the *Wedge* into its individual *3Dfaces*. The inclined plane is *Erased* (Fig. 40-69).

Figure 40-69 ——————————

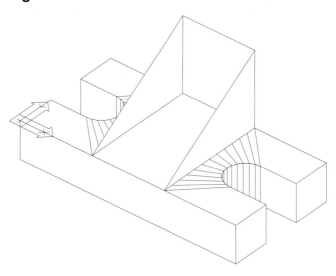

11. A new *UCS* is created to enable the construction of a new inclined surface with the hole. The *3point* option would be effective in this case.

Figure 40-70 ——————————

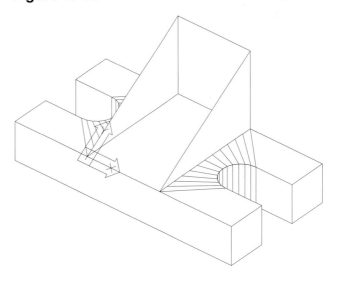

12. Wireframe elements (four *Arcs* and two *Lines*) are created on the wireframe layer as a foundation for construction of the new inclined surface and hole (Fig. 40-71).

Figure 40-71 ——————————

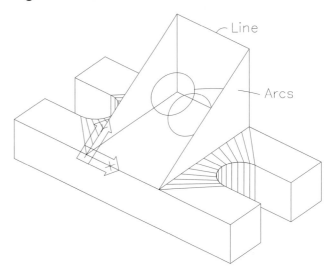

13. An inclined surface and hole are created on the surface model layer by applying the technique used in Surface Model Example 1 (or *Region* can be used). (Wireframe elements are not required for *3Dfaces*.)

Figure 40-72

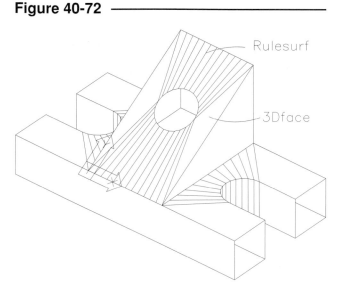

14. The model is completed by using *Rulesurf* to create the hole surface (sides), and a *Circle* establishes the bottom of the hole. Keep in mind that the *Rulesurf* must be defined by the original *Arcs*, not by the previous *Rulesurf* of the inclined plane (Fig. 40-73).

Figure 40-73

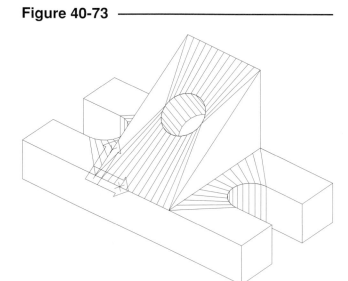

15. Figure 40-74 illustrates the completed surface model after performing a *Render*.

Figure 40-74

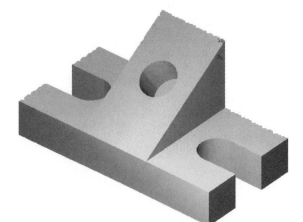

USING *THICKNESS* AND *ELEVATION*

THICKNESS

Pull-down Menu	COMMAND (TYPE)	ALIAS (TYPE)	Short-cut	Screen (side) Menu	Tablet Menu
Format Thickness	THICKNESS	TH	...	FORMAT Thicknes	V,3

ELEVATION

Pull-down Menu	COMMAND (TYPE)	ALIAS (TYPE)	Short-cut	Screen (side) Menu	Tablet Menu
...	ELEVATION	...	...	...	...

Thickness and *Elevation* were introduced with Version 2.1 (1985) as AutoCAD's first capability for creating 3D objects. *Thickness* and *Elevation* are still useful for creating simple surfaces. Most 2D objects (*Line, Circle, Arc, Point, Pline, Rectangle, Polygon*, etc.) can be assigned *Thickness* or *Elevation* property. Only newer (Release 13) objects such as NURBS curves (*Spline, Ellipse*), *Xline, Ray*, and *Mlines* cannot possess *Thickness* but <u>can</u> possess *Elevation*. *Thickness* and *Elevation* are usually assigned before using a 2D command; however, they can also be changed later using *Change* or *Ddmodify*.

Thickness is a value representing a <u>Z dimension</u> for a 2D object. The new object is created using the typical draw command (*Line, Circle,* or *Arc,* etc.), but the resulting object has a uniform Z dimension as assigned. For example, a two unit high cylinder can be created by setting *Thickness* to 2 and then using the *Circle* command (Fig. 40-75).

Thickness is <u>always perpendicular</u> to the XY plane of the object. Until later versions of AutoCAD, this posed a major limitation with the use of *Thickness*. This limited cylinders, for example, to having a vertical orientation <u>only</u>; no cylinders could be created with a horizontal orientation. The term "2 ½-D" was used to describe this capability of CAD packages during this phase of development.

Today, with the ability to create User Coordinate Systems (UCSs), *Thickness* can be used to create surfaces in any orientation. Therefore, the effectiveness of *Thickness* and *Elevation* commands is increased. Figure 40-76 displays a cylinder oriented horizontally, achieved by creating a UCS having a vertical XY plane and a horizontal Z dimension. UCSs can be created to achieve any orientation for using *Thickness*.

Figure 40-75 ─────────

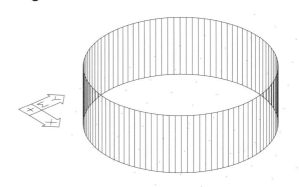

Figure 40-76 ─────────

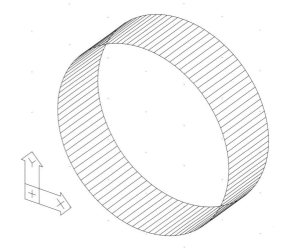

With a *Thickness* set to a value greater than **0**, commands normally used to create 2D objects can be used to create 3D surfaces having a uniform Z dimension perpendicular to the current XY plane. Figure 40-77 shows a *Line, Arc,* and *Point* possessing *Thickness* property. Note that a *Point* object with *Thickness* creates a line perpendicular to the current XY plane.

Figure 40-77

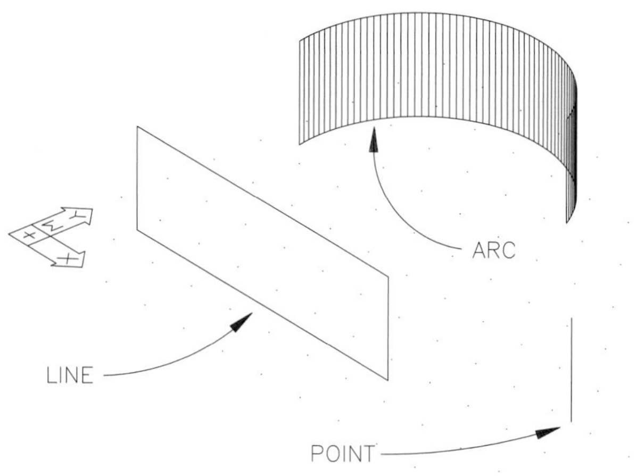

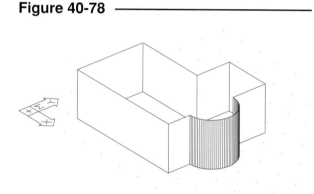

The ability to create *Plines* with *Thickness* offers many possibilities for developing complex surface models (Fig. 40-78). (Keep in mind that 2D shapes created with *Line* and *Arc* objects can be converted to *Plines* with *Pedit*.)

Figure 40-78

Using the *Hide* command with surfaces created with *Thickness* enhances your visibility. Figures 40-78 and 40-79 show examples of *Hide*. Notice that *Circles* created with *Thickness* have a "top." Other closed shapes created with *Line* or *Pline* do <u>not</u> have a "top."

Assigning a value for *Thickness* causes all subsequently created objects to have the assigned *Thickness* property. *Thickness* must be set back to **0** to create normal 2D objects.

Elevation controls the Z dimension of the "base plane" for subsequently created objects. Normally, unless an explicit Z value is given, 2D objects (with or without *Thickness*) are created on the XY plane or at an *Elevation* of **0**. Changing the value of the *Elevation* changes the "base plane," or the elevation, of the construction plane for newly created objects.

Figure 40-79

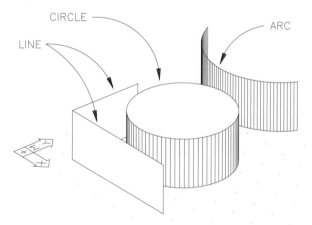

Figure 40-80 displays the same objects as in the previous figure except the *Circle* has an *Elevation* of **2** (level with the top edges of the other objects). It may be necessary to use *Hide* or several *Vpoints* to "see" the relationship of objects with different *Elevations*.

Like *Thickness*, *Elevation* should be set back to 0 to create geometry normally on the XY plane.

Figure 40-80

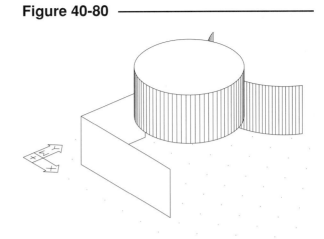

Changing *Thickness* and *Elevation*

Because *Thickness* and *Elevation* are properties that 2D objects may possess (like *Color* and *Linetype*), they can be changed or assigned <u>retroactively</u>. *Elevation* can only be changed retroactively using the *Change* command with the *Properties* option (see Chapter 16). *Thickness* can be changed using the *Modify (object)* dialog box.

DDMODIFY

Using the *Ddmodify* command (see Chapter 16) produces a dialog box providing the ability to change the *Thickness* of <u>existing</u> objects (Fig. 40-81).

As an alternative to setting *Thickness* before creating surfaces, the 2D geometry can first be created and then changed to 3D surfaces by using this dialog box to retroactively set *Thickness*.

Figure 40-81

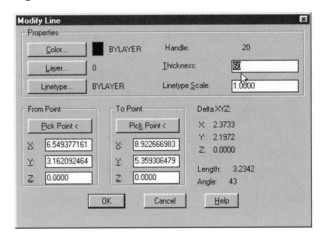

Figure 40-82 displays 2D *Pline* shapes. The *Modify (object)* dialog box can be used to change the *Thickness* of the shape to create a 3D surface model.

Figure 40-82

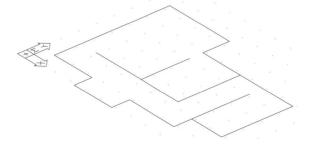

Figure 40-83 shows the same *Pline* shapes as in the previous figure after the *Thickness* property was changed to generate the 3D surface model.

If you want to change an object's *Elevation*, the *Change* command can be used. The command syntax is as follows:

Figure 40-83

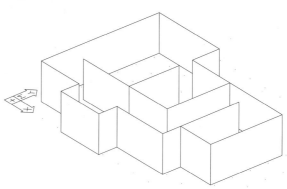

> Command: *change*
> Select objects: **PICK**
> Select objects: **Enter**
> Properties/<Change point>: *p*
> Change what property
> (Color/Elev/LAyer/LType/ltScale/Thickness) ? *e*
> New elevation <0.0000>: (**value**)
> Change what property (Color/Elev/LAyer/LType/ltScale/Thickness) ? **Enter**
> Command:

CHAPTER EXERCISES

1–3. Create a surface model of each of the objects in Figures 40-84 through 40-86. These are the same objects you worked with in Chapter 37, Wireframe Modeling. You can use each of the Chapter 37 Exercises wireframes (**WFEX1**, **WFEX2**, and **WFEX3**) as a "skeleton" on which to attach surfaces. Make sure you create a *New Layer* for the surfaces. Use each dimension marker in the figure to equal one unit in AutoCAD. Locate the lower-left corner of the model at coordinate **0,0,0**. Assign the names **SURFEX1**, **SURFEX2**, and **SURFEX3**. Make a plot of each model with hidden lines removed.

Figure 40-84

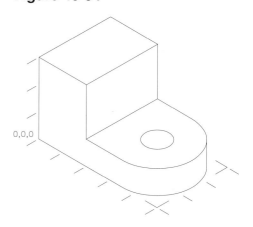

0,0,0

Figure 40-85

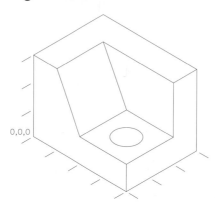

0,0,0

Figure 40-86

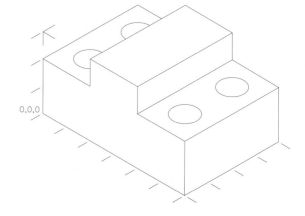

0,0,0

4. *Open* the **WFEX4** drawing that you created in Chapter 37 Exercises. *SaveAs* **SURFEX4**. For this exercise, create a surface model using a combination of regions and polygon meshes. Use the existing wireframe geometry. Convert the *Lines*, *Arcs*, and *Circles* representing <u>planar</u> surfaces to closed *Plines*, then to regions. Use polygon meshes for the curved surfaces. *Save* the drawing. Generate a pictorial *Vpoint* and align a *UCS* with the *View*. *Plot* the drawing with hidden lines removed.

Figure 40-87

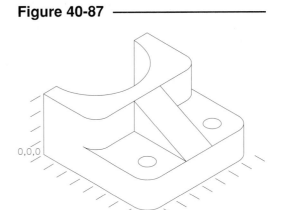

5. Create a surface model of the bar guide shown in Figure 40-88. You may want to use the **BGUID-WF** drawing that you created in Chapter 37 Exercises as a wireframe foundation. Make a *New Layer* for the surface model. Align a *UCS* with the *View* and create a title block and border. *Plot* the model with hidden lines removed. *Save* the surface model as **BGD-SURF**.

Figure 40-88

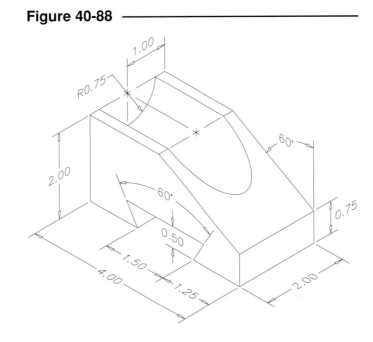

6. Create a surface model of the saddle in Figure 40-89. There is no previously created wireframe model, so you must begin a *New* drawing or use a *Template*. The fillets and rounds should be created using *Rulesurf* or *Tabsurf*. *Revsurf* must be used to create the complex fillets and rounds at the corners. Align a *UCS* with the *View* and create a title block and border. Plot the drawing with hidden lines removed. *Save* the drawing as **SDL-SURF**.

Figure 40-89

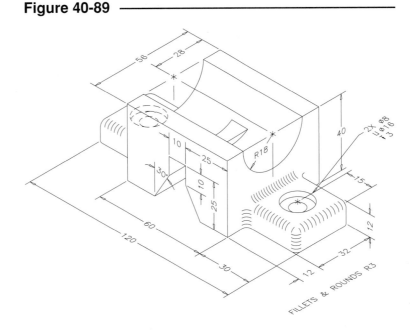

7. Begin a *New* drawing (or use a *Template*) for the pulley shown in Figure 40-90. This is the same object that you worked on in Chapter 38 Exercises. As a reminder, the vertical dimensions are diameters. Use *Revsurf* to create the curved (cylindrical) surfaces. Use *Regions* to create the vertical circular faces that would appear on the front and back (the surfaces containing the 4 counterbored holes or the keyway). Display the completed surface model from a pictorial *Vpoint*. Align a *UCS* with the *View* and create a border and title block. *Save* the drawing as **PUL-SURF**. *Plot* the model with the *Hide Lines* option.

Figure 40-90

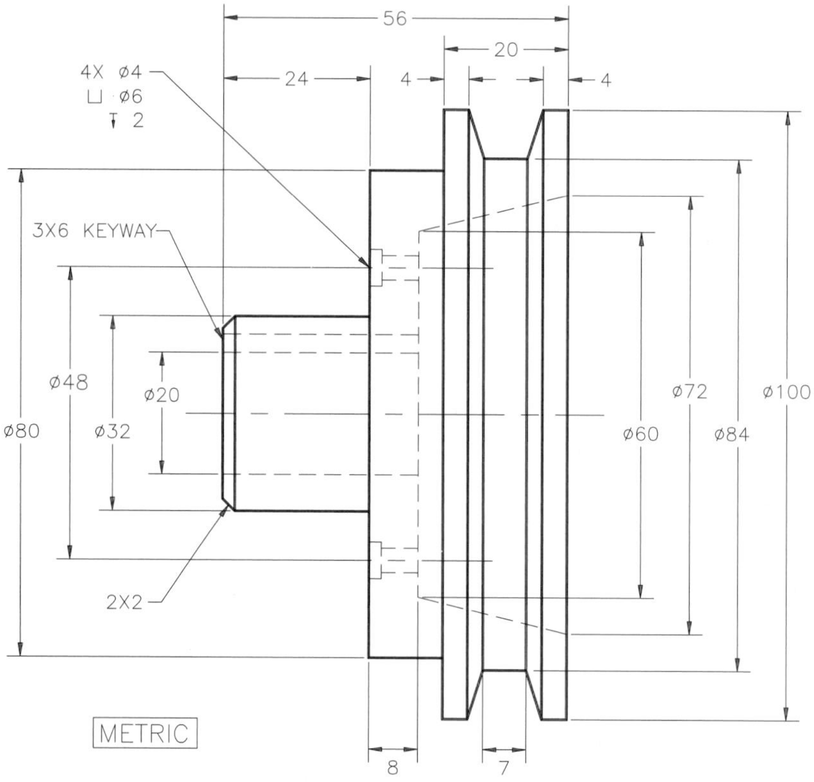

41

RENDERING

Chapter Objectives

After completing this chapter you should:

1. know the typical steps for rendering a surface or solid model using Render;

2. be able to create a variety of backgrounds for a rendering with *Fog* and *Background*;

3. be able to select and insert landscape images such as trees and shrubs;

4. understand and be able to create and set parameters for the four types of lighting: *Ambient Light*, *Point Light*, *Distant Light*, and *Spotlight*;

5. be able to select *Materials* from the *Materials Library*, *Import* them into the drawing, and *Attach* them to objects, colors, or layers;

6. be able to create *New* and *Modify* existing material properties of *Color*, *Ambient*, *Reflection*, *Roughness*, and other attributes;

7. be able to map raster images to objects and adjust the image orientation with *Setuv*;

8. be able to specify parameters and create a rendering using the *Render* command;

9. be able to use *Saveimg* to save and *Replay* to replay rendered images.

CONCEPTS

AutoCAD provides the ability to enhance the visualization of a <u>surface</u> or a <u>solid</u> model with the *Hide, Shade,* and *Render* commands. Because the default representation for surface and solid models is wireframe representation, it is difficult for the viewer to sense the three-dimensional properties of the model in this mode. *Hide, Shade,* and *Render* change the appearance of a model from wireframe to "solid." *Hide* calculates which surfaces obscure others and displays only the visible ones, and *Shade* adds the object color to the model. *Render* adds another level of realism to surface and solid models by giving you control of backgrounds, lights, colors, materials, and other effects.

Shade (discussed in Chapter 35, 3D Viewing and Display) is a simple and quick method that uses a default light location and fills the surfaces of the 3D model with the object color. You have control of only two variables, *SHADEDIF* and *SHADEDGE,* that affect the results of the shaded model.

Render, on the other hand, involves the use of *Light* sources and the application of *Materials* for the 3D model. You control the type, intensity, color, shadows, and positioning of lights in 3D space. Render also provides a variety of materials for attachment to objects, colors, or layers and allows you to adjust the properties of the material. With Render you can add and adjust *Background, Fog, Landscape* images, and images mapped onto surfaces of objects.

Render is intended to be used as a visualization tool for 3D models. Visualization of a design before it has been constructed or manufactured can be of significant value. Presentation of a model with light sources and materials can reveal aspects of a design that may otherwise be possible only after construction or manufacturing. For example, a rendered architectural setting may give the designer and client previews of color, lighting, or spatial relationships. Or a rendered model may allow the engineer to preview the functional relationship of mechanical parts in a manner otherwise impossible with wireframe representation. Thus, this "realistic" presentation, made possible by *Render,* can shorten the design feedback cycle and improve communication among designers, engineers, architects, contractors, and clients.

With AutoCAD Render, you can render to a viewport, to a separate rendering Window, or to a file. The *Render* command uses the *Lights* and *Materials* you assign and calculates the display based on the parameters that you specify. If no *Lights* or *Materials* are created for the drawing, a default light and material are used to render the current viewport. If you have created several *Lights, Materials,* and *Views,* you can use the *Scene* command to specify a particular named *View* and a combination of *Lights.*

R14 In AutoCAD Release 14, Render has been enhanced to provide "photo-realistic" rendering capabilities. The ability to calculate shadows and reflection of color and light from one surface to another supplies this "photo-realistic" quality.

The rendered screen image created with the *Render* command can be saved as a TIFF, TGA, and/or a BMP file and replayed at a later time. The *Render* and *Rendering Preferences* dialog boxes provide control for format and quality of the rendered image.

Rendering commands can be accessed by typing or selecting from the *View* screen pull-down menu (Fig. 41-1) and the *VIEW2* screen menu.

Figure 41-1

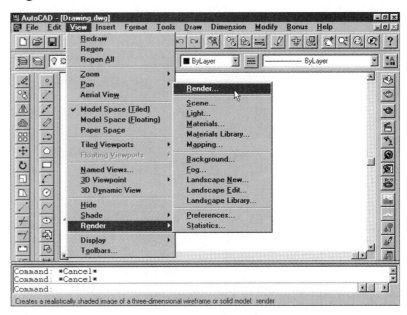

A *Render* toolbar (Fig. 41-2) is available by accessing the *Toolbar* dialog box. Optionally, several rendering commands can be selected from the digitizing menu.

Figure 41-2

Creating a Default Rendering

Creating a realistic rendering often takes more time that creating the 3D model. Although many parameters, adjustments, and steps can be used to create a realistic rendering (discussed later), you can use the *Render* command to create a rendering without specifying any background, lights, or materials.

Figure 41-3

To create a rendering, you must first create a 3D model, then use *Vpoint* or *Dview* (by any method) to generate a representative view of the function or feature of the model you want to display. For example, Figure 41-3 displays the support bracket from the desired viewpoint in wireframe representation.

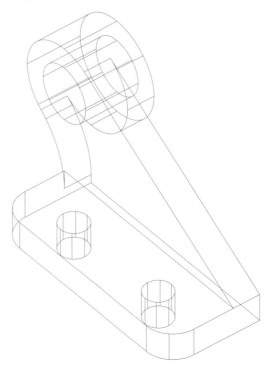

Next use the *Render* command which produces the *Render* dialog box (Fig. 41-4). Using all the defaults, press the *Render* button to generate a default rendering. AutoCAD creates the rendering using the current background color, the current object color, and a virtual "over-the-shoulder" *Distant* light source (Fig. 41-5).

Figure 41-4 ————————————————

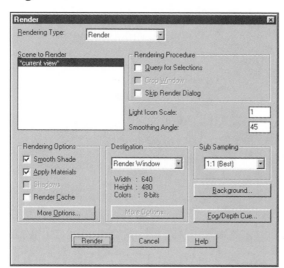

Figure 41-5 ————————————————

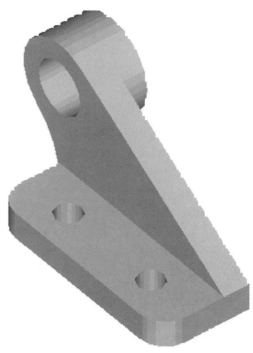

Typical Steps for Using Render

Although a rendering of a 3D surface or solid model can be created easily using default settings, the following steps are normally taken to create a rendering with the desired effect:

1. As with using *Shade,* you must first create or open an existing 3D model. Both surface and solid models can be rendered.

2. Use *Vpoint* or *Dview* to specify the desired viewing position. Perspective generations (created with *Dview Distance*) are allowed. If you want to create several renderings from different viewpoints, use the *Save* option of *View* to save each display to a named view so it can be *Restored* at a later time.

3. Set the *Background* for the object to be rendered. You can use a solid color or gradient color as the background. Optionally, specify an image (.BMP, .TIF, or other file) as the background to merge with the objects to be rendered. You can select from supplied *Landscape* images if needed and even adjust a *Fog* appearance.

4. Use the *Lights* dialog box series to create one or more lights. Parameters such as light type, intensity, color, and location (position in space) can be specified. A light source representing the Sun can be specified by geographic location and time of day. You can specify if the *Lights* cast shadows.

5. Use the *Materials* dialog box series to select the desired material(s) for the object(s). The material properties such as specification of color, reflective qualities, and other attributes can be adjusted. You can *Preview* the material as you adjust the parameters.

6. If you have created several *Views* and *Lights,* use the *Scenes* dialog box to select the specific combination of *Lights* and the desired *View* for the rendered scene.

7. Specify the desired format and quality for the rendering by the *Render* or *Rendering Preferences* dialog box. Several options are available for the rendering.

8. Use the *Render* command to create the rendered image on the screen.

9. Because rendering normally involves a process of continual adjustment to *Light* and *Material* parameters, any or all of steps 2 through 8 can be repeated to make the desired adjustments. Don't expect to achieve exactly what you want the first time.

10. When the final adjustments have been made, use the *Render* or *Rendering Preferences* dialog box to specify parameters such as *Photo Raytrace, Shadows,* and other rendering quality settings to create a realistic image.

11. Use the *Render* command again to create the rendered display or file. Usually, you should render to the screen until you have just the image you want; then save to a file if desired.

12. If you rendered to the screen, you can then save the rendered image to a .BMP, .TIF, or .TGA file using the *Save Image* dialog box, or save the image as a .BMP file from the rendering window. Alternately, you can write the rendering directly to a file without generating the render to the display.

Color Systems

There are several methods for choosing and assigning colors in the Render for such applications as background colors, material colors, and light colors. Because these systems appear throughout the many Render dialog boxes, these concepts are presented here.

Color for a specific application (background, material, or light) can be assigned by one or more of the four systems described below. These systems appear in several dialog boxes. In each case, current color adjustments are displayed in a square color box or *Preview* image tile appearing in the dialog box.

RGB

Figure 41-6

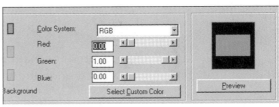

The *RGB* system determines the color by the amount of *Red, Green,* and *Blue* components (Fig. 41-6). Values for each can be entered into the edit boxes or changed using the slider bars. View the "color box" to see the changes in real time or press *Preview* to see the image tile. Pure red, green, or blue hues can be achieved by moving only that color to the right, or entering 1.0 in the edit box for that value and 0 for the other two.

Moving all of the *Red, Green,* and *Blue* color bars to the left produces black, and moving all to the right produces white (although gray may appear in the color box).

HLS

Figure 41-7

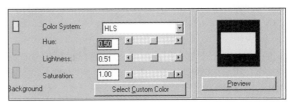

The slider bars adjust the *Hue, Lightness,* and *Saturation* visible in the small boxes (Fig. 41-7). The mix of *HLS* appears in the "color box."

Hue	Controls the color (red, blue, yellow, etc.).
Lightness	Controls the value (white to black). White is all the way to the right (although the color box may appear gray), or a value of 1.0, and black is to the left, or 0.
Saturation	Controls the purity of the color (mix of *Hue* and *Lightness*). Pure color is to the right, or a value of 1.0, and pure gray scale is to the left, or 0.

A pure hue is achieved by a *Lightness* of .5 and *Saturation* of 1.

Custom Color

The *Custom Color* button produces the *Color* dialog box (Fig. 41-8). The desired color can be PICKed from the selections in the color spectrum area. The value (gray scale) of the entire set of colors in the spectrum can be changed by using the slider bar on the right: up for lighter colors, down for darker. *Basic Colors* can be selected from boxes on the left and *Custom Colors* can be created by selecting the *Add to Custom Colors* button. RGB values or HLS values can be entered in the edit boxes on the lower-right.

Figure 41-8

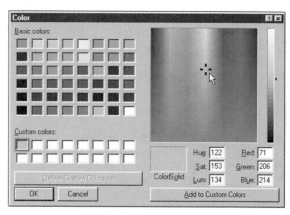

Select from ACI...

Selecting this option invokes the 256-color ACI (AutoCAD Color Index) palette (Fig. 41-9). Select the desired color from the palette or enter the name (for colors 1-16) or number in the edit box.

Figure 41-9

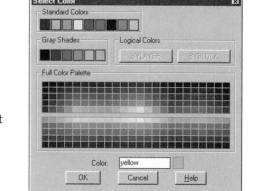

Photorealism, Raytracing, and Shadows

Release 14 offers enhancements that allow Render to create images that are very photo-realistic: raytracing and the ability to calculate shadows. These features increase the scene's realism but can increase rendering time.

Photo Real

This option of *Render* is called a "scanline renderer" because it "paints" one horizontal row of pixels at a time. It can display transparent materials and bitmapped images. Bitmapped images are images from an external file that can be planar (as for a background) or mapped to geometry such as a sphere, box, or other shape. *Photo Real* can also generate volumetric and mapped shadows (see explanation next).

Photo Raytrace

Raytracing is the process of calculating or following the paths of light rays as they are generated from a source and cast shadows or bounce off objects and illuminate other objects. Raytracing is needed to calculate reflections of light and color.

Volumetric Shadows

This type of shadow is found by calculating the volume of space cast by the shadow of an object, then generating a shadow based on this volume. These shadows have a sharp edge between the shadow and the illuminated area. Volumetric shadows are sensitive to transparent and translucent objects and calculate a shadow based on the color (light to dark) of the object.

Shadow Maps

Shadow mapping allows you to adjust the "fuzziness" or softness of the edge of a shadow (as opposed to a hard or sharp edge). You can also adjust the size or accuracy of the shadow map that is generated—the greater the size (in pixels), the more accurate. Shadow maps do not display the color cast by translucent objects.

Raytraced Shadows

Raytraced shadows are generated by tracing the path of light rays generated by a light source. These shadows have hard edges but can transmit color when passing through translucent objects.

Rendering Time

Although Release 14 Render is faster than earlier versions for simple rendering, shadows always increase rendering time. Shadow maps are particularly time consuming. Consider turning *Shadows* on only after setting background effects and adjusting lights.

Now that you know the concepts and general procedure for selecting colors and creating a rendered image, background, lighting, and material concepts and commands are presented next. Examples of the dialog boxes used to achieve the desired effects are included in the following information.

CREATING A BACKGROUND

Once the 3D model is created, you can use the *Background* command to specify the background you want to appear in the rendering. There are five choices for the background:

1. the current *AutoCAD Background* (in the Drawing Editor),
2. a *Solid* color background,
3. a *Gradient* color blended from two or three color bands,
4. an *Image* in the form of a .TIF, .GIF, .JPG, .BMP, .TGA, or .PCX file,
5. a previous rendering to *Merge* with the current rendering.

When you specify a background and generate a rendering, the object(s) in the rendering is automatically displayed "in front of" the background color or scene.

BACKGROUND

Pull-down Menu	COMMAND (TYPE)	ALIAS (TYPE)	Short-cut	Screen (side) Menu	Tablet Menu
View *Render >* *Background...*	*BACKGROUND*	...	...	*VIEW 2* *Backgrnd*	*Q,2*

Use the *Background* command to set up and adjust the background for the rendering. The background consists of colors or an image that appear behind the rendered geometry. Consider the background as a 2D plane behind the geometry always oriented normal to the viewing angle.

Background produces the *Background* dialog box (Fig. 41-10). At the top of the dialog box are four options: *Solid, Gradient, Image,* and *Merge.* The *AutoCAD Background* option is located on the middle left. The *HLS, RGB,* or *Custom Color* systems (in the central area) can be used to set background colors.

Figure 41-10

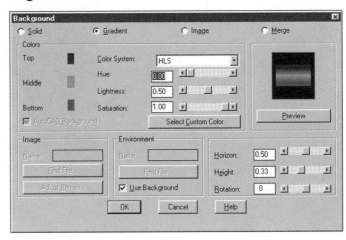

AutoCAD Background
Select this option to use the current color in the Drawing Editor as the render background. This option is accessible only if the *Solid* radio button is pressed.

Solid

Select *Solid* and remove the check from *AutoCAD Background* to create a specific color background. This action enables the *RGB* and *HLS* color systems. The selected color appears in the *Top* color box in "real time" as you make adjustments. Press *Preview* to see a large image of the solid color.

Gradient

This background type produces two or three colors that blend across the background (Fig. 41-11). Select the *Top, Middle,* and *Bottom* color boxes to set and view each color as you change it with one of the three available color systems.

When *Gradient* is selected, the *Horizon, Height,* and *Rotation* adjustments are enabled.

Horizon

This setting determines where the *Middle* color is located in the mix. A value of 50 puts the *Middle* color in the middle, a lower value places the *Middle* color near the bottom (when *Rotation* is 0), and a higher value moves the *Middle* color near the top (when *Rotation* is 0).

Height

Use this adjustment to specify the width of the band of *Middle* color. The greater the value, the wider the band; the smaller the value, the narrower the band. To use only two gradient colors (*Top* and *Bottom*), set *Height* to 0.

Rotation

You can change the orientation of the bands of color from horizontal to any other angle. For example, enter 45 to achieve a background with bands of color running diagonally across the display (Fig. 41-12).

Figure 41-11

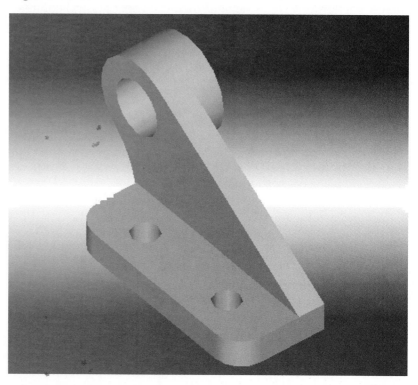

Figure 41-12

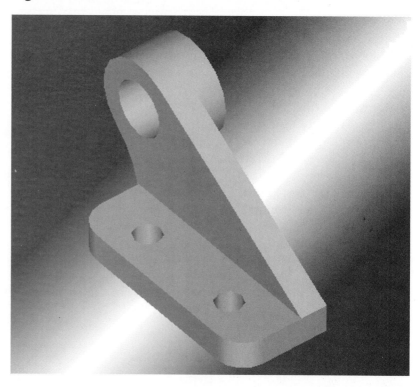

Image

Selecting the *Image* button enables the *Image* section of the *Background* dialog box (Fig. 41-13) and disables the other sections. With this option you can insert an image to use as the background for your rendering. Any .TIF, .GIF, .JPG, .BMP, .TGA, or .PCX file can be used as the image.

Figure 41-13

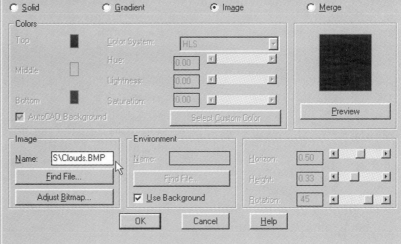

Several .BMP files are supplied with Windows, such as Clouds.bmp in the Windows directory (Fig. 41-14).

Figure 41-14

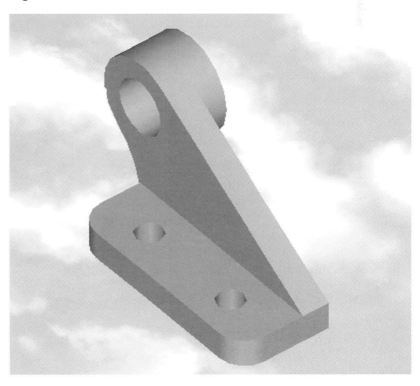

Environment

You can create additional reflection and refraction effects on the rendered objects by defining the *Environment*. This option causes the rendered objects to reflect or mirror the background image. For *Environment* to have an effect, *Materials* selected for the rendered objects must be reflective and the *Rendering Type* must be *Photo Real* or *Photo Raytrace*. You can use the selected *Solid, Gradient, Image,* or *Merge* background (check *Use Background*), or you can select a specific image such as a .TIF, .GIF, .JPG, .BMP, .TGA, or .PCX file (remove the check from *Use Background*, then enter a *Name* or pick *Find File…*).

Adjust Bitmap...

Select *Adjust Bitmap...* to produce the *Adjust Background Bitmap Placement* dialog box (Fig. 41-15). When an image is used for the background, this dialog box provides numerous tools for you to adjust the placement and proportions of the image.

Figure 41-15 —————————

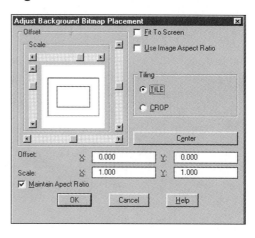

Fit to Screen

Select this option to automatically adjust the aspect ratio (width to height) of the image to fit the proportion of the screen. If *Use Image Aspect Ratio* is <u>also</u> checked, the image's larger dimension is fit to match the screen; therefore, the other dimension may not fill the screen. *Fit to Screen* disables all other options.

Use Image Aspect Ratio

This option prevents the width-to-height ratio of the image from being altered. You can still change *Offset* and *Scale*.

Offset and Scale

These values can be changed only when *Fit to Screen* is not checked. Use the slider bars or the edit boxes to change values. The red rectangle in the image tile represents the screen (render window) and the two-colored (black and magenta) rectangle represents the selected background image. Use *Offset* to move the image up, down, left, or right. Use *Scale* to make the selected image larger or smaller (this feature is misleading in that a larger value creates a smaller image and vice versa). If an image is *Offset* or *Scaled* so it is smaller than the entire screen area, the area outside the background image is controlled by *Tile* or *Crop*.

Center

Press *Center* to center the selected image file on the screen and reset the *Offset* values to 0.

Maintain Aspect Ratio

A check in this box locks the *X* and *Y* values proportionally. With this box checked, the image width-to-height ratio is maintained even if *Scale* values are altered. If you want to stretch the image, remove the check and change the *X* and *Y Scale* values independently. If *X* and *Y* values are changed (so the image proportions are distorted), you can check *Maintain Aspect Ratio* again to lock the new proportions while using *Scale*.

Figure 41-16 —————————

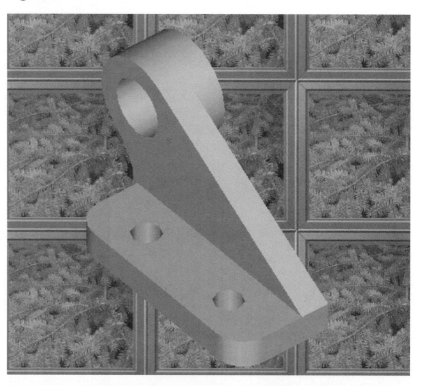

Tile and Crop

When the selected image is smaller (or *Scaled* to be smaller) than the screen or render window, these options have an effect. *Tile* causes the image to be repeated so it fills the screen (Fig. 41-16). *Crop* causes the rendered screen to display only one instance of the background image and all areas outside the background are black (cropped).

Merge
Merge is a pre-Release 14 method of using an external image file as a background. To do this, use the *Replay* command to display a .BMP, .TGA, or .TIF image. The display is stored in the "frame buffer." Then select *Merge* and render the model geometry. You must have 24-bit color enabled on your video display for this option to be available. However, a better method is to use the *Image* option, which allows a wider variety of file types as a background and automatically merges the image with the model geometry.

FOG

Pull-down Menu	COMMAND (TYPE)	ALIAS (TYPE)	Short-cut	Screen (side) Menu	Tablet Menu
View *Render >* *Fog…*	*FOG*	…	…	*VIEW 2* *Fog*	*P,2*

Using the *Fog* command displays the *Fog/Depth Cue* dialog box (Fig. 41-17). *Fog* adds a visual effect that simulates distance information in the rendering. You can specify any color for this effect. The amount of fog (color) applied to the geometry is based on the distance different parts of the geometry are from the camera (viewpoint). For most applications, the farther the geometry, the more fog applied. The terms "fog" and "depth cueing" are use to describe two extremes of this one effect—that is, fog is white, and depth cueing is black.

Figure 41-17

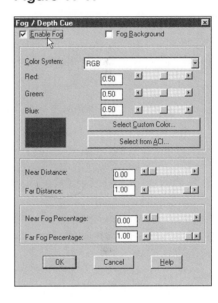

Enable Fog
Check this option to apply fog to the rendering. *Enable Fog* must be checked to enable all other options in the dialog box. Fog is applied only to the model geometry for *Photo Real* and *Photo Raytrace* rendering options.

Fog Background
Check this box to apply fog to the background image as well as the model geometry. Fog is applied to the background for any rendering option (*Render, Photo Real,* or *Photo Raytrace*).

Color System and Color Controls
Use these settings to adjust the color you want to use for fog. Select from the *RGB, HLS, Custom Color,* or *ACI* color systems (see Color Systems earlier this chapter).

Near Distance and Far Distance
Use these sliders or edit boxes to set the values to define where the fog begins and ends. The values are percentages of the distance from the camera to the background or back clipping plane.

Near Fog Percentage and Far Fog Percentage

These adjustments set the density of the fog for the near and far distances (ranging from 0 to 100 percent). For example, to make the model geometry appear less visible in the distance, set the *Near Fog Percentage* low and the *Far Fog Percentage* high (Fig. 41-18).

Figure 41-18 ───────────

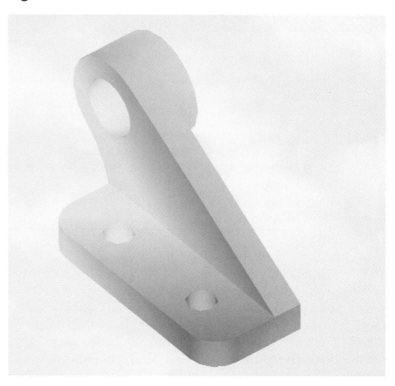

LSNEW

Pull-down Menu	COMMAND (TYPE)	ALIAS (TYPE)	Short-cut	Screen (side) Menu	Tablet Menu
View *Render >* *Landscape New...*	*LSNEW*	...	...	*VIEW 2* *Lsnew*	...

Lsnew produces the *Landscape New* dialog box (Fig. 41-19). This feature allows you to add photo-realistic images, such as trees, bushes, people, and signs to your renderings. Related commands *Lsedit* and *Lslib* allow you to modify landscape objects and maintain landscape object libraries, respectively.

A landscape object has a bitmap image mapped onto it. You can manipulate the placement of the landscape object in the drawing directly (with grips, for example) or you can use the *View Aligned* option in the *Landscape New* dialog box to align the object. Each landscape object has grips at the top, bottom, and at each corner. Use the base grip to move the entire object, top grips to adjust height, and bottom corner grips to scale and rotate it. Landscape objects in the drawing (unrendered) appear as planar objects such as a triangle, rectangle, or two intersecting triangles, depending on the alignment and single or double face specifications.

Figure 41-19 ───────────

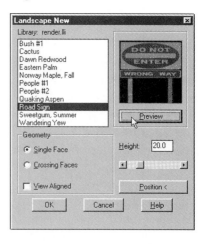

Library

The central area lists the AutoCAD-supplied landscape objects (in the RENDER.LLI file). Select one you want to insert into the drawing.

Height

Use this edit box or slider to specify the height of the object in the current drawing units. *Height* is always oriented as the positive Z direction in the current USC.

Position

Use this button to temporarily exit the dialog box and PICK a location in the drawing to place the landscape object. The default position is 0,0,0 of the current UCS.

Geometry

This section specifies the quality of the object and alignment of the object in the drawing.

Single Face or Crossing Faces

A *Single Face* object is faster to render but less realistic. *Crossing Face* objects result in better images, especially for *Photo Raytrace* renderings.

View Aligned

With this box checked, the object is positioned to be normal to the current view. The position of the landscape object changes to be normally aligned if the view changes. Use this option for trees and bushes. If you want planar objects such as road signs to retain the orientation you specify, remove this check before positioning them.

Landscape objects are planar objects that appear in the drawing depending on these settings (Fig. 41-20). A *Single Face, View Aligned* object appears as a triangle. You cannot rotate it with grips. A *Single Face* positioned object appears as a rectangle and can be rotated with grips. A *Crossing Face* object always appears as intersecting triangles. These objects cannot be rotated when *View Aligned* but can be when positioned (fixed).

The landscape objects appear as normal trees, bushes, people, and signs in the rendered image (Fig 41-21).

Figure 41-20

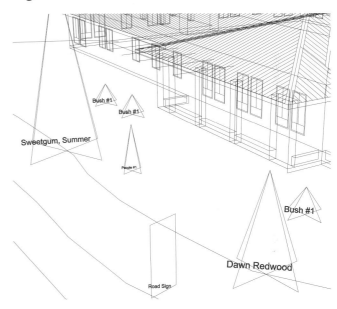

Figure 41-21

LSEDIT

Pull-down Menu	COMMAND (TYPE)	ALIAS (TYPE)	Short-cut	Screen (side) Menu	Tablet Menu
View *Render >* *Landscape Edit...*	*LSEDIT*	...	...	*VIEW 2* *Lsedit*	...

You can position landscape objects when you create them with *Lsnew* or when you manipulate a land-scape object directly in the drawing (with grips, for example). Alternately, you can use *Lsedit* to reposi-tion a selected object in the same manner as when you created the object. The *Lsedit* command produces the *Landscape Edit* dialog box, which is identical to the *Landscape New* dialog box (see Fig. 41-19), except you can only change options for the selected object—not create new objects.

LSLIB

Pull-down Menu	COMMAND (TYPE)	ALIAS (TYPE)	Short-cut	Screen (side) Menu	Tablet Menu
View *Render >* *Landscape Library...*	*LSLIB*	...	...	*VIEW 2* *Lslib*	...

The *Lslib* command produces the *Landscape Library* dialog box (Fig. 41-22). Use this tool to manage libraries of landscape objects. Only a few objects are supplied with AutoCAD. These objects can be modified and other objects can be created or purchased, then managed using this dialog box.

Figure 41-22 ————

Library:
The central area displays objects in the current library. The current *Library* file is listed above.

Modify
Select an object, then select this option to produce the *Landscape Library Edit* dialog box (see Fig. 41-23). All landscape objects are defined by an image file and an opacity map (to specify see-through areas of the image).

New
If you have an image to import into the library, use this option to produce the *New Landscape Library* dialog box. This dialog is identical to the *Landscape Library Edit* dialog except the edit boxes are blank. Enter or *Find* an image file and opacity map file for your new object.

Delete
Select an item to delete from the list, then select this button. The object is deleted from the library, but the referenced image and opacity map files are not deleted.

Open
Use this option to locate and open another library (.LLI file). The contents of the library appear in the list.

Save

When changes are made to the current library that you want to keep, click this button to save the .LLI file under the current name or create a new name.

Landscape Library Edit **Dialog Box**

This dialog box (Fig. 41-23) appears when you use the *Modify* option from the *Landscape Library* dialog box (Fig. 41-22). The information about the selected object is displayed in the edit boxes. In the *Default Geometry* section, you can modify the *Single* or *Crossing Faces* parameter and change the *View Aligned* assignment for the selected object. If you want the named object to reference another image file or opacity map file, enter or *Find* the desired file name. Change the name of the object in the *Name* edit box if desired.

Figure 41-23

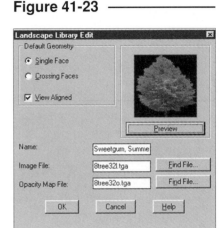

CREATING LIGHTS FOR THE MODEL

AutoCAD recognizes four types of light: *Ambient Light, Point Light, Distant Light,* and *Spotlight*. *Point Lights, Distant Lights,* and *Spotlights* can be positioned anywhere in space and adjusted for color and intensity. *Distant Lights* and *Spotlights* can be set to shine in a specific direction, whereas *Point Lights* shine in all directions. *Ambient Light* has no location and affects all surfaces equally. *Ambient Light* can be controlled only for color and intensity. A feature new with Release 14 is the ability for lights to cast shadows. *Point Lights, Distant Lights,* and *Spotlights* can cast shadows. *Ambient Light* cannot cast shadows.

Ambient Light

Ambient Light is the overall lighting of surfaces. Each surface is illuminated with an equal amount of ambient light, regardless of the surface's position. Thus, you cannot distinguish the adjoining faces of a box or curvature of a sphere when this is the only type of light in the model. Figure 41-24 illustrates the rendering of a box and sphere with <u>only</u> ambient light.

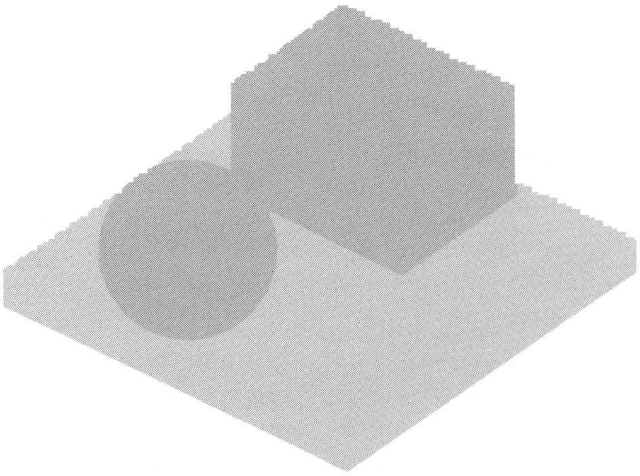

Adjusting ambient light would be like using a dimmer switch to darken or lighten a room except that the light has no source or direction. Ambient light is the most unrealistic type of light, but it is needed to display "background" surfaces that may not be illuminated by *Point* or *Distant* lights. You do not create or position an *Ambient Light* in the model; it is ever-present unless its intensity has been set to zero.

Figure 41-24

Distant Light

Figure 41-25

A *Distant Light* shines in one direction and has parallel rays. It is used to simulate sunlight. Because *Distant Light* represents the sun, you specify only a direction vector, not a light location. The vector determines the direction of the parallel beams of light which extend to infinity. Similar to sunlight, the light intensity does not fall off (grow dimmer) for surfaces farther away from the "source." Illumination of a surface is determined only by the angle between the surface and the light beams—surfaces that are perpendicular to the beams are more brightly illuminated.

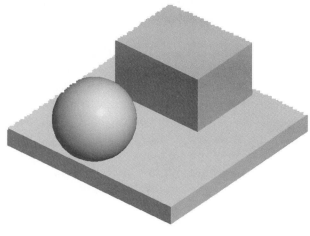

Figure 41-25 illustrates a box and sphere illuminated by <u>only</u> one *Distant Light* shining from above and to the right. Notice how each surface is illuminated differently because of the angle of the surface to the direction vector, yet each individual flat surface is illuminated evenly across the entire surface.

Point Light

Figure 41-26

A *Point Light* is inserted at the location you specify, but no direction vector is needed since this light shines in all directions. Point light would be typical of the light emitted from a light bulb. Point light falls off (grows dimmer) as the distance between the light source and surfaces increases, so surfaces close to the light are brighter than those at a distance. The box in Figure 41-26 is illuminated by <u>only</u> one *Point Light* above and slightly to the right side. Notice how the light falls off the farther the surfaces are from the light source.

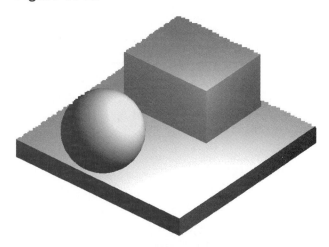

Spotlight

Figure 41-27

Spotlights have characteristics much like a spotlight used for stage lighting. You can specify the light's *Location* in space as well as the point that it shines toward (*Target*). Like point lights, the light falls off as the illuminated surfaces become farther from the light *Location*. Similar to stage spotlights, the beams of light from the source (*Location*) to the *Target* form a "cone" of light. Figure 41-27 shows the box and sphere illuminated by a *Spotlight* shining from directly above the objects. Notice the circular illuminated area formed by the cone of light. The angle of a *Spotlight's* cone of light can be adjusted for full illumination and for light fall-off.

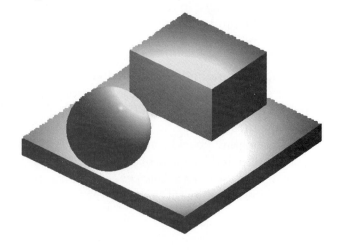

For *Point Lights, Distant Lights,* and *Spotlights,* the brightness of a rendered surface depends on the following parameters:

1. The *Intensity* of the light can be specified. The higher the *Intensity,* the brighter the rendered surfaces appear.

2. The *Color* of the light can be specified. The lighter the *Color,* the brighter the rendered surfaces appear.

3. The angle between the light source and the surface affects the brightness of the surface. A surface that shines the brightest is one in which the light is shining perpendicular to the surface. The more angled the surface is from perpendicular, the darker it appears.

4. The distance between the light and the surface affects brightness (*Point Lights* and *Spotlights* only).

5. When surfaces have a high *Reflective* attribute, the brightness of the reflected light is determined by the angle formed by the light source, object surface, and observer's viewpoint.

6. A surface's brightness can also be affected by a shadow or light reflected from another surface (for *Raytraced* displays).

Figure 41-28 ——————————

For dramatic effects, only one light might be used; however, for most renderings several lights or a combination of types of lights is used. Typically, a mix of *Ambient Light* and one or two other lights is used. Figure 41-28 illustrates the box and sphere illuminated by a combination of *Point, Distant,* and *Ambient* lighting.

Now that you understand the types of lights and the effects of each type on a rendered object, examine the commands that let you insert and control the lights.

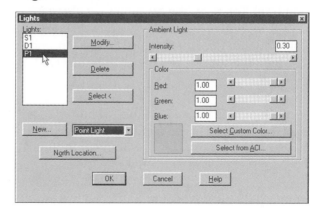

LIGHT

Pull-down Menu	COMMAND (TYPE)	ALIAS (TYPE)	Short-cut	Screen (side) Menu	Tablet Menu
View *Render >* *Light...*	*LIGHT*	...	...	*VIEW 2* *Light*	*O,1*

When you invoke the *Light* command by any method, the *Lights* dialog box appears, as shown in Figure 41-29.

Figure 41-29 ——————————

New...
To create a new light, select the type of light (*Distant Light, Point Light,* or *Spotlight*) from the drop-down list and PICK the *New* tile. The appropriate dialog box for the type of light appears and allows you to set the light's parameters. The light types and related dialog boxes are discussed later.

Lights:

The *Lights:* section in the upper-left corner displays any existing (previously created) lights. Using the related buttons beside the list, you can *Modify...*, *Delete*, or *Select* (from the drawing) an existing light. Modifying a light offers the <u>same dialog box</u> and options used to create a new light.

Ambient Light

The *Ambient Light* section (upper-right) is where the percentage of ambient light is set by using the slider bar or by entering a value in the edit box. The setting is <u>global</u> and affects all surfaces equally. Generally a percentage of .30 is a good starting value.

You can specify the *Color* of light that is emitted by the ambient source by using the slider bars in the *Lights* dialog box (*RGB* color system), by using the *Custom Color* system, or by using the *ACI* color system.

Light Colors

All lights can be assigned a color using either the *RGB* color system, the *Custom Color* system, or the *ACI* color system (see Color Systems earlier this chapter). The selected light color is the color that the light emits. Although each <u>object</u> in the model can have a color (determined by the object's assigned material), the light color casts a hue on the object's surfaces. Normally, white or bright colors are used for lights.

Creating New and Modifying Existing Lights

To insert new lights in your drawing, select the type of light (*Point Light, Distant Light,* or *Spotlight*) from the drop-down list in the *Lights* dialog box (Fig. 41-29); then pick the *New...* button. A different dialog box titled *New (type) Light* appears depending on your light choice.

If you want to modify an existing light, select the light name from the *Lights:* list (in the *Lights* dialog box) and PICK the *Modify...* button. The dialog box that appears is titled *Modify (type) Light* but is otherwise identical to the dialog box used to create the new light.

New Point Light and *Modify Point Light* Dialog Boxes

Light Name
When creating a new light, assign a name for the light in the *Light Name:* edit box as a first step (Fig. 41-30). In this example, "P1" is assigned as the name for a new point light. Existing light names can be renamed by changing the name in the edit box and selecting *OK*.

Intensity
Intensity is a value that specifies the brightness of the light; the higher the value, the brighter the light. The intensity can be set with the slider bar or by entering a value in the edit box. The resulting brightness of a surface illuminated by a *Point* light is based on several variables: intensity, color, attenuation, distance to the surface, and angle between the surface and the light beams.

Figure 41-30

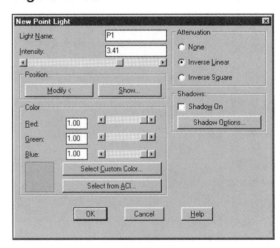

Position
When a light is created, AutoCAD automatically assigns a default location for the light based on the existing geometry. Selecting the *Show* tile displays the default *Location* of the light (Fig. 41-31). The *Location* is the coordinate position of the light in 3D space (*Target* is enabled only for *Spotlights*). You can change the position by selecting the *Modify* tile, then entering the desired coordinates or PICKing a new location (use *OSNAP* to PICK in 3D space).

Figure 41-31

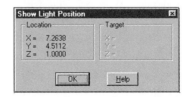

Color

The color of a *Point* light (as any other light type) can be assigned using the *RGB, Custom Color,* or *ACI* color system (see previous explanation in Color Systems).

Attenuation

Point Lights possess a fall-off quality similar to actual lighting. That is, the farther the light beams travel from the source (*Location*), the dimmer the light becomes. *Point Light Attenuation* has three options.

None

This option forces the light emitted from a point light to keep full intensity no matter how far the rendered surfaces are from the source. This is similar to a *Distant Light*.

Inverse Linear

This option sets an inverse linear relation between the distance and the illumination. The brightness of a rendered surface is inversely proportional to the distance from the source. Put another way, there is a 1 to 1 relationship between distance and darkness. Generally, *Inverse Linear* attenuation is the easiest to work with and appears realistic for most AutoCAD renderings.

Inverse Square

This is the greatest degree of light fall-off. The amount of light reaching a surface is inversely proportional to the square of the distance. In other words, if the rendered surfaces are 2, 3, and 4 units from the light source, the illumination for each will be $1/2^2, 1/3^2$, and $1/4^2$ (1/4, 1/9, and 1/16) as strong. This option simulates the physical law explaining light fall-off.

Shadows

Check the *Shadows On* checkbox if you want Render to calculate shadows cast by this light. (*Photo Real* or *Photo Raytrace* and the [global] *Shadows* checkbox in the *Render* dialog box must also be selected for shadows to be displayed in the model.) Pressing the *Shadow Options* produces a dialog box with the following options (Fig. 41-32).

Figure 41-32 ─────────

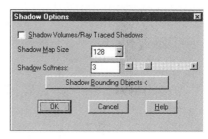

Shadow Volumes/Ray Traced Shadows

When this box is checked, *Photo Real* renderer produces volumetric shadows and *Photo Raytrace* produces raytraced shadows for this light. The other options are disabled.

Shadow Map Size

This setting controls the size in pixels of one side of the shadow map. The possible range is 64 to 4096. The larger the map size, the more accurate the shadows but the longer it takes to calculate them.

Shadow Softness

This value determines the "fuzziness" of the edges of shadows for shadow-mapped shadows. The value can range from 1 to 10 and represents the number of pixels along the edge of the shadow that are blended from the illuminated to the non-illuminated area. A value of 2 to 4 usually gives the best results.

Shadow Bounding Objects

You can reduce rendering time by selecting a set of objects that determine a bounding box. The bounding box is used to clip the shadow maps.

New Distant Light and Modify Distant Light Dialog Boxes

Distant Light simulates sunlight. With distant lights, parallel beams of light illuminate the solid or surface objects. Distance lights have no location or target, only a direction vector. Distance light has no attenuation, so the beams of light maintain the same intensity regardless of the location of the surfaces. The direction vector is specified by using either the *Azimuth* and *Altitude* or *Light Source Vector* sections of the *Distant Light* dialog box (Fig. 41-33) or the *Sun Angle Calculator*. The model in the Drawing Editor is oriented in space such that the XY plane of the WCS represents horizontal (Earth's surface), and north is a positive Y direction.

Figure 41-33

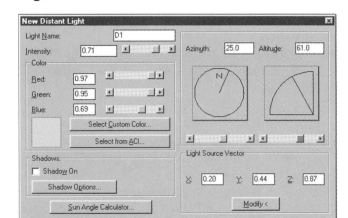

Light Name, Intensity, Color, and Shadows

The *Name, Intensity, Color,* and *Shadows* sections in the dialog boxes operate identically to the same sections in the other dialog boxes (see *Point Light*).

Azimuth

Azimuth is the bearing in degrees from north in a clockwise direction; for example, due east is 90 degrees. PICK in the image tile, use the slider bars, or enter a value in the edit box to make your selection.

Altitude

Altitude is the slope of the direction vector specified in degrees from horizontal; for example, 90 degrees specifies a light source from directly above (high noon) and 20 to 30 degrees simulates sunlight in the early morning or late afternoon. PICK in the image tile, use the slider bars, or enter a value in the edit box to make your selection.

Light Source Vector

An alternative to the *Azimuth* and *Altitude* method, the *Light Source Vector* edit boxes specify the location of a "source" related to 0,0,0 of the WCS. You can enter values in the edit boxes.

Sun Angle Calculator

Since a *Distant Light* is used to simulate sunlight, use the *Sun Angle Calculator* to determine the location of the sun with respect to the model by specifying a date, time, and time zone (Fig. 41-34). Making these adjustments automatically computes the *Azimuth* and *Altitude* of the *Point Light*. The location of the model on Earth can be specified if you know the *Longitude* and *Latitude,* or you can select a *Geographic Location*.

Figure 41-34

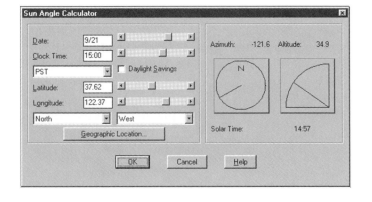

Geographic Location

This dialog allows you to select a city from the list or PICK a location on the map to determine the location on Earth of the model (Fig. 41-35). The drop-down list allows you to select from the seven continents. Selecting from the map or city list automatically enters the related values in the *Longitude* and *Latitude* edit boxes.

Figure 41-35

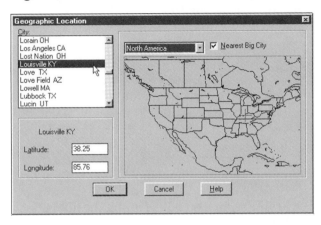

New Spotlight and *Modify Spotlight* Dialog Boxes

A *Spotlight* has characteristics similar to a spotlight used for stage lighting. Because you can specify a *Location* and a *Target* for the *Spotlight*, any surface or area of a surface or solid model can be illuminated from any direction. In addition, the angle of the "cone" of light (formed as the light beams travel from the source) can be adjusted. Spotlights also possess attenuation, color, and intensity. All options are available from the *New Spotlight* (Fig. 41-36) and *Modify Spotlight* dialog boxes.

Light Name, Intensity, Color, Attenuation, **and** *Shadows*
These options operate identically to the same options appearing in the *Distant* and *Point Light* dialog boxes. See *Distant Light* dialog boxes and *Point Light* dialog boxes.

Figure 41-36

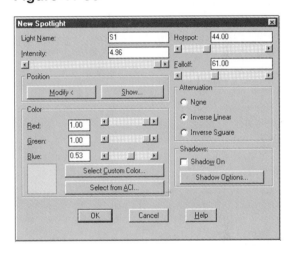

Hotspot **and** *Falloff*
These two sections determine characteristics of the cone of light emitted from a *Spotlight*. The *Hotspot* is the area of the model that receives full illumination (Fig. 41-37). The size of this area is determined by the angular value entered in the edit box or appearing there as a result of moving the slider bars. The light begins to fall off just outside the *Hotspot* until approaching no illumination at the *Falloff* cone. The area outside the *Falloff* cone receives no illumination. The *Falloff* angle must be equal to or greater than the *Hotspot* angle.

Figure 41-37

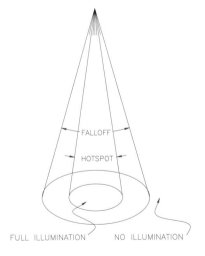

Position

You can indicate the *Location* and *Target* for the *Spotlight* in this cluster. A default *Location* and *Target* are calculated by AutoCAD when you create the light and those coordinates can be viewed by selecting the *Show...* tile (see Fig. 41-31). To specify another *Location* and *Target,* select *Modify <*. Enter coordinate values or PICK points (use *OSNAP* when PICKing in 3D space).

You can use the *Move* command to reposition the lights; however, it is recommended that you use the *Location* and *Target* options in the light dialog boxes. Changing a *Distance Light* location with *Move* also moves the *Target* position. A light can be deleted by *Erasing* its icon or by using the *Delete* option in the *Lights* dialog box.

Light Icons

For any light, an icon representing the light type and position appears in the drawing at the light *Location* (Figs. 41-38 and 41-39). These icons are actually *Block* insertions that appear in the drawing but not in renderings. The light icons are inserted on layer ASHADE, which is automatically created by AutoCAD when you place the first light. This layer can be *Frozen* to prevent the icons from appearing in drawings or plots.

Figure 41-38 ───────────────

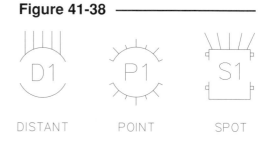

DISTANT   POINT   SPOT

Figure 41-39 ───────────────

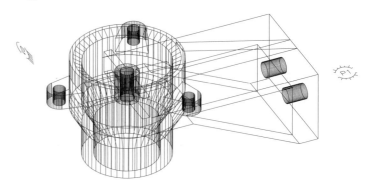

USING MATERIALS IN THE MODEL

When the background has been set and the lights have been inserted and adjusted for position, color, attenuation, and intensity, you are ready to select *Materials* and attach them to the model. Because the default template drawing has no materials loaded, one or more materials must be selected from the *Material Library*. Once the desired materials have been imported into the drawing, you must attach them to components of the model. Different materials can be attached to different components of the model. A material can be attached to an object, layer, or ACI (AutoCAD Color Index number). If a material is attached to a layer or ACI, every object on the layer or having the ACI assumes the material. Materials can be created and modified in a variety of ways to attain a particular color and reflective qualities.

MATLIB

	Pull-down Menu	COMMAND (TYPE)	ALIAS (TYPE)	Short-cut	Screen (side) Menu	Tablet Menu
	View *Render >* *Materials Library...*	*MATLIB*	...	...	VIEW 2 *Matlib*	Q,1

The *Matlib* command summons the dialog box that
provides access to the AutoCAD-supplied materials
(Fig. 41-40). The names in the *Library List:* (right side)
represent the choices from the RENDER.MLI file. A
material is brought into the drawing by highlighting it
and selecting the *Import* button. The selected mate-
rial(s) then appear in the *Materials List* (left side) representing all of the materials currently in the drawing
and ready for attachment to the model components.
Importing materials increases file size, so import only
those that are needed. This dialog box can also be
summoned from the *Materials* dialog box (see next).

Figure 41-40

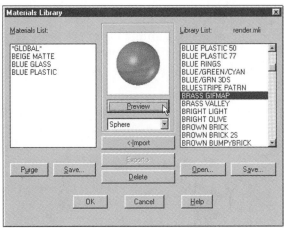

Preview
You can highlight <u>one material at a time</u> (from either
list) and select the *Preview* button to see how the material appears attached to the sphere or cube. Select
either the *Sphere* or *Cube* for preview from the drop-down list.

Import
Import brings material descriptions into the drawing from the RENDER.MLI or other .MLI files. After
materials are imported, select *OK*; then use the *Rmat* command (*Materials* dialog box) to attach materials
to objects, layers, or ACI numbers. If you try to import a name that exists in the drawing, the *Reconcile
Imported Material Names* dialog box appears, asking if you want to overwrite the existing material in the
drawing by the same name.

Export
Use this button to add selected materials from the *Materials List* to the *Library List*. This is useful when
you have created a new material or modified an existing material in the drawing and want to save it in
the material library for use in other drawings.

Delete
This option deletes selected materials from the *Materials List* or the *Library List*. Deleting materials from
the *Library List* does not delete them from the RENDER.MLI file.

Purge (Materials List:)
Use this button to purge all unattached materials from the material list. Unattached materials are those
that have been imported but have not been attached to objects, layers, or ACI numbers using the *Rmat*
command (*Materials* dialog box).

Save (Materials List:)

The materials listed on this side can be saved to a .MLI file for use in other drawings. Assign the desired file name to save in the *Library File* dialog box that appears.

Open (Library List:)

Material library (.MLI) files other than the default RENDER.MLI can be opened for material selection and used in the current drawing. Other .MLI files could be those you have created or those supplied with other software.

Save (Library List:)

The materials listed in the library list can be saved to a .MLI file for use in other drawings. If you have created new or modified existing materials, they can be saved in the RENDER.MLI or other .MLI file. Assign the desired file name to save in the subsequent *Library File* dialog box.

RMAT

Pull-down Menu	COMMAND (TYPE)	ALIAS (TYPE)	Short-cut	Screen (side) Menu	Tablet Menu
View *Render >* *Materials...*	RMAT	...	...	VIEW 2 Rmat	P,1

After materials have been imported from the materials library (see *Matlib*), this dialog box is used to attach selected materials to AutoCAD objects, layers, or ACI numbers. You can also gain access to the tools for creating new and modifying existing materials from this dialog box (Fig. 41-41).

Materials

This list represents the materials that are available for attachment. The list includes the materials imported previously using the *Materials Library* dialog box and a material named *GLOBAL*, which is the default material used for objects when you *Render* without attaching materials.

Preview

Use this button to see how the selected material (one at a time) appears on the sphere or cube. Select either a *Sphere* or *Cube* to use in the preview image tile from the drop-down list.

Materials Library...

This button invokes the *Materials Library* dialog box for importation of other materials (see *Matlib*).

Select <

This option temporarily removes the dialog box to allow selection of an object. When an object is selected, the *Materials Library* dialog box reappears with the selected object's material name and attachment method used displayed at the bottom of the dialog box.

Attach <

Use this button to attach the highlighted material from the *Materials:* list to an object. The dialog box is temporarily removed to allow selection of an object. When the *Render* command is used, the object is rendered with the attached material.

Figure 41-41

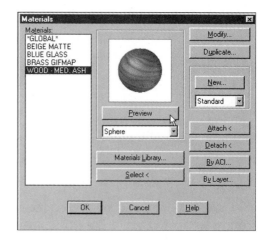

Detach

The *Detach* option allows you to unattach an assigned material by selecting an object.

By ACI...

You can attach a material to an *ACI* (AutoCAD Color Index number) using this button. With this method of attachment, every object in the drawing that has the selected color number displays the attached material when the *Render* command is used. The *Attach by AutoCAD Color Index* dialog box appears (Fig. 41-42).

To attach a material, you must select <u>both</u> the desired material from the list on the left and the desired color or number from the list on the right. Then PICK the *Attach* button and the material name appears next to the color in the *Select ACI:* list (see Fig. 41-42). If you want to *Detach* a material from a color number, select the choice from the *Select ACI:* list only; then PICK *Detach*.

Figure 41-42

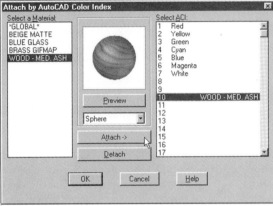

By Layer...

Using this option produces the *Attach by Layer* dialog box (not shown), which looks and operates identically to the *Attach by AutoCAD Color Index* dialog box. To attach a material, select <u>both</u> the desired material and the desired layer; then PICK *Attach.* Attached materials appear in the list with the layer name.

Modify...

Select a material from the list and choose the *Modify...* button to produce the *Modify Standard Material* dialog box. See Creating New and Modifying Existing Materials.

Duplicate...

If you want to create a new material but use an existing material as a starting point, choose this method. The highlighted name from the *Materials List:* is used as a template. The *New Standard Material* dialog box appears. See Creating New and Modifying Existing Materials.

New...

This button produces the *New Standard Material* dialog box and enables you to create a new material "from scratch." You can select from four template materials (in the drop-down list), *Standard, Granite, Marble,* and *Wood.* See Creating New and Modifying Existing Materials, next.

CREATING NEW AND MODIFYING EXISTING MATERIALS

Render allows you to control seven attributes that define a material: *Color/Pattern, Ambient, Reflection, Roughness, Transparency, Refraction,* and *Bump Map.* For many attributes, the *Value* (intensity) and *Color* can be adjusted. Other attributes such as *Transparency, Refraction,* and *Bump Map* have special adjustments. These adjustments are made in the *Modify Standard Material* and *New Standard Material* dialog boxes (Fig. 41-50). The attributes are first explained here.

Color/Pattern

Color/Pattern is the global color of the material. The color (hue) and the value (intensity) can be adjusted. Technically, *Color* represents the light that is reflected in a diffuse manner. Diffuse light affects all surfaces equally, not accounting for highlights (highlights are controlled by the *Reflection* attribute). Setting a low *Color* value makes the surfaces appear dark and dull (Fig. 41-43, left sphere). Increasing the *Color* value makes the surface reflect more light; therefore, the assigned color appears brighter (right sphere). Texture maps can be used in combination with this attribute. When a texture map is applied to the object it appears to be "painted" on the object's surface (see Image Mapping).

Figure 41-43

Ambient

Ambient light is background light that illuminates all surfaces. Because most of the surfaces are illuminated by *Point Lights, Distant Lights,* and *Spotlights,* adjusting the *Ambient* attribute controls the color and intensity of material "in the shadows." A low *Ambient* setting makes surfaces in the "shadows" appear dark and adds contrast (Fig. 41-44, left sphere), while a high *Ambient* setting increases the <u>overall lighting</u> and makes the "shadows" brighter (right sphere).

Figure 41-44

Reflection

This attribute controls the light that is reflected from *Point Lights, Distant Lights,* and *Spotlights. Reflection* accounts for the highlights on an object (technically called specular reflection). Controlling the value of *Reflection* increases or decreases the amount of reflected light. Use a low *Reflection* value to make the surfaces appear flat or matte, like blotting paper or soft material (Fig. 41-45, left sphere). Use a high *Reflection* value to increase light contrast on surfaces from *Point Lights, Distant Lights,* and *Spotlights* (right sphere). Reflection maps can be used in combination with this attribute.

Figure 41-45

Roughness

Roughness is a function of *Reflection*—if *Reflection* value is 0, *Roughness* has no effect. *Roughness* does not affect the amount of highlight, only the <u>size</u> of the highlight. A smaller *Roughness* value produces a small shiny spot on a sphere. Use a small *Roughness* value to produce a hard, shiny material (Fig. 41-46, left sphere). If a surface has a high *Roughness* value, its reflection is spread in a less perfect, larger cone. The larger the *Roughness* value, the larger the cone of reflection and the larger the highlight (right sphere).

Figure 41-46

Transparency
With this attribute you can make an object appear translucent or transparent. The *Value* adjusts the amount of transparency: from opaque (0.0), to translucent (0.1-0.9), to transparent (1.0). Do not use this attribute if you want materials to appear opaque (solid), as in Figure 41-47, left sphere. Use a medium-high *Value* to make an object appear translucent (right sphere). Opacity maps can be used in combination with this attribute (see Image Mapping). *Photo Real* must be selected in the *Render* dialog box for *Transparency* to be calculated.

Figure 41-47

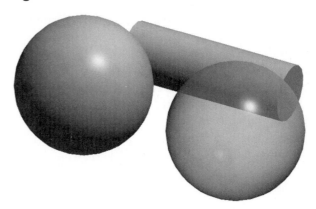

Refraction
This attribute is a function of *Transparency*, so this option can be used only when a positive *Transparency* value is set. Refraction is the bending of light waves as they pass through translucent or transparent material. Note the appearance of the cylinder through the right sphere in Figure 41-48 compared to the previous figure. *Refraction* calculates the effect of the light rays from the curvature of translucent objects. This feature requires the *Rendering Type* to be set to *Photo Raytrace* in the *Render* dialog box.

Figure 41-48

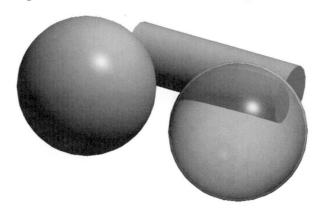

Bump Map
This attribute allows you to select a bump map to be applied to the object's surface. Bump maps create an embossed or bas-relief effect. For example, notice how the Autodesk "calipers" logo appears to be embossed on the surface of the right sphere (Fig. 41-49).

Figure 41-49

New Standard Material and *Modify Standard Material* Dialog Box

To make adjustments to a material's attributes, select either *Modify...*, *Duplicate...*, or *New...* from the *Materials* dialog box (Fig. 41-41). In any case, the resulting dialog box is the same; however, the title of the dialog box differs. For example, if you selected *Modify...*, the *Modify Standard Material* title is used (Fig. 41-50); otherwise, the *New Standard Material* title is used. Options in the dialog box are described here.

Figure 41-50

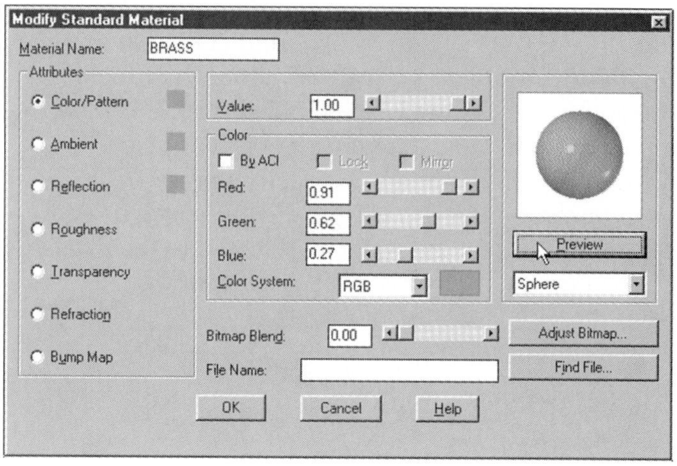

Material Name:

If you are modifying an existing material, the name appears here. You can rename an existing material, but it does not create a duplicate material. (Choose the *Duplicate...* button in the *Materials* dialog box to keep the original material and create a new one.)

Attributes

Only one of these radio buttons can be depressed. The selected *Color/Pattern, Ambient, Reflection, Roughness, Transparency, Refraction,* or *Bump Map* button determines which adjustments can be made in the *Value, Color,* and *Bitmap Blend* sections. See previous discussion of attributes.

Value

The magnitude or intensity of the selected attribute is adjusted by using the slider bar or entering a value in the edit box.

Color

This cluster enables you to set the color for each material attribute three ways: *RGB* system, *HLS* system, and the *ACI* pallet. These methods are identical to those used for setting light colors explained previously (see Color Systems).

If *By ACI* is checked in the *Color* cluster, the material color is set to match the object's drawing color (assigned *BYLAYER* or as an object-specific setting). In this case, the color is fixed to the object's drawing color and cannot be changed. Removing the "X" in the checkbox enables the *RGB* and *HLS* systems.

Lock

This is a useful checkbox to use if you want the *Ambient* and *Reflection* attribute colors to remain the same as the main *Color*. In this way, you can make adjustments to the attribute *Values* while ensuring that all reflection colors are not changed. Remove the check from *Lock* if you want *Ambient* and *Reflection* colors to be different than the object's global color.

Mirror

This feature (enabled only for the *Reflection* attribute) allows you to create mirrored reflections. The *Photo Raytrace* renderer must be used for this effect. If an image file is specified for the reflection map, the resulting reflection includes the background and the reflection map.

Bitmap Blend

See Image Mapping for a discussion of mapping concepts and this dialog box section.

Dialog Boxes for *New Granite, Marble,* and *Wood Material*

When you create a new material, you can select from four AutoCAD-supplied materials to use as a template: *Standard, Granite, Marble,* or *Wood.* Select from the drop-down list in the *Materials* dialog box before selecting the *New* button (Fig. 41-51, right side).

Figure 41-51

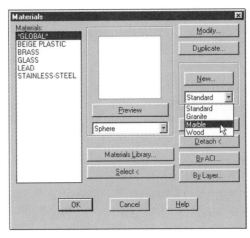

Selecting the *Standard* material template, then *New...* produces the *Create New Standard Material* dialog box which is identical to the *Modify Standard Material* dialog box except for the title (see Fig. 41-50). This dialog box and all related options (except the *Bitmap Blend* section) were previously discussed.

Create New Granite Material Dialog Box

If you want to create a new granite material, select *Granite* from the drop-down list (Fig. 41-51), then select *New.* The *Create New Granite Material* dialog box appears (Fig. 41-52). A template that simulates granite material appears.

Figure 41-52

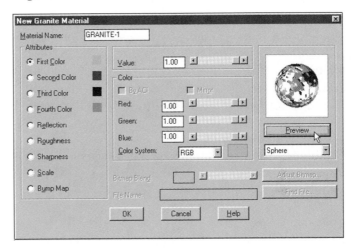

Material Name, Value, Color, Preview, Bump Map, and *Bitmap Blend*

These options have identical functions to those options in the *Create New Standard Material* dialog box. See the previous section for information related to these options. See Image Mapping (next) for information on *Bitmap Blend.*

First, Second, Third, and *Fourth Color*

Granite can have four colors present. Select the desired button to change the color. The *Value* specifies the amount of each color in the granite pattern. To use fewer than four colors, enter a *Value* of zero for one or more colors. The color systems are the same as for *Standard* material.

Reflection and *Roughness*

Reflection adjusts the material's highlight (specular) *Color* and *Value.* In other words, *Reflection* is the amount of light reflected or the intensity of the highlight.

Roughness is a function of *Reflection.* *Roughness* adjusts the <u>size</u> of the highlight. A small *Roughness* value makes the material (granite) appear hard and shiny (like polished granite), whereas a high *Roughness* value makes the material appear rough and abrasive.

Sharpness

This setting defines the sharpness of the edges of color on the stone. A *Sharpness* of 0 makes the colors blurred, or blended together. A *Sharpness* of 1.0 makes the colors distinct with sharp edges between different colors.

Scale

Use this option to scale the pattern relative to the objects you attach it to. Larger values create a bigger pattern while smaller values create a tighter pattern.

Create New Marble Material **Dialog Box**

Figure 41-53

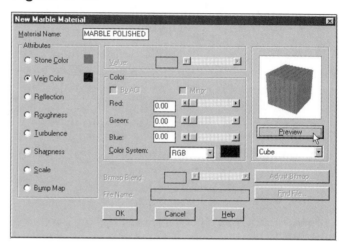

Select *Marble* from the drop-down list in the *Materials* dialog box (Fig. 41-51), then select *New* to produce the *Create New Marble Material* dialog box (Fig. 41-53). The template material is marble.

Material Name, Value, Color, Bump Map, **and** *Bitmap Blend*

These options have identical functions to the same options in the *Create New Standard Material* dialog box. See the previous sections for information related to these options. See Image Mapping (next) for information on *Bitmap Blend*.

Reflection, Roughness, Scale, **and** *Sharpness*

These options are the same as discussed in the previous section, *Create New Granite Material* Dialog Box.

Preview

It may be helpful when creating marble to view both the *Sphere* and the *Cube* in the *Preview* window. The *Cube* gives a clear indication of the "vein" characteristics in the stone.

Stone Color **and** *Vein Color*

Marble has two color characteristics: the color of the body of the marble and the color of the long "veins" or "swirls" that run through the stone color. Usually a white or black color is typical for the veins in marble.

Turbulence

This feature adjusts the amount of "swirl" and color in the veins of the stone. Higher values produce more vein color and more swirling, but higher values increase rendering time. Values between 1 and 10 are recommended.

Create New Wood Material **Dialog Box**

Figure 41-54

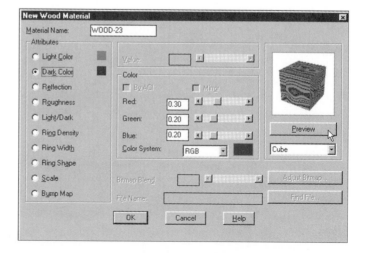

Selecting *Wood* from the drop-down list in the *Materials* dialog box allows you to create a new wood pattern from a supplied template. The *Create New Wood Material* dialog box has the following settings (Fig. 41-54).

Material Name, Value, Color, Preview, Bump Map, Reflection, Roughness, Scale, **and** *Bitmap Blend*
These options have functions identical those options in the *Create New Standard Material* and other material dialog boxes. The *Bitmap Blend* section is discussed later in Image Mapping.

Light Color **and** *Dark Color*
To visually describe wood, use two colors to define the main body of wood and the grain. *Light Color* is generally used as the color for the body of the wood (soft, wide grain), and *Dark Color* is used for the thin grain in the wood. Generally shades of brown are used.

Light/Dark
This adjustment controls the ratio of light color to dark color, that is, the size of the light grain in proportion to the dark grain. A low *Light/Dark Ratio* (adjusted by the *Value* slider) creates more dark grain color, while a high value produces more light color grain.

Ring Density
This adjustment varies the density of the grain, or viewing from the end of a "log," the number of rings. Larger values create more rings and a denser grain.

Ring Width
You can change the variation in the widths of grain by adjusting the value of *Ring Width*. A value of 0 produces uniform width for all grain (or rings, if viewed from the end). A high value creates grain (rings) of different widths.

Ring Shape
Use *Ring Shape* to control circularity of the rings of grain (when viewed from the end of a "log"). A value of 0 gives perfectly circular rings, whereas a high value gives distorted (not round) rings. Use a high value to create more "swirls" in the wood grain.

IMAGE MAPPING

Image mapping is the process of projecting a 2D image onto the surface of a 3D object. In AutoCAD Release 14, the image used for projecting can be any image in BMP, TGA, TIFF, and JPEG file formats (Fig. 41-55, for example, is the CAMOFLAB.TGA file in the AutoCAD R14\Textures directory.)

Figure 41-55

Another type of mapping accomplished with the *Setuv* command allows you to adjust the <u>orientation</u> of an image or of an AutoCAD material. *Setuv* can be used to change the orientation of <u>materials</u> as well as mapped <u>images</u> on objects. See *Setuv* at the end of this section.

Image mapping can be accomplished by using the *Rmat* command and *Create Material* or *Modify Material* dialog boxes (including those for creating new or modifying existing *Standard, Granite, Marble,* or *Wood* material templates) to attach any of the four types of maps. See Figure 41-50, *Modify Standard Material* dialog box.

You must render using the *Photo Real* or *Photo Raytrace* renderer for mapped images to be visible in the resulting rendering.

You can use four kinds of image maps in AutoCAD Release 14: texture maps, reflection maps, opacity maps, and bump maps.

R14

Texture Maps

A texture map is the projection of an image (such as a tile pattern) onto an object (such as a table or floor). Texture maps define colors for the surface of the object, as if the bitmap image were "painted" onto the object. For example, you could apply an image of a brick pattern to a horizontal flat surface to create the appearance of a brick floor. Texture maps conform to the attached object's surface geometry and therefore are affected by light and shadow. These capabilities can result in some very realistic renderings.

Figure 41-56 displays the CAMOFLB.TGA image mapped onto the right sphere. You can attach a texture map using the *Modify Standard Materials* dialog box. This example was accomplished by using the *Modify Standard Materials* dialog box with the *Color/Pattern* attribute on and assigning the image file in the *Bitmap Blend* section (see Fig. 41-50).

Figure 41-56

Reflection Maps

A reflection map (also referred to as an "environment map") simulates the effect of a scene reflected on the surface of an object. For example, an image used as a background (maybe a sky scene) can be reflected on an object with a highly reflective surface (such as chrome-plated metal). For reflection maps to render well, the object's material should have high reflection and low roughness, and the image file used for the reflection bitmap should have a high resolution.

Figure 41-57 displays the pattern from the right sphere "mapped" on the left sphere. This was accomplished by selecting the *Mirror* checkbox and setting a high *Reflective* and low *Roughness* value to the material for the left sphere in the *Modify Standard Materials* dialog box. Although the pattern is not actually attached (mapped) to the left sphere, it is considered an environmental map since it affects the appearance of the left sphere.

Figure 41-57

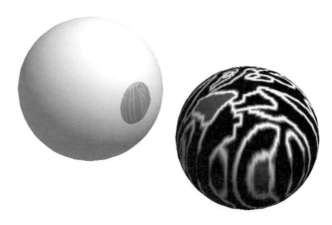

Opacity Maps

Opacity maps create the illusion of opaque and transparent areas on the attached object. (Since most objects are naturally opaque, think of an opacity map as a "transparency" map.) AutoCAD uses the color (lightness or darkness) of the mapped image to determine opacity or transparency. Pure white areas in an opacity map are translated as completely opaque, while pure black areas are treated as transparent. For example, if an image of a white square with a black circle were mapped onto a surface, the white area would be seen as solid (opaque) and the black area would appear as a hole in the surface. If an opacity map is a color image (not black and white), only the equivalent gray-scale values of the colors are used for the opacity calculation. Opacity maps do not necessarily change the color of the opaque areas of the mapped object.

Figure 41-58

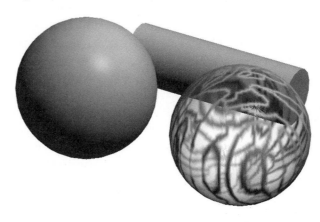

Figure 41-58 displays the previous image attached as an opacity map to the right sphere. This effect was created by setting a high *Transparency* attribute value in the *Modify Standard Materials* dialog box and using the *Bitmap Blend* section to map the CAMOFLAB.TGA image (with the *Transparency* attribute on.)

Bump Maps

Bump maps are similar to opacity maps in that the value (gray-scale) of the image is translated to change the surface appearance of attached object. However, in this case the brightness values of a bump map image are translated into apparent changes in the height of the surface of an object. For example, consider mapping an image of white text with a black background onto a surface. The resulting rendering would give the text (white image) the appearance of being raised (or embossed) against a flat (black) background. The surface has not actually changed, although the <u>appearance</u> of the surface (due to the map) has changed. Bump mapping increases rendering time significantly.

Figure 41-59

The bump map in Figure 41-59 (right sphere) was created by selecting the *Bump Map* attribute, then using the *Bitmap Blend* section to specify and adjust the image. Use a small *Bitmap Blend* value (from 0.1 to 0.02).

R14

Combination Maps

You can apply maps in combination. For example, you can apply a wood grain bitmap as both a bump map and a texture map on a paneled wall to give the wall both the "feel" and the color of wood. In addition, an opacity map (a white surface with black dots) could be used to create the illusion of worm holes in the wood.

All maps have a blend value that specifies how much they affect the rendering. For example, a texture map with a blend value less than the maximum (1.0) allows some of the material's surface colors to show through. Lower blend values reduce the effect of the mapped image.

Adjust Material Bitmap Placement **Dialog Box**

Invoke this dialog box (Fig. 41-60) by selecting the *Adjust Bitmap...* button in the *New Standard Material* or *Modify Standard Material* dialog box. This box is essentially the same as the *Adjust Background Bitmap Placement* dialog box used to place a background image with the *Background* command (see Figure 41-15).

Figure 41-60

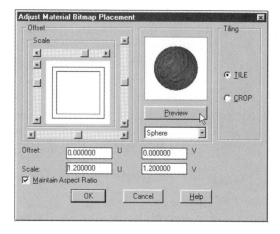

Offset **and** *Scale*

The image in the upper-left corner indicates the extents of the object (red box) and the bitmap image (black and magenta rectangle). Initially the two rectangles are the same size, indicating a mapping scale of 1:1. *Offset* values range from 1 to -1 in either dimension (U or V). Adjust the *Offset* and *Scale* by moving the slider bars or entering values in the edit boxes.

Maintain Aspect Ratio

This checkbox keeps the U and V dimensions together. Remove the check to scale the image more or less in one dimension.

Tile **and** *Crop*

If the size of the image you use is smaller than the objects, you can choose to create either a tiled or a cropped map.

A tiled map repeats the pattern indefinitely, fitting together like "tiles," until the entire object is covered. This effect is used in Microsoft Windows when a small image is tiled repeatedly to produce a pattern on your desktop. Tiling is adjustable to obtain different tiling values along the mapping axes, U and V.

With cropped projection, you fix one image in a specific location on an object. The rest of the object is rendered with the original colors of the material. This effect is sometimes known as a "decal" map because fixed-size image is attached to the object's surface.

Setting Mapping Coordinates

If you want to change the orientation of an AutoCAD <u>material</u> or of a mapped <u>image</u> with respect to the object, you can change the mapping coordinates with the *Setuv* command. Normally, when a material such as wood or an image such as a checkerboard is attached to an object, AutoCAD automatically sets the orientation of the material or image on the objects. As you know from the previous discussion, you can adjust the position of an image only with the *Adjust Material Bitmap Placement* dialog box. In contrast, *Setuv* allows you to change the entire orientation of both materials and mapped images.

For example, assume you used the *Rmat* command to assign wood material to a 3D box to simulate a block of wood. Using the default AutoCAD material orientation of the material with respect to the box, the object would render with the results shown in Figure 41-61.

Figure 41-61

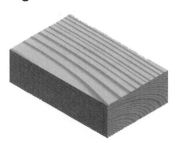

Let's assume that you wanted the block of wood to appear as if it were cut diagonally across the grain. You could use the *Setuv* command to change the orientation of the material to achieve a rendering such as that in Figure 41-62.

Figure 41-62

Mapping coordinates are also referred to as *UV* coordinates, hence the command *Setuv*. The letters *UV* are used because the coordinates are independent of the *XY* coordinates used to describe the AutoCAD geometry; the coordinates are related to the mapped objects, not to any traditional coordinate system.

SETUV

Pull-down Menu	COMMAND (TYPE)	ALIAS (TYPE)	Short-cut	Screen (side) Menu	Tablet Menu
View *Render >* *Mapping…*	*SETUV*	…	…	*VIEW 2* *Mapping*	*R,1*

Setuv allows you to change the orientation of an attached <u>material</u> or mapped <u>image</u> with respect to the object. *Setuv* sets the mapping coordinates (U, V, and W). See previous discussion, Setting Mapping Coordinates.

When you use *Setuv,* AutoCAD first prompts you to select objects. When you have selected the AutoCAD objects to which you want to assign mapping coordinates, AutoCAD displays the *Mapping* dialog box (Fig. 41-63). You can use the *Preview* box in the *Mapping* and *Adjust Projection* dialog boxes to experiment to find the best result before you do a full rendering.

Figure 41-63

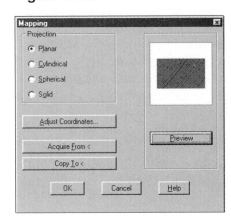

Projection
Select one of the buttons to specify *Planar, Cylindrical, Spherical,* or *Solid* projection. Normally, select the projection type that matches the overall shape of the object for which you want to specify coordinates.

Adjust Coordinates
Selecting this button displays a dialog box for adjusting projection coordinates relative to the object. The dialog box that is displayed is based on your choice of *Projection: Planar, Cylindrical, Spherical,* or *Solid.*

Acquire From

Use this option to return to the drawing so you can select one object to "copy" coordinates from. The object should already have mapping coordinates assigned to it—these become the current mapping settings. You can accept the acquired mapping coordinates as is or use *Adjust Coordinates* to modify them.

Copy To

This option allows you to copy the current mapping coordinates to any number of objects. When you finish the selection, then choose *OK,* the current mapping coordinate settings are applied to the objects you selected.

Adjust Planar Projection **Dialog Box**

When you select *Planar Projection,* then *Adjust Coordinates,* this dialog box appears (Fig. 41-64). Use this dialog box to specify the plane used for projecting the bitmap onto the object. *Planar* projection maps the material or image onto the object with a one-to-one correspondence, as if you were projecting the image from a slide projector onto the surface. This does not distort the texture; it just scales the image to fit the object.

Figure 41-64

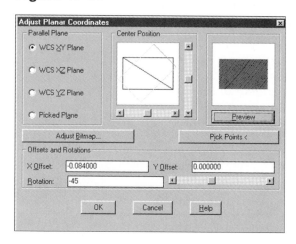

Parallel Plane

This section is used to select one of the three perpendicular reference planes of the WCS or a plane you pick using *Pick Points.* If you want to select a plane, use *Pick Points,* then select *Picked Plane.*

Center Position

Shows a diagram of normal to the object's current parallel plane (red image). The current projection square for the material or image is in blue (the top of the projection square is indicated by a small blue tick mark, and the square's left edge is shown in green). You can move the material/image projection square by moving the sliders.

Adjust Bitmap

Choose this button to adjust the material/image offset and scaling. This dialog box operates similar to those for *Adjust Material Bitmap Placement* (see Fig. 41-60) and *Adjust Background Bitmap Placement.*

Pick Points

Selecting this button temporarily displays the Drawing Editor so you can specify a projection plane by picking points on an object. You must select three points: the lower-left, lower-right, and upper-left corners of the plane.

Offset **and** *Rotation*

You can change the *X* and *Y* offsets of the map by entering new values in the boxes. You can change the rotation by entering a new value or by moving the slider.

Adjust Cylindrical Projection Dialog Box

Selecting the *Cylindrical* projection type, then *Adjust Coordinates* displays the *Adjust Cylindrical Projection* dialog box (Fig. 41-65). Here you can define the axis of the cylindrical coordinate system and the wrap line for the bitmap projection. *Cylindrical* projection maps an image or material onto an object with a cylindrical projection; the horizontal edges are wrapped together but not the top and bottom edges. The height of the image is scaled along the cylinder's axis.

Figure 41-65

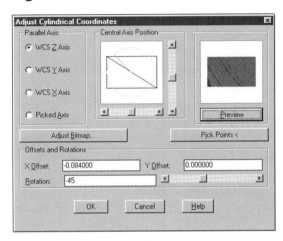

Parallel Axis
Use this option to select one of the three perpendicular axes of the WCS or an axis you pick using the *Pick Points* button.

Central Axis Position
This section shows a view of the selected objects' meshes (red image) on a plane perpendicular to the current axis. It also shows the projection axis with a blue circle (the center of the circle is the projection axis). A green radius shows the wrap line.

By default, the axis is placed at the center of the object's extents in both directions, with the base of the cylinder at the minimum vertical extent and its top at the maximum vertical extent. The wrap line is placed toward minimum Z—or minimum Y when the Z axis is parallel to the cylindrical coordinate system axis.

Pick Points
Use this option to select a projection cylinder by picking points in the drawing. You must select three points: the center-bottom of the mapping cylinder, the center top of the cylinder, and the direction toward the seam (wrap line).

Offset and *Rotation, Adjust Bitmap*
These options operate similarly to the other dialog boxes (see *Adjust Planar Projection* Dialog Box, previous).

Adjust Spherical Projection Dialog Box

This dialog box is used to define the polar axis of the spherical coordinate system and the wrap line for the bitmap projection. *Spherical* projection wraps the texture both horizontally and vertically. The top edge of the texture map is compressed to a point at the "north pole" of the sphere as is the bottom edge at the "south pole."

The *Adjust Spherical Projection* dialog box has the same options and operates similarly to the *Adjust Cylindrical Projection* dialog box (see Fig. 41-65).

Adjust *UVW Coordinates* Dialog Box

Figure 41-66

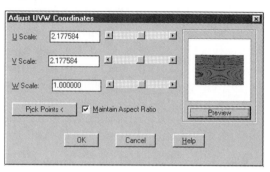

Use this dialog box (Fig. 41-66) to adjust the three coordinates to shift a solid, 3D material: marble, granite, or wood. Because solid materials are three-dimensional, they have three mapping coordinates, U, V, and W, and can be applied from any angle. You do not always need to specify the mapping coordinates for these materials, but you can. For example, you might want to change the material's orientation for a particular rendering or skew a pattern along one dimension.

U Scale, V Scale, and *W Scale*

The material/image scale changes the scale of the material relative to the object mapping coordinates in any dimension (U, V, or W). Adjust the scale by using the three sliders or by entering a value in the edit boxes. The scale is the number of times the image fits onto the object in the U, V, or W direction. Select *Preview* to see the results of the new scale.

Maintain Aspect Ratio

The *Maintain Aspect Ratio* check box locks the U, V, and W scaling dimensions together. When *Maintain Aspect Ratio* is turned on, moving one slider moves the other, and a value you enter in one box changes the other. Remove the check to change the scale in any one, two, or three independent directions to distort the image or material for particular effects.

USING *SCENES*

When the *Background* has been set (if desired), *Lights* have been inserted, and *Materials* have been attached and adjusted, you can use the *Render* command to calculate the rendered view and display it in the viewport or rendering window, or write the image to a file (whichever is configured). If no *Scene* has been specified, AutoCAD uses the current view and all lights in the rendering. However, you can choose to use *Scene* to specify a particular view or lighting combination. *Scenes* are helpful if you want to create several renderings of a model using different viewpoints and different light combinations.

SCENE

Pull-down Menu	COMMAND (TYPE)	ALIAS (TYPE)	Short-cut	Screen (side) Menu	Tablet Menu
View *Render >* *Scene...*	*SCENE*	...	...	VIEW 2 *Scene*	*N,1*

If you want to render objects from one of several named *Views* or use different light combinations for a viewpoint, you can use the *Scene* command to select the *View* and *Light* combinations for each render. You can save particular *Views* and lighting configurations to a *Scene* with a name that you assign. This action allows you to create multiple renderings without having to recreate the configuration each time. *Scenes* do not have icons that are inserted into the drawing like *Light* icons.

Figure 41-67

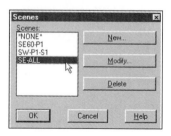

Invoking the *Scene* command by any method displays the *Scenes* dialog box (Fig. 41-67). Selecting the *NONE* entry renders the current viewpoint and uses all lights inserted in the drawing. If no lights are inserted, AutoCAD uses one from over the shoulder like a default rendering.

New...

Selecting the *New...* option allows you to create a new *Scene* for the drawing and produces the *New Scene* dialog box (see Fig. 41-68).

Modify...

This option produces the *Modify Scene* dialog box where you can modify existing *Scenes*. This dialog box is the same as the *New Scene* dialog.

Delete

Delete removes the highlighted scene from the drawing.

Creating New and Modifying Existing Scenes

The *New Scene* and the *Modify Scene* dialog boxes allow you to select from a list of existing named *Views* and inserted *Lights*. First, assign the desired name for the scene by entering it in the edit box in the upper-right (Fig. 41-68). The name is limited to eight characters, no spaces.

Figure 41-68 ───────

The *Views* column lists all previously created views. The *CURRENT* view can be selected if no *Views* have been previously assigned. You can only have <u>one</u> *View* in a *Scene*. Selecting an existing view makes it the current view for the drawing as well as that used for the subsequent *Render*.

You can also select *Lights* from the list to be included for the rendering of the scene. Any number of lights can be selected for a scene. Select multiple lights by holding down Ctrl while selecting. Highlighted lights are included in the scene. There is no limit to the number of *Scenes* that can be created.

RENDERING AND SPECIFYING PREFERENCES

Finally, after setting a *Background*, inserting *Lights*, attaching and modifying *Materials*, you are ready to set the rendering preferences and create the rendering. Using the *Render* command is the easiest step in creating a rendering because Render does the work.

Use the *Render* command to set the preferences and create the rendering. If you want to set preferences without creating a rendering, you can use *Rpref* to produce the *Rendering Preferences* dialog box (see next).

RENDER						
	Pull-down Menu	**COMMAND (TYPE)**	**ALIAS (TYPE)**	**Short-cut**	**Screen (side) Menu**	**Tablet Menu**
	View *Render >* *Render...*	*RENDER*	*RR*	...	*VIEW 2* *Render*	*M,1*

Using this command causes Render to calculate and display the rendered scene to the configured device. The rendering is generated according to the settings in the *Render* or *Rendering Preferences* dialog box.

Render uses the *Lights* and *View* selected in the *Scene* dialog. If no *Scene* or selection set is specified, the current view is rendered with all *Lights*. If there are no *Lights* in the drawing, Render assumes an "over-the-shoulder" *Distant Light* source with an intensity of 1.

After issuing the *Render* command, the *Render* dialog box appears (Fig. 41-69) unless the *Skip Render Dialog* option is checked in the *Rendering Preferences* dialog box. This dialog box offers the same options as the *Rendering Preferences* dialog box.

Figure 41-69

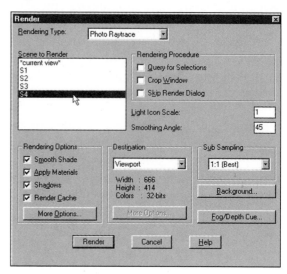

After setting the preferences or accepting the defaults, select the *Render* button to generate the rendering. Rendering can take considerable time, depending on the complexity of the scene, your computer system capabilities, and the settings in the *Render* or *Rendering Preferences* dialog box. Once you have selected the *Render* button to initiate the rendering process, you can stop (cancel) the rendering by pressing Esc.

It is a good habit to determine or check the settings in this dialog box immediately before using the *Render* command. Since rendering takes time, taking a minute to make sure the desired options are set beforehand can save wasted rendering time.

Render Dialog Box

Rendering Type
This drop-down box lists *Render, Photo Real,* and *Photo Raytrace*. Other rendering devices can be installed.

Render
This is the default option. *Render* creates the fastest rendering. Use this option to save time when you first begin adjusting *Background, Lights,* and *Materials* for the model.

Photo Real
This of *Render* can display transparent materials and bitmapped images. Mapped images are images from an external file that can be planar (as for a background) or mapped to geometry such as a sphere, box, or other shape. *Photo Real* can also generate volumetric and mapped shadows.

Photo Raytrace
Photo Raytrace produces the highest quality (most realistic) renderings. Raytracing is the process of calculating or following the paths of light rays as they are generated from a source and cast shadows or bounce off objects and illuminate other objects. Raytracing is needed to calculate reflections of light and color.

Scene to Render
This section lists all scenes previously created, including the current view. Select any scene from the list to render.

Rendering Procedure

Query for Selections
When you render, you are prompted to specify a selection set to render if this button is selected. This is an excellent way to save time when you need to render only one object or set of objects in the view as a test, such as when you are adjusting *Material* for one object.

Crop Window

When this option is checked and you then select *Render*, AutoCAD prompts you to select an area (window) on the screen to render. This is also a good way to save time when working on only one section of the model.

Skip Render Dialog

Normally, the *Render* dialog box appears when you use the *Render* command. Checking this box skips this procedure. In this case, you can use the *Rendering Preferences* dialog box to set your preferences.

Rendering Options

Smooth Shading

If *Smooth Shading* is checked, Render automatically smoothes out the multifaceted appearance of polygon meshes. With *Smooth Shading* off, the mesh edges appear in the final rendering. If an edge defines a corner of greater than 45 degrees, it is not affected by *Smooth Shading*.

Apply Materials

This option applies your materials when the rendering is calculated. If *Apply Materials* is not checked, Render uses the color, ambient, and reflection settings for the default **GLOBAL** material for the rendering.

Shadows

This is the global setting for calculating shadows. Each *Light* also has a *Shadows* checkbox. For a *Light* to calculate shadows, this button as well as the individual *Light's* button must be checked. *Photo Real* and *Photo Raytracing* must be used to calculate shadows.

More Options…

Selecting this tile produces a dialog box with advanced options. The dialog box that appears depends on whether you have *Render, Photo Real* or *Photo Raytrace* selected. See *AutoCAD Rendering Options* Dialog Box at the end of this section.

Destination

Viewport

If this option is checked, Render calculates the rendering and displays it in the Drawing Editor. If you are using tiled or paper space viewports, the rendering is in the current viewport.

Render Window

Selecting this option creates the render in a separate Window, not in the Drawing Editor.

File

If this option is checked, the rendering is calculated and written to the file but is not displayed. You can specify the type of file to render to by selecting *More Options…* in this cluster, which produces the *File Output Configuration* dialog box. See Saving a Rendered Image to a File in Saving, Printing, and Replaying Renderings.

If you specified that Render create a file, the rendering is calculated and written to the specified file or device but is not displayed on the screen (see Saving, Printing, and Replaying Renderings). Alternately, if the *Viewport* or *Rendering Window* is the selected *Destination* and you decide to save the rendered screen display after creating the rendering, you can use *Saveimg* (see *Saveimg*) to save the image as a .BMP, .TGA, or .TIF, or you can use *Save* from the Render Window to save the image as a .BMP file.

Lights Icon Scale
Enter a value to scale the *Light* icons that are inserted into the drawing. Any value larger than 0 is valid.

Smoothing Angle
This value determines the angle at which smoothing is applied. For any angles on the object less than the value specified, AutoCAD renders as if no edge exists (see *Smooth Shading*).

Sub Sampling
Use this option to reduce rendering time and image quality without dropping the effects (such as shadows). This feature speeds up rendering by rendering only a sample of the pixels. Select an option from the drop-down list ranging from 1:1 (slowest, best quality) to 8:1 (fastest, lowest quality).

Background...
This option invokes the *Background* dialog box (see the *Background* command).

Fog/Depth Cue...
Selecting this button produces the *Fog/Depth Cue* dialog box (see the *Fog* command).

Render Options Dialog Box

The *Rendering Options* dialog box (Fig. 41-70) appears when you select *More Options...* in the *Rendering Options* section of the *Rendering* dialog box, and <u>*Render*</u> is the selected *Rendering Type*.

Figure 41-70

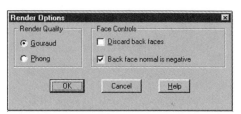

Render Quality

Gouraud
This method calculates light intensity at each vertex and interpolates the intensities between vertices. As a result, a sphere, for example, may have a diamond-shaped highlight.

Phong
This creates the most realistic type of rendering with higher-quality highlights. Phong calculates light intensity at each pixel.

Face Controls

Discard Back Faces
This setting prevents Render from reading the back faces of surfaces when rendering, which can save considerable time calculating the rendering (see *Back Face Normal Is Negative*).

Back Face Normal Is Negative
Checking this setting reverses which faces Render considers as back faces. Normally, a positive normal vector points outward from the interior of the object toward the viewer. When constructing surfaces such as a *3Dface,* specifying vertices in a counterclockwise direction creates a positive normal vector toward the observer. Turning off this setting reverses the direction AutoCAD considers back faces. If surfaces in a surface model rendering are black and *Discard Back Faces* is checked, reverse this setting or turn off *Discard Back Faces.*

Photo Real and *Photo Raytrace Render Options* Dialog Boxes

These dialog boxes appear when either <u>*Photo Real*</u> or
<u>*Photo Raytrace*</u> is selected and you select *More
Options...* in the *Rendering Options* section of the
Render or *Rendering Preferences* dialog box. The *Photo
Real* options are the same as the *Photo Raytrace* options,
excluding *Adaptive Sampling* and *Ray Tree Depth*. Only
the *Photo Raytrace Rendering Options* dialog box is
shown here (Fig. 41-71).

Figure 41-71 ————————————————

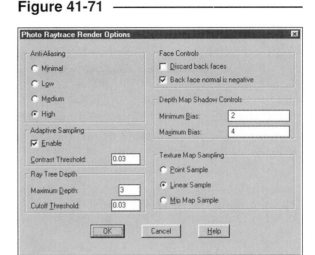

Anti-Aliasing
This section offers four levels of anti-aliasing resolution. Aliasing is the jagged edges on a diagonal edge
caused by rows and columns of pixels. The range (*Minimal* to *High*) controls how many multiple
samples are rendered per pixel. The trade-off in quality is the increased rendering time.

Adaptive Sampling
Anti-aliasing can be accelerated with *Adaptive Sampling* enabled. This feature prevents anti-aliasing
from samples with low color contrast. The *Contrast Threshold* value determines the level of contrast
(from 0 to 1.0) before anti-aliasing takes place. Higher values increase rendering time at the expense of
image quality.

Ray Tree Depth
This option controls the depth of the tree to track reflected and refracted rays. Greater values for
Maximum Depth give better images (10 is the recommended maximum value). The *Cutoff Threshold* value
(percentage) determines how strong the rays must be for each "bounce" before they are terminated.
Higher values create a greater tree depth.

Depth Map Shadow Controls
Use this option to prevent "detached" shadows (from the object). Set *Minimum Bias* from 2 to 20 and
Maximum Bias from 4 to 10.

Texture Map Sampling
When a texture map is projected onto an object smaller than itself, use this section to control how the
image is sampled. *Point Sample* causes render to use the closest pixel, *Linear Sample* is a linear average of
pixels, and *Mip Map Sample* takes a triangular sample.

Face Controls
See *Render Options* Dialog Box.

RPREF

	COMMAND (TYPE)	ALIAS (TYPE)	Short-cut	Screen (side) Menu	Tablet Menu
Pull-down Menu					
View Render > Preferences...	*RPREF*	*RPR*	...	*VIEW 2 Rpref*	*R,2*

The *Rpref* command invokes the *Rendering Preferences* dialog box (not shown). This box is identical to the *Render* dialog box except you cannot generate a rendering from this box (there is no *Render* button). Use this box when you want to make changes to the rendering preferences, then press *OK* to save the changes. (The *Render* dialog box provides no *OK* button to save changes; *Render* and *Cancel* are the only two methods to leave the *Render* dialog box.)

Because the options in the *Rendering Preferences* dialog are identical to those in the *Render* dialog box (other than the *Render* button), see the previous sections for information on these options.

Rendering Example

A solid model of an intake housing and machinist's file are used here as an example for creating a *Background,* inserting *Lights,* creating *Materials,* using a *Bump Map,* setting rendering preferences, and creating a rendering (refer to Typical Steps for Using Render).

1. After the geometry is completed, *Dview* or *Vpoint* is used to provide an appropriate viewing orientation for the rendering. If desired, the viewpoint can be saved to a named *View* to be used later for defining *Scenes*. Figure 41-72 displays the solid models ready to begin the rendering process. Notice that the drawing includes 3 objects: the intake housing, a machinist's file, and a surface (only the back edge of the *Box* appears) that is used to simulate a table top, which will provide a surface for shadows and reflection for the intake housing and file.

Figure 41-72 ──────────────────────────────────────

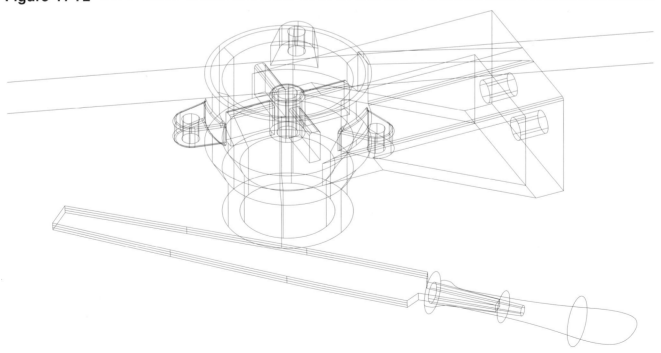

2. To begin the rendering setup, a *Background* is created (Fig. 41-73). Using the *Gradient* option, three colors are selected. A 90-degree *Rotation* is specified for the bands of color to give the illusion of vertically oriented background objects and to provide directional contrast to the horizontal orientation of the objects to be rendered. A quick, "rough" rendering is generated for checking the background using the *Render* option of *Rendering Type* (not *Photo Real* or *Photo Raytrace*). The rendering is generated with no lights or materials specified (using the "over-the-shoulder" light and GLOBAL material). The *Query for Selections* option is checked in the *Render* dialog box to allow selec-

Figure 41-73

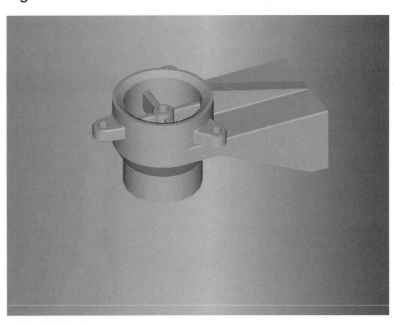

tion of only the intake housing to speed rendering time. All other options in the *Render* dialog box are set to the minimum, fastest, or default values.

3. Next, the *Lights* dialog box series is used to insert a light in the drawing (Fig. 41-74). For this example, the first light is a *Point Light* positioned in the upper right. Setting the *Point Light* color to a cool color (very light shade of purple or blue) can be used to reinforce the appearance of metallic surfaces. *Shadows On* is selected for the light when created, but the (global) *Shadows* option (in the *Render* dialog box) is off for this preliminary work. The light position illuminates the front, top, and right sides of the intake housing. Notice how the point light illumination falls off along the top surface. Since only one light is used, the top and right side sur-

Figure 41-74

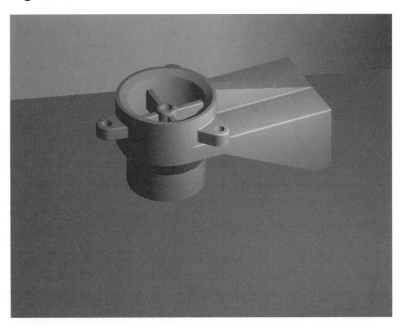

faces appear very bright, while the back of the cylinder has no light. Also notice the "floating" effect of the model due to the lack of shadows.

A material is selected and modified to represent a smooth, reflective metal for the intake housing (see Fig. 41-74). In the *Modify Standard Material* dialog box, the *Reflection* attribute is set with a high value and *Roughness* is given a low value. A hard, dark, and reflective material is selected for the table top. Only the intake housing and table top are selected for the rendering (use *Query for Selection*). *Photo Real* is set as the *Rendering Type* to enhance the light reflective characteristics of the materials. Generating a *Render* (Fig. 41-74) displays the results of the light position and intensity as well as material characteristics. You can make several adjustments to *Light* and *Material* settings, then generate "rough" renderings until the desired effect is obtained.

4. Next, materials are assigned to the file and handle (each created as a separate model). The handle has a supplied wood material and the metal portion of the file has a standard metallic material attached (Fig. 41-75). The *Photo Real* renderer is used, but *Shadows* are toggled off to speed rendering time when checking and adjusting material characteristics for the file and handle.

Figure 41-75 ————————————

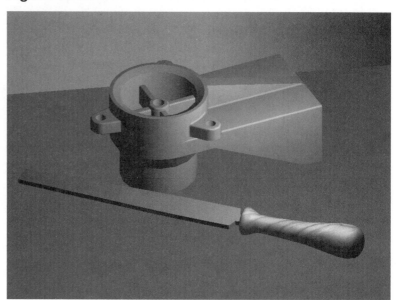

5. A rendering with only one point light can create drastic contrast from one side of the scene to the other—some portions of the model are well illuminated, while others are dark. Therefore, a second *Light* is inserted into the drawing to provide more even and realistic illumination over the surfaces of the model. This second light is a *Distant* light positioned on the left side and having a direction set to illuminate the left side of the models (Fig. 41-76). The new light compensates for the previously dark left side of the models; however, with two or more new lights, a scene can have a bright, washed-out effect. *Intensity* for the lights should be adjusted to

Figure 41-76 ————————————

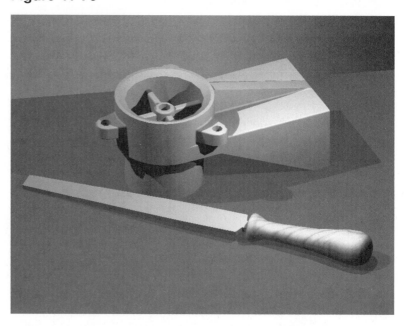

provide the desired illumination. Make several *Intensity, Location,* and *Color* adjustments, if necessary, to achieve the desired effects. A useful technique is to use a slightly different color for the two lights; for example, with the *Distant Light*, use a warm tone (light yellow).

The *Shadows* checkbox is selected in the *Render* dialog box to generate shadows (both lights were created with *Shadow On* in the light dialog boxes). Shadows can make a dramatic difference in the realism of the rendered image. Note that the objects no longer "float," but now appear to be sitting on the table. Also note both sets of shadows, one from each light source.

6. For the final adjustments, an image file (of a grid pattern) is used as a *Bump Map* to provide the serrated edges on the surface of the file model. The *Setuv* command is then used to position (rotate) the bitmapped image at the desired angle with respect to the body of the file model.

    For this final rendering (Fig. 41-77), *Photo Raytrace* is the desired *Rendering Type* so reflections can be generated. The table surface material *Reflection* attribute is modified by toggling *Mirror* in the *Modify Standard Material* dialog box. With these options, rendering time increases considerably.

    Also note that a perspective view is used in this rendering. This view can be accomplished by using *Dview Distance* to create the perspective view initially, but turning the perspective *Off* for preliminary renderings, then restoring perspective for the final rendering. Alternately, create two similar *Views*—one with and one without perspective.

    The *Shadows* options can be adjusted for the final rendering in the individual *Light's* dialog boxes. For example, you may want to generate shadow maps, which are necessary for "soft" shadows. Figure 41-77 displays *Shadow Volumes/Raytraced Shadows*.

    Other rendering options can be set at this time. *Anti-aliasing* can be set to *Medium* or *High* to minimize the "stair-step" effect of the table edge (this problem occurs with model edges that are "almost" horizontal or vertical).

    Don't expect to achieve excellent results when you create your first few renderings. Many options are available; therefore, a surprising amount of time can be spent making adjustments and generating "rough" renderings. Please consider using all time-saving techniques except for "final" renderings.

Figure 41-77

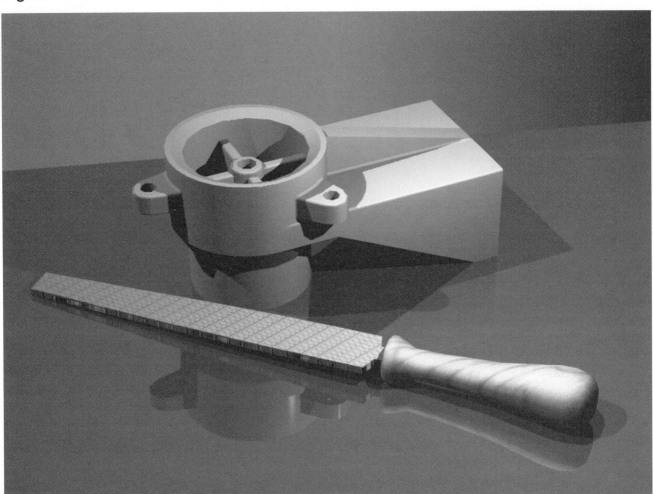

SAVING, PRINTING, AND REPLAYING RENDERINGS

Renderings can be saved to files in a number of ways. If you have used the *Render* command to generate a rendered display in the Drawing Editor, then want to save the image to a .BMP, .TGA, or .TIF file, use *Saveimg*. If you created the rendering in the Render Window, then you can save the image to a .BMP file only. If you want to render directly to a file (without rendering to the display first), the *File* option of the *Destination* cluster of the *Render* or *Rendering Preferences* dialog box can be checked. To replay an image from a file, use *Replay*. If you want to print a rendering, first select *Render Window* as the *Destination*, then select *Print*.

SAVEIMG

Pull-down Menu	COMMAND (TYPE)	ALIAS (TYPE)	Short-cut	Screen (side) Menu	Tablet Menu
Tools *Display Image >* *Save...*	*SAVEIMG*	...	...	*TOOLS 1* *Saveimg*	*V,7*

The *Saveimg* command allows you to save the rendered image in the Drawing Editor (you selected *Viewport* as the rendering *Destination*) to a .BMP, .TIF, or .TGA file format. The image can be displayed at a later time with the *Replay* command. No matter how long it took for *Render* to calculate and display the original rendering, when it is saved to a .BMP, .TIF, or .TGA file, it can be *Replayed* in a matter of seconds.

If you want to create a rendering and save it to a file directly (without first rendering to the screen), you can select *File* as the *Destination* in the *Rendering Preferences* dialog box (see Saving a Rendered Image to a File, next).

The *Saveimg* command invoked by any method produces the *Save Image* dialog box where you can select the file format and the portion of the image you want to save (Fig. 41-78).

Figure 41-78

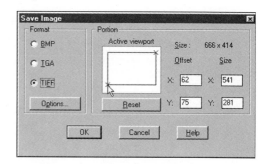

Format
BMP
A BMP file is a Windows bitmap raster file format.
TGA
This is the Truevision v2.0 format (.TGA file extension). It is a 32-bit format (16 million colors).
TIFF
Also a 32-bit format, this option is a Tagged Image File Format (.TIF file extension).

Options
Depending on the type of file (TIF and TGA only), file compression is available.

Offset
This option sets the lower-left corner for the image to save. The X and Y values are in screen pixels. You can also PICK in the *Active viewport* image tile to set *Size* and *Offset*.

Size
This option sets the upper-right corner for the image in pixels.

Reset
This option resets the *Offset* and *Size* values to the full-screen defaults.

Saving a Rendered Image to a File

If your choice for *Destination* is *File* (in the *Render* or *Rendering Preferences* dialog box) you can render directly to the selected raster file format without rendering to the screen. In this case, when *Render* is invoked, the screen display keeps its previous drawing or image and the calculated rendering is written directly to the designated file.

When *File* is selected as the *Destination* in either the *Render* dialog box (see *Render*, Fig. 41-69) or in the *Rendering Preferences* dialog box, you can choose the file format for the rendering to be written by selecting *More Options...* button. The *File Output Configuration* dialog box appears (Fig. 41-79).

Figure 41-79 ───────────

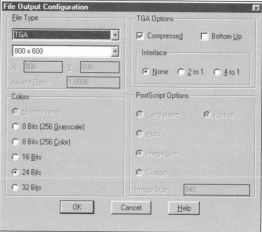

File Type
This option specifies the type of output file and rendering resolution. Supported option types are *BMP, TGA, PCX, TIFF* (.TIF), and *PostScript* (.EPS). Screen resolution is the number of columns and rows of pixels created for the image.

Aspect Ratio
This area sets the aspect ratio (ratio of horizontal/vertical).

Colors
There are several options for the number of colors based on the file type.

Options
Compression turns on file compression if supported by the selected file type. *Bottom Up* starts the scan lines from the bottom left instead of the top left.

Interlace
None turns off interlacing. Checking *2 to 1* or *4 to 1* sets those interlacing modes on.

PostScript Options
These options are available only if *PostScript* is selected as a file type.

Rendering and Printing from a Rendering Window

If you have selected *Render Window* as the *Destination* in either the *Render* dialog box or in the *Rendering Preferences* dialog box, your renderings will appear in a separate window (Fig. 41-80). You can toggle between the rendering window and AutoCAD using the Windows Task Bar or the Alt+Tab key sequence.

Figure 41-80 ───────────

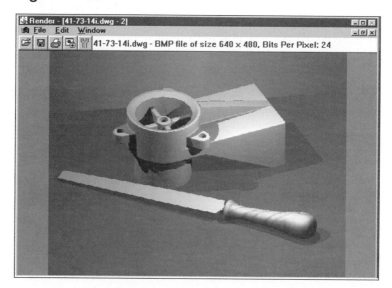

Several options are here that are not available by other means. For example, several options for printing a rendered image are available by selecting either the print icon or *Print* from the *File* pull-down menu. If you have configured AutoCAD to use the configured Windows printer, the *Print* dialog box appears (Fig. 41-81).

Figure 41-81 ────

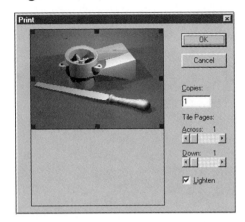

Printing is allowed across several pages by using the *Tile Pages* slider bars. You can also drag the position squares at the corners of the image to change or reproportion the size or reposition the image on the selected sheet(s).

Selecting the options icon or *Option* from the *File* pull-down menu produces the *Windows Render Options* dialog box (Fig. 41-82). Here you can change the resolution (*Size in Pixels*) and *Color Depth* of the bitmap image.

Figure 41-82 ──

	Pull-down Menu	COMMAND (TYPE)	ALIAS (TYPE)	Short-cut	Screen (side) Menu	Tablet Menu
REPLAY	*Tools* *Display Image >* *View...*	*REPLAY*	...	...	*TOOLS 1* *Replay*	*V,8*

The *Replay* command opens the *Replay* dialog box (not shown) where you specify the file you want to replay. You can replay any .BMP, .TGA, or .TIF files. Enter the desired file format to replay in the *List Files of Type:* or *Pattern:* edit box to make the existing files appear in the files list. After you select the file to replay, the *Image Specifications* dialog box appears.

In the *Image Specifications* dialog box (Fig. 41-83), you can specify that the entire image or a portion of the image be displayed. You can also offset the image in the screen area.

Figure 41-83 ────

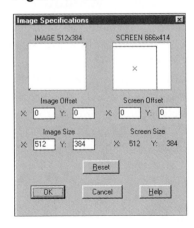

R13

The *Image* area of the dialog box displays the full image size (listed below in the *Image Size* edit boxes). You can display the full image or PICK the lower-left and upper-right corners in this box to define a portion of the image to display. Optionally, you can enter the lower-left corner X,Y values (in pixels) in the *Image Offset* edit box and the image size (in pixels) in the *Image Size* edit box.

The *Screen* image box allows you to PICK the location on the screen for the image to be replayed. Just PICK a location in this image tile to specify the new location for the <u>center</u> of the image. Alternately, the lower-left corner position can be specified by entering X,Y values (in pixels) in the *Screen Offset* edit boxes. When AutoCAD displays the image, the size is limited to the size (in pixels) of the window in which it is displayed.

MISCELLANEOUS RENDERING TIPS AND FEATURES

STATS

Pull-down Menu	COMMAND (TYPE)	ALIAS (TYPE)	Short-cut	Screen (side) Menu	Tablet Menu
View *Render >* *Statistics...*	*STATS*	...	...	*VIEW 2* *Stats*	...

The *Stats* command reports information about the last rendering that you created. The *Statistics* dialog box appears (Fig. 41-84) giving information such as time spent rendering and the status of all options used (as set in the *Render* or *Rendering Preferences* dialog box).

The data collected by *Stats* can be saved to a file that you designate in ASCII format. Any file extension can be assigned. In many cases it is a good idea to save this information along with the rendering file (.BMP, .TGA, .TIF, etc.) to keep a record of the settings used to create the image.

Figure 41-84

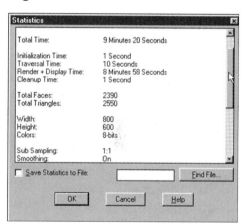

VIEWRES and FACETRES

The value you set with the *VIEWRES* command controls the accuracy of the display of circles, arcs, and ellipses. To increase performance while you draw, set a low value for *VIEWRES*. However, to make sure you get a good-quality rendering, raise the value before rendering drawings that contain arcs or circles. (See Solid Modeling Display Variables in Chapter 39.)

The *FACETRES* system variable controls the smoothness of meshed solids as well as the smoothness of shaded and rendered curved solids. It is linked to the value set by the *VIEWRES* command. When *FACETRES* is set to 1, there is a one-to-one correspondence between the viewing resolution of circles, arcs, and ellipses, and the tessellation of solid objects. When *FACETRES* is set to two, the tessellation will be twice that set by *VIEWRES*, and so on. When you raise and lower the value of *VIEWRES*, objects affected by both *VIEWRES* and *FACETRES* are affected. When you raise and lower the value of *FACETRES*, only solid objects are affected.

Drawing Outward-Facing Surfaces for Surface Models

When you prepare surface models for rendering, consider these factors. AutoCAD uses the "normal" on each face to determine which is a front face and which is a back face. A normal is a vector that is perpendicular to each polygon face on your model and points outward from the surface. If you draw the face (such as a *3Dface*) by PICKing counterclockwise, the normals point outward; if you draw the face clockwise, the normals point inward. You should draw all faces consistently. Mixing methods produces unexpected rendering results. AutoCAD calculates all the normals in the drawing during rendering.

If all surface model faces are drawn consistently, all normals should point outward from (or inward toward) the model. If problems still result in rendering, you can select *Back Face Normal Is Negative* in the *Rendering Preferences* dialog box (see *Rpref*).

If all faces are created consistently, you can save time in the rendering process by turning on *Discard Back Faces* in the *Render* or *Rendering Preferences* dialog box. This action discards the faces with normals pointing away from your viewpoint because they wouldn't be visible from that viewpoint. The time saved is proportional to the number of faces discarded and the total number of faces.

In cases when you are rendering transparent objects or when you can see two sides of an object (such as looking inside an open container), you may want to leave back faces in the rendering (do not check *Discard Back Faces*).

Unloading Render

When you first use any rendering command in an AutoCAD session, the AutoCAD loads Render (an ARX application) into memory. This may cause a slow-down in performance if you have a system with limited memory.

If you are finished using Render for a particular drawing session and intend to continue drawing, you can improve system performance by unloading Render. To do this follow the command sequence listed below:

```
Command: arx
?/Load/Unload/Commands/Options: u
Unload ARX file: render
- LsObj was just informed that C:\PROGRA~1\render.arx was unloaded.
render successfully unloaded.
Command:
```

After issuing these commands, some memory in your computer is freed. Rendering commands are still available as before, but when you issue a rendering command again, *Render* will have to reload.

R14

CHAPTER EXERCISES

1. *Render, Lights, Materials, Saveimg, Replay*

   Open the **PULLY-SL** drawing that you created in Chapter 38 Exercises. Refer to the "Typical Steps for Using Render" in this chapter to create a rendering. As an aid, follow these brief suggestions.

Create a *Vpoint* showing the best view of the pulley. Create a *Distant Light* located to the left side and slightly above the pulley. Create a *Point Light* located to the right side and above. Define and *Attach* a *Material* representing steel. Modify the material to have high *Reflective* and low *Roughness* values. Using the *Render* option of *Rendering Type*, perform several renders and adjust the *Light* positions and intensity until you find the desired results. Experiment also with the *Material* properties. When you have the best settings, perform a *Phong Render* (use *More Options…*) to achieve results similar to those shown in Figure 41-85. Finally, use *Saveimg* to save the rendering as **PULLY-REND.TIF**. Use *Replay* to ensure the image is saved. Use *SaveAs* and rename the drawing **PULLY-REND**.

Figure 41-85 ———————————————

2. *Lights, Materials, Saveimg, Replay, Phong Render*

   Open the **SADL-SL** model that you created in Chapter 38 Exercises. For this rendering, use *Dview* to generate a useful viewpoint and a small amount of perspective (*Distance*). Set up two lights—a cool-colored *Distant Light* with an *Azimuth* of **-130** and an *Altitude* of **50**, and a warm-colored *Point Light* at **300,-200,300**. *Attach* the *Blue Metallic Material* to represent metal. *Modify* the material to have a high *Reflective* value (**.80**) and a low *Roughness* value (**.25**).

Create several *Gouraud Renderings* to adjust the light intensities. The *Point Light* should provide slightly more light (at the near end of the model), and the *Distant Light* should provide slightly less (on the far side). The varying light intensities with the color effects give the proper impression of distance. Also make necessary adjustments to the *Material* parameters. When you have a good combination, compare your image to Figure 41-86 and create a *Phong Render*. Save the rendering to a **.TIF** file with the file name **SADL-REND**. Use *SaveAs* and rename the drawing **SADL-REND**.

Figure 41-86 ———————————————

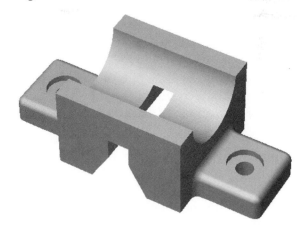

3. *Background, Lights, Materials, Saveimg, Replay, Shadows, Photo Real*

Open the **PULLY-REND** drawing again. Save the existing *Viewpoint* (from the previous rendering, exercise 1) to a named *View*. Create another *Vpoint* to show the back of the part. Try to show several of the through holes in the view. Save the new *Vpoint* to a named *View*. Use *Background* and create a *Gradient* background with three colors. Use the *Rotation* option to bring the bands of color into a vertical orientation. Create a few quick *Renders* to check and adjust the background.

Create one more *Light* with a *Location* of your choosing. Turn *Shadows On* in each of the *Light's* dialog boxes, but leave *Shadows* off in the *Render* dialog box while making preliminary adjustments. In order to create the new rendering, create a *Scene* by selecting the new *View*, the new *Light*, and one of the *Lights* previously created for the first rendering. Experiment to see which of the original two lights gives the best results. Adjust the light *Intensities* and *Colors* if necessary. When you have your best settings, turn *Shadows* on and create a *Photo Real* rendering. Use *Saving* and name the rendering **PULLY-R-BK.TIF**. Save the drawing as **PULLY-R-BK**.

4. *Spotlight, Shadows, New Standard Material, Photo Raytrace*

Open **SADL-REND** again. Change *VIEWRES* to **1000** and *FACETRES* to **1.0**. Make a solid *Box* using the *Center* option (center located at **0,-70,0**) with *Length* **600**, *Width* **300**, and *Height* **-20**. Then use *Rotate* with the *Last* selection option and rotate **20** degrees. This *Box* should serve as a base or "table top" for the saddle to rest on and provide a background for the *Spotlight* you will soon create. Create a *New Standard Material* with high *Reflection* (**.60**) and low *Roughness* (**.25**) and select a dark *Color* with a low *Value* (**.20**). *Attach* the new material to the "table." Create a *Background* for the scene using the *Gradient* option, or use the *Image* option and find a suitable image in the Textures or Windows folders. Create a few "rough" *Photo Real* renderings with *Shadows* off to test and adjust the background.

Next, insert a *Spotlight* with a *Target* of **60,0,0** and a *Location* of **200,-200,300**. Assign a warm color to the *Spotlight*. Turn *Shadows On* for the light and set the *Shadow Map Size* to **1024**. Adjust the *Intensity* of the *Point Light* to **0** so that the *Spotlight* provides the primary illumination from this viewing direction. Perform a few *Photo Real* renderings with *Shadows* off to test the angles of the *Hotspot* and *Falloff*. Make adjustments to those values so the saddle is fully illuminated but the light falls off just beyond the saddle. Perform several *Photo Real* renderings to test and adjust the lighting parameters. Any of several parameters can be adjusted—*Intensities, Colors,* and light *Locations*.

Turn on *Shadows* for all lights and globally for the scene. *Modify* the material you created for the table by selecting the *Reflective* attribute and toggling on *Mirror*. Perform a *Photo Raytrace* rendering. Try to achieve a reflective effect on the table top similar to that shown in Figure 41-87. Use *SaveAs* and rename the drawing **SADDL-REND2**.

Optional (time permitting): Create several renderings and use *Saving* to assign the names **SADL-REND2, SADL-REND3,** and so on. Use *Replay* to compare and determine the best renderings.

Figure 41-87 ——————————

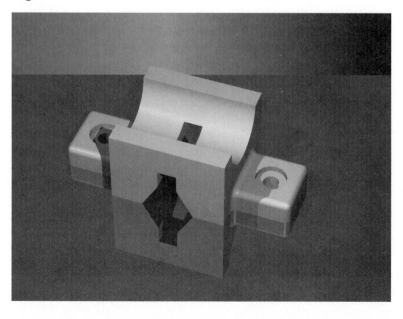

5. **Image Mapping,** *Setuv*

Create a machinist's file with handle similar to that used previously as a rendering example earlier in the chapter. Use the dimensions shown in Figure 41-88 to create half of the file using *Pline* and *Spline*, then *Mirror* the shape along the longitudinal axis. Create a *Region* from the resulting shape, then use *Extrude* with a height of **3/16** to create a solid model of the file.

Figure 41-88 ────────────────────────────────────

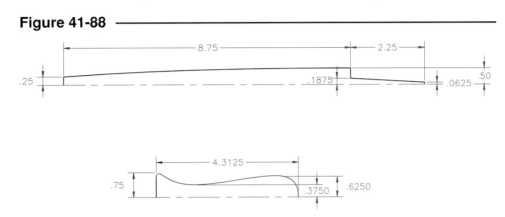

Create the handle solid model. First, draw the shape shown in Figure 41-88 with *Pline* and *Spline*, then use *Region* to create one closed shape. Make the solid by using *Revolve.* Use *Move* to attach the handle to the file. Set *VIEWRES* to **1000** and *FACETRES* to **1.0**. Save the drawing as *FILE*.

Next, attach appropriate wood and metallic materials to the two solids. Create one *Point Light* and one *Distant Light,* then adjust the intensities and colors. Make adjustments to the materials and lights as needed by creating several rough *Renders*.

Modify the file material by using the *Bump Map* attribute. Locate the **GRID.TGA** image file in the **AutoCAD R14/Textures** directory and set *Bitmap Blend* to **.02**. Use *Setuv* to change the *Rotation* angle to **30** and the *Scale* to **2.0**. Set the *Rendering Type* to *Photo Raytrace* and create a rendering. Make adjustments as necessary. Try to achieve a realistic file such as that shown in Figure 41-89. *Save* the drawing and use *Saveimg* to save the rendering as **FILE.TIF**.

Figure 41-89 ────────────────────────────────────

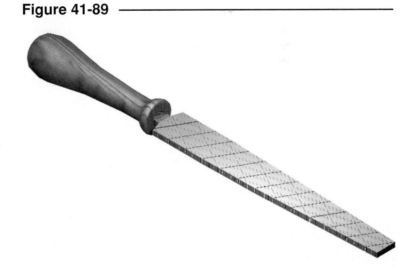

6. *Lsnew, Lsedit*

   Open the **CAMPUS.DWG** from the **Samples** directory. Rename the file to **CAMPUS2** when you use *SaveAs* to save it to your working directory. *Zoom* in to any area (see Fig. 41-21 for an example), then insert several landscape objects using *Lsnew.* Use *Lsedit* and **grips** to adjust the position and location of the images. Generate several preliminary *Renders* while you create new and adjust existing *Lights*. Use *Saveimg* to save your best rendering as **CAMPUS2.TIF.**

7. *Setuv*

   Create a scene similar to that shown in Figure 41-90. Use *Spheres* for the balls; create the glass by *Revolving* a profile; create the rack by *Extruding* two shapes and *Subtracting* the inner shape; and create the pool cue by *Revolve, Cone,* or *Extrude* with a taper angle. Create appropriate *Lights* and *Attach* appropriate materials. *Save* the drawing and assign the name **POOL.**

   Here is the challenge: Use AutoCAD to create several TIFF images (stripes and numbers) to attach to the balls. Use *Modify Material* with the *Color/Pattern* attribute to locate and assign the images. Use *Setuv* to adjust the orientation of the mapped images.

8. *Open* the **ADJMT-SL** or other solid or surface model that you would like to render. There are several good possibilities with the AutoCAD sample drawings (in the SAMPLE subdirectory), especially the **CHEVY.DWG**. Achieving skill with Render requires time and experimentation. Experiment with different *Backgrounds, Lights, Materials,* and *Scenes,* and *Images*. Have fun.

Figure 41-90

42

CREATING 2D DRAWINGS FROM 3D MODELS

Chapter Objectives

After completing this chapter you should:

1. know the suggested procedure for using the *Standard Engineering* option of *Mvsetup*;

2. be able to use *Mvsetup* to create viewports showing the standard engineering "views" of an existing 3D model;

3. be able to use *Solview* to create viewports and layers for use with *Soldraw*;

4. be able to use *Soldraw* to project 3D solids onto 2D planes, complete with hidden and visible lines;

5. be able to use layers created with *Soldraw* to complete a dimensioned 2D multiview drawing;

6. be able to create a 2D or 3D profile, complete with hidden lines, from any view of a solid model using *Solprof*;

7. be able to use *Solprof* in conjunction with *Mvsetup* to create a 2D drawing from a solid model.

CONCEPTS

This chapter discusses the use of AutoCAD features that assist in the creation of a 2D drawing from a 3D model. Several features in AutoCAD Release 14 provide a means for creating a 2D drawing from a 3D model: *Mvsetup, Solview, Soldraw,* and *Solprof. Mvsetup* can be used with any type of 3D model—wireframe, surface, or solid. *Solview, Soldraw,* and *Solprof* are used specifically with AutoCAD solid models for converting 3D geometry to 2D drawings.

Mvsetup is an automated routine that can be invoked by typing at the command prompt. It operates with <u>any type</u> of 3D model. This routine automatically sets up paper space viewports. It also has a series of options that allow alternate arrangements for the viewports. If you use the *Standard Engineering* option, *Mvsetup* creates the viewports and automatically places the 3D model correctly with a top, front, right side, and isometric viewpoints. Although the viewports are set up and the 3D geometry is arranged for you, other operations are required to create a 2D drawing complete with hidden lines and dimensions with correct visibility within viewports.

The *Solview* and *Soldraw* commands used together are far more automated and powerful than any other alternative for creating 2D drawings from AutoCAD solid models. *Solview* is similar to *Mvsetup* in that it automatically sets up the views in paper space viewports. It differs from *Mvsetup,* however, because you can select the desired views and their placement, including section and auxiliary views. *Solview* also creates layers for use with *Soldraw.* The *Soldraw* command is used as the next step to project the geometry (of AutoCAD solid models only) onto a 2D plane and automatically uses the newly created layering scheme for creating the 2D geometry. These new layers are complete with hidden lines and correct viewport-specific visibility settings. Dimensions can be drawn as an additional "manual" step on the layers provided by *Solview.*

Solprof (solid profile) creates a profile of an AutoCAD solid model. The profiled geometry can be used as a wireframe or projected onto a 2D plane. *Solprof* creates new layers for the profile geometry, complete with correct hidden lines. *Solprof* can be used in conjunction with *Mvsetup* to create standard engineering views. If you use the *Standard Engineering* option, *Mvsetup* creates the viewports and automatically places the 3D model correctly with top, front, right side, and isometric views. *Solprof* can then be used as a second operation to <u>profile</u> the 3D model (convert the 3D model in each view into 2D line drawings).

In summary, using *Solview* and *Soldraw* is the superior method for creating 2D drawings from AutoCAD ACIS solid models. It is the most universal method and accomplishes the most for you automatically, including setting up layers for dimensioning. *Mvsetup* can be used with wireframe or surface models to semi-automatically set up standard engineering views, but it cannot project the model onto a 2D plane with hidden and visible lines.

AutoCAD Designer

AutoCAD Designer is a separate software package that runs in conjunction with (and requires) AutoCAD. This product is intended particularly for mechanical applications and includes capabilities for constructing 3D models and 2D drawings.

AutoCAD Designer is a parametric-based solid modeler, meaning that parametric relationships rather than dimensional values can be specified for geometric features. In this way, when you change a dimension, the related features also change. The model construction method is features based rather than Boolean based. With features modeling, typical manufacturing operation terms are used to create geometry. For example, you may use the *Hole* command instead of creating a *Cylinder* and *Subtracting* it. In addition, Designer automatically creates a 2D drawing with dimensions from the 3D model. The 2D and 3D geometries are bidirectionally linked so that if you make a (dimensional) change to one, the other is automatically updated.

USING *MVSETUP* FOR STANDARD ENGINEERING DRAWINGS

Mvsetup is a program that assists you in setting up paper space viewports. The *Standard Engineering* option can be used to set up a 3D model in viewports, each viewport having a different standard view. The fundamental concepts and details of <u>other</u> options of *Mvsetup* are discussed in Chapter 33, Advanced Paper Space Viewports. Refer to Chapter 33 if you need more information on *Mvsetup* or guidelines for using paper space viewports, since only the *Standard Engineering* option of *Mvsetup* is discussed in this chapter.

MVSETUP

Pull-down Menu	COMMAND (TYPE)	ALIAS (TYPE)	Short-cut	Screen (side) Menu	Tablet Menu
...	*MVSETUP*	...		...	...

This section discusses the use of *Mvsetup* using the *Standard Engineering* option—the option for creating engineering drawings from 3D models. This option of *Mvsetup* operates with <u>any type</u> of 3D model. It automatically creates four viewports and displays the model from four *Vpoints* (front, top, side, and isometric views).

The *Mvsetup* (multiview setup) command is actually a routine that is invoked by typing *Mvsetup*. For the application of *Mvsetup* for creating standard engineering paper space viewports, the sequence is given here.

Typical Steps for Using the *Standard Engineering* Option of *Mvsetup*

1. Create the 3D part geometry in model space.

2. Create a layer for viewports and a layer for the title block (named VPORTS and TITLE, for example). Set the viewports layer current.

3. Invoke *Mvsetup* and use *Options* to set the *Mvsetup* preferences.

4. Use *Title block* to *Insert* or *Xref* one of many AutoCAD-supplied or user-supplied borders and title blocks.

5. Use the *Create* option to make the paper space viewports. Select the *Standard Engineering* option from the list.

6. Use *Scale viewports* to set the scale factor (*Zoom XP* factor) for the model geometry displayed in the viewports.

7. The model geometry that appears in one viewport can be aligned with the model in adjacent viewports if necessary using the *Align* option.

8. From this point, other AutoCAD commands must be used to create dimensions for the views or to convert some lines to "invisible" lines.

Mvsetup **Example**

Figure 42-1

To illustrate these steps, a 3D wireframe
model (WFEX3 drawing from Chapter 37
Exercises) is used as an example.

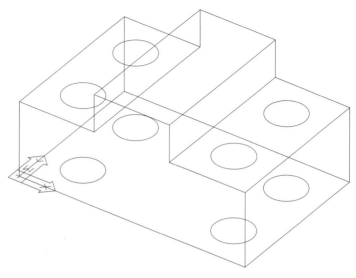

1. Create the 3D part geometry in model
   space. The wireframe is shown here as it
   exists in model space viewed from an isometric-type *Vpoint*.

2. Create a VPORTS layer for viewports and
   a TITLE layer for the title block. Set the
   VPORTS layer current.

3. Invoke *Mvsetup* and use *Options* to set the
   Mvsetup preferences. For the example,
   the *Layer* option is used to specify the
   layer for the title block. Additionally, the
   Limits option is used to automatically set
   paper space *Limits* equal to the border extents.

> Command: **mvsetup**
> Initializing... Enable Paper/Modelspace? <Y>: **Enter**
> Entering Paper space. Use MVIEW to insert Model space viewports.
> Regenerating paperspace.
> Align/Create/Scale viewports/Options/Title block/Undo: **o**
> Choose option to set—Layer/LImits/Units/Xref: **L**
> Layer name for title block or . for current layer: **title**
> Choose option to set—Layer/LImits/Units/Xref: **Li**
> Set drawing limits? <N>: **y**
> Choose option to set—Layer/LImits/Units/Xref: **Enter**
> Align/Create/Scale viewports/Options/Title block/Undo:

4. The *Title* block option is used to insert a title block and border in paper space.

> Align/Create/Scale viewports/Options/Title block/Undo: **t**
> Delete objects/Origin/Undo/<Insert title block>: **Enter**
> Available title block options:
> 0:   None
> 1:   ISO A4 Size(mm)
> 2:   ISO A3 Size(mm)

(Thirteen title block options are displayed here. See Chapter 33 for details on this option.)

> Add/Delete/Redisplay/<Number of entry to load>: **8**
> Align/Create/Scale viewports/Options/Title block/Undo:

The previously selected options produce a title block and border appearing in paper space, as shown in Figure 42-2. The 3D model is not visible because viewports have not yet been created.

Figure 42-2

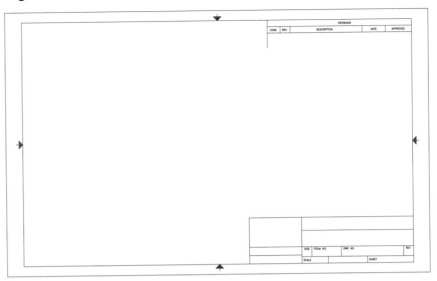

5. Use the *Create* option of *Mvsetup* to create the desired viewport configuration:

> Align/Create/Scale viewports/Options/Title block/Undo: **c**
> Delete objects/Undo/<Create viewports>: **Enter**
> Available Mview viewport layout options:
> 0:   None
> 1:   Single
> 2:   Std. Engineering
> 3:   Array of Viewports
> Redisplay/<Number of entry to load>: **2**
> Bounding area for viewports. First point: **PICK**
> Other point: **PICK**
> Distance between viewports in X. <0.0>: **Enter**
> Distance between viewports in Y. <0.0>: **Enter**

Select the "*2: Std. Engineering*" option. PICK two corners to define the bounding area for the four new viewports. The action produces four new viewports and automatically defines a *Vpoint* for each viewport. The resulting drawing displays the standard engineering front, top, and right side views of the 3D model in addition to an isometric-type view, as shown in Figure 42-3.

Figure 42-3

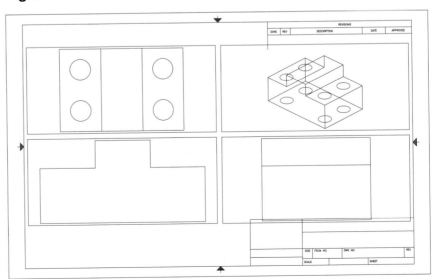

6. Notice in Figure 42-3 that the geometry in each viewport is displayed at its maximum size, as if a *Zoom Extents* were used. The *Scale viewports* option is used to set a scale for the geometry in each viewport:

    Align/Create/Scale viewports/Options/Title block/Undo: **s**
    Select the viewports to scale:
    Select objects: **PICK** the <u>viewport objects (borders)</u>
    Select objects: **Enter**
    Set zoom scale factors for viewports. Interactively/<Uniform>: **u**
    Enter the ratio of paper space units to model space units...
    Number of paper space units. <1.0>: **Enter** or (**value**)
    Number of model space units. <1.0>: **Enter** or (**value**)
    Align/Create/Scale viewports/Options/Title block/Undo:

Make sure you select the viewport objects (borders) at the "Select viewports to scale:" prompt. The *Uniform* option ensures that the 3D model will be scaled to the same proportion in each viewport. The ratio of paper space units to model space units is like a *Zoom XP* factor; for example, a ratio of 1 paper space unit to 2 model space units is equivalent to a *Zoom 1/2XP*.

The above action produces the scaling of the 3D model space units relative to paper space units, as shown in Figure 42-4.

Figure 42-4 ———————————————————

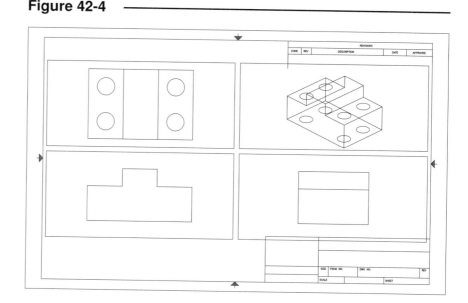

The *Align* option of *Mvsetup* can be used to orthogonally align the views if they are not automatically aligned by the *Scale viewports* option. The *Align* option allows you to *OSNAP* to an object in one view to accomplish a *Horizontal* or *Vertical* alignment with an object in another viewport. The *Horizontal* option can be used to align the model appearing in the front and side views and the *Vertical* option is used to select model geometry for alignment of the top and front views.

An additional step is required for this example in order to display the isometric view as desired within the viewport. *Zoom Extents, Zoom XP,* and *Pan* can be used to locate and size the model appropriately.

Finally, the VPORTS layer created for inserting the viewports can be turned *Off* or *Frozen* to display the views without the viewport borders, as shown in Figure 42-5.

Figure 42-5

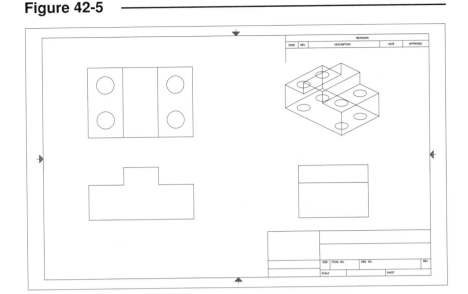

This is as far as *Mvsetup* goes. As you notice, the result is not a complete conventional multiview drawing because there are no dimensions nor are there hidden lines representing the invisible edges of the object. Since the views are actually different *Vpoints* of a 3D model, hidden lines are not possible for solid models using *Mvsetup*, but hidden lines can be created for wireframe models.

Creating Hidden Lines for a 3D Wireframe Shown with *Mvsetup*

If a complete 2D drawing is to be constructed from the 3D wireframe model, some lines from the model must be converted to "invisible" lines, and/or other lines may have to be added. It is not possible to convert edges of a solid or surface model to a hidden linetype, but edges of a wireframe model can be converted. Because a wireframe model is composed of simple objects such as *Line*, *Arc*, and *Circle*, these objects can easily be changed to a hidden linetype or to a layer with a hidden linetype assigned. Probably the simplest way to create the "invisible" lines is to use *Ddmodify*, select the desired objects, and change them to a layer having a hidden linetype assigned.

Better yet, it is desirable to create a separate hidden layer for each view (viewport). In this way, you have control over each view's hidden-layer visibility. For example, you may want to turn off just the isometric viewport's hidden lines or change that layer's linetype to continuous. Creating separate layers for each viewport's hidden linetypes may take more effort but affords you the most flexibility for displaying the drawing. Refer to Chapter 33, Advanced Paper Space Viewports, for more information on viewport-specific layer visibility.

Creating Dimensions for a 3D Model Shown with *Mvsetup*

If you use *Mvsetup* or any other method of creating 2D "views" of a 3D model in paper space viewports, dimensions can be created for each view. This is accomplished by creating the dimensions for each view on a separate layer and making each dimensioning layer visible only in the appropriate view (viewport). For example, assume you have used *Mvsetup* to create the top, front, and right side views of a model. The strategy for dimensioning is as follows:

1. Create three new layers: DIM-TOP, DIM-FRONT, and DIM-SIDE.

2. If not existing, create and save three UCSs named TOP, FRONT, and SIDE, each with its XY plane parallel to the matching view.

3. Starting with the front viewpoint, make the FRONT UCS active and make layer DIM-FRONT the current layer. Create the dimensions for the front view on this UCS and layer.

4. Make layer DIM-FRONT frozen for all <u>other</u> viewports so the dimensions for the front view appear only in the front viewport.

5. Moving to the TOP viewport, set its matching UCS and layer current. Create the dimensions for the view. Make layer DIM-TOP frozen for all other viewports.

6. Do the same for the remaining side view.

Don't forget any other procedures necessary for a complete 2D drawing, such as turning off the viewport layer and inserting title block and border.

USING *SOLVIEW* AND *SOLDRAW* TO CREATE MULTIVIEW DRAWINGS

The *Solview* command creates new layers and new views in paper space viewports for the existing 3D model space geometry. The *Soldraw* command creates new 2D objects on the new layers. *Soldraw* projects the model geometry onto a 2D plane with the appropriate continuous or hidden linetype. *Soldraw* can only operate with viewports that have been created with *Solview* (not with those created with *Mvsetup*). *Solview* and *Soldraw* operate for solid models only. The typical steps for using *Solview* and *Soldraw* to create a 2D drawing from a 3D model are described here.

Typical Steps for Using *Solview* and *Soldraw*

1. Create the part geometry in model space. Set up a UCS parallel to the desired profile (front) plane of the object. Also ensure that the HIDDEN linetype is loaded in the drawing.

2. Enable paper space (change *TILEMODE* to 0) and set up paper space *Limits* to the desired paper size. Set paper space *Snap* and *Grid* if desired.

3. Type *Solview* at the command prompt, select the *Setup View* icon button, or select *Solids > Setup > View* from the *Draw* pull-down menu.

4. Use the *UCS* option to create the profile (front) view. You can select the location and scale for the view.

5. Use the *Ortho* option to create the other principal orthographic views. Usually a top and/or side view is needed.

6. If a section or auxiliary view is desired, use the *Section* or *Auxiliary* option to create the view in the desired location.

7. Invoke *Soldraw* at the command prompt. *Soldraw* is used to project the "views" to a 2D plane, thus creating new 2D geometry on the appropriate layers (created by *Solview*).

8. To display the new 2D geometry, *Freeze* the model layer(s) and VPORTS layer.

9. Create a new layer named TITLE and draw, *Insert,* or *Xref* a title block and border in paper space.

10. You can then create dimensions for the 2D drawing on the "*-DIM" layers prepared by *Solview*. Use the normal commands for creating and editing dimensions.

SOLVIEW

Pull-down Menu	COMMAND (TYPE)	ALIAS (TYPE)	Short-cut	Screen (side) Menu	Tablet Menu
Draw *Solids >* *Setup >* *View*	SOLVIEW	...	...	DRAW 2 SOLIDS *Solview*	...

Solview creates paper space viewports and automatically specifies the correct viewing angle (*Vpoint*) for each *viewport*. You select which views you want and the location for each. *Solview* also prepares new layers for subsequent use with the *Soldraw* command:

        Command: **solview**
        Ucs/Ortho/Auxiliary/Section/<eXit>:

Ucs

The *Ucs* option creates a <u>view</u> normal (perpendicular) to the XY plane of a User Coordinate System. The *Ucs* option is generally the best way to create the first viewport from which other viewports can be created. All other *Solview* options require an existing viewport. You can select and readjust the view's location. You can also specify the size of the viewport.

Ortho

This option creates a principal orthographic view (top, side, bottom) from an existing viewport. New *Ortho* viewports are created by selecting an <u>edge</u> of an existing viewport object to project from. You can select and readjust the view's location and set the size of the viewport.

Auxiliary

This option is used to create an auxiliary view from an existing view. You are switched to model space to PICK two points to define the inclined plane used for the auxiliary projection. *OSNAPs* should be used for selection.

Section

This option creates a section view. *Solview* uses *Slice* to create the section at the cutting plane you define. The 3D geometry behind the cutting plane remains visible. Similar to the *Auxiliary* option, you PICK two points to define a cutting plane.

Solview creates the layers that *Soldraw* uses to place the visible lines, hidden lines, and section hatching for <u>each view</u>. *Solview* also creates layers for dimensioning that are set for visibility per viewport. Because of the possible complexity of the drawing, *Solview* places the visible lines, hidden lines, dimensions, and section hatching for each view on <u>separate layers</u>. This provides you with complete visibility control. The following naming convention is used for the layering scheme.

<u>Layer Name</u>	<u>Object Type</u>
view name–VIS	Visible lines
view name–HID	Hidden lines
view name–DIM	Dimensions
view name–HAT	Hatch patterns

(The *view name* is the original name that you specified for the view when it was created.)

Solview also creates a layer called VPORTS for the viewport objects. This layer should be reserved exclusively for the use of *Solview* and *Soldraw*. Do not draw or alter information on this layer.

SOLDRAW

Pull-down Menu	COMMAND (TYPE)	ALIAS (TYPE)	Short-cut	Screen (side) Menu	Tablet Menu
Draw *Solids >* *Setup >* *Drawing*	*SOLDRAW*	...	...	DRAW 2 SOLIDS *Soldraw*	...

Soldraw is generally used immediately after *Solview*. The *Soldraw* command uses the 3D model in each viewport created by the *Solview* command and projects the profiles onto a 2D plane. (*Soldraw* actually uses the *Solprof* command.) New 2D objects are created as profiles and sections in the viewports. The new 2D objects are drawn in the appropriate *Continuous* or *Hidden* linetypes. If sectional views are included, hatch patterns are drawn using the current values of the *HPNAME*, *HPSCALE*, and *HPANG* variables. You can use *Soldraw* only with previously created *Solview* viewports.

Soldraw is very easy to use because all you have to do is select the viewports that you want to be affected by *Soldraw*. There are no options for *Soldraw*:

```
Command: soldraw
Viewports to draw...
Select objects: PICK viewport objects
Command:
```

Solview and *Soldraw* Example

As the steps (given previously) for using *Solview* and *Soldraw* are explained here, an example AutoCAD ACIS solid model is used to illustrate the process (Fig. 42-6). The AutoCAD ACIS solid model is the angle brace shown in previous chapter exercises.

1. The first step to use *Solview* and *Soldraw* is to create a 3D model. When the geometry is complete, create a UCS parallel to the desired front view of your model, as shown in Figure 42-6. Your first view selected using *Solview* should be the profile (front) view.

2. The second step is to enable paper space and set *Limits* to the intended plotting paper size. If desired, also set *Snap* and *Grid* increments for paper space. (Optional: Insert a title block and border.)

3. Invoke *Solview* at the Command: prompt or select the *Solids View* icon button.

Figure 42-6

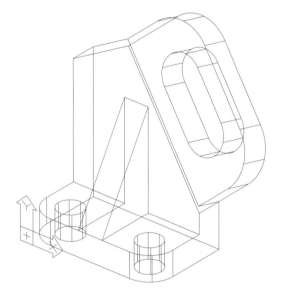

4. The *Current UCS* option is used to create a paper space viewport displaying the profile (front) view (see Fig. 42-7). The *UCS* option creates a view normal (perpendicular) to the current UCS XY plane.

> Command: **solview**
> Ucs/Ortho/Auxiliary/<eXit>: **UCS**
> Named/World/?/<Current>: **Enter** (to use current UCS)
> Enter view scale<1>: **Enter** or (**value**)
> View center: **PICK**
> View center: **PICK** (to readjust if necessary)
> View center: **Enter**
> Clip first corner: **PICK**
> Clip other corner: **PICK**
> View name: **front**
> Ucs/Ortho/Auxiliary/Section/<eXit>:

The "Enter view scale<1>:" value is the same value as a *Zoom XP* factor. That is, enter the proportion of model geometry units to paper space units.

PICK the desired "View Center" to specify the desired location of the view. The model geometry appears <u>after</u> PICKing the view center. The "View Center" can be readjusted if necessary.

At the "Clip first corner:" and "Clip other corner:" prompts, select the desired paper space viewport corners as indicated in Figure 42-7. Make sure you allow sufficient room for dimensions or other drawing objects you may want to include later. (It is OK for the viewports to overlap.) Finally, you must name the view. In this case, "FRONT" is used as the name of the new viewport.

Figure 42-7

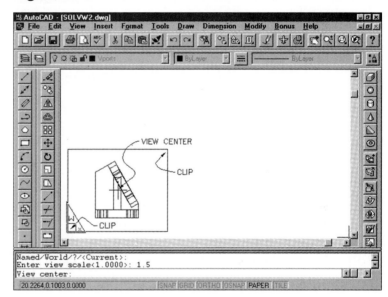

5. After the front view is created, you can create other views with the *Ortho* option. This option creates a new viewport orthographically aligned with the <u>edge</u> of the viewport object that you PICK:

> Ucs/Ortho/Auxiliary/Section/<eXit>: **o**
> Pick side of viewport to project: **PICK** the desired edge of viewport
> View center: **PICK**
> View center: **PICK** (to adjust if necessary)
> View center: **Enter**
> Clip first corner: **PICK**
> Clip other corner: **PICK**
> View name: **top**
> Ucs/Ortho/Auxiliary/Section/<eXit>:

As shown in Figure 42-8, PICK the edge of the viewport object representing the viewing direction for the new *Ortho* viewport. In other words, PICK the top edge of the front viewport to produce a top view. Notice the *Midpoint OSNAP* option is automatically invoked when you PICK the edge of the viewport. Next, PICK the view center for the top view (as before when the front viewport was established).

Figure 42-8

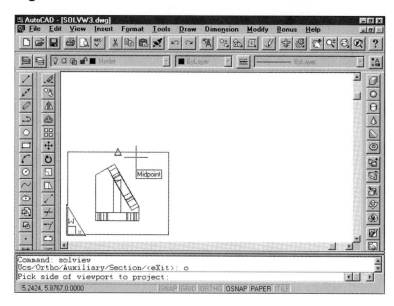

The new view then appears (Fig. 42-9). Clip the corners to specify the size for the viewport. Assign a name for this viewport such as "TOP."

You can continue creating viewports with the *Ortho* option until the necessary principal views are established. For example, a right side view could be created by using the *Ortho* option, then PICKing the right edge of the front viewport object.

Figure 42-9

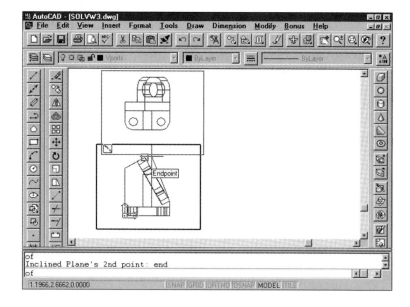

6. In this example, an auxiliary view is required to provide the necessary visual and dimensional information for the drawing. *Solview* is used with the *Auxiliary* option to create the viewport.

For this procedure you must select two points to define the auxiliary surface. You are automatically switched to model space for PICKing the surface's first and second points (*OSNAP*s should be used). For the example, the auxiliary surface is established as the upper right inclined plane as shown in Figure 42-10.

Figure 42-10

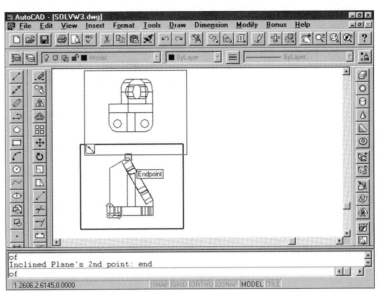

Next, PICK a point to define which side of the auxiliary surface to view the object from:

```
Ucs/Ortho/Auxiliary/Section/<eXit>: a
Inclines Plane's 1st point: PICK
Inclined Plane's 2nd point: PICK
Side to view from: PICK (on desired side of cutting plane)
Enter view scale<1>: Enter or (value)
View center: PICK
View center: Enter
Clip first corner: PICK
Clip other corner: PICK
View name: aux
Ucs/Ortho/Auxiliary/Section/<eXit>:
```

The resulting viewport with auxiliary view is shown in Figure 42-11.

The model geometry visible in the viewports is the existing 3D model. The model has not been changed in any way by *Solview*. However, as well as the viewports that are new, *Solview* creates new layers complete with viewport-specific layer visibility.

Figure 42-11

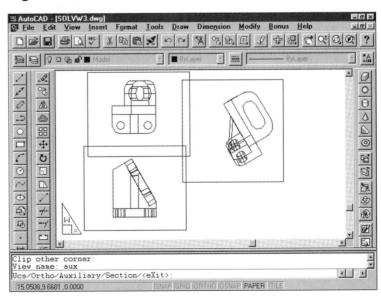

For our example, listing the layers reveals the work that was done by *Solview*. The following list shows <u>all</u> and <u>only</u> layers created by the previous options of *Solview*:

Layer Name	State	Color	Linetype
FRONT-DIM	On 7	(white)	CONTINUOUS
FRONT-HID	On 7	(white)	HIDDEN
FRONT-VIS	On 7	(white)	CONTINUOUS
AUX-DIM	On 7	(white)	CONTINUOUS
AUX-HID	On 7	(white)	HIDDEN
AUX-VIS	On 7	(white)	CONTINUOUS
TOP-DIM	On 7	(white)	CONTINUOUS
TOP-HID	On 7	(white)	HIDDEN
TOP-VIS	On 7	(white)	CONTINUOUS
VPORTS	On 7	(white)	CONTINUOUS

Current layer: VPORTS

Solview uses the name that you specify for the views as prefixes for the layer names. The layer visibility has automatically been set so that layers are visible only in the associated viewports (TOP-* layers are only visible in the top view, for example).

7. Invoke the *Soldraw* command. Now that the views have been established by the *Solview* command, the model geometry is ready for projection onto 2D planes.

```
Command: soldraw
Select viewports to draw:
Select objects: PICK
Select objects: Enter
```

The viewport objects are PICKed in response to the "Select viewports to draw:" prompt. Normally you would <u>select all viewports</u>. PICK the viewport <u>objects</u> (borders) with the pickbox or Crossing Window, as shown in Figure 42-12. Because much computation is involved, *Soldraw* may take some time, depending on the complexity of the model and the speed of your computer system. When *Soldraw* has finished, the results may not be apparent, because even though the layering scheme is set up correctly, all of the layers are still visible including the 3D model layer(s).

Figure 42-12

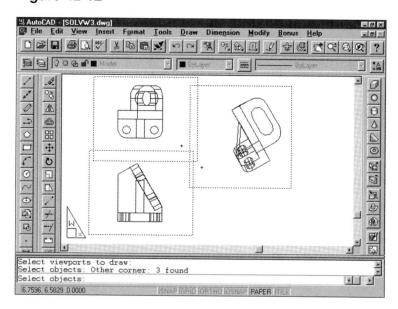

8. Use the *Layer Control* drop-down list or the *Layer* tab of the *Layer & Linetype Properties* dialog box to control the visibility of the layers. *Freeze* the MODEL layer (or whatever layer you used for the model geometry). Also, *Freeze* the VPORTS layer to prevent the viewport borders from displaying, as shown in Figure 42-13.

Figure 42-13 ——————————

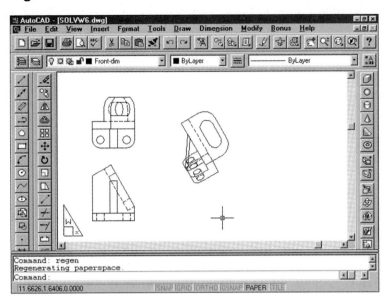

9. If not done previously, create a new layer (called TITLE or BORDER, for example) and draw, *Insert*, or *Xref* a title block and border in paper space.

10. If you want to create dimensions for the new 2D views, use the normal dimensioning commands. Dimensioning is not as "automatic" as using *Solview* and *Soldraw*; however, all the layers for dimensions have been created and viewport-specific visibility is preset by *Solview*. Set the appropriate layers *Current* and create the related associated dimensions in model space. Remember to create or set a UCS normal to each view for creating dimensions related to that view.

USING *SOLPROF* WITH AutoCAD SOLID MODELS

SOLPROF

Pull-down Menu	COMMAND (TYPE)	ALIAS (TYPE)	Short-cut	Screen (side) Menu	Tablet Menu
Draw *Solids >* *Setup >* *Profile*	*SOLPROF*	...	...	DRAW 2 SOLIDS *Solprof*	...

Solprof is a command that creates a profile of AutoCAD solids. *Solprof* generates a separate profile containing only the edges of straight and curved surfaces of a solid as seen from a particular viewpoint (parallel projection only—perspectives cannot be profiled). The "profile" can be projected to a plane or can be generated as a wireframe model. The profile is generated on two new <u>layers that are automatically created</u>, one with hidden lines and one with visible lines. *Solprof* determines which profile edges should be projected as hidden or as visible lines and draws them on the appropriate layer. *Solprof* can be used in conjunction with *Mvsetup* to create standard engineering 2D drawings.

Solprof must operate <u>within a paper space viewport</u>. Set *TILEMODE* to 0, make a paper space viewport with *Mview,* and activate model space with the *Mspace* command. Then use *Solprof* and select the solids that you want profiled. Remember to <u>select the solids, not the viewport objects</u> as with *Soldraw.*

```
Command: solprof
Select objects: PICK (Select the solid or solids that you want profiled.)
Select objects: Enter
Display hidden profiles lines on separate layer? <Y>:
Project profile lines onto a plane? <Y>:
Delete tangential edges? <Y>:
```

Accepting the *Yes* default for "Display hidden profiles on a separate layer?" prompt causes *Solprof* to create two *Blocks,* one with hidden lines and one with the linetype of the existing solid. Two new layers are created for the *Block* insertions:

PV-*nx*   Visible profile layer
PH-*nx*   Hidden profile layer

(Where PV designates the Profile Visible objects, PH designates the Profile Hidden objects, and *nx* represents a number and letter AutoCAD randomly assigns to the viewports. This letter and number combination is called the "viewport handle." You can *List* the viewport object to display its handle.)

The *Hidden* linetype must be loaded into the drawing before *Solprof* can use it, or you can assign the HIDDEN linetype to layer PH-* after using *Solprof.* Answering *No* to the "Display hidden profiles on a separate layer?" prompt causes *Solprof* to create one *Block* drawn as visible lines with the linetype of the existing solid.

The next prompt, "Project profile lines onto a plane? <Y>:," allows you to create a 3D wireframe or a 2D profile. Answering *Yes* causes AutoCAD to project the profile onto a plane parallel to the screen, similar to the *UCS View* option. In other words, the solid as it appears from the current viewpoint is transformed to a 2D projection. This option is particularly useful for creating a 2D "view" from a solid model *Vpoint.* Answering *No* creates a 3D wireframe of the model.

The last prompt allows you to delete "tangential edges." These edges are lines between two tangent faces. Answering *No* to this prompt creates a line between a planar surface and a tangent curved surface. You should answer *Yes* to delete tangential edges for most applications.

When *Solprof* has been used, the new objects may not be visible because *Solprof* does not change the layer visibility. To see the new *Solprof* layers, use the *Layer* tab of the *Layer & Linetype Properties* dialog box or the *Layer Control* drop-down list to *Freeze* the original model layer(s). The new *Solprof* layers can then be viewed and edited if you choose.

Solprof **Example**

As an example, consider the solid model of the angle brace shown in Figure 42-14. In this case, the model has been completed and is displayed in model space. Also assume that the *Hidden* linetype has been loaded into the drawing.

Figure 42-14

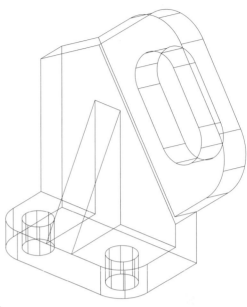

First, change *TILEMODE* to 0 to activate paper space. Use *Mview* to create a paper space viewport. Next, type *Mspace* to activate model space and, if needed, use *Vpoint* to achieve the desired view of the model. In this case, an isometric-type viewpoint is desired.

Invoke *Solprof* while model space is active (the cursor is <u>in</u> the viewport) and select the solid. All of the defaults are accepted. The resulting new layers are created as shown:

> PH-1DD   *Hidden* linetype
> PV-1DD   *Continuous* linetype

Layer visibility is <u>automatically</u> set to display the profile geometry only in the specific viewport (*Frozen for New* viewports).

Using the *Layer* command, the model layer is *Frozen* to reveal the two profile layers. The resulting profile geometry is displayed in its paper space viewport in Figure 42-15.

Figure 42-15

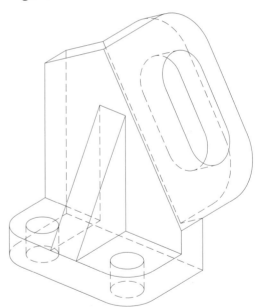

This procedure can also be used to generate a 2D multiview drawing from an AutoCAD ACIS solid model. The procedure involves creating paper space viewports (with *Mview*) for each desired "view"—top, front, side, etc. Alternately, *Mvsetup* can be used to lay out the views for you. Next, *Solprof* is used to create the views, complete with hidden and visible lines, for each viewport (see the following example).

Keep in mind the power of using *Solprof* to automatically create a wireframe model from a solid model (answering *No* to, "Project profile lines onto a plane?"). In many cases when a wireframe is desired, it may be easier to create a solid model using the relatively simple CSG techniques, then use *Solprof* to automatically create the wireframe model from the solid. This method can be used when the model consists of multiple curved surfaces and complex lines of intersection (edges) between surfaces.

Using *Solprof* with *Mvsetup*

The procedure for creating 2D drawings from AutoCAD ACIS solid models using *Solprof* can be automated by using *Mvsetup* in conjunction with *Solprof*. *Mvsetup* is a routine that automatically creates paper space viewports with standard engineering views (see *Mvsetup,* earlier this chapter). *Solprof* is then used to create the hidden and visible lines for each "view."

For example, consider this case to illustrate *Mvsetup*. Given the adjustable slide solid model shown in Figure 42-16, assume the same procedure was followed to create the layout as that illustrated in Figures 42-1 through 42-5 earlier this chapter (for a wireframe model).

Figure 42-16 ————————————————

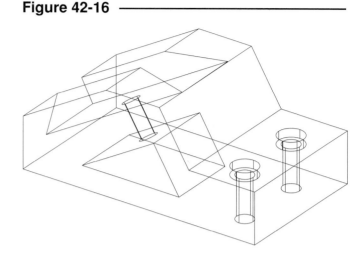

However, in this case *Mvsetup* is used with a <u>solid</u> model (adjustable slide) to create the paper space viewports using the "standard engineering" option, resulting in the setup shown in Figure 42-17. The drawing at this point (as far as *Mvsetup* goes) displays four *Vpoints* of the same solid model.

Figure 42-17 ————————————————

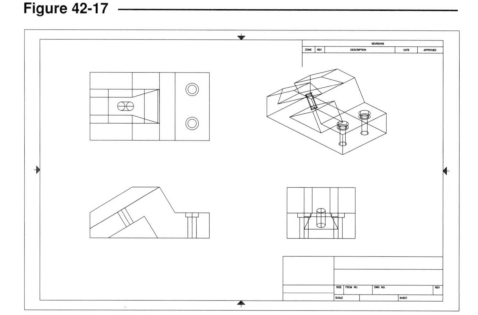

Solprof can be used next. The *Mspace* command is used to activate any viewport. Invoke *Solprof* and select the geometry in the current viewport to profile:

```
Command: solprof
Select objects: PICK (Select the solid or solids that you want profiled.)
Select objects: Enter
Display hidden profiles lines on separate layer? <Y>: Enter
Project profile lines onto a plane? <Y>: Enter
Delete tangential edges? <Y>: Enter
```

This procedure is followed for each viewport. *Solprof* creates the profiles for each viewport. Remember that *Solprof* creates two layers for each profile (or for each viewport), one for hidden lines and one for visible lines. For this example, the following layers exist after the profile geometry is created. (Only PH-* and PV-* are created by *Solprof*.)

Layer Name	State	Color	Linetype
0	On	7 (white)	CONTINUOUS
MODEL	On	7 (white)	CONTINUOUS
PH-A61	On C N	7 (white)	HIDDEN
PH-A62	On C N	7 (white)	HIDDEN
PH-A63	On C N	7 (white)	HIDDEN
PH-A64	On C N	7 (white)	HIDDEN
PV-A61	On C N	7 (white)	CONTINUOUS
PV-A62	On C N	7 (white)	CONTINUOUS
PV-A63	On C N	7 (white)	CONTINUOUS
PV-A64	On C N	7 (white)	CONTINUOUS
TITLE	On	7 (white)	CONTINUOUS
VPORTS	Frozen	7 (white)	CONTINUOUS

Notice that *Solprof* automatically determines the viewport-specific layer visibility. Each of the profile layers (PV-* and PH-*) is *Frozen* for the *Current* and *New* viewports. In other words, only the front view profile is visible in the front viewport, only the top view profile is visible in the top viewport, and so on.

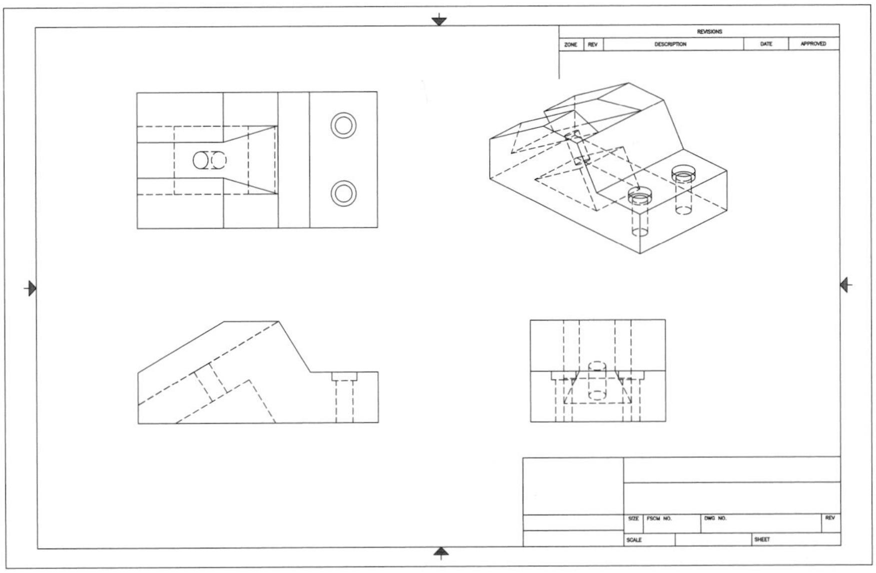

Freezing the MODEL layer reveals the profile geometry (PV-* and PH-* layers). Figure 42-18 displays the drawing after *Solprof* has been used in each viewport and the MODEL layer has been *Frozen*. The drawing is now ready for adding centerlines, dimensions, and annotation.

If you want to dimension a drawing, such as that in Figure 42-18, it is possible to do so by using the normal dimensioning commands. However, additional work is required to create layers for dimensions so they appear

Figure 42-18

only in the top, front, and side views. The top, front, and side view dimensions each must have their own layers. Next, viewport-specific visibility must be set for each layer. In other words, only the dimensions related to the top view can be visible in the top viewport, and only dimensions related to the front view can be visible in the front viewport, and so on. This can be quite tedious. Remember that *Solview* creates viewport-specific dimensioning layers automatically and is the best method for creating 2D drawings from solid models.

CHAPTER EXERCISES

1. **Using** *Mvsetup* **with a Solid Model**

Open the **BGUID-SL** drawing that you created in Chapter 38 Exercises. Use *SaveAs* to assign a new name, **BGUD-MVS**. Create two new *Layers* named **VPORTS** and **TITLE**. Make **VPORTS** the *Current* layer. Use *Mvsetup* with the following *Options*: *Layer, Limits,* and *Title*. Insert the *B-size* sheet title block. Then use the *Create* option for *Standard Engineering* setup. *Scale* the geometry in each viewport *Uniformly* to a factor of **1**. The drawing should look like Figure 42-19. *Save* the drawing.

Figure 42-19 ───────────────

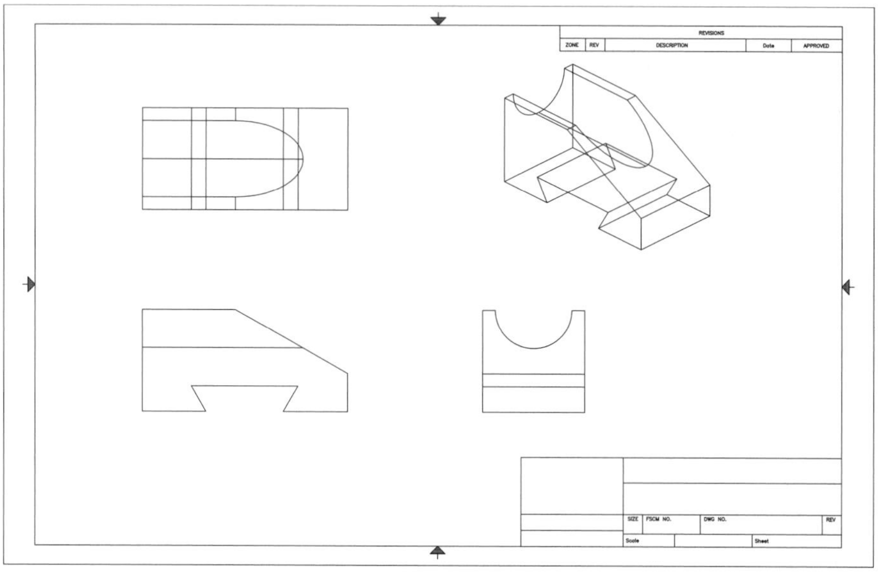

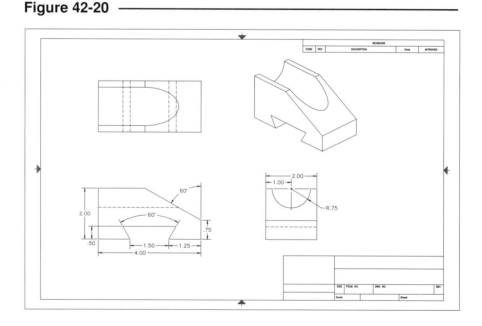

2. **Using** *Mvsetup* **with a Wireframe Model**

A. *Open* the **BGUID-WF** drawing that you created in Chapter 36 Exercises. Use *SaveAs* to assign a new name, **BGUD-MV2**. Make a *Layer* for the viewports and one for the title block. Set the viewports layer *Current*. Use *Mvsetup* with the same procedure as in the previous exercise.

Figure 42-20 ───────────────

B. Convert the necessary lines to "invisible" lines by creating four new *Layers* for hidden lines, one for each viewport. Assign the *Hidden* linetype to each new layer. Use the *Freeze in Current Viewport* and *Freeze in New Viewports* options to assign the correct viewport-specific layer visibility. Using *Ddmodify*, change the appropriate lines to the matching layers.

C. *Freeze* the invisible edges of the wireframe model in the isometric viewport. *Save* the drawing. The completed drawing should look like Figure 42-20.

3. **Creating Dimensions for the New 2D Drawing**

Open the **BGUD-MV2** drawing if not already open. Use *SaveAs* to save the drawing as **BGUD- MV3**. Create dimensions for the views. Follow the steps given in the chapter to create the appropriate UCSs, create the layers, and draw the dimensions as shown in Figure 42-21. Don't forget to set the correct visibility settings for each **DIM** layer. *Save* the drawing and *Plot* to scale.

Figure 42-21 ─────────────────────

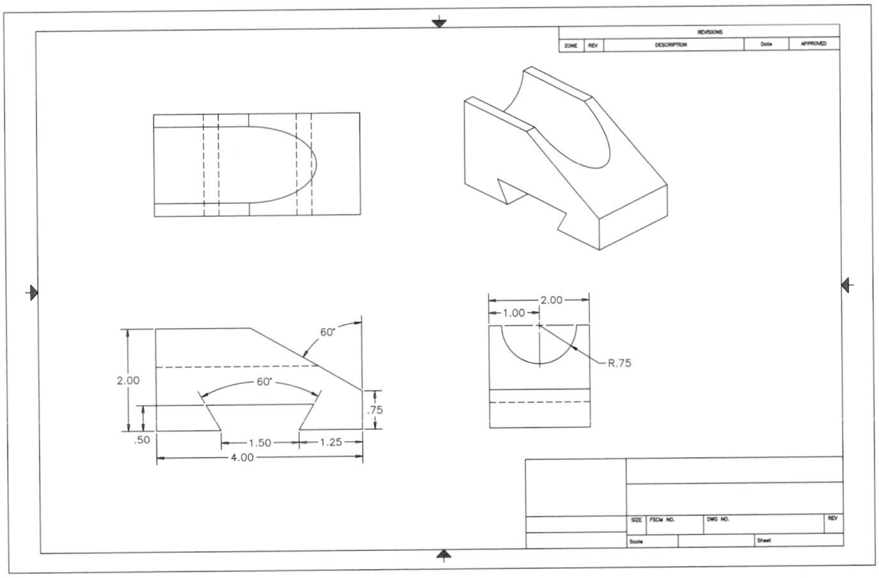

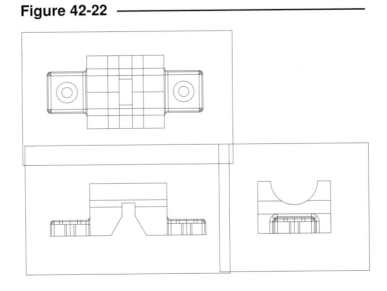

4. **Creating a 2D Drawing Using** *Solview* **and** *Soldraw*

A. *Open* the **SADL-SL** drawing from the Chapter 38 Exercises. Set up a *UCS* parallel with the front profile of the model (to create the front view as shown in Fig. 42-22). Change *TILE-MODE* to **0** and set *Limits* in paper space to a metric A size (**279,216**). Next, make sure the *Hidden Linetype* is loaded and *LTSCALE* is set to **13**. Use *SaveAs* to give the name **SADDLE-SV**.

Figure 42-22 ─────────────────

B. Use *Solview* with the *UCS* option to create the **FRONT** view. A *Scale* factor of **1** can be used. Use the *Ortho* option for the **TOP** and **RIGHT** views. If the views do not align orthogonally, use *Zoom Extents*, then *Zoom 1XP* in each viewport, or use *Mvsetup* with the *Align* option. Your results should appear like Figure 42-22. *Save* the drawing (as **SADDLE-SV**.)

C. Next, use *Soldraw* to create the visible and hidden line geometry. When finished, *Freeze* the layer that the solid model is on. Use *Pspace* and insert or draw a title block and border. If the model does not fit within the border, use *Move* to move the viewport objects accordingly (make sure you *Move* two viewports together to ensure alignment). The drawing should appear like Figure 42-23. Use *SaveAs* and change the name to **SADDLE-SD**.

D. Next, dimension the drawing and add centerlines. Remember to use the layers that *Solview* created for the dimensions (one for each viewport). *Save* the drawing when you are finished (as **SADDLE-SD**). Make a *Plot* to scale.

Figure 42-23 ───────────

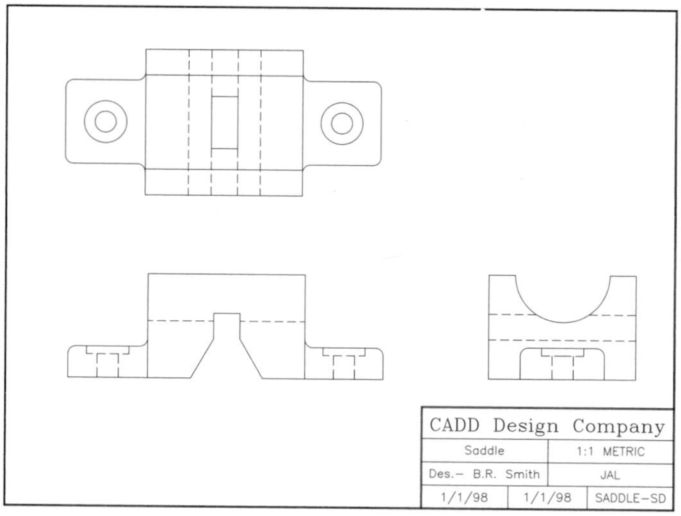

5. Create a multiview drawing from the solid model you created of the angle brace (**AGLBR-SL** drawing from Chapter 38). Use *Solview* and *Soldraw* to create the multiviews. Use the *Auxiliary* option to create an auxiliary view. Create the dimensions on the **DIM** layers. *Save* the drawing as **AGLBR-SV**.

6. Use the solid model of the **SWIVEL** from Chapter 38 Exercises. Set up paper space for a B-size sheet and plan to print or plot the drawing at full size (1 = 1). Use *Solview* and *Soldraw* to create a front, top, and auxiliary view in a manner similar to that used in the previous exercises (4 and 5). Dimension the drawing (use Fig. 38-83 as a guide). Use *SaveAs* to assign the new name **SWIVEL-SD**.

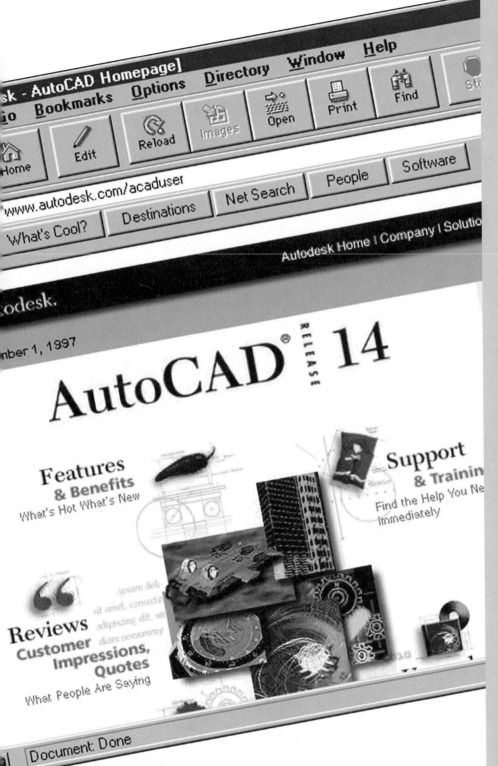

43

INTERNET TOOLS

Chapter Objectives

After completing this chapter you should:

1. understand the difference between a raster file and a vector file;

2. be able to use *Browser* to launch your web browser from within AutoCAD;

3. be able to obtain and use the *WHIP!*™ Plug-In for Netscape Navigator™ or Microsoft Internet Explorer™ to view DWF files;

4. be able to use the *WHIP!* menu to *Zoom*, *Pan*, display *Named Views*, *Highlight URLs*, and *Print* DWF images;

5. be able to convert an AutoCAD drawing (DWG) to a Drawing Web Format (DWF) file for others to view over the Internet or intranets;

6. be able to use the Internet Utilities to attach, detach, list, and select URLs in an AutoCAD drawing, and to open, insert, and save DWG files via the Internet;

7. be able to use HTML codes to embed a DWF file in a web page.

CONCEPTS

A new standard file format for 2D CAD drawings was established just prior to AutoCAD Release 14 to provide a method to view drawings over the Internet and within intranets. The DWF (Drawing Web Format) file is a compact vector-based file. In AutoCAD Release 14 you can convert a DWG file to a DWF file. You can even embed Web addresses (Uniform Resource Locators, or URLs) within your AutoCAD drawings. These URLs become hyperlinks to other Web pages when your drawing is saved as a DWF file. Using the *WHIP!*™ (Windows High Performance) Plug-In for Netscape Navigator™ and Microsoft Internet Explorer™, you can use zoom and pan tools to view and maneuver through complex, detailed drawings (DWFs) over the Internet. The DWF file format, *WHIP!* Plug-In for Internet browsers, and AutoCAD's Internet Utilities provide new opportunities for project teams, clients, and customers to collaborate over the Internet or within an intranet using only basic web-browsing tools. The following components make up AutoCAD's Internet capabilities.

1. AutoCAD Release 14 built-in commands:

    Dwfout   converts a DWG file to a DWF file
    Browser   launches the system Internet browser (such as Netscape Navigator or Microsoft Internet Explorer)

2. *WHIP!* Plug-In for Netscape or Microsoft Internet Explorer

3. AutoCAD Internet Utilities: additional commands for attaching and managing URLs saved with the DWF and for saving, opening, and inserting DWG files via the Internet

With the new DWF file format, the *WHIP!* Plug-In, and AutoCAD's Internet Utilities, drawings can be made accessible over the World Wide Web or within an intranet for a variety of applications. For example, an engineering architectural, construction, or design firm could post DWF images on its web page, providing immediate viewing accessibility to associates or clients in the field. The images could be construction plans for viewing from laptops on the construction site or requests-for-bid drawings displayed for interested clients. Mechanical parts drawings or instructional assembly information could be available on the manufacturer's pages for viewing by consumers or clients. Parts catalogs or consumer goods catalogs with detailed graphical information could be available to the general public or available only to subscribers using password protection. Architectural or design projects in process can be posted for collaboration between team members in remote locations. Maps, public building layouts, directional information, and other diagrams can be made viewable to the public.

When designers, engineers, or other colleagues want to exchange drawing (DWG) files, the DWF format and the *WHIP!* Plug-In can expedite the process. This is possible because the "parent" DWG file used to create the DWF can be copied across the Internet if it is located in the same remote directory along with the DWF image file. The *WHIP!* Plug-In provides a *SaveAs...* option that copies the related DWG file to the local (viewing) system. *WHIP!* also allows users to drag-and-drop an image into your browser for viewing. If users drag-and-drop a DWF image into an open AutoCAD session, the related DWG file (if made available) is copied and loaded into AutoCAD. This technique allows users to easily view drawings (via DWF) before they decide to copy and use the drawings (DWG).

RASTER AND VECTOR FILE FORMATS

Raster Files: The Common WWW Standard

Figure 43-1

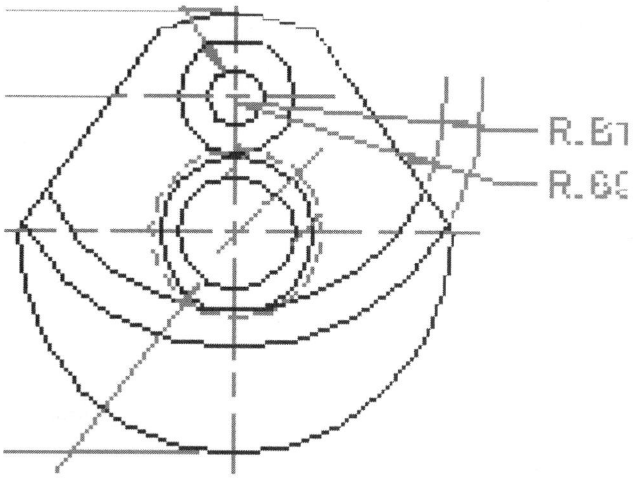

The common World Wide Web graphic standard uses a raster image format such as GIF or JPG. These files can be highly compressed, making the information displayed in the images quickly transportable over data transmission lines and devices used for the Internet. However, because raster images are pixel-based, or composed of tiny "dots," viewing detail is not possible—zooming in only enlarges the "dots." Figure 43-1 displays a raster image that has been enlarged.

Raster images, such as a "screen grab" for example, are defined in the data file as a map of pixels—rows and columns of picture elements either in black and white or in colors. Raster images are sometimes referred to as "bitmaps," since the file retains the map of each dot (placement within the rows and columns) with its color information. Examples of raster image file formats (file name extensions) are given in the following table.

RASTER FILES

File Format (Extension)	Application/Creator
BMP	Windows and OS/2
CLP	Windows Clipboard
DIB	Windows BMP
GIF	CompuServe
ICO	Windows Icon
IFF	Amiga ILBM
JPG	JPEG
MAC	MacPaint
MSP	Microsoft Paint
NIF	Navy Image File
PBM	Portable Bitmap
PCX	PC Paintbrush
PSD	Photoshop
RAS	Sun Raster
SGI	Silicon Graphics RGB
TGA	Truevision
TIF	TIFF
XBM	X-Windows Bitmap
XPM	X-Windows Pixel Map
XWD	X-Windows Dump

Vector Files: The CAD Data Standard

Raster images are of little use to professionals who require drawings with detailed information. Instead, the engineering, architecture, design, and construction industries use vector images to retain data with great precision and detail. CAD programs, which are the primary graphics creation and storage media for these industries, generate vector data. Figures 43-2 and 43-3 illustrate the same vector image (a CAD drawing) in a "zoomed out" and "zoomed in" display.

Figure 43-2

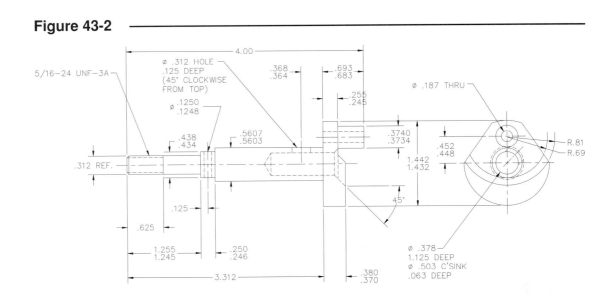

Figure 43-3

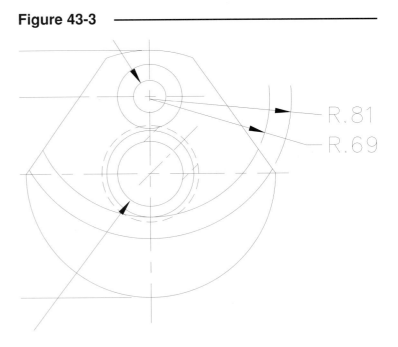

Because a vector file defines a line, for example, as a vector between two endpoints, only one dimension is retained—length. "Zooming in" does not create a second dimension (thickness or width) for the line. The thickness of the displayed line on a computer monitor is always one pixel, the size of which is determined by the display device. The following table lists some examples of vector file formats.

VECTOR FILES

File Format (Extension)	Application/Creator
CDR	CorelDRAW
CGM	Computer Graphics Metafile
CMX	Corel Metafile Exchange
DGN	MicroStation
DRSW	Micrografx Draw
DWG	AutoCAD
DXF	AutoCAD
GEM	GEM Metafile
IGS	IGES
MCS	MathCAD
P10	Tektronix Plot 10
PCL	HP Laser Jet II
PGL	HP 7475A Plotter
PIC	Lotus
PRT	CADKEY
WRL	VRML

Before the new DWF standard was introduced, the World Wide Web provided inadequate usefulness for the engineering, architecture, design, and construction industries. Raster files are unable to retain the necessary detailed information, while the data-rich vector files generated by CAD programs are too large to be "viewable" over the web. A CAD file's size is partially due to the non-graphic information contained in the file, such as layers and other named dependent objects, procedures used in creating the files, drawing and editing preferences, symbol (*Block*) definitions, and so on. Although CAD files can be transmitted via the Internet from one location to another and then loaded into the associated CAD program, they cannot be "viewed" practically over the web from a remote location.

The DWF File Format

The DWF file format developed by Autodesk provides a medium for storing only the necessary display information of a CAD-generated drawing file. The DWF file format is a 16-, 20-, or 32-bit vector format for 2D graphics. The main advantage of this file format is size, usually about 10–40% of the parent CAD file. This compact vector file facilitates faster transport of images over the web, thus making it possible to view a vector image stored on a web server, with zoom and pan capabilities, from a remote location such as your computer at home or from a laptop on the construction site.

Only basic web-browsing tools plus the *WHIP!* Plug-In are needed to view the images—not the original CAD software. Because the DWF files cannot be edited, drawing security and integrity are maintained. The complete data-rich CAD file can be attached to the DWF for transport over the Internet if desired.

Now that you are familiar with the concepts of how the DWF format can be created and viewed, let's examine the commands and utilities needed for these operations. The information is presented in three sections: Viewing DWF Files, Creating DWF Files, and Publishing DWF Files.

VIEWING DWF FILES

Although DWF files are created from AutoCAD drawings, you cannot view the DWF files from within AutoCAD. To view DWF files over the Internet you need:

1. Internet access through your company, school, or an "Internet service provider" (ISP);
2. a popular Web browser; and
3. the *WHIP!* Plug-In.

The *WHIP!* Plug-In operates with two manufacturer's products—Netscape Navigator™ (or Communicator™) and Microsoft Internet Explorer™. Internet Explorer is included with the Microsoft Windows products (Windows 95™ or Windows NT™ 4.0). Netscape Navigator and Communicator are available from most software supply sources or can be downloaded from "www.netscape.com." The software is normally free for educational institutions (downloading to an .EDU web address).

See your network administrator or information systems administrator in your school or company for Internet access. Otherwise, contact your local computer store or computer services provider for information on local Internet service providers.

Obtaining the *WHIP!* Plug-In

The *WHIP!* Plug-In for your browser should be downloaded from the Autodesk web site. After you have obtained a web browser, open the browser (or use the *Browser* command from within AutoCAD to launch the browser) and enter the following address:

http://www.autodesk.com/products/acadr14/features/

At this site you will find information about the *WHIP!* Plug-In and how to download it directly to your system. If you are using Netscape, print out the information given about installing the *WHIP!* Plug-In while you are at the site. The latest plug-in; at the time of this writing compatible with AutoCAD Release 14 is *WHIP!* Release 2.0 (for Netscape Communicator [Navigator 4.0] or Navigator 3.0x) or *WHIP!* Release 2.0 ActiveX Control (for Internet Explorer 3.02). Downloading this file (WHIP2.EXE) for Netscape may take from 8 minutes (T1 access) to 20 minutes (28.8 modem), and for Internet Explorer may take from 1 to 8 minutes, respectively.

Using the *WHIP!* Plug-In

Once the *WHIP!* Plug-In is installed, you can view DWF files. Several methods can be used to view DWF files:

1. Log onto a web site that has DWF files posted for viewing.
2. Use the *File* pull-down menu in the browser to locate and open a DWF file.
3. Drag-and-drop the file from File Manager or Explorer into the browser.
4. Double-click on the DWF file in File Manager or Explorer to launch the browser and view the file.

Several sample DWF files are posted on the Autodesk site for you to view and test your Plug-in. The DWF sample drawings can be viewed by entering the following address:

http://www.autodesk.com/products/autocad/whip/whpsites.htm

(Notice there is only one "i" in "whpsites.htm")

A DWF file <u>does not have to be posted on a web page</u> for you to view it, as long as you have access to the file on your system or via any network connection. For example, once you create a DWF file from within AutoCAD (using the *Dwfout* command), it can be immediately viewed from within your browser. In this case, use the *File* pull-down menu in Netscape or Internet Explorer to locate and load the DWF file.

You can also drag-and-drop the file (file name or icon) from Windows File Manager, Explorer, or other source into the open browser program. The file is automatically opened for viewing.

Once a DWF file is loaded in Netscape Navigator or Microsoft Internet Explorer, the *WHIP!* Plug-In provides real time *Zoom* and *Pan* capabilities. Right-clicking (on the mouse) invokes the *WHIP!* menu displaying zoom, pan, and other options (Fig. 43-4). Initially, a web page displaying a DWF image normally gives a full view of the image, similar to a *Zoom Extents*. The *WHIP!* menu provides the following options. (All menu options do not appear unless the open DWF includes the features. For example, a DWF without named *Views* causes the *Named Views...* option not to appear in the *WHIP!* menu.)

Figure 43-4

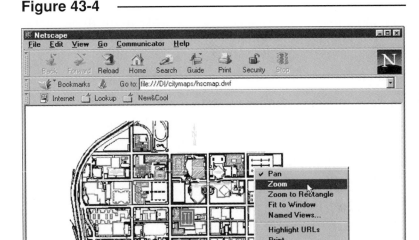

Pan
By default you are in real-time *Pan* mode and the pointer becomes a "hand" icon. Press the left mouse button to drag the image around on the screen in a manner identical to the *Real Time* option of *Pan* in AutoCAD.

Zoom
The cursor changes to a magnifying glass. Real-time zoom is accomplished by holding down the left mouse button while moving the mouse up (for zoom in) or down (for zoom out), like the *Real Time* option *Zoom* in AutoCAD. In this way genuine detail can be viewed, similar to capabilities in the original CAD program with the associated CAD drawing. Figure 43-5 displays detail within the same DWF image displayed previously.

Figure 43-5

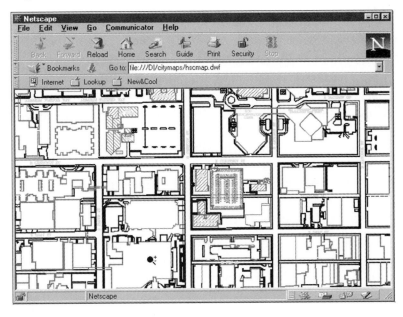

Zoom to Rectangle
The cursor changes to a cross and allows you drag a Window (rectangle) around the area you want to zoom.

Fit to Window

The view is automatically zoomed in or out so the image fills the available Window in the browser.

Named Views...

If the AutoCAD drawing from which the DWF image was created contained named *Views*, those views are recorded in the DWF image and can be viewed with this option. A dialog box appears with the list of named views for you to select from.

Highlight URLs

Toggling this option on causes all the URLs in the image to be highlighted with a "blinking" box over each URL area (see Uniform Resource Locators next). This allows you to see where URLs are located before using *Pan* or *Zoom*. To remove the highlighted boxes, remove the check (toggle). You can hold down the Shift key as a shortcut for highlighting URLs.

Print...

Selecting *Print* causes the current view to be sent to the system printer. You can also print the HTML file along with the embedded DWF using the browser's *Print* button.

SaveAs...

You can save the DWF file to a local drive in one of three formats: DWF, DWG, or BMP. If DWG is selected, *WHIP!* copies the related DWG file used to create the DWF provided that the DWG file is located in the same directory as the DWF. You can also use drag-and-drop to bring a related DWG to your local system if the DWG is located in the remote directory where the DWF is stored.

This is a very powerful feature that clients and colleagues can use to exchange drawings. You can post the DWF file for viewing because it is small and can be easily viewed. If you locate the related "parent" DWG file (used to create the DWF) in the same directory as the DWF, it can be copied (across the Internet) to the viewer's system using the *SaveAs...* feature. The drag-and-drop feature (see Drag-and-Drop into AutoCAD) and the Internet Utilities command, *Openurl,* can also be used to copy DWG files via Internet but cannot be used to save the files in BMP or DWF format.

Back

This accomplishes the same action as using the *Back* button in the browser.

Forward

This is equivalent to the browser's *Forward* button.

Drag-and-Drop into AutoCAD

If you drag-and-drop the DWF image <u>from your browser into an open AutoCAD Release 14 session</u>, a special code is invoked to grab the "parent" DWG file (if available in the directory) and bring it into AutoCAD for viewing and editing. To open a DWG in AutoCAD using drag-and-drop:

1. Press and hold the Ctrl and Shift keys simultaneously.
2. Click on the DWF image and drag it into AutoCAD.
3. Release the mouse button and then release the Ctrl and Shift keys.

Uniform Resource Locators (URLs)

Uniform Resource Locators (URLs) are Internet addresses, such as "http://www.autodesk.com." URLs can be attached to objects or to areas in a DWF image. This feature offers particularly creative possibilities. With the ability to attach URLs, a DWF image can "contain" additional graphical or textual information.

When viewing a DWF image, passing the mouse pointer over a URL-linked object displays a small hand image. The related URL address is also listed at the bottom of the browser, similar to the Help strings that appear in AutoCAD. Clicking on the linked object automatically links to the new URL, which causes the browser to display the text and images at the new address. The URL could be another DWF image or textual data (HTML file). For example, clicking on a door, window, or fixture in an architectural layout could in turn display a table of text information listing price, manufacturer, specifications, etc. Or, clicking on one component of a mechanical assembly could in turn display a detail drawing of the selected part. It would also be possible to view a map, select a building, and see written information about the business, or link to the web site of the business.

The Internet Utilities must be used to attach a URL to a DWF file. Use *Attachurl* in the "parent" drawing to assign a URL to specific objects or to an area in a drawing, then use *Dwfout* to create the DWF file with the attached URLs. See the next section, Creating DWF Files.

BROWSER

Pull-down Menu	COMMAND (TYPE)	ALIAS (TYPE)	Short-cut	Screen (side) Menu	Tablet Menu
Help *Connect to Internet*	*BROWSER*	...	...	...	...

AutoCAD Release 14 includes a command to launch your web browser. An icon button on the *Standard* toolbar is available to invoke the *Browser* command.

The information about your browser (location and name of the executable file) is stored in the system registry. Therefore, when you use *Browser*, AutoCAD automatically locates and launches your current Web browser from within AutoCAD. This action makes it possible to view DWF files after you have created them, view other DWF files on the Internet, and drag-and-drop files from the Internet directly into your AutoCAD session.

When you use *Browser*, your registered web browser is launched. The AutoCAD User homepage at the Autodesk web site is located first by default (Fig. 43-6).

Figure 43-6

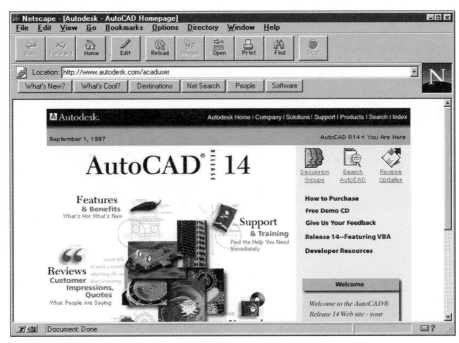

INETLOCATION

Pull-down Menu	COMMAND (TYPE)	ALIAS (TYPE)	Short-cut	Screen (side) Menu	Tablet Menu
...	*INETLOCATION*	...	...	...	...

The initial Internet address that is located when *Browser* is used can be specified by using the *INETLOCATION* system variable or by using the *Preferences* dialog box. The default *INETLOCATION* in AutoCAD Release 14 is:

> http://www.autodesk.com/acaduser

You can specify any other Internet address. In the dialog box (Fig. 43-7), expand the *Menu, Help, Log, and Miscellaneous File Names* section. Expand *Default Internet Location*, highlight the address, and select the *Remove* button. Enter the desired new address.

Figure 43-7

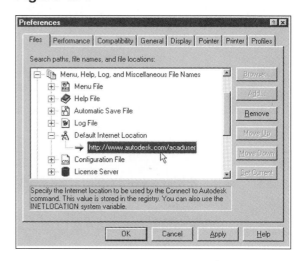

CREATING DWF FILES

Although the DWF is an open standard, only AutoCAD drawings (DWG files) currently are convertible to the DWF format. Producers of CAD programs other than Autodesk should soon follow suit to enable their proprietary CAD drawing files to be convertible and viewable as DWF drawings. It is likely that the DWF format will emerge as the standard for web-viewable drawings, based on AutoCAD's large user base and the existence of the DWG and DXF as industrywide CAD drawing standard formats.

AutoCAD provides several commands to assist you to create DWFs, attach and manage Uniform Resource Locators (URLs, or web addresses) associated with drawing objects, and to furnish other utilities. *Dwfout* and *Browser* are the only Internet-related commands "built into" AutoCAD. The other commands are accessible by installing the bonus features (use "Full" installation). Additional information about these commands cannot be found in the official AutoCAD Release 14 documentation, but can be read or printed from the Autodesk web site (see address in the previous section, Obtaining the *WHIP!* Plug-In) or found in *Inethelp* (available after installing Internet Utilities). The commands discussed in this section are as follows:

> *Dwfout, Attachurl, Detachurl, Listurl, Selecturl, Openurl, Inserturl, Saveurl, Inetcfg,* and *Inethelp*

These commands are used <u>in AutoCAD</u>, not in the browser while viewing a DWF. In other words, use the commands in this section in AutoCAD when preparing a drawing file (DWG) to generate a DWF.

The *Dwfout* command creates a DWF file from the current drawing file. The other commands (ending in "url") are used to manage URLs, or Internet addresses. With these commands you have the ability to attach URLs to objects in a DWF drawing. Therefore, objects in a viewed DWF can be "clicked on" to take the viewer to another address. The address could be another drawing, text information, or a related link. Other utilities allow you to open, insert, and save AutoCAD drawings from a web site. The ability to attach URLs to a drawing and manipulate related drawings over the World Wide Web gives you incredible power to use the web to graphically communicate ideas and transfer information with colleagues, clients, customers, and the general public.

The Internet Utilities commands in this section (except *Dwfout* and *Browser*) are included on the Release 14 CD-ROM and available <u>only if you install the Internet components</u> of AutoCAD Release 14. The Internet Utilities are automatically installed with AutoCAD if you choose the "Full" installation. If you installed only the "Typical" components, you can install just the Internet Utilities at any time by inserting the Release 14 CD-ROM, starting Setup, selecting the "ADD" option, then checking Internet components.

Once the Internet components are installed, the *Internet Utilities* toolbar is accessible by using *MENULOAD* or selecting *Toolbars* from the *Tools* pull-down menu. In the *Toolbars* dialog box, select *inet* as the *Menu Group*, then *Internet Utilities* to produce the toolbar (Fig. 43-8).

Figure 43-8

DWFOUT

Pull-down Menu	COMMAND (TYPE)	ALIAS (TYPE)	Short-cut	Screen (side) Menu	Tablet Menu
...	*DWFOUT*	...	...	...	...

The *Dwfout* command creates a (separate) DWF file from the current open drawing. The DWF file is generally approximately 10–40% the size of the original parent CAD drawing (DWG) file, depending on the precision of the file created (16-, 20-, or 32-bit) and the ratio of the number of objects in the display to the amount of other non-graphic data in the CAD file.

Typically, the *Dwfout* command is used when you have finished the drawing and are ready to create a separate DWF file for presentation on the Internet or an intranet. If you intend to attach any URLs to objects in the drawing (*Attachurl*), complete this work before using *Dwfout*.

When the *Dwfout* command is used, all named *Views* in the original DWG file are recorded in the resulting DWF file. A view named INITIAL is created in the DWF for the current display when the *Dwfout* command is used. When viewing a DWF file in your browser, you can select which view to display by right-clicking to produce the *WHIP!* menu (see Fig. 43-4 and the section on Viewing DWF Files). When using *Dwfout*, the resulting DWF file that is created has the same background color as in the current session.

If you are creating a DWF to post on a web page and intend to make the related DWG file available to the public for copying across the Internet, the DWG file must be *Saved* and named before creating the DWF because the "parent" DWG name (without the path) is stored in the DWF image when *Dwfout* is used. The file names do not have to match, but the related DWG file should not be renamed.

File names are case sensitive in UNIX-based Internet environments. Therefore, special precautions must be used if the files are located in or moved to UNIX servers.

Typing *Dwfout* produces the *Create DWF File* dialog box directly (Fig. 43-9). Select the directory location and enter the desired file name. The drawing name is used as the default name for the DWF file.

Figure 43-9

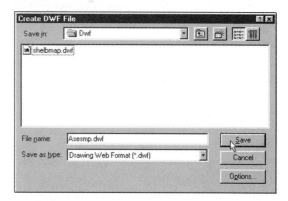

Alternately, you can select *Export* from the *File* pull-down menu to produce the *Export Data* dialog box. In this case you must use the drop-down list to indicate *Drawing Web Format (*.dwf)* as the file type to export (Fig. 43-10).

Figure 43-10

Options...
Selecting the *Options...* button in either dialog box described above produces the *DWF Export Options* dialog with the following choices (Fig. 43-11):

Figure 43-11

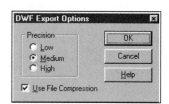

Precision	This section saves the file in 16-bit (*Low*), 20-bit (*Medium*), and 32-bit (*High*) format. Higher precision results in larger file size. *Low Precision* can produce a file about 40% smaller than *High* and as low as 10% of the original DWG file.
File Compression	This option reduces the size of binary format files (cannot be used on ASCII format files). File compression is recommended for all applications except when another application cannot read the compressed format.

AutoCAD 13 Version C4 for Windows 95 or NT can also be used to create DWF files. An additional *WHIP!* display driver especially for Release 13 is required to create DWF files from within AutoCAD Release 13 (DWG) drawings. See the Autodesk web site for more information and to download the needed driver.

ATTACHURL

Pull-down Menu	COMMAND (TYPE)	ALIAS (TYPE)	Short-cut	Screen (side) Menu	Tablet Menu
...	*ATTACHURL*	...	...	...	...

The *Attachurl* feature offers particularly creative possibilities. *Attachurl* allows you to assign an Internet address to objects or areas in the drawing (DWG). When viewing the resulting DWF in a browser, a small "hand" icon appears when the pointer is passed over the object. Clicking on the object activates the browser to search for and display the new URL. The attached URL is a "hyperlink" to another address, which could be another web site, DWF image, or additional explanatory text data.

Invoke the *Attachurl* command to select an object or to define an area, then enter an Internet address to be associated with that object or area of the resulting DWF (when *Dwfout* is later used). To attach the URL to an object, follow the default prompts:

```
Command: attachurl
URL by (Area/<Objects>): Enter
Select objects: PICK
Select objects: Enter
Enter URL:  (enter Internet address or path and file name)
```

If you are specifying a URL for a remote web site or FTP site, enter the address prefixed by "http://" or "ftp://," respectively. If you are loading a file (such as a DWF file or HTML file) from a local network or local PC, use any of the following formats:

```
file:///drive:/pathname/filename.dwf (or .htm)
file:///drive|/pathname/filename.dwf (or .htm)
file://\\localPC\pathname\filename.dwf (or .htm)
file:////localPC/pathname/filename.dwf (or .htm)
file://localhost/drive:/pathname/filename.dwf (or .htm)
file://localhost/drive|/pathname/filename.dwf (or .htm)
```

If one file calls another file and all files are located in the same directory, the drive and directory specifications are not necessary. For example, DWF files that reference other DWFs (or HTMLs that call other HTMLs) do not have to include the drive and directory in the URL reference when all files are located together.

You can also attach the URL to an area. This option is useful if you want one URL to be accessible from a large area of the drawing (the "hand" appears instead of the pointer when viewing the DWF image). This option is also helpful if you want to attach a URL to a specific location in a drawing where no objects exist. Avoid creating overlapping areas, since either URL may be activated when PICKing the overlapped areas.

```
Command: attachurl
URL by (Area/<Objects>): A
First corner: PICK
Other corner: PICK
Enter URL:  (enter Internet address)
```

AutoCAD creates a *Pline* object in the drawing to define the URL-attached area and places it on a new layer named *URLLAYER*. The newly created *Pline* is visible in the drawing but will not be visible in the DWF image. You can turn the layer *On* or *Off*, but remember to turn the layer *On* before the *Dwfout* command is used or the desired address may not be attached. The *Freeze/Thaw* state and properties of the layer should be managed only by AutoCAD.

NOTES: At the time of this writing, *Attachurl* does not provide support for drawings created in paper space. URLs cannot be attached to *Rays* or *Xlines*. *Attachurl* is designed to be used with objects in the WCS (World Coordinate System); attaching URLs to objects on other UCSs may cause unexpected results.

DETACHURL

Pull-down Menu	COMMAND (TYPE)	ALIAS (TYPE)	Short-cut	Screen (side) Menu	Tablet Menu
...	DETACHURL	...	...	...	...

This command reverses the action of *Attachurl*. You can select one or more objects in the drawing to remove the URL or URLs associated with them:

    Command: **detachurl**
    Select objects: **PICK**
    Select objects: **Enter**
    Command:

If a URL has been attached by the *Area* option, AutoCAD also deletes the *Pline* defining the area and gives the following prompt:

    Command: **detachurl**
    Select objects: **PICK**
    Select objects: **Enter**
    DetachURL, deleting the Area

LISTURL

Pull-down Menu	COMMAND (TYPE)	ALIAS (TYPE)	Short-cut	Screen (side) Menu	Tablet Menu
...	LISTURL	...	...	...	...

Listurl reports the Internet address (URL) that is attached to an object or area. When you invoke *Listurl*, you are prompted to "Select objects:," so you must know which objects have URLs attached. (Use *Selecturl* to highlight objects with attached URLs.)

    Command: **listurl**
    Select objects: **PICK**
    Select objects: **Enter**
    URL for selected object is: http://www.xxxxxx...

SELECTURL

Pull-down Menu	COMMAND (TYPE)	ALIAS (TYPE)	Short-cut	Screen (side) Menu	Tablet Menu
...	*SELECTURL*	...	...	...	...

Use *Selecturl* to "select" or highlight all objects or areas in the drawing that have attached URLs. This feature is useful for displaying all objects and areas that have attached URLs. The highlighting action essentially enables grips on the objects (Fig. 43-12, chair [object] and sofa [area]). Press Escape twice to cancel the grips.

Figure 43-12

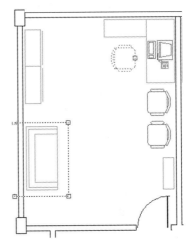

OPENURL

Pull-down Menu	COMMAND (TYPE)	ALIAS (TYPE)	Short-cut	Screen (side) Menu	Tablet Menu
...	*OPENURL*	...	...	...	...

Use this command to *Open* a drawing in AutoCAD from an Internet location, assuming you are currently connected to the Internet through your browser. Invoking *Openurl* produces the *Open Drawing from URL* dialog box (Fig. 43-13). Specify the desired Internet address in the edit box, then select the *Open* button. You can enter a URL in the following formats:

Figure 43-13

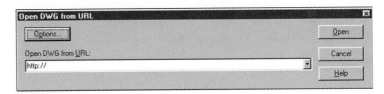

    http://servername/pathname/filename.dwg
    ftp://servername/pathname/filename.dwg
    file:///drive:/pathname/filename.dwg
    file:///drive|/pathname/filename.dwg
    file://\\localPC\pathname\filename.dwg
    file:////localPC/pathname/filename.dwg
    file://localhost/drive:/pathname/filename.dwg
    file://localhost/drive|/pathname/filename.dwg

This method accomplishes the same action as opening a DWG file by using drag-and-drop to bring its related DWF file from your browser into an open AutoCAD session. The *Options* button produces the *Internet Configuration* dialog box (see *Inetcfg*).

INSERTURL

Pull-down Menu	COMMAND (TYPE)	ALIAS (TYPE)	Short-cut	Screen (side) Menu	Tablet Menu
...	SELECTURL	...	...	...	...

Inserturl is similar to *Openurl* except the DWG is inserted as a *Block* into the current drawing. Using *Inserturl* produces the *Insert DWG from URL* dialog box, which is essentially the same as the *Open DWG from URL* dialog box (see Fig. 43-13). Specify the desired Internet address in the edit box (use any of the formats given in *Openurl*), then select the *Open* button. The *Options* button produces the *Internet Configuration* dialog box (see *Inetcfg*).

When you use *Inserturl*, the files that you download from the Internet are placed in your system's temporary directory (usually C:\Temp). These files are not automatically deleted, so you should periodically remove the unwanted files from your temporary directory.

Drag-and-Drop to *Insert* a Drawing

Similar to the drag-and-drop feature for *Opening* a drawing from a URL, you can use drag-and-drop to *Insert* a drawing into the current drawing from a URL. Your browser must be open and connected to the desired site, and you must have AutoCAD running with the desired drawing *Open* that you want to *Insert* the new (remote) drawing into. This function operates only when the DWG file used to create the DWF file exists in the same directory as the DWF file at the time you drag-and-drop. To *Insert* a DWG as a *Block* into an open AutoCAD drawing using drag-and-drop:

1. Press and hold the Ctrl key.
2. Click on the DWF image and drag it into AutoCAD.
3. Release the mouse button and then release the Ctrl key.

SAVEURL

Pull-down Menu	COMMAND (TYPE)	ALIAS (TYPE)	Short-cut	Screen (side) Menu	Tablet Menu
...	SAVEURL	...	...	...	...

This command can be used to save the current drawing to an FTP address. The FTP address can be on a local or remote server, assuming you are currently connected to the web. If the related DWF is saved in the same directory, saving the DWG file to that directory makes the DWG file accessible to the public when

Figure 43-14

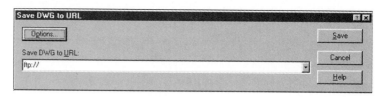

remote AutoCAD users visit the web site and drag-and-drop the related DWF from their browser into AutoCAD or use *Openurl*. *Saveurl* produces the *Save DWG to URL* dialog box (Fig. 43-14). Enter the desired FTP location to save the drawing in the edit box in the following format:

ftp://servername/pathname/filename.dwg

The *Options* button produces the *Internet Configuration* dialog box.

INETCFG

Pull-down Menu	COMMAND (TYPE)	ALIAS (TYPE)	Short-cut	Screen (side) Menu	Tablet Menu
...	*INETCFG*	...	...	...	...

Inetcfg produces the *Internet Configuration* dialog box (Fig. 43-15). This dialog box allows you to specify a *User name:* and *Password:* when logging in to an FTP site (when required) or to view a secure HTTP site (when required). Generally, this information is not needed to view "public" sites but login user names and passwords can be helpful for ensuring only a select group of designers, engineers, colleagues, etc. have access to specific AutoCAD drawings.

If you log in to a secured FTP server that contains files you want access to open, insert, or save to the server, turn *Anonymous Login* off and enter your user name and password for that server. If you attempt to access a site requiring password and access is denied, AutoCAD displays the *User Authentication* dialog box, which prompts you to enter your user name and password.

Figure 43-15

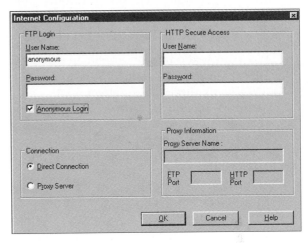

You can also specify if you want to use a *Direct Connection* to the site or specify a *Proxy Server* and port addresses. If you have a direct connection to the Internet through an ISP (Internet service provider), select *Direct Connection*. If you are connected to the Internet through a proxy server, select *Proxy Server*. If you don't know if you are using a proxy server, contact your network system administrator or look at the proxy settings in your browser. Proxy servers are used as a gateway between a company's internal intranet and the external Internet. This "firewall" machine provides access to the outside world for people inside the firewall and provides protection from unauthorized visitors to the company's intranet.

INETHELP

Pull-down Menu	COMMAND (TYPE)	ALIAS (TYPE)	Short-cut	Screen (side) Menu	Tablet Menu
...	*INETHELP*	...	...	...	...

This feature produces separate help windows similar to AutoCAD's built-in Help screens; however, the built-in AutoCAD Help does not contain information on the Internet Utilities (or other bonus features). *Inethelp* provides information online similar to the pages in this chapter.

PUBLISHING DWF FILES

Once the DWF files have been generated, they can be included in a web page using HTML codes. HTML (Hypertext Markup Language) is the platform-generic open standard set of codes used for formatting text and images to be displayed within web pages. HTML files are in ASCII format (no word processing codes). You can use any text editor (like Microsoft Notepad) or word processor (like Microsoft Word or Corel WordPerfect) to create HTML files as long as you save the file in "Text Only" or ASCII format.

The instructions below and on the following pages explain how to embed a DWF file into an HTML document by adding special tags using an ASCII text editor. Because the required format is very specific, this section is reprinted from AutoCAD documentation available at the Autodesk web site and in AutoCAD's *Inethelp*. If you need more information on this subject, access the following addresses:

> http://www.autodesk.com/products/acadr14/features/r2-prod.htm
> http://www.autodesk.com/products/autocad/whip/whip.htm.

Using a Text Editor to Add a DWF File to a Web Page

1. Start your text editor and open an existing HTML document.
2. Copy the following code and paste it into your HTML document in the location you want the DWF file to appear in the document:

```
<object
classid ="clsid:B2BE75F3-9197-11CF-ABF4-08000996E931"
codebase = "ftp://ftp.autodesk.com/pub/autocad/plugin/whip.cab#version=2,0,0,0"
width=400
height=300>
<param name="Filename" value="drawingname.dwf">
<param name="View" value="10000,20000 30000,40000">
<param name="Namedview" value="viewname">
<embed name= drawingname src="drawingname.dwf"
pluginspage=http://www.autodesk.com/products/autocad/whip/whip.htm
width=400
height=300
view="10000,20000 30000,40000"
namedview="kitchen">
</object>
```

3. Modify the italicized text as described in Description of Special HTML Tags.
4. Save your HTML document.
5. Start your browser and open your HTML document to verify its correctness.
6. Copy the HTML document and its associated DWF files to the proper location on your Web server.

Description of Special HTML Tags

The special HTML code consists of four HTML tags. Each tag is made up of text enclosed in brackets, <like this>. The italicized text should be changed to suit your usage, and the regular text must be entered as shown. The following four sections describe how to modify these tags to suit your application.

Although some of these tags apply only to Netscape Navigator or Microsoft Internet Explorer, it is recommended that you always include all four tags so your DWF images can be viewed by someone using either browser.

The "object" Tag

This optional tag is ignored by Netscape Navigator and is included only for compatibility with Microsoft Internet Explorer. If you decide to use this optional tag, you must also include the second, third, and fourth tags. If you choose not to use this tag, you must remove the second and fourth tags.

```
<object
classid ="clsid:B2BE75F3-9197-11CF-ABF4-08000996E931"
codebase="ftp://ftp.autodesk.com/pub/autocad/plugin/whip.cab#version=2,0,0,0"
width=400
height=300>
```

Change the width and height values to the size (in pixels) that you want the DWF image to appear on your Web page. These values must match the values specified in the third tag. The other options in this tag (classid and codebase) are used to identify the *WHIP!* Plug-In and tell Microsoft Internet Explorer where to find the latest version of the plug-in.

The "param" Tag

This is an optional tag that is ignored by Netscape Navigator and only included for Microsoft Internet Explorer. This tag requires the presence of the first and fourth tags.

```
<param name="Filename" value="drawingname.dwf">
<param name="View" value="10000,20000 30000,40000">
<param name="Namedview" value="viewname">
```

Change the drawingname string to the name of the DWF file you are inserting into your HTML document. This parameter actually specifies a URL and can be either a relative URL or an absolute URL. If the DWF file is in the same directory as the HTML file, this parameter is just the name of the DWF file. Here are some examples of legal relative URLs for this parameter:

my_drawing.dwf	specifies that the DWF file is in the same directory as the HTML file
drawings/my_drawing.dwf	specifies that the DWF is in a subdirectory named drawings, which is below the location of the HTML file
../drawings/my_drawing.dwf	specifies that the DWF file is in a sibling directory to the location of the HTML file

Examples of legal absolute URLs for this parameter:

```
http://www.acme.com/drawings/my_drawing.dwf
ftp://ftp.acme.com/our_valut/my_drawing.dwf
file:///C|/official/my_drawing.dwf
```

Relative URLs are recommended since the respective files can be moved more easily. The View and Namedview parameters are optional. You can use either View or Namedview, but not both. View defines the coordinates of the display area that you want to be the initial view of the DWF image. Namedview defines the named view to be displayed as the initial view of the DWF image. All named views that are present in the DWG file when the *Dwfout* command is run are recorded in the resulting DWF file. For convenience, if the named view INITIAL is not already specified in the DWG, it is automatically created in the DWF file. The INITIAL view for the DWF matches the view of the DWG when the *Dwfout* command is run. When viewing a DWF, the available named views can be accessed from the *WHIP!* menu by clicking the right mouse button.

The "embed" Tag

This tag is required and is the only tag (of the four) that Netscape Navigator uses.

```
<embed name= drawingname src=" drawingname.dwf"
pluginspage=http://www.autodesk.com/products/autocad/whip/whip.htm
width=400
height=300
view="10000,20000 30000,40000"
namedview="kitchen">
```

Change the width and height to match the width and height values set in the "object" tag. The drawing-name string is a name parameter that lets you give each DWF in an HTML document a unique name that can be referenced by Java and JavaScript. This can be changed as desired so that each DWF in the HTML document has a unique name. This name is not a filename or a URL, and is usually a short one word identifier. If your DWF filename were linkrod.dwf, for example, you could use linkrod as the drawing's name, or just drawing.

The drawingname.dwf string is a URL that references the desired DWF file. This parameter must be the same as the URL parameter in the "param" tag.

The pluginspage parameter is used to tell Netscape where the current version of the _WHIP!_ Plug-In can be found. This parameter should not be changed.

The view and namedview parameter values must match those defined by the param tag.

The Ending "object" Tag

This is an optional tag that is ignored by Netscape Navigator and only included for Microsoft Internet Explorer. This tag requires the presence of the first and second tags.

Registering a MIME Type

To publish DWF files, your Internet server must be able to recognize the DWF file type. Ask your web-master to add the MIME type "drawing/x-dwf" (with the .dwf extension) to your Internet server machine so that it is registered with your Internet server software. If you use an Internet service provider (ISP), contact them to register the "drawing/x-dwf" MIME type on their servers. If you need more information on this procedure, connect to the following address:

http://www.autodesk.com/products/acadr14/features/usedwf.htm

CHAPTER EXERCISES

These exercises require Netscape Navigator 3.0x (or higher) or Microsoft Internet Explorer to be installed and operational on your computer system. Internet access is needed to download the _WHIP!_ Plug-In for Netscape Navigator or the _WHIP!_ ActiveX Control plug-in for Microsoft Internet Explorer (Exercise 1) and to view sample DWF files (Exercises 2–5). If the _WHIP!_ Plug-In is installed, Internet access is not needed for Exercises 3, 4, and 5. Exercises 4 and 5 require the AutoCAD Internet Utilities to be installed.

1. **Obtain the _WHIP!_ Plug-In**

   A. Open the browser on your system (Netscape Navigator 3.0x or higher or Microsoft Internet Explorer). If you need to initialize your Internet connection by activating Dial Up Networking or other program, do so. When you have an Internet connection, access the Autodesk web site by entering the following address at the location edit box:

      http://www.autodesk.com/products/acadr14/features/

   B. Locate the section giving information about downloading the **_WHIP!_ Release 2** (or higher) Plug-In you need (the _WHIP!_ Plug-In for Netscape Navigator or the _WHIP!_ ActiveX Control plug-in for Microsoft Internet Explorer). Find and print the **Installation Instructions**, then proceed with the **download**.

C. When the download is complete, follow the steps to complete the installation (for Netscape Navigator). (If you are using MS Internet Explorer, installation is automatic.)

2. **View Sample DWF Images**

A. While your Internet connection is active, use your web browser to view sample DWF images. Enter the following address in the location edit box in your browser:

   http://www.autodesk.com/products/autocad/whip/whpsites.htm

   These files were created with *WHIP!* Release 1 (for AutoCAD Release 13), so some of the options are not available (*Named Views, Highlight URLs...,* and *Saveas...*).

B. Select the **Neutrino Property Management** site and select the **Site Plan**. When the DWF image appears, *Pan* the image about in the window. Next, press the **right mouse** button to display the *WHIP!* menu. (Only options that apply to the specific DWF file are available from the menu.) Use *Zoom, Pan, Zoom to Window,* and the other options to answer the following questions:

   a. What is the name of the road passing horizontally through the property?
   b. How many elevators (yellow color) are in the red building near Skyline Drive?
   c. How many URLs are attached to the image?

C. Next visit the **Architectural Project Site**. Locate and display the **Plan View** of the project posted for bid. Use the *WHIP!* menu (**right click**) to answer the following questions:

   a. What is the name of the project posted for bid?
   b. What is in the center of the circular section of the building?
   c. What color are the plumbing fixtures?

   Next use *SaveAs...* in the *WHIP!* menu to save the image as a **DWF** file in your working directory.

   d. What is the size (in KBs) of the DWF file?

D. Finally, view the **Zephyr Tools** catalog. Select the **Workbench Equipment**. Use the *WHIP!* menu to examine the vice drawing and answer the following questions:

   a. How many screws are shown on the vice (brown in color)?
   b. How many components make up the vice assembly, including the screws (yellow objects represent hidden features, not components)? Don't miss the two small green pins on the bottom.
   c. How many "feet" does the vice base have?

3. *Dwfout, Browser*

A. *Open* the **ASESMP** sample drawing from the **AutoCAD R14/Sample** directory. If the drawing is not on your system, it can be copied from the AutoCAD Release 14 CD-ROM.

B. Use *Zoom* to zoom in to display only rooms **101** and **102** (upper-left). Next, use the *Dwfout* command in AutoCAD. Specify your normal working directory as the target directory location for the DWF file in the *Create DWF File* dialog box. Accept the default name, **ASESMP.DWF**.

C. Invoke the *Browser* command by selecting the button on the *Standard* toolbar. When the browser opens, use the *File* pull-down menu (in the browser) to locate and load the **ASESMP.DWF** file. Look for a *File...*, *Open...*, *Open File...*, or *Open Page...* option. Also, make sure you indicate to list *Drawing Web Format files (*.dwf)*. (If you are currently connected to the Internet, you do not have to wait for the browser to load the Autodesk page.)

D. When the DWF image appears, it should display only rooms 101 and 102. Use the *WHIP!* menu to *Fit to Window*. Next, use the right-click menu and select the *Named Views...* option. In the dialog box, select **INITIAL** to restore the original view (rooms 101 and 102).

The following exercises require the Internet Utilities to be installed in AutoCAD. If these utilities are not yet installed, you can insert the AutoCAD Release 14 CD-ROM, invoke *Setup*, then select the *ADD* option of installation. From the list of components, place a check by *Internet* and proceed with the installation. After installation, use *Toolbar* in AutoCAD to load the *inet* Menu Group and the *Internet Utilities* toolbar.

4. *Attachurl*. In this exercise you will create two DWF images. Each image will contain a URL that will link to the other image when viewed in the browser.

A. *Open* the **ASESMP** drawing again. *Zoom Extents* to display the entire drawing. Use *Mtext* with text *Height* of **24** to create a paragraph with the following text near the bottom of the drawing:

Click here to view the OFF-URL drawing

B. Use *Attachurl* with the *Area* option. Create the area rectangle around the *Mtext* object. Specify the URL as "**file:///Off-url.dwf**" (the drive and location of your working directory are not necessary if all related DWF files are in the same directory). The text and area box should appear in the drawing similar to Figure 43-16. Use *SaveAs*, name the drawing **ASE-URL**, and locate it in your working directory. Finally, use *Dwfout*, accept the default name (ASE-URL), and save the DWF in your working directory.

Figure 43-16 ———————————————————————————

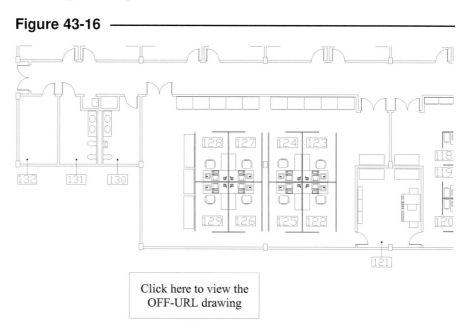

Click here to view the
OFF-URL drawing

C. *Open* the **OFF-DIM** drawing you worked on last in Chapter 29 Exercises. *Freeze* the **Title** layer so the title block and border are not visible. Create an *Mtext* object with the following text near the bottom of the drawing. Use a *Height* of **24**.

> Click here to view the ASE-URL drawing

Use *Attachurl* with the *Area* option. Create the box around the *Mtext* object. Specify the URL as "**file///Ase-url.dwf**". The text and area box should appear in the drawing similar to Figure 43-17. Use *SaveAs*, name the drawing **OFF-URL**, and locate it in your working directory. Finally, use *Dwfout*, accept the default name (OFF-URL), and save the DWF in your working directory.

Figure 43-17

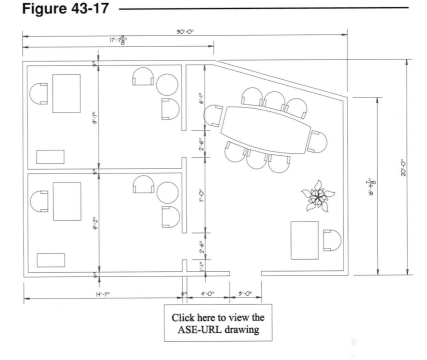

D. Open your **browser** by any method. Use the *File* pull-down menu to locate and load one of the DWF files you just created. You should be able to click on the embedded URL and the browser should automatically load the other DWF.

E. If this scheme does not work, you should do some "debugging." Make sure that the URL (path and file name) in each drawing is listed correctly. If not, use *Detachurl*, then *Attachurl* and specify the correct URL. Remember to *Save* the drawing, then create a new DWF using *Dwfout* in each case.

5. *Attachurl, Dwfout, Listurl, Detachurl*. In this exercise you will use the same procedure as in Exercise 4 to "link" several DWF files. The DWF images are composed of one assembly image and several component images. You will attach several URLs to the assembly image so that clicking on any one component links to the related detail drawing of that component, and each component image has one URL that links back to the assembly.

A. Locate the following drawings (Old Name) that you worked with in Chapters 29 and 30 and copy them to your working directory. All drawings except SADDL-DM are *Xrefs* that make up the ASSY drawing.

Old Name	New Name
ASSY.DWG	**ASSY-URL.DWG**
SHAFT.DWG	**SHAFT-URL.DWG**
SLEEVE.DWG	**SLEEVE-URL.DWG**
BOLT.DWG	**BOLT-URL.DWG**
PUL-SEC.DWG	**PUL-URL.DWG**
SADDLE.DWG	
SADDL-DM.DWG	**SADDL-URL.DWG**

B. *Open* the **ASSY.DWG**. Invoke the *External Reference* dialog box and *Bind* all the drawings. Create an *Mtext* object with *Height* of **4** or **5** to state "**Click on any component to view the related detail drawing**." The ASSY-URL drawing should appear as that in Figure 43-18. Use *SaveAs* to rename ASSY.DWG to **ASSY-URL.DWG**.

Figure 43-18 ———————

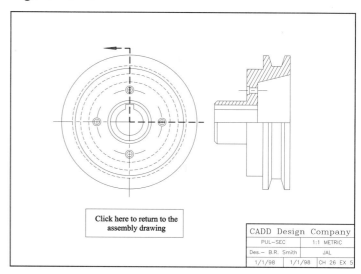

Click any component
to view the related
detail drawing

C. *Open* each of the other drawings except SADDLE.DWG (the SADDLE drawing will not be used again). Follow these steps <u>for each of the component drawings</u>:

a. Create an *Mtext* object with *Height* of **4** or **5** stating, "**Click here to return to the assembly drawing**."
b. Use *Attachurl* to attach a URL by *Area* to the *Mtext* object. The URL should give the file name for ASSY-URL.DWF such as "**file:///Assy-url.dwf**."
c. Then, use *SaveAs* to rename each drawing as indicated under "New Name" listed previously.
d. Finally, use *Dwfout* for each drawing to create the related DWF.

When this step is complete, the PUL-URL drawing should appear as in Figure 43-19 and the SADDL-URL drawing should appear as in Figure 43-20, for example.

Figure 43-19 ———————

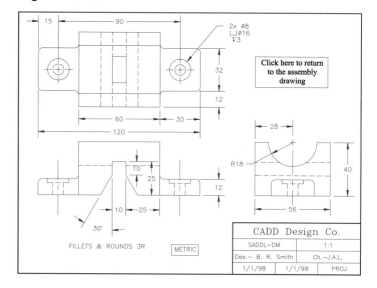

Click here to return to the
assembly drawing

CADD Design Company

PUL−SEC	1:1 METRIC
Des.− B.R. Smith	JAL
1/1/98	1/1/98	CH 26 EX 5

Figure 43-20 ———————

15   90   2x ø8
⌴ø16
⊤3

32

Click here to return
to the assembly
drawing

12

60   30

120

28

TO

25

R18

40

12

10   25

56

30°

FILLETS & ROUNDS 3R   METRIC

CADD Design Co.

SADDL−DM	1:1
Des.− B. R. Smith	Ch.−J.A.L.
1/1/98	1/1/98	PROJ

D. *Open* the **ASSY-URL** drawing and use *Attachurl* by the *Object* method and specify the correct URL for each related component (attach a URL to each component). For example, the URL attached to the shaft component would be "**file:///Shaft-url.dwf.**"

F. Finally, open your browser and test your work. Open **ASSY-URL.DWF** first, then click on any component. This action should "link" to the related DWF detail drawing. Check each component DWF to ensure the link to return to the ASSY-URL image operates correctly.

G. Don't expect everything to work flawlessly the first time. You will most likely have to "debug" the URLs. Do this by using *Listurl*, *Detachurl*, *Attachurl*, then *Dwfout* again.

H. <u>Optional</u>: Create dimensions, borders, and title blocks for any drawings that are lacking these features (but do not create dimensions on the ASSY_URL drawing). *Save* each drawing when finished, then create a new DWF image.

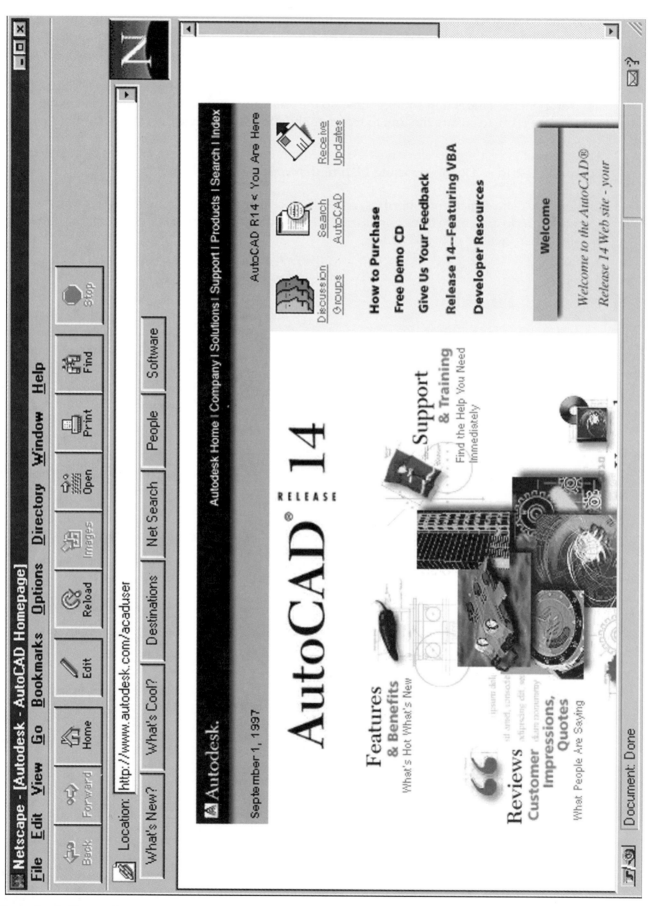

Autodesk Home Page (September 1997), "http://www.autodesk.com/acaduser"

44

MISCELLANEOUS COMMANDS AND FEATURES

Chapter Objectives

After completing this chapter you should:

1. be able to manage named objects and files using wildcards;

2. be able to control the display with *REGENAUTO*, *BLIPMODE*, *DRAGMODE*, and *Fill*;

3. be able to reinitialize I/O ports and the ACAD.PGP file using *Reinit* and be able to check the drawing for errors using *Audit*;

4. know how to use *Multiple* to automatically repeat commands;

5. be able to control dialog boxes with the *FILEDIA*, *CMDDIA*, and *MAXSORT* system variables;

6. be able to create and view slides using *Mslide* and *Vslide*;

7. know how to create and run a *Script* for viewing slide shows;

8. know how to use *Cal* to invoke the geometry calculator for computing arithmetic expressions, locating points, and creating and editing geometry.

CONCEPTS

This chapter discusses several unrelated AutoCAD commands and features that are helpful for intermediate-level users. The following commands, variables, and features are covered in this chapter:

> Miscellaneous Features
> > Wildcards
> > *REGENAUTO*
> > *BLIPMODE*
> > *DRAGMODE*
> > *Fill*
> > *Multiple*
> > *FILEDIA*
> > *CMDDIA*
> > *MAXSORT*
> > *Reinit*
> > *Audit*

> Using Slides and Scripts
> > *Mslide*
> > *Vslide*
> > Creating and Using Scripts
> > Creating a Silde Show
> > *Script*

> The Geometry Calculator
> > *Cal*
> > Calculator Functions and Examples

MISCELLANEOUS FEATURES

AutoCAD keeps objects in a drawing file other than those we know as graphical entities (*Line, Circle, Arc,* etc.). These are called "<u>named objects</u>." The named objects are:

> *Blocks*
> *Dimension Styles*
> *Layers*
> *Linetypes*
> *Text Styles*
> *User Coordinate Systems*
> *Views*
> *Viewport* configurations

Wildcard characters, the *Rename* command, and the *Purge* command can be used to access or alter named objects.

Using Wildcards

Whenever AutoCAD prompts for a list of names, such as file names, system variables, *Block* names, *Layer* names, or other named objects, any of the wildcards in the following wildcard list can be used to access those names. These wildcards help you specify a select group of named objects from a long list without having to repeatedly type or enter the complete spelling for each name in the list.

For example, the asterisk (*) is a common wildcard that is used to represent any alphanumeric string (group of letters or numbers). You may choose to specify a list of layers all beginning with the letters "DIM" and ending with any string. Using the *-Layer* command and entering "DIM*" in response to the *?* option yields a list of only the layer names beginning with DIM.

Commands That Accept Wildcards

Several AutoCAD commands prompt you for a name or list of names or display a list of names to match your specification. Any command that has a ? option provides a list. Wildcards can be used with any of these commands and with many dialog boxes:

Attedit
Block
Dimstyle
Insert
-Layer
-Linetype
Load
Rename
Setvar
-Style
UCS
View
Vplayer
Xref

Valid Wildcards That Can Be Used in AutoCAD

The following list defines valid wildcard characters that can be used in AutoCAD:

Character	Definition
# (pound)	Matches any numeric digit
@ (at)	Matches any alpha character
. (period)	Matches any nonalphanumeric character
* (asterisk)	Matches any string, including the null string. It can be used anywhere in the search pattern—at the beginning, middle, or end of the string
? (question mark)	Matches any single character
~ (tilde)	Matches anything but the pattern
[...]	Matches any one of the characters enclosed
[~...]	Matches any character not enclosed
- (hyphen)	Used with brackets to specify a range for one character
' (reverse quote)	Escapes special characters (reads next character literally)

Below are listed one or more examples for each application of wildcard patterns:

Pattern	Will match or include . . .	But not . . .
ABC	Only ABC	
~ABC	Anything but ABC	
A?C	Any 3-character sequence beginning with A and ending with C	AC, ABCD, AXXC, or XABC
AB?	ABA, AB3, ABZ, etc.	AB, ABCE, or XAB
?BC	ABC, 3BC, XBC, etc.	AB, ABCD, BC, or XXBC
A*C	AAC, AC, ABC, AX3C, etc.	XA or ABCD
A*	Anything starting with A	XAAA
*AB	Anything ending with AB	ABX
AB	AB anywhere in string	AXB
~*AB*	All strings without AB	AB, ABX, XAB, or XABX
'*AB	*AB	AB, XAB, or *ABC
[AB]C	AC or BC	ABC or XAC
[~AB]C	XC or YC	AC, BC, or XXC
[A-J]C	AC, BC, JC, etc.	ABC, AJC, or MC
[~A-J]C	Any character not in the range A–J, followed by C	AC, BC, or JC

Wildcard Examples

For example, assume that you have a drawing of a two-story residential floor plan with the following *Layers* and related settings:

Layer name	State	Color	Linetype
0	On	7 (white)	CONTINUOUS
1-ELEC-DIM	On	7 (white)	CONTINUOUS
1-ELEC-LAY	On	7 (white)	CONTINUOUS
1-ELEC-TXT	On	7 (white)	CONTINUOUS
1-FLPN-DIM	On	7 (white)	CONTINUOUS
1-FLPN-LAY	On	7 (white)	CONTINUOUS
1-FLPN-TXT	On	7 (white)	CONTINUOUS
1-HVAC-DIM	On	7 (white)	CONTINUOUS
1-HVAC-LAY	On	7 (white)	CONTINUOUS
1-HVAC-TXT	On	7 (white)	CONTINUOUS
2-ELEC-DIM	On	7 (white)	CONTINUOUS
2-ELEC-LAY	On	7 (white)	CONTINUOUS
2-ELEC-TXT	On	7 (white)	CONTINUOUS
2-FLPN-DIM	On	7 (white)	CONTINUOUS
2-FLPN-LAY	On	7 (white)	CONTINUOUS

2-FLPN-TXT	On	7 (white)	CONTINUOUS
2-HVAC-DIM	On	7 (white)	CONTINUOUS
2-HVAC-LAY	On	7 (white)	CONTINUOUS
2-HVAC-TXT	On	7 (white)	CONTINUOUS

If you wanted to turn *Off* all the layers related to the second floor (names beginning with "2"), you can use the asterisk (*) wildcard as follows:

```
Command: -Layer
?/Make/Set/New/ON/OFF/Color/Ltype/Freeze/Thaw/LOck/Unlock: off
Layer name(s) to turn Off: 2*
?/Make/Set/New/ON/OFF/Color/Ltype/Freeze/Thaw/LOck/Unlock: ?
Layer name(s) to list <*>:
```

Layer name	State	Color	Linetype
0	On	7 (white)	CONTINUOUS
1-ELEC-DIM	On	7 (white)	CONTINUOUS
1-ELEC-LAY	On	7 (white)	CONTINUOUS
1-ELEC-TXT	On	7 (white)	CONTINUOUS
1-FLPN-DIM	On	7 (white)	CONTINUOUS
1-FLPN-LAY	On	7 (white)	CONTINUOUS
1-FLPN-TXT	On	7 (white)	CONTINUOUS
1-HVAC-DIM	On	7 (white)	CONTINUOUS
1-HVAC-LAY	On	7 (white)	CONTINUOUS
1-HVAC-TXT	On	7 (white)	CONTINUOUS
2-ELEC-DIM	Off	7 (white)	CONTINUOUS
2-ELEC-LAY	Off	7 (white)	CONTINUOUS
2-ELEC-TXT	Off	7 (white)	CONTINUOUS
2-FLPN-DIM	Off	7 (white)	CONTINUOUS
2-FLPN-LAY	Off	7 (white)	CONTINUOUS
2-FLPN-TXT	Off	7 (white)	CONTINUOUS
2-HVAC-DIM	Off	7 (white)	CONTINUOUS
2-HVAC-LAY	Off	7 (white)	CONTINUOUS
2-HVAC-TXT	Off	7 (white)	CONTINUOUS

You may want to *Freeze* all layers (both floors) related to the electrical layout (having ELEC in the layer name). (Assume all the layers are *On* again.) You could use the question mark (?) to represent any floor number and an asterisk (*) for any string after ELEC as follows.

```
Command: -Layer
?/Make/Set/New/ON/OFF/Color/Ltype/Freeze/Thaw/LOck/Unlock: fr
Layer name(s) to Freeze: ?-elec*
?/Make/Set/New/ON/OFF/Color/Ltype/Freeze/Thaw/LOck/Unlock: ?
Layer name(s) to list <*>:
```

Layer name	State	Color	Linetype
0	On	7 (white)	CONTINUOUS
1-ELEC-DIM	Frozen	7 (white)	CONTINUOUS
1-ELEC-LAY	Frozen	7 (white)	CONTINUOUS
1-ELEC-TXT	Frozen	7 (white)	CONTINUOUS
1-FLPN-DIM	On	7 (white)	CONTINUOUS

1-FLPN-LAY	On	7 (white)	CONTINUOUS
1-FLPN-TXT	On	7 (white)	CONTINUOUS
1-HVAC-DIM	On	7 (white)	CONTINUOUS
1-HVAC-LAY	On	7 (white)	CONTINUOUS
1-HVAC-TXT	On	7 (white)	CONTINUOUS
2-ELEC-DIM	Frozen	7 (white)	CONTINUOUS
2-ELEC-LAY	Frozen	7 (white)	CONTINUOUS
2-ELEC-TXT	Frozen	7 (white)	CONTINUOUS
2-FLPN-DIM	On	7 (white)	CONTINUOUS
2-FLPN-LAY	On	7 (white)	CONTINUOUS
2-FLPN-TXT	On	7 (white)	CONTINUOUS
2-HVAC-DIM	On	7 (white)	CONTINUOUS
2-HVAC-LAY	On	7 (white)	CONTINUOUS
2-HVAC-TXT	On	7 (white)	CONTINUOUS

You may want to use only the layout layers (names ending with LAY) and *Freeze* all the other layers. (Assume all layers are *On* and *Thawed*.) The tilde (~) character can be used to match anything but the pattern given. The tilde (~) character is translated as "anything except."

Command: **-layer**
?/Make/Set/New/ON/OFF/Color/Ltype/Freeze/Thaw/LOck/Unlock: **fr**
Layer name(s) to Freeze: **~*lay**
?/Make/Set/New/ON/OFF/Color/Ltype/Freeze/Thaw/LOck/Unlock: **?**
Layer name(s) to list <*>:

Layer name	State	Color	Linetype
0	On	7 (white)	CONTINUOUS
1-ELEC-DIM	Frozen	7 (white)	CONTINUOUS
1-ELEC-LAY	On	7 (white)	CONTINUOUS
1-ELEC-TXT	Frozen	7 (white)	CONTINUOUS
1-FLPN-DIM	Frozen	7 (white)	CONTINUOUS
1-FLPN-LAY	On	7 (white)	CONTINUOUS
1-FLPN-TXT	Frozen	7 (white)	CONTINUOUS
1-HVAC-DIM	Frozen	7 (white)	CONTINUOUS
1-HVAC-LAY	On	7 (white)	CONTINUOUS
1-HVAC-TXT	Frozen	7 (white)	CONTINUOUS
2-ELEC-DIM	Frozen	7 (white)	CONTINUOUS
2-ELEC-LAY	On	7 (white)	CONTINUOUS
2-ELEC-TXT	Frozen	7 (white)	CONTINUOUS
2-FLPN-DIM	Frozen	7 (white)	CONTINUOUS
2-FLPN-LAY	On	7 (white)	CONTINUOUS
2-FLPN-TXT	Frozen	7 (white)	CONTINUOUS
2-HVAC-DIM	Frozen	7 (white)	CONTINUOUS
2-HVAC-LAY	On	7 (white)	CONTINUOUS
2-HVAC-TXT	Frozen	7 (white)	CONTINUOUS

Remember that wildcards can be used with many AutoCAD commands. For example, you may want to load a certain set of *Linetypes*, perhaps all the HIDDEN variations. The following sequence could be used:

Command: **-linetype**
?/Create/Load/Set: **l**
Linetype(s) to load: **hid***

Linetype HIDDEN loaded.
Linetype HIDDEN2 loaded.
Linetype HIDDENX2 loaded.

Or you may want to load all of the linetypes except the "X2" variations. This syntax could be used:

Command: **-linetype**
?/Create/Load/Set: **l**
Linetype(s) to load: **~*x2**

Linetype BORDER loaded.
Linetype BORDER2 loaded.
Linetype CENTER loaded.
Linetype CENTER2 loaded.
Linetype DASHDOT loaded.
Linetype DASHDOT2 loaded.
Linetype DASHED loaded.
Linetype DASHED2 loaded.
etc.

RENAME

The *Rename* and *Ddrename* commands allow you to rename <u>any named object</u> that is part of the current drawing. See Chapter 21, *Blocks*.

PURGE

Purge allows you to selectively delete any named object that is not referenced in the drawing. In other words, if the drawing has any named objects defined but not appearing in the drawing, they can be deleted with *Purge*. See Chapter 21, *Blocks*.

MAXSORT

The *MAXSORT* system variable controls the maximum number of named objects or files that are alphabetically sorted when a list is produced in a dialog box or command line listing. The default setting is 200, meaning that a maximum of 200 items are sorted. If the list contains more than the number specified in *MAXSORT*, no sorting is done.

If *MAXSORT* is set to 0, no alphabetical sorting is done and the list is displayed in the order that the items were created. It may be helpful in some cases to set *MAXSORT* to 0 if you want items to be displayed in the order they were created. For example, if *MAXSORT* is set to 0, the display of layers may appear, as shown in Figure 44-1. Change the variable by typing "maxsort" at the Command: prompt or by using the *Setvar* command.

Figure 44-1

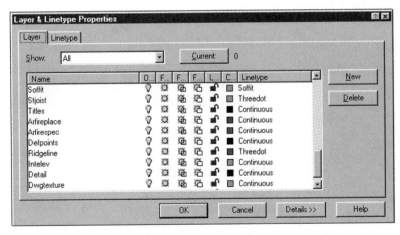

Keep in mind that display of <u>external files</u> in AutoCAD for Windows may appear in alphabetical order, even though *MAXSORT* is set to 0. The Windows sorting overrides the *MAXSORT* setting for external files only but should not affect named objects (internal). The setting for *MAXSORT* is saved in the system registry.

REGENAUTO

Pull-down Menu	COMMAND (TYPE)	ALIAS (TYPE)	Short-cut	Screen (side) Menu	Tablet Menu
...	*REGENAUTO*	...	...	...	...

When changes are made to a drawing that affect the appearance of the objects on the screen, the drawing is generally regenerated automatically. A regeneration can be somewhat time consuming if you are working on an extremely complex drawing or are using a relatively slow computer. In some cases, you may be performing several operations that cause regenerations between intermediate steps that you feel are unnecessary.

Regenauto allows you to control whether automatic regenerations are performed. *Regenauto* is *On* by default, but for special situations you may want to turn the automatic regenerations *Off* temporarily. The setting of *Regenauto* is stored in the *REGENMODE* variable (1=*On* and 0=*Off*). If *Regenauto* is *Off* and a regeneration is needed, AutoCAD prompts:

> About to regen—proceed? <Y>

If you prompt with a *No* response, the regeneration will be aborted.

BLIPMODE

Pull-down Menu	COMMAND (TYPE)	ALIAS (TYPE)	Short-cut	Screen (side) Menu	Tablet Menu
Tools *Drawing Aids...*	*BLIPMODE*	...	...	...	...

"Blips" are small markers that can be made to appear whenever you PICK a point. Most users prefer that the blips do not appear; therefore, Autodesk decided to set *Blipmode* to *Off* in Release 14. The *BLIPMODE* variable is saved with the drawing file.

Releases of AutoCAD previous to Release 14 have a *Blipmode* setting of *On* by default. In the case that you work with drawings created in Release 13 or earlier, it is likely that the blips may appear when you PICK points. If you prefer, you can turn *Blips Off*. The command format is this:

> Command: **blipmode**
> ON/OFF <ON>: (option)

The setting is saved with the drawing file in the *BLIPMODE* system variable. The variable can also be accessed using the *Setvar* command (see Appendix A).

DRAGMODE

Pull-down Menu	COMMAND (TYPE)	ALIAS (TYPE)	Short-cut	Screen (side) Menu	Tablet Menu
...	*DRAGMODE*	...	...	...	...

AutoCAD dynamically displays some objects such as *Circles, Blocks, Polygons*, etc., as you draw or insert them. This dynamic feature is called "dragging." If you are using a slow computer or for certain operations such as complex *Block* insertions, you may wish to turn dragging off.

The *DRAGMODE* system variable provides three positions for the dynamic dragging feature. The command format is as follows:

> Command: **dragmode**
> ON/OFF/Auto <Auto>: (option)

By default, *DRAGMODE* is set to *Auto*. In this position, dragging is automatically displayed whenever possible, based on the ability of the command in use to support it. Commands that support this feature issue a request for dragging when the command is used. This request is automatically filled when *DRAGMODE* is in the *Auto* position.

When *DRAGMODE* is *On*, you must enter *"drag"* at the command prompt when the request is issued if you want to enable dragging. If *DRAGMODE* is *Off*, dragging is not displayed and all requests are automatically ignored. The *DRAGMODE* variable setting is stored with the current drawing file.

FILL

Pull-down Menu	COMMAND (TYPE)	ALIAS (TYPE)	Short-cut	Screen (side) Menu	Tablet Menu
...	FILL	...	...	...	...

The *Fill* command controls the display of objects that are filled with solid color. Commands that create solid filled objects are *Donut*, *Solid*, and *Pline* (with *width*), *Bhatch*, and text commands using TrueType fonts. *Fill* can be toggled *On* or *Off* to display or plot the objects with or without solid color. The command format produces this prompt:

Command: *fill*
ON/OFF <On>: (option)
Command:

Figure 44-2 displays several objects with *Fill On* and *Off*. The default position for the *Fill* command is *On*. If *Fill* is changed, the drawing must be regenerated to display the effects of the change. The setting for *Fill* is stored in the *FILLMODE* system variable.

Since solid filling a drawing with many wide *Plines*, *Donuts*, and *Solids* may be time consuming to regenerate, it may be useful to turn *Fill Off* temporarily. *Fill Off* would be <u>especially</u> helpful for speeding up test plots.

Figure 44-2

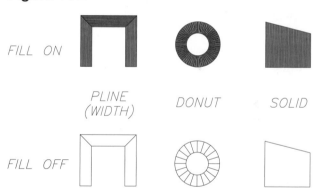

See also Chapters 18 and 26 for using *Fill* with solid-filled TrueType fonts and solid hatch patterns.

REINIT

Pull-down Menu	COMMAND (TYPE)	ALIAS (TYPE)	Short-cut	Screen (side) Menu	Tablet Menu
...	REINIT	...	...	...	...

Many software programs require you to exit and restart the program in order for reinitialization to be performed. AutoCAD provides the *Reinit* command to prevent having to exit AutoCAD to accomplish this task.

The *Reinit* command reinitializes the input/output ports of the computer (to the digitizer) and reinitializes the AutoCAD Program Parameters (ACAD.PGP) file. Reinitialization may be necessary if you physically switch the port cable from the digitizer or if it loses power temporarily. If you edit the ACAD.PGP file, it must be reinitialized before you can use the new changes in AutoCAD (see Chapter 45 for information on the ACAD.PGP file).

Invoking the *Reinit* command causes the *Re-initialization* dialog box to appear (Fig. 44-3). You may check one of the boxes. PICKing the *OK* tile causes an immediate reinitialization.

Figure 44-3

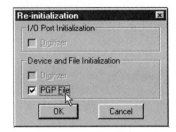

If you have configured the digitizer as the "Current System Pointing Device," AutoCAD uses the Windows mouse driver; therefore, the "Digitizer" checkbox is disabled. If your pointing device is disabled and you need to use this dialog box, use TAB to move to the desired options, the space bar to check the boxes, and Enter to execute an *OK*.

MULTIPLE

Pull-down Menu	COMMAND (TYPE)	ALIAS (TYPE)	Short-cut	Screen (side) Menu	Tablet Menu
...	MULTIPLE	...	...	...	...

Multiple is not a command, but rather a <u>command modifier</u> (or adjective) since it is used in conjunction with another command. Entering *"multiple"* at the Command: prompt <u>as a prefix</u> to almost any command causes the command to automatically repeat until you stop the sequence with Escape. For example, if you wanted to use the *Insert* command repetitively, you enter the following:

```
Command: multiple
Multiple command: insert
Block name (or ?):
```

The *insert* command would repeat until you stop it with Escape. This is helpful for inserting multiple *Blocks*.

The multiple modifier repeats only the command, not the command options. For example, if you wanted to repeatedly use *Circle* with the *Tangent, Tangent, Radius* option, the *TTR* would have to be entered each time the *Circle* command was automatically repeated.

AUDIT

Pull-down Menu	COMMAND (TYPE)	ALIAS (TYPE)	Short-cut	Screen (side) Menu	Tablet Menu
File Drawing Utilites > Audit	*AUDIT*	...	...	*FILE Audit*	*Y,24*

The *Audit* command is AutoCAD's diagnostic utility for checking drawing files and correcting errors. *Audit* will examine the current drawing. You can decide whether or not AutoCAD should fix any errors if they are found. The command may yield a display something like this:

```
Command: audit
Fix any errors detected? <N> y
10    Blocks audited
15    Blocks audited
Pass 1 100   objects audited
Pass 1 200   objects audited
Pass 1 282   objects audited
Pass 2 100   objects audited
Pass 2 200   objects audited
Pass 2 282   objects audited
Pass 3 100   objects audited
```

Pass 3 200   objects audited
Pass 3 282   objects audited
Total errors found 0 fixed 0
Command:

If errors are detected and you requested them to be fixed, the last line of the report indicates the number of errors found and the number fixed. If *Audit* cannot fix the errors, try the *Recover* command (see Chapter 2).

You can create an ASCII report file by changing the *AUDITCTL* system variable to 1. In this case, when the *Audit* command is used, AutoCAD automatically writes the report out to disk in the current directory using the current drawing file name and an .ADT file extension. The default setting for *AUDITCTL* is 0 (off).

FILEDIA

This system variable enables and disables FILE-related DIAlog boxes. File-related dialog boxes are those that are used for reading or writing files. For example, when you request to *Open* a file, the *Select File* dialog box appears (Fig. 44-4).

Figure 44-4 ——————————————————————

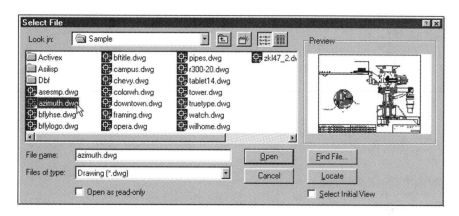

The settings for *FILEDIA* are as follows:

FILEDIA = 1   File dialog boxes appear when a file command is invoked (default setting)
FILEDIA = 0   File dialog boxes are disabled and command prompts are used

The *FILEDIA* variable can be set to 0 to present the command line prompt shown below instead of the *Select File* dialog box:

Command: **open**
Enter name of drawing:

If file dialog boxes are disabled, they can be invoked at the command line by entering a tilde (~) symbol as a prefix to the command. For example, when *FILEDIA* is set to 0, invoking the *Open* command by any method produces the command prompt; however, typing "~open" produces the *Select File* dialog box.

Disabling the dialog boxes can be helpful for running a script file. Since dialog boxes require user input with a pointing device, some scripts require the command line interface to be used instead of dialog boxes.

The *FILEDIA* variable setting is saved in the system registry.

R14

CMDDIA

The *CMDDIA* variable controls the display of dialog boxes for the *Plot* command and for external databases. For example, if *CMDDIA* is set to 0, the *Print / Plot Configuration* dialog box does not appear when the *Plot* command is used, but the command prompts appear instead:

```
Command: plot
What to plot—Display, Extents, Limits, View, or Window <D>:
Device Name:  HP LaserJet 4 Plus/4M Plus
Output Port: LPT1:
Driver Version:      4.0
Use the Control Panel to make permanent changes to a printer's configuration
Plot device is System Printer ADI 4.3—by Autodesk, Inc
Description: Default System Printer
Plot optimization level = 0
Plot will NOT be written to a selected file
Sizes are in Inches and the style is landscape
Plot origin is at (0.00,0.00)
Plotting area is 7.93 wide by 10.48 high (MAX size)
Plot is rotated 90 degrees
Area fill will NOT be adjusted for pen width
Hidden lines will NOT be removed
Plot will be scaled to fit available area
0. No changes, proceed to Plot
1. Merge partial configuration from .pcp file
2. Replace configuration from .pc2 file
3. Save partial configuration as .pcp file
4. Save configuration as .pc2 file
5. Detailed plot configuration

Enter choice, 0-5 <0>:
Effective plotting area:  6.51 wide by 10.48 high
Plot complete
Command:
```

R14 Using the command line interface is helpful for plotting from a script file since user input with a pointing device (required for dialog boxes) is not possible in a script. Script files can be written for batch plotting. The *CMDDIA* variable is saved in the system registry.

ATTDIA

The *ATTDIA* variable controls the display of the *Enter Attributes* dialog box. The default setting is 0 (off). Ensure the setting is off if you are using a script to enter attributes. The variable is saved in the drawing file. (See Chapter 22, Block Attributes, for more information.)

USING SLIDES AND SCRIPTS

A slide is a "snapshot" of an AutoCAD drawing that can be made and saved to a file using the *Mslide* command. The resulting slide file contains only one image with no other drawing information. Slide files can later be viewed in AutoCAD using the *Vslide* command. A prepared series of slides can be presented at a later time, much like a slide show. This section also explains how to create a self-running slide show from your slide files by creating and running a *Script*.

MSLIDE

Pull-down Menu	COMMAND (TYPE)	ALIAS (TYPE)	Short-cut	Screen (side) Menu	Tablet Menu
...	*MSLIDE*	...	...	...	...

Mslide is short for "make slide." A slide is a "screen capture" of the current AutoCAD display. The image that is saved is composed of the objects that appear in the drawing editor when the *Mslide* command is issued (excluding the menus, tool bars, grid, cursor, and command line). If tiled or floating viewports are active, the current viewport's image is captured. When a slide is made, the resulting image is saved as a file in the directory of your choice with an extension of .SLD.

Mslide is simple to use. Assume that you are working on a drawing—for example, Figure 44-5—and want to take a "snapshot" of the image using *Mslide*. Invoke the command by any method:

Command: **mslide** (The *Create Slide File* dialog box appears. Assign a name and directory.)
Command:

Figure 44-5

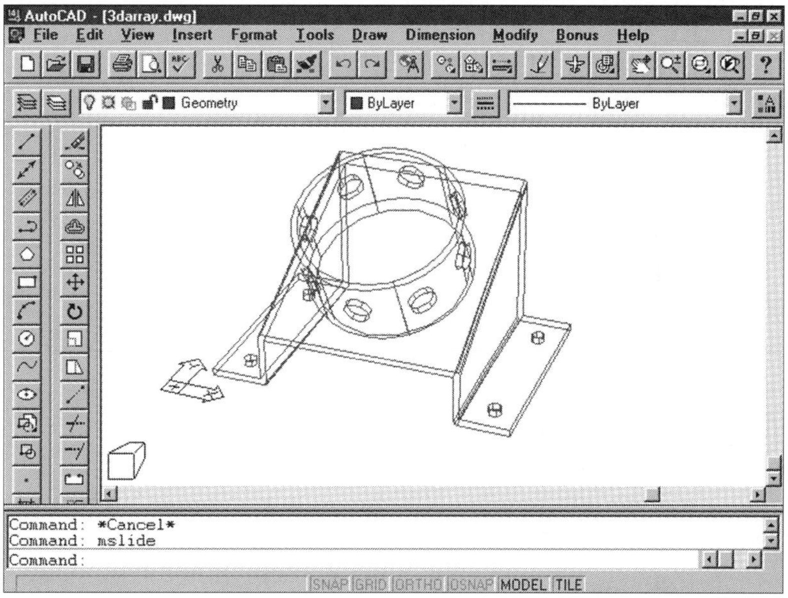

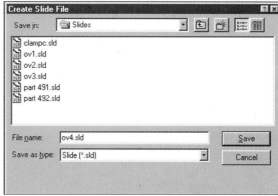

When the *Create Slide File* dialog box appears (Fig. 44-6), assign a name and directory for the slide. (If the *FILEDIA* variable is set to 0, the command line format is activated instead of the dialog box.) Only the file name is required because AutoCAD automatically assigns the .SLD file extension. The drawing image is saved without the grid, cursor, menus, toolbars, etc. Only the drawing image is saved in the .SLD; no other drawing information is saved. Therefore, an .SLD file is much smaller (file size in bytes) than the corresponding .DWG file. Use the *Vslide* command to view the slide again.

Figure 44-6

AutoCAD .SLD files are proprietary, meaning the file format was developed by AutoCAD to be used in AutoCAD. The .SLD file format is not widely used like a .TIF, .GIF, .JPG, etc. but can be viewed or converted by some software. Slide files are also a convenient format for saving an image created with the AutoCAD *Shade* command (see Chapter 35, 3D Viewing and Display).

VSLIDE

Pull-down Menu	COMMAND (TYPE)	ALIAS (TYPE)	Short-cut	Screen (side) Menu	Tablet Menu
...	VSLIDE	...	...	...	...

Vslide is short for "view slide." Use this command to view previously made AutoCAD slide (.SLD) files. Assuming previously created slide files are accessible, invoke *Vslide* by any method:

Command: **vslide** (The *Select Slide File* dialog box appears. Select the desired directory and slide name.)
Command:

When the *Select Slide File* dialog appears (Fig. 44-7), enter or select the desired directory and slide name. The selected slide then appears in the Drawing Editor (if viewports are active, it appears in the current viewport). Using the slide from the previous example (see *Mslide*), the slide image is "projected" in the Drawing Editor (Fig 44-8).

A slide file may be viewed during any AutoCAD session. If a drawing is in progress, the slide file is projected on top of the current drawing, much like a photographic slide is displayed on a projection screen or wall. Because a slide file contains only a description of the display, the slide file image cannot be edited. You cannot use *Pan* or *Zoom* to modify the display of the slide. In fact, any AutoCAD display command issued (*Redraw*, *Pan*, *Zoom*, etc.) will cause the slide to disappear and act upon the current drawing, not the slide. After a slide has been viewed, use *Redraw* to refresh the current drawing before using draw or edit commands. Because .SLD files are generally much smaller and contain less information than .DWG files, viewing a slide is much faster than loading a drawing.

Slide files are helpful for keeping a visual record of drawings, since the .SLD file is much smaller and faster loading than the corresponding .DWG file. The most common use for a series of slides is giving a presentation or a demonstration. Using prepared slides to present a complex drawing is much faster and more reliable than using *Pan* and *Zoom*. A script file (*.SCR) file can be created to "run" a series of slide files—a scripted slide show (see Creating and Using Scripts).

Figure 44-7 —————————

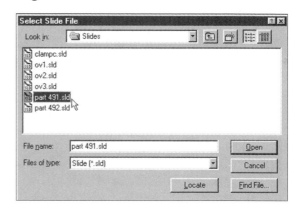

Figure 44-8 —————————

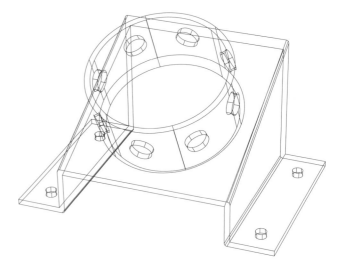

Creating and Using Scripts

Scripts are text files that contain AutoCAD commands. Similar to a batch routine written in DOS, a script file contains a user-specified listing of the desired AutoCAD commands. To run a script, use the *Script* command. AutoCAD reads and operates each command and option in the sequence it is listed in the script file. A script file contains essentially the same text that you would enter at the command line during a drawing session. When a script is executed, each line of the script is echoed at the command line.

Script files can be used for a number of purposes. If you want to give a presentation, you can create a self-running display of a series of slides by creating a script file. For this purpose, the script file repeats the *Vslide* command and the slide names. Since a script can contain any AutoCAD commands, you could make scripts to execute a series of frequently used commands—like a macro (this topic is discussed in Chapter 45, Basic Customization). You could even recreate an entire drawing by listing every command used to create the drawing in the script file.

There are three basic steps to creating and running a script:

1. In AutoCAD, determine the commands, options, and responses that you want to include in the script file. This can involve a "practice run" to determine each step. You may have to write down each command, option, and response. The *Logfileon* and *Logfileoff* commands can help with this process (see Chapter 45).

    If you are creating a slide show, the slides must be created first. A practice run of the show using *Vslide* can be helpful to ensure you know the correct slide sequence.

2. Use a text editor such as Windows Notepad or other program that can save an ASCII text file. Include all the desired commands. Save the file as an .SCR file extension.

3   In AutoCAD, use the *Script* command to locate, load, and run the script.

Creating a Slide Show

Assume that you have created the slides and know the sequence that you want to display them. You are ready to create the script file. Use any text editor or word processor that can create text files in ASCII format (without internal word processing codes). The script file should be saved using any descriptive file name but should have a .SCR file extension.

Begin the script file (in your text editor) by entering the *Vslide* command. When the *Vslide* command is issued in AutoCAD, the "Slide file:" prompt appears. Therefore, your next line of text (in the script file) should be the name of the desired slide file (SLIDE1, for example) without the .SLD extension. Upper- or lowercase letters can be used. For example, the script file at this point might look like this:

```
VSLIDE
SLIDE1
```

Each line or space in a script file is equivalent to one AutoCAD command, option, or response by the user. In other words, place a <u>space</u> or begin a <u>new line</u> for each time you would press <u>Enter</u> in AutoCAD. The file would be translated as:

Command: **VSLIDE** Enter
Slide file: **SLIDE1** Enter

You probably want to allow the slide to stay visible for a specific amount of time—for example, 5 seconds—then move on to the next slide. You may also want the slide show to be self-repeating automatically for use during presentations or an open house. A few script-controlling commands are given here that will enable you to more effectively present the slide show:

DELAY   With the *Delay* command, you can enter the number of milliseconds you wish to have the current slide displayed. For example, a 5 second display of the slide would be accomplished by entering "*Delay 5000.*" When entering the delay time, consider that the slide will remain displayed while the next *Vslide* command is issued, as well as during the time that it takes your system to retrieve and display the next slide.

RSCRIPT   Typically entered as the last line of a script file, this command loops to the first line of the script and repeats the script indefinitely (until you interrupt with Backspace or Escape). Make sure you include an <u>extra space or line</u> after the *Rscript* line (to act as an Enter).

RESUME   This command cannot be used in the script file, but must be typed at the keyboard during a script execution. You can interrupt a script, then type *Resume* to pick up with the script again. See *Script* next for details on using *Resume*.

An example script file is given here for displaying two slides (SLIDE1 and SLIDE2), then the drawing in the background, then repeating the show:

```
VSLIDE
SLIDE1
DELAY 2000
VSLIDE
SLIDE2
DELAY 500
REDRAW
DELAY 1000
RSCRIPT
```

In the script above, SLIDE1 displays for 2 seconds, SLIDE2 displays for .5 seconds, the *Redraw* causes the drawing (in the background) to appear and display for 1 second, and then the entire script repeats.

The drive and directory location of the slide files is significant. The slide files should be located in the path designated by the *Support File Search Path* section of the *Files* tab in the *Preferences* dialog box (Fig. 44-9). Otherwise, the path should be given in the script file before each slide name as shown below:

Figure 44-9

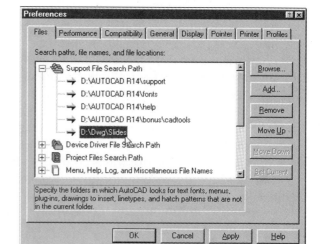

```
VSLIDE
C:\DWGS\IMAGES\SLIDE1
DELAY 2000
VSLIDE
C:\DWGS\IMAGES\SLIDE2
DELAY 500
REDRAW
DELAY 1000
RSCRIPT
```

Once the script is created and saved as an .SCR file, return to AutoCAD and test it by using the *Script* command.

SCRIPT

	COMMAND (TYPE)	ALIAS (TYPE)	Short-cut	Screen (side) Menu	Tablet Menu
Pull-down Menu					
Tools *Run Script...*	*SCRIPT*	*SCR*	*...*	*TOOLS 1* *Script*	*V,9*

The *Script* command allows you to locate, load, and run a script file. Invoking the *Script* command produces the *Select Script File* dialog box (Fig. 44-10):

> Command: **script** (The *Select Script File* dialog box appears. Select the desired directory and script name.)
> Command:

In the dialog box, the desired directory and file can be selected. Notice the default file extension of .SCR. (If the *FILEDIA* variable is set to 0, the dialog box does not appear and the "Script file:" prompt appears at the Command line instead.)

Figure 44-10

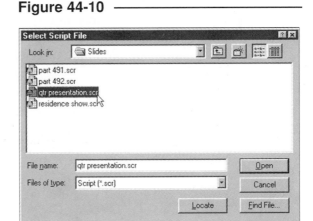

When the file is selected, the script executes. There is nothing more to do at this point other than watch the script (or begin your verbal presentation). If you are giving a verbal presentation, here are some other ideas:

1. Text slides can be created as an introduction, conclusion, and topic or section heading. Just use the *Mtext*, *Dtext*, or *Text* command to create the desired text screen in AutoCAD; then create a slide with *Mslide*.

2. To temporarily interrupt the script and view a particular slide, press the Backspace or Escape key. This enables you to discuss one slide indefinitely or stop to answer questions. Typing *Resume* causes the script to pick up again from the stopping point.

   Remember that any AutoCAD drawing can be in the background while the slide show is projected over the drawing. The Backspace or *Resume* feature allows you to cut from the slides and return to the drawing for editing, *Pans*, or *Zooms*, then return to the slide show. You may want to have a drawing loaded in the background (an entire view of the drawing, for example) and show slides "on top" (close-ups, etc.).

NOTE: The *FILEDIA* variable does not have to be set to 0 before running a slide show script. Even though the *Vslide* command normally invokes a dialog box, the command line format is automatically activated when using *Vslide* from a script. With scripts using other commands that activate dialog boxes, it may be helpful to set *FILEDIA* to 0.

THE GEOMETRY CALCULATOR

The geometry calculator can be used as a standard 10-key calculator and can also be used to calculate the location of points in a drawing. It can be used transparently (within a command) to calculate and pass values to the current command. The real power of the geometry calculator is the ability to be used with existing geometry utilizing *OSNAPs*. The geometry calculator is an ADS application that was introduced with Release 12.

The geometric calculator provides the following capabilities:

1. It can be used as a standard calculator to return values for arithmetic expressions.

2. The calculator can be used transparently to compute numbers, distances, angles, and vector directions as input for a command.

3. The calculator can be used transparently to pass coordinate data to the current command. This feature simplifies creating new geometry and editing existing geometry by enabling you to locate points that are not easily accessible by other means.

4. *OSNAP*s can be used in the expressions to further enhance your ability to use existing geometry (coordinates) in calculations.

5. Built-in functions are available as shortcuts for using *OSNAP*s, creating geometry, and making mathematical calculations and conversions.

6. The calculator can be used as a programmable calculator because of its ability to store variables and interact with AutoLISP defined functions and expressions.

CAL

Pull-down Menu	COMMAND (TYPE)	ALIAS (TYPE)	Short-cut	Screen (side) Menu	Tablet Menu
...	*CAL*	...	...	...	...

The calculator prompts you for the entry of an "expression." The calculator uses the command line interface as shown in this example (divide 12 by 5):

```
Command: cal
Initializing...>> Expression: 12/5
2.4
Command:
```

The expression can contain arithmetic expressions, vector points, coordinates, and variables. When you enter the expression, the geometry calculator recognizes the following standard operation priorities:

1. Expressions in parentheses are considered first, beginning with the innermost.
2. Exponents are calculated first, any multiplication or division is calculated next, and addition or subtraction is calculated last.
3. Operators of equal precedence (for example, several multiplications) are taken from left to right.

These expressions can be applied to existing geometry using standard *OSNAP* functions or used in any AutoCAD command where points, vectors, and numbers are expected. The calculator supports many of the standard numeric functions found on typical scientific calculators.

Numeric Operators

Operator	Operation
()	group expressions
^	exponentiation
*,/	multiplication and division
+,-	addition and subtraction

A few examples of simple mathematical expressions using numeric operators are given here:

(5+6)*2	=	22
2^2*(14/2)	=	28
((6/2)-1)*PI	=	6.28319

Format for Feet and Inches

The default unit of measure for distances is one inch; however, feet and inches can be entered as follows:

feet'-inches" or feet'inches"

Expressions using feet and inch notation convert to real numbers as in the following examples:

3′	evaluates to	3*12=36.0
3.75″	evaluates to	3*12+ 9= 45.0
4′-5″	evaluates to	4*12+ 5= 53.0
6.5′3″	evaluates to	6*12+ 6+ 3=81.0

Format for Angles

The default unit of measure for angles is decimal degrees and is entered like this:

*deg**d**min'sec"*

Enter a number followed by an **r** to designate radians or **g** to designate grads. Examples are as follows:

45d15′22″
0d15′22″
1.55r
21g

Points and Vectors

Both points and vectors can be used in expressions. A point or vector is a set of real numbers enclosed in brackets [ ]. You can omit from your input string coordinates with a value of zero and the comma(s) immediately preceding the right bracket. Some examples of this syntax are shown below:

[2,5,8]	=	2,5,8
[1,2]	=	1,2,0
[,4]	=	0,0,4
[ ]	=	0,0,0

A point defines a location in space. A vector defines a direction in space. You can add a vector to a point to obtain another point. (In the examples later in this chapter, the notation p1,p2 is used to designate points and v1,v2 is used to designate vectors.)

Point Formats

The calculator also supports the standard input format for rectangular, polar, relative, and other input:

Coordinate entry method	Format
Relative rectangular	[@X,Y,Z}
Relative polar	[@dist<angle]
Relative cylindrical	[@dist<angle,Z]
Relative spherical	[@dist<angle<angle]
WCS (overrides UCS)	use the * (asterisk) prefix inside brackets: [*X,Y,Z]

Standard Numeric Functions

The geometric calculator supports these standard numeric functions:

Function	Description
sin(*angle*)	Sine of the angle.
cos(*angle*)	Cosine of the angle.
tang(*angle*)	Tangent of the angle.
asin(*real*)	Arcsine of the number. The number must be between -1 and 1.
acos(*real*)	Arccosine of the number. The number must be between -1 and 1.
atan(*real*)	Arctangent of the number.
ln(*real*)	Natural log of the number.
log(*real*)	Base-10 log of the number.
exp(*real*)	Natural exponent of the number.
exp10(*real*)	Base-10 exponent of the number.
sqr(*real*)	Square of a number.
sqrt(*real*)	Square root of a number. The number must be non-negative.
abs(*real*)	Absolute value of the number.
round(*real*)	Number rounded to the nearest integer.
trunc(*real*)	Integer potion of the number.
r2d(*angle*)	Angles in radians converted to degrees. For example, r2d(pi) converts the constant pi to 180 degrees.
d2r(*angle*)	Angles in degrees converted to radians. For example, d2r (180) converts 180 degrees to radians and returns the value of the constant pi.
pi	The constant pi.

OSNAP Modes

You can include *OSNAP* modes as part of an arithmetic expression. When *Cal* encounters an *OSNAP* mode, the aperature appears and you are prompted to "Select entity for (*OPTION*) snap:". The selected *OSNAP* point coordinate values are passed to the expression for evaluation. To use *OSNAP* modes, enter the three-character name in the expression:

CAL OSNAP Mode	AutoCAD *OSNAP* Mode
END	*Endpoint*
INS	*Insert*
INT	*Intersection*
MID	*Midpoint*
CEN	*Center*
NEA	*Nearest*

CAL OSNAP Mode	AutoCAD *OSNAP* Mode
NOD	*Node*
QUA	*Quadrant*
PER	*Perpendicular*
TAN	*Tangent*

Built-in Functions

The Geometry Calculator has a number of "built-in" functions that provide a variety of utilities to be used for conversions and inquiries of existing geometry.

Function	Description
abs(v)	Calculates the length of vector v, a non-negative real number.
abs([1,2,4])	Calculates the length of vector [1,2,4].
ang(p)	Angle between the X axis and vector p.
ang(p1, p2)	Angle between the X axis and line (p1, p2).
ang(apex, p1, p2)	Angle between the lines (apex, p1) and (apex, p2) projected onto the XY plane.
ang(apex, p1, p2, p)	Angle between the lines (apex, p1) and (apex, p2).
cvunit(1,inch,cm)	Converts the value 1 from inches to centimeters.
cur	Sets the value of the variable *LASTPOINT*.
cur(p)	Pick a point using the graphics cursor with a rubberband line from point p.
dist(p1, p2)	Distance between points p1 and p2.
dpl(p, p1, p2)	Distance between point p and the line (p1, p2).
dpp(p, p1, p2, p3)	Distance between point p and the plane defined by p1, p2 & p3.
ill(p1, p2, p3, p4)	Intersection of lines (p1, p2) and (p3, p4).
ilp(p1,p2,p3,p4,p5)	Intersection of line (p1, p2) and plane defined by p3, p4 & p5.
nor	Unit vector normal to a circle, arc, or arc polyline segment.
nor(v)	Unit vector in the XY plane and normal to the vector v.
nor(p1, p2)	Unit vector in the XY plane and normal to the line (p1, p2).
nor(p1, p2, p3)	Unit vector normal to plane defined by p1, p2, & p3.
pld(pl, p2, units)	Point on the line (p1, p2) that is drawing units away from the point p1.
plt(p1, p2, t)	Point on line (p1, p2) that is t segments from p1. A segment = distance from p1 to p2.
rad	Radius of the selected object.
rot(p, origin, ang)	Rotates point p through angle ang using a line parallel to the Z axis passing through origin as the axis of rotation.

Function	Description
rot(p1, p2, ang)	Rotates point p through angle ang using line (p1, p2) as the axis of rotation.
u2w(p1)	Converts point p1 expressed in WCS to the current UCS.
vec(p1, p2)	Vector from point p1 to point p2.
vec1(p1, p2)	Unit vector from point p1 to point p2.
w2u(p1)	Converts point p1 expressed in WCS to the current UCS.

Shortcut Functions

In addition to computing numbers and points, you can use the calculator to create and edit geometry in your drawing. This is one of the most powerful features of *CAL*. The following table lists the geometric functions supported by the calculator. Using these shortcuts prevents you from having to enter the three-letter *OSNAP* mode and parentheses in the expression.

Function	Shortcut for	Description
dee	dist(end, end)	Distance between two endpoints.
ille	ill(end, end, end, end)	Intersection of two lines defined by four endpoints.
mee	(end+end)/2	Midpoint between two endpoints.
nee	nor(end, end)	Unit vector in the XY plane and normal to two endpoints.
vee	vec(end, end)	Vector from two endpoints.
vee1	vec1(end, end)	Unit vector from two endpoints.

Point Filter Functions

You can use these fuctions to filter the X,Y, and Z components of a point or vector coordinate.

Function	Description
xyof($p1$)	X and Y components of a point. The Z component is set to 0.0.
xzof($p1$)	X and Z components of a point. The Y component is set to 0.0.
yzof($p1$)	Y and Z components of a point. The X component is set to 0.0.
xof($p1$)	X component of a point. The Y and Z components are set to 0.0.
yof($p1$)	Y component of a point. The X and Z components are set to 0.0.
zof($p1$)	Z component of a point. The X and Y components are set to 0.0.
rxof($p1$)	X component of a point.
ryof($p1$)	Y component of a point.
rzof($p1$)	Z component of a point.

Geometry Calculator Examples

Point, Vector, and *OSNAP*s Example

The following example places the beginning of a *Line* [1,-.5] units over from the selected *Quadrant* of a *Circle*, then adds two more calculated vectors to create a wedge (Fig. 44-11). Don't forget to enter an apostrophe (') to use *Cal* transparently if you are typing the command.

Figure 44-11

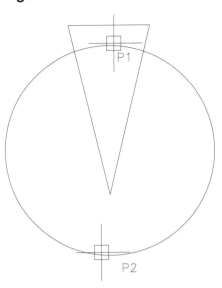

```
Command: Line
From point: 'cal
>>Expression: qua+[-1,.5]
>>Select entity for QUA snap: PICK
(PICK the top quadrant on the circle, P1.)
```

To continue with the line from the previous sequence, a vector is added to a current point to obtain the next point on the line.

```
To point: 'cal
>> Expression: cur+[,1.5]
>> Enter a point: qua
of PICK (PICK the bottom quadrant of the circle, P2.)
```

To finish the wedge, the next point of the vector is placed in the same fashion as the first point.

```
To point: 'cal
>> Expression: qua+[1,.5]
>> Select entity for QUA snap: PICK (PICK the top quadrant of the circle, P1 again.)
To point: c (To close the wedge shape.)
```

Numeric Functions and Built-in Functions Example

The following example displays the use of numeric functions and built-in functions in finding the circumference and area of a *Circle* (no figure).

To find the circumference of a *Circle*, first create the variable (R1) which collects the circle radius. Entering any variable name (R1 in this case) followed by an equal (=) sign stores the value or point to an AutoLISP variable:

```
Command: 'cal
>> Expression: R1=rad
>> Select circle, arc or polyline segment for RAD function: PICK (PICK anywhere on a Circle.)
2.58261
```

The value is stored to variable R1. Then enter the radius in a numeric function to determine the circumference:

```
Command: 'cal
>> Expression: 2*pi*R1
16.227
```

Now use the radius variable (R1) in a numeric function format to determine the area of a circle:

```
Command: 'cal
>> Expression: pi*sqr(R1)
20.9541
```

OSNAP Examples

In the following example, the calculator returns the point (Fig. 44-12, point A) halfway between the center of the circle and the endpoint of the object:

```
Command: Line
From point: 'cal
>> Expression: (cen+end)/2
>> Select entity for CEN snap: PICK
(PICK any point on the circle, P1).
>> Select entity for END snap: PICK
(PICK the upper-right corner of the
rectangle, P2.)
```

Figure 44-12

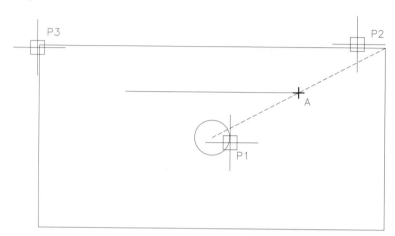

AutoCAD finds the point halfway between the selected points (A) and passes the coordinate to the *Line* command as the "From point:". Do the same to determine the "To point:" of the *Line*:

```
To point: 'cal
>> Expression: (cen+end)/2
>> Select entity for CEN snap: PICK  (PICK the circle, P1 again.)
>> Select entity for END snap: PICK  (PICK the upper-left corner of the rectangle, P3.)
To point: Enter
```

The *Line* is drawn with its endpoints halfway between the *OSNAP*s.

This example uses *END* and *CEN* to calculate the centroid (Fig. 44-13, point B) defined by 3 points (2 endpoints and the center of a circle):

```
Command: Line
From point: mid
of PICK (PICK the bottom hor-
izontal line of the rectangle, P4.)
To point: 'cal
>> Expression:
(end+cen+end)/3
>> Select entity for END snap:
PICK (PICK the lower-left
corner of the rectangle, P5.)
```

Figure 44-13

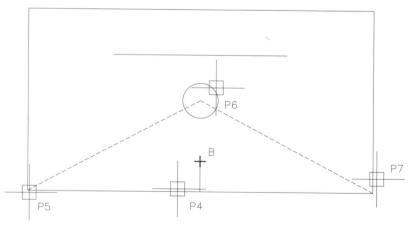

```
>> Select entity for CEN snap: PICK  (PICK the circle, P6.)
>> Select entity for END snap: PICK  (PICK the lower-right corner of the rectangle, P7.)
To point: Enter
```

For the next drawing example, *Cal* prompts for an object, then returns a point that is 1 unit to the left of the selected object's midpoint. In the second sequence, the point is placed 1 unit to the right of the selected object's midpoint.

Figure 44-14

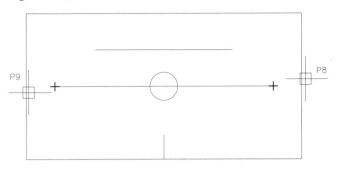

Command: *Line*
From point: **'cal**
>> Expression: **mid+[-1,]**
>> Select entity for MID snap: **PICK** (PICK the right vertical line of the rectangle, P8.)

To point: **'cal**
>> Expression: **mid+[1,]**
>> Select entity for MID snap: **PICK** (PICK the left vertical line of the rectangle, P9.)
To point: **Enter**
Command:

Shortcut Function Examples

The calculator shortcut functions are used to add text and circles to the previous drawing. In the first example, middle justified text is placed at the intersection of two lines defined by endpoints (*ille*) (Fig. 44-15).

Figure 44-15

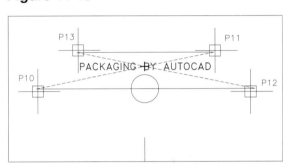

Command: *dtext*
Justify/Style/<Start point>: *j*
Align/Fit/Center/Middle/Right/TL/TC/
TR/ML/MC/MR/BL/BC/BR: *m*
Middle point: **'cal**
>> Expression: **ille**
>> Select one endpoint for ILLE:First line: **PICK** (PICK the left end of the long horizontal line, P10.)
>> Select another endpoint for ILLE:First line: **PICK** (PICK the right end of the short horizontal line, P11.)
>> Select one endpoint for ILLE:Second line: **PICK** (PICK the right end of the long horizontal line, P12.)
>> Select another endpoint for ILLE:Second line: **PICK** (PICK the left end of the short horizontal line, P13.)
Height <0.2000>: **Enter**
Rotation angle <0>: **Enter**
Text: **PACKAGING BY AUTOCAD**
Text: **Enter**

Use another built-in function to return the midpoint between two endpoints (*mee*) for the center of a circle (Fig. 44-16):

Figure 44-16

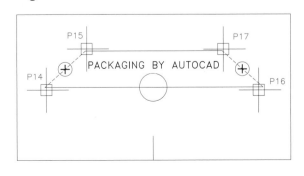

Command: *Circle*
3P/2P/TTR/<Center point>: **'CAL**
>> Expression: **mee**
>> Select one endpoint for MEE: **PICK** (PICK the left end of the long horizontal line, P14.)
>> Select another endpoint for MEE: **PICK** (PICK the left end of the short horizontal line, P15.)
Diameter/<Radius> <0.5000>: **.25**

Command: CIRCLE 3P/2P/TTR/<Center point>: **`CAL**
>> Expression: **mee**
>> Select one endpoint for MEE: **PICK** (PICK the right end of the long horizontal line, P16.)
>> Select another endpoint for MEE: **PICK** (PICK the right end of the short horizontal line, P17.)
Diameter/<Radius> <0.2500>: **Enter**
Command:

The Geometry Calculator has tremendous power, but it can be beneficial only if you know how and when to use it. Review the function tables (particularly the Built-in and Shortcut Functions) and practice with the expressions. Many of the functions can save drawing time and allow you to construct and edit more efficiently than by other methods.

CHAPTER EXERCISES

1. **Wildcards**

   In this exercise, you will use wildcards to manage layers for a complex drawing. *Open* the AutoCAD **WILHOME** drawing that is located in the \AutoCAD R14\Sample directory. Use the *Layer & Linetype Properties* dialog box to view the layers. Examine the layer names; then *Cancel*.

   A. Type the *-Layer* command. *Freeze* all of the layers whose names begin with the letters **AR**. (Hint: Enter **AR*** at the Layer name(s) to Freeze: prompt.) Do you notice the change in the drawing? Use the *Layer Control* drop-down box or list the layers using the *?* option to view the *State* of the layers.

   B. *Thaw* all layers (use the asterisk ***). Now *Freeze* all of the layers <u>except</u> the architectural layers (those beginning with **AR**). (Hint: Enter **~AR*** at the "Layer name(s) to Freeze:" prompt.) Do only the architectural features appear in the drawing now? Examine the list of layers and their *State*. *Thaw* all layers again.

   C. Next, *Freeze* all of the layers except the wall layers (layers that have a "WALL" string). (Hint: Enter **~*WALL*** at the "Layer name(s) to Freeze:" prompt.) Do only the grids appear in the drawing now? Examine the list of layers and their *State*. *(Layer 0 cannot be Frozen* because it is the Current layer.) *Thaw* all layers again and do <u>not</u> *Save* changes.

2. *Multiple* modifier

   Open the **OFF-ATT2** drawing and prepare to *Insert* several more furniture *Blocks* into the office. Type *Multiple* at the Command: prompt, then Enter, then *Insert* and Enter. Proceed to *Insert* the *Blocks*. Notice that you do not have to repeatedly enter the *Insert* command. Now try the *Multiple* modifier with the *Line* command. Remember this for use with other repetitive commands. (Do not *Save* the drawing.)

3. *Audit*

   With the OFF-ATT2 drawing open, use the *Audit* command to report any errors. If errors exist, use *Audit* again to fix the errors.

4. *BLIPMODE, DRAGMODE*

A. Begin a *New* drawing using your **ASHEET** prototype. Turn *BLIPMODE On*. Experiment with this setting by drawing a few *Lines* and *Circles*. Use *Erase*. Notice that you have to *Redraw* often when *BLIPMODE* is *On*.

B. Now change the setting of *DRAGMODE* to *Off*. Draw a few *Circles*, a *Polygon*, an *Ellipse*, and then **dimension** your figures. Do you prefer *DRAGMODE On* or *Off*?

C. Since the *BLIPMODE* and *DRAGMODE* system variables are saved with the drawing file, you can make the settings in the template drawings so they will preset to your preference when you begin a *New* drawing (if you prefer other than the default settings). If you prefer *BLIPMODE On* or *DRAGMODE Off*, *Exit* the **ASHEET** and *Discard Changes*. Use *Open* to open each of the template drawings (**ASHEET, BSHEET,** and **CSHEET**), change the desired setting, and *Save*.

5. *Fill, Regen*

Open the **AZIMUTH** sample drawing (from the C:\Program Files\AutoCAD R14\Sample directory). When the drawing has loaded, type *Mspace* to make model space active. Then type *Regen* and check the amount of time it takes for the drawing to regenerate (use your stopwatch or count the seconds). Now turn *Fill Off*. Also turn *Qtext On*. Make another *Regen* as before to check the time. Is there any improvement?

Imagine the time required for plotting this drawing using a pen plotter. *Fill* and *Qtext* can be set to save time for test plots. If you have a pen plotter, make a plot of the drawing with *Fill* and *Qtext* set in each position. How many minutes are saved with *Qtext On* and *Fill Off*? (Do <u>not</u> *Save* the changes to AZIMUTH).

6. *Mslide, Vslide*

A. Open your choice of the following sample drawings: **ASESMP, AZIMUTH, CAMPUS, OPERA,** or **WILHOME**. (Assuming the default installation was used for AutoCAD, the drawings are located in the C:\Program Files\AutoCAD R14\Sample directory.) Make eight to ten slides of the drawing, showing several detailed areas (using *Zoom* and *Pan*, etc.) and at least one slide showing the entire drawing. Experiment with *Freezing* layers and include several slides with some layers frozen. Assign descriptive names for the slides so you can determine the subject of each slide by its name. Strive to describe as much as possible about the drawing in the limited number of slides.

B. When the slides of the drawing are complete, prepare at least two text slides to use as informative material such as an introduction, conclusion, or topic headings (make the text in one or more separate drawings, then use *Mslide*).

C   Finally, use *Vslide* to view your slides and determine the optimum sequence to describe the drawing. Write down the order of the slides that you choose.

7. *Script*

   Use **Windows Notepad** or other text editor to create a script file to run your slide show. Include appropriate *delays* to allow viewing time or time for you to verbally describe the slides. Make the script self-repeating. Name the script **XXXSHOW.SCR** (where XXX represents your initials). In AutoCAD, use *Script* to run the show to "debug" it and optimize the delays.

8. **Geometry Calculator**

   A. This exercise utilizes several *Cal* functions. Begin by drawing the shape shown in Figure 44-17 using a *Pline*. Do not draw the dimensions. *Save* the drawing as **CAL-EX**.

   Figure 44-17

   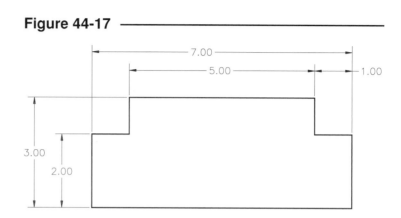

   B. Draw the *Circle* in the center of the *Pline* shape with a **.5** unit *radius* (Fig. 44-18). The center of the *Circle* is located halfway between the *Midpoints* of the top and bottom horizontal line segments. Use the *Calculator* transparently to place the *Center* of the *Circle*:

   Figure 44-18

   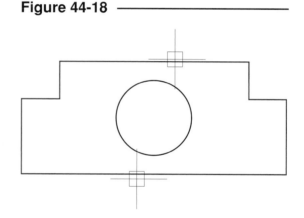

   Hint: Command: **circle**
          CIRCLE 3P/2P/TTR/<Center point>: **'cal**
          > Expression: **(mid+mid)/2**

   C. Draw the small *Circle* (Fig. 44-19) with a **.1** *radius*. Its *Center* is located **.25** units to the right and **.25** units up from the lower-left corner of the *Pline* shape. Use *Cal* transparently to place the *Circle*:

   Figure 44-19

   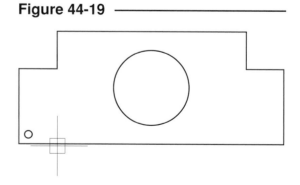

   Hint: Command: **circle**
          3P/2P/TTR/<Center point>: **'cal**
          >> Expression: **end+[.25,.25]**

D. Create the *Circle* in the upper-right corner (Fig. 44-20, P1). It has a *Radius* of **1.5** times the existing small *Circle*. Store the existing small circle radius to a variable named **R1**. Then create the new *Circle* with the *Center* located **.5** units to the left and down from the upper-right corner:

Hint 1: >> Expression: **R1=rad**
Hint 2: >> Expression: **end+[-.5,-.5]**

Figure 44-20

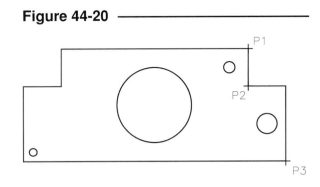

E. The last *Circle* (on the far right) is located midway between two endpoints (P2 and P3). It has a radius of **1.8** times the size of the last circle that you created. Use a shortcut function and variable (if necessary) to create the *Circle*. *Save* the drawing.

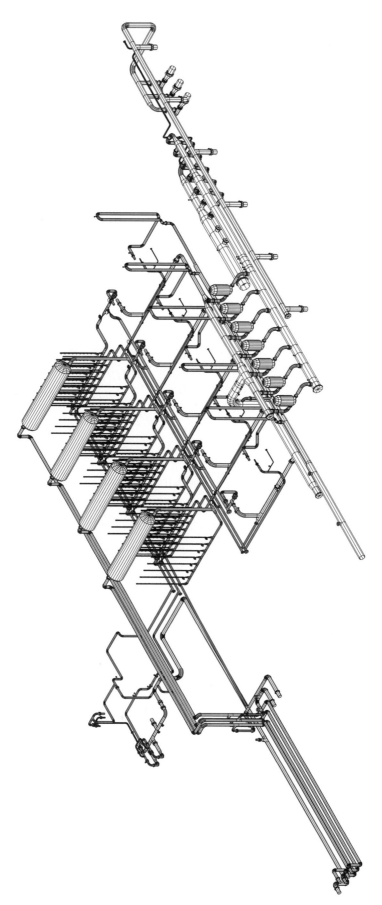

PIPES.DWG Courtesy of Autodesk, Inc. (Release 14 Sample Drawing)

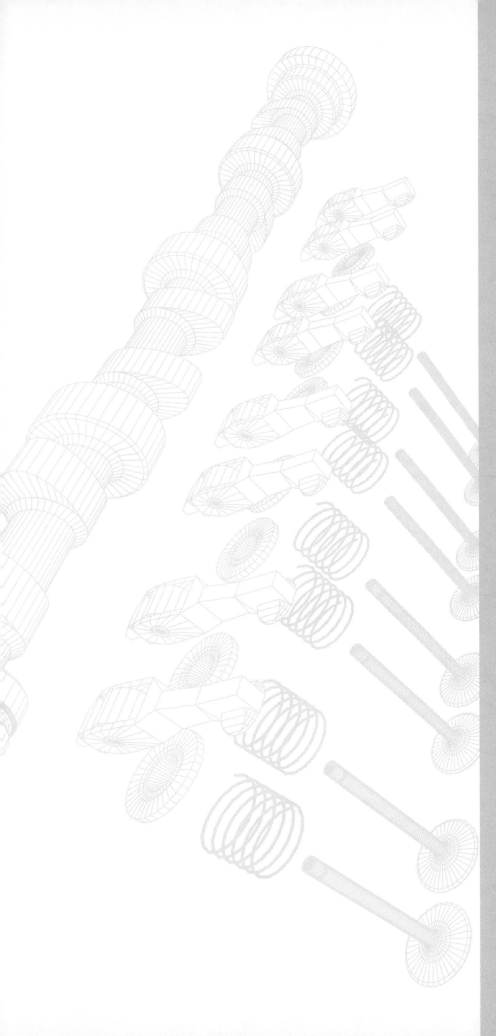

45

BASIC CUSTOMIZATION

Chapter Objectives

After completing this chapter you should:

1. be able to assign which pull-down menus appear on the menu bar;

2. be able to customize the AutoCAD for Windows existing toolbars and create new toolbars;

3. be able to create new and modify existing tool button icons and assign them to perform special tasks;

4. be able to create your own simple and complex linetypes;

5. be able to customize the ACAD.PGP file to create your own command aliases;

6. be able to create script files for accomplishing repetitive tasks automatically.

CONCEPTS

This chapter is offered as an introduction to customizing AutoCAD. Autodesk has written AutoCAD with an "open" architecture that allows you to customize the way the program operates to suit your particular needs. You have the ability to change any of the menus and toolbars as well as create new linetypes.

The topics that are best for beginning your customization experience are introduced in this chapter. They are:

> Customizing which pull-down menus are visible on the menu bar
> Customizing toolbars and icon buttons
> Creating your own simple and complex linetypes
> Customizing the ACAD.PGP file and creating command aliases
> Creating script files to use for automated activities

In addition to the basic customization capabilities covered in this chapter, one other widely used customization feature is the ability to customize the menus. You can modify existing and create new pull-down menus, screen menus, and digitizing tablet menus. See Chapter 46, Menu Customization.

PARTIAL MENU LOADING AND CUSTOMIZING THE MENU BAR

Since Release 13, AutoCAD has the ability to load "partial menus." These are menu files that can "overlay" an existing menu file. A menu file can contain all the information for a complete set of pull-down, screen, toolbar, and digitizing menus. This section describes the *Menuload* command used for partial loading and for customizing the existing menu bar. See Chapter 46, Menu Customization, for more information on menu loading.

MENULOAD

Pull-down Menu	COMMAND (TYPE)	ALIAS (TYPE)	Short-cut	Screen (side) Menu	Tablet Menu
Tools *Customize Menus...*	*MENULOAD*	...	...	*TOOLS 2* *Menuload*	*Y,9*

The *Menuload* command invokes the *Menu Customization* dialog box (Fig. 45-1). The *Menu Bar* tab allows you to customize the visibility of particular pull-down menus. You can also use the *Menu Groups* tab to load (or unload) one or more menu groups as an "overlay" to the existing menu.

In the *Menu Bar* tab, you can remove menu bars (individual pull-down menus) that currently appear on screen. For example, you may not be dimensioning and therefore want to remove the *Dimensioning* menu bar. To do this, select the desired *Menu Bar* from the list on the right, then select *Remove*. The selected bar disappears from the top of the AutoCAD screen.

Figure 45-1 ————

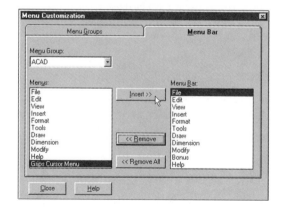

You can also add menus from the list on the left to the existing group of menus bars. Do this by making a selection from the *Menus:* list, then press *Insert*. For example, you can add the *Grips Cursor Menu* to the menu bar on top of the screen (see Fig. 45-1). There is no option within this interface for positioning the order of the menu bars.

You can use the *Menu Groups* tab to load partial menus to overlay the existing menus (Fig. 45-2). A partial menu can contain a complete set of pull-down, screen, toolbar, and digitizing menus, so using this tab installs all of those components. The menu loading feature is mainly used for your own special (customized) menus or third party (add-on products) menus. See Chapter 46, Menu Customization, for examples.

Figure 45-2

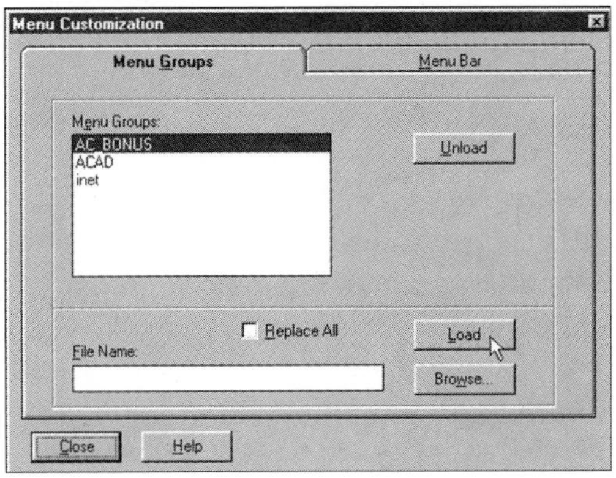

You can use this tab to unload partial menus, as well. For example, if you installed AutoCAD using the "Full Installation," the bonus menu and the Internet utilities menu were installed. Either of those menus can be unloaded if you do not intend to use those features for the drawing session.

Partial menu loading and unloading is effective only for the current drawing session (until you *Exit* AutoCAD). Starting AutoCAD again loads all the original configured menus.

CUSTOMIZING TOOLBARS

In AutoCAD, you have the ability to customize toolbars. By doing so, you can create toolbars that have your most frequently used commands or even create buttons (tools) that invoke "new" commands, that is, choices that activate a command, command option, system variable, or series of commands not previously offered in the existing toolbars or menus. Using the series of dialog boxes that provide for tool and toolbar customization, you can accomplish these actions:

Modify existing toolbar groups
Create new toolbar groups
Modify existing tool buttons
Create new tool buttons

First, you should gain experience with modifying existing and creating new toolbars, then try creating new icon buttons.

Modifying Existing and Creating New Toolbar Groups

In AutoCAD Release 14, you can modify existing toolbars or create your own toolbar by placing the tools you want into a group. For example, you can make a new toolbar that contains your most frequently used solid modeling commands <u>and</u> the Boolean tools, thus allowing you to turn off the *Solids* and *Modify II* toolbars. The *Toolbar* command provides access to a series of dialog boxes that allow you to specify which tools to include in an existing or new toolbar. This section explains the options of the *Toolbars* and related dialog boxes and suggests steps for modifying and creating toolbars.

TOOLBAR

Pull-down Menu	COMMAND (TYPE)	ALIAS (TYPE)	Short-cut	Screen (side) Menu	Tablet Menu
View Toolbar...	TOOLBAR	...	TO	VIEW 2 Toolbar	R,3

Using the *Toolbar* command produces the *Toolbars* dialog box (Fig. 45-3). You can access the *Toolbars* dialog box by the methods shown in the command table above or by right-clicking on any tool. The options are presented below.

Figure 45-3 ————

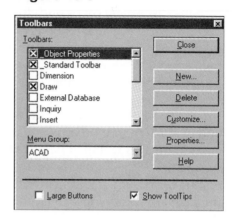

Toolbars
This list gives the existing toolbar names. If you want to activate an existing toolbar group, check the name of the toolbar.

Close
The *Close* option closes the *Toolbars* dialog box. If you have edited an existing toolbar or created a new one, AutoCAD automatically compiles the menu structure so that the toolbar can be used immediately.

New
New opens the *New Toolbar* dialog box and allows you to create a new toolbar (Fig. 45-4). Enter the name, or title, of the new toolbar you want to create in the *Toolbar Name* edit box. Once a name has been entered, choose *OK* to have the new toolbar appear in the drawing and the name added to the *Toolbars* dialog box list. When a new toolbar is created, a small box with a title bar but no tools in it appears near the top of the AutoCAD window.

Figure 45-4 ————

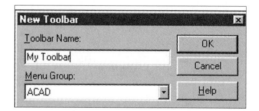

Delete
The *Delete* option deletes the toolbar name currently highlighted in the list. You are prompted for a confirmation for the delete, at which point you can choose *Yes* or *No*.

Customize
This option produces the *Customize Toolbars* dialog box (Fig 45-5). Here, you choose the tools (buttons) to be added to or removed from any toolbar that is displayed in the Drawing Editor. To add a tool to an open toolbar, simply drag the tool from the current category and drop it into the toolbar. Remove an existing tool from a toolbar by dragging it from the toolbar and releasing it in the graphics area of the Drawing Editor. The toolbar must be visible in order to add or remove a tool. If the toolbar is "hidden," use the *Properties* option to display the toolbar.

Figure 45-5 ————

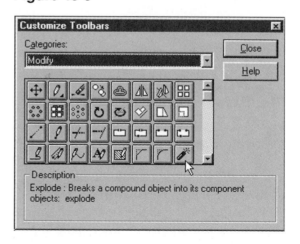

To locate the tools you want to work with, use the *Categories* drop-down list to select the group of tools you need. There are sixteen categories that contain all the tools found in the Release 14 toolbars. The available categories are:

Attribute	*Modify*
Custom	*Object Properties*
Custom Flyout	*Object Snap*
Dimensioning	*Render*
Draw	*Solids*
External Database	*Standard*
External Reference	*Surfaces*
Miscellaneous	*UCS*

After selecting the category, choose a tool and "drag and drop" it into a toolbar. Choose *Close* to close the *Customize Toolbars* dialog box when you are finished editing a toolbar. When you choose *Close* from the *Toolbars* dialog box, AutoCAD compiles the menu structure so that you can use the edited toolbar.

Properties

The *Toolbar Properties* dialog box (Fig. 45-6) appears with this selection and allows you to specify the name and help message for a toolbar. The *Name* edit box allows you to specify a name for a new toolbar or edit an existing toolbar name. The *Help* field enables you to type information about the toolbar that appears when accessed from the *Help* command. The *Alias* is automatically created by AutoCAD when a toolbar is made. The alias cannot be edited from this dialog box. Select *Apply* to apply the changes you have made. To close the *Toolbar Properties* dialog box, choose the small "X" (once, not a double-click) in the upper-right corner of the dialog box.

Figure 45-6

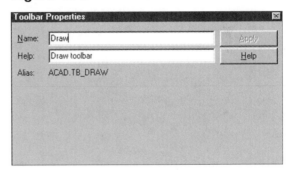

Help

The *Help* option opens the AutoCAD Help program, which gives help regarding the *TBCONFIG* command.

Large Buttons

This option doubles the number of pixels (from 16 x 16 to 32 x 32) used to display the tool buttons. Users with difficulty seeing the icons or who use a very high-resolution display should find the use of *Large Buttons* less straining on the eyes.

Show Tooltips

A "tooltip" is the small text string displayed in a yellow box when you rest the pointer on a tool button. This toggle enables you to turn on or off the display of tooltips.

Steps for Modifying an Existing Toolbar

1. Type *Toolbar*, right-click on any tool, or use any other method (listed in the *Toolbar* command table in this book) to invoke the *Toolbars* dialog box.

2. Ensure the toolbar you want to modify is visible. If that toolbar is not visible on the screen, select the checkbox next to its name to make the toolbar appear.

3.   Choose the *Customize* button to open the *Customize Toolbars* dialog box.

4.   When the *Customize Toolbars* dialog box is open, you can:

   A.   Remove any tool from a visible toolbar by dragging it out of the toolbar group and dropping it into the graphics area of the Drawing Editor.

   B.   Move any tool from one group to another by dragging and dropping.

   C.   Add a tool from the *Customize Toolbars* dialog box to a visible toolbar group. See Step 5.

5.   To add a tool to a visible group (without removing it from another group), select the desired tool's group from the *Categories* drop-down list. This displays the available tool icons in the toolbar group ("category"). Drag the desired tool(s) from the dialog box and drop into the desired visible toolbar group. In Figure 45-7, the Boolean tools (*Union, Subtract, Intersect*) are being added to the Solids toolbar group.

Figure 45-7

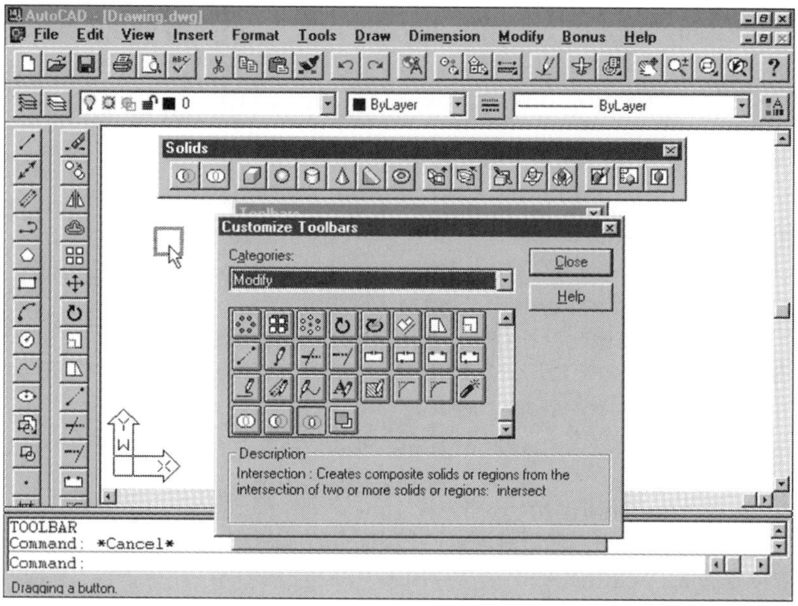

6.   *Close* the *Customize Toolbars* dialog box; then *Close* the *Toolbars* dialog box. AutoCAD automatically compiles the ACAD.MNR and ACAD.MNS menu files. This enables the modified toolbars to be used immediately.

Steps for Creating a New Toolbar Group

1.   Invoke the *Toolbars* dialog box by any method.

2.   Choose *New* to open the *New Toolbar* dialog box. In the *Toolbar Name* field, enter the new name, for example, "My Toolbar". PICK the *OK* button. When the dialog box closes, your new toolbar name (ACAD.My Toolbar) is highlighted in the toolbar listing. The new empty toolbar appears near the top of the AutoCAD drawing window.

3. Drag the new toolbar down into an open drawing area; then choose *Customize* from the *Toolbars* dialog box. From the *Customize* dialog box, drag and drop any tool icons onto the new toolbar (Fig. 45-8). Use the *Categories:* list to highlight toolbar groups and display any tool icon that you need. (The tooltip does not appear on tools in the *Customize* dialog box, only on tools that have been placed in a toolbar.)

4. *Close* the *Customize Toolbars* dialog box; then *Close* the *Toolbars* dialog box. AutoCAD automatically compiles the ACAD.MNR and ACAD.MNS menu files.

Figure 45-8

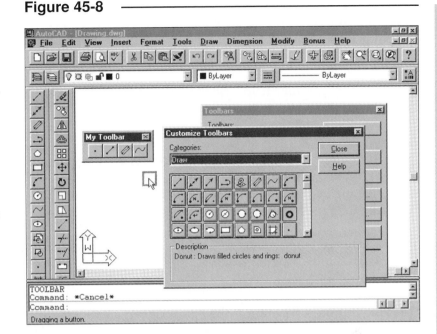

5. If you want to customize the help message for your new toolbar, open the *Toolbars* dialog box and highlight the new toolbar name from the *Toolbars:* list. Select the *Properties* button to produce the *Toolbar Properties* dialog box where the *Help* edit box is available. After entering the desired help massage, select *Apply,* then *Close* all the dialog boxes.

Note: If you are dragging a tool from a category in the *Customize Toolbars* dialog box and release the tool in the graphics area prior to dropping it into an existing toolbar, a new toolbar is created with the name "Toolbar*n*" where *n* is a number. To close the toolbar, pick the toolbar's close "X" in the upper-right corner. Delete it from the *Toolbars* dialog box listing by scrolling down to "ACAD.Toolbar*n*", then select the *Delete* button. *Close* the *Customize Toolbars* dialog box.

Customizing Existing and Creating New Buttons

Not only can you modify existing and create new toolbar groups in AutoCAD, you can customize individual buttons. Existing buttons can be modified and new buttons can be created. You can modify or create the image that appears on the icon and assign the command(s) that are activated when you PICK the button. The *Button Properties* dialog box provides the interface for "drawing" the button image and assigning the associated command.

The *Button Properties* dialog box (Fig. 45-9) is invoked by double right-clicking on any tool button visible in the Drawing Editor. Alternately, you can open the *Customize Toolbars* dialog box, then right-click (once) on any visible tool button. The areas of the *Button Properties* dialog box are described as follows.

Figure 45-9

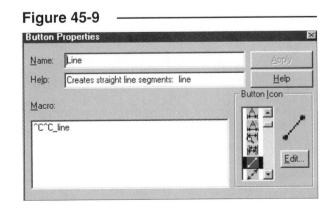

Name

This edit box defines the name of the button which also serves as the "tool tip" string that appears when the pointer rests on the button.

Help

This field contains the "help string" that appears at the bottom of the screen over the Status bar when the pointer rests on the button.

Macro

Enter any command, command and option, series of commands, system variable changes, etc., in this edit box. Similar to menu customization, all commands, options, or system variables must contain the exact spelling as you would enter at the command prompt. Spaces in the string are treated as Returns (pressing the Enter key).

Button Icon

Use this area and slider bar to select any of the AutoCAD-supplied icons for editing.

Edit

Selecting this button produces the *Button Editor* (Fig. 45-10). The *Button Editor* is like a miniature "paint" program that you can use to modify existing or create new button images. From the top of the editor, select a pencil for sketching, a line drawing instrument, a circle and ellipse drawing instrument, or an eraser. The color pallet appears on the right. Activate the *Grid* to "see" the pixels. You can *Clear* the entire button face, *Open* an existing button (.BMP file), or *Undo* the last drawing sequence. The drawing procedure is quite simple and fun. Use *Save* to save your work to a .BMP file before you *Close* the editor, or you can *Close* without saving. Use *SaveAs* to rename the button image to a .BMP file. After you *Close*, select *Apply* from the *Button Properties* dialog box to force the selected tool (that you originally selected from the Drawing Editor) to assume the new button face.

Figure 45-10 ――――――――――

Steps for Changing Existing and Creating New Buttons

Changing Existing Buttons

1. Double right-click (click twice quickly) on any existing tool button (except a flyout button) visible in the Drawing Editor. The *Toolbars* and the *Button Properties* dialog boxes appear.

    Alternate Method: Invoke the *Toolbars* dialog box by any method. Then right-click (once) on any existing tool button visible in the Drawing Editor. The *Button Properties* dialog box appears.

2. If it is necessary to change the appearance of the icon, select *Edit...* to produce the *Button Editor*. Make the necessary changes, *Save* the changes, and select *Apply* to have the new icon appear on the button in the Drawing Editor.

3. Make any desired changes in the *Name:* and *Help:* edit boxes.

4. Enter the new action desired for the button in the *Macro:* area. You can alter the existing action or delete the existing action and create a new series of commands or variable changes. Select *Apply*.

5. Close the *Button Properties* dialog box; then *Close* the *Toolbars* dialog box. AutoCAD automatically compiles the ACAD.MNR and ACAD.MNS menu files.

Creating New Buttons

1. If you want the new button(s) to reside in a new toolbar group, create the new toolbar group first (see Steps for Creating a New Toolbar Group). If you want the new button to reside in an existing group, make sure the toolbar is visible in the Drawing Editor.

2. Invoke the *Toolbars* dialog box by any method. Then select the *Customize* tile to make the *Customize Toolbars* dialog box appear.

3. In the *Customize Toolbars* dialog box, select *Custom* from the *Categories* drop-down list. Drag and drop the empty button (not the flyout button) into the desired toolbar (into the Drawing Editor). *Close* the *Customize Toolbars* dialog box.

4. Right-click (once) on the new empty button (in the Drawing Editor). The *Button Properties* dialog box appears.

5. Select *Edit...* to produce the *Button Editor*. Design the new button, *Save* the changes, and *Close* the *Button Editor*.

6. Make the desired entries in the *Name:* and *Help:* edit boxes.

7. Enter the new action desired for the button in the *Macro:* area. Select *Apply* to have the new icon appear on the button in the Drawing Editor.

8. Close the *Button Properties* dialog box, and then *Close* the *Toolbars* dialog box. AutoCAD automatically compiles the ACAD.MNR and ACAD.MNS menu files.

An Example Custom Toolbar Group and Buttons

Assume you work for a civil engineering firm and utilize *Point* objects often in your work. You decide to create a new toolbar group to change the display of the *Point* objects with <u>one</u> PICK rather than accessing the *Point Style* dialog box from the *Format* pull-down menu, PICKing the desired style, PICKing the *OK* button, and invoking a *Regen* (five PICKs required). To create a custom toolbar, you must create new button icons, assign the buttons to set the *PDMODE* system variable to the appropriate value, and then invoke a *Regen*.

Figure 45-11 illustrates the custom toolbar group under construction. The new toolbar group (in the Drawing Editor, left side) is titled *My Toolbar*. Only four buttons have been created at this time, with one under construction for changing the *Point* style to a circle with a cross (*PDMODE* 34). The other two buttons change *PDMODE* to values of 2 and 3.

Note the contents of the *Button Properties* dialog box (Fig. 45-11). The *Button Editor* (*Edit...*) was used previously to create the new icon (circle and two crossing lines). The *Name:* edit box contains the "tool tip" text (note the pointer on the left resting on the button). The *Help:* edit box

Figure 45-11

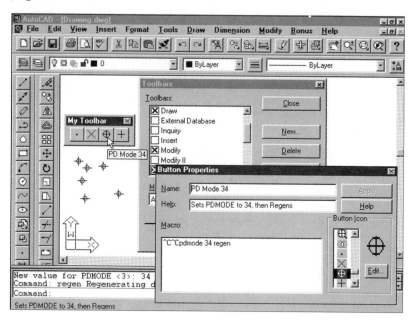

contains the "help string" that appears at the Status line (not shown during construction). The *Macro:* area contains two *Cancels* (^C^C), the *PDMODE* system variable call, a space to force a Return (like pressing Enter), the desired value (34), another space (Enter), and the *Regen* command.

When this toolbar group is complete, selecting any one of the buttons causes the *Point* objects in the drawing to immediately display the new point style.

Using the *Flyout Properties* Dialog Box

Flyout tools have a small right triangle in the lower-right corner and cause several related buttons to "fly out." If you right-click on a flyout tool while the *Toolbars* dialog box is open (or double right-click anytime on a flyout tool), the *Flyout Properties* dialog box appears (Fig. 45-12).

The *Flyout Properties* dialog box allows you to assign which toolbar you want to operate as a flyout group. In other words, with the *Flyout Properties* dialog box you attach an entire <u>existing toolbar</u> to one flyout button.

You can modify an existing flyout button (one with the small triangle in the lower-right corner) by attaching a different toolbar to it. For example, you could double-click on one of the *Zoom* flyouts, then assign another toolbar from the list to operate in its place. However, normally you would create a <u>new</u> button (from the *Custom* group) and assign an existing toolbar to it. The options in the *Flyout Properties* toolbar are the following.

Figure 45-12

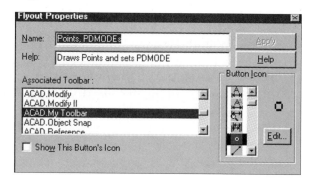

Name:
Enter the string for the "tool tip" that appears when the pointer rests on the button. (This option is only effective if the *Show This Button's Icon* option is checked.)

Help:
Enter the "help string" that appears on the Status line when the pointer rests on the button. (This option is only effective if the *Show This Button's Icon* option is checked.)

Associated Toolbar:
Select the existing flyout group (from the list) that you want to "fly out" when the button is selected.

Edit...
Selecting this produces the *Button Editor*. (See Customizing Existing and Creating New Buttons.)

Show This Button's Icon
If you have made your own icon for the flyout group using the *Button Editor*, checking this box forces your icon to display on the button. If this is not checked, one of the buttons from the associated flyout (toolbar) group appears—that is, the last button used.

Apply
This applies the icon and entries of the other fields. You must *Apply*, then close the *Flyout Properties* dialog box and the *Toolbars* dialog box to use the new flyouts properly.

Steps for Assigning an Existing Toolbar Group to a New Flyout Button

1. Make sure the toolbar (to attach the flyouts) is visible in the Drawing Editor.

2. Invoke the *Toolbars* dialog box by any method. Then select the *Customize* tile to make the *Customize Toolbars* dialog box appear.

3. In the *Customize Toolbars* dialog box, select *Custom* from the *Categories* drop-down list. Drag and drop the blank flyout button into the desired toolbar (into the Drawing Editor). *Close* the *Customize Toolbars* dialog box.

4. Right-click (once) on the new empty flyout button (in the Drawing Editor). The *Flyout Properties* dialog box appears.

5. Select *Edit...* to produce the *Button Editor*. Design the new button, *Save* the changes, and *Close* the *Button Editor*.

6. Make the desired entries in the *Name:* and *Help:* edit boxes.

7. Select the desired flyout group to attach to the button from the *Associated Toolbar* list. Check the *Show This Button's Icon* box to use your new icon. Select *Apply* to have the new icon appear on the button in the Drawing Editor.

8. *Close* the *Flyout Properties* dialog box, and then *Close* the *Toolbars* dialog box. AutoCAD automatically compiles the ACAD.MNR and ACAD.MNS menu files.

Example: New Flyout Group

Using *My Toolbar* from the previous
example, assume you wanted to add
this toolbar (*My Toolbar*) as a flyout
group to the existing *Draw* toolbar.
First, invoke the *Toolbars* and
Customize Toolbars dialog boxes and
drag the "blank" flyout button from
the *Customize* group and drop it into
the *Draw* toolbar. (See Steps 1, 2,
and 3.)

Next, produce the *Flyout Properties*
dialog box by right-clicking on the
new blank flyout button. You can
create a new icon at this point and
select *Show This Button's Icon*
(Fig. 45-13) or accept the defaults to
force one of the existing buttons (last
button used) to appear. Next, assign
My Toolbar as the toolbar to associate with (attach to) the
button. Enter the *Name* and *Help* strings in the edit boxes if
you want to *Show This Button's Icon*. Select *Apply*, then close
the dialog box. The resulting toolbar group appears as a
flyout group on the *Draw* toolbar (Fig. 45-14).

Figure 45-13 ───────────────

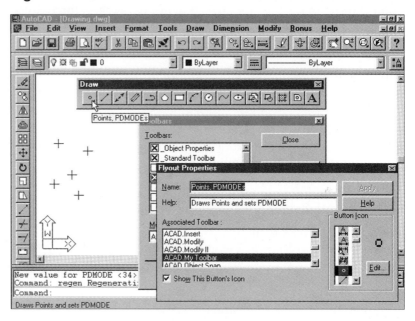

Figure 45-14 ───────────────

CREATING CUSTOM LINETYPES

AutoCAD has a wide variety of linetype definitions from which you can choose (Chapter 12, Layers,
Linetypes, and Colors). These definitions are stored in an ASCII file called ACAD.LIN. The ACAD.LIN
file in Release 14 describes both simple and complex linetypes. Simple linetypes can be composed of
line segments, dots, and spaces only. Complex linetypes include text or graphical shapes within the definition of the linetype pattern. However, you are not limited to these linetype definitions. New linetypes can be easily created that contain the line spacing, text, and shapes that you need.

Linetypes are created by using a text editor or word processor to describe the components of the linetype. The DOS Editor and the Windows Notepad are sufficient text editors that create ASCII files (only
alphanumeric characters without word processing codes) and can be used for this purpose. New linetype definitions are stored in ASCII files that you name and assign with an .LIN file extension. The file
names should indicate the nature of the linetype definitions. For example, ELECT.LIN can include linetype definitions for electrical layouts, or CIVIL.LIN can contain linetype definitions for the civil engineering discipline. This scheme promotes better file maintenance than appending custom linetypes to
the ACAD.LIN file.

Creating Simple Linetypes

A linetype definition is created by typing the necessary characters using a text editor and saving as an .LIN file. Examine the basic components of a simple linetype definition below. Each element of the line-type definition is separated by a comma (with no spaces). A linetype definition includes the following items:

1. The <u>name</u> of the linetype with a prefix of *

2. A <u>description</u> of the linetype (either text or symbols)

3. The <u>alignment</u> field

4. Numerical <u>distances</u> for dashes, dots, and spaces

5. The <u>shape</u> or <u>text</u> creation elements (contained in square brackets [ ])

A Simple Linetype Example

An example of creating a line that looks like line definition "EXAMPLE1"in the figure is the following:

```
*example1,a sample linetype ____ . . _____ . __
A,1.0,-0.25,0,-0.1,0,-0.25,1.5,-0.25,0,-0.25,0.5,-0.5
```

The above definition can be analyzed as follows:

1. The name of the linetype is "EXAMPLE1."

2. A description of the linetype is a "sample" of the linetype that you create with the keyboard characters: ____ . . _____ . __ (Note that this "graphical" description is created using the underscore and period characters.)

3. A   This is the alignment field.

4. 1.0   A positive value designates a line segment. This segment has a 1 unit length.

5. -0.25   A negative value denotes a gap or blank space. This gap has a length of 0.25.

6. 0   A value of 0 creates a dot.

7. repeat of the previous elements.

Figure 45-15 shows the "EXAMPLE1" linetype defined above.

Figure 45-15 ─────────────────────────────────

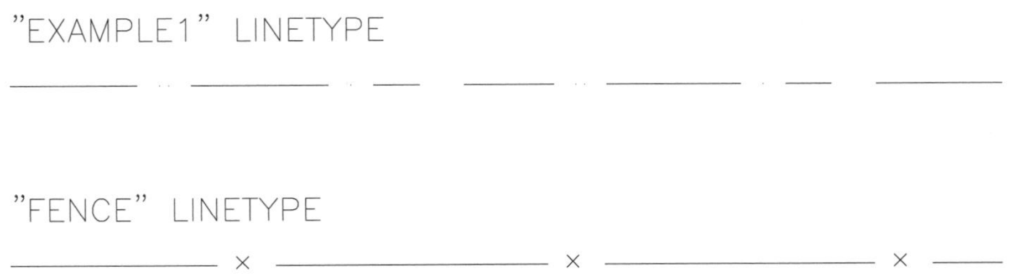

"EXAMPLE1" LINETYPE

──────── ·· ──────── ── ─── ──────── ·· ──────── ── ───

"FENCE" LINETYPE

─────────── × ─────────── × ─────────── × ───

Creating a Complex Linetype

In a complex linetype, two additional fields are possible in the definition and are placed in brackets [ ]. These fields are the Shape field and the String field. Each field has the following form:

Shapes: [shape description, shape file, scale, rotation, Xoffset, Yoffset]
Strings: ["text string," text style name, scale, rotation, Xoffset, Yoffset]

For linetypes containing shapes, the shape description and the compiled file name (.SHX) are provided in the first two places of the field. In comparison, a linetype containing a text string has the text string in quotes (" ") and the defined text style name in its first two places. If the text style does not exist, the current text style is used. The last four places of string and shape fields are the same. They are described as follows:

Scale	S=*value*	This is the scale factor by which the height of the text or shape definition is multiplied.
Rotation	R=*value* or A=*value*	A rotation angle for the text or shape is given here. The "A" designator is for an absolute rotational value, regardless of the the angle of the current line segment. The "R" designator is to specify a rotational value relative to the angle of the current line segment.
Xoffset	X= *value*	You can specify an X offset value from the end of the last element with this value.
Yoffset	Y=*value*	Specify a Y offset value from the end of the last element with this value.

Complex Linetype Examples

```
*PL,Property Line ----PL----PL----
A,0.5,-0.5,[PLSYM,SYMBOLS.SHX,S=1.5,R=0,X=-0.1,Y=-0.1],-0.5,0.5,1

*Fence,Fence line ----x----x----x----
A,1.5,-0.25,["x",STANDARD,S=0.2,R=0.0,X=0.05,Y=-0.1],-0.4,1.5
```

See Figure 45-15 for a sample of the "Fence" linetype defined above. For more examples of simple and complex linetypes, use the DOS editor or the Windows Notepad to open the ACAD.LIN file. It is suggested that you make <u>new</u> files when you create your first linetypes rather than appending to the ACAD.LIN file.

CUSTOMIZING THE ACAD.PGP FILE

The ACAD.PGP (AutoCAD Program Parameter) file is an ASCII text file that provides two main features. It provides a link between AutoCAD and other programs by specifying what external programs can be accessed from within AutoCAD, and it defines command aliases—the one- or two-letter shortcuts for commands. In short, the ACAD.PGP file defines what commands can be typed at the AutoCAD Command: prompt that are not native to AutoCAD.

Because the ACAD.PGP file is an ASCII file, you can edit the file to add your own command aliases and to define which external commands you want to use from within AutoCAD. Use any text editor, such as DOS EDIT, Windows Notepad, or Wordpad, to view and modify the file. The AutoCAD Release 14 ACAD.PGP file follows:

```
;   AutoCAD Program Parameters File For AutoCAD Release 14
;   External Command and Command Alias Definitions

;   Copyright (C) 1997 by Autodesk, Inc.

;   Each time you open a new or existing drawing, AutoCAD searches
;   the support path and reads the first acad.pgp file that it finds.

;   While AutoCAD is running, you can invoke other programs or utilities,
;   such Windows system commands, utilities, and applications.
;   You define external commands by specifying a command name to be used
;   from the AutoCAD command prompt and an executable command string
;   that is passed to the operating system.

;   You can abbreviate frequently used AutoCAD commands by defining
;   aliases for them in the command alias section of acad.pgp.
;   You can create a command alias for any AutoCAD command,
;   device driver command, or external command.

;   Recommendation: back up this file before editing it.

;   There is a bonus application for editing command aliases as well as
;   a sample acad.pgp file with many more command aliases.
;   See the bonus\cadtools folder for more details.


;   External command format:
;   <Command name>,[<DOS request>],<Bit flag>,[*]<Prompt>,

;   The bits of the bit flag have the following meanings:
;   First bit (1): if set, don't wait for the application to finish
;   Second bit (2): if set, run the application minimized
;   Third bit (4): if set, run the application "hidden"
;   Bits 2 and 4 are mutually exclusive; if both are specified only the 2 bit
is used.
;   The most useful values are likely to be 0 (start the application and wait
;   for it to finish), 1 (start the application and don't wait), 3 (minimize
and don't
;   wait), and 5 (hide and don't wait). Values of 2 and 4 should normally be
avoided,
;   as they make AutoCAD unavailable until the application has completed.

;   Examples of external commands for command windows

CATALOG,    DIR /W,        0,File specification: ,
DEL,        DEL,           0,File to delete: ,
DIR,        DIR,           0,File specification: ,
EDIT,       START EDIT,    1,File to edit: ,
SH,         ,              1,*OS Command: ,
SHELL,      ,              1,*OS Command: ,
```

```
START,      START,               1,*Application to start: ,
TYPE,       TYPE,                0,File to list: ,

; Examples of external commands for Windows
; See also the STARTAPP AutoLISP function for an alternative method.

EXPLORER,   START EXPLORER, 1,,
NOTEPAD,    START NOTEPAD,  1,*File to edit: ,
PBRUSH,     START PBRUSH,   1,,


; Command alias format:
;   <Alias>,*<Full command name>

;  The following are guidelines for creating new command aliases.
;  1.  Try the first character of the command, then try the first two,
;         then the first three.
;  2.  Ignore "DD" at the beginning of a command.
;  3.  Abbreviate the following prefixes:
;         Examples: 3 for 3D, A for ASE, D for Dim, I for Image, R for render.
;  4.  Once an alias is defined, add suffixes for related aliases:
;         Examples: R for Redraw, RA for Redrawall, L for Line, LT for
Linetype.
;  5.  An alias should reduce a command by at least two characters.
;  6.  Commands with a control key equivalent, status bar button, or function
key
;         do not require a command alias.
;         Examples: Use Control-N, -O, -P, and -S for New, Open, Print, and
Save
;  7.  Use a hyphen to differentiate between command line and dialog box
commands.
;  8.  Exceptions to the rules include AA for Area, T for Mtext, X for
Explode.

;  Sample aliases for AutoCAD commands
;  These examples include most frequently used commands.

3A,         *3DARRAY
3F,         *3DFACE
3P,         *3DPOLY
A,          *ARC
AA,         *AREA
AL,         *ALIGN
AP,         *APPLOAD
AR,         *ARRAY
AAD,        *ASEADMIN
AEX,        *ASEEXPORT
ALI,        *ASELINKS
ASQ,        *ASESQLED
ARO,        *ASEROWS
ASE,        *ASESELECT
AT,         *DDATTDEF
-AT,        *ATTDEF
ATE,        *DDATTE
-ATE,       *ATTEDIT
B,          *BMAKE
-B,         *BLOCK
```

```
BH,        *BHATCH
BO,        *BOUNDARY
-BO,       *-BOUNDARY
BR,        *BREAK
C,         *CIRCLE
CH,        *DDCHPROP
-CH,       *CHANGE
CHA,       *CHAMFER
COL,       *DDCOLOR
CO,        *COPY
D,         *DDIM
DAL,       *DIMALIGNED
DAN,       *DIMANGULAR
DBA,       *DIMBASELINE
DCE,       *DIMCENTER
DCO,       *DIMCONTINUE
DDI,       *DIMDIAMETER
DED,       *DIMEDIT
DI,        *DIST
DIV,       *DIVIDE
DLI,       *DIMLINEAR
DO,        *DONUT
DOR,       *DIMORDINATE
DOV,       *DIMOVERRIDE
DR,        *DRAWORDER
DRA,       *DIMRADIUS
DST,       *DIMSTYLE
DT,        *DTEXT
DV,        *DVIEW
E,         *ERASE
ED,        *DDEDIT
EL,        *ELLIPSE
EX,        *EXTEND
EXIT,      *QUIT
EXP,       *EXPORT
EXT,       *EXTRUDE
F,         *FILLET
FI,        *FILTER
G,         *GROUP
-G,        *GROUP
GR,        *DDGRIPS
H,         *BHATCH
-H,        *HATCH
HE,        *HATCHEDIT
HI,        *HIDE
I,         *DDINSERT
-I,        *INSERT
IAD,       *IMAGEADJUST
IAT,       *IMAGEATTACH
ICL,       *IMAGECLIP
IM,        *IMAGE
-IM,       *-IMAGE
IMP,       *IMPORT
IN,        *INTERSECT
INF,       *INTERFERE
IO,        *INSERTOBJ
L,         *LINE
```

```
LA,        *LAYER
-LA,       *-LAYER
LE,        *LEADER
LEN,       *LENGTHEN
LI,        *LIST
LS,        *LIST
LT,        *LINETYPE
-LT,       *-LINETYPE
LTS,       *LTSCALE
M,         *MOVE
MA,        *MATCHPROP
ME,        *MEASURE
MI,        *MIRROR
ML,        *MLINE
MO,        *DDMODIFY
MS,        *MSPACE
MT,        *MTEXT
MV,        *MVIEW
O,         *OFFSET
OS,        *DDOSNAP
-OS,       *-OSNAP
P,         *PAN
-P,        *-PAN
PA,        *PASTESPEC
PE,        *PEDIT
PL,        *PLINE
PO,        *POINT
POL,       *POLYGON
PR,        *PREFERENCES
PRE,       *PREVIEW
PRINT,     *PLOT
PS,        *PSPACE
PU,        *PURGE
R,         *REDRAW
RA,        *REDRAWALL
RE,        *REGEN
REA,       *REGENALL
REC,       *RECTANGLE
REG,       *REGION
REN,       *DDRENAME
-REN,      *RENAME
REV,       *REVOLVE
RM,        *DDRMODES
RO,        *ROTATE
RPR,       *RPREF
RR,        *RENDER
S,         *STRETCH
SC,        *SCALE
SCR,       *SCRIPT
SE,        *DDSELECT
SEC,       *SECTION
SET,       *SETVAR
SHA,       *SHADE
SL,        *SLICE
SN,        *SNAP
SO,        *SOLID
```

R14

```
SP,         *SPELL
SPL,        *SPLINE
SPE,        *SPLINEDIT
ST,         *STYLE
SU,         *SUBTRACT
T,          *MTEXT
-T,         *-MTEXT
TA,         *TABLET
TH,         *THICKNESS
TI,         *TILEMODE
TO,         *TOOLBAR
TOL,        *TOLERANCE
TOR,        *TORUS
TR,         *TRIM
UC,         *DDUCS
UCP,        *DDUCSP
UN,         *DDUNITS
-UN,        *UNITS
UNI,        *UNION
V,          *DDVIEW
-V,         *VIEW
VP,         *DDVPOINT
-VP,        *VPOINT
W,          *WBLOCK
WE,         *WEDGE
X,          *EXPLODE
XA,         *XATTACH
XB,         *XBIND
-XB,        *-XBIND
XC,         *XCLIP
XL,         *XLINE
XR,         *XREF
-XR,        *-XREF
Z,          *ZOOM
```

; The following are alternative aliases and aliases as supplied in AutoCAD
Release 13.

```
AV,         *DSVIEWER
CP,         *COPY
DIMALI,     *DIMALIGNED
DIMANG,     *DIMANGULAR
DIMBASE,    *DIMBASELINE
DIMCONT,    *DIMCONTINUE
DIMDIA,     *DIMDIAMETER
DIMED,      *DIMEDIT
DIMTED,     *DIMTEDIT
DIMLIN,     *DIMLINEAR
DIMORD,     *DIMORDINATE
DIMRAD,     *DIMRADIUS
DIMSTY,     *DIMSTYLE
DIMOVER,    *DIMOVERRIDE
LEAD,       *LEADER
TM,         *TILEMODE
```

As you can see the ACAD.PGP file in Release 14 is quite long—about 5 times the length of the Release 13 ACAD.PGP. The main reason is that for the <u>first time in AutoCAD's history almost all commands have an alias defined</u>, whereas previous releases only gave 15 or so useful command aliases. Autodesk's intention is for you to modify the file and to define the external commands and command aliases for personal productivity, although most of that work has been done for you in this release.

All lines of this file that begin with a semicolon (;) are remarks only. These lines (more than a page) give useful information about modifying the file, the format for specifying external commands, and the format for specifying command aliases. Only lines without the semicolon (;) are read by the AutoCAD program. Defining external commands and command aliases are explained next.

Specifying External Commands

The External Commands section of the ACAD.PGP file determines what commands or other programs can be accessed from <u>within AutoCAD</u> by typing the command at the AutoCAD Command: prompt. By examining this section of the file, you can see that the following DOS commands are already usable from within AutoCAD:

```
CATALOG,    DIR /W,        0,File specification: ,
DEL,        DEL,           0,File to delete: ,
DIR,        DIR,           0,File specification: ,
EDIT,       START EDIT,    1,File to edit: ,
SH,         ,              1,*OS Command: ,
SHELL,      ,              1,*OS Command: ,
START,      START,         1,*Application to start: ,
TYPE,       TYPE,          0,File to list: ,
```

The following Windows commands are also usable from within AutoCAD:

```
EXPLORER,   START EXPLORER, 1,,
NOTEPAD,    START NOTEPAD,  1,*File to edit: ,
PBRUSH,     START PBRUSH,   1,,
```

For example, type DIR from the Command: prompt to list the contents of the current directory. Also note the *Shell* command is defined here (see Chapter 2 for information on using *Shell*). If you prefer using Windows Explorer for file management, just type "Explorer" at the AutoCAD command prompt to open it.

Note that Notepad can be opened from within AutoCAD by typing "Notepad" at the command prompt. In addition, EDIT, the Microsoft DOS text editor, can be activated by typing EDIT at the command line (in this case, EDIT appears in a window while AutoCAD runs "in the background"). These programs make <u>customizing ACAD.PGP possible from within AutoCAD.</u>

NOTE: If you edit ACAD.PGP while AutoCAD is running, use the *Reinit* command after editing to reinitialize ACAD.PGP to make the new changes usable (see Chapter 44 for information on *Reinit*).

If you use another text editor—Norton Editor, for example—you can change the ACAD.PGP file to allow "NE" to be typed at the command prompt. Do it by changing this:

```
EDIT,       START EDIT,    1,File to edit: ,
```

to this:

```
NE,        START NE,        1,File to edit: ,
```

Ensure you follow the format guidelines specified in the first page of the file:

```
<Command name>,[<DOS request>],<Bit flag>,[*]<Prompt>,
```

You should also use spaces (press the space bar) instead of tabs in the ACAD.PGP file.

Defining Command Aliases

Autodesk provided only 16 command aliases in Release 13. Autodesk intended for you to define your own command aliases. After questioning users, Autodesk decided to define a command alias for almost all commands in Release 14. However, you can add more aliases, delete the defined aliases that you don't use, or alter the aliases that AutoCAD provides. The command alias section begins with these entries:

```
3A,        *3DARRAY
3F,        *3DFACE
3P,        *3DPOLY
A,         *ARC
AA,        *AREA
AL,        *ALIGN
and so on...
```

Because most commands already have an alias defined, it is probably more productive to learn the existing aliases rather than create new ones. There may be cases where it would be helpful to redefine existing aliases, however. For example, you may use *Trim* often but not *Mtext* (*TR* is the current alias for the *Trim* command, and *T* _and_ *MT* for *Mtext*). In this case you could delete the *T* alias for *Mtext* (and just use *MT* for *Mtext*) and change *Trim* to *T*. To do this, follow the format specified at the beginning of the aliases section (give the alias and a comma, space over, and then give the AutoCAD command prefaced by an asterisk). First, find and delete this:

```
T,        *MTEXT
```

and change this:

```
TR,        *TRIM
```

to this:

```
T,        *TRIM
```

You may want to modify some of the other defined aliases. For example, you could change the alias C to invoke *Copy* rather than *Circle*. An alternative to *Circle* could be CI.

You can make comments anywhere in the ACAD.PGP by typing a semicolon (;) at the beginning of the line. This is helpful for making note of the changes that you have made.

One last suggestion: make a copy of the ACAD.PGP on diskette or in another directory before you begin experimenting, just in case.

CREATING SCRIPT FILES

A script file is an ASCII text file that contains a series of AutoCAD commands to be performed in succession. A script file is an external file that you create with a text editor and assign an .SCR extension. The script file is composed of commands in the sequence that you would normally use just as if you entered the commands at the AutoCAD command prompt. The *Script* command is used to tell AutoCAD to run the specified script file. When AutoCAD "runs" a script file, each command in the file is executed in the order that it is listed.

Script files have many uses, such as presenting a slide show, creating drawing geometry, or invoking a series of AutoCAD commands you use often—like a "macro." Script files are commonly used for repetitive tasks such as setting up a drawing. In fact, script "macros" may be incorporated directly into a menu file, thereby eliminating the need for an external .SCR file. However, in this chapter, only independent script files are addressed.

Chapter 44 discusses the *Script* command and the steps for creating scripts to present slide shows. A script for viewing slides uses the *Vslide* and *Delay* commands repetitively. If you are not familiar with the procedure for creating and running a script for this fundamental application, please review Using Slides and Scripts in Chapter 44.

Scripts for Drawing Setups

As you become familiar with AutoCAD, you realize several steps are repeated each time you set up a new drawing. For example, several commands are used and variables are set to prepare the drawing for geometry construction and for eventual plotting. Drawing setup is a good candidate for script application. Script files can be created to perform those repetitive drawing setup steps for you. Although an interactive LISP routine would provide a more efficient and powerful method for this task, creating a script is relatively simple and does not require much knowledge or effort beyond that of the AutoCAD commands that would typically be used. Keep in mind that template drawings that have many of the initial settings can be used as an alternative to creating a script for drawing setup. However, one advantage of a script (in general) is that it can be executed in any drawing and at any time during the drawing process.

For example, assume that you often set up drawings for plotting at a 1/4" = 1'-0" scale on a 24" x 36" sheet of paper. This involves setting *Limits, Grid, Snap, LTSCALE, DIMSCALE*, and *Scaling* an existing title block and border. Instead of executing each of these operations each time "manually," create a script to automate the process. Here are the steps:

1. Open your text editor from within AutoCAD or before starting AutoCAD. Creating scripts while AutoCAD is running is the most efficient method for writing and "debugging" scripts. In AutoCAD, use EDIT or Notepad (or other text editor specified in the ACAD.PGP) to open your text editor.

2. Create the ASCII text file with the commands that you would normally use for setting up the drawing. Remember (from Chapter 44) that a space or a new line in the script is read as a Return (pressing the Enter key) while the script executes in AutoCAD. The script file may be written like this:

```
LIMITS 0,0 1728,1152
GRID 48
SNAP 12
LTSCALE 48
SCALE ALL  0,0 48
DIMSCALE 48
DIMSTYLE SAVE STANDARD Y
```

3. Save the file with an .SCR extension. A good name for this file is SETUP48.SCR. Consider locating the file in a directory in the AutoCAD environment (as specified in the *Support Files Search Path* section of the *Files* tab in the *Preferences* dialog box) or in a location with other scripts.

4. Return to AutoCAD and use the *Script* command to test the script. Don't expect the script to operate the first time if this is your first experience writing scripts. It is common to have "bugs" in the script. For example, an extra space in the file can cause the script to get out of the intended sequence. An extra space in the script is hard to locate (since it is not visible), particularly using DOS EDIT. A good text editor allows the cursor to move only to places occupied by a character (including a space) and should be used for script writing.

Examining the commands in SETUP48.SCR, first, set the *Limits* using inch values. *Scale All* is used to scale the existing border and title block. The "DIMSTYLE SAVE STANDARD Y" line saves the *DIM-SCALE* setting to the STANDARD dimension style. (If a dimension style other than STANDARD exists, the new *DIMSCALE* setting should be incorporated also into these dimension styles.) Keep in mind that scripts operate as if the commands were entered at the command prompt, not as you might operate using dialog boxes.

Many other possibilities exist for drawing setup scripts. For example, you could set *Units* (command) and the *DIMUNITS* variable. You may want to create a script for each plot scale and sheet size. It depends on your particular applications, hence the term "customization."

Possible applications for scripts are numerous. In addition to drawing setups, scripts can be used for geometry creation. It is possible to write scripts to create particular geometry "in real time" as the script executes. For example, repetitively drawn shapes could be created by scripts as an alternative to inserting *Blocks*. This idea is not generally used, but it may have advantages for certain applications. There are endless possibilities for script applications. Generally, anytime you find yourself performing the same tasks or series of commands repetitively, remember that a script can automate that process.

Special Script Commands

Creating scripts is a very straightforward procedure because the script file contains the AutoCAD commands that you would normally use at the command prompt for the particular operation. A few commands are intended for use specifically with scripts and other commands can be used as an aid to create or run script files. Those commands are listed next. Many of these commands are discussed in detail in Using Slides and Scripts, Chapter 44.

SCRIPT	Use this command in AutoCAD to instruct AutoCAD to load and execute a script. The *Select Script File* dialog box appears, prompting you for the desired script file (see Chapter 44).
DELAY	Include this command in a script file to force a delay (specified in milliseconds) until the next command in the script is executed. *Delay* is commonly used in script files that display slide shows (see Chapter 44).
RSCRIPT	*Rscript* is short for "Repeat script." This command can be entered as the last line of a script file to make the script repeat until interrupted. *Rscript* is helpful for self-running slide shows (see Chapter 44).
RESUME	If you want to interrupt a script with the Break or backspace key, type *Resume* to resume the script from the point at which it was interrupted. This command has no purpose in the script file itself but is intended to be typed at the keyboard (see Chapter 44).

LOGFILEON/
LOGFILEOFF

These commands are used to turn on and off the log file creation utility in AutoCAD. When *Logfileon* is used, the conversation that appears at the command prompt (command history) is written to an external text file until *Logfileoff* is used (explained next in this chapter).

LOGFILEON/
LOGFILEOFF

Pull-down Menu	COMMAND (TYPE)	ALIAS (TYPE)	Short-cut	Screen (side) Menu	Tablet Menu
Tools *Preferences...* *General*	LOGFILEON or LOGFILEOFF	...	...	TOOLS 2 *Prefernc* *General*	...

Logfileon and *Logfileoff* are used to turn on and off the creation of the AutoCAD log file. The log file is a text file that contains the command history—the text that appears at the command line. When *Logfileon* is invoked, nothing appears to happen; however, any commands, prompts, or other text that appear at the command prompt from that time on are copied and written to the AutoCAD log file. Use *Logfileoff* to close the log file and discontinue recording the command history.

If the log file is on, AutoCAD continues to write to the log file, and each AutoCAD session in the log is separated by a dashed line. If you turn *Logfileoff* and *Logfileon* again, the new session is appended to the existing file. The default log file name is ACAD.LOG. The name and path of the file can be changed in the *Preferences* dialog box.

The ACAD.LOG file is an ASCII text file that can be edited. The file has no direct link to AutoCAD and can be deleted without consequence to AutoCAD operation. The log file cannot be accessed externally until *Logfileoff* is used or AutoCAD is exited.

A sample log file is shown here:

```
[ AutoCAD - Wed Aug 06 10:16:29 1997 ]-----------------------------
Command: _line From point:
To point:
To point:
To point:
Command:
Command: _circle 3P/2P/TTR/<Center point>: Diameter/<Radius>:
Command:
Command: _copy
Select objects: 1 found
Select objects: 1 found
Select objects: 1 found
Select objects:
<Base point or displacement>/Multiple: Second point of displacement:
Command:
Command: _trim
Select cutting edges: (Projmode = UCS, Edgemode = No extend)
Select objects: 1 found
Select objects: 1 found
Select objects:
<Select object to trim>/Project/Edge/Undo:
<Select object to trim>/Project/Edge/Undo:
<Select object to trim>/Project/Edge/Undo:
Command:
Command: _qsave
Command: logfileoff
```

The log file is especially helpful when you are writing scripts. When you create a script file, the commands and options listed in the file must be typed in the exact sequence that would occur if you were actually using AutoCAD. In order to write the script correctly, you have three choices: (1) remember every command and prompt exactly as it would occur in AutoCAD; (2) go through a "dry run" in AutoCAD and write down every step; (3) turn *Logfileon*; then execute a "dry run" to have all the steps and resulting text recorded.

In other words, turn *Logfileon* to create a rough script file. Then turn *Logfileoff*, open the ACAD.LOG, and edit out the AutoCAD prompts such as "Command:" and "Select objects:," etc. Finally, save the file to the desired name with an .SCR extension.

If you want to change the name or location of the log file that is created when you use *Logfileon,* use the *LOG-FILENAME* system variable or the *Files* tab of the *Preferences* dialog box (Fig. 45-16). In the dialog box, expand the *Menu, Help, Log, and Miscellaneous File Names* tree and enter the desired log file name and location under the *Log File* section.

Figure 45-16

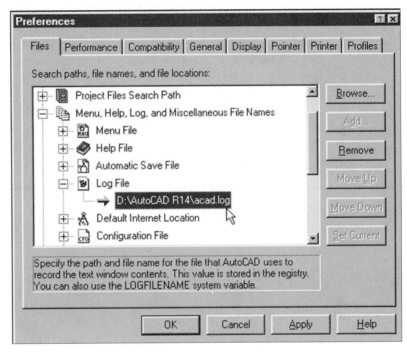

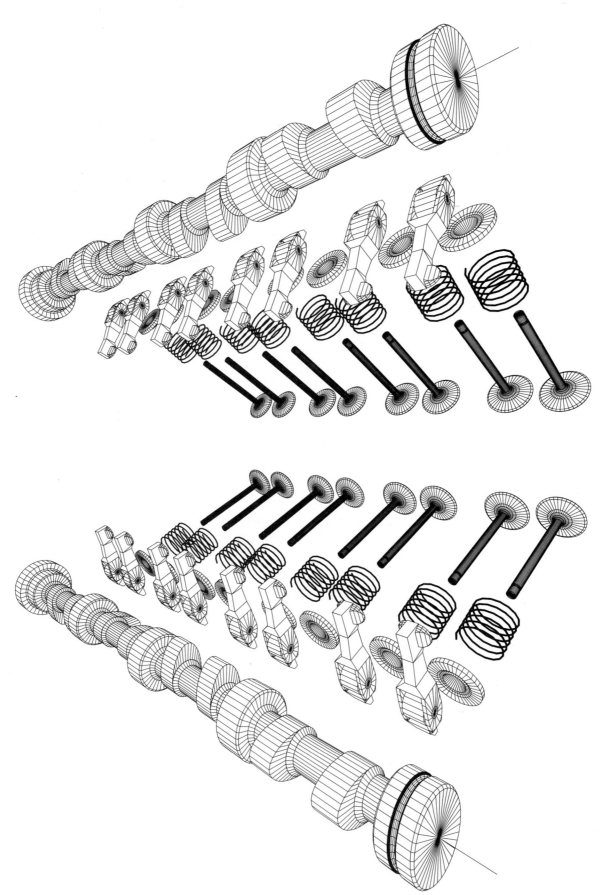

CAMASSY.DWG Courtesy of Autodesk, Inc. (Release 14 Sample Drawing)

46

MENU CUSTOMIZATION

```
**POP1
*FILE
ID_MnFile    [&File]
ID_New       [&New...\tCtrl+N]^C^C_new
ID_Open      [&Open...\tCtrl+O]^C^C_op
             [--]
ID_Save      [&Save\tCtrl+S]^C^C_qsav
ID_Saveas    [Save &As...]^C^C_save
ID_Export    [&Export...]^C^C_export
             [--]
ID_Config    [P&rinter Setup...]^C^C_
ID_Preview   [Print Pre&view]^C^C_
ID_Print     [&Print...\tCtrl+P]^C^C_p
             [--]
ID_MnDrawing [->Drawing &Utilitie
ID_Audit     [&Audit]^C^C_audit
ID_Recover   [&Recover...]^C^C_
             [--]
ID_MnPurge   [->&Purge]
ID_PurgeAll  [&All]^C^C_purg
             [--]
ID_PurgeLay  [&Layers]^C^C
ID_PurgeLin  [Li&netypes]^C
ID_PurgeTxt  [&Text Styles]
ID_PurgeDim  [&Dimension
ID_PurgeMln  [&Multiline S
ID_PurgeBlk  [&Blocks]^C
ID_PurgeShp  [<-<-&Shap
ID_SendMail  [Sen&d...]
             [--]
ID_MRU       [Drawing Histo
             [--]
             [E&xit]^C^C_q
```

Chapter Objectives

After completing this chapter you should:

1. understand the menu structure;

2. be able to use name_tags, item labels, and menu macros;

3. be able to use special characters in menu macros;

4. be able to use line terminators;

5. understand the menu sections;

6. understand menu swapping;

7. understand the base menu versus partial menus.

CONCEPTS

One of the special features of AutoCAD is the ability to customize the menu system. AutoCAD enables you to change the menu system by providing the primary menu, ACAD.MNU, as an ASCII text file. You can use a word processor or text editor to change the entries in the ACAD.MNU file to include your own combinations of commands, command options, and system variable settings. The AutoCAD menu file is a detailed list of the various commands available to you. The menu file is organized in sections; each section controls a separate function, such as the pointing device buttons, the pull-down menus, the screen menu, the digitizing tablet menu, and the toolbars.

The ability to modify the existing menu file and to create new menu files gives you a surprising amount of power to control the way AutoCAD operates. However, customizing the menus requires knowledge of the menu organization and special codes (characters) that are understood by AutoCAD for performing specific tasks. Although specific knowledge is required for customizing the AutoCAD menus, it is easy to learn how to control the menus by making small changes to the existing menus or creating simple menus "from scratch."

A menu file is an ASCII text file that executes commands based on the tasks you select on the menu. By customizing part or all of the menu system, you can tailor AutoCAD to work for you. Customizing menus is a means to automate repetitive tasks and maintain consistency on drawings throughout an organization. Customizing the menus for your specific tasks can make you much more productive.

MENU FILE TYPES AND MENU LOADING

Menu File Types

A menu file is usually a group of files working together to define the appearance and functionality of the particular menu you have loaded with your drawing file. A number of file extensions are associated with the basic AutoCAD menu. The following table defines the various file extensions that may be associated with the menu:

File Extension	Description
MNU	Menu template file: This ASCII file contains the AutoCAD command strings and menu macros.
MNL	Menu LISP file: This ASCII file contains AutoLISP expressions that are used by the menu file and are loaded into memory when a menu file with the same name is loaded.
MNC	Compiled menu file: This binary file contains the command strings and menu syntax that defines the appearance and functionality of the menu.
MNR	Menu resource file: This binary file contains the bitmaps used by the menu.
MNS	Menu source file: The source menu file is created by AutoCAD from the MNU file.
MND	Menu definition files: These files must be compiled with the menu compiler utility file MC.EXE. These are special menu source files that contain macros.
DLL	Dynamic link libraries: Bitmap resource DLLs are used to store the bitmaps for toolbars, pull-down, and cursor menus. A DLL's filename must be the same as the menu file name it is associated with. All resources must be named not index numbered. In addition, the DLLs must be located in the same directory as the menu file that it is associated with.

Before you begin creating custom menus, you should make a backup of the ACAD.MNU and the ACAD.MNL files. You can then make alterations in these files to create your own custom menu files. Each new menu you create should have its own MNL file (using the same filename) to evaluate and load the necessary LISP routines to interact with the other menu files.

Loading Menu Files

MENU

Pull-down Menu	COMMAND (TYPE)	ALIAS (TYPE)	Short-cut	Screen (side) Menu	Tablet Menu
...	*MENU*	...	...	...	...

The *Menu* command loads a menu file. Normally this command is used to load a base menu, complete with all menu areas such as pull-down menus, toolbars, cursor menu, image tile menus, and screen menus. The *Menu* command produces the *Select Menu File* dialog box (Fig. 46-1). Enter or select a menu file name. To have no menu displayed, select *Type It* and enter a period (.) on the command line.

Figure 46-1

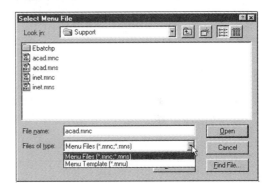

The name of the most recently loaded menu file is stored in the system registry in Release 14. The name of this menu is also stored with the drawing file in the *MENUNAME* system variable for backward compatibility. The last menu used is loaded each time you restart AutoCAD. In order to speed up initial drawing loads, beginning with R14, AutoCAD no longer reloads the menu file each time you open a drawing file in any one AutoCAD session.

Each time you restart AutoCAD or load a specific menu file with the *MENU* command, AutoCAD systematically searches for the menu file based on the library search path (see Library Search Path).

MENULOAD

Pull-down Menu	COMMAND (TYPE)	ALIAS (TYPE)	Short-cut	Screen (side) Menu	Tablet Menu
Tools Customize Menus...	*MENULOAD*	...	...	TOOLS 2 Menuload	Y,9

The *Menuload* command is used to load additional menus (called "partial menus") or add and remove individual pull-down menus from the menu bar. The *Menuload* command produces the *Menu Customization* dialog box (Fig. 46-2). Use the *Menu Groups* section of this dialog box to *Load*, *Unload*, and *Browse* partial menu files. The *Menu Bar* tab is used to select pull-down menus to *Insert* into or *Remove* from the menu bar at the top of your screen. When *FILEDIA* is set to 0, the command line issues the following prompt:

Figure 46-2

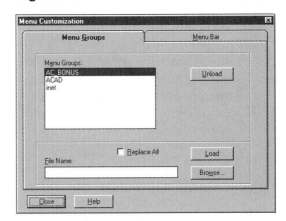

```
Enter name of menu file to load:
```

MENUUNLOAD

Pull-down Menu	COMMAND (TYPE)	ALIAS (TYPE)	Short-cut	Screen (side) Menu	Tablet Menu
Tools *Customize Menus...*	*MENUUNLOAD*	...	...	*TOOLS 2* *Menuload*	*Y,9*

The *Menuunload* command also produces the *Menu Customization* dialog box (see Fig. 46-2). Here you may also unload partial menus and menu bars. When *FILEDIA* is set to 0, the command line issues the following prompt:

```
Enter name of the MENUGROUP to unload:
```

Library Search Path

When AutoCAD attempts to locate a menu specified by the system registry, *MENUNAME* variable, or the *Menu* or *Menuload* commands, the following search path order is used:

1. Current directory

2. Directory that contains the current drawing file

3. Directories listed in the Support path (for more information, see Support File Search Path in the *Installation Guide*)

4. Directory that contains the AutoCAD program files

NOTE: Two or more of these directories may be the same, depending on your drawing environment.

Steps for Loading Menu Files

The default AutoCAD menu files are named "ACAD" with the appropriate file extension, for example, ACAD.MNU, ACAD.MNS, etc. You can use any name (8 character limit) when you create a new custom menu. Assuming the menu file you are loading is called ACAD menu, AutoCAD looks for the menu using the following steps. Figure 46-3 visually describes the process.

Figure 46-3

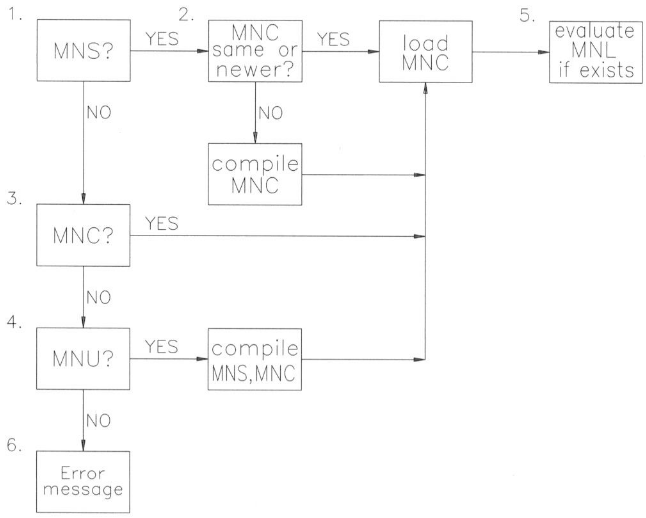

1. AutoCAD looks for the ACAD.MNS (menu source) file following the library search procedure shown above. If found, it proceeds to step 2; otherwise, it continues with step 3.

2. After locating the ACAD.MNS (menu source) file, AutoCAD looks for the ACAD.MNC (compiled menu) file in the same directory. If it finds the ACAD.MNC (compiled menu) file with the same or later date and time as the ACAD.MNS (menu source) file, AutoCAD loads the ACAD.MNC (compiled menu) file. Otherwise, AutoCAD compiles the ACAD.MNS (menu source) file, generating a new ACAD.MNC (compiled menu) file in the same directory, and loads that file. Next AutoCAD goes to step 4.

3. If AutoCAD doesn't find the ACAD.MNS (menu source) file during step 1, it looks for the ACAD.MNC (compiled menu) file, following the library search procedure shown above. If AutoCAD finds the ACAD.MNC (compiled menu) file, it loads that file and continues with step 5; otherwise, it moves to step 6.

4. If AutoCAD fails to locate either the ACAD.MNS (menu source) or the ACAD.MNC (compiled menu) file, it searches the library path for the ACAD.MNU (menu template) file. If the ACAD.MNU file is found, it compiles ACAD.MNC (compiled menu) and ACAD.MNS (menu source) files, then loads the ACAD.MNC (compiled menu) file and continues with step 5; otherwise, it moves to step 6.

5. When AutoCAD is finished locating, compiling, and loading the ACAD.MNU (menu template) file, AutoCAD looks for a ACAD.MNL (menu LISP) file, following the library search procedure shown above. If AutoCAD finds a ACAD.MNL (menu LISP) file, it evaluates the AutoLISP expressions within the ACAD.MNL (menu LISP) file. The ACAD.MNL file is optional.

6. If AutoCAD doesn't find any of the ACAD menu files, an error message is displayed and you are prompted for another menu file name. You need to specify the full path and file name for AutoCAD to locate the ACAD menu.

Each time you load the standard AutoCAD menu (ACAD.MNU), the standard AutoCAD LISP menu file (ACAD.MNL) is also evaluated. AutoCAD then compiles an ACAD.MNS file and an ACAD.MNC file and generates the ACAD.MNR, which contains the bitmaps used by the menu. The ACAD.MNU file is an ASCII file that contains comments and special formatting.

Sample Menu Files

Excerpt from the AutoCAD Standard Menu Template File (ACAD.MNU)
Listed below is the first section of the standard AutoCAD ACAD.MNU menu template file:

```
//
//    AutoCAD Menu - Release 14.0
//    28 April 1997
//
//    Copyright (C) 1986, 1987, 1988, 1989, 1990, 1991, 1992, 1994, 1996, 1997
//    by Autodesk, Inc.

....
//    NOTE:  AutoCAD looks for an ".mnl" (Menu Lisp) file whose name is
//           the same as that of the menu file, and loads it if
//           found.  If you modify this menu and change its name, you
//           should copy acad.mnl to <yourname>.mnl, since the menu
//           relies on AutoLISP routines found there.
//
//    Default AutoCAD NAMESPACE declaration:
//
***MENUGROUP = ACAD
//
//    Begin AutoCAD Digitizer Button Menus
//
***BUTTONS1
// Simple + button
// if a grip is hot bring up the Grips Cursor Menu (POP 17), else send a carriage return
$M=$(if,$(eq,$(substr,$(getvar,cmdnames),1,5),GRIP_),$P0=ACAD.GRIPS $P0=*);
$P0=SNAP $p0=*
^C^C
^B
```

A pair of forward slashes (//) indicates comments in a menu file. The ACAD.MNS file is initially the same as the ACAD.MNU without the comments or special formatting. The first 35 lines of the standard AutoCAD menu (ACAD.MNU) are comments. When AutoCAD compiles the ACAD.MNS file, only the first three and the last three lines of the file are kept as comments. AutoCAD modifies the ACAD.MNS file each time you make changes to the menu using the AutoCAD interface (such as creating custom toolbars).

Whenever you modify the toolbars (using the AutoCAD interface) and you want to keep the new changes, you should cut and paste the modified sections of the MNS file to the MNU file prior to deleting the MNS file or specifically reloading the MNU file.

Excerpt from the AutoCAD Standard Menu Source File (ACAD.MNS)
Listed below is a section of the standard AutoCAD ACAD.MNS menu source file:

```
//
//   AutoCAD menu file - R:\NTEXE\RELEASE\support\acad.mnu
//

***MENUGROUP=ACAD

***BUTTONS1
$M=$(if,$(eq,$(substr,$(getvar,cmdnames),1,5),GRIP_),$P0=ACAD.GRIPS $P0=*);
$P0=SNAP $p0=*
^C^C
^B
```

The MNU and the MNS files contain the initial positioning of the toolbars. Changes to the status (docked/floating and show/hide) are stored in the system registry. When you modify the MNU file after the MNS file is generated, you must specifically load the MNU to generate new MNS and MNR files to make your changes recognized.

MENU STRUCTURE

Menu Sections

The menu is made up of several sections. Section labels are in the format of ***section_name*. The following table shows the standard AutoCAD menu section labels and their associated menu areas. This table is reprinted from the AutoCAD Release 14 *Customization Guide*.

Menu Section	Use of Label
***MENUGROUP	Menu file group name
***BUTTONS1–4	Digitizer or other Non-System Pointing Device menu
***AUX1–4	System Pointing Device menus
***POP0	Cursor menu
***POP1–16	Pull-down menus
***POP17	Grips Cursor menu
***TOOLBARS	Toolbar menu
***IMAGE	Image Tile menus
***SCREEN	Screen menus
***TABLET	Tablet menu area
***HELPSTRINGS	Text that is displayed in the Status bar when your cursor passes over a pull-down menu item, cursor menu item, or a toolbar icon
***ACCELERATORS	Accelerator key definitions

Each section of the menu controls the way AutoCAD responds to that particular device or area of the Drawing Editor. When you make a menu selection in AutoCAD from one of the menu areas, AutoCAD responds by following the instructions (commands) listed in that particular line of the menu.

Menu Item Syntax

A menu item is generally one line in a menu that performs one action in AutoCAD when selected from the menu. In general, menu item syntax is consistent across all sections of the menu that allow menu items. Each menu item normally is comprised of these elements: a name tag, a label, and a menu macro. A menu item is generally on one line in the following format.

name_tag   label   menu_macro

Name Tags

A name tag is generally the first component listed in a menu item. It consists of a combination of alphanumeric characters and an underscore (_) character. The name_tag should be a unique string within the menu file. The functionality provided by the name tags are:

Name_tags link menu items and toolbar buttons to their associated Help strings which are displayed on the Status bar line.

Name_tags link accelerator keys to the corresponding POP menu macros.

Name_tags enable menu items to switch from one state to another. Menu items can be enabled/disabled and checked/unchecked with menu macros and AutoLISP.

Name tags are not allowed in the BUTTONS, AUX, and IMAGE sections of the menu. They serve no purpose in the SCREEN and TABLET menu sections.

Item Labels

Item labels define the words that are displayed in AutoCAD for a pull-down or screen menu. Each menu section uses a different format for item labels. In the menu file, item labels are enclosed in brackets [label]. Menu sections that have no interface for displaying information to the user, such as the BUTTONS, AUX, and TABLET sections, do not require item labels. You may wish to use item labels in the BUTTONS, AUX, and TABLET sections for programming notes, but they serve no other purpose. The use of each of the menu section item labels is shown in the following table. This table is reprinted from the AutoCAD Release 14 *Customization Guide*.

Menu Section	Use of Item Label
BUTTONS*n*, AUX*n*, and POP*n*	Defines the content and formatting of pull-down and cursor menu selections.
TOOLBARS	Defines the toolbar name, status (floating or docked and hidden or visible), and position. Also, it defines each button and its properties.
IMAGE	Defines the text and image displayed in the image tile menus.
SCREEN	Defines the text that is displayed in the screen menus.
HELPSTRINGS	Defines the Status line help related to POP and TOOLBAR menu items.
ACCELERATORS	Associates keyboard action with menu macros.

The proper syntax for each of the labels is described in the particular menu section descriptions.

Menu Macros

Menu macros are single commands or a string of commands in a menu that complete tasks for you. When you include menu macros in the menu item, you must use the proper syntax, otherwise you may get unexpected results. Every character in a menu macro is significant. Even blank spaces serve a vital purpose and must be located properly. With each revision and upgrade of AutoCAD, command syntax might change and may require you to make minor changes to your customized menus. Sometimes with new releases, command names change. One example of this is the LWPOLYLINE object, or "optimized polyline," which is new with Release 14.

Menu macros often involve using special characters. The special characters, which AutoCAD recognizes, are shown in the following table. The table is reprinted from the AutoCAD Release 14 *Customization Guide*.

Special Characters for Menu Macros

Character	Description
;	Issues Enter
^M	Issues Enter
^I	Issues Tab
Spacebar	Enters a space; blank space between command sequences in a menu item is equivalent to pressing the Spacebar
\	Pauses for user input
–	Translates AutoCAD commands and key words that follow
+	Continues menu macro to the next line (if last character)
=*	Displays the current top level image, pull-down, or cursor menu
*^C^C	Prefix for a repeating item
$	Special character code that loads a menu section or introduces a conditional DIESEL macro expression ($M=)
^B	Toggles *Snap* on or off (Ctrl+B)
^C	Cancels command (Esc)
^D	Toggles *Coords* on or off (Ctrl+D)
^E	Sets the next isometric plane (Ctrl+E)
^G	Toggles *Grid* on or off (Ctrl+G)
^H	Issues a backspace
^O	Toggles *Ortho* on or off (Ctrl+O)
^P	Toggles *MENUECHO* on or off
^Q	Echoes all prompts, status listings, and input to the printer (Ctrl +Q)
^T	Toggles *Tablet* on or off (Ctrl+T)
^V	Changes current viewport (Ctrl+V)
^Z	Null character that suppresses the automatic addition of Spacebar at the end of a menu item

Interpreting a Macro
Built into AutoCAD is the ability to add a space at the end of a menu macro:

    ID_Line   [&Line]^C^C_line(blank space)

AutoCAD interrupts the menu macro shown above as:

    name_tag   =   ID_Line

    item label   =   Line (with the letter L underlined for use as an accelerator key combination)

    ^C^C   =   (cancel any currently active command)

    underscore (_) =   automatic translation for foreign language support

    line   =   Standard AutoCAD *Line* command

    blank   =   AutoCAD automatically adds a blank space at the end of a menu macro line

AutoCAD automatically adds a blank space (like pressing the Spacebar) at the end of a line when no other menu macro item terminator is at the end of the line.

Special Menu Macro Line Terminators
Some of the special menu macros line terminators are the semi colon (;), the backslash (\), all of the control characters, and a plus sign (+). These terminators appear at the end of a menu macro line. AutoCAD does not add a space after any of these line terminators.

    ;   =   A semicolon is interpreted as pressing the Enter key on the keyboard.

The command to execute pressing the Enter key would simply be   *ID_Enter   [&Enter];*

    \   =   A backslash causes the menu macro to pause for user input.

A sample of good use of the backslash is in the **TABLET3 section of the AutoCAD menu where you can reconfigure your digitizing tablet simply by picking the corners of each area of the tablet menu. The following section of code is taken from the ACAD.MNU.

    **TABLET 3

    ...
    [Config]_cfg 4 \\\25 9 \\\11 9 \\\9 13 \\\25 7 _y _y

Each one of the backslashes in the menu macro is a pause for you to select one of the corners on the particular tablet menu or screen pointing area. Note the number of rows and the number of columns are automatically input for you. The first "*y*" is in response to the prompt: "Do you want to respecify the Fixed Screen Pointing Area?" The second "*y*" is in response to the prompt: "Do you want to specify the Floating Screen Pointing Area?"

NOTE: Since the backslash (\) pauses for user input, whenever you need to specify a directory path you should use a forward slash (/) for any and all directory path delimiters in a menu file (for example, /directory/filename).

When you use the backslash to pause for user input, continuation of the macro will be delayed under the following conditions:

1. When input of a point is expected, the user may set Object Snap modes prior to entering the data to the prompt for a point.
2. When *X/Y/Z* point filters are used, the macro is suspended until the entire point is entered.
3. For the *Select* command only, the menu item does not resume until object selection is completed.
4. When the user responds with a transparent command, the menu macro remains suspended until the transparent command is completed and the original requested input is received.
5. When the user responds by choosing another menu item to supply options, the menu macro remains suspended until the options portion is processed and the original requested data is received.
6. When the user responds by selecting a completely different command, most likely the menu macro will be terminated.

Canceling a Command

Canceling a command is controlled by two Ctrl+Cs (^C^C) or the Escape key (Esc). Typically, a ^C^C is at the beginning of all menu macros to make sure that no previous incomplete commands are active when a new command is executed. Two Ctrl+Cs (^C^C) are common in the AutoCAD standard menu. The reasoning is that it takes two Cancels to exit the *DIM* command; therefore, it has became standard practice to use a pair of ^C^C to verify all previous commands have been terminated.

Menu Echoes and Prompts

Menu echoes and prompts are controlled by Ctrl+P (^P). When you enter commands at the keyboard, the characters you type are echoed to the Command: prompt line. The same is true when the commands are read from a menu. AutoCAD uses the special character Ctrl+P (^P) to toggle the menu echoing feature On or Off to the command line. The echoing of commands can also be controlled by the system variable *MENUECHO*.

Foreign-Language Support

The underscore (_) is the special character to enable foreign-language support for a menu. The standard AutoCAD menu makes extensive use of an underscore placed immediately before the command names to enable automatic translation between languages of AutoCAD. The portion of the *File* pull-down menu shown below has two examples: the underscore preceding the word *new* and the underscore preceding the word *open*.

```
***POP1
**FILE
ID_MnFile    [&File]
ID_New       [&New...\tCtrl + N]^C^C_new
ID_Open      [&Open...\tCtrl + O]^C^C_open
```

Control Characters in Menu Macro Items

You can enter control characters to the command string by placing items on the menu that send ASCII control characters to the system. An example of this is the Ctrl+C (^C). This sequence issues a Cancel or Escape command to AutoCAD. When entered through the keyboard the Ctrl+C (^C) is a *Copyclip* as defined by the Windows system interface.

AutoCAD recognizes the following non-alphabetic control characters:

^@	(ASCII code 0)
^[	(ASCII code 27)
^\	(ASCII code 28)
^]	(ASCII code 29)
^^	(ASCII code 30)
^_	(ASCII code 31)

The caret character (^) maps to the Ctrl key on the keyboard when used in a menu macro. You can use this feature in your menu macros to toggle settings *On* or *Off*:

```
[GridFlip]^G
[OrthoFlip]^O
[SnapFlip]^B
```

The caret is also used to cancel a command, that is, ^C^C. Remember it takes two ^Cs to exit a *DIM* command:

```
[*Cancel*]^C^C
```

You may need to send one or more characters to the command prompt line, but do not want AutoCAD to send them as final input. You can use the special characters Ctrl+H to accomplish this delay for additional user input. Ctrl+H is the ASCII code for a Backspace character. AutoCAD uses a numeric keypad in the following two examples:

```
[1]1x^H
[2]2x^H
[3]3x^H
```

When you choose one of these items, the appropriate digit is entered. Another character follows (the letter x in this case), and that character is removed by ^H (backspace). Each of these menu items ends with a control character, and AutoCAD does not add a space or Enter to such items. Thus, you can choose [2], [2], [3], [1] to construct the input 2231. Press Enter to enter the completed number.

The following sample menu items perform the function as shown above using the backslash:

```
[1]1\
[2]2\
[3]3\
```

The second method is easier to implement and in most cases produces the same result. A command issued while a menu pause is active might not function as expected, so the first method is recommended.

Support for Long Menu Macros

You can add support for long menu macros by using the Plus (+) symbol. When you have a menu macro that extends beyond one line, you can use the Plus (+) symbol as the last character in the line to continue the macros to the next line. There is no limit to the number of lines to use for a single macro. Line continuations are not preserved in the AutoCAD MNS file. The following example of a line continuation shows the toolbar *Copy* command in the **TB_MODIFY menu section of the AutoCAD menu:

```
**TB_MODIFY
ID_TbModify
ID_Erase
ID_Copy    [_Button("Copy Object",......., ICON_24_COPYOB)]$M=$(if,$(eq,+
$(substr,$(getvar,cmdnames),1,4),grip),_copy,^C^C_copy)
```

Repeating Menu Macros

Sometimes it is useful to allow the user to repeat a menu macro. To have a menu macro repeat until the user intentionally cancels the macro, put an asterisk (*) immediately following the item label. When used in combination with the *Single* option of object selection, it makes for a very friendly repeating command. The *Move'em* menu macro below enables the user to move one object or a selection set of objects to a new location in the drawing. The command automatically repeats until the user intentionally cancels the command:

```
[Move'em]*^C^Cmove Single
```

The same syntax can be beneficial for many other commands like *Copy, Erase, Stretch* (use with *Single Crossing*), *Rotate, Scale, Trim,* and *Extend.*

Menu Groups

The term ***MENUGROUP refers to a label. The MENUGROUP section contains no menu items. The MENUGROUP label must precede all menu sections in the menu. Typically, a menu has one MENUGROUP. The AutoCAD menu has one MENUGROUP that labels all other menus in the AutoCAD menu with the following syntax:

```
***MENUGROUP=ACAD
```

A simple way to refer to the POP2 (*Edit*) menu of the standard AutoCAD menu would be by MENU-GROUP=ACAD and the MENUNAME=POP2 or by MENUGROUP=ACAD and MENUALIAS=EDIT.

Now that AutoCAD uses a MENUGROUP label, a name_tag must be unique only within the menu file in which the name_tag is defined.

Menu group names combined with POP menu names or aliases provide the following functionality:

> Interactive loading and unloading of partial menus
> (see the *Menuload* command)

> Control of pull-down menu display and layout from menu macros or AutoLISP
> (see Display Control of Menu Item Labels in the AutoCAD *Customization Guide*)

Menu Swapping

AutoCAD supports menu swapping of the same type menu (AUX*n* for another, BUTTONS*n* for another, etc.) in the menu sections of an active BUTTONS, AUX, POP, SCREEN, or TABLET menu. The resulting menu content can be that from another section or submenu in the base menu, or it can come from a currently loaded partial menu (see Buttons and Auxiliary Menus).

BUTTON AND AUXILIARY MENUS

Whether you are using the Button (***BUTTONS*n*) or the Auxiliary (***AUX*n*) menus depends on your pointing device. The system mouse uses the Auxiliary menus. A digitizer puck or any other input device uses the Buttons menus. The format of the two menus is identical:

Menu_Section	=	Required
Section_Alias	=	Optional, one or more allowed
Section_Label	=	Not allowed
Name_tags	=	Not allowed
Labels	=	Not required but are allowed and used for short comments
Menu Macros	=	Needed to execute commands

Each line in the BUTTONS*n* or AUX*n* menu corresponds to a button on your puck/input device or on the mouse. A three-button mouse typically has button # 1 on the left, button # 2 on the right, and button # 3 in the middle. Button # 1 is always the PICK button and cannot be reassigned to another menu macro; therefore, no line in the menu defines the button # 1 function.

Since a mouse is considered a system pointing device, it uses the AUX*n* menus. Button # 1 (left) is the PICK button. Button # 2 (right) would execute the menu macro one line below the menu section label ***AUX1. Button # 3 would execute the menu macro on the second line below the menu section label ***AUX1.

Using a Three-Button Mouse as an Input Device

// AUX1 are executed when you press the buttons on your system input device:

```
***AUX1                  // The menu section label
[Button # 2];            // Press Button # 2 and send an Enter
[Button # 3]^C^C         // Press Button # 3 and send an Escape/Cancel
[Button # 4]^B
[Button # 5]^O
[Button # 6]^G
[Button # 7]^D
[Button # 8]^E
[Button # 9]^T
```

Using a Four-Button Digitizer Puck as an Input Device

// BUTTONS1 are executed when you press the buttons on your input device:

```
***BUTTONS1              // The menu section label
[Button # 2 ];           // Press Button # 2 and send an Enter
[Button # 3 ]$P0=SNAP $p0=*   // Press Button # 3 and call the OSNAP Pop menu
[Button # 4 ]^C^C        // Press Button # 4 and send an Escape/Cancel
[Button # 5 ]^B
[Button # 6 ]^O
[Button # 7 ]^G
[Button # 8 ]^D
[Button # 9 ]^E
[Button # 10]^T
```

AutoCAD ships with menu sections labeled BUTTONS1–4 and AUX1–4. To access the BUTTONS2–4 or AUX2–4 menu areas, use the key combinations shown in the table below.

Menu Sections and Associated Button Combinations to Execute the Menu Macro

Menu Section	Key/Button Combination
AUX1 and BUTTONS1	Button
AUX2 and BUTTONS2	Shift + Button
AUX3 and BUTTONS3	Ctrl + Button
AUX4 and BUTTONS4	Ctrl + Shift 1 Button

Under the Key/Button Combination column, the word "Button" refers to any single button on your input device (e.g., button #1 or button # 2 or button # *n*). Your pointing device can recognize as many lines in the BUTTONS or AUX menus as the number of assignable buttons.

Button and Auxiliary Menu Swapping

Button and Auxiliary menus allow menu swapping. BUTTON and AUX menu swapping gives you access to more commands at your fingertips by allowing you to dynamically swap the commands assigned to your input device. BUTTON and AUX menus recognize what are called "menu section aliases. " The format for a menu section alias is **alias. In previous versions of AutoCAD, the double asterisk (**) designated menu subsections, which are no longer supported. The following sample of code shows two methods to swap the Auxiliary menus in and out of ***AUX2. The ***AUX3 section swaps the menu sections ***AUX2 and ***AUX5 by using menu section aliases. The ***AUX4 section swaps the ***AUX2 and ***AUX5 menu sections by using the menu section names.

```
***MENUGROUP=ACAD

***AUX1
// Simple button
// if a grip is hot bring up the Grips Cursor Menu (POP 17), else send a carriage return
$M=$(if,$(eq,$(substr,$(getvar,cmdnames),1,5),GRIP_),$P0=ACAD.GRIPS $P0=*);
$P0=SNAP $p0=*
^C^C
^B
^O
^G


***AUX2
**DEFAULT
// Shift + button
[Shift+Right]^C^Cline
[Shift+Middle]^C^Cline \


***AUX3
[CTRL+Right]^C^C^P(menucmd "a2=acad.video") ^P      // Alias for ***AUX5
[CTRL+Middle]^C^C^P(menucmd "a2=acad.default") ^P   // Alias for ***AUX2
//$P0=SNAP $p0=*


***AUX4
[CTRL+Shift+right]^C^C(menucmd "A2=acad.aux5")       // Uses menu section name
[CTRL+Shift+Middle]^C^C(menucmd "A2=acad.aux2")      // Uses menu section name


***AUX5
**VIDEO
[Button #2 Right]^C^C'_zoom ;
[Button #3 Middle]^C^C'_pan
$P0=SNAP $p0=*
```

Since all of these menu sections are in the ***MENUGROUP called ACAD, you do not have to list the MENUGROUP name in the commands. The MENUGROUP name is in the macros because you can swap menu sections from the same MENUGROUP or a different MENUGROUP. Only like menu sections can be swapped (BUTTON1 with BUTTON*n* or AUX1 with AUX*n*). Another reason for including the MENUGROUP name is because when you or someone else needs to revise the menu in the future, it will be much easier to understand the macros. You can add as many uniquely named ***BUTTON and ***AUX sections as you wish in a given menu. One thing to keep in mind is that each added section takes up additional memory and system resources. Excessive menu sections in a single menu file could require more resources than some systems have available, which could lead to unpredictable results.

Special Use of the Backslash by BUTTON and AUX Menus

AutoCAD receives both the button number and the cursor screen coordinate values when you select a button on a multi-button pointing device. You can use this to your advantage when writing menu macros. The *Line* command shown previously in ***AUX2 is an example of how you can use the backslash (\) to your advantage. The syntax "[Shift+Right]^C^Cline" issues a *Line* command and waits for the user to select the beginning point of the line (From point:). The "[Shift+Middle]^C^Cline\" issues a *Line* command and uses the coordinates from the cursor location on the screen as the first point (From point:) of the *Line* command. By using the backslash to capture coordinates and pass them to the system, you can make the menu macros work for you.

PULL-DOWN AND CURSOR MENUS

Cursor Menus

The default Cursor (***POP0) menu allows quick access to the Object Snap modes and switches to the Grips Cursor (***POP17) menu when a grip is "hot" (see Chapter 23, Grip Editing). The Cursor menu allows a maximum of 499 menu items. If the Cursor menu exceeds this limit, the extra items are ignored. The Cursor menu is displayed at or near the cursor location on the graphics screen when you press button 2 through button n on your input device. If the Cursor menu is taller than the available graphics screen space, it is truncated to fit on the graphics screen. The cursor menus title is not displayed, but a title is required. Access to the Cursor menu is through the $P0=* menu command. When the Cursor menu is active, the menu bar is disabled. You must define at least one active Pull-down menu (***POPn) in order for the Cursor menu to function.

Pull-down Menus

The Pull-down (***POP1–***POPn) menus are the menus along the top of your graphics screen. This area is generally called the Menu Bar. Each Pull-down menu is allowed a maximum of 999 menu items. If the menu items in the menu exceed the maximum items allowed, the extra items are ignored. Pull-down menu items are always "pulled-down" from the menu bar. If the Pull-down menu is taller than the available graphics screen space, it is truncated to fit on the graphics screen. Access to the Pull-down menus is through the Pn$=* menu command.

Creating Pull-down and Cursor Menus

The cursor (***POP0) and the Pull-down (***POP1-POPn) menu sections control the pull-down menus. These menus are know as cascading menus. Presentation of these menus is generally in a logical order. When AutoCAD loads a menu file it uses the ***POP1-***POP16 menu sections to construct the menu bar at the top of the graphics screen. The ***POP16+n (where n =>1) menu sections are loaded, but are not active on the menu bar. You can use the *Menuload* command to load any additional Pull-down menus. If no **POPn sections are defined, AutoCAD inserts the default *File* and *Help* menus.

Special Characters for Pull-down and Cursor Menus

Character	Description
--	Item label that expands to become a separator line in the pull-down and cursor menus (when used with no other characters).
+	Continues the macro to the next line (if last character).
->	Label prefix that indicates that the pull-down or cursor menu item has a submenu (cascading menu).
<-	Label prefix that indicates that the pull-down or cursor menu item is the last item in the submenu (cascading menu).
<-<-	Label prefix that indicates that the pull-down or cursor menu item is the last item in the submenu and terminates the parent menu. (One <- is required to terminate each parent menu.)
$(	Enables the pull-down or cursor menu item label to evaluate a DIESEL string macro if $(are the first characters.
~	Label prefix that disables (grays out) a menu item.
!.	Label prefix that marks a menu item with a check mark.
&	An ampersand placed directly before a character specifies that character as the menu accelerator key in a pull-down or cursor menu label (the character following the "&" appears with an underscore). For example, S&ample displays as Sample.
/c	Specifies the menu accelerator key in a pull-down or cursor menu label. For example, /aSample displays with an underscore under the "a" in Sample.
\t	Specifies that all label text to the right of these characters is pushed to the right side of the menu.

NOTE: The only non-alphanumeric characters that can be used as the first character in a menu label are those listed above. Non-alphanumeric characters not listed in the previous table are reserved for future use as special menu characters.

Pull-down and Cursor Menu Syntax

Menu_Section	=	Required
Section_Alias	=	Optional, One or more allowed
Section_Label	=	Required (For Pull-down menus this is the heading shown in the menu bar. Cursor menus do not display this required label).
Name_tags	=	Not required, but adds functionality (See Name Tags)
Labels	=	Command Label the user sees on the menu
Menu Macros	=	Needed to execute commands

The following example illustrates the syntax that is used to create a pull-down or cursor menu:

```
***POP6
**TOOLS
ID_MnTools      [&Tools]                          // No Menu Macro Associated with Label
ID_Spell        [&Spelling]^C^C_spell
ID_MnOrder      [_->Display &Order]
ID_DrawordeF    [Bring to &Front]^C^C^P(ai_draworder "_f") ^P
ID_DrawordeB    [Send to &Back]^C^C^P(ai_draworder "_b") ^P
                [--]
ID_DrawordeA    [Bring &Above Object]^C^C^P(ai_draworder "_a") ^P
ID_DrawordeU    [<-Send &Under Object]^C^C^P(ai_draworder "_u") ^P
```

In the example above, ***POP6 is the *Menu_Section* label, **TOOLS is the *Section_Alias*, and ID_MnTools is the *Name_tag* for the *Section_Label*. The [&Tools] entry is the heading shown in the menu bar. The rest of the lines are of the following format:

Name_Tag	Label	Menu Macro
ID_Spell	[&Spelling]	^C^C_spell

The name_tags in the Pull-down menu can be used to enable and disable the entire menu or various commands within the menu. The label [&Spell] causes *Spell* to be displayed as the menu bar heading, and the letter S is underscored to indicate that Alt+S is the accelerator key for this menu item. As shown in the sample above, menu macros cannot be associated with Pull-down menu labels.

Cascading Submenus

The special characters used in Pull-down and Cursor menu labels to indicate the beginning and end of submenus are _->, and <_. These characters control the cascading of the menus and terminate all parent menus. They are always the first characters in the item label. Review of the **TOOLS menu above shows that a submenu begins with the label [_->Display &Order] and ends with the label [<-Send &Under Object].

You must have one returning special character (<_) for each and every submenu level you create; there-fore, you may need to have multiple special characters (<_<_) on one menu item label as shown below.

```
ID_PurgeShp     [<_<_&Shapes]^C^C_purge _sh
```

Create Menu Label Separator Lines with Two Hyphens

In order to make reading of menus easier, AutoCAD uses the following format and two hyphens to create menu item separators [--]. The line created by the hyphens extends the width of the menu. You cannot select separator lines from the menu. Any menu macro assigned to a separator line is ignored.

Base and Partial Pull-down Menus

Release 14 supports the theory of base (e.g., AutoCAD menu) and partial menus (e.g., your menu) working in combination. What this means to you is that you can create a menu with only your menu macros and commands, then use it in combination with any other menu that is supported by AutoCAD. In the past, if you wanted a complete working menu, you had to create your commands and cut and paste them to a copy of the AutoCAD menu. Now you can add or remove your partial menus anytime you wish without affecting the standard AutoCAD menu.

You can insert Pull-down menus with the following syntax:

    $Pn=+menugroup.menuname

In the following example, the menu group name is MYGROUP and the menu name is MYPOP; therefore, the code to insert the partial menu on the standard AutoCAD menu bar between the MODIFY and HELP menu sections would be:

    $P9=+mygroup.mypop

The same thing can be done with the AutoLISP function *menucmd*:

    (menucmd "P9=+mygroup.mypop")

Similar syntax can be used to remove MYPOP from the menu bar:

    $P9=.

The syntax $P9 =- in a menu macro removes whatever menu is in the ninth position on the menu bar. A safer method to remove the expected menu from the menu bar is to use an AutoLISP command to do the job:

    (menucmd "Gmygroup.mypop=-")

The above command searches for the menu group MYGROUP and the menu MYPOP and removes only this menu if it is found; otherwise, it takes no action.

AutoCAD uses the ***POP1 through ***POP16 for menu swapping. You may wish to always have your menus placed in the same location on the menu bar. One method that works is to always place your partial menus with the same statement using a P*n* larger than the location where your menu will be placed. In the following command, any partial menus are loaded in the farthest position to the right on the menu bar.

Place partial menu on menu bar:	(menucmd "P100 =+mygroup.mypop")
Remove specific partial menu from menu bar:	(menucmd "Gmygroup.mypop=-")

By using a combination of the two commands above you can safely place and remove partial menus to and from the menu bar.

IMAGE TILE MENUS

Image tile menus make use of individual slides and slide libraries to allow you to see an image of the items you are selecting. AutoCAD displays image tiles in groups of 20, along with a scrolling list box containing the associated slide names or label text. Image tile submenus may be of unlimited length. When the image tile submenu contains more than 20 images, AutoCAD provides *Next* and *Previous* buttons to enable you to scroll through the images.

One use for image tile menus is for placing blocks. You may not know the name of a *Block* you need to place in the drawing file, but often you will know what the item looks like. By combining an image tile menu of the *Blocks* you often place in drawing files with the menu macros to *Insert* the *Blocks*, block insertion becomes easy.

One example of the use of the image tile menus can be found by selecting the *Draw* pull-down, then *Surfaces,* then *3D Surfaces....* This selection displays an image tile menu titled *3D Objects* (Fig. 46-4). When you select one of the images, the related command is invoked.

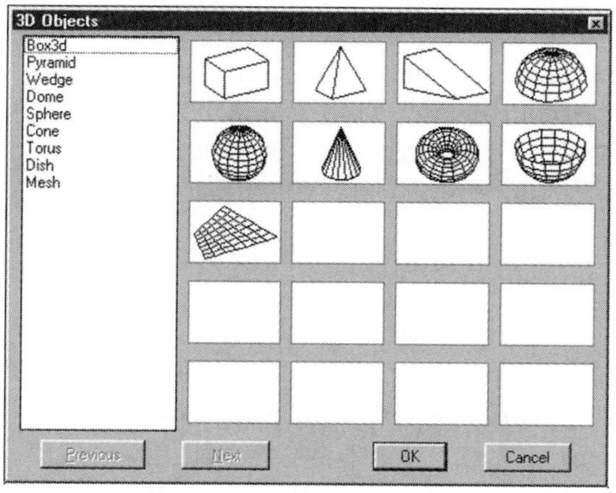

Slide Preparation

Any slide generated by AutoCAD can be used for an image tile (see Chapter 44, Slides and Script Files, for more information). In order to present the best image and for slides to be consistent, you should follow these guidelines:

Figure 46-4

Keep it simple	Images should be easily recognizable. If the image is to represent a complex object, you may wish to create a simple form of the object to use for the image tile.
Fill the image box	It is easier to recognize the images that fill the image box. Images are displayed with an aspect ratio of 1.5:1 (1.5 units wide by 1 unit high). If you create your slide in a floating viewport that has this aspect ratio, you will be assured the images will look the same on your slides.
Shade solids	Solid filled areas like wide *Plines* and filled solids display as outlines unless you issue a *Shade* command before making your slides.

To set up a correctly proportioned viewport for making a slide, start with a drawing with no views:

```
Command: tilemode
New value for TILEMODE <1> 0
Entering Paper space. Use MVIEW to insert Model space viewports.
Command: mview
ON/OFF/Hideplot/Fit/2/3/4/Restore/<First Point>: 0,0
Other corner: 3,2
Command: zoom
All/Center/Dynamic/Extents/Previous/Scale(X/XP)/Window/<Realtime>: e
Command: mspace
```

You can create consistent slides for your image tile menu by issuing the commands from a script file. Open an ASCII text editor (for example, Windows Notepad) and type the following lines as shown:

```
undo
mark
tilemode
0
mview
0,0
3,2
zoom
e
mspace
mslide

undo
back
```

The line after the command *Mslide* is blank to save the slide file using the same name as the drawing file with an extension of SLD. Save this text file to a name (MAKESLD.SCR, for example). Copy the script file to your AutoCAD support directory. Start AutoCAD and open a drawing. To create a slide, type the command *Script* at the command prompt. At the prompt for a script file to load, type in the name of your script file: MAKESLD. AutoCAD issues the commands and creates a slide file with the same name as your drawing file, then places it in the same directory where your drawing file is located. Your drawing file will be returned to the state it was before you ran the script file.

The format for the image tile menu section is:

Menu_Section	=	Required
Section_Alias	=	Optional, one or more allowed
Section_Label	=	Required
Name_tags	=	Not allowed
Labels	=	Not required, but are used for the text in the list box
Menu Macros	=	Needed to execute commands

Image tile item labels that appear in the list box may be the slide file name or text labels. The image displayed in each image tile box may be a single slide, part of a slide library, a combination of both, or can even be from different slide libraries. The slide file name in the menu macros should be the same as the slide file name used at the command line for the *Vslide* command.

A number of image labeling options are available. Often the scrolling list box has the name of each slide in the list. You can input up to 19 characters for labels in lieu of using the slide file names.

Image Tile Menu Labeling Options

[sldname]	The sldname is displayed in the list box and the sldname image is displayed.
[sldname,labeltext]	Labeltext is displayed in the list box and the sldname image is displayed.
[sldlib(sldname)]	Sldname is displayed in the list box and the sldname image from sldlib is displayed.
[sldlib(sldname,labeltext)]	Labeltext is displayed in the list box and the sldname image from sldlib is displayed.
[blank]	When you use the word blank as an image label, a separator line is displayed in the list box and a blank image is displayed.
[(space)labeltext]	When the item label begins with a space, the text supplied as labeltext is displayed in the list box and no image is displayed. This allows you to include simple commands (words) like EXIT in the image tile without having to create slides.

One image tile menu can display another image tile menu. Since image tile menus are called sequentially, and not nested, there is no limit to the complexity of the menus you can create. The asterisk (*) character, for repeating menu macros, is not supported in the image tile menus.

Sample of an Image Tile Menu from the AutoCAD Standard Menu

```
***IMAGE
**IMAGE_3DOBJECTS
[3D Objects]
[acad(Box3d,Box3d)]^C^Cai_box
[acad(Pyramid,Pyramid)]^C^Cai_pyramid
[acad(Wedge,Wedge)]^C^Cai_wedge
[acad(Dome,Dome)]^C^Cai_dome
```

You must first load an image tile menu before you can display it:

```
$I=menugroup.menuname
```

You can then display the currently loaded image tiles menu by:

```
$I=*
```

The syntax to load and display the image menu YOURBLKS from a partially loaded menu group YGROUP is:

```
$I=ygroup.yourblks $I=*
```

Below is an example of an image tile menu used to create office furniture layouts using individual slides located in a directory already listed in the search path:

```
***MENUGROUP=OFFICE

***IMAGE
**IFURN
[Furniture]
[Chair]^C^C_insert chair
[ArmChair]^C^C_insert armchair
[Desk]^C^C_insert desk
```

Another example follows of an image tile menu used to create office furniture layouts using slides complied in a slide library called FURN.SLB located in a directory already listed in the search path:

```
***MENUGROUP=OFFICE

***IMAGE
**IFURN
[Furniture]
[furn(Chair)]^C^C_insert chair
[furn(armchair]^C^C_insert armchair
[furn(Desk)]^C^C_insert desk
```

Image Tile Menus Support Pull-down Partial Menus

As of Release 13 c3, image tile menus are supported in pull-down partial menus. The format for loading a partial image tile menu from a pull-down menu is:

    [Menu Item Label]^C^C$I=menugroup.menuname

The following menu macro loads and displays a partial image tile menu called IFURN, which is in the menu group file called OFFICE:

    [Furniture]^C^C$i=office.ifurn $i=*

SCREEN MENUS

In early versions of AutoCAD, the screen menu was the main menu interface for AutoCAD users. The screen menus section controls the text listing of commands along the right side of your graphics screen. When you initially install AutoCAD Release 14, the screen menu is disabled. You can enable the Screen menu by selecting the *Tools* pull-down menu, then *Preferences,* the *Display* tab, and check *Display AutoCAD screen menu in drawing window.*

The format for the screen menus section of the menu is:

Menu_Section Name	=	Required
Submenu Sections	=	Used extensively
Start Number	=	*Optional*
Section_Label	=	Not allowed
Name_tags	=	Not required and serve no purpose
Labels	=	Command label the user sees in the screen menu
Menu Macros	=	Needed to execute commands

The following information is from the beginning of the screen menus section of the ACAD.MNU file.

```
//
//    AutoCAD Screen Menus
//
//    There are two types of screen menus: command menus and options menus
//    Command menus provide access to the lists of AutoCAD commands.
//    Options menus provide access to the options available for individual commands.
//
//    There are 22 lines between menu titles.  This is one method for assuring that each time
//    that a menu is called that it overwrites the previous menu.
//
//    The organization of the command menus generally follows the organization of the pull-down
//    menus.
//    A command has a screen menu item only if it has a pull-down menu item.
//    Command menus have, as much as possible, the same name as the equivalent pull-down menu.
//    Command menus have, as much as possible, the same items in the same order as the pull-down
//    menus.
//
```

```
//    Command menus generally use the command name while pull-down menus offer a more
//    descriptive title.
//    Items in command menus that call other command menus are in upper case.
//
//    Command menu names start with a number and are located after the special menus.
//
***SCREEN                        Menu Section
**S                              Submenu
[AutoCAD ]^C^C^P(ai_rootmenus) ^P
[* * * * ]$S=ACAD.OSNAP
[FILE    ]$S=ACAD.01_FILE
[EDIT    ]$S=ACAD.02_EDIT
[VIEW 1  ]$S=ACAD.03_VIEW1
[VIEW 2  ]$S=ACAD.04_VIEW2
[INSERT  ]$S=ACAD.05_INSERT
[FORMAT  ]$S=ACAD.06_FORMAT
[TOOLS 1 ]$S=ACAD.07_TOOLS1
[TOOLS 2 ]$S=ACAD.08_TOOLS2
[DRAW 1  ]$S=ACAD.09_DRAW1
[DRAW 2  ]$S=ACAD.10_DRAW2
[DIMNSION]$S=ACAD.11_DIMENSION
[MODIFY1 ]$S=ACAD.12_MODIFY1
[MODIFY2 ]$S=ACAD.13_MODIFY2

[HELP    ]$S=ACAD.14_HELP




[ASSIST ]$S=ACAD.ASSIST
[LAST   ]$S=ACAD.
```

Screen submenu labels use the format:

> Format for submenu label: **menuname [startnum]
>
> Sample submenu label: **01_FILE 3

The maximum length of a screen menu name is 33 characters. It can include the special characters dollar sign ($), hyphen (-), and underscore (_). A submenu label must be on a line by itself and cannot have embedded blanks. You can specify a starting line for your submenu using the optional start number. The "3" in the sample submenu label indicates the 01_FILE submenu will begin on line 3 of the screen menu area. When the word FILE is selected on the screen menu, the *FILE* menu, which begins on line 3 of the graphics screen, overwrites the AutoCAD main screen menu. The main menu selections (*AutoCAD)* and the second line (* * * *) stay in their respective locations on the graphics screen.

If a label is provided, the first eight characters of the label are displayed in the appropriate screen menu box. Any additional characters can serve as comments.

NOTE: The maximum number of menu items depends on your system. You can retrieve the number of screen menu boxes with the *SCREENBOXES* system variable.

Your graphics screen has a fixed number of lines for you to work with when developing screen submenus. A good rule is not to exceed 22 items in a single screen submenu. This allows the menu to be presented properly on both low and high resolution monitors. AutoCAD starts all screen submenus at or below line 3 to retain the first 2 lines (*AutoCAD* and * * * *) of the screen menu. The menu macro associated with the first line recalls the AutoCAD main screen menu. The second line of the screen menu calls the OSNAP submenu. When you select a submenu item on the screen menu, AutoCAD generally replaces lines 3 through 24 of the screen menu with the referenced screen submenu. If you create submenus that exceed the visible area of your display monitor, the menu items that are below the bottom viewing edge of the screen are inaccessible.

Control Submenu Swapping with the *MENUCTL* System Variable

You can control the automatic swapping of screen submenus with the *MENUCTL* system variable. When *MENUCTL* is set to 1 (On), the screen menus swap corresponding to selections made on the menus.

TABLET MENUS

There are up to four configurable tablet menu areas. The following tablet menu areas are defined in the standard AutoCAD menu.

***TABLET1 = 25 columns x 9 rows	Reserved for your customization
***TABLET2 = 11 columns x 9 rows	Commands defined by AutoCAD
***TABLET3 = 9 columns x 13 rows	Commands defined by AutoCAD
***TABLET4 = 25 columns x 7 rows	Commands defined by AutoCAD

The format for the tablet menus section of the menu is:

Menu_Section Name	=	Required
SectionsAlias	=	Optional
Section_Label	=	Not allowed
Name_tags	=	Optional, but serve no purpose
Labels	=	Optional, may be used for comments
Menu Macros	=	Needed to execute commands

Tablet area 1 [A-1]–[I-25] has been reserved for you to customize. This means there are 225 command lines on the standard tablet area 1 for you to customize. If you need more boxes, you may define up to a maximum of 32,766 items for each of the tablet areas.

The tablet boxes are ordered from left to right across the rows of the tablet, then top to bottom; therefore, A-1 (row A, column 1) is the upper-left box in the tablet 1 area and I-25 is the lower-right box in the tablet 1 area. All of the boxes are of equal size in a particular tablet area.

In the standard AutoCAD menu file, the 225 labels following the ***TABLET1 Menu section name and the **TABLET1STD section alias correspond to the 225 template grid boxes in tablet area 1 on your digitizing tablet:

```
//    Begin AutoCAD Tablet Menus
//
//    This is the TABLET1 menu.  You may put your own
//    macros and menu items here in these spaces.
//    All of the "blank" line items actually contain a
//    backslash so that no command is issued when you pick any
//    of them from the tablet.  Remove them if you want an Enter
```

```
//   to happen when they are selected, or place your own
//   macros in their place.

***TABLET1
**TABLET1STD
[A-1]\
[A-2]\
[A-3]\
```

Modifying any lines following box [I-25] is not recommended.

STATUS LINE HELP

In the standard AutoCAD menu each command associated with the pull-down and toolbar menus echoes a Help message to the Status line. These messages occur when items are highlighted (when you pass your cursor over the various commands). When an item is highlighted, the name_tag for that item is queried for a matching entry in the ***HELPSTRINGS section of the menu. When a match occurs, the Help string message contained within the label is displayed on the Status line of the graphics screen.

The menu has a special section for entering HELPSTRINGS. A sample of the ACAD.MNU menu ***HELPSTRINGS sections follows:

```
//   IDs are listed in alphabetical order.
//   Helps strings end with a colon, two spaces and the name of the command.
//

     ***HELPSTRINGS
ID_3darray   [Creates a three-dimensional array:  3darray]
ID_3dface    [Creates a three-dimensional face:  3dface]
ID_3dmesh    [Creates a free-form polygon mesh:  3dmesh]
```

The format for entering HELPSTRINGS is:

name_tag [Help String: Command name]

By adding Help string messages to your commands as you create custom menus, you provide first line Help support.

Help strings can be extremely useful when you are searching toolbar buttons for a particular AutoCAD command.

ACCELERATOR KEYS

Two methods of user-defined accelerator keys are supported by AutoCAD. The format for the first method maps a key sequence to an existing menu item:

```
name_tag [Modifier(s)+accelerator key]

// Map CTRL+Z key combination to the menu item UNDO
ID_U   [CONTROL+"Z"]
```

The format for the second method maps a key sequence to a command string:

    [Modifier(s)+accelerator key] command string

    // Toggle Orthomode
    [CONTROL+"L"]^O

The ***ACCELERATOR menu section from the ACAD.MNU is shown below:

    //
    //   Keyboard Accelerators
    //
    //   If a keyboard accelerator is preceded by an ID string that references a menu item
    //   in a pull-down menu, then the keyboard accelerator will run the command referenced
    //   by that menu item.
    //
    ***ACCELERATORS
    // Toggle PICKADD
    [CONTROL+"K"]$M=$(if,$(and,$(getvar,pickadd),1),'_pickadd 0,'_pickadd 1)
    // Toggle Orthomode
    [CONTROL+"L"]^O
    // Next Viewport
    [CONTROL+"R"]^V
    // ID_Spell   ["\"F7\""]   Item is commented out
    // ID_PanRealti   ["\"F11\""]   Item is commented out
    // ID_ZoomRealt   ["\"F12\""]   Item is commented out
    ID_Copyclip   [CONTROL+"C"]
    ID_New   [CONTROL+"N"]
    ID_Open   [CONTROL+"O"]
    ID_Print   [CONTROL+"P"]
    ID_Save   [CONTROL+"S"]
    ID_Pasteclip   [CONTROL+"V"]
    ID_Cutclip   [CONTROL+"X"]
    ID_Redo   [CONTROL+"Y"]
    ID_U   [CONTROL+"Z"]

The valid modifiers are the Control and/or Shift keys. AutoCAD does not distinguish between the left and right sides of the keyboard; therefore, the left Control or Shift keys are interpreted the same as the right Control or Shift keys. You can use Control+Shift key combinations to define accelerator keys.

Special Virtual Keys

The following table is reprinted from the AutoCAD Release 14 *Customization Guide*. (Functions keys must be enclosed in quotation marks.)

String	Description	Exceptions
"F1"	F1 key	Although the F1 key can be assigned a menu macro, this is discouraged because this key is generally associated with Help. Using a modifier with this key is acceptable.
"F2"	F2 key	Unmodified, this key toggles the state of the text window
"F3"	F3 key	Unmodified, this key runs *OSNAP*
"F4"	F4 key	Unmodified, this key toggles *TABMODE*
"F5"	F5 key	Unmodified, this key toggles *ISOPLANE*

"F6"	F6 key	Unmodified, this key toggles *COORDS*
"F7"	F7 key	Unmodified, this key toggles *GRIDMODE*
"F8"	F8 key	Unmodified, this key toggles *ORTHOMODE*
"F9"	F9 key	Unmodified, this key toggles *SNAPMODE*
"F11"	F11 key	None
"F12"	F12 key	None
"INSERT"	INS key	None
"DELETE"	DEL key	None
"ESCAPE"	ESC key	Although the Esc key can be assigned a menu macro, this is discouraged because this key is generally associated with Cancel. Ctrl+Escape and Ctrl+Shift+Escape cannot be assigned a menu macro; these sequences are controlled by Windows. Using the Shift modifier with this key is acceptable.
"UP"	UPARROW key	Must be used with the Control modifier
"DOWN"	DOWNARROW key	Must be used with the Control modifier
"LEFT"	LEFTARROW key	Must be used with the Control modifier
"RIGHT"	RIGHTARROW key	Must be used with the Control modifier
"NUMPAD0"	0 key	None
"NUMPAD1"	1 key	None
"NUMPAD2"	2 key	None
"NUMPAD3"	3 key	None
"NUMPAD4"	4 key	None
"NUMPAD5"	5 key	None
"NUMPAD6"	6 key	None
"NUMPAD7"	7 key	None
"NUMPAD8"	8 key	None
"NUMPAD9"	9 key	None

NOTE: F10 is used by the Windows operating system as an alternative to Alt and is therefore not user-configurable.

SUMMARY OF CHANGES TO THE MENUS OF RELEASE 14

1. Submenus are not allowed in the menu sections:

    ***BUTTONS
    ***AUX
    ***POP
    ***TABLET

    Any old submenus in these sections should be converted to full menus.

2. Since the plain keys are used for command line recall and editing features, only the Control and Shift versions of the following numeric pad keys are available as accelerator keys:

    UP
    DOWN
    LEFT
    RIGHT
    HOME
    END

R14

3. AutoCAD is shipped with one standard base menu and two optional partial menus that are installed with a full installation:

    BASE MENU   = \ACADR14\SUPPORT\ACAD.MNU   (Standard ACAD Menu)
    PARTIAL MENU = \ACADR14\SUPPORT\INET.MNU   (Internet Tools Menu)
    PARTIAL MENU = \ACADR14\BONUS\CADTOOLS\BONUS.MNU (Bonus Tools Menu)

4. Changes in the ways menus load:

    When you start AutoCAD and open a drawing file, the last menu used is loaded. AutoCAD keeps track of the last menu used and loads that menu along with any partial menus that were loaded. In addition, AutoCAD no longer reloads menus when you change drawings.

5. Right-click menus

    New and revised context-sensitive right-click (right mouse button) menus are:

    Real Time Pan and Zoom functions
    Plot Preview functions
    Grip editing functions when a grip is "hot"
    OLE object functions
    Command line paste functions

6. Toolbar customization

    A more direct interface for access and customization of tool bars is provided in Release 14. (See Chapter 45, Basic Customization).

CHAPTER EXERCISES

1. **Buttons and Auxiliary Menus**

    Using a copy of the standard AutoCAD menu, create Buttons or Auxiliary menus to do the following:

    Ctrl+Button # 2 executes a *Zoom Real Time* command
    Shift+Button # 2 executes a *Zoom Extents* command
    Ctrl+Button # 3 executes a *Copy* command
    Shift+Button # 3 executes a *Move* command
    Assign any menu macro you wish to the combination Ctrl+Shift+Button # 2
    Assign any menu macro you wish to the combination Ctrl+Shift+Button # 3

    (HINT: Remember Button # 2 on a three-button mouse is normally the right button.)

2.   **Partial Pull-down Menus**

Create a pull-down menu titled **Layouts**. In **Layouts,** create menu macros for the following AutoCAD commands:

<u>Draw</u>   <u>Edit</u>

Mline   *Copy*
Circle   *Erase*
Dtext   *Move*
Mtext   *Rotate*
Rectangle

3.   **Partial Image Tile Menus**

A.   ***Open*** the **OFFICE** drawing that you created in Chapter 21 Exercises. Use the ***Wblock*** command to create .DWG files of each of the following *Blocks* from the drawing:

CHAIR   DESK   FILECAB   TABLE

B.   Create a slide of each of the furniture items (*Wblocks*). Next, create a slide library of the slides and name the file **FURN.SLB.** (HINT: Use the file \ACADR14\SUPPORT\SLIDELIB.EXE to create the slide library). The slide library file (FURN.SLB) <u>must</u> be in your AutoCAD Release 14 Support directory to function correctly as image tiles.

C.   Add a menu item to the pull-down menu you created in Exercise 2 (this chapter) called **FURNITURE** and have that menu item display the four furniture slides in an Image Tile menu titled **Office Furniture.** Once the images are displayed, you should be able to select any of the office furniture objects and insert them in your drawing at a scale of 1:1. You should not be prompted for the X and Y scale factors—only the rotation angle and insertion point—as you insert the objects into the drawing.

4.   **Partial Tablet Menu**

Create *Block* insertion menu macros for the four *Blocks* in Exercise 3. Add the menu macros to the drawing tablet menu area 1 in the following locations:

[A-5] CHAIR   [B-5] DESK   [B-6] FILECAB   [C-5] TABLE

Be prepared to turn in your menu macros in ASCII text format and a plot of the TABLET1 menu area you would use to *Insert* the *Blocks* into a drawing file.

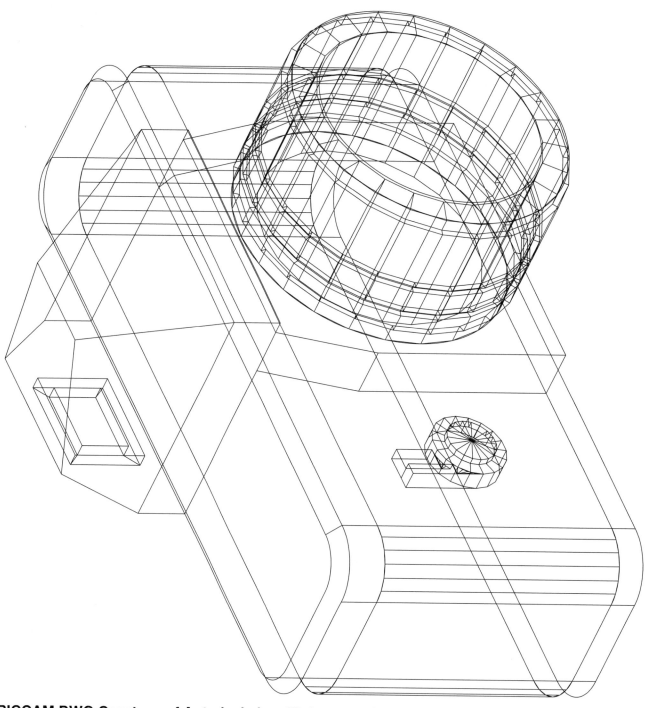

RICCAM.DWG Courtesy of Autodesk, Inc. (Release 14 Sample Drawing)

47

BATCH PLOT UTILITY AND BONUS TOOLS

Chapter Objectives

After completing this chapter you should:

1. be able to use the Batch Plot Utility to select and plot several drawings unattended;

2. be able to specify which layers and what drawing areas to plot with the batch utility;

3. be able to generate text "stamps" and log files when plotting drawings in batch;

4. be able to install the Bonus Tools from the Release 14 CD-ROM, load the Bonus menus, and invoke the Bonus toolbars;

5. be able to use the Bonus Text commands;

6. be able to use the Bonus Layer commands;

7. be able to use the Bonus Standard toolbar commands;

8. be able to use miscellaneous Bonus commands from the pull-down menus and the command line.

CONCEPTS

AutoCAD Release 14 is shipped with many useful bonus tools and a batch plotting feature. The Batch Plot Utility is installed automatically (if the *Typical* or *Full* installation is selected). The Bonus Tools are installed only if you choose the *Full* option when you first install AutoCAD or use the Setup utility at a later time and select the Bonus Tools for installation.

The Batch Plot Utility is very useful in an office or laboratory where many drawings must be printed or plotted. This utility allows you to specify multiple drawings to be printed or plotted unattended. For example, a set of drawings can be set up to be printed at night so they are ready for you when you return in the morning.

Bonus Tools include several utility commands that have a specialized or streamlined usefulness. These commands were not fully quality tested in time to make it in the documented Release 14 command set. However, many very powerful and useful features are available, including layer utilities and a Layer Manager, text utilities, modify commands, draw commands, and miscellaneous tools.

BATCH PLOT UTILITY

AutoCAD Release 14's Batch Plot Utility is useful in offices or laboratories where many prints or plots are made because a set of drawings can be assigned for plotting as a "batch" without human intervention. For other applications, such as when a number of plots are required as a complete project set, the entire set of drawings can be saved to a list to simplify plotting the same set again at a later time. Other specifications about the drawings to be plotted can be saved, such as which layers should be included in the plot, drawing areas to plot, time and date stamps, and the print/plot devices to use. A plot test can be used as a "dry run" to locate any potential problems before running the batch. If errors occur during an unattended plot run, an error log is kept so problems can be tracked.

The Batch Plot Utility requires AutoCAD for operation. Invoking the Batch Plot Utility in turn starts its own dedicated AutoCAD session used just for plotting.

You can start the Batch Plot Utility by three methods:

1. Use the Windows Start Menu, choose Programs, then locate and select Batch Plot Utility in the AutoCAD R14 program group.

2. Click the *Batch Plot Utility* shortcut icon in the AutoCAD R14 program group (Fig. 47-1).

3. Use Windows Explorer to locate the EBATCHP.EXE file in the AutoCAD R14\SUPPORT\EBATCHP folder.

Figure 47-1

Once the Batch Plot Utility is started (with its own AutoCAD session), the *AutoCAD Extended Batch Plot Utility* dialog box appears (Fig. 47-2). Five tabs in this dialog box allow you to specify parameters with respect to these features: *File, Layers, Plot Area, Logging,* and *Plot Stamping*.

Figure 47-2

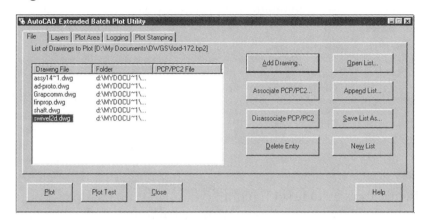

File

In this tabbed section you indicate which drawings you want to include in the "batch" of drawings to plot. Selected drawings appear in the *List of Drawings to Plot*.

Add Drawing...
Use this button to select drawings to include in the list.

Save List As...
When a group of drawings has been indicated for plotting, you can save the list for future use (with *Open List…*). The file is saved as an ASCII file with a .BP2 file extension.

Open List...
Use this button to select a list of drawings (previously saved) to include in the list of drawings to plot.

Append List...
This option appends (adds) a previously saved list (.BP2 file contents) to the *List of Drawings to Plot* list. Using this method, several "sets" of drawings can be plotted in batch.

New List...
Use this button to clear the *List of Drawings to Plot* list. This action may be necessary in preparation for creating a new list or selecting a saved list for plotting.

Associate PCP/PC2...
Use this button to select and associate a PCP or PC2 file with the highlighted drawing(s) in the list. PCP and PC2 files specify settings previously made in the *Pen Parameters* section of the *Print/Plot Configuration* dialog box.

A .PCP (partial plot configuration) file stores the pen settings (made in *Pen Parameters* section of the *Print/Plot Configuration* dialog box), but the information is device independent, so a .PCP file can be used for multiple print/plot devices. A .PC2 (complete plot configuration) file is a device-complete file. The file keeps the *Pen Parameter* information and also keeps the name and all other driver and configuration specifications for a <u>particular device</u>, so when a .PC2 file is loaded, it automatically <u>installs and config-</u><u>ures the device and makes it the current plotter or printer</u>. The .PC2 files are useful for batch plotting because different printers and plotters can be controlled while AutoCAD is unattended. See Chapter 14, Plotting, for information on creating .PCP and PC2 files.

Disassociate PCP/PC2...
Select this button to unattach a previously associated .PCP or .PC2 file to the highlighted drawing(s).

Layers

This tab of the AutoCAD Extended Batch Plot Utility dialog box allows you to indicate which layers should be included in <u>each</u> of the drawings to plot (Fig. 47-3). When you select a drawing in the *List of Drawings to Plot,* the drawing is loaded into AutoCAD and its list of layers appears in the box on the right. The selected drawing's layer settings (*On, Off, Thawed, Frozen*) are read and displayed as *On* or *Off* in the *Plotting On/Off* column (a drawing's *Frozen* layers are displayed as *Off* for plotting).

Figure 47-3

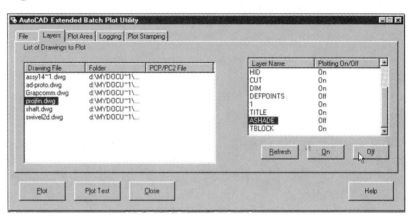

Layer Name **Column**
Lists the layers in the selected drawing from the *List of Drawings to Plot.*

On/Off **Column**
Lists the state of the layers as read by the current settings of the highlighted drawing or as changed using the *On* or *Off* buttons.

On/Off
Use these buttons to specify if you want the highlighted layer to be plotted (*On*) or not (*Off*). The drawing's layer settings are changed only temporarily while the drawing is plotted. The settings you specify in this dialog box do not permanently change the settings in the drawing file.

Plot Area

The *Plot Area* tab (Fig. 47-4) allows you to specify what area of the selected drawing (from the *List of Drawings to Plot*) you want to appear in the plot. You can also specify the plot scale for each file.

Figure 47-4

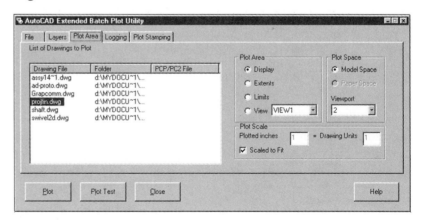

Plot Area **Section**
In this section, select to plot the *Display, Extents, Limits,* or *View* (if *Views* have been saved in the drawing). These options are identical to those in the *Print/Plot Configuration* dialog box (see Chapter 14, Plotting).

Plot Space **Section**

Here you indicate if you want to plot from *Model Space* or *Paper Space* (if paper space is enabled in the selected drawing). If viewports are used in the drawing, you can select from available viewports to plot in the *Viewport* drop-down list.

Plot Scale **Section**

This section operates identically to the same section in the *Print/Plot Configuration* dialog box (see Chapter 14, Plotting). Enter the desired plot scale in the edit boxes or select *Scaled to Fit* for each drawing in the list.

Logging

To locate problems that may have occurred during an unattended batch plot, a log file can be checked. First, you must *Enable Logging* before running the batch to plot. The log is kept for all drawings in the batch to plot. All options selected for the plot as well as the current plotting device are listed in the log file. An example log is shown below:

Figure 47-5

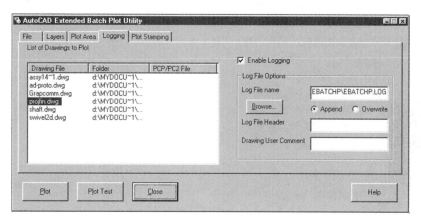

```
+++ Batch Plot Session started at
09-16-1997, 16:33:17 by jim
*** Starting batch plot of <d:\MYDOCU~1\DWGS\Grapcomm.dwg> at 09-16-1997, 16:33:18
    Using default layers in drawing
    Setting drawing to plot display
    Setting drawing to model space, viewport #-1
    Setting up Plot Stamping
    Plot Device: Plotting to current plotting device: Default System Printer
    DONE: batch plot of <d:\MYDOCU~1\DWGS\Grapcomm.dwg> at 09-16-1997, 16:33:23
*** Starting batch plot of <d:\MYDOCU~1\DWGS\swivel.dwg> at 09-16-1997, 16:33:23
    Using default layers in drawing
    Setting drawing to plot display
    Setting drawing to model space, viewport #-1
    Setting up Plot Stamping
    Plot Device: Plotting to current plotting device: Default System Printer
    DONE: batch plot of <d:\MYDOCU~1\DWGS\swivel.dwg> at 09-16-1997, 16:33:32
```

Log File Options

In this section specify the *Log File* name and path you want to use (the default is "AutoCAD R14\Support\Ebatchp\Ebatchp.log"). Alternately, use the *Browse* button to locate a path. Select *Append* to add a new log file to a previous one or select *Overwrite* to have a new log file replace the last one (of the same name). A *Log File Header* can be specified—this places any text you want at the beginning of each new log file (helpful when you use the *Append* option). An additional *Drawing User Comment* can be added if desired.

R14

Plot Stamping

This tab provides options for you to have an informational text "stamp" plotted on each drawing (Fig. 47-6). For example, the drawing name, date, and time of plot can be added to each plot for tracking and verification purposes.

Figure 47-6

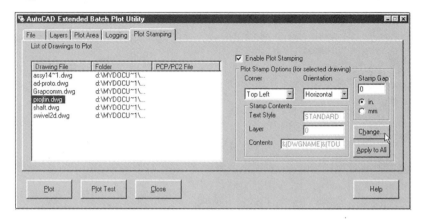

Enable Plot Stamping

This option enables plot stamping for the entire *List of Drawings to Plot.* Although you can specify different options for each drawing, all or none of the drawings get stamped.

Plot Stamp Options

Set these options for <u>each</u> drawing. *Corner* specifies the location on the drawing for the stamp to appear. Select from *Vertical* or *Horizontal Orientation. Stamp Gap* is the gap from the edge of the maximum plot area for the stamp to appear.

Apply to All

If you want to apply the same stamp contents to every drawing, select this button. Each drawing will be plotted with the text and options shown in the *Plot Stamp Options* section.

Stamp Contents, Change...

The *Stamp Contents* section indicates the current settings for *Text Style, Layer,* and text *Contents* of the stamp. Select *Change...* to change these contents. Selecting *Change...* produces the *Plot Stamp Contents* dialog box.

Plot Stamp Contents

Selecting the *Change...* button (see above) produces this dialog box (Fig. 47-7). Here you can select from several *Text Styles* and assign the *Layers* that the plotting utility uses to temporarily place the text. Several *Pre-defined Fields* are available for your selection. The *Insert* button inserts the highlighted predefined choice into the edit box. If you want to insert text from a file, select the *File* button to locate a .TXT (ASCII text) file.

Figure 47-7

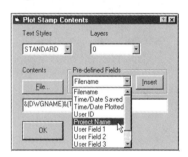

Plot Test

Selecting the *Plot Test* button causes AutoCAD to make a "dry run" of all the drawings listed for batch plotting. In this test, each drawing is checked to ensure all component parts are located (Xrefs, linetypes, fonts, etc.) and to ensure that the selected settings are operable. Warning messages appear if a problem (or potential problem) exists. For example, if an attached Xref, raster file, or font is missing, AutoCAD reports it. A sample *Plot Test Results* dialog box is displayed in Figure 47-8.

Figure 47-8

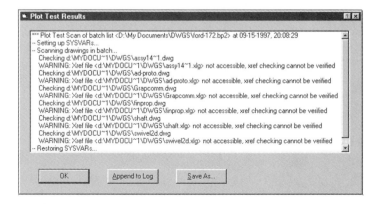

In Figure 47-8, a common warning is displayed. The message "WARNING: Xref file <path/filename> not accessible, xref checking cannot be verified" means that when Plot Test attempted to check for missing Xrefs in the drawing, it could not create an XLG file. If the drawing does not contain Xrefs, <u>this message does not indicate a problem.</u>

Plot

When you are ready to make the plots, select the *Plot* button. As AutoCAD progresses through the plot batch, the *Batch Plot Progress* box gives the current status (Fig. 47-9).

Figure 47-9 ────────

BONUS TOOLS

Installing the Bonus Tools from the CD-ROM

To install the Bonus Tools when you install AutoCAD for the first time:

1. Insert the AutoCAD Release 14 CD-ROM.

2. The Windows Setup utility should start automatically. If not, use Windows Explorer to find the Setup.exe file on the CD-ROM. Double-click on Setup.exe.

3. In the *Setup Choices* dialog box (Fig. 47-10), select *Install*.

4. In the *Setup Type* dialog box (Fig. 47-11), select *Full*.

Figure 47-10 ────────

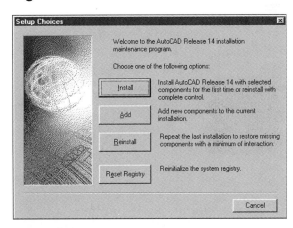

Figure 47-11 ────────

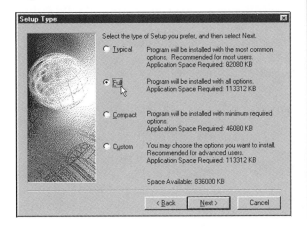

To install the Bonus Tools after AutoCAD is installed on your system:

1. Insert the AutoCAD Release 14 CD-ROM.
2. Use Windows Explorer to find the Setup.exe file on the CD-ROM. Double-click on Setup.exe.
3. In the *Setup Choices* dialog box (Fig. 47-10), select *Add*.
4. In the *Custom Components* dialog box (Fig. 47-12), select *Bonus*.

Figure 47-12

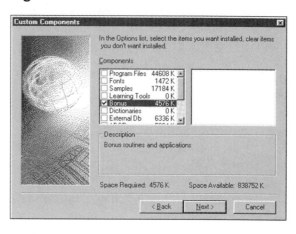

Installing the Bonus Menu and Toolbars

You may have to install the Bonus menu files if you installed the Bonus Tools after you installed AutoCAD. If the Bonus Tools were installed when you first installed AutoCAD, you should not have to perform this step. The Bonus Tools menu is located in the **AutoCAD R14\Bonus\Cadtools** folder.

To Load the Bonus Tools Menu

To load the Bonus Tools menu, use the *Menuload* command (which is used to load a "partial menu"). *Menuload* invokes the *Menu Customization* dialog box (Fig. 47-13). In the *Menu Groups* tab, select the *Browse…* button to locate AC_BONUS.MNU. The *Load* button loads the menu and "overlays" it on the AutoCAD standard menu. Use the *Menu Bar* tab to ensure the *Bonus* pull-down menu is installed.

Figure 47-13

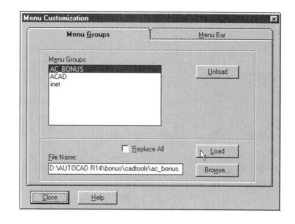

To Load the Bonus Tools Toolbars

To bring the *Bonus Menu* toolbars onto the screen use the *Toolbar* command or right-click on any tool to invoke the *Toolbars* dialog box (Fig. 47-14). At the bottom of the box (in the *Menu Group* area), select *AC_BONUS*, then check each of the toolbars you want bring to the screen. Three toolbars are available: *Bonus Standard Toolbar, Bonus Text Tools,* and *Bonus Layer Tools.*

Bonus Layer Tools Commands

Several useful layer utilities are available in this set. The productivity you can gain from these bonus commands is well worth the trouble of loading these menus and toolbars. These commands operate as an extension to the standard layer commands, allowing you to accomplish in one command what might otherwise take two or three commands or what might require multiple object or dialog box selections.

Figure 47-14

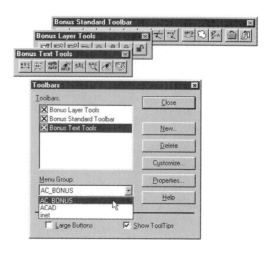

LMAN

	COMMAND (TYPE)	ALIAS (TYPE)	Short-cut	Screen (side) Menu	Tablet Menu
Pull-down Menu					
Bonus Layers> Layer Manager...	*LMAN*	...	...	...	...

Lman is short for *Layer Manager*. *Layer Manager* allows you to save, restore, and edit layer settings. With many drawings you may have certain combinations of layers that are frozen, thawed, on, or off for particular operations. For example, you may require a certain combination of layers to be *On* for making a plot, while a different set of layers may be *On* for working with construction lines (*Xlines, Rays*) or for creating dimensions and text. Instead of using the *Layer* command <u>each time</u> you make a plot and <u>each time</u> you create and edit *Xlines*, dimensions, and text, you can *Save* these settings to be *Restored* at any later time.

Lman produces the *Layer Manager: Save and Restore Layer Settings* dialog box (Fig. 47-15). The central area lists the *Saved Layer states* (empty when you first begin). The options operate as described here.

Figure 47-15

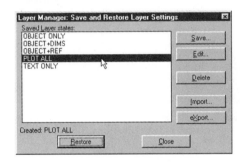

Save...
Using *Save...* invokes the *Layer State Name* dialog box (Fig. 47-16). Here you can assign a name to the current layer settings. Normally, you would want to specify the desired *On, Off, Freeze, Thaw* state for each layer before using this option. However, once a layer state is created, you can use *Edit...* to change the *On, Off, Freeze, Thaw* state of each layer. The layer states are saved in the drawing file, so you can *Restore* the layer states whenever needed. Spaces and other characters are allowed in the name.

Figure 47-16

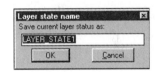

Restore
Highlight a name from the list, then use this option to restore the selected *Saved Layer state*. AutoCAD automatically sets the *On, Off, Freeze, Thaw* settings to those saved under the assigned name.

Edit...
This option allows you to change the layer settings for any named state. First, select the desired name from the list, then press *Edit...*. The *Layer* tab of the *Layer & Linetype Properties* dialog box appears for you to make the desired changes.

Delete
This button deletes a highlighted *Saved Layer state* from the list.

Import...
You can import previously saved layer states from a file. Layer states are saved (with *Export...*) with a .LAY file extension. Importing works well when you have several drawings with the same layer names and functions (as when similar drawings are created from the same template drawing). Layer states can be saved in one drawing, then *Exported* to a file and *Imported* to the other drawings.

R14

Export...
Once you have created the desired layer states, you can save them to a file (with a .LAY extension). The layer states can be imported into other drawings with *Import....*

LAYMCH

	Pull-down Menu	COMMAND (TYPE)	ALIAS (TYPE)	Short-cut	Screen (side) Menu	Tablet Menu
	Bonus Layers> Layer Match	*LAYMCH*	...	...	...	...

Laymch allows you to change the layer of selected objects to the layer of the selected destination object. This command performs the same function as the *Matchprop* command (see Chapter 12), but only with respect to layers (*Matchprop* can change other properties such as color, linetype, or linetype scale).

Using *Laymch* produces the following prompt:

```
Command: laymch
Select objects to be changed:
Select objects: PICK
Select objects: PICK
Select objects: Enter
2 found.
Type name/Select entity on destination layer: PICK
2 objects changed to layer 0.
Command:
```

You can use the *Type name* option to enter the name of the destination layer:

```
Type name/Select entity on destination layer: t
Enter layer name: text
2 objects changed to layer text.
```

LAYCUR

	Pull-down Menu	COMMAND (TYPE)	ALIAS (TYPE)	Short-cut	Screen (side) Menu	Tablet Menu
	Bonus Layers> Change to Current	*LAYCUR*	...	...	...	...

Laycur changes the layer of selected objects to the current layer. The following prompt is issued:

```
Command: laycur
Select objects to be CHANGED to the current layer:
Select objects: PICK
Select objects: PICK
Select objects: Enter
2 found.
2 objects changed to layer TEXT (the current layer).
Command:
```

R14

LAYISO

	COMMAND (TYPE)	ALIAS (TYPE)	Short-cut	Screen (side) Menu	Tablet Menu
Pull-down Menu					
Bonus *Layers>* *Layer Isolate*	*LAYISO*	...	...	...	...

Use this bonus command if you want to turn *Off* <u>all layers except</u> the layer(s) of the selected object(s). Several layers can be selected to be "isolated." If more than one layer is isolated, the last one selected becomes the current layer. The following prompt is used:

```
Command: layiso
Select object(s) on the layer(s) to be ISOLATED:
Select objects: PICK
Select objects: PICK
Select objects: Enter
2 layers have been isolated. Layer TEXT is current.
Command:
```

LAYFRZ

	COMMAND (TYPE)	ALIAS (TYPE)	Short-cut	Screen (side) Menu	Tablet Menu
Pull-down Menu					
Bonus *Layers>* *Layer Freeze*	*LAYFRZ*	...	...	...	...

This utility *Freezes* the layer(s) of the selected object(s). More than one object (layer) can be selected for *Freezing* with this command, but you select only one object at a time, then the prompt repeats. You cannot *Freeze* the current layer. AutoCAD gives the following prompt:

```
Command: layfrz
Options/Undo/<Pick an object on the layer to be FROZEN>: PICK
Layer TEXT has been frozen.
Options/Undo/<Pick an object on the layer to be FROZEN>: PICK
Layer GEOMETRY has been frozen.
Options/Undo/<Pick an object on the layer to be FROZEN>: Enter
Command:
```

You can use *Undo* to undo the last layer (object) selected. Type *O* for *Options,* which issues the following prompt:

```
Options/Undo/<Pick an object on the layer to be FROZEN>: o
Entity level nesting/No nesting/<Block level nesting>:
```

With the *No Nesting* option, only layers in the current drawing (not Xref layers or nested Block layers) are *Frozen. Entity level nesting* and *Block level nesting* automatically *Freeze Xrefs* or *Blocks* down to the level of the selected object (*Block* or *Xref*).

LAYTHW

Pull-down Menu	COMMAND (TYPE)	ALIAS (TYPE)	Short-cut	Screen (side) Menu'	Tablet Menu
Bonus Layers> Thaw All Layers	*LAYTHW*	...	...	...	...

Use this command to <u>*Thaw* all layers</u> in the drawing. Layers that are *Frozen* <u>and</u> *Off* are only *Thawed* by this command so they still remain invisible until turned *On*:

    Command: **laythw**
    All layers have been thawed.
    Command:

LAYOFF

Pull-down Menu	COMMAND (TYPE)	ALIAS (TYPE)	Short-cut	Screen (side) Menu	Tablet Menu
Bonus Layers> Layer Off	*LAYOFF*	...	...	...	...

Layoff is used to turn *Off* the layer(s) of the selected object(s). Multiple objects can be selected (one at a time) to turn off their layers. The *Options* are the same as those for *Layfrz*:

    Command: **layoff**
    Options/Undo/<Pick an object on the layer to be turned OFF>: **PICK**
    Layer ASESMP | 1-WALL has been turned off.
    Options/Undo/<Pick an object on the layer to be turned OFF>: **Enter**
    Command:

LAYON

Pull-down Menu	COMMAND (TYPE)	ALIAS (TYPE)	Short-cut	Screen (side) Menu	Tablet Menu
Bonus Layers> Turn All Layers On	*LAYON*	...	...	...	...

This command is a quick way to turn <u>all layers on</u>. Layers that are *Frozen* are not affected; only layers that are *Off* can be turned *On* with this command:

    Command: **layon**
    Warning: layer 0 is frozen. Will not display until thawed.
    All layers have been turned on.
    Command:

LAYLCK

	Pull-down Menu	COMMAND (TYPE)	ALIAS (TYPE)	Short-cut	Screen (side) Menu	Tablet Menu
	Bonus Layers> Layer Lock	*LAYLCK*	...	...	...	...

Laylck locks the layers of selected objects. There are no options for this command. If objects are selected that are part of an *Xref* or *Block,* the layer that was current when the *Xref* or *Block* was *Attached* or *Inserted* is locked.

    Command: **laylck**
    Pick an object on the layer to be LOCKED: **PICK**
    Layer TEXT has been locked.

LAYULK

	Pull-down Menu	COMMAND (TYPE)	ALIAS (TYPE)	Short-cut	Screen (side) Menu	Tablet Menu
	Bonus Layers> Layer Unlock	*LAYULK*	...	...	...	...

Layulk has the opposite function of *Laylck;* that is, it unlocks the layers of selected objects. If the selected objects are part of an *Xref* or *Block,* the layer that was current when the *Xref* or *Block* was *Attached* or *Inserted* is unlocked. There are no options for this command.

    Command: **layulk**
    Pick an object on the layer to be UNLOCKED: **PICK**
    Layer TEXT has been unlocked.
    Command:

Bonus Text Tools Commands

Several outstanding text utilities are included in this set. For applications that insert and manipulate text, these commands are truly a bonus. Some of the functions of these commands can be accomplished otherwise only by writing custom AutoLISP routines, not by using standard AutoCAD commands.

ARCTEXT

	Pull-down Menu	COMMAND (TYPE)	ALIAS (TYPE)	Short-cut	Screen (side) Menu	Tablet Menu
	Bonus Text> Arc Aligned Text	*ARCTEXT*	...	...	...	...

Figure 47-17 ─────────

There is no other easy method to use in AutoCAD to create text along an arc. The built-in AutoCAD commands allow creation of a line of text only along a straight line. With this command, you can "attach" text to an existing *Arc* (Fig. 47-17). You can use the same command to edit existing *ArcAlignedText* objects, as noted by the prompt:

    Command: **arctext**
    Select an Arc or an ArcAlignedText: **PICK** (existing *Arc* or existing
    ArcAlignedText)
    Command:

After selecting an arc to use for text alignment or selecting an existing *ArcAlignedText* object, AutoCAD produces the *ArcAlignedText Workshop* dialog box (Fig. 47-18). Here you can control the typical text features (style, font, height, width factor, etc.) as well as many special features.

Figure 47-18

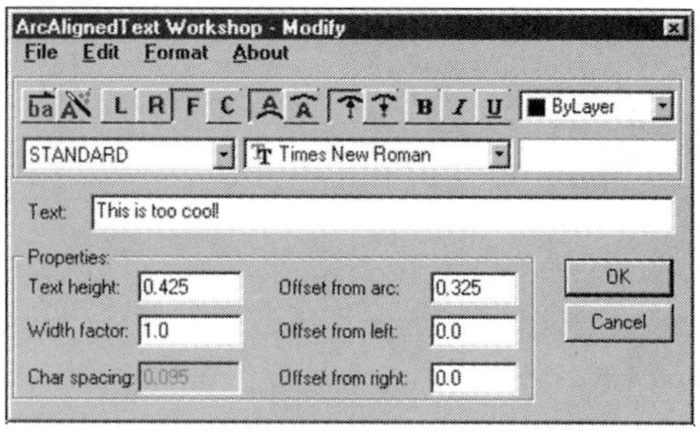

The following options are available in the *ArcAlignedText Workshop* dialog box.

Reverse Text
This option reverses the text so it reads "backwards."

Alignment
Left, Right, Fit, and *Center* methods can be used. If *Fit* is not selected, you can specify the *Offset from left* and *Offset from right*.

Position
The text can be created on the *Convex* or *Concave* side of the arc. You can also set a value for *Offset from arc*.

Outward from the center/Inward to the center
This controls which direction the "top" of the letters point. Using Figure 47-17 as an example, if *Inward to the center* were selected, the top of the letters would point downward resulting in upside-down text.

Typeface
Bold, Italic, and *Underline* can be selected.

Other Options
Other typical text options are available, such as text *Height, Width factor, Style,* and font file.

Existing *ArcAlignedText* objects can be modified with grips but keep the concentric orientation with the aligned arc.

BURST

Pull-down Menu	COMMAND (TYPE)	ALIAS (TYPE)	Short-cut	Screen (side) Menu	Tablet Menu
Bonus *Text>* *Explode Attributes to...*	BURST	...	...	...	...

Use *Burst* with attributed text (text combined with a *Block*) like you would use *Explode* with an unattributed *Block. Burst* actually explodes text attributes but converts the attributes to *Dtext* objects. Normally, if you select attributes to *Explode,* the *Block* is exploded and the text is changed back to text, indicating the original attribute definitions—that is, tags, values, and prompts. (See Chapter 22, Block Attributes.) Using *Burst* may cause the newly created *Dtext* objects to be misaligned from their previous orientation.

```
Command: burst
BURST loaded, Type BURST to explode blocked attributes into Dtext entities
Select objects: PICK
Select objects: Enter
Command:
```

R14

CHT

Pull-down Menu	COMMAND (TYPE)	ALIAS (TYPE)	Short-cut	Screen (side) Menu	Tablet Menu
Bonus Text> Change Text	CHT	...	...	...	...

This utility command is similar to the option of the *Change* command (see Chapter 16) that allows you to change text characteristics (*Height, Rotation,* etc.), except that you can select multiple text objects to change rather than one object at a time. All selected text objects then acquire the new characteristics. This feature is helpful for changing text "globally" across a drawing.

```
Command: cht
Select annotation objects to change.
Select objects: PICK
Select objects: PICK
Select objects: Enter
2 annotation objects found.
Height/Justification/Location/Rotation/Style/Text/Undo/Width:
```

FIND

Pull-down Menu	COMMAND (TYPE)	ALIAS (TYPE)	Short-cut	Screen (side) Menu	Tablet Menu
Bonus Text> Find and Replace Text	FIND	...	...	...	...

This utility operates similarly to the *Find and Replace* option of *Mtext;* however, the *Mtext* option allows the operation only in selected *Mtext* objects, whereas *Find* allows you to find and replace text objects <u>globally throughout the drawing</u>. Using this command produces the *Find and Replace* dialog box (Fig. 47-19). Enter the desired text strings in the *Find:* and *Replace With:* edit boxes. If you are concerned with an exact upper- and lower-case match, select *Case Sensitive. Global Change* automatically makes the change throughout the current drawing, whereas without this box checked you are prompted to select which text items in the drawing to use for the selection set to search.

Figure 47-19 ——————

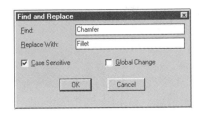

If *Global Change* is not checked and you select objects, another *Find and Replace* dialog box appears with the several options. The *Replace* option replaces the individual objects one at a time. *Auto* replaces <u>all</u> objects matching the criteria <u>in the selection set</u>. *Skip* does not change the highlighted text object but proceeds to the next text object in the selection set matching the criteria.

Figure 47-20 ——————

R14

GATTE

Pull-down Menu	COMMAND (TYPE)	ALIAS (TYPE)	Short-cut	Screen (side) Menu	Tablet Menu
Bonus *Text>* *Global Attribute Edit*	*GATTE*	...	...	...	...

Gatte makes global changes to attribute values. *Gatte* allows you to make changes to all insertions of the same *Block* in a drawing with one operation. For example, assume you wanted to change the PART_NO attribute for all insertions of the CAP (capacitor) block in the schematic shown in Figure 47-21. Use *Gatte* and follow the command prompt below to change all PART_NO attribute values to "C-47XXX:"

Figure 47-21

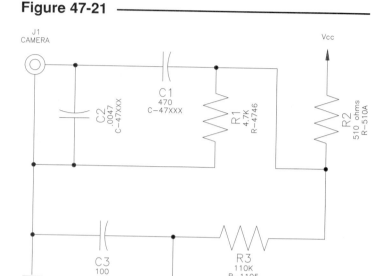

```
Command: gatte
Block name/<select block or attribute>:
PICK (select only one attribute)
Block: CAP  Attribute tag: PART_NO
New Text: C-47XXX
Number of inserts in drawing = 3  Process
all of them? <Yes>/No:  Enter
Please wait…
3 attributes changed.
Command:
```

The attribute values for all insertions of the *Block* are updated (three *Blocks* on left, Fig. 47-21).

TEXTFIT

Pull-down Menu	COMMAND (TYPE)	ALIAS (TYPE)	Short-cut	Screen (side) Menu	Tablet Menu
Bonus *Text>* *Text Fit*	*TEXTFIT*	...	...	...	...

Textfit fits text between two specified points. Using this command gives the same results as using the *Fit* option of text alignment; however, this Bonus command acts on <u>existing</u> text:

```
Command: textfit
Initializing… TEXTFIT loaded.  Type TEXTFIT to stretch/shrink text between points
    Type TFHELP for help
Select Text to stretch/shrink:  PICK
Starting Point/<Pick new ending point>:  PICK
Command:
```

Using the default option you need to PICK only the new ending point and the line of text is changed using the existing start point but specifying a new endpoint. You can also change the start point and endpoint, as shown below:

    Command: *textfit*
    Select Text to stretch/shrink: **PICK**
    Starting Point/<Pick new ending point>: *s*
    Pick new starting point: **PICK**
    ending point: **PICK**
    Command:

TEXTMASK

Pull-down Menu	COMMAND (TYPE)	ALIAS (TYPE)	Short-cut	Screen (side) Menu	Tablet Menu
Bonus Text> Text Mask	*TEXTMASK*	...	...	...	...

When text and other objects are drawn "on top of" each other (so they occupy the same space), you can use *Textmask* to make the text appear "in front of" other objects by "masking" the other objects. For example, using *Textmask* creates text that masks other drawing objects as shown in Figure 47-22. Use the following command syntax:

Figure 47-22

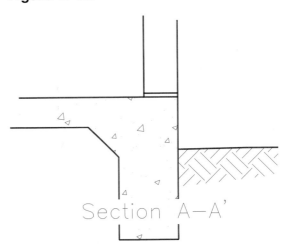

    Command: *textmask*
    Initializing... Loading WIPEOUT for use with
    TEXTMASK...
    Enter offset factor relative to text height <0.35>: **Enter**
    Select Text to MASK...
    Select objects: **PICK**
    Select objects: **Enter**
    Wipeout created.
    Command:

Offset factor is the distance around the text for the imaginary box that is used for "masking" (trimming) the objects behind the text.

TXTEXP

Pull-down Menu	COMMAND (TYPE)	ALIAS (TYPE)	Short-cut	Screen (side) Menu	Tablet Menu
Bonus *Text>* *Explode Text*	*TXTEXP*	...	...	...	...

Textexp explodes text. The resulting objects are *Plines*. As you probably know, a line or paragraph of text is one object; therefore, without this command you cannot explode text. Exploded text is helpful if you need to alter or manipulate the individual text characters. For example, to convert text into 3D solids, use *Textexp,* convert each letter to a *Region,* then *Extrude* the text (Fig. 47-23):

Figure 47-23

```
Command: txtexp
Initializing... Select text to be EXPLODED:
Select objects: PICK
Select objects: Enter
1 found.
1 text object(s) have been exploded to lines.
The line objects have been placed on layer 0.
Command:
```

Bonus Standard Tools Commands

The Standard Bonus toolbar contains a number of commands useful principally for modifying objects. A few commands are included for drawing and for other utilities. If you use pull-down menus, these commands are spread through the *Modify, Draw,* and *Tools* sections of the *Bonus* pull-down menu.

EXCHPROP

Pull-down Menu	COMMAND (TYPE)	ALIAS (TYPE)	Short-cut	Screen (side) Menu	Tablet Menu
Bonus *Modify>* *Extend Change* *Properties*	*EXCHPROP*	...	...	...	...

The *Exchprop* utility is an extended change properties (*Ddchprop*). *Exchprop* produces the *Change Properties* dialog box shown in Figure 47-24. This dialog box is similar to the *Ddchprop* dialog box with the added sections for *Linetype Scale, Thickness, Polyline,* and *Text/Mtext/Attdef.* Multiple items can be selected; for example, to change the *Linetype Scale* property for all objects selected.

Figure 47-24

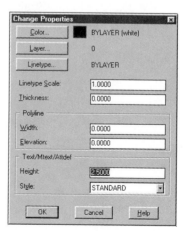

MSTRETCH

Pull-down Menu	COMMAND (TYPE)	ALIAS (TYPE)	Short-cut	Screen (side) Menu	Tablet Menu
Bonus Modify> Multiple Entity Stretch	*MSTRETCH*	...	...	...	...

The typical *Stretch* command requires that you select objects to stretch with one *Crossing Window* or *Window Polygon*. Any objects that are not within the selection Window must be stretched with another use of the command. In contrast, *Mstretch* allows you to create multiple selection Windows before stretching the objects:

```
Command: mstretch
Define crossing windows or crossing polygons...
CP(crossing polygon)/<Crossing First point>: PICK
Other corner: PICK
CP/Undo/<Crossing First point>: PICK
Other corner: PICK
CP/Undo/<Crossing First point>:  Enter
Done defining windows for stretch...
Remove objects/<Base point>: PICK
Second base point: PICK
Command:
```

MOCORO

Pull-down Menu	COMMAND (TYPE)	ALIAS (TYPE)	Short-cut	Screen (side) Menu	Tablet Menu
Bonus Modify> Move Copy Rotate	*MOCORO*	...	...	...	...

Mocoro allows you to perform a *Move, Copy, Rotate,* or *Scale,* or any combination of these operations in one command. This utility is especially useful for those who set *GRIPS* to 0 (off), but occasionally need these capabilities without using Grips:

```
Command: mocoro
Select objects: PICK
Select objects: Enter
Base point: PICK
Move/Copy/Rotate/Scale/Base pt/Undo/<eXit>: m
Second point of displacement: PICK
Move/Copy/Rotate/Scale/Base pt/Undo/<eXit>: r
Second point or rotation angle: PICK
Move/Copy/Rotate/Scale/Base pt/Undo/<eXit>: c
Second point of displacement/Undo/<eXit>: PICK
Second point of displacement/Undo/<eXit>: Enter
Move/Copy/Rotate/Scale/Base pt/Undo/<eXit>: Enter
Command:
```

EXTRIM

Pull-down Menu	COMMAND (TYPE)	ALIAS (TYPE)	Short-cut	Screen (side) Menu	Tablet Menu
Bonus Modify> Cookie Cutter Trim	EXTRIM	...	...	...	...

Extrim allows you to use an existing *Circle*, closed *Pline*, *Arc*, or *Line* as a "trimming edge." Even though the normal *Trim* command allows closed *Plines*, this utility automates the process by prompting only for "Side to Trim" and automatically trims all objects outside or inside the closed boundary that cross the trimming edge object. In other words, *Extrim* allows you to trim like using a "cookie cutter" when closed objects are used as trimming edges.

```
Command: extrim
Pick a POLYLINE, LINE, CIRCLE, or ARC for cutting edge..
Select objects: PICK
Command:
Pick the side to trim on: PICK
Command:
```

CLIPIT

Pull-down Menu	COMMAND (TYPE)	ALIAS (TYPE)	Short-cut	Screen (side) Menu	Tablet Menu
Bonus Modify> Extended Clip	CLIPIT	...	...	...	...

Release 14 has the unique feature of clipping *Xrefed* drawings by specifying a polygonal or rectangular *Pline* (no arc segments) as the clipping boundary. *Clipit* extends that capability by allowing you to use an existing *Circle*, *Arc*, or *Pline* as a clipping edge for an *Xref*, *Block*, or *Wipeout* (see *Wipeout*):

```
Command: clipit
Pick a POLYLINE/CIRCLE/ARC for clipping edge..
Select objects: PICK
Command:
Pick an IMAGE, a WIPEOUT, or an XREF/BLOCK to clip...
Select objects: PICK
Command:
Enter max error distance for resolution of arcs <0.0200>: Enter
Specify clipping boundary: PICK
Command:
```

MPEDIT

Pull-down Menu	COMMAND (TYPE)	ALIAS (TYPE)	Short-cut	Screen (side) Menu	Tablet Menu
Bonus Modify> Multiple Pedit	MPEDIT	...	...	...	...

Mpedit is similar to *Pedit* with the additional capability to select multiple *Plines* for editing. All selected *Plines* are updated as a group. For example, you can change the *width* of all *Plines* in a drawing globally with this feature rather than having to change one *Pline* at a time:

Command: *mpedit*
Select objects: **PICK**
Select objects: **PICK**
Select objects: **PICK**
Select objects: **PICK**
Select objects: **Enter**
Convert Lines and Arcs to polylines? <Yes>: **Enter**
Open/Close/Width/Fit/Spline/Decurve/Ltype gen/eXit <X>:

Although you may select only *Plines,* you will still get the "Convert Lines and Arcs to polylines? <Yes>:" prompt. All typical *Pedit* options are available (note the last prompt line).

NCOPY

Pull-down Menu	COMMAND (TYPE)	ALIAS (TYPE)	Short-cut	Screen (side) Menu	Tablet Menu
Bonus Modify> Copy Nested Entities	NCOPY	...	...	...	...

Ncopy allows you to copy objects that are nested within *Xrefs* or *Blocks.* With built-in AutoCAD commands this is possible to do for *Xrefs* (using *Xbind*), but the process is more complex than *Ncopy.* *Ncopy* allows you to select any object within an *Xref* or *Block* to use for copying. *Ncopy* operates similarly to the *Copy* command:

Command: *ncopy*
Select nested objects to copy: **PICK**
Select nested objects to copy: **Enter**
Select objects: <Base point or displacement>/Multiple: **PICK**
Second point of displacement: **PICK**
Command:

BTRIM

Pull-down Menu	COMMAND (TYPE)	ALIAS (TYPE)	Short-cut	Screen (side) Menu	Tablet Menu
Bonus Modify> Trim to Block Entities	BTRIM	...	...	...	...

Btrim allows you to trim objects using nested *Block* or *Xref* objects as cutting edges. The command is simple to use since it operates identically to the *Trim* command. If you select an object within a clipped *Xref,* the entire object (outside the clipping boundary) is highlighted.

Command: *btrim*
Select cutting edges: **PICK**
Select cutting edges: **Enter**
Select objects:
<Select object to trim>/Project/Edge/Undo: **PICK**
<Select object to trim>/Project/Edge/Undo: **Enter**
Command:

Btrim has the same options as *Trim* (see *PROJECTMODE* and *EDGEMODE,* Chapter 9).

BEXTEND

	Pull-down Menu	COMMAND (TYPE)	ALIAS (TYPE)	Short-cut	Screen (side) Menu	Tablet Menu
	Bonus Modify> Extend to Block Entities	*BEXTEND*	...	...	...	...

Bextend allows you to extend objects using nested *Block* or *Xref* objects as boundary edges. This utility (like the *Btrim* utility) is similar to the built-in AutoCAD counterpart but with the added capability of extending objects to *Block* and *Xref* objects. *Bextend* operates identically to the *Extend* command:

```
Command: bextend
Select objects for extend: PICK
Select objects for extend: Enter
Select objects:
<Select object to extend>/Project/Edge/Undo: PICK
<Select object to extend>/Project/Edge/Undo: Enter
Command:
```

WIPEOUT

	Pull-down Menu	COMMAND (TYPE)	ALIAS (TYPE)	Short-cut	Screen (side) Menu	Tablet Menu
	Bonus Draw> Wipeout	*WIPEOUT*	...	...	...	...

Wipeout creates an opaque object that is used to "hide" other objects. The *Wipeout* object is created from an existing closed *Pline*. All objects behind the *Wipeout* become "invisible." For example, assume a *Pline* shape was created so as to cross other objects. You could use *Wipeout* to make the shape appear to be "on top of" the other objects (Fig. 47-25, left):

Figure 47-25

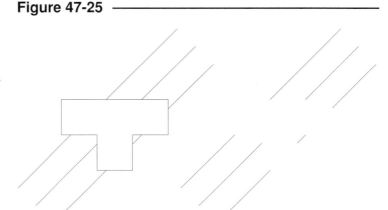

```
Command: wipeout
Frame/New <New>: Enter
Select a polyline: PICK
Erase polyline? Yes/No <No>: Enter
Wipeout created.
Command:
```

You can create an "invisible" *Wipeout* by selecting the *Erase Polyline* option when you create the *Wipeout*, then use the *Frame Off* option to make the *Wipeout* object disappear (Fig. 47-25, right). A *Wipeout* is an AutoCAD object. The *List* command reports the object as a *Wipeout* object.

REVCLOUD

	Pull-down Menu	COMMAND (TYPE)	ALIAS (TYPE)	Short-cut	Screen (side) Menu	Tablet Menu
	Bonus *Draw>* *Revision Cloud*	*REVCLOUD*	...	...	...	...

Use this command to draw a revision cloud on the current layer. Revision clouds are the customary method of indicating an area of an architectural drawing that contains a revision to the original design. Revision clouds are created on a separate layer so they can be controlled for plotting.

Revcloud is an automated routine that simplifies drawing revision clouds. All you have to do is pick a point and move the cursor in a counter-clockwise (circular) direction and bring the cursor around to the start point to complete the cloud (Fig. 47-26). The cloud automatically closes and the command ends:

Figure 47-26

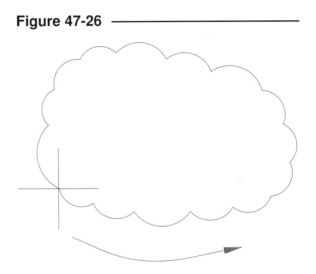

```
Command: revcloud
Arc length set at .375
Arc length/<Pick cloud starting point>: a
Arc length <10.000>: 20
Pick cloud start point: PICK
Guide cursor along cloud path... (move the cursor in a
counter-clockwise circular direction)
Cloud finished.
Command:
```

Use the *Arc length* option to specify a value in drawing units for the arc segments.

PACK

	Pull-down Menu	COMMAND (TYPE)	ALIAS (TYPE)	Short-cut	Screen (side) Menu	Tablet Menu
	Bonus *Tools>* *Pack'n Go...*	*PACK*	...	...	...	...

The *Pack* command is useful for preparing drawings to be sent to clients, vendors, colleagues, etc. *Pack* ensures that the drawing and all its dependencies (related files) such as Xrefs, images, and fonts are "packed" and ready to go.

Pack produces the *Pack & Go* dialog box (Fig. 47-27). Here you can use the *List View* or *Tree View* options (upper-left corner) to display the list of files. The *Tree View* option displays a hierarchical structure to the parent drawing and attached *Xrefs, Images,* selected *Fonts,* etc. (see Fig. 47-27).

Figure 47-27 ————————

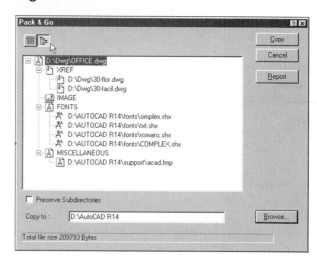

Use the *Report* button in the *Pack & Go* dialog box to produce a written description of the dependencies (Fig. 47-28). The report can be printed and sent as hard copy along with the files you send to your clients.

Figure 47-28 ————————

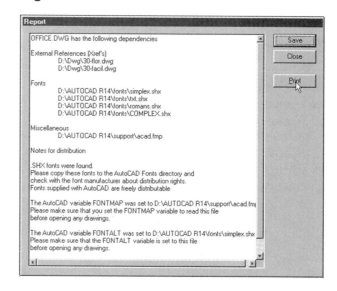

XLIST

Pull-down Menu	COMMAND (TYPE)	ALIAS (TYPE)	Short-cut	Screen (side) Menu	Tablet Menu
Bonus Tools> List Xref/Block Entities	*XLIST*	...	...	...	...

The *Xlist* command displays properties of objects that are nested in *Xrefs* or *Blocks*. This is helpful since the *List* command reports information about only the top level object (*Xref* or *Block*), not information on nested objects. Using *Xlist* produces the following prompt:

Command: **xlist**
Select nested xref or block object to list: **PICK**

Figure 47-29 ————————

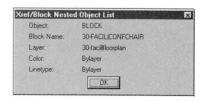

After selecting an object within an *Xref* or *Block, Xlist* produces the *Xref/Block Nested Object List* dialog box (Fig. 47-29). This dialog box lists properties of the selected *Block* or *Xref.*

QLEADER

	Pull-down Menu	COMMAND (TYPE)	ALIAS (TYPE)	Short-cut	Screen (side) Menu	Tablet Menu
	Bonus *Draw>* *Leader Tools>* *Quick Leader*	*QLEADER*	...	...	...	...

Qleader (quick leader) creates a leader with or without text. *Qleader* is similar to the *Leader* command except the prompts for text insertion are streamlined. If you insert text, *Qleader* creates an *Mtext* object (by default) but without producing the *Mtext* dialog box. Many other options are available with the *Qleader* utility.

Command: **qleader**
First Leader point or press Enter to set Options:
Next Leader point: **PICK**
Next Leader point: **PICK**
Enter Leader text: **Enter** (or enter text)
Command:

Typing "E" for "Enter to set Options:" produces the *Quick Leader Options* dialog box (Fig. 47-30). There are four tabs for setting options: *Annotation/Format*, *Points*, *Angles*, and *Attachment*.

Annotation/Format
Annotation
This section specifies the type of text objects created (*None, Copy, Mtext,* or *Tolerance*) and the related prompts when you use *Qleader*. Your selection stays in effect until changed. *Copy an Object* prompts you to select an existing annotation object, while *Tolerance* produces the *Tolerance* dialog box when you use *Qleader*. Use *Block Reference* to use a *Block* with the leader.

Annotation Memory
Here you can select which text *Qleader* uses as the default <inside brackets> text.

Format
Spline allows you to create a *Spline* object leader.

Points
Set the maximum number of points for the leader in the *Points* tab (Fig. 47-31). If set to *No Limit*, *Qleader* repeatedly prompts for "Next Leader point."

Figure 47-30 ───────

Figure 47-31 ───────

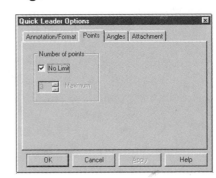

Angles

The *Angles* tab (Fig. 47-32) allows you to specify (force) an angle for the *Qleader* to be drawn. For example, selecting 45 causes the *Qleader* to be drawn always at 45 degrees.

Figure 47-32

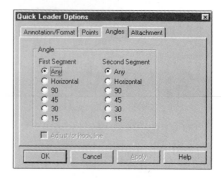

Attachment

The Attachment tab (Fig. 47-33) is used for setting the attachment point when *Qlattach* is used (see *Qlattach* in Additional Bonus Tools Commands, Pull-down Menus).

Figure 47-33

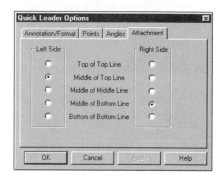

Additional Bonus Tools Commands, Pull-down Menus

Commands in this section are included in the *Bonus* pull-down menus (under *Modify, Draw* or *Tools*) but are <u>not</u> included in any of the Bonus toolbars.

QLATTACH

Pull-down Menu	COMMAND (TYPE)	ALIAS (TYPE)	Short-cut	Screen (side) Menu	Tablet Menu
Bonus *Draw>* *Leader Tools>* *Attach Leader to...*	QLATTACH	...	...	...	...

Use *Qlattach* to attach an existing leader to an annotation (*Text, Dtext,* or *Mtext* object). *Qleader* automatically changes the endpoint of the leader and/or moves the annotation to align and attach to the annotation as set in the *Attachment* tab of the *Quick Leader Options* dialog box (see Fig. 47-33):

```
Command: qlattach
Select Leader: PICK
Select Annotation: PICK
Command:
```

QLATTACHSET/QLDETACHSET

These two commands have been known to cause AutoCAD Fatal Errors and should not be used.

ALIASEDIT

Pull-down Menu	COMMAND (TYPE)	ALIAS (TYPE)	Short-cut	Screen (side) Menu	Tablet Menu
Bonus Tools> Command Alias Editor	*ALIASEDIT*	...	...	...	...

Aliasedit allows you to edit the existing command aliases so you can create your own aliases. *Aliasedit* produces the *AutoCAD Alias Editor* (Fig. 47-34). This editor automates making changes to the ACAD.PGP file (see Chapter 45, Basic Customization). Highlight the desired alias, select *Edit*, then enter the new alias in the *Edit Command Alias* dialog box (not shown). You can *Add* or *Remove* aliases.

The *Shell* tab allows you to edit the "External Commands" section of the ACAD.PGP (not shown). This tab operates similarly to the *Command Alias* tab options.

Figure 47-34 ———

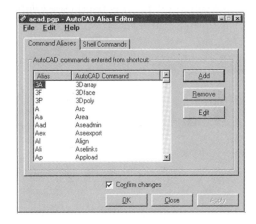

BONUSPOPUP

Pull-down Menu	COMMAND (TYPE)	ALIAS (TYPE)	Short-cut	Screen (side) Menu	Tablet Menu
Bonus Tools> Popup Menu	*BONUSPOPUP*	...	...	...	...

The *Bonuspopup* utility enables the capability of pressing Ctrl+right-click to produce one of the pull-down menus on the screen as a cursor menu. For example, when you use Shift+right-click, the *Osnap* menu appears. With the *Bonuspopup* utility you can dictate which of the normal pull-down menus appears (like the *Osnap* menu) when you press Ctrl+right-click.

After using *Bonuspopup* to load this utility, you can press Alt+right-click to produce the *Pick a Pop-Up Menu* dialog box (Fig. 47-35). Here you specify which pull-down menu appears as a cursor menu when you Ctrl+right-click.

Figure 47-35 ——

CONVERTPLINES

Pull-down Menu	COMMAND (TYPE)	ALIAS (TYPE)	Short-cut	Screen (side) Menu	Tablet Menu
Bonus Tools> Pline Converter	*CONVERTPLINES*	...	...	...	...

Use this utility as an alternative to the *Convert* command to convert the pre-Release 14 *Plines* to the Release 14 "lightweight" *Plines*. This is a streamlined version of *Convert,* since it globally and unconditionally converts all *Plines* in the current drawing to lightweight *Plines*:

    Command: **convertplines**
    ***Warning: This will convert all polylines to lightweight unconditionally.
    this will remove all xdata on existing polylines and
    may cause third party applications reliant on this data to fail.
    Type "YES" <all caps> to proceed: **YES**
    20 Polylines converted to lwPolyline
    Command:

DIMEX

Pull-down Menu	COMMAND (TYPE)	ALIAS (TYPE)	Short-cut	Screen (side) Menu	Tablet Menu
Bonus Tools> Dimstyle Export...	*DIMEX*	...	...	...	...

This is a very useful utility for exporting dimension styles from the current drawing to a file for importation into another drawing with *Dimim* (dimension import). *Dimex* produces the *Dimension Style Export* dialog box (Fig. 47-36). Select a dimension style to export, then select *OK*. The dimension style is saved to a file with a .DIM file extension.

Figure 47-36 ——————

DIMIM

Pull-down Menu	COMMAND (TYPE)	ALIAS (TYPE)	Short-cut	Screen (side) Menu	Tablet Menu
Bonus Tools> Dimstyle Import...	*DIMIM*	...	...	...	...

Use *Dimim* to import previously exported dimension styles (see *Dimex*). *Dimim* invokes the *Dimension Style Import* dialog box (Fig. 47-37). Use *Browse...* to locate the desired *.DIM file.

Figure 47-37 ——————

GETSEL

Pull-down Menu	COMMAND (TYPE)	ALIAS (TYPE)	Short-cut	Screen (side) Menu	Tablet Menu
Bonus Tools> Get Selection Set	GETSEL	...	...	...	...

Getsel allows you to PICK specific objects by object type or by layer and use them as the current selection set. You can select the desired layer (or all layers) or the desired object type (or all object types). This utility is a fast alternative to using the *Filter* command to locate and find specific object types. *Getsel* issues the following prompt:

    Command: **getsel**
    Select Object on layer to Select from <*>: **PICK**
    Select type of entity you want <*>: **PICK**
    Collecting all LWPOLYLINE items on layer GEOMETRY...
    12 items have been placed in the active selection set.
    Command:

SYSVDLG

Pull-down Menu	COMMAND (TYPE)	ALIAS (TYPE)	Short-cut	Screen (side) Menu	Tablet Menu
Bonus Tools> System Variable Editor	SYSVDLG	...	...	...	...

You can edit values stored in system variables using a dialog box interface with *Sysvdlg*. *Sysvdlg* invokes the *System Variables* dialog box (Fig. 47-38). Here you can select any number of system variables to change.

Figure 47-38

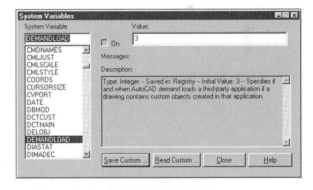

Save Custom.../Read Custom...
The power of this utility is its capability for saving the entire set of system variables settings to a file with an .SVF file extension. This feature allows you also to import previously saved .SVF files into a drawing. Importing .SVF files is analogous to restoring a *Dimstyle*, except all system variable settings can be restored!

XDATA

Pull-down Menu	COMMAND (TYPE)	ALIAS (TYPE)	Short-cut	Screen (side) Menu	Tablet Menu
Bonus Tools> Xdata Attachment	XDATA	...	...	...	...

Extended entity data is additional information that can be attached to any AutoCAD object. Extended entity data provides an internal level of information "attached" to each individual object. This AutoCAD feature is extremely powerful for customization applications.

Xdata is a Bonus utility to simplify attaching extended entity data to an object. Use *Xdata* to produce the following prompt:

```
Command: Xdata
Select object: PICK
Application name: (name)
(NAME) new application.
3Real/DIR/DISP/DIST/Hand/Int/LAyer/LOng/Pos/Real/SCale/STr/<eXit>: (option)
New xdata appended.
Command:
```

The prompt displays the areas of information that can be attached to an object: *3 Real numbers, 3D world space direction, 3D world space displacement, Distance, Database handle, 16-bit integer, Layer, 32-bit long integer, 3D world space position, Real number, Scale factor,* and *ASCII string.*

XDLIST

Pull-down Menu	COMMAND (TYPE)	ALIAS (TYPE)	Short-cut	Screen (side) Menu	Tablet Menu
Bonus Tools> *List Entity Xdata*	*XDLIST*	...	...	...	...

Xdlist lists extended entity data attached to a nested object. An example is listed below:

```
Command: xdlist
Select object: PICK
Application name <*>: (name)
* Registered Application Name: (NAME)
* Code 1002, Starting or ending brace: {
* Code 1000, ASCII string:
* Code 1003, Layer name: 0
* Code 1042, Scale factor: 0.5000
* Code 1000, ASCII string: sample object
* Code 1002, Starting or ending brace: }
Object has 16347 bytes of Xdata space available.
```

XPLODE

Pull-down Menu	COMMAND (TYPE)	ALIAS (TYPE)	Short-cut	Screen (side) Menu	Tablet Menu
Bonus Modify> *Extended Explode*	*XPLODE*	...	...	...	...

The *Xplode* utility performs an *Explode* with the additional capability of allowing you to control properties of the exploded object when the explode is performed. For example, you can control which *Layer* the objects are exploded onto, as shown in the following prompt:

Command: *xplode*
Select objects to XPlode.
Select objects: **PICK**
Select objects: **Enter**
1 objects found.
All/Color/LAyer/LType/Inherit from parent block/<Explode>: **LA**
XPlode onto what layer? <0>: **DIM**
Object exploded onto layer DIM.
Command:

Additional Bonus Tools Commands, Command Line Entry Only

Commands in this section are <u>not included</u> in the Bonus toolbars or *Bonus* pull-down menus (under *Modify, Draw,* or *Tools*). All of these commands <u>must be typed</u>.

ATTLST

This utility command lists all attributes in a *Block* insertion. Only one *Block* can be selected. The listing gives coordinate data, attribute tags, and attribute values. For example, selecting the "Tblock" block produces the following list:

Command: *ATTLST*
Select object: **PICK**
(-3.88095 1.25 0.0) "COMPANY": "CADD Design Company"
(-3.88214 0.5 0.0) "DES_NAME": "Des.- B.R. Smith"
(6.1581 6.88 0.0) "PROJ_TITLE": "Adjustable Mount"
(8.56667 6.88 0.0) "SCALE": "1/2"=1""
(8.53571 6.5 0.0) "CHK_NAME": "Chk.-JRS"
(6.2969 6.12 0.0) "COMP_DATE": "1/1/98"
(7.6069 6.13 0.0) "CHK_DATE": "1/1/98"
(8.80929 6.13 0.0) "PROJ_NO": "42B-ADJM"
Block TBLOCK has 2 constant attributes.
Block TBLOCK has 6 variable attributes.
Command:

CATTL

This command is similar to *Attlst,* except it lists only the *Constant* attributes in a *Block* insertion. Only one *Block* can be selected.

Command: *cattl*
Block name < >: *TBLOCK*
(-3.88095 1.25 0.0) "COMPANY": "CADD Design Company"
(-3.88214 0.5 0.0) "DES_NAME": "Des.- B.R. Smith"
Block TBLOCK has 2 constant attributes.
Command:

BLKTBL

Use *Blktbl* to produce a report (on the command line) of the current drawing's block table (all stored block definitions). Two *Blocks* are listed in this example:

```
Command: blktbl
((0 . "BLOCK") (2 . "TBLOCK") (70 . 2) (10 0.0 0.0 0.0) (-2 . <Entity name: 2b90668>))
((0 . "BLOCK") (2 . "CHAIR") (70 . 0) (10 0.0 0.0 0.0) (-2 . <Entity name: 2b90750>))
Command:
```

BLKLST

This utility lists the object types that make up a *Block* definition. Only one *Block* at a time can be selected from the drawing. The "Tblock" (title block) block was selected for this example:

```
Command: blklst
Block name <TBLOCK>:
"LWPOLYLINE"
"LINE"
"LINE"
"LINE"
"ATTDEF"
"ATTDEF"
"ATTDEF"
"ATTDEF"
Command:
```

BLOCK?

This utility lists coordinate and other information about objects composing a *Block* definition as shown in the following example:

```
Command: block?
Block name/<Return to select>: PICK
An entity type/<Return for all>: Enter
Press ESC to exit or any key to continue.

((0 . "BLOCK") (2 . "TBLOCK2") (70 . 2) (10 0.0 0.0 0.0) (-2 . <Entity name:
2b90668>))

((-1 . <Entity name: 2b90668>) (0 . "LWPOLYLINE") (5 . "65") (100 .
"AcDbEntity") (67 . 0) (8 . "0") (100 . "AcDbPolyline") (90 . 4) (70 . 0) (43 .
0.02) (38 . 0.0) (39 . 0.0) (10 -4.0 0.0) (40 . 0.02) (41 . 0.02) (42 . 0.0)
(10 -4.0 0.0) (40 . 0.02) (41 . 0.02) (42 . 0.0) (10 -4.0 1.625) (40 . 0.02)
(41 . 0.02) (42 . 0.0) (10 0.0 1.625) (40 . 0.02) (41 . 0.02) (42 . 0.0) (210
0.0 0.0 1.0))

((-1 . <Entity name: 2b90698>) (0 . "LINE") (5 . "6B") (100 . "AcDbEntity") (67
. 0) (8 . "0") (100 . "AcDbLine") (10 -4.0 0.375 0.0) (11 0.0 0.375 0.0) (210
0.0 0.0 1.0))
```

COUNT

The *Count* Bonus command counts and lists the number of *Block* insertions in each drawing or selection set. You can count all *Blocks* in a drawing or specify a selection Window. The following example is the *Count* results for the example schematic drawing used in Chapter 22:

```
Command: count
Press <CR> to select entire drawing or,
Select objects: Enter

Counting block insertions...Block  Count
----------
CAP ...... 3
RES ...... 3

Command:
```

CROSSREF

You can search a *Block* definition for references to a specific *Linetype, Style, Dimstyle, Mlinestyle,* or *Layer* with this utility command. For example, the following *Crossref* lists only one block on the specified layer:

```
Command: crossref
Cross reference Block/LType/Style/Dimstyle/Mlinestyle/<Layer>: Enter
Name of LAYER to cross reference: text
Scanning blocks for references to layer TEXT.
Scanning block: CAP
No nested references to layer TEXT found.
Command:
```

SSX

SSX is a powerful utility that allows you to find objects of similar type (*Circle, Arc, Line,* etc.) for the selection set by selecting only one object. You can also specify a certain criteria. This feature is simpler than using *Filter* but can be almost as powerful. Enter *SSX* at the command prompt to initialize the routine, then enter "(SSX)" at any "Select objects:" prompt to use it. For example, you can select all similar objects ("Flatfile" block in this case) by selecting only one:

```
Command: move
Select objects: (ssx)
Select object/<None>: PICK
Filter: ((0 . "INSERT") (2 . "FLATFILE") (8 . "FURNITURE") (210 0.0 0.0 1.0))
>>Block name/Color/Entity/Flag/LAyer/LType/Pick/Style/Thickness/Vector: Enter
10 found. 10 found
Select objects: Enter
Base point or displacement: PICK
Second point of displacement: PICK
Command:
```

PQCHECK

If you write AutoLISP programs, this utility is helpful for finding mismatched parentheses and closing quotes.

ASCPOINT

Ascpoint generates a continuous string of lines from an ASCII data file. This utility is useful for generating topology drawings from survey data.

A,B,C,D

APPENDICES

Contents

APPENDIX A

System Variables

Variable	Characteristic	Description
ACADPREFIX	(Read-only) Type: String Not saved	Stores the directory path, if any, specified by the ACAD environment variable, with path separators appended if necessary.
ACADVER	(Read-only) Type: String Not saved	Stores the AutoCAD version number, which can have values like 14 or 14a. This variable differs from the DXF file $ACADVER header variable, which contains the drawing database level number.
ACISOUTVER	Type: Integer Saved in: Drawing Initial value: 16	Controls the ACIS version of SAT files created using the *ACISOUT* command. Currently, *ACISOUT* supports only a value of 16 (for ACIS version 1.6). Future releases of AutoCAD will support additional ACIS versions.
AFLAGS	Type: Integer Not saved Initial value: 0	Sets attribute flags for *ATTDEF* bit-code. It is the sum of the following: 0 No attribute mode selected 1 Invisible 2 Constant 4 Verify 8 Preset
ANGBASE	Type: Real Saved in: Drawing Initial value: 0.0000	Sets the base angle to 0 with respect to the current UCS.
ANGDIR	Type: Integer Saved in: Drawing Initial value: 0	Sets the positive angle direction from angle 0 with respect to the current UCS. 0 Counter-clockwise 1 Clockwise
APBOX	Type: Integer Saved in: Registry Initial value: 1	Turns the AutoSnap aperture box on or off. The aperture box is displayed in the center of the cursor when you snap to an object. This option is available only when Marker, Magnet, or Snaptip is selected. 0 Aperture box is not displayed 1 Aperture box is displayed
APERTURE	Type: Integer Saved in: Registry Initial value: 10	Sets *APERTURE* command object snap target height, in pixels.
AREA	(Read-only) Type: Real Not saved	Stores the last area computed by *AREA, LIST,* or *DBLIST.*
ATTDIA	Type: Integer Saved in: Drawing Initial value: 0	Controls whether *INSERT* uses a dialog box for attribute value entry. 0 Issues prompts on the command line 1 Uses a dialog box

R14

R14

Variable	Characteristic	Description
ATTMODE	Type: Integer Saved in: Drawing Initial value: 1	Controls display of attributes. 0 Off makes all attributes invisible 1 Normal retains current visibility of each attribute: visible attributes are displayed; invisible attributes are not 2 On makes all attributes visible
ATTREQ	Type: Integer Saved in: Drawing Initial value: 1	Determines whether *INSERT* uses default attribute settings during insertion of blocks. 0 Assumes the defaults for the values of all attributes 1 Turns on prompts or dialog box for attribute values, as specified at *ATTDIA*
AUDITCTL	Type: Integer Saved in: Registry Initial value: 0	Controls whether *AUDIT* creates an ADT file (audit report). 0 Prevents writing of ADT files 1 Writes ADT files
AUNITS	Type: Integer Saved in: Drawing Initial value: 0	Sets units for angles. 0 Decimal degrees 1 Degrees/minutes/seconds 2 Gradians 3 Radians 4 Surveyor's units
AUPREC	Type: Integer Saved in: Drawing Initial value: 0	Sets number of decimal places for angular units displayed on the Status line or when listing an object.
AUTOSNAP	Type: Integer Saved in: Registry Initial value: 7	Controls the display of the AutoSnap marker and Snaptip and turns the AutoSnap magnet on or off. It is the sum of the following bit values: 0 Turns off the marker, Snaptip, and magnet 1 Turns on the marker 2 Turns on the Snaptip 4 Turns on the magnet
BACKZ	(Read-only) Type: Real Saved in: Drawing	Stores the back clipping plane offset from the target plane for the current viewport, in drawing units. Meaningful only if the back clipping bit in VIEWMODE is on. The distance of the back clipping plane from the camera point can be found by subtracting *BACKZ* from the camera-to-target distance.
BLIPMODE	Type: Integer Saved in: Drawing Initial value: 0	Controls whether marker blips are visible. 0 Turns off marker blips 1 Turns on marker blips
CDATE	(Read-only) Type: Real Not saved	Sets calendar date and time.

R14

Variable	Characteristic	Description
CECOLOR	Type: String Saved in: Drawing Initial value: "BY-LAYER"	Sets the color of new objects.
CELTSCALE	Type: Real Saved in: Drawing Initial value: 1.0000	Sets the current object linetype scaling factor. This sets the linetype scaling for new objects relative to the *LTSCALE* setting. A line created with *CELTSCALE*=2 in a drawing with *LTSCALE* set to 0.5 would appear the same as a line created with *CELTSCALE*=1 in a drawing with *LTSCALE*=1.
CELTYPE	Type: String Saved in: Drawing Initial value: "BY-LAYER"	Sets the linetype of new objects.
CHAMFERA	Type: Real Saved in: Drawing Initial value: 0.5000	Sets the first chamfer distance.
CHAMFERB	Type: Real Saved in: Drawing Initial value: 0.5000	Sets the second chamfer distance.
CHAMFERC	Type: Real Saved in: Drawing Initial value: 1.0000	Sets the first chamfer length.
CHAMFERD	Type: Real Saved in: Drawing Initial value: 0.0000	Sets the first chamfer angle.
CHAMMODE	Type: Integer Not saved Initial value: 0	Sets the input method by which AutoCAD creates chamfers. 0 Requires two chamfer distances 1 Requires one chamfer distance and an angle
CIRCLERAD	Type: Real Not saved Initial value: 0.0000	Sets the default circle radius. A zero sets no default.
CLAYER	Type: String Saved in: Drawing Initial value: 0	Sets the current layer.
CMDACTIVE	(Read-only) Type: Integer Not saved	Stores the bit code that indicates whether an ordinary command, transparent command, script, or dialog box is active. It is the sum of the following: 1 Ordinary command is active 2 Ordinary command and a transparent command are active 4 Script is active 8 Dialog box is active 16 AutoLISP or *DDE* command is active

Variable	Characteristic	Description
CMDDIA	Type: Integer Saved in: Registry Initial value: 1	Controls whether dialog boxes are turned on for *PLOT* and external database commands. 0 Turns off dialog boxes 1 Turns on dialog boxes
CMDECHO	Type: Integer Not saved Initial value: 1	Controls whether AutoCAD echoes prompts and input during the AutoLISP **(command)** function. 0 Turns off echoing 1 turns on echoing
CMDNAMES	(Read-only) Type: String Not saved	Displays the name of the currently active command and transparent command. For example, *LINE'ZOOM* indicates that the *ZOOM* command is being used transparently during the *LINE* command. This variable is designed for use with programming interfaces such as AutoLISP, DIESEL, and ActiveX Automation.
CMLJUST	Type: Integer Saved in: Drawing Initial value: 0	Specifies multiline justification. 0 Top 1 Middle 2 Bottom
CMLSCALE	Type: Real Saved in: Drawing Initial value: 1.0000	Controls the overall width of a multiline. A scale factor of 2.0 produces a multiline that is twice as wide as the style definition. A zero scale factor collapses the multiline into a single line. A negative scale factor flips the order of the offset lines (that is, the smallest or most negative is placed on top when the multiline is drawn from left to right).
CMLSTYLE	Type: String Saved in: Drawing Initial value: "STANDARD"	Sets the multiline style that AutoCAD uses to draw the multiline.
COORDS	Type: Integer Saved in: Drawing Initial value: 1	Controls when coordinates are updated on the Status line. 0 Coordinate display is updated as you specify points with the pointing device 1 Display of absolute coordinates is continuously updated 2 Display of absolute coordinates is continuously updated, and distance and angle from last point are displayed when a distance or angle is requested.
CURSORSIZE	Type: Integer Saved in: Registry Initial value: 5	Determines the size of the cursor as a percentage of the screen size. Valid settings range from 1 to 100 percent. When set to 100, the cursor are full-screen and the ends of the cursor are never visible. When less than 100, the ends of the cursor may be visible when the cursor is moved to one edge of the screen.

R13

R14

Variable	Characteristic	Description
CVPORT	Type: Integer Saved in: Drawing Initial value: 2	Sets the identification number of the current viewport. You can change this value, thereby changing the current viewport, if the following conditions are met: • The identification number you specify is that of an active viewport. • A command in progress has not locked cursor movement to that viewport. • Tablet mode is off.
DATE	(Read-only) Type: Real Not saved	Stores the current date and time represented as a Julian date and fraction in a real number: <Julian date>.<Fraction> For example, on January 29, 1993, at 2:29:35 in the afternoon, the *DATE* variable would contain 2446460.603877364. Your computer clock provides the date and time. The time is represented as a fraction of a day. To compute differences in time, subtract the times returned by *DATE*. To extract the seconds since midnight from the value returned by *DATE*, use AutoLISP expressions: **(setq s (getvar "DATE"))** **(setq seconds (* 86400.0 – s (fix s)))** The *DATE* system variable returns a true Julian date only if the system clock is set to UTC/Zulu (Greenwich Mean Time). *TDCREATE* and *TDUPDATE* have the same format as *DATE,* but their values represent the creation time and last update time of the current drawing.
DBMOD	(Read-only) Type: Integer Not saved	Indicates the drawing modification status using bit code. It is the sum of the following: 1 Object database modified 4 Database variable modified 8 Window modified 16 View modified
DCTCUST	Type: String Saved in: Registry Initial value: ""	Displays the path and file name of the current custom spelling dictionary.
DCTMAIN	Type: String Saved in: Registry Initial value: *varies by country*	Displays the file name of the current main spelling dictionary. The full path is not shown, because this file is expected to reside in the *support* directory. You can specify a default main spelling dictionary using *SETVAR*. When prompted for a new value for *DCTMAIN,* you can enter one of the keywords below. *Keyword Language Name* enu American English ena Australian English

Variable	Characteristic	Description
DCTMAIN (continued)		ens — British English (ise) enz — British English (ize) ca — Catalan cs — Czech da — Danish nl — Dutch (primary) nls — Dutch (secondary) fi — Finnish fr — French (unaccented capitals) fra — French (accented capitals) de — German (Scharfes s) ded — German (Dopple s) it — Italian no — Norwegian (Bokmal) non — Norwegian (Nynorsk) pt — Portuguese (Iberian) ptb — Portuguese (Brazilian) ru — Russian (infrequent io) rui — Russian (frequent io) es — Spanish (unaccented capitals) esa — Spanish (accented capitals) sv — Swedish
DELOBJ	Type: Integer Saved in: Drawing Initial value: 1	Controls whether objects used to create other objects are retained or deleted from the drawing database. 0 Objects are retained 1 Objects are deleted
DEMANDLOAD	Type: Integer Saved in: Registry Initial value: 3	Specifies if and when AutoCAD demand loads a third-party application if a drawing contains custom objects created in that application. 0 Turns off demand loading. 1 Demand loads the source application when you open a drawing that contains custom objects. This setting does not demand load the application when you invoke one of the application's commands. 2 Demand loads the source application when you invoke one of the application's commands. This setting does not demand load the application when you open a drawing that contains custom objects. 3 Demand loads the source application when you open a drawing that contains custom objects or when you invoke one of the application's commands.
DIASTAT	(Read-only) Type: Integer Not saved	Stores the exit method of the most recently used dialog box. 0 Cancel 1 OK

R13

R14

Variable	Characteristic	Description
DIMADEC	Type: Integer Saved in: Drawing Initial value: -1	Controls the number of places of precision displayed for angular dimension text. The default value (-1) uses the *DIMDEC* system variable setting to generate places of precision. When *DIMADEC* is set to a value from 0 to 8, angular dimensions display precisions that differ from their linear dimension values. -1 Angular dimension is drawn using the number of decimal places corresponding to the *DIMDEC* setting 0-8 Angular dimension is drawn using the number of decimal places corresponding to the *DIMADEC* setting
DIMALT	Type: Switch Saved in: Drawing Initial value: Off	Controls use of alternate units in dimensions. See also *DIMALTD, DIMALTF, DIMALTZ, DIMALTTZ, DIMALTTD,* and *DIMAPOST.* Off Disables alternate units On Enables alternate units
DIMALTD	Type: Integer Saved in: Drawing Initial value: 2	Controls the number of decimal places in alternate units.
DIMALTF	Type: Real Saved in: Drawing Initial value: 25.4000	Controls scale factor in alternate units. If *DIMALT* is turned on, *DIMALTF* multiplies linear dimensions by a factor to produce a value in an alternate system of measurement. The initial value represents the number of millimeters in an inch.
DIMALTTD	Type: Integer Saved in: Drawing Initial value: 2	Sets the number of decimal places for the tolerance values in the alternate units of a dimension. *DIMALTTD* sets this value when you enter it on the command line or set it under Alternate Units in the *DDIM* Annotation dialog box.
DIMALTTZ	Type: Integer Saved in: Drawing Initial value: 0	Toggles suppression of zeros in tolerance values. 0 Suppresses zero feet and precisely zero inches. 1 Includes zero feet and precisely zero inches. 2 Includes zero feet and suppresses zero inches. 3 Includes zero inches and suppresses zero feet. To the preceding values, add: 4 Suppresses leading zeros. 8 Suppresses trailing zeros. *DIMALTTZ* sets this value when you enter it on the command line or set it under Alternate Units in the Annotation dialog box.
DIMALTU	Type: Integer Saved in: Drawing Initial value: 2	Sets the units format for alternate units of all dimension style family members except angular. 1 Scientific 2 Decimal

R14

R13

Variable	Characteristic	Description
DIMALTU (continued)		3 Engineering 4 Architectural (stacked) 5 Fractional (stacked) 6 Architectural (stacked) 7 Fractional 8 Windows Desktop (decimal format using Control Panel settings for decimal separator and number group systems) *DIMALTU* sets this value when you enter it on the command line or set it under Alternate Units in the Annotation dialog box.
DIMALTZ	Type: Integer Saved in: Drawing Initial value: 0	Controls the suppression of zeros for alternate unit dimension values. *DIMALTZ* stores this value when you enter it on the command line or set it under Alternate Units in the Annotation dialog box. *DIMALTZ* values 0-3 affect feet-and-inch dimensions only. 0 Suppresses zero feet and precisely zero inches 1 Includes zero feet and precisely zero inches 2 Includes zero feet and suppresses zero inches 3 Includes zero inches and suppresses zero feet 4 Suppresses leading zeroes in decimal dimensions (for example, 0.5000 becomes .5000) 8 Suppresses trailing zeroes in decimal dimensions (for example, 12.5000 becomes 12.5) 12 Suppresses both leading and trailing zeroes (for example, 0.5000 becomes .5)
DIMAPOST	Type: String Saved in: Drawing Initial value: ""	Specifies a text prefix or suffix (or both) to the alternate dimension measurement for all types of dimensions except angular. For instance, if the current units are Architectural, *DIMALT* is on, *DIMALTF* is 25.4 (the number of millimeters per inch), *DIMALTD* is 2, and *DIMAPOST* is set to "mm," a distance of 10 units would be displayed as 10"[254.00mm]. To turn off an established prefix or suffix (or both), set it to a single period (.).
DIMASO	Type: Switch Saved in: Drawing Initial value: On	Controls the associative of dimension objects. Off Creates no association between the various elements of the dimension. The lines, arcs, arrowheads, and text of a dimension are drawn as separate objects. On Creates an association between the elements of the dimension. The elements are formed into a single object. If the definition point on the object moves, the dimension value is updated. The *DIMASO* value is not stored in a dimension style.

R13

Variable	Characteristic	Description
DIMASZ	Type: Real Saved in: Drawing Initial value: 0.1800	Controls the size of dimension line and leader line arrowheads. Also controls the size of hook lines. Multiples of the arrowhead size determine whether dimension lines and text should fit between the extension lines. *DIMASZ* is also used to scale arrowhead blocks if set by *DIMBLK*. *DIMASZ* has no effect when *DIMTSZ* is other than zero.
DIMAUNIT	Type: Integer Saved in: drawing Initial value: 0	Sets the angle format for angular dimension. 0 Decimal degrees 1 Degrees/minutes/seconds 2 Gradians 3 Radians 4 Surveyor's units *DIMAUNIT* sets this value when entered on the command line or set from the Primary Units area of the Annotation dialog box.
DIMBLK	Type: String Saved in: Drawing Initial value: ""	Sets the name of a block to be drawn instead of the normal arrowhead at the ends of dimension lines or leader lines. To turn off an established block name, set it to a single period (.).
DIMBLK1	Type: String Saved in: Drawing Initial value: ""	If *DIMSAH* is on, specifies a user-defined arrowhead block for the first end of the dimension line. This variable contains the name of a previously defined block. To turn off an established block name, set it to a single period (.).
DIMBLK2	Type: String Saved in: Drawing Initial value: ""	If *DIMSAH* is on, specifies a user-defined arrowhead block for the second end of the dimension line. This variable contains the name of a previously defined block. To turn off an established block name, set it to a single period (.).
DIMCEN	Type: Real Saved in: Drawing Initial value: 0.0900	Controls drawing of circle or arc center marks and centerlines by *DIMCENTER, DIMDIAMETER,* and *DIMRADIUS*. 0 No center marks or lines are drawn <0 Centerlines are drawn >0 Center marks are drawn The absolute value specifies the size of the mark portion of the centerline. *DIMDIAMETER* and *DIMRADIUS* draw the center mark or line only if the dimension line is placed outside the circle or arc.

Variable	Characteristic	Description
DIMCLRD	Type: Integer Saved in: Drawing Initial value: 0	Assigns colors to dimension lines, arrowheads, and dimension leader lines. Also controls the color of leader lines created with *LEADER*. The color can be any valid color number. Integer equivalents for *BYBLOCK* and *BYLAYER* are 0 and 256, respectively.
DIMCLRE	Type: Integer Saved in: Drawing Initial value: 0	Assigns colors to extension lines of dimensions. The color can be any valid color number. See *DIMCLRD*.
DIMCLRT	Type: Integer Saved in: Drawing Initial value: 0	Assigns colors to dimension text. The color can be any valid color. See *DIMCLRD*.
DIMDEC	Type: Integer Saved in: Drawing Initial value: 4	Sets the number of decimal places displayed for the primary units of a dimension. When the dimension family member is Parent, the precision is based on the units or angle format you have selected. *DIMDEC* stores this value when you enter it on the command line or set it under Primary Units in the Annotation dialog box.
DIMDLE	Type: Real Saved in: Drawing Initial value: 0.0000	Sets the distance the dimension line extends beyond the extension line when oblique strokes are drawn instead of arrowheads.
DIMDLI	Type: Real Saved in: Drawing Initial value: 0.3800	Controls the spacing of the dimension lines in baseline dimensions. Each dimension line is offset from the previous line by this amount, if necessary, to avoid drawing over it. Changes made with *DIMDLI* are not applied to existing dimensions.
DIMEXE	Type: Real Saved in: Drawing Initial value: 0.1800	Specifies how far to extend the extension line beyond the dimension line.
DIMEXO	Type: Real Saved in: Drawing Initial value: 0.0625	Specifies how far extension lines are offset from origin points. If you point directly at the corners of an object to be dimensioned, the extension lines do not touch the object.
DIMFIT	Type: Integer Saved in: Drawing Initial value: 3	Controls the placement of text and arrowheads inside or outside extension lines based on the available space between the extension lines. When space is available, AutoCAD always places text and arrowheads between the extension lines. Otherwise, the *DIMFIT* setting affects the placement of text and arrowheads as follows: 0 Text and arrows 1 Text only 2 Arrows only 3 Best fit

Variable	Characteristic	Description
DIMFIT (continued)		4 Leader 5 No leader *DIMFIT* sets this value when you enter it on the command line or set it under Fit in the Format dialog box.
DIMGAP	Type: Real Saved in: Drawing Initial value: 0.0900	Sets the distance around the dimension text when the dimension line breaks to accommodate dimension text. Also sets the gap between annotation and a hook line created with *LEADER*. A negative *DIMGAP* value creates basic dimensioning—dimension text with a box around it. AutoCAD also uses *DIMGAP* as the minimum length for pieces of the dimension line. When calculating the default position for the dimension text, it positions the text inside the extension lines only if doing so breaks the dimension lines into two segments at least as long as *DIMGAP*. Text placed above or below the dimension line is moved inside only if there is room for the arrowheads, dimension text, and a margin between them at least as large as *DIMGAP*: 2 * (*DIMASZ* + *DIMGAP*). *DIMGAP* also sets the gap between a tolerance symbol and its feature control frame.
DIMJUST	Type: Integer Saved in: Drawing Initial value: 0	Controls the horizontal position of dimension text. 0 Centered along the dimension line between the extension lines 1 Next to the first extension line 2 Next to the second extension line 3 Above and aligned with the first extension line 4 Above and aligned with the second extension line
DIMLFAC	Type: Real Saved in: Drawing Initial value: 1.0000	Sets a global scale factor for linear dimensioning measurements. All linear distances measured by dimensioning (including radii, diameters, and coordinates) are multiplied by the *DIMLFAC* setting before being converted to dimension text. *DIMLFAC* has no effect on angular dimensions, and it is not applied to the values held in *DIMTM*, *DIMTP*, or *DIMRND*. If you are creating a dimension in paper space and *DIMLFAC* is not zero, AutoCAD multiplies the distance measured by the absolute value of *DIMLFAC*. In model space, negative values for *DIMLFAC* are ignored, and the value 1.0 is used instead.

R13

R13

Variable	Characteristic	Description
DIMLFAC (continued)		AutoCAD computes a value for *DIMLFAC* if you try to change *DIMLFAC* from the Dim prompt while in paper space and you select the *Viewport* option. Dim: **dimlfac** Current value <1.0000> New value (Viewport): **v** Select viewport to set scale: AutoCAD calculates the scaling of model space to paper space and assigns the negative of this value to *DIMLFAC*.
DIMLIM	Type: Switch Saved in: Drawing Initial value: Off	Generates dimension limits as the default text. Setting *DIMLIM* on forces *DIMTOL* to be off. Off Dimension limits are not generated as default text On Dimension limits are generated as default text
DIMPOST	Type: String Saved in: Drawing Initial value: ""	Specifies a text prefix or suffix (or both) to the dimension measurement. For example, to establish a suffix for millimeters, set *DIMPOST* to mm; a distance of 19.2 units would be displayed as 19.2mm. If tolerances are turned on, the suffix is applied to the tolerances as well as to the main dimension. Use <> to indicate placement of the text in relation to the dimension value. For example, enter **<>mm** to display a 5.0 millimeter radial dimension as "5.0mm." If you entered **mm<>,** the dimension would be displayed as "mm5.0." Use the <> mechanism for angular dimensions.
DIMRND	Type: Real Saved in: Drawing Initial value: 0.0000	Rounds all dimensioning distances to the specified value. For instance, if *DIMRND* is set to 0.25, all distances round to the nearest 0.25 unit. If you set *DIMRND* to 1.0, all distances round to the nearest integer. Note that the number of digits edited after the decimal point depends on the precision set by *DIMDEC*. *DIMRND* does not apply to angular dimensions.
DIMSAH	Type: Switch Saved in: Drawing Initial value: Off	Controls use of user-defined arrowhead blocks at the ends of the dimension line. Off Normal arrowheads or user-defined arrowhead blocks set by *DIMBLK* are used On User-defined arrowhead blocks are used *DIMBLK1* and *DIMBLK2* specify different user-defined arrowhead blocks for each end of the dimension line.

Variable	Characteristic	Description
DIMSCALE	Type: Real Saved in: Drawing Initial value: 1.0000	Sets the overall scale factor applied to dimensioning variables that specify sizes, distances, or offsets. Also affects the scale of leader objects created with *LEADER*. 0.0 AutoCAD computes a reasonable default value based on the scaling between the current model space viewport and paper space. If you are in paper space, or in model space and not using the paper space feature, the scale factor is 1.0. >0 AutoCAD computes a scale factor that leads text sizes, arrowhead sizes, and other scaled distances to plot at their face value. *DIMSCALE* does not affect tolerances or measured lengths, coordinates, or angles.
DIMSD1	Type: Switch Saved in: Drawing Initial value: Off	Controls suppression of the first dimension line. When turned on, suppresses the display of the dimension line and arrowhead between the first extension line and the text.
DIMSD2	Type: Switch Saved in: Drawing Initial value: Off	Controls suppression of the second dimension line. When turned on, suppresses the display of the dimension line and arrowhead between the second extension line and the text.
DIMSE1	Type: Switch Saved in: Drawing Initial value: Off	Suppresses display of the first extension line. Off Extension line is not suppressed On Extension line is suppressed
DIMSE2	Type: Switch Saved in: Drawing Initial value: Off	Suppresses display of the second extension line. Off Extension line is not suppressed On Extension line is suppressed
DIMSHO	Type: Switch Saved in: Drawing Initial value: On	Controls redefinition of dimension objects while dragging. Associative dimensions recompute dynamically as they are dragged. On some computers, dynamic dragging can be very slow, so you can set *DIMSHO* to off to drag the original image instead. The *DIMSHO* value is not stored in a dimension style.
DIMSOXD	Type: Switch Saved in: Drawing Initial value: Off	Suppresses drawing of dimension lines outside the extension lines. Off Dimensions lines are not suppressed On Dimension lines are suppressed If the dimension lines would be outside the extension lines and *DIMTIX* is on, setting *DIMSOXD* to on suppresses the dimension line. If *DIMTIX* is off, *DIMSOXD* has not effect.
DIMSTYLE	(Read-only) Type: String Saved in: Drawing	Sets the current dimension style by name. To change the dimension style, use *DDIM* or *DIMSTYLE*.

R13

Variable	Characteristic	Description
DIMTAD	Type: Integer Saved in: Drawing Initial value: 0	Controls the vertical position of text in relation to the dimension line. 0 Centers the dimension text between the extension lines. 1 Places the dimension text above the dimension line except when the dimension line is not horizontal and text inside the extension lines is forced horizontal (*DIMTIH* = 1). The distance from the dimension line to the baseline of the lowest line of text is the current *DIMGAP* value. 2 Places the dimension text on the side of the dimension line farthest away from the defining points. 3 Places the dimension text to conform to Japanese Industrial Standards (JIS).
DIMTDEC	Type: Integer Saved in: Drawing Initial value: 4	Sets the number of decimal places to display in tolerance values for the primary units in a dimension. *DIMTDEC* stores this value when you enter it on the command line or set it under Primary Units in the Annotation dialog box.
DIMTFAC	Type: Real Saved in: Drawing Initial value: 1.0000	Specifies the height of tolerance text and the relative size of stacked fractions. Use *DIMTFAC* for plus and minus tolerance strings when *DIMTOL* is on and *DIMTM* is not equal to *DIMTP* or when *DIMLIM* is on.
DIMTIH	Type: Switch Saved in: Drawing Initial value: On	Controls the position of dimension text inside the extension lines for all dimension types except ordinate. Off Aligns text with the dimension line On Draws text horizontally
DIMTIX	Type: Switch Saved in: Drawing Initial value: Off	Draws text between extension lines. Off The result varies with the type of dimension. For linear and angular dimensions, AutoCAD places text inside the extension lines if there is sufficient room. For radius and diameter dimensions that don't fit inside the circle or arc, *DIMTIX* has no effect and always forces the text outside the circle or arc. On Draws dimension text between the extension lines even if AutoCAD would ordinarily place it outside those lines.
DIMTM	Type: Real Saved in: Drawing Initial value: 0.0000	When *DIMTOL* or *DIMLIM* is on, *DIMTM* sets the minimum (or lower) tolerance limit for dimension text. AutoCAD accepts signed values for *DIMTM*. If *DIMTOL* is on and *DIMTP* and *DIMTM* are set to the same value, AutoCAD draws a tolerance value.

Variable	Characteristic	Description
DIMTM (continued)		If *DIMTM* and *DIMTP* values differ, the upper tolerance is drawn above the lower, and a plus sign is added to the *DIMTP* value if it is positive. For *DIMTM*, AutoCAD uses the negative of the value you enter (adding a minus sign if you specify a positive number and a plus sign if you specify a negative number). No sign is added to a value of zero.
DIMTOFL	Type: Switch Saved in: Drawing Initial value: Off	Controls whether a dimension line is drawn between the extension lines even when the text is placed outside. For radius and diameter dimensions (when *DIMTIX* is off), it draws a dimension line and arrowheads inside the circle or arc and places the text and leader outside. Off Does not draw dimensions lines between the measured points when arrowheads are placed outside the measured points On Draws dimension lines between the measured points even when arrowheads are placed outside the measured points
DIMTOH	Type: Switch Saved in: Drawing Initial value: On	When turned on, controls the position of dimension text outside the extension lines. Off Aligns text with the dimension line On Draws text horizontally
DIMTOL	Type: Switch Saved in: Drawing Initial value: Off	Appends tolerances to dimension text. Setting *DIMTOL* on forces *DIMLIM* off.
DIMTOLJ	Type: Integer Saved in: Drawing Initial value: 1	Sets the vertical justification for tolerance values relative to the nominal dimension text. 0 Bottom 1 Middle 2 Top
DIMTP	Type: Real Saved in: Drawing Initial value: 0.0000	When *DIMTOL* or *DIMLIM* is on, sets the maximum (or upper) tolerance limit for dimension text. AutoCAD accepts signed values for *DIMTP*. If *DIMTOL* is on and *DIMTP* and *DIMTM* are set to the same value, AutoCAD draws a tolerance value. If *DIMTM* and *DIMTP* values differ, the upper tolerance is drawn above the lower and a plus sign is added to the *DIMTP* value if it is positive.
DIMTSZ	Type: Real Saved in: Drawing Initial value: 0.0000	Specifies the size of oblique strokes drawn instead of arrowheads for linear, radius, and diameter dimensioning. 0 Draws arrowheads >0 Draws oblique strokes instead of arrowheads The size of the oblique strokes is determined by this value multiplied by the *DIMSCALE* value.

R13

Variable	Characteristic	Description
DIMTVP	Type: Real Saved in: Drawing Initial value: 0.0000	Controls the vertical position of dimension text above or below the dimension line. AutoCAD uses the *DIMTVP* value when *DIMTAD* is off. The magnitude of the vertical offset of text is the product of the text height and *DIMTVP*. Setting *DIMTVP* to 1.0 is equivalent to setting *DIMTAD* to on. AutoCAD splits the dimension line to accommodate the text only if the absolute value of *DIMTVP* is less than 0.7.
DIMTXSTY	Type: String Saved in: Drawing Initial value: "STANDARD"	Specifies the text style of the dimension.
DIMTXT	Type: Real Saved in: Drawing Initial value: 0.1800	Specifies the height of dimension text, unless the current text style has a fixed height.
DIMTZIN	Type: Integer Saved in: Drawing Initial value: 0	Controls the suppression of zeros in tolerance values. *DIMTZIN* stores this value when you enter it on the command line or set it under Primary Units in the Annotation dialog box. *DIMTZIN* values 0-3 affect feet-and-inch dimensions only. 0 Suppresses zero feet and precisely zero inches 1 Includes zero feet and precisely zero inches 2 Includes zero feet and suppresses zero inches 3 Includes zero inches and suppresses zero feet 4 Suppresses leading zeroes in decimal dimensions (for example, 0.5000 becomes .5000) 8 Suppresses trailing zeroes in decimal dimensions (for example, 12.5000 becomes 12.5) 12 Suppresses both leading and trailing zeroes (for example, 0.5000 becomes .5)
DIMUNIT	Type: Integer Saved in: Drawing Initial value: 2	Sets the units format for all dimension style family members except angular. 1 Scientific 2 Decimal 3 Engineering 4 Architectural (stacked) 5 Fractional (stacked) 6 Architectural 7 Fractional 8 Windows Desktop (decimal format using Control Panel settings for decimal separator and number grouping symbols)
DIMUPT	Type: Switch Saved in: Drawing Initial value: Off	Controls options for user-positioned text. Off Cursor controls only the dimension line location On Cursor controls both the text position and the dimension line location

Variable	Characteristic	Description
DIMZIN	Type: Integer Saved in: Drawing Initial value: 0	Controls the suppression of zeroes in the primary unit value. *DIMZIN* stores this value when you enter it on the command line or set it under Primary Units in the Annotation dialog box. *DIMZIN* values 0-3 affect feet-and-inch dimensions only. 0 Suppresses zero feet and precisely zero inches 1 Includes zero feet and precisely zero inches 2 Includes zero feet and suppresses zero inches 3 Includes zero inches and suppresses zero feet 4 Suppresses leading zeroes in decimal dimensions (for example, 0.5000 becomes .5000) 8 Suppresses trailing zeroes in decimal dimensions (for example, 12.5000 becomes 12.5) 12 Suppresses both leading and trailing zeroes (for example, 0.5000 becomes .5) *DIMZIN* also affects real-to-string conversions performed by the AutoLISP **(rtos)** and **(angtos)** functions.
DISPSILH	Type: Integer Saved in: Drawing Initial value: 0	Controls display of silhouette curves of solid objects in wireframe mode. Also controls whether mesh is drawn or suppressed when a solid object is hidden. 0 Off 1 On
DISTANCE	(Read-only) Type: Real Not saved	Stores the distance computed by *DIST*.
DONUTID	Type: Real Not saved Initial value: 0.5000	Sets the default for the inside diameter of a donut.
DONUTOD	Type: Real Not saved Initial value: 1.0000	Sets the default for the outside diameter of a donut. It must be nonzero. If *DONUTID* is larger than *DONUTOD*, the two values are swapped by the next command.
DRAGMODE	Type: Integer Saved in: Drawing Initial value: 2	Controls display of objects being dragged. 0 Does not display an outline of the object as you drag it 1 Displays the outline of the object as you drag it only if you enter **drag** on the command line after selecting the object to drag 2 Auto; always displays an outline of the object as you drag it
DRAGP1	Type: Integer Saved in: Registry Initial value: 10	Sets regen-drag input sampling rate.
DRAGP2	Type: Integer Saved in: Registry Initial value: 25	Sets fast-drag input sampling rate.

Variable	Characteristic	Description
DWGCODEPAGE	(Read-only) Type: String Saved in: Drawing	Stores the same value as *SYSCODEPAGE* (for compatibility).
DWGNAME	(Read-only) Type: String Not saved	Stores the drawing name as entered by the user. If the drawing has not been named yet, *DWGNAME* defaults to "Drawing.dwg." If the user specified a drive/directory prefix, it is stored in the *DWGPREFIX* variable.
DWGPREFIX	(Read-only) Type: String Not saved	Stores the drive/directory prefix for the drawing.
DWGTITLED	(Read-only) Type: Integer Not saved	Indicates whether the current drawing has been named. 0 Drawing has not been named 1 Drawing has been named
EDGEMODE	Type: Integer Saved in: Registry Initial value: 0	Controls how *TRIM* and *EXTEND* determine cutting and boundary edges. 0 Uses the selected edge without an extension 1 Extends or trims the selected object to an imaginary extension of the cutting or boundary edge Line, arc, elliptical arc, ray, and polyline are objects eligible for natural extension. The natural extension of a line or ray is an unbounded line (xline); an arc is a circle; and an elliptical arc is an ellipse. A polyline is broken down into its line and arc components, which are extended to their natural boundaries.
ELEVATION	Type: Real Saved in: Drawing Initial value: 0.0000	Stores the current elevation relative to the current UCS for the current space.
EXPERT	Type: Integer Not saved Initial value: 0	Controls whether certain prompts are issued. 0 Issues all prompts normally. 1 Suppresses "About to regen, proceed?" and "Really want to turn the current layer off?" 2 Suppresses the preceding prompts and "Block already defined. Redefine it?" (*BLOCK*) and "A drawing with this name already exists. Overwriteit?" (*SAVE* or *WBLOCK*). 3 Suppresses the preceding prompts and those issued by *LINETYPE* if you try to load a linetype that's already loaded or create a new linetype in a file that already defines it. 4 Suppresses the preceding prompts and those issued by *UCS Save* and *VPORTS Save* if the name you supply already exists.

R13

Variable	Characteristic	Description
EXPERT (continued)		5 Suppresses the preceding prompts and those issued by the *DIMSTYLE Save* option and *DIMOVERRIDE* if the dimension style name you supply already exists (the entries are redefined). When a prompt is suppressed by *EXPERT*, the operation in question is performed as though you entered **y** at the prompt. The setting of *EXPERT* can affect scripts, menu macros, AutoLISP, and the command functions.
EXPLMODE	Type: Integer Not saved Initial value: 1	Controls whether *EXPLODE* supports non-uniformly scaled (NUS) blocks. 0 Does not explode NUS blocks 1 Explodes NUS blocks
EXTMAX	(Read-only) Type: 3D Point Saved in: Drawing	Stores the upper-right point of the drawing extents. Expands outward as new objects are drawn, shrinks only with *ZOOM All* or *ZOOM Extents*. Reported in World coordinates for the current space.
EXTMIN	(Read-only) Type: 3D Point Saved in: Drawing	Stores the lower-left point of the drawing extents. Expands outward as new objects are drawn, shrinks only with *ZOOM All* or *ZOOM Extents*. Reported in World coordinates for the current space.
FACETRES	Type: Real Saved in: Drawing Initial value: 0.5	Further adjusts the smoothness of shaded and rendered objects and objects with hidden lines removed. Valid values are from 0.01 to 10.0.
FILEDIA	Type: Integer Saved in: Registry Initial value: 1	Suppresses display of the file dialog boxes. 0 Dialog boxes are not displayed. You can still request a file dialog box to appear by entering a tilde (~) in response to the command's prompt. The same is true for AutoLISP and ADS functions. 1 File dialog boxes are displayed. However, if a script or AutoLISP/ADS program is active, AutoCAD displays an ordinary prompt.
FILLETRAD	Type: Real Saved in: Drawing Initial value: 0.5000	Stores the current fillet radius.
FILLMODE	Type: Integer Saved in: Drawing Initial value: 1	Specifies whether multilines, traces, solids, solid-fill hatches, and wide polylines are filled in. 0 Objects are not filled in 1 Objects are filled
FONTALT	Type: String Saved in: Registry Initial value: "simple.shx"	Specifies the alternate font to be used when the specified font file cannot be located. If an alternate font is not specified, AutoCAD displays the Alternate Font dialog box. The dialog box is displayed in the following cases:

Variable	Characteristic	Description
FONTALT (continued)		1. A Release 13 drawing is opened; *FONTALT* is not set or not found; and a TrueType, SHX, or PostScript font is not found for a defined text style. 2. A Release 14 drawing is opened, *FONTALT* is not set or not found, and a SHX or PostScript font is not found for a defined text style. For missing TrueType fonts in Release 14 drawings, AutoCAD automatically substitutes the closest TrueType font available. 3. The Browse button is pressed in the Preferences dialog box when you specify an alternate font. AutoCAD validates the alternate font specified for *FONTALT*. If the font name or font file name is not found, the message "Font not found" is displayed. Enter either a TrueType font name (for example, Times New Roman Bold) or a TrueType file name (for example *timebd.ttf*). When a TrueType file name is entered for *FONTALT*, AutoCAD returns the font name in place of the file name) if the font is registered with the operating system.
FONTMAP	Type: String Saved in: Registry Initial value: "acad.fmp"	Specifies the font mapping file to be used. A font mapping file contains one font mapping per line; the original font used in the drawing and the font to be substituted for it are separated by a semicolon (;). For example, to substitute the Times TrueType font for the Romans font, the line in the mapping file would read as follows: **romanc.shx;times.ttf** If *FONTMAP* does not point to a font mapping file, if the FMP file is not found, or if the font file name specified in the FMP file is not found, AutoCAD uses the font defined in the style. If the font in the style is not found, AutoCAD substitutes the font according to substitution rules.
FRONTZ	(Read-only) Type: Real Saved in: Drawing	Stores the front clipping plane offset from the target plane for the current viewport, in drawing units. Meaningful only if the front clipping bit in *VIEWMODE* is on and the front-clip-not-at-eye bit is also on. The distance of the front clipping plane from the camera point is found by subtracting *FRONTZ* from the camera-to-target distance.
GRIDMODE	Type: Integer Saved in: Drawing Initial value: 0	Specifies whether the grid is turned on or off. 0 Turns the grid off 1 Turns the grid on

R13

Variable	Characteristic	Description
GRIDUNIT	Type: 2D point Saved in: Drawing Initial value: 0.5000, 0.5000	Specifies the grid spacing (*X* and *Y*) for the current viewport.
GRIPBLOCK	Type: Integer Saved in: Registry Initial value: 0	Controls the assignment of grips in blocks. 0 Assigns a grip only to the insertion point of the block 1 Assigns grips to objects within the block
GRIPCOLOR	Type: Integer Saved in: Registry Initial value: 5	Controls the color of nonselected grips (drawn as box outlines). The valid range is 1-255.
GRIPHOT	Type: Integer Saved in: Registry Initial value: 1	Controls the color of selected grips (drawn as filled boxes). The valid range is 1-255.
GRIPS	Type: Integer Saved in: Registry Initial value: 1	Controls use of selection set grips for the Stretch, Move, Rotate, Scale, and Mirror grip modes. 0 Turns off grips 1 Turns on grips To adjust the size of the grips and the effective selection area used by the cursor when you snap to a grip, use the *GRIPSIZE* system variable.
GRIPSIZE	Type: Integer Saved in: Registry Initial value: 3	Sets the size of the box drawn to display the grip, in pixels. The valid range is 1-255.
HANDLES	(Read-only) Type: Integer Saved in: Drawing Initial value: On	Reports whether object handles can be accessed by applications. Off Handles are not available On Handles are available
HIGHLIGHT	Type: Integer Not saved Initial value: 1	Controls object highlighting; does not affect objects selected with grips. 0 Turns off object selection highlighting 1 Turns on object selection highlighting
HPANG	Type: Real Not saved Initial value: 0	Specifies the hatch pattern angle.
HPBOUND	Type: Integer Saved in: Registry Initial value: 1	Controls the object type created by *BHATCH* and *BOUNDARY*. 0 Creates a region 1 Creates a polyline
HPDOUBLE	Type: Integer Not saved Initial value: 0	Specifies hatch pattern doubling for user-defined patterns. 0 Turns off hatch pattern doubling 1 Turns on hatch pattern doubling
HPNAME	Type: String Not saved Initial value: "ANSI31"	Sets a default hatch pattern name of up to 34 characters, no spaces allowed. Returns "" if there is no default. Enter a period (.) to set no default.

R13

Variable	Characteristic	Description
HPSCALE	Type: Real Not saved Initial value: 1.0000	Specifies the hatch pattern scale factor; must be nonzero.
HPSPACE	Type: Real Not saved Initial value: 1.0000	Specifies the hatch pattern line spacing for user-defined simple patterns; must be nonzero.
INDEXCTL	Type: Integer Saved in: Drawing Initial value: 0	Controls whether layer and spatial indexes are created and saved in drawing files. 0 No indexes are created 1 Layer index is created 2 Spatial index is created 3 Layer and spatial indexes are created
INETLOCATION	Type: Real Saved in: Registry Initial value: "www.autodesk. com/acaduser"	Stores the Internet location used by *BROWSER*.
INSBASE	Type: 3D point Saved in: Drawing Initial value: 0.0000, 0.0000,0.0000	Stores insertion basepoint set by *BASE*, expressed as a UCS coordinate for the current space.
INSNAME	Type: String Not saved Initial value: ""	Sets a default block name for *DDINSERT* or *INSERT*. The name must conform to symbol naming conventions. Returns "" if no default is set. Enter a period (.) to set no default.
ISAVEBAK	Type: Integer Saved in: Registry Initial value: 1	Improves the speed of incremental saves, especially for large drawings. *ISAVEBAK* controls the creation of a backup file (BAK). In Windows, copying the file data to create a BAK file for large drawings takes a major portion of the incremental save time. 0 No BAK file is created (even for a full save) 1 A BAK file is created **WARNING:** In some cases (such as a power failure in the middle of a save), it's possible that drawing data can be lost.
ISAVEPERCENT	Type: Integer Saved in: Registry Initial value: 50	Determines the amount of wasted space tolerated in a drawing file. The value of *ISAVEPERCENT* is an integer between 0 and 100. The default value of 50 means that the estimate of wasted space within the file does not exceed 50% of the total file size. Wasted space is eliminated by periodic full saves. When the estimate exceeds 50%, the next save will be a full save. This resets the wasted space estimate to 0. If *ISAVEPERCENT* is set to 0, every save is a full save.

R14

R13

Variable	Characteristic	Description
ISOLINES	Type: Integer Saved in: Drawing Initial value: 4	Specifies the number of isolines per surface on objects. Valid integer values are from 0 to 2047.
LASTANGLE	(Read-only) Type: Real Not saved	Stores the end angle of the last arc entered relative to the XY plane of the current UCS for the current space.
LASTPOINT	Type: 3Dpoint Saved in: Drawing Initial value: 0.0000, 0.0000,0.0000	Stores the last point entered, expressed as a UCS coordinate for the current space; referenced by the at symbol (@) during keyboard entry.
LASTPROMPT	(Read-only) Type: String Not saved Initial value: ""	Stores the last string echoed to the command line. This string is identical to the last line seen at the command line and includes any user input.
LENSLENGTH	(Read-only) Type: Real Saved in: Drawing	Stores the length of the lens (in millimeters) used in perspective viewing for the current viewport.
LIMCHECK	Type: Integer Saved in: Drawing Initial value: 0	Controls creation of objects outside the drawing limits. 0 Objects can be created outside the limits 1 Objects cannot be created outside the limits
LIMMAX	Type: 2D point Saved in: Drawing Initial value: 12.0000,9.0000	Stores the upper-right drawing limits for the current space, expressed as World coordinate.
LIMMIN	Type: 2D point Saved in: Drawing Initial value: 0.0000, 0.0000	Stores the lower-left drawing limits for the current space, expressed as a World coordinate.
LISPINIT	Type: Integer Saved in: Registry Initial value: 1	Specifies whether AutoLISP-defined functions and variables are preserved when you open a new drawing or whether they are valid in the current drawing session only. 0 AutoLISP functions and variables are preserved from drawing to drawing 1 AutoLISP functions and variables are valid in current drawing only (Release 13 behavior)
LOCALE	(Read-only) Type: String Not saved Initial value: "en"	Displays the ISO language code of the current AutoCAD version you are running.
LOGFILEMODE	Type: Integer Saved in: Registry Initial value: 0	Specifies whether the contents of the text window are written to a log file. 0 Log file is not maintained 1 Log file is maintained

Margin tabs: R13, R14, R14, R13, R14

Variable	Characteristic	Description
LOGFILENAME	Type: String Saved in: Registry Initial value: "C:\ACADR14\ acad.log"	Specifies the path for the log file. The log file is created if you select Maintain a Log File on the General tab in the Preferences dialog box. **NOTE:** The initial value varies depending on where you installed AutoCAD.
LOGINNAME	(Read-only) Type: String Not saved	Displays the user's name as configured or as input when AutoCAD is loaded. The maximum length for a login name is 30 characters.
LTSCALE	Type: Real Saved in: Drawing Initial value: 1.0000	Sets the global linetype scale factor (cannot equal zero).
LUNITS	Type: Integer Saved in: Drawing Initial value: 2	Sets linear units. 1 Scientific 2 Decimal 3 Engineering 4 Architectural 5 Fractional
LUPREC	Type: Integer Saved in: Drawing Initial value: 4	Sets the number of decimal places displayed for linear units in commands (except dimensioning commands), variables, and output. This includes coordinate display, *LIST, DIST, ID, AREA,* and *DDMODIFY.*
MAXACTVP	Type: Integer Saved in: Drawing Initial value: 48	Sets the maximum number of viewports that can be active at one time.
MAXOBJMEM	Type: Integer Not saved Initial value: 0	Controls the object pager and specifies how much virtual memory AutoCAD allows the drawing to use before it starts paging the drawing out to disk into the object pager's swap files. The default value of 0 turns off the object pager. The object pager is also turned off when this variable is set to a negative value or the value 2,147,483,647. When *MAXOBJMEM* is set to any other value, the object pager is turned on and the specified value is used as the upper limit for the object pager's virtual memory. The *ACADMAXOBJMEM* environment variable also controls the object pager. **WARNING:** If you reboot your system without exiting AutoCAD, these swap files will not be deleted, so you should delete them. Do not delete them from within AutoCAD.
MAXSORT	Type: Integer Saved in: Registry Initial value: 200	Sets the maximum number of symbol names or block names sorted by listing commands. If the total number of items exceeds this value, no items are sorted.

R14

R13

Variable	Characteristic	Description
MEASUREMENT	Type: Integer Saved in: Drawing Initial value: 0	Sets drawing units as English or metric. Specifically, *MEASUREMENT* controls which hatch pattern and linetype file an existing drawing uses when it is opened. 0 English; AutoCAD uses the hatch pattern file and linetype file designated by the ANSIHatch and ANSILinetype registry settings 1 Metric; AutoCAD uses the hatch pattern file and linetype file designated by the ISOHatch and ISOLinetype registry settings The drawing units for new drawings are controlled by the *MEASUREINIT* registry variable (*MEASUREINIT* uses the same values as *MEASUREMENT*). The *MEASUREMENT* setting of a drawing always overrides the *MEASUREINIT* registry setting.
MENUCTL	Type: Integer Saved in: Registry Initial value: 1	Controls the page switching of the screen menu. 0 Screen menu does not switch pages in response to keyboard command entry 1 Screen menu does switch pages in response to keyboard command entry
MENUECHO	Type: Integer Not saved Initial value: 0	Sets menu echo and prompt control bits. It is the sum of the following: 1 Suppresses echo of menu items (^P in a menu item toggles echoing) 2 Suppresses display of system prompts during menu 4 Disables ^P toggle of menu echoing 8 Displays input/output strings; debugging aid for DIESEL macros
MENUNAME	(Read-only) Type: String Saved in: Application header	Stores the *MENUGROUP* name. If the current primary menu has no *MENUGROUP* name, the menu file includes the path if the file location is not specified in AutoCAD's environment setting.
MIRRTEXT	Type: Integer Saved in: Drawing Initial value: 1	Controls how *MIRROR* reflects text. 0 Retains text direction 1 Mirrors the text
MODEMACRO	Type: String Not saved Initial value: ""	Displays a text string on the Status line, such as the name of the current drawing, time/date stamp, or special modes. Use *MODEMACRO* to display a string of text, or use special text strings written in the DIESEL macro language to have AutoCAD evaluate the macro from time to time and base the Status line on user-selected conditions.
MTEXTED	Type: String Saved in: Registry Initial value: "Internal"	Sets the name of the program to use for editing multiline text objects. The default editor is the multiline text editor. If you enter a period (.), the system's default text editor is used.

R14

R13

Variable	Characteristic	Description
OFFSETDIST	Type: Real Not saved Initial value: 1.0000	Sets the default offset distance. <0 Offsets an object through a specified point >0 Sets the default offset distance
OLEHIDE	Type: Integer Saved in: Registry Initial value: 0	Controls the display of OLE objects in AutoCAD. 0 All OLE objects are visible 1 OLE objects are visible in paper space only 2 OLE objects are visible in model space only 3 No OLE objects are visible *OLEHIDE* affects both screen display and printing.
ORTHOMODE	Type: Integer Saved in: Drawing Initial value: 0	Constrains cursor movement to the perpendicular. When *ORTHOMODE* is turned on, the cursor can move only horizontally or vertically relative to the UCS and the current grid rotation angle. 0 Turns off Ortho mode 1 Turns on Ortho mode
OSMODE	Type: Integer Saved in: Drawing Initial value: 0	Sets running object snap modes using the following bit-codes. 0 *None* 1 *Endpoint* 2 *Midpoint* 4 *Center* 8 *Node* 16 *Quadrant* 32 *Intersection* 64 *Insertion* 128 *Perpendicular* 256 *Tangent* 512 *Nearest* 1024 *Quick* 2048 *Apparent Intersection* To specify more than one object snap, enter the sum of their values. For example, entering **3** specifies the *Endpoint* (bit code 1) and *Midpoint* (bit code 2) object snaps. Entering **4095** specifies all object snaps.
OSNAPCOORD	Type: Integer Saved in: Registry Initial value: 2	Controls whether coordinates entered on the command line override running object snaps. 0 Running object snap settings override keyboard coordinate entry 1 Keyboard entry overrides object snap settings 2 Keyboard entry overrides object snap settings except in scripts
PDMODE	Type: Integer Saved in: Drawing Initial value: 0	Controls how point objects are displayed. For information about values to enter, see *POINT*.

R14

R14

Variable	Characteristic	Description
PDSIZE	Type: Real Saved in: Drawing Initial value: 0.0000	Sets the display size for point objects. 0 Creates a point at 5% of the graphics area height >0 Specifies an absolute size <0 Specifies a percentage of the viewport size
PELLIPSE	Type: Integer Saved in: Drawing Initial value: 0	Controls the ellipse type created with *ELLIPSE*. 0 Creates a true ellipse object 1 Creates a polyline representation of an ellipse
PERIMETER	(Read-only) Type: Real Not saved	Stores the last perimeter value computed by *AREA, LIST,* or *DBLIST*.
PFACEVMAX	(Read-only) Type: Integer Not saved	Sets the maximum number of vertices per face.
PICKADD	Type: Integer Saved in: Registry Initial value: 1	Controls additive selection of objects. 0 Turns off *PICKADD*. The objects most recently selected, either individually or by windowing, become the selection set. Previously selected objects are removed from the selection set. Add more objects to the selection set by holding down Shift while selecting. 1 Turns on *PICKADD*. Each object selected, either individually or by windowing, is added to the current selection set. To remove objects from the set, hold down Shift while selecting.
PICKAUTO	Type: Integer Saved in: Registry Initial value: 1	Controls automatic windowing at the Select Objects prompt. 0 Turns off *PICKAUTO* 1 Draws a selection window (for either Window or Crossing selection) automatically at the Select Objects prompt
PICKBOX	Type: Integer Saved in: Registry Initial value: 3	Sets object selection target height, in pixels.
PICKDRAG	Type: Integer Saved in: Registry Initial value: 0	Controls the method of drawing a selection window. 0 Draws the selection window using two points. Click the pointing device at one corner and then at the other corner. 1 Draws the selection window using dragging. Click at one corner, hold down the pick button on the pointing device, drag, and release the button at the other corner.
PICKFIRST	Type: Integer Saved in: Registry Initial value: 1	Controls whether you select objects before (noun-verb selection) or after you issue a command. 0 Turns off *PICKFIRST* 1 Turns on *PICKFIRST*

R13

Variable	Characteristic	Description
PICKSTYLE	Type: Integer Saved in: Drawing Initial value: 1	Controls use of group selection and associative hatch selection. 0 No group selection or associative hatch selection 1 Group selection 2 Associative hatch selection 3 Group selection and associative hatch selection
PLATFORM	(Read-only) Type: String Not saved	Indicates which platform of AutoCAD is in use. One of the following strings may appear: "Microsoft Windows NT Version 3.51 (x86)" "Microsoft Windows NT Version 4.00 (x86)" "Microsoft Windows Version 4.00 (x86)"
PLINEGEN	Type: Integer Saved in: Drawing Initial value: 0	Sets how linetype patterns are generated around the vertices of a two-dimensional polyline. Does not apply to polylines with tapered segments. 0 Polylines are generated to start and end with a dash at each vertex 1 Generates the linetype in a continuous pattern around the vertices of the polyline
PLINETYPE	Type: Integer Saved in: Registry Initial value: 2	Specifies whether AutoCAD uses optimized 2D polylines. *PLINETYPE* controls both the creation of new polylines with the *PLINE* command and the conversion of existing polylines in drawings from previous releases. 0 Polylines in older drawings are not converted on open; *PLINE* creates old-format polylines 1 Polylines in older drawings are not converted on open; *PLINE* creates optimized polylines 2 Polylines in older drawings are converted on open; *PLINE* creates optimized polylines For more information on the two formats, see *CONVERT*. *PLINETYPE* also controls the polyline type created with the following commands: *BOUNDARY* (when object type is set to Polyline), *DONUT, ELLIPSE* (when *PELLIPSE* is set to 1), *PEDIT* (when selecting a line or arc), *POLYGON*, and *SKETCH* (when *SKPOLY* is set to 1).
PLINEWID	Type: Real Saved in: Drawing Initial value: 0.0000	Stores the default polyline width.
PLOTID	Type: String Saved in: Registry Initial value: ""	Changes the default plotter, based on its assigned description, and retains the text string of the current plotter description. Change to another configured plotter by entering its full or partial description. For example, if you have a plotter named "Draft Plotter," enter **draft**. Applications can use *PLOTTER* to step through the available plotter descriptions retained by *PLOTID* and thus control the default plotter.

R13

R14

Variable	Characteristic	Description
PLOTROTMODE	Type: Integer Saved in: Registry Initial value: 1	Controls the orientation of plots. 0 Rotates the effective plotting area so that the corner with the Rotation icon aligns with the paper at the lower-left for 0, top-left for 90, top-right for 180, and lower-right for 270 1 Aligns the lower-left corner of the effective plotting area with the lower-left corner of the paper
PLOTTER	Type: Integer Saved in: Registry Initial value: 0	Changes the default plotter, based on its assigned integer, and retains an integer number that AutoCAD assigns for each configured plotter. This number can be in the range of 0 up to the number of configured plotters. You may configure up to 29 plotters. Change to another configured plotter by entering its valid, assigned number. For example, if you configured four plotters, the valid numbers are 0 through 3. **NOTE:** AutoCAD does not permanently assign a number to a given plotter. If you delete a plotter configuration, AutoCAD assigns new numbers to each of the remaining configured plotters. AutoCAD also updates the value of *PLOTTER*.
POLYSIDES	Type: Integer Not saved Initial value: 4	Sets the default number of sides for *POLYGON*. The range is 3-1024.
POPUPS	(Read-only) Type: Integer Not saved	Displays the status of the currently configured display driver. 0 Does not support dialog boxes, the menu bar, pull-down menus, and icon menus 1 Supports these features
PROJECTNAME	Type: String Saved in: Drawing Initial value: ""	Assigns a project name to the current drawing. The project name points to a section in the registry which can contain one or more search paths for each project name defined. Used when an xref or image is not found in its original hard-coded path. Project names and their search directories are created through the Files tab of the Preferences dialog box. Project names make it easier for users to manage xrefs and images when drawings are exchanged between customers, or if users have different drive mappings to the same location on a server. If the xref or image is not found at the hard-coded path, the project paths associated with the project name are searched. The search order is hard-coded path, then project name search path, then AutoCAD search path.

R13

R14

Variable	Characteristic	Description
PROJMODE	Type: Integer Saved in: Registry Initial value: 1	Sets the current Projection mode for trimming or extending. 0 True 3D mode (no projection) 1 Project to the *XY* plane of the current UCS 2 Project to the current view plane
PROXYGRAPHICS	Type: Integer Saved in: Drawing Initial value: 1	Specifies whether images of proxy objects are saved in the drawing. 0 Image is not saved with the drawing; a bounding box is displayed instead 1 Image is saved with the drawing
PROXYNOTICE	Type: Integer Saved in: Registry Initial value: 1	Displays a notice when a proxy is created. A proxy is created when you open a drawing containing custom objects created by an application that is not present. A proxy is also created when you issue a command that unloads a custom object's parent application. 0 No proxy warning is displayed 1 Proxy warning is displayed
PROXYSHOW	Type: Integer Saved in: Registry Initial value: 1	Controls the display of proxy objects in a drawing. 0 Proxy objects are not displayed 1 Graphic images are displayed for all proxy objects 2 Only the bounding box is displayed for all proxy objects
PSLTSCALE	Type: Integer Saved in: Drawing Initial value: 1	Controls paper space linetype scaling. 0 No special linetype scaling. Linetype dash lengths are based on the drawing units of the space (model or paper) in which the objects were created, scaled by the global *LTSCALE* factor. 1 Viewport scaling governs linetype scaling. If *TILEMODE* is set to 0, dash lengths are based on paper space drawing units, even for objects in model space. In this mode, viewports can have varying magnifications, yet display linetypes identically. For a specific linetype, the dash lengths of a line in a viewport are the same as the dash lengths of a line in paper space. You can still control the dash lengths with *LTSCALE*.
PSPROLOG	Type: String Saved in: Registry Initial value: ""	Assigns a name for a prolog section to be read from the *acad.psf* file when you are using *PSOUT*.
PSQUALITY	Type: Integer Saved in: Registry Initial value: 75	Controls the rendering quality of PostScript images and whether they are drawn as filled objects or as outlines. 0 Turns off PostScript image generation <0 Sets the number of pixels per AutoCAD drawing unit for the PostScript resolution

Variable	Characteristic	Description
PSQUALITY (continued)		>0 Sets the number of pixels per drawing unit but uses the absolute value; causes AutoCAD to show the PostScript paths as outlines and does not fill them
QTEXTMODE	Type: Integer Saved in: Drawing Initial value: 0	Controls how text is displayed. 0 Turns off Quick Text mode; displays characters 1 Turns on Quick Text mode; displays a box in place of text
RASTERPREVIEW	Type: Integer Saved in: Registry Initial value: 1	Controls whether BMP preview images are saved with the drawing. 0 No preview image is created 1 Preview image created
REGENMODE	Type: Integer Saved in: Drawing Initial value: 1	Controls automatic regeneration of the drawing. 0 Turns off *REGENAUTO* 1 Turns on *REGENAUTO*
RE-INIT	Type: Integer Not saved Initial value: 0	Reinitializes the digitizer, digitizer port, and *acad.pgp* file using the following bit-codes: 1 Digitizer I/O port reinitialization 4 Digitizer reinitialization 16 PGP file reinitialization (reload) To specify more than one reinitialization, enter the sum of their values, for example, **5** to specify both digitizer port (1) and digitizer reinitialization (4).
RTDISPLAY	Type: Integer Saved in: Registry Initial value: 1	Controls the display of raster images during real-time zoom or pan. 0 Displays raster image content 1 Displays raster image outline only *RTDISPLAY* is saved in the current profile.
SAVEFILE	(Read-only) Type: String Saved in: Registry Initial value: "auto.sv$"	Stores current auto-save file name.
SAVENAME	(Read-only) Type: String Not saved Initial value: ""	Stores the file name and directory path of the current drawing once you save it.
SAVETIME	Type: Integer Saved in: Registry Initial value: 120	Sets the automatic save interval, in minutes. 0 Turns off automatic saving >0 Automatically saves the drawing at intervals specified by the nonzero integer The *SAVETIME* timer starts as soon as you make a change to a drawing. It is reset and restarted by a manual *SAVE*, *SAVEAS*, or *QSAVE*. The current drawing is saved to *auto.sv$*.

R13

R14

Variable	Characteristic	Description
SCREENBOXES	(Read-only) Type: Integer Saved in: Registry	Stores the number of boxes in the screen menu area of the graphics area. If the screen menu is turned off, *SCREENBOXES* is zero. On platforms that permit the AutoCAD graphics area to be resized or the screen menu to be reconfigured during an editing session, the value of this variable might change during the editing session.
SCREENMODE	(Read-only) Type: Integer Saved in: Registry	Stores a bit-code indicating the graphics/text state of the AutoCAD display. It is the sum of the following bit values: 0 Text screen is displayed 1 Graphics area is displayed 2 Dual-screen display is configured
SCREENSIZE	(Read-only) Type: 2D point Not saved	Stores current viewport size in pixels (*X* and *Y*).
SHADEDGE	Type: Integer Saved in: Drawing Initial value: 3	Controls shading of edges in rendering. 0 Faces shaded, edges not highlighted 1 Faces shaded, edges drawn in background color 2 Faces not filled, edges in object color 3 Faces in object color, edges in background color
SHADEDIF	Type: Integer Saved in: Drawing Initial value: 70	Sets the ratio of diffuse reflective light to ambient light (in percentage of diffuse reflective light).
SHPNAME	Type: String Not saved Initial value: ""	Sets a default shape name; must conform to symbol naming conventions. If no default is set, it returns ""; enter a period (.) to set no default.
SKETCHINC	Type: Real Saved in: Drawing Initial value: 0.1000	Sets the record increment for *SKETCH*.
SKPOLY	Type: Integer Saved in: Registry Initial value: 0	Determines whether *SKETCH* generates lines or polylines.
SNAPANG	Type: Real Saved in: Drawing Initial value: 0	Sets the snap and grid rotation angle for the current viewport. The angle you specify is relative to the current UCS. **NOTE:** Changes to this variable are not reflected in the grid until the display is refreshed. AutoCAD does *not* automatically redraw when variables are changed.
SNAPBASE	Type: 2Dpoint Saved in: Drawing Initial value: 0.0000, 0.0000	Sets the snap and grid origin point for the current viewport relative to the current UCS. **NOTE:** Changes to this variable are not reflected in the grid until the display is refreshed. AutoCAD does *not* automatically redraw when variables are changed.

Variable	Characteristic	Description
SNAPISOPAIR	Type: Integer Saved in: Drawing Initial value: 0	Controls the isometric plane for the current viewport. 0 Left 1 Top 2 Right
SNAPMODE	Type: Integer Saved in: Drawing Initial value: 0	Turns Snap mode on and off. 0 Snap off 1 Snap on for the current viewport.
SNAPSTYL	Type: Integer Saved in: Drawing Initial value: 0	Sets snap style for the current viewport. 0 Standard 1 Isometric
SNAPUNIT	Type: 2D Saved in: Drawing Initial value: 0.5000, 0.5000	Sets the snap spacing for the current viewport. If the *SNAPSTYL* system variable is set to 1, AutoCAD automatically adjusts the *X* value of *SNAPUNIT* to accommodate the isometric snap. **NOTE:** Changes to this system variable are not reflected in the grid until the display is refreshed. AutoCAD does *not* automatically redraw when system variables are changed.
SORTENTS	Type: Integer Saved in: Drawing Initial value: 96	Controls *DDSELECT* object sort order operations. *SORTENTS* uses the following bit codes: 0 Disables *SORTENTS* 1 Sorts for object selection 2 Sorts for object snap 4 Sorts for redraws 8 Sorts for *MSLIDE* slide creation 16 Sorts for *REGEN*s 32 Sorts for plotting 64 Sorts for PostScript output To select more than one, enter the sum of their codes. For example, enter **3** to specify sorting for both object selection and object snap. The initial value of 96 enables sorting for plotting and PostScript output only. Setting additional sorting options can result in slower regeneration and redrawing times.
SPLFRAME	Type: Integer Saved in: Drawing Initial value: 0	Controls display of splines and spline-fit polylines. 0 Does not display the control polygon for splines and spline-fit polylines. Displays the fit surface of a polygon mesh, not the defining mesh. Does not display the invisible edges of 3D faces orpolyface meshes. 1 Displays the control polygon for splines and spline-fit polylines. Only the defining mesh of a surface-fit polygon mesh is displayed (not the fit surface). Invisible edges of 3D faces or polyface meshes are displayed.

Variable	Characteristic	Description
SPLINESEGS	Type: Integer Saved in: Drawing Initial value: 8	Sets the number of line segments to be generated for each spline-fit polyline generated by *PEDIT* Spline.
SPLINETYPE	Type: Integer Saved in: Drawing Initial value: 6	Sets the type of curve generated by *PEDIT* Spline. 5 Quadratic B-spline 6 Cubic B-spline
SURFTAB1	Type: Integer Saved in: Drawing Initial value: 6	Sets the number of tabulations to be generated for *RULESURF* and *TABSURF*. Also sets the mesh density in the *M* direction for *REVSURF* and *EDGESURF*.
SURFTAB2	Type: Integer Saved in: Drawing Initial value: 6	Sets the mesh density in the *N* direction for *REVSURF* and *EDGESURF*.
SURFTYPE	Type: Integer Saved in: Drawing Initial value: 6	Controls the type of surface fitting to be performed by *PEDIT* Smooth. 5 Quadratic B-spline surface 6 Cubic B-spline surface 8 Bezier surface
SURFU	Type: Integer Saved in: Drawing Initial value: 6	Sets the surface density in the *M* direction.
SURFV	Type: Integer Saved in: Drawing Initial value: 6	Sets the surface density in the *N* direction.
SYSCODEPAGE	(Read-only) Type: String Not saved	Indicates the system code page specified in the *acad.xmx* file. Codes are as follows: ascii dos860 dos932 iso8859-7 big5 dos861 gb2312 iso8859-8 dos437 dos863 iso8859-1 iso8859-9 dos850 dos864 iso8859-2 johab dos852 dos865 iso8859-3 ksc5601 dos855 dos866 iso8859-4 mac-roman dos857 dos869 iso8859-6
TABMODE	Type: Integer Not saved Initial value: 0	Controls use of the tablet. 0 Turns off Tablet mode 1 Turns on Tablet mode
TARGET	(Read-only) Type: 3D point Saved in: Drawing	Stores a location (as a UCS coordinate) of the target point for the current viewport.
TDCREATE	(Read-only) Type: Real Saved in: Drawing	Stores the time and date the drawing was created.
TDINDWG	(Read-only) Type: Real Saved in: Drawing	Stores the total editing time.

Variable	Characteristic	Description
TDUPDATE	(Read-only) Type: Real Saved in: Drawing	Stores the time and date of the last update/save.
TDUSRTIMER	(Read-only) Type: Real Saved in: Drawing	Stores the user-elapsed timer.
TEMPPREFIX	(Read-only) Type: String Not saved	Contains the directory name (if any) configured for placement of temporary files, with a path separator appended.
TEXTEVAL	Type: Integer Not saved Initial value: 0	Controls the method of evaluation of text strings. 0 All responses to prompts for text strings and attribute values are taken literally 1 Text starting with an opening parenthesis [(] or an exclamation mark (!) is evaluated as an AutoLISP expression, as for nontextual input **NOTE:** *DTEXT* takes all input literally regard less of the setting of *TEXTEVAL*.
TEXTFILL	Type: Integer Saved in: Registry Initial value: 1	Controls the filling of TrueType fonts while plotting, exporting with *PSOUT*, and rendering. 0 Outputs text as outlines 1 Outputs text as filled images.
TEXTQLTY	Type: Integer Saved in: Drawing Initial value: 50	Sets the resolution of TrueType fonts while plotting, exporting with *PSOUT*, and rendering. Values represent dots per inch. Lower values decrease resolution and increase plotting speed. Higher values increase resolution and decrease plotting speed.
TEXTSIZE	Type: Real Saved in: Drawing Initial value: 0.2000	Sets the default height for new text objects drawn with the current text style (has no effect if the style has a fixed height).
TEXTSTYLE	Type: String Saved in: Drawing Initial value: "STANDARD"	Sets the name of the current text style.
THICKNESS	Type: Real Saved in: Drawing Initial value: 1	Sets the current 3D solid thickness.
TILEMODE	Type: Integer Saved in: drawing Initial value: 1	Controls access to paper space and the behavior of viewports. 0 Turns on paper space and viewport object (uses *MVIEW*). AutoCAD clears the graphics area and prompts you to create one or more viewports. 1 Turns on Release 10 Compatibility mode (uses *VPORTS*). AutoCAD restores the most recently active tiled-viewport configuration. Paper space objects—including viewport objects—are not displayed, and *MVIEW, MSPACE, PSPACE,* and *VPLAYER* cannot be used.

R13

Variable	Characteristic	Description
TOOLTIPS	Type: Integer Saved in: Registry Initial value: 1	Controls the display of tooltips. 0 Turns off display of tooltips 1 Turns on display of tooltips
TRACEWID	Type: Real Saved in: Drawing Initial value: 0.0500	Sets the default trace width.
TREEDEPTH	Type: Integer Saved in: Drawing Initial value: 3020	Specifies the maximum depth, that is, the number of times the tree-structured spatial index may divide into branches. 0 Suppresses the spatial index entirely, eliminating the performance improvements it provides in working with large drawings. This setting assures that objects are always processed in database order, making it unnecessary ever to set the *SORTENTS* system variable. >0 Turns on *TREEDEPTH*. An integer of up to four digits is valid. The first two digits refer to model space, and the second two digits refer to paper space. <0 Treats model space objets as two-dimensional (Z coordinates are ignored), as is always the case with paper space objects. Such a setting is appropriate for 2D drawings and makes more efficient use of memory without loss of performance. **NOTE:** You cannot use *TREEDEPTH* transparently.
TREEMAX	Type: Integer Saved in: Registry Initial value: 10000000	Limits memory consumption during drawing regeneration by limiting the number of nodes in the spatial index (oct-tree). By imposing a fixed limit with *TREEMAX*, you can load drawings created on systems with more memory than your system and with a larger *TREEDEPTH* than your system can handle. These drawings, if left unchecked, have an oct-tree large enough to eventually consume more memory than is available to your computer. *TREEMAX* also provides a safeguard against experimentation with inappropriately high *TREEDEPTH* values. The initial default for *TREEMAX* is 10000000 (10 million), a value high enough to effectively disable *TREEMAX* as a control for *TREEDEPTH*. The value to which you should set *TREEMAX* depends on your system's available RAM. You get about 15,000 oct-tree nodes per megabyte of RAM.

R13

R13

Variable	Characteristic	Description
TREEMAX (continued)		If you want an oct-tree to use up to, but no more than, 2 megabytes of RAM, set *TREEMAX* to 30000 (2×15,000). If AutoCAD runs out of memory allocating oct-tree nodes, restart AutoCAD, set *TREEMAX* to a smaller number, and try loading the drawing again. AutoCAD might occasionally run into the limit you set with *TREEMAX*. Follow the resulting prompt instructions. Your ability to increase *TREEMAX* depends on your computer's available memory.
TRIMMODE	Type: Integer Saved in: Registry Initial value: 1	Controls whether AutoCAD trims selected edges for chamfers and fillets. 0 Leaves selected edges intact. 1 Trims selected edges to the endpoints of chamfer lines and fillet arcs
UCSFOLLOW	Type: Integer Saved in: Drawing Initial value: 0	Generates a plan view whenever you change from one UCS to another. You can set *UCSFOLLOW* separately for each viewport. If *UCSFOLLOW* is on for a particular viewport, AutoCAD generates a plan view in that viewport whenever you change coordinate systems. Once the new UCS has been established, you can use *VPOINT, DVIEW, PLAN,* or *VIEW* to change the view of the drawing. It will change to a plan view again the next time you change coordinate systems. 0 UCS does *not* affect the view 1 Any UCS change causes a change to plan view of the new UCS in the current viewport The setting of *UCSFOLLOW* is maintained separately for paper space and model space and can be accessed in either, but the setting is ignored while in paper space (it is always treated as if set to 0). Although you can define a non-World UCS in paper space, the view remains in plan view to the World Coordinate System.
UCSICON	Type: Integer Saved in: Drawing Initial value: 1	Displays the user coordinate system icon for the current viewport using bit code. It is the sum of the following: 0 No icon displayed 1 On; icon is displayed 2 Origin; if icon is displayed, the icon floats to the UCS origin if possible
UCSNAME	(Read-only) Type: String Saved in: Drawing	Stores the name of the current coordinate system for the current space. Returns a null string if the current UCS is unnamed.
UCSORG	(Read-only) Type: 3D point Saved in: Drawing	Stores the origin point of the current coordinate system for the current space. This value is always stored as a World coordinate.

Variable	Characteristic	Description
UCSXDIR	(Read-only) Type: 3D point Saved in: Drawing	Stores the *X* direction of the current UCS for the current space.
UCSYDIR	(Read-only) Type: 3D point Saved in: Drawing	Stores the *Y* direction of the current UCS for the current space.
UNDOCTL	(Read-only) Type: Integer Not saved	Stores a bit code indicating the state of the *UNDO* feature. It is the sum of the following values: 0 *UNDO* is turned off 1 *UNDO* is turned on 2 Only one command can be undone 4 Turns on the Auto option 8 A group is currently active
UNDOMARKS	(Read-only) Type: Integer Not saved	Stores the number of marks that have been placed in the *UNDO* control stream by the Mark option. The Mark and Back options are not available if a group is currently active.
UNITMODE	Type: Integer Saved in: Drawing Initial value: 0	Controls the display format for units. 0 Display fractional, feet and inches, and surveyor's angles as previously set 1 Displays fractional, feet and inches, and surveyor's angles in input format
USERI1-5	Type: Integer Saved in: Drawing Initial value: 0	*USERI1, USERI2, USERI3, USERI4,* and *USERI5* are used for storage and retrieval of integer values.
USERR1-5	Type: Real Saved in: Drawing Initial value: 0.0000	*USERR1, USERR2, USERR3, USERR4,* and *USERR5* are used for storage and retrieval of real numbers.
USERS1-5	Type: String Not saved Initial value: ""	*USERS1, USERS2, USERS3, USERS4,* and *USERS5* are used for storage and retrieval of text string data.
VIEWCTR	(Read-only) Type: 3D point Saved in: Drawing	Stores the center of view in the current viewport, expressed as a UCS coordinate.
VIEWDIR	(Read-only) Type: 3D vector Saved in: Drawing	Stores the viewing direction in the current viewport expressed in UCS coordinates. This describes the camera point as a 3D offset from the target point.
VIEWMODE	(Read-only) Type: Integer Saved in: Drawing	Controls Viewing mode for the current viewport using bit code. The value is the sum of the following: 0 Turned off 1 Perspective view active 2 Front clipping on 4 Back clipping on 8 UCS Follow mode on

R14

Variable	Characteristic	Description
VIEWMODE (continued)		16 Front clip not at eye. If on, the front clip distance (*FRONTZ*) determines the front clipping plane. If off, *FRONTZ* is ignored, and the front clipping plane is set to pass through the camera point (vectors behind the camera are not displayed). This flag is ignored if the front clipping bit (2) is off.
VIEWSIZE	(Read-only) Type: Real Saved in: Drawing	Stores the height of the view in the current viewport, expressed in drawing units.
VIEWTWIST	(Read-only) Type: Real Saved in: Drawing	Stores the view twist angle for the current viewport.
VISRETAIN	Type: Integer Saved in: Drawing Initial value: 1	Controls visibility of layers in xref files. 0 The xref layer definition in the current drawing takes precedence over these settings: On/Off, Freeze/Thaw, color and linetype settings for xref-dependent layers. 1 On/Off, Freeze/Thaw, color, and linetype settings for xref-dependent layers in the current drawing take precedence over the xref layer definition.
VSMAX	(Read-only) Type: 3D point Saved in: Drawing	Stores the upper-right corner of the current viewport's virtual screen, expressed as a UCS coordinate.
VSMIN	(Read-only) Type: 3D point Saved in: Drawing	Stores the lower-left corner of the current viewport's virtual screen, expressed as a UCS coordinate.
WORLDUCS	(Read-only) Type: Integer Not saved	Indicates whether the UCS is the same as the World Coordinate System. 0 Current UCS is different from the World Coordinate System 1 Current UCS is the same as the World Coordinate System
WORLDVIEW	Type: Integer Saved in: Drawing Initial value: 1	Controls whether the UCS changes to the WCS during *DVIEW* or *VPOINT*. 0 Current UCS remains unchanged 1 Current UCS is changed to the WCS for the duration of *DVIEW* or *VPOINT*; *DVIEW* and *VPOINT* command input is relative to the current UCS
XCLIPFRAME	Type: Integer Saved in: Drawing Initial value: 0	Controls visibility of xref clipping boundaries. 0 Clipping boundary is not visible 1 Clipping boundary is visible

Variable	Characteristic	Description
XLOADCTL	Type: Integer Saved in: Registry Initial value: 1	Turns xref demand loading on and off and controls whether it opens the original drawing or a copy. 0 Turns off demand loading; entire drawing is loaded 1 Turns on demand loading; reference file is kept open 2 Turns on demand loading; a copy of the reference file is opened When XLOADCTL is set to 2, the reference copy is stored in the AutoCAD temporary files directory (defined by *PREFERENCES*) or in a user-specified directory.
XLOADPATH	Type: String Saved in: Registry Initial value: ""	Creates a path for storing temporary copies of demand-loaded xref files. For more information, see *XLOADCTL.*
XREFCTL	Type: Integer Saved in: Registry Initial value: 0	Controls whether AutoCAD writes external reference log (XLG) files. 0 Xref log (XLG) files are not written 1 Xref log (XLG) files are written

Source: AutoCAD *Command Reference*

R14

APPENDIX B

AutoCAD Release 14 Command Alias List Sorted by <u>Command</u>

Command	Alias	Command	Alias
3DARRAY	3A	DDVPOINT	VP
3DFACE	3F	DIMALIGNED	DAL
3DPOLY	3P	DIMALIGNED	DIMALI
ALIGN	AL	DIMANGULAR	DAN
APPLOAD	AP	DIMANGULAR	DIMANG
ARC	A	DIMBASELINE	DBA
AREA	AA	DIMBASELINE	DIMBASE
ARRAY	AR	DIMCENTER	DCE
ASEADMIN	AAD	DIMCONTINUE	DCO
ASEEXPORT	AEX	DIMCONTINUE	DIMCONT
ASELINKS	ALI	DIMDIAMETER	DDI
ASEROWS	ARO	DIMDIAMETER	DIMDIA
ASESELECT	ASE	DIMEDIT	DED
ASESQLED	ASQ	DIMEDIT	DIMED
ATTDEF	-AT	DIMLINEAR	DIMLIN
ATTEDIT	-ATE	DIMLINEAR	DLI
BHATCH	BH	DIMORDINATE	DIMORD
BHATCH	H	DIMORDINATE	DOR
BLOCK	-B	DIMOVERRIDE	DIMOVER
BMAKE	B	DIMOVERRIDE	DOV
-BOUNDARY	-BO	DIMRADIUS	DIMRAD
BOUNDARY	BO	DIMRADIUS	DRA
BREAK	BR	DIMSTYLE	DIMSTY
CHAMFER	CHA	DIMSTYLE	DST
CHANGE	-CH	DIMTEDIT	DIMTED
CIRCLE	C	DIST	DI
COPY	CO	DIVIDE	DIV
COPY	CP	DONUT	DO
DDATTDEF	AT	DRAWORDER	DR
DDATTE	ATE	DSVIEWER	AV
DDCHPROP	CH	DTEXT	DT
DDCOLOR	COL	DVIEW	DV
DDEDIT	ED	ELLIPSE	EL
DDGRIPS	GR	ERASE	E
DDIM	D	EXPLODE	X
DDINSERT	I	EXPORT	EXP
DDMODIFY	MO	EXTEND	EX
DDOSNAP	OS	EXTRUDE	EXT
DDRENAME	REN	FILLET	F
DDRMODES	RM	FILTER	FI
DDSELECT	SE	GROUP	-G
DDUCS	UC	GROUP	G
DDUCSP	UCP	HATCH	-H
DDUNITS	UN	HATCHEDIT	HE
DDVIEW	V	HIDE	HI

Command	Alias	Command	Alias
-IMAGE	-IM	REDRAW	R
IMAGE	IM	REDRAWALL	RA
IMAGEADJUST	IAD	REGEN	RE
IMAGEATTACH	IAT	REGENALL	REA
IMAGECLIP	ICL	REGION	REG
IMPORT	IMP	RENAME	-REN
INSERT	-I	RENDER	RR
INSERTOBJ	IO	REVOLVE	REV
INTERFERE	INF	ROTATE	RO
INTERSECT	IN	RPREF	RPR
-LAYER	-LA	SCALE	SC
LAYER	LA	SCRIPT	SCR
LEADER	LE	SECTION	SEC
LEADER	LEAD	SETVAR	SET
LENGTHEN	LEN	SHADE	SHA
LINE	L	SLICE	SL
-LINETYPE	-LT	SNAP	SN
LINETYPE	LT	SOLID	SO
LIST	LI	SPELL	SP
LIST	LS	SPLINE	SPL
LTSCALE	LTS	SPLINEDIT	SPE
MATCHPROP	MA	STRETCH	S
MEASURE	ME	STYLE	ST
MIRROR	MI	SUBTRACT	SU
MLINE	ML	TABLET	TA
MOVE	M	THICKNESS	TH
MSPACE	MS	TILEMODE	TI
-MTEXT	-T	TILEMODE	TM
MTEXT	MT	TOLERANCE	TOL
MTEXT	T	TOOLBAR	TO
MVIEW	MV	TORUS	TOR
OFFSET	O	TRIM	TR
-OSNAP	-OS	UNION	UNI
-PAN	-P	UNITS	-UN
PAN	P	VIEW	-V
PASTESPEC	PA	VPOINT	-VP
PEDIT	PE	WBLOCK	W
PLINE	PL	WEDGE	WE
PLOT	PRINT	XATTACH	XA
POINT	PO	-XBIND	-XB
POLYGON	POL	XBIND	XB
PREFERENCES	PR	XCLIP	XC
PREVIEW	PRE	XLINE	XL
PSPACE	PS	-XREF	-XR
PURGE	PU	XREF	XR
QUIT	EXIT	ZOOM	Z
RECTANGLE	REC		

APPENDIX C

AutoCAD Release 14 Command Alias List Sorted by <u>Alias</u>

Alias	Command	Alias	Command
3A	*3DARRAY*	*DIMDIA*	*DIMDIAMETER*
3F	*3DFACE*	*DIMED*	*DIMEDIT*
3P	*3DPOLY*	*DIMLIN*	*DIMLINEAR*
A	*ARC*	*DIMORD*	*DIMORDINATE*
AA	*AREA*	*DIMOVER*	*DIMOVERRIDE*
AAD	*ASEADMIN*	*DIMRAD*	*DIMRADIUS*
AEX	*ASEEXPORT*	*DIMSTY*	*DIMSTYLE*
AL	*ALIGN*	*DIMTED*	*DIMTEDIT*
ALI	*ASELINKS*	*DIV*	*DIVIDE*
AP	*APPLOAD*	*DLI*	*DIMLINEAR*
AR	*ARRAY*	*DO*	*DONUT*
ARO	*ASEROWS*	*DOR*	*DIMORDINATE*
ASE	*ASESELECT*	*DOV*	*DIMOVERRIDE*
ASQ	*ASESQLED*	*DR*	*DRAWORDER*
-AT	*ATTDEF*	*DRA*	*DIMRADIUS*
AT	*DDATTDEF*	*DST*	*DIMSTYLE*
-ATE	*ATTEDIT*	*DT*	*DTEXT*
ATE	*DDATTE*	*DV*	*DVIEW*
AV	*DSVIEWER*	*E*	*ERASE*
-B	*BLOCK*	*ED*	*DDEDIT*
B	*BMAKE*	*EL*	*ELLIPSE*
BH	*BHATCH*	*EX*	*EXTEND*
-BO	*-BOUNDARY*	*EXIT*	*QUIT*
BO	*BOUNDARY*	*EXP*	*EXPORT*
BR	*BREAK*	*EXT*	*EXTRUDE*
C	*CIRCLE*	*F*	*FILLET*
-CH	*CHANGE*	*FI*	*FILTER*
CH	*DDCHPROP*	*-G*	*GROUP*
CHA	*CHAMFER*	*G*	*GROUP*
CO	*COPY*	*GR*	*DDGRIPS*
COL	*DDCOLOR*	*H*	*BHATCH*
CP	*COPY*	*-H*	*HATCH*
D	*DDIM*	*HE*	*HATCHEDIT*
DAL	*DIMALIGNED*	*HI*	*HIDE*
DAN	*DIMANGULAR*	*I*	*DDINSERT*
DBA	*DIMBASELINE*	*-I*	*INSERT*
DCE	*DIMCENTER*	*IAD*	*IMAGEADJUST*
DCO	*DIMCONTINUE*	*IAT*	*IMAGEATTACH*
DDI	*DIMDIAMETER*	*ICL*	*IMAGECLIP*
DED	*DIMEDIT*	*-IM*	*-IMAGE*
DI	*DIST*	*IM*	*IMAGE*
DIMALI	*DIMALIGNED*	*IMP*	*IMPORT*
DIMANG	*DIMANGULAR*	*IN*	*INTERSECT*
DIMBASE	*DIMBASELINE*	*INF*	*INTERFERE*
DIMCONT	*DIMCONTINUE*	*IO*	*INSERTOBJ*

Alias	Command	Alias	Command
L	LINE	RR	RENDER
-LA	-LAYER	S	STRETCH
LA	LAYER	SC	SCALE
LE	LEADER	SCR	SCRIPT
LEAD	LEADER	SE	DDSELECT
LEN	LENGTHEN	SEC	SECTION
LI	LIST	SET	SETVAR
LS	LIST	SHA	SHADE
-LT	-LINETYPE	SL	SLICE
LT	LINETYPE	SN	SNAP
LTS	LTSCALE	SO	SOLID
M	MOVE	SP	SPELL
MA	MATCHPROP	SPE	SPLINEDIT
ME	MEASURE	SPL	SPLINE
MI	MIRROR	ST	STYLE
ML	MLINE	SU	SUBTRACT
MO	DDMODIFY	-T	-MTEXT
MS	MSPACE	T	MTEXT
MT	MTEXT	TA	TABLET
MV	MVIEW	TH	THICKNESS
O	OFFSET	TI	TILEMODE
OS	DDOSNAP	TM	TILEMODE
-OS	-OSNAP	TO	TOOLBAR
-P	-PAN	TOL	TOLERANCE
P	PAN	TOR	TORUS
PA	PASTESPEC	TR	TRIM
PE	PEDIT	UC	DDUCS
PL	PLINE	UCP	DDUCSP
PO	POINT	UN	DDUNITS
POL	POLYGON	-UN	UNITS
PR	PREFERENCES	UNI	UNION
PRE	PREVIEW	V	DDVIEW
PRINT	PLOT	-V	VIEW
PS	PSPACE	VP	DDVPOINT
PU	PURGE	-VP	VPOINT
R	REDRAW	W	WBLOCK
RA	REDRAWALL	WE	WEDGE
RE	REGEN	X	EXPLODE
REA	REGENALL	XA	XATTACH
REC	RECTANGLE	-XB	-XBIND
REG	REGION	XB	XBIND
REN	DDRENAME	XC	XCLIP
-REN	RENAME	XL	XLINE
REV	REVOLVE	-XR	-XREF
RM	DDRMODES	XR	XREF
RO	ROTATE	Z	ZOOM
RPR	RPREF		

APPENDIX D

Buttons and Special Keys

Mouse and Digitizing Puck Buttons

Depending on the type of mouse or digitizing puck used for cursor control, a different number of buttons is available. In any case, the buttons perform the following tasks:

#1 (left mouse)	**PICK**	Used to select commands or point to locations on screen.
#2 (right mouse)	**Enter**	Generally, performs the same action as the Enter or Return key on the keyboard. When some dialog boxes are visible, right-clicking produces a menu of choices affecting the active dialog box.
#3 (center mouse)	*OSNAP*	Activates the cursor *OSNAP* menu.
#4	**Cancel**	Cancels a command.

Function (F) Keys

Function keys in AutoCAD offer a quick method of turning on or off (toggling) drawing aids.

F1	*Help*	Opens a help window providing written explanations on commands and variables.
F2	*Flipscreen*	Activates a text window showing the previous command line activity (command history).
F3	*Osnap Toggle*	If Running Osnaps are set, toggling this tile temporarily turns the Running Osnaps off so that a point can be picked without using Osnaps. If no Running Osnaps are set, F3 produces the *Osnap Settings* dialog box (discussed in Chapter 7).
F4	*Tablet*	Turns the *TABMODE* variable on or off. If *TABMODE* is on, the digitizing tablet can be used to digitize an existing paper drawing into AutoCAD.
F5	*Isoplane*	When using an *Isometric* style *SNAP* and *GRID* setting, toggles the cursor (with *ORTHO* on) to draw on one of three isometric planes.
F6	*Coords*	Toggles the coordinate display between cursor tracking mode and off. If used transparently (during a command in operation), displays a polar coordinate format.
F7	*GRID*	Turns the *GRID* on or off (see Drawing Aids).
F8	*ORTHO*	Turns *ORTHO* on or off (see Drawing Aids).
F9	*SNAP*	Turns *SNAP* on or off (see Drawing Aids).
F10	*Status*	Turns the Status line on or off.

R13

R14

Control Key Sequences (Accelerator Keys)

Accelerator keys (holding down the Ctrl key and pressing another key simultaneously) invoke regular AutoCAD commands or produce special functions. Several have the same duties as F3 through F9.

Drawing Aids

Ctrl+F (F3)	*Osnap Toggle*	If Running Osnaps are set, pressing Ctrl+F temporarily turns the Running Osnaps off so that a point can be picked without using Osnaps. If no Running Object Snaps are set, Ctrl+F produces the *Osnap Settings* dialog box. This dialog box is used to turn on and off Running Object Snaps (discussed in Chapter 7).
Ctrl+T (F4)	*Tablet*	Turns the *TABMODE* variable on or off. If *TABMODE* is on, the digitizing tablet can be used to digitize an existing paper drawing into AutoCAD.
Ctrl+E (F5)	*Isoplane*	When using an *Isometric* style *SNAP* and *GRID* setting, toggles the cursor (with *ORTHO* on) to draw on one of three isometric planes.
Ctrl+D (F6)	*Coords*	Toggles the Coordinate Display between cursor tracking mode and off. If used during a command operation, can be toggled to a polar coordinate format.
Ctrl+G (F7)	*GRID*	Turns the *GRID* on or off (see Drawing Aids).
Ctrl+L (F8)	*ORTHO*	Turns *ORTHO* on or off (see Drawing Aids).
Crtl+B (F9)	*SNAP*	Turns *SNAP* on or off (see Drawing Aids).

Windows Copy, Cut, Paste (see Chapter 31)

Ctrl+C	*Copyclip*	Copies the highlighted objects to the Windows clipboard.
Ctrl+X	*Cutclip*	Cuts the highlighted objects from the drawing and copies them to the Windows clipboard.
Ctrl+V	*Pasteclip*	Pastes the clipboard contents into the AutoCAD drawing as a Block Reference.

File Operations (see Chapter 2)

Ctrl+O	*Open*	Invokes the *Open* command to open an existing drawing.
Ctrl+N	*New*	Invokes the *New* command to start a new drawing.
Ctrl+S	*Qsave*	Performs a quick save or produces the *SaveAs* dialog box if the file is not yet named.

Other Control Key Sequences

Ctrl+Z	*Undo*	Undoes the last command (see Chapter 5).
Ctrl+Y	*Redo*	Invokes the *Redo* command (see Chapter 5).
Ctrl+P	*Print/Plot*	Produces the *Print/Plot Configuration* dialog box for creating and controlling prints and plots (see Chapter 14).
Ctrl+A	*Group*	Toggles selectable *Groups* on or off (see Chapter 20).
Ctrl+J	*Enter*	Has the same function as Enter.
Ctrl+K	*PICKADD*	Changes the setting (0 or 1) of the *PICKADD* variable (see Chapter 20).

R14

R13

Special Key Functions

Esc		The Escape key cancels a command, menu, or dialog box or interrupts processing of plotting or hatching.
Spacebar		In AutoCAD, the Spacebar performs the same action as the Enter key or #2 button. Only when you are entering text into a drawing does the Spacebar create a space.
Enter		If Enter, Spacebar, or #2 button is pressed when no command is in use (the open Command: prompt is visible), the last command used is invoked again.
Up Arrow		The up arrow key can be used to recall previous command line entries, similar to the DOSKEY feature. Pressing the up arrow places the previous command at the Command: prompt, but does not execute it. You can use the up arrow repeatedly to find a previously used command, then press Enter to execute it.
Down Arrow		After the up arrow is used to recall previously used commands in reverse order, the down arrow cycles back through those commands in forward order. The up arrow and down arrow keys are used together to locate particular command line entries.

INDEX

CHAPTER EXERCISE INDEX